PHASE RELATIONS, HIGH-PRESSURE TERRANES, P-T-OMETRY, AND PLATE PUSHING

PHASE RELATIONS, HIGH-PRESSURE TERRANES, P-T-OMETRY, AND PLATE PUSHING

A Tribute to W. G. Ernst
International Book Series, Vol. 9

J. G. Liou and Mark Cloos, editors

Bellwether Publishing, Ltd. for the
THE GEOLOGICAL SOCIETY OF AMERICA

Copyright © 2006 by Bellwether Publishing, Ltd.

All rights reserved. Printed in the United States of America. No part of this publication may be reproduced, stored in a retrieval system, or transmitted, in any form or by any means, electronic, mechanical, photocopying, recording, or otherwise, without prior written permission of the publisher.

Published by

    Bellwether Publishing, Ltd.
    8640 Guilford Road, Suite 200
    Columbia, MD 21046-3163
    http://www.bellpub.com
    for the Geological Society of America

Exclusively marketed throughout the world by:

    Geological Society of America
    P. O. Box 9140
    Boulder, CO 80301-9140
    email: pubs@geosociety.org
    or order from the GSA Bookstore on the World Wide Web at
    http://www.geosociety.org/bookstore.

Library of Congress Control Number: 2006920682
ISBN 978-0-9665869-9-2

Printed by The Sheridan Press, Hanover, Pennsylvania.

PHASE RELATIONS, HIGH-PRESSURE TERRANES, P-T-OMETRY, AND PLATE PUSHING: A TRIBUTE TO W. G. ERNST

J. G. Liou and Mark Cloos, editors

Contents

Editors' Preface
   —J. G. Liou and Mark Cloos 1

General

Psychology of a Changing Paradigm: 40+ Years of High-Pressure Metamorphism
   —Harry W. Green, II 5

Blueschists and Blue Amphiboles: How Much Subduction Do They Need
   —Walter V. Maresch and Taras V. Gerya 23

Involvement of Crustal Material in Delamination of the Lithosphere
after Continent-Continent Collision
   —Hans-Joachim Massonne 38

Himalayas

Global UHP Metamorphism and Continental Subduction/Collision: The Himalayan Model
   —J. G. Liou, T. Tsujimori, R. Y. Zhang, I. Katayama, and S. Maruyama 53

U-Pb Zircon Dating of Regional Deformation in the Lower Crust of the Kohistan Arc
   —Hiroshi Yamamoto, Katsura Kobayashi, Eizo Nakamura, Yoshiyuki Kaneko,
   and Allah Bakhsh Kausar 80

Kazakhstan

Cathodoluminescence of Microdiamond in Dolomite Marble from the Kokchetav Massif—
Additional Evidence for Two-Stage Growth of Diamond
   —Nobuhiro Yoshioka and Yoshihide Ogasawara 95

Diamond Formation in UHP Dolomite Marbles and Garnet-Pyroxene Rocks of the
Kokchetav Massif, Northern Kazakhstan: Natural and Experimental Evidence
   —V. S. Shatsky, Y. N. Pal'yanov, A. G. Sokol, A. A. Tomilenko, and N. V. Sobolev 108

The Role of Fluid in Diamond-Free Dolomitic Marble from the Kokchetav Massif
   —Yoshihide Ogasawara and Kazumasa Aoki 120

Western China

A New HP/LT Metamorphic Terrane in the Northern Altyn Tagh, Western China
—Jianxin Zhang, Fancong Meng, and Jingsui Yang 139

Two Ultrahigh-Pressure Metamorphic Events in the Central Orogenic Belt of China:
Evidence from U-Pb Dating of Coesite-Bearing Zircons
—Jingsui Yang, Fulai Liu, Caiali Wu, Zhiqin Xu, Rending Shi, Songyong Chen,
Etienne Deloule, and Joseph L. Wooden 155

A New Caledonian Khondalite Series in West Kunlun, China: Age Constraints and
Tectonic Significance
—Zhiqin Xu, Xuexiang Qi, Fulai Liu, Jingsui Yang, Lingsen Zeng, and Cailai Wu 172

Eastern China

Tracing the Boundary between UHP and HP Metamorphic Belts in the Southwestern
Sulu Terrane, Eastern China: Evidence from Mineral Inclusions in Zircons
from Metamorphic Rocks
—Fulai Liu, Zhiqin Xu, and J. G. Liou 187

Fluid Composition and Evolution Attending UHP Metamorphism: Study of Fluid
Inclusions from Drill Cores, Southern Sulu Belt, Eastern China
—Zeming Zhang, Kun Shen, Yilin Xiao, Alfons M. van den Kerkhof,
Jochen Hoefs, and J. G. Liou 204

Late Mesozoic–Eocene Mantle Replacement beneath the Eastern North China Craton:
Evidence from the Paleozoic and Cenozoic Peridotite Xenoliths
—Jianping Zheng, W. L. Griffin, Suzanne Y. O'Reilly, J. G. Liou,
R. Y. Zhang, and Fengxiang Lu 217

Geochronological Constraints on the Paleoproterozoic Evolution of the North China Craton:
SHRIMP Zircon Ages of Different Types of Mafic Dikes
—Peng Peng, Mingguo Zhai, Huafeng Zhang, and Jinghui Guo 233

Fluid-Rock Interaction in UHP Phengite-Kyanite-Epidote Eclogite from the
Sulu Orogen, Eastern China
—S. Ferrando, M. L. Frezzotti, L. Dallai, and R. Compagnoni 250

^{40}Ar-^{39}Ar Thermochronological Constraints on the Exhumation of Ultrahigh-Pressure
Metamorphic Rocks in the Sulu Terrane, Eastern China
—Li-Hung Lin, Pei-Ling Wang, Ching-Hua Lo, Chin-Ho Tsai, and Bor-ming Jahn 275

Limiting Effects of UHP Metamorphism on Length Scales of Oxygen, Hydrogen, and
Argon Isotope Exchange: An Example from the Qinglongshan UHP Eclogites, Sulu
Terrain, China
—M. A. Cosca, D. Giorgis, D. Rumble, and J. G. Liou 290

SHRIMP U-Pb Dating of Zircon from the Xugou UHP Eclogite, Sulu Terrane, Eastern China
—Ruixuan Zhao, Juhn G. Liou, Ru Y. Zhang, and Joseph L. Wooden 324

Tectonic Division of the Sulu Ultrahigh-Pressure Region and the Nature of Its Boundary
with the North China Block
  —Mingguo Zhai and Wenjun Liu 334

Petrogenesis of UHP Metamorphic Crustal and Mantle Rocks from the Chinese
Continental Scientific Drilling Pre-pilot Hole 1, Sulu Belt, Eastern China
  —Zeming Zhang, Yilin Xiao, Jochen Hoefs, Zhiqin Xu, and J. G. Liou 350

Petrologic Study of Ultrahigh-Pressure Metamorphic Cores from 100 to 2000 m Depth
in the Main Hole of the Chinese Continental Scientific Drilling Project, Eastern China
  —Shangguo Su, Juhn G. Liou, Zhendong You, Fenghua Liang, and Zeming Zhang 368

Low-Grade Metamorphic Rocks in the Dabie-Sulu Orogenic Belt: A Passive-Margin
Accretionary Wedge Deformed during Continent Subduction
  —Yong-Fei Zheng, Jian-Bo Zhou, Yuan-Bao Wu, and Zhi Xie 384

Taiwan

Isotopic Composition of Carbonaceous Material in Metamorphic Rocks from the
Mountain Belt of Taiwan
  —Tzen-Fu Yui 407

South Korea

Metamorphic Evolution of the Ogcheon Belt, Korea: A Review and New Age Constraints
  —Moonsup Cho and Hyeoncheol Kim 425

Ridge Subduction–Related Jurassic Plutonism in and around the Okcheon
Metamorphic Belt, South Korea, and Implications for Northeast Asian Tectonics
  —S. W. Kim, C. W. Oh, S. G. Choi, I.-C. Ryu, and T. Itaya 442

Metamorphic Evolution of the Southwest Okcheon Metamorphic Belt in South Korea
and Its Regional Tectonic Implications
  —S. W. Kim, C. W. Oh, H. Hyodo, T. Itaya, and J. G. Liou 464

Japan

Eclogite-Facies Mineral Inclusions in Clinozoisite from Paleozoic Blueschist, Central Chugoku
Mountains, Southwest Japan: Evidence of Regional Eclogite-Facies Metamorphism
  —T. Tsujimori and J. G. Liou 493

Accretionary Complex Origin of the Mafic-Ultramafic Bodies of the Sanbagawa Belt,
Central Shikoku, Japan
  —Masaru Terabayashi, Kazuaki Okamoto, Hiroshi Yamamoto, Yoshiyuki Kaneko,
    Tsutomu Ota, Shigenori Maruyama, Ikuo Katayama, Tsuyoshi Komiya,
    Akira Ishikawa, Ryo Anma, Hiroaki Ozawa, Brian F. Windley, and J. G. Liou 511

Exhumation Tectonics of the Sanbagawa High-Pressure Metamorphic Belt, Southwest
Japan—Constraints from the Upper and Lower Boundary Faults
  —Hideki Masago, Kazuaki Okamoto, and Masaru Terabayashi 527

U-Pb Dating of Large Zircons in Low-Temperature Jadeitite from the Osayama
Serpentinite Mélange, Southwest Japan: Insights into the Timing of Serpentinization
—T. Tsujimori, J. G. Liou, J. Wooden, and T. Miyamoto 540

Western North America

Fast Cooling and Exhumation of the Valhalla Metamorphic Core Complex,
Southeastern British Columbia
—Frank S. Spear 553

Multi-stage Origin of the Coast Range Ophiolite, California: Implications for the
Life Cycle of Supra-subduction Zone Ophiolites
—J. W. Shervais, David L. Kimbrough, Paul Renne, Barry B. Hanan, Benita Murchey,
Cameron A. Snow, Marchell M. Zoglman Schuman, and Joe Beaman 570

Field and Isotopic Evidence for Fluid Mobility in the Franciscan Complex:
Forearc Paleohydrogeology to Depths of 30 Kilometers
—Seth J. Sadofsky and Gray E. Bebout 597

Aluminous Xenoliths in Miocene Andesite, Central California Coast Ranges:
Magma-Crust Interaction in a Subduction-Transform Transitional Setting
—Ellen P. Metzger, W. G. Ernst, and Dennis Sorg 633

Dominican Republic

UHP Magma Paragenesis, Garnet Peridotite, and Garnet Clinopyroxenite:
An Example from the Dominican Republic
—Richard N. Abbott, Jr., Grenville Draper, and Shantanu Keshav 653

Editors' Preface

A symposium entitled "Phase Relations, High-Pressure Terranes, P-T-ometry, and Plate Pushing" was held at the 2003 National Meeting of the Geological Society of America (GSA) as a tribute to W. G. Ernst. Over a three-day period, 109 presentations were made in four oral sessions and a poster session. Three special issues are planned for these papers to honor this scientific giant by providing a showcase of his diverse interests. This volume represents a collection of papers published in *International Geology Review*, most of which were presented at the GSA symposium.

Gary Ernst was born in St. Louis, Missouri and moved to St. Paul, Minnesota at the age of nine. He received a B.A. degree in geology from Carleton College in 1953 where he completed an honors thesis on the St. Peter Sandstone. He went to the University of Minnesota where he received an M.S. degree in 1955, working on the petrology of the Endion sill near Duluth, Minnesota. Both his B.A. honors thesis and M.S. thesis were published in *American Mineralogist* and the first issue of *Journal of Petrology*, respectively. These two studies, one focused on the petrology of sedimentary rocks and the other on igneous rocks, reflect the breath of his career-long interests, the most prominent of which center on his papers concerning metamorphism and tectonics of convergent plate margins. Ernst received a Ph.D. from Johns Hopkins University in 1959, during which time he began his long association with the Carnegie Institute of Washington. He joined the faculty at the University of California, Los Angeles in 1960 and remained there until 1989. Gary quickly rose to the rank of Professor and served two four-year stints as Chair of the Department. He moved to Stanford University to become Dean of the School of Earth Sciences, a position he held until 1994. Gary continues to teach despite having formally retired in September 2004.

Gary Ernst's research career has been prodigious. He has been a lead author on more than 120 scientific papers, co-author (with more than 85 colleagues!) on nearly as many more. He is author of six books and an editor of 14 volumes of papers. And the numbers keep growing. Gary's research approach involves an exceptionally broad range of geography, tools, and techniques. All of it has been firmly rooted in the rock record and field basis. Since the beginning of his career, he has undertaken geologic mapping to document the field relations of rock assemblages. For these and other places of interest, he collects samples to perform the geochemical characterization needed to address the scientific questions at hand. He has published papers and geologic maps based on his own field work in the California Coast Ranges, White-Inyo and Klamath mountains of California, European Alps, New Zealand, Russia, Kazakhstan, Taiwan, Japan, and China.

His analytical approach is equally diverse, as he has alone, and with co-authors, published studies centered on major- and trace-element analysis of bulk rocks by XRF and INAA methods as well as mineral composition and element partitioning relations from electron microprobe, micro-Raman, and Mössbauer analyses. With others, he has published on stable and radiogenic isotopic analysis and various forms of geochronology. For most of his career he maintained a program of experimental geochemistry using apparatus for mineral synthesis and for determining phase equilibria and element partitioning relationships at the high pressures and temperatures of the middle crust to upper mantle.

Several themes for Gary's research can be identified, each of which is the subject of a dozen or so papers. These include: (1) the mineralogy and petrology of amphiboles; (2) blueschist-facies subduction-zone metamorphism and plate tectonics; (3) the petrology of ultramafic and eclogitic rocks of the western Alps; (4) the metamorphic petrology and ophiolites of Japan and Taiwan; (5) the geology of the Coast Ranges, Klamaths, and White-Inyo Mountains of California; and (6) the petrology and tectonic significance of ultrahigh-pressure metamorphic rocks in the Maksyutov Complex of Ural Mountains, the Kokchetav Massif of northern Kazakhstan, and the Dabie-Sulu terrane of east-central China. His recent interests include geologic remote sensing and vegetation, and medical geology.

Gary's experimental work on sodic amphiboles led him to field work on blueschists in the Franciscan Complex of California. His lab and field studies were done with a theoretician's eye toward its megatectonic significance. In the early 1960s, a few others suspected that blueschist-facies metamorphic rocks were the product of unusually high-pressure/low-temperature conditions. Gary proved that blueschist belts mark sites of ancient and active subduction—a key piece of evidence supporting the theory of plate tectonics. This fundamental scientific advance, widely cited in introductory and petrology textbooks, was the primary basis for his election to the National Academy of Sciences in 1976.

Ophiolitic rocks are associated with blueschists in subduction zones. Ernst was thus drawn to work on these fragments of ocean crust and mantle in various parts of the Earth. From a study in the Italian and Swiss Alps in the mid-1970s, he concluded that the mineral chemistry of certain ultramafic and eclogitic rocks required their equilibration at pressures approaching 40 kbar. At the time, this was an astonishing conclusion, for it required that these rocks rose to the surface from depths of about 110 km—that is, from near the base of the lithosphere. It has not only been proven correct, but is now hailed as a precursor of the mid-1980s discovery by other workers of ultra-high-pressure rock terranes in the Alps, and later in China, Norway, Russia, and elsewhere.

Gary's interest in education is far greater than that of most world-class researchers. He is an inspiring classroom teacher. He regularly taught undergraduate and graduate student courses, as well as summer field geology courses at both UCLA and Stanford. He wrote graduate-level texts such as *Amphiboles* in 1968, and *Petrologic Phase Equilibria* in 1976. He has completed several undergraduate textbooks including *Earth Materials* in 1967, and *The Dynamic Planet* in 1990. In addition, Gary has edited many research and undergraduate texts including *Petrology and Plate Tectonics*, *Subduction-Zone Metamorphism*, *Earth Systems: Processes and Issues*, *Frontiers in Geochemistry*, and *Ultrahigh-Pressure Metamorphism*, to name a few. In 1997, a W. G. Ernst Scholarship was established in his honor at UCLA. In 1999, he was appointed as the first holder of the Benjamin M. Page Chair at Stanford University.

Gary's vision and remarkably broad knowledge have enabled him to help steer students and post-doctoral fellows to work on important research problems (14 M.S., 35 Ph.D., 31 postdoctoral fellows and about 20 visiting scientists from Japan, China, South Korea, Switzerland, Slovakia, India, and Pakistan). This is demonstrated by his success rate in producing first-class researchers and teachers. Many of his Ph.D. students, post-doctoral fellows, and research scientists hold academic positions at prestigious institutions around the world, and several others hold staff positions in major research organizations. One of his lesser-known attributes is that he never added his name as a co-author to a student publication unless he contributed substantially to the work. In this way, he is a part of many more geoscience discoveries than his vita will ever show.

Gary's service to our scientific communities has been extraordinary. For decades he made frequent cross-country trips to Washington, D.C. to serve on numerous panels and committees of the National Research Council/National Academy of Sciences. He has been a Trustee of the Carnegie Institution of Washington since 1990. He has served on Advisory Committees for more than 20 major U.S. universities. He has been an efficient editor for several scientific journals, the latest one being *International Geology Review*. At UCLA, he edited seven Rubey Volumes ranging from *Geotectonic Development of California* and *The Environment of the Deep Sea* through *Energy* to *Metamorphism and Tectonics*. At Stanford, he honored the contributions of several distinguished scientists through two-day symposia and a scientific volume respectively for Clarence Hall, Larry Taylor, Ben Page, Konrad Krauskopf, George Thompson, and Bob Coleman, produced as the Bellwether/GSA International Book Series. He was elected President of the Mineralogical Society of America in 1980, and the Geological Society of America in 1985. He was awarded the Mineralogical Society of America Award for 1969, was the first recipient of the Geological Society of Japan Medal in 1998, and received the Penrose Medal of the GSA for 2004.

W. G. Ernst is an extraordinary geoscientist who has made groundbreaking research contributions over the wide spectrum of mineralogy, metamorphic petrology, geochemistry, and tectonics. He has taught thousands of students the wonders of geology and converted some of them to our profession. He formally supervised the theses and dissertations of about 50 graduate students and nearly as many post-doctoral scholars. This volume includes papers from many of his former students, postdocs, and friends. Our science has moved forward because of Gary Ernst—via ideas and inspiration. And it has been made better because he is also one of the most compassionate persons we know.

—J. G. LIOU
Department of Geological and
Environmental Sciences,
Stanford University

—MARK CLOOS
Department of Geological Sciences,
University of Texas at Austin

GENERAL

Psychology of a Changing Paradigm: 40+ Years of High-Pressure Metamorphism

HARRY W. GREEN, II[1]

*Institute of Geophysics and Planetary Physics and Department of Earth Sciences,
University of California, Riverside, California 92521*

Abstract

The story of ultrahigh-pressure metamorphism (UHPM) is a confused mixture of surprising, sometimes spectacular, discoveries and emotional reactions. Surprisingly, the process has been a repeating cycle of disbelief followed by confirmation, with little evidence that the community response in a given cycle has learned from previous cycles. To this writer's understanding, it began in the early 1960s, before the plate-tectonic revolution, when W. Gary Ernst determined experimentally that the blue amphibole, glaucophane, is stable only at low temperatures and elevated pressures. The implication that glaucophane-bearing rocks, such as those of the Franciscan Complex of northern California, could have been rapidly carried to high pressure and returned to the surface equally rapidly was rejected as impossible. The advent of plate tectonics provided a conceptual answer to the conundrum, but resistance remained, and the concept of tectonic overpressure was invented to explain away the depths implied by experimental results. Despite both theoretical and practical demonstration that this concept is invalid, the concept continues to resurrected to this day. In the 1970s, the development of the technique of thermobarometry brought a welcome ability to infer the pressure/temperature from which kimberlite xenoliths had been brought to the surface. However, application of the technique to garnet peridotites in the field, again led by Ernst, brought resistance from those for whom depths exceeding 120 km did not fit readily into their previous interpretations. Subsequent discoveries of coesite in the 1980s and diamonds in the 1990s in continental-collision terranes met with similar disbelief initially, and microstructural evidence for exhumation from depths of hundreds of km in the same terranes experienced the same reception. Piece by piece, the general community has now largely accepted these discoveries, but a new claim of rocks perhaps from the lower mantle is experiencing renewal of the cycle. This paper presents some highlights of the discoveries during this period, and attempts to understand some of the human aspects of this slow but not-so-quiet revolution.

Introduction

THE PROCESS BY which science advances is a conservative one. That is, each step of the way involves documentation that the new information is reproducible and consistent with the existing body of knowledge. For most day-to-day operations of science, this approach is reasonable and successful; inaccurate observations and incorrect deductions are weeded out, and accurate observations and deductions are added to the fabric of knowledge. However, when new observations or arguments do not fit into the existing paradigm, there is a strong tendency for them to be rejected by the scientific community. The larger the shift of paradigm required to accommodate the new information, the more harsh and emotional is likely to be the rejection. This article provides a brief chronicle of the history of ultrahigh-pressure metamorphism. The shift in paradigm directly involved is not a particularly large one, but it is embedded in the much larger shift involved in the acceptance of plate tectonics by the Earth Science community; it shows that a segment of a scientific community (in this case metamorphic petrology) can be left behind during a scientific revolution because the underlying assumption upon which the old paradigm for that segment was based was not recognized as a holdover from the fixist views of the 19th century and the first two-thirds of the 20th century.

---

[1]Email: harry.green@ucr.edu

FIG. 1. Gary Ernst (right) in the field in the Franciscan complex of California circa 1962 with Robert Coleman (left) and Arden Albee. Photo by Y. Seki.

The Beginning: Franciscan Paradox

In the early 1960s, revolution was in the air in the Earth Sciences. Periodic reversal of Earth's magnetic field had been discovered but was not yet fully accepted (e.g., Opdyke and Runcorn, 1956). Nevertheless, application of paleomagnetism by the early leaders in the field was pointing (Collinson and Runcorn, 1960) to similar continental movements in the past, as had been envisioned by Wegener (1912, 1923) and others based on physiography of the globe and geological "fits" across ocean basins (c.f. Holmes, 1965). Harry Hess had advanced the theory of sea-floor spreading to explain ocean ridges (see review in Hess and Revelle, 1963), but the work of Vine and Matthews (1963), Tuzo Wilson (1963, 1965), and the full explosion of plate tectonics (Isaacs et al., 1968; Le Pichon, 1968; Morgan, 1968) were still to come.

A young assistant professor at UCLA, W. G. Ernst (Fig. 1), was conducting experiments on the stability relations of glaucophane, a sodium amphibole that is a relatively rare mineral but is prominent in blueschists of the Franciscan Complex of central and northern California. This formation was of considerable interest because of its chaotic internal structure, including randomly mixed "knockers" (with sizes from meters to hundreds of meters or more), of widely varying lithology and metamorphic grade in a very fine-grained matrix. During 1961–1963, the present author was employed as an undergraduate assistant, separating glaucophane and other minerals from the rocks with heavy liquids (ah, the headaches!!). Ernst found, to his consternation, that the stability of glaucophane was confined to low temperatures and, for that time, high pressures (Ernst, 1963). As a consequence, the straightforward explanation of the metamorphic history of glaucophane-bearing rocks was that: (1) somehow they were transported to sufficient depth to explain the pressures implied by his experiments (0.7–0.8 GPa); and (2) they then were returned to the surface sufficiently rapidly that they never warmed to the temperatures expected for such depths. Had they not been returned to the surface quickly, the glaucophane would have broken down to chlorite and other minerals and the evidence of their rapid descent would have been lost (Ernst, 1965). Thus, the simple experimental result that glaucophane stability is limited to low temperatures but elevated pressures implied not only uncomfortably high pressures but also uncomfortably high descent and ascent rates, all of which violated the then-current paradigm.

The reaction of the geological community to these laboratory results was predictable: Either the experimental results had to be incorrect or misinterpreted, or their application to Earth was flawed. The implied depths were unacceptable because surficial geology was viewed as being essentially separated from Earth's interior by isostasy—continental rocks are light and float on mantle rocks that are more dense. The descent and ascent rates were unacceptable because they seemed much too rapid to be consistent with geological processes. At that time, metamorphosed sedimentary rocks were thought to be subjected to elevated pressures by burial beneath more sediments and to be exhumed from depth by erosion; both of these processes were too slow to allow the temperatures to remain low enough for glaucophane to be grown stably and then preserved by rapid cooling. Demonstration that the conditions of experimentation could not be significantly in error was relatively easy to achieve. Nevertheless, the inferred geological implications of the results were unacceptable. It was realized that if the error lay in the pressure (depth) implication, then the burial and exhumation rates would automatically be resolved.

At the same time (late 1950s, early 1960s), another, seemingly unrelated, subject was of interest to geologists concerned with metamorphism. In the late 19th century, thermodynamic understanding of chemical equilibrium in heterogeneous systems had been put on a firm footing by J. Willard Gibbs (e.g. Gibbs, 1948). However, one of the loose ends of Gibbs's work was the question of what happens to thermodynamic equilibrium under conditions of nonhydrostatic stress. Gibbs himself showed that his concept of chemical potential was not a single-valued global function under nonhydrostatic stress, and this question had been revisited but not resolved by a number of scientists in both geology and physics during the first half of the 20th century. There was a flurry of activity in this subject in the geological literature of the 1950s and early 1960s, culminating in a confrontation in print between W. B. Kamb (1961a, 1961b) and G. J. F. MacDonald (1961) in which MacDonald's theory of nonhydrostatic thermodynamics was shown to be flawed and Kamb's vindicated. We shall return to this subject below; suffice it to say for now that interpretation of the thermodynamic state under nonhydrostatic stress in the early 1960s was not well understood.

The basic problem is that under hydrostatic (isotropic) stress, pressure is a scalar quantity and Gibbs's famous equation for the free energy of a chemical component can be expressed as:

$$\mu = U + PV - TS, \qquad (1)$$

where μ is the Gibbs (free) energy per mole or the chemical potential, U is the partial molar internal energy of the component, P is pressure, V is the partial molar volume of the component, T is the absolute temperature, and S is the partial molar entropy of the component. The critical question in the present context is "What happens to this equation when the scalar parameter P representing pressure (and requiring one number to describe it) becomes a nonhydrostatic stress and therefore a second-rank tensor (requiring six numbers to describe its magnitude and orientation)?" In the early 1960s, it was thought by many in the geological community that one of the possible answers to this question was that one could generalize pressure in the following way:

$$P = (\sigma_1 + \sigma_2 + \sigma_3)/3, \qquad (2)$$

where σ_1, σ_2, and σ_3 are the three principal stresses; hence the effective pressure is assumed to be the average of the three principal stresses.

It was well known that Earth is a dynamic body and that, in general, the pressure at depth during mountain building must be nonhydrostatic. It was therefore argued that if the horizontal stresses were sufficiently high during tectonic deformation, then the apparent pressure as indicated by the minerals present could be different from that calculated assuming:

$$P = \rho g h, \qquad (3)$$

where ρ is the density of the overlying rock, g is the acceleration of gravity, and h is the height of the overlying rock. Therefore, if P were given by (2) rather than by (3) as assumed by Ernst, and if the horizontal stresses were sufficiently high, the vertical stress could be low, and the depth of burial could have been much less than that implied by Ernst. Voilá: the experimental results could be correct but their application to Earth was invalid. The difference between the pressure one would calculate from (2) and that calculated from (3) was termed "tectonic overpressure" (see Brace et al., 1970 and references therein). Whew!! The paradigm was saved.

Undaunted, Ernst enlisted one of the leading rock mechanics experts of the day, W. F. Brace, and in Brace's laboratory at MIT they demonstrated that in order to explain away Ernst's implications by this argument, the strength of Franciscan rocks during prograde metamorphism would have had to be greater than the *room temperature crushing strength* of the rocks in the laboratory. In short, they showed that quantitatively the argument was grossly inadequate. Their diplomatically worded conclusion was that to explain the presence of blueschists (glaucophane-bearing rocks) by tectonic overpressure, "It would be necessary to have rapid, geologically unrealistic strain rates, for fluid to be absent [or almost absent], and to have temperatures insufficient for the facies to develop" (Brace et al., 1970). Despite this absolutely clear repudiation of the concept of tectonic overpressure on a practical basis, the concept has reappeared over and over again to this day as the evidence for greater and greater depths of equilibration of rocks has been assembled.

The reader should note the date of the Brace et al. (1970) paper. That paper appeared two years after the plate tectonics revolution burst forth undeniably onto the world stage (1968), carrying with it the easy answer to the dilemma presented to the metamorphic community by Ernst (1963). The easy answer is that Ernst was correct; plate tectonic theory easily explained the Franciscan Complex as sediments carried down ("subducted," in modern terminology) to previously unbelievable depths (20–50 km) along the interface between the Pacific and North American plates where they were metamorphosed at high pressures and low temperatures, and then returned back up that interface at comparable rates. In the process of exhumation they were chaotically deformed, mixing rocks of a wide variety of metamorphic grades, and plastered onto the western margin of North America.

Tectonic Overpressure: A Failed Concept

Advances in nonhydrostatic thermodynamics over the last 40 years have shown that under nonhydrostatic stress, stability of a phase is determined not by the total stress state but by the local normal stress across a specific interface. This has been verified both theoretically (e.g., Kamb, 1961a; Green, 1970, 1980, 1986) and experimentally (Vaughan et al., 1984; Green, 1986). The experimental work demonstrated that polycrystalline specimens of Mg_2GeO_4 olivine subjected to nonhydrostatic stress within the stability field of the high-pressure spinel polymorph undergo the phase transformation preferentially on interfaces of higher normal stress. In particular, under the special conditions in which the stress state straddled the phase boundary as determined under hydrostatic stress, those surfaces whose normal stress placed them in the spinel stability field experienced nucleation and growth of the spinel phase, whereas those surfaces whose normal stress placed them in the olivine field experienced recrystallization to new crystals of the olivine phase (Fig. 2). Thus, the very idea of tectonic overpressure is faulty because there is no such thing as global equilibrium under nonhydrostatic stress. As a consequence, stable growth of glaucophane (or any other high-pressure mineral) can occur only on interfaces whose normal stress reaches the glaucophane stability field, and any relaxation of that normal stress would cause the glaucophane to be consumed unless the temperature was reduced quickly to stop reaction. Coupled with current knowledge that under metamorphic conditions rocks cannot support large stresses except at very high strain rates (essentially the conclusion reached by Brace et al., 1970), it follows that under natural conditions nonhydrostatic stress can perturb growth conditions of minerals only marginally. This conclusion remains valid despite the various attempts that have been made in the more recent past to revive the concept of tectonic overpressure. No amount of theoretical calculations can make rocks in a metamorphic environment stronger, nor can they remove the experimental observations of Figure 2.

Thermobarometry

In the early 1970s, a revolutionary paper by F. R. (Joe) Boyd (1973) showed that the partitioning of elements between pyroxenes of garnet-bearing peridotite in high-pressure experiments is significantly pressure dependent. As a consequence, comparison of the chemical composition of peridotite xenolith minerals from diamond-bearing kimberlite pipes with phases synthesized in experiments conducted under known conditions can enable calculation of the pressure and temperature from which each xenolith was plucked. Analysis of a suite of xenoliths from the same pipe allows determination of a mineralogical cross section of the mantle and the associated thermal profile at the time of eruption. The result showed that kimberlite magmas and many of their xenoliths come from depths greater than approximately 200 km (e.g., Green and Gueguen, 1974), fully consistent with the presence of diamonds (that would be stable only at pressures in excess of 4 GPa = depths exceeding approximately 120 km at mantle temperatures).

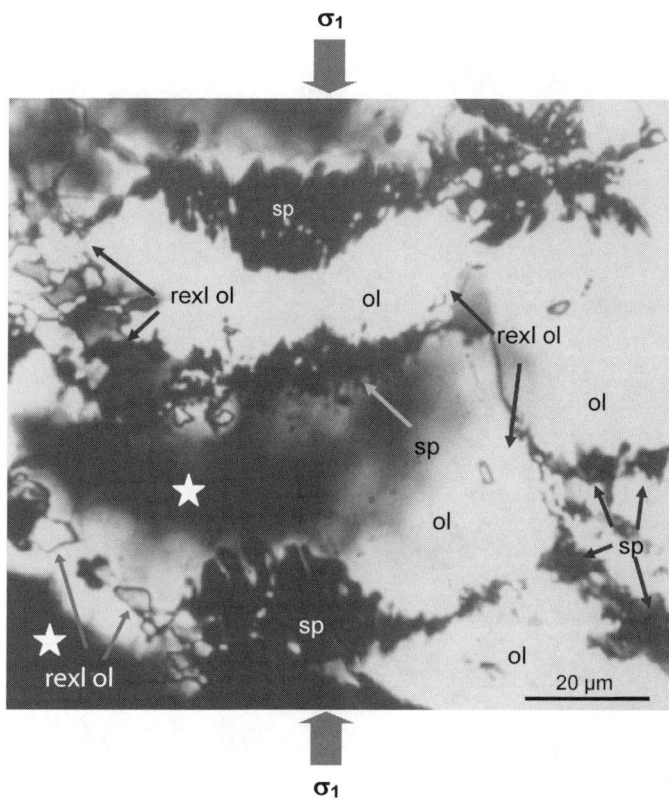

FIG. 2. Optical micrograph of a very thin section (0–2 μm thick) of a polycrystal of Mg_2GeO_4 olivine (ol) undergoing transformation to the high-pressure spinel polymorph under nonhydrostatic stress. Conditions of the experiment were such that the normal stress on grain boundaries approximately perpendicular to the maximum principal stress, σ_1 (N-S in photo), was in the spinel stability field and the normal stress on grain boundaries approximately perpendicular to the minimum principal stress, σ_3 (E-W in photo), was in the olivine stability field. Note that the first set of grain boundaries are populated with newly grown spinel crystals (sp) that in many cases can be seen to be growing fastest parallel to σ_1, whereas the second set of grain boundaries show recrystallization of olivine (rexl ol). Stars denote holes in the thin section produced during ion milling. Modified after Figure 4 of Vaughan et al. (1984).

Once again, the pioneer in applications to field geology was Ernst, who applied this "pyroxene geotherm" technique to the Alpe Arami garnet peridotite of the Swiss Alps, with the astounding result that it had last equilibrated at conditions of ~4.0 GPa (Ernst, 1978), implying that this alpine peridotite in a continental collision environment had somehow been incorporated into the continental crust and brought to the surface without losing its "memory" of depths approximating that for diamond stability. Once again, conventional wisdom found these results unacceptable; the conclusion was that this new technique must still contain flaws; detractors were confident that subsequent refinement of thermobarometry would surely yield lower pressures for these rocks. In fact, the contrary happened; successive reanalyses of this massif yielded ~5.0 GPa (Medaris and Carswell, 1990) and ~5.9 GPa (Paquin and Alther, 2001a). Only the results of Nimis and Trommsdorff (2001a, 2001b) yielded lower pressures than determined by Ernst (1978), and the results of that study have been criticized for selection of data that effectively eliminated the high-pressure signal (e.g., Paquin and Alther, 2001b). We will return below to discussion of the depth from which the Alpe Arami massif has been transported to the surface but, in the interest of following approximately a time line, we will first turn to another startling discovery, nearby in the Italian Alps at Dora Maira and shortly thereafter in Norway.

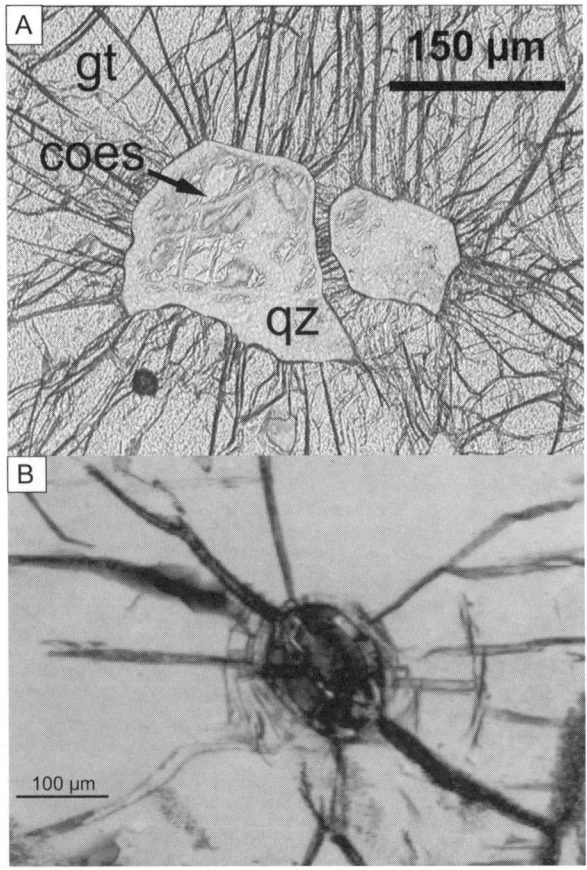

FIG. 3. Photomicrographs of thin sections showing relict coesite enclosed in garnet from Dora Maira, Italy (A) and omphacite from the Western Gneiss Region, Norway (B). Both images also show retrograde quartz and radial fractures in the host mineral generated by expansion as a consequence of partial transformation to quartz during exhumation of the UPHM rocks. Image A courtesy of C. Chopin; image B courtesy of D. Smith.

Coesite in Continental Rocks?

Natural occurrences of coesite, a high-pressure polymorph of SiO_2, were thought to be restricted to meteorite impact craters (Chao et al., 1960), where they were created by transient high pressures during the shock of impact, and to eclogite xenoliths plucked from environments of static high pressure >3 GPa in the mantle during kimberlite eruptions (Smyth and Hatton, 1977). That image changed suddenly with the discovery of coesite in metamorphosed rocks of continental affinity in the Dora Maira massif (Chopin, 1984), followed very shortly thereafter by similar discovery in eclogite of the Western Gneiss Region of Norway (Smith, 1984). The discovery at Dora Maira (Fig. 3A) was well documented and, for once in this saga, acceptance of the obvious implication of high pressures came reasonably quickly. In Norway, however, others were unable to confirm Smith's discovery (Fig. 3B) and, for reasons baffling to this author, the clear implications of high pressures presented by the unambiguous images in Smith (1984) and subsequent papers by the same author were essentially ignored for 13 years until Alice Wain (1997) found an extensive outcrop area of eclogite displaying coesite inclusions in both garnet and pyroxene, by which time many other pieces of evidence of very high pressures had been found, including diamond farther north in these same gneisses (next section). In the meantime, coesite was first described in the Dabie/SuLu region of Eastern China (Wang et al., 1989) and has since been shown to be rare but ubiquitous in those rocks. Most importantly, in both Norway (Wain, 1997) and China (Ye et al., 2000), coesite has been documented from the quartzofeldspathic country gneisses as well as the mafic eclogites, demonstrating that the coesite is recording bulk metamorphism of the terrane, not simply incorporation of coesite-bearing eclogites into lower-pressure continental material. In the classical UHPM terrane of northern Kazakhstan, coesite was even found exsolving from titanite in marble (Ogasawara et al, 2002). Thus, in this century the ultrahigh-pressure metamorphism problem becomes not "How did these high-pressure rocks get incorporated into the crust?," but rather, "How did continental rocks and sediments get carried to depths greater than 100 km and returned to the surface?"

Diamonds in Rocks of Continental Affinity

Returning to the timeline of discovery, once again the world of metamorphic petrology was startled by documentation for the first time in the Western literature of microdiamonds (Figs. 4A and 4B) in metasediments from the Kokchetav massif, northern Kazakhstan (Sobolev and Shatsky, 1990). In this case, the diamonds were so abundant that, despite their small size, their host rocks became ores for exploitation for abrasives. Like Chopin's coesite, there was no denying the existence of the Kazakhstan diamonds, establishing a clear benchmark that rocks of continental composition had been subducted to more than ~120 km and returned to the surface sufficiently rapidly that the diamonds were not totally consumed by back-reaction to graphite.

Other discoveries of diamonds in continental rocks soon followed in China (Xu et al., 1992) and Norway (Fig. 4C; Dobrzhinetskaya et al., 1995), but neither was generally believed; they both were attributed to contamination, not because there was reason to suspect such a problem, but because it is easy to assert and very difficult to disprove. Indeed, in the case of Norwegian diamonds, the infrared spectra clearly showed them to be fundamentally different from kimberlitic or synthetic diamonds, the only reasonable potential source of contamination, but to have similarities to the metamorphic diamonds of Kokchetav. Both occurrences have since been confirmed in thin section (van Roermund et al., 2002; Yang et al., 2003) (Figs. 4D and 4E). Other diamond localities now known or suspected include Sulawasi (Parkinson and Katayama, 1999), the Urals (Bostick et al., 2003) and the Greek Rhodope (Mposkos and Kostopoiulos, 2001). Most spectacularly, a Kokchetav look-alike has been found in the Erzgebirge of Germany. Of course, it was summarily rejected when first presented at the 1999 Kimberlite Conference (Massonne, 1999), but quickly demonstrated to be real (Fig. 4F; Nasdala and Massonne, 2000).

Onward and Downward

The Alps

By the early 1990s, it might seem that the theme of disbelief of indicators of high pressure followed by new discoveries indicating even higher pressures should have begun to settle down. Basically, blueschists were by that time fully accepted as markers of fossil subduction zones worldwide, coesite was accepted in the Dora Maira, and metamorphic diamonds were accepted in northern Kazakhstan. However, complete rejection of coesite and diamond continued unabated in Norway and the discovery of coesite and diamond in China were viewed with great skepticism. Dora Maira and Kokchetav were viewed as isolated curiosities, not the keys to a major new discovery that extensive subduction of continental material can occur during continental collision, followed by rapid return to the surface, essentially up the same subduction zone. Thus, the possibility of subduction to 100–120 km, as indicated by coesite and diamond, was generally accepted, but heavy resistance continued in specific locations where established researchers were reluctant to abandon previously developed interpretations.

Stimulated by a presentation of Stephen Haggerty (University of Massachusetts) during a session of the Spring 1994 meeting of the American Geophysical Union in Baltimore, MD, the present author and his colleagues began an investigation of specimens of the Alpe Arami peridotite (collected in 1973 but never examined in detail) that displayed garnet/pyroxene microstructures similar to those Dr. Haggerty had suggested might indicate very great depth of origin. We already knew that, of the two generations of olivine present in the rocks, the older generation displayed an extremely rare (at that time thought to be unique) lattice preferred orientation (LPO) defined by a maximum of [100] normal to the foliation (Möckel, 1969; Buiskool Toxopeus, 1976, 1977). This LPO had no explanation at that time,

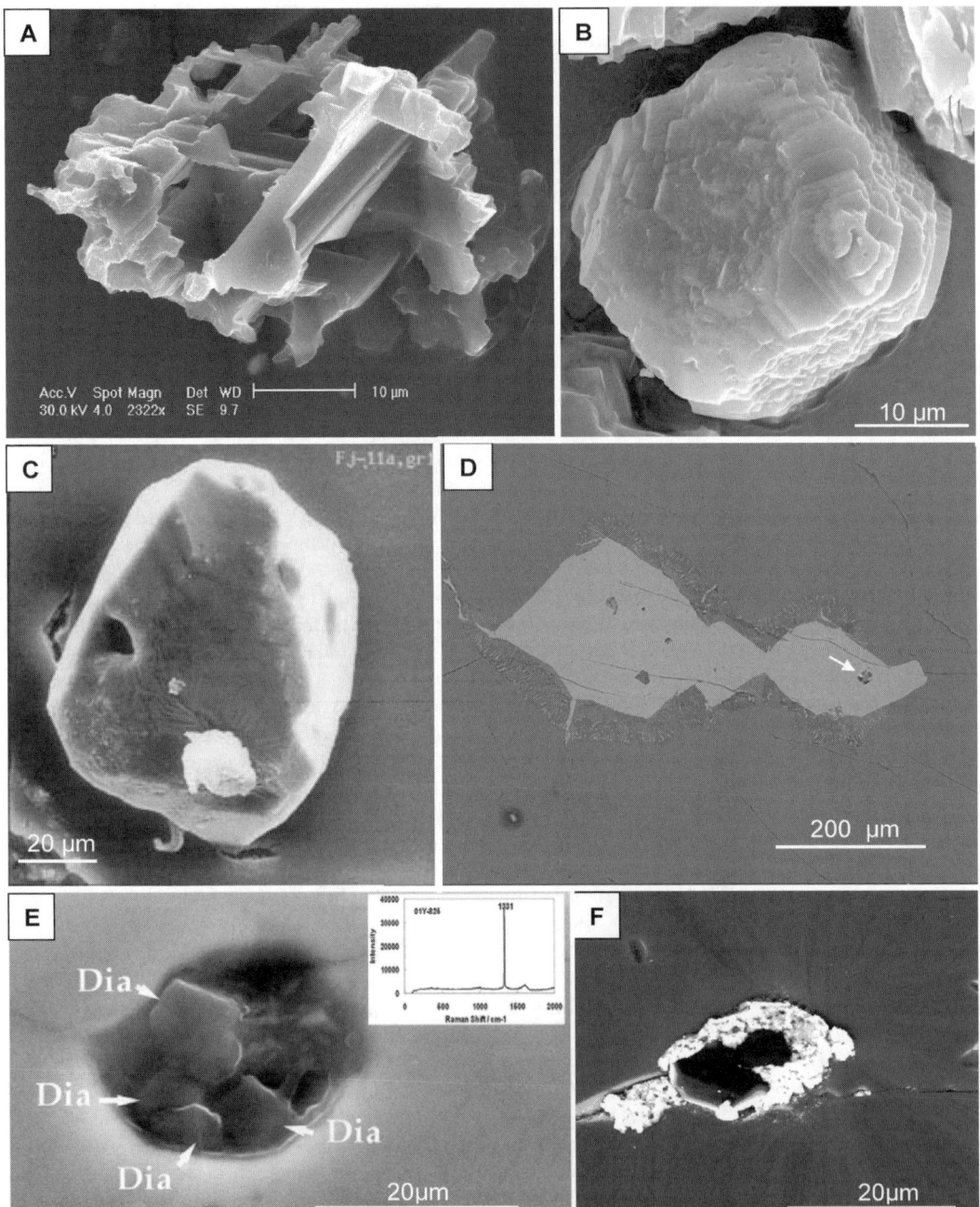

FIG. 4. Scanning electron micrographs of microdiamonds and their microstructural settings. A. Skeletal diamond composed of thin {111} plates making up a "house of cards" structure from metasediments; Kokchetav massif, Kazakhstan (see Dobrzhinetskaya et al., 2001). B. Cuboid diamond also composed of octahedral plates, Kokchetav massif, Kazakhstan. C. Microdiamond from metasediments, Fjortoft island, Norway (see Dobrzhinetskaya et al., 1995). D. Spinel within garnet containing a microdiamond-bearing polyphase inclusion (arrow) from peridotite, Fjortoft Island, Norway (see van Roermund et al., 2003). E. Crystal pocket containing diamond and other minerals from Qinling, China; insert shows Raman peak characteristic of diamond (see Yang et al., 2003). F. Microdiamond from Seidenbach Reservoir, Erzgebirge, Germany. The diamond stands out from the polished surface; white material around diamond is metal dug out of the polishing wheel by the diamond. Radial streaks around diamond are left by polishing process (see Massonne, 1999; Nasdala and Massonne, 2000). Images A, B, C, and F are courtesy of L. F. Dobrzhinetskaya; image D is courtesy of H. van Roermund; image E is courtesy of J. Yang.

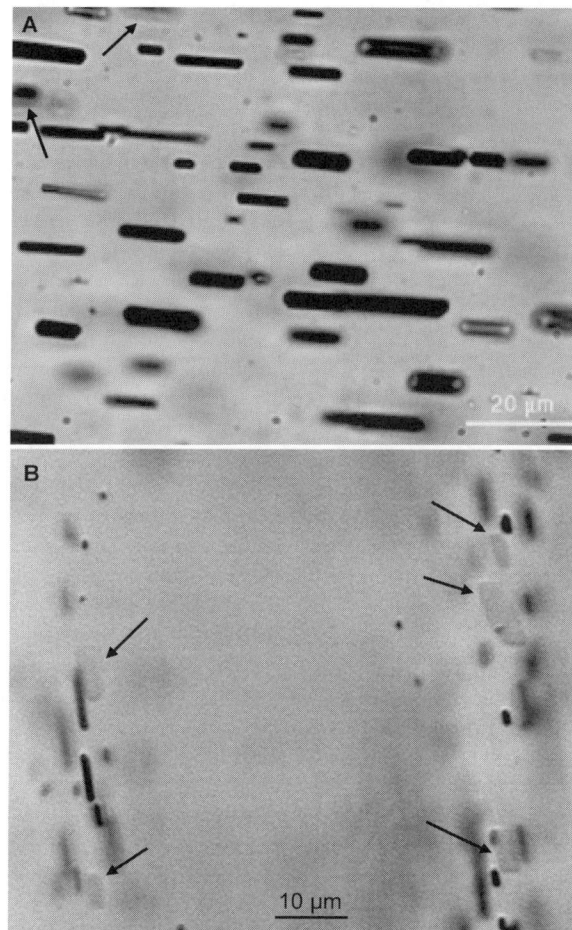

FIG. 5. Ilmenite and chromite precipitates in first-generation Alpe Arami olivine. A. Precipitates are commonly spatially random; ilmenite rods are oriented parallel to [010] of olivine; chromite platelets (difficult to see in this image) are oriented parallel to (100) of olivine. Two chromites that are in focus and attached to ilmenite rods are indicated by arrows (see Dobrzhinetskaya et al., 1996; Green et al., 1997). B. Precipitates are in a few cases concentrated on (001) planes, but exhibit the same topotaxy as in A. In this image, virtually all ilmenite rods have a chromite platelet attached; arrows point out spinels without ilmenite (see Bozhilov et al., 2003). Figure 5A is modified after Figure 1 of Green et al. (1997); Figure 5B is modified after Figure 3C of Bozhilov et al. (2003).

but had been inferred to have developed at great depth (den Tex, 1971). To our surprise, we discovered that this generation of olivine also contained oriented inclusions of ilmenite, up to >1% by volume in abundance (Fig. 5A). Our investigations determined that the ilmenite is topotactically related to the olivine, strongly suggesting an origin by exsolution from the olivine and therefore implying a TiO_2 content of the olivine before exsolution of >0.6 wt. %. Further examination determined that the same olivine grains also contain topotactically oriented chromite inclusions (Fig. 5B) with an estimated volume ratio to ilmenite of about 1:4 or 1:5. The observations can be quantitatively explained by the following chemical reaction:

$$Fe_5(Cr,Fe,Al)_2\Delta Ti_4O_{16} \rightarrow 4FeTiO_3 + Fe(Cr,Fe,Al)_2O_4,$$
defect olivine $\rightarrow$ ilmenite + chromite,

where Δ denotes an octahedral vacancy. The second generation of olivine was normal in every way; it contained no inclusions, displayed a normal LPO

([010] normal to foliation), and was associated with interstitial ilmenite. It clearly formed by recrystallization of the earlier, ilmenite-bearing, generation under conditions of low ilmenite solubility during exhumation of the massif.

The strong indication that the ilmenite + chromite originated by exsolution created a considerable conceptual problem because olivine, although accepting Fe substitution for Mg in any proportion, accommodates virtually no other elements into solution except very small amounts of Mn and Ca. In particular, it essentially does not accept trivalent and tetravalent cations (e.g. Cr and Ti) into its structure. There are no published analyses of olivine with more than ~500 ppm Ti, with the highest concentrations from high-pressure mantle xenoliths (Hervig et al., 1986). We concluded that the most plausible conditions of origin were at depths greater than represented by the mantle xenolith suite, meaning depths of ~300 km or more (Dobrzhinetskaya et al., 1996).

A firestorm of scientific criticism greeted publication of that *Science* paper (and its discussion worldwide in the popular press). A new circle of disbelief concerning UHPM was beginning. First, the experimental implications of glaucophane in blueschists were rejected because of the implications of depth and rates of burial and exhumation, implications explained by formulation of plate tectonics theory. Then coesite and diamond had been hard to accept because of their much greater depth implications, but once the presence of these minerals was demonstrated to be correct, only local pockets of entrenched resistance remained (discounting the arguments of tectonic overpressure which continue to be resurrected to this day). However, now a major jump in the depth from which rocks could be exhumed was being proposed, not based on a high-pressure polymorph of known minimum depth implications, but based on chemical evidence without pre-existing experimental demonstration that the hypothesis was at least possible.

Realizing that such skepticism would be a reasonable response of the community, especially given the history of UHPM denial, even before publication of this paper we had secured funding to test whether this radical suggestion could be correct. We initiated a high-pressure research program to investigate whether there are any conditions under which olivine can dissolve > 0.6 wt% TiO_2 and, if so, to map out those conditions. We determined (to our own surprise) that there are conditions where the solubility of TiO_2 in olivine is sufficient to explain the observations in Alpe Arami, but only at pressures of 9–12 GPa, depending on temperature (Dobrzhinetskaya et al., 2000). The explanation appears to be that the solubility of TiO_2 in olivine is significantly enhanced when the pressure becomes sufficiently high to stabilize Ti in the tetrahedral site of the olivine crystal structure. However, as had occurred earlier concerning Smith's discovery of coesite in Norway, the criticism had taken on a life of its own and demonstration that the intepretation was viable seemingly had no effect at all on the scientific community. In response, we reinvestigated several aspects of the study and showed that our original interpretation of the amount of TiO_2 previously in solution was correct and that the high-pressure interpretation is the only one consistent with all of the observations (Bozhilov et al., 2003). In particular, we showed that an alternative interpretation that the ilmenite could have arisen from dehydration of titanian clinohumite (Risold et al., 2001) is incompatible with the presence of co-precipitated ilmenite + chromite.

Moreover, as the firestorm over Alpe Arami olivine was expanding, we were already finding that diopside inclusions in garnet of these same rocks contain exsolution lamellae of clinoenstatite (technically pigeonite, because of small Ca content) displaying antiphase domains that indicate the originally precipitated pyroxene had the C2/c space group (Fig. 6). There are two C2/c clinopyroxene phases that could fulfill this observation: high-temperature clinoenstatite and high-pressure clinoenstatite. From the totality of our observations, we concluded that the C2/c precursor had to be the high-pressure variety, providing an independent constraint of greater than ~ 8 GPa (~250 km) for origin of the Alpe Arami peridotite (Bozhilov et al., 1999). The alternative interpretation, that the C2/c precursor could have been high-temperature clinoenstatite metastably exsolved in the orthopyroxene stability field (Arlt et al., 2000), is inadequate because: (1) it would require temperatures at least 200°C higher than the Alpe Arami exhumation curve derived by others (Brenker and Brey, 1997); (2) it is incompatible with the crystallographic orientation of the lamellae within diopside; (3) existing phase diagrams show that low-pressure exsolution of clinoenstatite from diopside is inconsistent with the very high Mg content of the lamellae. Finally, recent discovery (Liu et al., 2005) of essentially identical C2/c clinoenstatite/pigeonite exsolution lamellae

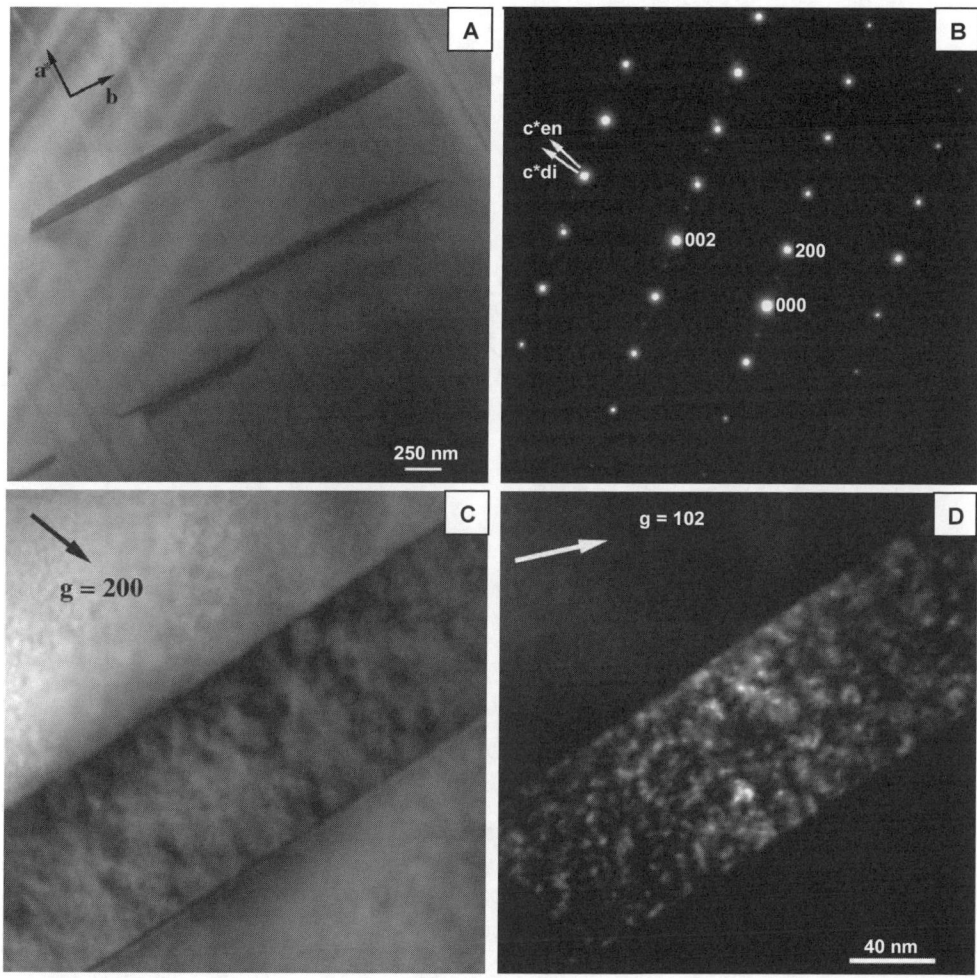

FIG. 6. Exsolution of high-pressure clinoenstatite/pigeonite from diopside. A. Low-magnification bright-field image of lamellae. B. Selected-area diffraction pattern showing spots of both diopside and (low) clinoenstatite. C. Dark-field image of a single lamella showing strain mottling due to inversion from the denser C2/c form. D. Dark-field image of same lamella, same area, under different diffraction conditions, showing antiphase domains (bright) recording the former presence of the C2/c form.

from diopside in the Dabie UHPM belt of eastern China, a terrane with a much colder exhumation curve than Alpe Arami, confirms that such exsolution lamellae in garnet peridotite represent high-pressure clinoenstatite/pigeonite (in addition to establishing a higher than thus-far-accepted depth of exhumation for that eastern China UHPM belt). Lastly, the conditions necessary to account for the TiO_2-in-olivine observations are incompatible with the conditions necessary to explain away the high-pressure C2/c pyroxene, and vice-versa, yet both can be found repeatedly in the same thin section.

In summary, exsolution of oxides in olivine and of high-pressure pyroxenes in diopside, coupled with experimental determination of the P/T dependence of solubility of TiO_2 in olivine and the thermobarometry of others, demonstrate that the Alpe Arami garnet peridotite carries memory of a previous equilibration at a depth in excess of ~300 km (Fig. 7). The body was transported to the surface during the later stages of the Alpine collision between Europe and Africa, presumably buoyantly uplifted by the quartzofeldspathic Lepontine gneisses in which it is found (but which so far have

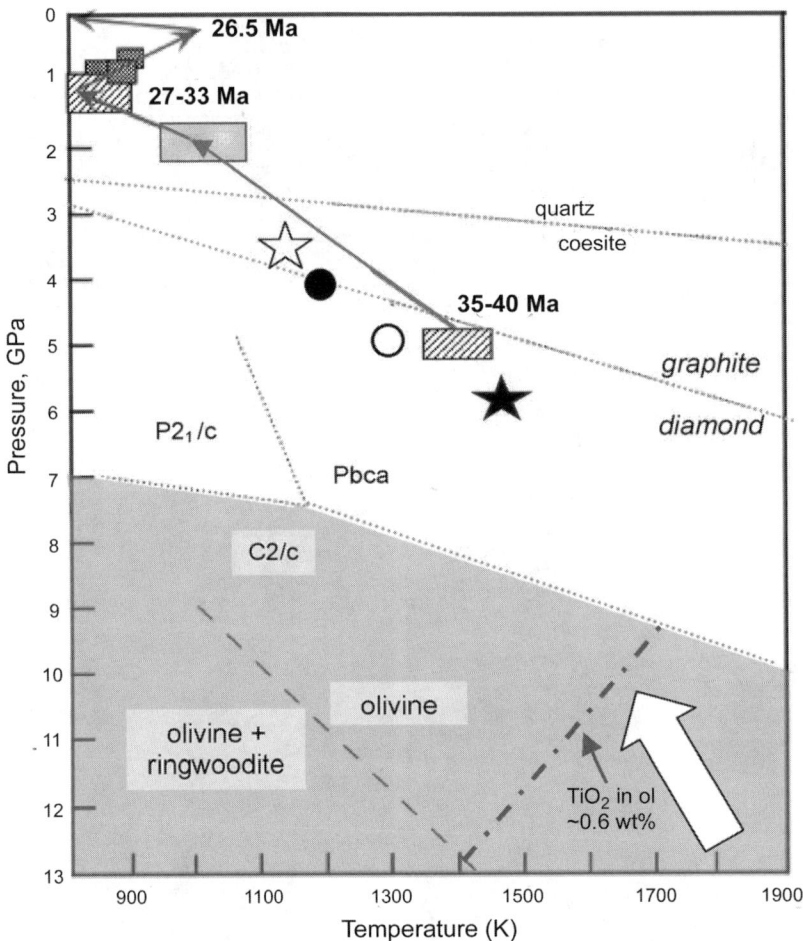

FIG. 7. Exhumation path for the Alpe Arami peridotite. Five thermobarometric studies have been published on the Alpe Arami peridotite. The maximum pressure/temperature points calculated by Ernst (1978), Medaris and Carswell (1990), Paquin and Alther (2001a), and Nimis and Trommsdorff (2001a) are shown by a solid circle, open circle, solid star, and open star, respectively. A more ambitious study by Brenker and Brey (1997) derived several points along the exhumation path by careful examination of microstructural relationships and associated chemical inhomogeneities (hatchured boxes connected by arrows). Three points along this path are labeled with dates estimated by these authors to correspond with points they derived. In addition, we plot (1) the quartz/coesite and graphite/diamond phase boundaries as well as the equilibrium boundaries between the three principal phases of $Mg_{0.9}Fe_{0.1}SiO_3$ (i.e., low clinoenstatite, space group $P2_1/c$; orthoenstatite, space group Pbca; high-pressure clinoenstatite/pigeonite, space group C2/c); (2) the reaction boundary of olivine to olivine + ringwoodite (dashed line); (3) the approximate set of conditions under which olivine can dissolve a minimum of 0.6 wt% TiO_2 (dot-dashed line). The combination of the observations on olivine and diopside described here require that the exhumation path for the Alpe Arami peridotite passes through the shaded region to the right of the dot-dashed line (block arrow).

not yielded direct evidence of very high pressures). Whether this was a single-stage exhumation directly from > 300 km along the Alpine subduction zone or whether a more complicated history is involved is still not known.

The Caledonides

While the controversy concerning Alpe Arami was in full bloom, another remarkable discovery was made in the Western Gneiss Region of Norway. The

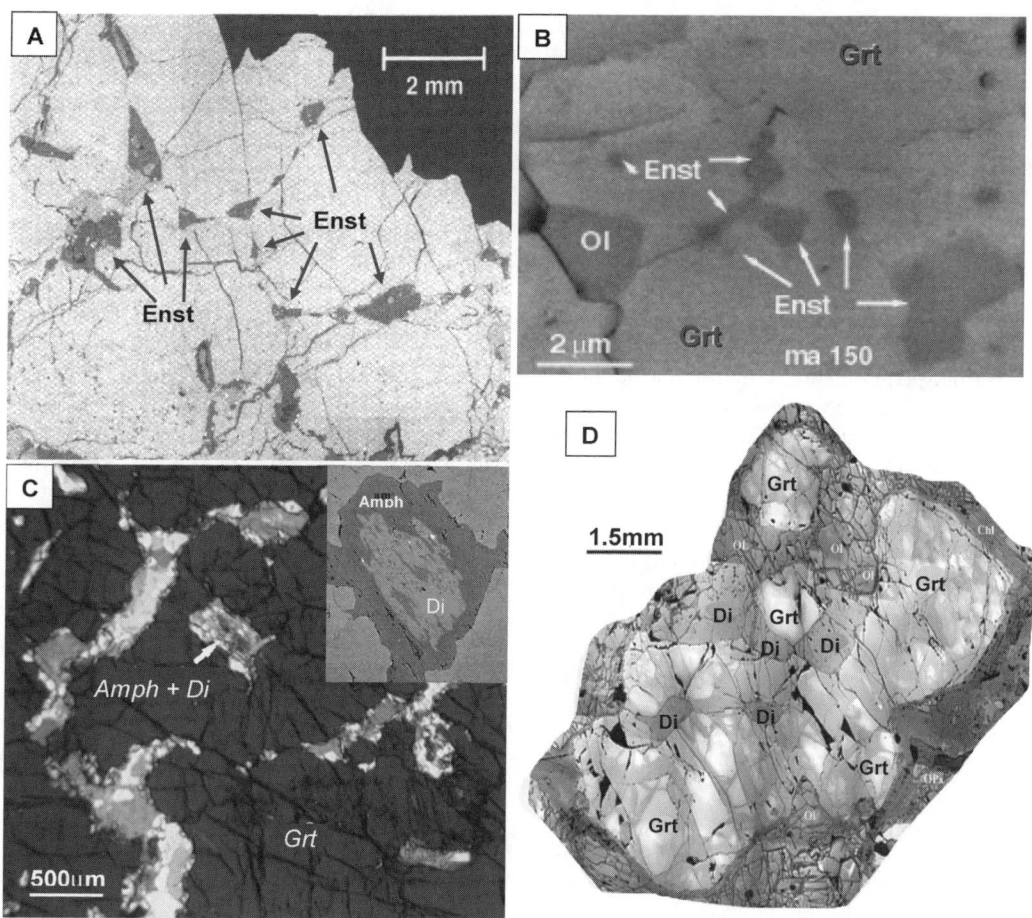

FIG. 8. Precipitation of pyroxenes in garnet. A. Grain-boundary precipitates of enstatite (enst) in garnet nodule of peridotite, Western Gneiss Region, Norway. Such nodules also display typical exsolution lamellae of enstatite in the centers of garnets greater than 4 mm diameter (see van Roermund and Drury, 1998). B. Similar precipitates produced in an experiment first equilibrated at 8 GPa, in which all enstatite dissolved and then held at the same temperature and 5 GPa to exsolve enstatite (Dobrzhinetskaya et al., 2004). Note difference of 10^3 between scale bars in A and B. C. Optical micrograph (crossed polarizers) of sodic diopside (now mostly altered to amphibole) along grain boundaries in garnet nodule of Alpe Arami peridotite, Switzerland. Insert is SEM image of a diopside relic in amphibole showing exsolution lamellae of quartz and an earlier amphibole of different composition (Green and coworkers, unpubl. results). D. Diopside along grain boundaries in garnet nodule of Sulu peridotite, China (Dobrzhinetskaya and Green, unpubl. results).

garnet harzburgites abundant in that terrane in many places contain large (>10 mm) purple garnets. Microscopic examination of such garnets showed them to be polycrystalline. Individual garnets with diameters >4 mm were found to contain in their cores abundant exsolution lamellae of enstatite, with minor exsolution of diopside; the rims of these grains were clear of pyroxene, as were the entirety of smaller garnets, but all of the triple junctions between garnets (and many grain boundaries) con-

tained enstatite crystals (Fig. 8A). This classic exsolution microstructure showed that these garnets had contained a significant majoritic component, implying depths in excess of 200 km (van Roermund and Drury, 1998). This microstructure had been reported previously in xenoliths from kimberlite (Haggerty and Sautter, 1990; Sautter et al., 1991) and is consistent with the experimentally well-documented progressive dissolution of pyroxenes into garnet with increasing pressure (e.g., Ringwood,

1991). We have tested this concept in the laboratory by first dissolving significant enstatite into garnet at 8 GPa and then re-equilibrating the specimen at 5 GPa, precipitating enstatite along grain boundaries (Fig. 8B; see also Dobrzhinetskaya et al., 2004). Other polycrystalline garnets in peridotites worldwide show inclusions of diopside along grain boundaries, suggesting much higher pressures, because diopside goes into garnet in significant amounts only at higher pressures than enstatite. Examples from Alpe Arami, Switzerland and the SuLu belt of eastern China are given in Figures 8C and 8D. We have also verified experimentally that majoritic garnet grown at 14 GPa exsolves diopside at grain boundaries at pressures of 10–12 GPa (not shown). Future work is necessary, however, to demonstrate that the experimental studies are fully consistent with observations on natural peridotites.

Van Roermund and Drury (1998) attributed the exsolution of pyroxene in garnet in Norway to a pre-Cambrian diapiric event unrelated to Caledonian subduction, with subsequent storage in the upper mantle until the peridotites were exhumed along with the Caledonian UHPM rocks. The previous widely accepted experimental work and presence of the microstructure in xenoliths, coupled with an interpretation placing the exsolution event prior to the Caledonian orogeny, made acceptance of this discovery almost instantaneous, despite the only recently terminated denial of coesite and continuing denial of diamond in this terrane. Similar microstructures involving interstitial diopsides in polycrystalline garnets from other terranes, including Alpe Arami and eastern China, are suggestive of even greater depths, but sorting out the meaning of those microstructures remains for future work.

More recently (van Roermund et al, 2002), microdiamonds have been discovered within similar peridotites on the same island (Fjortoft) where the original Norwegian diamond discovery had been made in Caledonian metasediments (Dobrzhinetskaya et al., 1995). Hence that controversy has also been put to rest.

Eastern China

Thus far, only brief mention has been made of the Chinese UHPM belt of the Dabie Mountains and the Sulu terrane. Although less prominent in the controversies in the literature, this belt has steadily increased in importance for several reasons. Firstly, it is by far the largest UHPM terrane; it is now known to be comprised of at least two UHPM events separated by ~200 m.y. in age and extending more than 4000 km across China in an approximately E-W orientation (Yang et al., 2003), with coesite and diamond occurrences sprinkled along its length. The Chinese belt has another very important feature. Through systematic extraction and analysis of zircons from the continental country rocks in which the eclogites and peridotites are embedded, it has been abundantly and unequivocally demonstrated that these rocks all contain coesite (Ye et al., 2000). Thus, it is clear that the mafic and ultramafic rocks have not been exhumed from great depth and tectonically emplaced within the continental gneisses in which they are found, but that the continental gneisses were also subducted to and exhumed from great depth. Only within zircons have these rocks retained their "memory" of the voyage, because the bulk rocks are less refractory and have efficiently back-reacted to lower-pressure assemblages during the return trip. The nonreactive zircons, with their broad stability range, have served as almost perfect environmental capsules, and withstood the exhumation without breaking down or allowing their contents to be back-reacted. As mentioned above, discovery of coesite in the country gneisses of Norway (Wain, 1997) demonstrated similar bulk subduction and exhumation. In both cases, recognition that the country gneisses were also deeply subducted then provides a ready explanation for the UHPM mafic and ultramafic rocks within them. They were simply emplaced together after exhumation. The strikingly similar deformation microstructures and migmatization common in the quartzofeldspathic gneisses of other UHPM terranes strongly suggests that these gneisses, too, have experienced the two-way street of subduction during continental collision, despite the current lack of positive confirmation within them. As has become the pattern, this argument is controversial in every place where it has not been proven (e.g., in the western Alps), yet no alternative reasonable explanation of how the UHPM rocks in these terranes could have been transported to the surface and emplaced in their present positions has been offered.

The Chinese UHPM belt also has some other specific characteristics that make it particularly interesting. First, oxygen isotope ratios are extremely light (Rumble et al., 2002), implying arctic subduction or perhaps a manifestation of the extraordinary glacial conditions in the late Precam-

FIG. 9. As if clairvoyant, W. G. Ernst flashes the victory sign (or perhaps signifying peace with his critics) ca. 1978. Shown also is B. Clark Burchfiel. Photo by G. Davis.

brian (interpreted as "snowball earth" by some authors). Second, extraordinary amounts of magnetite are found in olivine of some localities, seemingly in exsolution relationship to the olivine. It has been suggested (Zhang et al., 1999) that this microstructure could imply extensive solid solution between magnetite and the post-olivine phases of the mantle transition zone (wadsleyite with a spineloid structure or ringwoodite with a true spinel structure), implying the greatest depth of exhumation yet proposed. The remarkably fresh nature of many of the eclogites and peridotites (e.g., Fig. 8D) and the 5 km drill hole in progress in Sulu (Xu et al., 2004) allows the prospect of more exciting discoveries to come.

Lastly, chromite deposits of the Luobosa ophiolite of Tibet contain diamonds and moissanite (SiC), and have been argued to contain minerals from the mantle transition zone, the lower mantle, and perhaps even the core-mantle boundary (Bai et al., 2004). This deposit appears to be igneous, and therefore different from UHP metamorphism. However the pattern of denial so consistently applied to UHPM discoveries has fallen very hard on these descriptions. Perhaps rightly so, but stay tuned....

Conclusions and Psychoanalysis

I conclude that the saga of ultrahigh-pressure metamorphism is probably in its young adulthood or, perhaps, still only at the end of its childhood. It has been a relatively slow revolution, with its principal punctuations being Ernst (1963), Ernst (1978), Chopin (1984) and Smith (1984), Sobolev and Shatsky (1990), and Dobrzhinetskaya et al. (1996). In each case, a powerful new set of observations was added to the fabric of metamorphic petrology: The first two were due to initial applications of new experimental advances, the third and fourth were field-based (but with experimental backing required for the stability range of the high pressure polymorphs discovered), and the fifth, based on microstructures, required subsequent experimental verification to be viable.

Was the slow and grudging acceptance of each forward step in this new field of metamorphic petrology a drama of human resistance to a changing paradigm, or is this just a good example of science working in the conservative way it must? I see strong examples of both in the sometimes intellectual, sometimes emotional, debates that have ensued for

40 years. It will be great fun to watch over the next 10 years as new gems are added to the crown of UHPM.

Lastly, I come to the subject of this volume. I have tried to show that here we are commemorating more than a strikingly successful scientific career (with more to come, we hope), but also the birth and childhood of a new scientific discipline that Gary helped to initiate. It is quite astonishing that the person who began it all has also made major individual contributions at each step along the way (Fig. 9). Congratulations, Gary!!!—but I think my headaches are coming back.

Acknowledgments

I thank Gary Ernst for many years of inspiration and many of those cited here for their perseverance in the face of emotional and financial discouragement. I also thank Y. Seki for the image in Figure 1, Claude Chopin for Figure 3A; David Smith for Figure 3B; Larissa Dobrzhinetskaya for Figures 4A, 4B, 4C, 4F, and 8D; Herman van Roermund for Figures 4D and 8A; and Jingsui Yang for Figure 4E. I also thank Larissa Dobrzhinetskaya and Junfeng Zhang for helpful reviews. My own work discussed in this paper was supported by various grants from the U.S. National Science Foundation.

REFERENCES

Arlt, Th., Kunz, M., Stolz, J., Armbruster, Th., and Angel, R. J., 2000, P-T-X data on P21/c-clinopyroxenes and their displacive phase transitions: Contributions to Mineralogy and Petrology, v. 138, p. 35–45.

Bai, W., Robinson, P., Fang, Q.-S., Malpas, J., Yang, J.-S., Zhou, M.-F., Hu, X.-F., 2004, Silicon spinel—an ultrahigh pressure mineral recovered from podiform chromitites of the Luobusa ophiolite, Tibet: 32nd International Geological Congress, Florence, Italy, Abs. Vol., pt. 1, p. 720 (abs. #153-6).

Bostick, B. C., Jones, R. E., Ernst, W. G., Chen, C., Leech, M. L., and Beane, R. J., 2003, Low-temperature microdiamond aggregates in the Maksyutov Metamorphic Complex, South Ural Mountains, Russia: American Mineralogist, v. 88, p. 1709–1717.

Boyd, F. R., 1973, A pyroxene geotherm: Geochimica et Cosmochimica Acta, v. 37, p. 2533–2546.

Bozhilov, K. N., Dobrzhinetskaya, L. F., and Green, II, H. W., 2003, Quantitative 3D measurement of ilmenite abundance in Alpe Arami olivine: Confirmation of high pressure origin: American Mineralogist, v. 88, p. 596–603.

Bozhilov, K. N., Green, H. W., II, and Dobrzhinetskaya, L., 1999, Clinoenstatite in Alpe Arami peridotite: Additional evidence for very high pressure: Science, v. 284, p. 128–132.

Brace, W. F., Ernst, W. G., and Kallberg, R. W., 1970, An experimental study of tectonic overpressure in Franciscan rocks: Geological Society of America Bulletin, v. 81, p. 1325–1338.

Brenker, F. E., and Brey, G. P., 1997, Reconstruction of the exhumation path of the Alpe Arami garnet-peridotite body from depths exceeding 160 km: Journal of Metamorphic Geology, v. 15, p. 581–592.

Buiskool Toxopeus, J. M. A., 1976, Petrofabrics, microstructures, and dislocation substructures of olivine in peridotite mylonite (Alpe Arami, Switzerland): Leidse Geol. Med., v. 51, p. 1–36.

Buiskool Toxopeus, J. M. A., 1977, Fabric development of olivine in a peridotite mylonite: Tectonophysics, 1977, v. 39, p. 55–71.

Chao, E. C.-T., Shoemaker, E. M. J., and Madsen, B. M., 1960, First natural occurrence of coesite [Arizona]," Science, v. 132, p. 220–222.

Chopin, C., 1984, Coesite and pure pyrope in high-grade blueschists of the western Alps: A first record and some consequences: Contributions to Mineralogy and Petrology, v. 86, p. 107–118.

Collinson, D. W., and Runcorn, S. K., 1960, Polar wandering and continental drift—evidence from paelomagnetic observations in the United States: Geological Society of America Bulletin, v. 71, p. 915–958.

den Tex, E., 1971, Age, origin, and emplacement of some alpine peridotites in the light of recent petrofabric researches: Fortschrift Mineralogie, v. 48, p. 69–74.

Dobrzhinetskaya, L. F., Bozhilov, K. N., and Green, H. W., II, 2000, The solubility of TiO_2 in olivine: Implications for the mantle wedge environment: Chemical Geology, v. 163, p. 325–338.

Dobrzhinetskaya, L. F., Eide, E. A., Larsen, R. B., Sturt, B. A., Tronnes, R. G., Smith, D. C., Taylor, W. R., and Posukhova, T. V., 1995, Microdiamond in high-grade metamorphic rocks of the Western Gneiss Region, Norway: Geology v. 23, p. 597–600.

Dobrzhinetskaya, L.F., Green, H. W., II, Mitchell, T., and Dickerson, R. M., 2001, Metamorphic diamonds: Mechanism of growth and inclusion of oxides: Geology, v. 29, p. 263–266.

Dobrzhinetskaya, L. G., Green, H. W., II, Renfro, A. P., Bozhilov, K. N., Spengler, D., and van Roermund, H. L. M., 2004, Precipitation of pyroxenes and olivine from majoritic garnet: Simulation of peridotite exhumation from great depth: Terra Nova, v. 16, p. 325–330.

Dobrzhinetskaya, L. F., Green, H. W., II, and Wang, S., 1996, Alpe Arami: A peridotite massif from depths of more than 300 kilometers: Science, v. 271, p. 1841–1845.

Ernst, W. G., 1963, Petrogenesis of glaucophane schists: Journal of Petrology, v. 4, p. 1–30.

Ernst, W. G., 1965, Mineral paragenesis in Franciscan metamorphic rocks, Panoche Pass, California: Geological Society of America Bulletin, v. 76, p. 879–914.

Ernst, W. G., 1978, Petrochemical study of lherzolitic rocks from the Western Alps: Journal of Petrology, v. 19, p. 341–392.

Gibbs, J. W., 1948, On the equilibrium of heterogeneous substances, in Collected works of J. Willard Gibbs: New Haven, CT, Yale University Press, p. 55–371.

Green, H. W., II, 1970, Diffusional flow in polycrystalline materials: Journal of Applied Physics, v. 41, p. 3899–3902.

Green, H. W., II, 1980, On the thermodynamics of nonhydrostatically stressed solids: Philosophical Magazine, v. A41, p. 637–647.

Green, H. W., II, 1986 Phase transformation under stress and volume transfer creep, in Hobbs, B. E., and Heard, H. C., eds., Mineral and rock deformation: Laboratory studies—the Paterson volume: Washington, DC, American Geophysical Union, Geophysical Monograph, v. 36, p. 201–211.

Green, H. W., II, Dobrzhinetskaya, L., and Bozhilov, K., 1997, Determining the origin of ultra-high pressure lherzolites (response): Science, v. 278, p. 704–707.

Green, H. W., II, and Gueguen, Y., 1974 Origin of kimberlite pipes by diapiric upwelling in the upper mantle: Nature, v. 249, p. 617–620.

Haggerty, S. E., and Sautter, V., 1990, Ultra-deep (>300 km) ultramafic, upper mantle xenoliths: Science v. 248, p. 993–996.

Hervig, R. L., Smith, J. V., and Dawson, J. B., 1986, Lherzolite xenoliths in kimberlites and basalts: Petrogenetic and crystallochemical significance of some minor and trace elements in olivine, pyroxenes, garnets, and spinel: Royal Society of Edinburgh Transactions, Earth Sciences, v. 77, p. 181–201.

Hess, H. H., and Revelle, R., 1963, The origins of the continents, oceans, and atmosphere, in The scientific endeavor—centennial celebration of the National Academy of Sciences: New York, NY, Rockefeller Institute Press.

Holmes, A., 1965, Principles of physical geology, rev. ed.: New York, NY, Ronald Press.

Isaacs, B., Oliver, J., and Sykes, L. R., 1968, Seismology and the new global tectonics: Journal of Geophysical Research, v. 73, p. 5855–5899.

Kamb, W. B., 1961a, The thermodynamic theory of nonhydrostatically stressed solids: Journal of Geophysical Research, v. 66, p. 259–271.

Kamb, W. B., 1961b, Author's reply to discussions of the paper "The thermodynamic theory of nonhydrostatically stressed solids," Journal of Geophysical Research, v. 66, p. 3985–3988.

Le Pichon, X., 1968, Sea-floor spreading and continental drift: Journal of Geophysical Research, v. 73, p. 3661–3697.

Liu, X.-W., Jin, Z.-M., Green, H. W., II, and Qu, J., 2005, Clinoenstatite exsolution in diopside of garnet lherzolite from Bixiling massif of Dabie mountains and its geological significance (submitted).

MacDonald, G. J. F., 1961, Discussion of paper by W. Barclay Kamb, "The thermodynamic theory of nonhydrostaticallly stressed solids," Journal of Geophysical Research, v. 66, p. 2599.

Massonne, H.-J., 1999, A new occurrence of microdiamonds in quartzofeldspathic rocks of the Saxonian Erzgebirge, Germany, and their metamorphic evolution, in Nixon, P. H., ed., Proceedings of VIIth International Kimberlite Conferrence–1998. Cape Town, South Africa, p. 533–539.

Medaris, L. G., and Carswell, D. A., 1990, The petrogenesis of Mg-Cr garnet peridotites in European metamorphic belts, in Carswell, D. A., ed., Eclogite facies rocks: Glasgow-London, UK, Blackie, p. 260–290.

Möckel, J. R., 1969, Structural petrology of the garnet peridotite of Alpe Arami (Ticino, Switzerland). Leidse Geol. Med., v. 42, p. 61–130.

Morgan, W. J., 1968, Rises, trenches, great faults and crustal blocks: Journal of Geophysical Research, v. 73, p. 1959–1982.

Mposkos, E. D., and Kostopoulos, D. K., 2001, Diamond, former coesite, and supersilicic garnet in metasedimentary rocks from the Greek Rhodope: A new ultrahigh-pressure metamorphic province established: Earth and Planetary Science Letters, v. 192, p. 497–506.

Nasdala, L., and Massonne, H.-J., 2000, Microdiamonds from the Saxonian Erzgebirge, Germany: In situ micro-Raman characterisation: European Journal of Mineralogy, v. 12, p. 495–498.

Nimis, P., and Trommsdorff, V., 2001a, Revised thermobarometry of Alpe Arami and other garnet peridotites from the Central Alps: Journal of Petrology, v. 42, p. 103–115.

Nimis, P., and Trommsdorff, V., 2001b, Comment on "New constraints on the P-T evolution of the Alpe Arami garnet peridotite body (Central Alps, Switzerland)" by Paquin & Altherr (2001): Journal of Petrology, v. 42, p. 1773–1779.

Ogasawara, Y., Fukasawa, K., and Maruyama, S., 2002, Coesite exsolution from supersilicic titanite in UHP marble from the Kokchetav Masssif, northern Kazakhstan: American Mineralogist, v. 87, p. 454–4361.

Opdyke, N. D., and Runcorn, S. K., 1956, New evidence for reversal of the geomagnetic field near the Pliocene–Pleistocene boundary: Science, v. 123, p. 1126–1127.

Paquin, J., and Altherr, R., 2001a, New constraints on the P-T evolution of the Alpe Arami garnet peridotite body

(Central Alps, Switzerland): Journal of Petrology, v. 42, p. 1119–1140.

Paquin, J., and Altherr, R., 2001b, "New constraints on the P-T evolution of the Alpe Arami garnet peridotite body (Central Alps, Switzerland)": Reply to comment by Nimis & Trommsdorff (2001): Journal of Petrology, v. 42, p. 1781–1787.

Parkinson, Ch., and Katayama, I., 1999, Metamorphic diamond and coesite from Sulawesi, Indonesia: Evidence of deep subduction at the SE Sundaland margin: EOS (Transactions of the American Geophysical Union), v. 80, no. 46, p. 1181.

Ringwood, A. E., 1991, Phase transformations and their bearing on the constitution and dynamics of the mantle: Geochimica et Cosmochimica Acta, v. 55, p. 2083–2110.

Risold, A.-C., Trommsdorff, V., and Grobéty, B., 2001, Genesis of ilmenite rods and palisades along humite-type defects in olivine from Alpe Arami: Contributions to Mineralogy and Petrology, v. 140, p. 619–628.

Rumble, D., Giorgis, D., Ireland, T., Zhang, Z. M., Xu, H. F., Yui, T. F., Yang, J. S., Xu, Z. Q., and Liou, J. G., 2002, Low delta O-18 zircons, U-Pb dating, and the age of the Qinglongshan oxygen and hydrogen isotope anomaly near Donghai in Jiangsu Province, China: Geochimica et Cosmochimica Acta, v. 66, p. 2299–2306.

Sautter, V., Haggerty, S. E., and Field S., 1991, Ultra-deep (>300km) ultramafic xenolith: New petrologic evidence from the transition zone: Science, v. 252, p. 827–830.

Smith, D. C., 1984, Coesite in clinopyroxene in the Caledonides and its implications for geodynamics: Nature, v. 310, p. 641–644.

Smyth, J. R., and Hatton, C. J., 1977, A coesite-sanidine grospydite from the Roberts-Victor kimberlite: Earth and Planetary Science Letters, v. 34, p. 284–290.

Sobolev, N., and Shatsky, V., 1990, Diamond inclusions in garnets from metamorphic rocks: A new environment of diamond formation: Nature, v. 343, p. 742–746.

van Roermund, H. M., and Drury, M. R., 1998, Ultra-high pressure (P>6GPa) garnet peridotites in Western Norway: Exhumation of mantle rocks from >185 km depth: Terra Nova, v. 10, p. 295–301.

van Roermund, H. L. M., Carswell, D. A., Drury, M. R., and Heijboer, T. C., 2002, Microdiamonds in megacrystic garnet websterite pod from Bardane on the island of Fjortoft, western Norway: Evidence for diamond formation in mantle rocks during deep continental subduction. Geology, v. 30, p. 959–962.

Vaughan, P. J., Green, H. W., II, and Coe, R. S., 1984, Anisotropic growth in the olivine-spinel transformation of Mg_2GeO_4 under nonhydrostatic stress: Tectonophysics, v. 108, p. 299–322.

Vine, F. J., and Matthews, D. H., 1963, Magnetic anomalies over oceanic ridges: Nature, v. 199, p. 947–949.

Wain, A., 1997, New evidence for coesite in eclogite and gneisses: Defining an ultrahigh-pressure province in the Western Gneiss region of Norway: Geology, v. 25, p. 927–930.

Wang, X., Liou, J. G., and Mao, H. K., 1989, Coesite-bearing eclogites from the Dabie Mountains in central China: Geology, v. 17, p. 1085–1088.

Wegener, A., 1912, Die Entestehung der kontinente: Geologische Rundschau, v. 3, p. 276–292.

Wegener, A., 1923, The origin of continents and oceans (translated from the third German edition by J. G. A. Skerl). New York, NY, E.P. Dutton & Co.

Wilson, J. T., 1963, A possible origin of the Hawaiian Islands: Canadian Journal of Physics, v. 41, p. 863–870.

Wilson, J. T., 1965, A new class of faults and their bearing on continental drift: Nature, v. 207, p. 343–347.

Xu, S., Okay, A. L., Sengor, A., Su, W., Liu, Y., and Jiang, L., 1992, Diamond from the Dabie Shan metamorphic rocks and its implication for tectonic setting: Science, v. 256, p. 80–82.

Xu, Z., 2004, 32nd International Geological Congress, Florence, Italy, Abs. Vol., pt. 1, p. 723 (abs. #153-17).

Yang, J., Xu, Z., Dobrzhinetskaya, L. F., Green, H. W., II, Pei, X., Shi, R., Wu, C., Wooden, J. L., Zhang, J., Wan, Y., and Li, H., 2003, Discovery of metamorphic diamonds in central China: an indication of a >4000-km-long zone of deep subduction resulting from multiple continental collisions: Terra Nova [doi: 10.1046/j.1365-3121.2003.00511.x].

Ye, K., Yao, Y. P., Katayama, I., Cong, B. L., Wang, Q. C., and Maruyama, S., 2000, Large areal extent of ultrahigh-pressure (UHP) metamorphism in the Sulu ultrahigh-pressure terrane of East China: New implications from coesite and omphacite inclusions in zircon of granitic gneiss: Lithos, v. 52, p. 157–164.

Zhang R. Y., Shu, J. F., Mao, H. K., and Liou, J. G., 1999, Magnetite lamellae in olivine and clinohumite from Dabie UHP ultramafic rocks, central China: American Mineralogist, v. 84, p. 564–569.

Blueschists and Blue Amphiboles: How Much Subduction Do They Need?

WALTER V. MARESCH[1]

Institut für Geologie, Mineralogie, und Geophysik, Ruhr-Universität Bochum, 44780 Bochum, Germany

AND TARAS V. GERYA

Geologisches Institut, ETH - Zürich, CH-8092 Zürich, Switzerland; Institut für Geologie, Mineralogie, und Geophysik, Ruhr-Universität Bochum, 44780 Bochum, Germany; and Institute of Experimental Mineralogy, Russian Academy of Sciences, Chernogolovka, Moscow District, 142432, Russia

Abstract

We focus on blueschist formation in evolving, nascent intra-oceanic subduction zones where cessation of subduction, e.g., by collision with continental margins, preserves the first-formed blueschist products. The Caribbean area provides a case study for a tectonic regime that has led to obducted blueschist occurrences without obvious coeval volcanism, raising the question of the minimum subduction duration and amount of convergence necessary to produce them. Systematic numerical modeling shows that for slab ages of 20–100 Ma and subduction rates of 2–14 cm/yr blueschist-facies conditions can be attained in only 0.25 to 3 m.y., with amounts of total convergence ranging between 35 and 75 km. Because of the geometrical interplay between the subducting slab "nose" and the evolving array of isotherms, younger and hotter slabs can lead to earlier blueschist formation than older, cooler slabs. There is a distinct optimum for slabs between 40 and 60 m.y. In principle, Andean-type, continental margin models yield almost identical values. Depending on the specific exhumation scenario, the time required for exhumation must in general be added to the minimum life span of the subduction zone.

Introduction

IT IS A PLEASURE and privilege to contribute to a compilation of papers dedicated to honoring the impact of W. Gary Ernst's career on the Earth sciences. Gary has accompanied, steered, and decidedly influenced studies on the subduction process ever since the "early days" about 40 years ago, when his studies in metamorphic and experimental petrology (e.g., Ernst, 1963) emphasized the necessity for high dP/dT geotherms during the orogenic process. Such ideas were not necessarily popular in all quarters, and it is a further credit to Ernst's influence and perseverance that he steadily and patiently collected, presented, and honed the scientific arguments that contributed to a convincing subduction model. Since these early days, subduction has come to be viewed as a "factory" in a twofold sense of the word. On the one hand, "factory" describes in a fitting way the interconnected processes that produce crust (i.e., metamorphic and magmatic rocks: Eiler, 2004; Hacker et al., 2003; Tatsumi and Kogiso, 2003) from raw materials such as sediments, oceanic crust, and mantle lithosphere, and, on the other hand, the voluminous literature on the subject now being churned out at an unprecedented rate on this key element of System Earth's processes.

In this contribution we propose to take modern, state-of-the-art methodology and return to the first harbingers of high-pressure, subduction-related metamorphism. Given that blueschists and the relatively aluminous and magnesian blue amphiboles they contain are classical indicators for the sites of fossil subduction zones, and should be the first evidence to develop in a maturing subduction zone, we propose to use numerical methods to quantify how much time and how much subduction are actually needed to produce the first blueschists after the birth of a subduction zone.

In the present study we focus on intra-oceanic subduction zones for two reasons. As summarized by Leat and Larter (2003), a significant 40% of modern subduction zones on Earth are intra-oceanic. These

---

[1]Corresponding author; email: walter.maresch@rub.de.

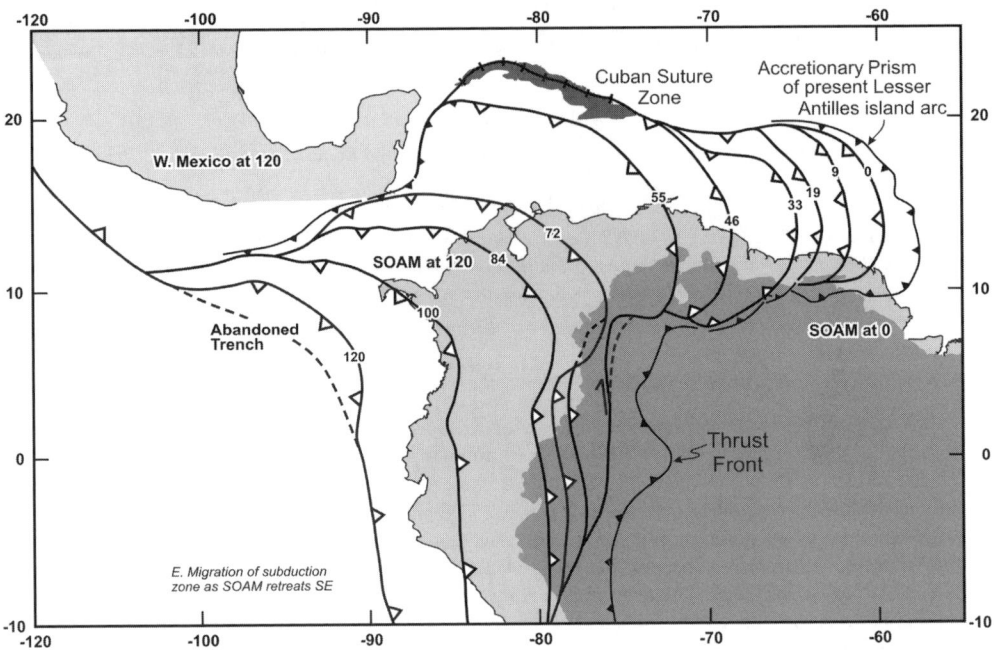

FIG. 1. Relative eastward advance of the Caribbean plate from the Pacific with respect to North America (proxied by western Mexico in this representation) and South America (SOAM) with time (all ages are in Ma), as derived from the detailed plate tectonic model of Pindell and Kennan (2001). The bow of the plate is the location of an intra-oceanic subduction zone system called the "Great Arc" by Burke (1988). The widening gap between North and South America is indicated by the relative positions of South America at 120 Ma and 0 Ma. Complex interaction between the oceanic Great Arc and North as well as South America has led to obduction and suturing of blueschists and other high-pressure rock sequences to the northern and southern continental margins of the Caribbean (Pindell et al., 2005).

subduction systems are generally simpler than those at continental margins, and therefore easier to understand (e.g., Leat and Larter, 2003). They also commonly have a shorter history of subduction, and they tend to be more mobile. On the other hand, these modern intra-oceanic arcs (most of which are found along the Southwest Pacific rim) have received much less attention, because of their largely inaccessible character in remote parts of the world or well below the surface of the sea. It is therefore of great interest to study older examples that have collided with and were obducted onto continental margins (e.g., Clift et al., 2003), with concomitant shutdown of the "subduction factory" and emphasis on the preservation of prograde features. The present study draws on both of these aspects. The numerical approach employed uses a conceptual model patterned after intra-oceanic subduction zones to address questions that have arisen from studies of arc systems that have swept through the Caribbean area during the last 120 m.y.

The Caribbean Example: Definition of the Problem

In many respects, the history of the Caribbean area suggests strong similarities with the systems of island arcs now characterizing the Southwest Pacific, or with the Tethys realm before closure (e.g., Maresch, 1974; Burke, 1988; Meschede and Frisch, 1998; Kerr et al., 1999; Pindell and Kennan, 2001; Pindell et al., 2005, among many others). Island arcs in such settings are probably very mobile, constantly evolving and individually may be short lived. They need not reach the dynamic equilibrium we usually expect in textbook examples. In keeping with the scenario of rapidly evolving systems of intra-oceanic island arcs interacting with each other and the continental margins of North and South America, there are diverging ideas on the actual details of Caribbean plate tectonic models.

The summary and synthesis of Pindell et al. (2005) suggest that the model outlined in Figure 1 fits the presently available data on the geological history of the circum-Caribbean margins well. This model serves as a template for Figure 1, which schematically shows the inferred progress of a Pacific plate moving relatively eastward between North and South America as these rift apart, overriding and subducting first newly formed Protocaribbean and then Atlantic oceanic lithosphere. Burke (1988) coined the term "Great Arc" to describe the system of intra-oceanic arcs marking the front of the eastwardly progressing Pacific lobe, whose present eastward position is indicated by the accretionary prism of the Lesser Antilles intra-oceanic arc. Because the "Great Arc" must lengthen and change shape during its progress, subduction zones and related island arcs must be continuously evolving. It is also postulated that at ~120 Ma the east-dipping subduction zone bridging the gap between North and South America must have reversed direction to allow eastward migration (Fig. 1). Progressive, very oblique eastward collision along the northern and southern margins of the Caribbean led to obduction of the high-pressure metamorphic rocks now found from Guatemala and Cuba in the NW Caribbean via Jamaica, the Dominican Republic, and Venezuela to Colombia in northwestern South America (see detailed summary by Pindell et al., 2005). One of the largest examples is the blueschist-dominated Villa de Cura complex in northern Venezuela (e.g., Smith et al., 1999; Pindell et al., 2005), an obducted block 250 km by 30 km in exposed areal extent.

Clearly, it is of general interest to gauge the amount of time that elapsed between the birth of a particular subduction zone segment and its shut-down by obduction and suturing. Because of the highly oblique angle of collision along the southern margin, obduction was generally followed by strike-slip shearing with almost no post-collisional volcanism. Rapid exhumation of high-pressure rocks is also postulated to have occurred by arc-parallel extension prior to obduction (Avé Lallemant and Guth, 1990; Avé Lallemant, 1997). In the north, a southward jump of the collision zone has decoupled and preserved the suture zone well, as for instance in Cuba. Obduction in such an arc-continent collision can be expected to take place in less than 2–3 Ma (e.g., Clift et al., 2003). No further subduction was available to exhume the high-pressure rocks, as required in more mature situations with return flow in so-called subduction channels (e.g., Hsu, 1971; Cloos, 1982; Cloos and Shreve, 1988a, 1988b; Shreve and Cloos, 1986). The current plate tectonic model of Pindell and Kennan (2001) also incorporates ideas on early back-arc spreading in northwestern South America (Fig. 2), before relative eastward movement of the Pacific plate closed this spreading "wedge." Such models (e.g., Pindell and Erikson, 1994; Pindell et al., 2005) attempt to relate isolated occurrences of blueschist in the northern Andes well inland of the Pacific to closure of such a backarc basin. This raises another interesting question. How wide must a backarc basin be to produce blueschists after closure and subduction of such young oceanic crust?

Numerical Modeling: Design and Implementation

Initial and boundary conditions of the 2-D model

We use a newly designed regional 2-D model that takes into account the process of hydration of the mantle wedge by the fluid released from a kinematically described subducting plate (e.g., Gerya et al., 2002, 2004; Gerya and Yuen, 2003a). Figure 3 shows the initial (panel a) and boundary (panel b) conditions and the hydration model used (panel c). We assume that dehydration of the subducting slab liberates an upward migration of aqueous fluid, resulting in the hydration of the mantle wedge near the slab (see panel c). This hydration leads to a sharp decrease in density and viscosity of the mantle rocks, creating favorable conditions for the development of a weak hydrated zone along the plate interface (Gerya et al., 2002; Gerya and Yuen, 2003a).

We use a simplified layered structure of subducting oceanic crust of 8 km thickness (Fig. 3A), composed of sedimentary (1 km), basaltic (2 km), and gabbroic (5 km) layers characterized by different physical properties (Table 1). The position of the incipient subduction zone (Fig. 3A) is imposed using a weak, 8 km thick layer composed of hydrated peridotite (e.g., Regenauer-Lieb et al., 2001). During modeling of the subduction process, this weak zone is spontaneously replaced by weak, subducted crustal rocks and hydrated mantle, implying a decoupling along the plate interface. The initial temperature field in both the subducting and overriding plates (Fig. 3A) is defined by an oceanic geotherm $T_0(z)$ based on a cooling half-space model with a specified age.

The kinematic boundary conditions (Fig. 3B) correspond to the corner flow model (e.g., Gerya and Yuen, 2003a) and simulate asthenospheric mantle flow at temperatures exceeding 1000°C in the

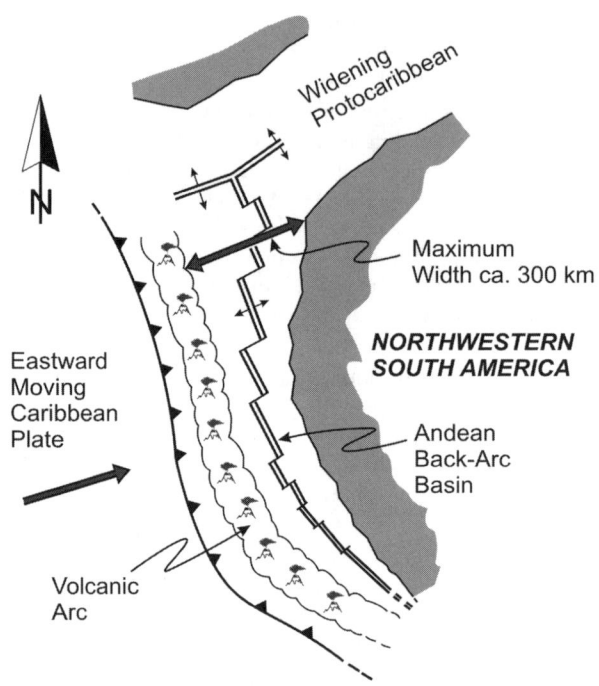

FIG. 2. Backarc basin suggested for northwest South America in the plate tectonic model of Pindell and Kennan (2001) during the early Aptian (119 Ma). Can closure of such a narrow basin and subduction of this young warm crust lead to the isolated belts of blueschists now found in the northern Andes?

mantle wedge. To insure a smooth continuity of the temperature field across the lower boundary of the truncated regional model, we use an infinite-like boundary condition involving vanishing changes of the vertical Lagrangian heat flux, q_z, with depth ($\partial q_z/\partial z = 0$; Gerya and Yuen, 2003a).

The upper surface is calculated dynamically at each time-step like a free surface. To account for changes in the topography, we use a layer with a lower viscosity (10^{18} Pa·s), whose initial thickness is 8 km on top of the oceanic crust. The density of this layer is taken to be 1 kg/m^3 (air) at $z < 4$ km and 1000 kg/m^3 (sea water) at $z > 4$ km. An interface between this layer and the top of the oceanic crust is considered to be the erosion/sedimentation surface, which evolves according to the following transport equation, solved at each time step:

$$\partial z_{es}/\partial t = v_z - v_x \partial z_{es}/\partial x - v_s + v_e, \quad (1)$$

where z_{es} is the vertical position of the surface as a function of the horizontal distance, x; v_z and v_x are the vertical and horizontal components of the material velocity vector at the surface; v_s and v_e are, respectively, sedimentation and erosion rates corresponding to the relation

$v_s = 0$ mm/a, $v_e = 1$ mm/a when $z < 4$ km,
$v_s = 0.3$ mm/a, $v_e = 0$ mm/a when $z > 4$ km.

Dehydration and hydration models

To account for the effects of the hydration process (Fig. 3C) in a viscous medium, we describe the vertical displacement of the hydration front (i.e., the interface between hydrated and dry mantle rock) with respect to the upper surface of the subducting slab (in Eulerian coordinates) by the following transport equation (Gerya et al., 2002), which is solved at each time-step:

$$\partial z_h/\partial t = v_z - v_x \partial z_h/\partial x - v_h, \quad (2)$$

where z_h is the vertical location of the hydration front as a function of the horizontal distance, x, measured from the trench; v_h is the hydration rate; and v_z and v_x are the vertical and horizontal components of the material velocity vector at the front.

It has been shown (e.g., Peacock, 1987) that the availability of water controls the progress of mantle

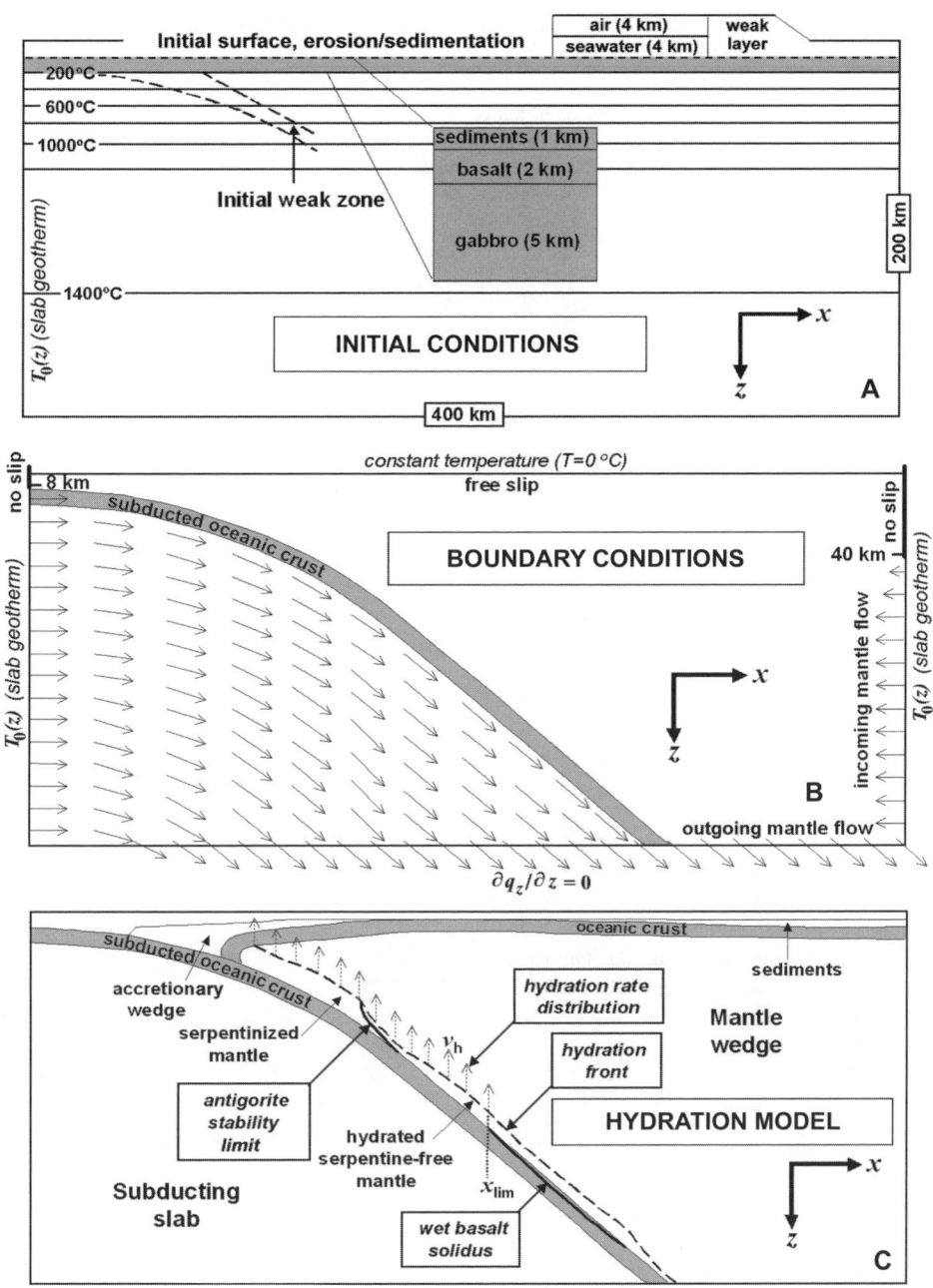

FIG. 3. Design of (A) initial and (B) boundary conditions, as well as (C) hydration model of the mantle wedge as used in our 2-D numerical experiments. See text for details.

hydration. We assume, therefore, that spatial changes in the hydration rate along the hydration front should mainly depend on spatial changes in the rate of fluid release along the surface of the subducting plate. Assuming a continuous (Schmidt and Poli, 1998) dehydration of the subducting slab, we also account for the vertical displacement of the hydration front (Fig. 3) in the deforming mantle

TABLE 1. Material Properties Used in 2-D Numerical Experiments[1]

Material	ρ, kg/m^3	Thermal conductivity, W/(m·K)	Rheology	H_r, μW/m^3
Sedimentary rocks	2700	$0.64 + \frac{807}{T + 77}$	Wet quartzite flow law, $\lambda_{10}=0.95$	2
Basaltic crust	3100	$1.18 + \frac{474}{T + 77}$	Constant viscosity 10^{19} Pa·s	0.25
Gabbroic crust	As above	as above	Plagioclase (An$_{75}$) flow law, $\lambda = 0.95$	As above
Serpentinized mantle	3000	$0.73 + \frac{1293}{T + 77}$	Constant viscosity 10^{19} Pa·s	0.022
Hydrated mantle beyond the serpentine stability field[2]	3300	As above	Wet olivine flow law, $\lambda_{10} = 0.95$	As above
Dry mantle	3300	As above	Dry olivine flow law, $\lambda_{10} = 0$	As above
References	Turcotte and Schubert, 1982	Clauser and Huenges, 1995	Gerya et al., 2002; Ranalli, 1995	Turcotte and Schubert, 1982

[1]λ = pore fluid pressure coefficient (see Eq. 5 in text); p = density (see Eq. 9 in text); H_r = radioactive heat production (see Eq. 10 in text).
[2]The serpentine stability field (Schmidt and Poli, 1998): $T > 751 + 0.18P - 0.000031P^2$ at P < 2100 MPa, $T > 1013 - 0.0018P - 0.0000039P^2$ at P>2100 MPa.

wedge with respect to the upper interface of the subducting slab. Hydration leads to a sharp decrease in density and viscosity of the mantle rocks, creating favorable conditions for corner flow in a wedge-shaped subduction channel. The hydration rate, v_h, is approximated (Gerya et al., 2002) by a linear function of the horizontal distance, x, measured from the trench (Fig. 3):

$$v_h/v_s = A[1 - Bx/x_{lim}] \text{ when } x < x_{lim}, \quad (3)$$

$$v_h/v_s = 0 \text{ when } x > x_{lim},$$

where v_s is the subduction rate; and x_{lim} is the limiting horizontal distance from the trench beyond which fluid release from the subducting plate is negligible (Gerya et al., 2002). The nondimensional parameter B may vary from –1 to +1, characterizing either the increase (B < 0) or the decrease (B > 0) in hydration rate with depth. The parameter A is a nondimensional intensity in the hydration process of the mantle wedge. Typical values of the parameter A range between 0.01 and 0.30 as a function of (1) the water contents in the hydrated peridotite and (2) the total amount of water released from the subducting plate within the limiting distance x_{lim} (Gerya et al., 2002). In the present study we set A = 0.035 and B = 0.

Rheological model

For materials of the crust and mantle, we employ a composite viscous rheology that encompasses the variability of the lithological phases (Table 1), temperature, and strain rate. The effective dislocation creep viscosity of rocks depending on stress and temperature is defined in terms of deformation invariants according to Ranalli (1995):

$$\eta_{disl} = (\dot{\varepsilon}_{II})^{(1-n)/2n} F (A_D)^{-1/n} \exp(E/nRT), \quad (4)$$

where $\dot{\varepsilon}_{II} = \frac{1}{2}\dot{\varepsilon}_{ij}\dot{\varepsilon}_{ij}$ is the second invariant of the strain-rate tensor, with dimension s^{-2}; A_D, E, and n are experimentally determined flow law parameters (Ranalli, 1995). F is a dimensionless coefficient depending on the type of experiment on which the flow law is based (e.g., $F = 2^{(1-n)/n}/3^{(1+n)/2n}$ for triaxial compression and $F = 2^{(1-2n)/n}$ for simple shear).

The strength of solid rock in the brittle field is implemented as a limiting maximum viscosity

$$\eta_{max} = \sigma_{yield}/(4\dot{\varepsilon}_{II})^{1/2}, \quad (5)$$

where $\sigma_{yield} = (N_1 P_{lith} + N_2)(1-M)$ is the yield stress; P_{lith} is the lithostatic pressure, MPa; N_1 and N_2 are empirical constants, MPa: $N_1 = 0.85$, $N_2 = 0 < \sigma_{yield} < 200$ MPa and $N_1 = 0.6$, $N_2 = 60$ when $\sigma_{yield} > 200$ MPa (Brace and Kohlstedt, 1980); and M is the pore fluid pressure coefficient (Table 1).

Finally, an effective creep viscosity η is calculated according to a formula (e.g., Schott and Schmeling, 1998) providing smooth transitions between different regimes of deformation

$$1/\eta = 1/\eta_{disl} + 1/\eta_{max}. \quad (6)$$

Mathematical modeling and numerical implementation

We consider 2-D creeping flow wherein buoyant forces are included. The conservation of mass is approximated by the incompressible continuity equation

$$\partial v_x/\partial x + \partial v_z/\partial z = 0. \quad (7)$$

The 2-D Stokes equations take the form:

$$\partial \sigma_{xx}/\partial x + \partial \sigma_{xz}/\partial z = 0 \quad (8)$$

$$\partial \sigma_{zz}/\partial z + \partial \sigma_{xz}/\partial x = \partial P/\partial z - g\rho. \quad (9)$$

We employ viscous rheological constitutive relationships between the stress and strain rate, whose coefficient η represents the effective viscosity, which depends on the composition, temperature, pressure, and strain rate:

$$\sigma_{xx} = 2\eta\varepsilon_{xx}, \sigma_{xz} = 2\eta\varepsilon_{xz}, \sigma_{zz} = 2\eta\varepsilon_{zz},$$
$$\varepsilon_{xx} = \partial v_x/\partial x, \varepsilon_{xz} = $$
$$\tfrac{1}{2}(\partial v_x/\partial z + \partial v_z/\partial x), \varepsilon_{zz} = \partial v_z/\partial z.$$

We adopt (Gerya and Yuen, 2003b) a Lagrangian frame of reference in which the temperature equation with a temperature-dependent thermal conductivity $k(T)$ (Table 1) takes the form

$$\rho Cp(D\,T/D\,t) = \partial q_x/\partial x + \partial q_z/\partial z + H_r + H_s \quad (10)$$
$$q_x = k(T) \times (\partial T/\partial x), q_z = k(T) \times (\partial T/\partial z),$$
$$H_r = constant, H_s = \sigma_{xx}\varepsilon_{xx} + \sigma_{zz}\varepsilon_{zz} + 2\sigma_{xz}\varepsilon_{xz},$$

where $D/D\,t$ represents the substantive time derivative.

The notations in equations 7–10 represent as follows: x and z are, respectively, the horizontal and vertical coordinates in m, v_x and v_z are components of the velocity vector $\underline{v}$ in m·s^{-1}; t is time in s; σ_{xx}, σ_{xz}, and σ_{zz} are components of the viscous deviatoric stress tensor in Pa; ε_{xx}, ε_{xz}, and ε_{zz} are components of the strain rate tensor in s^{-1}; P is the pressure in Pa; T is the temperature in K; q_x and q_z are the horizontal and vertical heat fluxes, respectively, in W·m^{-2}; η is the effective viscosity in Pa·s; ρ is the density in kg·m^{-3}; $g = 9.81$ m·s^{-2} is the gravitational acceleration; k is the thermal conductivity in W·m^{-1}·K^{-1}; the isobaric heat capacity Cp is set equal to 1000 J·kg^{-1}·K$_{-1}$ (Table 1); and H_r and H_s denote, respectively, radioactive and shear heat production in W·m^{-3}.

We employ a recently developed 2-D code (I2VIS) based on finite differences with a moving marker technique allowing for the accurate conservative solution of the governing equations on a rectangular, fully staggered Eulerian grid for multiphase viscoplastic flow. A detailed description of the numerical method as well as algorithmic tests are provided by Gerya and Yuen (2003b).

Numerical Modeling: Choosing the Specific Parameters for This Study

For reasons of basic reproducible simplicity, and in keeping with the focus on intra-oceanic subduction zones outlined above, we chose a basic subduction zone model initiating in oceanic lithosphere, thus leading to a convergence of oceanic plate vs. oceanic plate. In this way, the model could be built on a well-defined thermal structure of the oceanic lithosphere as a function of its age. Slab dip was considered to be uncritical at the relatively shallow depths of 15–20 km at which blueschists can be expected to become stable, and should be dictated by the possible bending radius of the oceanic lithosphere. Assuming this to be about 200 km, we set the slab dip at a constant 45°. With these assumptions, we selected as variables the age of the oceanic lithosphere to constrain the thermal structure, and the rate of subduction. In total, 25 combinations of initial slab age (20 to 100 Ma) and subduction rate (2 to 14 cm/yr) were modeled to obtain a systematic data set. The model uses a 400 × 200 km 2-D box with 500,000 markers, implying a resolution of 400 m for the lithological field.

To define the P-T conditions of the blueschist facies we accepted the suggestion of Peacock (1992), which incorporates the calculations of Evans (1990) for the definition of the epidote-blueschist facies (Fig. 4). Blue amphiboles in this context are the (Mg,Al)–rich sodic amphiboles glaucophane and ferroglaucophane (Leake et al., 1997), as well as "crossite" as originally defined by Miyashiro (1957). The term "blueschist" is used in its broadest sense, incorporating all isofacial rocks metamorphosed in the P-T field outlined in Figure 4.

Results of the Calculations

The results of the model calculations are summarized in Figures 5–8. Figures 5 and 6 show selected typical examples of the self-organizational evolution of the 2-D distribution of sediment, crust, mantle, hydrated mantle, etc. of the subduction zone model as defined in Figure 3. The panels in Figure 5 indicate the changes with time of this cross-section as a function of the rate of subduction at a constant age of the lithosphere of 60 Ma. The cross-sections are grouped according to the same amount of subduction, and therefore represent different times elapsed after the initiation of subduction (see label for each cross-section). In Figure 6, the subduction rate is kept constant at 8 cm/yr, and the cross-sections are compared for three ages of lithosphere. Thus these cross-sections are grouped together for comparable stages in development both in time and amount of subduction.

As expected, the fastest subduction rate leads to the earliest appearance of blueschist-facies conditions (Fig. 5). Note however, that these conditions are first reached at different locations in the subducting oceanic crust. For slow subduction of 2 cm/yr, blueschist-facies conditions first appear in the lower part of the crustal section, i.e., where the isotherms are relatively flat, whereas at subduction rates of 8 cm/yr and more these conditions are first reached in the upper and frontal parts of the slab where the isotherms are steep. In Figure 6, it is again apparent that blueschist-facies conditions are first reached at different parts of the "nose" of the subducting basaltic crust. For young crust, the blueschist facies is first developed in the lower parts marked by flat isotherms, whereas in older crust blueschists form in the upper and frontal sections.

All data obtained for the modeled 25 combinations of initial slab age and subduction rate are summarized in Figure 7, where the data are contoured

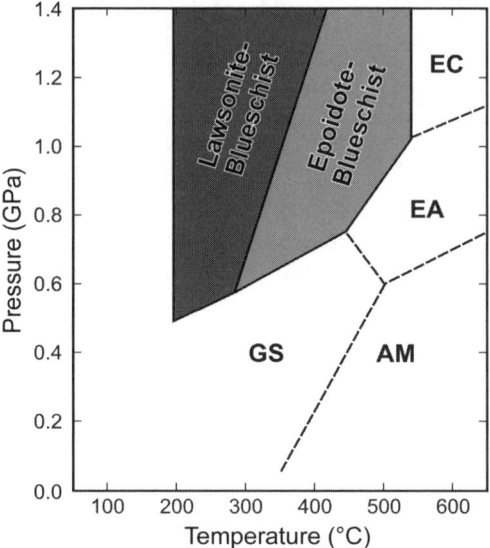

FIG. 4. Pressure-temperature template for blueschist-facies conditions adopted in numerical modeling (Evans, 1990; Peacock, 1992). Abbreviations: GS = greenschist facies; AM = amphibolite facies; EA = edpidote-amphibolite facies; EC = eclogite facies.

both in terms of the time needed after subduction initiation and the convergence necessary to attain blueschist conditions. These diagrams underline the observation made above that time and convergence needed are not simple functions of slab age and convergence rate.

Discussion

Numerical results

This numerical study leads to some logical, some unexpected, and also some seemingly non-intuitive results (Fig. 7). It is certainly not surprising that the highest subduction rates should lead to the fastest attainment of blueschist-facies conditions in the subducted slab. On the other hand, we do find it surprising that, for a wide range of slab ages and subduction rates, blueschist-facies conditions can be attained in less than 500,000 years, and for an amount of convergence/subduction of less than 50 km. Even for very low subduction rates of 2 cm/yr, blueschist-facies conditions can be attained within 2–3 m.y. The amount of required convergence is only moderately dependent on subduction rate (Fig. 7), and mainly a function of slab age. Figure 7 indicates that for all values of slab age between 20 and

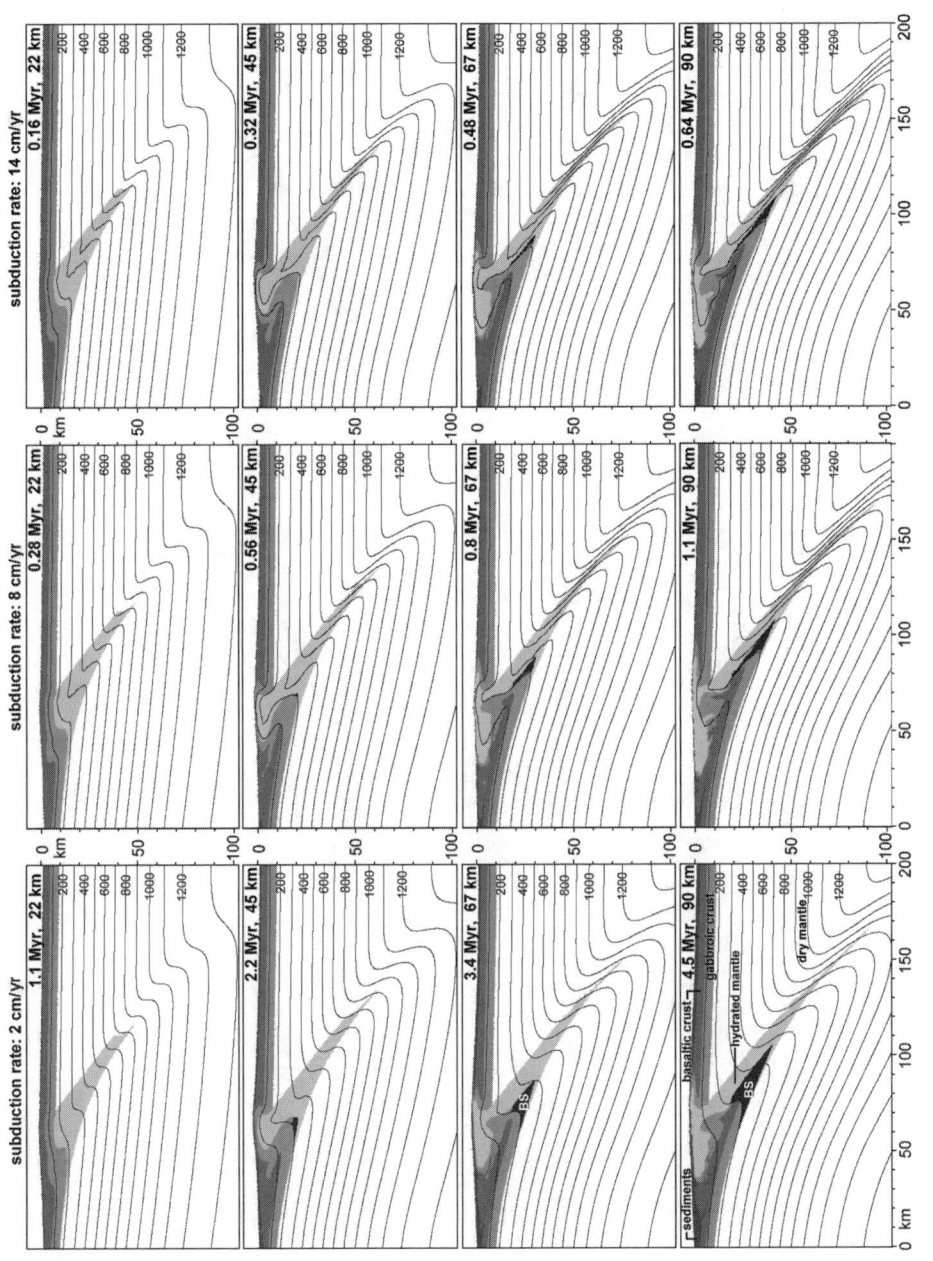

FIG. 5. Sequences of selected panels comparing the modeled self-organizational development of an intra-oceanic subduction zone for three different subduction rates (2, 8, and 14 cm/yr) and the same initial age of 60 Ma for the oceanic lithosphere. Note that the panels are grouped according to the amount of subduction and therefore represent different times elapsed after subduction initiation. These representations have been simplified for clarity (cf., Fig. 3) and show half the experimental box used in the modeling.

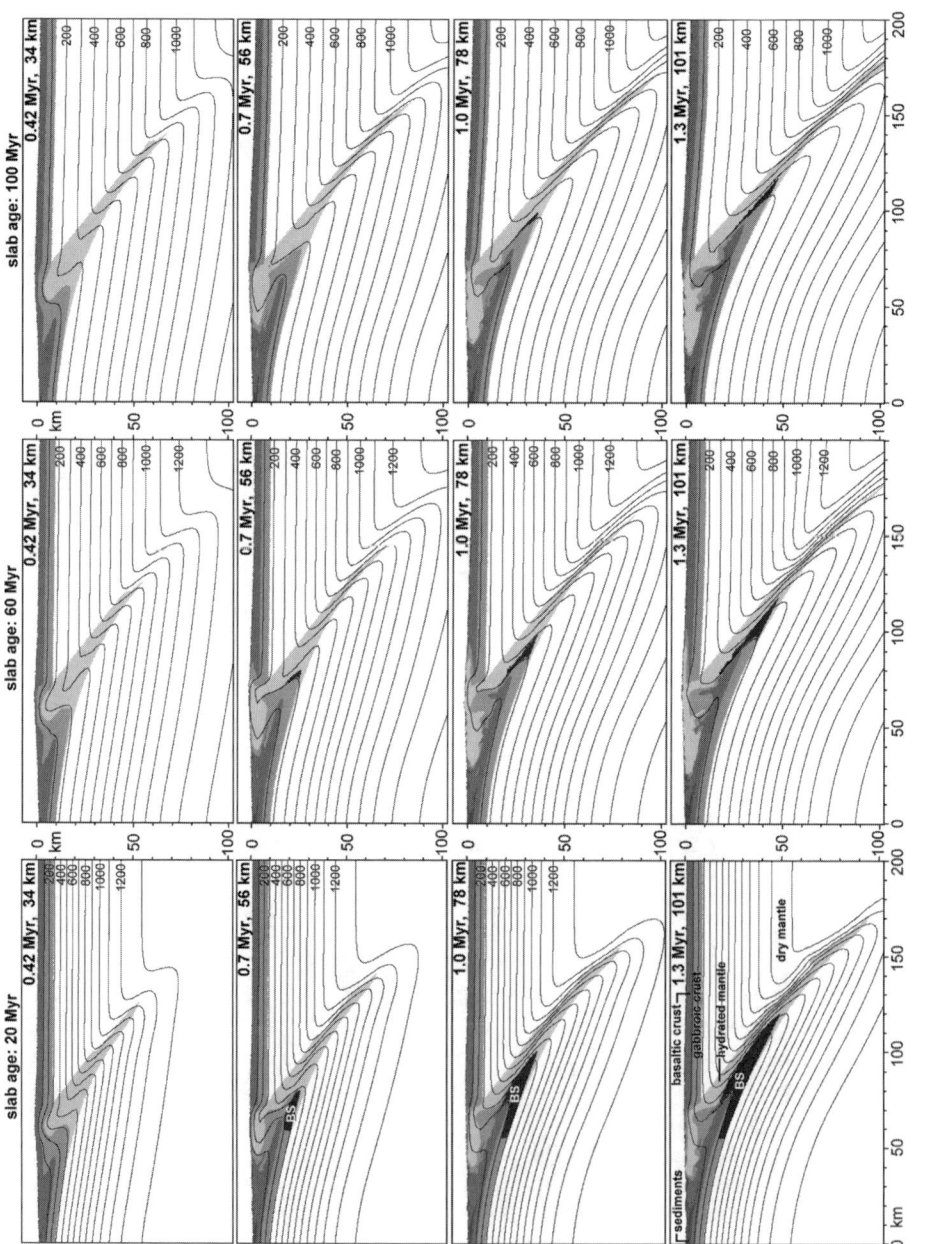

FIG. 6. Sequences of selected panels comparing the modeled self-organizational development of an intra-oceanic subduction zone for three different initial ages of oceanic lithosphere (20, 60, 100 Ma) at a constant subduction rate of 8 cm/yr. The groups of panels represent equal amounts of subduction and the same time elapsed after subduction initiation. These representations have been simplified for clarity (cf., Fig. 3) and show half the experimental box used in the modeling.

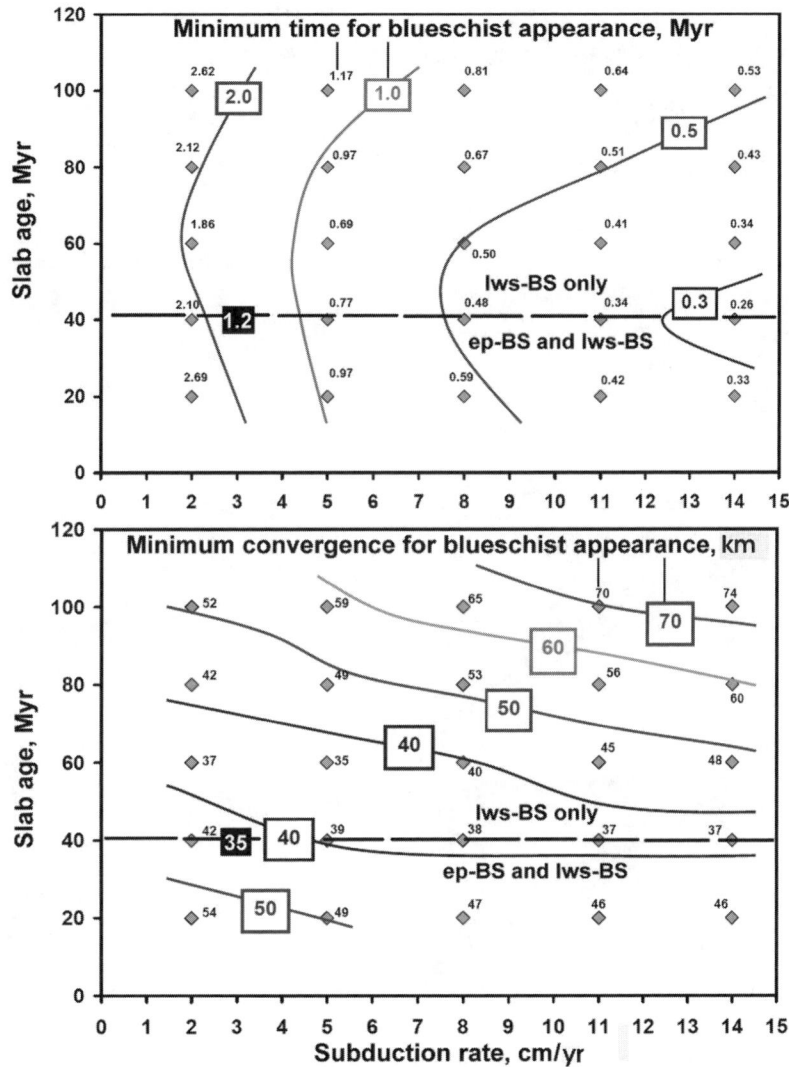

FIG. 7. Summary diagram of slab age vs. subduction rate contoured for equal time elapsed after subduction initiation (upper panel) and minimum convergence needed (lower panel) to reach blueschist-facies conditions. The two data points at 40 Ma and 3 cm/yr are taken from an Andean-type subduction model by Stöckhert and Gerya (2005).

100 Ma, as well as all rates of convergence between 2 and 14 cm/yr, less than 80 km of convergence are required to reach blueschist-facies conditions. Thus, in the context of Caribbean-, Tethyan-, or western Pacific–type oceanic island-arc scenarios, blueschist formation without any coeval volcanic activity is distinctly possible in many cases of ephemeral subduction.

Surprising, and indeed seemingly non-intuitive, is the result that younger, hotter slabs can lead to earlier blueschist formation than older, cooler slabs. In fact, Figure 7 indicates that there is a distinct optimum for slabs between approximately 40 and 60 Ma of age. This result is a logical consequence of the interplay between the geometry of the subducting slab "nose" and the evolving array of isotherms. The subducting slab can enter the blueschist P-T field from the low-temperature side (i.e., already deep enough but not yet hot enough), or from the low-pressure side (i.e., already hot enough but not

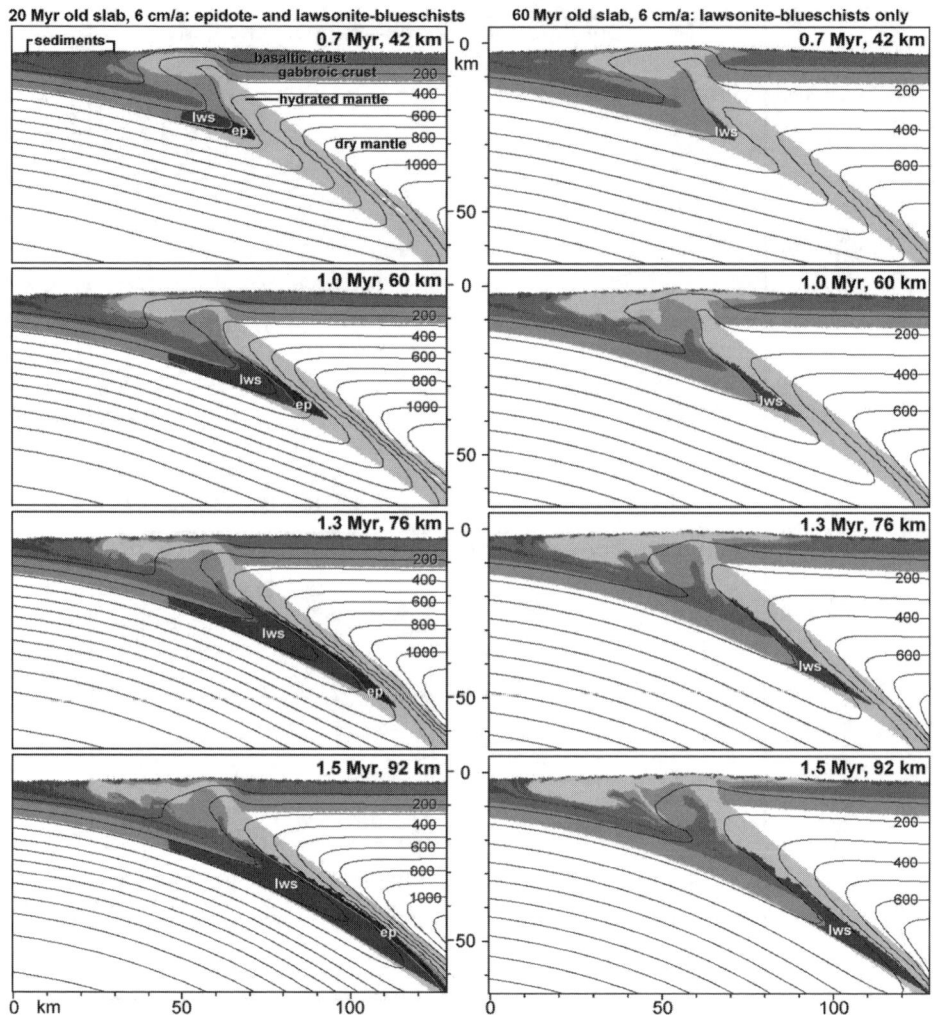

FIG. 8. Sequences of selected panels indicating the development of epidote- vs. lawsonite-blueschist–facies conditions as a function of slab age.

yet deep enough). Between these two geometries there will exist an optimized situation where the slab "nose" will directly "impinge" on the blueschist P-T field at approximately 200°C and 0.5 GPa (Fig. 4). Lawsonite-blueschists are the first to form for slabs older than 40 Ma (Figs. 7 and 8). Both epidote- and lawsonite-blueschists form at approximately the same time for the younger slabs (Figs. 7 and 8) that enter the blueschist field from the low-pressure side (Fig. 4), but with time, as the subduction zone matures, lawsonite-blueschist conditions dominate in the subducting slab. In the context of the present intra-oceanic model, it appears that epidote-blueschists play only a minor role in early blueschist formation.

General significance

In gauging the significance of the "minima" in time and convergence calculated in this study, it is important to note that we have ignored the influence of reaction kinetics. Even if equilibration in some massive, water-poor rock types may be slow, we assume that amphibole-rich blueschists (i.e., usually strongly deformed, water-rich metavolcanics and metasediments) should probably be a very reactive system. As noted many years ago by Ernst

(1971), the commonly observed systematic distribution in space and time of high-pressure, low-temperature metamorphic facies in subducted slabs (see also Yardley, 1989; Spear, 1993), and the commonly observed prograde zoning from Fe-rich to (Mg,Al)–rich sodic amphiboles (i.e., "crossite" to glaucophane) suggest that large-scale overstepping of the P-T boundaries in Figure 4 should not be a general feature.

Although we have focused here on an intra-oceanic model system characteristic of the Caribbean area (see review by Pindell et al., 2005), where interaction with, and obduction onto, continental margins can shut down the subduction process and allow separation of the prograde path from various possible steady-state exhumation scenarios, the systematic results obtained here can in principle be corroborated by analogous modeling of subduction at a continental margin. Figure 7 includes two data points derived from an Andean-type, continental-margin model calculated by Stöckhert and Gerya (2005). The values do not differ significantly from the results calculated in the present, more systematic study. However, such calculated "minima" become meaningful only when augmented by information on the exhumation path, which is dependent on the specific exhumation scenario involved. Disregarding various possible ad hoc tectonic scenarios (e.g. extension, exhumation during continental collision, etc.) for which no systematic predictions are feasible, it is nevertheless possible to draw on recent quantitative studies of exhumation rates in subduction zones.

Assuming forced return flow in so-called subduction channels (Hsu, 1971; Cloos, 1982; Shreve and Cloos, 1986; Cloos and Shreve, 1988a, 1988b), Gerya and Stöckhert (2002), Gerya et al. (2002), and Stöckhert and Gerya (2005) have numerically investigated the exhumation process for a continental-margin setting with a conceptual model analogous to Figure 3. Maximum exhumation rates are found to vary between about one-sixth of the vertical component of the given subduction rate when assuming uniform power-law rheology for the subduction channel medium (Gerya and Stöckhert, 2002), and up to three times the vertical component of the subduction rate when exhumation is localized in discrete parts of this channel (Stöckhert and Gerya, 2005). Such results are basically in accord with studies based on diffusion rates in mineral grains (e.g., Perchuk et al., 1999) or on geochronology (Rubatto and Hermann, 2001). One caveat to consider, however, is that the subduction zone must mature to a critical stage before return flow from depths greater than about 20 km is possible (Gerya et al., 2002). Thus only very low-grade blueschists could be recycled in the very early stages of subduction. Gerya et al. (2002) suggested that a mature stage with intense return flow from greater depths should be possible after about 10 m.y. of subduction. With the data now available, a case-to-case appraisal for a given subduction-zone scenario now appears possible.

With regard to the Caribbean setting that prompted this study, we may conclude that extremely short-lived subduction can rapidly produce rocks of the blueschist facies without any correlatable volcanic activity. This may not necessarily simplify paleotectonic interpretation, but it will allow a more realistic appraisal of the significance of the various blueschist occurrences exposed in the circum-Caribbean area (e.g., Pindell et al., 2005). Although the closure of a young, narrow backarc basin as in Figure 2 cannot be considered to be a purely intra-oceanic process, the above discussion indicates that rapid blueschist formation, as in the numerical model studied here, is possible. This can lead logically to intramontane occurrences of blueschist far from the presumed trace of major subduction activity.

Acknowledgments

We thank W. Gary Ernst for many decades of thought-provoking scientific input. Thanks also to Lorcan Kennan for providing Figure 1 and Frau Renate Lehmann for help with the drafting. Funding by the Deutsche Forschungsgemeinschaft (German Science Foundation) within the scope of the Sonderforschungsbereich SFB 526 (Collaborative Research Center) in Bochum: "Rheology of the Earth—from the upper crust into the subduction zone" is gratefully acknowledged. TVG was also generously supported by the Alexander von Humboldt Foundation, the Russian Foundation for Basic Research (grants # 03-05-64633 and #1645-2003-5), and by ETH Research Grant TH-12/04-1.

REFERENCES

Avé Lallemant, H. G., 1997, Transpression, displacement partitioning, and exhumation in the eastern Caribbean/South American plate boundary zone: Tectonics, v. 16, p. 272–289.

Avé Lallamant, H. G., and Guth, L. R., 1990, Role of extensional tectonics in exhumation of eclogites and blueschists in an oblique subduction zone setting: Northeast Venezuela: Geology, v. 18, p. 950–953.

Brace, W. F., and Kohlstedt, D. L., 1980, Limits on lithospheric stress imposed by laboratory experiments: Journal of Geophysical Research, v. 85, p. 6248–6252.

Burke, K., 1988, Tectonic evolution of the Caribbean: Annual Review of Earth and Planetary Sciences, v. 16, p. 210–230.

Clauser, C., and Huenges, E., 1995, Thermal conductivity of rocks and minerals, in Ahrens, T. J., ed., Rock physics and phase relations: Washington, DC, American Geophysical Union, AGU Reference Shelf 3, 1995, p. 105–126.

Clift, P. D., Schouten, H., and Draut, A. E., 2003, A general model of arc-continent collision and subduction polarity reversal from Taiwan and the Irish Caledonides, in Larter, R. D. and Leat, P. T., eds., Intra-oceanic subduction systems: Tectonic and magmatic processes: Geological Society of London Special Publications, v. 219, p. 1–17.

Cloos, M., 1982, Flow melanges: Numerical modelling and geologic constraints on their origin in the Franciscan subduction complex, California: Geological Society of America Bulletin, v. 93, p. 330–345.

Cloos, M., and Shreve, R. L., 1988a, Subduction-channel model of prism accretion, melange formation, sediment subduction, and subduction erosion at convergent plate margins, 1. Background and description: Pure and Applied Geophysics, v. 128, p. 455–500.

Cloos, M., and Shreve, R.L., 1988b, Subduction-channel model of prism accretion, melange formation, sediment subduction, and subduction erosion at convergent plate margins, 2, Implications and discussion: Pure and Applied Geophysics, v. 128, p. 501–545.

Eiler, J., ed., 2004, Inside the subduction factory: American Geophysical Union, Geophysical Monographs Series, v. 138, 324 p.

Ernst, W. G., 1963, Petrogenesis of glaucophane schists: Journal of Petrology, v. 4, p. 1–30.

Ernst, W. G., 1971, Metamorphic zonations on presumably subducted lithospheric plates from Japan, California and the Alps: Contributions to Mineralogy and Petrology, v. 34, p. 43–59.

Evans, B. W.,1990, Phase relations of epidote-blueschists: Lithos, v. 25, p. 3–23.

Gerya, T. V., and Stöckhert, B., 2002, Exhumation rates of high pressure metamorphic rocks in subduction channels: the effect of rheology: Geophysical Research Letters, v. 29, Article No. 1261

Gerya, T. V., Stoeckhert, B., and Perchuk, A. L., 2002, Exhumation of high-pressure metamorphic rocks in a subduction channel: A numerical simulation: Tectonics, v. 21, p. 6-1 to 6-19.

Gerya, T. V., and Yuen, D. A., 2003a, Rayleigh-Taylor instabilities from hydration and melting propel "cold plumes" at subduction zones: Earth and Planetary Science Letters, v. 212, p. 47–62.

Gerya, T. V., and Yuen, D. A., 2003b, Characteristics-based marker-in-cell method with conservative finite-difference schemes for modeling geological flows with strongly variable transport properties: Physics of the Earth and Planetary Interiors, v. 140, p. 293–318.

Gerya, T. V., Yuen, D. A., and Sevre, E. O. D., 2004, Dynamical causes for incipient magma chambers above slabs: Geology, v. 32, p.89–92.

Hacker, B. R., Abers, G. A., and Peacock, S. M., 2003, Subduction factory 1. Theoretical mineralogy, density, seismic wave speeds, and H_2O content: Journal of Geophysical Research, v. 108 [art. no. 2029].

Hsu, K. J., 1971, Franciscan melanges as a model for eugeosynclinal sedimentation and underthrusting tectonics: Journal of Geophysical Research, v. 76, p. 1162–1170.

Kerr, A. C., Iturralde Vinent, M. A., Saunders, A. D., Babbs, T. L., and Tarney, J., 1999, A new plate tectonic model of the Caribbean: Implications from a geochemical reconnaissance of Cuban Mesozoic volcanic rocks: Geological Society of America Bulletin, v. 111, p. 1581–1599.

Leake, B. E., Woolley, A. R., Arps, C. E. P., Birch, W. D., Gilbert, M. C., Grice, J. D., Hawthorne, F. C., Kato, A., Kisch, H. J., Krivovichev, V. G., Linthout, K., Laird, J., Mandarino, J., Maresch, W. V., Nickel, E. H., Rock, N. M. P., Schumacher, J. C., Smith, D. C., Stephenson, N. C. N., Ungaretti, L., Whittaker, E. J. W., and Youzhi, G., 1997, Nomenclature of amphiboles: Report of the subcommittee on amphiboles of the International Mineralogical Association Commission on New Minerals and Mineral Names: European Journal of Mineralogy, v. 9, p. 623–651.

Leat, P. T., and Larter, R. D., 2003, Intra-oceanic subduction systems: Introduction, in Larter, R. D., and Leat, P. T., eds., Intra-oceanic subduction systems: Tectonic and magmatic processes: Geological Society of London, Special Publications, v. 219, p. 1–17.

Maresch, W. V., 1974, The plate tectonic origin of the Caribbean mountain system of northern South America: Discussion and proposal: Geological Society of America Bulletin, v. 85, p. 669–682.

Meschede, M., and Frisch, M., 1998, A plate-tectonic model for the Mesozoic and early Cenozoic history of the Caribbean Plate: Tectonophysics, v. 296, p. 269–291.

Miyashiro, A., 1957, The chemistry, optics and genesis of alkali amphiboles: Journal of the Faculty of Science, University of Tokyo, v. 11, p. 57–83.

Peacock, S. M., 1987, Serpentinization and infiltration metasomatism in the Trinity peridotite, Klamath province, northern California: Implications for subduction zones: Contributions to Mineralogy and Petrology, v. 95, p. 55–70.

Peacock, S.M., 1992, Blueschist-facies metamorphism, shear heating, and P-T-t paths in subduction shear zones: Journal of Geophysical Research, v. 97, p. 17,693–17,707.

Perchuk, A. L., Philippot, P., Erdmer, P., and Fialin, M., 1999, Rates of thermal equilibration at the onset of subduction deduced from diffusion modeling of eclogitic garnets, Yukon-Tanana terrain: Geology, v. 27, p. 531–534.

Pindell, J. L., and Erikson, J. P., 1994, The Mesozoic passive margin of northern South America, in Salfity, J. A, ed., Cretaceous tectonics in the Andes: Wiesbaden, Germany, Vieweg Publishing, International Monograph Series, Earth Evolution Sciences, p. 1–60.

Pindell, J. L., and Kennan, L. J. G., 2001, Kinematic evolution of the Gulf of Mexico and the Caribbean, in Fillon, R., ed., Transactions, GCSSEPM Foundation 21st Annual "Bob F. Perkins" Research Conference on "Petroleum Systems of Deep-Water Basins," Houston, Texas December 2–5, 2001, p. 193–220.

Pindell, J. L., Kennan, L., Maresch, W. V., Stanek, K.-P., Draper, G., and Higgs, R., 2005, Plate kinematics and crustal dynamics of circum-Caribbean arc-continent interactions: Tectonic controls on basin development in Proto-Caribbean margins, in Avé Lallemant, H. G., and Sisson, V., eds., Caribbean/South American plate interactions, Venezuela: Geological Society of America Special Paper, in press.

Ranalli, G., 1995, Rheology of the Earth, 2nd ed.: London, UK, Chapman and Hall, 413 p.

Regenauer-Lieb, K., Yuen, D. A., and Branlund, J., 2001, The initiation of subduction: Criticality by addition of water?: Science, v. 294, p. 578–580.

Rubatto, D., and Hermann, J., 2001, Exhumation as fast as subduction: Geology, v. 29, p. 3–6.

Schmidt, M. W., and Poli, S., 1998, Experimentally based water budgets for dehydrating slabs and consequences for arc magma generation: Earth and Planetary Science Letters, v. 163, p. 361–379.

Schott, B., and Schmeling, H., 1998, Delamination and detachment of a lithospheric root: Tectonophysics, v. 296, p. 225–247.

Shreve, R. L., and Cloos, M., 1986, Dynamics of sediment subduction, melange formation, and prism accretion: Journal of Geophysical Research, v. 91, p. 10,229–10,245.

Smith, C. A., Sisson, V. B., Avé Lallemant, H. G., and Copeland, P., 1999, Two contrasting pressure-temperature-time paths in the Villa de Cura blueschist belt, Venezuela: Possible evidence for Late Cretaceous initiation of subduction in the Caribbean: Geological Society of America Bulletin, v. 111, p. 831–848.

Spear, F. S., 1993, Metamorphic phase equilibria and pressure-temperature-time paths: Mineralogical Society of America Special Publication, Washington, DC, 799 p.

Stöckhert, B., and Gerya, T. V., 2005, Pre-collisional high pressure metamorphism and nappe tectonics at active continental margins: A numerical simulation: Terra Nova, in press

Tatsumi, Y., and Kogiso, T., 2003, The subduction factory: Its role in the evolution of the Earth's crust and mantle, in Larter, R. D. and Leat, P. T., eds., Intra-oceanic subduction systems: Tectonic and magmatic processes: Geological Society of London, Special Publications, v. 219, p. 55–80.

Turcotte, D. L., and Schubert, G., 1982, Geodynamics: Applications of continuum physics to geological problems: New York, NY, John Wiley, 430 p.

Yardley, B. W. D., 1989, An introduction to metamorphic petrology: Englewood Cliffs, NJ, Longman Earth Sciences Series, 248 p.

Involvement of Crustal Material in Delamination of the Lithosphere after Continent-Continent Collision

HANS-JOACHIM MASSONNE[1]

Institut für Mineralogie und Kristallchemie, Universität Stuttgart, Azenbergstr. 18, D-70174 Stuttgart, Germany

Abstract

Deep submersion of crustal material into the mantle, evidenced for the Variscan orogen at about 340 Ma, is explained by delamination of the continental lithosphere. The cause for such delamination is thickening of the continental crust and the underlying mantle lithosphere due to the collision of Gondwana and Laurussia. Scientific information on this continent-continent collision is scrutinized to test if the evolutionary model for the Variscan orogeny proposed here is compatible with the vast geoscientific data. It is then hypothesized that the evolution of an orogen, resulting from continent-continent collision since the late Neoproterozoic, can, in general, be described by six stages: (1) approach of continental plates; (2) beginning of continent-continent collision; (3) lateral extension of continental crust that was thickened to 60 km or more, for tens of million years; (4) delamination of the continental lithosphere involving material from the continental crust; (5) continuation of continent-continent collision without significant thickening of continental crust lasting again several tens of million years; and (6) cessation of continent-continent collision. These stages are characterized by the formation of different sedimentary, igneous, and metamorphic rock types. For instance, the origin and the beginning of exhumation of high- and ultrahigh-pressure rocks are related to stages 2 and 4. Thus, the corresponding rocks are good relative time markers.

The above model was applied to the Cretaceous to Quaternary evolution of the orogenic chains in the Mediterranean region. It is concluded that this region underwent extended lithospheric delamination (stage 4) in the late Paleogene. Finally, the consequences of lithospheric delamination with continental crust involved are discussed. It is speculated that this process has resulted in an enormous loss of continental crust over the last 600 m.y. This loss could have also significantly influenced the climatic and biologic evolution on Earth.

Introduction

DELAMINATION OF THE CONTINENTAL lithosphere after continent-continent collision is a process that was predicted by geophysical considerations and modeling experiments (e.g., Houseman et al., 1981; Platt and England, 1994; Schott and Schmeling, 1998). The collision of continental plates results in thickening of both continental crust and underlying lithospheric mantle. As the thickened lithospheric and, thus, relatively cold mantle material tends to be somewhat denser than chemically equivalent but hotter mantle material, the lithospheric root under thickened continental crust can delaminate to sink deep into the mantle (Schott and Schmeling, 1998). The lithospheric mantle below the Tibetan Plateau, resulting from crustal thickening through collision of the Indian plate with Eurasia, is believed to delaminate currently according to seismological studies (Kosarev et al., 1999).

The question arises if the delaminated material consists only of ultramafic mantle rocks or of parts of the overlying continental crust in addition to the ultramafic material. The possibility of the involvement of continental crust in the delamination process has been recently discussed. For instance, numerical modeling experiments by Willner et al. (2002) indicate that this process occurs when the continental crust and underlying lithosphere are immensely thickened (120 km) and the density of the mantle lithosphere is ≥ 3.4 g/cm^3. Wittenberg et al. (2000) concluded from geochemical and seismological investigations, which, for instance, indicated a deficit of mafic material in the Variscan crust in central Europe compared with ordinary continental crust, that there was a loss of portions of the lower crust, consisting mainly of metabasites, during a delamination event. In contrast to indirect evidence for continental crust involved in a lithospheric

---

[1]Email: h-j.massonne@mineralogie.uni-stuttgart.de

delamination process, more direct evidence comes from ultrahigh-pressure (UHP) rocks of continental affinity. Massonne (2003a) assumed that diamondiferous quartzofeldspathic rocks from the same Variscan area (Bohemian Massif in the northeastern Variscan orogen) investigated by Wittenberg et al. (2000) provide evidence of lithospheric delamination. In this paper, the evolution of the continental crust of the Variscan orogen is reviewed in the context of a delamination process. In addition, consequences of lithospheric delamination in a global context are identified.

Evolution of the Variscan Orogen

Large areas of Variscan basement, such as the Iberian Massif, the Armorican Massif, the Massif Central, and the Bohemian Massif, occur in western and central Europe. The formerly western portion of the Late Paleozoic Variscan orogen is exposed on the North American continent (Alleghenides in the eastern United States and in southeastern Canada). However, broad areas of the denuded Variscan orogen are covered by Permian and Mesozoic rocks. Moreover, the southern portion of this orogen was involved in the Alpine orogeny. Nevertheless, fragments of weakly Alpine-overprinted Variscan rocks are abundant in the Alpine and related orogenic chains in the Mediterranean region.

It has become a common view that the Variscan orogen resulted from the collision of Gondwana and Laurussia (Laurentia + Baltica). However, different paleogeographic arrangements of these continental plates have been proposed for the time span of some tens of million years before the collision. In addition, some authors (e.g., Tait et al., 2000) assumed microplates, such as Armorica, broke off from Gondwana (as today's Madagascar from the African plate) and drifted separately to collide with Laurussia before the main collision of Laurussia and Gondwana. In this context, the existence of a single oceanic basin (e.g., Rheic ocean as termed by Robardet et al., 1990) or two and more basins (e.g., Stampfli et al., 2002) separating Gondwana and Laurussia, the size of the(se) oceanic basin(s) (e.g., narrow: McKerrow et al., 2000; wide: Tait et al., 2000) and the time of final closure, ranging from Lower Devonian to Early Carboniferous times, are also matters of debate.

Most authors, presenting paleogeographic reconstructions for the Variscan orogeny, have assumed that the oceanic crust between the microplates and continental plates was subducted, accounting also for the minor but widespread occurrences of high-pressure (HP) metamorphic rocks, such as eclogites, in the Variscan terranes. After subduction, the terranes amalgamated according to the principle of terrane accretion (see, for instance, examples in Condie, 1989) to form more extended continental plates. In general, it is neglected in the corresponding paleogeographic reconstructions that the continental crust of the colliding plates can overlap during a long-lasting collisional event. Thus, the process of amalgamation (simple docking of plates) cannot explain the widespread occurrence of HP (P > 10 kbar) gneisses, typically rich in potassic white mica and poor (or lacking) in biotite (unless formed late as retrograde product), in the Variscan crust (see Massonne, 2003a). These rocks point to crustal thickening over an extended area due to an underplating process during the Variscan orogeny. A few geodynamic models related to this orogeny, for instance, presented by Matte (1998), take this into account.

This author proposed subduction of one continental plate beneath the other. Firstly, the subducted continental plate is drawn into the depths by the adherent oceanic plate that finally breaks off to sink deeper into the mantle according to the slab-breakoff model by Davies and von Blanckenburg (1995). Then, buoyancy forces cause an ascent of the subducted continental plate. After the initiation of this process in the Early Devonian, when Gondwana and Laurussia collided, continental subduction was repeated several times to form an imbricated megastructure (O'Brien, 2000), relatively narrow (a few hundred km across at most) with crustal slices (Laurussia) subducted to the south. This structure was located at the northern boundary of the Variscan orogen (Fig. 1) and could thus explain the occurrence of UHP rocks of continental affinity including diamondiferous quartzofeldspathic rocks (Massonne, 2003b) in the northwestern Bohemian Massif. However, the geodynamic model proposed by Massonne (2003a) and Massonne and O'Brien (2003), comparing the Variscan orogen with the Tertiary collisional orogen between India and Asia (Himalaya and Tibetan plateau), is preferred here, inasmuch as it also accounts for the clear bimodal distribution of metamorphic ages for various types of medium- to high-pressure rocks. This age distribution, noted by Willner et al. (2000), shows a clear peak between 390 and 400 Ma (U-Pb, Sm-Nd ages) with cooling ages (K-Ar, Ar-Ar) about

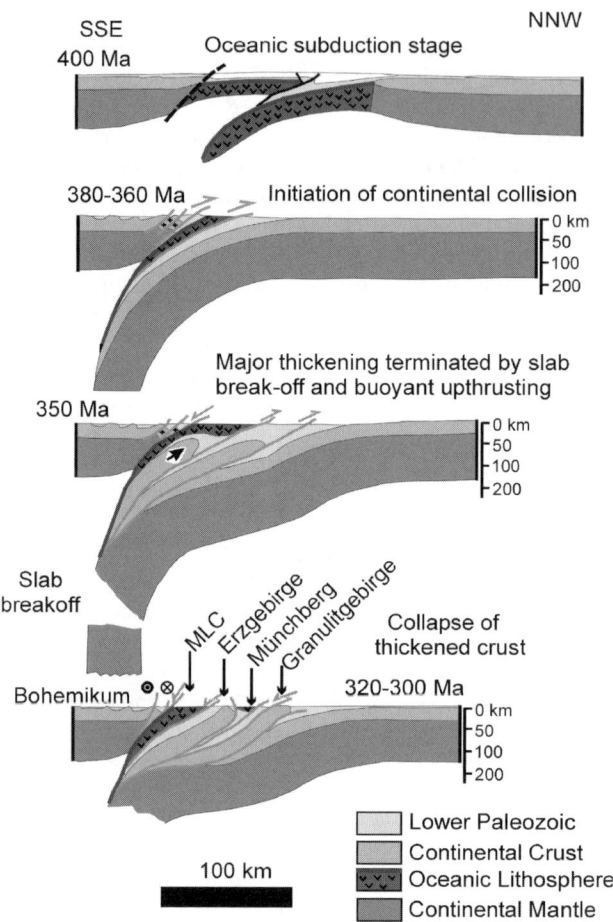

FIG. 1. The subduction-collision model by Matte (1998) redrawn by O'Brien (2000), considering the geological situation of the Bohemian Massif in the north-eastern Variscan orogen. MLC = Mariánské Lázne Complex.

15 m.y. younger. The second peak appears around 340 Ma. Cooling ages are on average only a few m.y. younger.

The first HP event was related to the beginning of the collision of Gondwana and Laurussia (O'Brien, 2000; Massonne, 2003a; Massonne and O'Brien, 2003). This continent-continent collision was responsible for the exhumation of portions of subducted oceanic crust, transformed to eclogites, back to the surface. The corresponding process is not well understood but at least in the case of the India-Asia collision, eclogites were found in the Himalaya with ages (~ 50 Ma) coinciding with the age of the contact of India with Asia (Kohistan arc). The characteristics of these eclogites (see Massonne and O'Brien, 2003), which contain relics of coesite, are: (1) peak P-T conditions close to 30 kbar, 700°C; (2) amphibole porphyroblasts formed during a late eclogite stage; (3) cooling ages that are 10 to 15 m.y. younger than the peak metamorphic age; and (4) occurrence as lenses embedded, for instance, in medium-temperature, medium-pressure metapelites. These characteristics also apply to Lower Devonian eclogites of the Variscan orogen, for instance, exposed in the northwestern Bohemian Massif (Massonne and O'Brien, 2003). In addition, such eclogites have geochemical MORB signatures (Bernard-Griffiths and Cornichet, 1985; Stosch and Lugmair, 1990) expected for subducted oceanic crust. After the initial event, the area of continental crust, thickened to 60 km and more, steadily increased to an area of several million km^2 (see Fig. 2). The upper plate

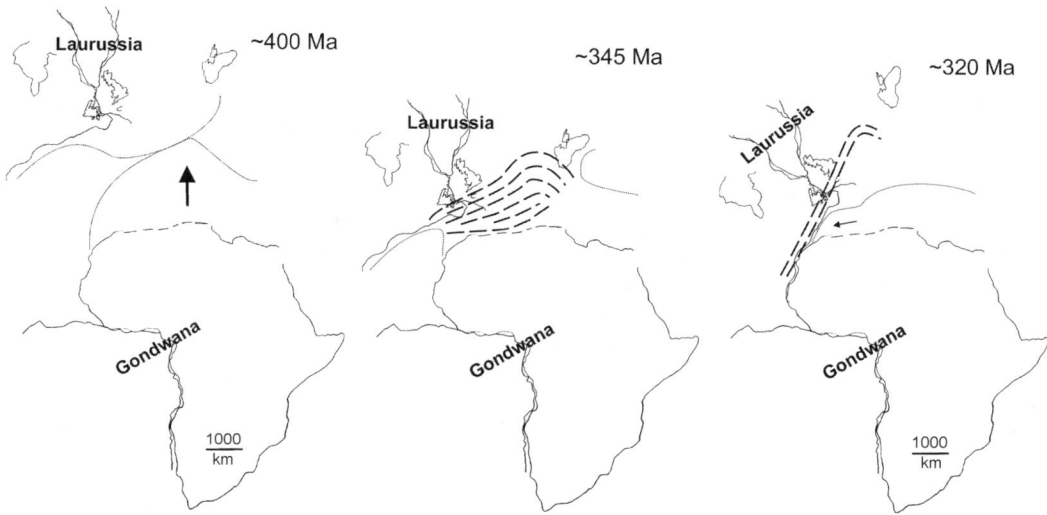

FIG. 2. Collision of Gondwana and Laurussia presented by the position of these continental plates relative to each other in the Early Devonian (left hand side), Early Carboniferous, and Late Carboniferous (right hand side), according to Massonne (2003a). The positions of the plates refer to paleomagnetic data (see Tait et al., 2000). For orientation the current shapes of the continental plates, except portions overprinted by the Alpine orogeny (see northern Africa), and other typical geographic shapes (Hudson Bay, Newfoundland, Great Britain, Black Sea) are shown. The estimated shapes of the continental plates at the corresponding times are displayed by dotted lines (northern margin of Africa refers to pre-Alpine times). Thick dashed lines mark orogenic zones (= thickened continental crust). Arrows point to the relative plate motions of Gondwana.

during this process was very likely Gondwana (see Massonne, 2003a) because crustal fragments, which are believed to have been part of this plate, for instance, gneisses and granitoids in the Lugicum of the Bohemian Massif, were not metamorphosed or, at least, never experienced metamorphism under HP conditions (> 8 kbar) during Variscan times (see, e.g., Linnemann et al., 2000).

Crustal thickening continued until an event took place about 340 Ma, leading to the second formation of HP and UHP rocks (Massonne, 2003a; Massonne and O'Brien, 2003). During the intermediate time period of 50–55 m.y., evidence for sedimentation is rare, except in marginal areas of the Variscan orogen, and igneous rocks related to magmatic arcs do not occur. In contrast to that, sedimentation and volcanism as evidence for Silurian to Early Devonian subduction of oceanic crust exist (see also above). For instance, MORB-derived low-grade metamorphic rocks in the northern part of the Variscan orogen in central Europe (e.g., Kryza, 1995; Meisl, 1995), probably part of accretionary wedges, and calc-alkaline volcanism, which can be related to a magmatic arc, also are known from the North American portions of the Variscan orogen (see summaries in Keppie, 1989; Neugebauer, 1989).

The second HP event, around 340 Ma after extensive crustal and lithospheric thickening, was caused by lithospheric delamination. In contrast to the first HP event related to the subduction of oceanic crust that was mainly transformed to eclogites, various lithologies experienced HP and even UHP conditions. Currently exposed sections of basement from the lower portion of the thickened crust, characterized by diverse HP granulites and migmatitic gneisses, contain lensoid bodies of serpentinized garnet peridotites, eclogites, and UHP felsic rocks that are spatially closely related (see review by Massonne and O'Brien, 2003). Slices of garnet peridotites, representing deeper mantle (e.g., Brueckner and Medaris, 1998) could have been introduced into the base of the continental crust when hot mantle ascended to replace cold lithospheric mantle that had sunk into the mantle through the delamination process. Eclogites, typically hotter (> 800°C) than those formed during subduction of oceanic crust in the Middle Devonian, could have been part of the lower crust before the delamination process started. This crustal origin is suggested by the variability of the geochemical signature of the eclogites, that can be assigned to various protoliths, for instance, of within-plate magmatism or marly sediments (Massonne and Czambor, 2003).

For felsic rocks, the origin from the lower portion of thickened crust is supported by the study of diamondiferous quartzofeldspathic rock (Massonne, 2003b) from the Saxonian Erzgebirge, situated in a northwestern part of the Bohemian Massif. After this rock sank deep (possibly 250 km) into the mantle, it became partially molten (Massonne, 2003b) causing the subsequent fast ascent of the rock. After ascent, the diamondiferous rock was emplaced at the base of still-thickened continental crust. The thickness of the continental crust may have been reduced by at least 10 km compared to the thickness before delamination because a pressure of 18–20 kbar (≥60 km) was determined for the pre-UHP stage of the diamondiferous rock and, thus, prior to its descent into the mantle (Massonne and Nasdala, 2003) and about 15 kbar (~50 km) after emplacement. This would be evidence for thinning of the continental crust by the delamination process, which would also explain why basins were newly formed despite continuous collision. In these, up to several km-thick sequences of Viséan and Upper Carboniferous sediments were deposited. Moreover, both sedimentary and metamorphic ages can be contemporaneous (e.g., Franke and Engel, 1986). Nearly at the same time, large volumes of S-type granitoid magma, the formation of which was possibly triggered by ascending hot mantle material causing an increase of the geothermal gradient, intruded the Variscan crust (see stage 5 of Fig. 3). Nappe tectonics (e.g., Matte and Burg, 1981; Schulmann et al., 1991) as well as major fault and shear zones (e.g., Rajlich, 1990; Krohe, 1996) also contributed to a collage of many small basement blocks of different age and metamorphic evolution, and sedimentary rocks, which are predominantly only anchimetamorphic, currently exposed at the surface side by side. This collage-like assembly is discernible already in the Bohemian Massif.

The collision of Gondwana and Laurussia did not seem to have ceased immediately after the delamination process, but the speed of plate convergence might have been significantly diminished. In addition, it was reported from various portions of the Variscan orogen (e.g., Conti et al., 2001; Edel, 2001; Konopásek et al., 2001) that the direction of maximum strain had changed from NNW-SSE to close to an E-W direction (see Fig. 2) during a late stage of the Variscan orogeny. The change of the stress regime could indicate that the movement of (Paleo)Asia (the northern portion of the present Asian plate), which had already resulted in the formation of the Urals by collision with Laurussia, began to dominate the plate motions of Laurussia and Gondwana. Despite continuous orogenic events (Hercynian phase) in the Late Carboniferous, crustal thickening did not reach 60 to 70 km again. Instead, a maximum range of 40 to 50 km crustal thickness was estimated by adding emplacement depths of non-deformed Late Carboniferous S-type granitoids, resulting in values around 15 km for currently exposed rocks (Massonne, 1984), and the present thickness of the mid-European crust (28–30 km). Finally, the movement of the continental plates, Gondwana, Laurussia, and (Paleo)Asia, toward each other ended at the beginning of the Permian as generally accepted (see, e.g., Condie, 1989). These irrecoverably amalgamated plates, however, started to separate again by the opening of the Atlantic Ocean about 100 m.y. later.

General Characteristics of Continent-Continent Collision

The attempt is made here to derive a general scheme for the main geodynamic processes and their consequences during a continent-continent collision lasting tens of million years, and also to show where the delamination process settled within an orogenic development. For that purpose, the evolution of the Late Paleozoic collision of Gondwana and Laurussia outlined above and the current and Tertiary situation of the Himalayan orogen (+ Tibetan plateau) is considered. The following stages and corresponding characteristics are presented in a chronological sequence for continental plates drifting towards each other (see also Fig. 3).

Stage 1. Approach of continental plates. The oceanic lithosphere between the approaching continental plates is continuously subducted. As a consequence, a magmatic arc occurs at one of the continental plates parallel to its margin. This magmatic arc remains active until the end of stage 1. Flysch sedimentation occurs at the margin of the continental plate overlying the subduction zone, and starts at the previously passive margin of the other continent when both continental plates are almost in contact.

Stage 2. Beginning of continent-continent collision. Magmatic activity and flysch sedimentation of stage 1 have ceased. A distinct and discrete time marker for the beginning of stage 2 is the metamorphic age of eclogites representing the previously subducted oceanic crust. These eclogites were subjected to relatively low metamorphic temperatures at UHP or at least near-UHP conditions. They are exhumed during stage 2 (see the above examples such as the 50 Ma old Himalayan eclogites) because the underplated leading wedge of the continent, which follows the subducted oceanic crust, squeezes out upper portions of the formerly developed subduction channel, containing ascended eclogitic material (see results of numerical modeling on the mass flows in subduction channel environments by Gerya et al., 2002) as well as portions of lower crust from the overlying continental plate.

Stage 3. Lateral extension of thickened crust. The formation of thickened continental crust, simplified in Figure 3 by the overlap of the continental crust of both colliding plates, was initiated during stage 2. During a long period of time, possibly tens of million years, the area with 60 and more km thick continental crust can increase dramatically. There is no magmatic arc. Sedimentary basins are rare except along the margins of thickened continental crust.

Stage 4. Delamination of the continental lithosphere involving material from the continental crust. After thickened continental crust and lithospheric mantle had spread over an extended area, this metastable state starts to turn to a stable state by delamination of the lithosphere. This may not be a single cataclysmic event. Instead, repeated delamination of different portions of the lithosphere within a limited time span could be similar to a single event. Lower portions of the thickened crust are involved in the delamination process. The corresponding rocks are metabasic, metasedimentary, and possibly felsic meta-igneous rocks as well; many of them are subjected to UHP metamorphism. Unknown but possibly minute quantities of these rocks return to the Earth's surface either by buoyancy forces alone or through an anatectic stage as partially molten rocks. Metabasic rocks, eclogites, involved in the delamination differ from those of stage 2 by higher metamorphic temperatures. As a response of the sinking of the continental lithosphere deep into the mantle, hot mantle from the asthenosphere ascends (Fig. 3). Fragments of these rocks, as well as ascending crustal material involved in the delamination, are introduced into the lower crust (after delamination). The local crustal involvement in the lithospheric delamination should depend on the quantity of dense, i.e., mainly basic, material in the corresponding crustal portions. This involvement can be so intense that only relatively thin crust (20–30 km thick) remains in segments of the formerly equally thickened crust, so that basins suddenly evolve that can be filled with km-thick sequences of sediments.

Stage 5. Continuation of continent-continent collision. Despite the evolution of basins and ranges after the lithospheric delamination due to differentially thickened (or thinned) crust, continent-continent collision continues with reduced convergence rates of the colliding plates and/or with different convergence directions inasmuch as the plate motions are the result of the global force balance on the plates. Portions of the orogenic crust could be somewhat thickened again, but many young sedimentary basins are closed. The formation of large volumes of S-type granitoid magmas and their penetration into high crustal levels are characteristic phenomena of stage 5. The corresponding anatexis of the lower crust is due to hot mantle which had replaced the cold lithospheric mantle below the orogenic crust during stage 4.

Stage 6. Cessation of continent-continent collision. The waning stages of the orogeny resulting from continent-continent collision finally ceased a few tens of million years after lithospheric delamination. Denudation of the orogen is the final act.

As previously demonstrated for the Variscan orogen, area-wide geochronological, geochemical, petrotectonic, and geophysical data are required to reconstruct the geodynamic evolution of a collisional orogen. For a number of other Phanerozoic orogens formed by continent-continent collision, the data quantity is significantly less. Under these circumstances, it is difficult to evaluate if the above orogenic scheme is applicable to all these orogens. Even in case of the well-studied and young Alpine orogenic chains (Tethysides) in the Mediterranean region, resulting from the collision of the African plate and the European part of the Eurasian plate, the application of this scheme would result in a controversial view of the corresponding orogenic development. At present, several other evolutionary models have been proposed (e.g., Avigad et al., 1997; Stampfli et al., 1998; Jolivet and Faccenna, 2000) that differ significantly from a model postulating crustal thickening over an extended area (hatched area in Fig. 4) in a time interval of several tens of million years. Nevertheless, it can be

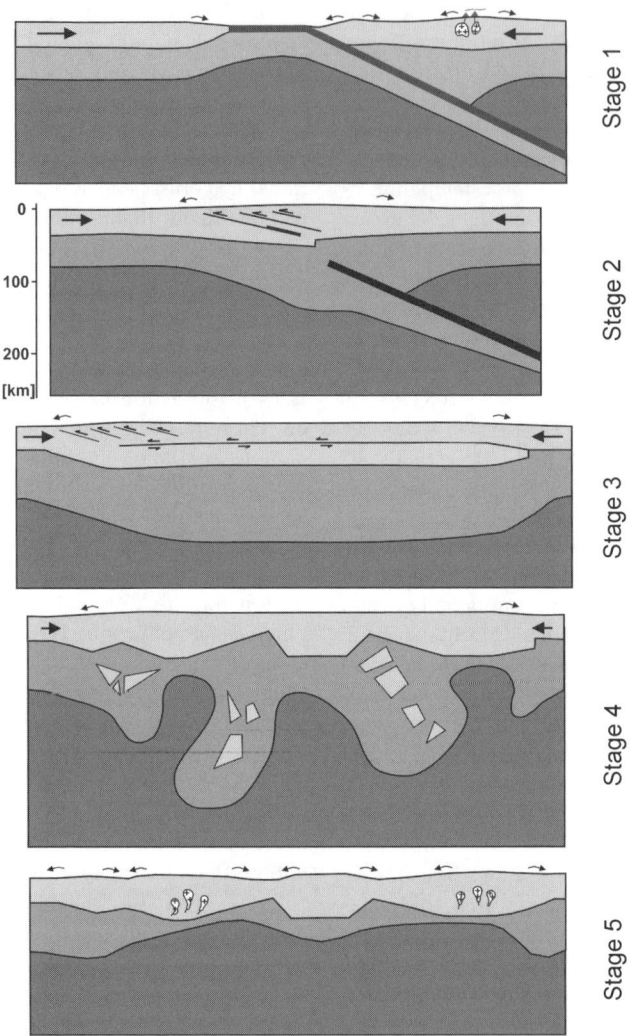

FIG. 3. Model for continent-continent collision with lithospheric delamination (stage 4) shown in a chronological sequence from top to bottom. The five stages displayed correspond to orogenic events outlined in the text. Black = oceanic crust; light = continental crust; light grey = lithospheric mantle; dark grey = asthenospheric mantle. Bubbles filled with crosses = intermediate to acidic magmas. Thick arrows mark the motions of the continental plates. The lengths of these arrows indicate the relative velocity of the plates. Small curved arrows mark transport of sediments into basins. Half arrows indicate directions of overthrusting.

demonstrated that the characteristics of the above stages 1 to 5 can be found in the Tethysides as well.

There is no doubt since the discovery of the HP nature of metabasites in the European Alps (most are low-temperature eclogites; e.g., Ernst, 1973) that subduction of oceanic crust occurred in such a way that the African and Eurasian plates could approach each either (stage 1). Important time markers for the close of this process and, thus, the inferred collision of the plates (stage 2) are the ages of eclogite-facies metamorphism and the exhumation of eclogites. According to Hunziker et al. (1992), this so-called Eo-Alpine event occurred in Late Cretaceous time (however, see also Froitzheim et al., 1996). Eclogites of the Austrian Alps can even be somewhat older (e.g., Neubauer et al., 1999). Evidence for stage 3 (crustal thickening) is rare. Nevertheless, rocks of the lower portion of the

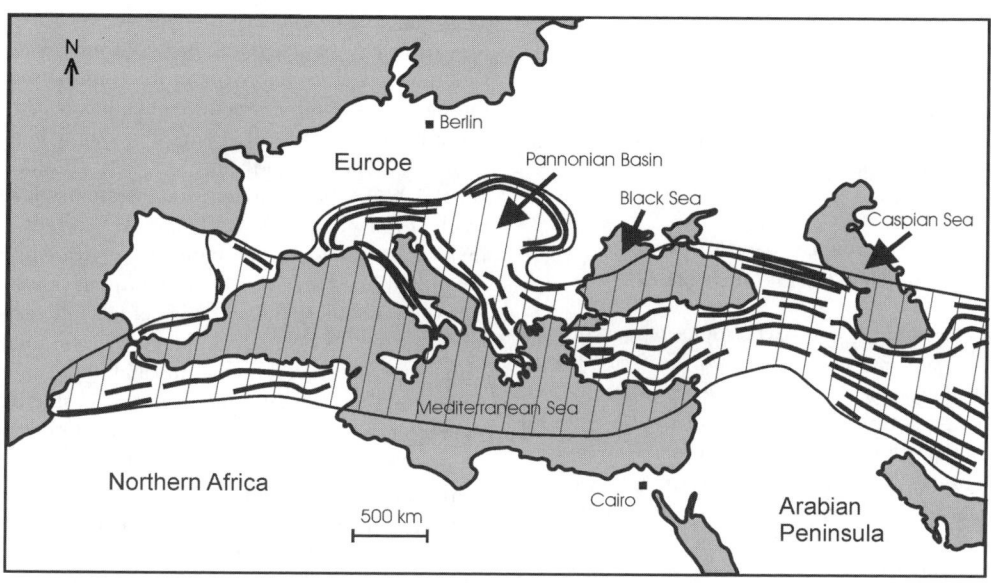

FIG. 4. Orogenic region of the Tethysides (marked by the hatched area). At present, this region can be subdivided into mountain chains (thick lines) and several basins (Pannonian Basin or those [grey] with surface below sea level). The single thick arrow shows relative motion of the Anatolian plate caused by the Arabian plate indenting into the Tethysides. Perhaps this also resulted in the deflection of the Carpathians and Balkanides southeast of the Pannonian Basin.

crust, which resulted from stacking or duplicating continental crust, occur, such as metapelites and metagranitoids of the Sesia zone in the European Alps, which experienced HP metamorphism (19.2 kbar and 605°C according to Massonne, 1995) proven by the appearance of jadeitic pyroxene in felsic rocks. These P-T conditions were attained at about 65 Ma (Duchêne et al., 1997) and, thus, during the hypothesized stage 3 for the Tethysides. Another episode of the formation of HP and UHP rocks is related to Paleogene time. UHP rocks of the European Alps, consisting of various lithologies—for instance, felsic and mafic rocks (eclogites) with coesite and coesite pseudomorphs in the western Alps (see summary in Compagnoni and Rolfo, 2003) or mafic and ultramafic rocks in the central Alps (e.g., Dobrzhinetskaya et al., 1996)—are as young as 35 Ma (see summary in Rubatto et al., 2003). This age is related here to stage 4, the delamination of the lithosphere, which resulted in the burial of rocks from the continental crust as deep as 100 km or more.

However, in the westernmost portions of the Tethysides, the HP rocks and associated garnet peridotites with pseudomorphs after diamond (Pearson et al., 1995) in the Betic Cordillera and the Moroccan extension, the Rif Mountains, seem to be as young as Early Miocene (Sánchez-Rodríguez and Gebauer, 2000). The Alboran Sea between these ranges is believed to be thinned continental crust that was previously thickened (Platt and Vissers, 1989). Zeck (1996) proposed a lithospheric delamination process to explain this. Other basins of the Mediterranean region formed in the Late Eocene or Oligocene according to the oldest sediments that fill these basins (see, e.g., summaries in Horváth and Tari, 1999; Séranne, 1999). This would indicate that the proposed delamination involved large portions of continental crust, resulting in a significant reduction of the thickness of continental crust in some areas, and started indeed around 35 Ma. Currently, the Mediterranean region would represent stage 5 of the above model for an orogenic evolution due to continent-continent collision. Other regions on Earth could also pass through this orogenic stage at present. A possible region is in southeast Asia where areas of thin continental crust (e.g., basins of the Sunda Sea and the South China Sea) and those with thickened continental crust (Malayan Peninsula, southern Indochina, and islands of northwestern Indonesia) are closely related. Cretaceous UHP metamorphic rocks, formed possibly at stage 2 of the above scheme, have been reported by Parkinson et al. (1998). However, a recent geochronological study on a garnet lherzolite from central Sulawesi (Kadarusman et al., 2001) gave a Late Oligocene age. This age could possibly be related to delamination (stage 4) in that region.

Global Consequences of Crustal Delamination

If we consider that UHP rocks of continental protoliths are not older than 600 Ma (see Chopin, 2003), the process of lithospheric delamination, involving continental crust, could play a role not earlier than the Late Neoproterozoic. However, in Phanerozoic time it might be that this process was common for colliding continental plates. What could be the consequences of lithospheric delamination involving continental crust with respect to the evolution of the continental crust on Earth, in general? First, we have to consider which portions of the crust can be involved in the delamination. Most likely are portions of the lower crust, which is believed to consist of a high proportion of granulite-facies metabasic rocks. At pressures of ~15 kbar (~ 50 km Earth's depth) these rocks transform to eclogites that are denser even than the lithospheric mantle.

Density calculations show that some quartz-poor metapelitic rocks can also be denser than the mantle at HP (Massonne and Willner, 2004). A significant step toward high densities of metapelitic rocks is caused by the formation of Na-rich clinopyroxene from plagioclase. The high densities of metabasic and metapelitic rocks at depths greater than 50 km could account for the involvement of continental crust in lithospheric delamination since Phanerozoic times if we assume that thickening of continental crust did not exceed 50 km in the Precambrian. Once this magic limit of thickened continental crust is exceeded, the lowermost crust can become so dense that this portion of the continental crust can sink together with the thickened mantle lithosphere into the deeper mantle. Once the process of delamination of continental crust has started, the downgoing continental material can pull adherent lighter material beyond the 15 kbar limit. Then, plagioclase in the lighter material is transformed to omphacite-jadeite, resulting in a significant density increase, as mentioned above. Hence, this process, like subduction of oceanic crust, can steadily run for a while by itself.

A major consequence of the above process is the loss of (basic to intermediate) continental crust, which would presumably be recycled in the mantle through every continent-continent collision. Thus, it could be that the so-called enriched mantle component (e.g., Hawkesworth et al., 1984; Zindler and Hart, 1986), characterized, for instance, by relatively high contents of Rb and U, reflects the involvement of crustal material in lithospheric delamination. However, once continent-continent collision, including delamination of the lithosphere, has occurred, this collisional process in conjunction with extensive crustal thickening and, thus, another delamination of the lithosphere would not start again soon after, to be possibly repeated several times.

In case of the collision of the African plate (formerly part of Gondwana) and the western part of Eurasia, lithospheric delamination happened in the Late Paleozoic, as outlined above, but then there was a pause of at least 200 m.y. until a very similar continent-continent collision started again, resulting in the formation of the Tethysides. Surprisingly, a pause of similar time span can be assigned to orogenies in today's China. For instance, UHP metamorphism occurred in the Dabie-Sulu terrane as a result of the collision of the Sino-Korean and Yangtze cratons at about 220 Ma (see summary in Jahn et al., 2003). Because felsic, mafic, and ultramafic lithologies are involved in UHP metamorphism, it might be that this age represents that of stage 4 of the corresponding orogeny. The same process (stage 4 again) happened at ~490 Ma during a very similar collisional situation, inasmuch as corresponding UHP rocks occur in the area of the Qilian Massif–Qaidam Basin (Yang et al., 2002; Song et al., 2003), which is more or less the western extension of the Dabie-Sulu terrane.

Despite these orogenic pauses, the loss of a significant volume of continental crust since the Late Neoproterozoic must be considered, although a quantification of this loss would be highly speculative at present. For a rough estimate, it is assumed here that the process of lithospheric delamination (with involvement of continental crust) occurred only 10 times during the last 600 m.y. In addition, a mean area equivalent to the present thickened continental crust in the regions of the Himalaya, Tibetan Plateau, Tien Shan, and Nan Shan (about 4 million km^2) would have been subjected to this process. Furthermore, it is assumed that on average the complete lower crust (15 to 18 km thick) of the lower plate (India of this example) would have been recycled into the mantle by a delamination process, then an equivalent of about two-thirds of the continental crust of the African plate (~30 million km^2, 33 km thick) would have been lost during the last 600 m.y. However, such events could have occurred more often than 10 times since the Late Neoproterozoic and the mean thickness of delaminated continental crust could be larger than assumed above, as is suggested by the extended basins (= thin continental crust) in the Mediterranean area (see Fig. 4). The consequence would be that amounts of continental crust even clearly larger than an equivalent of two-thirds of the continental crust of the African plate have been recycled since the Late Neoproterozoic. Under these circumstances, one conclusion must be that it is almost impossible to reconstruct former constellations of crustal plates, for instance, those of the ancestral supercratons (see, e.g., Bleeker, 2003).

In addition, as continent-continent collision will occur in the future as well, and, thus, lithospheric delamination with continental crust involved, it is likely that a possible new supercontinent (Amasia?) would be the last one that would appear on Earth. Moreover, the loss of continental crust implies that the ratio of continental crust/oceanic crust was higher in Precambrian time. The area related to deep oceanic basins could have been equal or even minor compared to the area covered by continental masses (continental shelf included) before the onset of lithospheric delamination with continental crust involved. Instead of extended deep oceans as today's Pacific Ocean, shallow oceans were abundant, as wide areas of the continents were covered by water. Thus, the loss of continental crust would have had a significant influence on Earth's climate and on biological evolution as well.

Finally, in order to emphasize the difference between subduction of an oceanic plate into the mantle along a convergence zone, and burial of large portions (see Fig. 3 and above estimation) of continental crust by lithospheric delamination, it is proposed not to use the term "subduction" for the latter process. Instead, an old but nearly forgotten term "subfluence" (see, e.g., Kraus, 1951) is suggested here. For instance, Behr (1978) referred to subfluence as a geodynamic process related to deep burial of crustal rocks from all sides toward a center deep in the mantle, in contradistinction to the burial of a nearly planar megasegment (that is subducted oceanic crust), bounded by the Benioff zone. This author thus explains the formation of high-temperature and HP Variscan rocks, including their metamorphic fabric on a large scale, for the entire Bohemian Massif. Because of the similarity to the suggested lithospheric delamination with involvement of continental crust, the above proposal, subfluence, is justified.

Acknowledgments

I am indebted to M. Cloos and J.G. Liou for constructive comments on an earlier version of the manuscript.

REFERENCES

Avigad, D., Garfunkel, Z., Jolivet, L., and Azanon, J. M., 1997, Back arc extension and denudation of Mediterranean eclogites: Tectonics, v. 16, p. 924–941.

Behr, H.-J., 1978, Subfluenz-Prozesse im Grundgebirgs-Stockwerk Mitteleuropas: Zeitschrift der deutschen Geologischen Gesellschaft, v. 129, p. 283–318.

Bernard-Griffiths, J., and Cornichet, J., 1985, Origin of eclogites from southern Brittany, France: A Sm-Nd isotopic and REE study: Chemical Geology, v. 52, p. 185–201.

Bleeker, W., 2003, The late Archean record: A puzzle in ca. 35 pieces: Lithos, v. 71, p. 99–134.

Brueckner, H. K., and Medaris, L. G., Jr., 1998, A tale of two orogens: The contrasting T-P-t history and geochemical evolution of mantle in high- and ultrahigh-pressure metamorphic terranes of the Norwegian Caledonides and the Czech Variszides: Schweizerische Mineralogische und Petrographische Mitteilungen, v. 78, p. 293–307.

Chopin, C., 2003, Ultrahigh pressure metamorphism: Tracing continental crust into the mantle: Earth and Planetary Science Letters, v. 212, p. 1–14.

Compagnoni, R., and Rolfo, F., 2003, UHPM Units in the Western Alps, *in* Carswell, D. A., and Compagnoni, R., eds., Ultrahigh pressure metamorphism: EMU Notes in Mineralogy, v. 5, p. 13–49.

Condie, K. C., 1989, Plate tectonics and crustal evolution: Oxford, UK, Pergamon Press, 476 p.

Conti, P., Carmignani, L., and Funedda, A., 2001, Change of nappe transport direction during the Variscan collisional evolution of central-southern Sardinia (Italy): Tectonophysics, v. 332, p. 255–273.

Davies, J. H., and von Blanckenburg, F., 1995, Slab breakoff: A model of lithospheric detachment and its test in the magmatism and deformation of collisional orogens: Earth and Planetary Science Letters, v. 129, p. 85–102.

Dobrzhinetskaya, L., Green, H. W., II, and Wang, S., 1996, Alpe Arami, a peridotite massif from depths of more than 300 kilometers: Science, v. 271, p. 1841–1845.

Duchêne, S., Blichert-Toft, J., Luais, B., Télouk, P., Lardeaux, J.-M., and Albarède, F., 1997, The Lu-Hf dating of garnets and the ages of the Alpine high-pressure metamorphism: Nature, v. 387, p. 586–589.

Edel, J. B., 2001, The rotations of the Variscides during the Carboniferous collision: Paleomagnetic constraints from the Vosges and the Massif Central (France): Tectonophysics, v. 332, p. 69–92.

Ernst, W. G., 1973, Interpretative synthesis of metamorphism in the Alps: Geological Society of America Bulletin, v. 84, p. 2053–2088.

Franke, W., and Engel, W., 1986, Synorogenic sedimentation in the Variscan belt of Europe: Bulletin de Société Géologique de France, v. 1986 II, p. 25–33.

Froitzheim, N., Schmid, S. M., and Frey, M., 1996, Mesozoic palaeogeography and the timing of eclogite facies metamorphism in the Alps: A working hypothesis: Eclogae Geologica Helvetiae, v. 89, p. 51–110.

Gerya, T. V., Stöckhert, B., and Perchuk, A. L., 2002, Exhumation of high-pressure metamorphic rocks in a subduction channel: A numerical simulation: Tectonics, v. 21, p. 1–19.

Hawkesworth, C. J., Rogers, N. W., van Calsteren, P. W. C., and Menzies, M. A., 1984, Mantle enrichment processes: Nature, v. 311, p. 331–335.

Horváth, F., and Tari, G., 1999, IBS Pannonian Basin project: A review of the main results and their bearing on hydrocarbon exploration, in Durand, B., Jolivet, L., Horváth, F., and Séranne, M., eds., The Mediterranean basins: Tertiary extension within the Alpine orogen: Geological Society of London Special Publications, v. 156, p. 195–213.

Houseman, G. A., McKenzie, D. P., and Molnar, P., 1981, Convective instability of a thickened boundary layer and its relevance for the thermal evolution of the continental crust: Journal of Geophysical Research, v. 86, p. 6115–6132.

Hunziker, J. C., Desmond, J., and Hurford, A. J., 1992, Thirty-two years of geochronological work in the Central and Western Alps: A review on seven maps: Mémoir de Géologique, v. 13, 59 p.

Jahn, B.-M., Rumble, D., and Liou, J. G., 2003, Geochemistry and isotope tracer study of UHP metamorphic rocks, in Carswell, D. A., and Compagnoni, R., eds., Ultrahigh pressure metamorphism: EMU Notes in Mineralogy, v. 5, p. 365–414.

Jolivet, L., and Faccenna, C., 2000, Mediterranean extension and the Africa-Eurasia collision: Tectonics, v. 19, p. 1095–1106.

Kadarusman, A., Brueckner, H., Yurimoto, H., Parkinson, C. D., and Maruyama, S., 2001, Geochemistry and Sm-Nd dating of garnet peridotites from central Sulawesi, and its implication to the Neogene collision complex in Eastern Indonesia [abs.]: EOS (Transactions of the American Geophysical Union), v. 82, no. 47, abstract T5D-08.

Keppie, J. D., 1989, Northern Appalachian terranes and their accretionary history: Geological Society of America Special Paper, v. 230, p. 159–192.

Konopásek, J., Schulmann, K., and Lexa, O., 2001, Structural evolution of the central part of the Krusne Hory (Erzgebirge) Mountains in the Czech Republic—evidence for changing stress regime during Variscan compression: Journal of Structural Geology, v. 23, p. 1373–1392.

Kosarev, G., Kind, R., Sobolev, S. V., Yuan, X., Hanka, W., and Oreshin, S., 1999, Seismic evidence for a detached Indian lithospheric mantle beneath Tibet: Science, v. 283, p. 1306–1309.

Kraus, E. C., 1951, Vergleichende Baugeschichte der Gebirge: Berlin, Germany, Akademie Verlag, 587 p.

Krohe, A., 1996, Variscan tectonics of central Europe: Postaccretionary intraplate deformation of weak continental lithosphere: Tectonics, v. 15, p. 1364–1388.

Kryza, R., 1995, VI. Western Sudetes, D. Igneous activity, in Dallmeyer, R. D., Franke, W., and Weber, K., eds., Pre-Permian Geology of Central and Eastern Europe: Berlin, Germany, Springer, p. 341–350.

Linnemann, U., Gehmlich, M., Tichomirowa, M., Buschmann, B., Nasdala, L., Jonas, P., Lützner, H., and Bombach, K., 2000, From Cadomian subduction to Early Palaeozoic rifting: The evolution of Saxo-Thuringia at the margin of Gondwana in the light of single zircon geochronology and basin development (Central European Variscides, Germany), in Franke, W., Haak, V., Oncken, O., and Tanner, D., eds., Orogenic processes: Quantification and modelling in the Variscan belt: Geological Society of London, Special Publications, v. 179, p. 131–153.

Massonne, H.-J., 1984, Bestimmung von Intrusionstiefen variszischer Granite Mitteleuropas und Neuschottlands anhand der Chemie ihrer Hellglimmer: Fortschritte der Mineralogie, v. 62, Beiheft 1, p. 147–149.

Massonne, H.-J., 1995, Is the concept of "in situ" metamorphism applicable to deeply buried continental crust with lenses of eclogites and garnet peridotites?: Chinese Science Bulletin, v. 40 (suppl.), p. 145–147.

Massonne, H.-J., 2003a, Paläozoische Hochdruck-und Ultrahochdruck-Metamorphite in Mitteleuropa und ihre Beziehungen zur variszischen Orogenese: Zeitschrift für geologische Wissenschaften, v. 31, p. 239–249.

Massonne, H.-J., 2003b, A comparison of the evolution of diamondiferous quartz-rich rocks from the Saxonian Erzgebirge and the Kokchetav Massif: Are so-called diamondiferous gneisses magmatic rocks?: Earth and Planetary Science Letters, v. 216, p. 345–362.

Massonne, H.-J., and Czambor, A., 2003, Protoliths of eclogites from the Variscan Erzgebirge [abs.]: Norges geologiske undersøkelse, Report 2003.055, Eclogite Field Symposium, p. 93–94.

Massonne, H.-J., and Nasdala, L., 2003, Characterization of an early metamorphic stage through inclusions in zircon of a diamondiferous quartzofeldspathic rock from the Erzgebirge, Germany: American Mineralogist, v. 88, p. 883–889.

Massonne, H.-J., and O'Brien, P.J., 2003, The Bohemian Massif and the NW Himalaya, in Carswell, D. A., and Compagnoni, R., eds., Ultrahigh pressure metamorphism: EMU Notes in Mineralogy, v. 5, p. 145–187.

Massonne, H.-J., and Willner, A.P., 2004, Densities of psammopelitic rocks at ultrahigh pressure (UHP) conditions [abs.]: 32nd International Geological Congress, Florence, Italy, Abstract Volume, p. 536.

Matte, P., 1998, Continental subduction and exhumation of HP rocks in Paleozoic orogenic belts: Uralides and Variscides: Geologiska Föreningens Stockholm Förhandlingar, v. 120, p. 209–222.

Matte, P., and Burg, J. P., 1981, Sutures, thrusts and nappes in the Variscan Arc of western Europe: Plate tectonic implications, in McClay, K. R., and Price, N. J., eds., Thrust and nappe tectonics: Geological Society of London Special Publications, v. 9, p. 353–358.

McKerrow, W. S., Mac Niocaill, C., Ahlberg, P. E., Clayton, G., Cleal, C. J., and Eagar, R. M. C., 2000, The Late Palaeozoic relations between Gondwana and Laurussia, in Franke, W., Haak, V., Oncken, O., and Tanner, D., eds., Orogenic processes: Quantification and modelling in the Variscan belt: Geological Society of London, Special Publications, v. 179, p. 9–20.

Meisl, S., 1995, III.C Metamorphic units (Northern Phyllite Zone), III.C.3 Igneous activity, in Dallmeyer, R. D., Franke, W., and Weber, K., eds., Pre-Permian geology of central and eastern Europe: Berlin, Germany, Springer, p. 118–131.

Neubauer, F., Dallmeyer, R. D., and Takasu, A., 1999, Conditions of eclogite formation and age of retrogression within the Sieggraben unit, Eastern Alps: Implications for Alpine-Carpathian tectonics: Schweizerische Mineralogische und Petrographische Mitteilungen, v. 79, p. 297–307.

Neugebauer, J., 1989, The Iapetus model: A plate tectonic concept for the Variscan belt of Europe: Tectonophysics, v. 169, p. 229–256.

O'Brien, P. J., 2000, The fundamental Variscan problem: High-temperature metamorphism at different depths and high-pressure metamorphism at different temperatures, in Franke, W., Haak, V., Oncken, O., and Tanner, D., eds., Orogenic processes: Quantification and modelling in the Variscan belt: Geological Society of London, Special Publications, v. 179, p. 369–386.

Parkinson, C. D., Miyazaki, K. Wakita, A. J., Barber, A., and Carswell, D. A., 1998, An overview and tectonic synthesis of the very high pressure and associated rocks of Sulawesi, Java and Kalimantan, Indonesia: The Island Arc, v. 7, p. 184–200.

Pearson, D. G., Davies, G. R., and Nixon, P. H., 1995, Orogenic ultramafic rocks of UHP (diamond facies) origin, in Coleman, R. G. and Wang, X., eds., Ultrahigh pressure metamorphism: Cambridge, UK, Cambridge University Press, p. 456–510.

Platt, J. P., and England, P., 1994, Convective removal of lithosphere beneath mountain belts: thermal and mechanical consequences: American Journal of Science, v. 294, p. 307–336.

Platt, J. P., and Vissers, R. L. M., 1989, Extensional collapse of thickened continental lithosphere: A working hypothesis for the Alboran Sea and Gibraltar arc: Geology, v. 17, p. 540–543.

Rajlich, P., 1990, Variscan shearing tectonics in the Bohemian massif: Mineralia Slovaka, v. 22, p. 33–40.

Robardet, M., Paris, F., and Racheboeuf, P. R., 1990, Palaeogeographic evolution of southwestern Europe during Early Palaeozoic times, in McKerrow, W. S. and Scotese, C. R., eds., Paleozoic paleogeography and biogeography: Geological Society of London Memoirs, v. 12, p. 411–419.

Rubatto, D., Liati, A., and Gebauer, D., 2003, Dating UHP metamorphism, in Carswell, D. A., and Compagnoni, R., eds., Ultrahigh pressure metamorphism: EMU Notes in Mineralogy, v. 5, p. 341–363.

Sánchez-Rodríguez, L., and Gebauer, D., 2000, Mesozoic formation of pyroxenites and gabbros in the Ronda area (southern Spain), followed by Early Miocene subduction metamorphism and emplacement into the middle crust: U-Pb sensitive high resolution ion microprobe dating of zircon: Tectonophysics, v. 316, p. 19–44.

Schott, B., and Schmeling, H., 1998, Delamination and detachment of a lithospheric root: Tectonophysics, v. 296, p. 225–247.

Schulmann, K., Ledru, P., Autran, A., Melka, R., Lardeaux, J. M., Urban, M., and Lobkowicz, M., 1991, Evolution of nappes in the eastern margin of the Bohemian Massif: A kinematic interpretation: Geologische Rundschau, v. 80, p. 73–92.

Séranne, M., 1999, The Gulf of Lion continental margin (NW Mediterranean) revisited by IBS: An overview, in Durand, B., Jolivet, L., Horváth, F., and Séranne, M., eds., The Mediterranean basins: Tertiary extension within the Alpine orogen: Geological Society of London Special Publications, v. 156, p. 15–36.

Song, S. G., Yang, J. S., Xu, Z. Q., Liou, J. G., and Shi, R. D., 2003, Metamorphic evolution of the coesite-bearing ultrahigh-pressure terrane in the North Qaidam, northern Tibet, NW China: Journal of Metamorphic Geology, v. 21, p. 631–644.

Stampfli, G. M., Mosar, J., Marchant, R., Marquer, D., Baudin, T., and Borel, G., 1998, Subduction and obduction processes in the western Alps: Tectonophysisc, v. 296, p. 159–204.

Stampfli, G. M., von Raumer, J., and Borel, G. D., 2002, Paleozoic evolution of pre-Variscan terranes: From

Gondwana to the Variscan collision, *in* Martínez Catalán, J. R., Hatcher, R. D., Jr., Arenas, R., and Díaz García, F., eds., Variscan-Appalachian dynamics: The building of the Late Paleozoic basement: Geological Society of America, Special Papers, v. 364, p. 263–280.

Stosch, H.-G., and Lugmair, G. W., 1990, Geochemistry and evolution of MORB-type eclogites from the Münchberg Massif, southern Germany: Earth and Planetary Science Letters, v. 99, p. 230–249.

Tait, J., Schätz, M., Bachtadse, V., and Soffel, H., 2000, Palaeomagnetism and Palaeozoic palaeogeography of Gondwana and European terranes, *in* Franke, W., Haak, V., Oncken, O., and Tanner, D., eds., Orogenic processes: Quantification and modelling in the Variscan belt: Geological Society London, Special Publications, v. 179, p. 21–34.

Willner, A. P., Krohe, A., and Maresch, W. V., 2000, Interrelated P-T-t-d paths in the Variscan Erzgebirge dome (Saxony, Germany): Constraints on the rapid exhumation of high-pressure rocks from the root zone of a collisional orogen: International Geology Review, v. 42, p. 64–85.

Willner, A. P., Sebazungu, E., Gerya, T. V., Maresch, W. V., and Krohe, A., 2002, Numerical modelling of PT-paths related to rapid exhumation of high-pressure rocks from the crustal root in the Variscan Erzgebirge Dome (Saxony/Germany): Journal of Geodynamics, v. 33, p. 281–314.

Wittenberg, A., Vellmer, C., Kern, H., and Mengel, K., 2000, The Variscan lower continental crust: Evidence for crustal delamination from geochemical and petrophysical investigations, *in* Franke, W., Haak, V., Oncken, O., and Tanner, D., eds., Orogenic processes: Quantification and modelling in the Variscan belt: Geological Society of London, Special Publications, v. 179, p. 401–414.

Yang, J., Xu, Z., Zhang, J., Song, S., Wu, C., Shi, R., Li, H., and Brunel, M., 2002, Early Palaeozoic North Qaidam UHP metamorphic belt on the north-eastern Tibetan plateau and a paired subduction model: Terra Nova, v. 14, 397–404.

Zeck, H. P., 1996, Betic-Rif orogeny: Subduction of Mesozoic Tethys lithosphere under E-ward drifting Iberia, slab detachment shortly before 22 Ma, and subsequent uplift and extensional tectonics: Tectonophysics, v. 254, p. 1–16.

Zindler, A., and Hart, S. R., 1986, Chemical geodynamics: Annual Reviews of Earth and Planetary Sciences, v. 14, p. 493–571.

HIMALAYAS

Global UHP Metamorphism and Continental Subduction/Collision: The Himalayan Model

J. G. LIOU,[1]

Department of Geological and Environmental Sciences, Stanford University, Stanford, California 94305-2115

T. TSUJIMORI,

Department of Geological and Environmental Sciences, Stanford University, Stanford, California 94305-2115 and Research Institute of Natural Sciences, Okayama University of Science, Okayama 700-0005, Japan

R. Y. ZHANG,

Department of Geological and Environmental Sciences, Stanford University, Stanford, California 94305-2115

I. KATAYAMA,

Department of Geology and Geophysics, Yale University, New Haven, CT 06520 and Department of Earth and Planetary Sciences, Tokyo Institute of Technology, Meguro-ku, Tokyo 152-8551, Japan

AND S. MARUYAMA

Department of Earth and Planetary Sciences, Tokyo Institute of Technology, Meguro-ku, Tokyo 152-8551, Japan

Abstract

Continental crust (density ~2.8 g·cm^{-3}) resists subduction into the earth's mantle (~3.3 g·cm^{-3}) because of buoyancy. However, more than 20 recognized ultrahigh-pressure (UHP) terranes have been documented; these occurrences demonstrate that not only is continental crust subducted to depths as great as 150 km, but also that some supracrustal rocks were then exhumed to the earth's surface. UHP terranes are composed of mainly supracrustal rocks that contain minor amounts of minerals such as coesite or diamond, indicative of P > 2.5 GPa. In general, quartzofeldspathic units are thoroughly back reacted, and only mafic eclogite lenses and boudins retain scattered UHP phases. These index minerals are restricted to micron-scale inclusions in chemically and mechanically resistant zircon, garnet, and a few other strong container minerals, and are difficult to identify by conventional petrologic studies. The continental rocks were subjected to UHP metamorphism at T ranging from ~700 to 950°C and P > 2.8 to 5.0 GPa, corresponding to depths of ~100 to 150 km. These UHP units were subsequently exhumed to crustal depths and subjected to intense hydration and amphibolite-facies overprint. Widespread Barrovian-type metamorphism in many collisional orogens may mask an earlier, higher-pressure metamorphic history. We suspect that coesite-bearing UHP rocks were once generated in the majority of exhumed collisional orogens.

The recent finding of coesite inclusions in rare Himalayan eclogites and country rock gneisses is a typical example. We use the Himalayan model to illustrate UHP metamorphism and subduction of continental crustal rocks to mantle depths and later Barrovian-type overprint during exhumation. Himalayan UHP eclogites and adjacent gneisses were formed at mantle depths > 100 km at 46 to 52 Ma. These rocks were exhumed to crustal depths and subjected to Barrovian amphibolite- to granulite-facies metamorphism; associated magmatism occurred at 30 to 15 Ma. The Himalayan metamorphic belt was domally uplifted and the mountain-building process initiated since 11 Ma, when underthrusting of the Indian tectosphere beneath the Lesser Himalayas occurred.

Introduction

ULTRAHIGH-PRESSURE (UHP) metamorphism refers to the metamorphism of crustal rocks (both continental and oceanic) brought to pressures high enough to crystallize index minerals such as coesite at a minimum P > 2.7 GPa at T > 600°C and/or diamond. Prior to the initial discoveries of coesite in supracrustal rocks (Chopin, 1984; Smith, 1984), coesite and diamond were thought to occur only in meteorite impact craters and mantle xenoliths. Figure 1 shows the relevant P-T conditions defining

---

[1]Corresponding author; email: liou@pangea.stanford.edu

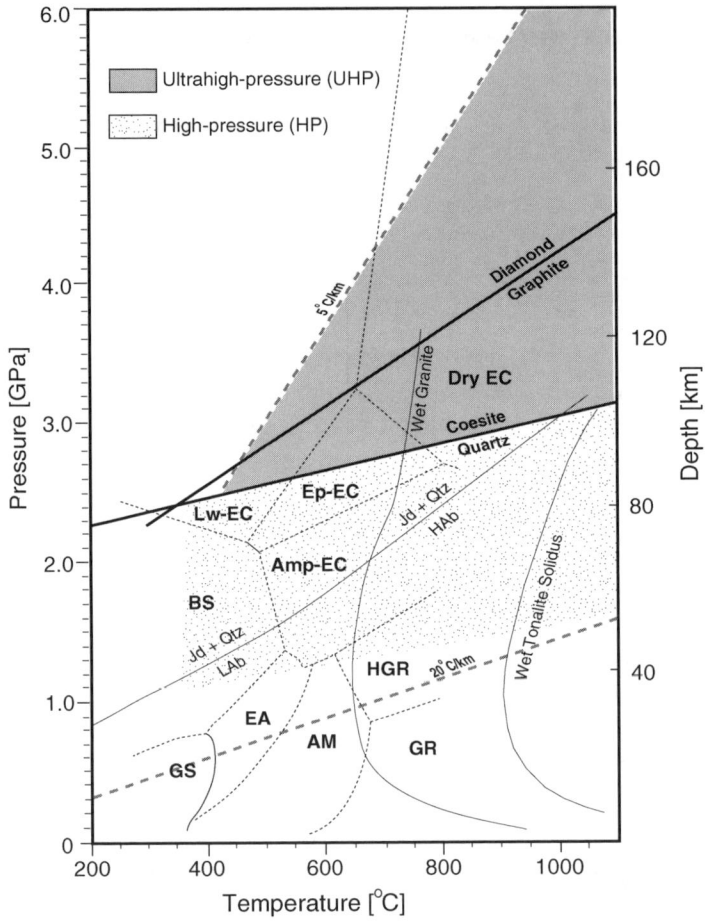

FIG. 1. P-T regimes assigned to various metamorphic types: (1) ultrahigh-P; (2) high-P; and (3) low-P. Geotherms of 5°C/km and 20°C/km are indicated. Stabilities of diamond, coesite, glaucophane, jadeite + quartz, aragonite, kyanite, sillimanite, andalusite, paragonite, and the minimum meltings of granite and tonalite are shown. P-T boundaries of various metamorphic facies [granulite, amphibolite, epidote amphibolite, greenschist, and subgreenschist facies] and subdivision of the eclogite field into amphibole (Amp) eclogite, epidote (Ep) eclogite, lawsonite (Lw) eclogite, and dry eclogite are indicated (for abbreviations, see Liou et al., 2000).

UHP, high-pressure (HP), and low-pressure (LP) metamorphism; in addition, geotherms of about 5°C·km⁻¹ (extreme high P/T) and 20°C·km⁻¹ (old plates) are illustrated.

The objective of the present article is to present a summary to the literature of continental crust subduction and UHP metamorphism and illustrate recent petrochemical research of various UHP rocks. To honor our teacher and mentor, W. G. Ernst, on the occasion of his retirement, we have prepared this review article to honor his contributions to the studies of both HP and UHP subduction-zone metamorphism and tectonics. We emphasize the interactions between metamorphism and tectonics in continental collision belts using the Himalayan orogen as an example. Distribution and mineralogical and petrological characteristics of other UHP terranes in the world are also described. Similar reviews for recent mineralogical-petrochemical-tectonic studies of global UHP rocks are included in Chopin (2003), Zheng et al. (2003), Rumble et al. (2003) and various chapters of the book *Ultrahigh-Pressure Metamorphism* published by the European Mineralogical Union and edited by Carswell and Compagnoni (2003).

The stabilities of coesite and other UHP minerals in metamorphic regime require abnormally low geothermal gradients, less than approximately 7°C·km^{-1}. Such environments can be attained only by the subduction of old, cold, oceanic crust–capped lithosphere ± pelagic sediments or ancient continental crust. Eclogites with compositions of mid-oceanic ridge basalt (MORB) + H_2O have been experimentally studied (e.g., Schmidt and Poli, 1998; Okamoto and Maruyama, 1999); the results shown in Figure 1 subdivide P-T regimes for the eclogite facies into those for amphibole eclogite, epidote eclogite, lawsonite eclogite, and dry eclogite. UHP and HP metamorphism can be separated conveniently by the P-T boundary for the quartz-coesite equilibrium. Eclogitic rocks formed at pressures greater or less than the transition boundary have been called "hot" or "cold" eclogite, respectively. The equilibrium boundary for the graphite-diamond transition further subdivides UHP regimes into diamond grade and coesite grade; for appropriate compositions and f_{O2} and X_{CO2} conditions, microdiamond ± coesite occur in diamond-grade UHP rocks.

We have classified global HP and UHP belts into two types according to protoliths (Maruyama et al., 1996). In active margins of the Pacific type, protoliths for HP metamorphism consist of rock assemblages forming an accretionary complex, including bedded cherts, MORB-origin greenstones, seamount fragments, and enclosing trench turbidites. On the other hand, protoliths of the collision-type HP-UHP rocks include continental basement complexes and the overlying sediment + volcanoclastic rocks of a variety of tectonic settings. Figure 2 is a schematic diagram showing the contrasting tectonic settings for accumulation of A-type protoliths in a rifted continental margin, and B-type protoliths in an active continental margin.

Most UHP rocks have A-type protoliths and are characterized by occurrences of continental shelf carbonates, bimodal volcanics, peraluminous sediments, and granite-gneiss basement rocks. Eclogites and garnet peridotites, although volumetrically small, are widespread in all recognized UHP terranes of the world. They are the most significant components inasmuch as they preserve most completely UHP index minerals such as coesite and microdiamond. These mafic-ultramafic rocks may have originated from different tectonic settings and were subjected to similar UHP metamorphism, deformation, and retrogression along with the enclosing granitic gneisses and supracrustal rocks.

Because of better preservation of the effect of UHP metamorphism compared with adjacent gneissic rocks, eclogites and garnet peridotites have received the most intensive study; accordingly, they provide important petrochemical and isotopic constraints on tectonic models of continental subduction, collision, and exhumation for UHP terranes.

Global Distribution of UHP Metamorphic Terranes

Occurrences of UHP rocks have been increasingly recognized and reviewed extensively (e.g. Liou, 1999, 2002; Chopin, 2003, Carswell and Compagnoni, 2003). Thus far, more than 20 UHP terranes, shown in Figure 3, have been documented. These UHP terranes lie within major continental plate collision belts and extend for several hundred km or more; many are in Eurasia, but a few are in Africa, South America, or Antarctica. They share common structural and lithological characteristics. (1) UHP records are preserved mainly in eclogites and garnet peridotites enclosed as pods and slabs within gneissic units. A few of these rocks contain minute inclusions of coesite in zircon, garnet, and omphacite, and microdiamonds in garnet and zircon. (2) Most lithologies are continental in chemical composition. (3) Exhumed UHP units are now present in the upper continental crust as thin, sub-horizontal slabs, bounded by normal faults on top, and reverse faults on bottom, and sandwiched in amongst HP or lower-grade metamorphic units. (4) Coeval island-arc volcanic and plutonic rocks do not occur, whereas post-collisional or late-stage anorogenic granitic plutons are common in some occurrences.

In this section, classical (Ernst and Liou, 2000), less intensively studied (Liou, 1999), and recently recognized UHP terranes are described below. Lithological and tectonic characteristics including index minerals, P-T conditions, size, and peak metamorphic ages of these UHP terranes are summarized in Table 1.

Classical UHP Terranes

Several classical UHP terranes including (1) the Dora Maira massif of the Western Alps (Chopin, 1984), (2) the Western Gneiss Region of Norway (Smith, 1984), (3) the Dabie-Sulu terrane of east-central China (Wang et al., 1989), (4) the Kokchetav massif of northern Kazakhstan (Sobolev and Shatsky, 1990), and (5) the Bohemian massif

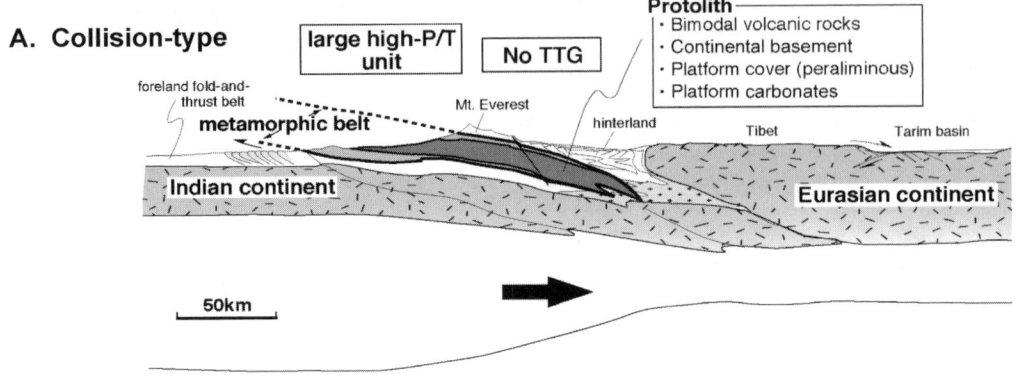

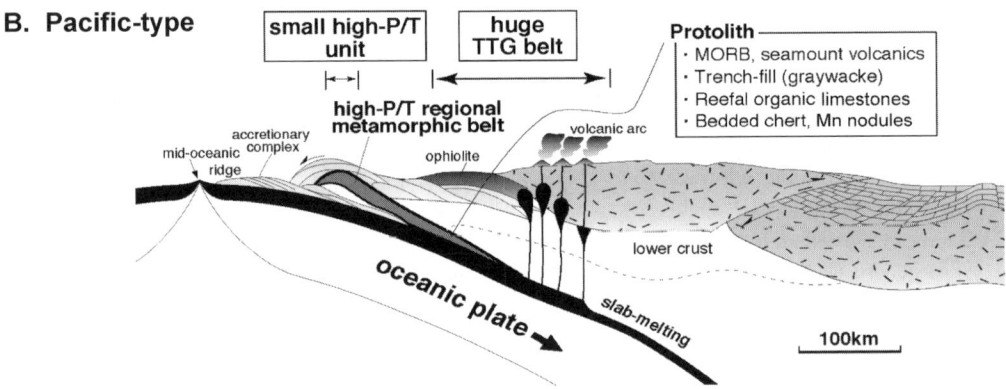

FIG. 2. Schematic cross-section contrasting the tectonic setting of generation and exhumation of (A) Collision (A)-type and (B) Pacific (B)-type HP-UHP metamorphic belts. Characteristics of the Pacific type include subduction of oceanic lithosphere and development of an accretionary complex, forearc basin, huge tonalite-trondhjeimite-granitoid (TTG) belt, and volcanic arc. Collision-type belts, which are associated with continental collision, lack those characteristics, and the high-P belt is generally of a much large extent (after Maruyama et al., 1996).

(Bakun-Czubarow, 1991) have been well studied and extensively reviewed (e.g. Ernst and Liou, 2000). Except for the Dora Maira massif, occurrences of microdiamond in these UHP terranes have been reported; petrochemical characteristics of these terranes are summarized below.

Dora Maira massif, Western Alps. The Dora Maira massif consists of Late Paleozoic and older continental basement rocks, metamorphosed under UHP and HP conditions as a result of Mesozoic–Cenozoic convergence of the European and African plates (for a recent review, see Compagnoni and Rolfo, 2003). The HP + UHP belt is a stack of four major units separated by low-angle faults. These units have been affected by Cenozoic HP metamorphism and later re-equilibration at lower pressures. The UHP unit consists of dark- and light-colored eclogites, pyrope quartzite with phengite schist and jadeite-rich rock inclusions, as well as orthogneiss country rocks and undeformed metagranites. Boudins of pyrope quartzite contain UHP relics such as coesite and other UHP minerals. This entity was subdivided into two metamorphic subunits: (a) a coesite-bearing, UHP subunit contains kyanite-eclogites with relict coesite and coesite pseudomorphs, and has estimated peak P-T conditions of 3.7 GPa and 790°C, within the diamond stability field (Schertl et al., 1991); and (b) a lower-P eclogite subunit was metamorphosed at 1.5 GPa and 500°C. The time of UHP/HP metamorphism has been considered to have culminated between 90 to 125 Ma, and retrograded between 35 to 41 Ma. However, recent

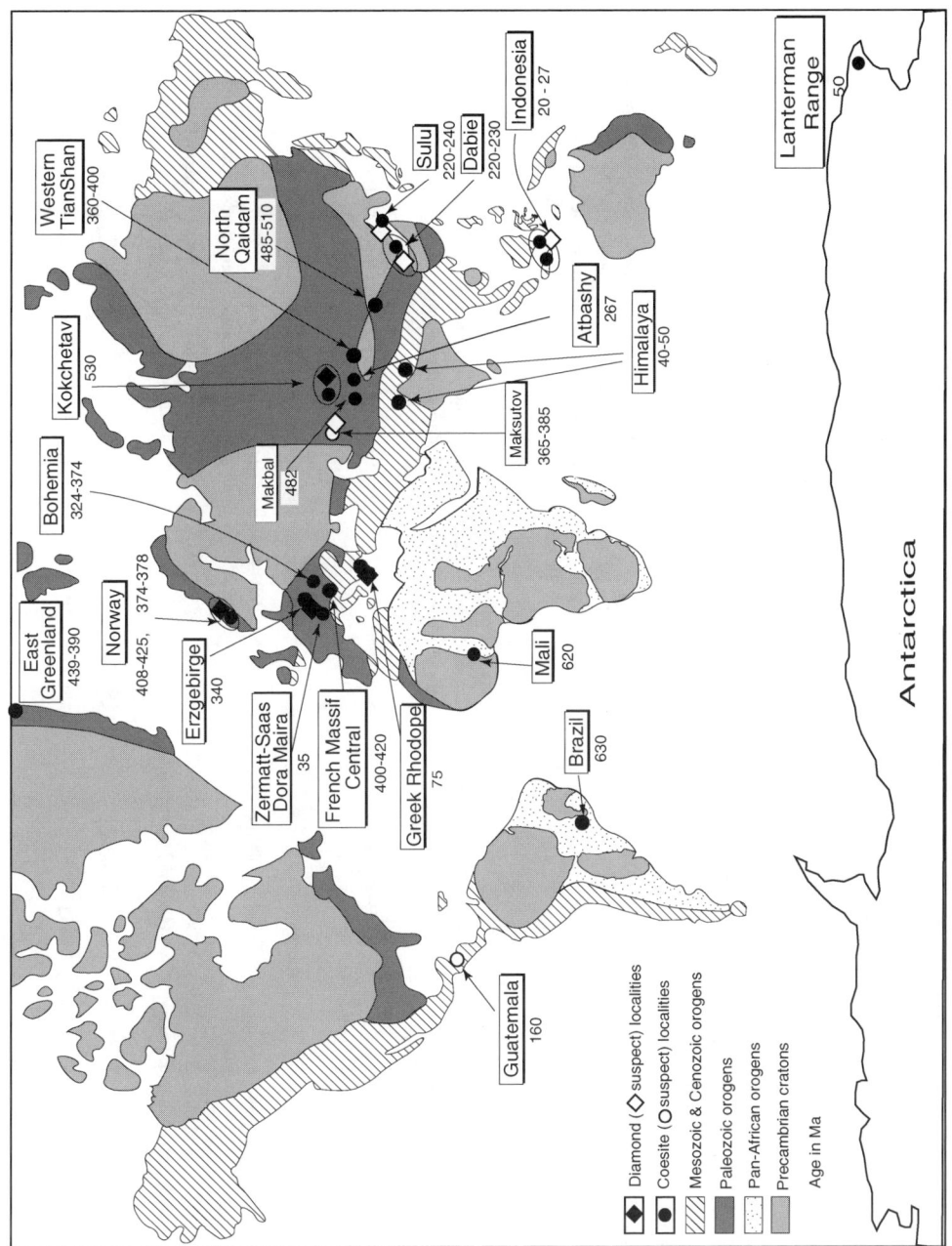

FIG. 3. Distribution and peak metamorphic age of recognized UHP terranes in the world (modified after Liou et al., 2000).

TABLE 1. Summary of Petrochemical Features of Recognized UHP Terranes in the World

UHP terrane	UHP rock type	Ages	P-T estimates	Special features
			Classical terranes	
Dora Maira Massif, Western Alps, Italy	Eclogite Pyrope quartzite Pelites Whiteschist	35.4 ± 1.0 Ma (peak) 30 Ma (retrograde)	37 kbar, 790°C	1. Pyrope whiteschist contains Mg-Al silicates 2. No garnet peridotite
West Gneiss Region, Norway	Eclogite Garnet peridotite Diamond-bearing gneiss	408–425 Ma (peak) 374–378 Ma (retrograde)	P > 28 kbar T > 790°C	1. Rare diamond in country rock gneiss and garnet peridotite 2. Majoritic garnet in peridotite formed at mantle, ~1700 Ma
Dabie-Sulu terrane, East-Central China	Eclogite Garnet peridotite Jadeite quartzite Minor whiteschist	220–240 Ma (peak)	Eclogite: 26–40 kbar, 700–850°C Garnet peridotite: 50–60 kbar, 700–850°C	1. Abundant coesite in garnet, omphacite, kyanite, zoisite, zircon 2. Rare diamond in eclogite 3. Eclogite and garnet peridotite contain UHP hydrous phases 4. Dual origin for garnet peridotite 5. Majoritic garnet in eclogite
Kokchetav Massif, northern Kazakhstan	Eclogite Diamond-bearing gneiss and marble Whiteschist	537±9 Ma (peak) 507±9 Ma (retrograde)	Diamond grade: 40–60 kbar, 920–1000°C Coesite grade: 36 kbar, 750–800°C	1. Microdiamond is abundant in some garnet gneiss and dolomite marble 2. Coesite inclusions in whiteschist, eclogite, and garnet gneiss 3. Rare garnet peridotite
Bohemian Massif	Eclogite Garnet peridotite	379–395 Ma	>30 kbar, 700–800°C	Garnet peridotite yields much higher P than coesite-eclogite
			Less intensively studied terranes	
Zermatt-Saas area, Switzerland	Eclogite Mn-quartzite	52 ± 18 Ma (peak)	26–30 kbar, 590–630°C	1. Protoliths are oceanic affinity 2. Coesite inclusion in the mantle of garnet where garnet core has quartz
Saxonian Erzgebirge, Germany	Eclogite Diamond-bearing gneiss	360 ± 7 Ma (peak) 348–355 Ma (retrograde)	>42 kbar, 900–1000°C	1. Inclusions of microdiamond in zircon, kyanite, and garnet from gneiss 2. Rutile with α-PbO2 structure
Mali, Africa	Eclogite	620 Ma (peak)	>27 kbar, 700–750°C	1. Oldest age of UHP metamorphism
Makbal, western Kyrgyzstan	Eclogite Garnet-bearing pelitic schist	482 Ma	>26 kbar, 610–680°C	1. Inclusions of coesite pseudomorph in garnet from eclogite and schist

Locality	Rock type	Age	P-T	Evidence
Atbashy, Kazakhstan	Eclogite	270 Ma	> 25 kbar, 660°C	1. Inclusions of coesite in garnet
Central Indonesia, SW Sulawesi	Eclogite Garnet peridotite	115–120 Ma	>27 kbar, 720–760°C 33–35 kbar, 1150°C	1. Inclusions of coesite in zircon and coesite pseudomorph in jadeite
		Recently recognized terranes		
French Alps	Eclogite Metapelites	400–420 Ma (peak) 360–380 Ma (retrograde)	> 28 kbar, 750°C	1. Inclusions of coesite in garnet and omphacite
Northeast Greenland	Eclogite	400–440 Ma	24–28 kbar, 820°C	1. Inclusions of coesite pseudomorph in garnet
Southeast Brazil	Eclogite	630 Ma	> 27 kbar, 800°C	1. Inclusions of coesite in zircon
Chuacus Complex, northern Guatemala	Eclogite Banded gneiss	48–72 Ma (retrograde)	20–30 kbar, 700–800°C	1. Coesite pseudomorph in garnet and kyanite
Rhodope Massif, Greece	Eclogite Orthogneiss Pelitic gneiss Ultramafic rocks	30–42 Ma (peak) 8–12 Ma (retrograde)	> 30 kbar, ~1200°C	1. Microdiamond inclusion in garnet porphyroblast 2. Silica, rutile and apatite rods in garnet
Lanterman Range, Antarctica	Eclogite Felsic gneiss	500 Ma (peak) 486–490 Ma (retrograde)	> 29 kbar, > 850°C	1. Inclusions of coesite and its pseudomorph in garnet
Western Tian-Shan, northwest China	Eclogite Marble	< 310 ± 5 Ma (peak)	26–27 kbar, 500–600°C	1. Inclusions of coesite and its pseudomorph in garnet. 2. Possible coexistence of magnesite + aragonite
Qaidam, western China	Eclogite Gneiss Garnet peridotite	495 ± 7 Ma (peak) 467 ± 1 Ma (retrograde)	> 27 kbar, 700°C	1. Inclusions of coesite in zircon from paragneiss
Altun Mountains, western China	Eclogite Grt-lherzolite	504 ± 5 Ma (peak)	28–32 kbar, 820–850°°C	1. Coesite pseudomorph in garnet
North Qinling Mtns., Central China	Eclogite Gneiss	507 ± 38 Ma 400 ± 16 Ma	> 26 kbar, 590–760°C	1. Coesite in garnet 2. Microdiamond in zircons from eclogite and gneiss
Kaghan valley, Pakistan Himalayas	Eclogite Gneiss	46 ± 1 Ma (peak) 44 ± 1 Ma (retrograde)	27–29 kbar, 690–750°C	1. Inclusions of coesite and its pseudomorph in omphacite and zircon from eclogite and in zircon from gneiss
Tso Morari, Indian Himalayas	Eclogite Gneiss	55 Ma (peak) 47 Ma and 30 Ma (retrograde)	> 28 kbar, 700–800°C	1. Inclusions of coesite and its pseudomorph in garnet from eclogite

SHRIMP dating of zircons extracted from gneiss, schist, pyrope inclusion, and pyrope quartzite indicates the occurrence of 240–275 Ma old zircon cores and newly formed 35 Ma rims for some zircons (Gebauer et al., 1997). Thus, UHP metamorphism must have taken place during Oligocene time, with an estimated exhumation rate of 2 to 2.4 cm·yr^{-1}.

Western Gneiss Region, Norway. The Western Gneiss Region lies within an Early Paleozoic collision zone; the gneissic unit, about 300 km long and 150 km wide, consists of interlayered pelite and migmatite, marble, quartzite, and amphibolite, with tectonic inclusions of gabbro and peridotite (for a recent review see Carswell and Cuthbert, 2003). The gneissic unit exhibits mainly amphibolite-facies assemblages, but relics of HP assemblages also occur. Eclogite boudins are widespread. Coesite was first reported from Grytting (Smith, 1984); several new localities of coesite and coesite pseudomorphs have been recognized recently in both eclogites and the adjacent gneissic rocks (Wain, 1997; Cuthbert et al., 2000; Wain et al., 2000; Carswell et al., 2003). Microdiamond grains 20–50 microns across have been described from residues separated from two gneisses (Dobrzhinetskaya et al., 1995); the associated kyanite-eclogites contain inclusions of coesite pseudomorphs in garnet. Relict majoritic garnets showing exsolution of pyroxene lamellae in peridotite bodies have been recently discovered on the islands of Otrogy and Flemsoy (van Roermund and Drury, 1998; van Roermund et al., 2000). These peridotite bodies originally must have had an even higher-pressure, deeper-mantle origin, probably within a rising mantle diapir at ~1700 Ma. In these garnet peridotites, diamond inclusions in sympletetic spinel after garnet have been discovered (van Roermund et al., 2002). Radiometric ages for eclogites of the Western Gneiss Region are 408–425 Ma for the peak metamorphism, and 374–378 Ma for the retrogressive mid-crustal metamorphic overprinting (for review, see Carswell et al., 1999). Carswell et al. (2003) provide new petrographic evidence and a review of the latest radiometric age data, and conclude that the UHP metamorphism of eclogites occurred at 400–410 Ma, significantly younger than the previous, widely accepted age of 425 Ma. A two-stage exhumation process is suggested: an initial exhumation to about 35 km depth by about 395 Ma at a mean rate of about 10 m·Ma^{-1}, and subsequent exhumation to 8–10 km by about 375 Ma at a much slower rate of about 1.3 m·Ma^{-1}.

The suture zone occupied by UHP/HP rocks of the Western Gneiss Region reflects collision of the eclogite-bearing Greenland sialic crust-capped lithospheric plate with Fennoscandia during the Caledonian orogeny (Gilotti, 1993; Krogh and Carswell, 1995; Brueckner and Medaris, 1998). Inclusions of coesite pseudomorph in garnet and omphacite from kyanite eclogites have been recently documented in the East Greenland eclogite province (Gilotti and Ravna, 2002); this new UHP/HP terrane is remote, but has excellent exposures and requires close examination.

Dabie-Sulu terrane of east-central China. The Dabie-Sulu terrane occupies a Triassic collision zone between the Sino-Korean and Yangtze plates (for a recent review, see Hirajima and Nakamura, 2003). The central UHP coesite-bearing eclogite belt (P = 2.7–5 GPa) is flanked to the north by a migmatite zone (P < 2 GPa), and to the south by a belt of blueschist, epidote amphibolite, and eclogite-facies rocks (P = 0.5–1.2 GPa). Prevalent rock types include felsic gneiss, orthogneiss, marble, and quartzite. Protolith ages range from Proterozoic to Ordovician, but the majority of reported ages are Neoproterozoic (Li et al., 1993; Hacker et al., 2000). Blocks, boudins, and layers of eclogites and garnet peridotites occur as enclaves in gneisses in the UHP unit. The Dabie-Sulu UHP rocks show several characteristics. (1) Widespread coesite with hydrous minerals such as talc, zoisite/epidote, and phengite occur in eclogites. (2) Two distinct types of mantle- and crustal-derived garnet peridotites are present. (3) Abundant exsolution textures were identified in UHP minerals from garnet peridotite and eclogite. (4) Diamond separates from both eclogite and garnet peridotite have been reported from few restricted regions (e.g., Xu et al., 1992, 2003). Some of these findings are not confirmed by thin-section observation and extensive studies of mineral inclusions in zircon (for review, see Liou et al., 2002). (5) Some garnet peridotites record much higher pressure than associated coesite-bearing eclogites, reaching 5–6 GPa (Yang et al., 1993; Zhang R. Y. et al., 2000, 2003). (6) Various isotopic age dating methods for UHP minerals and rocks from Dabie-Sulu terrane gives 220–240 Ma (Li et al., 1993; Hacker et al., 2000).

Eclogites in the Huwan shear zone from the northwestern Dabie Mountains have SHRIMP U/Pb dates of 309 ± 3 Ma (Sun et al., 2002) suggesting a discrete Carboniferous subduction/collision between North and South China. Farther to the west in North Qinling, inclusions of microdiamond were discovered in zircons from both eclogite and its gneissic country rocks (Yang et al., 2002a); SHRIMP U/Pb dating of zircons from granitic gneiss yielded 507 ± 38 Ma for metamorphic rims and 1200–1800 Ma for relic magmatic or old cores, suggesting an Early Paleozoic UHP metamorphic event. These new data suggest that episodic subduction and collision of continents occurred during stages of closures of paleo Tethys prior to the well-documented major Triassic continent-continent collision for the Dabie-Sulu terrane.

Kokchetav massif. The Kokchetav massif, northern Kazakhstan, is the type site for the UHP diamond-eclogite regional metamorphic facies (Sobolev and Shatsky, 1990; Parkinson et al., 2002; Shatsky and Sobolev, 2003). Diamonds occur as minute inclusions in garnet, diopside, and zircon porphyroblasts in marble, pyroxene-carbonate-garnet rock, and garnet-biotite gneiss and schist. Inclusions of coesite and coesite pseudomorphs occur in garnet and zircon from eclogite and diamond-bearing gneiss. The UHP/HP unit extends NW-SE for at least 80 km, and is about 17 km wide. The UHP slab is structurally overlain by a weak- to low-grade metamorphic unit and is underlain by a low-P metamorphic andalusite-staurolite facies unit (Kaneko et al., 2000). The UHP slab is composed of felsic gneiss with locally abundant eclogite lenses and minor orthogneiss, metacarbonate, and rare garnet peridotite, with a range of Proterozoic protolith ages. A well-constrained P-T-time path deduced from mineral inclusions and SHRIMP geochronology of zircons is illustrated in Figure 4. Zoned zircons from diamond-bearing gneiss yield the following ages: a quartz-bearing inherited core with 1100–1400 Ma as the protolith age, coesite-bearing mantle with 537 ± 9 Ma for UHP metamorphism, and the plagioclase-bearing rim with 507 ± 8 Ma for amphibolite-facies overprint (Katayama et al., 2001). Many diamond-grade marbles contain clinopyroxene with exsolution lamellae of quartz, K-feldspar, phengite, and phlogopite (Katayama et al., 2002; Zhu and Ogasawara, 2002), and titanite with coesite lamellae (Ogasawara et al., 2002); these occurrences suggest that UHP metamorphism occurred at P > 6 GPa and 1000°C.

C^{-13} isotopic data of diamond suggest biogenic sources; inclusions of water, carbonate, and nanometric oxide in diamonds indicate that metamorphic microdiamonds in this and other UHP terranes were precipitated from C-O-H supercritical fluids (e.g.,

Stockhert et al., 2001). Ogasawara et al. (2002) further suggested a two-stage growth mechanism for microdiamonds. Masago et al. (2003) reported negative ^{18}O values (-3.9 per mil) for minerals in eclogite and whiteschist of the Kokchetav Massif, similar to the Chinese UHP rocks first described by Yui et al. (1995). The Kokchetav Massif is the second recognized UHP region that preserves a significant effect of the interaction of cold meteoric water with the protolith prior to subduction.

Saxonian Erzgebirge, Germany. The Erzgebirge Crystalline Complex (ECC) at the northern margin of the Bohemian massif is in fault contact with a low-grade Paleozoic sequence at its margins (for a recent review, see Massonne and O'Brien, 2003). The Variscan ECC consists of abundant gneisses that include numerous lenses of eclogites. The "Gneiss-Eclogite Unit" at the core of this massif contains UHP rocks typified by inclusions of coesite pseudomorph in garnets from eclogite (Schmadicke, 1991) and inclusions of microdiamonds in garnet, kyanite, and zircon from gneisses (Massonne, 1999). Thus far, the diamondiferous gneisses seem to occur only in a 1 km long strip near the eastern shore of the Saidenbach Reservoir. Diamond micro-inclusions occur exclusively in garnet, kyanite, and zircon of several gneissic rocks. Besides completely preserved diamonds with grain sizes between 1 to 25 μm, partially graphitized diamonds and graphite pseudomorphs after diamond occur. Such microdiamond inclusions in garnet and zircon are similar to those in Kokchetav diamondiferous biotite gneiss. Some of these inclusions contain additional fluid-bearing phases including apatite, phengitic mica, and possible fluid inclusions; the morphologies of inclusions and mineral associations suggest that microdiamonds from both the Erzgebirge and the Kokchetav were crystallized from supercritical fluids under UHP conditions (Stockhert et al., 2001; Hwang et al., 2001; Dobrzhnetskaya et al., 2001, 2003).

Schmadicke et al. (1995) reported Sm-Nd isochrons for garnet-Cpx-WR at 360 ± 7 Ma for eclogite and 353 ± 6 Ma for garnet pyroxenite from Erzgebirge UHP rocks. $^{40}Ar/^{39}Ar$ spectra of phengite from two eclogite samples give plateau ages of 348 ± 2 and 355 ± 2 Ma. Diamond-bearing zircons yield SHRIMP U/Pb dates at 336 ± 2 Ma (Massonne, 2001). Such similarity in ages for garnet peridotites and eclogite enclosed in gneiss suggests a coeval Variscan UHP in-situ metamorphism of the ECC around 340–360 Ma.

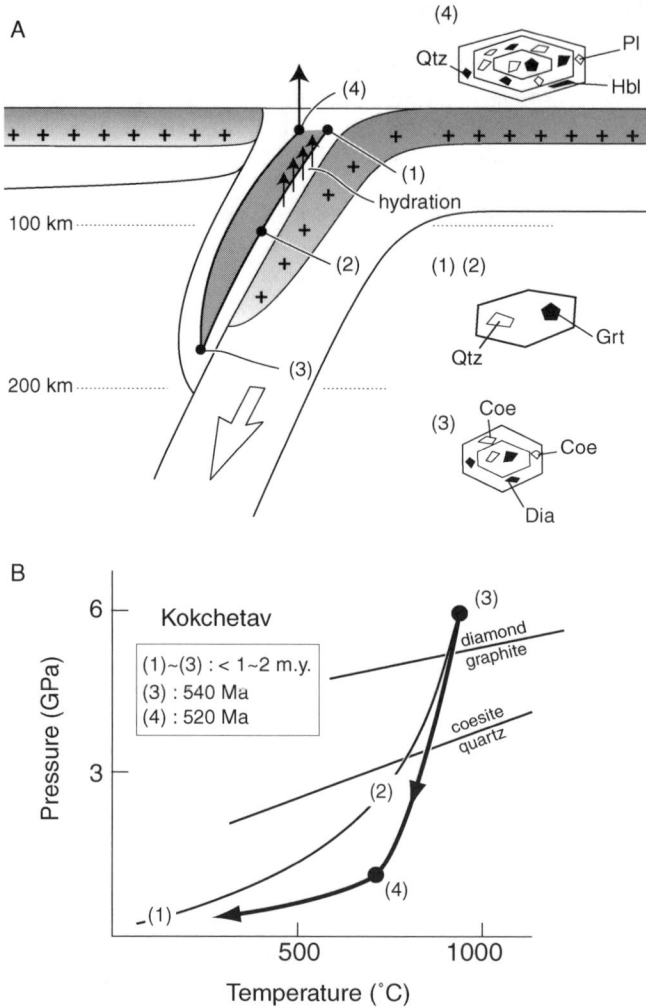

FIG. 4. A. Schematic diagram showing the growth of detrital zircon (stage 1) through subduction (stage 2 and peak stage 3) and exhumation (stage 4) of supracrustal rocks. Mineral inclusions in various stages of zircon growth are shown. B. SHRIMP U-Pb dating of zoned zircons from a Kokchetav diamond-bearing gneiss yields a well constrained P-T-time path (for details, see Katayama et al., 2001).

Less Intensively Studied UHP terranes

Several UHP terranes described below, except for the Maksyvtov Complex, have received less intensive study mainly due to remoteness and difficult access, despite their long recognition.

Maksyutov Complex, Southern Urals. The Maksyutov Complex is an elongated (10 × 120 km) N-S belt between the Russian and Siberian cratons in the southern Ural Mountains (Coleman and Wang, 1995; Dobretsov et al., 1996), and contains blocky graphitized diamond(?) in pelitic schist (Leech and Ernst, 1998). The complex is the locality from which quartz aggregates within garnet exhibiting radial cracks were first recognized as coesite-pseudomorphs by Chesnokov and Popov (1965), nearly 20 years prior to the independent finding of coesite inclusions by Chopin (1984) and Smith (1984). The eclogite-bearing unit contains boudins of eclogite, layers of eclogitic gneiss, and rare ultramafic bodies within host metasedimentary mica schist and quartzite.

Some Maksyutov eclogites were recrystallized at 594–637°C with a minimum pressure from 1.5 to 1.7 GPa (Beane et al., 1995; Leech and Ernst, 1998) or up to 2.7 GPa if coesite pseudomorphs (Chesnokov and Popov, 1965; Dobretsov and Doretsova, 1988) are indeed present. However, subsequent studies have not revealed the occurrence of inclusions of coesite or coesite pseudomorph in garnet and zircon (Beane et al., 1995; Shatsky et al., 1995). Leech and Ernst (1998) described unusual graphite aggregates (up to 13 mm edge length) that may be diamond pseudomorphs. This suggestion was based on the blocky morphology of the cuboid graphite aggregates cross-cutting rock foliation, pressure shadows, biogenic isotopic signatures, and other spectroscopic characters similar to those graphitized diamonds from the Beni Bousera massif. Recent study of several microdiamond inclusions in garnet has yielded further suggestive possibility that the eclogitic unit of the Maksyutov Complex experienced diamond-grade UHP metamorphism (Bostick et al., 2003). The UHP/HP metamorphism with peak metamorphic ages of 370–374 Ma (Shatsky et al., 1995; Beane and Connelly, 1998) may be related to the collision of the Russian platform with a fragment of the Siberian craton during Devonian time, but was affected by the Triassic collision of the Kazakhstan block with the Russian platform (e.g., Matte, 1998).

Zermatt-Saas area, Western Alps. In contrast to most other UHP terranes in which the protolith is continental crust, eclogites and metasediments at Lago di Cignana of the Western Alps were derived from oceanic crust and constitute the ophiolite sequence of the Zermatt-Saas zone. Inclusions of coesite and coesite pseudomorphs occur in tourmaline and garnet of manganiferous quartzite, and in omphacite and garnet of the underlying mafic eclogite (Reinecke, 1991). The metasediments are mainly garnet-phengite-quartz schist, with variable amounts of garnet-clinopyroxene quartzite, and piedmontite-phengite-quartz schist. Eclogites and retrograded eclogites with preserved UHP mineral parageneses consist of garnet, omphacite, glaucophane, zoisite, phengite, and dolomite. Zoned garnets preserve prograde quartz crystals in their cores and coesite inclusions in mantles and rims. Coesite inclusions are much more common in omphacite than in garnet. UHP conditions were estimated at 2.6–3.0 GPa and 590–630°C. Sm-Nd isotopic analyses of eclogites yield UHP metamorphism at 52 ± 18 Ma (Bowtell et al., 1994), consistent with a Tertiary age of eclogitization recently established for Alpine metamorphism in the Western Alps (Gebauer et al., 1997).

Thus far, the recognized areal extent of the coesite-bearing rocks is ~2 km². The apparent thickness of the UHP slice does not exceed a few hundred meters. This area is the best example for subduction of oceanic lithosphere to mantle depth, and later exhumation as a fragmented tectonic slab. Similarly, the Allalin metagabbro, part of the oceanic Piemonte zone, contains magnesite + talc + kyanite + garnet recrystallized at ~600°C and 3.5 GPa, with an estimated geothermal gradient of 5–6°C·km^{-1}; however, inclusions of coesite or coesite pseudomorphs have not been reported from the Allalin metagabbro.

Mali, Africa. Eclogitic rocks in the Pan-African collision zone cropping out in northern Mali contain inclusions of coesite in omphacite from mafic nodules within a calc-silicate layer enclosed in impure marble, and coesite pseudomorphs in garnets from the surrounding eclogitic metasediments (Caby, 1994). These UHP eclogitic mica schists have P-T estimates of 700–750°C and > 2.7 GPa; phengite yields a well-defined plateau ^{39}Ar/^{40}Ar age of 1045 ± 9 Ma. Recent Sm/Nd ages of 620 Ma were obtained by Jahn et al. (2001) for peak UHP metamorphism. This may be the oldest UHP terrane in the world.

The coesite-bearing UHP eclogitic unit about 3 km thick lies within an internal nappe as a flat, thin slice bounded on the top by a thick passive-margin shaly formation, and on the bottom by low-grade greenschist-facies phyllites. Caby (1994) suggested that a large portion of the terrigenous metasediments of passive-continental-margin affinity of the West African plate was subducted eastward to mantle depths (>90 km) beneath an oceanic domain. Collision of the West African plate with an island arc resulted in a low-angle, westward subhorizontal extrusion of this subducted slice of sialic materials to its present position. The coesite-bearing eclogite unit may also crop out 1500 km to the south in Togo, where kyanite-bearing eclogites occur as part of the passive-margin assemblage.

Makbal (480 Ma) and Atbashy (270 Ma), Kazakhstan. The Tian-Shan is divided into three mountain ranges—northern, middle, and southern. Within the Kyrgyzstan Tian-Shan, UHP eclogite localities with coesite pseudomorph inclusions in garnet and omphacite from the northern and southern Tian-Shan represent Caledonian and Hercynian orogenic belts, respectively (Tagiri and Bakirov, 1990; Tagiri et al., 1995). The Makbal Formation in the western Kyrgyz Ridge of the northern Tian-Shan consists of quartzose schist alternating with pelitic schists intercalated with thin layers or lenses of marble, eclogite, and amphibolitized eclogite. Tagiri and Bakirov (1990) found inclusions of coesite pseudomorphs in garnet from a garnet-chloritoid-talc schist with a peak metamorphic assemblage of coesite + almandine + chloritoid + talc + phengite (Si = 3.42 p.f.u) + phlogopite + rutile. Makbal coesite-grade eclogite has a paragonite K/Ar age of 480 Ma.

The eclogite-bearing complex of Atbashy Ridge from the southern Tian-Shan is composed of pelitic and siliceous schists alternating with thin UHP eclogitic layers, and is unconformably overlain by Upper Paleozoic molasse and silicic volcanic rocks. Inclusions of coesite pseudomorphs occur in omphacite cores and in garnet. Eclogites from the Atbashy area preserve prograde and retrograde paths with a peak metamorphic condition at 660°C, 2.5 GPa. The Atbashy coesite-eclogites yielded a Rb/Sr mineral isochron age of 270 Ma (Tagiri et al., 1995).

Central Indonesia UHP terrane. The pre-Tertiary basement of the Indonesian region comprises a variety of imbricate terranes, mélange, ophiolite, and variably metamorphosed accretionary complexes; some contain HP to UHP metamorphic rocks resulting from the collision of an Australia-derived continent with Eurasia (Parkinson, 2003). HP rocks including eclogites and garnet peridotites are widely distributed in Cretaceous accretionary complexes. Many of these rocks occur as imbricate slices of carbonate, quartzite, and pelitic schist of shallow-marine or continental-margin parentage, interthrust with subordinate mafic schist and serpentinite; some yield mica K/Ar ages of 110–120 Ma (Parkinson, 2003).

HP and UHP rocks are sporadically exposed as tectonic blocks throughout the Cretaceous accretionary complexes. They include eclogite, garnet-glaucophane rock (P = 1.8–2.4 GPa, T = 580–620°C), and jadeite-garnet quartzite (P > 2.7 GPa, T = 720–760°C) in Bantimala, southwest Sulawesi, eclogite and garnet granulite in west-central Sulawesi, eclogite and jadeite-glaucophane-quartz rock (P ~2.2 GPa, T ~530°C) in central Java, Mg chloritoid–bearing whiteschists (P ~1.8 GPa) in southeast Kalimantan, garnet lherzolites in east-central Sulawesi (P = 2.2–2.8 GPa, T = 1000–1100°C), west-central Sulawesi (P = 1.6–2.0 GPa, T = 1050–1100°C), and garnet pyroxenite (P ~2 GPa, T ~ 850°C) in Sabah, northeast Borneo. Evidence for UHP rocks include: (1) inclusions of coesite in zircon and coesite pseudomorphs in jadeite from jadeite quartzite and eclogite of the Bantimala Complex of south Sulawesi (Parkinson, 2003); and (2) P-T estimates of peak-stage recrystallization at 2.7–3.5 GPa and 1000–1100°C for most garnet peridotites (Kardarusman and Parkinson, 2000). Many of these rocks were probably recrystallized in a N-dipping subduction zone at the margin of the Sundaland craton in the Early Cretaceous. Exhumation may have been facilitated by the collision of a Gondwana continental fragment with the Sundaland margin at ~120–115 Ma.

Recently Recognized UHP Terranes

Several new UHP terranes were recently identified, inasmuch as they contain partially preserved trace index minerals in strong containers such as zircon or garnet. In fact, zircon has been considered to be the best container and many new terranes were discovered through positive identification of inclusions of coesite or coesite pseudomorphs, or diamond in zircons. Detailed examination of mineral inclusions in core, mantle, and rims of zircon separates from eclogites and their enclosing country-rock gneisses have yielded both prograde and retrograde P-T-time paths for various UHP terranes mentioned above (see Katayama et al., 2001 for details; also see the section on the Himalayan eclogite for an example).

The latest recognized UHP terranes include the following: (1) coesite in the French Massif Central (Lardeaux et al., 2001); (2) coesite and diamond from the Greek Rhodope metamorphic province (Mposkos and Kostopoulos, 2001); (3) coesite pseudomorphs in the Northeast Greenland Eclogite Province (Gilotti and Ravna, 2002); (4) coesite in granulite-facies overprinting eclogites of Southeast Brazil (Parkinson et al., 2001); (5) coesite in Himalayan eclogite from the Upper Kaghan Valley, Pakistan (O'Brien et al., 2001; Kaneko et al., 2003) and coesite from the Tso-Morari crystalline complex of India (Mukherjee et al., 2003); (6) coesite inclusions in gneissic rocks from the North Qaidam belt, western China (Yang et al., 2001; Song et al., 2003); (7) diamond inclusions in garnets of eclogite and gneiss from North Qinling (Yang et al., 2002a); (8) inclusions of quartz pseudomorphs and minor coesite in garnets of mafic eclogites from the Lanterman Range in Antarctica (Ghiribelli et al., 2002); (9)

inclusions of calcite pseudomorphs after aragonite + magnesite in dolomite in metapelites from the western Tian-Shan inferred as metamorphism in a UHP region of dolomite decomposition (Zhang L. et al., 2002); and (10) inclusion of possible coesite pseudomorphs in garnet, lamellar inclusions in garnet, kyanite, and high sodium content of garnet from eclogites of the Chuacus Complex of north-central Guatemala (Solari et al., 2003). The results of petrochemical and geochronological data for these new terranes are summarized below. The Himalayan UHP terrane in the Upper Kaghan Valley and the Tso-Morari complex are used to illustrate the processes of continental subduction and collision in a later section.

French Massif Central. The French Massif Central in the western part of the Variscan Belt has experienced Late Silurian–Early Devonian to Late Carboniferous orogenic events (e.g., Matte, 1986). High-pressure metamorphic rocks occur mainly in the upper gneissic unit within the leptyno-amphibolite group. The latter consists of an association of metagreywacke, mica schist, metabasalt, leptynite, metagranite, and peridotite. Eclogites are associated with silicic and mafic HP granulites (Lardeaux et al., 2001) and also with spinel and/or garnet lherzolites (Gardien et al., 1990). Inclusions of coesite and coesite pseudomorphs occur in garnets of kyanite-bearing eclogite (Lardeaux et al., 2001). Only two coesite grains are preserved as relics; most coesite grains are completely transformed into polycrystalline radial quartz (palisade texture) or into polygonal quartz surrounded by radial cracks. Metamorphic temperatures were estimated to be 740-780°C, and subsequent amphibolite-facies overprinting at 750°C at 1.5–1.7 GPa during decompression. UHP metamorphism occurred between 420 and 400 Ma (Paquette et al., 1995); $^{40}Ar/^{39}Ar$ data from amphibole separates from retrogressed eclogites yielded 339 ± 4 Ma (Costa et al., 1993).

Northeast Greenland Caledonides. The first evidence for UHP metamorphism in the Greenland Caledonites was reported from kyanite eclogites and associated host gneisses on an island in Jokelbugt (Gilotti and Ravna, 2002). Inclusions of quartz pseudomorphs exhibiting palisade structure and radiating fractures occur in garnet and omphacite of eclogites and in garnet of the host gneisses. P-T estimates of peak-stage metamorphism of eclogite at ~972°C and 3.6 GPa lie well within the coesite stability field. SHRIMP U/Pb dates of 403 ± 5 Ma for zircon rims from HP metapsammite and 404 ± 4 Ma for zircons from anatectic melt derived from metapelite are coeval with the estimated ages of exhumation of UHP terranes in the Scandinavian Caledonides described in a previous section (McClelland and Gilotti, 2003).

Neoproterozoic nappes in Southeast Brazil. Neoproterozoic nappes of Southeast Brazil consist mainly of coarse-grained kyanite-garnet granulite with intercalations of impure quartzite, calc-silicate rock, and minor lenses and sills of mafic-ultramafic rocks, including eclogite. Early eclogitic phases were strongly obliterated during granulite-facies overprinting during a major Pan-African continent collision at 640–630 Ma (Campos Neto and Caby, 1999). Rare ultramafic lenses consist of phlogopite-orthopyroxene rocks and garnet clinopyroxenite. Cores of granulite boudins retain eclogitic assemblages; some quartz inclusions within the mantle regions of zoned garnet porphyroblasts are surrounded by intense radial fractures, suggestive of the former presence of coesite. In fact, coarse-grained zircons in some granulites contain numerous micro-inclusions of K-feldspar + kyanite + coesite, as confirmed by Raman spectroscopy (Parkinson et al., 2001). Many exsolution microstructures observed in other UHP terranes also occur. This area represents the first coesite-grade UHP rocks in the Americas, and it can be correlated with the Mali UHP terrane of equatorial Africa (Caby, 1994). It has been suggested that eclogitic assemblages and probable UHPM components may be present in several other correlative nappe systems at the margins of the Sao Francisco craton in Brazil.

Chuacus Complex, north-central Guatemala. HP eclogites within banded gneisses, schists, and migmatites of the Chuacus Complex contain relict textures and mineralogical features that suggest UHP metamorphism (Solari et al., 2003). These include inclusions of possible coesite pseudomorphs in garnet and kyanite, lamellar inclusions of rutile/ilmenite in garnet, high sodium content of some garnets (up to 1200 ppm), and preliminary P-T estimates of 700–800°C, and 2–3 GPa. Although K-Ar dating of mica and hornblende yield ages of 48–72 Ma possible for amphibolite facies retrogression, regional and structural data support a pre-Mesozoic age for the multi-stage HP-UHP recrystallization. Such a discovery opens up a new interpretation regarding the tectonic evolution of Maya-Chortis-Oaxaquia microcontinental blocks forming the structural backbone of Mesoamerica.

Greek Rhodope. The Greek Rhodope Metamorphic Province (RMP) at the border between Greece and Bulgaria represents a syn-metamorphic nappe system of Alpine age that was formed during the Cretaceous to Mid-Tertiary collision of Apulia and paleo-Europe. The Kimi nappe complex comprises crustal eclogites, leucocratic orthogneisses, pelitic gneisses, and volumetrically minor mantle-derived ultramafic rocks; these units contain many mineralogical indicators of UHP metamorphism (Mposkos and Kostopoulos, 2001). The ultramafic lithologies include garnet spinel–bearing lherzolite and layers of spinel-garnet clinopyroxenite, garnet pyroxenite, and clinopyroxene garnetite. These rocks were subjected to UHP recrystallization at P-T conditions of ~1200°C and >3.0 GPa. The pelitic gneiss contains microdiamond inclusions in garnet porphyroblasts as well as rods or needles of silica, rutile, biotite, and apatite in sodic garnet, suggesting the prior occurrence of majoritic garnet. The growth of such supersilicic garnet was suggested to be at a pressure of ~7 GPa. Inclusions of polycrystalline quartz aggregates surrounded by radial cracks occur in eclogitic garnet. Exsolution lamellae of quartz in eclogitic clinopyroxene also indicate an UHP precursor of supersilicic clinopyroxene.

Lanterman Range, Antarctica. The Lanterman Range in northern Victoria Land of Antarctica consists of several metamorphic complexes. The Gateway Hills Metamorphic Complex is a thin discontinuous belt more than 50 km long, consisting of mafic and ultramafic rocks including lenses and pods of eclogite within felsic gneiss. Some less foliated eclogites contain inclusions of coesite and coesite pseudomorphs in garnet porphyroblasts as confirmed by *in situ* Raman spectroscopy, as well as radial fractures around the inclusions (Ghiribelli et al., 2002). The eclogites were subjected to UHP metamorphism at T > 850°C and P > 2.9 GPa, then amphibolite-facies overprinting during isothermal decompression. The UHP event is dated by Sm/Nd and ^{238}U/^{206}Pb data at 500 Ma, ^{40}Ar/^{39}Ar ages of 490–486 Ma for Ca-amphibole (amphibolite-facies overprinting) suggest rapid cooling, with an average exhumation rate of 3–4 km_Ma^{-1} for the UHP terrane (Di Vincenzo and Palmeri, 2001).

Western Tian-Shan, China. Caledonian eclogites of the western Tian-Shan, China have been classified into three types (Zhang L. et al., 2002). Type I eclogite pods are layered with mafic blueschists and Type II eclogites preserved pillow structures, whereas Type III eclogites are banded with marbles. This 200 km long eclogite-bearing belt extends westward to connect with the Atbashy UHP belt in Kazakhstan (Tagiri et al. 1995). Reported UHP evidence includes inclusions of coesite pseudomorphs in garnets from Type I and III eclogites, quartz exsolution lamellae in omphacite from Type II eclogites, and relict magnesite within matrix dolomite in Type III eclogites (Zhang, J. X. et al., 2002; Zhang L. et al., 2002). P-T estimates of the peak UHP stage of metamorphism are T = 500–600°C and P = 2.6–2.7 GPa (Wei et al., 2003).

North Qaidam Mountains, western China. Many lenses and layers of garnet peridotite and eclogite occur in an amphibolite-facies terrane in the North Qaidam Mountains along the northeastern rim of the Qaidam Basin. This belt comprises mainly felsic gneiss, quartz schist, garnet-bearing gneiss, amphibolite, and minor eclogite + garnet peridotite; it extends westward for more than 1000 km to the Altun Mountains and was displaced by the left-lateral Altyn-Tagh fault. At the eastern end of this UHP belt at Dulan, occurrences of coesite inclusions in zircon separates from paragneiss, inclusions of coesite pseudomorph in garnet and omphacite of eclogites, quartz rods in eclogitic omphacite, and P-T estimates of eclogites (P = 2.87–3.17 GPa, and T = 631–687°C) demonstrate that this is a typical UHP metamorphic terrane (Yang et al., 2001; Song et al., 2003a, 2003b). The Dulan eclogites have an Sm-Nd mineral isochron age of 497 ± 87 Ma and a SHRIMP U/Pb zircon age of 495 Ma (Yang et al., 2002b).

Altun Mountains, western China. This UHP terrane, about 200 km long in the southern margin of the Altun Mountains, is the western extension of the North Qaidam belt. Eclogitic lenses and blocks of various sizes together with garnet-lherzolite and clinopyroxenite are enclosed within an Early Proterozoic sequence of amphibolite, amphibole schist, garnet-bearing granitic gneiss, and pelitic schist. Garnet lherzolite contains magnesite that reacted to form dolomite; P-T estimates of the peak magnesite-bearing assemblage yield P > 3.6 GPa and 850°C (Liu et al., 2001). Garnet porphyroblasts include exsolution lamellae of clinopyroxene and rutile similar to that of eclogite reported by Ye et al. (2000) as majoritic garnet. The associated eclogites contain inclusions of coesite pseudomorphs in garnet with characteristic radial fractures; this together with the exsolved quartz rods in omphacite and P-T estimates of 820–850°C and 2.8 to 3.2 GPa provide additional evidence for UHP (Zhang J. X. et al.,

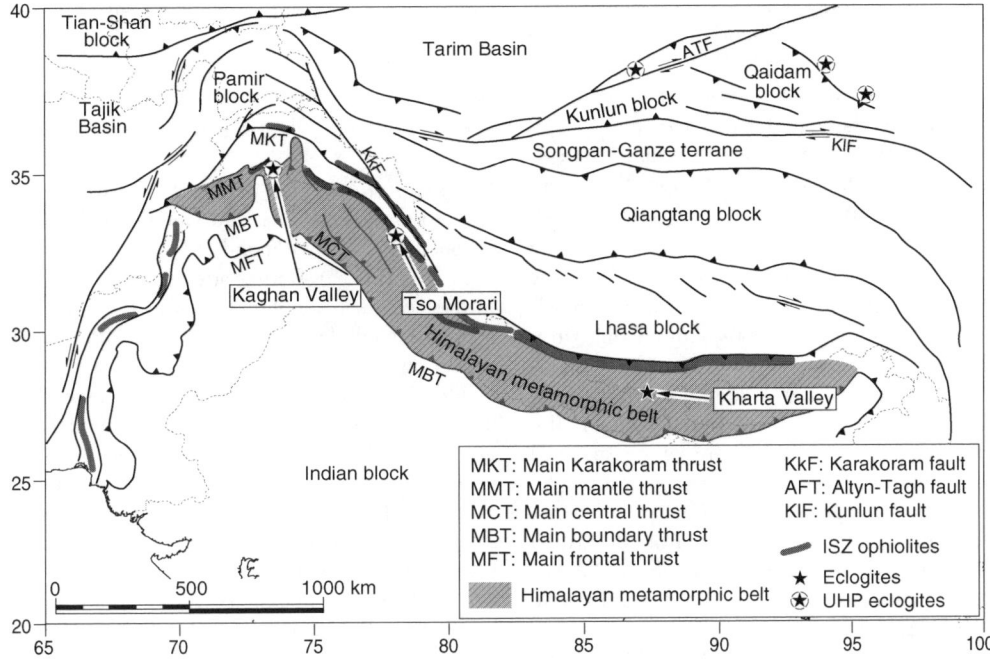

FIG. 5. A simplified tectonic map of the western Himalaya, Hindu Kush, Pamir, Karakoran, and western Tibet modified after Figure 3 of Searle et al. (2001). This map shows the locations of eclogite and UHP eclogites from both Himalayan and North Qaidam, Indus Suture Zone (ISZ) ophiolites and major structures.

2001). An Sm/Nd mineral isochron for eclogitic garnet, omphacite, and the whole rock yields 500 ± 10 Ma, whereas an U/Pb isochron for zircon separates from eclogite gives an age of 504 ± 5 Ma (Zhang J. X. et al., 2000).

North Qinling Mountains, central China. Lenses and blocks of eclogitic rocks in northern Qinling are enclosed in amphibolite-facies gneiss and garnet-bearing quartz + phengite mica schist. Inclusions of coesite in eclogitic garnet (Hu et al., 1995) and microdiamonds in zircon separates from both eclogite and gneiss (Yang et al., 2002a) have been recently reported. Although individual outcrops are no more than a few meters wide, the belt of eclogites extends more than 10 km. P-T estimates of amphibolitized, coesite-bearing eclogite yield T = 590–758°C and P > 2.6 GPa. An Sm-Nd isochron for garnet, omphacite, rutile, amphibole, and whole-rock yields 400 ± 16 Ma. Microdiamonds were recently discovered as inclusions in zircon from both eclogite and its gneissic country rocks (Yang et al., 2002a); SHRIMP U/Pb dating of zircons from granitic gneiss yields 507 ± 38 Ma for peak UHP metamorphism (Yang et al., 2002a). These data indicate that two discrete collision events occurred between the North China and Yangtze cratons: (1) an Early Paleozoic North Qinling UHP suturing; and (2) a Triassic suturing along the Dabie-Sulu UHP-HP belt. The Early Paleozoic belt extends westward to the North Qaidam–Altun Mountains for more than 4000 km (Yang et al., 2002b).

Himalayan UHP Eclogites

The Himalayan orogen has long been considered a classic locality of continental collision between the Indian and Eurasian continents (Fig. 5). Several eclogite-facies HP rocks in the Himalayas have been reported since the late 1980s (e.g., Chaudary and Ghazanfar, 1987; Pognante and Spencer, 1991; de Sigoyer et al., 1997; Lombardo and Rolfo, 2000). Inclusions of coesite were recently discovered in omphacite and garnet in several Himalayan eclogites from the upper Kaghan Valley, Pakistan (O'Brien et al., 2001; Kaneko et al., 2003). Subsequent findings of additional coesite localities in the Tso Morari area (Sachan et al., 2001; Mukherjee et al., 2003) have established the presence of a UHP belt in the Himalayas, an active ongoing collision since the Eocene (Massonne and O'Brien, 2003). However, UHP relics are volumetrically very minor constituents of the metamorphic belts dominated mostly by later Barrovian overprintings. In these rocks, hydration related to Barrovian-zone metamorphism has resulted in remarkably pervasive recrystallization during the later exhumation; this event has almost entirely masked the preceding UHP record.

The finding of HP-UHP evidence has led to new insights into the Himalayan tectonic model. England and Houseman (1986) showed that, in a model of homogeneous thickening of the Himalayan-Tibet boundary, the maximum crustal thickness should be less than 70 km. However, the occurrences of coesite-bearing eclogites and gneisses in the Himalayan orogen cannot be explained by a homogeneous thickening of the crust, but rather by continental subduction as described below. According to available geochronological data, (e.g., Treloar et al., 2003) the exhumation rate of the Himalayan UHP rocks has been estimated to range from ~1 cm·yr^{-1} (O'Brien et al., 2001) to >3 cm·yr^{-1} (Treloar et al., 2003; Massone and O'Brien, 2003). These estimates are higher than those calculated from other UHP-HP terranes. This requires tectonic emplacement of the UHP-HP Himalayan slab into shallow crust, following isostatic rebound due to slab decoupling between the Indian continent and the oceanic crust initiated during the Middle Eocene. UHP rocks from both the Kaghan Valley and Tso-Mori are described below.

Kaghan Valley. The Kaghan nappe, in the northwestern Himalayan syntaxis, lies between the Indus suture (also called main mantle thrust: MMT) to the north and the main central thrust (MCT) to the southwest (Pognante and Spear, 1991; Spencer, 1993) (Fig. 5). The Cretaceous Kohistan island-arc sequence lies north of the Indus suture. It developed as a result of intra-oceanic subduction. The main central thrust separates an underlying Salkhala unit from an overlying higher Himalayan crystalline (HHC) unit. The Kaghan nappe consists mainly of granitic gneiss and paragneiss, with minor intercalated layers of amphibolite, marble, and quartzite. Eclogites are preserved in the cores of mafic boudins (less than 1 m thick) in gneiss, and are significantly overprinted by amphibolite assemblages. They contain fine-grained (< 1 mm) garnet and omphacite, together with aggregates of quartz, chains of rutile rimmed by titanite, and mm-sized phengitic micas overgrown by randomly oriented amphiboles up to 1 cm in length. Coesite and coesite pseudomorphs occur as inclusions in omphacite, and show the characteristic palisade-quartz aggregate texture; fractures radiating from these inclusions are characteristically developed in the host pyroxene.

Garnet pyroxene–phengite barometry was applied to the eclogites, and yielded peak P-T conditions of 2.7–2.9 GPa and 690–750°C (O'Brien et al., 2001) (Fig. 6). Retrograde textures are common in the eclogites, including symplectic intergrowths of augite, amphibole, and plagioclase after omphacite; thermobarometric analyses of these assemblages yielded 1.0–1.3 GPa and 600–710°C (Kaneko et al., 2003). On the other hand, the dominant granitic gneisses and metapelites of the higher Himalayan crystalline unit contain typical Barrovian-zone amphibolite-facies assemblages, represented by kyanite- and staurolite-bearing garnetiferous metapelites (Treloar, 1995). Extensive overprinting has mostly obliterated the peak assemblages; matrix minerals of the gneisses yielded P-T conditions of 0.7–1.1 GPa and 600–700°C (Treloar, 1995), consistent with retrogression of the eclogites. However, relict coesite is preserved as inclusions in zircon; the inclusions are ovoid and approximately 10 µm in diameter (Fig. 7A). This discovery of coesite-bearing zircons from Himalayan quartzofeldspathic gneisses demonstrates subduction of Indian continental crust to the depths of UHP conditions (Kaneko et al., 2003).

Geochronological studies using Sm/Nd (garnet-omphacite), Rb/Sr (phengite), and U/Pb (rutile, zircon) methods suggest that the eclogite-facies event took place at 40–50 Ma (Tonarini et al., 1993; Spencer and Gebauer, 1996). $^{40}Ar/^{39}Ar$ cooling ages for hornblende and mica from gneisses were reported to be ~43 Ma and ~25 Ma, respectively (Chamberlain et al., 1991). Recent SHRIMP U/Pb dates of zoned zircons with mineral inclusions formed at different stages indicate the ages of non-UHP mineral-bearing mantle domains of zircon and UHP mineral-bearing rims are at about 50 and 46 Ma, respectively (Kaneko et al. 2003). A new U-Pb age of 44 ± 1.1 Ma was obtained for rutile from Kaghan coesite-bearing eclogite (Treloar et al., 2003); this together with available data implies that exhumation of the UHP rocks from mantle depth to mid-crustal level occurs within a few million years.

Tso Morari. The Tso Morari crystalline dome is located between the Indus suture zone on the north, bordered by the Zildat detachment fault, and the

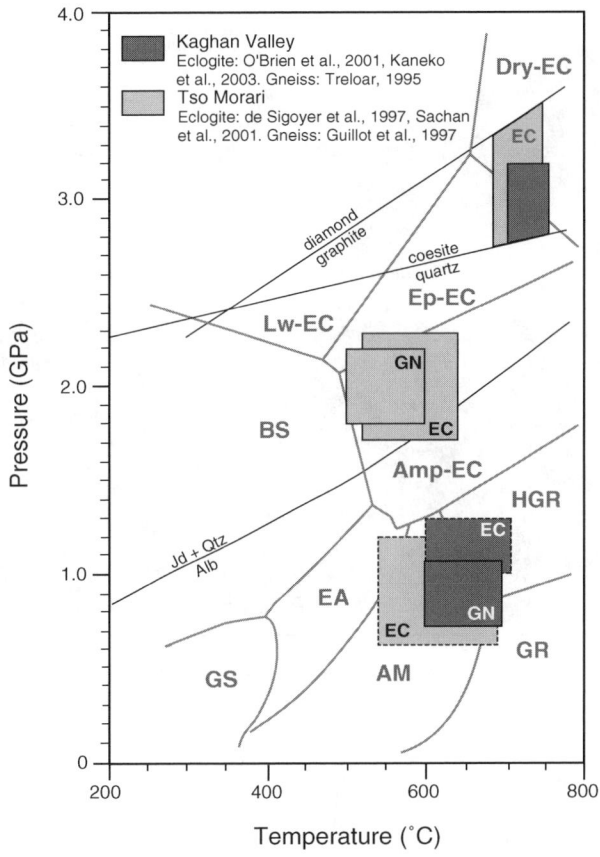

FIG. 6. P-T estimates of peak- and retrograde-stage metamorphism for coesite-bearing eclogites and the country-rock gneisses from the Kaghan Valley and Tso Morari.

Zanskar sedimentary unit on the south (de Sigoyer et al. 1997). The dome represents an internal massif of the northern part of the higher Himalaya. The Tso Morari area has been considered a typical Indian Tethyan margin because of a stratigraphic succession similar to the Nimaling antiform in the core of the dome; a metamorphic basement consists of Cambro-Ordovician augen gneisses (Stutz, 1988). This complex is covered by a metasedimentary series of Cambrian to Devonian quartzite, schist, and conglomerate, and is overlain by Lower Carboniferous to Triassic marble and metapelite. Eclogites occur as lenses or irregular bodies within the basement and cover gneisses, and are strongly overprinted by amphibolite-facies assemblages. de Sigoyer et al. (1997) reported peak P-T conditions for the eclogites at 1.7–2.3 GPa and 520–640°C (Fig. 7). However, relics of coesite 30–80 μm in size were recently found as inclusions in garnet (Sachan et al., 2001; Mukherjee et al., 2003) typified by radial fractures (Figs. 7B and 7C); for the SiO_2 inclusions, a characteristic Raman spectrum is centered at 523 cm^{-1} (Fig. 7D). The country rock metasediments with jadeite + chloritoid + paragonite + garnet assemblages have peak P-T estimates of 1.8–2.2 GPa and 500–600°C, and record retrogression at the eclogite-blueschist facies transition of 1.3–1.8 GPa and 490–590°C (Guillot et al., 1997).

The geochronology of the Tso Morari eclogites has been extensively studied and subdivided into three different stages (de Signoyer et al., 2000). Eclogitization at ~55 Ma was obtained by Lu-Hf and Sm-Nd methods on garnet, omphacite, glaucophane, and the whole rock. An amphibolite-facies overprint occurred at 47 Ma, judging by Rb/Sr dating on phengite, apatite, and whole rock from metapelites.

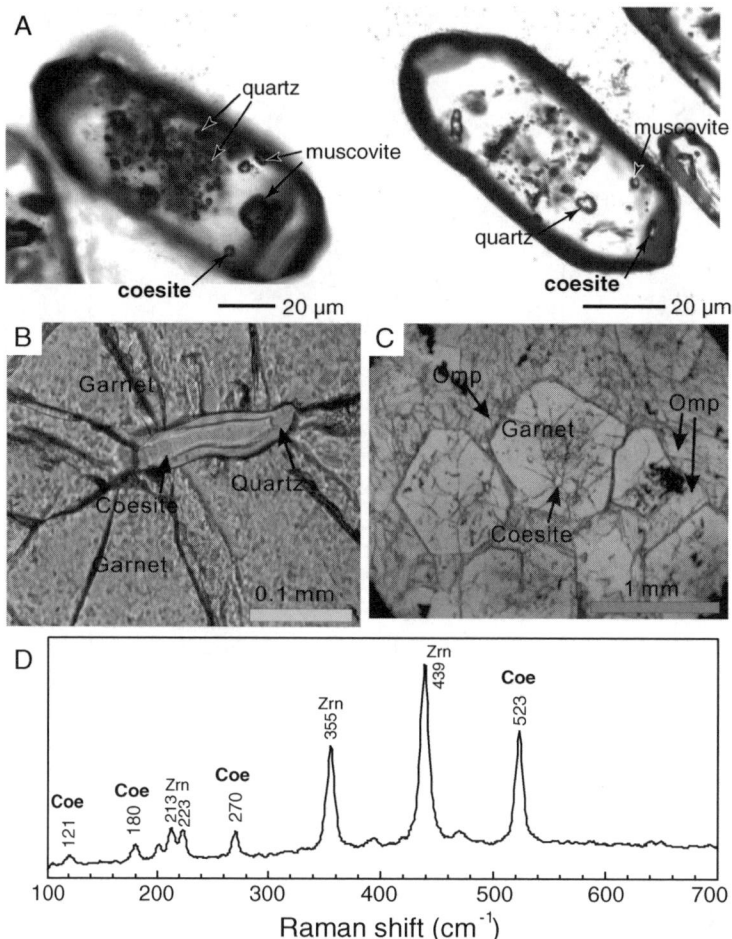

FIG. 7. A. Inclusions of low-P phases (quartz and muscovite) at core and coesite at rim from zircon separates of gneissic rocks from Kaghan Valley (Kaneko et al., 2003). B. Photomicrograph of coesite inclusion in garnet from Tso Morari eclogite showing thin palisade quartz around coesite and radial fractures in host garnet (from Fig. 2 of Mukherjee et al., 2003). C. Inclusion of coesite in eclogitic garnet from Tso Morari (from Fig. 2 of Mukherjee et al., 2003). D. Laser Raman spectra of coesite inclusion in zircon separates of Figure 7A.

^{40}Ar/^{39}Ar ages of biotite and muscovite at ~30 Ma suggest that the Tso Morari unit rose to upper crustal levels and recrystallized at the end of the exhumation (de Signoyer et al., 2000). Recent SHRIMP U/Pb dating of zoned zircons from country gneissic rocks to eclogite from the Tso Morari yields 48 ± 1 Ma for the peak eclogite-facies metamorphism and protolith ages of 700 ± 6 Ma to 1668 ± 14 Ma (Leech et al., 2003), consistent with the 46 ± 1 Ma UHP ages for Kaghan eclogite (Kaneko et al., 2003). Diffusion modeling of garnet overgrowth composition steps by O'Brien and Sachan (2000) yields an exhumation rate of 23–45 mm·yr^{-1} from the UHP stage to low greenschist-facies stage occuring within about 3 Ma (45–48 Ma) (Massone and O'Brien, 2003). The similarity in the tectonic setting, lithologies, and UHP ages suggest they belong to a single UHP belt.

Continental Subduction And Collision (the Himalayan Model)

Continental subduction and UHP metamorphism at 45 to 52 Ma

The Himalayan orogeny was controlled principally by subduction of the Indian continental crust beneath the Asian continental lithosphere; this may

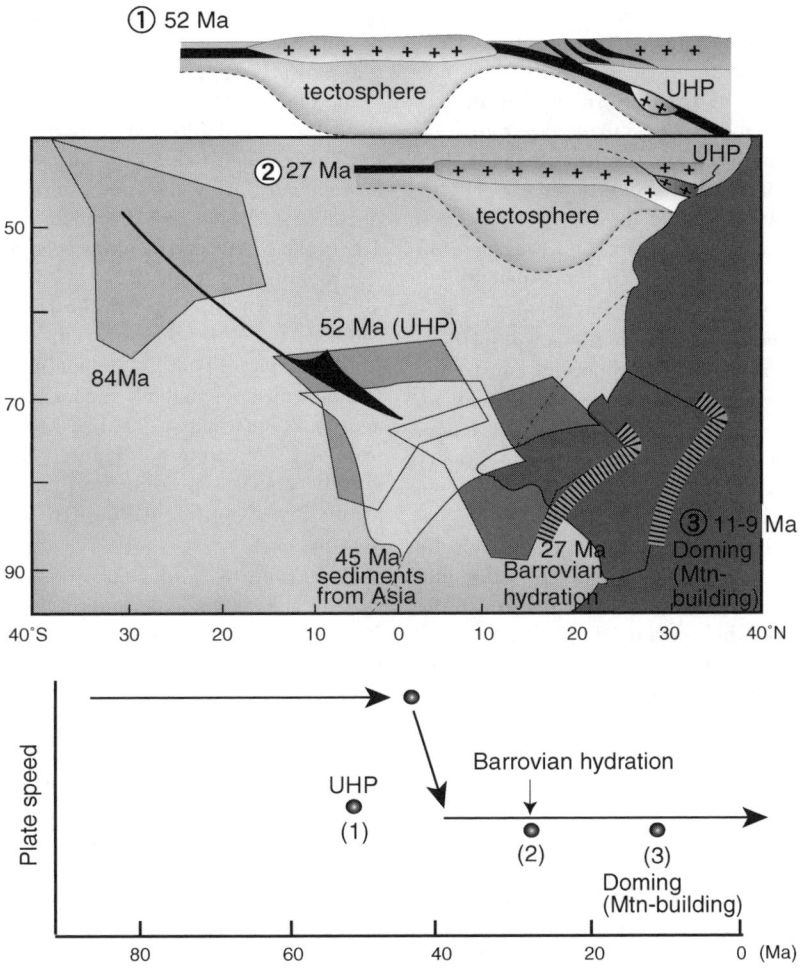

FIG. 8. Subduction and collision of the Indian and Eurasian continents showing the subduction of microcontinent for formation of UHP rocks at 52 Ma (1) and exhumation of UHP rocks from the Himalayan metamorphic belt. Exhumation of the UHP slab includes (2) wedge extrusion from mantle depth to crustal level at >27 Ma and (3) a doming event at 11–9 Ma. Paleogeographic positions of the Indian continent at 84, 52, 45, 27, and 11–9 Ma are shown together with the rate estimates of northeastward motion of the Indian plate. The doming event was responsible for the mountain-building processes for the High Himalayan Mountain chain.

provide a present-day analog for UHP metamorphism. So far, numerous tectonic models for the Himalayan orogeny have been proposed (for a review, see Hodges, 2000); most contributions have focused on the dynamics of unusual crustal thickening (up to 75 km) that resulted in a regional Barrovian-zone metamorphic sequence. In general, the closure of Neo-Tethys to form the Indus suture was followed by the India-Asia continental collision at 52–50 Ma (Fig. 8). Both Asian and Indian margins were thickened and metamorphosed, and the intra-oceanic convergence generated a continental suture (e.g., Rowley, 1996). During the infant stage of the Himalayan orogeny (~80–50 Ma), a Pacific-type arc-trench system was produced, and huge, multiple volcanic/plutonic calc-alkaline arcs formed, cored by the Kohistan-Ladakh and Trans-Himalayan batholiths. Pacific-type subduction formed a Cretaceous accretionary complex along the southern margin of the Kohistan-Ladakh arc; the accretionary complex contains 80 Ma blueschists (e.g., Anczkiewicz et al., 2000). Granitic arc-magmatism continued until 50 Ma at Ladakh (Weinberg and Dunlap, 2000), when collision was initiated at the western syntax and propagated to the SE. Since then, subduction of the Indian continent has occurred, and the India-Asia convergence rate drastically decreased from 19 to 5 cm·yr^{-1} at about Middle Eocene time (e.g. Klootwijk et al., 1992). Continental collision closed the intervening ocean and terminated arc magmatism (Fig. 8); hence 45 Ma lacustrine deposits on the Indian subcontinent contain detritus derived from the Asian mainland.

Recent seismologic studies of the ongoing Himalayan orogeny have provided important constraints on its tectonic evolution. In the western Himalaya, the Hindu Kush region is seismically most active, with abundant intermediate-depth seismicity (Pegler and Das, 1998). The seismic zone extends to a depth of 100-250 km, and steepens with increasing depth. A seismic reflection profile through the region shows the presence of subducted continental crust and a trapped oceanic basin at depth. On the basis of the available seismic data, Searle et al. (2001) proposed a continental subduction model for the formation of UHP metamorphic rocks beneath the Hindu Kush. Their model suggests that a cold lower crust of the Indian block comprising Precambrian granulite-facies basement and a cover sequence of Paleozoic–Lower Mesozoic supracontinental sediments was subducting beneath the Hindu Kush; subsequently the crustal rocks underwent UHP metamorphism at about 150 km depth and ~800–900°C at 4–6 GPa. Kaneko et al. (2003) obtained a sinking rate of 1.2–1.6 cm·yr^{-1} for the downgoing Indian crust to ~110 km depth, based on SHRIMP U/Pb ages of zoned zircon in UHP rocks from the Kagan Valley. These data suggest a time-integrated subduction angle of 18–22°, similar to the estimated angle of the present Moho between the Himalaya foreland and southern edge of the Tethyan Himalaya (Zhao et al., 1993). The timing of UHP metamorphism at 46 Ma (Kaneko et al., 2003) is synchronous with that of the rapid cooling (Weinberg and Dunlap, 2000) of the youngest plutons of the Ladakh batholith.

Barrovian metamorphic overprint at >30–15 Ma

The Himalayan orogeny involves crustal thickening that resulted in a regional Barrovian-type kyanite-sillimanite series metamorphism. This metamorphism is well documented in metapelites of the Higher Himalaya (including the Zanskar Himalayan series) and the north Himalayan gneiss domes. Most Oligocene kyanite-grade rocks were metamorphosed at P = 0.6–1.0 GPa and T = 450–600°C; Early Miocene sillimanite-grade rocks were formed at higher T (650–770°C) but lower P (0.4–0.7 GPa). The U/Pb and Sm/Nd geochronology in the Zansker and Lahoul regions yields ~37–29 Ma for regional Barrovian-type kyanite-grade metamorphism (e.g., Vance and Harris, 1999; Walker et al., 1999). The ^{40}Ar/^{39}Ar cooling ages of muscovite in the metapelites yield 22–21 Ma (Walker et al., 1999) and overlap those in the Himalayan leucogranite (24–17 Ma: Searle et al., 1988). Monazite from sillimanite-bearing metapelites of the greater Himalayan series in the Everest region yielded U/Pb ages of 32 Ma (Simpson et al., 2000); monazite U/Pb ages of 23 Ma from sillimanite-cordierite–bearing granitic gneiss represent the timing of cordierite-grade low-P metamorphism related to crustal melting.

The kyanite- and staurolite-bearing mineral assemblage occurs as a retrograde phase of the Tso Morari eclogitic metapelites (Guillot et al., 1997) in the largest gneiss dome of the Himalayan chain. ^{40}Ar/^{39}Ar plateau ages of 31–29 Ma from low-Si muscovite and biotite of retrograded metapelites represent thermal relaxation at upper crustal levels (de Sigoyer et al., 2000). A series of gneiss domes (north Himalayan gneiss domes) that are mantled by Barrovian-type kyanite-grade metamorphic rocks can be traced for at least 2000 km; their cooling age is similar to the Barrovian overprint of the Tso Morari UHP eclogites.

Domal uplift and mountain-building since 11 Ma

The uplift of the higher Himalayas apparently occurred in the Late Miocene, as documented by sedimentary flooding demonstrated on ODP Leg 116, by on-land geology, and by climate change in central Asia. Drilling of the distal Bengal fan in the central Indian Ocean revealed that the sedimentation rate drastically increased at 10.9 Ma (Amano and Taira, 1992). Paleontological evidence suggests that glaciation began in the higher Himalayas at about 7 Ma (Lakhanpal et al., 1983). The loess-paleosol sequence of the Chinese Loess Plateau is known to have formed by eolian transport of particles from the inland deserts of northwestern China and possibly, Central Asia. In recent years, uplift of the Himalayan-Tibetan Plateau (Manabe and Broccoli, 1990; An et al., 2001) and changes in land-sea distribution (Ramstein et al., 1997) have been considered principal driving forces for long-term

Cenozoic climatic change. A very surprising result has just been published in Nature (Guo et al., 2003), which recounts the discovery of a 22 Ma old loess deposit in Gansu Province, and the onset of Asian desertification has been linked to the uplift of the Tibetan Plateau.

The doming event at 15–11 Ma is also suggested by mica ^{40}Ar/^{39}Ar thermochronology of the Kangmar dome (Maluski et al., 1988; Lee et al., 2000). This doming event raised the Himalayan Mountain chain more than 5 km in elevation; active erosion has continued and thick flysch sequences have accumulated in the Bengal fan. The major driving force for the domal uplift is not well understood. The Miocene underthrusting of the Indian tectosphere beneath the Lesser Himalayas may have elevated the Higher Himalaya.

These data clearly show a two-step exhumation process for the Himalayan UHP and high-grade metamorphic terrane. (1) Emplacement of the Himalayan high-grade metamorphic rocks at shallow crustal levels occurred at > 30 to 20 Ma, when no topographic mountain building was apparent, and (2) Himalayan mountain building began at 10.9 Ma, when extensive surface erosion commenced. Figure 8 schematically shows such events together with the paleogeography of the Indian and Eurasian continents since 80 Ma.

Acknowledgments

We submit this review article for the retirement celebration of W. G. Ernst, who has documented subduction-zone metamorphism for both Pacific-type and Alpine-type blueschist terranes. This manuscript was initiated in 2000–2001 while S. Maruyama held a sabbatical position as a Cox Professor at Stanford; both T. Tsujimori and I. Katayama were supported by a JSPS Research Fellowship for Young Scientists, and a Grant-in-Aid for JSPS Fellows. The paper represents a product of the U.S.–Japan project supported by the NSF Global Partnership Fellowship INT-9820171, and the NSF US-China project EAR 0003355. We thank Y. Kaneko and Himanshu Sachan for preprints and photomicrograph of coesite inclusions in UHP rocks respectively from the Kaghan Valley of Pakistan and Tso-Mori. We appreciate a constructive review of the manuscript by our mentor W. G. Ernst.

REFERENCES

Amano, K., and Taira, A., 1992, Two-phase uplift of Higher Himalayas since 17 Ma: Geology, v. 20, p. 391–394.

An, Z. S., Kutzbach, J. E., Prell, W. L., and Porter, S. C., 2001, Evolution of Asian monsoons and phased uplift of the Himalaya-Tibetan Plateau since Late Miocene times: Nature, v. 411, p. 62–66.

Ankiewicz, R., Burg, J. P., Villa, I. M., and Meier, M., 2000, Late Cretaceous blueschist metamorphism in the Indus suture zone, Shangla region, Pakistan Himalaya: Tectonophysics, v. 324, p. 111–134.

Bakun-Czubarow, N., 1991, On the possibility of occurrence of quartz pseudomorphs after coesite in the eclogite-granulite rock series of the Zlote Mountains in the Sudetes, SW Poland: Archiwum Mineralogy, v. 48, p. 3–25.

Beane, R. J., and Connelly, J. N., 1998, ^{40}Ar/^{39}Ar, U-Pb, and Sm-Nd constraints on the timing of metamorphic events in the Maksyutov Complex, southern Ural Mountains: Journal of the Geological Society of London, v. 157, p. 811–822.

Beane, R. J., Liou, J. G., Coleman, R. G., and Leech, M. L., 1995, Petrology and retrograde P-T path for eclogites of the Maksyutov complex, southern Ural Mountains, Russia: The Island Arc, v. 4, p. 254–266.

Bostick, B., Jones, R. E., Ernst, W. G., Chen, C., Leech, M. L., and Beane, R. J., 2003, Low-temperature microdiamond aggregates in the Maksyutov metamorphic complex, south Ural Mountains, Russia: American Mineralogist, v. 88, p. 1709–1717.

Bowtell, S. A., Cliff, R. A., and Barnicoat, A. C., 1994, Sm-Nd isotopic evidence on the age of eclogitization in the Zermatt-Saas ophiolite: Journal of Metamorphic Geology, v. 12, p. 187–196.

Brueckner, H. K., and Medaris, L. G., Jr., 1998, A tale of two orogens: The contrasting T-P-t history and geochemical evolution of mantle in high- and ultra-high-pressure metamorphic terranes of the Norwegian Caledonides and the Czech Variscides: Schweizerische Mineralogische und Petrographische Mitteilungen, v. 78, p. 293–307.

Caby, R., 1994, Precambrian coesite from N. Mali: First record and implications for plate tectonics in the trans-Saharan segment of the Pan-African belt: European Journal of Mineralogy, v. 6, p. 235–244.

Carswell, D. A., Brueckner, H. K., Cuthbert, S. J., Mehta, K., and O'Brien, P. J., 2003, The timing of stabilisation and the exhumation rate for ultra-high pressure rocks in the Western Gneiss Region: Journal of Metamorphic Geology, v. 21, p. 601–612.

Carswell, D. A., and Compagnoni, R., eds., 2003, Ultra-high pressure metamorphism: European Mineralogical Union, Notes in Mineralogy, v. 5, p. 51–74.

Carswell, D. A., and Cuthbert, S. J., 2003, Reviews of representative UHPM terranes: The Western Gneiss

Region of Norway: *in* Carswell, D. A., and Compagnoni, R., eds., Ultra-high pressure metamorphism: European Mineralogical Union, Notes in Mineralogy, v. 5, p. 51–74.

Carswell, D. A., Cuthbert, S. J., and Krogh Ravna, E. J., 1999, Ultrahigh-pressure metamorphism in the western Gneiss Region of the Norwegian Caledonides: International Geology Review, v. 41, p. 955–966.

Campos Neto, M. da C., and Caby, R., 1999, Tectonic constraint on Neoproterozoic high-pressure metamorphism and nappe system south of the San Francisco craton, southeast Brazil: Precambrian Research, v. 97, p. 3–26.

Chamberlain, C. P., Zeitler, P. K., and Erickson, E., 1991, Constraints on the tectonic evolution of the northwestern Himalaya from geochronologic and petrologic studies of Babusar Pass, Pakistan: Journal of Geology, v. 99, p. 829–849.

Chaudary, M. N., and Ghazanfar, M., 1987, Geology, structure and geomorphology of upper Kaghan valley, Northwestern Himalaya, Pakistan: Geological Bulletin, University of Punjab, v. 22, p. 13–57.

Chesnokov, B. V., and Popov, V. A., 1965, Increasing volume of quartz grains in eclogites of the south Urals: Doklady Akademii Nauk, v. 162, p. 176–178 (in Russian).

Chopin, C., 1984, Coesite and pure pyrope in high-grade blueschists of the Western Alps: A first record and some consequences: Contributions to Mineralogy and Petrology, v. 86, p. 107–118.

———, 2003, Ultrahigh-pressure metamorphism: Tracing continental crust into the mantle: Earth and Planetary Science Letters, v. 212, p. 1–14.

Coleman, R. G., and Wang, X., 1995, Ultrahigh pressure metamorphism: Cambridge, UK, Cambridge University Press.

Compagnoni, R., and Rolfo, F., 2003, Reviews of representative UHPM terranes: The Western Alps, *in* Carswell, D. A., and Compagnoni, R., eds., Ultra-high pressure metamorphism: European Mineralogical Union, Notes in Mineralogy, v. 5, p. 13–50.

Costa, S., Maluski, A., and Lardeaux, J. M., 1993, $^{40-39}$Ar chronology of Variscan tectono-metamorphic events in an exhumed crustal nappe: The Monts du Lyonnais complex (Massif Central France): Chemical Geology, v. 105, p. 339–359.

Cuthbert, S. J., Carswell, D. A., Krogh-Ravna, E. J., and Wain, A., 2000, Eclogites and eclogites in the Western Gneiss Region, Norwegian Caledonides: Lithos, v. 52, p. 165–195.

de Sigoyer, J., Chavagnac, V., Blichert-Toft, J., Villa, I. M., Luais, B., Guillot, S., Cosca, M., and Mascle G., 2000, Dating the Indian continental subduction and collisional thickening in the northwest Himalaya: Multichronology of the Tso Morari eclogites: Geology, v. 28, p. 487–490.

de Sigoyer, J., Guillot, S., Laudeaux, J. M., and Mascle, G., 1997, Glaucophane-bearing eclogites in the Tso Morari dome, eastern Ladakh, NW Himalaya: European Journal of Mineralogy, v. 9, p. 1073–1083.

Di Vincenzo, G., and Palmeri, R., 2001, An ^{40}Ar-^{39}Ar investigation of high-pressure metamorphism and the retrogressive history of mafic eclogites from the Lanterman Range, Antarctica: Evidence against a simple temperature control on argon transport in amphibole: Contributions to Mineralogy and Petrology, v. 141, p. 15–35.

Dobretsov, N. L., and Dobretsova, L. V., 1988, New mineralogic data on the Maksyutov eclogite-glaucophane schist complex, southern Urals: Doklady Akademii Nauk, v. 300, p. 111–116.

Dobretsov, N. L., Shatsky, V. S., Coleman, R. G., Lennykh, V. I., Valizer, P. M., Liou, J. G., Zhang, R. Y., and Beane, R. J., 1996, Tectonic setting and petrology of ultrahigh-pressure metamorphic rocks in the Maksyutov Complex, Ural Mountains, Russia: International Geology Review, v. 38, p. 136–160.

Dobrzhinetskaya, L. F., Eide, E. A., Larsen, R. B., Sturt, B. A., Tronnes, R. G., Smith, D. C., Taylor, W. R., and Posukhova, T. V., 1995, Microdiamond in high-grade metamorphic rocks of the Western Gneiss region, Norway: Geology, v. 23, p. 597–600.

Dobrzhinetskaya, L. F., Green, H. W., Mitchell, T. E., and Dickerson, R. M., 2001, Metamorphic diamonds: Mechanism of growth and inclusion of oxides: Geology, v. 29, p. 263–266.

Dobrzhinetskaya, L. F., Green, H. W., Weschler, M., Darus, M., Wang, Y. C., Massonne, H., and Stöckert, B., 2003, Focused ion beam technique and transmission electron microscope studies of microdiamonds from the Saxonian Erzgebirge, Germany: Earth and Planetary Science Letter, in press.

England, P. C. and Houseman, G. A., 1986, Finite strain calculations of continental deformation. II. Application to the India-Asia plate collision: Journal of Geophysical Research, v. 91, p. 3664–3676.

Ernst, W. G., and Liou, J. G., eds., 2000, Ultrahigh-pressure metamorphism and geodynamics in collision-type orogenic belts: Geological Society of America, International Book Series, v. 4, 293 p.

Gardien, V., Tegyey, M., Lardeaux, J. M., Misseri, M., and Dufour, E., 1990, Crustal-mantle relationship in French Variscan chain: The example of the southern Monts du Lyonnais unit (eastern French Massif Central): Journal of Metamorphic Geology, v. 8, p. 477–492.

Gebauer, D., Schertl, H. P., Brix, M., and Schreyer, W., 1997, 35 Ma old ultrahigh-pressure metamorphism and evidence for very rapid exhumation in the Dora Maira Massif, western Alps: Lithos, v. 41, p. 5–24.

Ghiribelli, B., Frezzotti, M.-L., and Palmeri, R., 2002, Coesite in eclogites of the Lanterman range, Antarctica: Evidence from textural and Raman studies: European Journal of Mineralogy, v. 14, p. 355–360.

Gilotti, J. A., 1993, Discovery of a medium-temperature eclogite province in the Caledonides of North-East Greenland: Geology, v. 21, p. 523–526.

Gilotti, J. A., and Ravna, E. J. K., 2002, First evidence for ultrahigh-pressure metamorphism in the Northeast Greenland Caledonides: Geology, v. 30, p. 551–554.

Guillot, S., de Sigoyer, J., Lardeaux, J. M., and Mascle, G., 1997, Eclogitic metasediments from the Tso Morari area, Ladakh, Himalaya.: Evidence for continental subduction during India-Asia convergence: Contributions to Mineralogy and Petrology, v. 128, p. 197–212.

Guo, Z. T., Ruddiman, W. F., Hao, Q. Z., Wu, H. B., Qiao, Y. S., Zhu, R. X., Peng, S. Z., Wei, J. J., Yuan, B. Y. and Liu, T. S., 2003, Onset of Asian desertification by 22 Myr ago inferred from loess deposits in China: Nature, v. 416, p. 159–162.

Hacker, B. R., Ratschbacher, L., Webb, L., McWilliams, M. O., Ireland, T., Calvert, A., Dong, S., Wenk, H.-R. and Chateigner, D., 2000, Exhumation of ultrahigh-pressure continental crust in East-central China: Late Triassic–Early Jurassic tectonic unroofing: Journal of Geophysical Research, v. 105, p. 13,339–13,364.

Hirajima, T., and Nakamura, D., 2003, Reviews of representative UHPM terranes: The Dabie Shan and Sulu region of China, in Carswell, D. A., and Compagnoni, R., eds., Ultra-high pressure metamorphism: European Mineralogical Union, Notes in Mineralogy, v. 5, p. 105–144.

Hodges, K. V., 2000, Tectonics of the Himalaya and southern Tibet from two perspectives: Geological Society of America Bulletin, v. 112, p. 324–350.

Hu, N., Zhao, D., Xu, B., and Wang, T., 1995, Petrography and metamorphic study on high-ultrahigh pressure eclogite from Guanpo area, northern Qinling Mountain: Journal of Mineralogy and Petrology, v. 15, p. 1–9 (in Chinese).

Hwang, S. L., Shen, P., Chu, H. T., Yui, T. F., and Lin, C. C., 2001, Genesis of microdiamonds from melt and associated multiphase inclusions in garnet of ultrahigh-pressure gneiss from Erzegebirge, Germany: Earth and Planetary Science Letters, v. 188, p. 9–15.

Jahn, B. M., Caby, R., and Monie, P., 2001, The oldest UHP eclogites of the World: Age of UHP metamorphism, nature of protoliths, and tectonic implications: Chemical Geology, v. 178, p. 143–158.

Kaneko, Y., Katayama, I., Yamamoto, H., Misawa, K., Ishikawa, M., Rehman, H. U., Kausar, A. B., and Shiraishi, K., 2003, Timing of Himalayan ultrahigh-pressure metamorphism: Sinking rate and subduction angle of the Indian continental crust beneath Asia: Journal of Metamorphic Geology, v. 21, p. 589–599.

Kaneko, Y., Maruyama, S., Terabayashi, M., Yamamoto, H., Ishikawa, M., Anma, R., Parkinson, C. D., Ota, T., Nakajima, Y., Katayama, I., Yamamoto, J., and Yamauchi, K., 2000, Geology of the Kokchetav UHP-HP metamorphic belt, Northern Kazakhstan: The Island Arc, v. 9, p. 264–283.

Kardarusman, A., and Parkinson, C. D., 2000, Petrology and P-T evolution of garnet peridotites from central Silawesi, Indonesia: Journal of Metamorphic Geology, v. 18, p. 193–209.

Katayama, I., Maruyama, S., Parkinson, C. D., Terada, K., and Sano, Y., 2001, Ion micro-probe U-Pb zircon geochronology of peak and retrograde stages of ultrahigh-pressure metamorphic rocks from the Kokchetav massif, northern Kazakhstan: Earth and Planetary Science Letters, v. 188, p. 185–198.

Katayama, I., Ohta, M., and Ogasawara, Y., 2002, Mineral in zircon from diamond-bearing marble in the Kokchetav massif, northern Kazakhstan: European Journal of Mineralogy, v. 14, p. 1103–1108.

Krogh, E. J., and Carswell, D. A., 1995, HP and UHP eclogites and garnet peridotites in the Scandinavian Caledonides, in Coleman, R. G., and Wang, X., eds., Ultrahigh pressure metamorphism: Cambridge, UK, Cambridge University Press, p. 244–298.

Klootwijk, C., Gee, J., Peirce, J., Smith, G., and McFadden, P., 1992, An early India-Asia contact: Paleomagnetic constraints from the Ninetyeast Ridge, ODP Leg 121: Geology, v. 20, p. 395–398.

Lakhanpal, R. N., Sah, S. C. D., Kewal, K., Sharma, K. K., and Guleria, J. S., 1983, Occurrence of Livistona in the Hemis conglomerate horizon of Ladakh, in Thakur, V. C., and Sharma, K. K., eds., Geology of Indus suture zone of Ladakh: New Delhi, India, Hindustan Book Publishing Corporation, p. 179–185.

Lardeaux, J. M., Ledru, P., Daniel, I., and Duchene, S., 2001, The Variscan French Massif Central—a new addition to the ultrahigh pressure metamorphic "club": Exhumation processes and geodynamic consequences: Tectonophysics, v. 332, p. 143–167.

Lee, J., Hacker, B. R., Dinklage, W. S., Gans, P. B., Calvert, A., Wang, Y., Wan, J., and Chen, W., 2000, Evolution of the Kangmar Dome, southern Tibet: Structural, petrologic, and thermochronologic constraints: Tectonics, v. 19, p. 872–895.

Leech, M. L., and Ernst, W. G., 1998, Graphite pseudomorphs after diamond? A carbon isotope and spectroscopic study of graphite cuboids from the Maksyutov complex, south Ural Mountains, Russia: Geochimica et Cosmochimica Acta, v. 62, p. 2143–2154.

Leech, M. L., Singh, S., Jain, A. K., and Manickavasagam, R. M., 2003, New U-Pb SHRIMP ages for the UHP Tso-Morari crystallines, eastern Ladakh, India [abs.]: Geological Society of America, Abstracts with Programs, v. 34, p. 637.

Li, S., Chen, Y., Cong, B., Zhang, Z., Zhang, R., Liou, D., Hart, S. R., and Ge, N., 1993, Collision of the North China and Yangtze blocks and formation of coesite-bearing eclogites: Timing and processes: Chemical Geology, v. 109, p. 70–89.

Liou, J. G., 1999, Petrotectonic summary of less-intensively studied UHP regions: International Geology Review, v. 41, p. 571–586.

Liou, J. G., and Zhang, R. Y., 2002, Ultrahigh-pressure metamorphic rocks: Encyclopedia of Physical sciences and technology, third ed., v. 17, p. 227–244: Tarzana, CA, Academia Press.

Liou, J. G., Zhang, R. Y. Katayama, I., Maruyama, S., and Ernst, W. G., 2002, Petrotectonic characterization of the Kokchetav Massif and the Dabie-Sulu terranes—Ultrahigh-P metamorphism in the so-called P-T Forbidden-Zone: Western Pacific Earth Sciences, v. 2, p. 119–148.

Liu, F., Xu, Z., Katayama, I., Yang, J. S., Maruyama, S., and Liou, J. G., 2001, Mineral inclusions in zircons of para- and orthogneiss from pre-pilot drillhole CCSD-PP1, Chinese Continental Scientific Drilling Project: Lithos, v. 59, p. 199–215.

Lombardo, B., and Rolfo, F., 2000, Two contrasting eclogite types in the Himalaya: implications for the Himalayan orogeny: Journal of Geodynamics, v. 30, p. 37–60.

Maluski, H., Matte, P., and Brunel, M., 1988, Argon 39-argon 40 dating of metamorphic and plutonic events in the North and High Himalaya belts (southern Tibet-China): Tectonics, v. 7, p. 299–326.

Manabe, S., and Broccoli, A.J., 1990, Mountains and arid climates of middle latitudes: Science, v. 247, p. 192–195.

Maruyama, S., Liou, J. G., and Terabayashi, M., 1996, Blueschists and eclogites of the world, and their exhumation: International Geology Review, v. 38, p. 485–594.

Masago, H., Rumble, D., Ernst W. G., Parkinson, C., and Maruyama, S., 2003, O^{18} depletion in eclogites from the Kokchetav massif, northern Kazakhstan: Journal of Metamorphic Geology, v. 21, p. 579–587.

Massonne, H. J., 1999, A new occurrence of microdiamonds in quartzofeldspathic rocks of the Saxonian Erzgebirge, Germany, and their metamorphic evolution: Proceedings of 7th International Kimberlite Conference, Capetown, v. 2, p. 533–539.

———, 2001, First find of coesite in the ultrahigh-pressure metamorphic region of the Central Erzgebirge, Germany: European Journal of Mineralogy, v. 13, p. 565–570.

Massonne, H. J., and O'Brien, P., 2003, Reviews of representative UHPM terranes: The Bohemian Massif and the NW Himalaya: in Carswell, D. A. and Compagnoni, R. (eds.), Ultra-high pressure metamorphism: European Mineralogical Union, Notes in Mineralogy, v. 5, p. 145–188.

Matte, P., 1986, Tectonics and plate tectonics model for the Variscan belt of Europe: Tectonophysics, v. 126, p. 329–374.

Matte, P., 1998, Continental subduction and exhumation of HP rocks in Paleozoic orogenic belts: Uralides and Variscides: Geologiska Foreningen i Stockholm Forhandlingar, v. 20, p. 209–222.

McClelland, W. C., and Gilotti, J. A., 2003, Late-stage extensional exhumation of high-pressure granulites in the Greenland Caledonides: Geology, v. 31, p. 259–262.

Mposkos, E. D., and Kostopoulos, D. K., 2001, Diamond, former coesite, and supersilicic garnet in metasedimentary rocks from the Greek Rhodope: A new ultrahigh-pressure metamorphic province established: Earth and Planetary Science Letters, v. 192, p. 497–506.

Mukherjee, B. K., Sachan, H. K., Ogasawara, Y., Muko, A., and Yoshioka, N., 2003, Carbonate-bearing UHPM rocks from the Tso-Morari region, Ladakh, India: Petrological implications: International Geology Review, v. 45, p. 49–69.

O'Brien, P. J., and Sachan, H. K., 2000, Diffusion modelling in garnet from Tso Morari eclogite and implications for exhumation models: Earth Science Frontier (China University of Geoscience, Beijing), v. 7, p. 25–27.

O'Brien, P. J., Zotov, N., Law, R., Khan, M. A., and Jan, M. Q., 2001, Coesite in Himalayan eclogite and implications for models of India-Asia collision: Geology, v. 29, p. 435–438.

Ogasawara, Y., Fukasawa, K., and Maruyama, S., 2002, Coesite exsolution from supersilicic titanite in UHP marble from the Kokchetav Massif, northern Kazakhstan: American Mineralogist, v. 87, p. 454–461.

Ogasawara, Y., Ohta, M., Fukasawa, K., Katayama, I., and Maruyama, S., 2000, Diamond-bearing and diamond-free metacarbonate rocks from Kumdy-Kol in the Kokchetav massif, northern Kazakhstan: The Island Arc, v. 9, p. 400–416.

Okamoto, K., and Maruyama, S., 1999, The high-pressure synthesis of lawsonite in the MORB + H_2O system: American Mineralogist, v. 84, p. 362–373.

Paquette, J. L., Monchoux, P., and Couturier, M., 1995, Geochemical and isotopic study of a norite-eclogite transition in the European Variscan belt. Implications for U-Pb zircon systematics in metabasic rocks: Geochimica et Cosmochimica Acta, v. 59, p. 1611–1622.

Parkinson, C. D., 2003, Coesite-bearing quartzites from Sulawesi, Indonesia: A first record of UHP metamorphism in SE Asia: Tectonophysics, in press.

Parkinson, C. D., Katayama, I. Liou, J. G., and Maruyama, S. eds., 2002, The diamond-bearing Kokchetav Massif, Kazakhstan: Petrochemistry and tectonic evolution of an unique ultrahigh-pressure metamorphic terrane: Tokyo, Japan, Universal Academy Press, Inc. 527 p.

Parkinson, C. D., Motoki, A., Onishi, C. T., and Maruyama, S., 2001, Ultrahigh-pressure pyrope-kyanite granulites and associated eclogites in Neoproterozoic Nappes of Southeast Brazil: UHPM Workshop 2001, Waseda University, p. 87–90.

Pegler, G., and Das, S., 1998, An enhanced image of the Pamir–Hindu Kush seismic zone from relocated earthquake hypocentres: Geophysical Journal International, v. 134, p. 573–595.

Pognante, U., and Spear, D. A., 1991, First record of eclogites from the High Himalayan belt, Kaghan Valley (northern Pakistan): European Journal of Mineralogy, v. 3, p. 613–618.

Ramstein, G., Fluteau, F., Besse, J., and Joussaume, S., 1997, Effect of orogeny, plate motion, and land-sea distribution on Eurasian climate change over the past 30 million years: Nature, v. 386, p. 788–795.

Reinecke, T., 1991, Very-high-pressure metamorphism and uplift of coesite bearing metasediments from the Zermatt-Saas zone, Western Alps: European Journal of Mineralogy, v. 3, p. 7–17.

Rowley, D. Y., 1996, Age of initiation of collision between India and Asia: A review of stratigraphic data: Earth and Planetary Science Letters, v. 145, p. 1–13.

Rumble, D., Liou, J. G., and Jahn, B. M., 2003, Continental crust subduction and UHP metamorphism, in Rudnick, R. L., ed., The crust: Oxford, UK, Elsevier, Treatise on Geochemistry, v. 3 (Holland, H. D., and Turekian, K. K., eds.), p. 293–319.

Sachan, H., Mucherjee, B. K., Ishida, H., Muko, A., Yoshioka, N., Ogasawara, Y., and Maruyama, S., 2001, New discovery of coesite from the Indian Himalaya, in Fluid/slab/mantle interactions and ultrahigh-P minerals: UHPM Workshop 2001, Waseda, Japan, p. 124–128.

Schertl, H. P., Schreyer, W., and Chopin, C., 1991, The pyrope-coesite rocks and their country rocks at Parigi, Dora Maira massif, western Alps: Detailed petrography, mineral chemistry, and PT-path: Contributions to Mineralogy and Petrology, v. 108, p. 1–21.

Searle, M. P., Cooper, D. J. W., and Rex, A. J., 1988, Collision tectonics of the Ladakh-Zanskar Himalaya, in Shackleton, R. M., Dewey, J. F., and Windley, B. F., eds., Tectonic evolution of the Himalayas and Tibet: London, UK, The Royal Society, p. 117–149.

Searle, M., Hacker, B. R., Bilham, R., 2001, The Hindu Kush seismic zone as a paradigm for the creation of ultrahigh-pressure diamond and coesite-bearing rocks: Journal of Geology, v. 109, p. 143–154.

Schmadicke, E., 1991, Quartz pseudomorphs after coesite in eclogites from the Saxonian Erzgebirge: European Journal of Mineralogy, v. 3, p. 231–238.

Schmadicke, E., Mezger, K., Cosca, M. A., and Okrusch, M., 1995, Variscan Sm-Nd and Ar-Ar ages of eclogite facies rocks from the Erzgebirge, Bohemian Massif: Journal of Metamorphic Geology, v. 13, p. 537–552.

Schmidt, M. W., and Poli, S., 1998, Experimentally based water budgets for dehydrating slabs and consequences for arc magma generation: Earth and Planetary Science Letters, v. 163, p. 361–379.

Shatsky, V. S., and Sobolev, N. V., 2003, Reviews of representative UHPM terranes: The Kokchetav massif, Kazakhstan, in Carswell, D. A., and Compagnoni, R. eds., Ultra-high pressure metamorphism: European Mineralogical Union, Notes in Mineralogy, v. 5, p. 75–104.

Shatsky, V. S., Sobolev, N. V., and Vavilov, M. A., 1995, Diamond-bearing metamorphic rocks of the Kokchetav massif, N. Kazakhstan, in Coleman, R. G., and Wang, X., eds., Ultrahigh pressure metamorphism: Cambridge, UK, Cambridge University Press, p. 427–455.

Simpson, R. L., Parrish, R. R., Searle, M. P., and Waters, D. J., 2000, Two episodes of monazite crystallization during metamorphism and crustal melting in the Everest region of the Nepalese Himalaya: Geology, v. 28, p. 403–406.

Smith, D. C., 1984, Coesite in clinopyroxene in the Caledonides and its implications for geodynamics: Nature, v. 310, p. 641–644.

Sobolev, N. V., and Shatsky, V. S., 1990, Diamond inclusions in garnets from metamorphic rocks: A new environment for diamond formation: Nature, v. 343, p. 742–746.

Solari, L., Ortega, F. Sole-vinas, J., Gomez-Tuena, A., Ortega-Obregon, C., Reher-Salas, M., Martens, U., and Moran, S., 2003, Petrologic evidence for possible ultrahigh pressure metamorphism in the Chuacus Complex of north-central Guatemala [abs.]: Geological Society of America, Abstracts with Programs, v. 34, p. 639.

Song, S. G., Yang, J. S., Liou, J. G., Wu, C. L., Shi, R. D., and Xu, Z. Q., 2003a, Petrology, geochemistry and isotopic ages of eclogites from the Dulan UHPM terrane, the North Qaidam, NW China: Lithos, v. 70, p. 195–211.

Song, S. G., Yang, J. S., Xu, Z. Q., Liou, J. G., and Shi, R. D., 2003b, Metamorphic evolution of the coesite-bearing ultrahigh-pressure terrane in the North Qaidam, Northern Tibet, NW China: Journal of Metamorphic Geology, v. 21, p. 631–644.

Spencer, D. A., 1993, Tectonics of the Higher- and Tethyan Himalaya, upper Kaghan Valley, NW Himalaya, Pakistan: Implications of an early, high-pressure, eclogite-facies metamorphism to the Himalayan belt: Unpubl. Ph.D. dissertation, ETH Zurich, Switzerland, 1050 p.

Spencer, D. A., and Gebauer, D., 1996, SHRIMP evidence for a Permian age and a 44 Ma metamorphic age for the Himalayan eclogites, Upper Kaghan, Pakistan: Implications for the subduction of Tethys and the subdivision terminology of the NW Himalaya [ext. abs.]: 11th Himalaya-Karakoram-Tibet Workshop, Flagstaff, Arizona, USA, Abstracts, p. 147–150.

Stöckhert, B., Duyster, J., Trepmann, C. and Massonne, H. J., 2001, Microdiamond daughter crystals precipitated from supercritical COH + silicate fluids included in garnet, Erzgebirge, Germany: Geology, v. 29, p. 391–394.

Stutz, E., 1988, Geologie de la chaine du Nyimaling aux confines du Ladakh et du Rupshu, NW-Himalaya, Indes—evolution paleogeographique et tectonique d'un segment de la marge nord-indienne: Memorie de Geologie, Lausanne, no. 3, 149 p.

Sun, W. D., Williams, I. S., and Li, S. G., 2002, Carboniferous and Triassic eclogites in the western Dabie Mountains, east-central China. Evidence for protracted convergence of the North and South China Blocks: Journal of Metamorphic Geology, v. 20, p. 873–886.

Tagiri, M., and Bakirov, A., 1990, Quartz pseudomorph after coesite in garnet from a garnet-chloritoid-talc schist, northern Tien-Shan, Kirghiz, USSR: Proceedings of the Japan Academy, v. 66, p. 135–139.

Tagiri, M., Yano, T., Bakirov, A., Nakajima, T., and Uchiumi, S., 1995, Mineral parageneses and metamorphic P-T paths of ultrahigh-pressure eclogites from Kyrghyzstan Tien-Shan: The Island Arc, v. 4, p. 280–292.

Tonarini, S., Villa, I. M., Oberli, F., Meier, M., Spencer, D. A., Pognante, U., and Ramsay, J. R., 1993, Eocene age of eclogite metamorphism in Pakistan Himalaya: Implications for India-Eurasia collision: Terra Nova, v. 5, p. 13–20.

Treloar, P. J., 1995, Pressure-temperature-time paths and the relationship between collision, deformation, and metamorphism in the north-west Himalaya: Geological Journal, v. 30, p. 333–348.

Treloar, P., O'Brien, P. J., Parrish, R. R., and Khan, M. A., 2003, Exhumation of early Tertiary coesite-bearing eclogites from the Pakistan Himalaya: Journal of the Geological Society of London, v. 160, p. 367–376.

van Roermund, H. L. M., Carswell, D. A., Drury, M. R., and Heijboer, T. C., 2002, Microdiamonds in a megacrystic garnet websterite pod from Bardane on the island of Fjortoft, western Norway: Evidence for diamond formation in mantle rocks during deep continental subduction: Geology, v. 30, p. 959–962.

van Roermund, H. L. M., and Drury, M. R., 1998, Ultrahigh pressure (P > 6 Gpa) garnet peridotites in western Norway: Exhumation of mantle rocks from > 185 km depth: Terra Nova, v. 10, p. 295–301.

van Roermund, H. L. M., Drury, M. R., Barnhoom, A., and Ronde, A., 2000, Super-silicic garnet microstructures from an orogenic garnet peridotite, evidence for an ultra-deep, > 6 GPa Origin: Journal of Metamorphic Geology, v. 18, p. 135–148.

Vance, D., and Harris, N. B. W., 1999, The timing of prograde metamorphism in the Zanskar Himalaya: Geology, v. 27, p. 395–398.

Wain, A., 1997, New evidence for coesite in eclogite and gneisses: Defining an ultrahigh-pressure province in the Western Gneiss Region of Norway: Geology, v. 25, p. 927–930.

Wain, A., Waters, D., Jephcoat, A. and Olijynk, H., 2000, The high-pressure to ultrahigh-pressure eclogite transition in the Western Gneiss Region, Norway: European Journal of Mineralogy, v. 12, p. 667–687.

Walker, J., Martin, M. W., Bowring, S. A., Searle, M. P., Waters, D. J., and Hodges, K. V., 1999, Metamorphism, melting, and extension: Age constraints from the High Himalayan slab of southeast Zanskar and northwest Lahaul: Journal of Geology, v. 107, p. 473–495.

Wang, X., Liou, J. G., and Mao, H. K., 1989, Coesite-bearing eclogite from the Dabie Mountains in central China: Geology, v. 17, p. 1085–1088.

Wei, C. J., Powell, R., and Zhang, L. F., 2003, Eclogites from the south Tianshan, NW China: Petrological characteristic and calculated mineral equilibria in the Na_2O-CaO-FeO-MgO-Al_2O_3-SiO_2-H_2O system: Journal of Metamorphic Geology, v. 21, p. 163–180.

Weinberg, R. F., and Dunlap, W., 2000, Growth and deformation of the Ladakh Batholith, NW Himalayas: Implications for timing of continental collision and origin of calc-akaline batholiths: Journal of Geology, v. 108, p. 303–320.

Xu, S., Liu, Y., Chen, G., Compagnoni, R., Rolfo, F., He, M., and Liu, H., 2003, New findings of microdiamonds in eclogites from Dabie-Sulu region in central-eastern China: Chinese Science Bulletin, v. 48, p. 988–994.

Xu, W., Okay, A. I., Ji, S., Sengor, A. M. C., Su, W., and Jiang, L., 1992, Diamond from the Dabie Shan metamorphic rocks and its implication for tectonic setting: Science, v. 256, p. 80–82.

Yang, J. J., Godard, G., Kienast, J. R., Lu, Y., and Sun, J., 1993, Ultrahigh-pressure, 60 kbar magnesite-bearing garnet peridotite from Northeastern Jiangsu, China: Journal of Geology, v. 101, p. 541–554.

Yang, J., Xu, Z., Pei, X., Shi, R., Wu, C., Zhang, J., Li, H., Meng, F., and Rong, H., 2002a, Discovery of diamond in North Qiling: Evidence for a giant UHPM belt across central China and Recognition of Paleozoic and Mesozoic dual deep subduction between North China and Yangtze plates: Acta Geologica Sinica, v. 76, p. 484–495 (in Chinese).

Yang, J. S., Xu, Z., Song, S., Zhang, J., Shi, R., Li, H., and Brunel, M., 2001, Discovery of coesite in the North Qaidam Early Paleozoic ultrahigh pressure, UHP metamorphic belt, NW China: Comptes Rendus de l'Academie des Sciences, Paris, Sciences de la Terre et des Planets, v. 333, p. 719–724.

Yang, J. S., Xu, Z., Zhang, J., Chu, C., Zhang, R. and Liou, J. G., 2001, Tectonic significance of early Paleozoic high-pressure rocks in Altun-Qaidam-Qilian Mountains, northwest China, in Hendrix, M. S., and Davis, G. A., eds., Paleozoic and Mesozoic tectonic evolution of central Asia: From continental assembly to intracontinental deformation: Geological Society of America Memoir, v. 194, p. 151–170.

Yang, J. S., Xu, Z. Zhang, J., Song, S., Wu, C., Shi, R., Li, H., and Brunel, M., 2002b, Early Palaeozoic North Qaidam UHP metamorphic belt on the north-eastern

Tibetan Plateau and a paired subduction model: Terra Nova, v. 14, p. 397–404.

Ye, K., Cong, B., and Ye, D., 2000, The possible subduction of continental material to depths greater than 200 km: Nature, v. 407, p. 734–736.

Yui, T. F., Rumble, D., and Lo, C. H., 1995, Unusually low d^{18}O ultrahigh-pressure metamorphic rocks from Su-Lu terrane, China: Geochimica et Cosmochimica Acta, v. 59, p. 2859–2864.

Zhang, J. X., Yang, J. S., Xu, Z. Q., Meng, F. C., Li, H. B., and Shi, R., 2002, Evidence for UHP metamorphism of eclogite from the Altun Mountains: Chinese Science Bulletin, v. 47, p. 751–755.

Zhang, J. X., Zhang, Z. M., Xu, Z. Q., Yang, J. S., and Cui, J. W., 2001, Petrology and geochronology of eclogites from the western segment of the Altyn Tagh, northwestern China: Lithos, v. 56, p. 187–206.

Zhang, L., Ellis, D. J., Arculus, R. J., and Jiang, W., 2003, "Forbidden zone" subduction of sediments to 150? km depth—the reaction of dolomite to magnesite + aragonite in the UHPM metapelites from western Tianshan, China: Journal of Metamorphic Geology, v. 21, p. 523–529.

Zhang, L., Ellis, D. J., and Jiang, W., 2002, Ultrahigh-pressure metamorphism in western Tianshan, China: Part I. Evidence from inclusions of coesite pseudomorphs in garnet and from quartz exsolution lamellae in omphacite in eclogites; Part II. Evidence from magnesite in eclogite: American Mineralogist, v. 87, p. 853–860; 860–866.

Zhang, R. Y., Liou, J. G., and Yang, J. S., 2000, Petrochemical constraints for dual origin of garnet peridotites of the Dabie-Sulu UHP terrane, China: Journal of Metamorphic Geology, v. 18, p. 149–166.

Zhang, R. Y., Liou, J. G., Yang, J., and Ye, K., 2003, Ultrahigh-pressure metamorphism in the forbidden zone: The Xugou garnet peridotite, Sulu terrane, eastern China: Journal of Metamorphic Geology, v. 21, p. 539–550.

Zheng, Y. F., Fu, B., Gong, B., and Li, L., 2003, Stable isotope geochemistry of ultrahigh-pressure metamorphic rocks from the Dabie-Sulu orogen in China: Implications for geodynamics and fluid regime: Earth Science Review, v. 62, p. 105–161.

Zhao, W., Nelson, K. D., and Team, P. I., 1993, Deep seismic reflection evidence for continental underthrusting beneath southern Tibet: Nature, v. 366, p. 557–559.

Zhu, Y.-F., and Ogasawara, Y., 2002, Phlogopite and coesite exsolution from super-silicic clinopyroxene: International Geology Review, v. 44, p. 831–836.

U-Pb Zircon Dating of Regional Deformation in the Lower Crust of the Kohistan Arc

HIROSHI YAMAMOTO,[1]

Department of Earth and Environmental Sciences, Faculty of Science, Kagoshima University, Kagoshima 890-0065, Japan

KATSURA KOBAYASHI, EIZO NAKAMURA,

The Pheasant Memorial Laboratory, Institute for Study of the Earth's Interior, Okayama University, Misasa 682-0193, Japan

YOSHIYUKI KANEKO,[2]

National Institute of Advanced Industrial Science and Technology, Geological Survey of Japan, Tsukuba 305-8567, Japan

AND ALLAH BAKHSH KAUSAR

Geoscience Laboratory, Geological Survey of Pakistan, Shahzad Town, P. O. Box 1461, Islamabad 44000, Pakistan

Abstract

Felsic veins several cm thick to bodies several hundreds of m wide are sporadically distributed in the Kamila amphibolites of the Kohistan arc sequence. Some small dikes (less than one m thick) were deformed together with ductile deformation of wall-rock amphibolites; other small undeformed dikes crosscut foliation in the amphibolites. Large granitic intrusive bodies include gneissose and massive varieties. Ion-microprobe U-Pb dating of zircon grains reveal ages as follows: three deformed dikes, 107.7 ± 1.8 Ma, 94.0 ± 1.9 Ma, 81.0 ± 1.6 Ma; a single gneissose granitic body, 89.5 ± 4.2 Ma; three undeformed dikes, 82.0 ± 2.0 Ma, 80.6 ± 4.5 Ma, 75.8 ± 1.7 Ma; a massive granitic body, 75.7 ± 2.2 Ma; and a foliated amphibolite xenolith in the massive granitic body, 110.7 ± 4.9 Ma. Ages of the deformed dikes, the gneissose granitic body, and the amphibolite xenolith indicate that ductile deformation in the lower crust of the Kohistan arc took place episodically or successively during the period from 107 to 81 Ma. Ages of the undeformed dikes and the massive granitic body indicate that ductile deformation ceased prior to 80 Ma. Collision between Asia and the Kohistan arc elsewhere was inferred to be before 80 Ma. We conclude that the ductile deformation events did not extend beyond the time of Asia-Kohistan collision.

Introduction

HIGH-GRADE METAMORPHIC terranes undergo various extents of ductile deformation. Although it is generally difficult, proper dating of a regional deformation event is of great importance for a better understanding of the tectonic evolution of a metamorphic terrane. In this paper, we present new evidence for the timing of ductile deformation in deep-seated rocks of the Kohistan arc.

The Kohistan sequence, composed of volcanic, sedimentary, high-grade metamafic, ultramafic, and granitic rocks, is exposed in northern Pakistan (Fig. 1). It is sandwiched between Asian and Indian continental crust and believed to be an arc amalgamated with Asia and India (Tahirkheli et al., 1979; Bard et al., 1980; Bard, 1983; Coward et al., 1987; Treloar et al., 1996). The high-grade metamafic lower-crustal rocks in the lower crust of the Kohistan arc underwent ductile deformation probably due to subduction of the Tethyan lithosphere, the Asia-Kohistan collision, and/or the Asia-India collision. It has been suggested that ductile deformation in the lower crust occurred before 83 Ma or between 100 and 83 Ma (Treloar et al., 1990; Arbaret et al., 2000). However these estimated dates are not well supported by available chronological evidence. Ages of metamorphic rocks can be dated by various isotopic systems but those ages cannot be unambiguously related to a specific deformation event.

However, felsic intrusions in the southern part of the Kohistan arc sequence are distributed in the metamafic rocks. Some of these rocks underwent

---

[1]Corresponding author; email: hyam@sci.kagoshima-u.ac.jp
[2]Present address: Graduate School of Environment and Information Sciences, Yokohama National University, Yokohama 240-8501, Japan.

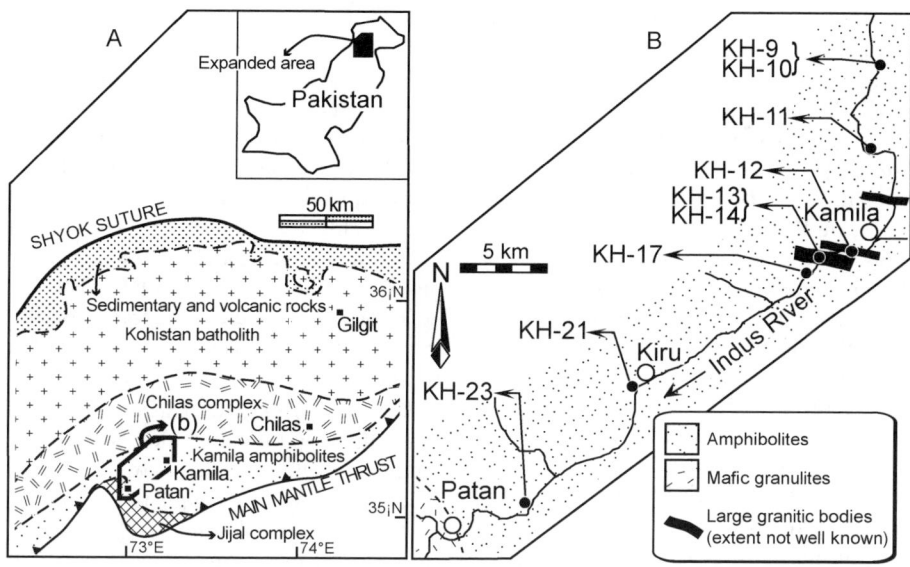

FIG. 1. A. Regional setting of the Kohistan arc sequence. Lithological boundaries are adopted from Coward et al., (1986) and Treloar, et al. (1996). B. Lithological route map of the southern part of the Kohistan sequence along the Indus River.

ductile deformation and others are apparently undeformed, thus post-kinematic. We conducted ion-microprobe U-Pb dating of zircons from these felsic intrusions. The timing of the ductile deformation event was bracketed by the ages of deformed and undeformed intrusions. Our new U-Pb data in general support the earlier work of Treloar et al. (1990) and Arbaret et al. (2002).

Geological Setting

The northernmost portion of the Kohistan arc sequence is comprised of Jurassic to Cretaceous sedimentary and volcanic rocks. These sedimentary and volcanic rocks are separated from Upper Paleozoic metasediments of the Asian plate by the Shyok suture (Coward et al., 1982, 1986; Pudsey 1986; Searle et al., 1989; Treloar et al., 1996; Rex et al., 1998).

Volcanic rocks and sediments of the Kohistan sequence were intruded by gabbroic rocks of the Chilas complex (Coward et al., 1987). The gabbroic rocks were re-equilibrated under granulite-facies metamorphic conditions of 750–850°C at 6–8 kbar (Jan and Howie, 1980; Bard, 1983). These rocks were subsequently intruded by tonalite and diorite during Late Cretaceous to Miocene time. The intrusions form a vast complex of plutons in the middle to northern part of the Kohistan sequence, known as the Kohistan batholith (e.g. Coward et al., 1986). The southern margin of the Chilas complex is bounded by a zone of retrograde hydration and shear deformation. To the south of the Chilas complex, gabbroic amphibolite, banded amphibolite, amphibole schist, and pelitic schist of the Kamila amphibolites are associated with intrusions of hornblendite and felsic rocks (Coward et al., 1987). Ductile shear zones and isoclinal folds are developed in the Kamila amphibolites (Arbaret et al., 2000; Jan, 1988). A highly sheared part of the Kamila amphibolites is called the "Kamila shear zone" (Coward et al., 1987; Treloar et al., 1990); it is considered to have been formed by southward thrusting of the main island arc (Chilas complex) over its forearc region (Treloar et al., 1990) or due to arc-related strain localization during subduction of the Tethyan lithosphere beneath the Kohistan arc (Arbaret et al., 2000).

Metagabbroic rocks of the Kamila amphibolites are separated from mafic granulites of the Jijal complex by a high-strain zone (Arbaret et al., 2000; Coward et al., 1982). The Jijal complex is composed of garnet-bearing gabbroic granulites and ultramafic rocks. Pressure-temperature conditions for garnet-bearing granulites are estimated at 670 to 790°C at and 12 to 14 kbar (Jan and Howie, 1981) and 697 to 949°C at and 11 to 17 kbar (Yamamoto, 1993). A wider rage of P-T conditions (713–1134°C, at 15–21 kbar) for this rock type is has been reported by Ringuette et al. (1999). The metamorphic pressure conditions of the garnet-bearing granulites in the Jijal complex are higher than those of the Chilas complex. Ultramafic rocks on the southern margin of the Jijal complex are bounded by the main mantle thrust (MMT) where the Kohistan arc sequence is juxtaposed against Indian continental crust. South of the MMT, between the Indus and the Swat rivers, there is a tectonic wedge up to 10–15 km wide containing high P/T metamorphic rocks such as glaucophane schists is exposed (Coward et al., 1982; Shams, 1980). The high-P/T metamorphism was related to northward subduction of the Tethyan lithosphere beneath the Kohistan arc, or to one of the deformation events within the Kohistan arc (Coward et al., 1987).

The Kamila amphibolites are well exposed from Patan to Kamila and to the north along the Indus River (Fig. 1B). Felsic rocks are sporadically distributed in the Kamila amphibolites. The felsic rocks range in their thickness from several cm thick veins to a few km wide bodies. Two varieties of small-scale felsic dike (less than one m thick) occur; some were deformed together with ductile deformation of the wall-rock amphibolites, whereas others crosscut foliation in the amphibolites. Large gneissose and massive granitic bodies consisting of sheet-like intrusions lie to the south and north of Kamila (Fig. 1B). Macroscopically, these granitic bodies extend are largely concordant with the general E-W trend of foliation and compositional banding in the Kamila amphibolites, although at smaller scales, the massive variety crosscuts the fabrics in the host rocks.

Sample Descriptions

Deformed felsic rock samples

KH-11: anorthositic dike. An anorthositic dike (Fig. 2A) up to 5 cm thick is tightly folded with its axial plane parallel to the foliation in the wall-rock amphibolite (N75° E, 58° N). The sample is mostly composed of medium-grained plagioclase (>90%) with accessory epidote, quartz, rutile, zircon, and opaque minerals. Most of the plagioclase is saussuritized.

KH-12: gneissose tonalite. Sample KH-12 was taken from gneissose tonalite in a complex of sheet-like granitic bodies (Fig. 2B). The apparent total width of this complex is about 500 m. The sample is composed of medium-grained plagioclase (0.5–2 mm) and fine- to coarse-grained quartz (up to 3 mm), epidote, muscovite with accessory of garnet, biotite, chlorite, and zircon. The gneissosity is defined by the sub-parallel arrangement of tabular muscovite, short-prismatic plagioclase, and elongated aggregate of quartz. Large-elongated quartz grains (1–3 mm in length) exhibit intense undulose extinction.

KH-17: tonalitic dike. The dike (Fig. 2C) is about 25 cm thick and tightly folded, with its axial plane parallel to the foliation in the wall-rock amphibolite (N75° W, 68° N). The sample is composed of fine- to medium-grained plagioclase (up to 2 mm) and fine- to medium-grained quartz (up to 1.5 mm), medium-grained garnet (1–2 mm), biotite, hornblende, and epidote with accessory muscovite, zircon, and opaque minerals.

KH-21: tonalitic dike. The dike is less than 25 cm thick and extends for more than 3 m in length (Fig. 2D). Strike and dip of the dike is N83° W, 50° N, which is parallel to the foliation in the wall-rock amphibolite. The sample consists of fine-grained quartz (up to 1 mm), fine- to medium-grained plagioclase (0.2–5 mm), muscovite, and biotite with accessory epidote and zircon. The foliation in the tonalitic dike is defined by the planar arrangement of rectangular-shaped muscovite and biotite and is parallel to that of amphibolite.

Undeformed felsic rock samples

KH-9: trondhjemitic dike. The dike is about 15 cm thick and extends for more than 10 m (Fig. 2E). Strike and dip of the dike is N7° E, 66° E, which crosscuts foliation in the wall-rock amphibolite (E-W, 80° S). The sample consists of medium- to coarse-grained plagioclase (1–10 mm), fine- to medium- grained quartz (up to 2 mm), muscovite, epidote, and rare chlorite and zircon.

KH-10: tonalitic dike. The location is the same as KH-9. The dike is about 8 cm thick and extends for more than 5 m (Fig. 2F). Strike and dip of the dike is N79° W, 11° S, which crosscuts foliation in the wall-rock amphibolite. The sample consists of medium- to coarse-grained plagioclase (2–10 mm), medium-grained garnet (2–5 mm), fine-grained hornblende, quartz, epidote, opaque minerals, and

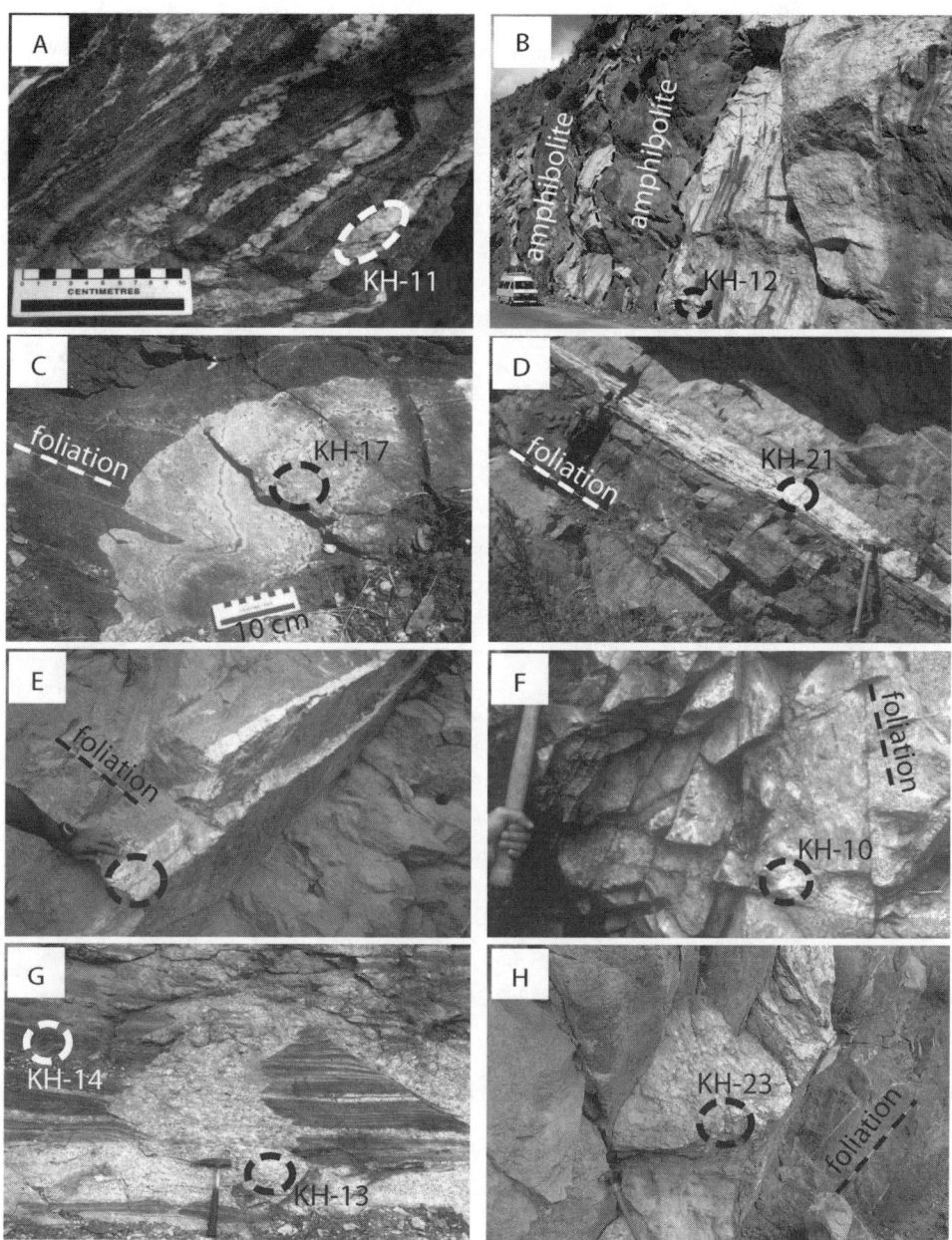

FIG. 2. Photographs showing the occurrence of felsic rocks. Positions of samples are indicated by thick dashed circles with sample numbers.

a few zircon grains. Plagioclase is altered into saussurite to various extents.

KH-13: massive trondhjemite. Sample KH-13 was taken from a massive trondhjemitic body in the central part of the granitic complex about 700 m wide, and is composed of coarse-grained plagioclase (5–10 mm) and fine- to medium- grained quartz (up to 2 mm) with accessory epidote, muscovite, and zircon. This body includes numerous amphibolite xenoliths (Fig. 2G).

KH-23: tonalitic dike. The dike (Fig. 2H) is about 20 cm thick and extends more than 10 m in length. Strike and dip of the dike is N8° W, 65° W, which crosscuts foliation in the wall-rock amphibolite (N50° W, 70° S). The sample consists of coarse-grained plagioclase (5–15 mm), fine- to medium-grained quartz (0.5–3 mm), and biotite with accessory garnet, epidote, and zircon.

Amphibolite xenolith

KH-14: banded amphibolite. The location is the same as KH-13. The sample was taken from one of the xenoliths in the massive trondhjemite (Fig. 2G) and is composed of fine- to medium-grained hornblende (0.5–2.5 mm), plagioclase (0.2–4 mm), quartz (up to 2 mm), epidote (up to 1 mm), rutile (up to 1 mm), and opaque minerals (up to 0.5 mm) with accessory muscovite, chlorite, and zircon. Most of the plagioclase is saussuritized. The foliation of banded amphibolite is defined by the planar arrangement of elongate-tabular and/or short-prismatic amphibole and epidote.

Zircons and Analytical Techniques

Non-magnetic heavy minerals were separated from each rock sample. Zircon grains were hand-picked from heavy minerals and embedded in epoxy resin disks and polished. Zircon grains from felsic rocks are euhedral (acicular or prismatic), colorless or pale pink, and 100–400 µm in length. Some zircons contain quartz and plagioclase as inclusions. Backscattered electron (BSE) and cathode luminescence (CL) imaging capability of a scanning electron microscope reveal concentric-oscillatory or sector zones in the zircon grains (Fig. 3). Some homogeneous zircons were observed even in the BSE-CL images. Boundaries of growth domains are straight and generally euhedral. Therefore the zircon separates are presumably magmatic in origin. Zircon grains from amphibolite xenolith are subhedral (rounded short prismatic) or anhedral, colorless or pale pink, and 50–200 µm in length. Some zircons contain inclusions of plagioclase, apatite, and titanite. Concentric-oscillatory or sector zones were observed in BSE-CL images. Boundaries of oscillatory growth domains are euhedral. Some anhedral grains include euhedral or irregular-shaped domains and those boundaries are discordant to the external grain shapes. This observation implies postcrystallization metamorphic overgrowth.

The zircon separates with even surfaces were selected and analyzed using a high-resolution secondary ion mass spectrometer (HR-SIMS), CAMECA ims-1270, at the Pheasant Memorial Laboratory, Institute for study of the Earth's Interior, Okayama University. Gem-quality zircon from Sri Lanka (560 Ma) was used as standard for HR-SIMS analyses. U-Pb isotopic data for the standard zircon have been described in Usui et al. (2002). A mass-filtered O^{2-} primary ion beam of 5–15 nA was focused on a 10–30 µm ellipsoidal spot. Sites for ion bombarding were selected to avoid inclusions, cracks, and, if possible, boundaries of growth domains. Mass resolution power was about 5000 (measured at 90% peak height). Individual spot analysis consisted of 55 cycles of the isotopic spectrum. Concordia age calculation follows the method of Ludwig (1998) using Isoplot/Ex (Ludwig, 2000). Results of the HR-SIMS U-Pb analyses were summarized in Table 1 and Figures 4–6. All ages quoted in the text and figures are U/Pb concordia ages calculated from several spot analyses on one to five zircon grains of each sample. The errors for ages are quoted in 2σ including errors in the decay constant.

U-Pb Ages

Deformed felsic rock samples

Four zircon grains from sample KH-11 were analyzed. U-Pb data of nine spots were combined, giving a concordia age of 81.0 ± 1.6 Ma. Analyses at six spots from three zircon grains of KH-12 yield a concordia age of 89.5 ± 3.3 Ma. Concordant U-Pb data of eight spot analyses from four zircon grains of KH-17 yield an age of 94.0 ± 1.9 Ma. Six spot data from four zircon grains of KH-21 yield a concordia age of 107.7 ± 1.8 Ma.

Undeformed felsic rock samples

Eight spot analyses from four zircon grains of KH-9 were combined, giving a concordia age of 75.8 ± 1.7 Ma. Three spot analyses from one zircon fraction of KH-10 yield a concordia age of 80.6 ± 4.5 Ma. Five spots from three zircon grains of KH-13 were analyzed. The U-Pb data yield a concordia age of 75.7 ± 1.4 Ma. Six spot analyses from two zircon fractions of KH-23 yield a concordia age of 82.0 ± 2.0 Ma.

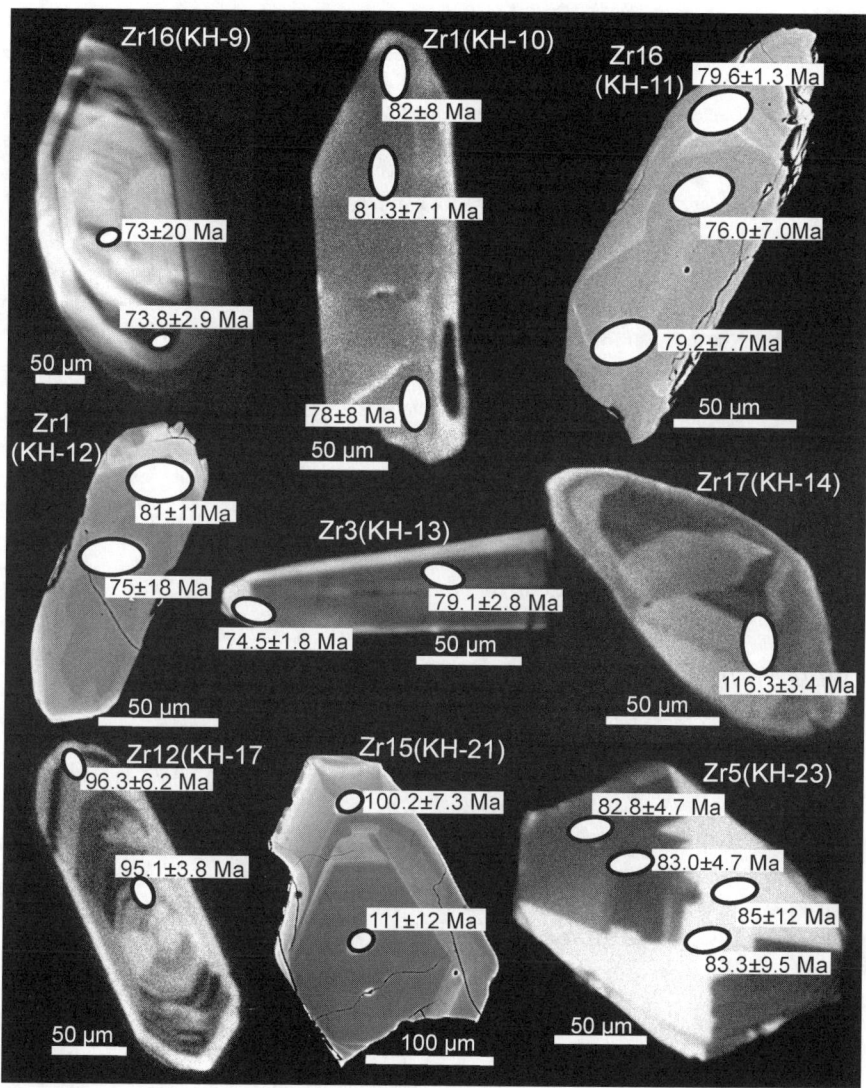

FIG. 3. Electron microscope images (CL and BSE) of selected zircon fractions. Oval marks with ages indicate spots of HR-SIMS analysis. CL images: Zr16(KH-9); Zr1(KH-10); Zr3(KH-13); Zr17(KH-14); Zr12(KH-17); Zr5(KH-23). BSE images: Zr16(KH-11); Zr1(KH-12); Zr15(KH-21).

Amphibolite xenolith

Four spot analyses from four zircon grains of amphibolite xenolith (KH-14) yield a concordia age of 110.7 ± 4.9 Ma.

Tectonic Implications

A minimum 83 Ma age for shearing within the Kamila shear zone has been proposed based on ^{40}Ar-^{39}Ar amphibole ages and closure temperature (~500°C) of Ar diffusion in amphibole (Treloar et al., 1989, 1990). The other minimum age determined by ^{40}Ar-^{39}Ar amphibole dating has been reported to be 76 ± 4 Ma (Wartho et al., 1996). Their ^{40}Ar-^{39}Ar age spectra are commonly U-shaped, suggesting excess ^{40}Ar incorporation in minerals (Damon and Kulp, 1958; Lanphere and Dalrymple, 1971; 1976; Kaneoka, 1974). The presence of excess ^{40}Ar casts doubt on the explanation of their

TABLE 1. U–Pb Isotopic Results of Zircons

Grain no.	$^{207}Pb/^{206}Pb$	Error, ±2σ	$^{207}Pb/^{235}U$	Error, ±2σ	$^{206}Pb/^{238}U$	Error, ±2σ	$^{206}Pb/^{204}Pb$	Age, Ma	Error, ±2σ
				KH-9					
Zr10	0.048479	0.004260	0.082805	0.062574	0.011083	0.002034	333	71	13
Zr10	0.048606	0.002017	0.077168	0.028504	0.011189	0.001144	1455	71.3	6.9
Zr12	0.053248	0.003147	0.086460	0.045966	0.011438	0.001283	651	72.7	7.8
Zr12	0.046799	0.000769	0.081466	0.012263	0.012477	0.000528	5591	80.0	2.9
Zr16	0.059661	0.006654	0.104306	0.101315	0.011647	0.003228	205	73	20
Zr16	0.047218	0.000752	0.074800	0.010732	0.011491	0.000593	10748	73.8	2.9
Zr26	0.045431	0.002478	0.075218	0.034055	0.011480	0.001666	534	74	10
Zr26	0.046762	0.001217	0.076976	0.019389	0.011594	0.000954	2134	74.1	5.3
				KH-10					
Zr1	0.048256	0.001739	0.083779	0.025395	0.012700	0.001181	1530	81.3	7.1
Zr1	0.050005	0.002552	0.090002	0.037845	0.012851	0.001267	1165	82.0	8.0
Zr1	0.049889	0.002238	0.087281	0.031765	0.012206	0.001327	1163	78.0	8.4
				KH-11					
Zr1	0.051224	0.001952	0.079474	0.027565	0.011063	0.001107	569	70.2	6.7
Zr1	0.049449	0.001136	0.081051	0.016354	0.012005	0.000728	2331	76.8	4.6
Zr15	0.057668	0.002621	0.101230	0.044392	0.012451	0.001686	147	79	10
Zr15	0.047438	0.000634	0.089717	0.011731	0.013653	0.000656	5279	87.5	3.9
Zr15	0.047156	0.000485	0.085408	0.009266	0.013035	0.000387	7047	83.5	1.1
Zr16	0.051181	0.001995	0.088845	0.030625	0.012043	0.001242	701	76.0	7.0
Zr16	0.050546	0.001519	0.087514	0.021575	0.012447	0.001236	1062	79.2	7.7
Zr16	0.047700	0.000425	0.081895	0.006335	0.012438	0.000355	15937	79.6	1.3
Zr17	0.048070	0.000902	0.081545	0.013067	0.012307	0.001002	4997	78.7	6.1
				KH-12					
Zr1	0.051242	0.001141	0.084635	0.020622	0.011915	0.002767	2400	75	18
Zr1	0.048271	0.000311	0.084104	0.012133	0.012608	0.001660	49877	81	11
Zr2	0.051550	0.001871	0.086951	0.023528	0.012284	0.002795	643	78.0	18
Zr2	0.050022	0.001176	0.101248	0.020631	0.014514	0.001048	3582	91.8	5.8
Zr6	0.053052	0.001924	0.111593	0.035176	0.014704	0.001680	1405	92.0	9.8
Zr6	0.048190	0.000455	0.095628	0.008954	0.014338	0.000888	17384	91.4	5.3
				KH-13					
Zr1	0.047703	0.002630	0.072794	0.032742	0.010729	0.001642	864	68.5	9.9
Zr2	0.047511	0.000454	0.078680	0.045104	0.010137	0.002146	454	63	12
Zr3	0.047481	0.000751	0.077451	0.010992	0.011660	0.000370	8569	79.1	2.8
Zr3	0.047754	0.000870	0.081956	0.012702	0.012379	0.000606	8876	74.5	1.8
Zr4	0.048557	0.001144	0.086327	0.017920	0.012744	0.001408	3907	80.4	7.4

Table continues

TABLE 1. Continued

Grain no.	$^{207}Pb/^{206}Pb$	Error, ±2σ	$^{207}Pb/^{235}U$	Error, ±2σ	$^{206}Pb/^{238}U$	Error, ±2σ	$^{206}Pb/^{204}Pb$	Age, Ma	Error, ±2σ
KH-14									
Zr3	0.048299	0.000702	0.108875	0.016376	0.016319	0.001150	10975	104.0	4.9
Zr10	0.048541	0.001144	0.118441	0.025861	0.017509	0.001684	3255	112	10
Zr13	0.048038	0.000849	0.111806	0.017737	0.016657	0.000931	5899	106.1	4.4
Zr17	0.047865	0.000833	0.120176	0.018992	0.018173	0.000758	8310	116.3	3.4
KH-17									
Zr3	0.046863	0.000655	0.097482	0.013671	0.015018	0.001104	12453	97.0	5.7
Zr3	0.048396	0.000965	0.097782	0.015845	0.014624	0.000991	6371	93.3	5.5
Zr4	0.047495	0.001294	0.092105	0.025266	0.013829	0.001441	1846	88.3	7.8
Zr4	0.047444	0.000600	0.097368	0.012167	0.014913	0.000812	9006	95.9	4.1
Zr8	0.046716	0.000914	0.089987	0.014751	0.013802	0.001530	6458	88.3	9.7
Zr8	0.048572	0.001439	0.092983	0.024325	0.013890	0.001127	2300	88.6	5.9
Zr12	0.047628	0.000926	0.099587	0.019608	0.014912	0.000910	3943	95.1	3.8
Zr12	0.048041	0.000451	0.100960	0.012639	0.015170	0.001305	24652	96.3	6.2
KH-21									
Zr4	0.059952	0.003655	0.152279	0.090932	0.017034	0.002644	331	106	15
Zr4	0.048802	0.000314	0.115173	0.006936	0.017067	0.000483	35706	108.0	2.0
Zr6	0.054634	0.003555	0.136982	0.068540	0.017297	0.002198	2400	109	13
Zr15	0.052303	0.002029	0.127006	0.044658	0.017397	0.001884	983	111	12
Zr15	0.048651	0.000698	0.104446	0.014994	0.015690	0.001233	9025	100.2	7.3
Zr17	0.048385	0.001673	0.115152	0.031026	0.017115	0.001013	2244	109.4	6.4
KH-23									
Zr5	0.048389	0.001295	0.086948	0.021838	0.013000	0.000903	2134	83.0	4.7
Zr5	0.053070	0.003606	0.098511	0.060996	0.013164	0.001668	432	83.3	9.5
Zr5	0.046951	0.003072	0.087366	0.047760	0.013321	0.001901	440	85	12
Zr5	0.045642	0.001666	0.089888	0.025095	0.013015	0.000847	2081	82.8	4.7
Zr9	0.048624	0.001120	0.082375	0.016746	0.012573	0.000713	3380	81.0	3.0
Zr9	0.049637	0.001393	0.082868	0.023433	0.012882	0.001393	3133	83.2	6.8

^{40}Ar-^{39}Ar data. The K-Ar system is not very suitable for the dating of deep crustal rocks. Furthermore, the closure temperature of Ar diffusion in amphibole is not well constrained. Our U-Pb results give a more rigorous interpretation of the Kamila shear zone. On the basis of U-Pb analyses of zircon grains, the age of the amphibolite xenolith in this study (110.7 ± 4.9 Ma) represents the date of crystallization of the magmatic protolith or subsequent amphibolite-facies metamorphism. The zircon ages of the felsic intrusive rocks range between 107.7 ± 1.8 Ma and 75.7 ± 1.4 Ma. These ages are taken as the dates of magmatic emplacement of felsic rocks into the amphibolites. The new ages of the felsic rocks are consistent with the zircon age of the amphibolite xenolith.

The ages of deformed felsic dikes and the gneissose trondhjemitic body range from 107.7 ± 1.8 Ma (KH-21) to 81.0 ± 1.6 Ma (KH-11). Schaltegger et al. (2002) reported a U-Pb zircon age of metamorphosed, foliated diorite stock near Kiru (Fig. 1B) as 91.8 ± 1.4 Ma. Nakajima et al. (2004) reported an

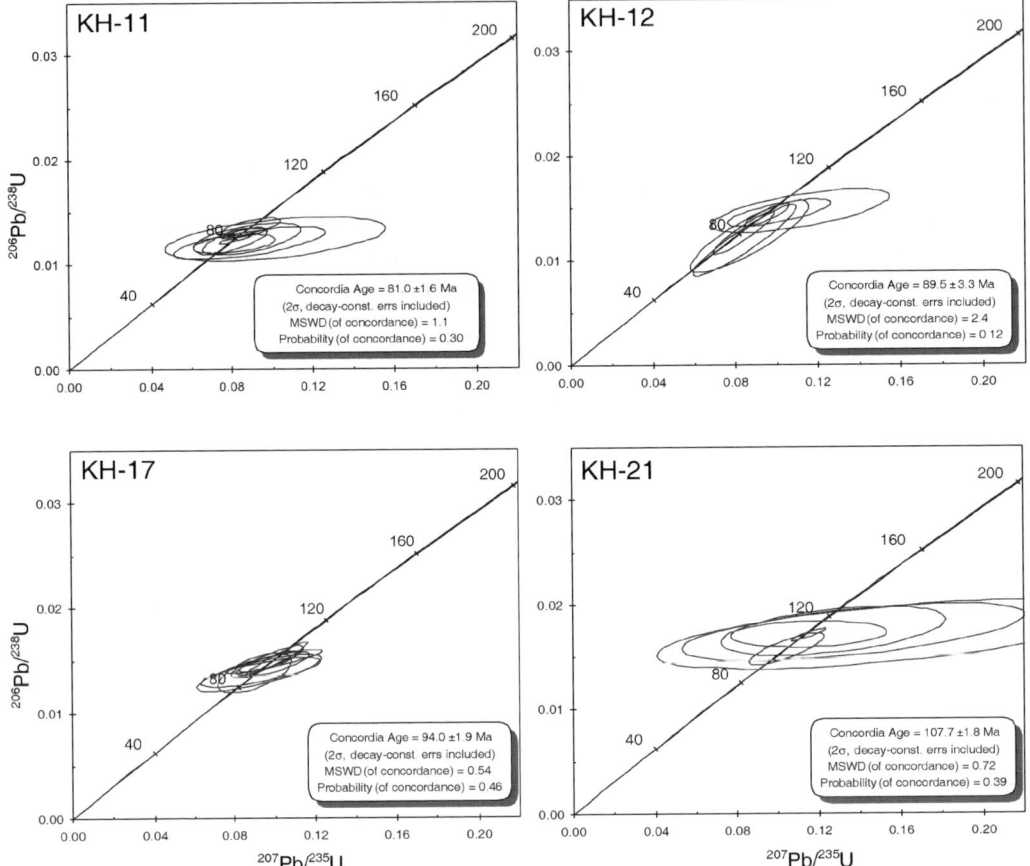

FIG. 4. U-Pb concordia plots for data on deformed felsic rocks (KH-11, KH-12, KH-17, KH-21). Error ellipses are 2σ.

U-Pb zircon age of a weakly foliated tonalite sheet extending concordant with the structure of the host amphibolite near Kamila as 98 Ma. These ages are within the range of deformed felsic rocks in this study. These ages of deformed and foliated felsic rocks indicate that ductile deformations that formed the Kamila shear zone in the lower crust of the Kohistan arc took place episodically or successively after ~107 Ma, and the latest ductile deformation event after ~81 Ma.

The zircon ages of undeformed dikes and massive trondhjemite are 82.0 ± 2.0 Ma (KH-23), 80.6 ± 4.5 Ma (KH-10), 75.8 ± 1.7 Ma (KH-9), and 75.7 ± 1.4 Ma (KH-13). These ages of undeformed felsic rocks suggest that the Kamila shear zone was no longer active after ~80 Ma. The age of KH-23 is older than the youngest deformed dike (KH-11). This contradiction can be interpreted from the regional age trend. The location of KH-23 is near the southern margin of the Kamila shear zone and 28 km from KH-11, which is close to the northernmost part of the shear zone. The deformed felsic dikes in the southern part of study area (KH-17 and KH-21) are older than the more northerly deformed felsic dike (KH-11). Similarly, the undeformed felsic dike in the south (KH-23) is older than the undeformed felsic dikes in the north (KH-9 and KH-10). We infer that ductile deformations in the Kamila shear zone were initiated and terminated earlier in the southern part (structural low level) in comparison with those in the northernmost part.

The tectonic accretion of the Kohistan arc to the Asian continent has been inferred to occur between 102 ± 12 and ~80 Ma based on Rb-Sr and K-Ar ages

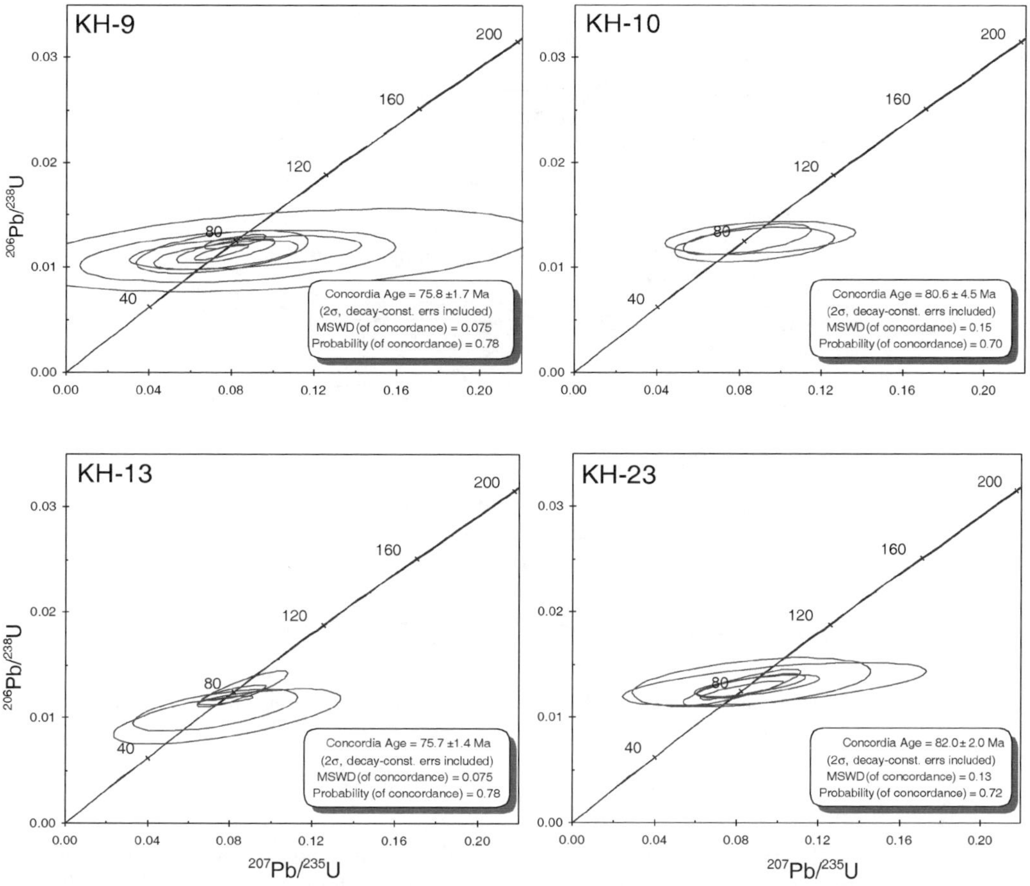

FIG. 5. U-Pb concordia plots for data on undeformed felsic rocks (KH-9, KH-10, KH-13, KH-23). Error ellipses are 2σ.

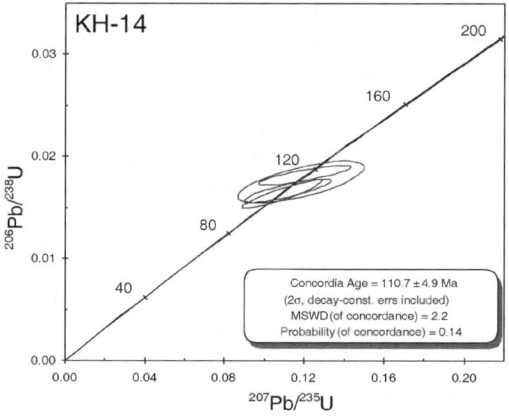

FIG. 6. U-Pb concordia plots for data on amphibolite xenolith in the massive trondhjemitic body (KH-14). Error ellipses are 2σ.

of syn-collisional leucogranites and post-collisional basic mafic dikes intruded into the suture zones between Asia and Kohistan (Petterson and Windley, 1985, 1992; Treloar et al., 1989). The initiation of the India-Asia collision (that is, the closure of the Tethys) is constrained at 65 Ma or later from paleomagnetic study (Klootwijk et al., 1992). The ductile deformation events in the Kamila shear zone largely coincided with the Asia-Kohistan collision and were certainly before the Asia-India collision (Fig. 7). It is supposed that the deformation due to the Asia-India collision was propagated to the south and accommodated by a thrust system in the northern margin of the Indian plate (Coward and Butler, 1985; Coward, et al., 1988). The lower-crustal section of the Kohistan arc did not undergo ductile deformation throughout the collision.

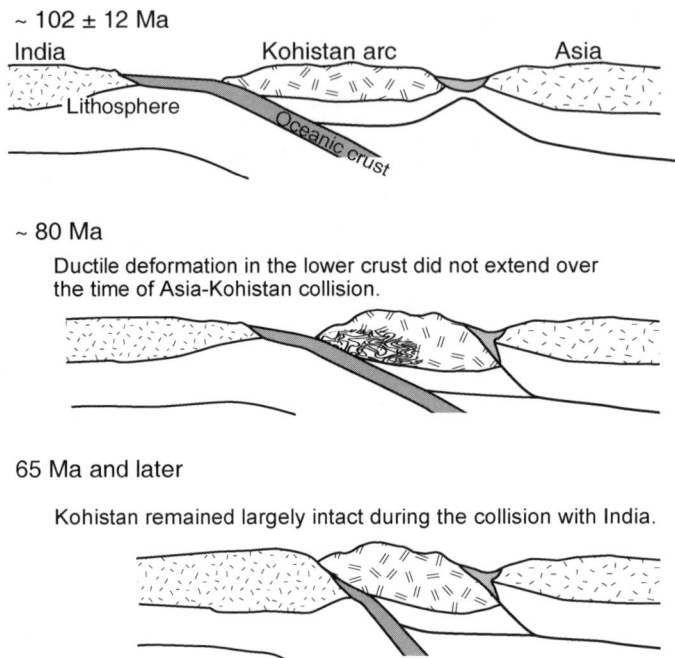

FIG. 7. Schematic interpretation of the ductile deformation related to the Asia-Kohistan collision. Diagram is not to scale.

A U-Pb zircon age of a gabbroic body near Patan ("the Sarangar gabbro" described by Arbaret et al., 2000) is reported to be 98.9 ± 0.4 Ma. (Schaltegger et al., 2002). The sample was taken from undeformed gabbro preserved between amphibolite-facies shear zones. Garnet-bearing gabbroic granulite of the Jijal complex is exposed to the west of the Sarangar gabbro. The distribution pattern of amphibolite-facies shear zones in the Sarangar gabbro is concordant with that in the gabbroic granulite of the Jijal complex (Arbaret et al., 2000). Thus the Sarangar gabbro was kinematically combined with the Jijal complex before the deformation under amphibolite-facies conditions. Sm-Nd isochron ages of 90.9 ± 4.6 Ma, 94.0 ± 4.7 Ma and 95.7 ± 2.7 Ma were determined for the garnet granulite of the Jijal complex (Yamamoto and Nakamura, 1996, 2000; Anczkiewicz and Vance, 2000). A relict noritic mineral assemblage (orthopyroxene, clinopyroxene, and plagioclase) is preserved within irregularly shaped patches distributed in the northeastern part of the Jijal complex (Yamamoto and Yoshino, 1998). A Sm-Nd isochron age of 118 ± 12 Ma was obtained for a sample from a noritic patch (Yamamoto and Nakamura, 2000). These ages combined with the previous petrological studies indicate that the gabbroic rocks around Patan were crystallized at ~100 Ma or earlier (Yamamoto and Yoshino, 1998; Yamamoto and Nakamura, 2000), followed by high-pressure granulite-facies metamorphism crystallizing garnet at P-T conditions up to ~1100°C at ~20 kbar (Jan and Howie, 1981; Yamamoto, 1993; Ringuette et al., 1999) before about 94 Ma (Anczkiewicz and Vance, 2000; Yamamoto and Nakamura, 2000). Based on the pressure-temperature history of granulites in the southern Kohistan arc, crustal thickening of the arc is considered to have been caused by loading of mafic magma at intermediate depths of the pre-existing gabbroic crust (Yoshino et al., 1998). Considering the 94 ± 1.9 Ma (KH-17) and younger ages of deformed felsic rocks of this study and amphibolite-facies conditions of ductile deformation in the Kamila shear zone (Jan, 1988; Treloar et al., 1990), we suggest that shearing activity was mainly after the crustal thickening due to magma loading in the Kohistan arc. Strain in the Kamila shear zone did not contribute much to the crustal thickening process.

Acknowledgments

The authors express our great esteem for the achievements of W. G. Ernst in the broad field of Earth Science. We are grateful to Dr. T. Usui for his technical assistance in BSE-CL imaging and sample preparation. This paper was improved by the constructive review of J. G. Liou. The field survey was financially supported by a Sasakawa Scientific Research Grant from The Japan Science Society and a Grant-in-Aid for Scientific Research (KAKENHI) from the Japan Society for the Promotion of Science (No. 12740284 and 13373005). U-Pb dating was carried out under the Visiting Researcher's Program of the Institute for Study of the Earth's Interior, Okayama University.

REFERENCES

Anczkiewicz, R., and Vance, D., 2000, Isotopic constraints on the evolution of metamorphic conditions in the Jijal-Patan complex and the Kamila Belt of the Kohistan arc, Pakistan Himalaya, in Khan, M. A., Treloar, P. J., Searle, M. P., and Jan, M. Q., eds., Tectonics of the Nanga Parbat syntaxis and the western Himalaya: Geological Society of London Special Publication 170, p. 321–331.

Arbaret, L., Burg, J. P., Zeilinger, G., Chaudhry, N., Hussain, S., and Dawood, H., 2000, Pre-collisional anastomosing shear zones in the Kohistan arc, NW Pakistan, in Khan, M. A., Treloar, P. J., Searle, M. P., and Jan, M. Q., eds., Tectonics of the Nanga Parbat syntaxis and the western Himalaya: Geological Society of London Special Publication 170, p. 295–311.

Bard, J. P., 1983, Metamorphism of an obducted island arc: Example of the Kohistan sequence (Pakistan) in the Himalayan collided range: Earth and Planetary Science Letters, v. 65, p. 133–144.

Bard, J. P., Maluski, H., Matte, P., and Proust, F., 1980, The Kohistan sequence: Crust and mantle of an obducted island arc: Geological Bulletin, University of Peshawar, v. 13, p. 87–93.

Coward, M. P., and Butler, R. W. H., 1985, Thrust tectonic and the deep structure of Pakistan Himalayas: Geology, v. 13, p. 417–420.

Coward, M. P., Butler, R. W. H., Chambers, A. F., Graham, R. H., Izatt, C. N., Khan, M. A. Knipe, R. J., Prior, D. J., and Treloar, P. J., 1988, Folding and imbrication of the Indian crust during Himalayan collision: Philosophical Transactions of the Royal Society of London, v. A326, p. 89–116.

Coward, M. P., Butler, R. W. H., Khan, M. A., and Knipe, R. J., 1987, The tectonic history of Kohistan and its implications for Himalayan structure: Journal of Geological Society of London, v. 144, p. 377–391.

Coward, M. P., Jan, M. Q., Rex, D. C., Tarney, J., Thirlwall, M., and Windley, B. F., 1982, Geo-tectonic framework of the Himalaya of N. Pakistan: Journal of the Geological Society of London, v. 139, p. 299–308.

Coward, M. P., Windley, B. F., Broughton, R. D., Luff, I. W., Petterson, M. G., Pudsey, C. J., Rex, D. C., and Khan, M. A., 1986, Collision tectonics in the NW Himalayas, in Coward M. P., and Ries, A. C., eds., Collision tectonics: Geological Society of London Special Publication 19, p. 203–219.

Damon, P. E., and Kulp, J. L., 1958, Excess helium and argon in beryl and other minerals: American Mineralogist, v. 43, p. 433–459.

Jan M. Q., 1988, Geochemistry of amphibolites from the southern part of the Kohistan arc, N. Pakistan: Mineralogical Magazine, v. 52, p. 147–159.

Jan, M. Q., and Howie, R. A., 1980, Ortho- and clinopyroxenes from the pyroxene granulites of Swat Kohistan, northern Pakistan: Mineralogical Magazine, v. 43, p. 715–726.

Jan, M. Q., and Howie, R. A., 1981, The mineralogy and geochemistry of the metamorphosed basic and ultrabasic rocks of the Jijal complex, Kohistan, NW Pakistan: Journal of Petrology, v. 22, p. 85–126.

Kaneoka, I., 1974, Investigations of excess argon in ultramafic rocks from the Kola Peninsula by the $^{40}Ar/^{39}Ar$ method: Earth and Planetary Science Letters, v. 22, p. 145–156.

Klootwijk, C. T., Gee, J. S., Pierce, J. W., Smith, G. M., and McFadden, P. L., 1992, An early India-Asian contact: Paleomagetic constraints from Ninetyeast ridge, ODP Leg 121: Geology, v. 20, p. 395–398.

Lanphere, M. A., and Dalrymple, G. B., 1971, A test of the $^{40}Ar/^{39}Ar$ age spectrum technique on some terrestrial materials: Earth and Planetary Science Letters, v. 12, p. 359–372.

Lanphere, M. A., and Dalrymple, G. B., 1976, Identification of excess ^{40}Ar by the $^{40}Ar/^{39}Ar$ technique: Earth and Planetary Science Letters, v. 32, p. 141–148.

Ludwig, K. R., 1998, On the treatment of concordant uranium-lead ages: Geochimica et Cosmochimica Acta, v. 62, p. 665–676.

Ludwig, K. R., 2000, Isoplot/Ex version 2.3: Berkeley, CA, Berkeley Geochronology Center Special Publication No. 1a, 53 p.

Nakajima, T., Williams, I. S., Hyodo, H., Miyazaki, K., Kono, Y., Kausar, A. B., Khan, S. R., and Shirahase, T., 2004, The lower crustal Dasu Tonalite and its implications for the formation-reformation-exhumation history of the Kohistan arc crust: Himalayan Journal of Sciences, v. 2, p. 211.

Petterson, M. G., and Windley, B. F., 1985, Rb-Sr dating of the Kohistan arc-batholith in the Trans-Himalaya of

N. Pakistan and tectonic implications: Earth and Planetary Science Letters, v. 74, p. 45–57.

Petterson, M. G., and Windley, B. F., 1992, Field relations, geochemistry and petrogenesis of the Cretaceous basaltic Jutal dykes, Kohistan, northern Pakistan: Journal of the Geological Society of London, v. 149, p. 299–308.

Pudsey, C. J., 1986, The Northern Suture, Pakistan: Margin of a Cretaceous island arc: Geological Magazine, v. 123, p. 405–423.

Rex, A. J., Searle, M. P., Crawford, M. B., Prior, D. J., Rex, D. C., and Barnicoat, A., 1998, The geochemical and tectonic evolution of the central Karakoram, north Pakistan: Philosophical Transactions of the Royal Society of London, v. A326, p. 229–255.

Ringuette, L., Martignole, J., and Windley, B. F. 1999, Magmatic crystallization, isobaric cooling, and decompression of the garnet-bearing assemblages of the Jijal sequence (Kohistan terrane, western Himalayas): Geology, v. 27, p. 139–142.

Schaltegger, U., Zeilinger, G., Frank, M., and Burg, J. P., 2002, Multiple mantle sources during island arc magmatism: U-Pb and Hf isotopic evidence from the Kohistan arc complex, Pakistan: Terra Nova, v. 14, p. 461–468.

Searle, M. P., Rex, A. J., Tirrul, R., Rex, D. C., Barnicoat, A., and Windley, B. F., 1989, Metamorphic, magmatic, and tectonic evolution of the Central Karakoram in the Biafo-Baltoro-Hushe regions of N. Pakistan: Geological Society of America Special Paper 232, p. 47–73.

Shams, F. A., 1980, Origin of the Shangla blueschists, Swat Himalaya, Pakistan: Geological Bulletin University of Peshawar, v. 13, p. 67–70.

Tahirkheli, R. A. K., Mattauer, M., Proust, F., and Tapponnier, P., 1979, The India Eurasia suture zone in northern Pakistan: Synthesis and interpretation of recent data at plate scale, in Farah, A., and DeJong, K. A., eds., Geodynamics of Pakistan: Quetta, Pakistan, Geological Survey of Pakistan, p. 125–130.

Treloar, P. J., Brodie, K. H., Coward, M. P., Jan, M. Q., Khan, M. A., Knipe, R. J., Rex, D. C., and Williams, M. P., 1990, The evolution of the Kamila shear zone, Kohistan, Pakistan, in Salisbury, M. H., and Fountain, D. M., eds., Exposed cross-sections of the continental crust: Dordrecht, Netherlands, Kluwer Academic Publishers, p. 175–214.

Treloar, P. J., Petterson, M. G., Jan, M. Q., and Sullivan, M. A., 1996. A re-evaluation of the stratigraphy and evolution of the Kohistan arc sequence, Pakistan Himalaya: Implications for magmatic and tectonic arc-building processes: Journal of the Geological Society of London, v. 153, p. 681–693.

Treloar, P. J., Rex, D. C., Guise, P. G., Coward, M. P., Searle, M. P., Windley, B. F., Petterson, M. G., Jan, M. Q., and Luff, I. W., 1989, K-Ar and Ar-Ar geochronology of the Himalayan collision in NW Pakistan: Constraints on the timing of suturing, deformation, metamorphism and uplift: Tectonics, v. 8, p. 881–909.

Usui, T., Kobayashi, K., and Nakamura, E., 2002, U-Pb isotope systematics of micro-zircon inclusions: Implications for the age and origin of eclogite xenolith from the Colorado Plateau: Proceedings of the Japan Academy, Series B, v. 78, p. 51–56.

Wartho, J. A., Rex, D. C., and Guise, P. G., 1996, Excess argon in amphiboles linked to greenschist alteration in the Kamila Amphibolite Belt, Kohistan island arc system, northern Pakistan: insights from $^{40}Ar/^{39}Ar$ step-heating and acid leaching experiments: Geological Magazine, v. 133, p. 595–609.

Yamamoto, H., 1993, Contrasting metamorphic P-T-time paths of the Kohistan granulites and tectonics of the western Himalayas: Journal of the Geological Society of London, v. 150, p. 843–856.

Yamamoto, H., and Nakamura, E., 1996, Sm-Nd dating of garnet granulites from the Kohistan complex, northern Pakistan: Journal of the Geological Society of London, v. 153, p. 965–969.

Yamamoto, H., and Nakamura, E., 2000, Timing of magmatic and metamorphic events in the Jijal complex of the Kohistan arc deduced from Sm-Nd dating of mafic granulites, in Khan, M. A., Treloar, P. J., Searle, M. P., and Jan, M. Q., eds., Tectonics of the Nanga Parbat Syntaxis and the western Himalaya: Geological Society of London Special Publication 170, p. 313–319.

Yamamoto, H., and Yoshino, T., 1998, Superposition of replacements in the mafic granulites of the Jijal complex of the Kohistan arc, northern Pakistan: Dehydration and rehydration within deep arc crust: Lithos, v. 43, p. 219–234.

Yoshino, T., Yamamoto, H., Okudaira, T., and Toriumi, M., 1998, Crustal thickening of the lower crust of the Kohistan arc (N. Pakistan) deduced from Al-zoning in clinopyroxene and plagioclase: Journal of Metamorphic Geology, v. 16, p. 729–748.

KAZAKHSTAN

Cathodoluminescence of Microdiamond in Dolomite Marble from the Kokchetav Massif—Additional Evidence for Two-Stage Growth of Diamond

NOBUHIRO YOSHIOKA AND YOSHIHIDE OGASAWARA[1]

Department of Earth Sciences, Waseda University, Shinjuku-ku, Tokyo 169-8050, Japan

Abstract

Abundant microdiamonds occur as inclusions in garnet and diopside in dolomite marble from the Kumdy-kol area of the Kokchetav Massif. These microdiamonds have been classified by previous work into three types (S type, R type, T type) according to their morphology. S-type diamond, the most abundant type in Kokchetav dolomite marble, was examined using microcathodoluminescence (CL) spectroscopy to detect the difference between core and rim. Cathodoluminescence was obtained as spectra and intensity images at the peak of the main CL band using two kinds of CL spectrometer combined with a scanning electron microscope (SEM). Strong broad bands of the CL spectra at 514 to 537 nm were detected at the rim of all S-type microdiamond grains. The core also has the same broad band, but its intensities are very weak. The weak broad band at about 393 nm was detected at the rim of some grains; another weak broad band at about 440–450 nm was also detected both in the core and the rim of several grains by the MP-32 spectrometer system. The intensity images of CL spectra at the peak wavelength (514–537 nm) of the main broad band showed a clear difference between core (weak) and rim (strong). These CL data indicate different geochemical environments for the growth of core and rim of S-type microdiamonds, and provide additional evidence for the two-stage growth of microdiamond in dolomite marble, as previously proposed.

Introduction

THE FIRST REPORT of metamorphic diamond from the Kokchetav Massif, northern Kazakhstan (Sobolev and Shatsky, 1990) changed our understanding of the process of subduction and exhumation within collisional orogenic belts. In the Kokchetav Massif, microdiamonds have been reported in feldspathic gneiss, garnet, pyroxenite, and metacarbonate rock (e.g., Sobolev and Shatsky, 1990; Ogasawara et al., 2000; Dobrzhinetskaya et al., 2001). Subsequently, metamorphic microdiamonds have been reported from other UHP terranes including: the Western Gneiss Region, Norway (Dobrzhinetskaya et al., 1995); the Saxonian Erzgebirge, Germany (Stöckhert et al., 2001); Sulawesi, Indonesia (Parkinson and Katayama, 1999); and the Greek Rhodope Metamorphic Province (Mposkos and Kostopoulos, 2001). Among these UHP terranes, the Kokchetav Massif is the most attractive because of the occurrence of abundant microdiamond; the highest domain of diamond content reaches 2700c/t (Yoshioka et al., 2001). Ogasawara et al. (2000) described diamond-bearing dolomite marble and diamond-free dolomitic marble at Kumdy-kol, and demonstrated that mineral assemblages and the presence/absence of microdiamond in these rocks depended on the fluid composition during unltrahigh-pressure (UHP) metamorphism. Recently, the role of fluid for diamond crystallization was emphasized both by experimental studies (e.g., Akaishi and Yamaoka, 2000; Kumar et al., 2000) and by investigation of UHP rocks (e.g., De Corte et al., 1998; Dobrzhinetskaya et al., 2001; Ogasawara and Ishida, 2001; Stöckhert et al., 2001, Ishida et al., 2003).

Ishida et al. (2003) classified microdiamonds in Kokchetav dolomite marble into three types (S type, R type, and T type) based on morphologies and other characteristics, and they concluded that S-type microdiamond consisting of a single-crystal core and a polycrystalline, fine-grained rim formed discontinuously in two different stages. These characteristics also suggested that the rim of S-type diamonds could have formed from a carbon-bearing fluid at the second stage. They also interpreted that R-type microdiamonds, lacking clear rims, formed at the first stage, and that the T type, similar to the

---

[1]Corresponding author; email:yoshi777@waseda.jp

rim of the S type, lacks a distinctly different rim and formed at the second stage.

In order to clarify the two-stage growth of microdiamond in dolomite marble, microcathodoluminescence spectroscopy was conducted on a sample of S-type microdiamonds. Cathodoluminescence is a traditional method to describe the morphological character of natural diamonds. Microcathodoluminescence study using a scanning electron microscope (SEM) is effective for examining very fine-grained diamond in order to understand heterogeneity at the micrometer scale.

Petrography of Diamond-Bearing Dolomite Marble

UHP rocks of the Kumdy-kol area consist mainly of gneiss and eclogite (e.g., Sobolev and Shatsky, 1990; Kaneko et al., 2000; Katayama et al., 2000; Okamoto et al., 2000) with minor amounts of impure carbonate rock and calc-silicate rocks (e.g., Ogasawara et al., 2000, 2002). These Kumdy-kol metamorphic rocks belong to Unit II of Kaneko et al. (2000). Two types of dolomite-bearing marbles occur: (1) diamond-bearing dolomite dominant marble; and (2) diamond-free marble that contains both Mg-calcite and dolomite (Ogasawara et al., 2000). The differences in mineral assemblages, including the presence or absence of diamond, can be explained by local heterogeneity of fluid compositions, particularly X_{CO2}, during UHPM (Ogasawara et al., 2000). Another interesting UHP carbonate rock is a calcite marble that contains titanite with coesite exsolution needles and plates. Excess Si contents of titanite prior to the coesite exsolution yield a minimum metamorphic pressure of 6 GPa (Ogasawara et al., 2002). These marbles were subjected to the same P-T conditions as the diamond-bearing units.

In the present study, we used a diamond-bearing dolomite marble, sample no. XX01, for microcathodoluminescence study; this sample contains an average concentration of microdiamonds among diamond-bearing dolomite marbles at Kumdy-kol. This dolomite marble shows granoblastic texture (Fig. 1), and contains dolomite (60%), garnet (10%), diopside (10%), and secondary phlogopite (15%). Dolomite occurs in the matrix and as inclusions in garnet and diopside; both garnet and diopside are chemically homogeneous. Dolomite contains graphite inclusions that are polycrystalline aggregates (5–30 μm in diameter). Diopside is anhedral and subhedral coarse-grained crystals 0.4 to 2.5 mm in the long dimension. All diopside grains contain phengite needles. A few grains of microdiamond and many aggregates of graphite are included in diopside. Garnet occurs as anhedral granular-shaped crystals (0.5 to 0.7 mm in diameter, rarely larger than several mm), and is rimmed by secondary phlogopite + calcite. Chemical compositions of garnet are: pyrope: 36.5 to 46.9 mol%, grossular: 38.3 to 48.4 mol%, and almandine: 8.8 to 15.1 mol%. Garnet contains abundant inclusions of pure calcite, low Mg-calcite, dolomite, microdiamond, graphite, and minor apatite, rutile, and phengite. Microdiamond occurs as inclusions, mainly in garnet and phlogopite after garnet, and as a few inclusions in diopside. Distribution of microdiamond in thin section scale and even in a single garnet grain is heterogeneous. The average garnet contains several tens of microdiamond grains (Fig. 2). The diamond grains range from 5 to 25 μm in diameter. Some garnet grains lack diamond.

Three Types of Microdiamonds

Variable diamond morphologies occur in the dolomite marble. Ishida et al. (2003) classified these diamonds into three types, on the basis of their morphologies under an optical microscope. The present study follows their classification.

S type

S-type diamond is the most abundant type in dolomite marble. Representative photomicrographs of S-type diamonds are shown in Figure 3. These diamonds consist of distinct cores and rims. The core is characterized by a translucent single crystal (Ishida et al., 2003) measuring 5 to 20 μm in diameter. The rims are very fine-grained subhedral to euhedral transparent crystals (about 1 to 5 mm); some show a platy form with well-developed crystal faces. The core is similar to an R-type diamond in terms of morphology, size, surface, and translucent appearance, and the rim is similar to those of T-type diamonds.

R type

R-type diamond is a single translucent crystal with rugged surface (5 to 20 μm in diameter); crystal forms are rounded (Fig. 4A) or cuboid. The rugged surface indicates weakly developed, very fine-

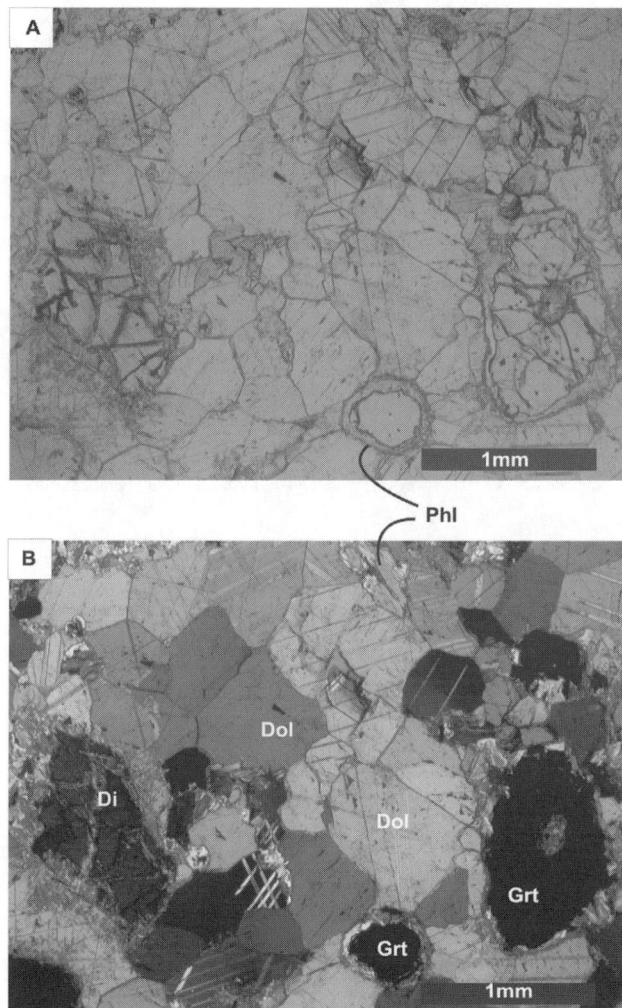

FIG. 1. Photomicrographs of a representative specimen of diamond-bearing dolomite marble (sample no. ZW46). A. Single polar. B. Crossed polars. Dolomite (Dol), garnet (Grt), and diopside (Di) show granoblastic textures. Garnet is rimmed by secondary phlogopite (Phl).

grained rim crystals that could not be confirmed as a single crystal by micro-Laue diffraction methods (Ishida et al., 2003).

T type

T-type microdiamonds are transparent crystals with smooth surfaces; their size is relatively small (1–7 μm) compared to the other two types (Fig. 4B). The shape is either angular or rounded. They seem to be a single crystal. In some cases, this type of microdiamond forms random aggregations (5–10 μm).

Results of CL Analysis

Analytical techniques

Micro-CL studies were conducted on S-type microdiamonds in UHP dolomite marble. The purpose of our micro-CL analysis was to detect differences between the cores and rims of S-type diamonds. CL spectra from μm-sized areas and CL images at the peak wavelengths were obtained employing two different CL instruments: a HITACHI S-4300SE SEM equipped with a CL spectro-

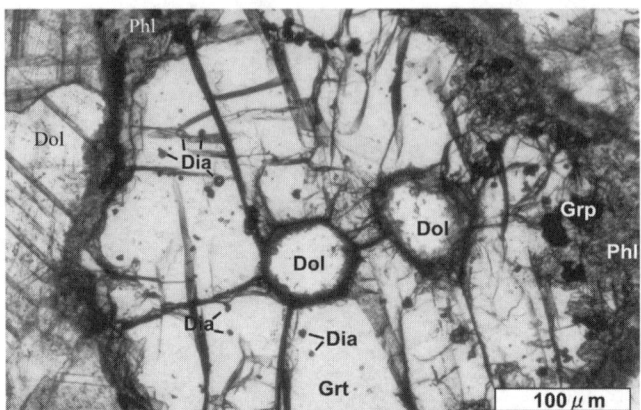

FIG. 2. Photomicrograph showing the distribution of microdiamond (Dia) in garnet (Grt) having the average concentration of microdiamond, sample no. XX01. Abbreviations: Dol = dolomite; Grp = graphite; Phl = phlogopite.

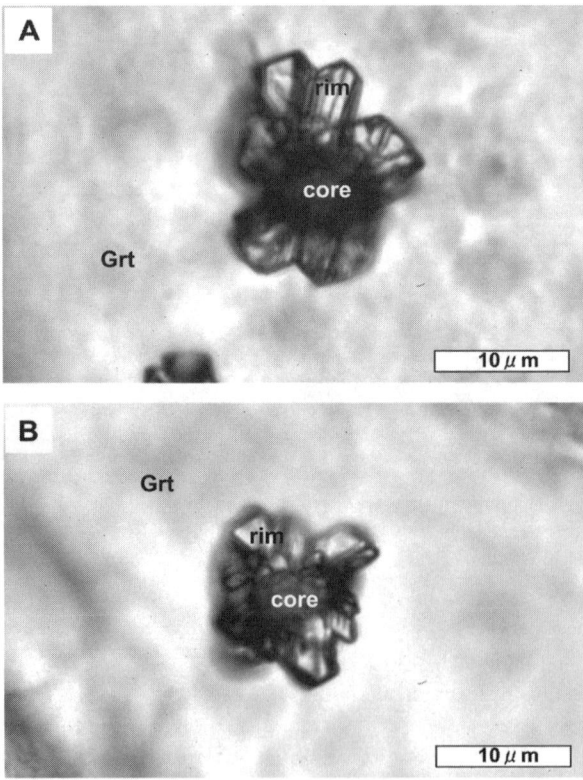

FIG. 3. Photomicrographs of S-type microdiamond in garnet (Grt) (sample no. XX01).

meter F-4500 and a HITACHI S-3000N SEM equipped with a HORIBA CL spectrometer MP-32. Accelerating voltages for both SEM were 15 kV. Samples for CL analysis were prepared as ordinary polished thin sections polished with diamond abrasives (no contamination by abrasives was confirmed under a microscope). Analyzed areas are microdiamond grains that were cut by the polished

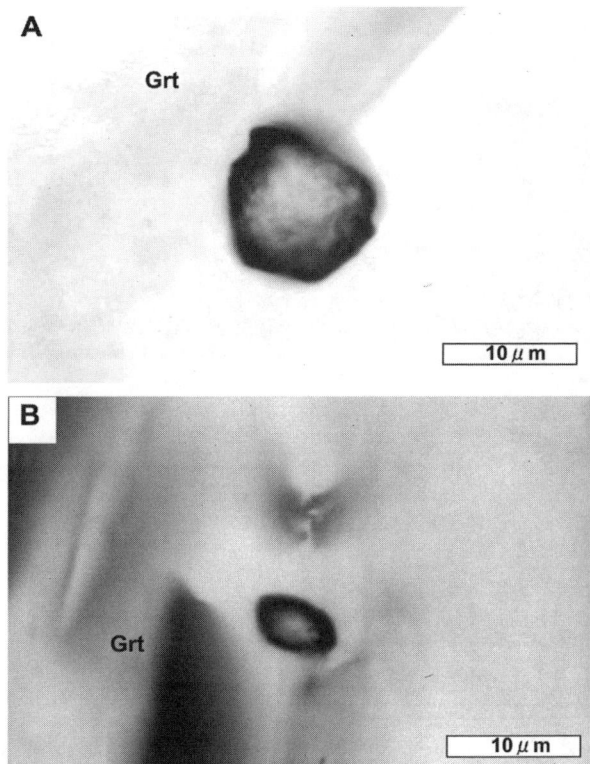

FIG. 4. Photomicrographs of R-type and T-type microdiamonds. A. R-type microdiamond in garnet (Grt), sample no. C43. B. T-type microdiamond in garnet (Grt), sample no. ZW46.

surface. For each analyzed grain, a spectrum by spot analysis was acquired for rim and core; a CL image was then taken for the area covering whole grain at the strongest peak wavelength. On the analysis using a MP-32 system, line scanning profiles of CL spectra were obtained.

Nine S-type microdiamond grains were analyzed with the F-4500 system and five grains with the MP-32 system. The peak locations of CL spectra by F-4500 and MP-32 systems were summarized in Table 1. A strong broad band of CL spectra was detected from microdiamond rims in the ranges from 514 to 526 nm by the F-4500 system, and from 525 to 537 nm by the MP-32 system. The weak or very weak broad band of the same ranges was detected from the cores of S-type microdiamonds. Representative results of CL analyses are shown in Figures 5 to 10 by sets of optical photomicrographs and SEM images, intensity images of CL peak spectra, and CL spectra. Details are described below.

TABLE 1. Peak Locations of CL Spectra of S-type microdiamonds in Kokchetav Dolomite Marble Detected with CL Spectrometer F-4500 and MP-32

Thin section no.	Grain no.	Peaks of CL spectra, nm
	F-4500	
XX01C	XX01C-1	395, 525
	XX01C-2	524
	XX01C-3	518
	XX01C-4	395, 517
	XX01C-5	523
	XX01C-11	523
	XX01C-14	520
	XX01C-15	514
	XX01C-17	526
	MP-32	
XX01C	XX01C-20	440, 537
	XX01C-21	440, 535
	XX01C-22	454, 529
	XX01C-23	453, 535
XX01	XX01-1-4	450, 525

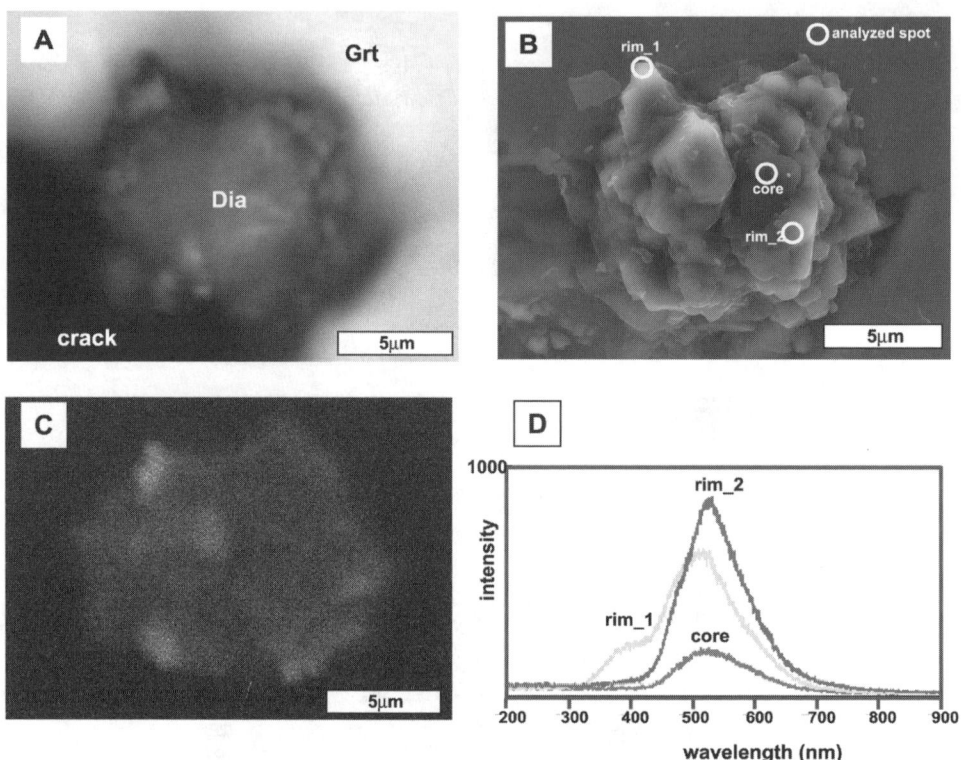

FIG. 5. A. Photomicrograph of S-type microdiamond in garnet, grain no. XX01C-1. B. SEM image of the same microdiamond in A. C. CL image at the wavelength 525 nm of the same microdiamond grain as in A. D. CL spectra of the microdiamond obtained at three spots indicated in B.

CL data with the F-4500 system

With the F-4500 system, nine grains were analyzed. Grain no. XX01C-1 (Fig. 5) is a typical S-type microdiamond with 2–3 μm thin rim surrounding a core of about 10μm diameter. This grain exhibits rather complex CL image features. The analyzed rim no. 2 surrounding the core above the polished surface shows a very strong spectrum of the peak at 525 nm. The adjacent core shows very weak spectra at the same peak position. The analyzed rim no. 1 shows a strong, broad peak at 525 nm, and also shows a weak peak at 395 nm. These results indicate that rimmed diamonds grown on the core vary in CL spectra. The intensity image at 525 nm (Fig. 5D) indicates the occurrence of very fine-grained diamond rims. Similar CL patterns were obtained from grain no. XX01C-4. Grain no. XX01C-03 (Fig. 6) is an S-type microdiamond with relatively coarse-grained rims (about 5 μm in longer dimension) surrounding a small core (< 10 μm in diameter). The rim shows a very strong broad band at 518 nm, and the core has a weak peak at the same range. Grain no. XX01C-05 (Fig. 7), having 2–4 μm crystalline rims on a relatively small core (10 < μm in diameter) shows simple CL features. The rim shows strong CL spectra at 523 nm, and the core has a very weak peak at the same wavelength.

CL data employing the MP-32 system

Using this system, line scanning CL spectra were obtained for 11 to 21 analyzed spots along a grain traverse. Grain no. XX01C-20 (Fig. 8) is a S-type microdiamond with a relatively large rim crystal (about 5 μm in longer dimension) surrounding a small core (10 <μm in diameter). Spectra (Fig. 8D) along the traverse shown in Figure 8B indicate that the rim has strong peak at 537 nm, but its intensity is weak in the core. The CL image at the peak of 537 nm clearly shows that the rim has strong luminescence. A weak peak was detected at about 440 nm. Grain no. XX01C-21 (Fig. 9) is a typical S type with

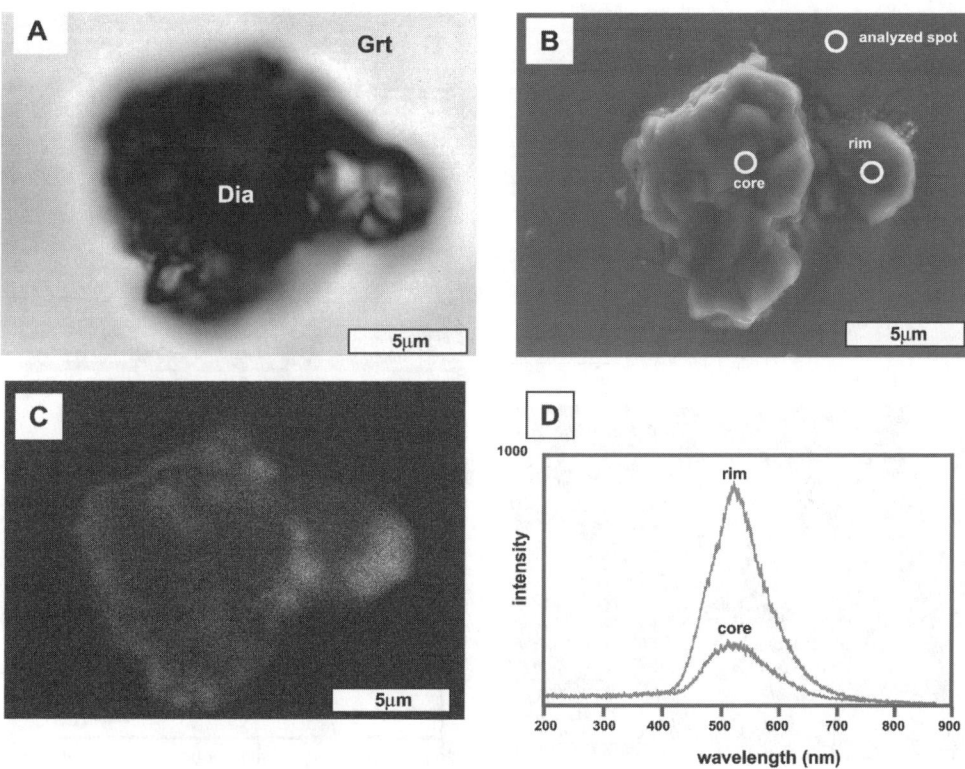

FIG. 6. A. Photomicrograph of S-type microdiamond in garnet, grain no. XX01C-3. B. SEM image of the same microdiamond in A. C. CL image at the wavelength 518 nm of the same microdiamond grain in A. D. CL spectra of the microdiamond obtained at two spots indicated in B.

relatively fine-grained rim diamond surrounding the core (about 10 μm in diameter). The traverse for CL spectra is located along the rim. The rim has relatively strong luminescence at 535 nm, and its intensity is weak in the core. This grain also shows a weak peak at about 440 nm. The CL image at a peak of 535 nm shows strong luminescence at the rim. Grain no. XX01C-22 (Fig. 10) is an S-type microdiamond having a relatively large rim crystal with well-developed crystal surface on a small core (10 <μm in diameter). The 529 nm peak was detected in the rim, and its intensity is rather strong compared with other grains. This is well demonstrated in the image at the peak of 529 nm (Fig. 10C). A rather strong peak at 454 nm (weaker than the peak at 529 nm) was detected along the traverse, including both the core and the rim.

The peak at about 520 nm detected with the F-4500 system is regarded as the same spectrum as the 530 nm peak obtained by the MP-32 system. The small difference in the peak position may reflect instrument conditions. Appearance or absence in the 440-454 nm spectrum could depend on contrasting machine characteristics of the MP-32 and F-4500 systems; the MP-32 system is much more sensitive than the F-4500 system. The intensity difference between core and rim in the range of 514–526 nm by F-4500 and 525–537 nm by MP-32 was clearly detected; the intensity of the rim in every case is stronger than the core.

Discussion

New evidence for two-stage growth

Ishida et al. (2003) examined microdiamonds in dolomite marble by micro-Laue diffraction and laser Raman spectra, together with microscopic observation. They demonstrated that: (1) the rims of S-type diamonds have different orientations than the cores; (2) the core is a single crystal; and (3) full width at half maximum (FWHM) of the Raman band of the rim diamond (1330 to 1335 cm^{-1}) is slightly smaller

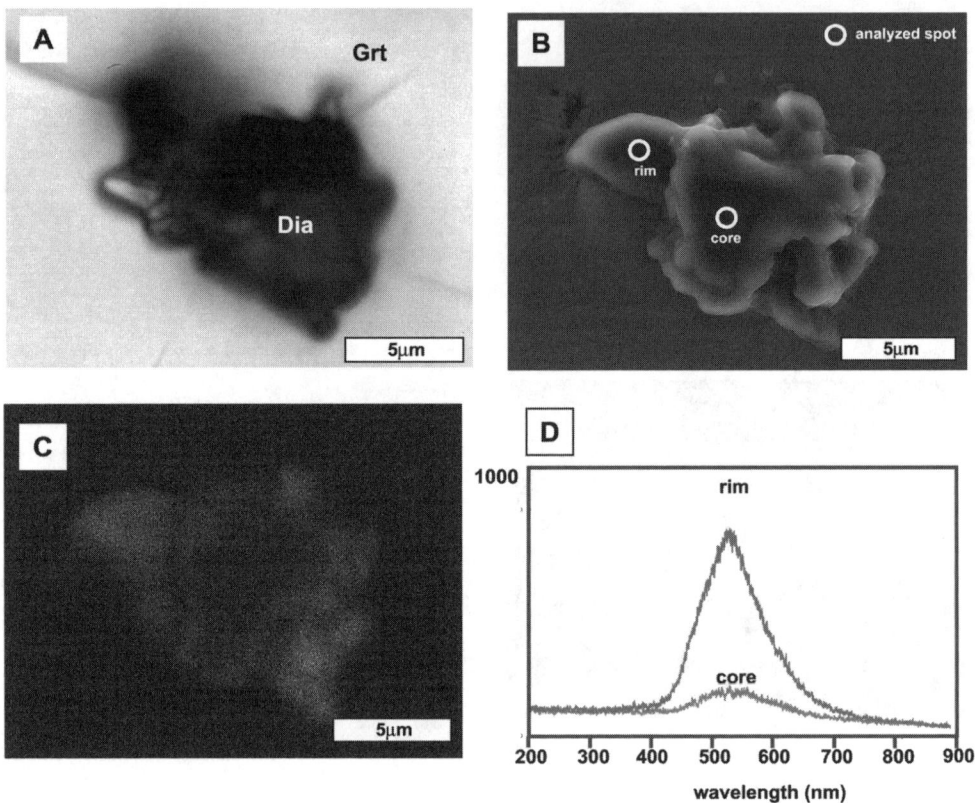

FIG. 7. A. Photomicrograph of S-type microdiamond in garnet, grain no. XX01C-5. B. SEM image of the same microdiamond in A. C. CL image at the wavelength 523 nm of the same microdiamond grain in A. D. CL spectra of the microdiamond obtained at two spots indicated in B.

than that of the core. They concluded that S-type diamonds crystallized discontinuously at two different stages: a first stage of core formation and a second stage of rim formation: the rim diamond was not formed by transformation from graphite, but was precipitated on the core as a seed crystal from a fluid during the second stage. R-type diamond is similar to the core part of S-type diamond, so Ishida et al. thought that R-type diamond formed at the first stage. T-type diamond was considered to have formed directly from the fluid without a seed crystal.

CL data obtained for S-type microdiamonds in the present study add new evidence for the two-stage growth of microdiamond. Apparent intensity differences in the CL spectra of peak positions of 514 to 537 nm were detected between the core (strong) and the rim (weak). Such differences may be due to different environments for core and rim formation of S-type microdiamonds. One of the possible causes of intensity difference in CL spectra may be nitrogen content. Although the cause of the intensity differences is not clear, our results of CL spectra of S-type microdiamonds strongly support a two-stage growth of microdiamond in dolomite marble, as previously proposed by Ishida et al. (2003), and illustrated in Figure 11. Graphite as a possible source of carbon for the first stage growth of diamonds (R-type and core of the S-type) has been supposed in this figure.

De Corte et al. (2002) used CL images to characterize the internal morphology and zoning pattern of microdiamonds from the Kokchetav Massif. They prepared polished sections of thermochemically extracted diamonds from marble. Some analogies in CL images are shown between their results and ours. Their images of analyzed microdiamonds display

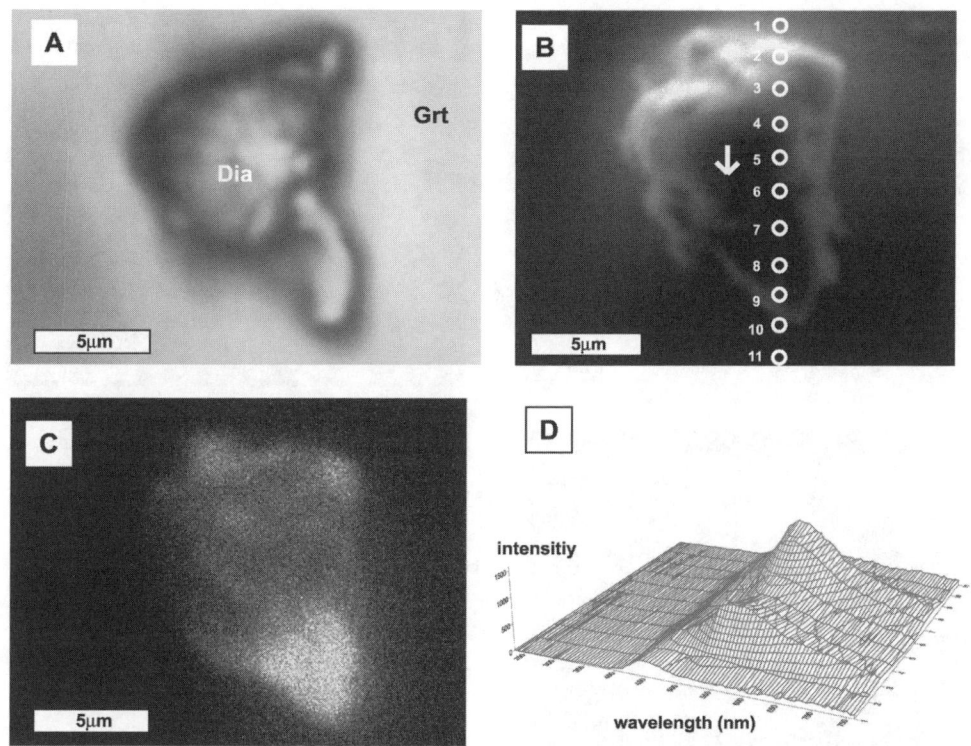

FIG. 8. A. Photomicrograph of S-type microdiamond in garnet, grain no. XX01C-20. B. SEM image of the same microdiamond in A. C. CL image at the wavelength 537 nm of the same microdiamond grain in A. D. CL spectra of the microdiamond along the line indicated in B.

pronounced banding contrasts; the core exhibits cubic sector zoning, whereas the rim is irregularly sectored. The rim is lighter than the core. Differences in intensity of CL emission and morphology between core and rim are consistent with our results. Inasmuch as their results do not yield CL spectra but simply show CL images, we cannot evaluate the peak spectra of their CL data. They also observed a switch of internal morphology from cubo-octahedral (the core) to cubic growth (external shape). They showed that cubic sectoring was the dominant internal growth feature in Kokchetav microdiamonds. They concluded that the process of diamond growth was complex, and influenced by the geochemical environment.

We did not detect banding patterns in S-type microdiamonds. The reason is that: (1) we did not polish the *in situ* surface of diamond, and therefore cross-sections of diamond were be exposed; (2) diamond is too small (10–15 μm) for sector zoning to be observed. De Corte et al. (2002) collected relatively large grains (20–52 μm) for their investigation.

Comparison with carbonado

Carbonado is a mysterious polycrystalline diamond found in alluvial placers only from the Central African Republic and Brazil. De et al. (2001) studied carbonado by combining CL and SIMS, and found a correlation between CL emission, $\delta^{13}C$ value, and nitrogen abundance. They obtained green luminescence in carbonado. The CL spectra of carbonado show three main peaks at 520, 580 and 665 nm (Magee and Taylor, 1999). The main broad band of CL spectra for S-type diamonds from Kokchetav obtained in this study has a peak ranging from 514 to 537 nm; these wavelengths correspond to a green color, which is not a common CL color in kimberlitic diamonds. This spectrum range corresponds to the 520 nm peak reported by Magee and Taylor (1999). De at al. (2001) concluded that

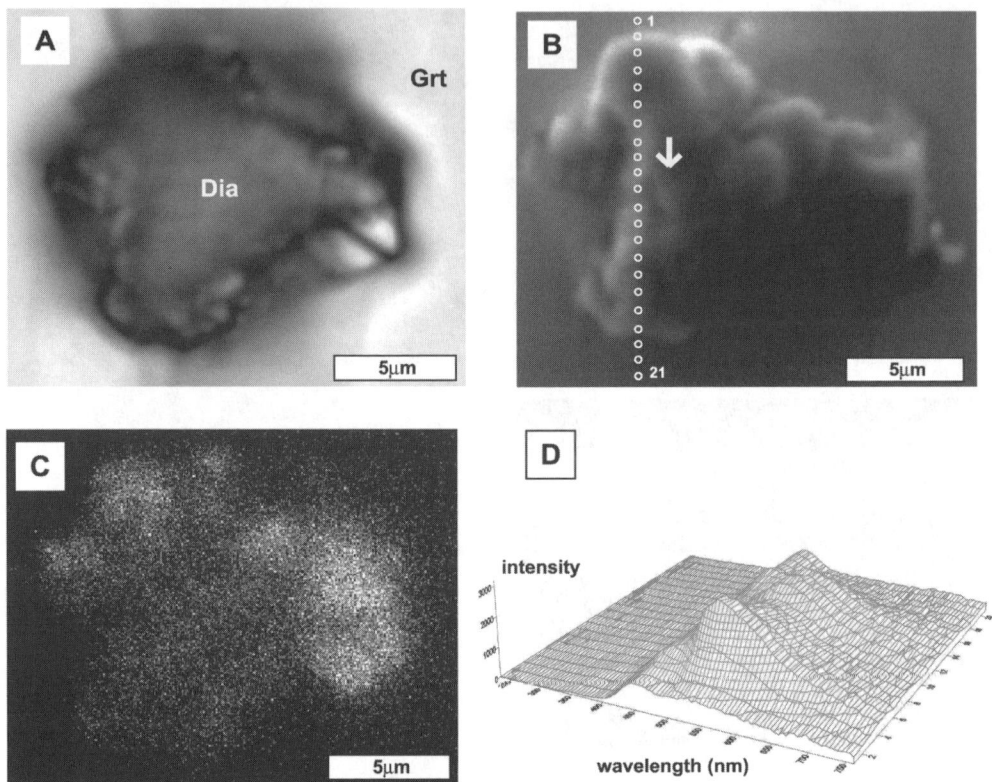

FIG. 9. A. Photomicrograph of S-type microdiamond in garnet, grain no. XX01C-21. B. SEM image of the same microdiamond in A. C. CL image at the wavelength 535 nm of the same microdiamond grain in A. D. CL spectra of the microdiamond along the line indicated in B.

carbonado grew in two steps; they did not demonstrate whether carbonado originated in the mantle, in the crust, or by meteorite impact.

Attribution of CL spectra

Many scientists have reported CL spectra for diamond using this technique in conjunction with FT/IR and SIMS to clarify special characteristics. Through this research, some CL spectra were interpreted as follows.

Bulanova (1995) interpreted the common blue CL color (430–490 nm) of most kimberlite-derived diamonds to reflect variations in the abundance of nitrogen, whereas the yellow color (550–590 nm) commonly reflects the presence of hydrogen. Lu et al. (2001) reported kimberlitic diamond from Fuxian in eastern China, and argued that variations in N abundance influence the CL color in the 430–590 nm range; bright blue CL has a high N content (>540 ppm), and dark green or greenish blue CL has moderate N contents (540–244 ppm).

Robins et al. (1989) assigned the wavelength at 439 nm to dislocations in chemical vapor deposition (CVD) diamond, and the 389 nm peak to either interstitial nitrogen or a N-C interstitial complex. Seo et al. (2002) also investigated CL emission of CVD diamond. They attributed the emission at 430 nm to be mainly due to growth-induced dislocations, and the peak at 538 nm to be caused by a vacancy trapped between two nitrogen atoms, N-V-N.

Inasmuch as it is difficult to explain two origins of the spectra without other instrumental analysis such as FT/IR and SIMS, we cannot contribute to an attribution of spectra only with our CL data. Nevertheless, the following possibilities could be considered for our S-type microdiamonds: (1) the spectral peak at 450 nm may be caused by dislocations in diamond crystals; (2) the intensity difference at the

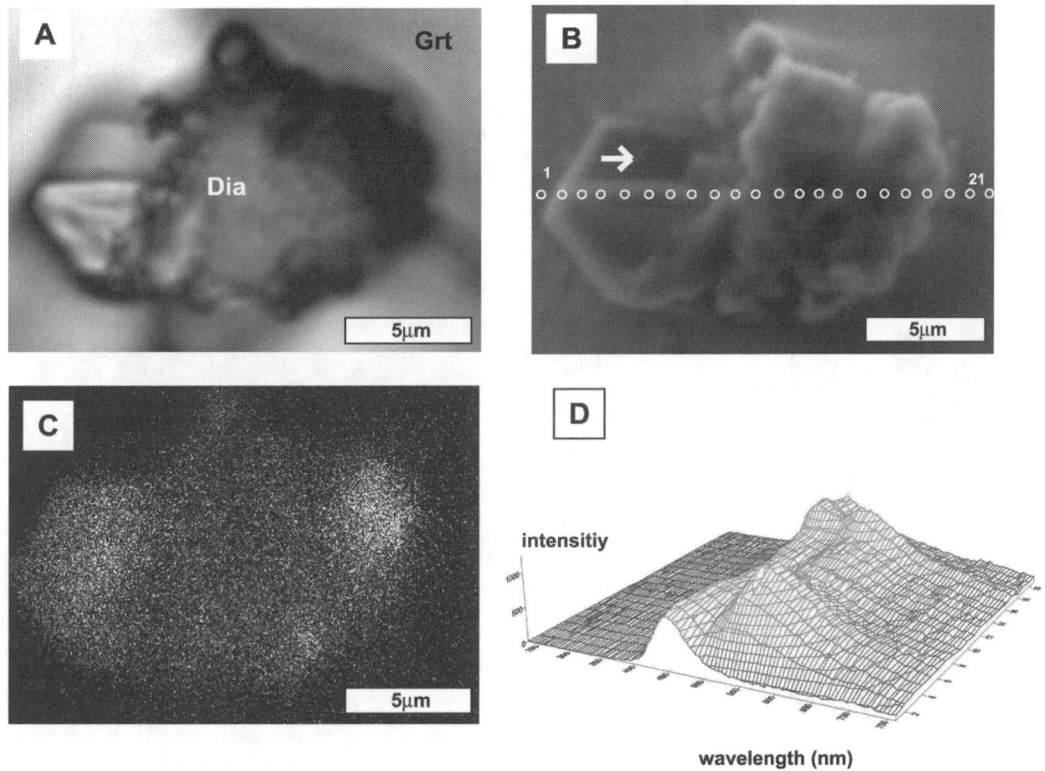

FIG. 10. A. Photomicrograph of S-type microdiamond in garnet, grain no. XX01C-22. B. SEM image of the same microdiamond in A. C. CL image at the wavelength 529 nm of the same microdiamond grain in A. D. CL spectra of the microdiamond along the line indicated in B.

514 to 537 nm peak may indicate considerable amounts of nitrogen. The attribution of the spectrum of the 395 nm peak is still unclear.

Conclusions

Cathodoluminescence of S-type microdiamonds in dolomite marble from the Kokchetav Massif was obtained as spectra and intensity images at the peak of the main CL band using two kinds of CL spectrometer combined with a SEM. The strong, broad bands of CL spectra at 514 to 537 nm were detected at the rim of all grains of S-type microdiamonds. The core also has the same broad bands, but its intensities are very weak. The weak, broad band at about 393 nm was detected at the rim of some grains; another weak broad band at about 440–450 nm also was detected both in core and rim of several grains by MP-32 system analysis. The intensity images of CL spectra at the peak wavelength (514–537 nm) of the main band showed a clear difference between core (weak) and rim (strong). These CL data signify that separate stages of diamond growth can be geochemically differentiated, and provide additional evidence for the two-stage growth of S-type microdiamonds in dolomite marble, as proposed previously (Ishida et al., 2003).

Acknowledgments

This paper was prepared to honor the retirement of W. G. Ernst, who has promoted many international cooperative projects. The authors appreciate the help of Akihiro Sato for CL analyses for this study. We also thank K. Yonemochi for his assistance in sample preparation. This study was financially supported by a JSPS Grant in Aid nos. 13640485 and 15204050, and a Waseda University Grant for Special Research Project no. 2001A-533.

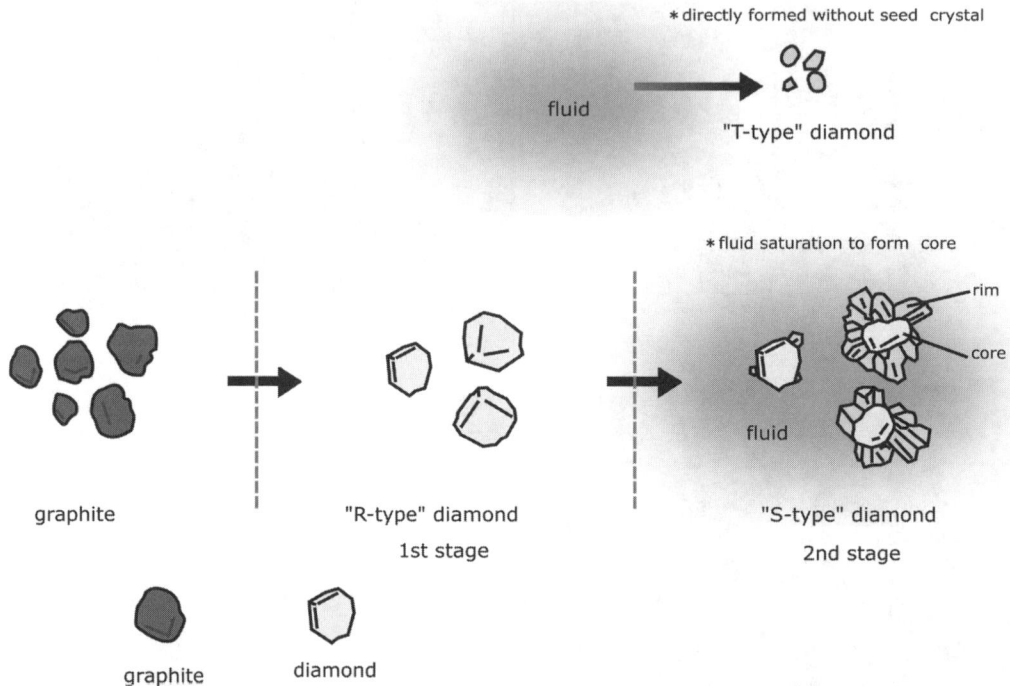

FIG. 11. Schematic illustration showing two-stage growth of microdiamond in dolomite marble as a function of fluid infiltration. Graphite as a possible source of carbon for the first stage growth of diamonds (R-type and the core of S-type) has been supposed.

REFERENCES

Akaishi, M., and Yamaoka, S., 2000, Crystallization of diamond from C-O-H fluids under high-pressure and high-temperature conditions: Journal of Crystal Growth, v. 209, p. 999–1003.

Bulanova, G., 1995, The formation of diamond: Journal of Exploration Geochemistry, v. 53, p. 1–23.

De, S., Heaney, P. J., Vicenzi, E. P., and Wang, J., 2001, Chemical heterogeneity in carbonado, an enigmatic polycrystalline diamond: Earth and Planetary Science Letters, v. 185, p. 315–330.

De Corte, K., Cartigny, P., Shatsky, V. S., Sobolev, N. V., and Javoy, M., 1998, Evidence of fluid inclusions in metamorphic microdiamonds from the Kokchetav Massif, northern Kazakhstan: Geochimica et Cosmochimica Acta, v. 62, p. 3765–2773.

De Corte, K., Trautman, R., Griffin, B., and De Paepe, P., 2002, Internal morphology of microdiamonds from UHPM rocks of the Kokchetav Massif, in Parkinson, C. D., Katayama, I., Liou, J. G., and Maruyama, S., eds., The diamond-bearing Kokchetav Massif, Kazakhstan: Tokyo, Japan, Universal Academy Press, p. 103–114.

Dobrzhinetskaya, L., Eide, E., Korneliussen, A., Larsen, R., Millege, J., Posukhova, T., Smith, D. S., Sturt, B., Taylor, W. R., and Tronnes, R., 1995, Diamond in metamorphic rocks of the Western Gneiss region in Norway: Geology, v. 23, p. 597–600.

Dobrzhinetskaya, L. F., Green, H. W., Mitchell, T. E., and Dickerson, R. M., 2001, Metamorphic diamonds: Mechanism of growth and inclusion of oxides: Geology, v. 29, p. 263–266.

Ishida, H., Ogasawara, Y., and Ohsumi, K., and Saito, A., 2003, Two-stage growth of microdiamonds in UHP dolomite marble from Kokchetav Massif, Kazakhstan: Journal of Metamorphic Geology, v. 21, p. 515–522.

Katayama, I., Zayachkovsky, A., and Maruyama, S., 2000, Prograde pressure-temperature records from inclusions in zircons from ultrahigh-pressure-high-pressure rocks of the Kokchetav Massif, northern Kazakhstan: The Island Arc, v. 9, p. 417–427.

Kaneko Y., Maruyama S., Terabayashi, H., Yamamoto, H., Ishikawa, M., Anma, R., Parkinson, C., Ota, T., Nakajima, Y., Katayama, I., Yamamoto, J. and Yamauchi, K., 2000, Geology of the Kokchetav UHP-HP metamorphic belt, Northern Kazakhstan: The Island Arc; v. 9, p. 264–283.

Kumar, M. D. S., Akaishi, M., and Yamaoka, S., 2000, Formation of diamond from supercritical CO_2-H_2O fluid at high pressure and high temperature: Journal of Crystal Growth, v. 213, p. 203–206.

Lu, F. X., Chen, M. H., and Zheng, J. P., 2001, Nitrogen distribution in diamonds from the kimberlite pipe no. 50 at Fuxian in Eastern China: A CL and FTIR study: Physics and Chemistry of the Earth, v. 26, p. 773–780.

Magee, C. W., and Taylor, W. R., 1999, Constrains from luminescence on the history and origin of carbonado, in Gurney, J. J., and Pascoe, M. D., eds., Seventh International Kimberlitic Conference, Cape Town, v. 2, p. 529–532.

Mposkos, E. D., and Kostopoulos, D. K., 2001, Diamond, former coesite, and supersilicic garnet in metasedimentary rocks from the Greek Rhodope: A new ultrahigh-pressure metamorphic province established: Earth and Planetary Science Letters, v. 192, 497–506.

Okamoto, K., Liou, J. G., and Ogasawara, Y., 2000, Petrology of the diamond-grade eclogite in the Kokchetav Massif, northern Kazakhstan: The Island Arc, v. 9, p. 379–399.

Ogasawara, Y., Fukasawa, K., and Maruyama, S., 2002, Coesite exsolution from supersilicic titanite in UHP calcite marble: American Mineralogist, v. 87, p. 454–461.

Ogasawara, Y., and Ishida, H., 2001, Diamond, fluid and carbonates—stability and genesis of metamorphic diamond [abs.], in Fluid/slab/mantle interactions and ultrahigh-P minerals: Abstracts of UHPM Workshop 2001 at Waseda University, p. 1–5.

Ogasawara, Y., Ohta, M., Fukasawa, K., Katayama, I., and Maruyama, S., 2000, Diamond-bearing and Diamond-free metacarbonate rocks from Kumdy-kol in the Kokchetav Massif, northern Kazakhstan: The Island Arc, v. 9, p. 400–416.

Parkinson, C., and Katayama, I., 1999, Metamorphic microdiamond and coesite from Sulawesi, Indonesia: Evidence of deep subduction at SE Sundaland margin: EOS (Transactions of the American Geophysical Union), v. 80, p. F1181.

Robins, L. H., Cook, L. P., Farabaugh, E. N., and Feldman, A, 1989, Cathode luminescence of defects in diamond films and particles grown by hot-filament chemical vapor deposition: Physical Review, v. 39, p. 13,367–13,377.

Seo, S. H., Park, C. K., and Park, J. S., 2002, Cathodoluminescence characteristics of polycrystalline diamond films grown by cyclic deposition method: Thin Solid Films, v. 420-421, p. 161–165.

Sobolev, N. V., and Shatsky, V. S., 1990, Diamond inclusions in garnets from metamorphic rocks: Nature, v. 343, p. 742–746.

Stöckhert, B., Duyster, J., Trepmann, C., and Massonne, H., 2001, Microdiamond daughter crystals precipitated from supercritical COH + silicate fluids included in garnet, Erzgebirge, Germany: Geology, v. 29, p. 391–394.

Yoshioka, N., Muko, A., and Ogasawara, Y., 2001, Extremely high diamond concentration in dolomite marble [abs.], in Fluid/slab/mantle interactions and ultrahigh-P minerals: Abstracts of UHPM Workshop 2001 at Waseda University, p. 51–55.

Diamond Formation in UHP Dolomite Marbles and Garnet-Pyroxene Rocks of the Kokchetav Massif, Northern Kazakhstan: Natural and Experimental Evidence

V. S. SHATSKY,[1] Y. N. PAL'YANOV, A. G. SOKOL, A. A. TOMILENKO, AND N. V. SOBOLEV

Institute of Mineralogy and Petrography, Siberian Branch of the Russian Academy of Sciences, Koptyug Pr., 3, Novosibirsk, 630090, Russia

Abstract

Based upon detailed studies of diamondiferous metamorphic rocks, many authors share the opinion that diamonds crystallize in the field of their thermodynamic stability. Nevertheless, some problems remain, and the most important questions are as follows: (1) What is the pressure under which diamond crystallized? (2) Does the composition of diamondiferous rocks correspond to the medium of diamond crystallization? (3) Why are microdiamonds irregularly distributed in dolomite marbles and garnet-pyroxene rocks? (4) What is the role of carbonates in diamond genesis? To answer these questions, we carried out petrographic and mineralogical studies and experimentally modeled the process of microdiamond crystallization in diamondiferous garnet-pyroxene rocks and dolomite marbles. Diamondiferous marbles and garnet-pyroxene rocks occur as layers and lenses in biotite gneisses of the Kumdy-Kol microdiamond deposit, northern Kazakhstan. Mineralogical and petrographical data demonstrate that pyroxene of diamondiferous rocks differs in composition from pyroxene of nondiamondiferous rocks. The pyroxene in diamondiferous garnet-pyroxene rocks and dolomite marbles is characterized by the presence of potassium and by occurrences of lamellae of K-feldspar and phengite as well as quartz needles. No potassium is found in the pyroxene from associated nondiamondiferous rocks.

Starting materials in experiments were diamondiferous marble and garnet-pyroxene rock. Experiments were carried out at P = 5.7 GPa and T = 1420° C, and at P = 7.0 GPa and 1700° C using a multi-anvil apparatus with a 300 mm outer diameter of the multi-anvil sphere. The following conclusions can be inferred from the data obtained. Unlike pyroxene in the starting specimens, the newly formed pyroxene is K-depleted, which indicates that the rocks used in the experiments differ in composition from the natural medium of diamond crystallization. The garnets synthesized in experiments with dolomite marble contain up to 4% majorite component, whereas the garnet from the initial rock contains no majorite. These data clearly show that the pressure under which dolomite marbles formed did not exceed 50 kbar. The experimental diamonds are all octahedra, whereas the diamonds in the starting samples were cubes. We believe that the main factor governing the morphology of diamond crystals is the composition of the medium of crystallization. The obtained data suggest that in dolomite marbles and garnet-pyroxene rocks, diamond crystallized from a carbonatite melt in equilibrium with a K-rich fluid.

Introduction

UNLIKE KIMBERLITES and lamproites, diamondiferous metamorphic rocks provide a unique opportunity for studying processes of diamond formation *in situ*. One of the best known localities of ultrahigh-pressure (UHP) diamond-bearing metamorphic rocks is the Kokchetav massif (Sobolev and Shatsky, 1990; Shatsky et al., 1995; Zhang et al., 1997). Four types of diamond-bearing rocks have been recognized in the Kumdy-Kol microdiamond deposit of the Kokchetav massif: garnet-biotite-gneiss, garnet-pyroxene-quartz rock, garnet-pyroxene rock, and garnet-pyroxene-carbonate rock (dolomite marble) (Shatsky et al., 1995). The first two rock types have undergone retrograde metamorphism, and as a result their primary mineral assemblages have been preserved only as inclusions in garnet and zircon. On the other hand, the garnet-pyroxene rock and dolomite marble experienced less retrogression. So we suggest that their rock compositions are close to those of the media of diamond crystallization.

---
[1]Corresponding author; email: shatsky@uiggm.nsc.ru

Extremely high contents of microdiamonds—up to several thousand carats per metric ton—are a special feature of those rocks. Several papers have recently reported data on the mineralogy of the dolomite marbles from this area and P-T estimates for their formation (Shatsky et al., 1995, 1999a; Ogasawara et al., 2000; Zhu and Ogasawara, 2002). It has been shown that the diamonds crystallized in the field of their thermodynamic stability. In addition, the physical properties and isotopic composition of microdiamonds from the garnet-pyroxene rocks and dolomite marbles have been studied (Shatsky et al., 1998; De Corte et al., 1998, 1999; Cartigny et al., 2001). Although much new data has been obtained, several important questions remain unanswered: (1) At which pressure were the diamonds crystallized? (2) Does the composition of the rocks correspond to that of the diamond environment? (3) Why are the microdiamonds unevenly distributed in the dolomite marbles and garnet-pyroxene rocks? (4) What is the significance of carbonate for the origin of metamorphic diamonds? To answer these questions, the authors have performed experimental modeling of microdiamond crystallization processes in diamondiferous garnet-pyroxene and pyroxene-carbonate rocks along with mineralogical and petrographic studies.

Petrography and Mineralogy of Diamondiferous Rocks

Diamondiferous marbles and garnet-pyroxene rocks occur as separate layers and lenses in the biotite gneisses of the Kumdy-Kol microdiamond deposit. Some varieties of marble and garnet-pyroxenite are banded.

Garnet-pyroxene rocks show alternation of pyroxene and garnet layers. Diamondiferous and diamond-free garnet-pyroxene rocks comprise garnet and clinopyroxene with subordinate carbonate and K-feldspar. Amphibole and chlorite are retrograde minerals. Two types of garnet pyroxene rocks are recognized based on the composition of garnet and clinopyroxene. The first type (samples 2-4, 5-2, and 88-1) contains Mg-rich clinopyroxene (12–17 wt% MgO) and grossular-pyrope-almandine garnet (Table 1). Orthoclase lamellae (1–20 μm) and silica needles are present in the pyroxene from the matrix and in clinopyroxene inclusions in garnet (Shatsky et al., 1985, 1995; Sobolev and Shatsky, 1987). If the pyroxenes were to show no exsolution, both pyroxenes from the matrix and the inclusions in garnet would have similar composition. Lamellae of high-Al titanite were found in clinopyroxene from these rocks. Most samples of this type of garnet-pyroxene rock contain diamonds included within garnet and clinopyroxene (Fig.1); however, diamond-free samples have been also noted. The lack of diamonds within a thin section cannot be taken to indicate that the sample as a whole does not contain diamonds.

In contrast, the second type of garnet-pyroxene rocks never contain diamonds (sample 92-50). These rocks comprise grossular-almandine garnet and Fe-rich Cpx (Table 1). The absence of K impurities in the clinopyroxene is the most characteristic feature.

Marbles containing dolomite, Mg-calcite, diopsidic pyroxene, and garnet in variable proportions are typical of the Kumdy-Kol microdiamond deposit. Microdiamonds are widely present as inclusions in garnet (Fig. 2), clinopyroxene, and zircon. The garnet from dolomites differs compositionally from that in other rock types (Table 1): it is characterized by high Ca# and Mg#. It should be noted that in all types of diamondiferous rocks, excluding garnet-kyanite-muscovite schist (Shatsky and Sobolev, 1993), the grossular and pyrope contents of garnets are high. Most compositional profiles across garnet grains show a homogeneous central portion, but a change in the chemical composition occurs in the outer, 100–125 μm wide rim (Shatsky et al., 1995). It should be noted that the composition of homogeneous garnets may vary from grain to grain within a thin section. Garnet inclusions in zircon from several marble samples have strongly variable compositions: Ca# [100Ca/(Ca + Mg + Fe + Mn)] and Mg# [100Mg/(Mg + Fe)] range from 27 to 49 and from 56.3 to 66.4, respectively. In contrast, garnets from the matrix have nearly uniform chemical composition (Shatsky et al., 1995). Garnet in the matrix contains inclusions of calcite and dolomite, with calcite inclusions dominating. Mg in calcite both from the matrix and from the inclusions in garnets, is low. These calcite inclusions in garnet are typically surrounded by radial cracks. Dolomite, magnesite, and magnesian calcite are present as inclusions in zircon; $MgCO_3$ contents in the magnesian calcite reach up to 23.5% (Shatsky et al., 1995).

Based on the clinopyroxene composition, two types of diamondiferous marbles also can be recognized. Cpx from the first type is similar to that from the first type of pyroxene-garnet rocks containing

TABLE 1. Representative Analyses of Natural Minerals from Marbles and Garnet-Pyroxene Rocks[1]

Sample	Phase	SiO_2	TiO_2	Al_2O_3	Cr_2O_3	MnO	FeO	MgO	CaO	Na_2O	K_2O	Total
92-99*	gt_c	41.8	0.25	22.6	0	0.66	6.59	10.9	18.1	0.02	0	100.92
	gt_r	42.1	0.27	22.7	0	0.64	6.69	11	17.9	0.02	0	101.32
	cpx	55.6	0.02	1.5	0	0.06	1.58	16.9	24.2	0.25	0.49	100.6
	car	0	0	0	0	0.27	2.08	21.4	31.4	0	0	55.15
	amf	47.8	0.57	12.5	0	0.1	5.67	16.6	12.8	0.48	1.39	97.91
	phl	39.7	1.58	15.9	0	0.06	5.43	22.5	0.03	0.09	8.95	94.24
98-200B*	gt_{1c}	40.6	0.61	21.5	0	0.82	5.77	10.5	19	0	0	98.8
	gt_{1r}	40.8	0.41	21.9	0.03	1.02	6.14	12.2	16.5	0	0	99
	gt_{2c}	40.5	0.41	22	0.02	1.01	6.18	12.4	16.2	0	0	98.76
	gt_{2r}	41	0.44	22	0.04	1.03	6.28	12.4	16.1	0	0	99.29
	inc in gt	0	0	0	0	0.15	1.27	21.4	33.9	0	0	56.72
	cpx inc	55	0.03	0.29	0	0.06	0.9	18	25.5	0.02	0.05	99.85
	cpx_{1c}	55	0.03	0.45	0	0.06	1.04	17.5	25.1	0.06	0.14	99.38
	cpx_{1r}	54.8	0.02	0.71	0	0.06	1.05	17.2	25.1	0.07	0.21	99.22
	cpx_{2c}	55	0.02	0.4	0	0.06	1.08	17.6	25.2	0.06	0.13	99.55
	cpx_{2r}	54.2	0.02	0.95	0	0.07	1.17	17.6	25	0.08	0.03	99.12
	car	0	0	0	0	0.18	1.17	22.9	34.7	0	0	58.95
94-253	gt_c	41.8	0.27	22.6	0.02	0.35	6.76	16.6	11.3	0.01	0	99.71
	gt_r	42.3	0.14	22.9	0.02	0.43	7.74	18.6	7.96	0	0	100.09
	inc_c	54.7	0.08	0.43	0	0	0.84	17.9	25.3	0.04	0	99.29
	inc_r	54.6	0.15	0.78	0	0.02	1.08	17.9	24.8	0.04	0.04	99.41
	cpx	53.4	0.36	2.11	0	0.05	1.2	17.3	24.7	0.02	0	99.14
	hum	37.5	3.24	0	0	0.28	6.83	50	0.11	0	0	97.96
94-275	cpx_1	53.9	0.22	1.86	0.02	0.05	2.06	16.4	24.9	0.12	0	99.48
	lam_1	40.3	0.28	21.5	0.01	0.67	10.3	10.1	15.4	0	0	98.5
	lam_2	40.3	0.22	21.6	0.04	0.54	8.55	12.1	13.7	0	0	97
	gt_{1c}	39	0.31	20.8	0.03	0.91	19.8	5.06	13.4	0.06	0	99.37
	gt_{1r}	40.1	0.31	21.7	0.03	0.63	14.6	8.24	14.2	0.03	0	99.84
	inc_{1c}	52.6	0.13	2.76	0.01	0.09	6.11	14.1	23	0.64	0.02	99.46
	inc_{1r}	53.1	0.11	1.93	0	0.09	5.84	14.4	23.3	0.57	0.03	99.37
	inc_2	38.6	0.73	17.3	0	0.01	4.32	22.6	0.04	0.06	9.35	93.06
	inc_3	0	0	0	0	0.21	1.1	3.53	58.7	0	0	63.54
	gt_{2c}	40.5	0.25	21.8	0.03	0.68	11	9.94	15.5	0	0.01	99.71
	gt_{2r}	41.4	0.12	23.3	0.03	0.98	12.2	11	12	0.08	0	101.11
	cpx_2	53.4	0.26	2.19	0	0.05	2.49	16.5	24.7	0.26	0.01	99.84
	lam	39.6	0.51	16.97	0.02	0.03	4.3	23.6	0.4	0.09	8.64	94.12
	amf	40.4	0.85	15.8	0	0.04	5.43	15.9	12.5	0.55	2.67	94.14
92-36	gt_{1c}	41.8	0.41	22.3	0.04	0.46	7.28	12.8	15.6	0	0	100.69
	gt_{1r}	41.3	0.38	22.2	0.03	0.45	7.68	13.2	14.4	0	0	99.64
	gt_2	41.2	0.44	21.6		0.47	6.85	11.8	17.2	0.03	0.01	99.6
	cpx	54.6	0.07	0.42		0.05	1.21	17.7	25.2	0.08	0.03	99.36
2-4*	cpx_c	54.4	0.02	1.37	0	0.06	4.63	14.7	23.6	0.47	0.49	99.74
	cpx_r	54.3	0.02	1.74	0	0.08	4.62	14.4	23.1	0.43	0.61	99.3
	gt_c	40	0.34	21.4	0	0.58	16.4	8.6	13.3	0.04	0	100.66
	gt_r	39.9	0.27	21.4	0	0.63	16.3	8.55	13.2	0.03	0	100.28
	amf	44.1	0.41	12.8	0	0.16	11	13.1	11.4	1.33	1.32	95.62
	sph	28.9	39.1	1.93	0	0	0.12	0.17	27.3	0.03	0	97.554
5-2*	cpx_1	54	0.03	1.87	0	0.16	4.85	13.5	24	0.68	0.05	99.14
	lam_1	64.4	0	16.8	0	0.02	0.42	0.63	1.98	1.31	12.7	98.26
	lam_2	34.5	20.2	11.6	0	0.02	0.85	1.17	24.7	0.1	1.4	94.54
	gt_{1c}	39.8	0.36	21.2	0	0.74	15.9	5.96	16.2	0.04	0	100.2
	gt_{1r}	39.4	0.36	21.1	0	0.76	16	5.68	16.4	0.04	0	99.74
	gt_{2c}	39.7	0.44	21.4	0	0.63	14.8	6.45	16.4	0.04	0.01	99.87
	gt_{2r}	39.2	0.26	21.1	0	1.32	17.9	4.74	15.4	0.03	0.03	99.98
	cpx_{2c}	53.6	0.06	2.79	0	0.17	5.76	12.4	23.1	1.05	0	98.93
	cpx_{2r}	53.1	0.04	1.46	0	0.43	7.25	12.4	24.3	0.23	0	99.21
K88-1	gt	40.5	0.57	21.2		0.63	8.38	8.24	19.9	0.03	0.01	99.46
	cpx	54.8	0.02	0.65		0.09	1.94	16.9	24.6	0.17	0.25	99.42
92-50	gt_c	37.9	0.13	19.9	0	1.35	13.6	0.31	25.9	0	0	99.09
	gt_r	38.7	0.13	20	0	1.2	13.9	0.5	25.5	0.01	0	99.94
	cpx_c	51.4	0	0.52	0.01	0.3	16.1	8.3	23.3	0.24	0.01	100.18
	cpx_r	51.9	0	0.53	0	0.29	16.1	8.39	23.5	0.29	0	101

[1] * = diamondiferous rocks; c = core; r = rim; 92-99, 98-200B, 94-253, 94-275, 92-36 = marbles; 2-4, 5-2, 88-1, 92-50 = garnet-pyroxene rocks.

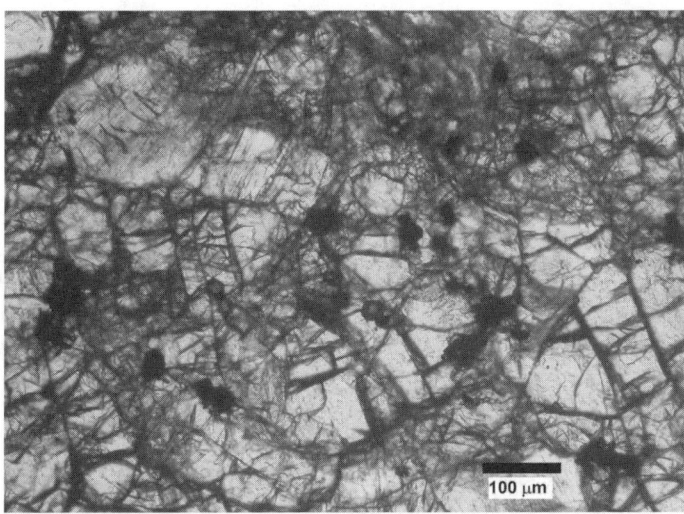

FIG. 1. Inclusions of diamond in garnet from garnet-pyroxene rock.

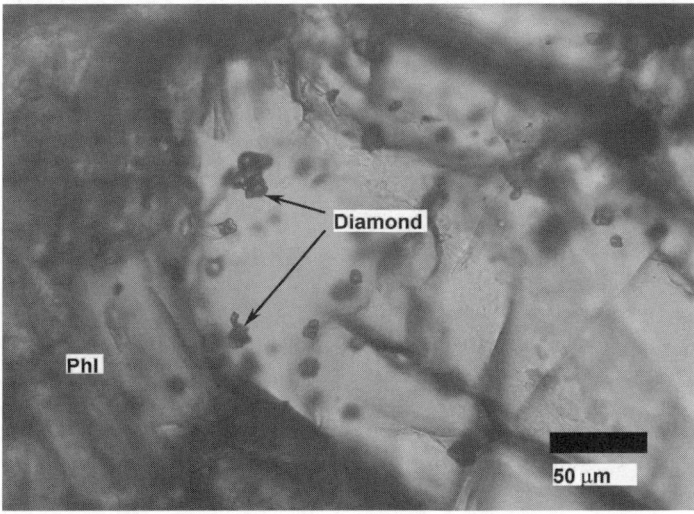

FIG. 2. Inclusions of diamond in garnet from dolomite marbles.

K-feldspar lamellae and quartz needles (sample 92-99). K-rich Cpx also occurs in the matrix and as an inclusion in garnet and zircon. The K_2O content of pyroxene inclusions in zircon is variable. Several samples of carbonate rocks contain two pyroxene generations. The first-generation pyroxene occurs as large grains, several millimeters in diameter. The grain cores include lamellae of K-feldspar and phengite. The second-generation, fine-grained pyroxene occurs in the matrix together with carbonate. The matrix pyroxene grains lack both lamellae and potassium.

Diamond Morphology

The cuboid is the predominant diamond form in garnet-pyroxene rocks and dolomite marbles. Crystal surfaces display a large number of pits of variable configuration. Octahedral faces may be found at the edges of the cuboids. Diamonds from the marbles characteristically show octahedral microfaces on their cuboidal surfaces, indicating a change in the environmental conditions during diamond growth.

The internal morphology of cuboids from the garnet-clinopyroxene rocks was studied by X-ray topography (Martovitskiy et al., 1987; Shatsky et al., 1998). The cuboids have a radiating structure formed by space-filling columnar growth, also known as "fibrous structure." An interesting feature of the internal structure of some crystals is the presence of a central core, which is seen as a slight blackening in the topograms.

Experimental Techniques

The experimental samples were diamondiferous dolomite marble (sample 92-99) 21.8% SiO_2, 0.23% TiO_2, 5.56% Al_2O_3, 3.52% Fe_2O_3, 0.17% MnO, 19.91% MgO, 21.12% CaO, Na_2O—n.d., 0.28% K_2O, 0.16 P_2O_5, LOI 26.9, total 99.68, and garnet-pyroxene (sample 2-4) rock 48.62% SiO_2, 0.23% TiO_2, 6.71% Al_2O_3, 9.35% Fe_2O_3, 0.17% MnO, 13.39% MgO, 19.20% CaO, 0.76% Na_2O, 0.13% K_2O, 0.16% P_2O_5, LOI 1.42 total 99.99 (Pal'yanov et al., 2001).

Experiments were carried out at P = 5.7 GPa and T = 1420°C, P = 7.0 GPa and 1700°C using a "split-sphere" multi-anvil apparatus with a 300 mm outer diameter of the multi-anvil sphere (Pal'yanov et al., 1997). In this work we used a cell developed specifically for the growth of large diamond crystals. This type of cell has the shape of a tetragonal prism 19 × 19 × 22 mm, and utilizes a graphite heater 9 mm in diameter and 15.8 mm in length. Pressure was calibrated at room temperature by the change in resistance of Bi at 2.55 GPa and of PbSe at 4.0 and 6.8 GPa, and at high temperatures by the graphite-diamond equilibrium (Kennedy and Kennedy, 1976). Temperature was measured in each experiment with a Pt6%Rh/Pt30%Rh thermocouple whose junction was placed inside the heater. The thermocouple was calibrated at 7 GPa by the melting of Ni and at 5.7 GPa by the melting of Ag, Au, and Al (Pal'yanov et al., 2002a). Based on previously published data (Pal'yanov et al., 2002a), we estimated a ±0.1 GPa and ±20°C accuracy of pressure and temperature measurements, respectively, at 5.7 GPa and 1420°C. For experiments at 7 GPa and 1700°C the estimated accuracies were ±0.2 GPa and ±40°C, respectively.

Cylinders 6 mm in diameter and 2 mm high were produced from the metamorphic rocks. The cylinders had special holes drilled for seed crystals. Then, the samples with 0.5 mm high graphite tablets together with natural and synthetic diamond seeds were placed into platinum ampoules and sealed by arc welding. After the experimental runs, several ampoules were unsealed directly in a gas-chromatographic unit in order to analyze the fluid component using a technique described by Sokol et al. (2000). The phase composition of the samples was analyzed with a DRON-3 X-ray diffractometer. Mineral major element compositions were obtained on a CAMEBAX-MICRO electron microprobe. Newly formed diamond and graphite crystals were analyzed with an OMARS-89 Raman spectrometer.

Experimental Results

The "garnet-pyroxene rock (sample 2-4) + graphite" system after a 41-hour long experiment at P = 5.7 GPa and T = 1420°C (experiment No. 1) experienced little or no change (Table 2). As before the experiment, the specimen is composed of clinopyroxene, garnet, and graphite. Microprobe analysis showed that clinopyroxene and garnet are compositionally unchanged (Table 2). Optical examination revealed no changes at the graphite-pyroxenite interface. No traces of growth or dissolution were observed on the faces {100} and {111} of diamond seeds. A small amount of a gas phase, released on opening the ampoule, is chiefly water-CO_2 with insignificant contents of N_2, CH_4, and H_2.

The experiments in the "dolomite marble (sample 92-99) + graphite" system led to radically different results. After the experiment (experiment No. 2) at P = 5.7 GPa and T = 1420°C, dolomite, clinopyroxene, garnet, graphite, coesite, and diamond were detected among run products. As evident from Table 2, the phase composition drastically differs from the initial one. The majority of the specimen is a fine-grained aggregate composed of quenched phases, dolomite, and clinopyroxene. A slight offset and broadening of dolomite X-ray peaks is indicative of its transformation into a dolomite solid solution. According to microprobe analysis, the dolomite solid solution is 3–4 wt% CaO richer than dolomite from marble (Table 2). Large (up to 500 micrometers) well-faceted crystals of garnet are localized in the cold part of the ampoule. The garnet crystals are colorless, transparent, and contain inclusions of two types: numerous black platy

TABLE 2. Representative Analyses of Minerals and Inclusions in Garnet-Pyroxene Rocks and Dolomite Marbles after Experiments[1]

	2-4 (run 1)				92-99 (run 2)						92-99 (run 3)		
	gt	Cpx	gt_{1c}	gt_{1r}	cpx	inc_1	inc_2	inc_3	gt_2	inc_4	gt_c	gt_r	inc_4	inc_4
SiO_2	40.2	54.2	43.9	44.3	56.4	3.84	3.16	12.7	45.6	7.21	43.9	43.8	5.6	14.4
TiO_2		2.7			1.67	0.18	0.26	0.16	0.03	0.17		0.05	0.03	0.07
Al_2O_3	21.5	0.01	23.2	23.3	0.15	1.41	0.71	0.49	23.6	0.89	23.2	22.6	0.05	0.17
MnO	0.59	4.2	0.26	0.36	0.99	0.22	0.2	0.19	0.28	0.16	0.26	0.25	0.23	0.16
FeO	15.4	14.7	2.12	2.02	20.9	1.75	1.52	1.29	1.98	1.09	1.71	1.65	1.32	1.24
MgO	8.26	22.1	24.6	24.1	19.7	18.3	16.9	20.1	23.2	18.1	24	24.3	20.5	20.5
CaO	13.5	0.53	5.05	5.14	0.13	31	31.4	29.3	5.1	31.4	6.32	6.53	30.4	29
Na_2O	0.07	0.57	0.03	0.02	0.06	0.64	0.07	0.12	0.02	0.1	0.02	0.02	0.05	0.03
K_2O						0.1	0.26	0.23		0.43			0.02	0.02
P_2O_5							0.08	0.08					0.08	0.1
Total	99.5	99	99.2	99.2	100	57.4	54.6	64.7	99.8	59.6	99.4	99.2	58.3	65.7
Si	3.04	1.99	3.05	3.07	2				3.13		3.05	3.05		
Ti									0			0		
Al	1.92	0.12	1.9	1.9	0.07				1.91		1.9	1.85		
Mn	0.04		0.02	0.02					0.02		0.02	0.02		
Fe	0.97	0.13	0.12	0.12	0.03				0.11		0.1	0.1		
Mg	0.93	0.81	2.54	2.49	1.11				2.37		2.48	2.52		
Ca	1.09	0.87	0.38	0.38	0.75				0.38		0.47	0.49		
Na	0.01	0.04	0.04	0	0.01				0		0	0		
K		0.03			0									

[1]c = core; r = rim; run 2: $inc_{1,2}$ = inclusions in garnet; inc_3 = inclusions in cpx; inc_4 = inclusion in coesite.

crystals of graphite up to 50 micrometers in size (identified by Raman spectroscopy) and single melt inclusions up to 80–100 micrometers in size. Melt inclusions are also present in pyroxene and newly formed coesite (Table 2). The melt inclusions represent fine-grained aggregates of recrystallized carbonatite melt. Compared to the initial composition of the dolomite marble, the melt has low Si and high Fe/Mg and K. In addition, the melt has a higher Ca/Mg value compared to the dolomite from the natural rock. The lowest Mg melt inclusions are found in coesite, where the influence of the Mg-rich matrix was minimal.

Newly formed diamond crystals (up to 100 micrometers in size) occur exclusively in the form of octahedra on the walls of the ampoule and immediately in the carbonate-silicate fine-grained aggregate at the contact with graphite. In the last case, the size of octahedral diamonds does not exceed 40 micrometers.

According to the X-ray data, the specimen of dolomite marble (92-99) after experiment (No. 3) at P = 7 GPa and T = 1700°C mainly consists of dolomite plus clinopyroxene, garnet, and diamond. As in the previous experiment, well-faceted crystals of garnet formed in the cold part of the ampoule. Abundant inclusions of diamond octahedra about 5–10 micrometers in size (Fig. 3) and their intergrowths are present as inclusions in garnets, and were identified by Raman spectroscopy. In addition, some garnets contained single-melt inclusions as well as melt inclusions containing diamond (Fig. 4). About 70% of the initial graphite was transformed to a fine-grained aggregate of diamonds, with octahedral crystals up to 150 micrometers in size. According to chromatographic analysis, the gaseous phase in the ampoule is composed chiefly of water with minor CO_2. The amount of the gaseous phase in this case is one and a half orders of magnitude higher than in the experiment with the garnet-pyroxene rock.

Discussion

Within the last decade, the significance of carbonates in mantle metasomatism has been widely discussed (Sweeney, 1994; Wyllie, 1995; Wang et al., 1996; Wood et al, 1996; Luth, 1999). Carbonate inclusions have been found in diamond cuboids and coated diamonds (Chrenco et al., 1967; Navon et al., 1988; Wamsley and Lang, 1992; Wang et.al., 1996; Schrauder and Navon, 1994) as well as in metamorphic diamonds (De Corte et al., 1998). The catalytic

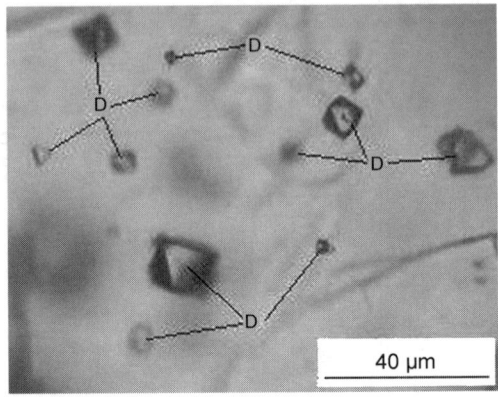

FIG. 3. Inclusions of diamond in newly formed garnet (experiment 3).

function of carbonates in promoting diamond crystallization has been studied in several experimental studies (Akaishi et al., 1990; Taniguchi et al., 1996; Litvin, 1998; Litvin et al., 1999; Pal'yanov et al., 1998, 1999, 2002 a,b).

The source of carbonates in the mantle is still under discussion. According to one hypothesis, carbonates can be carried to the mantle by subducted oceanic crust (Nishio et al., 1998). Another point of view considers that carbonates may be formed during carbonation of peridotites and eclogites by carbon dioxide carried from the lower lithosphere (Haggerty, 1999).

The occurrence of diamondiferous dolomite marbles in Kumdy-Kol proves the possibility of the subduction of carbonate rocks to the depth of diamond stability (Shatsky and Sobolev, 1995; Zhang et al., 1997; Ogasawara et al., 2000; Zhu and Ogasawara, 2002; Katayama et al., 2002). Our new experiments, combined with investigations of microdiamonds and mineralogical studies of garnet-pyroxene rock and dolomite marble, allow us to arrive at several conclusions.

The dolomite marble and garnet-pyroxene rocks contain grossular-rich garnet (up to 70% grossular component). As mentioned above, the composition of the garnet grossular content, which was newly formed in the experiments starting with dolomite marble, is different from that of the initial garnet. Yaxley and Green (1994) showed that the pyrope component in garnet, in equilibrium with the carbonate, increases with increasing temperature via the reaction:

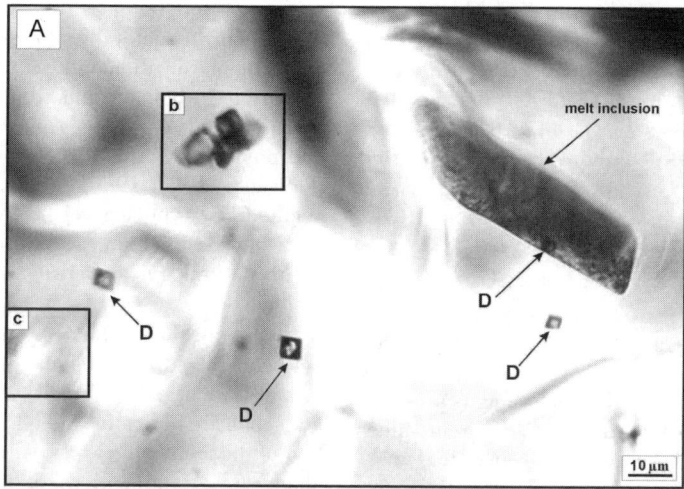

FIG. 4. Melt inclusion with diamond in newly formed garnet (experiment 3).

$$3Ca\ Mg(CO_3)_2 + Ca_3Al_2Si_3O_{12} =$$
$$6CaCO_3 + Mg_3Al_2Si_3O_{12}. \quad (1)$$

The grossular component contents are low in newly formed garnet (up to 13%) by comparison with garnet from the natural dolomite marble (53%). This suggests that the temperature of the experimental runs was significantly higher that the peak temperature of metamorphism.

It is particularly remarkable that the synthesized garnet contains majorite component (Table 2). Garnets from the Kumdy-Kol diamond-bearing rocks and eclogites lack a majorite component (Shatsky et al., 1995; Zhang et al., 1997). The synthesized garnet in association with carbonate contains up to 3% majorite component. Garnet in association with coesite contains up to 4% majorite component. We may assume that the majorite exsolved from the garnet solid solution during retrograde metamorphism of the diamond-bearing rocks,

but the garnet inclusions in zircon also contain no majorite components (Shatsky et al., 1995, Katayma et al., 2002). These experimental results show that the pressure of metamorphism was less than 50 kbar. So we doubt that the pressures during the metamorphism of diamond-bearing rocks exceeded 70 kbar (Ogasawara et al., 2000).

As mentioned above, coesite was detected among the run products. Coesite probably appears from the carbonation of pyroxene, which was experimentally studied by Luth (1995):

$$CaMgSi_2O_6 + 2\ CO_2 = CaMg(CO_3)_2 + 2\ SiO_2. \quad (2)$$

The curve for the grossular-carbonation reaction lies at a lower pressure for a given temperature than that of the diopside-carbonation reaction (Knoche et al., 1999). However the curve for garnet carbonation shifts to higher pressures with increasing pyrope content in garnet. As a result, garnet carbonation can begin after all clinopyroxene is exhausted. The coexistence of coesite, dolomite, diopside and diamond in the synthesized assemblages suggests that the oxidation state of the system was controlled by reaction (2). Ogasawara et al. (2000) reported that the stability of the assemblage Dol + Di + Dm (+Arg) in dolomite marble is determined by reaction (2) and reaction:

$$CaMgSi_2O_6 + 3\ CaMg(CO_3)_2 = 4\ CaCO_3 + 2\ Mg_2SiO_4 + 2\ CO_2. \quad (3)$$

Reaction (3) proceeds at a higher temperature than reaction (2). Ogasawara explained the absence of olivine in diamond-bearing rocks and its presence in diamond-free varieties as due to different values of CO_2 fugacity in those rocks. The curve for

$$C + O_2 = CO_2 \quad (4)$$

with decreasing X_{CO2} in the fluid shifts to lower oxygen fugacities. At the same values of oxygen fugacity, diamond is stable in rocks with a CO_2-rich fluid. Ogasawara et al. (2000) believed that low X_{CO2} was possibly responsible for garnet replacement by pyroxene-spinel symplectite in the diamond-free dolomite marble. Sobolev et al. (2001) studied a sample of banded carbonate rock with a 1 cm thick band containing garnet, pyroxene, dolomite, olivine, and clinohumite. Olivine is absent in other interlayers. Diamond inclusions occur in garnet from the garnet-pyroxene–rich interlayer adjacent to the olivine-bearing band. This garnet grain is rimmed by a pyroxene-spinel symplektite. Based on Ogasawara's model we should expect an extremely low fluid mobility at both prograde and retrograde stages of metamorphism. Although this suggestion is supported by an uneven distribution of diamonds in the rock, the occurrence of symplektite rims around garnet grains containing diamond inclusions contradicts the model of Ogasawara.

Shatsky et al. (1998, 1999b) showed that diamonds from garnet-pyroxene and carbonate rocks are cuboids with a radiating internal structure. The most recent models for diamond formation suggest that cubic diamonds are formed at lower temperatures than octahedra (Haggerty, 1999). Such a tendency was detected during experiments on diamond synthesis from metal-carbon systems (Giardini and Tydings, 1962). The diamonds we synthesized in the experiments with carbonate rocks are exclusively octahedra. If their morphology is a function of temperature, we would conclude that the experimental pressures and temperatures were different from those occurring naturally. However the results of the experiments on diamond synthesis in carbonate, fluid-rich carbonate, and fluid systems indicated that diamond morphology mainly depends on the composition of the medium of crystallization (Pal'yanov, 1998; 1999; 2002a). On this basis, the octahedral morphology of newly formed diamonds also may be explained by different compositions of the natural and experimental melts producing diamonds.

Although the problem of cubic diamonds has not been solved yet, abundant centers of diamond crystallization and the small size of the Kokchetav diamonds are indicative of high carbon supersaturation of the melt, which is possible only at an abrupt change of crystallization conditions. It should be noted that Shatsky et al. (1995) observed an octahedron and cuboid diamond in a single intergrowth in diamondiferous biotite gneisses.

Based on available models it is difficult to propose a path for such an abrupt change of P and T to create those variations in the degree of carbon supersaturation. A most probable mechanism is precipitation of diamond during fluid/rock interactions.

The isotope composition of diamond (δ^{13}‰ 10.2‰ to 10.5‰) is drastically different from that of carbonates ($\delta^{13}C$ +5.9‰ to +8.5‰) (Cartigny et al., 2001). Although the experimental data indicate fractionation of carbon isotopes during diamond formation as a result of decarbonation (Pal'yanov et al., 2002b), the amount of this fractionation cannot create the observed difference. This indicates that the source of carbon for diamond formation included not only carbonate minerals but organic carbon as well. Sedimentary organic matter has an isotope composition of $\delta^{13}C$ 25‰. Therefore, during the crystallization of diamonds, organic carbon interacted with H_2O-carbonate melts.

Graphite and diamond from the garnet-biotite gneisses have a lighter isotopic composition ($\delta^{13}C$ 17.5‰) compared to those from the pyroxene-carbonate rocks and dolomite marble (Pechnikov et al., 1993). Fluid from the silicate rocks interacted with the H_2O-CO_2 fluid of the carbonate rocks, resulting in a heavier isotopic composition of the diamonds. As previously mentioned, diamonds occur only in carbonate rocks with K-rich pyroxene or pyroxene with lamellae of K-feldspar and phengite. Therefore, we suggest that the natural diamonds crystallized under the interaction of carbonate rocks with K-rich fluids at Kumdy-Kol.

Our model for the formation of diamond takes into consideration the fact that carbonate and garnet-pyroxene rocks occur as interlayers and boudins in biotite-garnet gneisses. The model suggests an organic C–enriched metapelitic protolith for the biotite-garnet gneisses. Under UHP metamorphism of those rocks, a CH_4-K–rich fluid evidently was generated. The interaction of carbonate rocks with this fluid resulted in fluid oxidation and diamond precipitate via the reaction:

$$CH_4 + CO_2 = 2C + 2H_2O. \qquad (5)$$

K-rich pyroxene crystallized in equilibrium with that fluid. The precipitation of diamond raised f_{H2O} of the fluid. Taking into account the high temperature of UHP metamorphism, partial melting of dolomite marble and garnet-pyroxene rocks can occur.

In this connection, the low K_2O contents of the synthesized pyroxene also should be discussed. The bulk content of K_2O in pyroxene from carbonate rocks ranges from 0.5 to 1.5 wt%. The pyroxene used in the experiment contains 0.5 wt% in K_2O (sample 92-99). The melt/pyroxene partition coefficient for K exceeds 10 (Harlow, 1999), explaining the low K_2O content of the synthesized pyroxene. As mentioned above, the carbonatite melt inclusions contains about 0.7 wt% K_2O. Pyroxene in equilibrium with such a melt should contain less than 0.07 wt% K_2O, which is consistent with the obtained results. Therefore, we assume that the composition of the rock used in the experiment may be not absolutely the same as that of the fluid medium of pyroxene crystallization under UHP conditions.

Experimental data show that partial melting of carbonate-amphibole-bearing peridotite produces a H_2O-K_2O–rich melt (Wallace and Green, 1988). All K_2O partitions into the fluid coexisting with the carbonatite melt under high-pressure conditions (Wyllie, 1995). The most likely speculation is that hydrous fluid was removed from carbonate and garnet-pyroxene rocks during the partial melting. As mentioned above, some samples of carbonate rocks contain two pyroxene generations. The fine-grained pyroxene without potassium could crystallized from the intragranular carbonate-silicate melt after the removal of the K-rich fluid. The rim of porphyroblasts of K-rich pyroxene could continue their growth at the expense of that melt.

Thus, based on our new data, we suggest that diamond in the dolomite marble and pyroxene-carbonate rocks crystallized from a carbonatite melt in equilibrium with a K-rich fluid. The occurrence of diamond-bearing carbonate rocks supports the possibility of carbonate subduction to mantle depths in the diamond stability field. These deeply subducted carbonate and metapelites may be a potential source of high-K fluids, similar to those found in fluid inclusions in diamonds from kimberlites (Navon, 1999).

Acknowledgments

We thank D. G. Pearson for his critical reading of manuscript and useful comments. This research was partially supported by Grants 02-05-64632 and 03-05-65073 from the Russian Foundation for Basic Research and CRDF RG-2387-NO-02.

REFERENCES

Akaishi, M., Kanda, H., and Yamaoka, S., 1990, Synthesis of diamond from graphite-carbonate systems under very high temperature and pressure: Journal of Crystal Growth, v. 104, p. 578–581.

Cartigny, P., De Corte, K., Shatsky, V. S., Ader, M., De Paepe, P., Sobolev, N. V. and Javoy, M., 2001, The origin and formation of metamorphic microdiamonds from the Kokchetav massif, Kazakhstan: A nitrogen and carbon isotopic study: Chemical Geology, v. 176, p. 265–281.

Chrenko, R. M., McDonald, R. S., and Darrov, K. A., 1967, Infrared spectra of diamond coat: Nature, v. 213, p. 474–476.

De Corte, K., Cartigny, P., Shatsky, V. S., De Paepe, P., Sobolev, N. V., and Javoy, M., 1999, Characteristics of microdiamonds from UHPM rocks of the Kokchetav massif (Kazakhstan), in Gurney, J. J., Gurney, J. L., Pascoe, M. D., and Richardson, S. H., eds., VII International Kimberlite conference, v. 1: Cape Town, South Africa, Red Roof Design, p. 174–182.

De Corte, K., Cartigny, P., Shatsky, V. S., Javoy, M., and Sobolev, N. V., 1998, Evidence of inclusions in metamorphic microdiamonds from the Kokchetav Massif, northern Kazakhstan: Geochimica et Cosmochimica Acta, v. 62, p. 3765–3773.

Giardini, A. A., and Tydings, J. E., 1962, Diamond synthesis: Observations on the mechanism of formation: American Mineralogist, v. 11-12, p. 1393–1421.

Haggerty, S. E., 1999, A diamond trilogy: Superplumes, supercontinents, and supernovae: Science, v. 285, p. 851–860.

Harlow, G. E., 1999, Interpretation of Kcpx and CaEs components in clinopyroxene from diamond inclusions and mantle samples, in Gurney, J. J., Gurney, J. L., Pascoe, M. D., and Richardson, S. H., eds., VII International Kimberlite Conference, Cape Town, South Africa, Red Roof Design, p. 321–331.

Katayama, I., Ohta, M., and Ogasawara, Y., 2002, Mineral inclusions in zircons from diamond-bearing marble in the Kokchetav massif, northern Kazakhstan: European Journal of Mineralogy, v. 14, p. 1103–1108.

Kennedy, C. S., and Kennedy, G. C., 1976, The equilibrium boundary between graphite and diamond: Journal of Geophysical Research, v. 14, p. 2467–2470.

Knoche, R., Sweeney, R. J., and Luth, R. W., 1999, Carbonation and decarbonation of eclogites: The role of garnet: Contributions to Mineralogy and Petrology, v. 135, p. 332–339.

Litvin, Yu. A., 1998, Hot spots of mantle and experiment to 10 GPa: Alkaline reactions, lithosphere carbonatization, and new diamond-generating systems: Geologiya i Geofizika, v. 39, p. 1772–1779 (in Russian; English translation in Russian Geology and Geophysics, v. 39, p. 1760–1768.

Litvin, Yu. A., Aldushin, K. A., and Zharikov, V. A., 1999, Synthesis of diamond at 8.5–9.5 GPa in the system $K_2Ca(CO_3)_2$-$Na_2Ca(CO_3)_2$-C corresponding to the composition of fluid-carbonatitic inclusions in diamond from kimberlites: Doklady Akademii Nauk, v. 367, p. 529–532 (in Russian).

Luth, R. W., 1995, Experimental determination of the reaction dolomite + 2 coesite = diopside + 2 CO_2 to 6 GPa: Contributions to Mineralogy and Petrology, 1995, v. 122, p. 152–158.

Luth, R. W., 1999, Carbon and carbonates in mantle, in Mantle petrology: Field observation and high pressure experimentation: A tribute to Francis R. (Joe) Boyd: The Geochemical Society, Special Publication no. 6, p. 297–316.

Martovitskiy, V. P., Nadejdina, E. D., and Ekimova, T. E., 1987, Internal structure and morphology of non-kimberlitic diamonds: Mineralogicheskiy zhurnal, no. 9, p. 26–37 (In Russian).

Navon, O., 1999, Diamond formation in the Earth's mantle, in Gurney, J. J., Gurney, J. L., Pascoe, M. D., and Richardson, S. H., eds., VII International Kimberlite Conference, v. 2: Cape Town, South Africa, Red Roof Design, p. 546–554.

Navon, O., Hutcheon, I. D., Rossman, G. R., and Wasserburg, G. J., 1988, Mantle-derived fluids in diamond micro-inclusions: Nature, v. 335, p. 784–789.

Nishio, Y., Sasaki, S., Gamo, T., Hiyagon, H., and Sano, Y., 1998, Carbon and helium isotope systematics of North Fiji Basin basalt glasses: Carbon geochemical cycle in the subduction zone: Earth and Planetary Science Letters, v. 154, p. 127–138.

Ogasawara, Y., Ohta, M., Fukasava, K., Katayama, I., and Maruyama, S., 2000, Diamond-bearing and diamond-free metacarbonate rocks from Kumdy-Kol in the Kokchetav Massif, northern Kazakhstan: The Island Arc, v. 9, p. 400–416.

Pal'yanov, Yu. N., Khokhryakov, A. F., Borzdov, Yu. M., Sokol, A. G., Gusev, V. A., Rulov, G. M., and Sobolev, N. V., 1997, Growth conditions and real structure of synthetic diamond crystals: Geologiya i geofizika, v. 38, p. 882–918 (in Russian; English translation in Russian Geology and Geophysics, v. 38, p. 920–945).

Pal'yanov, Yu. N., Shatsky, V. S., Sokol, A. G., Tomilenko, A. A., and Sobolev, N. V., 2001, Crystallization of metamorphic diamond: An experimental modeling: Doklady Akademii Nauk, v. 381, p. 935–938 (in Russian).

Pal'yanov, Yu. N., Sokol, A. G., Borzdov, Yu. M., and Khokhryakov, A. F., 2002a, Fluid-bearing alkaline-carbonate melts as the medium for the formation of diamonds in the Earth's mantle: An experimental study: Lithos, v. 60, p. 145–159.

Pal'yanov, Yu. N., Sokol, A. G., Borzdov, Yu. M., Khokhryakov, A. F., and Sobolev, N. V., 1999, Diamond formation from mantle carbonate fluids: Nature, v. 400, p. 417–418.

Pal'yanov, Yu. N., Sokol, A. G., Borzdov, Yu. M., Khokhryakov, A. F., and Sobolev, N. V., 2002b, Diamond formation through carbonate-silicate interaction: American Mineralogist, v. 87, p. 1009–1013.

Pal'yanov, Yu. N., Sokol, A. G., Borzdov, Yu. M., and Sobolev, N. V., 1998, Experimental study of diamond crystallization in carbonate-carbon systems in connection with the problem of diamond genesis in magmatic and metamorphic rocks: Geologiya i geofizika, v. 39, p. 1780–1792 (in Russian; English translation in Russian Geology and Geophysics, v. 39, p. 1768–1779).

Pechnikov, V. A., Bobrov, V. A., and Podkuyko, Y. A., 1993, Isotopic compositions of diamond and accompanying graphite in North Kazakhstan metamorphic rocks: Geochemistry International, v. 30, p. 153–157.

Schrauder, M. and Navon, O., 1994, Hydrous and carbonatitic mantle fluids in fibrous diamonds from Jwaneng, Botswana: Geochimica et Cosmochimica Acta, v. 58, p. 761–771.

Shatsky, V. S., Jagoutz, E., Sobolev, N. V., Kozmenko, O. A., Parkhomenko, V. S., and Troesch, M., 1999a, Geochemistry and age of ultrahigh pressure metamorphic rocks from the Kokchetav massif (northern Kazakhstan): Contributions to Mineralogy and Petrology, v. 137, p. 185–205.

Shatsky, V. S., Rylov, G. M., Efimova, E. S., De Corte, K., and Sobolev, N. V., 1998, The morphology and real structure of microdiamonds from the Kokchetav massif metamorphic rocks, kimberlites and alluvial placers: Geologiya i geofizika, v. 39, p. 942–955 (in Russian; English translation in Russian Geology and Geophysics, v. 39, p. 949–961.

Shatsky, V. S., and Sobolev, N. V., 1993, Some specific features of the origin of diamonds in metamorphic rocks: Doklady Akademii Nauk, v. 331, p. 217–219 (in Russian).

Shatsky, V. S., Sobolev, N. V., and Stenina, N. G., 1985, Structural peculiarities of pyroxenes from eclogites: Terra Cognita, v. 5, p. 436–437.

Shatsky, V. S., Sobolev, N. V. and Vavilov, M. A., 1995, Diamond-bearing metamorphic rocks of the Kokchetav massif (northern Kazakhstan), in Coleman, R. G., and Wang, X., eds., Ultrahigh pressure metamorphism: Cambridge, UK, Cambridge University Press, p. 427–455.

Shatsky, V. S., Zedgenizov, D. A., Yefimova, E. S., Rylov, G. M., De Corte, K., and Sobolev, N. V., 1999b, A comparison of morphology and physical properties of microdiamonds from the mantle and crustal environments, in Gurney, J. J., Gurney, J. L., Pascoe, M. D., and Richardson, S. H., eds., VII International Kimberlite Conference, v. 2: Cape Town, South Africa, Red Roof Design, p. 757–763.

Sobolev, N. V., Schertl, H. P., Burchard, M., and Shatsky, V. S., 2001, Unusual pyrope-grossular-garnet and its paragenesis from diamond bearing carbonate-silicate rocks of the Kokchetav massif (northern Kazakhstan): Doklady Akademii Nauk, v. 380, p. 791–794 (in Russian).

Sobolev, N. V., and Shatsky, V. S., 1987, Inclusions of carbon minerals in the garnets from metamorphic rocks, Geologiya i geofizika, v. 7, p. 77–80 (in Russian; English translation in Russian Geology and Geophysics, v. 28, no. 7, p. 69–71).

Sobolev, N. V., and Shatsky, V. S., 1990, Diamond inclusions in garnets from metamorphic rocks: A new environment for diamond formation: Nature, v. 343, p. 742–746.

Sokol, A. G., Tomilenko, A. A., Pal'yanov, Yu. N., Borzdov, Yu. M., Pal'yanova, G. A., and Khokhryakov, A. F., 2000, Fluid regime of diamond crystallization in carbonate-carbon systems: European Journal of Mineralogy, v. 12, p. 367–375.

Sweeney, R. S., 1994, Carbonatite melt composition in the Earth's mantle: Earth and Planetary Science Letters, v. 128, p. 259–270.

Taniguchi, T., Dobson, D., Jones, A. P., Rabe, R., and Milledge, H. J., 1996, Synthesis of cubic diamond in the graphite-magnesium carbonate and graphite-$K_2Mg(CO_3)_2$ systems at high pressure of 9–10 GPa region: Journal of Materials Research, v. 11, p. 2622–2632.

Wallace, M. E., and Green, D. H., 1988, An experimental determination of primary carbonatite magma composition: Nature, v. 335, p. 343–346.

Walmsley, J. C., and Lang, A. R., 1992, On sub-micrometer inclusions in diamond coat: Crystallography0 and composition of ankerites and related rhombohedral carbonates: Mineralogical Magazine, v. 56, p. 533–543.

Wang, A., Pasteris, J. D., Meyer, H. O. A., and Dele-Duboi, M. L., 1996, Magnesite-bearing assemblage in natural diamond: Earth and Planetary Science Letters, v. 141, p. 293–306.

Wood, B. J., Pawley, A., and Frost, D. R., 1996, Water and carbon in the Earth's mantle: Philosophical Transactions: Mathematical, Physical, and Engineering Sciences, v. 354, p. 1459–1511.

Wyllie, P. J., 1995, Experimental petrology of upper mantle materials, processes, and products: Journal of Geodynamics, v. 20, p. 429–468.

Yaxley, G. M., and Green, D. H., 1994, Experimental demonstration of refractory carbonate-bearing eclogite and siliceous melt in subduction regime: Earth and Planetary Science Letters, v. 128, p. 313–325.

Zhang, R. Y., Liou, J. G., Ernst, W. G., Coleman, R. G., Sobolev, N. V., and Shatsky, V. S., 1997, Metamorphic evolution of diamond-bearing rocks from the Kokchetav massif, northern Kazakhstan: Journal of Metamorphic Geology, v. 15, p. 479–496.

Zhu, Y., and Ogasawara, Y., 2002, Carbon recycled into deep Earth: Evidence from dolomite dissociation in subduction-zone rocks: Geology, v. 30, p. 947–950.

The Role of Fluid for Diamond-Free UHP Dolomitic Marble from the Kokchetav Massif

YOSHIHIDE OGASAWARA[1] AND KAZUMASA AOKI

Department of Earth Sciences, Waseda University, Shinjuku-ku, Tokyo 169-8050, Japan

Abstract

A diamond-free, dolomite-bearing UHP marble showing a banded structure occurs at Kumdy-kol in the Kokchetav Massif, northern Kazakhstan. Three centimeter-scale zones occur in this marble (sample no. Y665): zone A = Ti-clinohumite–bearing dolomitic marble; zone B = dolomite marble; and zone C = dolomitic marble lacking Ti-clinohumite and forsterite. The occurrence of phlogopite and garnet lamellae in diopside in zone C suggests UHP metamorphism. Large-area chemical mapping covering about two-thirds of the sample surface was conducted. Ti-Kα X-ray images clearly indicate that TiO_2 is concentrated in zone A. Ca-Kα and Mg-Kα images showed that zones A and C have similar modal compositions of carbonates, but they have strong contrasts in TiO_2 contents and mineral paragenesis. Mineral assemblages in each zone were analyzed in terms of the CaO-(MgO, FeO)-TiO_2-SiO_2 compositional tetrahedron. In zone A, the aragonite-Ti-clinohumite tie-line was stable under UHP conditions, whereas it was unstable in zones B and C. This indicates that XCO_2 in zone A was lower than that in zones B and C. Infiltration of TiO_2-bearing aqueous fluid is implied on the basis of paragenetic relations, distribution of Ti-bearing phases and chemical zonation in Ti-clinohumite in zone A. The lack of microdiamond in this dolomite-bearing carbonate rock suggests extremely low-XCO_2 conditions, and implies the infiltration of an aqueous fluid during UHP metamorphism.

Introduction

SINCE THE FIRST report on microdiamond of crustal origin (Sobolev and Shatsky, 1990), the Kokchetav Massif of northern Kazakhstan has been the center of attention for many UHP petrologists. Particularly, metamorphic pressures greater than 5 GPa (e.g., Okamoto et al., 2000; Ogasawara et al., 2002) and extremely high concentrations of microdiamonds are nearly unique characteristics compared with other UHP terranes. In the Kokchetav UHP rocks, microdiamond is highly concentrated in some dolomite-bearing carbonate rocks; the highest concentration of microdiamond reaches 2700 carat/ton in some domains (Yoshioka et al., 2001). The strong contrast of the diamond occurrence between two types of dolomite-bearing UHP rocks has been described, and explained by compositional heterogeneity of the local fluid (such as XCO_2) during UHP metamorphism (Ogasawara et al., 2000).

A typical rock is diamond-free Ti-clinohumite-bearing dolomitic marble, of which mineral assemblages were stable at extremely low XCO_2 conditions under UHP. One immediate question is why this carbonate rock does not contain microdiamond, whereas the associated dolomite marble from the same locality is abundant in diamond. Considering such characteristic features of Kokchetav UHP marble, the relations between diamond stability and UHP fluid in carbonate systems had been discussed (Ogasawara, 2001; Ogasawara and Ishida, 2001). Although the occurrence of abundant microdiamonds has attracted attention, the lack of diamond in associated UHP marbles is equally important in order to evaluate the role of fluid for diamond formation in a deeply subducted carbonate system. One example of diamond-free UHP marbles is calcite marble that contains titanite with coesite exsolution needles; the Si content of the precursor titanite yields a minimum pressure > 6 GPa (Ogasawara et al., 2002).

In order to understand the phase relations of diamond and silicate phases in a carbonate system, the control of UHP fluid compositions and significance of the lack of diamond in dolomite-bearing carbonate rocks, we selected a dolomite-bearing marble (no. Y665) for detailed study. This rock exhibits a cm-scale banded structure and consists of layers of Ti-clinohumite–bearing dolomitic marble, diamond-free dolomite marble, and Ti-clinohumite– and

---

[1]Corresponding author; email: yoshi777@waseda.jp

diamond-free dolomitic marble. The purpose of this paper is to describe petrogeneses of these marble bands and to discuss heterogeneity of metamorphic fluid compositions in terms of XCO_2. The abbreviations of mineral names are after Kretz (1983).

Geological Setting

The Kokchetav Massif is located in northern Kazakhstan, Central Asia, with a NW-SE length of at least 80 km and a 17 km width (Dobretsov et al., 1995; Kaneko et al., 2000). The 530 ± 10 Ma terrane (Claoue-Long et al., 1991; Katayama et al., 2003) has been divided into four units (I, II, III, and IV) (Kaneko et al., 2000). Metamorphic rocks in Kumdy-kol belong to the highest grade, Unit II, and consist mainly of pelitic-psammitic gneiss, white schist, eclogite, and minor orthogneiss; Ti-clinohumite-bearing garnetiferous ultramafic rock; and metacarbonate rock (Kaneko et al., 2000; Ogasawara et al., 2000; Muko et al., 2002). Many lines of evidence indicate UHP metamorphism for the Kumdy-kol rocks. For example, diamond in gneiss and marble, inclusions of coesite in eclogitic minerals, and exsolution lamellae of coesite in titanite and of K_2O-bearing phases in diopside in marble (e.g., Ogasawara et al., 2000, 2002; Okamoto et al., 2000; Katayama et al., 2002) all suggest the peak metamorphic P > 5 GPa and temperature 900–1,000°C.

Analytical Conditions

Chemical compositions of constituent minerals were analyzed using two kinds of electron microprobes: JEOL JXA-733 combined with Voyager EDS system (Noran Instrument Co. Ltd.) and JEOL JXA-8900. Operating conditions of JXA-733 for spot analyses by EDS mode were: accelerating voltage of 15 kV, beam current of 0.4 nA and beam diameter of 10 µm, and those of JXA-8900 for spot analyses by WDS mode were: accelerating voltage of 15 kV, beam current of 20 nA, and beam diameter of 10 µm. The JXA-8900 also used for chemical mapping, and its operating conditions are the same as spot analyses except for beam diameters of 10 and 20 µm, and a counting interval of 0.05 sec for each spot. Fine-grained minerals such as exsolution lamellae were identified with a laser Raman spectrometers, HORIBA JOBIN YVON LabRam system, using a 633 nm line of He-Ne laser at 17 mW, and a laser spot size of 1 µm.

Petrography and Mineral Chemistry

A diamond-free dolomite-bearing marble sample (no. Y665) collected at a waste site in Kumdy-kol (Fig. 1) was used in this study. This sample (about 4 cm × 2.4 cm in size) consists of three layers defined by modal compositions; Ti-clinohumite–bearing dolomitic marble (zone A), dolomite marble (zone B); and Ti-clinohumite–free dolomitic marble (zone C) (Fig. 2)

Zone A

This zone is Ti-clinohumite–bearing dolomitic marble with a granoblastic texture (Fig. 3A), consisting of dolomite (20 vol%), diopside (15 vol%), Mg-calcite (15 vol%), forsterite (< 10 vol%), Ti-clinohumite (10 vol%), and symplectite after garnet (Di + Spl + Mg-Cal) (10 vol%), with minor amounts of pyrite (< 2 vol%), pyrrhotite (< 5 vol%), and talc (< 5 vol%). This zone is characterized by occurrences of Ti-clinohumite and forsterite, and is the same type as dolomitic marble described by Ogasawara et al. (2000). No lamellae were found in diopside of this zone. Some Mg-calcite grains contain coarse-grained lath-shaped exsolved dolomite. Mg-calcite grains are a retrograde product from dolomite and aragonite during exhumation (Ogasawara et al., 1998, 2000). Ti-clinohumite occurs mainly in the matrix and up to 200 µm in long dimension; some are present as inclusions in the cores of symplectite after garnet.

Representative microprobe analyses of major constituent minerals are listed in Tables 1, 2, and 3. Coarse-grained Ti-clinohumite has a distinct chemical zonation from core to rim (Fig. 4 and Table 3); TiO_2 and fluorine contents of the core range from 1.45 to 1.71 wt% and from 0.76 to 1.63 wt%, respectively, and those of the rim range from 2.04 to 3.28 wt% and 0.55 to 1.47 wt%. Ti-clinohumite is the only TiO_2-bearing phase in this zone.

Zone B

This zone is a dolomite marble with granoblastic texture, consisting mainly of dolomite (30 vol%), Mg-calcite (< 10 vol%), diopside (25 vol%), symplectite (10 vol%), and phlogopite (< 10 vol%) with minor amounts of pyrite (< 2 vol%), pyrrhotite (< 5 vol%), ilmenite (< 1 vol%), and talc (< 1 vol%) (Fig. 3B). This zone is characterized by an abundant dolomite, diopside-dolomite assemblage, and lack of Ti-clinohumite and forsterite, and is similar to dolomite marble described by Ogasawara et al. (2000), except for the following characteristics: lack

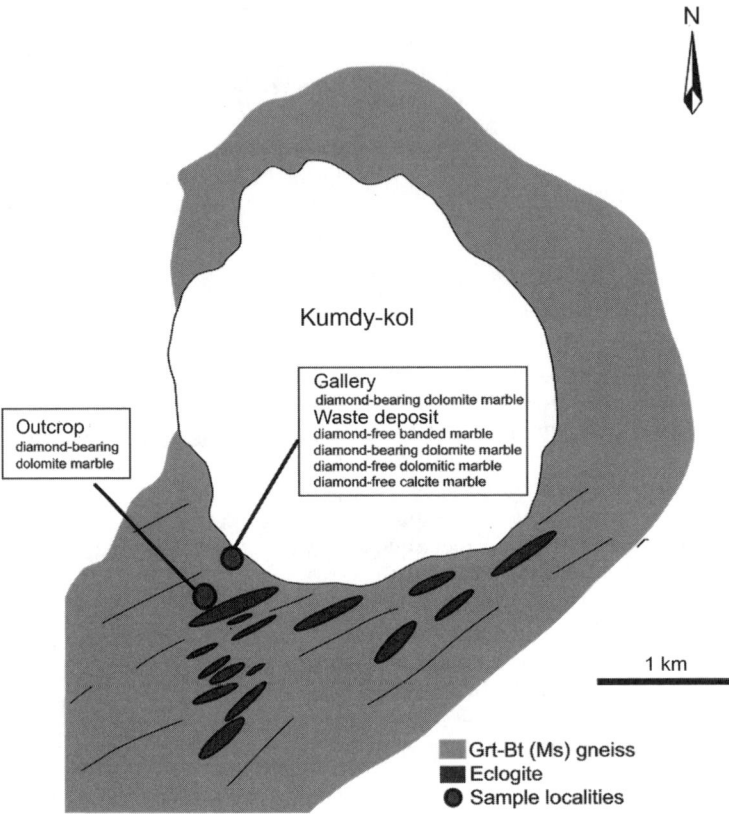

FIG. 1. Sample locality map of the marbles in Kumdy-kol.

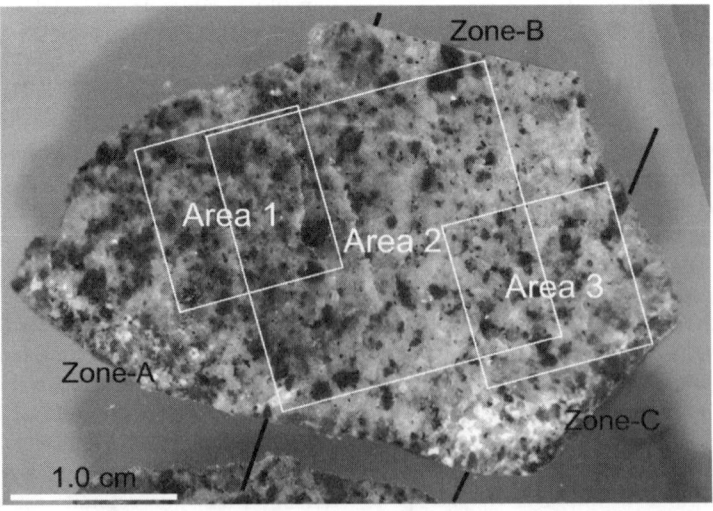

FIG. 2. Photograph of the polished surface of a diamond-free banded marble (sample no. Y665). Rectangles indicate the areas of chemical mapping (area 1 = 1.0 cm × 1.0 cm; area 2 = 1.7 cm × 1.7 cm; area 3 = 1.0 cm × 1.0 cm).

of diamond, presence of ilmenite instead of rutile, and no lamella in diopside. Some Mg-calcite grains contain coarse-grained exsolved dolomite of lath-like form. Representative microprobe analyses are listed in Tables 1, 2, and 3. Ilmenite grains occur only at the zone A side and contain 1 to 2 wt% of MgO (Table 3). No rutile occurs.

Zone C

This zone is a Ti-clinohumite–free dolomitic marble with granoblastic texture, consisting mainly of Mg-calcite (60 vol%), diopside (15 vol%), garnet (15 vol%), and pyrrhotite (5 vol%) with minor amounts of ilmenite (<1 vol%), phlogopite (<2 vol%), and pyrite (<2 vol%) (Fig. 3C). This zone is characterized by Mg-calcite with coarse-grained exsolved lath-like dolomite, diopside with exsolution lamellae of phlogopite and garnet, and lack of Ti-clinohumite and forsterite. Representative microprobe analyses are listed in Tables 1, 2, and 3.

Dolomitic marble of this zone is different from other dolomitic marbles described previously (Ogasawara et al., 2000). Both Ti-clinohumite and forsterite do not occur; no dolomite was found in matrix, and the dominant matrix carbonate is Mg-calcite. Coarse-grained, exsolved, lath-like dolomite from Mg-calcite indicates that precursor Mg-calcite contains high $MgCO_3$, and the presence of significant amounts of dolomite before Mg-calcite formation. Such Mg-calcite grains are regarded as aragonite + dolomite, stable under UHP conditions (Ogasawara et al., 1998, 2000).

Several diopside grains contain exsolved phlogopite lamellae (Fig. 5A) confirmed by laser Raman spectra characterized by 358.2 cm^{-1} and 682.5 cm^{-1} peaks (Fig. 5C). Coarse-grained exsolved phlogopite was analyzed with an electron microprobe (Table 2). One diopside grain contains both garnet and phlogopite lamellae (Fig. 5B). Garnet lamellae were confirmed by an electron microprobe (Table 3) and by laser Raman spectrometry with distinct 362.4 cm^{-1}, 552.5 cm^{-1} and 902.8 cm^{-1} peaks (Fig. 5D). Ilmenite is the only TiO_2-bearing phase in this zone.

Large Area Chemical Mapping

In order to clarify distributions and textural relations of minerals, characteristic X-ray mapping for eight elements (Ca, Mg, Fe, Si, K, Ti, Al, and F) was conducted in areas covering two-thirds of the surface of this banded sample. Three mapped areas are shown in Figure 2; area 1 (1.0 cm × 1.0 cm), area 2 (1.7 cm × 1.7 cm), and area 3 (1.0 cm × 1.0 cm).

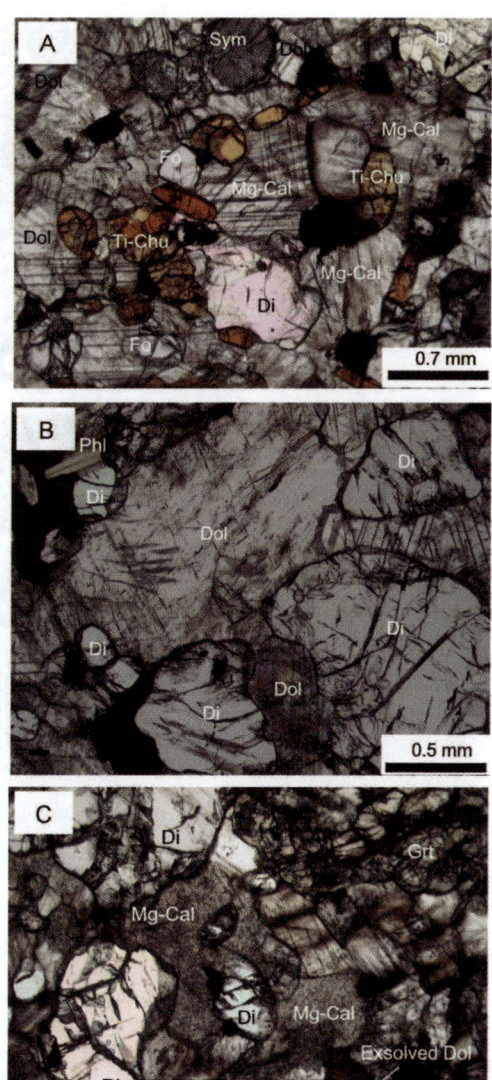

FIG. 3. Photomicrographs of each zone of diamond-free banded marble. A. Ti-clinohumite–bearing dolomitic marble (Zone A). B. Dolomite marble (Zone B). C. Ti-clinohumite–free dolomitic marble (Zone C)

Characteristic X-ray images of Ca-Kα, Ti-Kα, Mg-Kα, and Si-Kα are shown in Figure 6.

Ca-Kα

In Ca-Kα images, the green area corresponds to dolomite, orange and red areas to Mg-calcite, and light blue to diopside. Garnet and symplectite are

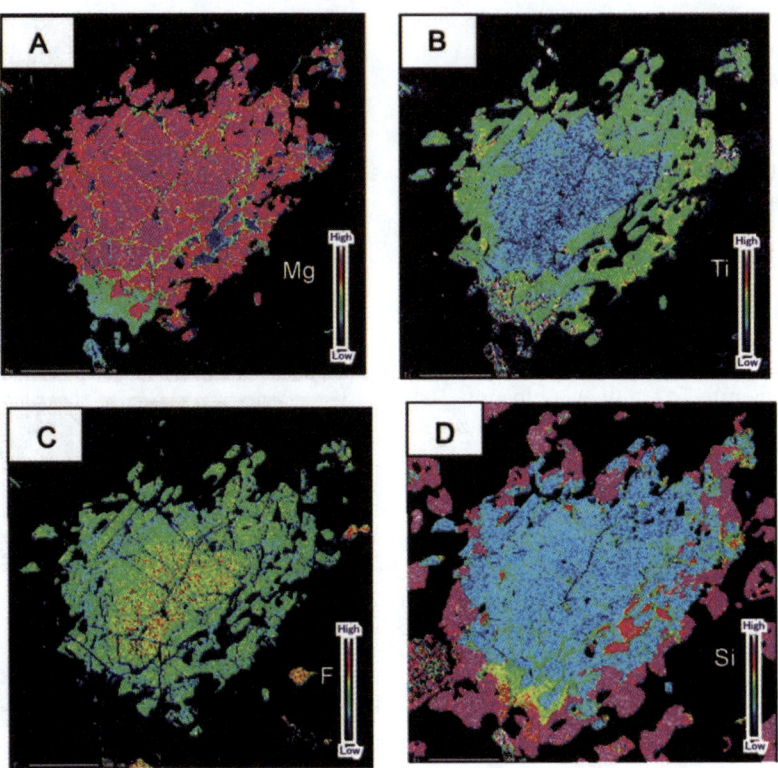

FIG. 4. Characteristic X-ray images of Ti-clinohumite in zone A. A. Mg-Kα. B. Ti-Kα. C. F-Kα. D. Si-Kα. Vertical color bars in the lower right corner of the figures show intensities of characteristic X-rays.

shown by dark blue (Fig. 6A). This map shows different modal contents of Mg-calcite and dolomite in each zone and the boundaries between three zones. The modal compositions of carbonates are: zone A = Mg-calcite> dolomite; zone B = dolomite >>Mg-calcite; and zone C = Mg-calcite only. The distribution of Mg-calcite is heterogeneous in zone B; this phase is restricted mainly along the zone C side.

Ti-Kα

This image clearly shows that TiO_2 is concentrated in zone A as Ti-clinohumite (shown by light blue and dark blue), and the different Ti-bearing phases between zones A, B, and C (Fig. 6B). Most Ti-clinohumite grains show increasing TiO_2 from core to rim (e.g., Fig. 4B). In contrast, zones B and C contain small amounts of ilmenite, shown as white. Ilmenite in zone B is limited only near the boundary with zone A.

Mg-Kα

This image shows the distribution of phlogopite (green), forsterite (red), Ti-clinohumite (red), diopside (blue), and dolomite (blue) (Fig. 6C). Ti-clinohumite and forsterite occur only in zone A. The distribution of phlogopite was heterogeneous; many grains are concentrated in some layers near the boundary between zone B and zone C.

Si-Kα

The Si-Kα image shows distributions of silicate minerals; diopside (red), Ti-clinohumite (light blue and dark blue), phlogopite (light blue), forsterite (green), symplectite (green), and garnet (light blue) (Fig. 6D). The distribution of diopside is homogeneous. Forsterite and symplectite are distributed only in zones A and B. Exsolution lamellae in some diopside grains of zone C are not recognized in the Si-Kα image because their grain sizes are too fine.

TABLE 1. Microprobe Analyses of Carbonate Minerals

Mineral:	Dolomite								Mg-calcite					
Zone:	A	A	B	B	A	A	C	C	A	A	B	B	C	C
Occurrence:	Matrix	Matrix	Matrix	Matrix	Lamella in Mg-cal	Lamella in Mg-cal	Lamella in Mg-cal	Lamella in Mg-cal	Matrix	Matrix	Matrix	Matrix	Matrix	Matrix
MgO	19.86	20.46	20.23	19.73	20.07	20.09	19.67	19.23	5.07	4.17	3.26	4.18	3.23	4.55
CaO	30.32	30.19	29.90	30.12	30.56	30.30	30.55	30.09	50.05	50.69	51.91	51.08	51.92	49.85
MnO	0.18	0.13	0.16	0.08	0.24	0.14	0.07	0.15	0.08	0.15	0.17	0.12	0.18	0.22
FeO*	0.71	0.84	1.03	1.17	0.83	0.91	1.30	1.87	0.31	0.32	0.44	0.41	0.62	0.72
CO_2**	46.29	46.02	46.17	46.28	45.98	46.11	46.03	46.16	44.22	44.30	44.10	44.10	44.03	44.30
Total	97.36	97.64	97.49	97.38	97.68	97.55	97.62	97.50	99.73	99.63	99.88	99.89	99.98	99.64
$XMgCO_3$	0.471	0.479	0.477	0.469	0.471	0.473	0.464	0.458	0.123	0.102	0.080	0.101	0.079	0.111
$XCaCO_3$	0.517	0.508	0.507	0.514	0.515	0.513	0.518	0.515	0.872	0.891	0.912	0.891	0.910	0.876
$XMnCO_3$	0.002	0.002	0.002	0.001	0.003	0.002	0.001	0.002	0.001	0.002	0.002	0.002	0.003	0.003
$XFeCO_3$	0.010	0.011	0.014	0.016	0.011	0.012	0.017	0.025	0.004	0.004	0.006	0.006	0.008	0.010

* Total Fe as FeO.
** Calculated on the basis of carbonate stoichiometry.

TABLE 2. Microprobe Analyses of Silicate Minerals

Mineral:	Diopside						Forsterite					Garnet				Phlogopite				
Zone:	A	A	B	B	C	C	A	A	A	C	C	B	C	C	C² Lamella in diopside	B	B	C	C	C² Lamella in diopside
Occurrence:	Matrix	Matrix	Matrix	Matrix	Matrix	Matrix	Matrix	Matrix	Matrix	Matrix	Matrix	Matrix	Matrix	Matrix		Matrix	Matrix	Matrix	Matrix	
SiO_2	54.76	54.83	54.55	54.58	54.51	55.01	41.26	41.43	41.41	40.09	40.30	40.37	40.3	40.87	39.52	40.03	40.60	40.2	38.3
TiO_2	0.07	0.03	0.13	0.13	0.08	0.02	0.01	0.04	0.04	0.34	0.28	0.27	0.2	0.52	0.51	0.56	0.56	0.9	0.9
Al_2O_3	0.35	0.35	0.83	0.59	0.34	0.52	0.00	0.01	0.00	21.50	21.56	21.74	20.7	16.17	16.95	16.99	16.36	18.0	16.2
Cr_2O_3	0.05	0.00	0.00	0.00	0.03	0.00	0.00	0.00	0.00	0.03	0.06	0.04	0.1	0.00	0.01	0.02	0.00	0.0	0.0
FeO^1	0.59	0.55	1.01	1.04	0.90	1.46	4.65	5.11	5.18	10.73	11.20	11.21	11.9	3.03	2.58	4.25	4.06	3.7	4.4
MnO	0.07	0.00	0.03	0.02	0.01	0.04	0.20	0.22	0.21	0.67	0.65	0.65	1.1	0.00	0.09	0.01	0.01	0.4	0.0
MgO	18.51	18.39	17.77	18.05	17.89	17.56	53.40	53.14	52.99	10.41	10.59	10.59	8.1	25.60	24.46	23.94	24.31	25.3	24.5
CaO	25.55	25.70	25.34	25.49	25.54	25.26	0.02	0.03	0.05	16.44	15.73	15.66	17.6	0.03	0.08	0.00	0.04	0.6	0.9
K_2O	0.01	0.04	0.13	0.01	0.00	0.03	0.00	0.02	0.00	0.00	0.02	0.01	0.0	10.23	9.68	9.93	9.56	9.2	9.5
Na_2O	0.00	0.05	0.08	0.05	0.05	0.12	0.00	0.00	0.02	0.00	0.00	0.01	0.0	0.21	0.21	0.56	0.24	0.5	0.8
Total	99.96	99.94	99.87	99.95	99.35	100.02	99.54	100.00	99.90	100.20	100.39	100.55	100.0	96.66	94.09	96.29	95.74	98.8	95.5

Number of atoms:

	O = 6						O = 4			O = 12				O = 22					
Si	1.981	1.985	1.978	1.978	1.987	1.993	0.993	0.994	0.995	2.981	2.989	2.988	3.029	5.670	5.611	5.606	5.690	5.473	5.459
Ti	0.002	0.001	0.004	0.004	0.002	0.000	0.000	0.001	0.001	0.019	0.015	0.015	0.013	0.054	0.054	0.059	0.059	0.089	0.099
Al	0.015	0.015	0.035	0.025	0.015	0.022	0.000	0.000	0.000	1.884	1.885	1.896	1.841	2.644	2.836	2.804	2.702	2.883	2.718
Cr	0.001	0.000	0.000	0.000	0.001	0.000	0.000	0.000	0.000	0.001	0.004	0.002	0.004	0.000	0.002	0.002	0.000	0.000	0.000
Fe^{2+}	0.018	0.017	0.031	0.031	0.027	0.044	0.094	0.103	0.104	0.667	0.695	0.694	0.748	0.352	0.306	0.498	0.476	0.422	0.517
Mn	0.002	0.000	0.001	0.001	0.000	0.001	0.004	0.004	0.004	0.042	0.041	0.041	0.071	0.000	0.010	0.000	0.001	0.041	0.000
Mg	0.998	0.992	0.961	0.975	0.972	0.948	1.916	1.901	1.898	1.154	1.171	1.168	0.909	5.295	5.177	4.998	5.079	5.153	5.189
Ca	0.990	0.997	0.985	0.990	0.997	0.981	0.000	0.001	0.001	1.310	1.250	1.242	1.418	0.004	0.012	0.000	0.006	0.080	0.133
K	0.000	0.000	0.006	0.000	0.000	0.001	0.000	0.000	0.000	0.000	0.000	0.001	0.000	1.811	1.753	1.774	1.709	1.588	1.729
Na	0.001	0.003	0.006	0.003	0.004	0.008	0.000	0.000	0.001	0.000	0.000	0.001	0.000	0.058	0.058	0.152	0.064	0.121	0.207

[1] Total Fe as FeO.
[2] Analyses with JXA-733.

TABLE 3. Microprobe Analyses of Ti-Clinohumite and Ilmenite

Mineral:	Ti-clinohumite						Ilmenite			
Zone:	A	A	A	A	A	A	B	C	C	C
Occurrence:	Matrix core	Matrix core	Matrix core	Matrix rim	Matrix rim	Matrix rim	Matrix	Matrix	Matrix	Matrix
SiO_2	36.85	37.11	36.59	35.84	36.84	36.03	0.26	0.22	0.05	0.02	0.06
TiO_2	1.62	1.71	1.53	3.28	2.23	2.46	51.00	51.65	48.46	49.76	49.61
Al_2O_3	0.03	0.03	0.00	0.02	0.02	0.01	0.04	0.02	0.02	0.03	0.02
Cr_2O_3	0.02	0.01	0.00	0.00	0.04	0.00	0.04	0.00	0.00	0.00	0.01
FeO[1]	6.03	5.98	6.28	6.03	5.57	6.41	42.78	41.23	48.17	46.99	47.31
MnO	0.23	0.17	0.21	0.26	0.18	0.25	2.28	2.24	2.35	1.82	2.46
MgO	50.90	50.89	50.66	48.78	49.95	50.35	1.75	1.46	0.19	0.17	0.17
CaO	0.01	0.01	0.02	0.00	0.18	0.00	0.16	0.17	0.00	0.08	0.05
K_2O	0.00	0.00	0.00	0.00	0.00	0.00	0.00	0.00	0.00	0.00	0.00
Na_2O	0.00	0.01	0.00	0.00	0.00	0.00	0.00	0.00	0.00	0.00	0.00
F	1.06	1.08	1.45	0.83	1.27	0.60	–	–	–	–	–
Total	96.75	97.00	96.74	95.04	96.28	96.11	98.31	97.01	99.24	98.85	99.69
Number of atoms:			Si = 4						O = 3		
Si	4.000	4.000	4.000	4.000	4.000	4.000	0.007	0.006	0.001	0.000	0.002
Ti	0.132	0.139	0.126	0.275	0.182	0.205	0.977	0.997	0.946	0.967	0.959
Al	0.004	0.004	0.000	0.002	0.003	0.001	0.001	0.001	0.001	0.001	0.001
Cr	0.001	0.001	0.000	0.000	0.004	0.000	0.001	0.000	0.000	0.000	0.000
Fe^{2+}	0.547	0.539	0.574	0.563	0.506	0.595	0.911	0.885	1.045	1.015	1.017
Mn	0.021	0.016	0.019	0.024	0.017	0.023	0.049	0.049	0.052	0.040	0.054
Mg	8.237	8.177	8.256	8.116	8.083	8.333	0.066	0.056	0.007	0.006	0.007
Ca	0.002	0.001	0.002	0.000	0.021	0.000	0.004	0.005	0.000	0.002	0.001
K	0.000	0.000	0.000	0.000	0.000	0.000	0.000	0.000	0.000	0.000	0.000
Na	0.000	0.002	0.000	0.000	0.000	0.000	0.000	0.000	0.000	0.000	0.000
F	0.365	0.368	0.501	0.294	0.436	0.211	–	–	–	–	–
O	17.079	17.019	17.103	17.257	17.001	17.364	–	–	–	–	–

[1]Total Fe as FeO.

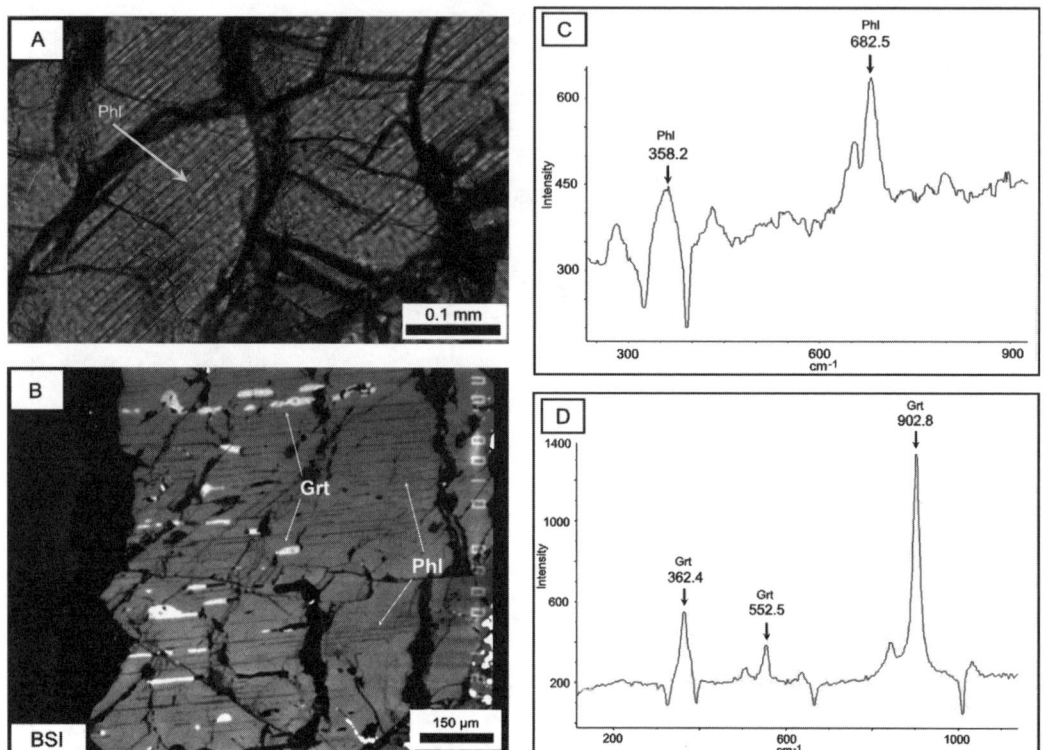

FIG. 5. Photomicrograph, backscattered electron image and laser Raman spectra of exsolution textures in diopside. A. Photomicrograph of diopside with phlogopite exsolution lamellae. B. Backscattered electron image of diopside with garnet and phlogopite exsolution lamellae. C. Raman spectra of phlogopite lamella. D. Raman spectra of garnet lamella.

Discussions

Evidence of UHP conditions

Exsolution textures of K_2O-bearing phases in clinopyroxenes have been reported in Kokchetav UHP rocks (e.g., Katayama et al., 2000; Okamoto et al., 2000; Ogasawara et al., 2000, 2002). Such texture indicates that precursor clinopyroxene contained a significant amount of K_2O under UHP conditions, consistent with the experimentally determined K_2O solubility in clinopyroxene (e.g., Harlow, 1997; Luth, 1997; Okamoto and Maruyama, 1998); it increases with pressure, and depends on bulk compositions and mineral assemblages. Diopside in zone C marble has phlogopite lamellae, but not in other zones A and B. This will be discussed later.

Exsolution lamellae of garnet in pyroxene have been reported in UHP rocks (e.g., Zhang et al., 1994; Zhang and Liou, 2000, 2003). Zhang and Liou (2003) described grossular-rich garnet exsolution from clinopyroxene in clinopyroxenite from the Sulu UHP terrane, China. They proposed two hypotheses for the precursor phase: (1) clinopyroxene, and (2) majoritic garnet. One diopside grain from zone C contains both garnet and phlogopite lamellae (Fig. 5B). Inasmuch as the garnet lamellae in our sample are thicker than phlogopite lamellae, and the host diopside surrounding garnet lamellae lacks phlogopite, garnet may have exsolved prior to the phlogopite exsolution. The garnet lamellae have similar compositions with matrix garnet, but have slightly lower pyrope contents. Although we cannot constrain P-T conditions for precursor clinopyroxene and for the two exsolution stages, the occurrence of these lamellae indicates UHP conditions.

Mineral assemblages in the model system CaO-(MgO, FeO)-TiO_2-SiO_2-CO_2-H_2O and heterogeneous XCO_2 conditions

Ogasawara et al. (1998) conducted a Schreinemakers' analysis of an invariant assemblage of

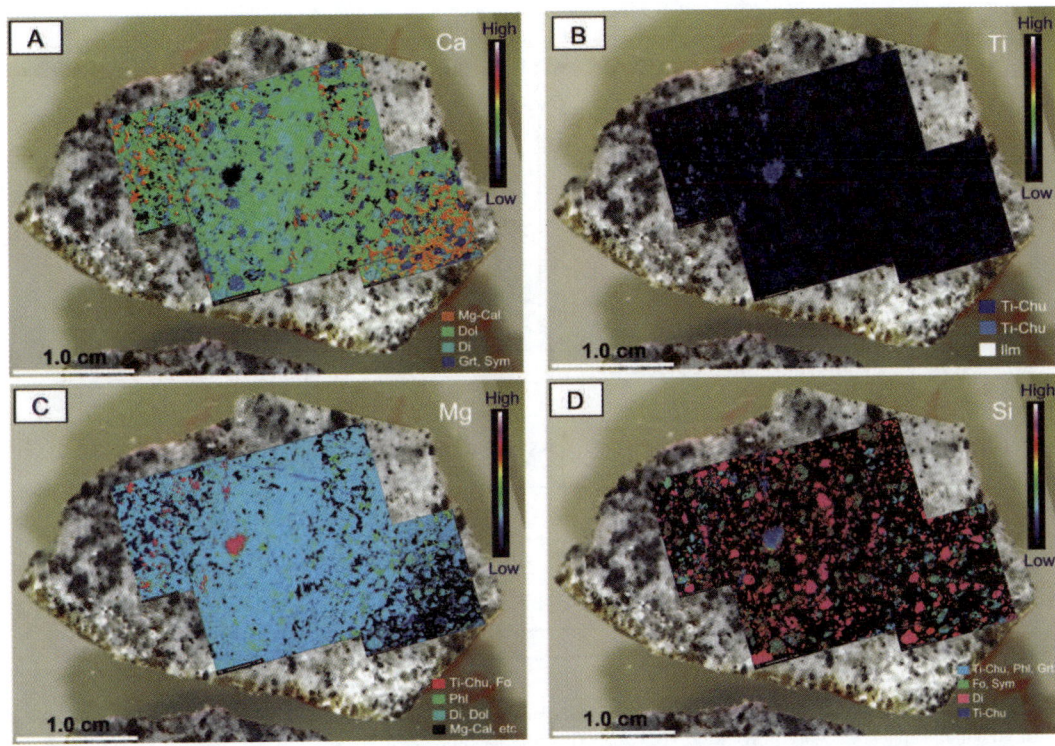

FIG. 6. Characteristic X-ray images of sample no. Y665. A. X-ray images of Ca-Kα. B. X-ray images of Ti-Kα. C. X-ray images for Mg-Kα. D. X-ray images for Si-Kα. Vertical color bars in the lower right corner of the figures show the intensities of characteristic X-rays.

Arg-Dol-Chu-Fo-Di in T-XCO_2 space for the model system $CaO-MgO-SiO_2-CO_2-H_2O$ in order to explain the parageneses of minerals in a Sulu UHP dolomitic marble (Fig. 7). In their analysis, TiO_2 was ignored, hence clinohumite was projected into the $CaO-MgO-SiO_2$ triangle and the aragonite-clinohumite tie-line was incompatible with the dolomite-diopside tie-line. Because clinohumite contains a significant amount of TiO_2, it should be plotted above the $CaO-MgO-SiO_2$ plane in the $CaO-MgO-TiO_2-SiO_2$ tetrahedron. Therefore, the aragonite-Ti-clinohumite tie-line becomes compatible with the dolomite-diopside join, and the four-phase assemblage Arg-Dol-Di-Ti-Chu becomes stable. Taking such a TiO_2 effect into account, we use the compositional tetrahedron of $CaO-(MgO, FeO)-TiO_2-SiO_2$ of Figure 8 to illustrate mineral assemblages for the three zoned marbles.

To simplify stability relations of constituent minerals in the banded marble, garnet was ignored from mineral assemblages in this system, inasmuch as garnet is a ubiquitous minor phase in all three zones. Compositions of six minerals (aragonite, dolomite, diopside, Ti-clinohumite, forsterite, and ilmenite) are plotted in the tetrahedron $CaO-(MgO, FeO)-TiO_2-SiO_2$ (Fig. 8). The TiO_2 component of Ti-clinohumite is emphasized in Figure 8.

Mineral assemblages in zones A, B, and C are interpreted in this model system. Zone A is characterized by the ubiquitous occurrence of Ti-clinohumite in entire zone A (Fig. 6B) and the stable coexistence of aragonite and Ti-clinohumite. However, it is difficult to determine whether the aragonite-forsterite or diopside-dolomite pair was stable. In zone A, the distribution of aragonite-forsterite is limited, and is less common compared with the diopside-dolomite pair. Textural relations indicate that aragonite-forsterite was (1) either stable in small domains or (2) occurs as a retrograde assemblage. However, no conclusive textural evidence for retrograde reaction was observed.

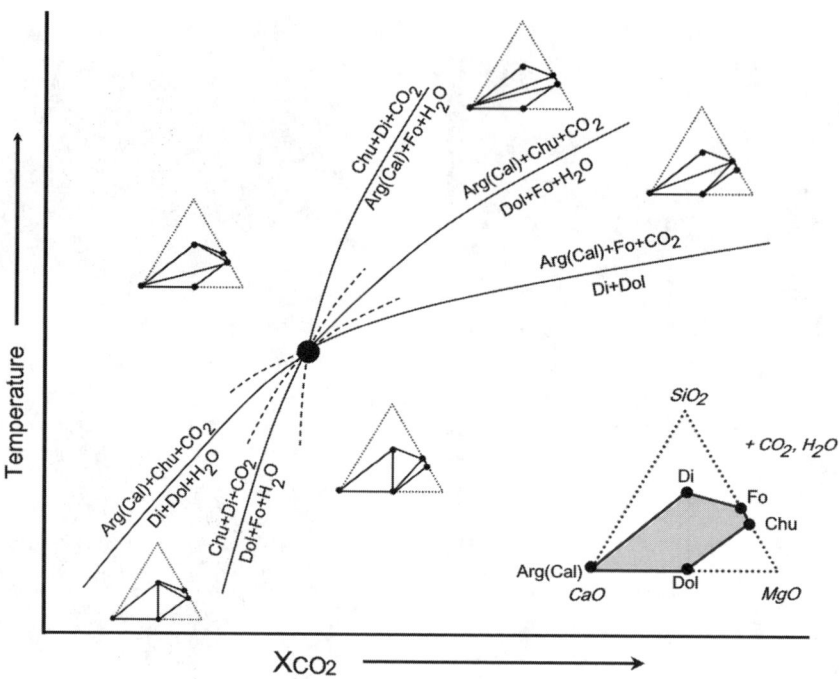

FIG. 7. Schematic T-XCO_2 diagram showing Schhreinemakers' bundle for the assemblage Arg (Cal) + Dol + Chu + Di + Fo in the system CaO-MgO-SiO_2-CO_2-H_2O (after Ogasawara et al., 1998).

If diopside-dolomite was stable instead of aragonite-forsterite together with aragonite-Ti-clinohumite, and the TiO_2-excess phase is ilmenite, the following four assemblages are possible in the tetrahedron: (1) Arg-Dol-Di-Ti-Chu; (2) Dol-Di-Fo-Ti-Chu; (3) Arg-Dol-Ti-Chu-Ilm; and (4) Arg-Di-Ti-Chu-Ilm. The ilmenite-bearing assemblages (3) and (4) do not appear in zone A. Forsterite is stable only in rocks with MgO-excess bulk compositions; aragonite does not appear. Therefore, if we assume that the diopside-dolomite pair was stable, only Arg-Dol-Di-Ti-Chu (Fig. 9A) was a stable assemblage in zone A, and this is consistent with the occurrence of the most abundant assemblage of Arg-Dol-Di-Ti-Chu in zone A. Based on the Schreinemakers' analysis in T-XCO_2 space for TiO_2-free system and the equilibrium T-XCO_2 conditions for Ti-Chu-free reactions (Fig. 7), the assemblage Arg-Dol-Di-Ti-Chu requires low-XCO_2 conditions, constrained by the following reaction (right-hand side was stable at low-XCO_2):

$$\text{Dol} + \text{Di} + TiO_2 \text{ (Ilm/fluid)} + H_2O = \text{Arg} + \text{Ti-Chu} + CO_2. \quad (1)$$

If we adopt the aragonite-forsterite pair as stable instead of diopside-dolomite in some portions of zone A, and XCO_2 values in these portions were different from other domains, the following assemblages are possible in the tetrahedron: (5) Arg-Dol-Fo-Ti-Chu, and (6) Arg-Di-Fo-Ti-Chu (Figs. 9B and 9C). Indeed, these two assemblages also occur in zone A. If we assume that XCO_2 was constant in zone A, the dolomite-diopside pair is not compatible with aragonite-forsterite. Hence, introduction of a compositionally heterogeneous fluid composition within zone A may be necessary to explain the observed assemblages. Alternatively, the formation of aragonite-forsterite assemblages in some domains of zone A during retrogression might be possible.

Zone B is characterized by the dolomite-diopside tie-line, the lack of Ti-clinohumite and forsterite, and the occurrence of ilmenite. The mineral assemblage Arg-Dol-Di-Ilm in the model tetrahedron (Fig. 9D) is stable at relatively high-XCO_2, compared with the aragonite-Ti-clinohumite pair constrained by reaction (1) in which the left-hand side assemblage was stable.

Zone C can be expressed by the same assemblage (Arg-Dol-Di-Ilm) as in zone B because Mg-

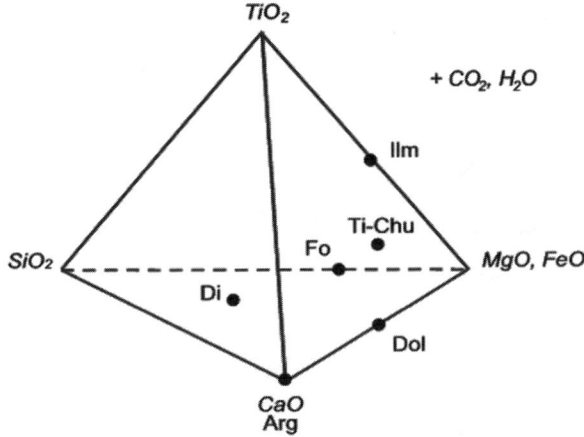

FIG. 8. Chemical compositions of six minerals projected onto the CaO-(MgO, FeO)-TiO_2-SiO_2 tetrahedron.

calcite (precursor assemblage aragonite + dolomite) with coarse-grained exsolved dolomite occurs in spite of the lack of matrix dolomite. The difference between zone B and C assemblages does not reflect a difference in tie-line relations of their mineral assemblages, but depends on the modal compositions of the carbonate (Fig. 6C).

As discussed above, the different mineral paragenesis in the studied zoned marble are controlled by two different tie-line relations. In zone A, aragonite-Ti-clinohumite is stable together with diopside-dolomite, whereas the pair aragonite-Ti-clinohumite is unstable in zones B and C. Such difference can be explained only by local difference in XCO_2. As shown in Figure 7, Ogasawara et al. (1998) pointed out that the aragonite-clinohumite pair is stable at very low XCO_2, and the diopside-dolomite association has a wide stability range in XCO_2 in the system CaO-MgO-SiO_2-CO_2-H_2O. Using this diagram, the zone A assemblage is stable at lower XCO_2 than mol fractions of carbon dioxide in zones B and C. This fact suggests the fluid heterogeneity at a cm scale due to different degrees of aqueous fluid infiltration during UHP metamorphism.

Lack of K_2O-bearing phase lamella in diopside in zones A and B

In Kokchetav UHP marbles, K_2O-bearing exsolved phases, including K-feldspar, phengite, and phlogopite in diopside, are common in diamond-bearing dolomite marble and calcite marble (Ogasawara et al., 2000, 2002). In the studied banded marble, only diopside of zone C contains phlogopite lamellae. These features have been described in other diamond-free dolomitic marbles from Kumdy-kol (Ogasawara et al., 2000). The question is why diopsides of diamond-free dolomitic marble lack K_2O-bearing lamellae, whereas all diopsides in diamond-bearing dolomite marble and diamond-free calcite marble exhibit K_2O-bearing lamellae textures. One possible explanation is very low bulk K_2O contents in zones A and B. According to their lamellae textures, hydroxyl also exsolved as phengite and phlogopite from diopside; these occurrences suggest significant amounts of hydroxyl are soluble in diopside together with K_2O. Based on P-T-XCO_2 stabilities of mineral assemblages, dolomitic marble requires extremely low-XCO_2 conditions compared with diamond-bearing dolomite marble (Ogasawara et al., 2000); XCO_2 is 0.01 in dolomitic marble and 0.1 in dolomite marble. Such XCO_2 conditions correspond to high H_2O activity in dolomitic marbles under UHP conditions, and suggest the possibility of leaching of K_2O from the marble to the fluid phase. However, high H_2O activity in dolomitic marble also suggests that hydroxyl still could be preserved in diopside.

Two-stage exsolution of K_2O-bearing phases has been reported in diopside in titanite-bearing calcite marble from Kumdy-kol; K-feldspar was exsolved during the first stage, and phengite at the second stage (Ogasawara et al., 2002). This indicates that hydroxyl was preserved in diopside at lower pressures than was the K_2O. Because the zone A assemblage formed at extremely low XCO_2 conditions,

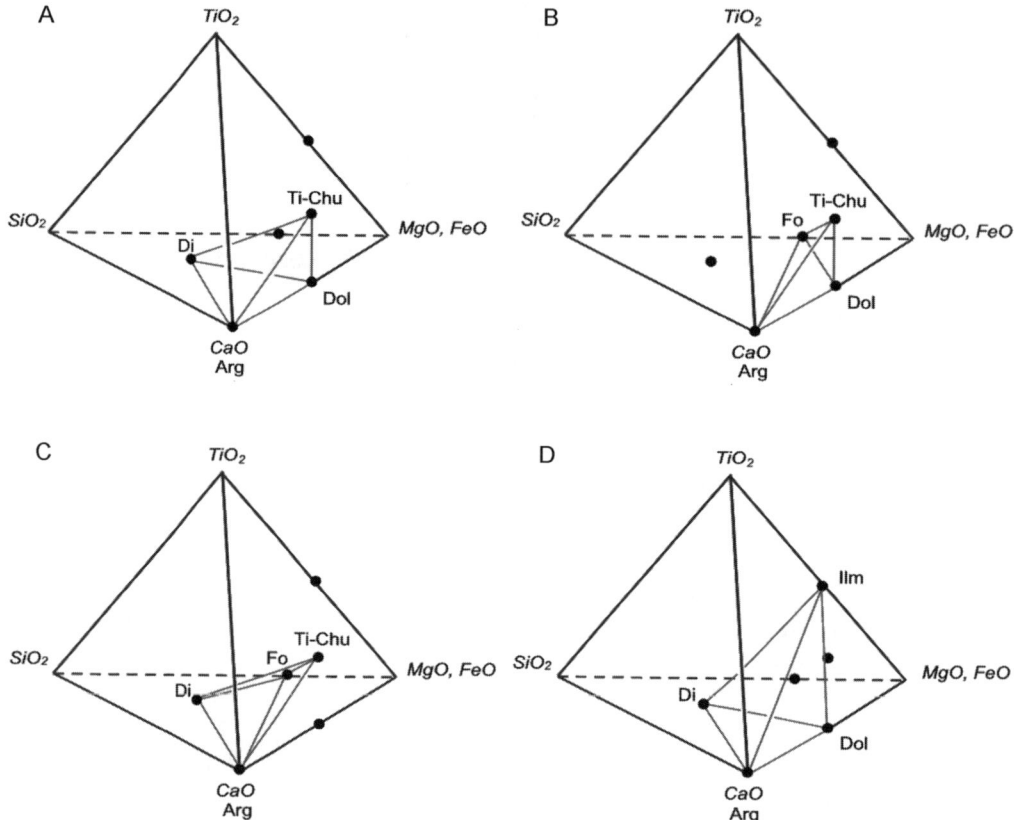

FIG. 9. Mineral assemblages projected onto the CaO-(MgO, FeO)-TiO$_2$-SiO$_2$ tetrahedron. A: Arg + Dol + Di + Ti-Chu. B: Arg + Dol + Fo + Ti-Chu. C: Arg + Di + Fo + Ti-Chu. D: Arg + Dol + Di + Ilm.

significant amounts of hydroxyl may be preserved in zone A diopside. Micro FT-IR spectra for such diopsides are expected to detect hydroxyl in diopside, as conducted by Katayama and Nakashima (2003).

Infiltration of aqueous fluid and transportation of TiO$_2$ during UHPM

One of the most important findings in UHP carbonate rocks is the dolomite-coesite assemblage from the Dabie Mountains, central China (Schertl and Okay, 1994; Zhang and Liou, 1996) as the lowest temperature assemblage generated in the coesite stability field during continental subduction that survived exhumation. Such an occurrence is an indicator of relatively dry conditions (low a_{H2O}) of the prograde stage of UHP metamorphism. Recently, Mukherjee et al. (2003) also reported magnesite-, aragonite/calcite-, and dolomite-bearing assemblages in coesite-bearing UHPM rocks from the Tso Morari complex, Ladakh, India.

In the Kokchetav Massif, carbonate rocks show strong fluid effects produced during UHP metamorphism. Occurrences of Ti-clinohumite–bearing assemblages in dolomitic marble, titanite-bearing marble, and the lack of coesite relics due to complete consumption of SiO$_2$ forming diopside in dolomite marble are a few examples. Katayama et al. (2002) reported coesite inclusion in zircon in diamond-bearing dolomite marble from Kumdy-kol (sample no. N21); coesite escaped diopside formation, and was preserved as inclusions trapped in zircon grain. These coesite inclusions are evidence that the diopside-forming reaction in dolomite marble was not completed in some domains even in the coesite stability field during prograde metamorphism, and this could have been caused by the weakness of aqueous fluid infiltration.

Our chemical mapping and microscopic observations have revealed a heterogeneous distribution of the Ti-bearing phases in this marble. As shown in Figure 6B, a Ti-bearing phase is concentrated in zone A as Ti-clinohumite (low-XCO_2 side). On the other hand, zones B and C (relatively high-XCO_2 side) contain less of the Ti-bearing phase. Such features could be explained by differences in bulk TiO_2 contents; however, another interpretation is that a TiO_2-bearing aqueous fluid was locally infiltrated in some layers to form zone A. Ayers and Watson (1993) experimentally determined TiO_2 solubility in aqueous fluid at 1.0–2.93 GPa, 800–1200°C, and showed that rutile precipitated from a fluid when activity of H_2O decreased. Their results support the latter interpretation, although the peak P condition at Kumdy-kol was much higher than their experimental pressures.

Ti-clinohumite is the only TiO_2-bearing phase as inclusions and in the matrix in zone A, and shows a distinct chemical zonation in TiO_2 content (Fig. 4B); the core is lower in TiO_2 than the rim. The formation of the TiO_2-rich rim of clinohumite cannot be explained by a bulk TiO_2 effect or by an isochemical process. Instead, the TiO_2 supply in zone A at the rim stage of Ti-clinohumite growth must be related to the infiltration of a TiO_2-bearing aqueous fluid.

Muko et al. (2002) described Ti-clinohumite–garnet peridotite at Kumdy-kol; Katayama et al. (2003) invoked metasomatism between the subducting slab and the mantle wedge due to the TiO_2-fluid infiltration for its formation at the 530 Ma UHP stage. Iizuka and Nakamura (1995) proposed metasomatism between the subducting slab and the mantle wedge in terms of TiO_2 behavior at 8 GPa. According to their experiment, TiO_2 is soluble in an aqueous fluid at UHP conditions because rutile becomes unstable in eclogite. This model provides one possible TiO_2 source for Ti-clinohumite–bearing dolomitic marble. Extremely low-XCO_2 conditions for Ti-clinohumite–bearing dolomitic marble is consistent with this model of mantle metasomatism by TiO_2-bearing aqueous fluid derived from the subducting slab. The aqueous fluid may have derived from dehydration of subducting eclogites and gneissose rocks associated with carbonate rocks and Ti-clinohumite garnet peridotite.

Significance of the lack of diamond

One of the critical characteristics of this banded sample is a lack of diamond in all zones. Microdiamond is common in dolomite marbles, as previously described (Yoshioka et al., 2001; Imamura et al., 2002; Ogasawara et al., 2000; Ishida et al., 2003). Almost all other dolomite marble samples from Kumdy-kol contain various amounts of microdiamond. Ishida et al. (2003) classified microdiamonds in Kokchetav dolomite marble into three types according to their morphologies and other characters; S-type (5–20 μm), R-type (5–20 μm), and T-type (1–7 μm). S-type microdiamond is characterized by a "star-shaped" morphology consisting of a single-crystal core and subhedral rim crystals. The R type has a translucent single crystal with a rugged surface, and is similar to the S-type core. The T type is a fine-grained transparent crystal with a smooth surface. They concluded that microdiamonds in dolomite marble were formed at two different stages according to the morphologies, micro-Laue diffraction, and laser Raman spectra of microdiamonds; the rim of S-type microdiamond was formed at the second stage. They suggested that fluid under UHP conditions was related to the rim formation of the S type. Composite inclusions of microdiamond with phengite in garnet also support a close relation of aqueous fluid to the formation of diamond. This two-stage growth has also been supported by cathodoluminescence of S-type microdiamond (Yoshioka and Ogasawara, 2005, in press). Several experimental studies demonstrated diamond formation from graphite and water system in C-H-O fluid mainly composed of H_2O and CO_2 under the diamond stability field (e.g., Kumar et al., 2000; Akaishi and Yamaoka, 2000), and natural diamond precipitated from supercritical C-H-O fluids in a carbonate system.

Kokchetav diamond-bearing carbonate rocks formed at higher XCO_2 condition than that in Ti-clinohumite–bearing dolomitic marble (zone A) and in dolomite marble of zone B and dolomitic marble of zone C. Why did dolomitic marble formed under extremely low-XCO_2 (~0.01) (such as aragonite-Ti-clinohumite stable) lack diamond, whereas dolomite marble formed at XCO_2 of 0.1 (Ogasawara et al., 2000) contains abundant diamond? Such a strong contrast indicates that the stability and formation of microdiamonds in a carbonate system is controlled by H_2O-rich fluid compositions. In extremely H_2O rich fluid, carbon species in the fluid could be lower than their solubility limits, and diamond could not have precipitated from such a a fluid. If fluid compositions were higher than $XCO_2 = 0.1$, carbon species could not be soluble in UHP aqueous fluid, and diamond was precipitated in dolomite marble.

Such an XCO_2 condition was too high to form the aragonite-Ti-clinohumite assemblage. In essence, microdiamonds were formed at low XCO_2 conditions (as 0.1), but were unstable at extremely low XCO_2 (as 0.01), at which aragonite-Ti-clinohumite was stable. Lack of diamond in zones B and C, in which aragonite-Ti-clinohumite and aragonite-forsterite pair were unstable, suggests that XCO_2 compositions in zones B and C marbles were located between those values of zone A and diamond-bearing dolomite marble. One possible source of carbon for rim microdiamond of the S type is soluble carbon species or a CO_2 component in the fluid.

Carbon isotopic compositions of S-type microdiamond are different, with heavier core and lighter rim; this supports the two-stage growth of S-type microdiamond (Imamura et al., 2003). However, such light carbon isotopic compositions for the rim of S-type diamond raise another mystery for the source of carbon for rim growth. At this moment, it is difficult to explain how such light ^{13}C compositions were possible under UHP conditions. The lack of diamond in UHP carbonates may indicate extremely low-XCO_2 conditions and be an indicator of aqueous fluid infiltration during UHP metamorphism.

Conclusions

Diamond-free dolomite-bearing marble occurs at Kumdy-kol in the Kokchetav UHP terrane, northern Kazakhstan. This marble consists of three different assemblage layers: Ti-clinohumite–bearing dolomitic marble (zone A); dolomite marble (zone B); and Ti-clinohumite–free dolomitic marble (zone C). UHP evidence is confirmed in zone C Ti-clinohumite–free dolomitic marble as phlogopite and garnet lamellae in diopside. Diopsides in the other two zones lack such exsolved lamellae.

The difference in mineral assemblages of the three zones can be explained by cm-scale local heterogeneity in fluid compositions (XCO_2) during UHP metamorphism. Ti-clinohumite-bearing marble indicates extremely low XCO_2, whereas zone B and C marbles require relatively high XCO_2. Large-area X-ray mapping strongly suggested that the source of TiO_2 in zone A could be aqueous fluid that infiltrated heterogeneously. The apparent TiO_2 zonation in Ti-clinohumite also suggests the infiltration of TiO_2-bearing aqueous fluid into some layers of the carbonate. Lack of microdiamond in dolomite-bearing carbonate rocks from the Kokchetav Massif is an indicator of extremely low XCO_2 fluid conditions during UHP metamorphism.

Acknowledgments

The paper was prepared to honor the retirement of W. G. Ernst, who has promoted many international cooperative projects. The authors thank J. G. Liou for his review and suggestions for improving this paper. This study was financially supported by the Grand in Aid of JSPS no. 13640485 and no. 15204050, and the Waseda University Grant for the Special Research Project no. 2001A-533.

REFERENCES

Akaishi, M., and Yamaoka, S., 2000, Crystallization of diamond from C-O-H fluids under high-pressure and high-temperature conditions: Journal of Crystal Growth, v. 209, p. 999–1003.

Ayers, J. C., and Watson, E. B., 1993, Rutile solubility and mobility in supercritical aqueous fluids: Contributions to Mineralogy and Petrology, v. 114, p. 321–330.

Claoue-Long, J. C., Sobolev, N. V., Shatsky, V. S., and Sobolev, A. V., 1991, Zircon response to diamond-pressure metamorphism in Kokchetav massif: Geology, v. 19, p. 710–713.

Dobretsov, N. L., Sobolev, N. V., Shatsky, V. S., Coleman, R. G., and Ernst, W. G., 1995, Diamondiferous paragenesis, northern Kazakhstan: The Island Arc, v. 4, p. 267–279.

Harlow, G. E., 1997, K in clinopyroxene at high pressure and temperature: An experimental study: American Mineralogist, v. 82, p. 259–269.

Iizuka, Y., and Nakamura, E., 1995, Experimental study of the slab-mantle interaction and implications for formation of titanoclinohumite at deep subduction zone: Proceedings of the Japan Academy, v. 71, p. 159–164.

Imamura, K., Ogasawara, Y., Yurimoto, Y., and Kusakabe, M., 2003, SIMS carbon isotope study of microdiamond in UHP dolomite marble from the Kokchetav Massif: EOS (Transactions of the American Geophysical Union), v. 84, p. F1532.

Imamura, K., Yoshioka, N., and Ogasawara, Y., 2002, Morphology and distribution of microdiamond in dolomite marble from Kumdy-kol: in Parkinson, C. D., Katayama, I., Liou, J. G., and Maruyama, S., eds., The diamond-bearing Kokchetav Massif, Kazakhstan: Tokyo, Japan, Universal Academy Press, p. 93–102.

Ishida, H., Ogasawara, Y., and Ohsumi, K., and Saito, A., 2003, Two-stage growth of microdiamonds in UHP dolomite marble from Kokchetav Massif, Kazakhstan: Journal of Metamorphic Geology, v. 21, p. 515–522.

Kaneko, Y., Maruyama, S., Terabayashi, H., Yamamoto, H., et al., 2000, Geology of the Kokchetav UHP-HP metamorphic belt, northern Kazakhstan: The Island Arc, v. 9, p. 264–283.

Katayama, I., Muko, A., Iizuka, T., Maruyama, S., Terada, K., Tsutsumi, Y., Sano, Y., Zhang, R. Y., and Liou, J. G., 2003, Dating of zircon from Ti-clinohumite-bearing garnet peridotite: Implication for timing metasomatism: Geology, v. 31, p. 713–716.

Katayama, I., and Nakashima, S., 2003, Hydroxyl in clinopyroxene from the deep subducted crust: Evidence for H_2O transport into the mantle: American Mineralogist, v. 88, p. 229–234.

Katayama, I., Ohta, M., and Ogasawara, Y., 2002, Mineral inclusions in zircon from diamond-bearing marble in the Kokchetav massif, northern Kazakhstan: European Journal of Mineralogy, v. 14, p. 1103–1108.

Katayama, I., Parkinson, C. D., Okamoto, K., Nakajima, Y., and Maruyama, S., 2000, Supersilicic clinopyroxene and silica exsolution in UHPM eclogite and pelitic gneiss from the Kokchetav massif, northern Kazakhstan: The Island Arc, v. 9, p. 417–427.

Kretz, R., 1983, Symbols for rock-forming minerals: American Mineralogist, v. 68, p. 277–279.

Kumar, M. D. S., Akaishi, M., and Yamaoka, S., 2000, Formation of diamond from supercritical H_2O-CO_2 fluid at high pressure and high temperature: Journal of Crystal Growth, v. 213, p. 203–206.

Luth, R. W., 1997, Experimental study of the system phlogopite-diopside from 3.5 to 17 GPa: American Mineralogist, v. 82, p. 1198–1209.

Mukherjee, B. K., Sachan, H. K., Ogasawara, Y., Muko, A., and Yoshioka, N., 2003, Carbonate-bearing UHPM rocks from the Tso-Morari region, Ladakh, India: Petrological implications: International Geology Review, v. 45, p. 49–69.

Muko, A., Okamoto, K., Yoshioka, N., Zhang, R. Y., Parkinson, C. D., Ogasawara, Y., and Liou, J. G., 2002, Petrogenesis of Ti-clinohumite-bearing garnetiferous ultramafic rocks from Kumdy-kol, in Parkinson, C. D., Katayama, I., Liou, J. G., and Maruyama, S., eds., The diamond-bearing Kokchetav massif, Kazakhstan. Tokyo, Japan, Universal Academy Press, p. 343–359.

Ogasawara, Y., 2001, Fluid control on metamorphic evolution of deeply subducted carbonates: The case of the Kokchetav UHP rocks [abs.]: Abstracts of the 11th Goldschmidt Conference, #3609.

Ogasawara, Y., Fukasawa, K., and Maruyama, S., 2002, Coesite exsolution from supersilicic titanite in UHP marble from the Kokchetav Massif, northern Kazakhstan: American Mineralogist, v. 87, p. 454–461.

Ogasawara, Y., and Ishida, H., 2001, Diamond, fluid and carbonates—stability and genesis of metamorphic diamond [abs.], in Fluid/slab/mantle interactions and ultrahigh-P minerals: Abstracts of UHPM Workshop 2001, Waseda University, p. 1–5.

Ogasawara, Y., Ohta, M., Fukasawa, K., Katayama, I., and Maruyama, S., 2000, Diamond-bearing and diamond-free metacarbonate rocks from Kumdy-kol in the Kokchetav Massif, northern Kazakhstan: The Island Arc, v. 9, p. 400–416.

Ogasawara, Y., Zhang, R. Y., and Liou, J. G., 1998, Petrogenesis of dolomitic marbles from Rongcheng in the Su-Lu ultrahigh-pressure metamorphic terrane, eastern China: The Island Arc, v. 7, p. 82–97.

Okamato, K., Liou, J. G., and Ogasawara, Y., 2000, Petrology of the diamond-grade eclogite in the Kokchetav Massif, northern Kazakhstan: The Island Arc, v. 9, p. 379–399.

Okamoto, K., and Maruyama, S., 1998, Multi-anvil re-equilibration experiments of a Dabie Shan ultrahigh-pressure eclogite within the diamond-stability fields: The Island Arc, v. 7, p. 52–69.

Schertl, H. P., and Okay, A. I., 1994, A coesite inclusion in dolomite in Dabie Shan, China: Petrological and rheological significance: European Journal of Mineralogy, v. 6, p. 995–1000.

Sobolev, N. V., and Shatsky, V. S., 1990, Diamond inclusions in garnets from metamorphic rocks: Nature, v. 343, p. 742–746.

Yoshioka, N., Muko, A., and Ogasawara, Y., 2001, Extremely high diamond concentration in dolomite marble [abs.], in Fluid/slab/mantle interactions and ultrahigh-P minerals: Abstracts of UHPM Workshop 2001, Waseda University, p. 51–55.

Yoshioka, N., and Ogasawara, Y., 2004, Cathodoluminescence of microdiamond in dolomite marble from the Kokchetav Massif—Another evidence for two-stage growth of diamond: International Geology Review (in press).

Zhang, R.Y., and Liou, J. G., 1996, Coesite inclusion in dolomite from eclogite in the southern Dabie Mountains, China: The significance of carbonate minerals in UHP rocks: American Mineralogist, v. 81, p. 181–186.

Zhang, R. Y., and Liou, J. G., 2000, Exsolution lamellae in minerals from ultrahigh-pressure rocks, in Ernst, W. G., Liou, J. G., eds., Ultrahigh-pressure metamorphism and geodynamics in collision-type orogenic belts: Columbia, MD, Bellwether Publishing, Geological Society of America, International Book Series, v. 4, p. 216–228.

Zhang, R. Y., and Liou, J. G., 2003, Clinopyroxenite from the Sulu ultrahigh-pressure terrane, eastern China: Origin and evolution of garnet exsolution in clinopyroxene: American Mineralogist, v. 88, p. 1591–1600.

Zhang, R. Y., Liou, J. G., and Cong, B. L., 1994, Petrogenesis of garnet-bearing ultramafic rocks and associated eclogites in the Su-Lu ultrahigh-pressure metamorphic terrane, China: Journal of Metamorphic Geology, v. 12, p. 169–186.

WESTERN CHINA

A New HP/LT Metamorphic Terrane in the Northern Altyn Tagh, Western China

JIANXIN ZHANG,[1] FANCONG MENG, AND JINGSUI YANG

Institute of Geology, Chinese Academy of Geological Sciences, Beijing 100037, People's Republic of China

Abstract

Field observations and petrological data suggest a new high-pressure/low temperature (HP/LT) metamorphic terrane in the northern Altyn Tagh. Newly recognized eclogites, blueschists, and metapelites constitute part of a coherent HP/LT metamorphic terrane that occurs as a tectonic slab in an ophiolitic mélange. Geobarometry and geothemometry show that peak metamorphic conditions of the eclogites were T = 430–540°C and P = 20–23 kbar. Petrographic data, P-T estimates, and amphibole compositions indicate distinctly different metamorphic histories for two outcrops about 1.2 km apart. For blueschists from outcrop I, the retrograde P-T path is a relatively straight line directed towards the greenschist facies, suggesting simultaneous cooling and decompression. In contrast, interlayered blueschist and eclogite in outcrop II underwent isothermal decompression from eclogite/blueschist facies to epidote-amphibole facies conditions during exhumation. This implies that rocks from these two outcrops experienced different tectono-thermal histories. Preliminary data regarding geological setting, lithologies, and P-T estimates reveal significant similarities between the HP/LT metamorphic terranes in the north Altyn Tagh and the north Qilian Mountains, and suggest that they may be parts of the same early Paleozoic HP/LT belt, subsequently displaced by the Altyn Tagh fault.

Introduction

BLUESCHISTS AND ECLOGITES are high-pressure/low-temperature (HP/LT) metamorphic rocks that exclusively form along active continental margins, and are considered as a result of oceanic plate subduction (e.g., Ernst, 1971; Maruyama et al., 1996). They constitute typical rocks of a HP/LT metamorphic terrane (belt). Previous studies in the North Altyn Tagh have led to the discovery of ophiolites (e.g., Sobel and Arnaud, 1999; Wu et al., 2001) and HP metapelites (Che et al., 1995). Similarities in mineral paragenesis to that of the north Qilian Mountains to the east of the Altyn Tagh has led some authors to propose a single HP/LT belt displaced by the Altyn Tagh fault (Yang et al., 2001; Yue et al., 2001, 2003; Zhang et al., 2001). However, thus far no mafic blueschist and eclogite have been recognized in the north Altyn Tagh. In recent field excursions, we discovered eclogites and blueschists in the North Altyn Tagh (Figs. 1 and 2). The blueschists, eclogites, and metapelites constitute part of a coherent HP/LT metamorphic terrane that occurs as a tectonic slab in an ophiolitic mélange (Fig. 2). As a large-scale strike-slip fault with a long tectonic history, identification of critical components that can serve as piercing points to reconstruct the slip history of the Altyn Tagh fault will be a significant contribution to improve our understanding of the tectonics of central Asia. Objectives of this contribution are: (1) to present mineralogical and petrological data for the newly discovered eclogites and blueschists from the North Altyn Tagh area; and (2) to draw implications for the reconstruction of western China tectonics in the light of such a discovery.

Geologic Setting

The Altyn Tagh, located at the northern margin of the Qinghai-Tibet Plateau, is bounded to the north by the Tarim Basin and to the south by the sinistral strike-slip Altyn Tagh fault (Fig. 1). This feature is one of the largest strike-slip fault systems in the world and has been considered a key element of the escape tectonic model for Eurasia-India continent-continent collision (e.g., Peltzer and Tapponnier, 1988). Major lithologic units of the Altyn Tagh include several terranes (Fig. 1): (1) The Archean Milan Group, composed of amphibolite to granulite-facies metamorphic complexes with an age of 2500–2800 Ma, have been considered as the exposed Tarim Percambrian cratonal basement rocks (BGMX,

---

[1]Corresponding author; email: zjx66@yeah.net

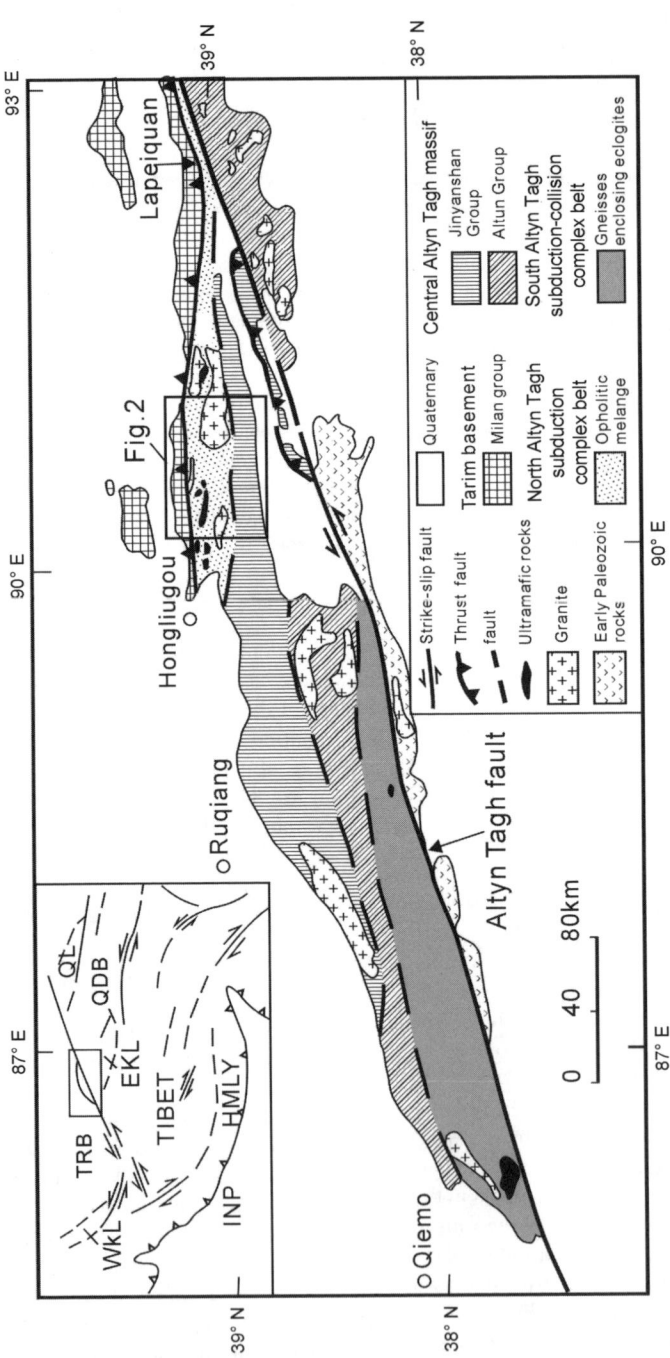

FIG. 1. Geological map showing the tectonic units in the Altyn Tagh area. Abbreviations: TRB = Tarim Basin; QL = Qilian Mountains; QDB = Qaidam Basin; HMLY = Himalaya Mountains; INP = India plate; WKL = Western Kunlun Mountains; EKL = Eastern Kunlun Mountains.

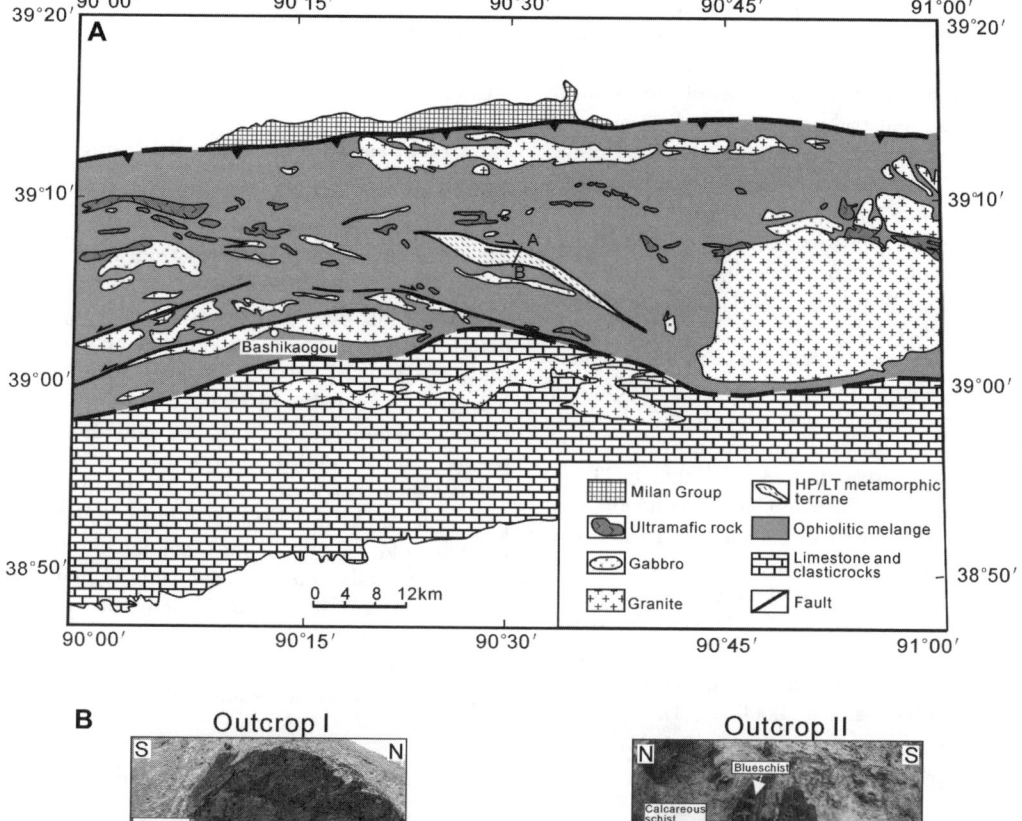

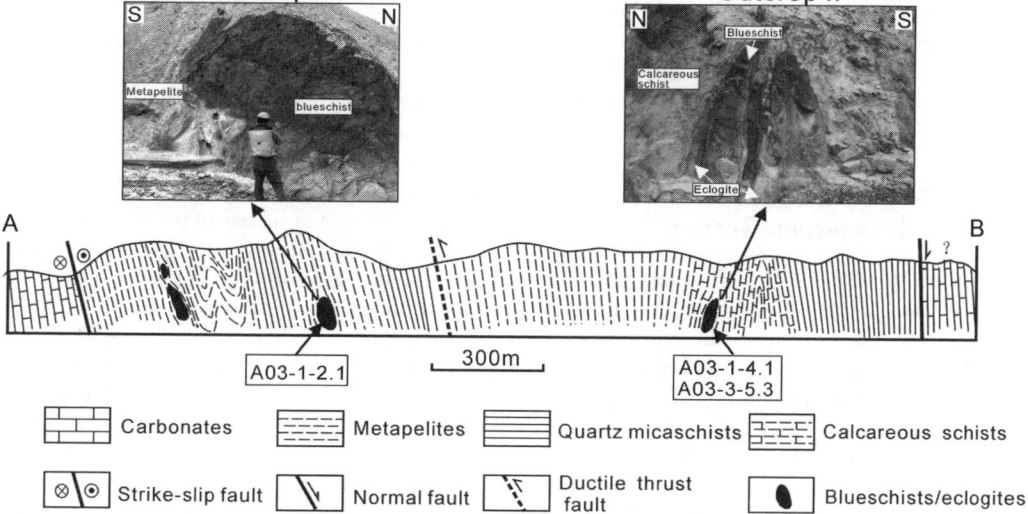

FIG. 2. A. Geological map showing distribution of the north Altyn Tagh HP/LT metamorphic terrane. B. Cross-section showing outcrops and sample locations for this study.

1993; Che and Sun, 1996). (2) The north Altyn Tagh subduction complex, occurs discontinuously along the northeastern part of the Altyn Tagh from Hongliugou to Lapeiquan; it consists of clastic and volcanic, as well as mafic and ultramafic intrusive rocks. Pillow lava and serpentinized ultramafic rocks suggest that they are part of an ophilolite suite. The newly recognized HP/LT metamorphic terrane occurs as a tectonic block or slab in the subduction complex (Fig. 2). (3) The central Altyn Tagh massif consists of the Altun and Jinyanshan groups. The Altun Group, which is made up of amphibolite facies quartzo-feldspathic gneisses, pelitic gneisses, marbles, and amphibolites, was considered previously as part of Precambrian cratonal basement rocks, and was mapped as lower Proterozoic (BGMX, 1993). However, recent radiometric data demonstrate that the Altun Group formed at 800–1000 Ma (e.g., Wan et al., 2001). The Jinyanshan Group is dominated by mid- to upper Proterozoic metasedimentary rocks that consist of thick carbonate-rich continental marginal sequences containing varying amounts of volcanics and clastics. The ages assigned to the Jinyanshan Group are based on stromtolites in carbonate rocks (BGMX, 1993). (4) The southern Altyn Tagh subduction-collision complex zone consists of blocks of eclogite, garnet-peridotite, and granulite within amphibolite-facies gneisses, and is exposed in the southwestern part of the Altyn Tagh (Zhang et al., 2001). The U/Pb dating on zircon separates from eclogites and granulites suggests the peak UHP metamorphism occurred at 450–500 Ma (Zhang et al., 2000, 2001).

These litho-tectonic units are bounded by nearly E-W–trending shear zones that are truncated by the ENE-WSW–trending Altyn Tagh fault. However, due to limited data, some boundaries between these units are not well constrained.

Occurrence of the HP/LT Metamorphic Terrane

The north Altyn Tagh HP/LT metamorphic terrane is an elongate slab 2–5 km wide and 40 km long, and is bounded by strike-slip and normal (?) faults (Fig. 2A). It consists of garnet-chloritoid–bearing micaschist (metapelite), quartz micaschist, calcareous schist, and metabasite. Metabasites are mainly eclogites and blueschists. In this study, we focus on two metabasite outcrops along a continuous expose section about 2.5 km long (Fig. 2B). Outcrop I consists of a large block of mafic blueschists rimmed by garnet actinolite schists within metapelites. Outcrop II, 1.2 km south of the outcrop I, consists of blueschist and eclogite. Eclogites constitute about 10–15% of outcrop II, and occur as layers or boudins within blueschists, or as layers along the contact between blueschists and interbedded calcareous micaschists. The two outcrops are separated by metapelites and quartz schists, which have experienced strong ductile deformation involving top-to-the-north shearing (Fig. 2B). High-pressure relict assemblages of Grt + Cld + Phe + Pg + Qtz in metapelites.

Petrography and Mineral Chemistry

Analytical methods

More than 30 samples of eclogites, blueschists, and metapelites were collected from the two outcrops. The estimated modal compositions are listed in Table 1. In the following sections, we present detailed descriptions of two representative blueschists (A03-1-2.1, A03-1-4.1), an eclogite (A03-3-5.3), and a metapelite (A03-1-2.2). Chemical compositions of mineral were analyzed at the Institute of Geology and Mineral Resources, Chinese Academy of Geological Sciences, using a JXA8800 electron probe with 20 kv accelerating voltage, 20 nA beam current, and a 5 µm beam diameter. Some fine-grained minerals were analyzed using 1 µm or 2 µm beam diameter. Representative mineral compositions are listed in Tables 2, 3, and 4. Mineral abbreviations are after Kretz (1983). Amphiboles are classified according to Leake et al. (1997).

Petrography

Blueschist (sample A03-1-2.1). In addition to garnet and glaucophane, sample A03-1-2.1 contains variable amounts of epidote, Ca-Na amphibole, actinolite, phengite, paragonite, chlorite, albite, quartz, carbonate, rutile, titanite and hematite (Table 1). The assemblage Grt + Gl + Phn + Pg + Qtz is dominant, and is interpreted to represent peak metamorphism. The sample has a medium- to fine-grained matrix with a foliated texture (Fig. 3A). Idioblastic garnets (0.5–2.0 mm) contain inclusions of epidote, quartz, winchite, and rutile, and are surrounded by aligned glaucophanes. Actinolite and epidote occur as porphyroblasts, contain relics of glaucophane inclusions, and are aligned at an angle to the foliation (Figs. 3C and 3E). Such features suggest late growth. Actinolite also rims glaucophane. Chlorite, albite and calcite mainly appear in fractures of the garnets.

Blueschist (sample A03-1-1.4). The dominating peak metamorphic assemblage in sample A03-1-4.1 is identical to that of sample A03-1-2.1: garnet, glaucophane, paragonite, and quartz. This assemblage defines a strong foliation (Fig. 3B). However, phengite is not present in the peak assemblage. Glaucophane is rimmed by Ca-Na amphibole (barroisite) or a symplectite of barroisite + albite (Fig. 3D). Fine-grained paragonite grew in textural

TABLE 1. Modes of Representative HP/LT Metamorphic Rocks from the North Altyn Tagh[1]

Sample:	A03-1-2.1	A03-1-2a	A03-1-2.2	A03-1-2.5	A03-1-4.1	A03-3-1.4	A03-1-5.1	A03-3-5.3
Rock type:	Blueschist	Grt-Act schist	Metapelite	Metapelite	Blueschist	Blueschist	Calcareous schist	Eclogite
Grt	20–25	20–22	12–15	3–5	20–25	10–15		15–17
Omp								25–30
Gl	27–30	<2			20–25	18–20		<2
Bar					5–7			7–10
Win	<1							
Cld			5–8	15–20				
Phn	<3	4	20–25	18–20		3	5–7	6–8
Pg	2–3	5	10–15	8–10	8–10	5	5	<2
Qtz	2–3	3–5	25–28	25–30	5	5–7	10–12	10
Act	3–5	30–35				10–15		
Ep	5–7	10–12	2–3	3	3–5	15–17		5
Cc	5	3	3	2	5		75	8
Chl	5	5	5–7	5	5	5	1	<2
Ab	5	5	2–3	3	5	8		<2
Others	Rt (2–3) Hem (Tr)	Ttn (2)	Tur (Tr) Hem (Tr)	Tur (Tr)	Ttn (3), Hem (Tr)	Ttn (3–4) Hem (Tr)	Ap (Tr)	Ttn (3–4) Hem (Tr)

[1]Vol%, determined by visual estimation; phengite and paragonite were differentiated by BSE images. Abbreviations: Cc = carbonates; Tr = traces.

equilibrium with glaucophane, whereas large paragonite porphyroblasts overgrew foliations composed of glaucophane and quartz (Fig. 3F). These indicate that paragonite formed during and after glaucophane. Chlorite and albite occurs along the fractures of garnets. In strongly retrograded domains, garnet is nearly entirely replaced by chlorite.

Eclogite (A03-3-5.3). Eclogite consists of garnet, omphacite, phengite, and quartz with varying amounts of glaucophane, Ca-Na amphibole (barroisite), epidote (zoisite or clinzoisite), paragonite, chlorite, albite, carbonate, and titanite. The alignment of omphacite, minor phengite, paragonite, and quartz defines a foliation in the matrix as well as within garnet (Fig. 4A). Inclusion trails within the garnet are straight and in continuation with foliation in the matrix. This implies that garnet may have continued to grow slightly after cessation of deformation. No rotation (i.e., coaxial progressive deformation with shortening direction normal to the foliation) can be inferred from inclusion trail geometry within the garnets during crystallization. However, the foliation deflection around garnet porphyroblasts also occurs locally. This suggests that garnet porphyroblasts generally were syntectonic, and garnet and omphacite may have grown synchronously under high-pressure conditions. Glaucophane occurs as inclusions in garnet cores, or within barroisite (Figs. 4B and 4C). Barroisite occurs either as (1) coarse-grained porphyroblasts containing glaucophane inclusions and obliquely cutting the foliation (Fig. 4C); or (2) fine-grained symplectites of Bar + Ab after omphacites or around garnets (Fig. 4D). Barroisite also occurs as rims around omphacites. Paragonite grew either parallel to—or at a small angle to—the matrix foliation, attesting to synchronous or slightly later growth as compared with the peak assemblage. Chlorite associated with albite mainly grew along fractures in garnets, omphacites, and barroisite prophyroblasts, suggesting that chlorite must have postdated growth of these minerals. These microscopic observations are used to infer a sequence of metamorphism that constitutes three stages: (1) garnet + omphacite + glaucophane + phengite + paragonite (peak stage); (2) garnet + epidote + barroisite + albite + paragonite (early retrogression); (3) chlorite + albite + epidote (late retrogression).

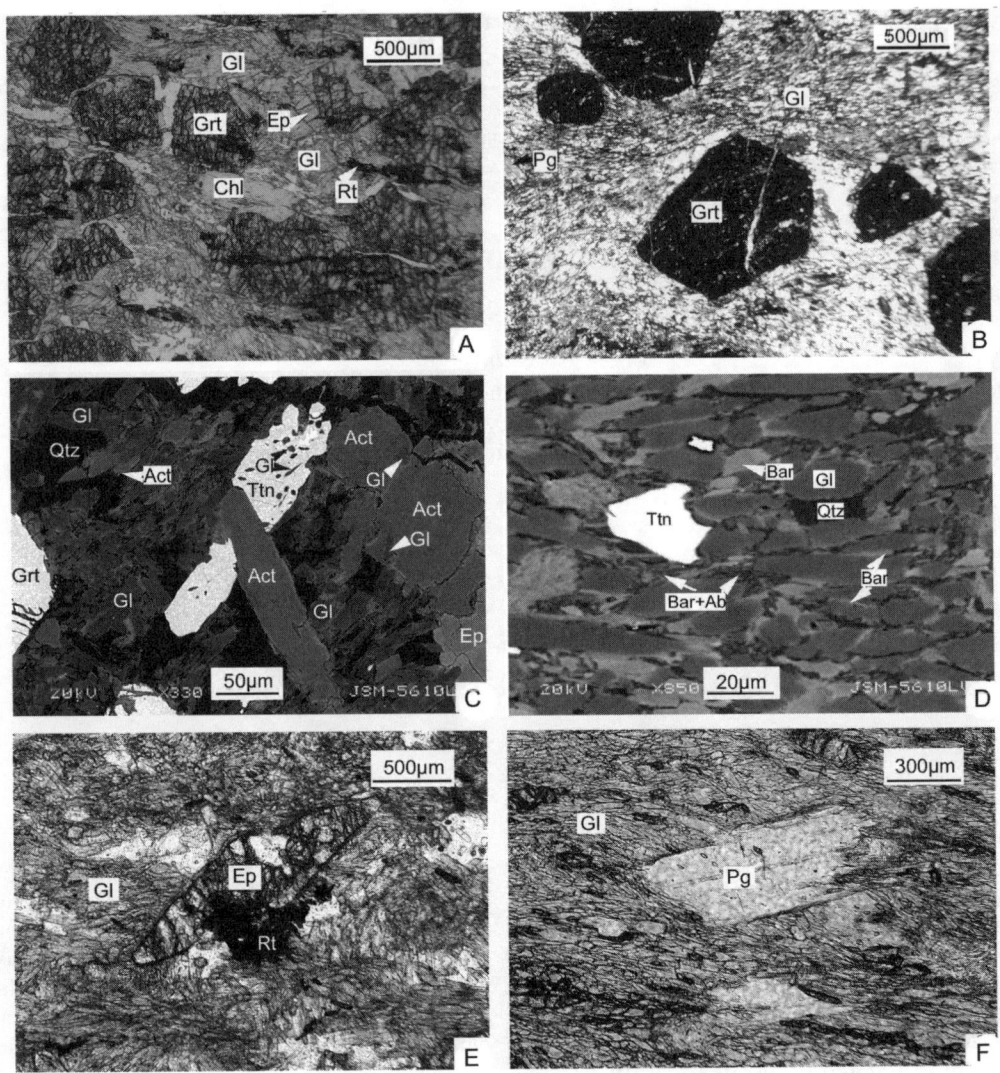

FIG. 3. Photomicrograghs of blueschists in the north Altyn Tagh. A. Idioblastic garnet porphyroblasts setting in a foliated matrix of Gl + Chl + Ab (A03-1-2.1, plane polarized light). B. Idioblastic garnet surrounded by oriented glaucophane and paragonite (A03-1-4.1, crossed polarized light). C. Actinolite porphyroblasts containing inclusion relics of glaucophane in a foliated matrix defined by glaucophane; glaucophane is also rimmed by actinolite (A03-1-2.1, backscattered electron image). D. Glaucophane rimmed by barroisite or barroisite + albite (A03-1-4.1, backscattered electron image). E. Epidote porphyroblasts cross-cutting the foliation defined mainly by glaucophane (A03-1-2.1, plane-polarized light). F. Late paragonite porphyroblasts overgrowing foliation defined by glaucophane (A03-1-4.1, plane- polarized light).

blasts in eclogites under high-pressure conditions. Garnet rims adjacent to the Bar + Ab symplectite exhibit a slight decrease in pyrope and grossular contents and an increase in almandine component. Garnet porphyroblasts of the metapelite have compositions of $Alm_{63-67}Prp_{8-10}Grs_{22-26}Sps_{1-3}$, to 3.45 p.f.u. (Table 2). In contrast, phengite in the eclogite has a Si content between 3.4 and 3.6 p.f.u. (Table 3, Fig. 9). Phengite in the metapelite contains 3.50–3.58 p.f.u. of Si (Table 4). Paragonite ranges in Si content from 3.05 to 3.15, and exhibits little variation in the different samples. Chloritoid is only

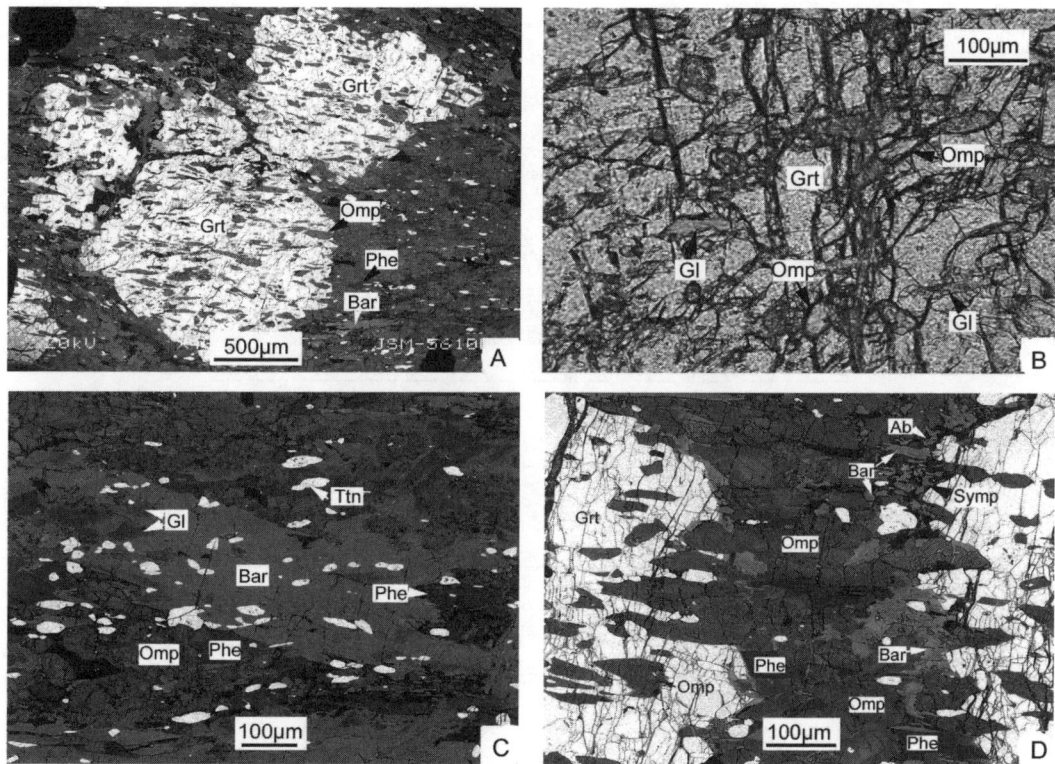

FIG. 4. Photomicrograghs of eclogites in the northern Altyn Tagh. A. Garnet porphyroblast in eclogite containing oriented inclusions of omphacites, phengites, and quartzes, extending to matrix defined well-developed foliations. Such texture suggests that the schistosity-defining minerals and garnet coexisted (backscattered electron image). B. Glaucophane occuring as inclusions in garnet cores, and oriented together with omphacite (plane-polarized light). C. Porphyroblasts of barroisite overgrowing foliations defined by oriented omphacite and phengite, and glaucophane also occuring as inclusions in barroisite (backscattered electron image). D. Barroisite-albite symplectite forming after omphacite near garnet rim (backscattered electron image).

present in the metapelite; analyzed Cld crystals are unzoned, with a Mg/(Mg+Fe^{2+}) ratio of 0.23 to 0.25. Epidotes in the eclogite and metapelite have lower Fe^{3+} contents than those in the blueschists. Chlorites from the eclogites and blueschists possess higher Mg/(Mg+Fe^{2+}) ratios than those in the metapelites. Where present, plagioclase is albitic (An content lower than 4).

P-T Estimations and Metamorphic Evolution

P-T conditions for the eclogite

The presence of paragonite coexisting with omphacite and the absence of kyanite constrain the peak metamorphic pressure to lie within the paragonite stability field. The upper pressure limit is defined by the reaction Pg = Jd + Ky + H$_2$O (Fig. 10). Microtextures indicate that albites formed by the breakdown of omphacites, so the lower pressure boundary for eclogite-facies recrystallization is estimated by the jadeite content of omphacite (Fig. 10).

Eclogite-facies recrystallization also was estimated by the Grt-Cpx-Phn geobarometry (Ravna and Terry, 2001) and the Grt-Cpx geothermometry (Ravna, 2000). Coexisting garnet, omphacite and phengite give equilibrium conditions of T = 430–540°C and P = 20–23 kbar, near the upper limit defined by Pg = Jd50 + Ky + H$_2$O (Fig. 10). Because of relatively constant garnet compositions, scatter in the temperature estimates is possibly due to variable omphacite compositions or to the uncertainties of Fe^{3+} contents in omphacite. The Fe^{3+} contents determined by charge balance may reflect analytical

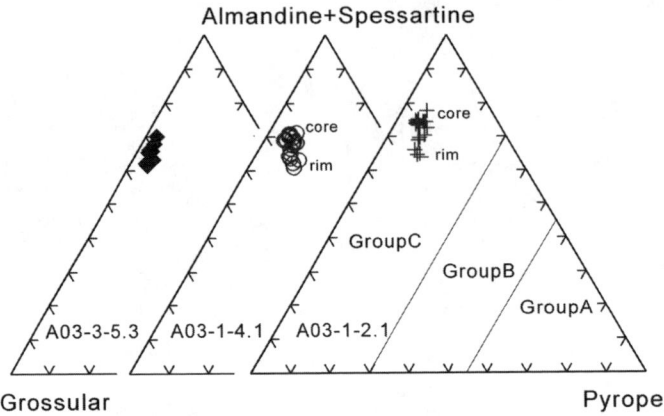

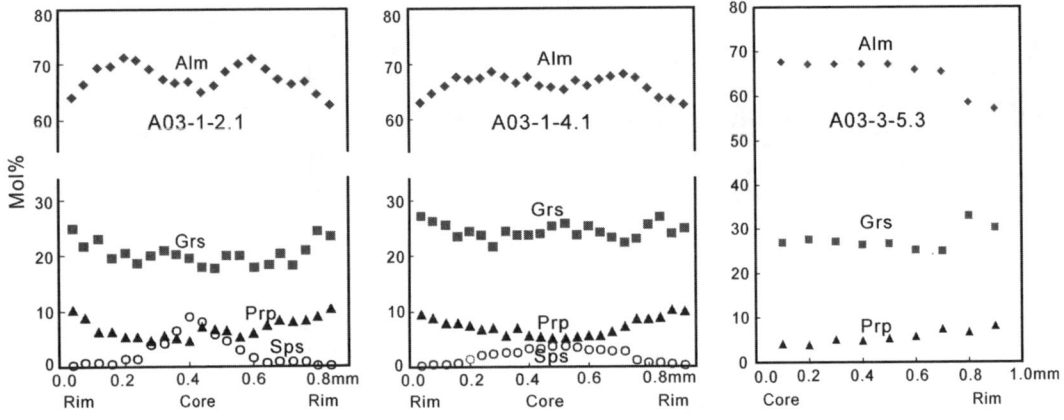

FIG. 5. Analyzed garnet compositions plotted on an Alm-Grs-Prp ternary diagram. Eclogite garnet classification after Coleman et al. (1965).

FIG. 6. Analyzed garnet compositional zoning profiles for blueschists and eclogite.

uncertainty rather than real proportions of ferric and ferrous iron.

In the early retrogression stage, the assemblage epidote + Ca-Na amphibole + albite was produced, characteristic of the epidote-amphibolite facies. In strongly retrograded domains of the eclogite, garnet rim and barroisite-albite symplectite compositions were chosen for geothermobarometric analysis of the early P-T conditions because local equilibrium between garnet and newly formed barroisite and plagioclase is hypothesized. P-T estimations based on the Grt-Bar-Ab-Qtz assemblage using THERMOCALC v3.1 yield P = 8–10 kbar and T = 450–500°C. The garnet-amphibole thermometry (Graham and Powell, 1984) yields slightly lower temperatures, ranging from 420 to 480°C. The garnet-amphibole-plagioclase-quartz geobarometry of Kohn and Spear (1990) gives a higher pressure range, 15–17 kbar, which is probably related to inappropriate compositional restrictions of the calibrated amphibole and plagioclase (Klemd et al., 2002).

Chlorite and albite along garnet fractures imply that eclogite underwent greenschist-facies metamorphism during later retrogression, but a suitable geothermobarometry does not exist for P-T estimation.

Constraints on metamorphic conditions employing amphibole compositions

The composition of amphiboles in mafic schists provides a good overall indication of the metamor-

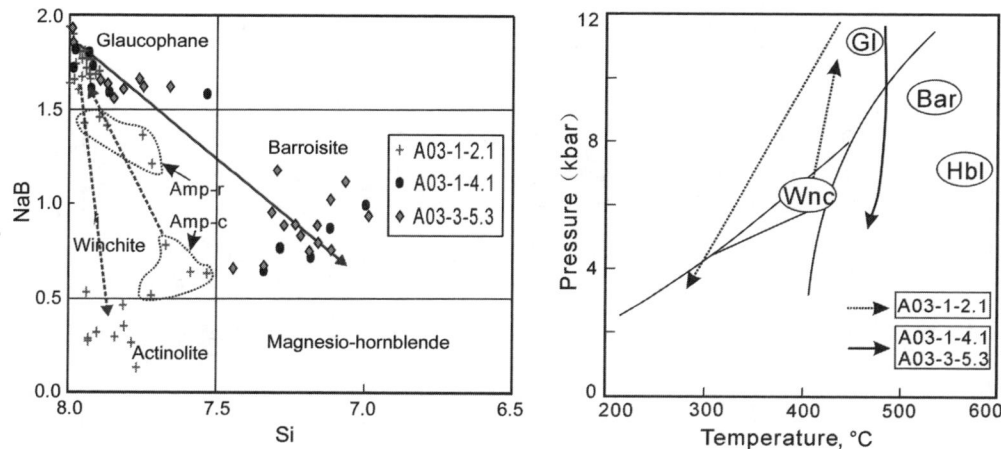

FIG. 7. A. Na_B vs. Si diagram for amphiboles in blueschists and eclogites. The amphibole nomenclature is after Leake et al. (1997). Fe^{3+}/Fe^{2+} ratios were calculated assuming a cation total of 13, excluding K, Na, Ca (O = 23). Amp-c and Amp-r represent the compositions of amphiboles as inclusions in garnet cores and garnet rims, respectively. Trends of compositional change of amphiboles are shown by dashed arrow (A03-1-2.1, blueschists) and lined arrow (A03-3-5.3, eclogites), respectively. B. Semi-quantitative phase relationships between Na-amphibole, winchite, actinolite, and barroisite in hematite-bearing white schist based on Ostsuki and Banno (1990). Qualitative P-T paths are shown by arrows.

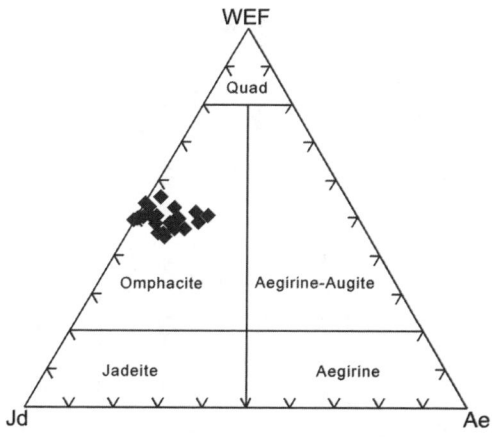

FIG. 8. Omphacite compostions plotted on WEF (Wo, En, Fs)–Jd–Ae ternary diagram. Clinopyroxene classification after Morimoto et al. (1988).

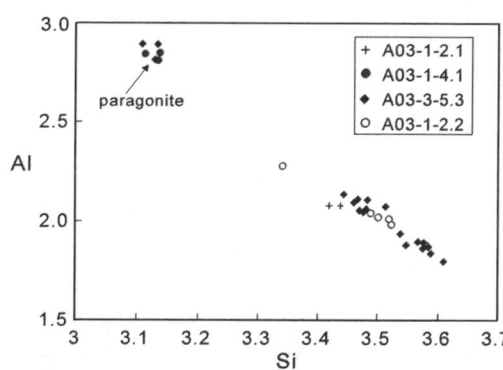

FIG. 9. Compositional variations of white micas are shown in Si p.f.u. versus Al p.f.u. plots. A03-1-2.2 is metepelite enclosing blueschist (A03-1-2.1).

phic grade (Otsuki and Banno, 1990; Matsumoto et al., 2003) (Fig. 7B). The P-T path can be constrained to some extent by compositional variation of the amphibole in hematite-bearing assemblages (e.g., Aoya et al., 2003; Matsumoto et al., 2003). All studied samples contain hematite. For sample A03-1-1.2 from outcrop I, amphibole inclusions in the core and rim of garnet as well as in matrix, have a composition ranging from Na_B-poor winchite through Na_B-rich winchite to glaucophane (Fig. 7). This suggests a prograde compositional variation in amphibole. During retrogression, glaucophane is replaced by actinolite (Figs. 4 and 7A). No replacement of glaucophane by barroisite or winchite was observed. In combination with semi-quantitative phase relationships between Na-amphibole, winchite, and actinolite in hematite-bearing assemblages (Otsuki and Banno, 1990), the amphibole compositions of sample A03-1-2.1 imply a lower geothermal gradient during the retrograde stage

TABLE 2. Representative Microprobe Analyses of Minerals from Blueschists in the North Altyn Tagh[1]

Sample no.:	A03-1-2.1		A03-1-4.1		A03-1-2.1			A03-1-4.1			A03-1-2.1	A03-1-4.1			A03-1-2.1	
	Garnet				Amphibole						White mica					
Mineral:	Grt-r	Grt-c	Grt-r	Grt-c	Gl	Gl	Wnc-r	Wnc-c	Act	Gl	Gl	Bar	Phn	Pg	Pg	Pg	Ep	Chl	Ab
SiO_2	37.96	37.46	38.06	37.78	58.68	58.94	57.54	53.70	55.52	58.36	58.43	49.82	50.86	49.72	49.34	49.06	38.70	26.59	68.68
TiO_2	0.06	0.19	0.05	0.17	0.02	0.02	0.04	0.08	0.03	0.02	0.03	0.12	0.22	0.07	0.04	0.08	0.17	0.04	0.02
Al_2O_3	20.65	19.93	20.91	20.46	11.40	10.21	8.88	4.44	2.76	10.20	11.78	10.21	26.11	37.87	38.24	37.80	27.53	20.03	19.24
Cr_2O_3	0.01	0.00	0.01	0.00	0.01	0.01	0.07	0.00	0.13	0.00	0.00	0.02	0.00	0.00	0.03	0.09	0.00	0.09	0.00
FeO	29.96	30.1	29.61	29.55	9.74	9.25	15.65	19.48	9.32	10.95	9.02	13.16	2.21	0.43	0.23	0.27	6.14	24.71	0.15
MnO	0.21	3.86	0.16	1.61	0.02	0.00	0.11	0.09	0.06	0.01	0.00	0.03	0.00	0.00	0.00	0.00	0.04	0.13	0.05
MgO	2.27	1.18	2.61	1.30	10.71	11.48	8.98	10.50	17.59	10.81	10.85	12.12	3.65	0.27	0.12	0.23	0.09	16.82	0.00
CaO	9.28	6.88	9.26	8.89	0.74	0.95	1.46	6.49	10.08	0.96	0.75	8.15	0.00	0.13	0.22	0.19	23.10	0.01	0.32
Na_2O	0.00	0.07	0.00	0.00	6.80	6.24	5.45	2.87	1.35	6.40	6.88	3.64	0.54	6.39	7.15	5.45	0.00	0.01	11.62
K_2O	0.00	0.00	0.00	0.00	0.00	0.01	0.04	0.02	0.08	0.02	0.02	0.24	10.50	1.30	0.52	1.18	0.00	0.01	0.03
Totals	100.4	99.67	100.67	99.76	98.12	97.11	98.22	97.67	96.92	97.73	97.76	97.51	94.09	96.18	95.89	94.35	95.79	88.30	100.11
O	12	12	12	12	23	23	23	23	23	23	23	23	22	22	22	22	12.5	14	8
Si	3.010	3.034	3.002	3.036	7.925	7.995	7.900	7.671	7.764	7.93	7.909	7.113	3.438	3.136	3.115	3.138	3.062	2.767	2.998
Ti	0.004	0.012	0.003	0.010	0.002	0.002	0.006	0.009	0.003	0.002	0.003	0.013	0.011	0.004	0.002	0.004	0.010	0.003	0.001
Al	1.928	1.901	1.942	1.936	1.814	1.631	1.426	0.747	0.000	3.449	3.139	3.117	2.078	2.812	2.843	2.847	2.680	2.580	0.990
Cr	0.001	0.000	0.001	0.000	0.001	0.001	0.008	0.000	0.014	0.000	0.000	0.002	0.000	0.000	0.002	0.005	0.000	0.007	0.000
Fe^{3+}	0.038	0.014	0.043	0.000	0.337	0.455	0.853	1.110	0.596	0.534	0.272	0.484	0.000	0.023	0.000	0.000	0.324	0.000	0.005
Fe^{2+}	1.948	2.025	1.910	1.986	0.763	0.595	0.944	1.217	0.494	0.711	0.749	1.088	0.125	0.026	0.012	0.015	0.046	2.139	0.000
Mg	0.268	0.142	0.307	0.156	2.156	2.321	1.838	2.236	3.667	2.190	2.189	2.580	0.368	0.000	0.012	0.022	0.011	2.609	0.003
Mn	0.014	0.265	0.011	0.110	0.002	0.000	0.013	0.011	0.007	0.001	0.000	0.004	0.000	0.009	0.015	0.000	0.003	0.010	0.000
Ca	0.788	0.597	0.782	0.766	0.107	0.138	0.215	0.993	1.510	0.140	0.109	1.247	0.071	0.782	0.875	0.676	1.960	0.001	0.015
Na	0.000	0.011	0.000	0.000	1.781	1.641	1.451	0.795	0.366	1.686	1.806	1.008	0.906	0.105	0.042	0.097	0.000	0.002	0.983
K	0.000	0.000	0.000	0.000	0.000	0.002	0.007	0.004	0.014	0.003	0.003	0.044	0.906	0.105	0.042	0.097	0.000	0.000	0.002
Totals	8.000	8.000	8.000	8.000	14.888	14.781	14.673	14.793	14.891	14.830	14.918	15.298	6.997	6.897	6.916	6.815	8.096	10.118	4.997

[1]Grt-c = garnet core; Grt-r = garnet rim; Wnc-r = winchite occurring as inclusions in garnet rim; Wnc-c = winchite occurring as inclusions in garnet core. Fe^{3+} of amphibole was calculated with total cations = 13 exclusive of K, Na, and Ca (O=23). Fe^{3+} contents for other minerals were calculated by charge balance.

TABLE 3. Representative Microprobe Analyses of Minerals from Eclogite in the North Altyn Tagh[1]

	Grt-c	Grt-r	Omp-r	Omp-r	Omp-c	Omp-c	Bar	Bar	Gl	Gl	Phn	Phn	Phn	Phn	Pg	Ep	Chl	Ab
SiO_2	38.08	38.19	56.75	56.33	56.46	55.62	50.44	49.02	57.64	54.8	54.72	52.09	53.10	52.96	49.77	39.00	31.32	68.07
TiO_2	0.14	0.06	0.02	0.05	0.05	0.10	0.16	0.09	0.04	0.09	0.2	0.21	0.21	0.21	0.04	0.07	0.03	0.01
Al_2O_3	20.78	20.57	11.01	10.46	10.54	8.90	9.70	10.40	10.13	10.22	23.15	23.42	26.68	23.43	38.00	29.72	15.48	18.92
Cr_2O_3	0.01	0.00	0.02	0.00	0.03	0.05	0.14	0.00	0.07	0.00	0.04	0.02	0.01	0.02	0.02	0.00	0.03	0.00
FeO	30.43	26.95	5.07	7.54	8.37	9.59	12.50	12.73	12.30	13.98	3.28	2.57	2.19	3.18	0.32	4.21	21.79	0.09
MnO	0.19	0.64	0.01	0.00	0.00	0.07	0.04	0.07	0.01	0.00	0.00	0.00	0.00	0.01	0.00	0.07	0.01	0.01
MgO	1.30	1.69	7.30	6.64	5.94	6.13	11.83	12.00	9.40	8.95	3.93	4.01	3.45	3.71	0.31	0.45	19.73	0.02
CaO	9.66	11.97	12.38	11.6	11.11	11.52	7.64	8.29	0.50	2.12	0.00	0.01	0.03	0.02	0.15	22.04	0.15	0.76
Na_2O	0.00	0.00	7.34	7.24	7.91	7.59	3.92	3.55	6.99	6.13	0.10	0.15	0.40	0.12	6.99	0.00	0.05	11.12
K_2O	0.00	0.00	0.02	0.04	0.00	0.00	0.25	0.20	0.04	0.07	11.55	11.48	10.56	10.91	0.69	0.23	0.08	0.02
Total	100.60	99.08	99.92	99.9	100.41	99.57	96.62	96.35	97.16	96.36	96.97	93.96	96.63	94.57	96.29	95.79	88.76	99.02
O	12	12	6	6	6	6	23	23	23	23	23	22	22	22	22	12.5	14	8
Si	3.029	3.028	2.020	2.022	2.016	2.014	7.268	7.086	7.975	7.749	3.609	3.547	3.480	3.575	3.130	3.049	3.760	3.002
Ti	0.008	0.004	0.001	0.001	0.001	0.003	0.017	0.010	0.004	0.010	0.010	0.011	0.011	0.011	0.002	0.004	0.002	0.000
Al	1.947	1.921	0.461	0.442	0.443	0.379	1.646	1.771	1.650	1.702	1.798	1.878	2.060	1.862	2.814	2.739	1.851	0.984
Cr	0.001	0.000	0.001	0.000	0.001	0.001	0.016	0.000	0.008	0.000	0.002	0.001	0.001	0.001	0.001	0.000	0.002	0.000
Fe^{3+}	0.000	0.009	0.004	0.015	0.068	0.118	0.267	0.438	0.354	0.446	0.000	0.000	0.000	0.000	0.000	0.245	0.000	0.003
Fe^{2+}	2.024	1.778	0.147	0.211	0.182	0.172	1.240	1.101	1.069	1.207	0.181	0.147	0.120	0.180	0.017	0.002	1.848	0.000
Mg	0.154	0.200	0.387	0.355	0.316	0.331	2.541	2.586	1.939	1.887	0.387	0.407	0.337	0.374	0.029	0.052	2.982	0.001
Mn	0.013	0.043	0.000	0.000	0.000	0.002	0.005	0.009	0.001	0.000	0.000	0.000	0.000	0.001	0.000	0.005	0.009	0.000
Ca	0.823	1.017	0.472	0.446	0.425	0.447	1.180	1.284	0.074	0.321	0.000	0.001	0.002	0.002	0.010	1.846	0.016	0.036
Na	0.000	0.000	0.506	0.504	0.548	0.533	1.095	0.995	1.882	1.681	0.013	0.020	0.051	0.016	0.853	0.000	0.010	0.951
K	0.000	0.000	0.001	0.002	0.000	0.000	0.046	0.037	0.007	0.013	0.972	0.998	0.883	0.940	0.056	0.023	0.010	0.001
Total	7.999	8.000	3.999	3.998	4.000	4.000	15.321	15.316	14.956	15.015	6.972	7.009	6.944	6.958	6.911	7.966	9.906	4.980

[1]Grt-c = garnet core; Grt-r = garnet rim; Omp-r = omphacite; Omp-c = omphacite core; Omp-r = omphacite. Fe^{3+} of amphibole was calculated with total cations = 13 exclusive of K, Na, and Ca (O=23). Fe^{3+} contents for other minerals were calculated by charge balance.

TABLE 4. Representative Microprobe Analyses of Minerals from Metapelite in the Northern Altyn Tagh

Mineral	Grt-r	Grt-c	Cld	Phn	Phn	Pg	Ep	Ab	Chl
SiO_2	38.58	37.64	25.41	53.12	54.63	48.39	40.35	67.57	25.11
TiO_2	0.09	0.10	0.01	0.16	0.15	0.05	0.23	0.00	0.03
Al_2O_2	21.18	20.59	40.80	26.00	24.38	37.2	28.89	18.81	21.86
Cr_2O_2	0.05	0.06	0.04	0.00	0.05	0.10	0.17	0.00	0.09
FeO	28.79	30.43	22.13	1.94	2.02	0.90	4.97	0.74	27.63
MnO	0.51	1.02	0.09	0.01	0.00	0.03	0.18	0.00	0.22
MgO	2.57	2.50	4.26	3.60	4.12	0.62	0.15	0.30	12.28
CaO	9.46	7.46	0.00	0.00	0.00	0.13	22.31	0.10	0.00
Na_2O	0.00	0.00	0.00	0.23	0.21	6.70	0.36	11.62	0.01
K_2O	0.00	0.00	0.00	11.32	11.31	1.38	0.01	0.01	0.01
Totals	101.24	99.81	92.74	96.39	96.88	95.50	97.62	99.15	87.24
O	12	12	6	11	11	11	12.5	8	14
Si	3.020	3.010	1.030	3.501	3.579	3.095	3.102	2.985	2.691
Ti	0.005	0.006	0.000	0.008	0.007	0.002	0.013	0.000	0.002
Al	1.955	1.941	1.950	2.020	1.883	2.805	2.619	0.980	2.762
Cr	0.003	0.004	0.001	0.000	0.003	0.005	0.010	0.000	0.008
Fe^{3+}	0.000	0.023	0.000	0.000	0.000	0.000	0.285	0.025	0.000
Fe^{2+}	1.885	2.010	0.750	0.107	0.111	0.048	0.003	0.000	2.476
Mn	0.034	0.069	0.003	0.001	0.000	0.002	0.012	0.000	0.020
Mg	0.300	0.298	0.258	0.354	0.402	0.059	0.017	0.020	1.961
Ca	0.794	0.64	0.000	0.000	0.000	0.009	1.838	0.005	0.000
Na	0.000	0.000	0.000	0.029	0.027	0.831	0.054	0.995	0.002
K	0.000	0.000	0.000	0.953	0.946	0.113	0.001	0.001	0.001
Totals	7.996	8.000	3.994	6.972	6.957	6.969	7.954	5.010	9.924

[1]Abbreviations: Grt-c = garnet core; Grt-r = garnet rim.

than that during prograde metamorphism (Fig. 7A). An anticlockwise P-T path is suggested. For sample A03-1-4.1 and sample A03-3-5.3 from outcrop II, no prograde amphibole was detected. The retrograde amphibole in symplectite or as overgrowths over peak assemblages is barroisite. This suggests that the retrograde P-T path continues from the eclogite/ blueschist facies to the epidote-amphibole facies, and is associated with isothermal decompression.

Discussion and Conclusions

Contrasting retrograde P-T paths for rocks from the two outcrops

Outcrop I is separated from outcrop II by a distance of about 1.2 km (Fig. 2). Petrograghic data, P-T estimations, and amphibole compositions indicate distinctly different metamorphic histories for these two rocks. For sample A03-1-2.1 from outcrop I, the retrograde P-T path is a relatively straight line directed toward the greenschist facies, and the thermal gradient during exhumation was lower than that during subduction (Figs. 7B and 10). Data presented in this study suggest an anticlockwise P-T path. A similar anticlockwise P-T path has been reported for Franciscan eclogites and high-grade blueschists, resulting from a decreasing geothermal gradient in the subduction zone due to refrigeration by a continuous supply of cold oceanic crust (Krogh et al., 1994).

In contrast, the decompression P-T path for the eclogite and blueschist from outcrop II is associated with an increase in the thermal gradient, which resulted in the formation of barroisite as the

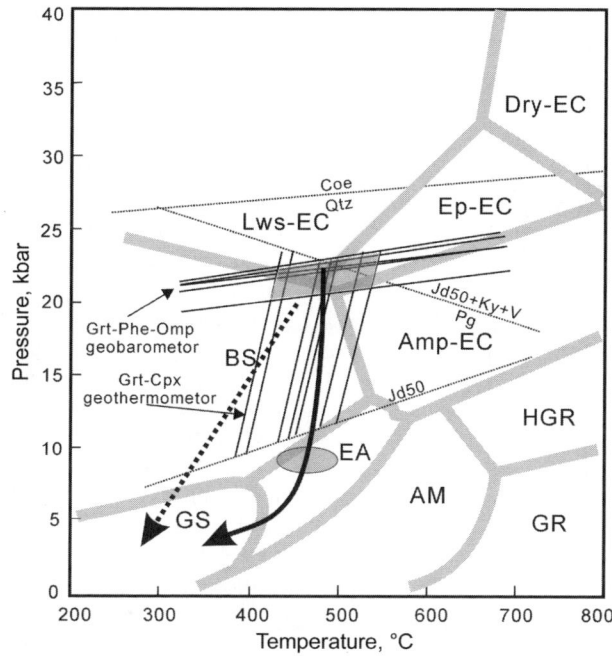

FIG. 10. Decompression P-T paths for eclogite (A03-3-5.3) and blueschist (A03-1-2.1). The dashed arrow represents the P-T path of blueschist from outcrop I, inferred qualitatively by amphibole compositions; the lined arrow represents the P-T path of eclogites from outcrop II. The univarient reaction Pg = Jd50 + Ky + H_2O is calculated with program THERMOCALC v3.1, with related data from Powell et al. (1998), assuming a water activity close to unity. Ab = Jd + Qtz is after Holland (1980). Metamorphic facies boundaries are taken from Liou et al. (1998).

retrograded amphibole. This suggests an isothermal decompression from eclogite/blueschist–facies to epidote/amphibole–facies conditions. The differences imply that although the two outcrops are only 1.2 km apart, the decompression thermal histories attending their exhumations were different. Considering the strong shear deformation between the two outcrops, a tectonic discontinuity between outcrops I and II is suggested. However, clarification of the relationship between these two outcrops requires additional structural and petrological data.

Relationship between eclogite and blueschist in outcrop II

Eclogites and blueschists occur as interlayers on a decimeter scale in outcrop II. Such interlayering is common in some HP/LT metamorphic belts (e.g., El-Shazly et al., 1997; Klemd et al., 2002). It has been ascribed to: (1) differences in P-T conditions in different layers (e.g., Ridley, 1984); (2) differences in the bulk-rock composition of the layers before metamorphism (e.g., Oh et al., 1991); (3) differences in the extent of retrograde overprinting due to fluid infiltration (e.g., Barrientos and Selverstone, 1993; Gao and Klemd, 2001). As far as the present rocks are concerned, due to lack of apparent structural breaks between the different layers, interlayering suggests that blueschist and eclogite underwent a common tectono-metamorphic history. The contrasting parageneses may be attributed to differences in bulk-rock compositions as well as the availability and composition of fluid during subduction and exhumation. This may support conclusions proposed by El-Shazly et al. (1990, 1997) for eclogites and blueschists from northeast Oman. Preliminary whole-rock major-element data also indicate that the eclogite and blueschist in outcrop II are chemically distinct (Table 5). Eclogite has higher CaO contents, which may control the occurrence of Cpx in peak metamorphic assemblages. However, more detailed petrological and geochemical data as well as fluid inclusion analyses are needed in order to further elucidate the relationship between interlayered eclogites and blueschists.

TABLE 5. Bulk-Rock Analyses of Major Elements for Blueschist and Eclogite from Outcrop II

Sample:	SiO_2	TiO_2	Al_2O_3	Fe_2O_3	FeO	MnO	MgO	CaO	Na_2O	K_2O	P_2O_5	H_2O^+	CO_2	LOI	Total
A03-1-4.1	49.93	1.81	14.77	2.38	12.57	0.28	6.55	5.56	3.15	0.18	0.27	2.28	0.69	0.69	100.45
A03-3-5.3	50.07	1.08	15.78	3.43	5.07	0.16	7.83	10.54	2.62	1.12	0.10	1.74	0.50	0.90	100.94

Tectonic implications: Comparsion with the HP/LT metamorphic belt in the north Qilian—a HP/LT metamorphic belt cut by the Altyn Tagh fault?

The eclogites and blueschists of the northern Altyn Tagh described in this study are the first reported, although HP metapelites have been previously recognized. They constitute part of a HP/LT metamorphic terrane consisting of metapelites, quartz micaschists, calcareous schists, and metabasites. The HP/LT terrane is associated with ophiolitic mélange, and lies within the northern Altyn Tagh subduction complex belt extending from Hongliugou to Lapeiquan (Fig. 1). In consideration of large-scale displacement along the Altyn Tagh fault, determination of the lateral continuation of the northern Altyn Tagh subduction complex belt including the distinctive HP/LT metamorphic terrane is very significant for the tectonic reconstruction of the Altyn Tagh–North Qilian orogen, and also helps to quantify the offset of the Altyn Tagh fault.

To the east of the Altyn Tagh, an early Paleozoic subduction complex belt occurs within the North Qilian Mountains (e.g., Xiao et al., 1978; Wu et al., 1993). Two HP/LT metamorphic terranes have been recognized in the North Qilian subduction complex belt (e.g., Wu et al., 1993). One of them contains blueschists and low-T glaucophane-bearing eclogites associated with an ophiolitic mélange, and constitutes the deeper parts of the North Qilian subduction belt (Zhang et al., 1998). Based on Grt-Omp-Phn geobarometry, P-T estimates for the North Qilian eclogites are T = 450–560°C and P = 19–23 kbar (Zhang et al., unpubl. data). These conditions are very similar to those computed for the north Altyn Tagh HP/LT metamorphic terrane. This suggests that the north Altyn Tagh may be the western extension of the North Qilian Mountains. However, because the northern Altyn Tagh HP/LT metamorphic terrane is recently discovered, no radiometric data constrain the ages of the eclogites and blueschists. A single ^{40}Ar-^{39}Ar plateau age of 575 ± 3 Ma on phengite from HP metapelite (Che et al., 1995) is different from the ages of 440–460 Ma for the North Qilian blueschists and eclogites (Zhang et al., 1997). However, phengite Ar/Ar ages often overestimate the time of HP metamorphism because of excess argon (Li et al., 1994). Detailed geochronological study is required to constrain the precise timing of the HP/LT metamorphism.

Another HP/LT metamorphic belt in the North Qilian Mountains consists of lower-grade blueschists characterized by the assemblage albite-lawsonite-glaucophane. A similar assemblage has not been recognized yet in the north Altyn Tagh. Therefore, further petrochemical study is required to constrain the correlation between the north Altyn Tagh and North Qilian HP/LT metamorphic terrane.

Acknowledgments

This paper is submitted in honor of W. G. Ernst for his contribution to the study of HP/LT metamorphism. This study was supported by the National Natural Science Foundation of China (40272095, 40472102), the Key Project of the Ministry of Land and Resources of China (20010201), and the Geological Survey Project of China (200313000058). Detailed reviews by Profesor J. G. Liou helped improve the manuscript substantially. Thanks also are given to Yu Jing for her assistance with the microprobe analyses.

REFERENCES

Aoya, M., Uehara, S., Matsumoto, M., Wallis, S. R., and Enami, M., 2003, Subduction-stage pressure-temperature path of eclogite from the sambagawa belt: Prophetic record for oceanic-ridge subduction: Geology, v. 31, p. 1045–1048.

Barrientos, X., and Selverstone, J., 1993, Infiltration vs. thermal overprinting of epidote blueschists, Ile de Groix, France: Geology, v. 21, p. 69–72.

BGMX (Bureau of Geology and Mineral Resources of Xinjiang Uygur Autonomous Region), 1993, Regional geology of Xingjiang Uygur Autonomous Region: Beijing, China, Geological Publishing, House, p. 315–318.

Che, Z., Liu, L., Liu, H., and Luo, J., 1995, The discovery and occurrence of high-pressure metapelitic rocks from Altun Mountain area, Xinjiang Autonomous Region: Chinese Science Bulletin, v. 40, p. 1298–1300.

Che, Z., and Sun, Y., 1996, The age of the Altun granulite facies complex and the basement of the Tarim basin: Regional Geology of China, v. 56, p. 51–57.

Coleman, R. G., Lee, D. E., Beatty, L. B., and Brannock, W. W., 1965, Eclogite and eclogites: Their differences and similarities: Geology Society of America Bulletin, v. 76, p. 483–508.

El-Shazly, A. K., Coleman, R. G., and Liou, J. G., 1990, Eclogites and blueschists from NE Oman: Petrology and P-t evolution: Journal of Geology, v. 31, p. 629–666.

El-Shazly, A. K., Worthing, M. A., and Liou, J. G., 1997, Interlayered eclogites, blueschists, and epidote amphibolites from NE Oman: A record of protolith compositional control and limited fluid infiltration: Journal of Petrology, v. 38, p. 1461–1487.

Ernst, W. G., 1971, Do mineral parageneses reflect unusually high pressure conditions of Fanciscan metamorphism?: American Journal of Sciences, v. 270, p. 81–108.

Gao, J., and Klemd, R., 2001, Primary fluids entrapped at blueschist to eclogites transition: Evidence from the Tianshan meta-subduction complex in northwestern China: Contributions to Mineralogy and Petrology, v. 142, p. 1–14.

Graham, C. M., and Powell, R., 1984, A garnet-hornblende geothermometer and application to the Pelona schists, southern California: Journal of Metamorphic Geology, v. 2, p. 13–22.

Holland, T. J. B., 1980, The reaction albite = jadeite + quartz determined experimentally in the range 600–1200°C: American Mineralogist, v. 65, p. 129–143.

Klemd, R., Schroter, F. C., Will, T. M., and Gao, J., 2002, P-T evolution of glaucophane-omphacite bearing HP/LT rocks in the western Tianshan Orogen, NW China: New evidence for "Alpine-type" tectonics: Journal of Metamorphic Geology, v. 20, p. 239–254.

Kohn, M. J., and Spear, F. S., 1990, Two new barometers for garnet amphibolites with applications to southeastern Vermont: American Mineralogist, v. 75, p. 89–96.

Kretz, R., 1983, Symbols for rock-forming minerals: Amercian Mineralogist, v. 68, p. 277–279.

Krogh, E. J., Oh, C. W., and Liou, J. G., 1994, Polyphase and anticlockwise P-T evolution for Francisican eclogites and blueschists from Jenner, California, USA: Journal of Metamorphic Geology, v. 12, p. 121–134.

Leake, B. E., Woolley, A. R., Arps, C. E. S., Birch, W. D., Gilbert, M. C., Grice, J. D., Hawthorne, F. C., Kato, A., Kisch, H. J., Krivovichev, V. G., Linthout, K., Laird, J., Mandarino J. A., Maresch, W. V., Nickel, E. H., Rock, N. M. S., Schumacher, J .C., Smith, D. C., Stephenson, N. C. N., Ungaretti, L., Whittaker, E. J. W., and Youzhi, G., 1997, Nomenclature of amphiboles: Report of the subcommittee on amphiboles of the International Mineralogical Association, Commission on New Minerals and Mineral Names: American Mineralogist, v. 82, p. 1019–1037.

Li, S., Wang, S., and Chen, Y., 1994, Excess argon in phengite from eclogite: Evidence from dating of eclogite minerals by Sm-Nd, Rb-Sr, and $^{40}Ar/^{39}Ar$ method: Chemical Geology, v. 112, p. 343–350.

Liou, J. G., Zhang, R. Y., and Ernst, W. G., 1998, High-pressure minerals from deeply subducted metamorphic rocks: Reviews in Mineralogy, v. 37, p. 33–96.

Maruyama, S., Liou, J. G., and Terabayashi, M., 1996, Blueschists and eclogites of the world and their exhumation: International Geology Review, v. 38, p. 485–594.

Matsumoto, M., Wallis, S., Aoya, M., Enami, M., Kawano, J., and Seto, Y., 2003, Petrological constraints on the formation conditions and retrograde P-T path of the

Kotsu eclogite unit, central Shikoku: Journal of Metamorphic Geology, v. 21, p. 363–176.

Morimoto, N., Fabries, J., ferguson, A. K., Ginzburg, I. V., Ross, M., Siefert, F. A., and Zussman, J., 1988, Nomenclature of pyroxenes: American Mineralogist, v. 73, p. 1123–1133.

Oh, C. W., Liou, J. G., and Maruyama, S., 1991, Low temperature eclogites and eclogitic schists in Mn-rich metabasites in War Creek, California; Mn and Fe effects on the transition between blueschist and eclogite: Journal of Petrology, v. 32, p. 275–301.

Otsuki, M., and Banno, S., 1990, Prograde and retrograde metamorphism of hemantite-bearing basic schist in Sanbagawa belt in central Shikoku: Journal of Metamorphic Geology, v. 8, p. 425–439.

Peltzer, G., and Tapponnier, P., 1988, Formation and evolution of strike-slip faults, rifts, and basins during the India-Asia collision: An experimental approach: Journal of Geophysical Research, v. 93, p. 119–133.

Powell, R., Holland, T. J. B., and Worlry, B., 1998, Calculating phase diagram involving solid solutions via nonlinear equations, with examples using THERMOCALC: Journal of Metamorphic Geology, v. 16, p. 327–342.

Ravna, E. K., 2000, The garnet-clinopyroxene Fe^{2+}-Mg geothermometer: An updated calibration: Journal of Metamorphic Geology, v. 18, p. 211–219.

Ravna, E. J. K., and Terry, M. P., 2001, Geothermobarometry of phengite–kyanite–quartz/coesite eclogites [abs.]: Eleventh Annual V. M. Goldschmidt Conference, abstract# 3145 p.

Ridley, J., 1984, Evidence of the temperature-dependent blueschist to eclogite transformation in high pressure metamorphism of metabasites: Journal of Petrology, v. 25, p. 852–870.

Sobel, E. R., and Arnaud, N., 1999, A possible middle Paleozoic suture in the Altyn Tagh, NW China: Tectonics, v. 18, p. 64–74.

Wan, Y., Xu, Z., Yang, J., and Zhang, J., 2001, Ages and compositions of the Precambrian high-grade basement of the Qilian terrane and adjacent areas: Acta Geologica Sinica, v. 75, p. 375–384.

Wu, J., Li, J., and Lan, C., 2001, Progress on studies of the ophiolite in the Altyn Tagh. Scientia Geologica Sinica, v. 36, p. 342–349.

Wu, H. Q., Feng, Y. M., and Song, S.G., 1993, Metamorphic and deformation of blueschist belts and their tectonic implications, North Qilian Mountains, China: Journal of Metamorphic Geology, v. 11, p. 523–536.

Xiao, X. C., Chen, G. M., and Zhu, Z. Z., 1978, A preliminary study on the tectonics of anicent ophiolites in Qilian Mountains, Northwest China: Acta Geologica Sinica, v. 52, p. 287–295.

Yang, J., Xu, Z., Zhang, J., Chu, C., Zhang, R., and Liou, J. G., 2001, Tectonic significance of early Paleozoic high-pressure rocks in Altun-Qaidam-Qilian Monutains, northern China: Geological Society of Amercia, Memoir 194, p. 151–170.

Yue, Y., Ritts, B. D., and Graham, S. A., 2001, Initiation and long-term slip history of the Altyn Tagh fault: International Geology Review, v. 43, p. 1097–1093.

Yue, Y., Ritts, B. D., Graham, S. A., Wooden, J. L., Gehrels, G. E., and Zhang, Z. C., 2003, Slowing extrusion tectonics: Lowered estimates of post–Early Miocene slip rate for the Altyn Tagh fault: Earth and Planetary Science Letters, v. 217, p. 111–122.

Zhang, J., Xu, Z., Chen, W., and Xu, H., 1997, A tentative discussion on the ages of the subduction-accretionary complex/volcanic arcs in the middle sector of North Qilian mountains: Atca Petrologica Mineralogica, v. 16, p. 112–119.

Zhang, J., Xu, Z., Xu, H. and Li, H., 1998, Framework of North Qilian Caledonian subduction-accretionary wedge and its deformation dynamics: Scientia Geologica Sinica, v. 33, p. 290–299.

Zhang, J., Zhang, Z., Xu, Z., Yang, J., and Cui, J., 2001, Petrology and geochronology of eclogites from the western segment of the Altyn Tagh, northwestern China: Lithos, v. 56, p. 187–206.

Zhang, J., Zhang, Z., Xu, Z, Yang, J., and Cui, J., 2000, Discovery of khondalite series from western segment of the Altyn Tagh and their petrological and geochronological studies: Science in China, series D, v. 43, p. 308–316.

Two Ultrahigh-Pressure Metamorphic Events Recognized in the Central Orogenic Belt of China: Evidence from the U-Pb Dating of Coesite-Bearing Zircons

JINGSUI YANG,[1] FULAI LIU, CAIALI WU, ZHIQIN XU, RENDING SHI, SONGYONG CHEN,

Key Laboratory for Continental Dynamics of MLR (KLCD), Institute of Geology, Chinese Academy of Geological Science, Beijing, China, 100037

ETIENNE DELOULE,

Centre de Researches Petrographiques et Geochimiques (CRPG), 15 rue Notre Dame des Pauvres, 54501 Vandoeuvre-lès-Nancy cedex, France

AND JOSEPH L. WOODEN

Department of Geological and Environmental Sciences, Stanford University, California 94305

Abstract

A ~4000 km long ultrahigh-pressure metamorphic (UHPM) belt in northern China has been documented on the basis of the discovery of coesite-bearing rocks in the Altun–North Qaidam terrane in the western Central Orogenic Belt (COB), and diamond-bearing rocks in Qinling in the central and Dabie-Sulu terrane in the east. New SIMS and SHRIMP U-Pb dates of zircons from coesite-bearing UHPM rocks indicate two UHPM events: one in the early Paleozoic and the other in the Triassic. Coesite-bearing zircons from a North Qaidam gneiss yielded UHP metamorphic ages of 452 ± 13.8 Ma and retrograde ages of 419 ± 6.7 Ma. A diamond-bearing gneiss from Qinling gave a lower intercept age of 502 ± 45 Ma, and an upper intercept age of 1545 ± 100 Ma, whereas a Qinling eclogite sample gave a lower intercept age of 493 ± 170 Ma and an upper intercept age of 1381 ± 82 Ma. The lower and upper intercept ages of the Qinling samples are interpreted as UHPM and protolith ages of the rocks, respectively. Coesite-bearing zircons from a Qinglongshan eclogite in the south Sulu belt yielded early Paleozoic UHPM ages of 441 ± 9 Ma, 449 ± 9 Ma, and 442 ± 9 Ma, whereas the core of a zircon containing plagioclase and apatite inclusions gave a protolith age of 761 ± 13 Ma. These age data suggest that the early Paleozoic UHPM rocks extend from west to east for about 4000 km across the COB, whereas the Triassic UHPM belt extends across the Dabie-Sulu region for about 1000 km.

Based on available geochronological and geochemical data, we suggest the following tectonic model for evolution of the COB. At about 1000 Ma, the area was amalgamated to form the Rodinian continent, which contained ophiolitic fragments of oceanic affinity. This part of Rodinia was then rifted at about 800–750 Ma to form an oceanic basin with a variety of MORB and intruded by granitic plutons. Closure of this ocean basin produced Neoproterozoic ophiolites and granitic gneisses. The UHPM rocks, along with subduction-related island-arc volcanics and granites of early Paleozoic age suggest a second cycle of rifting and subduction along the COB, whereas the Triassic UHPM rocks record a final subduction and collision event between the North China and South China blocks.

Introduction

THE CENTRAL OROGENIC BELT (COB) lies between the North China (NCB) and South China blocks (SCB). It extends about 4000 km across China and, from west to east, includes the Kunlun Mountains, Altun Mountains, Qilian Mountains, Qinling Mountains, and Dabie-Sulu Mountains. Previous studies have shown that the COB has a complex geological history. At least two periods of plate divergence and convergence are indicated by the occurrence of early Paleozoic and Triassic ophiolites and subduction-related magmatism (e.g., Yang J. S. et al., 1996; Meng and Zhang, 1999). The COB has become

---

[1]Corresponding author; email: yangjsui@ccsd.org.cn

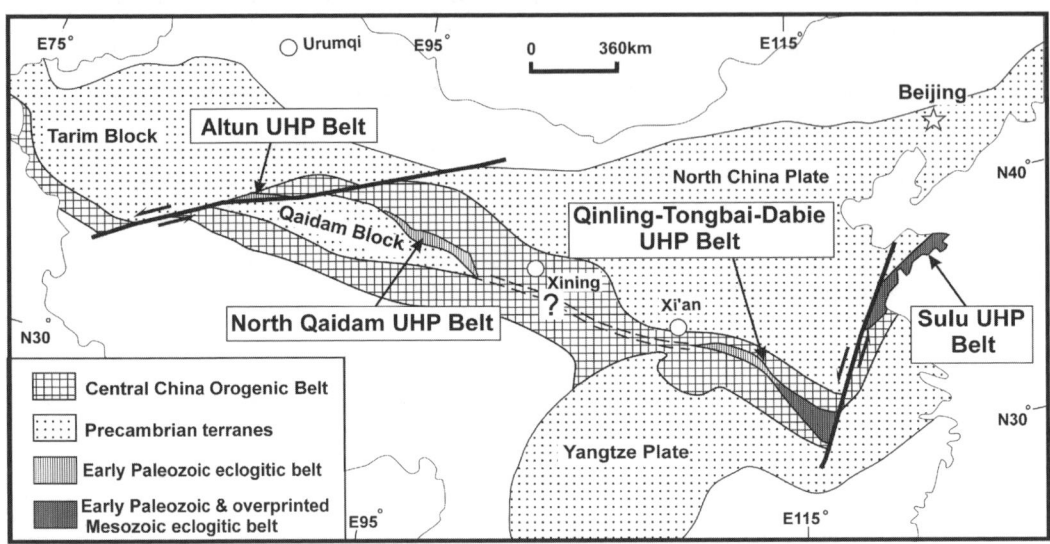

FIG. 1. Distribution of UHP terranes in the Central Orogenic Belt of China, and the two UHPM events recognized in this belt (modified after Yang J. S. et al., 2000b).

widely known because it contains the Dabie-Sulu ultrahigh pressure metamorphic (UHPM) belt in its eastern part. The Dabie-Sulu belt has been interpreted to represent a deep subduction and collision event between the NCB and SCB during the Triassic (Li S. G. et al., 1989, 1993, 2000; Ames et al., 1993, 1996; Hacker et al., 1998; Cong et al., 1994; Cong, 1996; Jahn, 1998, 1999). Recent studies indicate that an early Paleozoic UHPM belt exists in the Qilian-Altun region in the western COB, and Yang J. S. et al. (2000b, 2001a) thus proposed the existence of a huge UHPM belt across the COB from Altun in the west to Dabie-Sulu in the east. This interpretation has been further supported by the recent discovery of micro-diamonds in eclogite and gneiss in the middle part of the COB (Yang J. S. et al., 2003b; Fig. 1). This huge UHPM belt is probably over 4000 km long; some regions may have also experienced two UHPM events respectively in the early Paleozoic and Triassic. The spatial distribution of the rock records for these two UHPM events, their causal mechanisms, and tectonic implications are currently being studied. This paper reports new age data on coesite-bearing zircons from the North Qaidam gneiss and Sulu eclogite, as well as on zircons from the Qinling diamond-bearing gneiss and eclogite. These new data provide strong evidence for the existence of both early Paleozoic and Triassic UHPM events.

Recognition of Two UHPM Events across the COB

Coesite as a typical UHPM mineral that formed at depths >80 km was first discovered in China in eclogite at Changpu, Dabie Mountains (Xu, 1987). Since then, Dabie-Sulu UHPM rocks have become a focal point for UHP studies by research groups and individuals from many different countries. These studies led to a much better understanding of the nature of UHPM rocks and their petrotectonic significance. Many coesite- and diamond-bearing rocks have been discovered in different localities in the Dabie-Sulu UHPM terrane, suggesting that continental subduction reached depths >120 km (Xu S. T. et al., 1992, 2003; Lu Y. Z., 1998; Yang J. S. et al., 1999). Exsolution textures and mineral lamellae in eclogite and garnet peridotite suggest that some rocks have reached subduction depths >200 km (Zhang R. Y. et al., 1999; Yang J. J. and Jahn, 2000; Ye et al., 2000). Previous studies also suggest that the Sulu UHPM belt was offset from the Dabie belt up to 500 km by the Tanlu fault (Okay et al., 1989). In the Dabie-Sulu belt the peak UHP metamorphism has been dated at 240–220 Ma and is regarded as the result of continental subduction and collision between the NCB and SCB (Li S. G.. et al., 1989, 1993, 2000; Chavagnac and Jahn.. 1996). These studies indicate that Dabie-Sulu UHPM rocks are widely distributed, contain a variety of lithological types, and have complex metamorphic and deformational histories. A 5000 m deep hole currently being drilled in the Sulu UHPM terrane at Donghai, Jiangsu Province by the Chinese Continental Scientific Drilling Project aims to understand the formation and exhumation history of the UHPM rocks (Xu Z. Q. et al., 1998).

Recently, a new UHPM belt, over 350 km long, was discovered in North Qaidam, in the western COB (Fig. 1), and is characterized by the occurrence of abundant eclogite and sparse garnet peridotite in geissic rocks (Yang J. S. et al., 1998, 2000a, 2001b). Coesite inclusions in zircon separates from country rock gneiss, pseudomorphs of coesite in garnet, and exsolution lamellae of quartz in omphacite suggest that both eclogite and its country rock underwent UHP metamorphism (Yang J. S. et al., 2001a, 2002; Song S. G. et al., 2003). UHP metamorphism in the North Qaidam terrane occurred in the early Paleozoic (500–440 Ma), whereas exhumation started at 470–460 Ma and ended at about 400 Ma based on isotope age data obtained by SHRIMP U-Pb, Sm-Nd, and Ar-Ar methods (Zhang J. X. et al., 2000; Yang J. S. et al., 2002, Xu Z. Q. et al., 2003). Based on similarities in occurrence, composition, and metamorphic grade, the North Qaidam UHPM belt is correlated with the eclogite belt in South Altun, indicating that the latter has been offset at least 400 km by the Altun Tagh fault (Liu L. et al., 1996; Xu Z. Q. et al., 1999; Zhang J. X. et al., 2000, 2001; Yang J. S. et al., 2001b). Inclusions of coesite pseudomorphs and exsolution structures in the Altun eclogite show that these rocks underwent UHP metamorphism (Zhang J. X. et al., 2002), an interpretation supported by the recent discovery of UHP garnet peridotite in the South Altyn (Liu L. et al., 2002). The peak metamorphic age of these rocks is ~500 Ma based on TIMS U-Pb and Sm-Nd dating (Zhang J. X. et al., 2000, 2001).

Based on comparisons of UHP metamorphic rocks in North Qaidam, Qinling, and Dabie-Sulu, Yang J. S. et al. (2000b, 2001a) postulated the existence of a 4000 km long UHP belt of early Paleozoic age in central China, extending from Altun in the west to Sulu in the east. They suggested that the first subduction and collision event between the NCB and SCB occurred in the early Paleozoic and that the Triassic Dabie-Sulu UHPM represented a later separate discrete event. Microdiamonds were later found as inclusions in zircon from eclogites and their gneissic country rocks in Qinling, and the UHPM age of the Qinling gneiss was determined to be ~500 Ma by SHRIMP U-Pb dating (Yang J. S. et al., 2003b). This provides critical evidence for the existence of the 4000 km long UHP metamorphic belt across central China, and clarifies the distribution of the early Paleozoic and Triassic UHP rocks in the belt. Early Paleozoic UHPM rocks occur mainly in the western portion of the COB, including Altun, North Qaidam, and East Qinling, whereas Triassic UHPM rocks occur only in the eastern section in Dabie and Sulu.

However, there are still some different opinions about the timing of the UHPM events; some believe that the Sulu and Dabie UHP metamorphism occurred in the Neoproterozoic about 800–750 Ma (Cao et al., 1990; Wang et al., 1994; Kang et al., 1996; Chen et al., 2000; Song M. C. et al., 2003). Some disagreements stem from the complex history of the UHP rocks and variable protolith ages of eclogites and country rock gneisses. For example, some gneissic rocks have protolith ages of 800–700 Ma (Liu F. L. et al., 2002, 2003) and eclogites and ultramafic rocks in eastern Shandong Province have protolith ages of 1821 ± 19 Ma and 581 ± 44 Ma, respectively. However, coesite-bearing zircons in ultramafic rocks give a UHP metamorphic age of 221 ± 12 Ma (Yang J. S. et al., 2003a). Therefore, the most useful and direct method for determining accurate UHPM ages is to analyze UHPM mineral-bearing zircons by *in situ* U-Pb dating.

In Situ U-Pb Dating by SIMS of Coesite-Bearing Zircons from North Qaidam Gneiss

North Qaidam UHPM rocks consist of eclogite, garnet peridotite, and some associated gneissic country rocks. From west to east, they occur in three main localities: Daqaidam, Xitieshan, and North Dulan (Fig. 2). Some isotope ages related to the UHPM event have been previously reported. Zircons from eclogites in Daqaidam gave a 494.6 ± 6.5 Ma U-Pb age by TIMS, which is interpreted to represent the age of crystallization of metamorphic zircon during the UHPM event; phengite from same eclogite yielded an Ar-Ar plateau age of 466.7 ± 1.2 Ma and an isochron age of 465.9 ± 5.4 Ma, which are believed to represent cooling ages related to exhumation of the eclogites (Zhang J. X. et al., 2000). SHRIMP U-Pb dating of zircons from eclogite in Dulun yielded ages of 495–443 Ma (average of 473 ± 20 Ma), and the same sample yielded a 457.7 ± 3.3 Ma Sm-Nd whole rock-mineral age (Yang J. S. et al., 2002). Finally, zircons

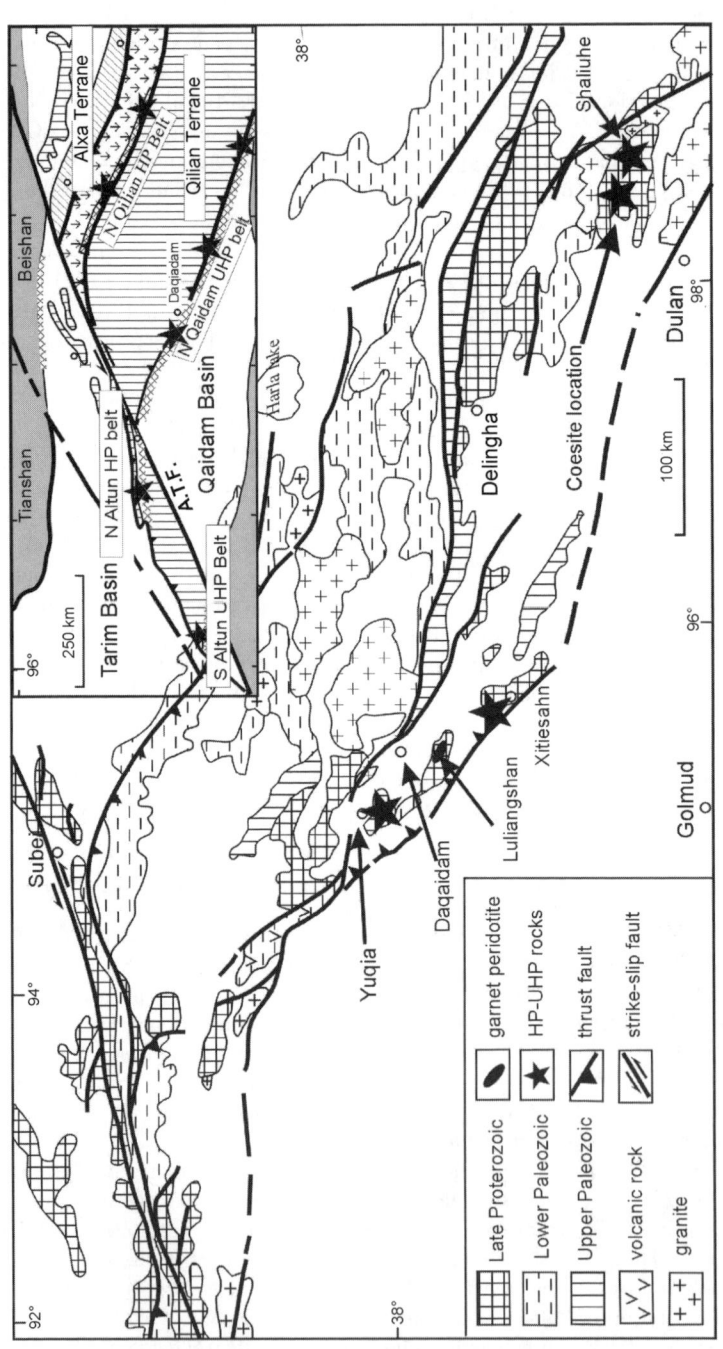

FIG. 2 Distribution of UHP metamorphic rocks in the North Qaidam Mountains.

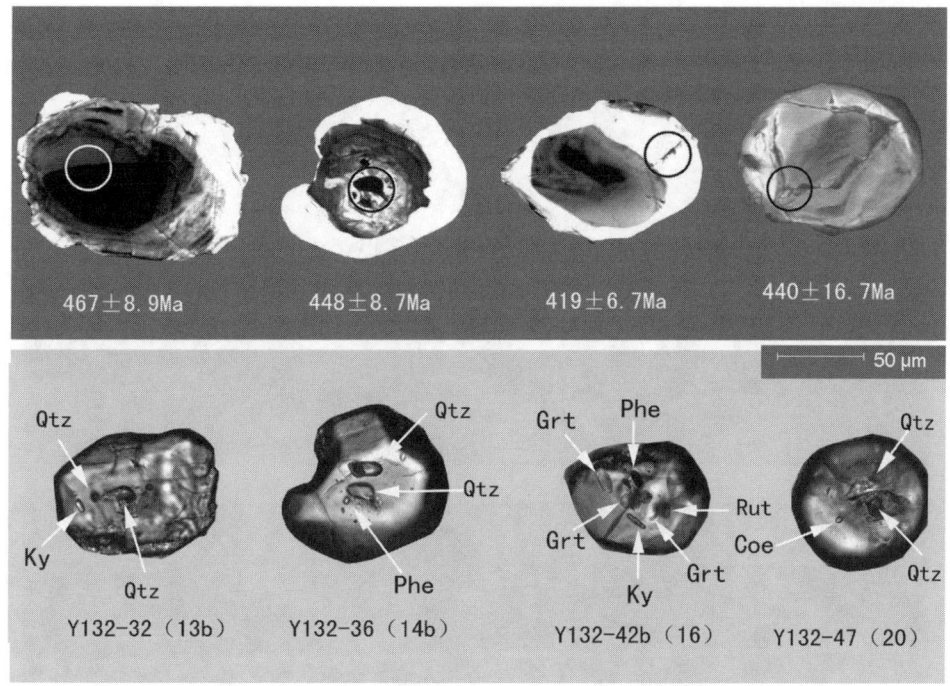

FIG. 3. Cathodoluminescence images of single zircons from coesite-bearing gneiss from Dulan (sample 01Y-132). Abbreviations: Qtz = quartz; Ky = kyanite; Phe = phengite; Grt = garnet; Ru = rutile; Coe = coesite.

from eclogite in Shaliuhe gave an age of 497 ± 10 Ma by TIMS U-Pb dating (Lu, 2003).

The above data indicate that the UHP metamorphism occurred in the Early Paleozoic at ~500–440 Ma. The formation age is not clearly separated from the retrograde age and thus cannot be used to estimate exhumation rate. In this study, we report new isotope ages obtained from coesite-bearing zircons in Dulan gneiss by U-Pb SIMS (CAMECA IMS-1270) dating performed in the Centre de Recherches Petrographiques et Geochemicques (CRPG-CRPG) in Nancy, France.

A coesite-bearing gneissic sample (00Y-132) from the North Belt of Dulan (Fig. 3) is strongly deformed, and consists mainly of garnet (5 wt%) + muscovite (30%) + biotite (10%) + quartz (> 50%) + titanite (rare). Garnets are elongated and rotated; most are partly replaced by biotite. Muscovite occurs as narrow sheets that define the foliation. No peak UHP minerals were observed—these have been replaced by assemblages of retrograde minerals. Most zircons from the gneissic sample are subrounded, with typical metamorphic features.

Cathodoluminescence (CL) images show well-developed zoning with distinct cores and rims (Fig. 3). UHP minerals, such as coesite, kyanite, phengite, garnet, and rutile, occur as inclusions in the zircon cores. Coesites show a strong peak at 521–525, and weak peaks at 272 and 181 cm^{-1} by laser Raman spectroscopy (Yang J. S. et al., 2001b). In situ dating of the zircons yielded 467 ± 8.9 Ma and 448 ± 6.7 Ma for the cores, whereas the coesite-bearing spot gave 440 ± 16.7 Ma. The average age of the three dates is 452 ± 13.8 Ma, which is taken as the age of UHP metamorphism. The well-developed rim on this grain yielded an age of 419 ± 6.7 Ma, which is believed to date the retrogression event. Both cores and rims show low Th/U ratios of <0.1, a typical feature of metamorphic zircons.

Sample 01Y-469 is another coesite-bearing gneissic rock hosting eclogite in the North Belt of Dulan. U-Pb dating was carried out by SHRIMP-RG at Stanford University. Some zircon grains contain a few inclusions of coesite, garnet, and rutile, whereas others contain graphite, apatite, and quartz (Fig. 4A). CL images show that some zircons are zoned

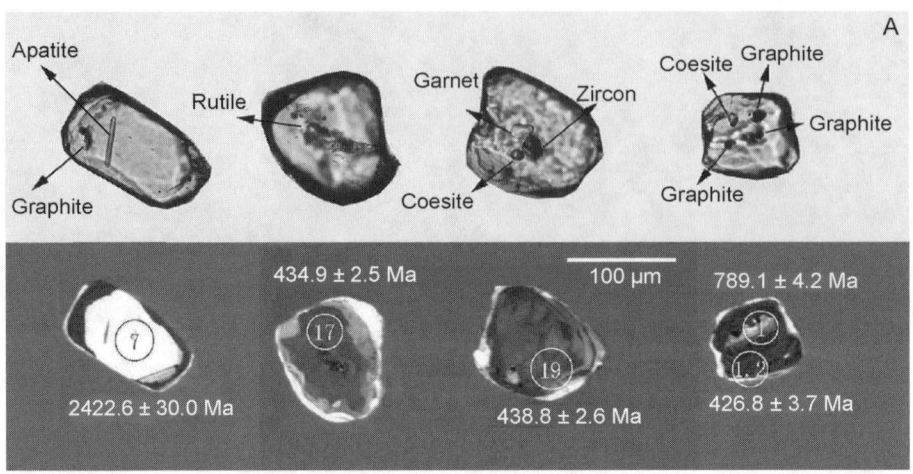
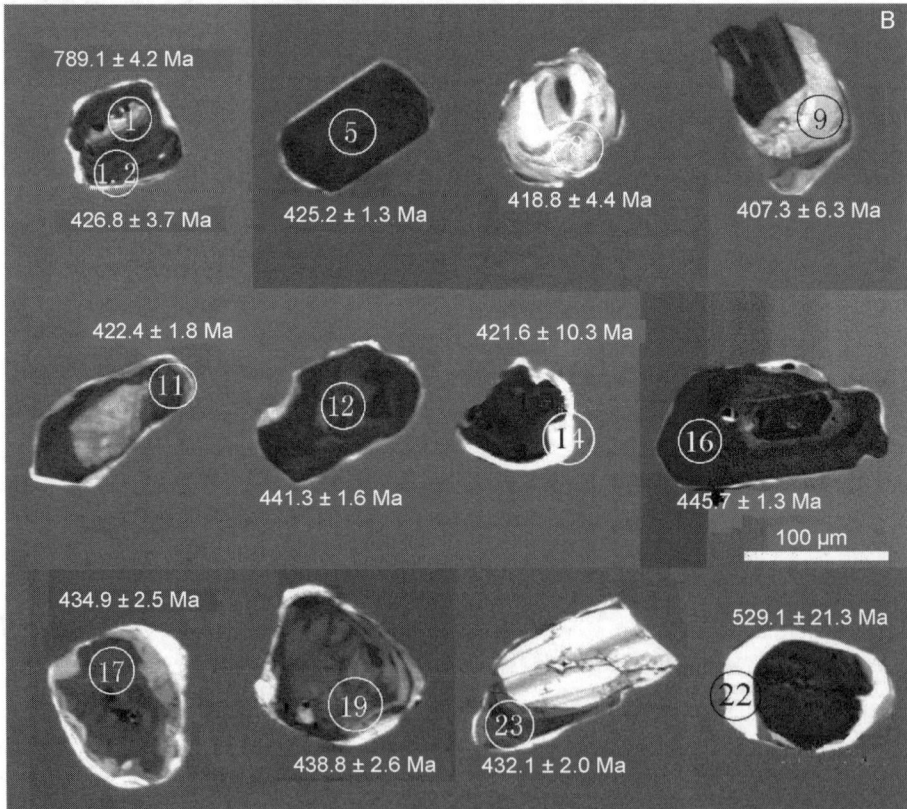

FIG. 4. A. Zircons with UHP mineral inclusions from coesite-bearing gneiss from Dulan. B–C. Cathodoluminescence images of single zircons. D. Average of 11 dates is 433.5 ± 6.6 Ma, identical to the average age (432.8 ± 3.2 Ma) of the two coesite-bearing zircons, and is taken to represent the UHPM age of the rock. E. Cores of all samples plot on a concordia line, suggesting that the different ages represent the original ages of zircons, which were most likely from fragments of different rocks.

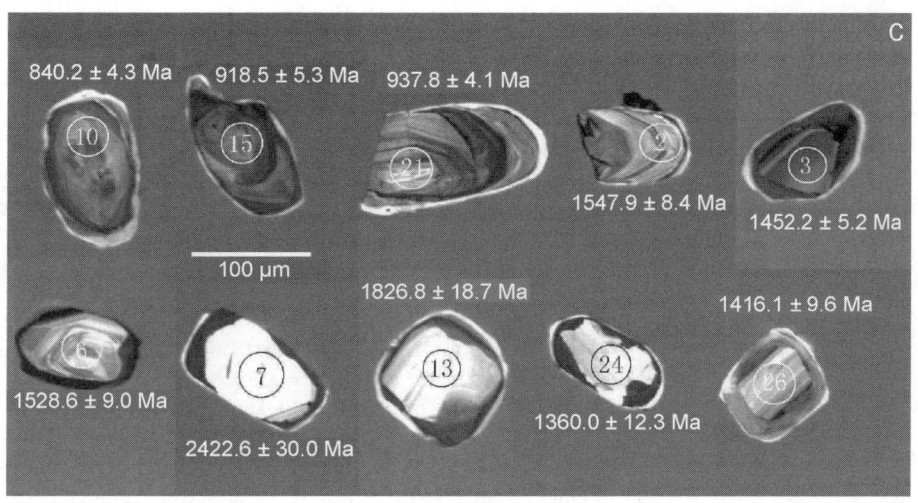

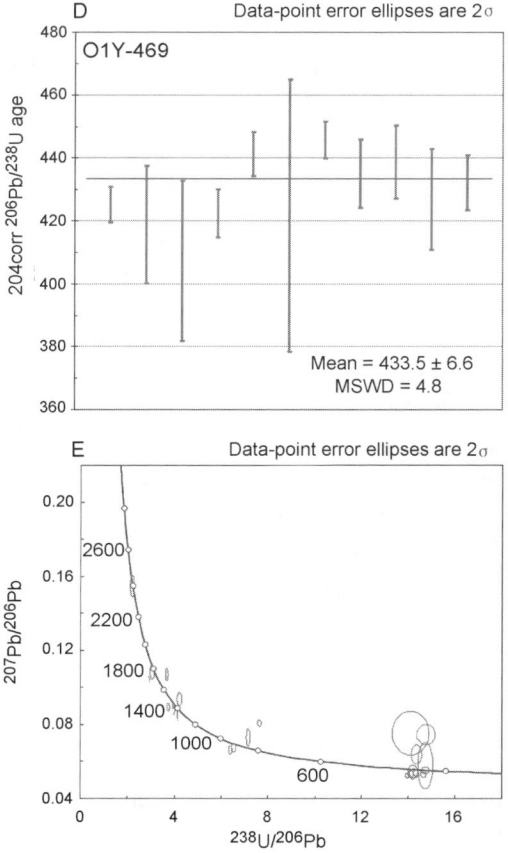

FIG. 4. Continued.

with distinct cores and rims (Figs. 4B–4C). The cores commonly show magmatic oscillatory zoning, and contain inclusions of graphite, quartz, and apatite. The rims contain inclusions of UHP minerals, such as coesite, garnet, and rutile, and are interpreted as metamorphic overgrowths. Some are unzoned, with or without coesite inclusions, but all have a low Th/U ratio, suggesting a metamorphic origin. Twenty-three measurements using SHRIMP U-Pb dating were carried out on 22 zircons, including two grains with coesite inclusions. One zircon with a coesite-bearing rim yielded an age of 426.8 ± 3.7 Ma (No. 1.2) for the rim, and 789.1 ± 4.2 Ma (No. 1.1) for the core, which contains graphite inclusions. One unzoned grain (No. 19) with coesite and garnet inclusions yielded an age of 438.8 ± 2.6 Ma. Eleven grains yielded a group of ages ranging from 407.3 ± 6.3 (No. 9) to 445.7 ± 1.3 Ma (No. 16). The average of 11 dates is 433.5 ± 6.6 Ma (MSWD = 4.8,) (Fig. 4D), which is the same as the average age (432.8 ± 3.2 Ma) of the two coesite-bearing zircons, and is taken to represent the UHPM age of the rock. However, the cores have a much larger age range from 840.2 ± 4.3 (No. 10) to 2422.6 ± 30 Ma (No. 7). In a concordia diagram, the cores all plot on a concordia line, suggesting that the different ages may represent the protolith ages of zircons, which were most likely from fragments of different rocks (Fig. 4E).

SHRIMP U-Pb Dating of Qinling UHP Rocks

Eclogites were first reported in Qinling in 1994 by Hu et al. (1995). Recently, metamorphic microdiamonds have been discovered as inclusions in zircons from both eclogite and its host gneiss (Yang J. S. et al., 2003b). The Qinling UHP rocks thus provide an important link between the Altun-Qilian and Dabie-Sulu UHP belts.

SHRIMP U-Pb dating of zircons from the gneissic country rock yielded an age of 507 ± 38 Ma for UHP metamorphic rims and prolith ages of 1200–1800 Ma for zircon cores (Yang J. S. et al., 2003b). Here we report two new isotopic ages by SHRIMP U-Pb dating, one from eclogite and the other from diamond-bearing gneiss.

SHRIMP U-Pb age of diamond-bearing gneiss

The diamond-bearing sample (01Y-827) is a well-foliated, coarse-grained (2–4 mm) paragneiss from Guangpo, North Qinling (Fig. 5), composed mainly of muscovite and quartz, with some albitic feldspar and a few garnets. Subrounded zircon grains are approximately 100 μm in size and contain numerous inclusions of diamond, graphite, apatite, rutile, and quartz.

Twenty-one ages from 15 grains were obtained by SHRIMP II in the Ion Probe Center of Beijing. CL images show that 6 grains are zoned, allowing for measurement of ages of both rim and core. Six metamorphic rims yielded ages ranging from 476 ± 12Ma to 573 ± 16 Ma, with an average of 511 ± 35 Ma (MSWD = 8.4 2σ) (Fig. 6A), which includes one unzoned zircon grain with an age of 467 ± 12 Ma. Core ages of the same grains range from 743 ± 21 Ma to 1559 ± 42 Ma. Th/U ratios of the rims are low (<0.1), whereas those of the cores are relatively high (>1.0). In a Tera-Wasserburg plot, the ages of the samples fall into two main groups, and yield a lower intercept of 502 ± 45 Ma and an upper intercept of 1545 ± 100 Ma (MSWD = 8.7 2σ) (Fig. 6B). The lower intercept age (502 ± 45 Ma) is close to the average age of the 7 rims, and is regarded as the age of the UHP metamorphism. The upper intercept age (1545 ± 100 Ma) is taken as the protolith age of the rock.

SHRIMP U-Pb dating of zircons from Qinling eclogite

Sample 01Y-826 is massive and fine-grained (0.5–2 mm) eclogite with a granular texture. It consists mainly of garnet and omphacite, the latter being mostly replaced by amphibole, feldspar, and quartz. About 100 zircon grains were recovered from a 10 kg sample. The zircons have variable sizes and form two distinct groups. Most are prismatic grains between 50 to 70 μm long, with a length/width ratio of (3–4):1. CL images show that some of these grains have magmatic cores surrounded by metamorphic rims, which are too narrow to be dated. Less abundant are short prismatic grains, up to 300 μm in size, with a length/width ratio of (1–2):1. This type of zircon has magmatic zoning and wide metamorphic rims. There are a few mineral inclusions in the zircons, including rutile, graphite, and quartz. No diamond inclusions were found despite the presence of diamond in the host gneiss (sample 0Y-827).

Sixteen SHRIMP U-Pb dates on 15 zircon grains from this sample were obtained at Stanford University. When plotted on a concordia diagram (Fig. 7), 16 spots yield a lower intercept age of 493 ± 170 Ma and an upper intercept age of 1381 ± 82 Ma (MSWD = 4.4, 2σ) (data not shown). However, these data are

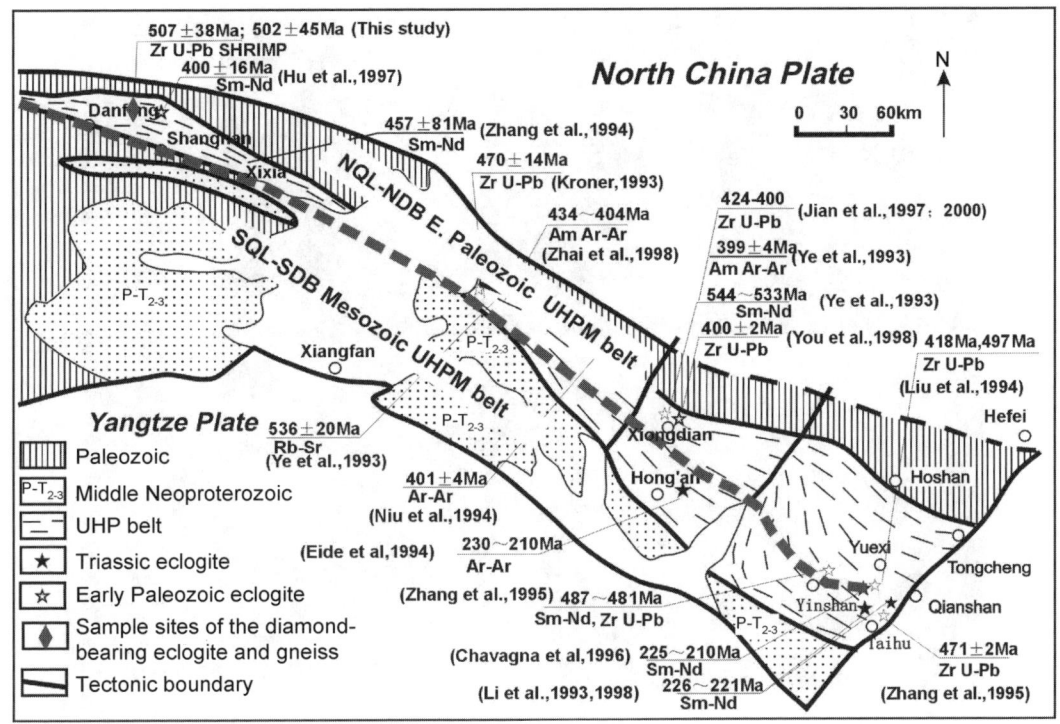

FIG. 5. Geologic sketch map of the East Qinling and Dabie regions (modified after Yang J. S. et al., 2000b).

considered preliminary because the small size of zircon grains allowed only one rim to be measured, resulting in a large uncertainties. However, the upper and lower intercept ages are similar to the ages of cores and rims of the two gneisses, and support an early Paleozoic age for the UHP metamorphism.

Early Paleozoic Coesite-Bearing Zircons in Sulu Eclogite

The Sulu UHP metamorphic belt is regarded as the eastern extension of the Dabie UHP metamorphic belt because both regions contain similar UHP metamorphics and host rocks formed under similar P-T conditions, and both are Triassic in age. Mantle-derived garnet peridotites are more common in Sulu than in Dabie, probably because the Sulu terrane has undergone greater uplift and erosion (Yang J. J., 1991; Zhang R. Y. et al., 1994, 2000; Xu Z. Q. et al., 1998; Ye et al., 2000; Liu F. L. et al., 2002).

An eclogite sample (QL-1) was collected from Qinglongshan in Donghai for dating (Fig. 8). At this site a body of fresh eclogite several hundred meters wide is exposed in a new road cut. Several types of eclogite are present, including phengite eclogite, phengite-kyanite-talc eclogite, quartz eclogite, and epidote-amphibole eclogite. Likewise, the eclogite blocks are hosted in a variety of gneiss, kyanite quartzite, and epidote-phengite schist.

Sample (QL-1) is a rutile-bearing phengite eclogite. Zircon separates from this sample are subrounded and have typical metamorphic morphologies. Laser Raman and CL images indicate that the zircons commonly have an old core and a younger metamorphic rim, and coesite occurs in both rim and core. Fifteen measurements from 10 zircon grains were carried out on the SHRIMP II in Institute of Geology, Chinese Academy of Geological Sciences, Beijing. The results yield three distinct age groups, Neoproterozoic, early Paleozoic, and Triassic. The Neoproterozoic ages are from old cores without UHP mineral inclusions, whereas the Paleozoic and Mesozoic ages are all from samples containing UHP mineral inclusions such as coesite, omphacite, rutile, and garnet. Early Paleozoic ages

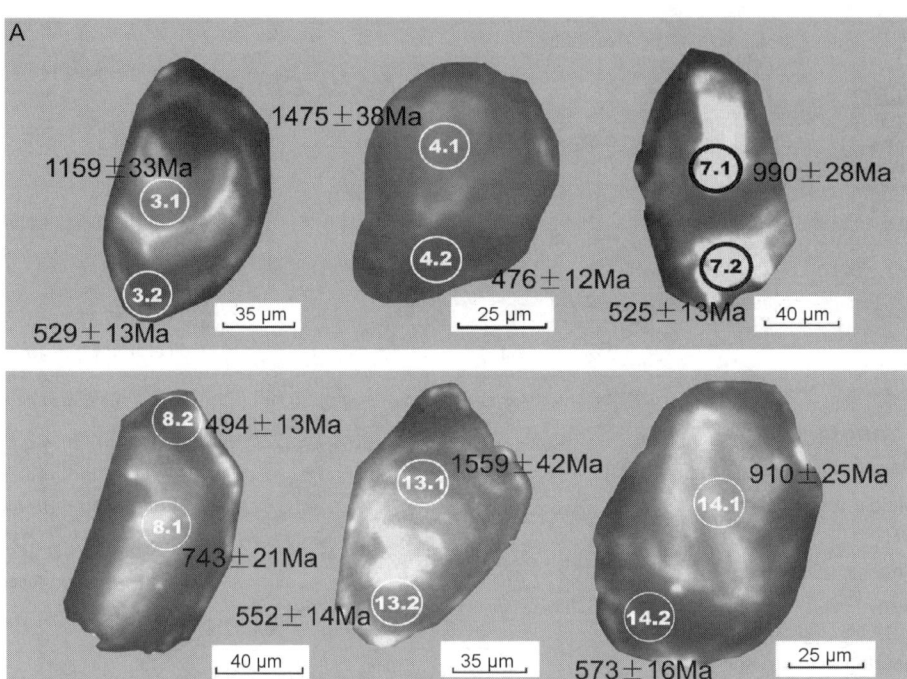

FIG. 6. SHRIMP U-Pb age and CL images of single zircons from a Qinling felsic gneiss (01Y-827). A. CL images reveal relic magmatic cores surrounded by UHP metamorphic rims. B. SHRIMP ages of diamond-bearing gneiss (01Y-827) plotted on a Tera-Wasserburg diagram. C. Ten zircon rims and cores from samples 00Y-17 and 01Y-827 show systematic variations in the Th/U ratios with $^{236}Pb/^{238}U$ ages (quoted from Yang et al., 2003b).

were obtained from rims on old cores as well as on zircons lacking distinct cores. Two zircons in the sample with Paleozoic ages are described below.

Zircon QL1-7 is rounded and has well-developed oscillatory zoning (Figs. 9A and 9B), but lacks a distinct core or rim. Mineral inclusions such as coesite, omphacite, and rutile were identified by Laser Raman, suggesting the zircon formed during UHP metamorphism. Two ages, 449 ± 9 Ma and 441 ± 9 Ma were obtained for this zircon, giving an average of 445 ± 9 Ma. The Th/U ratio is 0.60–0.64, which is obviously lower than that of old cores (1.34, see below), and is considered typical for metamorphic zircon. Although oscillatory zoning is typical of igneous zircon, it has been reported in metamorphic zircons from a Sesia-Lanzo quartz-jade vein in the western Alps (Rubatto et al., 1999). However, it is not clear whether that zircon formed from metamorphic fluids or was an inherited grain.

Zircon QL1-22 is also subrounded with distinct core and rim. In CL image, the core is bright whereas the rim has relatively low luminescence (Fig. 9D). The grain in Figure 9D has a core with well-developed magmatic zoning, which is characterized by high U (188 ppm) and Th (242 ppm) and a high Th/U ratio (1.34). These features, along with inclusions of plagioclase and apatite in the core (Fig. 9C), are typical for relict magmatic zircon. The core yielded a date of 761 ± 13 Ma, which is considered to be the protolith age.

The rim lacks oscillatory zoning, contains coesite and omphacite inclusions, has low contents of U (157 ppm) and Th (92 ppm), and a low Th/U ratio (0.61). The Th/U ratio is obviously lower than that of the core but similar to that of sample QL1-7 (0.60–0.64). The rim is typical of UHP metamorphic overgrowths and yielded an age of 442 ± 9 Ma.

Metamorphic zircon can form either by new growth from metamorphic fluids or by subsolidus recrystallization of a pre-existing magmatic or detrital crystal. Regardless of the process involved, Th/U ratios typically decrease during metamorphism (Rubatto et al., 1999). The Qinglongshan zircons have significantly lower Th/U ratios in the rim than

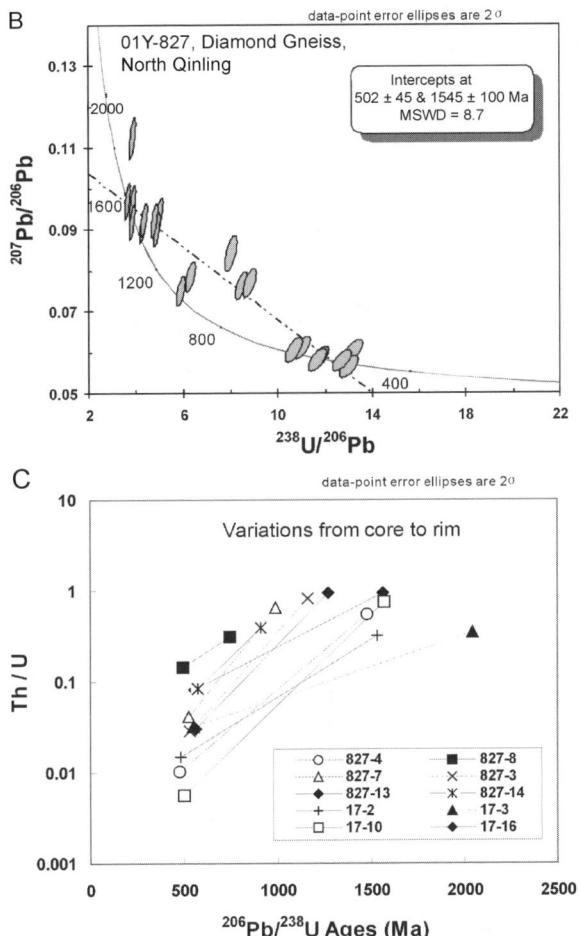

FIG. 6. *Continued.*

in the core, i.e., 0.6 and 1.34, respectively, but not as low (–0.1) as those reported from other UHP metamorphic zircons, such as those from the Kokchetav UHP massif (Katayama et al., 2001) and the Dabie UHP terrane (Hacker et al., 1998). These data suggest that zircons formed during UHP metamorphism may have a range of Th/U ratios, and that the relatively high values of the Qinglongshan samples were probably produced by melting of pre-existing magmatic crystals. A similar example has been reported from metamorphic gneiss formed by anatexis in the Kishar Formation, western Himalaya, where metamorphic zircons have high Th/U ratios and growth zoning (Dipietro and Isachsen, 2001). Another example is from the Altyn Tagh fault, where metamorphic zircons in veins formed by anatexis have high Th/U ratios (0.5–0.6) and growth zoning, both probably related to compositions of the primary zircon or the fluid (Li, H. B. et al., 2001).

Evidence of Triassic UHP Metamorphim of Age in the Dabie-Sulu Belt

A peak UHPM age of 240–200 Ma for the Dabie-Sulu UHPM rocks is generally accepted on the basis of various geochronological data (e.g., Li S. G. et al., 1989, 1993; Ames et al., 1996; Cong, 1996; Chavagnac and Jahn, 1996; Rowley et al., 1997; Hacker et al., 1998; Maruyama et al., 1998). However, although the metamorphic age of the belt has been considered to be Triassic by most authors, some doubts remain as to the age of the peak UHPM. For

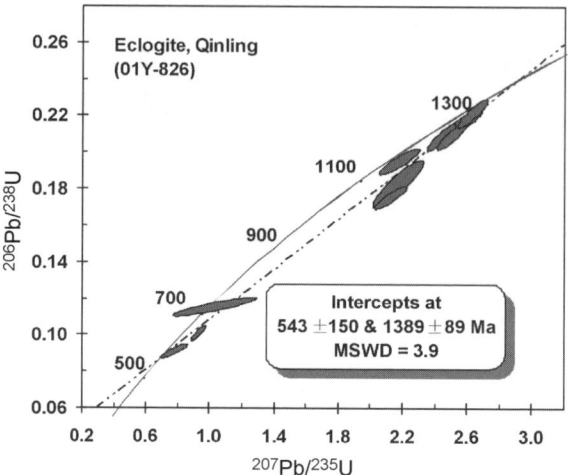

FIG. 7. SHRIMP age plots of eclogite (Sample 01Y-826) from Qinling.

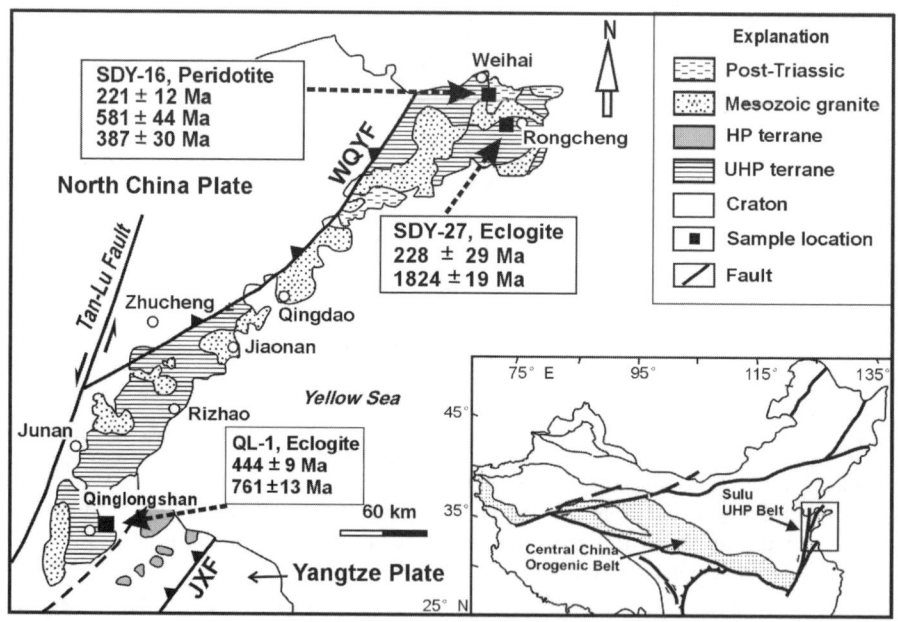

FIG. 8. Location of SHRIMP U-Pb dating samples (modified after Liu et al., 2001), and ages of SDY-16 and DY-27 from Yang J. S. et al. (2003a).

example, early Paleozoic ages of 487–420 Ma, based on Sm-Nd isochron and U-Pb isotope dating of single zircons, have been reported from eclogites from various localities in the Dabie Mountains (Zhang Z. et al., 1995; You et al., 1998; Jian et al., 2000). Moreover, eclogites in the Qinling orogenic belt are known to be of early Paleozoic age, and this terrane has traditionally been considered to be a western extension of the Dabie belt (e.g., Xu et al., 1998; Yang J. S. et al., 2000b; Zhang J. X. et al., 2001). Thus, two critical questions remain for the Dabie UHP belt: (1) Did the peak HP and UHP

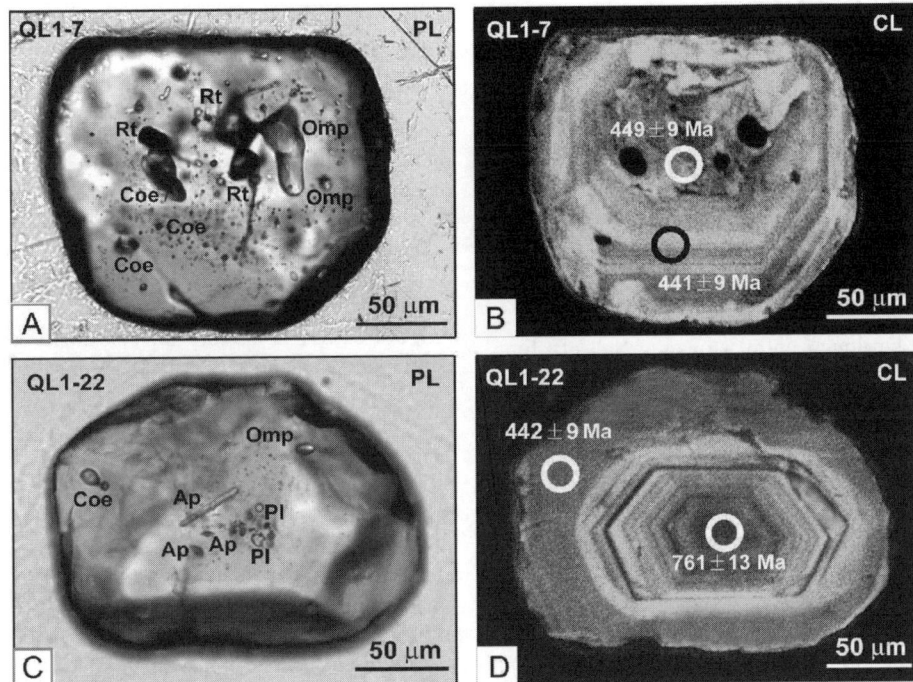

FIG. 9. CL images of coesite-bearing zircons and SHRIMP U-Pb ages of the Qinglongshan eclogite.

metamorphism occur in the Triassic or in the Paleozoic, or both? And (2), if the UHP rocks formed in a single Triassic event, what is the relationship of the Dabie HP/UHPM belt to the Qinling Paleozoic HP/UHPM belt?

Similar uncertainities exist for the age of the Sulu UHPM belt. Eclogite in Junnan with a U-Pb zircon age of ~265 Ma (Yang J. J., 1991) and eclogites in Jiaonan, Shandong Province with a U-Pb zircon age of ~ 217 Ma (Ames et al., 1996) have been reported. However, many supposedly Proterozoic UHPM ages have been reported from Jiaonan, southern Shandong. These include an Ar-Ar plateau age of 721 ± 13 Ma for phengite from eclogite (Cao et al., 1990), a $^{207}Pb/^{206}Pb$ age of 747 ± 13 Ma from a single zircon in eclogite (Wang et al., 1994), and U-Pb ages of 765–653 Ma from a single zircon in eclogite (Kang et al., 1996). Thus, the age of the peak UHP metamorphism in the Sulu UHP belt is uncertain, and the relationship between the Sulu and Dabie belts in terms of formational ages and tectonic environments is unclear.

Recently, a combined study involving laser Raman spectrometery (LR), CL imaging, and ion probe (SHRIMP) U-Pb in-situ dating provided accurate ages of UHPM from rocks collected from Weihai, in the northeastern part of the Sulu UHPM belt. Coesite and other UHP mineral inclusions were identified by LR in zircon separates from an amphibolized peridotite and an eclogite. CL images show distinct zoning in these zircons, and SHRIMP dating yielded ages representing different geological events. An age of 221 ± 12 Ma was obtained for coesite-bearing zircon from amphibolized peridotite, and an age of 228 ± 29 Ma was obtained from eclogite in the same region (Yang J. S. et al., 2003a). In addition, Triassic UHP metamorphic ages were also obtained by SHRIMP dating of coesite-bearing zircon from gneissic drill core samples from Hole PP2 of the Chinese Continental Scientific Drilling Project, i.e., a 234–220 Ma age from paragneiss and a 242–224 Ma age from orthogneiss (Liu F. L. et al., 2003, 2004).

Some new isotope ages from the Dabie UHPM rocks have also been recently reported. For example, three groups of ages were obtained from eclogite in Dabie. The oldest group (771–730 Ma) was interpreted as the protolith age, whereas two younger groups of 461–449 Ma and 216 Ma were interpreted as UHPM events (Gao et al., 2002), although no evidence was provided to support this idea. Similar problems exist in interpreting early Paleozoic or Neoproterozoic UHPM ages obtained by SHRIMP dating (e.g., Jian et al., 1997, 2000; Chen et al., 2000).

Two UHPM Events and Their Tectonic Implications

New age data in this study along with previuos study confirms that two UHPM events occurred in the 4000 km long Central Orogenic Belt of China. *In-situ* SHRIMP U-Pb dating of coesite-bearing zircon from North Qaidam gneiss yielded an age of 440 ± 16.7 Ma (452 ± 14 Ma on average), which represents the UHPM age of the rock. Diamond-bearing gneiss from Qinling yielded an age of ~500 Ma for UHP metamorphism, and associated eclogite was dated at 493 Ma. Two UHPM events were recognized in Qinglongshan eclogite of the Sulu belt: an early Paleozoic event at 450 Ma, and the widely recognized Triassic event at 240–220 Ma. Zircons formed during both events contain coesite inclusions. In a previous study, coesite was discovered in zircon from amphibolized peridotite in Weihai, Shandong Province, and coesite-bearing zircon yielded a metamorphic UHP age of 221 ± 12 Ma (Yang J. S. et al., 2003a). Thus, two UHPM events have been recognized in the Central Orogenic Belt, one of early Paleozoic age, and the other of Triassic age. Thus far, we have found no evidence for a Neoproterozoic UHP event.

Ultrahigh-pressure metamorphism in an orogen is believed to be due to continental subduction and collision between continents. The formation of huge mountain belts across central China is related to convergence, subduction, and collision between the North and South China continents and some microcontinents in between. Current data suggest that the COB of China is most likely an orogen with a complex history of multiple divergence and convergence. The Grenvillian orogenic event, which probably formed the Rodinian supercontinent, may have initiated tectonic and magmatic activities at about 1000 Ma in North Qaidam that created a large amounts of granite and mafic-ultramafic rocks (Zhang J. X. et al., 2000; Lu, 2003). An even earlier event may have occurred in the region because eclogite protoliths in Qinling and Shandong, in the middle and eastern parts of the COB, respectively, have been dated at 1800 Ma (Yang J. S. et al., 2003a, 2003b).

Rifting of the Rodinian supercontinent probably occurred about 800 Ma, as evidenced by several ophiolites and eclogite protoliths in North Qaidam dated at 800–750 Ma (Yang J. S. et al., 2003c). The occurrence of early Paleozoic ophiolites and UHP rocks along the COB suggests a discrete Caledonian plate tectonic event. Triassic rifting and collision are represented by the Dabie-Sulu UHP rocks of Triassic age. Thus, many petrotectonic assemblages of the COB record multiple events of rifting and convergence of continental blocks, and provide a unique natural laboratory for continuous study of continental dynamics.

Acknowledgments

The SHRIMP U-Pb dating was carried out at Stanford University and the Ion-probe Center of Beijing, and the SIMS U-Pb dating took place at the Centre de Researches Petrographiques et Geochimiques (CRPG) in Nancy, France. J. G. Liou, Maurice Brunel, Julie Schneider, and Yusheng Wan are thanked for assistance in this study and Paul Robinson for discussion and modification of the text. The study was financially supported by the Chinse National Key Project for Basic Research (2003CB716500), the China Geology Survey (200313000058), and NSFC (49732070).

REFERENCES

Ames, L., Titton, G. R., and Zhou, G., 1993, Timing of collision of the Sino-Korean and Yangtze cratons: U-Pb zircon dating of coesite-bearing eclogites: Geology, v. 21, p. 339–342.

Ames, L., Zhou, G., and Xiong B., 1996, Geochronology and geochemistry of ultrahigh-pressure metamorphism with implications for collision of the Sino-Korean and Yangtze cratons, central China: Tectonics, v. 15, p. 422–489.

Cao, G. Q., Wang, Z. B., and Zhang, C. J., 1990, Tectonic significance of the Jiaonan terrane, Shandong, and the its marginal Wulian-Rongchen fault: Journal of Shandong Geology, v. 6(1), p. 1–15 (in Chinese).

Chavagnac, V., and Jahn, B.-M., 1996, Coesite-bearing eclogites from the Bixiling Complex, Dabie Mountains, China: Sm-Nd ages, geochemical chracteristics, and tectonic implications: Chemical Geology, v. 133, p. 29–51.

Chen, Y. Q., Liu, D. Y., Williams, I. S., Jian, P., Zhuang, Y. X., and Gao, T. S., 2000, SHRIMP U-Pb dating of zircons of a dark-coloured eclogite and a garnet-bearing gneissic-granitic rock from Bixiling, Eastern Dabie

Area: Acta Geologica Sinica, v. 74(3), p. 193–205 (in Chinese with English abstract).

Cong, B. L., 1996, Ultrahigh-pressure metamorphic rocks in the Dabieshan-Sulu region of China: Beijing, China, Science Press, 224 p.

Cong, B. L., Wang, Q., Zhai, M., Zhang, R. and Ye, K., 1994, UHP metamorphic rocks on the Dabie-Su-Lu region, China: Their formation and exhumation: Island Arc, v. 3, p. 135–150.

DiPietro, J. A., and Isachsen, C. E., 2001, U-Pb zircon ages from the Indian plate in northwest Pakistan and their significance to Himalayan and pre-Himalayan geologic history: Tectonics, v. 20(4), p. 510–525.

Gao, S., Qiu, Y. M., Ling, W. L., et al., 2002, SHRIMP single zircon U-Pb geochronology of eclogites from Yingshan and Xiongdian: Earth Science Journal of China University of Geosciences, v. 27(5), p. 558–564 (in Chinese with English abstract).

Hacker, B. R, Ratschbacher, L., Webb, L., Ireland, T., Walker, D., and Dong, S., 1998, U/Pb zircon ages constrain the architecture of the ultrahigh-pressure Qinling-Dabie Orogen, China: Earth and Planetary Science Letters, v. 161, p. 215–230.

Hu, N. G., Zhao, D. L., Xu, B. Q., and Wang, T., 1995, Petrography and metamorphism study on high-ultrahigh pressure eclogite from Guanpo area, northern Qinling Mountains: Journal of Mineralogy and Petrology, v. 15(4), p. 1–9 (in Chinese with English abstract).

Jahn, B.-M., 1998, Geochemical and isotopic characteristics of UHP eclogites of the Dabie orogen: Implications for continental subduction and collisional tectonics, in Hacker, B., and Liou, J. G., eds., When continents collide: Geodynamics and geochemistry of ultrahigh-pressure rocks: Dordrecht, Netherlands, Kluwer Academic Publishing, p. 203–240.

Jahn, B.-M., 1999, Sm-Nd isotope tracer of UHP metamorphic rocks: implications for continental subduction and collisional tectonics: International Geology Review, v. 41, p. 859–885.

Jian, P., Liu, D., Yang, W., and Williams, I. S., 2000, Petrological study of zircons and SHRIMP dating of the Caledonian Xiongdian eclogite, northwestern Dabie mountains: Acta Geologica Sinica, v. 74, p. 259–264 (in Chinese with English abstract).

Jian, P., Yang, W., Li, Z., and Zhou, H., 1997, Caledonian eclogite in Xiongdian, west Dabie: Evidence in isotopic chronology: Acta Geologica Sinica, v. 71(2), p. 133–141 (in Chinese with English abstract).

Kang, W. G., Hu, K., Liang, W. T., Liu, X. C., and Liu, Y. Q., 1996, High pressure metamorphic belt of middle China: Beijing, China, Geological Publushing House, 110 p. (in Chinese).

Katayama, I., Maruyama, S., Parkinson, C. T., Terada, K., and Sano, Y., 2001, Ion micro-probe U-Pb zircon geochronology of peak and retrograde stages of ultrahigh-pressure metamorphic rocks from the Kokchetav massif, northern Kazakhstan: Earth and Planetary Science Letters, v. 188, p. 185–198.

Li, H. B., Yang, J. S., Xu, Z. Q., Wu, C. L., Wan, Y. S., Shi, R. D., Liou, J. G., Tapponnier, P., and Ireland, T. R., 2001, Geological and chronological evidence of Indo-Chinese strike-slip movement in the Altyn Tagh fault zone: Chinese Science Bulletin, v. 47(1), p. 27–32.

Li, S. G., Chen, Y., Cong, B., Zhang, Z., Zhang, R., Liou, D., Hart, S. R., and Ce, N., 1993, Collision of the North China and Yangtze blocks and formation of coesite-bearing eclogite: Timing and processes: Chemical Geology, v. 109, p. 70–89.

Li, S. G., Hart, S. R., Zheng, S. G., Liu, D. L., Zhang, G. W., and Guo, A. L., 1989, Timing of collision between north and south China blocks—the Sm-Nd isotopic age evidence: Science in China, v. 32, p. 1393–1400 (in Chinese with English abstract).

Li, S. G., Jagoutz, E., Chen, Y., and Li, Q., 2000, Sm-Nd and Rb-Sr isotope chrolonoly and cooling history of ultrahigh pressure metamorphicrocks and their country rocks at Shuanghe in the Dabie Mountains, central China: Geochimica et Cosmochimica Acta, v. 64, p. 1077–1093.

Liu, F. L., Xu, Z. Q., Liou, J. G., Katayama, I., Masago, H., Maruyama, S., and Yang, J. S., 2002, Ultrahigh-pressure mineral inclusions in zircons from gneissic core samples of the Chinese Continental Scientific Drilling Site in eastern China: European Journal of Mineralogy, v. 14, p. 499–512.

Liu, F. L., Xu, Z. Q., Liou, J. G., and Song, B., 2004, SHRIMP U-Pb ages of ultrahigh-pressure and retrograde metamorphism of gneisses, south-western Sulu terrane, eastern China: Journal of Metamorphic Geology, v. 22, p. 316–326.

Liu, F. L., Xu, Z. Q., and Song, B., 2003, Determination of UHP and retrograde metamorphic ages of the Sulu Terrane: Evidence from SHRIMP U-Pb dating on zircons of gneissic rocks: Acta Geologica Sinica, v. 77, p. 230–237 (in Chinese with English abstract).

Liu, F. L., Xu, Z. Q., Yang, J. S., Maruyama, S., Liou, J. G., Katayama, I., and Masago, H., 2001, Mineral inclusions of zircon and UHP metamorphic evidence from paragneiss and orthogneiss of pre-pilot drillhole CCSD-PP2 in north Jiangsu Province, China: Chinese Science Bulletin, v. 46(12), p. 1037–1042.

Liu, L., Che, Z. C., and Luo, J. H., 1996, Determination of eclogite in western Altyn and its geological significance: Chinese Science Bulletin, v. 41, p. 1485–1488.

Liu, L., Sun, Y., and Jiao, P. X., 2002, Discovery of ultrahigh pressure (>3.8Ga) garnet peridotite in Altun: Chinese Science Bulletin, v. 47(9), p. 657–662.

Lu, S. N., 2003, Preliminary study of Precambrian geology in the North Tibet–Qinghai Plateau: Beijing, China, Geological Publishing House, 125 p. (in Chinese).

Lu, Y. Z., 1998, Some ultra-high pressure metamorphic features of ultrabasic rocks in Northern Jiangsu: Jiangsu Geology, v. 22(1), p. 1–9 (in Chinese).

Maruyama, S., Tabata, H., Nutman, A. B., Morikawa, T., and Liou, J. G., 1998, SHRIMP U-Pb Geochronology of ultrahigh-pressure metamorphic rocks of the Dabie Mountains, Central China: Continental Dynamics, v. 3, p. 72–87.

Meng, Q. R., and Zhang, G. W., 1999, Timing of collision of the North and South China blocks: Controversy and reconciliation: Geology, v. 27, p. 123–126.

Okay, A. L., Xu, S., and Sengor, A. M. C., 1989, Coesite from the Dabie Shan eclogites, central China: European Journal of Mineralogy, v. 1, p. 595–598.

Rowley, D. B., Xue, F., Tucker, R. D., Peng, Z. X., Baker, J., and Davis, A., 1997, Ages of ultrahigh-pressure metamorphism and protolith orthogneisses from the eastern Dabie Shan: U/Pb zircon geochronology: Earth and Planetary Science Letters, v. 151, p. 191–203.

Rubatto, D., Gebauer, D., and Compagnoni, R., 1999, Dating of eclogite-facies zircons: The age of Alpine metamorphism in the Sesia-Lanzo Zone (Western Alps): Earth and Planetary Science Letters, v. 167, p. 141–158.

Song, M. C., Jin, Z. M., Wang, L. M., Zhang, X. D., and Li, Y. Y., 2003, New discovery of the contact relation between eclogite and country rocks in Guanshan, eastern Shandong and its enlightenment for chronology: Acta Geologica Sinica, v. 77(2), p. 239–244.

Song, S. G., Yang, J. S. Xu, Z. Q., Liou, J. G., and Shi, R. D., 2003, Metamorphic evolution of the coesite-bearing ultrahigh-pressure terrane in the North Qaidam, northern Tibet, NW China: Journal of Metamorphic Geology, v. 21, p. 631–644.

Wang, L. M., Song, B., Wu, H. X., et al., 1994, Age of the eclogite in Shandong province: Chinese Science Bulletin, v. 39, p. 1788–1791.

Xu, S. T., Liu, Y. C., Chen, G. B., Compagnoni, R., Rolfo, F., He, M. C., and Liu, H. F., 2003, New finding of micro-diamonds in eclogites from Dabie-Sulu region in central-eastern China: Chinese Science Bulletin, v. 48(10), p. 988–994.

Xu, S. T., Okay, A. I., Ji, S., Sengor, A. M. C., Su, W., Liu, Y. and Jiang, L., 1992, Diamond from the Dabie Shan metamorphic rocks and its implication for tectonic setting: Science, v. 256, p. 80–82.

Xu, Z. Q., 1987, Etude tectonique et microtectonique de la chaine paleozoique et triasique des qinlings (Chine): Montpellier, France, Academie de Montpellier Universite des sciences et techniques du Languedoc, 250 p.

Xu, Z. Q., Yang, J. S., Wu, C. L., Li, H. B., Zhang, J. X., Qi, X. X., Song, S. G., Wan, Y. S., Chen, W., and Qiu, H. J., 2003, Timing and mechanism of formation and exhumation of the Qaidam ultrahigh-pressure metamorphic belt: Acta Geologica Sinica, v. 77(2), p. 163–176 (in Chinese with English abstract).

Xu, Z. Q., Yang, J. S., and Zhang, J. X., 1999, Comparison of the tectonic units on both sides of the Altyn Tagh fault and the mechanism of lithospheric shearing: Acta Geologica Sinica, v. 73, p. 193–205 (in Chinese with English abstract).

Xu, Z. Q., Yang, W. C., Zhang, Z. M., and Yang, T. N., 1998, Scientific significance and site-selection researches of the first Chinese Continental Scientific Deep Drillhole: Continental Dynamics, v. 3(1-2), p. 1–13.

Yang, J. J., 1991, Eclogites, garnet, pyroxenites and related ultrabasics in Shandong and northern Jiangsu of east China: Beijing, China, Geological Publishing House, p. 26–52 (in Chinese).

Yang, J. J., and Jahn, B.-M., 2000, Deep subduction of mantle-derived garnet peridotites from the Su-Lu UHPM terrane in China: Journal of Metamorphic Geology, v. 18, p. 167–180.

Yang, J. S., Robinson, P. T., Jiang, C.-F., and Xu, Z.-Q., 1996, Ophiolites of the Kunlun Mountains, China and their tectonic implications: Tectonophysics, v. 258, p. 215–231.

Yang, J. S., Wooden, J. L., Wu, C. L., Liu, F. L., Xu, Z. Q., Shi, R. D., Katayama, I., Liou, J. G., and Maruyama, S., 2003a, SHRIMP U-Pb dating of coesite-bearing zircon from the ultrahigh-pressure metamorphic rocks, Sulu terrane, east China: Journal of Metamorphic Geology, v. 21, p. 551–560.

Yang, J. S., Xu, Z. Q., Bai, W. J., et al., 1999, Discovery of diamond in the eclogite of Sulu, potential diamond prospect in China: Earth Science Frontiers, v. 6(1), p. 69.

Yang, J. S., Xu, Z. Q., Dobrzhinetskaya, L. F., Green, H. D., II, Pei, X. Z., Shi, R. D., Wu, C. L., Wooden, J. L., Zhang, J. X., Wan, Y. S., and Li, H. B., 2003b, Discovery of metamorphic diamonds in central China: an indication of a >4000-km-long zone of deep subduction resulting from multiple continental collisions: Terra Nova, v. 15, p. 370–379.

Yang, J. S., Xu, Z. Q., Li, H. B., Wu, C. L., Cui, J. W., Zhang, J. X., and Chen W., 1998, Discovery of eclogite at the northern margin of Qaidam basin, NW China: Chinese Science Bulletin, v. 43, p. 1755–1760.

Yang, J. S., Xu, Z. Q., Li, H. B., Wu, C. L., Zhang, J. X., and Shi, R. D., 2000a, A convergent border at the southern margin of the Qilian terrain, NW China: Evidence from eclogite, garnet-peridotite, ophiolite, and S-type granite: Journal of the Geological Society of China (Taiwan), v. 43(1), p. 142–160.

Yang, J. S., Xu, Z. Q., Song, S. G., Wu, C. L., Shi, R. D., Zhang Jinxin, Wan, Y. S., Li, H. B., Jin, X. C., and Jalivet, M., 2000b, Discovery of eclogite in Dulan, Qinghai province and its significance for the HP-UHP metamorphic belt along the central orogenic belt of China: Acta Geologica Sinica, v. 74, p. 156–168 (in Chinese with English abstract).

Yang, J. S., Xu, Z. Q., Song, S. G., Zhang, J. X., Wu, C. L., Shi, R. D., Li, H. B., and Brunel, M., 2001a, Discovery of coesite in the North Qaidam Early Paleozoic ultrahigh pressure (UHP) metamorphic belt, NW China:

Compte Rendus de'l Academie des Sciences, Paris, v. 333, p. 719–724.

Yang, J. S., Xu, Z. Q., Zhang, J. X., Chu, C. Y., Zhang, R. Y., and Liou, J. G., 2001b, Tectonic significance of Caledonian high-pressure rocks in the Qilian-Qaidam-Altun mountains, NW China: From continental assembly to intracontinental deformation: Boulder, CO, Geological Society of America Memoir 194, p. 151–170.

Yang, J. S., Xu, Z. Q., Zhang, J. X., Song, S. G., Wu, C. L., Shi, R. D., Li, H. B. and Brunel, M., 2002, Early Palaeozoic North Qaidam UHP metamorphic belt on the north-eastern Tibetan plateau and a paired subduction model: Terra Nova, 14(5): 397–404.

Yang, J. S., Zhang J. X., Meng, F. C., Shi, R. D., Wu, C. L., Xu, Z. Q., Li, H. B., and Chen, S. Y., 2003c, Ultrahigh pressure eclogites of the North Qaidam and Altun Mountains, NW China and their protoliths: Earth Science Frontiers, v. 10(4), p. 291–314 (in Chinese with English abstract).

Ye, K., Cong, B. L., and Ye, D., 2000, The possible subduction of continental material to depths greater than 200 km: Nature, v. 407, p. 734–736.

You, Z. D., Han, Y. Q., Zhang, Z. M., et al., 1998, The high- and ultrahigh-pressure metamorphic belt of East Qinglin–Dabie: Wuhan, China, Publishing House of the Geological University of China, 157 p.

Zhang, J. X., Yang, J. S., Xu, Z. Q., Meng, F. C., Song, S. G., Li , H. B., and Shi, R. D., 2002, Evidence for the ultrahigh-pressure metamorphism in Altun eclogite: Chinese Science Bulletin, v. 47, p. 751–755.

Zhang, J. X., Yang, J. S., Xu, Z. Q., Zhang, Z. M., Li, H. B., and Chen, W., 2000, U-Pb and Ar-Ar ages of eclogites from the northern margin of the Qaidam basin, northwestern China: Journal of the Geological Society of China (Taiwan), v. 43(1), p. 161–169.

Zhang, J. X., Zhang, Z. M., and Xu, Z. Q., 2001, Petrology and geochronology of eclogites from the western segment of the Altyn Tagh, northwestern China: Lithos, v. 56(2–3), p. 187–206.

Zhang, R. Y., Liou, J. G., and Cong, B. L., 1994, Petrogenesis of garnet-bearing ultramafic rocks and associated eclogites in the Su-Lu ultrahigh-pressure metamorphic terrane, China: Journal of Metamorphic Geology, v. 12, p. 169–186.

Zhang, R. Y., Liou, J. G., Yang, J. S., and Yui, Z.-F., 2000, Petrochemical constrains for dual origin of garnet peridotites from the Dabie-Sulu UHP terrane, eastern-central China: Journal of Metamorphic Geology, v. 18, p. 149–166.

Zhang, R. Y., Shu, J. F., Mao, H. K., and Liou, J. G., 1999, Magnetite lamellae in olivine and clinohumite from Dabie UHP ultramafic rocks, central China: American Mineralogist, v. 84, p. 564–569.

Zhang, Z., You, Z., Han, Y., and Shang, L., 1995, Petrology, metamorphism and genesis of the Dabie-Sulu eclogite belt: Acta Geologica Sinica, v. 69, p. 306–324 (in Chinese with English abstract).

A New Caledonian Khondalite Series in West Kunlun, China: Age Constraints and Tectonic Significance

ZHIQIN XU,[1] XUEXIANG QI, FULAI LIU, JINGSUI YANG, LINGSEN ZENG, AND CAILAI WU

Key Laboratory for Continental Dynamics, MLR, Institute of Geology, Chinese Academy of Geological Sciences, Beijing 100037, China

Abstract

The Kangxiwar ductile strike-slip shear zone marks the southern boundary of the West Kunlun terrane, a large, nearly E-W trending metamorphic terrane in the western Qinghai Tibet Plateau region. This ductile shear zone is ~7 km wide, and consists of mylonitized khondalites. Protoliths of the khondalites were alumina-rich pelitic sedimentary and subordinate volcanic rocks. The pelitic khondalites have pronounced positive Th anomalies and subdued positive Ce and Zr anomalies, whereas the metavolcanic rocks have positive Nb and Zr anomalies. Both types of khondalite are LREE enriched, and show weak HREE depletions and moderate negative Eu anomalies. P-T conditions for the formation of the khondalites are estimated to be 6.8 kbar and 700°C. The khondalites formed in the Caledonian orogeny (428–445 Ma) and underwent strong shear deformation during the Indosinian (250–210 Ma). SHRIMP dating of detrital zircons in the khondalites suggests that they were derived from an older metamorphic basement, probably older than 644 Ma. The Kangxiwar khondalites are similar in their protoliths, trace- element and rare-earth element geochemistry, P-T conditions, and age of formation to those of the South Altyn Tagh khondalite series. This lateral correlation suggests that the West Kunlun and Altyn Tagh terranes were once contiguous, and provides evidence for the existence of a Caledonian orogenic belt in this region.

Introduction

THE NEARLY E-W–trending West Kunlun terrane is located in the northern part of the western Qinghai Tibet Plateau. It is a narrow terrane about 100 km wide, bounded on the north by the Tarim Basin, on the south by the Karakorum Mountains, and on the east by the Altyn Tagh fault. Geologic observations suggest that it can be divided into two subterranes South Kunlun and North Kunlun separated by the Küdi-Oytog suture. The Kangxiwar suture that separates the South Kunlun from the Tianshuihai terrane (the western extension of the Bayan Har terrane) defines the southern boundary of the West Kunlun terrane (Matte et al., 1996; Pan, 2000) (Fig. 1).

The oldest North Kunlun basement rocks have a zircon U-Pb age of ~2261 Ma and a Sm-Nd isochron age of ~2800 Ma. They are overlain by Precambrian schists, gneisses, and marbles with subordinate quartzites and mafic to intermediate volcanic rocks. Ordovician to Lower Silurian sedimentary sequences consist of deep-water sedimentary rocks and fan deposits that formed in continental shelf and continental slope settings. The Lower and Middle Devonian strata consist of terrestrial clastic and carbonate rocks, whereas Upper Devonian strata are marked by marine variegated clastic rocks intercalated with continental equivalents which contain plant fossils (XBGMR, 1985).

A granodiorite intrusion has ^{40}Ar-^{39}Ar ages of 449 and 474 Ma, and a zircon U-Pb age of 458 Ma (Matte et al., 1996), suggesting that the North Kunlun terrane experienced a Caledonian magmatic event. Such an inference is also supported by biotite Ar/Ar and whole-rock Rb-Sr ages from 384 to 480 Ma for calc-alkaline plutons in the North Kunlun (Xu et al., 2000). The Küdi-Oytog ophiolitic mélange, which lies in the Küdi-Oytog suture zone, consists of disrupted ophiolite rocks, siliceous pelites, and derivatives from arc volcanism. The ophiolite sequence consists of ultramafic rocks, mafic cumulates, mafic dike swarms, and tholeiitic basalts, which we regard as fragments of oceanic crust and uppermost mantle. The sequence as a whole represents rocks formed in either an island-arc or a continental arc environment (Jiang et al., 1992; Ding et al., 1996). The juxtaposition of ophiolitic material and an arc assemblage suggests that the West Kunlun terrane experienced Sinian to

---

[1]Corresponding author; email: xzq@ccsd.org.cn

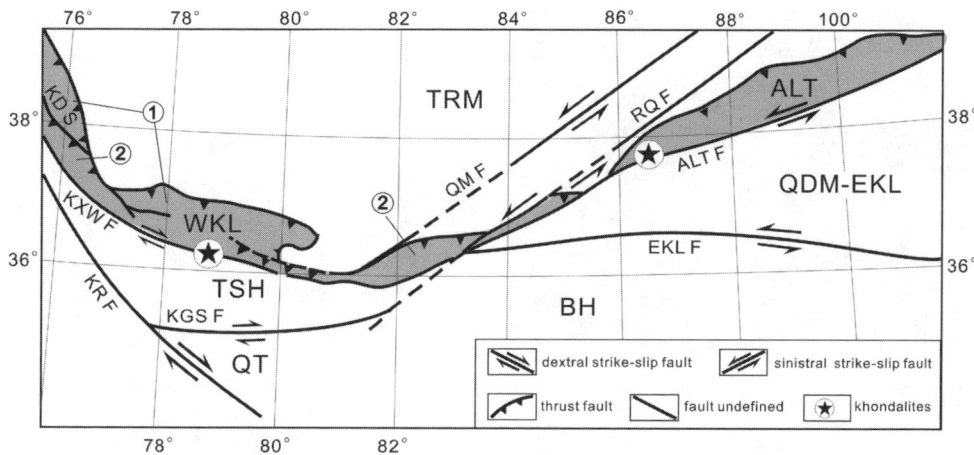

FIG. 1. Tectonic sketch map of the West Kunlun. Legend: 1 = thrust; 2 = sinistral strike-slip fault; 3 = dextral strike-slip fault; 4 = fault; 5 = khondalites. Abbreviations: WKL = West Kunlun terrane; ALT = Altyn terrane; TSH = Tianshuihai terrane; QDM + EKL = Qaidam–East Kunlun terrane; BH = Bayan Har terrane; QT = Qiantang terrane; ALT F = Altyn Tagh fault; EKL F = East Kunlun fault; KXW F = Kangxiwar fault; KR F = Karakunlun fault; KGS F = Konggushan fault; KD S = Küdi-Oytog suture; RQ F = Ruoqiang fault; QM F = Qiemo fault. Legend: circled number 1 = North Kunlun tectonic zone; circled number 2 = South Kunlun tectonic zone.

Early Paleozoic seafloor spreading, subduction, and finally collision (Ding et al., 1996).

The South Kunlun subterrane is a broad dome 15 km wide that contains Paleoproterozoic intermediate- to high-grade granitic gneiss in its core, overlain by low-amphibolite to greenschist facies schists, marbles, and metavolcanic rocks. These high-grade basement rocks were intruded by granitic plutons that fall into two age groups: a Variscan group with ages of 377 Ma (U/Pb zircon), 392 ± 35 Ma (whole-rock Rb-Sr), and 381 ± 4 Ma (biotite Ar/Ar) (Arnaud et al., 1991); and an Indosinian group with ages of 211 ± 8 Ma (whole-rock), 180 ± 10 Ma (K-feldspar and plagioclase Ar/Ar), and 180 ± 10 Ma (whole-rock Rb-Sr isochron) (Matte et al., 1996). These results suggest that: (1) the granitic gneisses underwent metamorphism sometime between 420 and 380 Ma; and (2) the metamorphic basement of the South Kunlun was involved in both Caledonian and Indosinian orogeneses.

The Kangxiwar fault is located along the southern margin of the South Kunlun subterrane, and extends eastward from Uzbel Pass, Kazakhstan through Bandinorth of Taxkorgan, Mazar, Sanshiliyingfang, Kangxiwar, and Muztag to Qong Muztag, where it is cut by the Altyn Tagh fault. The rocks of the South Kunlun subterrane are similar to those in the North Kunlun. Precambrian metamorphic basement is exposed in the northwestern part of the Tianshuihai terrane and is overlain by Lower Paleozoic (O-S) continental-margin slope flysch and siliceous rocks of abyssal basin facies, which are strongly foliated and folded. Middle Devonian strata were formed in a stable marine depositional environment, whereas Upper Devonian strata consist of continental variegated coarse clastic rocks (Ding et al., 1996). A very thick Permian-Triassic (P-T) flysch sequence consists of Paleo-Tethyan sedimentary rocks, which have been strongly foliated and folded and intruded by Indosinian granites (Xu et al., 2000; Matte et al., 1996). Most of the regional deformation and metamorphism in the Tianshuihai terrane is believed to be related to the Indosinian orogeny. The Taishuihai terrane and the Bayan Har-Songpan Garzê terrane together constitute a large-scale NW-trending Indosinian orogenic belt.

The Kangxiwar Khondalite Series

Khondalite-series rocks have recently been identified within the Kangxiwar shear zone. The Kangxiwar ductile shear zone is up to 7 km wide and consists of mylonites and mylonitized metamorphic rocks. Although the rocks in the shear zone have been subjected to strong brittle and ductile deformation, their compositions and textural characteristics are sufficiently preserved to allow identification of their protoliths. Such data are critical for recon-

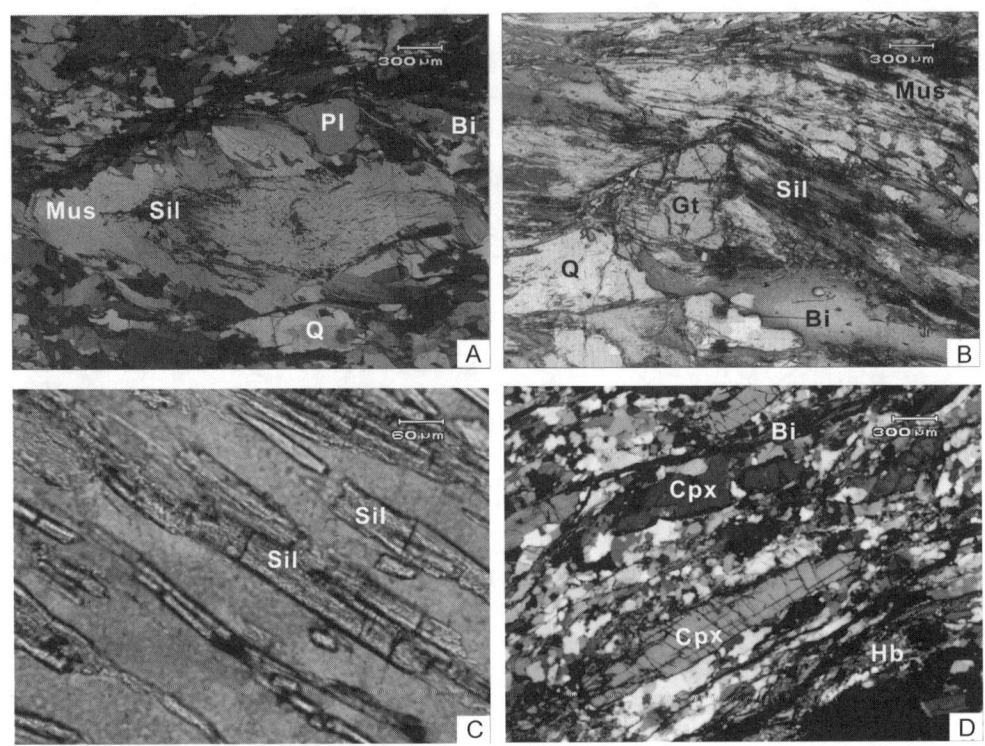

FIG. 2. Photomicrographs of the Kangxiwar khondalites. A–C. Sillimanite-garnet–rich gneiss. D. Pyroxene-amphibole-garnet-biotite monzonitic gneiss. Abbreviations: Sil = sillimanite; Gt = garnet; Bi = biotite; Mus = muscovite; Cpx = clinopyroxene; Q = quartz; Hb = hornblende; pl = plagioclase.

structing their environment of formation, and for tracing the deformational history of the shear zone. We have analyzed a suite of rocks from the shear zone in order to characterize major, trace-, and rare earth element compositions. Zircons were extracted from one of the samples for SHRIMP U/Pb dating. The radiometric age data, combined with the geochemical data, are used to constrain the tectonic history of the Kangxiwar khondalites. Field, petrologic, and geochemical studies (presented here) indicate that the protoliths of rocks in the Kangxiwar ductile shear zone were khondalites.

Khondalite series

Khondalites have long been considered to be alumina-rich, high-grade metamorphic rocks that formed during the early stages of continental crustal development. They consist dominantly of highly metamorphosed supracrustal rocks, potassic granites, and granulites that contain aluminous minerals such as sillimanite and garnet. Khondalites have been found on all continents and share a number of common features in the nature of their protoliths, metamorphism, magmatism, and ore potential (Narayanaswami, 1975; Banerji, 1982; Chacko et al., 1987; Lu et al., 1996). In China, most of the khondalites occur in Precambrian terranes in the northern and central parts of the country (Jiang, 1991; Lu and Jiang, 1992; Lu et al., 1992), although one occurrence of Phanerozoic (450 Ma) khondalite has been described from Qiemo, in the South Altyn Tagh Mountains (Zhang et al., 1999).

Field and petrographic observations show that the khondalites found along the Kangxiwar ductile shear zone have complicated lithologies dominated by aluminous gneisses (e.g., garnet-sillimanite rich gneisses and pyroxene-amphibole gneisses) with subordinate marbles (Fig. 2). Two types of gneiss as well as serpentinized marbles are present in the South Altyn Tagh mountains.

1. *Garnet-sillimanite-rich gneisses (khondalite series* sensu stricto*).* The garnet-sillimanite rich gneisses possess medium- to coarse-grained, nematoblastic to granoblastic textures and a gneissic

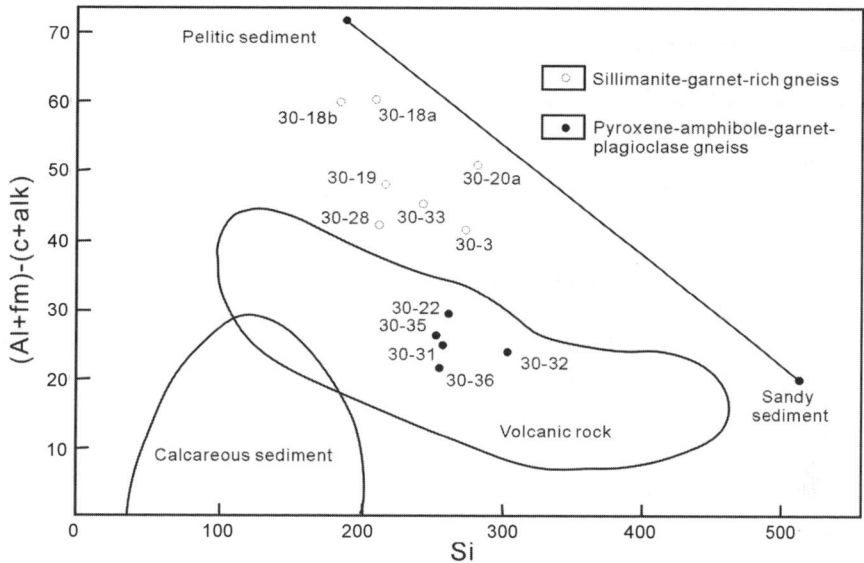

FIG. 3. Si-(Al+fm)-(c+alk) diagram of the Kangxiwar khondalites (after Simonen, 1953). Legend: 1 = sillimanite-garnet–rich gneisses; 2 = pyroxene-amphibole-garnet-plagioclase gneisses. Sample numbers are the same as in Table 1.

structure. They consist of sillimanite, biotite, almandine, graphite, quartz, K-feldspar, and plagioclase. Based on mineral compositions, this suite of rocks includes garnet-sillimanite-biotite-plagioclase gneisses, sillimanite-biotite-plagioclase gneisses, garnet-sillimanite-two-mica-plagioclase gneisses, and garnet-graphite-biotite-(two-mica) plagioclase gneisses.

2. *Pyroxene-amphibole-bearing garnet-biotite-plagioclase-K-feldspar gneisses*. These varieties occur along a ~50 m wide belt in the central part of the ductile shear zone. They have medium- to fine-grained, nematoblastic-granoblastic textures and a gneissic structure, or augen-mylonitic textures and a banded structure. The dominant minerals are diopside, zoisite, hornblende, with or without garnet, titanite, scapolite, biotite, quartz, microcline, plagioclase, and calcite.

3. *Serpentinized phlogopite-olivine marbles*. The marbles occur as lenses 1 to 2 m wide in the khondalite series in the northern part of the ductile shear zone. The marbles in the middle of the lenses have weakly deformed, weakly foliated, granular textures with a grain size of <1 mm to 1 cm, whereas those at the edges of the lenses have strongly deformed, pervasive-foliated mylonitic textures. The dominant minerals are forsterite, diopside, phlogopite, and hornblende, with minor euhedral-granular apatite. Some of the rocks are extensively serpentinized.

Geochemical characteristics of the khondalite series

Major, trace element, and REE analyses were carried out on 12 samples of the Kangxiwar khondalites at the Chinese National Research Center for Geoanalysis, Beijing. Major oxides and some trace elements were determined by standard XRF techniques, whereas the remainder of the trace elements and rare earth elements (REE) were determined by ICP-MS. The analyzed samples include garnet-sillimanite rich (GSR) gneisses (WKL30-18a, WKL30-18b, WKL30-22, WKL30-33, WKL30-3, WKL30-19, and WKL30-20a) and pyroxene-amphibole-garnet-plagioclase (PAGP) gneisses (WKL30-35, WKL30-36, WKL30-31, WKL30-32, and WKL30-28). As shown in the following discussion, the protolith of the garnet-sillimanite-rich gneisses was a peraluminous sedimentary rock, whereas that of the pyroxene-amphibole-garnet-plagioclase gneiss was a volcanic rock (Fig. 3). Analytic results are listed in Table 1 and are discussed in the following sections.

1. *Major element geochemistry*. The major element geochemistry of the khondalites provides a

TABLE 1. Chemical Composition of the Kangxiwar Khondalites[1]

Sample	30-18a	30-22	30-33	k30-3	30-18b	30-19	30-20a	30-28	30-31	30-32	30-35	30-36
SiO_2	60.93	66.85	59.09	65.49	62.63	65.62	71.44	65.67	69.96	72.07	69.91	69.2
TiO_2	0.84	0.43	0.7	0.74	0.87	0.88	0.58	0.67	0.46	0.44	0.45	0.38
Al_2O_3	19.58	10.66	18.07	15.71	17.43	15.97	13.35	15.78	11.73	10.32	9.7	12.06
Fe_2O_3	2.09	1.02	2.55	1.12	0.59	0.61	0.33	1.14	0.74	0.53	0.64	1.27
FeO	4.67	2.57	3.75	4.02	5.91	5.26	4.22	4.01	2.53	2.39	3.2	2.35
MnO	0.07	0.08	0.07	0.12	0.07	0.09	0.1	0.07	0.06	0.18	0.11	0.14
MgO	3.3	2.64	3.02	2.3	3.24	2.46	1.91	2.3	2.24	1.89	3.85	1.97
CaO	1.23	10.56	0.76	1.92	1.4	1.53	1.4	2.52	5.04	6.66	6.91	7.39
Na_2O	2.08	1.38	2.81	3.17	1.84	2.73	2.16	3.37	2.19	1.1	1.65	1.59
K_2O	3.61	1.39	6.73	2.86	3.5	3.12	2.73	2.95	2.76	1.36	1.01	1.51
P_2O_5	0.19	0.09	0.14	0.21	0.34	0.17	0.27	0.16	0.08	0.08	0.07	0.08
H_2O^+	1.39	0.72	1.74	1.22	1.5	1.28	1.35	0.6	0.84	1.08	0.68	0.62
CO_2	0.22	2.06	0.22	0.48	0.09	0.05	0.07	0.4	1.1	1.8	1.32	0.93
Total	100.2	100.45	99.65	99.36	99.41	99.77	99.91	99.64	99.73	99.9	99.5	99.49
Sr	156	175	55.2	207	153	197	181	113	185	153	204	267
Rb	130	36.1	176	93.3	134	113	90.8	93.5	77.7	58.6	33.7	52.8
Ba	549	441	777	459	508	480	460	314	420	249	204	267
Th	14.1	8.73	16.9	10.9	13.7	12.8	9.72	8.61	8.27	6.91	7.71	8.4
Nb	13.3	8.59	13.4	11.8	12.6	12.6	9.24	176	49.7	51.6	138	37.4
Zr	206	100	168	208	197	206	162	192	163	169	278	128
Sm	7.04	4.46	6.86	5.72	7.62	6.27	5.38	5.31	3.78	3.59	7.66	4.1
Y	24.7	22	30	28.8	32.6	23.2	26.1	23.1	18.2	17.5	33.5	19.4
Sc	15.4	9.33	14.6	12.5	13.8	12.3	10.3	9.98	6.47	6.23	8.2	7.74
La	38.4	39.8	30.4	31.4	29.3	22.2	22.9	28.6	22.3	21.5	43.8	22.6
Ce	78.8	82.2	62	73.3	68.8	54.8	45.8	57.7	43.5	40	85.8	45.6
Pr	9.14	9.24	7.32	9.26	8.36	6.69	5.44	6.73	5.02	4.77	10	5.32
Nd	34.8	34.7	27.9	34.8	31.3	24.7	21	26.5	19	17.9	37.8	19.9
Sm	7.04	6.86	5.72	7.62	6.27	5.38	4.46	5.31	3.78	3.59	7.66	4.1
Eu	1.3	1.27	1.39	1.44	1.51	1.42	0.89	1.12	0.81	0.81	1.61	0.8
Gd	5.38	5.45	4.76	6.97	5.38	4.95	3.71	4.96	3.61	3.4	7.51	3.78
Tb	0.9	0.94	0.83	1.16	0.86	0.87	0.67	0.82	0.55	0.55	1.2	0.6
Dy	5.11	5.64	5.29	6.71	4.78	5.28	4.09	4.64	3.4	3.3	7.23	3.75
Ho	0.94	1.13	1.11	1.27	0.93	1.06	0.85	0.91	0.72	0.69	1.49	0.74
Er	2.56	3.32	3.34	3.37	2.59	2.91	2.55	2.5	2.12	2.13	4.31	2.28
Tm	0.34	0.48	0.5	0.44	0.37	0.43	0.36	0.38	0.32	0.32	0.62	0.34
Yb	2.22	3.15	3.34	2.69	2.4	2.67	2.35	2.43	2.14	1.97	3.97	2.22
Lu	0.35	0.48	0.53	0.4	0.38	0.41	0.36	0.36	0.33	0.32	0.63	0.34
ΣREE	187.28	194.66	154.4	180.43	163.23	133.77	115.43	143	107.6	101.3	213.6	112.4
LREE/HREE	9.52	8.45	6.84	6.86	8.23	6.2	6.73	7.41	7.16	6.99	6.92	7
δEu	0.62	0.61	0.79	0.59	0.78	0.83	0.65	0.66	0.66	0.7	0.64	0.61
$(La/Sm)_N$	7.41	7.41	7.41	7.41	7.41	7.41	7.41	3.39	3.71	3.77	3.6	3.47
$(Gd/Yb)_N$	2.62	2.62	2.62	2.62	2.62	2.62	2.62	1.65	1.36	1.39	1.53	1.37

[1]Units for major elements are in wt %; trace elements and REE are in ppm.

good guide to their protoliths. The GSR gneisses have the following characteristics: (1) their SiO_2 contents are mostly between 59.09 and 66.85 wt%, with the exception of sample 30-20a that has 71.44 wt% SiO_2; (2) their K_2O, MnO, and TiO_2 contents are negatively correlated with SiO_2; (3) they have Al_2O_3 contents ranging from 10.66 to 19.58 wt%, with three samples (30-18a, 30-18b, and 30-33) having more than 17 wt% Al_2O_3. Such high Al_2O_3 is typical for peraluminous sedimentary rocks of the type formed in a stable continental margin environment. Samples of PAGP gneisses have: (1) Al_2O_3 contents ranging from 9.70 to 11.73 wt%, far lower than those of the GSR gneisses; (2) SiO_2 contents that vary between 65 and 72 wt%, similar to those in the GSR gneisses; and (3) high CaO of 5.04–7.39 wt%. Major element compositions suggest that they are calc-silicate rocks. When major element compositions of both types of rocks are converted into Niggli values and plotted in an Si (Al+fm) (c+alk) diagram, the GSR gneisses concentrate close to the pelitic end-member in the pelitic-psammitic sedimentary rock field, whereas all of the PAGP gneisses fall in the volcanic rock field (Fig. 3). Thus, protoliths of the analyzed samples are believed to be either pelitic sedimentary rocks or intermediate-silicic volcanic rocks. The association of pelitic sedimentary rocks with intermediate-silicic volcanic rocks is typical for rocks formed in an arc environment (Saleeby and Busby, 1993), and we suggest that protoliths of the Kangxiwar khondalites formed in the margin of an extinct continental arc.

2. *Trace element and REE geochemistry.* Concentrations of the large-ion lithophile elements (LILEs) such as K, Rb, and Ba, as well as Th, Nb, and Ce in the Kangxiwar khondalites are far higher than the MORB values given by Pearce et al. (1984). In contrast, the P_2O_5 contents and Sr, Zr, and Sm concentrations are highly variable, all lower than MORB values (Table 1). In primitive mantle normalized spider diagrams (Fig. 4), the GSR gneisses show pronounced positive Th, Ce, and Zr anomalies, reflecting a crustal source of the protoliths. The PAGP gneisses have no Th anomaly, but do have positive Nb and Zr anomalies, which distinguish them from the GSR gneisses. Both the trace element and major oxide compositions of these khondalites suggest that they originated from igneous rocks.

The total REE contents of the khondalites range from 115.43 to 187.28 ppm (Table 1). The LREE/HREE ratios vary from 2.34 to 3.63, indicating relative enrichment in LREE over HREE. Such enrichment in LREE is also consistent with the chondrite-normalized trace element distribution patterns (Fig. 4). Eu shows moderate depletion and Eu/Eu* is generally scattered around 0.6, with values up to 0.83. All of the samples have high $(La/Sm)_N$ ratios ranging from 2.59 to 3.65, and $(Gd/Yb)_N$ values exceeding 1.0. High $(La/Sm)_N$ ratios indicate enrichment of LREE over HREE, whereas near unity ratios of $(Gd/Yb)_N$ signify weak or no HREE fractionation. In summary, the rare earth element geochemistry shows that the khondalite series is characterized by relative enrichment in LREE, high degrees of REE fractionation, relatively low degrees of HREE fractionation, and moderate Eu depletion.

Estimates of P-T conditions for the formation of the khondalites

Six samples of fresh khondalite were selected for microprobe study at the Electron Microprobe Laboratory of the Institute of Mineral Resources, Chinese Academy of Geological Sciences (CAGS), Beijing. Operating conditions were an accelerating voltage of 20 kV and a beam current of 20 nA. The analyzed samples contain fresh biotite, muscovite, plagioclase, and garnet that are thought to represent equilibrium assemblages. The P-T conditions of metamorphism were calculated from the mineral data using ThermoCalc software developed by Powell and Holland (1994) and Holland and Powell (2001). The results show that peak metamorphism occurred at T = 668–729°C with a mean of 701°C and P = 6.6–7.1 kbar with a mean of 6.8 kbar.

Zircon SHRIMP U/Pb Ages of the Kangxiwar Khondalite Series

Sample 30-36, a PAGP gneiss, was collected from the khondalite series in the Kangxiwar ductile shear zone, West Kunlun. The rock is greyish white and has a mylonitic structure with well-developed foliation and stretching lineation. The dominant minerals are garnet, sillimanite, diopside, biotite, microcline, K-feldspar, and quartz. Zircon grains were separated at the Langfang Institute of the Geological Survey, Hebei Province, and euhedral crystals were handpicked under the microscope. The grains were then mounted and polished for laser Raman and SHRIMP analyses. The crystal shapes and internal structure of the zircons were observed with a laser Raman spectrometer and microscope at the Laboratory for Continental Dynamics of the Institute of Geology, CAGS. Cathodoluminescence

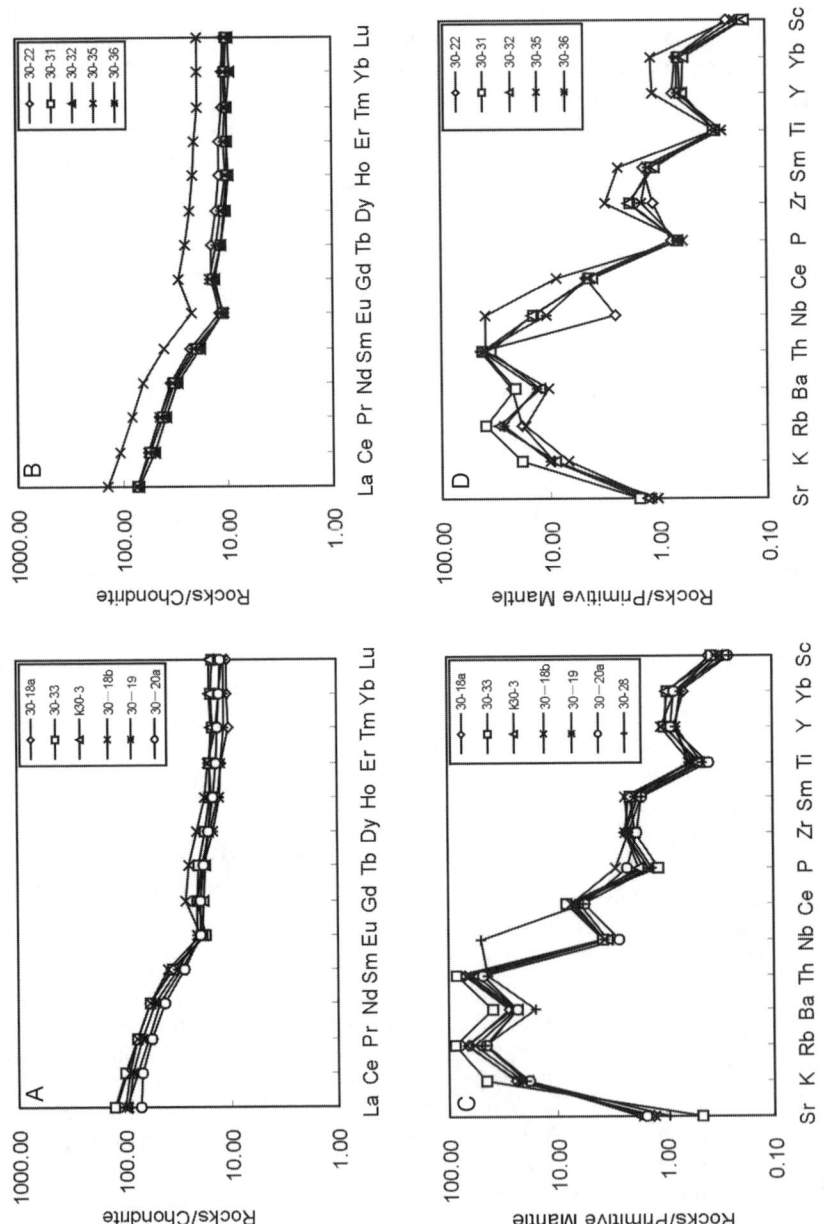

FIG. 4. Trace element spidergram and REE distribution patterns of the Kangxiwar khondalite series. A. Sillimanite-garnet–rich gneiss. Pyroxene-amphibole-garnet-plagioclase gneiss. C. Sillimanite-garnet–rich gneiss. D. Pyroxene-amphibole-garnet-plagioclase gneiss. Sample numbers are the same as in Table 1.

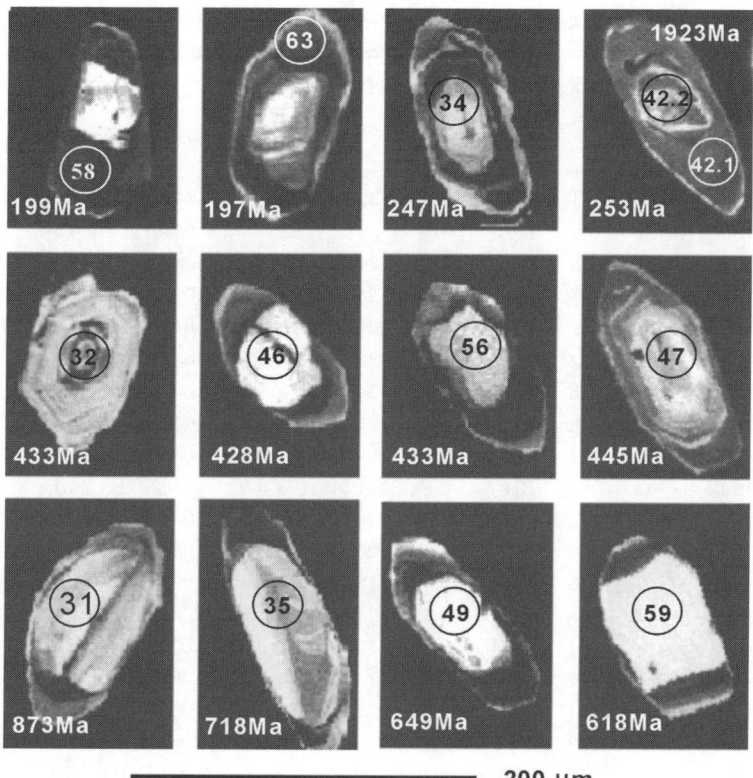

FIG. 5. Cathodoluminescence images of zircons from sample 30-36 of the Kangxiwar khondalite series. SHRIMP analytical spot numbers and the corresponding $^{206}Pb/^{238}U$ ages are shown in the images.

images were obtained at the Laboratory of Electron Microprobe of the Institute of Mineral Resources, CAGS.

Most zircons are prismatic with a length of 0.1–0.2 mm. The length/width ratio averages about 2.5:1 but may reach 3.5:1 in some grains. A few zircons have relatively complete crystal edges or faces, whereas others are rounded or granular in shape, with small sizes ranging from 0.05 to 0.1 mm. Some grains contain cracks parallel to the prismatic face, suggesting that these zircons were modified by late-stage geological processes after formation. Cathodoluminescence (CL) images show two different types of structures in the zircons—those with a distinct overgrowth around a primary relict core, and those showing obvious growth lamellae characteristic of magmatic crystallization (Fig. 5). The different shapes and geometries of the zircons in this sample indicate that they are the products of multi-stage geological processes.

The zircon U-Pb dating was carried out with the SHRIMP II ion microprobe at the Isotope Laboratory of the Institute of Geology, CAGS. Detailed analytical procedures have been described by Stern (1997). Ion microprobe analyses were made on 26 spots of different types of zircon (relict core, overgrowth rim, and growth lamellae). The analytical results are listed in Table 2, and plotted on a U-Pb concordia diagram (Fig. 6). Most of the data fall on the $^{207}Pb/^{235}U$–$^{206}Pb/^{238}U$ concordia, and the others points plot close to the concordia. Four age groups: 197–214 Ma, 245–256 Ma, 428–445 Ma, and 618–718 Ma can be recognized in the diagram (Fig. 6).

The first group includes three data points from the zircon overgrowths, which have ages of 197.1 Ma, 198.5 Ma, and 214.8 Ma, with a mean of 203 Ma. U and Th contents are relatively constant, varying from 462 to 679 ppm and from 8 to 51 ppm, respectively, giving Th/U ratios of less than 0.1. These features are similar to those of metamorphic

TABLE 2. SHRIMP U-Pb Data for Zircons from the Khondalite Series (sample 30-36)

Sample	Zircon domain	U	Th	206Pb*	Th/U	238Pb/206U		207Pb*/206Pb*		206Pb/238U	
30-36-63	Rim	679	51	18.4	0.08	32.2	± 3.5	0.0488	± 5.0	197.1	± 6.7
30-36-58	Rim	628	22	17	0.04	32	± 3.5	0.0536	± 4.5	198.5	± 6.8
30-36-30.2	Rim	462	8	13.8	0.02	29.5	± 3.5	0.0582	± 9.5	214.8	± 7.5
30-36-41	Rim	158	85	5.43	0.55	25.75	± 3.7	0.0528	± 12.0	245.6	± 8.8
30-36-34	Rim	351	182	12.4	0.54	25.64	± 3.7	0.042	± 24.0	246.6	± 9.0
30-36-50	Rim	669	2882	23.2	4.45	25.44	± 3.5	0.053	± 6.0	248.6	± 8.5
30-36-42	Core	530	5	18.5	0.01	24.91	± 3.5	0.0541	± 3.7	253.7	± 8.6
30-36-48	Rim	1224	478	43.2	0.4	24.65	± 3.4	0.0499	± 3.3	256.4	± 8.6
30-36-54	Rim	470	131	17.6	0.29	23.37	± 3.5	0.0532	± 6.3	270.1	± 9.2
30-36-33	Mantle	460	194	18.6	0.43	21.57	± 3.5	0.0562	± 4.1	292.2	± 9.9
30-36-37	Mantle	667	218	32.1	0.34	18.09	± 3.5	0.0652	± 2.9	347	± 12.0
30-36-38	Mantle	1031	364	43.7	0.36	20.55	± 3.5	0.0509	± 5.9	306	± 10.0
30-36-47.1	Core	248	155	15.4	0.65	13.98	± 3.5	0.054	± 4.7	445	± 15.0
30-36-46	Core	227	68	13.7	0.31	14.58	± 3.5	0.0636	± 6.8	428	± 15.0
30-36-32	Core	416	251	25.1	0.62	14.39	± 3.5	0.0578	± 3.2	433	± 15.0
30-36-56	Core	318	97	19.3	0.32	14.38	± 3.5	0.0546	± 5.5	433	± 15.0
30-36-67	Core	306	11	21.1	0.04	12.62	± 3.6	0.0668	± 5.2	492	± 17.0
30-36-44	Core	433	73	39.2	0.17	9.52	± 3.5	0.05962	± 1.6	644	± 21.0
30-36-49	Core	176	57	16.4	0.33	9.44	± 3.5	0.0647	± 5.3	649	± 22.0
30-36-59	Core	48	21	4.39	0.46	9.95	± 3.9	0.07	± 15.0	618	± 23.0
30-36-45	Core	562	25	56.4	0.05	8.65	± 3.4	0.1438	± 1.3	705	± 23.0
30-36-35	Core	235	21	24.7	0.09	8.49	± 3.5	0.0631	± 6.6	718	± 24.0
30-36-31	Core	232	246	29.2	1.09	6.89	± 3.5	0.0726	± 2.3	873	± 28.0
30-36-60	Core	70	50	15.1	0.74	4.06	± 3.6	0.1091	± 3.4	1419	± 46.0
30-36-42.2	Core	387	52	116	0.14	2.878	± 3.5	0.16671	± 0.50	1923	± 57.0

[1]Pb* = corrected for common Pb using ^{204}Pb. All errors are 1 sigma of standard deviation.

zircons (Claesson et al., 2000; Rubatto, 2002) (Fig. 5).

The second group includes five spots from concentrically zoned zircon, which have ages of 245.6 Ma, 246.6 Ma, 248.6 Ma, 252.7 Ma, and 256.4 Ma, with a mean of 250.2 Ma. These zircons have widely varying U and Th concentrations. Except for spot 30-36-42, which has a Th/U ratio of <0.1, all the grains have Th/U ratios >0.4. Because of growth zoning, these zircons are interpreted as having a magmatic or anatectic origin (Hanchar and Miller, 1993; Sue et al., 1999).

Zircons in the third group yielded four ages of 445 Ma, 428 Ma, 433 Ma, and 492 Ma, with an average of 435 Ma. A few individual data points fall below the concordia, suggesting a possible loss of Pb. The U/Th ratios of these grains are relatively constant, ranging from 0.31 to 0.65 (Fig. 7). In the cathodoluminescence images, these zircons have distinct growth zoning, implying that they may represent a growth stage related to an important regional tectonothermal event. The ages obtained from these zircons provide a key constraint on the timing of the Caledonian orogenic event.

Ages of 644, 649, 618, 705, 718, and 873 Ma, all of which fall on or close to the concordia, comprise the fourth group (Fig. 6). Their mean value is 667 Ma. Zircons from this group have highly variable U concentrations but relatively constant Th concentrations, with Th/U ratios varying between

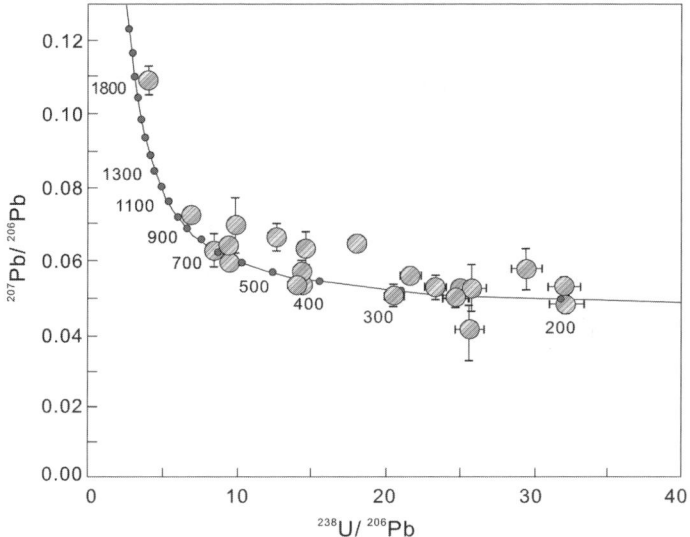

FIG. 6. Concordia diagram of SHRIMP U-Pb zircon ages for sample 30-36 of the Kangxiwar khondalites.

0.05 and 0.46 (Fig. 7). As shown in the CL images, all of these analyses were carried out on relict cores, and the ages probably represent the ages of the old metamorphic basement in the surrounding regions.

In addition to these four main groups of ages, two other clusters were obtained, one with ages of 873–1923 Ma and the other with ages of 292–347 Ma. Analytical spots of the first group are all from the relict cores of zircons, and these ages may reflect the presence of event older metamorphic basement components in the Kangxiwar khondalites. The second age group represents spots located in the transitional area between the relict cores and overgrowths. These are mixed ages and have no geological significance.

Discussion and Conclusions

A 7 km wide ductile shear zone at Kangxiwar in the southern part of the West Kunlun terrane contains a mylonitized khondalite series. The rocks are chiefly garnet-sillimanite–rich gneisses, pyroxene-amphibole–bearing garnet-biotite-plagioclase–K-feldspar gneisses, and serpentinized phlogopite-olivine marbles. The khondalite protoliths were aluminum-rich pelitic or pelitic-psammitic sedimentary rocks and interlayered intermediate to silicic volcanic rocks. Khondalites derived from pelitic sedimentary rocks have pronounced positive Th anomalies and relatively subdued positive Ce and Zr anomalies. In contrast, those derived from metavolcanic rocks have positive Nb and Zr anomalies. Both types are relatively enriched in LREE, relatively depleted in HREE, and moderately depleted in Eu. This khondalite series may have been derived from rocks that formed at an ancient rifted continental margin. Khondalites in the vicinity of Tura in the western segment of the Altyn Tagh terrane are composed of peraluminous gneisses with garnet-hornblende-two-pyroxene gneisses, whose protoliths were alumina-rich pelitic and pelitic-psammitic sedimentary rocks and basalts. This khondalite series is also thought to have formed in a continental-margin environment (Zhang et al., 1999). The Kangxiwar khondalite series is similar to that of the Altyn Tagh khondalite series.

The Kangxiwar khondalites have experienced granulite-facies metamorphism, as suggested by peak metamorphic temperatures and pressures of 800 ± 50°C and 6–9 kbar, respectively. However, relatively low metamorphic temperatures in the range of 650° to 750°C determined for a few samples, suggest that metamorphism may have been transitional between high-amphibolite facies and granulite facies conditions (Lu et al., 1996). Because no granulites have been found in the Kangxiwar khondalite series, the peak metamorphic P and T of 6.8 kbar and 700°C for the peraluminous gneisses imply that they have also undergone transitional metamorphism. Granulites in the Altyn Tagh

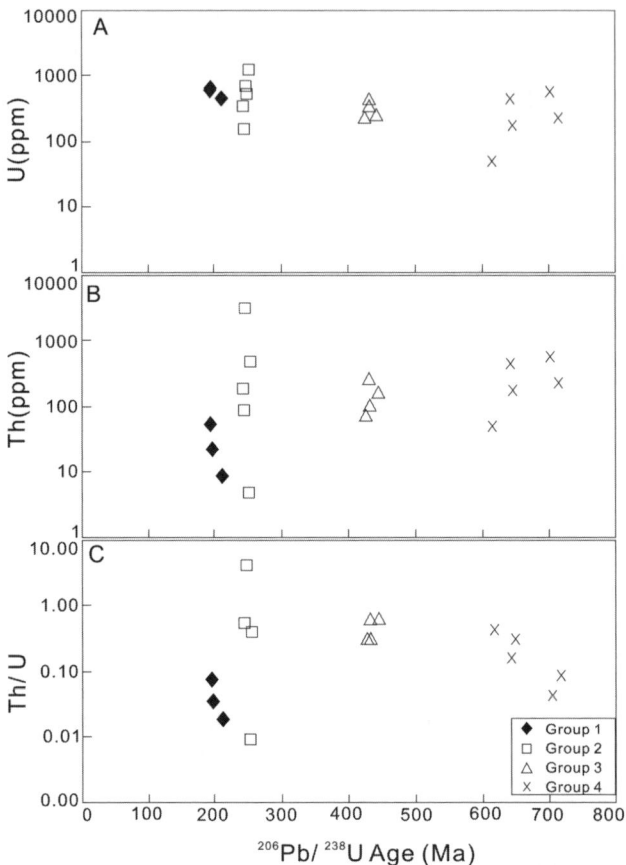

FIG. 7. Plots of U versus $^{206}Pb/^{238}U$ ages (A), Th versus $^{206}Pb/^{238}U$ ages (B), and Th/U versus $^{206}Pb/^{238}U$ ages (C) for zircons from sample 30-36, showing that these zircons can be divided into four age groups.

khondalite series have peak temperatures and pressures 700–850°C and 8–12 kbar, respectively (Zhang et al., 1999), similar to those of the Kangxiwar khondalites. Therefore, we conclude that the P-T conditions for the formation of the West Kunlun khondalites were comparable to those of the Altyn Tagh khondalites.

Zircon SHRIMP U/Pb dating of the Kangxiwar khondalites indicates that: (1) detrital zircons with ages from 644 to 873 Ma or older in the khondalites represent materials derived from pre-existing metamorphic terranes older that 644 Ma; (2) the khondalites were formed by high-grade metamorphism of pelitic sedimentary rocks and subordinate volcanic rocks at 428–445 Ma; and (3) the khondalites underwent strong shear strain in the Indosinian (250–210 Ma). Zhang et al. (1999) reported U-Pb and Pb-Pb ages of 447–462 Ma for metamorphic zircons, and upper intercept ages of 1027 and 2571 Ma for relict zircons from the Altyn Tagh khondalites. The older ages probably represent ancient source components in the protoliths for the Altyn Tagh khondalites. The sparse U-Pb zircon age data reported by Zhang et al. (1999) do not put a tight constraint on the potential old source components in the Altyn Tagh khondalites, as it is the case of the Kangxiwar khondalite. The available data indicate that the metamorphic ages of the two khondalite series are similar, and that the relict grains represent detrital zircons derived from an older metamorphic basement.

Khondalites in Kangxiwar and the Altyn Tagh are similar in age, chemical composition, protolith chemistry, and P-T conditions of formation (Zhang et al., 1999), and we suggest they may be originally part of the same terrane. SHRIMP dating on zircons collected from the ductile shear zones of the Altyn Tagh fault suggests that the main trace of the Altyn Tagh fault to the south of the Altyn Tagh terrane might have been initiated in Indosinian time (Li et al., 2001). Recent field, petrologic, and geochemical observations demonstrate excellent correlations between the Caledonian subduction complexes (e.g., the Northern Qilian subduction complex with the Northern Altyn Tagh, and the Northern Qaidam Mountain ultrahigh pressure metamorphic complex with the Southern Altyn Tagh) (Yang et al., 1998, 2000; Zhang et al., 2000a, 2000b). Such correlations on both sides of the main trace of the Altyn Tagh fault suggest an approximately 400 km left-lateral strike-slip motion along it (Xu et al., 1999). The ENE-WSW–trending Ruoqiang fault that is subparallel to main trace of the Altyn Tagh fault may be one of the abandoned branches of the paleo-Altyn Tagh fault and defines the northern boundary of the Altyn Tagh terrane (Fig. 1). It joins the Altyn Tagh fault at its southwestern end. Field observations and satellite images around this junction indicate that it is probably where the South Kunlun terrane meets the Altyn Tagh terrane. If the Altyn Tagh and South Kunlun terranes were once a coherent block and later offset by the Ruoqiang strike-slip fault, then about 150 to 200 km left-lateral strike-slip movement can be added to the total amount of slip on the Altyn Tagh fault system. Therefore, we suggest a total dextral slip of up to 600 km along the Altyn Tagh fault since the Indosinian.

The khondalite series has long been considered a peraluminous, high-grade metamorphic complex that developed in the early stages of crustal formation (Narayanaswami, 1975; Banerji, 1982; Chacko et al., 1987; Lu et al., 1996). Khondalites occur mainly in Early Precambrian terranes in the northern and central parts of China (Lu et al., 1992; Jiang, 1991). The khondalite series discovered at Kangxiwar of the West Kunlun and Qiemo of the South Altyn Tagh Mountains (Zhang et al., 1999) consist of high-grade metamorphic rocks located in the mountain root zones formed during the Caledonian, and thus have great significance for understanding the Caledonian orogeny and tectonic evolution of this region.

Acknowledgments

This study was supported by the "Terrane Boundaries and Lithospheric Shear Faults in the Qinghai-Tibet Plateau" section of the special project of the Ministry of Land and Resources "Collisional Orogeny of the Qinghai-Tibet Plateau and Its Effects." Thanks to Paul Robinson and J. G. Liou for their support and suggestions that greatly helped to improve the quality of this manuscript, and Fenghua Liang and Xiaowei Zhang for assistance in preparation of the figures.

REFERENCES

Arnaud, N., Xu, R., and Zhang, Y., 1991, Nouvelles donnees thermochronologique sur le batholite du kunlun et L'histoire thermique du Nord du Plateau Tibetain: Compte Rendus de l'Academie Sciences Paris, t. 312, Serie II, p. 905–911.

Banerji, P. K., 1982, The khondalites of Orissa, India—a case history of confusing terminology: Journal of the Geological Society of India, v. 23, p.155–159.

Chacko, T., Ravindra, K. G. R., and Newton, R. C., 1987, Metamorphic P-T conditions of Kerala (South India) khondalite belt: A granulite facies supracrustal terrain: Journal of Geology, v. 95, p. 343–358.

Claesson, V. S., Vetrin, T., and Bayanova, H. D., 2000, U-Pb zircon ages from a Devonian carbonatite dyke, Kola peninsula, Russia: A record of geological evolution from the Archean to the Palaeozoic: Lithos, v. 51, p. 95–108.

Ding, D. G., Wang, D. X., Liu, W. X., and Liu, W. X., 1996, The western Kunlun orogenic belt and basin: Beijing, China, Geological Publishing House (in Chinese).

Hanchar, J. M., and Miller, C. F., 1993, Zircon zonation patterns as revealed by cathodoluminescence and backscattered electron images: Chemical Geology, v. 110, p. 1–13.

Holland, T. J. B., and Powell, R., 2001, Calculation of phase relations involving haplogranitic melts using an internally consistent thermodynamic data set: Journal of Petrology, v. 42, p. 673–683.

Jiang, C. F., Yang, J. S., Feng, B. G., Zhu, Z. Z., Zhao, M., Chai, Y. C., Shi, X. D., Wang, H. D., and Hu, J. Q., 1992, Opening-closing tectonics of Kunlun mountains: Beijing, China, Geological Publishing House, p. 224 (in Chinese).

Jiang, J. S., 1991, Main-period areal metamorphism and evolution of khondalite series from the Mashan group: Acta Petrologica et Mineralogica, v. 11, no. 2, p. 97–110.

Li, H. B., Yang, J. S., Xu, Z. Q., Wu, C. L., Wang, Y. S., and Shi, R. D., 2001, Geological and geochronogy evidence of Indosinian strike-slip movement for the Altyn

Tagh fault: Chinese Science Bulletin, v. 46, no. 16, p. 1333–1338 (in Chinese with English abstract).

Lu, L. Z., and Jiang, J. S., 1992, Metamorphic evolution of the khondalic series from the early Proterozoic Kongling group, western Hubei Province, in Augustithis, S. S., ed., High-grade metamorphism: Athens, Greece, Theophrastus Publications, S.A., p. 265–296.

Lu, L. Z., Jin, S. Q., Xu, X. C., and Liu, F. L., 1992, Origin of an early Precambrian Khondalite series in southeastern inner Mongolia and its mineralization: Changchun, China, Jiling Science and Technology Press, 152 p. (in Chinese with English abstract).

Lu, L. Z., Xu, X. C., and Liu, F.L., 1996, Early Cambrian khondalite series in northern China: Changchun, China, Changchun Publishing House, 276 p. (in Chinese).

Matte, P. H., Tapponnier, P., Arnaud, N., Bourjot, L., Avouac, J. P., Vidal, P. H., Liu, Q., Pan, Y. S., and Wang, Y., 1996, Tectonics of western Tibet, between the Tarim and the Indus: Earth and Planetary Science Letters, v. 142, p. 311–330.

Narayanaswami, S., 1975, Proposal for charnockite-khondalite and Sargur-Nellore-Khamman-Bengpal- Deogarh-Pallahare-mahagiri rock groups—older than Dharwar type greenstone belts in the Peninsular Archaean: India Mineralogist, v. 16, p. 16–25.

Pan, Y. S., 2000, Geological evolution of the Karakorum-Kunlun Mountains: Beijing, China, Science Press, 525 p. (in Chinese with English abstract).

Pearce, J. A., Harris, N. B. W., and Tindle, A. G., 1984, Trace element discrimination diagrams for the tectonic interpretation of granitic rocks: Journal of Petrology, v. 25, p. 956–983.

Powell, R., and Holland, T. J. B., 1994, Optimal geothermometry and geobarometry: American Mineralogist, v. 79, p. 120–133.

Rubatto, D., 2002, Zircon trace element geochemistry: Partitioning with garnet and the link between U-Pb ages and metamorphism: Chemical Geology, v. 184, p. 123–138.

Saleeby, J. B., and Busby, C., 1993, Paleogeographic and tectonic setting of axial and western metamorphic framework rocks of the southern Sierra Nevada, California, in Dunn, G., and MacDougall, K., eds., Mesozoic paleogeography of western United States–II: Pacific Section SEPM, Book 71, p. 197–226.

Simonen, A., 1953, Stratigraphy and sedimentation of the Svecofennidic, early Archean supracrustal rocks in southwestern Finland: Bulletin de la Commission Géologique de Finlande, v. 160, p. 64.

Stern, R. A., 1997, The GSC Sensitive High Resolution Ion Microprobe (SHRIMP): Analytical techniques of zircon U-Th-Pb age determinations and performance evaluation, in Radiogenic age and isotopic studies: Geological Survey of Canada, Current Research, Report 10, p. 1–31.

Sue, K., David, S., and William, C., 1999, Identifying granite sources by SHRIMP U-Pb zircon geochronology: An application to the Lachlan fold belt: Contributions to Mineralogy and Petrology, v. 137, p. 323–341.

XBGMR (Xinjiang Bureau of Geology and Mineral Resources), 1985, 1:1 M Regional Geological Survey report on the Bulungkol-Qarlun area, West Kunlun Mountains: Beijing, China: Geological Publishing House, p. 14–89 (in Chinese).

Xu, R. H., Zhang Y. Q., Xie, Y. W., Vidal, P. H., Arnaud, N., Zhang, Q. D., and Zhao, D. M., 2000, Isotopic geochemistry of plutonic rocks, in Yusheng, P., ed., Geological evolution of the Karakoram and Kunlun mountains: Beijing, China, Scientific Publishing House, p. 324–392 (in Chinese).

Xu, Z. Q., Yang, S., Zhang, J. X, Jiang, M., Li, H. B, and Cui, J. W., 1999, A comparison between the tectonic units on the two sides of the Altun sinistral strike-slip fault and the mechanism of lithospheric shearing: Acta Geologica Sinica, v. 73, no. 3, p. 193–205 (in Chinese with English abstract).

Yang, J. S., Xu, Z. Q., Li, H. B., Wu, C. L., Cui, J. W., Zhang, J. X., and Chen, W., 1998, Discovery of eclogite at northern margin of Qaidam basin, NW China: Chinese Science Bulletin, v. 43, p. 1755–1760.

Yang, J. S., Xu, Z. Q., Li, H. B., Zhang, J. X., Wu, C. L., Zhang, J. X., and Shi, R. D., 2000, A Caledonian convergent border along the Qilian terraine, NW China: Evidence from eclogite, garnet-peridotite, ophiolite, and S-type granite: Journal of the Geological Society of China, v. 43, no. 1, p. 142–160.

Zhang, J. X., Xu, Z. Q., Yang, J. S., Li, H. B., and Wu, C. L., 2000a, The Altun–North Qaidam eclogite belt in western China—another HP-UHP metamorphic belt truncated by large-scale strike-slip fault in China: Earth Science Frontiers, v. 7 (suppl.), p. 254–255.

Zhang, J. X., Zhang, Z. M., Xu, Z. Q., Cui, J. W., and Yang, J. S., 1999, Discovery of khondalite series from the western segment of Altyn and their petrological and geochronological studies: Science in China (Series D), v. 29, no. 4, p. 298–305.

Zhang, J. X., Zhang, Z. M., Xu, Z. Q., Yang, J. S., and Cui, J. W., 2000b, Discovery of khondalite series from the western segment of Altyn and their petrological and geochronological studies: Science in China (Series D), v. 43, p. 308–316.

EASTERN CHINA

Tracing the Boundary between UHP and HP Metamorphic Belts in the Southwestern Sulu Terrane, Eastern China: Evidence from Mineral Inclusions in Zircons from Metamorphic Rocks

FULAI LIU,[1] ZHIQIN XU,

Institute of Geology, Chinese Academy of Geological Sciences, Beijing 100037, China

AND J. G. LIOU

Department of Geological and Environmental Sciences, Stanford University, Stanford, California 94305

Abstract

The southwestern Sulu terrane, eastern China is divided into four fault-bounded lithologic units. From northwest to southeast, they are: (1) orthogneiss unit (Unit I); (2) supracrustal rock unit (Unit II); (3) kyanite-bearing quartzite–marble unit (Unit III); and (4) paragneiss–schist unit (Unit IV). Inclusions of index minerals in zircon separates from more than 90 samples from these units were identified using laser Raman spectroscopy and electron microprobe analysis. Coesite and coesite-bearing ultrahigh-pressure (UHP) mineral assemblages occur in zircon separates from various amphibolite-facies metamorphic rocks in Units I and II. In contrast, aragonite- and phengite-bearing high-pressure (HP) mineral assemblages are preserved in zircons from many amphibolite- to greenschist-facies metamorphic rocks in Units III and IV. These together with previous studies yield P-T estimates of peak metamorphic recrystallization at 723°–852°C, and P > 2.8 GPa for UHP rocks, and 500°–600°C, and P = 1.2–2.5 GPa for HP rocks. The spatial distribution of these P-T estimates and recent surface mapping results constrain an exact boundary between UHP and HP belts in the southern Sulu terrane along a ductile shear zone in the Donghai region.

Introduction

IT IS WELL KNOWN that zircon, as an accessory rock-forming mineral, plays an important role in geochronologic dating. In recent years, zircon has also been considered an excellent container for the preservation of index ultrahigh-pressure (UHP) minerals, such as coesite and diamond (Sobolev et al., 1994; Chopin and Sobolv, 1995; Tabata et al., 1998; Parkinson and Katayama, 1999; Katayama et al., 2000). In the Sulu-Dabie terrane, identification of coesite inclusions in zircons from eclogite and eclogite-hosting gneisses (Tabata et al., 1998; Ye et al., 2000; Liu F. L. et al., 2001, 2002; Liu J. B. et al., 2001) concludes that both eclogites and their country rocks experienced *in situ* UHP metamorphism (Wang and Liou, 1992; Okay and Sengör, 1992; Zhang et al., 1995; Liou et al., 2000). However, more comprehensive assessment of the significance of these findings awaits a systematic study in a spatial distribution of mineral inclusions in zircon from Sulu-Dabie eclogite-hosting country-rocks. Furthermore, until now the potential of using zircon-hosted mineral inclusions to deduce a tectonic boundary between UHP and HP mineral assemblages has not been explored.

In the vicinity of Donghai, where the first Chinese Continental Scientific Drilling project is located, the Sulu terrane has been subdivided into UHP and HP belts separated by an inferred thrust (e.g., Zhang et al., 1995). Due to the extremely poor exposure of the region and extensive amphibolite- to greenschist-facies overprints, the boundary between the UHP and HP belts is tentative, and there has not been a valid method yet to map the boundary.

Zircons from numerous UHP and HP rocks in the Donghai region, southwestern Sulu terrane, were separated and mineral inclusions were identified and analyzed. Cathodoluminescence (CL) images of zircon were used to constrain zircon growth history. We have examined the distribution of various mineral inclusions in different zircon growth zones, and have evaluated whether it is possible to use index mineral inclusions to deduce the P-T conditions at which various zircon domains grew. In this paper, we report paragenesis and distribution of mineral inclusions in zircons separated from outcrops in Donghai and adjacent areas. We identified the index UHP and HP

---

[1]Corresponding author; email: liufulai@cags.net.cn

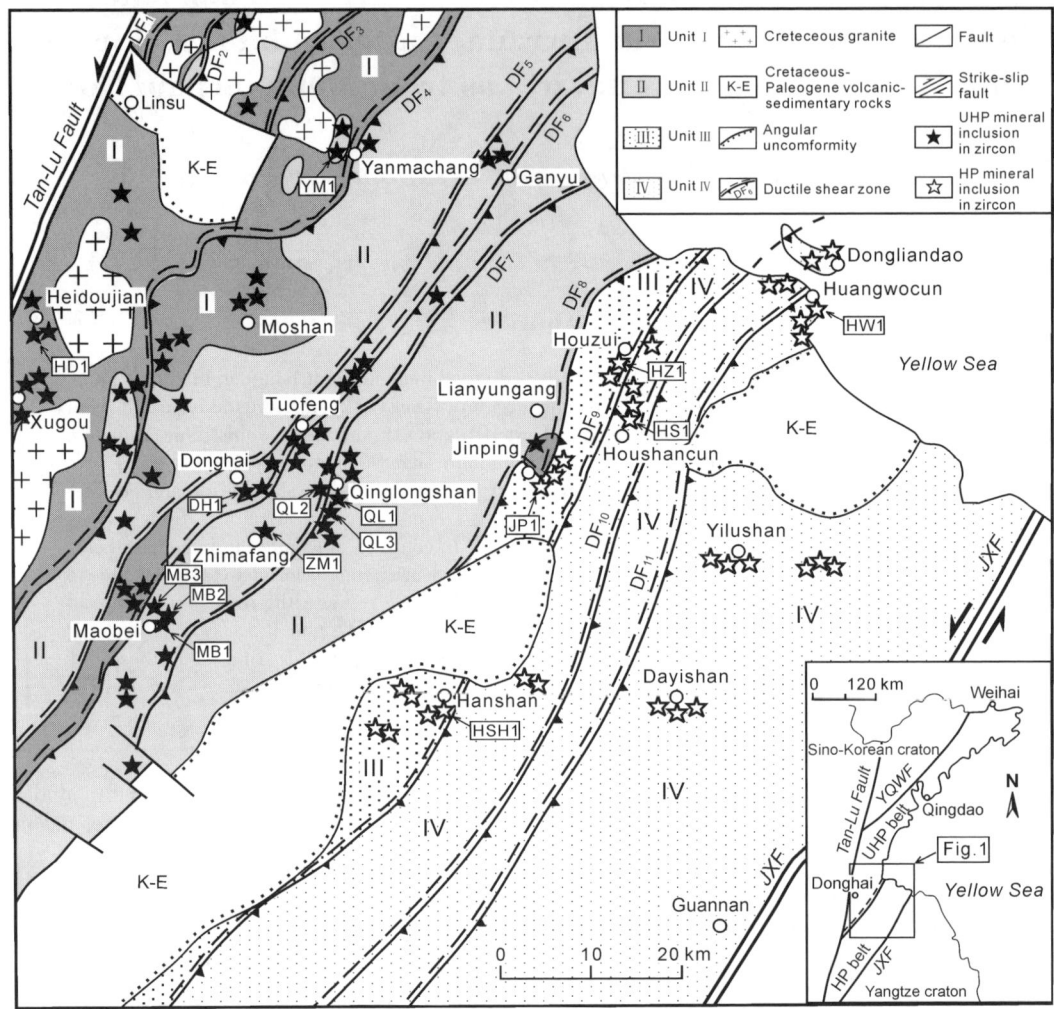

FIG. 1. Tectonic sketch map of the southwestern Sulu terrane, showing major lithotectonic units and the locations of zircon samples (modified after Xu et al., 2003). Abbreviations: YQWF = Yantai-Qingdao-Wulan fault; JXF = Jiashan-Xiangshui fault.

minerals as inclusions in zircons from amphibolite- to greenschist-facies metamorphic rocks, including paragneiss, orthogneiss, quartzite, marble, and amphibolite. The extent to which supracrustal rocks experienced UHP and HP metamorphism is now determined, and the exact boundary between the UHP and HP belts has been traced as well.

Geological Setting

The Triassic Qinling-Dabie collision zone between the Yangtze and Sino-Korean cratons in east-central China extends about 2000 km from the western Qinling to the Dabie Mountains. Its eastern extension, the Sulu region, has been displaced northward by the sinistral Tan-Lu fault. The Sulu UHP/HP belt, extending for about ~320 km from Weihai, northeastern Shandong Province, to Donghai, northern Jiangsu Province, is bounded by the Yantai-Qingdao-Wulian fault (YQWF) on the northwest and the Jiashan-Xiangshui fault (JXF) on the south (Fig. 1).

According to available petrological, petrographic, and geochemical characteristics of metamorphic rocks and mapping results (Xu et al., 2003), the southwestern Sulu terrane can be divided into four lithologic units: (1) orthogneiss unit (Unit I); (2) supracrustal rock unit (Unit II); (3) kyanite-bearing quartzite-marble unit (Unit III); and (4) paragneiss-schist unit (Unit IV) (Fig. 1). These four lithologic units were subjected to amphibolite- to greenschist-facies retrograde overprint and locally were intruded by Cretaceous granites. The units are bordered by shear zones consisting of mylonitic rocks. However, due to poor exposures, these boundaries are shown as dashed lines in Figure 1. Units I and II are considered to belong to the UHP belt, whereas Units III and IV lie in the HP belt.

Unit I is predominantly amphibolite-facies granitic gneiss, with some marble layers, and ultramafic rock and eclogite blocks. This unit was intruded by post-orogenic Mesozoic granite (Fig. 1). Unit II mainly consists of various supracrustal rocks, including epidote- and biotite-bearing two-feldspar gneiss, garnet-and biotite-bearing plagioclase gneiss, garnet- and amphibolite-bearing plagioclase gneiss, and phengite-and biotite-bearing two-feldspar gneiss. Many of these paragneisses contain thin layers of kyanite-bearing quartzite and kyanite-bearing phengite-quartz schist, as well as abundant blocks of eclogite and ultramafic rocks. In this unit, several small metamorphosed Neoproterozoic granitic intrusives into the supracrustal rocks crop out (Fig. 1). Unit III is predominantly comprised of kyanite- and topaz-bearing quartzite, kyanite- and phengite-bearing quartzite, phengite-bearing quartz schist, and phengite-bearing marble. Unit IV consists mainly of albite gneiss with variable amounts of biotite, epidote and amphibole, mica-bearing quartz schist, and rare blueschist (Zhang et al., 1995; Xu et al., 2003).

Sample Locations and Analytical Methods

Zircons were separated from 60 and 33 samples from the UHP and HP belts, respectively. These samples include orthogneiss, paragneiss, eclogite, amphibolite, quartzite, marble, and schist (see Fig. 1 for sample localities). After crushing and sieving approximately 5–30 kg of each sample, and processing through magnetic and heavy liquid separation, 100–300 zircon grains of ~50–400 μm diameter were hand-picked from the residue. The zircon grains were then mounted and polished in 25 mm epoxy discs. Discrete inclusions in zircons were identified using a JASCO NRS-2000 and a REN-ISHOW-1000 Laser Raman spectroscope with the 514.5 nm line of an Ar-ion laser, at the UHP laboratory of the Tokyo Institute of Technology (TIT), and the Open Laboratory of Continental Dynamics, Chinese Academy of Geological Sciences (CAGS), respectively. Distribution, textures, and inclusion minerals in host zircon were examined using an optical microscope, scanning electron microscope (SEM), and electron microprobe. The internal zoning patterns of zircon crystals were observed by CL image analysis at the Institute of Mineral Deposits, CAGS. Mineral abbreviations are after Kretz (1983).

Mineral Inclusions in Zircons from Metamorphic Rocks in UHP Belt

Zircon from paragneiss

Mineral inclusions in zircon separates from a few representative paragneisses were investigated in detail, including garnet-, epidote-, and biotite-bearing two-feldspar gneiss (QL1) and phengite- and biotite-bearing two-feldspar gneiss (QL2) from Qinglongshan, and garnet-, biotite-, and amphibolite-bearing plagioclase gneiss (MB1) in Maobei. Abundant mineral inclusions in zircons were identified; these include coesite, garnet, omphacite, jadeite, phengite, rutile, quartz, and apatite. Mineral inclusion assemblages vary significantly in different micro-domains of individual zircon grains (Figs. 2 and 3). Abundant coesite inclusions occur in zircons from all analyzed paragneiss samples. Most coesite-bearing zircons contain the following UHP mineral assemblages: Coe + Grt + Omp + Phe (Fig. 2A); Coe + Grt + Omp (Fig. 2C); Coe + Phe (Fig. 2E); Coe + Jd + Phe + Rt (Fig. 3A); Coe + Grt + Jd + Rt + Ap (Fig. 3C); Coe + Grt + Jd + Rt (Fig. 3D); and Coe + Jd + Rt (Fig. 3E). However, coesite, omphacite, and jadeite were not found in the matrix of these paragneisses. Omphacite inclusions are common in zircons from garnet-, epidote-, and biotite-bearing two-feldspar gneiss (QL1, Figs. 2A and 2C), whereas jadeite inclusions are preserved in zircons from garnet-, biotite-, and amphibole-bearing plagioclase gneiss (MB1) (Figs. 3A, 3C, 3D, and 3E). However, neither omphacite nor jadeite were identified in zircons from phengite- and biotite-bearing two-feldspar gneiss (QL2) (Fig. 2E). These features indicate that UHP mineral inclusion assemblages are not only related to P-T conditions, but also are controlled by host rock compositions (Liu F. L. et al., 2002).

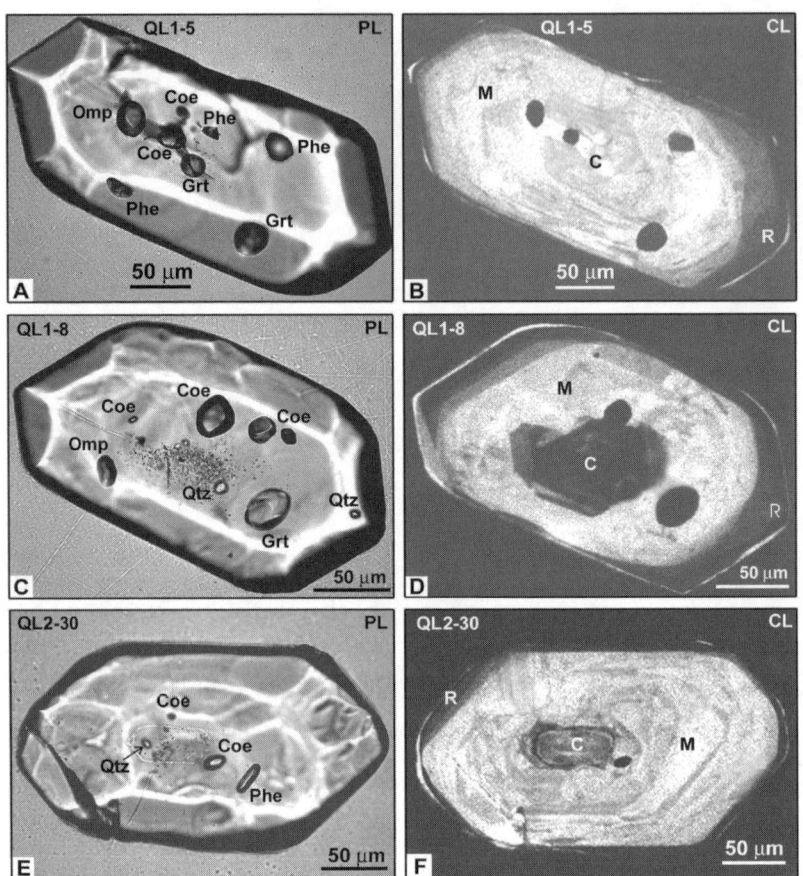

FIG. 2. Plane-polarized light (PL) images of mineral inclusions in zircons and CL images of host zircons from garnet-, epidote-, and biotite-bearing two-feldspar gneiss (QL1) and phengite- and biotite-bearing two-feldspar gneiss (QL2) in the Qinglongshan area. A. Zircon grain QL1-5 from QL1 sample contains Coe + Grt + Omp + Phe inclusions in the core and mantle. B. CL image of the same zircon as in Figure 2A showing core (C), mantle (M), and rim (R) relationships. C. Zircon grain QL1-8 from QL1 sample contains a Qtz inclusion in the core, Coe + Grt + Omp in the mantle, and a Qtz inclusion in the rim. D. CL image of the same zircon as in Figure 2C showing core (C), mantle (M), and rim (R) relationships. E. Zircon grain QL2-30 from QL2 sample contains a Qtz inclusion in the core and Coe + Phe in the mantle. F. CL image of the same zircon as in Figure 2E showing core (C), mantle (M), and rim (R) relationships.

Coesite and quartz inclusions commonly occur in different domains of a single zircon grain (Figs. 2C, 2E, 3D, and 3E). Cathodoluminescence (CL) images show that zircon crystals display distinct zoning (Figs. 2D, 2F, and 3F). For example, in zircons from QL1 and MB1, quartz inclusions occur in the cores of zircon (Figs. 2C and 3E) and in the rims (Fig. 2C), whereas Coe + Grt + Omp (Fig. 2C) and Coe + Jd + Rt (Fig. 3E) are found in the mantles. The intensity of CL images also reveals distinct zoning of zircon crystals; the low-luminescent cores in CL images are surrounded by relatively bright-luminescent mantles, and the mantles by relatively low-luminescent rims (Figs. 2D and 3F). In these samples, a few zircon grains show special patterns. For example, in zircon grains from QL1 and MB1, Coe + Grt + Omp + Phe (Fig. 2A), Coe + Jd + Phe + Rt (Fig. 3A), and Coe + Grt + Jd + Rt (Fig. 3D) occur in the core and mantle, whereas Qtz + Ab are trapped in the rim (Fig. 3D). The CL images show relatively bright-luminescent core-mantles surrounded by low-luminescent rims without a clear core-mantle boundary

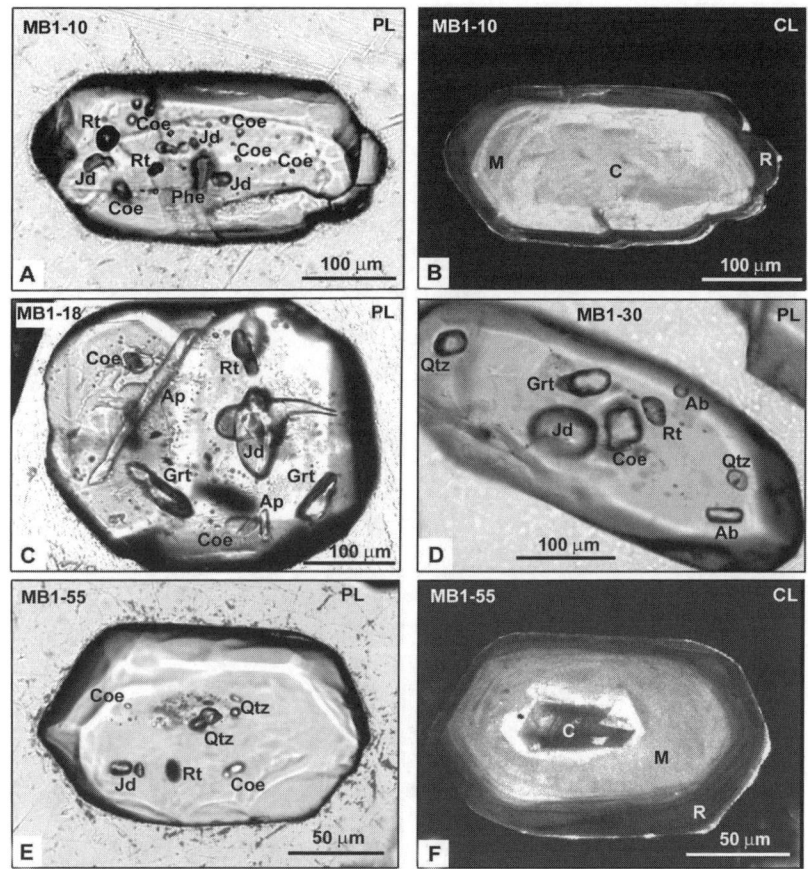

FIG. 3. Plane-polarized light (PL) images of mineral inclusions in zircons and CL images of host zircons from garnet-, biotite-, and amphibole-bearing plagioclase gneiss (MB1, 50.05 m depth in drillhole ZK2304) in the Maobei area. A. Zircon grain MB1-10 from MB1 sample contains Coe + Jd + Phe + Rt inclusions in the core and mantle. B. CL image of the same zircon as in Figure 3A showing core (C), mantle (M), and rim (R) relationships. C. Zircon grain MB1-18 from MB1 sample contains Coe + Jd + Grt + Rt + Ap inclusions in the core and mantle. D. Zircon grain MB1-30 from MB1 sample contains Coe + Grt + Jd + Rt in the core, and Qtz + Ab in the rim. E. Zircon grain MB1-55 from MB1 sample contains Qtz inclusions in the core, and Coe + Jd + Rt in the mantle. F. CL image of the same zircon as in Figure 3E showing core (C), mantle (M), and rim (R) relationships.

(Figs. 2B and 3B), indicating that these sections may not cut through the true cores of the zircon crystals.

Zircon from kyanite-bearing quartzite

UHP mineral inclusions in zircon separates from kyanite-bearing quartzite were also identified by laser Raman spectroscopy in the Qinglongshan-Hushan area (Fig. 1); these include coesite, kyanite, rutile, apatite, phengite, garnet, and omphacite (Figs. 4A–4C). The coesite-bearing zircons contain the following UHP inclusion assemblages: Coe + Ky + Rt + Ap (Fig. 4A); Coe + Ky + Grt + Omp + Phe + Ap (Fig. 4B); and Coe + Ky + Grt + Omp + Phe (Fig. 4C). The CL images reveal distinct zoning patterns in zircon crystals (Fig. 4D). In all zircon grains, coesite-bearing UHP assemblages occur in the core (C) and mantle (M) domains (Figs. 4A–4D), but were not found in the matrix of the same kyanite-bearing quartzite samples. These data indicate that the kyanite-bearing quartzite also experienced UHP metamorphism, and that zircons protected the UHP mineral inclusions from retrogressive recrystallization.

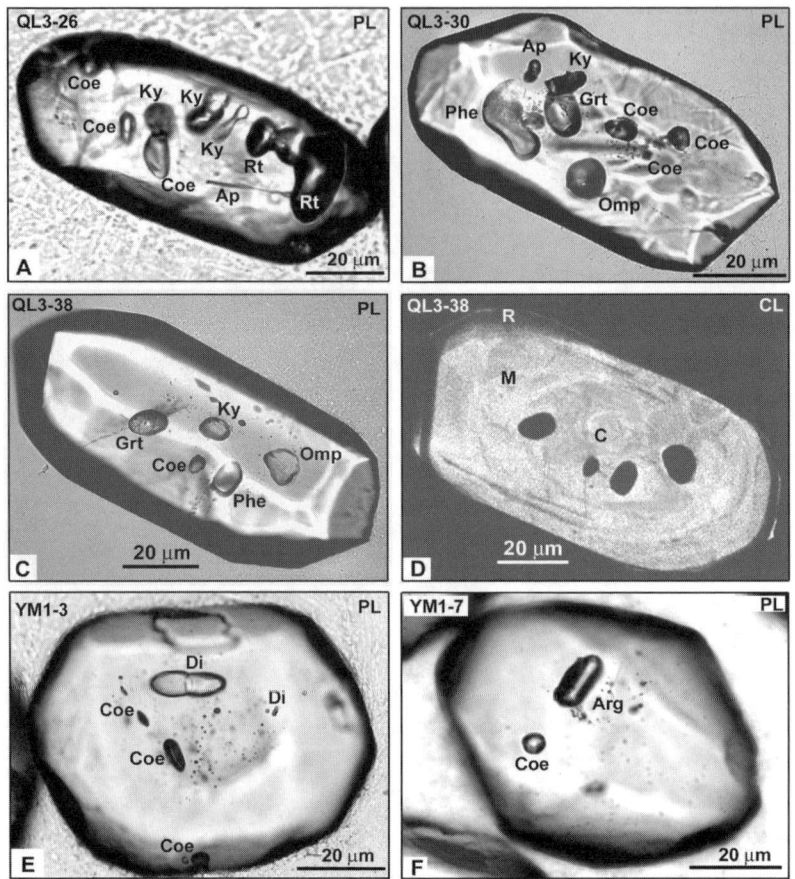

FIG. 4. Plane-polarized light (PL) images of mineral inclusions in zircons and CL images of host zircons from kyanite-bearing quartzite (QL3) at the Qinglongshan area, and diopside-bearing marble (YM1) from the Yanmachang area. A. Zircon grain QL3-26 from QL3 sample contains Coe + Ky + Rt + Ap inclusions in the core and mantle. B. Zircon grain QL3-30 from QL3 sample contains Coe + Ky + Grt + Omp + Phe + Ap inclusions in the core and mantle. C. Zircon grain QL3-38 from QL3 sample contains Coe + Ky + Grt + Omp + Phe inclusions in the core and mantle. D. CL image of the same zircon as in Figure 4C showing core (C), mantle (M), and rim (R) relationship. E. Zircon grain YM1-3 from YM1 sample contains Coe + Di inclusions in the core and mantle, and Coe inclusions in the rim. F. Zircon grain YM1-7 from YM1 sample contains Coe + Arg in the core and mantle.

Zircon from marble

Zircon is very fine-grained in marble, with grain size ranging from 40 to 80 μm. In each approximately 30 kg marble sample, only 10–30 zircon crystals were obtained. In general, most zircon grains are free in mineral inclusions; only a few crystals contain micro-coesite inclusions. With the example of diopside-bearing marble (YM1) from the Yanmachang area (Fig. 1), some zircons contain coesite + diopside inclusions in the core and mantle, and coesite inclusions in the rim (Fig. 4E), whereas others contain coesite + aragonite inclusions in the core and mantle domains (Fig. 4F). In addition, coesite-bearing UHP mineral inclusions were also identified in zircon separates from marbles from the north Linsu, Heidoujian, and Xugou areas (Fig. 1). These data indicate that all marble lenses in the orthogneiss unit (Unit I) were undoubtedly subjected to UHP metamorphism.

Zircon from orthogneiss

We have identified dusty and clear zircons from all orthogneiss samples in the orthogneiss unit (Unit

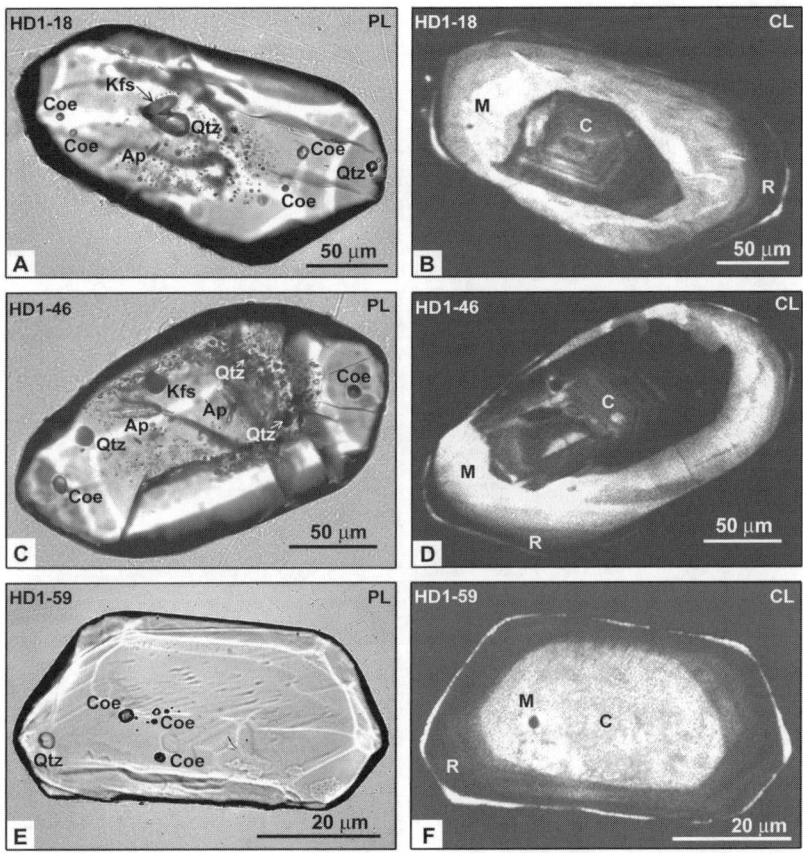

FIG. 5. Plane-polarized light (PL) images of mineral inclusions in zircons and CL images of host zircons from magnetite- and biotite-bearing orthogneiss (HD1) from the Heidoujian area. A. Zircon grain HD1-18 from HD1 sample contains Qtz + Kfs + Ap in the core, Coe in the mantle, and Qtz in the rim. B. CL image of the same zircon as in Figure 5A showing core (C), mantle (M), and rim (R) relationships. C. Zircon grain HD1-46 from HD1 sample contains Qtz + Kfs + Ap in the core, and Coe in the mantle. D. CL image of the same zircon as in Figure 5C showing core (C), mantle (M), and rim (R) relationships. E. Zircon grain HD1-59 from HD1 sample contains Coe inclusions in the core and mantle, and a Qtz inclusion in the rim. F. CL image of the same zircon as in Figure 5E showing core (C), mantle (M), and rim (R) relationships.

I) and supracrustal rock unit (Unit II) (Fig. 1); both zircon groups (Fig. 5) contain coesite inclusions. Dusty zircons from magnetite- and biotite-bearing orthogneiss (HD1) from the Heidoujian area (Fig. 1) preserve abundant mineral inclusions and contain many fractures (Figs. 5A and 5C). Raman spectroscopy analyses of these dusty zircons yield distinctive mineral assemblages in the cores, mantles, and rims. Abundant mineral inclusions in zircon cores consist mainly of Qtz + Kfs + Ap, and index UHP mineral inclusions in zircon mantles are characterized by coesite, whereas low-P mineral inclusions in zircon outmost rims are chiefly quartz (Figs. 5A and 5C). The CL images of these dusty zircons also consist of cores, mantles, and rims; relatively low-luminescent cores are surrounded by bright-luminescent mantles, and then by low-luminescent rims (Figs. 5B and 5D). Some zircons have euhedral cores, preserve obvious optical zoning (Fig. 5B), and represent growth zones of zircon crystallization from a magma.

Coesite inclusions were also identified in a few clear zircon grains from the same orthogneiss sample (HD1) (Figs. 5E and 5F). Compared to the dusty zircons (200–400 μm) in the same sample, the clear zircons are characterized by smaller sizes (50–80 μm), and by fewer inclusions including smaller

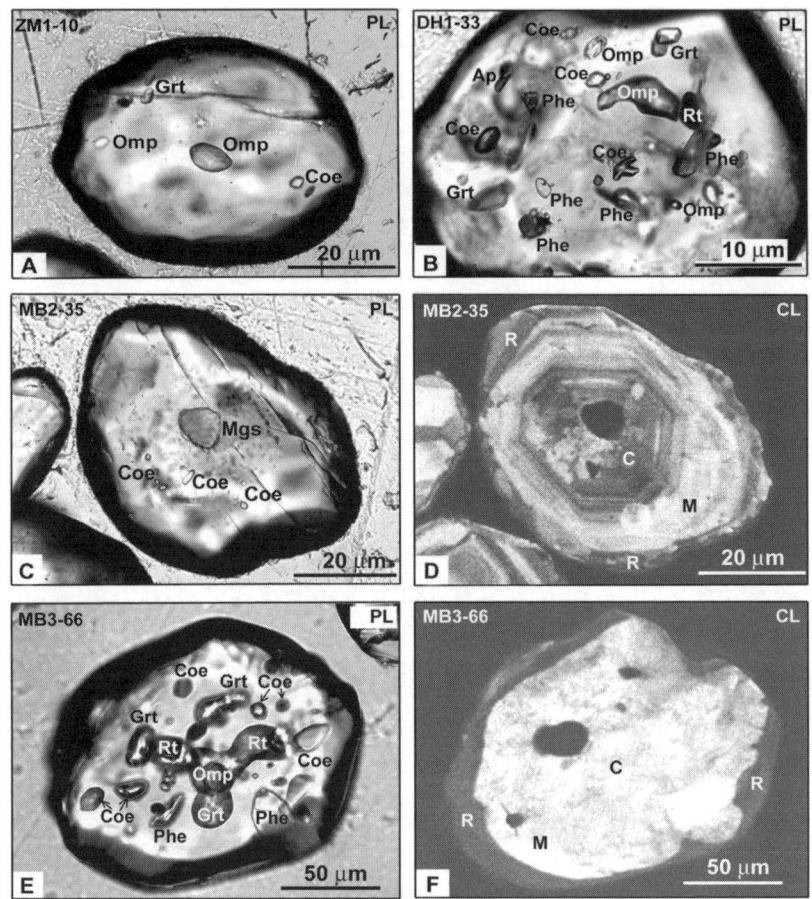

FIG. 6. Plane-polarized light (PL) images of mineral inclusions in zircons and CL images of host zircons in amphibolite from Zhimafang (ZM1, 290 m depth in drillhole CCSD-PP1), south Donghai (DH1) and Maobei (MB2, 958 m depth in drillhole CCSD-PP2), and from surface eclogite from Maobei (MB3). A. Zircon grain ZM1-10 from ZM1 sample contains Coe + Grt + Omp inclusions in the core and mantle. B. Zircon grain DH1-33 from DH1 sample contains Coe + Grt + Omp + Phe + Rt + Ap inclusions in the core and mantle. C. Zircon grain MB2-35 from MB2 sample contains Coe + Mgs inclusions in the core and Coe in the mantle, respectively. D. CL image of the same zircon as in Figure 6C showing core (C), mantle (M), and rim (R) relationships. E. Zircon grain MB3-66 from MB3 sample contains Coe + Grt + Omp + Phe + Rt inclusions in the core and mantle. F. CL image of the same zircon as in Figure 6E showing core (C), mantle (M), and rim (R) relationships.

ovoid coesite and quartz preserved in the core-mantle and rim, respectively (Fig. 5E). The relative CL image is characterized by bright-luminescent core-mantle overgrowths surrounded by low-luminescent outmost rims (Fig. 5F). These data indicate that the clear zircons may have crystallized only during UHP metamorphism.

Zircon from amphibolite and eclogite

Zircons separated from amphibolites contain abundant mineral inclusions, including coesite, quartz, garnet, omphacite, phengite, rutile, magnesite, and apatite. The coesite-bearing zircons preserve the following inclusion assemblages: Coe + Grt + Omp (Fig. 6A), Coe + Grt + Omp + Phe + Rt + Ap (Fig. 6B), and Coe + Mgs (Fig. 6C). CL images reveal that the zircons are characterized by low-luminescent cores (C), surrounded by bright-luminescent overgrowth mantles (M), and surrounded by irregular, narrow, low-luminescent outmost rims (R) (Fig. 6D). In most grains, coesite-bearing UHP assemblages occur in the core and mantle of zircons

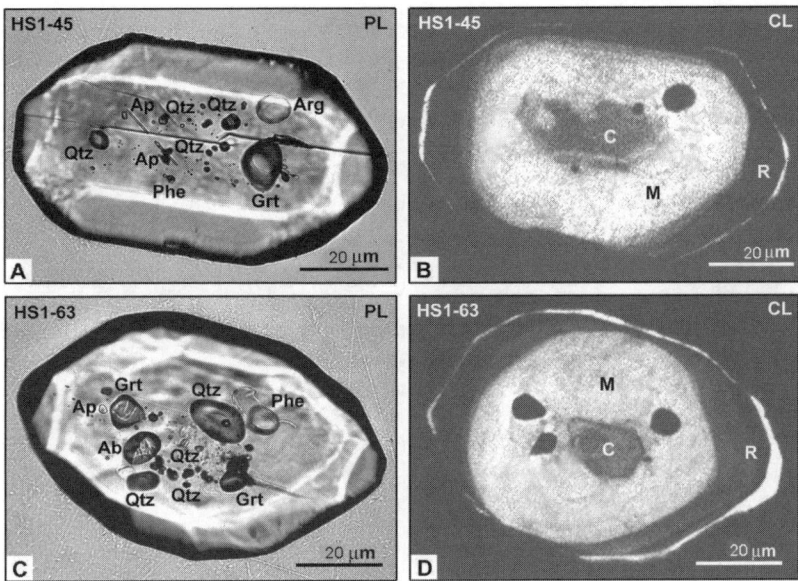

FIG. 7. Plane-polarized light (PL) images of mineral inclusions in zircons and CL images of host zircons from garnet- and biotite-bearing two-feldspar gneiss (HS1) from the Houshancun area. A. Zircon grain HS1-45 from HS1 sample contains Qtz + Ap inclusions in the core, and Arg + Grt + Phe + Qtz in the mantle. B. CL image of the same zircon as in Figure 7A showing core (C), mantle (M), and rim (R) relationships. C. Zircon grain HS1-63 from HS1 sample contains Qtz inclusions in the core, Grt + Phe + Ab + Qtz + Ap in the mantle. D. CL image of the same zircon as in Figure 7C showing core (C), mantle (M), and rim (R) relationship.

(Figs. 6A–6C), but are absent in the matrix of the same amphibolite samples (Liu F. L. et al., 2003). In addition, mineral inclusions hidden in zircons from eclogites were also detected by laser Raman spectroscopy and CL image analysis. Common UHP mineral assemblages are characterized by Coe + Grt + Omp + Phe + Rt (Fig. 6E), and CL images show relatively bright-luminescent core-mantles and low-luminescent rims (Fig. 6F). These characteristic UHP mineral inclusion assemblages preserved in zircons from eclogites are similar to those in zircons from amphibolite, and are also consistent with matrix assemblages of the same eclogites. These data indicate that the amphibolites undoubtedly experienced UHP metamorphism, and their peak-stage assemblages in the matrix were retrograded and completely replaced by amphibolite-facies assemblages related to the rapid exhumation of the Sulu terrane; zircons protected these index UHP mineral inclusions from retrogressive recrystallization.

Mineral Inclusions in Zircons from Metamorphic Rocks in the HP Belt

Zircons separated from paragneiss, orthogneiss, kyanite-bearing quartzite and marble in Units III and IV also contain abundant mineral inclusions (Figs. 7–10). However, coesite and other UHP minerals have not been identified; from all 33 samples in Units III and IV only a few index HP minerals (e.g., aragonite) were detected as inclusions in zircons (Fig. 1). The distribution of mineral inclusions and the CL images of zircons separated from paragneiss, kyanite-bearing quartzite, marble, and orthogneiss are described as follows.

Zircon from paragneiss

Many mineral inclusions in zircons from paragneiss samples in Units III and IV were identified; these include aragonite, garnet, phengite, quartz, albite, and apatite. Most common HP mineral inclusion assemblages in zircons are: Arg + Grt + Phe + Qtz (Fig. 7A) and Grt + Phe + Ab + Qtz + Ap (Fig.

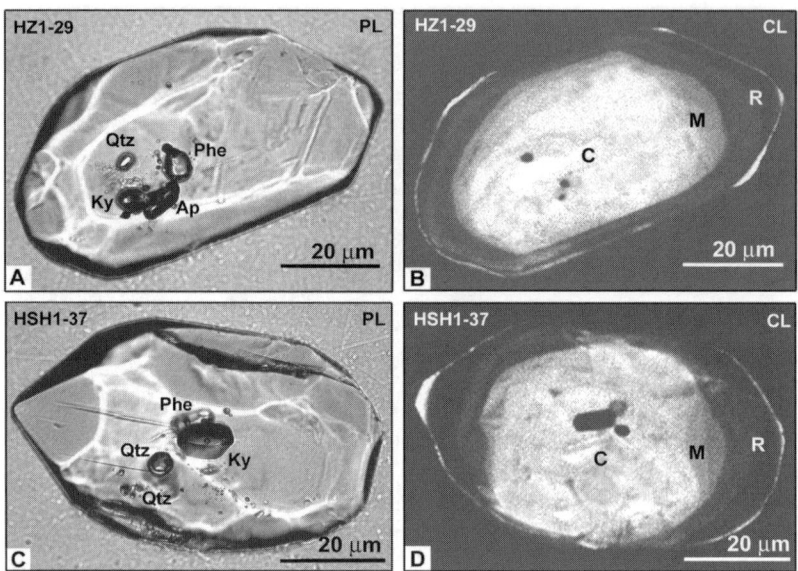

FIG. 8. Plane-polarized light (PL) images of mineral inclusions in zircons and CL images of host zircons in kyanite-bearing quartzite from the Houzui (HZ1) and Hanshan (HSH1) areas. A. Zircon grain HZ1-29 from HZ1 sample contains Ky + Phe + Qtz + Ap inclusions in the core and mantle. B. CL image of the same zircon as in Figure 8A showing core (C), mantle (M), and rim (R) relationships. C. Zircon grain HSH1-37 from HSH1 sample contains Ky + Phe + Qtz inclusions in the core and mantle. D. CL image of the same zircon as in Figure 8C showing core (C), mantle (M), and rim (R) relationships.

7C). Due to extensively amphibolite- to greenschist-facies retrograde metamorphism, aragonite-bearing HP mineral assemblages have not been found in the matrix of the analyzed paragneisses.

In some paragneiss samples, mineral inclusion assemblages differ greatly in different domains of the same zircon grains. For example, in zircons from garnet- and biotite-bearing two-feldspar gneiss (HS1) from Houshancun (Fig. 1), quartz and apatite inclusions occur in the core (C), and index HP mineral inclusions Arg + Grt + Phe + Qtz and Grt + Phe + Ab + Qtz + Ap are present in the mantle (M), whereas the rim (R) is free of mineral inclusions (Figs. 7A and 7C). The CL images are characterized by low-luminescent cores (C) surrounded by bright-luminescent mantle overgrowths (M), then surrounded by the low-luminescent outmost rims (R) (Figs. 7B and 7D). Such distribution of mineral inclusions in zoned zircons from paragneisses have recorded inherited (detrital) cores of protoliths, HP metamorphic mantles, and retrograde rim growth of zircons.

Zircon from kyanite-bearing quartzite

HP mineral inclusions were also identified in zircon grains from kyanite-bearing quartzite from Houzui (HZ1) and Hanshan (HSH1) (Fig. 8). Compared to zircons in paragneiss sample HS1, zircons in kyanite-bearing quartzite are characterized by smaller size (40-80 μm) and fewer inclusions including smaller ovoid kyanite, phengite, quartz, and apatite preserved in the core and mantle domains (Figs. 8A and 8C). In general, index HP mineral assemblages as inclusions preserved in zircons mainly consist of Ky + Phe + Qtz + Ap (Fig. 8A) and Ky + Phe + Qtz (Fig. 8C). The CL images reveal distinct zoning patterns in zircon crystals characterized by bright-luminescent core-mantle overgrowths surrounded by low-luminescent outmost rims (Figs. 8B and 8D). These data indicate that the zircons crystallized during HP metamorphism.

Zircon from marble

Zircons separated from marbles in Unit III are very fine, with grain size ranging from 30 to 70 μm.

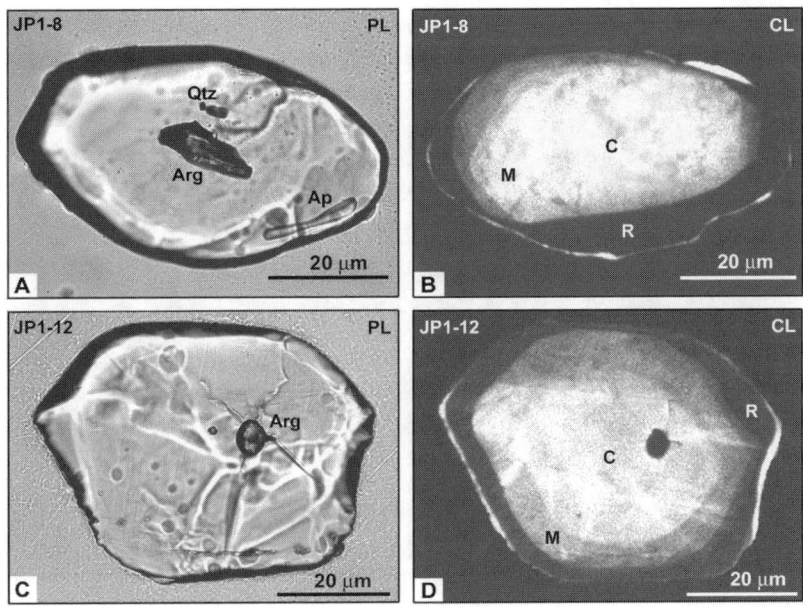

FIG. 9. Plane-polarized light (PL) images of mineral inclusions in zircons and CL images of host zircons in apatite-bearing marble (JP1) from the Jinping area. A. Zircon grain JP1-8 from JP1 sample contains Arg + Qtz in the core, and apatite in the rim. B. CL image of the same zircon as in Figure 9A showing core (C), mantle (M), and rim (R) relationships. C. Zircon grain JP1-12 from JP1 sample contains an Arg inclusion in the core. D. CL image of the same zircon as in Fig. 9C showing core (C), mantle (M), and rim (R) relationships.

In general, most zircon grains are free from mineral inclusions; only a few crystals contain micro-mineral inclusions. For example, some zircons of apatite-bearing marble (JP1) from Jinping (Fig. 1) contain aragonite + quartz inclusions in the cores and mantles, and apatite inclusions in the rims (Fig. 9A), whereas other zircon grains contain aragonite inclusions in the cores, and are free of mineral inclusions in the mantles and rims (Fig. 9C). The CL images are characterized by relatively bright-luminescent core-mantles, surrounded by low-luminescent rims (Figs. 9B and 9D). These data indicate that the marbles in the kyanite-bearing marble unit (Unit III) were also subjected to HP metamorphism.

Zircon from orthogneiss

Most zircon crystals separated from orthogneiss samples in Units III and IV are characterized by abundant mineral inclusions, including quartz, K-feldspar, phengite, and apatite. However, UHP mineral inclusions (e.g., coesite etc.) have not been found in all analyzed zircon grains. For example, in a magnetite- and biotite-bearing potash feldspar orthogneiss (HW1) from Huangwocun (Fig. 1), dusty zircons have anhedral to subhedral form and preserve abundant mineral inclusions (Figs. 10A and 10C). Cathodoluminescence images of these dusty zircons (Figs. 10B and 10D) are very different from those of the paragneiss mentioned above (Figs. 7B and 7D). In the cores (C), fine-grained mineral inclusions are mainly comprised of Qtz + Kfs + Ap (Fig. 10A) and Qtz + Kfs (Fig. 10C). In the CL images, the zircon cores exhibit euhedral crystal shapes, and preserve obvious optical zoning (Figs. 10B and 10D). These data suggest that the cores of zircons separated from orthogneiss samples inherited growth zones of primary magma crystallization prior to the HP event. However, HP mineral assemblages of Phe + Qtz (Fig. 10A) and Phe + Qtz + Ap (Fig. 10C) as inclusions are identified in the mantles of the same zircons, and CL images reveal irregular overgrowth zoning patterns in the mantles (Figs. 10B and 10D). Whereas the zircon rims are free of mineral inclusions, the CL images show relatively low-luminescent rims surrounding bright-luminescent mantles (Figs. 10B and 10D). These data conclude that zoned zircons separated from orthogneiss samples in Units III and IV record an igneous

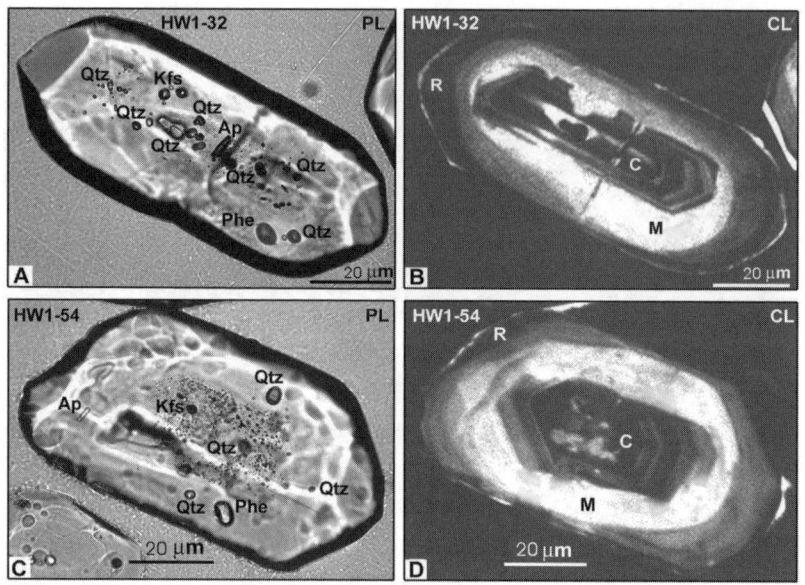

FIG. 10. Plane-polarized light (PL) images of mineral inclusions in zircons and CL images of host zircons from magnetite- and biotite-bearing potash feldspar orthogneiss (HW1) from the Huangwocun area. A. Zircon grain HW1-32 from HW1 sample contains Qtz + Kfs + Ap in the core, and Phe + Qtz in the mantle. B. CL image of the same zircon as in Figure 10A showing core (C), mantle (M), and rim (R) relationships. C. Zircon grain HW1-54 from HW1 sample contains Qtz + Kfs in the core, and Phe + Qtz + Ap in the mantle. D. CL image of the same zircon as in Fig. 10C showing core (C), mantle (M) and rim (R) relationship.

crystallization event in the inherited cores, HP metamorphism in the overgrowth mantle, and a retrogressive metamorphic event in the outermost rim, respectively.

Discussion and Conclusion

In the orthogneiss unit (Unit I) and supracrustal rock unit (Unit II) of the southwestern Sulu terrane (Fig. 1), many coesite-bearing UHP assemblages are preserved as inclusions in zircons. In contrast, zircon separates from rocks in the kyanite-bearing quartzite–marble unit (Unit III) and paragneiss-schist unit (Unit IV) (Fig. 1) contain index aragonite- and phengite-bearing HP mineral assemblages as inclusions. The spatial distribution of these index UHP and HP mineral assemblages as inclusions in zircons from northwestern Unit I to southeastern Unit IV are plotted in Figure 11. The size of the UHP slab provides a significant constraint on the subduction and exhumation model of the Sulu-Dabie terrane (Liou et al., 2000). In numerous previous studies, inclusions of coesite and/or quartz pseudomorphs after coesite, and other evidences of UHP metamorphism have been identified in many eclogites and ultramafic rocks (Wang et al., 1989; Okay et al., 1989; Yang J. J. et al., 1993; Zhang et al., 1995, 2003; Liou and Zhang, 1996; Yang, J. J., 2003; Yang J. S. et al., 2003). Moreover, our studies show that coesite occurs in zircons of amphibolite-facies paragneiss, orthogneiss, marble, kyanite-bearing quartzite, amphibolite, and eclogite in Units I and II of the southwestern Sulu terrane. Similarly, zircons separated from core samples, including paragneiss, orthogneiss, amphibolite, eclogite, phengite- and kyanite-bearing quartzite, and epidote-amphibole-biotite schist from pre-pilot drill-holes CCSD-PP1 (432 m depth) and CCSD-PP2 (1028 m depth) of the Chinese Continental Scientific Drilling Project also contain abundant coesite-bearing UHP mineral inclusions (Liu F. L. et al., 2001, 2002, 2003). Such consistent observations suggest that eclogite together with its country rocks experienced *in situ* UHP metamorphism.

Temperature estimates for the peak UHP metamorphic stage deduced from mineral inclusions in zircons of paragneiss, orthogneiss, kyanite-bearing quartzite, amphibolite, and eclogite in our studies

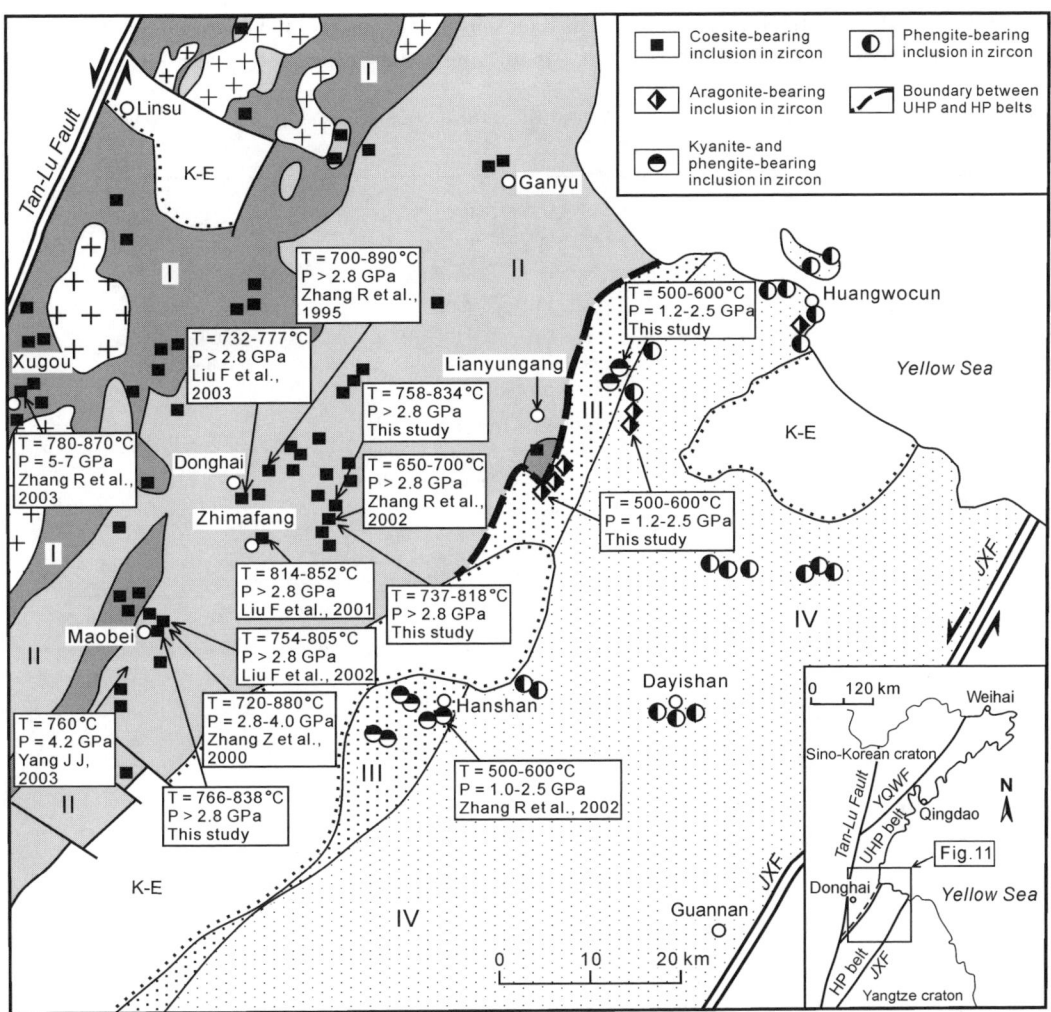

FIG. 11. Schematic lithologic map showing the distribution of identified index minerals as inclusions in zircons from UHP and HP rocks and available P-T estimates of the present and previous studies from selected localities in southwestern Sulu terrane. A deduced tectonic boundary between the UHP and HP belts is shown. The legends are same as those in Figure 1.

combined with previous data (Liu F. L. et al., 2001, 2002, 2003) range from 723° to 852°C, with a pressure > 2.8 GPa (Table 1). These supracrustal rocks have similar UHP P-T conditions as those estimates for surface and shallow drillhole eclogites (Enami et al., 1993; Zhang et al., 1995; Zhang and Liou, 1997; Zhang et al., 2000). This evidence suggests that the voluminous continental materials consisting mainly of granitic crustal materials of passive continental margins, including various sandstone, shale, limestone, and minor mafic-ultramafic intrusive and extrusive rocks, were subducted contemporaneously to mantle depths in excess of 100 km and metamorphosed, and then rapidly returned to the earth's upper crustal levels.

In the present studies, aragonite- and phengite-bearing index HP minerals as inclusions are preserved in many zircons from paragneiss, apatite-bearing marble, kyanite-bearing quartzite, and orthogneiss in both Units III and IV (Fig. 1), indicating that the various supracrustal rocks in Units III and IV undoubtedly experienced HP

TABLE 1. Index UHP and HP Minerals Identified as Inclusions in Zircons, and P-T Estimates of Present and Previous Studies for Both UHP and HP Rocks in Southwestern Sulu Terrane, Eastern China

Metamorphic belt or drillholes	Rock type	Index UHP or HP mineral inclusions	P-T conditions
UHP belt (Unit I and II; This study)	Paragneiss	Coe + Grt + Omp + Phe Coe + Grt + Omp Coe + Phe	T = 758 – 834 °C P > 2.8 GPa
		Coe + Jd + Phe + Rt Coe + Grt + Jd + Rt + Ap Coe + Grt + Jd + Rt Coe + Jd + Rt	T = 766 – 838 °C P > 2.8 GPa
	Kyanite-bearing quartzite	Coe + Ky + Rt + Ap Coe + Grt + Omp + Ky + Phe + Ap Coe + Grt + Omp + Ky + Phe	T = 737 – 818 °C P > 2.8 GPa
	Diopside-bearing Marble	Coe + Di Coe + Arg	
	Orthogneiss	Coe	
	Amphibolite	Coe + Grt + Omp + Phe + Rt + Ap	T = 732 – 777 °C (Liu F. L. et al., 2003) P > 2.8 GPa
	Eclogite	Coe + Grt + Omp + Phe + Rt	
Drillhole of PP1 in UHP belt (Liu F. L. et al., 2001)	Paragneiss	Coe + Grt + Omp Coe + Jd + Phe + Ap Coe + Grt	T = 814 – 852 °C P > 2.8 GPa
	Amphibolite	Coe + Grt + Omp	
Drillhole of PP2 in UHP belt (Liu F. L. et al., 2002, 2003)	Paragneiss	Coe + Grt + Omp + Rt Coe + Grt + Jd + Phe + Ap Coe + Phe + Rt + Ap	T = 754 – 805 °C P > 2.8 GPa
	Amphibolite	Coe + Grt + Omp + Rt Coe + Omp + Rt Coe + Mgs	T = 723 – 764 °C P > 2.8 GPa
HP belt (Unit III and IV; this study)	Paragneiss	Arg + Grt + Phe + Qtz Grt + Phe + Ab + Qtz + Ap	
	Kyanite-bearing quartzite	Ky + Phe + Qtz + Ap Ky + Phe + Qtz	T = 500 – 600 °C (Zhang et al., 2002) P = 1.2 – 2.5 GPa
	Apatite-bearing marble	Arg Arg + Qtz	
	Orthogneiss	Phe + Qtz Phe + Qtz + Ap	

metamorphism. Recently, Zhang et al. (2002) recognized hydroxyl-rich topaz in kyanite-bearing quartzite from Hanshan in the HP belt, and estimated the peak metamorphic temperatures of 500–600°C and pressures of 1.0–2.5 GPa for the assemblage. Our studies show that mineral assemblages preserved in zircons from the HP belt are characterized by Arg + Grt + Phe + Qtz (Fig. 7A) and Phe + Grt + Ab + Qtz + Ap (Fig. 7C) in paragneiss, and Arg ± Qtz in marble (Figs. 9A and 9C). Combining the experimental curves of aragonite = calcite (Holland and Powell, 1990), jadeite + quartz = albite (Holland, 1980), and coesite = quartz (Bohlen and Boettcher, 1982), pressure estimates for those aragonite-bearing inclusion assemblages range from 1.2 to 2.5 GPa, at 500–600°C, consistent with the estimates by Zhang et al. (2002).

Geochronologic data, including SHRIMP U-Pb dates of zircon, Sm-Nd mineral and whole-rock isochrons, yield 220–240 Ma for the peak UHP metamorphism of Proterozoic supracrustal rocks in the Sulu-Dabie UHP belt (Li et al., 1993; Ames et al., 1996; Hacker et al., 1998, 2000). Recently, Liu F. L. et al. (2004) conducted SHRIMP U-Pb dating of zoned zircons from drillhole CCSD-PP2 orthogneiss and paragneiss in order to constrain the geochronological evolution of the UHP belt. The results identify three discrete and meaningful stages: Neoproterozoic protolith age (> 680 Ma) for quartz-bearing cores, 231 ± 4 Ma for the UHP metamorphic event for coesite-bearing mantles, and 211 ± 4 Ma for the amphibolite-facies retrogressive overprint for quartz-bearing rims. These geochronological data suggest that Neoproterozoic supracrustal rocks in the Sulu UHP terrane experienced Middle Triassic subduction to mantle depths, then Late Triassic exhumation to mid-crustal levels. In our studies, the delicate micro-textures of zoned zircons, including inherited cores of protolith, HP overgrowth mantles, and amphibolite- to greenschist-facies outmost retrograde rims, also have been revealed by laser Raman spectroscopy and CL images in metamorphic rocks from Units III and IV (Figs. 7 and 10). Therefore, SHRIMP U-Pb microspot dating of these zircon cores, mantles, and rims is also essential for constraining the protolith age, HP metamorphic event, and late amphibolite- to greenschist-facies retrograde overprint in the HP belt of the Sulu terrane.

In all previous studies, because of very poor exposures and extensive amphibolite- to greenschist-facies retrograde overprints on both UHP and HP rocks in the Donghai region, the boundary between the UHP and HP belts of the southwestern Sulu terrane has remained unclear (Zhang et al., 1995, 2002). Index minerals identified as inclusions in zircons from both UHP and HP rocks in the region from the present and previous studies are summarized in Table 1 together with available P-T estimates from selected localities. These data are plotted in Figure 11 to illustrate the spatial distribution of index minerals and P-T estimates for both HP and UHP belts. The results clearly delineate the boundary between the UHP and HP belts in the southwestern Sulu terrane along ductile shear zone DF_8 in Figure 1. Detailed mapping and kinematic measurement indicate that shear zones in the studied region occur in mylonitized amphibolites and gneissic rocks; each shear zone is 3–20 km wide and is characterized by SE-NW shear strain. The shear strain and preferred orientation of quartz show that the shearing sense from SE to NW thrusting was transformed from NW to SE normal slipping, during the retrograde metamorphic process from amphibolite- to greenschist-facies metamorphism, related to exhumation of the Sulu terrane (Xu et al., 2003). Such deformation features show that Sulu UHP-HP slabs were exhumed by an extrusion mechanism and modified by post-exhumation extension.

Acknowledgments

We contribute this paper to honor Prof. W. G. Ernst for his contributions to the study of subduction-zone metamorphism and his support for the US-China project. We are grateful to Dr. R. Y. Zhang, Prof. Q. H. Shen, Z. M. Zhang, and J. S. Yang for discussion and constructive comments. Laser Raman spectroscopy of mineral inclusions in zircons was performed at the UHP laboratory of the Tokyo Institute of Technology (TIT) and the Open Laboratory of Continental Dynamics, Chinese Academy of Geological Sciences (CAGS). This study was financially supported by the National 973 Project of the Chinese Ministry of Science and Technology (Grant No. 2003CB716502), the Natural Science Foundation of China (Grant No. 40399143), and a joint Sino-American cooperative project supported by NSF EAR 0003355. We thank the above named individuals and institutes for support and help.

REFERENCES

Ames, L., Zhou, G., and Xiong, B., 1996, Geochronology and isotopic character of high-pressure metamorphism with implications for collision of the Sino-Korean and Yangtze cratons, central China: Tectonics, v. 15, p. 472–489.

Bohlen, S. R., and Boettcher, A. L., 1982, The quartz-coesite transformation: A pressure determination and effects of other components: Journal of Geophysical Research, v. 87, p. 7073–7078.

Chopin, C., and Sobolev, N. V., 1995, Principal mineralogic indicators of UHP in crustal rocks, in Coleman, R. G., and Wang, X. M., eds., Ultrahigh pressure metamorphism: Cambridge, UK, Cambridge University Press, p. 96–131.

Enami, M., Zang, Q. J., and Yin, Y. J., 1993, High-pressure eclogites in north Jiangsu-southern Shandong province, eastern China: Journal of Metamorphic Geology, v. 11, p. 589–603.

Hacker, B. R., Ratschbacher, L., Webb, L. E., McWilliams, M. O., Ireland, T. R., Calvet, A., Dong, S., Wenk, H. R., and Chateigner, D., 2000, Exhumation of ultrahigh-pressure continental crust in east central China: Journal of Geophysical Research, v. 105, p. 13,339–13,364.

Hacker, B. R., Ratschbacher, L., Webb, L. E., Ireland, T. R., Walker, D., and Dong, S., 1998, Orogen-scale architecture of the ultrahigh-pressure Dabie-Hong'an-Tongbai Shan, China: Earth and Planetary Science Letters, v. 161, p. 215–230.

Holland, T. J. B., 1980, The reaction albite = jadeite + quartz determined experimentally in the range 600-1200°C: American Mineralogist, v. 65, p. 125–134.

Holland, T. J. B., and Powell, R., 1990, An enlarged and updated internally consistent thermodynamic dataset with uncertainties and correlations: the system $K_2O – Na_2O – CaO – MgO – MnO – FeO – Fe_2O_3 – Al_2O_3 – TiO_2 – SiO_2 – C – H_2O – O_2$: Journal of Metamorphic Geology, v. 8, p. 89-124.

Katayama, I., Zayakhstan, A. A., and Maruyama, S., 2000, Progressive P-T records from zircon in Kokchetav UHP-HP rocks, northern Kazakhstan: The Island Arc, v. 9, p. 417–427.

Kretz, R., 1983, Symbols for rock-forming mineral: American Mineralogist, v. 68, p. 277–279.

Li, S., Chen, Y., Cong, B., Zhang, Z., Zhang, R. Y., Liu, D., Hart, S. R., and Ge, N., 1993, Collision of the North China and Yangtze blocks and formation of coesite-bearing eclogite: Timing and processes: Chemical Geology, v. 109, p. 70–89.

Liou, J. G., Hacker, B. R., and Zhang, R. Y., 2000, Into the forbidden zone: Science, v. 287, p. 1215–1216.

Liou, J. G., and Zhang, R. Y., 1996, Occurrence of intergranular coesite in Sulu ultrahigh-P rocks from China: Implications for fluid activity during exhumation: American Mineralogist, v. 81, p. 1217–1221.

Liu, F. L., Xu, Z. Q., Katayama, I., Yang, J. S., Maruyama, S., and Liou, J. G., 2001, Mineral inclusions in zircons of para- and orthogneiss from pre-pilot drillhole CCSD-PP1, Chinese Continental Scientific Drilling Project: Lithos, v. 59, p. 199–215.

Liu, F. L., Xu, Z. Q., Liou, J. G., Katayama, I., Masago, H., Maruyama, S., and Yang, J. S., 2002, Ultrahigh-pressure mineral inclusions in zircons from gneissic core samples of the Chinese Continental Scientific Drilling Site in eastern China: European Journal of Mineralogy, v. 14, p. 499–512.

Liu, F. L., Xu, Z. Q., Liou, J. G., and Song, B., 2004, SHRIMP U-Pb ages of ultrahigh-pressure and retrograde metamorphism of gneissic rocks, southwestern Sulu terrane, eastern China: Journal of Metamorphic Geology, in press.

Liu, F. L., Zhang, Z. M., Katayama, I., Xu, Z. Q., and Maruyama, S., 2003, Ultrahigh-pressure metamorphic records hidden in zircons from amphibolites in Sulu terrane, eastern China: The Island Arc, v. 12, p. 256–267.

Liu, J. B., Ye, K., Maruyama, S., Cong, B. L., and Fan, H. R., 2001, Mineral inclusions in zircon from gneisses in the ultrahigh-pressure zone of the Dabie Mountains, China: Journal of Geology, v. 109, p. 523–535.

Okay, A. I., and Sengör, A. M. C., 1992, Evidences for intra-continental thrust related exhumation of the ultra-high pressure rocks in China: Geology, v. 20, p. 411–414.

Okay, A. I., Xu, S., and Sengör, A. M. C., 1989, Coesite from the Dabie Mountains eclogite, central China: European Journal of Mineralogy, v. 1, p. 595–598.

Parkinson, C. D., and Katayama, I., 1999, Present day ultrahigh-pressure conditions of coesite inclusions in zircon and garnet: Evidence from laser Raman microspectroscopy: Geology, v. 27, p. 979–982.

Sobolev, N. V., Shatsky, V. S., Vavilov, S. V., and Goryainov, S. V., 1994, Zircon from ultrahigh pressure metamorphic rocks of folded regions as an unique containers of inclusions of diamond, coesite and coexisting minerals: Doklady Akademii Nauk, v. 334, p. 488–492 (in Russian).

Tabata, H., Yamauchi, K., Maruyama, S., and Liou, J. G., 1998, Tracing the extent of a UHP metamorphic terrane: Mineral-inclusion study of zircons in gneisses from the Dabieshan, in Hacker, B. R., and Liou, J. G., eds., When continents collide: Geodynamics and geochemistry of ultrahigh-pressure rocks: Dordrecht, Netherlands, Kluwer Academic Pulishers, p. 261–273.

Wang, X. M., and Liou, J. G., 1992, Regional ultrahigh-pressure coesite-bearing eclogitic terrane in central China: Evidence from country-rocks, gneiss, marble and metapelite: Geology, v. 20, p. 933–936.

Wang, X. M., Liou, J. G., and Mao, H. G., 1989, Coesite-bearing eclogites from the Dabie mountains in central China: Geology, v. 17, p. 1085–1088.

Xu, Z. Q., Zhang, Z. M., Liu, F. L., Yang, J. S., Li, H. B., Yang, T. N., Qiu, H. J., Li, T. F., Meng, F. C., Chen, S. Z., Tang, Z. M., and Chen, F. Y., 2003, Exhumation structure and mechanism of the Sulu ultrahigh-pressure metamorphic belt, central China: Acta Geologica Sinica, v. 77, p. 433–450 (in Chinese with English abstract).

Yang, J. J., 2003, Titanian clinohumite-garnet-pyroxene rock from the Su-Lu UHP metamorphic terrane, China: Chemical evolution and tectonic implications: Lithos, v. 70, p. 359–379.

Yang, J. J., Godard, G., Kienast, J. R., Lu, Y., and Sun, J., 1993, Ultrahigh-pressure (60 kbar) magnesite-bearing garnet peridotites from northeastern Jiangsu, China: Journal of Geology, v. 101, p. 541–554.

Yang, J. S., Wooden, C. L., Wu, C. L., Liu, F. L., Xu, Z. Q., Shi, R. D., Katayama, I., Liou, J. G., and Maruyama, S., 2003, SHRIMP U-Pb dating of coesite-bearing zircon from the ultrahigh-pressure metamorphic rocks, Sulu terrane, east China: Journal of Metamorphic Geology, v. 21, p. 551–560.

Ye, K., Yao, Y., Katayama, I., Cong, B. L., Wang, Q. C., and Maruyama, S., 2000, Large areal extent of ultrahigh-pressure metamorphism in the Sulu ultrahigh-pressure terrane of east China: New implications from coesite and omphacite inclusions in zircon of granitic gneiss: Lithos, v. 52, p. 157–164.

Zhang, R. Y., Hirajima, T., Banno, S., Cong, B. L., and Liou, J. G., 1995, Petrology of ultrahigh-pressure rocks from the southern Sulu region, eastern China: Journal of Metamorphic Geology, v. 13, p. 659–675.

Zhang, R. Y., and Liou, J. G., 1997, Partial transformation of gabbro to coesite-bearing eclogite from Yangkou, the Su-Lu terrane, eastern China: Journal of Metamorphic Geology, v. 15, p. 183–202.

Zhang, R. Y., Liou, J. G., and Shu, J. F., 2002, Hydroxyl-rich topaz in high-pressure and ultrahigh-pressure kyanite quartzites, with retrograde woodhouseite, from the Sulu terrane, eastern China: American Mineralogist, v. 87, p. 445–453.

Zhang, R. Y., Liou, J. G., Yang, J. S., and Ye, K., 2003, Ultrahigh-pressure metamorphism in the forbidden zone: The Xugou garnet peridotite, Sulu terrane, eastern China: Journal of Metamorphic Geology, v. 21, p. 539–550.

Zhang, Z. M., Xu, Z. Q., and Xu, H. F., 2000, Petrology of ultrahigh-pressure eclogites from the ZK703 drillhole in the Donghai, eastern China: Lithos, v. 52, p. 35–50.

Fluid Composition and Evolution Attending UHP Metamorphism: Study of Fluid Inclusions from Drill Cores, Southern Sulu Belt, Eastern China

ZEMING ZHANG,[1]

Institute of Geology, Chinese Academy of Geological Sciences, No. 26 Baiwanzhuang Road, Beijing, 100037, China

KUN SHEN,

Institute of Geological Sciences of Shandong, Jinan, 250013, China

YILIN XIAO, ALFONS M. VAN DEN KERKHOF, JOCHEN HOEFS,

Geoscience Center, University of Göttingen, Goldschmidtstrasse 1-3, D-37077 Göttingen, Germany

AND J. G. LIOU

Department of Geological and Environmental Sciences, Stanford University, Stanford, California 94305

Abstract

Rocks from the first pre-pilot hole of the Chinese Continental Scientific Drilling Project (CCSD-PPH1, 432 m), located in the eastern part of the Dabie-Sulu ultrahigh-pressure (UHP) metamorphic belt, have been subjected to a coesite-eclogite–facies metamorphic event, followed by an amphibolite-facies overprint. Primary fluid inclusions occur in garnet, omphacite, and apatite from eclogite; in kyanite and in topaz from quartzite; and in garnet, epidote, and apatite from paragneiss. Secondary fluid inclusions are present in all lithologies. Fluid inclusions are absent from ultramafic rocks. Based on fluid compositions and textural criteria we distinguished: (1) low-salinity aqueous-carbonic inclusions in topaz from quartzite, which may have originated from a supracrustal protolith; (2) primary $CaCl_2$-NaCl–rich brine inclusions in garnet and in omphacite from eclogite and in kyanite from quartzite, representing UHP metamorphic fluids; (3) high-salinity aqueous-carbonic inclusions in quartz from eclogite and quartzite, representing amphibolite-facies fluids; (4) aqueous fluids of low- and intermediate salinity trapped as primary inclusions in garnet, epidote (or allanite) and apatite from gneiss, or as secondary inclusions, representing amphibolite-facies and later retrograde fluids; (5) carbonic inclusions are distributed along transgranular fractures in quartz from quartzite, and probably represent the latest retrograde fluid. The diversity in fluid inclusion populations and compositions from different vertical depths suggests a closed fluid system without large-scale fluid migration during UHP metamorphism. However, the common low- and medium-salinity inclusions in most rock types suggests that a water-dominated fluid from an external source infiltrated into the rock system during amphibolite-facies metamorphism, resulting in extensive retrogression of the UHP rocks.

Introduction

SINCE THE DISCOVERY of coesite and diamond in metamorphic rocks from the Dabie-Sulu orogenic belt, significant progress concerning the role of fluids during UHP metamorphism have been made. Hydroxyl-bearing minerals, such as phengite, zoisite, talc, amphibole, and topaz were stable at the peak of UHP metamorphism (e.g., Liou et al., 1995;

R. Zhang et al., 1995, 2002; Z. M. Zhang et al., 2000, 2003). Extremely low $\delta^{18}O$ values in minerals from eclogites and associated gneisses suggest the interaction of their protoliths with surface water of low oxygen isotope composition (Yui et al., 1995; Zheng et al., 1996, 1998; Baker et al., 1997; Rumble and Yui, 1998; Rumble et al., 2002). Nominally anhydrous minerals in UHP rocks, such as garnet and omphacite, preserved significant amounts of fluid (Su et al., 2002). Fluid characteristics and evolution in Dabie-Sulu UHP rocks have been

---

[1]Corresponding author; email: zzm@ccsd.org.cn

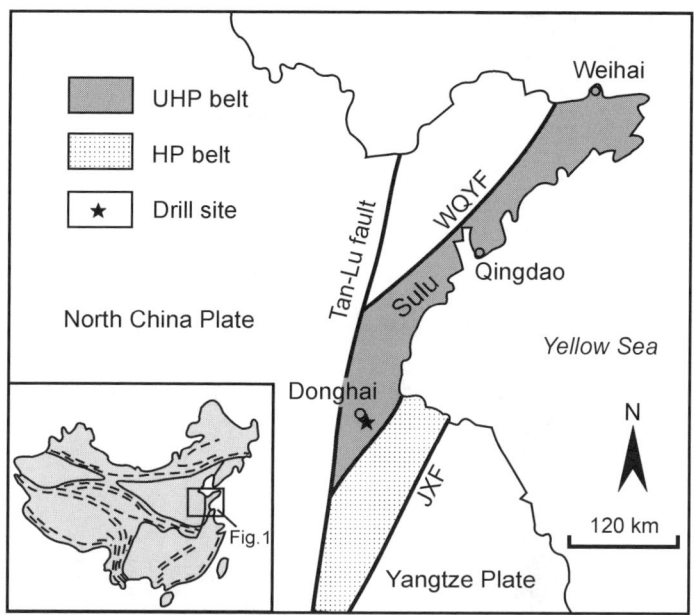

FIG. 1. Schematic geological map of the Sulu UHP metamorphic belt, showing the location of the CCSD-PPH1 (star). Abbreviations: WQYF = Wulian-Qingdao-Yantai fault; JXF = Jiashan-Xiangsui fault.

extensively discussed in the literature (e.g., Shen et al., 1996; You et al., 1996; Xiao et al., 2000, 2001; Franz et al., 2001; Fu et al., 2001, 2003; Zheng et al., 2003). Xiao et al. (2000) reported early Ca-rich brines which may have originated during prograde metamorphism; NaCl-rich fluids may have been trapped during the peak of UHP metamorphism. Fu et al. (2003) distinguished different generations of fluid inclusions in UHP rocks: pre- to syn-peak metamorphic fluids consisting of N_2, CH_4 and high-salinity brines, and post peak-metamorphic fluids rich in CO_2 and low-salinity water. These fluids reflect UHP metamorphism with variable degrees of fluid-rock interaction, and limited fluid mobility during subduction and exhumation. Inasmuch as these studies dealt with samples from discontinuous outcrops over a large area, systematic spatial and temporal variations of fluid compositions within the vertical sequence of UHP rocks could not be studied so far.

The CCSD drill site is located in Donghai County, south of the Sulu UHP belt, the eastern extension of the Dabie-Sulu UHP orogen between the Yangtze and North China plates (Fig. 1). Prior to the site selection for the CCSD main hole with a targeted depth of 5000 m, three pre-pilot holes

(PPH) have been drilled. The first pre-pilot hole, CCSD-PPH1, was drilled to 432 m depth, and revealed a sequence of garnet-peridotite (120 m thick), eclogite (10 m thick), gneiss, and minor quartzite. Petrological and petrochemical investigations show that: (1) felsic rocks, such as gneiss, schist, and quartzite, together with eclogite, have been subjected to *in situ* UHP metamorphism (Liu et al., 2001); (2) all UHP rocks, except garnet peridotite, show unusual low-oxygen isotopic values (Z. M. Zhang et al., 2004a), indicating Neoproterozoic water-rock interaction prior to Triassic subduction; and (3) garnet peridotite, derived from the mantle wedge above the subduction zone with a depleted-mantle geochemical signature, is now sandwiched within supracrustal rocks (Z. M. Zhang et al., 2004a). The continuous core samples from the CCSD-PPH1 drill hole provide a unique chance for studying the role of fluid during UHP metamorphism. In this paper, we report fluid inclusion data from the hole with special emphasis on the relation with lithology and depth. Mineral abbreviations are after Kretz (1983) with the following exceptions: Amp = amphibole; Coe = coesite; Cel = celadonite; Dia = diamond; and Phn = phengite.

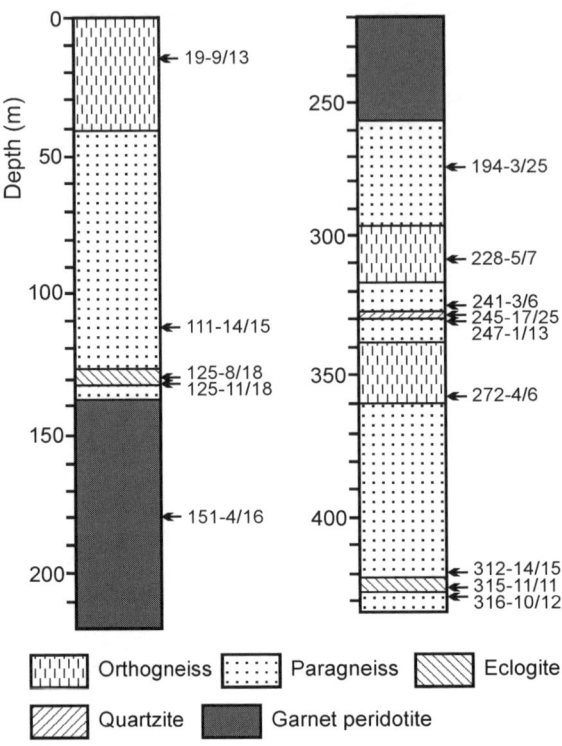

FIG. 2. Simplified lithological profile of the CCSD-PPH1, showing sampling locations.

Petrology of the Drill Core Samples

A simplified lithological profile of PPH1 and investigated samples are shown in Figure 2. Paragneiss and orthogneiss are the main lithologies and contain amphibolite-facies mineral assemblage Pl + Kfs + Qtz ± Grt ± Bt ± Ms ± Ep ± Amp. Symplectitic textures of biotite and plagioclase after phengite, and that of amphibole and plagioclase after garnet are common. Moreover, coesite inclusions in zircon suggest that the gneiss was subjected to UHP metamorphism before amphibolite-facies retrogression (Liu et al., 2001; Z. M. Zhang et al., 2004b). Kyanite quartzite occurs as a ~10–30 cm thick layer in paragneiss, and consists mainly of quartz, white mica, and kyanite with minor rutile, topaz, zircon, monazite, and pyrite. R. Zhang et al. (2002) assumed that topaz with oriented fine-grained kyanite inclusions formed from the reaction $2Al_2SiO_5$ (kyanite) + H_2O = $2Al_2SiO_4(OH)_2$ (topaz) at >3 GPa and ~650–700°C.

Ultramafic rocks between 137 and 257 m are mainly composed of garnet peridotite and dunite, which exhibit variable serpentinization. Fresh garnet peridotite displays a porphyroblastic texture with garnet and clinopyroxene porphyroblasts set in a matrix of garnet, clinopyroxene, olivine, orthopyroxene, and variable contents of phlogopite.

Two layers of fresh eclogite occur at 127–132 m and at 422–427 m. They contain the typical mineral paragenesis of Grt + Omp + Phn + Qtz (or Coe) + Rt + Amp + Zrn. Relic coesite and polycrystalline quartz pseudomorphs after coesite occur as inclusions in garnet and in omphacite. Thinner layers of retrograded eclogites including garnet-epidote amphibolite, epidote-biotite amphibolite, and epidote-biotite-amphibole schist occur within paragneiss.

Applying mineral thermobarometers, the peak metamorphic P-T conditions of coesite-bearing eclogites were estimated to be 2.7 to 3.5 GPa and 785–820°C, and amphibolite-facies retrogression around 540–710°C and 1.0 GPa. A retrograde P-T path characterized by rapid decompression and slight cooling is proposed (Z. M. Zhang, 2004b).

TABLE 1. Fluid Inclusion Types and Abundances in Different Rocks from CCSD- PPH1

Depth, m	Rock type	Representative sample and depth, m[1]	Fluid inclusion types[2]				
			I	II	III	IV	V
0–41	Orthogneiss	19-9/13 (16)	–[3]	–	–	++	–
41–127	Paragneiss	111-14/15 (112)	–	–	–	++++	–
127–132	Eclogite	125-8/18 (130) 125-11/18 (132)	–	++	++	+++	–
132–257	Peridotite	151-4/16 (180)	–	–	–	–	–
257–296	Paragneiss	194-3/25 (274)	–	–	–	+++	–
296–317	Paragneiss	228-5/7 (310)	–	–	–	++	–
317–328	Paragneiss	241-3/6-2 (326)	–	–	–	+++	–
328–330	Quartzite	245-17/25 (328)	++	++	–	++++	+
330–338	Paragneiss	247-1/13 (331)	+	–	–	+++	–
338–360	Orthogneiss	272-4/6 (358)	–	–	–	++	–
360–422	Paragneiss	312-14/15 (420)	–	–	–	+++	–
422–427	Eclogite	315-11/11 (425)	–	–	–	++	–
427–432	Paragneiss	316-10/12 (427)	–	–	–	+++	–

[1] Depth of samples is shown in parentheses.
[2] I = Low-salinity aqueous-carbonic inclusions; II = $CaCl_2$-rich brine inclusions; III = high-salinity aqueous carbonic inclusions; IV = aqueous inclusions of low and intermediate salinity; V = secondary carbonic inclusions.
[3] – = absent; + = present; ++ = common; +++ = abundant; ++++ = very abundant.

Analytical Methods

Doubly polished sections of drill core samples were prepared for fluid inclusion studies. Microthermometry was performed by using a Linkam THM600 heating/freezing stage at the Geoscience Centre, University of Göttingen, Germany, and at the State Key Laboratory for Ore Metallogeny of Nanjing University, China. The stages were calibrated by measuring the melting points of pure water inclusions (0°C), CO_2 inclusions (–56.6°C) and potassium dichromate (398°C). The accuracy of measured temperatures is about 0.2°C during cooling and better than 5°C during heating to 600°C. Fluid compositions and densities were calculated using the FLUIDS package developed by Bakker (2003). Laser Raman microanalysis was performed with a RENISHAW Raman spectrometer (RM100) at the Institute of Geology, Chinese Academy of Geological Sciences, for measuring gas compositions of fluid inclusions.

Petrography of Fluid Inclusions

Primary fluid inclusions have been recognized in garnet, omphacite, and apatite from eclogite; in kyanite, topaz, and monazite from quartzite; and in garnet, epidote (or allanite), and apatite from paragneiss. Secondary fluid inclusions are found in orthogneiss, eclogite, and quartzite. Fluid inclusions are absent from ultramafic rocks. Based on their composition, five types of fluid inclusions were distinguished, namely: (1) low-salinity aqueous-carbonic inclusions; (2) high-salinity aqueous inclusions; (3) high-salinity aqueous-carbonic inclusions; (4) aqueous inclusions with low to intermediate salinity; and (5) carbonic inclusions. The distribution of these inclusion types in the drill hole is shown in Table 1.

Mixed aqueous-carbonic inclusions contain two to three phases at room temperature. In general, the carbonic phases account for 10–35 vol%. The inclusions show elongated (tube-shaped) or negative crystal morphologies. They occur either randomly or clustered as primary inclusions oriented parallel to the crystallographic c-axis of kyanite and topaz, or as pseudosecondary inclusions along intragranular fractures in the cores of larger kyanite crystals from quartzite. Occasionally, two-phase aqueous-carbonic inclusions ≤5 μm are present in zircon inclusions in epidote from gneiss.

High-salinity aqueous inclusions can be subdivided into salt-undersaturated and salt-saturated types. Also amphibole and mica occur as daughter phases in the fluid inclusions. The salt-undersaturated inclusions show negative crystal shapes, and are isolated or form clusters in the cores of garnet grains from eclogite (Fig. 3A), and therefore are regarded as "primary." Salt-saturated aqueous inclusions contain one or more solids, typically a halite cube and in some cases also carbonate and opaque phases. They show negative crystal shapes and occur as isolated inclusions or in clusters in the cores of garnet (Figs. 3B and 3C). Some are distributed with preferential orientation parallel to the crystallographic c-axes of omphacite and kyanite crystals from eclogite and quartzite, respectively (Figs. 3D–3G), suggesting that they are primary in origin. The salt-saturated aqueous inclusions in kyanite from quartzite contain additional minerals like pyrite and zircon (Fig. 4A); these solids account for ~10–30 vol% of the total inclusions. Saturated fluid inclusions in garnet, which contain silicate crystals such as amphibole and mica are associated with monophase salt-undersaturated inclusions. Amphibole and mica are considered as reaction products of the Ca-bearing brine with the host garnet.

Highly saline aqueous-carbonic inclusions contain 30–50 vol% of CO_2. They normally contain a halite cube with a relatively constant phase ratio, and occur randomly in matrix quartz from eclogite (Fig. 3H). Irregular inclusions with higher H_2O/CO_2 ratios are probably the result of local mixing with low-salinity water.

Aqueous inclusions with low to intermediate salinity occur: (1) as tubular mono- and biphasic inclusions in kyanite from quartzite, which are similar to the so-called "exsolution inclusions" reported by Philippot et al. (1995); (2) as isolated or clustered inclusions in garnet, epidote (or allanite), and apatite from gneiss (Figs. 3I–3J); or (3) as secondary inclusions distributed along transgranular fractures in omphacite, garnet, epidote, and quartz from various lithologies.

Carbonic inclusions are mono- or biphasic at room temperature. They show rounded or negative crystal shape and occur in healed transgranular fractures in quartz from quartzite.

Relative Age of Fluid Inclusions

The trapping sequence of fluid inclusions is of utmost importance for the interpretation of fluid inclusion data. Due to rapid exhumation of the UHP rocks, most pre- and syn-peak fluid inclusions, especially in quartz, have been modified; only fluid inclusions in rigid host minerals may have survived (e.g., Andersen et al., 1989; Philippot et al., 1995; Xiao et al., 2000, 2001; Fu et al., 2001, 2003). Primary carbonic-aqueous inclusions found in topaz included in kyanite from quartzite, and in zircon included in epidote from gneiss, are assumed to be the earliest preserved fluids. They were probably trapped during pre- to syn- peak metamorphism and carry the fluid inherited from metasediments.

Primary brine inclusions occur in garnet and in omphacite from eclogite and in kyanite from quartzite, but are absent from gneiss, which has been overprinted by amphibolite-facies metamorphism. The coexistence of brine inclusions and carbonic-aqueous inclusions in kyanite suggests that they were trapped from immiscible CO_2-H_2O-salt-bearing fluids during UHP metamorphism.

Highly saline aqueous-carbonic inclusions in quartz were probably trapped during amphibolite-facies metamorphism inasmuch as the quartz contains amphibole inclusions. Low- to intermediate-salinity aqueous inclusions in healed transgranular fractures in omphacite and in quartz from eclogite and from quartzite must have been trapped after UHP metamorphism, whereas those in garnet, epidote (or allanite), and apatite from gneiss must have been trapped during the growth of these minerals at amphibolite-facies conditions. The carbonic inclusions together with aqueous inclusions occur along transgranular fractures in minerals from gneiss and quartzite and postdate amphibolite-facies metamorphism.

Microthermometry of Fluid Inclusions

Low-salinity aqueous-carbonic inclusions

These inclusions normally contain less than 10 mol% of CO_2. Final melting temperatures of solid CO_2 were measured between –57.5 and –56.9°C (Table 2). Clathrate dissociation temperatures range

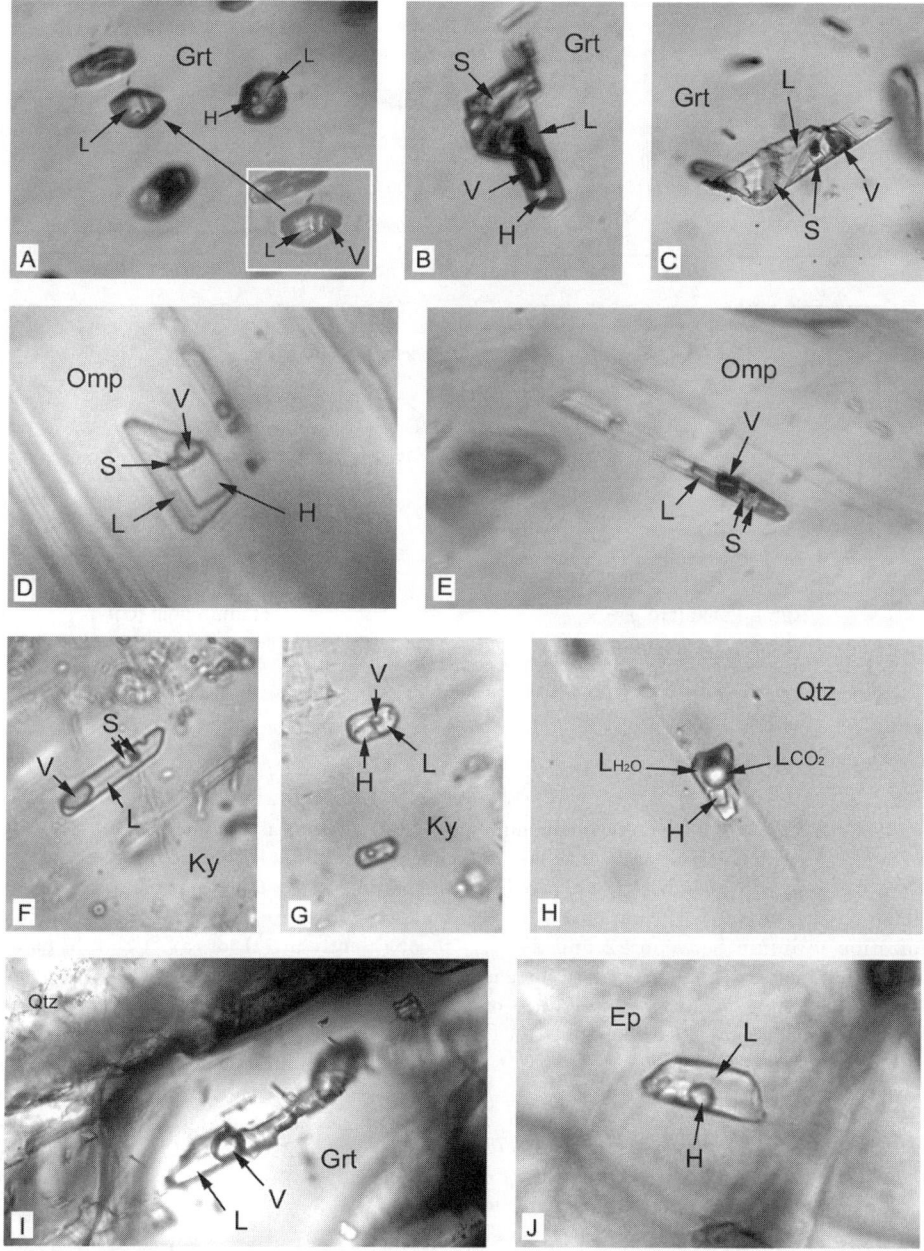

FIG. 3. Photomicrographs of fluid inclusions in eclogite, quartzite, and gneiss. A. High-salinity one-phase liquid (L) inclusions and a two-phase inclusion containing liquid (L) and halite (H) in the core of a garnet grain. The inset shows a bubble (V) nucleating in the monophase inclusion (~12 μm) during cooling. B and C. High-salinity aqueous inclusions occurring in the core of garnet contain vapor (V), liquid (L), and silicate (S) phases. The length of the inclusions is ~28 μm (B) and 29 μm (C). D. High-salinity inclusion (~19 μm) containing a bubble (V), a liquid (L), and two solids (H= halite) in omphacite. E. High-salinity inclusions containing a bubble, liquid, and two solids oriented along the c-axis of omphacite. The length of the inclusions is ~36 μm. F. High-salinity inclusion (24 μm) in kyanite with preferential orientation along the c-axis. G. High-salinity inclusions containing a bubble, liquid, and halite (H) in kyanite from quartzite. The biggest inclusion is ~8 μm. H. Aqueous-carbonic inclusion (~15 μm) containing liquid (L_{H2O}), liquid CO_2 (L_{CO2}), and halite (H) randomly distributed in quartz. I. Isolated 112 μm long inclusion with low or intermediate salinity in garnet from gneiss. J. Isolated 23 μm long inclusion in epidote from gneiss.

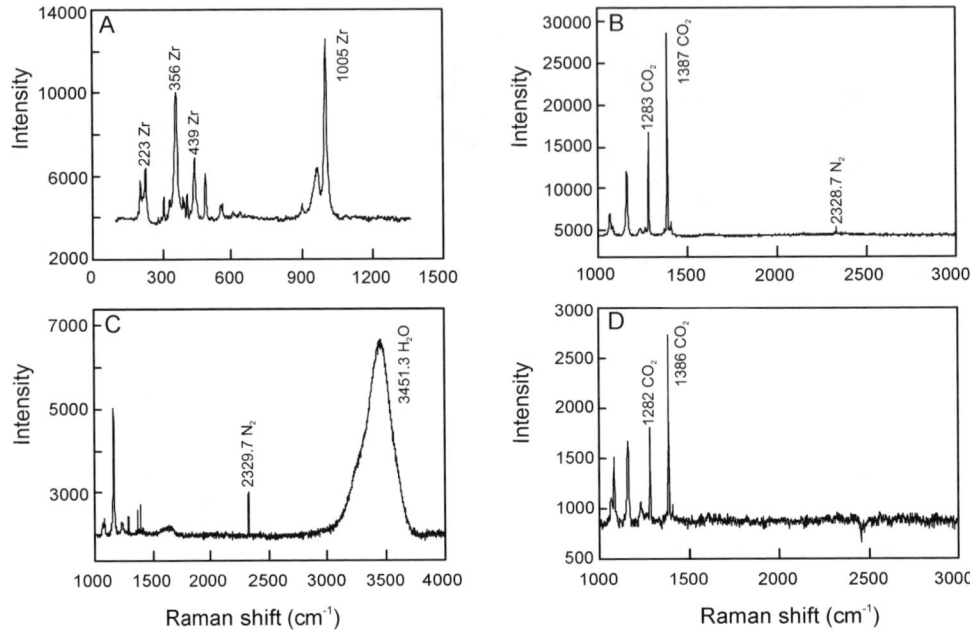

FIG. 4. Laser Raman spectra of (A) zircon daughter crystal in a high-salinity inclusion, (B) CO_2 and N_2 in a high-salinity aqueous-carbonic inclusion, (C) N_2 and H_2O in an aqueous-carbonic inclusion, and (D) carbonic inclusion.

between 1.2 and 8.7°C, with a frequency maximum between 7 and 8°C, which corresponds to salinities of 4–6 wt% NaCl-eq. Water-rich H_2O-CO_2 inclusions with high degrees of fill show partial CO_2 homogenization to liquid between 22 and 27°C, whereas CO_2-rich inclusions homogenize to the gas phase. The total homogenization temperature of aqueous-carbonic inclusions ranges between 193°C and 300°C.

High-salinity aqueous inclusions

The salt-undersaturated inclusions freeze at –75 to –80 °C. During subsequent warming eutectic melting was observed around –52°C, which points to Ca-Na-Cl–bearing brines. Ice melting between –27 °C and –22°C (Table 2) indicates salinities of 23 to 25 wt% NaCl-eq. In two inclusions, hydrohalite (NaCl · $2H_2O$) melting was observed at –15.6°C and at –8.5°C. In inclusions that are monophase at room temperature, a bubble may nucleate on quick warming during or after ice melting around –23°C (see the inset in Fig. 3A); during slow warming, the bubble shrinks until total homogenization at ≤–12.5°C.

The salt-saturated aqueous inclusions also show eutectic melting slightly below or equal to –52°C. Ice melting (metastable) was measured between –27 and –22°C. On further warming, hydrohalite melts incongruently at ~0 to +3.5°C, while halite starts growing. Halite dissolves between 180 and 230°C, or between 280 and ≥388°C for the inclusions in apatite and in omphacite, respectively. Birefringent crystals, which are considered to be calcite, do not melt. The saturated inclusions in kyanite show LV homogenization between 182°C and 260°C, with a mode between 200 and 220°C. On further heating, the inclusions decrepitate at ~450°C, before halite dissolution.

High-salinity aqueous-carbonic inclusions

During cooling, CO_2 solid melts between –58.0 and –56.9°C (Table 2), indicating minor amounts of other gas components. Laser Raman analysis confirms the presence of small amounts of N_2 (Figs. 4B and 4C). Halite dissolves between 180°C and 365°C; LV homogenization temperatures range between 300 and ≥450°C. Many inclusions decrepitate before total homogenization.

TABLE 2. Microthermometric Measurements of Fluid Inclusions in Eclogite, Quartzite, and Gneisses[1]

Type	Lithology	Host mineral	$Tmco_2$, °C[2]	Tmi, °C[2]	$Tmcl$, °C[2]	$Thco_2$, °C[2]	Th, °C[2]	Ts, °C[2]	Density, g/cm³	Salinity[3]
I	Quartzite	Kyanite	−57.5 to −56.9		−1.2–7.9	~22.5–28.2[4]	193–300		0.67–0.91	4.26–14.22
	Quartzite	Topaz			−6.1–8.7		193–242		0.89–1.04	2.59–7.2
	Gneiss	Zircon			4		225			
II	Eclogite	Garnet		−26.5 to −21.8			200–≥388	358	1.27–1.46	23.14–42.0
	Eclogite	Omphacite		−23.5 to −22.0			242–≥333	280–≥388	0.86–1.39	25.5–≥45.0
	Quartzite	Kyanite					182–260	≥450		38.0–≥60
III	Eclogite	Quartz	−58.0 to −56.9			−1.9–24.3	300–450	180–365	0.97–1.22	31.27–46.3
IVes[5]	Quartzite	Kyanite		−4.2 to −0.3			148–234		0.87–0.93	0.53–13.07
	Quartzite	Quartz		−4.9 to −3.2			117–254		0.87–0.98	5.26–7.73
	Eclogite	Omphacite		−11.2 to −7.9			221–276		0.87–0.97	11.58–15.17
	Eclogite	Quartz		−5.0 to −0.3			123–333		0.73–1.00	1.23–7.86
IVgp[5]	Gneiss	Garnet		−9.2 to −3.3			192–272		0.93–0.98	5.41–13.07
	Gneiss	Allanite		−20.6 to −7.0			181–304		0.89–1.04	10.49–22.8
	Gneiss	Epidote		−14.3 to −3.0			142–228		0.96–0.98	4.96–18.04
	Gneiss	Apatite		−3.9 to −3.4			100–250		0.91–0.99	5.56–6.30
IVgs[5]	Gneiss	Garnet		−2.0 to −1.3			185–215			2.24–3.39
	Gneiss	Epidote		−2.4 to −1.0			142–212			1.91–4.03
	Gneiss	Quartz		−2.5 to −0.3			69–289		0.89–0.99	0.53–4.18
V	Quartzite	Quartz	−56.8 to −56.6			5.7–27.5			0.67–0.89	

[1]For fluid inclusion types, see Table 1.
[2]$Tmco_2$ = last melting temperature of CO_2 solid; Tmi = last melting temperature of ice; $Tmcl$ = clathrate melting temperature; $Thco_2$ = homogenization temperature of CO_2 phase; Th = liquid-gas homogenization temperature; Ts = salt (halite) dissolving temperature.
[3]Salinity in wt% NaCl eq.
[4]Homogenization to vapor phase, others to liquid phase.
[5]Ives, IVgp, and IVgs refer to aqueous inclusions of low and intermediate salinity occurring as secondary in eclogite, and primary and secondary inclusions in gneiss, respectively.

Aqueous inclusions of lower and intermediate salinity

Most primary aqueous inclusions in gneiss show eutectic melting temperatures of around –21°C, indicating NaCl-bearing solutions. Lowest final melting temperatures were recorded for allanite (corresponding to 10 to 23 wt% NaCl-eq.); higher temperatures were recorded for garnet (5 to 13 wt%), epidote (3 to 18 wt%) and apatite (5 to 7 wt% NaCl-eq.) (Table 2 and Fig. 5). The inclusions homogenize to liquid between 110 and 300°C. Some inclusions in garnet, epidote and apatite contain an anisotropic solid (probably calcite). Exsolution inclusions in kyanite show final melting temperatures from –4.3 to –0.3°C and homogenize to liquid between 170 and 240°C.

Secondary inclusions in omphacite have Tm values between –11 and –8°C (Table 2); Th ranges from 221°C to 276°C. Tm and Th of secondary inclusions in quartz from eclogite fall in the range of –2.5 and –0.5°C and 60 and 220°C (rarely up to 300°C), respectively. Secondary inclusions along transgranular fractures in garnet, epidote (or allanite), and quartz from gneiss and quartzite generally melt between –2.5 and –0.5°C, indicating very low salinity. Homogenization temperatures range from 110 to about 300°C.

Carbonic inclusions

The melting of carbonic inclusions without visible water falls in a narrow range of –56.8 to –56.6°C, indicating nearly pure CO_2. This has been confirmed by Raman analysis (Fig. 4D). Most carbonic inclusions homogenize to liquid between 6 and 28°C, with a few homogenizing to gas around 25°C.

Discussion

Comparison of isochore and mineral thermobarometry calculations

Isochores corresponding to monophase, high-density brine inclusions lie well within eclogite-facies metamorphic conditions at 2.4 GPa and 820°C (IIa in Fig. 6), i.e., close to the coesite stability field. However, most brine inclusions (Th of 200 to 388°C) have lower densities. Their isochores run at P-T conditions that are much below UHP conditions (IIb in Fig. 6). This indicates that the density of these fluid inclusions must have been changed after trapping. Density changes of fluid inclusions are possibly the result of partial water leakage after

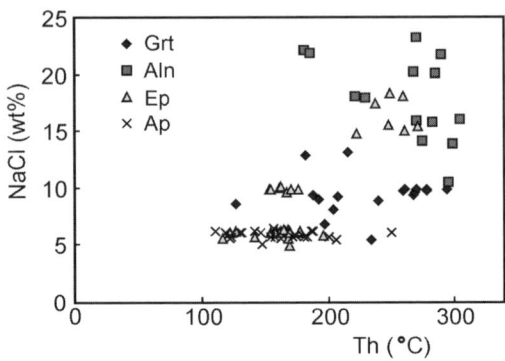

FIG. 5. Salinity-Th plot for primary aqueous inclusions (± CO_2) in garnet (Grt), allanite (Aln), epidote (Ep), and apatite (Ap) from gneiss samples.

trapping, and of reactions between the inclusion fluid and host minerals (Touret, 1981, 2001; Heinrich and Gottschalk, 1995; Franz et al., 2001). Partial leakage of fluid inclusions in eclogites has been reported also by Andersen et al. (1989), Philippot (1993), Philippot et al. (1995), Xiao et al. (2000, 2001) and Fu et al. (2003). These authors demonstrated that high-salinity fluid inclusions represent UHP metamorphism, but that fluid densities (and in part also compositions) were modified after trapping. Modification of fluid inclusion compositions has been recognized in natural samples (Andersen et al. 1989; Johnson and Hollister, 1995; Franz et al., 2001) and have been demonstrated also in experimental studies (e.g., Sterner and Bodnar, 1989; Bakker and Jansen, 1991). Amphibole and mica crystals in fluid inclusions are considered as reaction products of a $CaCl_2$-NaCl-H_2O fluid with the host garnet. The garnet reacts with water and Na^+ to form sodic amphibole, resulting in a change of fluid composition and a decrease of fluid density.

Isochores of secondary aqueous-carbonic inclusions in quartz of eclogite and primary inclusions in garnet and epidote from gneiss, pass through amphibolite-facies metamorphic conditions (III and IV in Fig. 6). Secondary carbonic inclusions have low densities, which correspond to trapping below amphibolite-facies conditions (V in Fig. 6).

Variation of fluid inclusion composition with lithology and fluid evolution

Several fluid inclusion generations were identified in UHP rocks from PPH1. The earliest aqueous-carbonic inclusions in topaz may have been

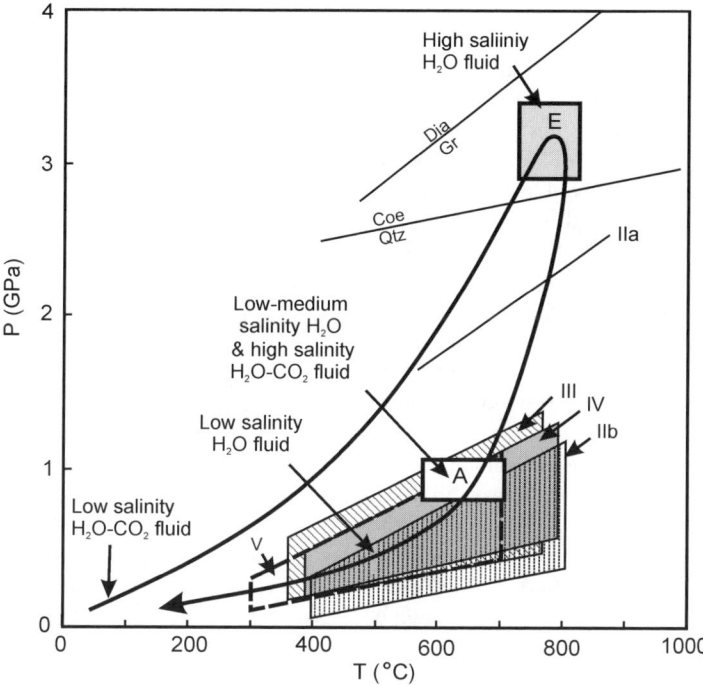

FIG. 6. P-T-fluid path for the UHP rocks from the CCSD-PPH1. Abbreviations: E = eclogite-facies stage; A = amphibolite-facies stage. Isochores are shown for monophase brine inclusions (IIa); $CaCl_2$-NaCl-rich brine inclusions (IIb); high-salinity aqueous carbonic inclusions (III); aqueous inclusions of low and intermediate salinity (IV); and secondary carbonic inclusions (V).

inherited from the protoliths (Fig. 6). Prograde dehydration reactions produced a water-rich fluid. R. Zhang et al. (2002) reported that topaz from the Sulu UHP belt may have formed from the reaction $2Al_2SiO_5$ (kyanite) + H_2O = $2Al_2SiO_4(OH)_2$ (topaz). This may indicate that surface water carried by sediments was subducted to great depth. Recycling of H_2O at convergent plate boundaries is also indicated by low $\delta^{18}O$ values of topaz-kyanite quartzite (R. Zhang et al., 2002) and of eclogite (Yui et al., 1995; Zheng et al., 1996, 1998; Rumble and Yui, 1998; Rumble et al., 2002; Fu et al., 2003). During prograde metamorphism, dehydration fluids were preserved in hydrous minerals, such as paragonite (phengite) and topaz in some quartzites. The low-salinity carbonic-aqueous inclusions in zircon are assumed to have formed before the UHP metamorphism.

Multicomponent brine inclusions in garnet and in omphacite from the eclogite and in kyanite from quartzite probably represent syn-UHP metamorphic fluids (Fig. 6), whereas CO_2 appeared not to be important at that stage. Philippot et al. (1995) and Philippot (1996) reported high-salinity fluid inclusions containing daughter minerals of halite, sylvite, carbonates, sulfates, and silicates in pyrope quartzite from the Dora-Maira Massif of the western Alps. Fu et al. (2003) also reported high-salinity brine inclusions in garnet from Sima eclogite (Dabie terrane) and suggested the presence of high-salinity fluids at pre- to syn-peak HP/UHP metamorphic conditions. In comparison with the peak-metamorphic fluids as recognized by previous workers, some primary fluid inclusions observed in this study show higher densities, indicating P-T conditions close to the coesite stability field.

H_2O-NaCl-CO_2 (± N_2) fluid inclusions with amphibole inclusions from eclogite may represent the amphibolite-facies fluid. During retrogression, the concentration of CO_2 and N_2 may have increased, and the water activity lowered because of the crystallization of hydrous minerals. This high-salinity fluid enriched in CO_2 and N_2 was trapped during quartz recrystallization.

Aqueous inclusions of varying salinity were trapped as primary inclusions in amphibolite-facies minerals like garnet, epidote (or allanite) and apatite (Fig. 6). Kerrick (1990) and Philippot et al. (1995) suggested that "exsolution" inclusions in kyanite formed as a result of decompression. Dehydration fluids were preserved in kyanite during prograde metamorphism through OH incorporation into the crystal structure at high pressure. Subsequent release of OH during decompression resulted in the formation of "exsolution inclusions." The low-salinity aqueous fluids trapped as secondary inclusions as well as all carbonic inclusions represent rock fluids after amphibolite-facies metamorphism (Fig. 6).

Fluid inclusion compositions vary among different rock types, but systematic variations with depth could not be demonstrated. This suggests the absence of a pervasive fluid flow. Water must have been introduced during exhumation. This fluid caused extensive hydration and serpentinization of the ultramafic rocks.

Fluid immiscibility at UHP conditions

The coexistence of low-salinity and high-salinity aqueous-carbonic inclusions in the same crystallographic planes of kyanite from quartzite suggests heterogeneous trapping of an H_2O-NaCl ($CaCl_2$)-CO_2 fluid during UHP metamorphism. The immiscibility gap of the H_2O-CO_2 system greatly expands at higher salt concentrations. The upper limit of immiscibility is 500 to 600°C for salinities of about 20 wt% NaCl (Sisson et al., 1981), about 800°C at 0.2 GPa for salinities of 35 wt% NaCl (Bowers and Helgeson, 1983), and >900°C for salinities of up to 50 wt% (Johnson, 1991). Shmulovich and Graham (2004) demonstrated a large miscibility gap in the ternary systems CO_2-H_2O-NaCl and CO_2-H_2O-$CaCl_2$ for temperatures and pressures as high as 800°C and 0.9 GPa. Increasing pressures may cause expansion of the immiscibility gap (e.g., Kerrick and Jacobs, 1981; Johnson, 1991). However, experimental data for the system CO_2-H_2O-salt are insufficient so far. Selverstone et al. (1992) reported both saline brines rich in silicate and oxide components and carbonic fluids from high-pressure (2.0 GPa) banded eclogite from the Tauern Window (Austria), which indicates the potential for generating immiscible fluid pairs during subduction. Our study suggests that fluid phase separation occurred during UHP metamorphism at >2.8 GPa.

Acknowledgments

This paper is dedicated to Prof. W. G. Ernst for his contribution in the study of subduction-zone metamorphism and his support of the Chinese Continental Scientific Drilling (CCSD) program. This work has been funded by a Sino-German and Sino-American project related to the CCSD. These projects were supported by the Major State Basic Research Development Program (2003CB716501), the National Natural Scientific Foundation of China (40399142), a project from the Ministry of Land and Resources of China (2002207), the National Science Foundation of Germany (DFG, Ho 375/22), and the U.S. National Science Foundation (EAR 0003355). We sincerely thank Xu Zhiqin, Yang Jingsui, Liu Fulai, You Zhendong, Jin Zhenmin, Zhang Ruyuan, D. Rumble, and scientists of the Geosciences Centre Göttingen for their help in the project. This manuscript has been critically reviewed and improved by I-Ming Chou and Doug Rumble. We thank the above-named individuals and institutes for their help and support.

REFERENCES

Andersen T., Burke, E. A. J., and Austrheim, H., 1989, Nitrogen-bearing, aqueous fluid inclusions in some eclogites from the Western Gneiss Region of the Norwegian Caledonides: Contribution to Mineralogy and Petrology, v. 103, p. 153–165.

Bakker, R. J., 2003, Package fluids I: Computer programs for analysis of fluid inclusion data and for modeling bulk fluid properties: Chemical Geology, v. 194, p. 3–23.

Bakker, R. J., and Jansen, J. B. H., 1991, Experimental post-entrapment water loss from synthetic CO_2-H_2O inclusions in natural quartz: Geochimica et Cosmochimica Acta, v. 55, p. 2215–2230.

Baker, J., Matthews, A., Mattey, D., Rowley, D., and Xue, F., 1997, Fluid-rock interactions during ultra-high pressure metamorphism, Dabie Shan, China: Geochimica et Cosmochimica Acta, v. 61, p. 1685–1696.

Bowers, T. S., and Helgeson, H. C., 1983, Calculation of the thermodynamic and geochemical consequences of non-ideal mixing in the system H_2O-CO_2-NaCl on phase relations in geologic systems: Equation of state for H_2O-CO_2-NaCl fluids at high pressures and temperatures: Geochimica et Cosmochimica Acta, v. 47, p. 1247–1275.

Franz, L., Romer, R. L., Klemd, R., Schmid, R., Oberhansli, Wanger, T., and Dong, S. W., 2001, Eclogite-facies quartz veins within metabasites of the Dabie

Shan (eastern China): Pressure-temperature-time-deformation path, composition of the fluid phase and fluid flow during exhumation of high-pressure rocks: Contributions to Mineralogy and Petrology, v. 141, p. 322–346.

Fu, B., Touret, J. L. R., and Zheng, Y. F., 2001, Fluid inclusions in coesite-bearing eclogites and jadeite quartzite at Shuanghe, Dabie Shan (China): Journal of Metamorphic Geology, v. 19, p. 529–545.

Fu, B., Touret, J. L. R., and Zheng, Y. F., 2003, Remnants of premetamorphic fluid and oxygen isotopic signatures in eclogites and garnet clinopyroxenite from the Dabie-Sulu terranes, eastern China: Journal of Metamorphic Geology, v. 21, p. 561–578.

Heinrich, W., and Gottschalk, M., 1995, Metamorphic reactions between fluid inclusions and mineral hosts. I. Progress of the reaction calcite+ quartz = wollastonite + CO_2 in natural wollastonite-hosted fluid inclusions: Contributions to Mineralogy and Petrology, v. 122, p. 51–61.

Johnson, E. L., 1991, Experimentally determined limits for H_2O-CO_2-NaCl immiscibility in granulites: Geology, v. 19, p. 925–928.

Johnson, E. L., and Hollister, L. S., 1995, Syndeformational fluid trapping in quartz: Determining the pressure-temperature conditions of deformation from fluid inclusions and the formation of pure CO_2 fluid inclusions during grain boundary migration: Journal of Metamorphic Geology, v. 13, p. 239–249.

Kerrick, D. M., 1990, Metamorphic reactions, in Ribbe, P. H., ed., The Al_2SiO_5 polymorphs: Washington, DC, Mineralogical Society of America, Reviews in Mineralogy, v. 22, 406 p.

Kerrick, D. M., and Jacobs, G. K., 1981, A modified Redlich-Kwong equation for H_2O, CO_2, and H_2O-CO_2, mixtures at elevated pressures and temperatures: American Journal of Science, v. 281, p. 735–767.

Kretz, R., 1983, Symbols for rock-forming minerals: American Mineralogist, v. 68, p. 277–279.

Liou, J. G., Zhang, R. Y., and Ernst, W. G., 1995, Occurrence of some hydrous and carbonate phases in ultrahigh-P rocks from east-central China-implications for the role of fluids in deep subduction zones: The Island Arc, v. 4, p. 362–375.

Liu, F., Xu, Z., Katayama, I., Yang, J., Maruyama, S., and Liou, J. G., 2001, Mineral inclusions in zircon of para- and orthogneiss from pre-pilot drillhole CCSD-PP1, Chinese continental scientific drilling project: Lithos, v. 59, p. 199–215.

Philippot, P., 1993, Fluid-melt-rock interaction in mafic eclogites and coesite-bearing metasediments: constraints on volatile recycling during subducion: Chemical Geology, v. 108, p. 93–112.

Philippot, P., 1996, The chemistry of high-pressure fluids (1 to 3 GPa): Natural observations vs. experimental constraints: Earth Science Frontiers, v. 3, p. 39–52.

Philippot, P., Chevallier, P., Chopin, C., and Dubessy, J., 1995, Fluid composition and evolution in the coesite-bearing rocks of the dora Maira Massif, Western Alps: Contributions to Mineralogy and Petrology, v. 121, p. 29–44.

Rumble, D., and Yui, T. F., 1998, The Qinglongshan oxygen and hydrogen isotope anomaly near Donghai in Jiangsu Province, China: Geochimica et Cosmochimica Acta, v. 62, p. 3307–3321.

Rumble, D., Giorgis, D., Ireland. T., Zhang, Z. M., Xu, H. F., Yui, T. F., Yang, J. S., Xu, Z. Q., and Liou, J. G., 2002, Low ^{18}O zircons, U-Pb dating, and the Qinglongshan oxygen and hydrogen isotope anomaly near Donghai in Jiangsu Province, China: Geochimica Cosmochimica Acta, v. 66, 2299–2306.

Selverstone, J., Franz, G., Thomas, S., and Gette, S., 1992, Fluid variability in 2 GPa eclogites as an indicator of fluid behavior during subduction: Contributions to Mineralogy and Petrology, v. 112, p. 341–357.

Shen, K., Xu, H. F., and Xu, Z. Q., 1996, Characteristics of metamorphic fluids in the eclogites and its country rocks from the ultrahigh-pressure metamorphic belt in Jiaonan terrain, Shandong (China), in Progress of research on the geology and mineral resources in Shandong Province (China): Jinan, China, Shandong Science and Technology Publishing House, p. 62–80 (in Chinese with English abstract).

Shmulovich, K. I., and Graham, C. M., 2004, An experimental study of phase equilibria in the systems H_2O-CO_2-$CaCl_2$ and H_2O-CO_2-NaCl at high pressures and temperatures (500–800°C, 0.5–0.9GPa): Geological and geophysical applications: Contributions to Mineralogy and Petrology, v. 146, p. 450–462.

Sisson, V. B., Crawford, M. L., and Thompson, P. H., 1981, CO_2-brine immiscibility at high temperatures: Evidence from calcareous metasedimentary rocks: Contributions to Mineralogy and Petrology, v. 78, 371–378.

Sterner, S. M., and Bodnar, R. J., 1989, Synthetic fluid inclusions-VII. Re-equilibration of fluid inclusions in quartz during laboratory-simulated metamorphic burial and uplift: Journal of Metamorphic Geology, v. 7, p. 243–260.

Su, W., You, Z. D., Cong, B. L., Ye, K., and Zhong, Z. Q., 2002, Cluster of water molecules in garnet from ultrahigh-pressure eclogite: Geology, v. 30, p. 611–614.

Touret, J. L. R., 1981, Fluid inclusions in high-grade metamorphic rocks, in Hollister, L. S., and Crawford, M. L. eds., Short course in fluid inclusions: Applications to petrology: Calgary, Canada, Mineralogical Association of Canada, p. 182–208.

Touret, J. L. R., 2001, Fluids in metamorphic rocks: Lithos, v. 55, p. 1–25.

Xiao, Y. L., Hoefs, J., van den Kerkhof, A. M., Fiebig, J., and Zheng, Y., 2000, Fluid history of UHP metamorphism in Dabie Shan, China: A fluid inclusion and oxygen isotope study on the coesite-bearing eclogite from Bixiling: Contributions to Mineralogy and Petrology, v. 139, p. 1–16.

Xiao, Y. L., Hoefs, J., van den Kerkhof, A. M., and Li, S. G., 2001, Geochemical constraints of the eclogite and granulite facies metamorphism as recognized in the Raobazhai complex from North Dabie Shan, China: Journal of Metamorphic Geology, v. 19, p. 3–19.

Yui, T., Rumble, D., and Lo, C., 1995, Unusually low $\delta^{18}O$ ultrahigh-pressure metamorphic rocks from Su-Lu terrane China: Geochimica et Cosmochimica Acta, v. 59, p. 2859–2864.

You, Z. D., Han, Y. J., Yang, W. R., Zhang, Z. M., Wei, B. Z., and Liu, R., 1996, The high-pressure and ultrahigh-pressure metamorphic belt in the east Qinling and Dabie Mountains, China: Wuhan, China, University of Geoscience Press.

Zhang, R., Hirajima, T., Banno, S., Cong, B., and Liou, J. G., 1995, Petrology of ultrahigh-pressure rocks from the southern Sulu region, eastern China: Journal of Metamorphic Geology, v. 13, p. 659–675.

Zhang, R., Liou, J., and Shu, J., 2002, Hydroxyl-rich topaz in high-pressure kyanite quartzites, with retrograde woodhouseite, from the Sulu terrane, eastern China: American Mineralogist, v. 87, p. 445–543.

Zhang, Z. M., Xiao, Y. L., Hoefs, J., Liou, J. G., Xu, Z., 2004a, Petrogenesis of UHP metamorphic crustal and mantle rocks from the Pre-Pilot Hole of the Chinese Continental Scientific Drilling Project in Sulu, eastern China: International Geology Review, in review.

Zhang, Z. M., Xiao, Y. L., Liu, F., Liou, J., and Hoefs, J., 2004b, Petrogenesis of UHP metamorphic rocks from Qinglongshan, Southern Sulu, East-Central China: Lithos, in review.

Zhang, Z. M., Xu, Z. Q., and Xu, H. F., 2000, Petrology of ultrahigh-pressure eclogite from the ZK703 drillhole in the Donghai, eastern China: Lithos, v. 52, p. 35–50.

Zhang, Z. M., Xu, Z. Q., and Xu, H. F., 2003, Petrology of the non-mafic UHP metamorphic rocks from a drill-hole in the Southern Sulu orogenic belt, eastern-central China: Acta Geologica Sinica, v. 77, p. 173–186.

Zheng, Y. F., Fu, B., Cong, B., and Li, S., 1996, Extreme $\delta^{18}O$ depletion in eclogite from the Su-Lu terrane in east China: European Journal of Mineralogy, v. 8, p. 317–323.

Zheng, Y. F., Fu, B., Cong, B., and Li, L., 2003, Stable isotope geochemistry of ultrahigh pressure metamorphic rocks from the Dabie-Sulu orogen in China: Implications for geodynamics and fluid regime: Earth-Science Review, v. 62, p. 105–161.

Zheng, Y. F., Fu, B., Li, Y., Xiao, Y. L., and Li, S., 1998, Oxygen and hydrogen isotope geochemistry of ultrahigh-pressure eclogites from the Dabie Mountains and the Sulu terrane: Earth and Planetary Science Letters, v. 155, p. 113–129.

Late Mesozoic–Eocene Mantle Replacement beneath the Eastern North China Craton: Evidence from the Paleozoic and Cenozoic Peridotite Xenoliths

JIANPING ZHENG,[1]

State Key Laboratory of Geoprocesses and Mineral Resources, Faculty of Earth Sciences, China University of Geosciences, Wuhan 430074, China; GEMOC ARC National Key Centre, Department of Earth and Planetary Sciences, Macquarie University, Sydney, NSW 2109, Australia; and Department of Geological and Environmental Sciences, Stanford University, Stanford, California 94305

W. L. GRIFFIN,

GEMOC ARC National Key Centre, Department of Earth and Planetary Sciences, Macquarie University, Sydney, NSW 2109, Australia; and CSIRO Exploration and Mining, North Ryde, NSW 2113, Australia

SUZANNE Y. O'REILLY,

GEMOC ARC National Key Centre, Department of Earth and Planetary Sciences, Macquarie University, Sydney, NSW 2109, Australia

J. G. LIOU, R. Y. ZHANG,

Department of Geological and Environmental Sciences, Stanford University, Stanford, California 94305

AND FENGXIANG LU

Faculty of Earth Sciences, China University of Geosciences, Wuhan 430074, China

Abstract

Xenolith-bearing Paleozoic kimberlites and Cenozoic basalts from the eastern part of the North China craton provide unusual insights into intraplate processes and Phanerozoic lithospheric evolution. Paleozoic peridotite xenoliths represent samples of ancient cratonic mantle; P-T estimates show that a thick (~230 km), cold (ca 40 mW/m^2) lithosphere existed beneath the craton during mid-Ordovician time. However, xenoliths from Tertiary basalts sample a thin (< 90 km), hot (mean geotherm ca 80 mW/m^2), compositionally heterogeneous lithosphere beneath the same area in Cenozoic time. Fertile, spinel-facies mantle makes up much of the Cenozoic lithosphere beneath the eastern North China craton, especially in regions along the translithospheric Tanlu fault. However, refractory spinel-facies xenoliths are found locally along the north-south gravity lineament in areas far away from the Tanlu fault. These refractory xenoliths are interpreted as derived from shallow relics of the cratonic mantle embedded in more fertile Cenozoic lithosphere. The increasing incidence of fine-grained, sheared microstructures in xenoliths from the north-south gravity lineament progressively toward the Tanlu fault suggests that the translithospheric fault system played an important role in the Mesozoic-Cenozoic replacement of pre-existing lithospheric mantle by more fertile material. Modification of cratonic mantle beneath the eastern North China craton involved irregular replacement of old lithosphere by cooling products of weakly depleted asthenosphere welling up along major shear systems. This lithosphere replacement was accompanied by an elevated geotherm and a shallower asthenosphere-lithosphere boundary.

Introduction

THE PRESERVATION OF cratonic roots appears to be common worldwide, due to their significant buoyancy (Griffin et al., 1998a) and high viscosity (Kelemen et al., 1998), but asthenosphere-lithosphere interaction can modify these roots and lead to their replacement by more fertile mantle material (e.g., Griffin et al., 1998b; Zheng et al., 1998; O'Reilly et al., 2001). Early Paleozoic diamond-bearing kimberlites (500–457 Ma; Lu et al., 1998) document the presence of a thick lithospheric root,

---

[1]Corresponding author; email: jpzheng@cug.edu.cn

Mengyin kimberlites are used to characterize the Paleozoic subcontinental lithospheric mantle (SCLM), whereas xenoliths from the Neogene Shanwang, Qixia, and Hebi basalts are used to characterize the Cenozoic SCLM. Locations of samples are shown in Figure 1. Shanwang and Qixia lie astride and east of the Tanlu fault zone, respectively. Hebi lies on the north-south gravity lineament.

Geological Setting

The North China craton (NCC) is divided into two regions by a major geophysical feature, the north-south gravity lineament (NSGL). The lineament extends from Guizhou in southwest China northward into Russia, and runs approximately parallel to, and about 400 km west of, the Tanlu fault, a major wrench fault system that cuts through the eastern NCC (Xu J. W.. et al., 1993). The NSGL coincides approximately with the Trans–North China orogen in the central NCC (Zhao et al., 2000) separating the Archean Liaolu and Ordos cratonic nuclei (Fig. 1). To the east of the NSGL, the Archean nucleus (≈ 3.6 Ga; Liu et al., 1992; Zheng et al., 2004a) is characterized by a thin lithosphere, high heat flow, and weak negative to positive regional Bouguer anomalies. To the west of the NSGL, the Ordos Archean nucleus is characterized by a thick lithosphere, low heat flow, and strong negative Bouguer gravity anomalies (Yuan, 1996a; Griffin et al., 1998b).

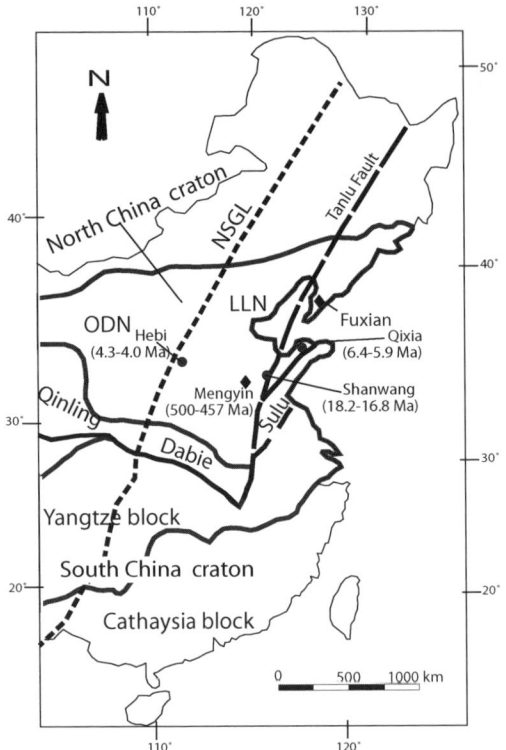

FIG. 1. Locality map and tectonic setting. Abbreviations: NSGL = north-south gravity lineament; LLN = Liaolu nucleus; ODN = Ordos nucleus.

The NCC underwent a series of tectonic events that may have included interactions between mantle plumes and continental lithosphere (Jahn, 1990; Shen and Gen, 1991; Wu and Gen, 1991; Zhao et al., 1998) after the formation of the initial nucleus (Liu et al., 1992; Zheng et al., 2004a), and was stabilized in the late Proterozoic (Ernst, 1988). This was followed by a period of tectonic and magmatic quiescence until the eruption of mid-Ordovician diamondiferous kimberlites in the Mengyin County, Shandong Province, and Fuxian County, Liaoning Province (Fig. 1). All Paleozoic xenoliths were collected from the largest pipe (Shengli No. 1) in Mengyin. This pipe has an elliptical exposure approximately 230 × 150 m, and is composed of diatreme-facies porphyritic kimberlites and subordinate hypabyssal-facies kimberlites. Rb-Sr and Sm-Nd dating yield ages of 500–457 Ma for the hypabyssal-facies kimberlites (Lu et al., 1998).

apparently of Archean age, during mid-Ordovician time beneath the eastern part of the North China craton (Griffin et al., 1992, 1998b; Lu and Zheng, 1996; Zheng and Lu, 1999). However, since late Mesozoic time, this region has been tectonically active, with widespread basaltic volcanism (e.g. Guo et al., 2001; Zhang et al., 2002, 2003), and the development of large sedimentary basins (Li et al., 1997). Seismic velocity (Vp) studies show that the present lithosphere of the region is ≤100 km thick on average (Yuan, 1996a). These observations indicate that lithospheric thinning and intense mantle-crust interactions occurred during Mesozoic–Cenozoic lithospheric evolution (Griffin et al., 1992, 1998b; Menzies et al., 1993; Xu et al., 1998, 2000). However, the mechanisms of such interactions are not well understood (Fan et al., 2001; Xu, 2001; Gao et al., 2002). Petrochemical comparisons of peridotite xenoliths from Paleozoic kimberlites and Neogene basalts can constrain such mechanisms. In this study, xenoliths from the Ordovician

The Neogene Shanwang and Qixia basalts occur astride and 80 km east of the Tanlu fault, respectively (Fig. 1). Three episodes of volcanism, including 15 lava flows (episodes 1, 2, and 3 comprising 10, 2, and 3 flows, respectively) took place in Shanwang. The mantle xenoliths described here were collected from lavas of episode 1 (18.2 ~ 16.8 Ma, K-Ar method; Jin, 1985). Two episodes of volcanism including 8 lava flows (episodes 1 and 2 comprising 2 and 6 flows, respectively) erupted at Qixia. The mantle xenoliths described here were collected from the flow of episode 1 (6.4–5.9 Ma, K-Ar method; Jin, 1985). Sr-Nd isotopic analyses of the basalts (Zhi et al., 1994) indicate the presence of depleted mantle beneath both areas. The Hebi area of Henan Province lies east of the NSGL (Fig. 1). Neogene basalts with abundant spinel peridotite xenoliths occur as NNW-oriented pipes and dikes with an eruption age of 4.3-4.0 Ma (K-Ar dating; Liu et al., 1990).

Peridotitic Petrography

The diverse microstructures of mantle-derived xenoliths in basalts and kimberlites, and their relationships to tectonic processes and magma genesis, are important in understanding mantle evolution (O'Reilly et al., 1989). Much of the early literature dealing with dominantly metamorphic microstructures was summarized by Harte (1977), who proposed a simple classification of olivine-rich xenoliths into major microstructural types—coarse grained, porphyroclastic, mosaic-porphyroclastic, and granuloblastic. Mercier and Nicolas (1975) and Harte (1977) noted that some xenoliths might have undergone more than one cycle of recrystallization, through coarse, porphyroclastic and granoblastic microstructures. The typical grain size for olivine (50–65 vol%) and orthopyroxene (20–35 vol%) in coarse-grained lherzolites is 2–5 mm, up to 10 mm. In porphyroclastic microstrustures, at least 10 percent of olivine occurs as coarse, commonly strained grains (porphyroclasts), set in a matrix of small, commonly strain-free recrystallized grains. Sheared microstructures used in the paper are defined as those in which porphyroclasts make up less than 10% of total olivine, with preferred orientation and/or foliation of olivines and "holly-leaf" spinels. Fine-grained microstructures in olivine-rich xenoliths have less than 5% of olivine occurring as porphyroclasts, and most mineral grains <1 mm have dominantly polygonal shapes.

Mantle xenoliths from Paleozoic Mengyin kimberlites include ~34% garnet harzburgite, and ~54% Cpx-poor garnet lherzolite (Fig. 2A); a few are pyroxenite, phlogopitite, or wehrlite. Most peridotitic xenoliths exhibit coarse-grained, porphyroclastic microstructures (Fig. 3A); a minor number show sheared, fine-grained microstructures (Fig. 3B). The peridotites are strongly serpentinized. Fresh relics of olivine, diopside (<5 vol%), enstatite, or garnet are only found in a few xenoliths. Dolomite and serpentinized olivine occur as inclusions in garnet. All garnets have reaction rims with fine-grained spinel. Pyroxenites are composed of 75–80% pyroxene, 14–16% garnet, 3–5% ilmenite, 1–2% rutile, and 1–2% apatite (Fig. 3C), or 80–85% pyroxene and 15–20% phlogopite (Fig. 3D). These pyroxenites have magmatic microstructures that vary widely from pegmatitic (= 15 mm) to coarse grained. All pyroxene is replaced by serpentine. Phlogopitites exhibit two types of microstructure: porphyroblastic (Fig. 3E) and magmatic (Fig. 3F). Phlogopitites with porphyroblastic textures are composed of 100% phlogopite including both porphyroblasts and matrix. Phlogopitites with magmatic textures have ~25% serpentinized olivine.

In contrast to the kimberlite-borne xenoliths, the basalt-borne xenoliths in the eastern NCC are almost all spinel lherzolite, and lack phlogopite-bearing rocks. These peridotites have microstructures and mineral compositions that vary with their locations relative to the Tanlu fault. Most Shanwang peridotites are fine-grained (Fig. 2B), or have sheared microstructures with strongly foliated olivines (Fig. 3G). They contain higher modal clinopyroxene than xenoliths from Qixia (5.6–19.5% vs. 1.2–12.8%), where mantle xenoliths with coarse-grained microstructures are found. Peridotites from Hebi contain less modal clinopyroxene (<5%), and comprise 30% harzburgite and 59% Cpx-poor lherzolite, mostly with coarse-grained microstructures (Fig. 3H).

Mineral Chemistry of Peridotites

$Mg^{\#}$ and $Cr^{\#}$ of minerals

The $Mg^{\#}$ values of the olivines in Mengyin xenoliths are high, with a mean of 92.5. Olivine in the Hebi xenoliths has similar $Mg^{\#}$ to Mengyin (Fig. 4A). However, the $Mg^{\#}$ values of olivines in xenoliths from Qixia, and especially in Shanwang, are lower than those in Mengyin, with mean values of 90.5 and 89.5, respectively.

Diopsides in Mengyin xenoliths have high $Mg^{\#}$ (92.0–94.5) and high $Cr^{\#}$ (20.1–48.5). The $Mg^{\#}$ and $Cr^{\#}$ of diopsides in Hebi xenoliths are slightly lower

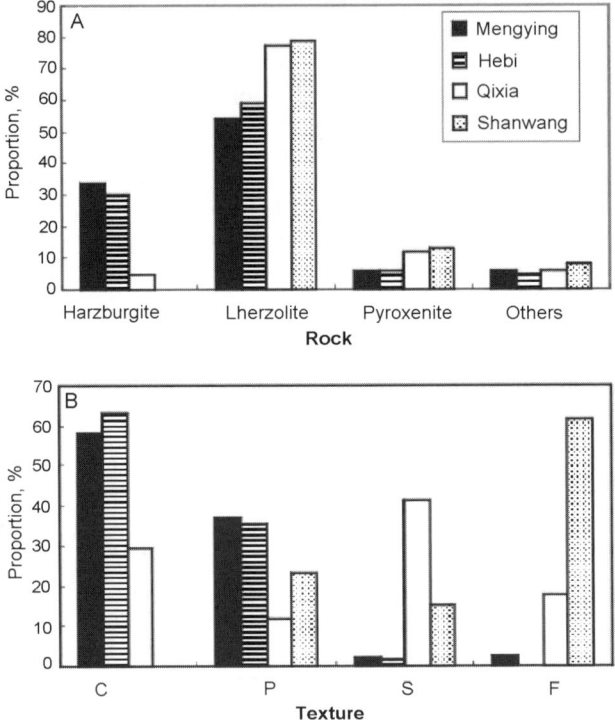

FIG. 2. Frequency distribution of types of mantle rocks (A) and textures (B). Abbreviations: C = coarse-grained; P = prophyroclastic; S = sheared; F = fine-grained.

than those in Mengyin. The $Mg^\#$ and $Cr^\#$ values of diopsides in xenoliths from Qixia, and especially from Shanwang, are much lower than those in Mengyin xenoliths (Fig. 4B). The patterns of relative differences in the $Cr^\#$ of chromites/spinels among the four localities are similar to the pattern shown by the $Cr^\#$ of diopsides (Fig. 4C).

Minor elements of diopside

The Mengyin diopsides contain 1.04–1.58 wt% Al_2O_3, 1.10–1.60 wt% Na_2O, 0.03–0.34 wt% TiO_2, and 2.5–11.7 Al^{VI}/Al^{IV} (Table 1). The Al_2O_3 contents and Al^{IV} are much lower than those of diopsides from the Cenozoic xenoliths. Relative to xenoliths from Hebi and Qixia, the Al_2O_3, Na_2O, and TiO_2 contents of Shanwang diopsides are high, and lie at the high end of the compositional trend defined by the Cenozoic xenoliths (Fig. 5).

CaO and Al_2O_3 of peridotites

The Mengyin peridotites have typical compositions of refractory Archean mantle, with CaO and Al_2O_3 contents much lower than the commonly accepted primitive upper mantle (McDonough and Sun, 1995), and in the refractory compositional range defined by low-T peridotite xenoliths from Kaapvaal craton (Fig. 6). The Hebi peridotites have similar compositions to those from Mengyin. However, most Qixia peridotites and all Shanwang peridotites are relatively fertile with high, but variable CaO and Al_2O_3 contents.

Lithospheric Thermal State

P-T conditions of the Paleozoic and Cenozoic xenoliths (Table 1) were estimated by single clinopyroxene thermobarometry of Nimis and Taylor (2000) and Mercier (1976), respectively, because there are no suitable and/or fresh mineral pairs. P-T estimates of the Mengyin peridotites suggest a thick (~230 km), cold (~45 mW/m²) lithosphere in Paleozoic time, whereas a significantly lower geotherm (≤40 mW/m²) has been derived from garnet concentrates (Griffin et al., 1998b) and silicate inclusions in diamonds (Lu and Zheng, 1996). In contrast, the Cenozoic lithosphere had a much higher geotherm.

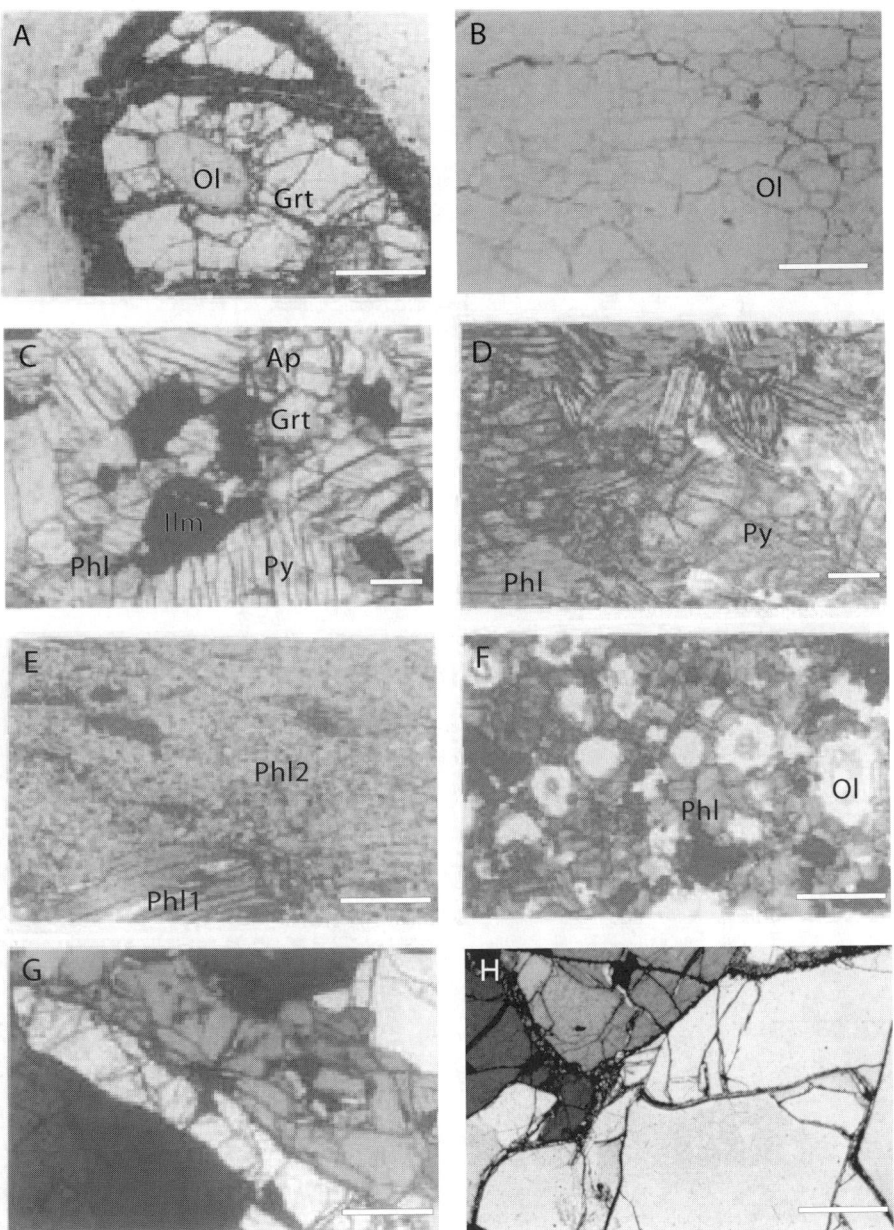

FIG. 3. Petrography of deep-seated xenoliths from the eastern North China craton. A–F are xenoliths from Paleozoic kimberlites. A. Porphyroclastic garnet in peridotite, the garnet (Grt) contains serpentinized olivine inclusions (Ol) and reaction rims with fine-grained spinel. B. Peridotite with fine-grained texture, composed of serpentinized olivine. C. Garnet pyroxenite, composed of altered pyroxene (Py) + garnet (up to 15 mm in size) + ilmenite (Ilm) + rutile (Rut) + apatite (Ap). D. Phlogopite-pyroxenite with magmatic and coarse-grained texture. E. Phogopitite with porphyroblastic texture (all of the porphyroblast and matrix are phlogopite). F. Olivine-phlogopitite with magmatic texture. G and H are peridotites from Cenozoic basalts. G. Shanwang peridotite with foliation defined by olivine. H. Hebi peridotite with coarse-grained, porphyroclastic textures. Scale bar = 2.5 mm.

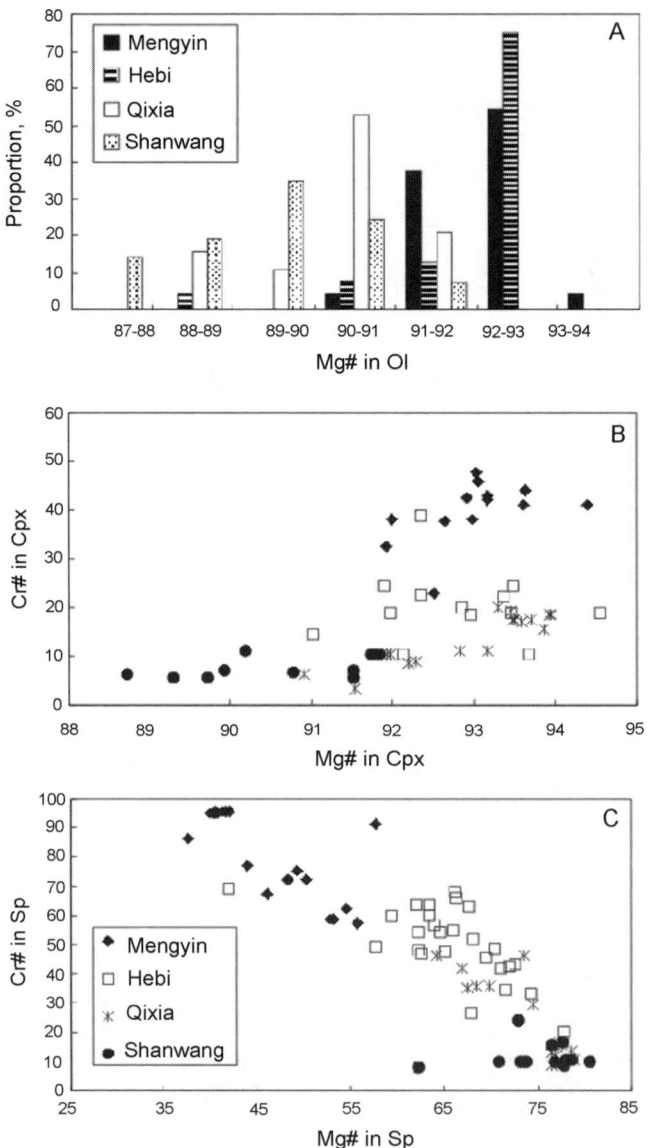

FIG. 4. Frequency distribution of Mg# in olivine (A) and plots of Mg# vs Cr# in diopside (B) and in spinel (C). Other data sources for Shanwang, Qixia, Hebi, and Mengyin are from Zheng et al. (1998, 2001) and Zheng (1999).

The single-pyroxene estimates spread between conductive model geotherms of 60–90 mW/m² (mean 80 mW/m²; Fig. 7). The spread of single-pyroxene estimates is consistent with more detailed geothermobarometry based on garnet websterite xenoliths, suggesting high convex geotherms reflecting advective heat transport (Xu et al., 1998; Chen et al., 2001). Both data sets suggest lithosphere with a thickness of ~ 85 km (Fig. 7).

Late Mesozoic-Eogene Mantle Replacement

The most significant feature of Archean mantle, which clearly distinguishes it from the mantle beneath younger terrains, is the presence of very depleted harzburgites with strongly subcalcic garnets (reflecting low Ca/Al) and high Opx/Ol ratios, reflecting low Mg/Si (Boyd, 1996). In general, subcalcic harzburgites are restricted to Archean

TABLE 1. Representative Electron Microprobe Analyses of Mantle Diopsides from the Eastern Part of the North China Craton (wt%)[1]

	Paleozoic													Cenozoic																							
Xenolith:	Interior of the North China craton													North-south gravity lineament									East of the Tanlu fault							Tanlu fault						
Setting:	Mengyin													Hebi									Qixia					Shanwang								
Locality:																																					
Rock:	GtH	GtH	GtH	GtH	GtH	GtH	GtH	GtH	GtH	GtH	GtH	GtH	GtH	SpH	SpH	SpH	SpH	SpH	SpH	SpH	SpL	SpL	SpL	SpL	SpL	SpL	SpL	SpL	SpL	SpL	SpL	SpL	SpL	SpL	SpL	SpL
Sample:	SD1	SD2	SD3	SD4	SD5	SD6	SD8	SD9	SD10	SD11	SD12	SD13	Hb4	Hb5	Hb6a	Hb10	Hb22	Hb30	Hb39	Hb47	Hb12	Hb45	Qx1	Qx7	QX8	Qx14	Qx17	SW4	SW12	SW14	SW36					
SiO$_2$	55.4	55.1	55.7	55.0	54.8	55.6	55.1	55.3	54.5	55.1	54.7	54.9	53.6	53.4	54.1	53.5	53.3	53.8	52.3	52.4	51.5	52.9	53.8	52.6	52.6	53.6	53.8	52.6	52.7	51.9	52.7
TiO$_2$	0.14	0.19	0.15	0.03	0.18	0.04	0.25	0.12	0.34	0.19	0.21	0.18	0.13	0.03	0.02	0.22	0.02	0.02	0.01	0.15	0.47	0.03	0.01	0.18	0.20	0.07	0.01	0.43	0.48	0.01	0.33
Al$_2$O$_3$	1.45	1.46	1.40	1.25	1.56	1.05	1.45	1.25	1.53	1.42	1.53	1.58	3.84	3.23	3.45	5.38	2.45	3.30	2.66	6.28	4.89	3.93	1.84	4.18	4.80	2.27	1.79	5.43	6.85	7.01	5.72
Cr$_2$O$_3$	1.53	1.30	1.84	1.42	0.67	1.05	1.29	1.52	1.63	1.24	1.63	1.11	1.69	1.11	1.11	1.88	1.19	1.14	1.14	1.09	1.24	0.67	0.58	0.78	0.91	0.62	0.61	0.94	0.72	0.63	0.98
FeO	2.44	2.96	2.41	2.27	2.77	1.99	2.48	2.54	2.52	2.62	2.41	2.90	2.53	2.23	2.35	2.46	2.22	2.33	2.19	2.42	2.70	2.20	2.13	2.07	2.12	2.04	2.05	2.47	2.69	3.11	2.43
MnO	0.07	0.18	0.07	0.11	0.14	0.09	0.08	0.16	0.08	0.06	0.07	0.14	0.09	0.07	0.09	0.08	0.06	0.06	0.04	0.11	0.05	0.04	0.08	0.06	0.10	0.07	0.07	0.05	0.09	0.08	0.01
MgO	18.5	18.9	17.9	18.5	19.0	18.7	18.2	18.9	18.4	18.3	18.3	18.3	17.1	17.8	17.1	15.8	17.9	17.2	17.3	15.9	15.3	18.3	17.8	15.9	15.4	17.5	17.8	15.6	14.9	14.6	15.1
CaO	18.3	18.2	18.1	19.8	19.0	20.2	18.9	18.3	18.5	19.0	19.1	19.0	18.4	20.8	18.9	17.9	21.6	20.5	21.4	19.8	22.5	21.4	24.0	22.4	22.1	23.8	23.8	20.8	19.4	19.9	20.6
Na$_2$O	1.52	1.48	1.61	1.26	1.46	1.11	1.53	1.49	1.64	1.33	1.40	1.32	1.78	0.72	1.62	2.65	0.49	1.11	0.76	1.62	0.67	0.23	0.12	1.06	1.09	0.25	0.06	1.48	2.11	2.01	1.65
K$_2$O	0.03	0.07	0.08	0.09	0.07	0.06	0.04	0.06	0.08	0.04	0.10	0.07	b.d.	b.d.	b.d.	b.d.	b.d.	b.d.	b.d.	b.d.	b.d.	b.d.	b.d.	b.d.	b.d.	b.d.	b.d.	b.d.	b.d.	b.d.	b.d.
Total	99.4	99.9	99.2	99.8	99.7	99.8	99.4	99.6	99.2	99.3	99.5	99.5	99.3	99.4	99.0	99.9	99.3	99.6	97.9	99.9	99.7	99.8	100.5	99.2	99.3	100.2	100.1	99.8	100.0	99.9	99.6
[O]=	6	6	6	6	6	6	6	6	6	6	6	6	6	6	6	6	6	6	6	6	6	6	6	6	6	6	6	6	6	6	6
Si	1.983	1.986	1.992	1.988	1.981	1.990	1.994	1.996	1.980	1.994	1.982	1.987	1.945	1.938	1.964	1.927	1.944	1.949	1.937	1.894	1.887	1.914	1.948	1.921	1.918	1.942	1.952	1.907	1.901	1.892	1.911
Al^{IV}	0.017	0.014	0.008	0.012	0.019	0.010	0.006	0.004	0.020	0.006	0.018	0.013	0.055	0.062	0.036	0.073	0.056	0.051	0.063	0.106	0.113	0.086	0.052	0.079	0.082	0.058	0.048	0.093	0.099	0.108	0.089
Al^{VI}	0.045	0.049	0.052	0.041	0.047	0.035	0.056	0.049	0.045	0.055	0.048	0.055	0.109	0.076	0.112	0.156	0.049	0.090	0.054	0.161	0.098	0.082	0.026	0.101	0.124	0.039	0.028	0.139	0.192	0.193	0.155
Ti	0.004	0.005	0.004	0.001	0.005	0.001	0.007	0.003	0.009	0.005	0.006	0.005	0.004	0.001	0.001	0.006	0.000	0.001	0.000	0.004	0.013	0.001	0.000	0.005	0.005	0.002	0.000	0.012	0.013	0.000	0.009
Cr	0.044	0.037	0.053	0.041	0.019	0.030	0.037	0.043	0.047	0.035	0.047	0.032	0.048	0.032	0.037	0.053	0.034	0.032	0.033	0.031	0.036	0.019	0.017	0.023	0.026	0.018	0.017	0.027	0.020	0.018	0.028
Fe	0.074	0.089	0.073	0.069	0.084	0.060	0.075	0.077	0.076	0.079	0.073	0.088	0.077	0.068	0.071	0.074	0.068	0.071	0.068	0.073	0.083	0.066	0.065	0.063	0.065	0.062	0.062	0.075	0.081	0.095	0.074
Mn	0.002	0.006	0.002	0.003	0.004	0.003	0.002	0.005	0.002	0.002	0.002	0.004	0.003	0.002	0.002	0.002	0.002	0.002	0.001	0.003	0.001	0.001	0.002	0.002	0.003	0.002	0.002	0.002	0.003	0.002	0.000
Mg	0.995	1.016	0.963	0.996	1.024	1.002	0.983	1.015	0.993	0.989	0.987	0.990	0.927	0.965	0.927	0.849	0.971	0.932	0.956	0.856	0.838	0.984	0.959	0.864	0.837	0.944	0.963	0.840	0.798	0.792	0.818
Ca	0.709	0.704	0.700	0.768	0.735	0.777	0.735	0.707	0.722	0.739	0.742	0.737	0.716	0.807	0.735	0.693	0.844	0.798	0.847	0.767	0.884	0.829	0.931	0.878	0.862	0.923	0.925	0.806	0.750	0.776	0.802
Na	0.106	0.103	0.113	0.088	0.102	0.078	0.107	0.104	0.115	0.093	0.098	0.093	0.125	0.051	0.114	0.185	0.034	0.078	0.055	0.113	0.048	0.016	0.008	0.075	0.077	0.018	0.004	0.104	0.147	0.142	0.116
K	0.001	0.003	0.004	0.004	0.003	0.003	0.002	0.003	0.004	0.002	0.005	0.003	0.000	0.000	0.000	0.000	0.000	0.000	0.000	0.000	0.000	0.000	0.000	0.000	0.000	0.000	0.000	0.000	0.000	0.000	0.000
Sum	3.98	4.01	3.96	4.01	4.02	3.99	4.00	4.01	4.01	4.00	4.01	4.01	4.01	4.00	4.00	4.02	4.00	4.00	4.01	4.01	4.00	4.00	4.01	4.01	4.00	4.01	4.00	4.00	4.00	4.02	4.00
Mg#	93.2	92.0	93.0	93.6	92.5	94.4	93.0	92.9	92.9	92.6	93.2	91.9	92.4	93.5	92.9	92.0	93.5	93.0	93.4	92.2	91.1	93.7	93.7	93.2	92.9	93.9	94.0	91.9	90.9	89.4	91.8
Cr#	42.3	38.2	47.8	44.2	22.9	41.0	38.3	45.8	42.5	37.7	42.8	32.6	23.4	19.3	20.6	19.5	25.2	18.9	22.9	10.8	15.0	10.5	18.1	11.5	11.6	16.0	19.1	10.7	6.8	5.9	10.6
P (kbar)	63	71	60	55	113	67	67	70	76	63	59	50	25	28	28	21	25	27	24	20	6	24	21	12	9	19	21	18	20	17	16
T(ºC)	1178	1222	1160	1104	1251	1099	1098	1150	1217	1165	1141	1120	1188	1121	1170	1151	1086	1105	1067	1072	866	1112	949	888	873	938	963	999	1039	1010	972
References[2]	(1)	(1)	(1)	(1)	(1)	(1)	(1)	(1)	(1)	(1)	(1)	(1)	(2)	(2)	(2)	(2)	(2)	(2)	(2)	(2)	(2)	(2)	(3)	(3)	(3)	(3)	(3)	(3)	(3)	(3)	(3)

[1]Abbreviations: GtH = garnet harzburgite; SpH = spinel harzburgite; SpL = spinel lherzolite; b.d. = below detection limit; for P and T estimates, see text; Al^{IV} = 2- Si; Al^{VI} = Al- Al^{IV}
[2]References and notes: (1) = this paper; (2) = Zheng et al., 2001; (3) = Zheng et al., 1998;

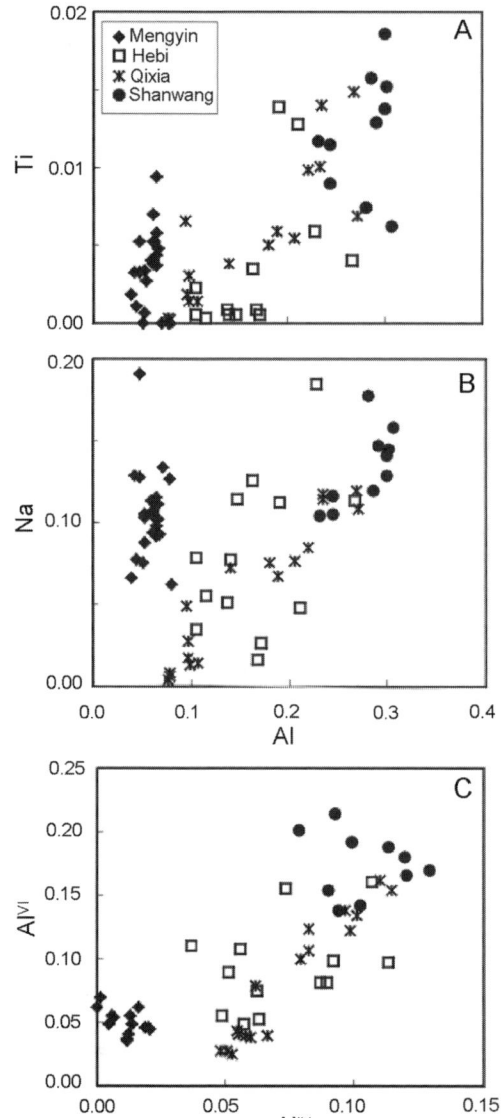

FIG. 5. Plots of Al vs Ti (A) and Na (B), and Al^{VI} vs Al^{IV} (C) in diopside. Data sources are the same as in Figure 3.

mantle, and the dominant lherzolites become progressively less depleted from Archean through Proterozoic to Phanerozoic time (Griffin et al., 1998a, 1999).

These differences are also seen in peridotite xenoliths from the eastern NCC. All of the Paleozoic peridotites, and most of the Hebi xenoliths, are harzburgite and Cpx-poor lherzolite, strongly depleted in CaO and Al_2O_3 contents relative to the primitive mantle (see Fig. 6), and closely similar to the Archean xenolith suite from the Kaapvaal craton (Fig. 8). The similarity in chemical characteristics between the Hebi spinel peridotites and the garnet peridotites from the Mengyin kimberlites, which are similar to xenoliths from the Archean Kaapvaal craton, suggests that the Hebi xenoliths represent the relict uppermost part of the cratonic mantle. Their spinel-facies mineralogy reflects both their shallow depths of origin and the high geotherm during Tertiary time, whereas the garnet-facies Mengyin samples were derived from greater depths attended by a cooler geotherm (Zheng et al., 2001).

In contrast, all Shanwang xenoliths, most Qixia xenoliths, and a few Hebi samples have high CaO, Al_2O_3 contents and lower $Mg^\#$-olivine/peridotite, and are typical of Phanerozoic rather than cratonic mantle (Griffin et al., 1998a, 1999). In terms of their Sr-Nd isotope compositions, the Paleozoic xenoliths are characteristic of an enriched mantle (EM2), whereas the Neogene xenoliths from Qixia and Shanwang are derived from the Phanerozoic depleted mantle (Fig. 9). The decrease in the proportion of refractory mantle (Figs. 4, 6, and 8) and increase of the fine-grained, sheared microstructures (Fig. 2) from Hebi through Qixia to Shanwang are ascribed to their spatial relationships to the Tanlu fault (Fig. 1). This major translithospheric structure evidently has played an important role in the Mesozoic–Cenozoic replacement of pre-existing Archean lithospheric mantle by more fertile material, allowing the upwelling of weakly depleted asthenosphere to shallow depths (Zheng et al., 1998, 2004b). This relatively undepleted material now makes up much of the subcontinental lithospheric mantle beneath the eastern NCC (Zheng et al., 1998; Xu et al., 1998, 2000; Gao et al., 2002; Wu et al., 2003), and represents the bulk of most members of Tertiary xenolith suites. Less common, more refractory xenoliths (e.g., Hebi; Zheng et al., 2001) may represent the shallow relics of older lithosphere.

A seismic tomography profile cross the eastern NCC along 36° N clearly reveals the presence of isolated high-Vp regions dispersed within a background of generally low-Vp material at depths of 100 km (Yuan, 1996a). Tomography (Fig. 10) suggests that lower-velocity material in the upper mantle has upwelled from depths greater than 150 km, and flowed laterally along weak zones in the mantle at about 60–130 km and along the Tanlu fault zone.

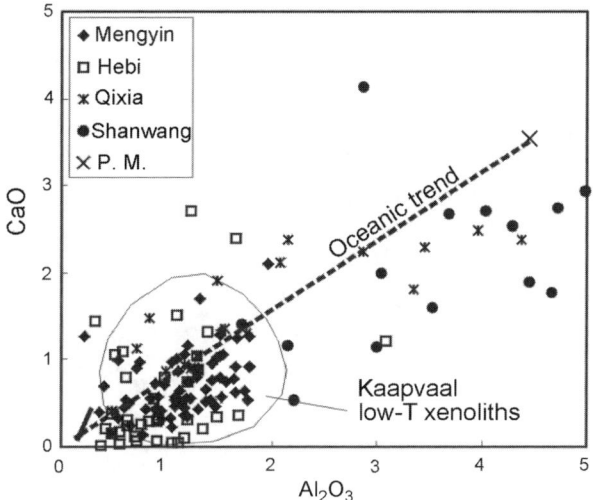

FIG. 6 CaO vs Al_2O_3 of peridotites. Abbreviations: P.M. = primitive mantle, from McDonough and Sun (1995); Kaapvaal Low-T xenoliths are from Boyd and Mertzman (1987), Cox et al. (1987), Nixon (1987), and Boyd et al. (1993).

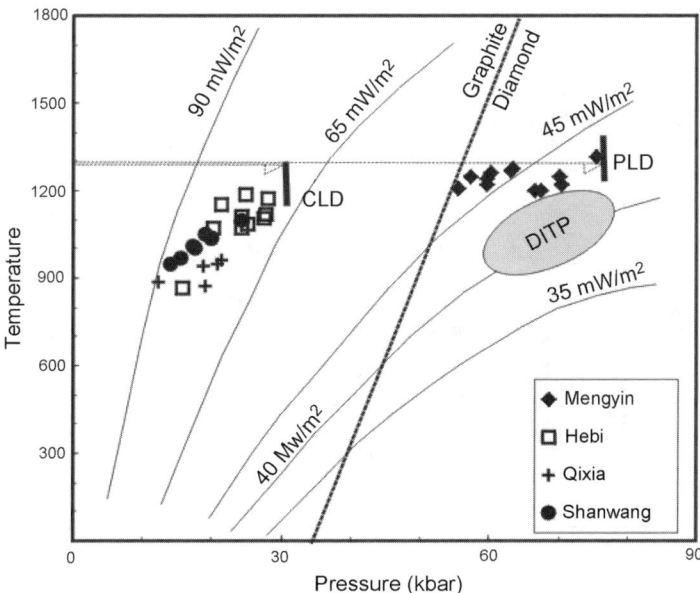

FIG. 7. Comparison between the Paleozoic and Cenozoic geotherms beneath the eastern North China craton. Abbreviations: DITP = P-T range obtained from diamond inclusions (Lu and Zheng, 1996; Zheng, 1999); the conductive geotherms are after Pollack and Chapman (1977); PLD = Paleozoic lithospheric thickness; CLD = Cenozoic lithospheric thickness.

Pegmatitic rocks associated with aborted magmatism (such as the pyroxenites, see Figs. 3C–3D), the shear zones (Xu et al., 1996), and the metasomatic zones (Menzies et al., 1993; Konzett, 1998) in the Paleozoic lithosphere have been interpreted as weak zones or channels for the transport

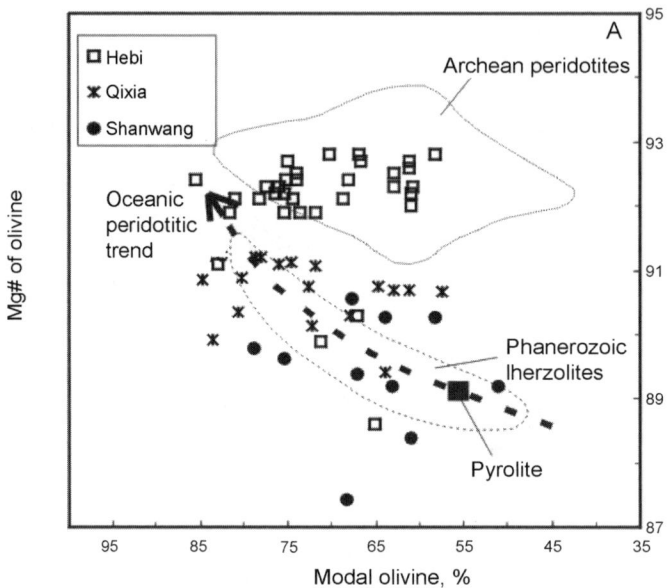

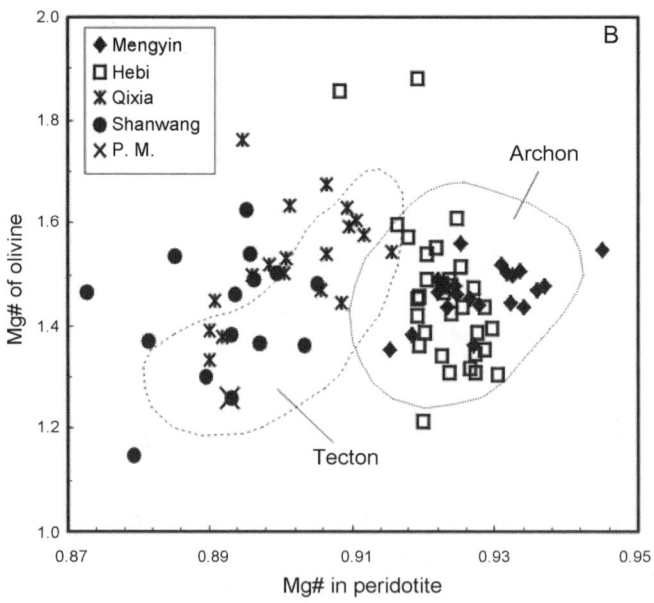

FIG. 8. Modal vs. Mg$^\#$ of olivine (A) and plots of Mg$^\#$ vs. Mg/Si of peridotites. Data sources are the same as in Figure 3. Oceanic trend from Boyd (1989); Archean and Phanerozoic fields are from Griffin et al. (1999); Archon and Tecton are defined as lithosphere domains that experienced their last major tectonothermal event more than 2.5 Ga and less than 1.0 Ga, as modified from the classification of Janse (1994) by Griffin et al (1998a).

of melts/fluids in mantle (Zheng, 1999). The penetration of the Tanlu fault into the lithospheric mantle during Mesozoic–Cenozoic time (Xu Y. et al., 1993) led to better connection between such zones, forming a reticulated pattern that provided paths for upwelling asthenospheric materials that replaced and eroded the pre-existing lithosphere (Fig. 11).

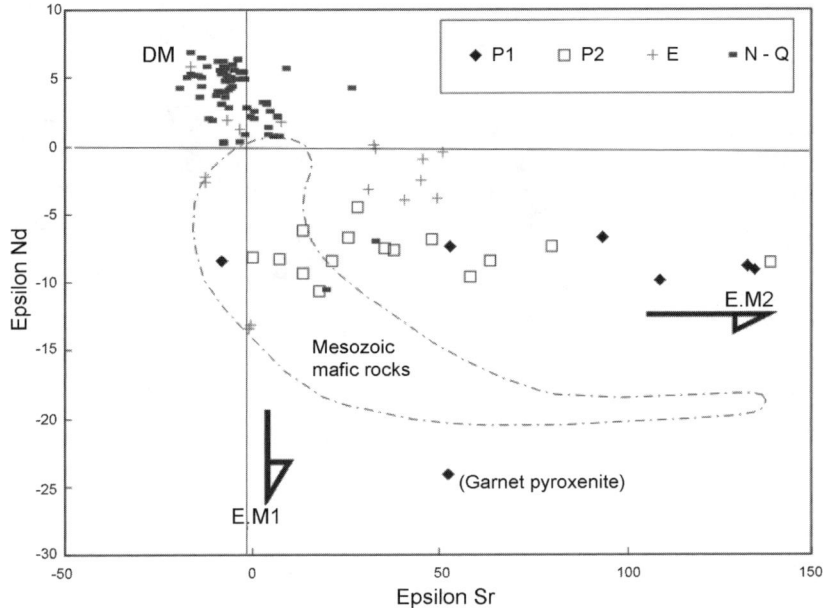

FIG. 9. Plots of εSr vs εNd of Paleozoic and Cenozoic mantle rocks from eastern North China craton. Data sources: P1 = peridotite xenoliths in the Paleozoic kimberlites (P2), from Zheng (1999); E = Eogene basalts from Jiyang basin in eastern Shandong Provinces (authors' unpubl. data); N–Q = Neogene–Quaternary basalts and their peridotite xenoliths from eastern Shandong Provinces (Zhi et al., 1994; Zheng, 1999; Fan et al., 2001); Mesozoic mafic magmatic rocks from Xu W. et al. (1993), Guo et al. (2001, 2003), and Zhang et al. (2002, 2003).

Recent studies on the Triassic, tectonically exhumed Sulu UHP peridotites show that they evolved from Paleozoic-like protoliths through interaction with crust-derived melts/fluids (Zheng et al., 2003b). The over-thickening of the lithosphere due to the collision of the North China and Yangtze cratons in the early Mesozoic (Li et al., 1993; Yin and Nie, 1993; Hacker et al., 2000), and the consequent exhumation (~ 205 Ma, Yang and Jahn, 2000) of the Dabie-Sulu UHP rocks, may have occurred in an extensional regime that promoted the early asthenospheric upwelling (Zheng et al., 2003a). The subduction of the Indian and Pacific plates would enhance the reaction between upwelling asthenosphere and the pre-existing lithosphere (Menzies et al., 1993; Griffin et al., 1992, 1998b; Zheng et al., 1998, 2001; Xu, 2001).

These processes coincided with the development of the large basins and strong crust-mantle interaction recorded in Sr-Nd isotopes of the late Mesozoic–Tertiary magmatic rocks (Guo et al., 2001; Zhang et al., 2002, 2003). The basins in eastern North China formed during two episodes: the first in Jurassic–Cretaceous time and second in Eocene time (Li et al., 1997). Two peaks of high heat flow, corresponding to these basin-subsidence episodes, have been calculated from coal reflectivity (Ro) measurements in the Songliao Basin (99 mW/m² at 115 Ma and 86.5 mW/m² at 57 Ma; Li, 1995). However, it should be noted that the Songliao Basin lies about 800–1000 km northeast of the studied areas, so similarities in tectonism at this time assume that the northeastern region of China was affected by similar and contemporaneous tectonic events. These thermal peaks are interpreted as resulting from the upwelling of asthenospheric material (Zheng et al., 2001). Basin subsidence would be a natural consequence of the replacement of buoyant Archean SCLM by denser fertile SCLM, followed by cooling (Poudjom Djomani et al., 2001).

The ancient Archean lithosphere, even if partially refertilized by interaction with melts from the rising asthenosphere, would remain relatively buoyant, and could not be delaminated (Poudjom Djomani et al., 2001). Both the seismic tomography (Fig. 10) and the xenolith suites indicate that lithosphere replacement has been effected by upwelling of asthenospheric material along major linear zones

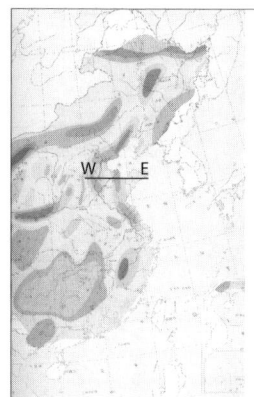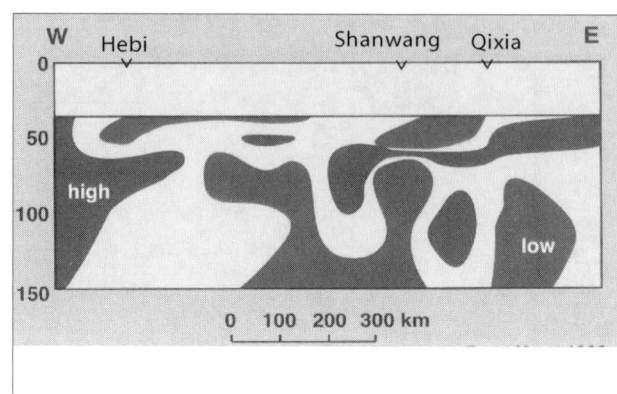

FIG. 10. Lithospheric structure from low-sensitivity layer depth on lithosphere (left) and seismic tomography (right). Modified from Yuan (1996a, 1996b).

cutting through the SCLM, such as the Tanlu fault zone, and spreading laterally near the crust-mantle boundary, and perhaps at other levels. This process would leave relicts of the Archean SCLM at different levels, scattered throughout the region beneath the eastern North China craton (Fig. 11). Therefore, the newly accreted, less refractory lithospheric mantle sampled by the Neogene basalts represents cooling products of such upwelled asthenosphere. Tomographic maps presented by Yuan (1996a) suggest that this process has involved moderate extension of the SCLM, opening sublinear channels for asthenospheric upwelling. Pre-existing translithospheric shear zones, such as the Tanlu fault and subsidiary structures, would provide natural zones of weakness that would serve as loci for such channels.

We therefore suggest that the ancient Archean SCLM has not so much been "eroded" or "thinned," but rather has been dispersed and modified by limited extension accompanied by asthenospheric upwelling. A similar mechanism has been proposed by Wang et al. (2003) to explain mixed Proterozoic and Phanerozoic Re-Os ages for xenoliths of the SCLM beneath the Taiwan Straits. Tertiary and especially Neogene basaltic eruptions in the North China craton are concentrated along the zones of asthenospheric upwelling (Yuan, 1996b), and this newly accreted, less refractory SCLM therefore dominates the basalt-borne xenolith suites. Only rare eruptions near the edge of the high-velocity zones shown by the seismic tomography (Fig. 10) sample relicts of the Archean mantle, as at Hebi. This model for the eastern part of the North China craton (Fig. 11) may be relevant to the behavior of the cratonic SCLM during episodes of major continental breakup.

Conclusions

1. The peridotite xenoliths from diamondiferous kimberlites in the eastern part of the North China craton are typical of refractory cratonic mantle, and reflect a thick (ca. 230 km), cold (ca. ~ 40 mW/m^2) lithosphere in the mid-Ordovician. In contrast, the xenoliths from basalts reflect a thin (< 90 km), hot (mean 80 mW/m^2 geotherm) and extremely heterogeneous lithosphere beneath the region in Neogene time, which suggests that the cratonic lithosphere was greatly thinned and modified during late Mesozoic–Eocene time.

2. Less common refractory xenoliths distributed in the area (e.g., Hebi) far from the Tanlu fault represent shallow relics of the cratonic mantle. However, seismic tomography suggests that other relics occur throughout the region, at both shallow and greater depths. The fertile SCLM peridotites are concentrated especially along the Tanlu fault and other translithospheric structures, which have played an important role in the replacement of the pre-existing lithospheric mantle by more fertile material. The newly accreted fertile SCLM may be over-represented in the younger basaltic xenolith suites, because basalts have preferentially erupted along zones of asthenospheric upwelling.

3. Modification of the ancient cratonic root of the North China craton by extension, rupture, and the upwelling of asthenospheric material along major lithospheric breaks may be a useful analog to

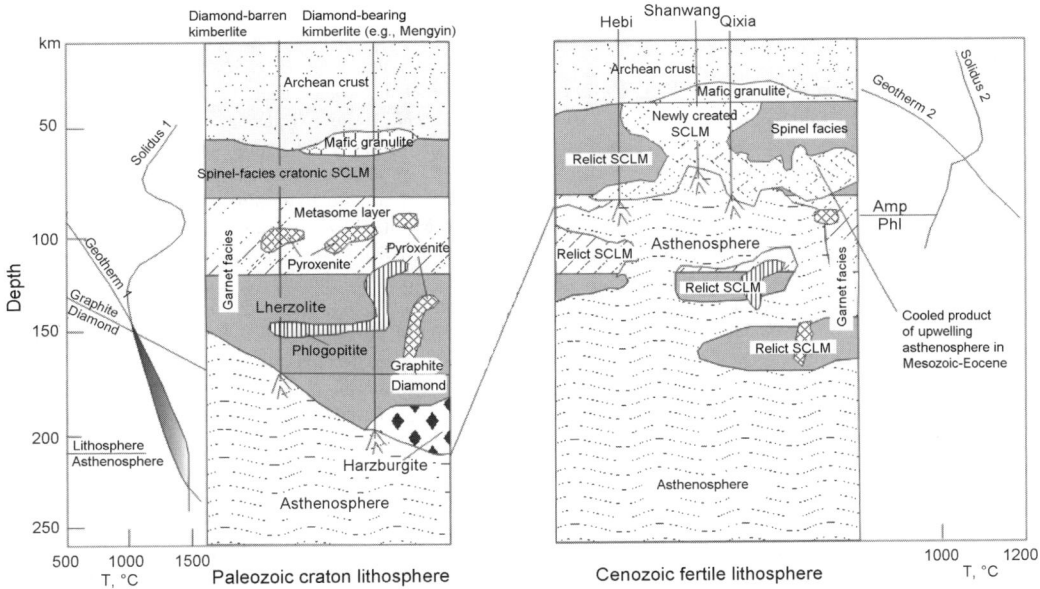

FIG. 11. Change of lithospheric/asthenospheric boundary during Paleozoic and Cenozoic beneath eastern North China craton. Paleozoic craton lithosphere (thick and cold) consists mainly of refractory harzburgite/Cpx-poor lherzolite with complex rock assemblages; in contrast, the Cenozoic lithosphere was highly heterogeneous, consisting mainly of fertile lherzolite (e.g., Shanwang and Qixia) except for few relics of refractory harzburgite/Cpx-poor lherzolite (e.g., Hebi). The fertile mantle with depleted Sr-Nd isotopes is the newly accreted lithospheric mantle from the cooling product of the upwelling asthenosphere during Mesozoic–Eocene time. Paleozoic and Cenozoic peridotites = CHO solidi adapted from Wyllie (1989) and Olafsson and Eggler (1983), respectively.

processes that have accompanied major continental breakup in the past. These processes may leave a "mixed" SCLM containing both ancient and modern components.

Acknowledgments

We dedicate this manuscript to Prof. W. G. Ernst for his support of many international projects and contribution to the study of subduction zone metamorphism and crust-mantle interactions. The first author would like to thank him for his encouragement and constructive instruction during his visit to Stanford in 2003–2004. The study was supported by the Chinese State Outstanding Youth Foundation (40425002), "973" project (G1999043303, 2003CB716500), Natural Science Foundation (40273001) and the EYTP, NSF EAR 0003355, and the ACILP AusAID Program. This is publication no. 384 from the ARC National Key Centre for Geochemical Evolution and Metallogeny of Continents [www.es.mq.edu.au/GEMOC/].

REFERENCES

Boyd, F. R., 1989, Composition and distinction between oceanic and cratonic lithosphere: Earth and Planetary Science Letters, v. 96, p. 15–26.

Boyd, F. R., 1996, Origin of peridotite xenoliths: Major element consideration, in Ranalli, G., Ricci Lucchi, F., Ricci, C. A., and Trommosdorff, T., eds., High pressure and high temperature research on lithosphere and mantle materials: Siena, Italy, University of Siena, p. 89–106.

Boyd, F. R., and Mertman, S. A., 1987, Composition and structure of the Kaapvaal lithosphere, southern Africa, in Mysen, B. O., ed., Magmatic processes: Physicochemical principles: The Geochemical Society, Special Publication 1, p. 13–24.

Boyd, F. R., Pearson, D. G., Nixon, P. H., and Mertzman, S. A., 1993, Low-calcium garnet harzburgites from southern Africa: Their relations to craton structure and diamond crystallization: Contributions to Mineralogy and Petrology, v. 113, p. 352–366.

Chen, S. H., O'Reilly, S. Y., Zhou, X. H., Griffin, W. L., Zhang, G. H., Sun, M., Feng, J. L., and Zhang, M., 2001, Thermal and petrological structure of the lithosphere beneath Hannuoba, Sino-Korean craton, China: Evidence from xenoliths: Lithos, v. 56, p. 267–301.

Cox, K. G., Smith, M. R., and Beswetherick, S., 1987, Textural studies of garnet lherzolites: Evidence of exsolution origin from high-temperature harzburgites, in Nixon, P. H., ed., Mantle xenoliths: New York, NY, Wiley, p. 537–550.

Ernst, W. G., 1988, Element partitioning and thermobarometry in polymetamorphic late Archean and Early-Mid Proterozoic rocks from eastern Liaoning and southern Jilin provinces, China: American Journal of Science, v. 288A, pp. 293–340.

Fan, W. M., Zhang, H. F., Baker, J., Jarvis, K. E., Mason, P. R. D., and Menzies, M. A., 2001, On and off the North China craton: Where is the Archean keel?: Journal of Petrology, v. 41, p. 933–950.

Gao, S., Rudnick, R. L., Carlson, R. W., McDonough, W. F., and Liu, Y. S., 2002, Re-Os evidence for replacement of ancient mantle lithosphere beneath the North China craton: Earth and Planetary Science Letters, v. 198, p. 307–322.

Griffin, W. L., O'Reilly, S. Y., and Ryan, C. G., 1992, Composition and thermal structure of the lithosphere beneath South Africa, Siberia and China: Proton microprobe studies, in International Symposium on Cenozoic volcanic rocks and deep-seated xenoliths of China and its environs: Beijing, p. 20.

Griffin, W. L., O'Reilly, S. Y., and Ryan, C. G., 1999, The composition and origin of sub-continental lithospheric mantle, in Fei, Y., Berka, C. M., and Mysen, B. O., eds., Mantle petrology: Field observations and high-pressure experimentation: A tribute to Francis R. (Joe) Boyd: The Geochemical Society, Special Publication 6, p. 13–45.

Griffin, W. L., O'Reilly, S. Y., Ryan, C. G., Gaul, O., and Ionov, D. I., 1998a, Secular variation in the composition of subcontinental lithospheric mantle: Geophysical and geodynamic implications, in Braun, J., Dooley, J. C., Goleby, B. R., van der Hilst, R. D., and Klootwijk, C. T., eds., Structure and evolution of the Australian continent: Washington, DC, American Geophysical Union, Geodynamics Series 26, p. 1–26.

Griffin, W. L., Zhang, A., O'Reilly, S. Y., and Ryan, C. G., 1998b, Phanerozoic evolution of the lithosphere beneath the Sino-Korean craton, in Flower, M., Chung, S. L., Lo, C. H., and Lee, Y. Y., eds., Mantle dynamics and plate interactions in eastern Asia: Washington, DC, American Geophysical Union, Geodynamics Series 27, p. 107–126.

Guo, F., Fan, W., Wang, Y., and Lin, G., 2001, Late Mesozoic mafic intrusive complexes in North China Block: Constraints on the nature of subcontinental lithospheric mantle: Physics and Chemistry of the Earth, v. 26, p. 759–772.

Guo, F., Fan, W., Wang, Y., and Lin, G., 2003, Geochemistry of late Mesozoic mafic magmatism in west Shandong Province, eastern China: Characterizing the lost lithospheric mantle beneath the North China block: Geochemical Journal, v. 37, p. 63–77.

Hacker, B. R., Ratschbacher, L., Webb, L. E., McWilliams, M., Ireland, T., Dong, S., Calvert, A., and Wenk, H. R., 2000, Exhumation of the ultrahigh-pressure continental crust in east-central China: Late Triassic–Early Jurassic extension: Journal of Geophysical Research, v. 105, p. 13,339–13,364.

Harte, B., 1977, Chemical variations in upper mantle nodules from southern African kimberlites: Journal of Petrology, v. 85, p. 279–288.

Jahn, B. M., 1990, Origin of high-pressure granulites: Geochemical constraints form Archean high-pressure granulite-facies rocks of the Sino-Korean craton, China, in Vielzeuf, D., and Vidal, Ph., eds., High-pressure granulites and crustal evolution: Boston, MA: Kluwer/NATO ASI Series, Series C: Mathematical and Physical Sciences, no. 311, p. 471–492.

Janse, A. J. A., 1994, Is Clifford's Rule still valid? Affirmative examples from around the world, in Meyer, H. O. A., and Leonardos, O., eds., Diamonds: Characterization, genesis, and exploration: Brazilia, Brazil, Dept. Nacional da Pord. Minera, CPRM Spec. Publ. 1A/93, p. 215–235.

Jin, L. Y., 1985, Xenoliths in Cenozoic basalts from Tanlu fault: Journal of the Changchun College of Geology, v. 3, p. 21–32 (in Chinese).

Kelemen, P. B., Hart, S. R., and Bernstein, S., 1998, Silica enrichment in the continental upper mantle via melt/rock reaction: Earth and Planetary Science Letters, v. 164, p. 387–406.

Konzett, J., 1998, The timing of MARID metasomatism in the Kaapvaal mantle: An ion probe study of zircons from MARID xenoliths: Earth and Planetary Science Letters, v. 160, p. 133–145.

Li, S. G., Xiao, T. L., Liou, D. L., Chen, Y. Z., Ge, N. J., Zhang, Z. Q., and Sun, S. S., 1993, Collision of the North China and Yangtze blocks and formation of coesite-bearing eclogites: Timing and processes: Chemical Geology, v. 109, p. 89–111.

Li, S. T., Lu, F. X., and Lin, C. S., 1997, Evolution of Mesozoic and Cenozoic basins in eastern China and their dynamic background: Wuhan, China, China University of Geosciences Press, p. 237 (in Chinese).

Li, Z., 1995, Evolution characters of mantle heat flow beneath the Songliao basin: Geotectonica et Metallogenia, v. 19, p. 108–112 (in Chinese).

Liu, D. Y., Nutman, A. P., Compston, W., Wu, J. S., and Shen, Q. H., 1992, Remnants of 3800 Ma crust in the Chinese part of the Sino-Korean craton: Geology, v. 20, p. 339–342.

Liu, R., Chen, W., Sun, J., and Li, D., 1990, The K-Ar age and tectonic environment of Cenozoic volcanic rock in China, in Liu, R., ed., The age and geochemistry of Cenozoic volcanic rock in China: Beijing, China, Seismology Publishing House, p. 1–43.

Lu, F. X., Wang, Y., Chen, M. H., and Zheng, J. P., 1998, Geochemical characteristics and emplacement ages of the Mengyin kimberlites, Shandong province, China: International Geology Review, v. 40, p. 998–1006.

Lu, F. X., and Zheng, J. P., 1996, Characteristics of Paleozoic lithosphere and deep processes in the North China platform, in Chi, J. S., and Lu, F. X., eds., Characteristics of kimberlites and Paleozoic lithosphere in the North China platform: Beijing, China: Science Press, p. 215–274.

McDonough, W. F., and Sun S. S., 1995, The composition of the Earth: Chemcal Geology, v. 120, p. 223–253.

Menzies, M. A., Fan, W., and Zhang, M., 1993, Paleozoic and Cenozoic lithoprobes and loss of >120 km of Archean lithosphere, Sino-Korean craton, China, in Prichard, H. M., Alabaster, T., Harris, N. B. W., and Neary, C. R., eds., Magmatic processes and plate tectonics: Geological Society Special Publication, v. 76, p. 71–81.

Mercier, J. C. C., 1976, Single-pyroxene thermobarometry and geobarometry: American Mineralogist, v. 61, p. 603–615.

Mercier, J. C. C., and Nicolas, A., 1975, Textures and fabrics of upper mantle peridotites as illustrated by xenoliths from basalts: Journal of Petrology, v. 16, p. 454–487.

Nimis, P., and Taylor, W. R., 2000, Single clinopyroxene thermobarometry for garnet peridotites. Part I. Calibration and test a Cr-in-Cpx barometer and an enstatite-in-Cpx thermometer: Contributions to Mineralogy and Petrology, v. 139, p. 541–554.

Nixon, P. H., 1987, Kimberlitic xenoliths and their cratonic setting, in Nixon, P. H., ed., Mantle xenoliths: Chichester, UK, John Wiley and Sons, p. 215–239.

O'Reilly, S. Y., Griffin, W. L., Poudjom, Y. H., and Morgan, P., 2001, Are lithospheres forever? Tracking changes in subcontinental lithospheric mantle through time: GSA Today, v. 11, p. 4–10.

O'Reilly, S. Y., Nicholls, N., and Griffin, W. L., 1989, Xenoliths and megacrysts of mantle origin. in Johnson, R. W., ed., Intraplate volcanism in eastern Australia and New Zealand: Cambridge, UK, Cambridge University Press, p. 254–274.

Olafsson, M., and Eggler, D. H., 1983, Phase relation of amphibole, amphibole, carbonate and phlogophite carbonate peridotite: Petrological constrains on the asthenosphere: Earth and Planetary Science Letters, v. 64, p. 305–315.

Pollack, H. N., and Chapman, D. S., 1977, On the regional variation of the heat flow, geotherms, and lithospheric thickness: Tectonophysics, v. 38, p. 279–296.

Poudjom Djomani, Y. H., O'Reilly, S. Y., Griffin, W. L., and Morgan, P., 2001, The density structure of subcontinental lithosphere: Constraints on delamination models: Earth and Planetary Science Letters, v. 184, p. 605–621.

Shen, Q. H., and Gen, Y. S., 1991, The evolution phase division of early Precambrian crust, in Wu, J. S., Shen, Q. H., and Liu, D. Y. et al., eds., The gravelly events taken place in early Precambrian in North China Craton: Beijing, China, Geological Publication House, p. 1–8.

Wang, K. L., O'Reilly, S. Y., Griffin, W. L., Chung, S. L., and Pearson, N. J., 2003, Proterozoic mantle lithosphere beneath the extended margin of the South China block: In situ Re-Os evidence: Geology, v. 31, p. 709–712.

Wu, F. Y., Walker, R. J., Ren, X. W., Sun, D. Y., and Zhou, X. H., 2003, Osmium isotopic constraints on the age of lithospheric mantle beneath northeastern China: Chemical Geology, v. 196, p. 107–129.

Wu, J. S., and Gen, Y. S., 1991, The initial cognition on formation and evolution of deep lithosphere beneath North China craton, in Wu, J. S., Shen, Q. H., and Liu, D. Y., eds., The gravelly events taken place in early Precambrian in North China Craton: Beijing, China: Geological Publication House, p. 98–108.

Wyllie, P. J., 1989, Transfer of subcratonic carbon into kimberlites and rare earth carbonatites: Geochemical Society Publication, v. 1, p. 107–119.

Xu, J. W., Ma, G. F., Tong, W. X., Zhu, G., and Lin, S. F., 1993, Displacement of the Tancheng-Lujiang wrench fault system and its geodynamic setting in the northwestern circum-Pacifc, in Xu, J. W., ed., The Tancheng-Lujiang wrench fault system: New York, NY, John Wiley and Sons, p. 51–76.

Xu, W., Chi, X., Yuan, C., Huang, Y., and Wang, W., 1993, Mesozoic diorite and its xenolith from the interior of the North China platform: Beijing, China, Geological Publication House, p. 125 (in Chinese).

Xu, X., O'Reilly, S. Y., Griffin, W. L., and Zhou, X., 1998, The nature of the Cenozoic lithosphere at Nushan, eastern China, in Flower, M., Chung, S. L., Lo, C. H., and Lee, Y. Y., eds., Mantle dynamics and plate interactions in east Asia: Washington, DC, American Geophysical Union, Geodynamics Series 27, p. 167–196.

Xu, X., O'Reilly, S. Y., Griffin, W. L., and Zhou, X., 2000, Genesis of young lithospheric mantle in southeastern China: A LAM-ICPMS trace element study: Journal of Petrology, v. 40, p. 111–148.

Xu, Y. G., 2001, Thermo-tectonic destruction of the Archean lithospheric keel beneath the Sino-Korean craton in China: Evidence, timing, and mechanism: Physics and Chemistry of the Earth, v. 26, p. 747–758.

Xu, Y. G., Menzies, M. A., Mattery, D. P., and Lowry, D., 1996, The nature of the lithospheric mantle near the Tancheng-Lujiang fault, China: An integration of texture, chemistry, and isotopes: Chemical Geology, v. 134, p. 67–81.

Yang, J. J., and Jahn, B. M., 2000, Deep subduction of mantle-derived garnet peridotites from the Su-Lu UHP metamorphic terrane in China: Journal of Metamorphic Geology, v. 18, p. 167–180.

Yin, A., and Nie, S. Y., 1993, An indentation model for the North and South China collision and the development of Tanlu and Honam fault systems, Eastern Asia: Tectonics, v. 12, p. 801–813.

Yuan, X. C., 1996a, Atlas of geophysics in China: Beijing, China, Geological Publishing House, p. 60.

Yuan, X. C., 1996b, Velocity structure of the Qinling lithosphere and mushroom cloud model: Science in China (Ser. D), v. 39, p. 235–244.

Zhang, H. F., Sun, M., Zhou, X. H., Fan, W. M., Zhai, M. G., and Yin, J. F., 2002, Mesozoic lithosphere destruction beneath the North China Craton: Evidence from major-, trace-element, and Sr-Nd-Pb isotope studies of Fangcheng basalts: Contributions to Mineralogy and Petrology, v. 144, p. 241–253.

Zhang, H. F., Sun, M., Zhou, X. H., Zhou, M. F., Fan, W. M., and Zheng, J. P., 2003, Secular evolution of the lithosphere beneath the eastern North China craton: Evidence from Mesozoic basalts and high-Mg andesites: Geochimica et Cosmochimica Acta, v. 15, p. 4373–4387.

Zhao, G. C., Cawood, P. A., Wilde, S. A., Sun, M., and Lu, L. Z., 2000, Metamorphism of basement rocks in the Central Zone of the North China craton: Implications for Paleoproterozoic tectonic evolution: Precambrian Research, v. 103, p. 55–88.

Zhao, G. C., Wilde, S. A., Cawood, P. A., and Lu, L. Z., 1998, Thermal evolution of Archean basement rocks from the eastern part of the North China craton and its bearing on tectonic setting: International Geology Review, v. 40, p. 706–721.

Zheng, J. P., 1999, Mesozoic–Cenozoic mantle replacement and lithospheric thinning beneath eastern China: Wuhan, China, China University of Geosciences Press, 126 p. (in Chinese).

Zheng, J. P., and Lu, F. X., 1999, Mantle xenoliths from kimberlites, Shandong and Liaoning: Paleozoic lithospheric mantle character and its heterogeneity: Acta Petrologica Sinica, v. 15, p. 65–74 (in Chinese).

Zheng, J. P., Griffin, W. L., O'Reilly, S. Y., Lu, F. X., Wang, C. Y., Zhang, M., Wang, F. Z., and Li, H. M., 2004a, 3.6 Ga lower crust in central China: New evidence on the assembly of the North China craton: Geology, v. 32, p. 229–232.

Zheng, J. P., Griffin, W. L., Zhang, R. Y., Liou, J. G., and O'Reilly, S. Y., 2003a, Comparison on geochemistry of minerals in garnet peridotites from Paleozoic and Neocene xenoliths and from Triassic UHP terrane to constrain the lithospheric evolution of east China [abs.]. American Geophysical Union, San Francisco, AGU Meeting Abstract.

Zheng, J. P., O'Reilly, S. Y., Griffin, W. L., Lu, F. X., and Zhang, M., 1998, Nature and evolution of Cenozoic lithospheric mantle beneath Shandong peninsula, Sino-Korean craton: International Geology Review, v. 40, p. 471–499.

Zheng, J. P., O'Reilly, S. Y., Griffin, W. L., Lu, F. X., Zhang, M., and Pearson, N. J., 2001, Relics of refractory mantle beneath the eastern North China block: Significance for lithosphere evolution: Lithos, v. 57, p. 43–66.

Zheng, J. P., O'Reilly, S. Y., Griffin, W. L., Zhang, M., and Lu, F. X., 2004b, Nature and evolution of Mesozoic–Cenozoic mantle beneath the Cathaysia block, SE China: Lithos, v. 74, p. 41–65.

Zheng, J. P., Zhang, R. Y., Liou, J. G., Griffin, W. L., and O'Reilly, S. Y., 2003b, A heterogeneous and metasomatic mantle recorded by trace elements of minerals from the Donghai garnet peridotite in the Sulu UHP terrane, China: Chemical Geology, in review.

Zhi, X. C., Chen, D. G., and Wei, C., 1994, The neodymium and strontium isotopic compositions of Cenozoic alkalic basalts from Penglai and Lingqu, Shandong Province: Geological Review, v. 40, p. 526–533 (in Chinese).

Geochronological Constraints on the Paleoproterozoic Evolution of the North China Craton: SHRIMP Zircon Ages of Different Types of Mafic Dikes

PENG PENG,[1] MINGGUO ZHAI, HUAFENG ZHANG, AND JINGHUI GUO

Institute of Geology and Geophysics, Chinese Academy of Sciences, Beijing 100029, China

Abstract

Widespread magmatic and metamorphic events during the interval 2350–1650 Ma suggest that the North China craton (NCC) may have been involved in the evolution of the supercontinent Columbia. Metamorphosed and unmetamorphosed dikes have been characterized in terms of their geochemistry and geochronology. Dike suite 1 in the northern Wutai-Fuping terrane comprises amphibolite-facies assemblages and has a SHRIMP U-Pb zircon crystallization age of 2147 ± 5 Ma. Dike suite 2, distributed in the northern part of the Huai'an-Fengzhen terrane, has a two-pyroxene granulite assemblage, and yields a SHRIMP metamorphic zircon age of 1929 ± 8 Ma. Dike suite 3 in the Sanggan structural zone between the two terranes is composed of garnet two-pyroxene granulites, and has a SHRIMP zircon age of 1973 ± 4 Ma for the cores and 1834 ± 5 Ma for the rims, defining the time of crystallization and peak metamorphism, respectively. Dike suites 1 and 2 were possibly emplaced close to a continental margin and an arc respectively; whereas dike suite 3 was most likely post-orogenic. Zircon grains from an unmetamorphosed mafic dike in the north-central NCC yields a SHRIMP crystallization age of 1778 ± 3 Ma. We suggest that the metamorphosed mafic dike suites probably resulted from the amalgamation of the NCC in the Columbia supercontinent between 2080 and 1980 Ma (~2000 Ma), whereas the unmetamorphosed mafic dike swarms probably was emplaced during its break-up at 1780–1750 Ma. The metamorphosed dikes were likely uplifted and exhumed during a plume-driven upwelling event during 1830–1750 Ma, causing intrusion of (unmetamorphosed) dikes throughout the NCC.

Introduction

MANY RECENT STUDIES show that the North China craton (NCC) was probably involved in the amalgamation and subsequent break-up of a supercontinent in the Paleoproterozoic (Lu et al., 2002; Rogers and Santosh, 2002; Wilde et al., 2002; Zhao et al., 2002a, 2002b; Zhai et al., 2003). This supercontinent, termed Columbia by Rogers and Santosh (2002), apparently drifted apart at about 1.8 Ga (Zhao et al., 2002c). However, constraints for the involvement of the NCC with Columbia are limited. The NCC was thought to be composed of microcontinental blocks (Bai et al., 1993; Wu et al., 1998; Zhai et al., 2000), although it has recently been suggested to consist of two main components, the eastern and western blocks, separated by an intervening central orogenic belt (Zhao et al., 2001; Kusky et al., 2001; Li et al., 2002; Kusky and Li, 2003; Zhai et al., 2003).

The time of amalgamation of the NCC is nevertheless controversial. Some researchers suggest that Archean blocks did not collide until the Paleoproterozoic, primarily based on high-pressure (HP) granulite-facies metamorphism dated at 1800–1900 Ma (Wu and Zhang, 1998; Guo and Zhai, 2001; Zhao et al., 2001, 2002a; Kröner et al., 2002, 2004; Wilde et al., 2002; Guo et al., 2004). In contrast, several studies suggest that the blocks may have collided at ~2.5 Ga and were involved in the Paleoproterozoic evolution as a uniform block (Kusky et al., 2001; Li et al., 2002; Kusky and Li, 2003; Zhai and Liu, 2003). The common presence of 2350–1650 Ma magmatic rocks in the NCC is important for understanding the evolution of the NCC in the Paleoproterozoic (Zhao et al., 1993; Zhai et al., 2003). Many geochronological studies have been performed on granites, rapakivi granites, anorthosites, mafic dikes, and volcanic rocks exposed in the NCC (e.g., Qian and Chen, 1987; Zhang et al., 1994; Halls et al., 2000; Zhao et al., 2002c; Wilde et al., 1997, 2002, 2004; Kröner et al.,

---

[1]Corresponding author; email: pengpengwj@mail.igcas.ac.cn

2002, 2004). We here present new age data for four dike suites and discuss the Paleoproterozoic crustal evolution of the NCC.

Geological Background of the Host Units

Figures 1A and 1B show the study area in the central NCC where three geological units have been subdivided, including, from south to north, the Wutai-Fuping terrane (WFT), the Sanggan structural zone (SSZ), and the Huai'an-Fengzhen terrane (HFT). The WFT is comprised of the Wutai and Fuping complexes. The Wutai Complex is characterized by greenschist- to amphibolite-facies rocks, including the Wutai and Hutuo groups. The Wutai Group is a mafic-dominated volcanic-sedimentary series, including several old pre-tectonic granitoid plutons, such as the Lanzhishan, Ekou, and Chechang-Beitai granites that yield SHRIMP zircon ages of 2537 ± 10, 2555 ± 6 to 2566 ± 13, and 2546 ± 3 to 2542 ± 7 Ma, respectively (Wilde et al., 1997; Kröner et al., 2002). The Hutuo Group comprises a thick medium to low greenschist facies metamorphic volcanic-sedimentary sequence (Cheng and Zhang, 1982); a felsic tuffaceous rock contains two SHRIMP zircon populations at 2180 ± 5 and 2087 ± 9 Ma, respectively (Wilde et al., 2004). The younger date is interpreted to be the age of the volcanism, whereas the older date is considered to be age of a crustal magmatic source, from which the Dawaliang granite was derived. The Fuping Complex is composed of upper amphibolite– to granulite-facies rocks. SHRIMP zircon ages reveal that the WFT is characterized by the emplacement of major granitoid bodies at ~2.50–2.48 Ga, deposition of supracrustal rocks in the Paleoproterozoic, intrusion of granitic bodies at ~2.1–2.0 Ga, and regional metamorphism at ~1.875–1.802 Ga (Guan et al., 2002; Zhao et al., 2002b).

The HFT can be divided into the Fengzhen khondalites and the Huai'an Complex. The Fengzhen khondalites are composed of paragneisses containing garnet and sillimanite. Guo et al. (2001a) suggested that sedimentary protoliths of the khondalite series formed before about 2.0 Ga, and were subjected to granulite-facies metamorphism at about 1.87 Ga. Some 1800–1900 Ma S-type garnet granites are linearly aligned in them (Zhai et al., 2003). The Huai'an Complex is dominated by 2.7 to 2.8 Ga grey gneisses and 2.5 Ga granitic gneisses, with less common lenses and thin-bedded supracrustal rocks (Zhai, 1996).

The NE-trending SSZ, separating the WFT and HFT, is also known as a granite and high-pressure belt/zone (Zhai et al., 1992; Guo et al, 1996; Zhai and Liu, 2003). It is about 400 km long and composed of numerous deformed granitic gneiss bodies and granite sills and veins with/without gneissosity, parallel to the regional deformation. The main granites in SSZ formed during 2510–2370 Ma; while the ductile deformation resulting in mylonitic textures were dated at 1900–1800 Ma (Zhang et al., 1994; Zhang, 1997; Zhai et al., 2003).

Mafic Dike Suites

Metamorphosed mafic dikes constitute three contrasting suites with several features in common. They mostly range from several meters to several hundreds of meters in width, and from several hundred meters to several kilometers in length. They are mainly meta-gabbroic or diabase rocks. Chemically, the rocks have a tholeiitic composition (Figs. 2A and 2B), fractionated rare earth element (REE) patterns exhibiting light REE enrichment, depletion in high field strength (HFS) elements, and variation in large ion lithophile (LIL) elements (Zhai et al., 2001; Huang et al., 2001; Peng et al., 2004 and unpubl. data).

Amphibolites in the northern WFT (dike suite 1)

Numerous deformed NE to ENE oriented dikes/ layers occur in the Wutai Complex (Fig. 1B, Table 1). They underwent amphibolite-facies metamorphism, which corresponds to the maximum metamorphic grade of the Wutai Complex. The dikes are composed mainly of amphibole and plagioclase, with minor biotite, chlorite, and quartz. Relict microtextures and geochemistry show that the protoliths were gabbros and diabases. In geochemistry discrimination diagrams (Figs. 2C and 2D), they plot approximately in the field of within-plate basalts. A provisional U-Pb SHRIMP zircon age of ~1900 Ma has been reported (unpubl. data, cited in Kröner et al., 2004).

Two-pyroxene granulites in the northern HFT (dike suite 2)

Lenses of metagabbroic dikes are abundant in the northern HFT, mainly in the garnet granite belt of the Fengzhen khondalites (Fig. 1B, Table 1). These lenses (original dikes) trend NE-ENE parallel to the regional gneissosity. The mafic-intermediate granulite lenses are composed mainly of

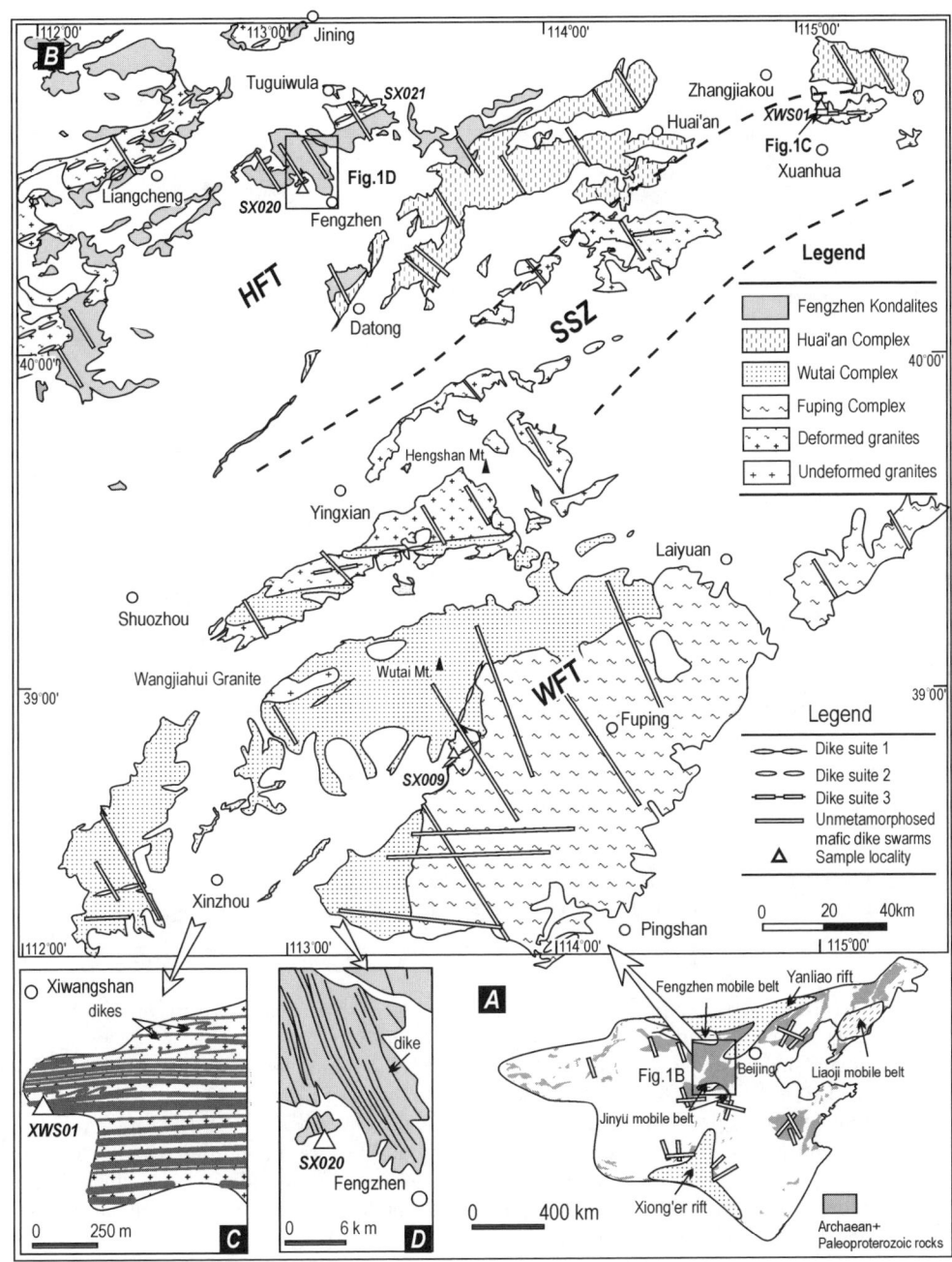

FIG. 1. A. Overview of the NCC. B. Simplified geological map of the study area, showing petrographical provinces and distribution of different dike suites. C. Field sketch map near Xiwangshan village. D. Geological map near Fengzhen county.

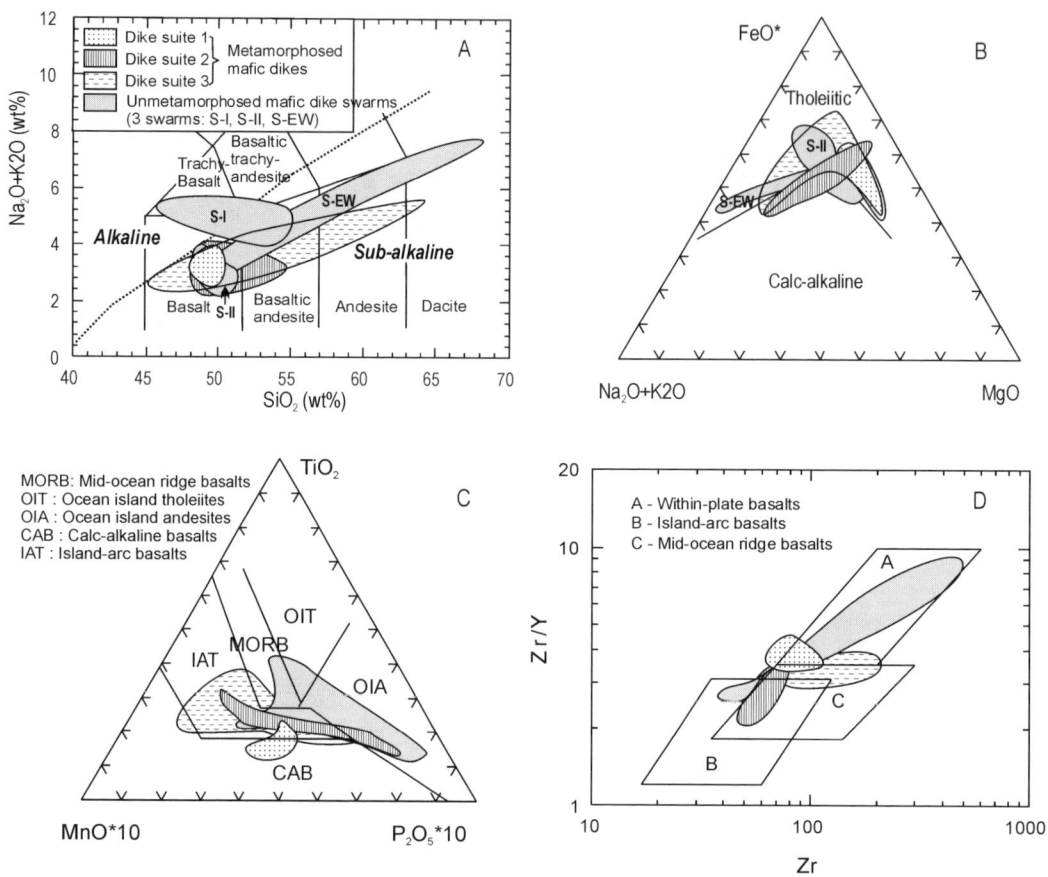

FIG. 2. A and B. Rock classification diagrams. C. TiO_2-$(MnO*10)$-(P_2O_5) discrimination diagram of Mullen (1983). D. Zr/Y-Zr discrimination diagram of Pearce and Cann (1973). Data source is the same as that in Table 1.

orthopyroxene, clinopyroxene, plagioclase, and minor quartz. These rocks underwent granulite-facies metamorphism under conditions of 800–900°C and 9.2–11 kbar (Zhai et al., 2003). Their $\varepsilon_{Nd}(t)$ values range from –1 to –6, showing a change of mantle evolution trend from relative depletion in the Neoarchean [$\varepsilon_{Nd}(t)$: 1–5] to more enrichment in the Paleoproterozoic; these metagabbros are thought to be related to extension and upwelling of the lower crust and mantle (Zhai and Liu, 2003). Discrimination diagrams (Figs. 2C and 2D) show that these dikes are possibly arc associated. Their T_{DM} ages are clustered between 2200 Ma and 1850 Ma (Zhai et al., 2003). A conventional multigrain chemical zircon U-Pb method age of 1921 ± 1 Ma, interpreted as the magmatic age was previously reported (Guo et al., 2001a).

Garnet two-pyroxene granulites in the SSZ (dike suite 3)

ENE-oriented lenses and dikes of HP granulites occur in the SSZ and are oriented parallel to the regional gneissosity. About 20 nearly E-trending vertical dikes were identified within a 1000 m section from 0.1 to 3 m wide and more than 1000 m long south of Xiwangshan town, Xuanhua county (Figs. 1B and 1C). The mineral assemblage of the granulite-facies dikes is garnet + orthopyroxene + clinopyroxene + hornblende + plagioclase; most of them are typical HP granulites (Guo et al., 2004). Discrimination diagrams of Figures 2C and 2D indicated island arc affinities. Two age groups, 2600–2400 and 1900–1800 Ma, have been reported for HP granulite-facies rocks (Guo et al., 1993, 2001a, 2001b, 2004; Shen et al., 1994; Zhai, 1996;

TABLE 1. Concise Characteristics of the Selected Dike Suites from the NCC[1]

Suites:	Dike suite 1		Dike suite 2		Dike suite 3		Unmetamorphosed mafic dikes			
Distributed in:	Northern WFT		Northern HFT		SSZ		Throughout NCC			
Orientation:	NE to ENE, locally cutting regional gneissosity		NE to ENE, parallel to regional gneissosity		E-W to ENE, parallel to regional gneissosity		Radiating NNW-NW to NE, and some NW and E-W, cutting prior structure lines			
Occurrence:	Deformed dikes and layers		Deformed lens and dikes		(Sub) Vertical dikes, or deformed lens (dikes)		(Sub)Vertical dikes, undeformed			
Metamorphism facies:	Amphibole facies		Granulite facies		MP-HP granulite facies		Unmetamorphosed			
Mineralogy:	Hbl + Pl + Chl ± Ilm ± Qtz		Opx + Cpx + Pl ± Qtz		Gt + Opx + Cpx + Hbl + Pl		Cpx + Pl ± Qtz, etc.			
	Average[2]	SX009	Average	SX021	Average	XWS01	S-I	S-II	S-EW	SX020
Chemical composition										
SiO_2	49.04	48.75	52.76	55.03	51.63	48.22	49.80	50.35	55.79	47.17
TiO_2	0.85	0.55	1.24	1.47	1.22	2.20	1.60	2.57	1.58	3.25
Al_2O_3	15.33	15.83	15.69	17.27	14.65	12.66	14.10	13.03	14.08	13.42
Fe_2O_3	13.59	13.07	12.07	9.89	13.33	20.11	14.21	15.28	10.65	15.54
MnO	0.18	0.17	0.16	0.12	0.20	0.27	0.20	0.20	0.15	0.20
MgO	7.66	7.98	5.23	4.15	4.83	4.51	6.07	4.07	3.92	4.62
CaO	8.04	7.58	7.88	6.84	8.42	9.10	9.77	7.40	6.19	7.96
Na_2O	2.39	2.89	2.61	2.35	2.68	2.24	2.22	2.67	2.37	2.48
K_2O	0.88	0.40	1.14	0.68	0.98	0.60	0.73	2.25	2.69	2.59
P_2O_5	0.17	0.14	0.41	0.63	0.17	0.37	0.17	1.01	0.49	1.86
LOI	1.76	2.52	0.41	0.68	0.32	−0.67	0.84	0.69	1.50	0.29
Total	99.87	99.86	99.72	99.11	98.42	99.61	99.79	99.43	99.41	99.39
CIPW norms										
Qtz	–	0	5.20	14.97	3.79	1.44	2.55	1.68	10.60	–
Or	5.37	2.46	6.87	4.12	5.98	3.6	13.65	4.42	16.41	15.68
Ab	20.84	25.39	22.48	20.36	23.37	19.22	23.15	19.22	20.66	21.45
An	29.36	30.14	28.22	31.01	25.8	23.03	17.32	26.96	20.41	18.21
Crn	–	–	–	1.60	–	–	–	–	–	–
Di wo	4.50	3.38	3.84	–	6.8	8.6	5.96	9.06	3.47	4.61
Di en	2.33	1.79	1.82	–	2.94	2.9	2.33	4.3	1.57	1.98
Di fs	2.04	1.49	1.96	–	3.85	5.97	3.7	4.63	1.87	2.63
Hy en	13.60	12.02	11.5	10.63	9.52	8.55	8.11	11.25	8.53	8.01
Hy fs	11.89	9.96	12.37	11.02	12.47	17.61	12.87	12.12	10.12	10.66
Ol fo	2.68	4.84	–	–	–	–	–	–	–	1.28
Ol fa	2.58	4.43	–	–	–	–	–	–	–	1.88
Mag	2.77	2.68	2.43	2.00	2.72	4.04	3.1	2.88	2.17	3.14
Ilm	1.67	1.09	2.40	2.86	2.39	4.24	5.01	3.11	3.1	6.32
Ap	0.38	0.32	0.91	1.41	0.38	0.82	2.26	0.38	1.1	4.15

[1]Mineral symbols as in Kretz (1983) except Hy = hypersthene.
[2]Average = average value of the corresponding suite, with 4, 4, 16, 11, 37, 10 analyzed samples for suite 1, 2, 3, S-I, S-II, and S-EW, respectively. Except for 14 data samples of suite 3 from Huang et al. (2001), all other data are from Peng et al. (2004) and the authors' own unpublished data.

Mao et al., 1999; Zhao, 2001; Kröner et al., 2004; S. Wilde, pers. commun., April 17, 2004). The ages of 2600–2400 Ma were mostly obtained by the Sm-Nd whole-rock isochron method, e.g., an age of 2402 ± 73 Ma (Shen et al., 1994) and 2647 ± 115 Ma (Guo et al., 1993); in contrast, the ages of 1900–1800 Ma include zircon U-Pb ages, e.g., a zircon U-Pb age of 1833 ± 23 Ma, a garnet Ar-Ar isochron age of 1862 ± 37 Ma (Guo et al., 2001b), and SHRIMP zircon ages of 1827 ± 4 (S. Wilde, unpubl. data, cited by Zhao, 2001) and 1817 ± 12 to 1819 ± 16 Ma (Guo et al., 2004). The two groups were interpreted as the ages of the source rocks and the peak metamorphism, respectively, by some researchers (e.g., Guo et al., 2001a, 2001b), two metamorphic events (e.g., Li et al., 1996), or multiple episodes of HP granulite-facies metamorphism by others (e.g., Zhong, 1999). A metamorphic SHRIMP U-Pb zircon age of 1819 ± 16 Ma for a dike was reported by Guo et al. (2004). They also noted a $^{207}Pb/^{206}Pb$ age of 2237 ± 17 Ma for the core of a zircon, interpreted as inheritance from igneous country rocks. However, the magmatic age of the dikes was not well constrained.

Unmetamorphosed mafic dike swarms throughout the NCC

Approximately 1.8 Ga unmetamorphosed mafic dikes are widespread in the NCC, with a density of several to tens of dikes per km (Figs. 1B, 1D, Table 1; Chen and Shi, 1983; Qian and Chen, 1987; Zhang et al., 1994; Hou et al., 1998, 2001; Halls et al., 2000; Li et al., 2001; Peng et al., 2004). The dikes are vertical to subvertical intrusions with chilled margins. Individual dikes are up to 60 km long and 0.5 to 100 m wide. They have NW-NNW and E-W orientations in the west and center, and NE-ENE and NW orientations in the east. They consist of clinopyroxene, plagioclase, and accessory hornblende, Fe-Ti oxides, biotite, alkali-feldspar, apatite, and quartz or olivine.

Unmetamorphosed dikes have both alkali and tholeiitic compositions (Figs. 2A and 2B), showing high contents of total REE and light to moderate light REE enrichment. They are relatively enriched in LIL (except for Sr) and are depleted in HFS elements. Geochemistry discrimination diagrams reveal that the dike swarms are of within-plate basalts, showing affinities for ocean island basalts/andesites (Figs. 2C and 2D). Three swarms of dikes including two NNW-trending swarms (S-I, S-II) and an E-W-trending one (S-EW) have been identified, i.e.: S-I comprising relatively low FeO(total)-TiO_2-P_2O_5 contents and tholeiite compositions; S-II with relatively high FeO(total)-TiO_2-P_2O_5 contents and alkaline–calc-alkaline compositions; and the third group including high-Fe tholeiitic basalts and andesites (Peng et al., 2004). These three swarms show distinct differentiation trends with different degrees of crustal assimilation. An anorogenic origin possibly associating with a plume centered near the south margin of the NCC has been suggested. Reported ages include a K-Ar whole-rock isochron age of 1760 Ma (Zhuang et al., 1997) and a new 1830 ± 17 Ma SHRIMP zircon U-Pb age (G.-T. Hou, pers. commun., June 1, 2004) from the Taishan area, Sm-Nd whole rock isochron ages of 1729–1759 Ma from northern Hebei (Li, 1999), scattered K-Ar early to middle Proterozoic ages from the central and western NCC (Qian and Chen, 1987; Zhang et al., 1994; Hou et al., 1998), and a single-zircon U/Pb age of 1769 ± 3 Ma (n = 5) for a dike from the central NCC (Li et al., 2001).

SHRIMP Zircon Analytical Techniques

Zircons from four dike samples were separated using standard heavy liquid and magnetic techniques, and then were handpicked under a binocular microscope. Zircon grains were mounted in epoxy resin with standard Temora zircon (TEM zircon, 417 Ma). The sample mount was polished to expose the centers of zircon grains, and then were gold coated. Transmission and reflection microscope images were taken to obtain information about the shape of grains and their position in the mount. Backscattered electron (BSE) and cathodoluminescence (CL) images were performed on the zircon grains at the electronic microprobe laboratory of the Institute of Mineral Resources, Chinese Academy of Geology Sciences (CAGS), to examine their internal structures; the textures provide important information regarding zircon genesis and growth history, and thus are useful for geochronological interpretation (e.g., Pidgeon, 1992). The analyses were performed on the SHRIMP II at the Beijing SHRIMP Center, Institute of Geology, CAGS. Operation and data processing followed Williams (1998). The radius of the ion target is about 20–30 μm. Mass resolution was about 5400 (1% peak height). Common Pb was corrected using the determined ^{204}Pb. All data were processed using the Squid and Isoplot programs (Ludwig, 1999). The data in Table 2 are given in 1-sigma errors with weighted mean at the 95% confidence level.

Samples and Results

Dike sample SX009 (amphibolite)

Sample SX009 was collected from a dike exposed near Hengling village, south of the Wutai Mountains (38°46' N, 113°37' E; Fig. 1B). Zircon grains from this sample are brownish and translucent, and mostly shorter than 150 μm (Fig. 3A). These zircons have U contents of 259–728 ppm, Th contents of 226–793 ppm, and Th/U ratios of 0.81–1.43 (Table 2, Fig. 4).

Fourteen SHRIMP analysis spots were measured. Except for spot 5, all analyses give concordant or nearly concordant ages; for instance, spots 2, 8, and 14 give concordant ages of 2141 ± 10 Ma, 2156 ± 12 Ma, and 2148 ± 8 Ma, respectively (Table 2). The 13 concordant analyses yield a weighted mean $^{207}Pb/^{206}Pb$ age of 2147 ± 5 Ma (MSWD = 0.75; Table 2, Fig. 5A).

Dike sample SX021 (two-pyroxene granulite)

Sample SX021 was collected from Xuwujia village, Tuguiwula county (40°44' N, 113°15' E; Fig. 1B). Zircon grains are round and multifaceted, typical of a metamorphic origin. The grain size is commonly larger than 100 μm (Fig. 3B). Analyzed zircons have U contents of 50–454 ppm, Th contents of 33–986 ppm, and Th/U ratios of 0.68–2.23 (Table 2, Fig. 4).

Fifteen analysis spots (excluding spot 6) on different zircon grains give concordant or nearly concordant U-Pb ages and these yield a weighted mean $^{207}Pb/^{206}Pb$ age of 1929 ± 8 Ma (MSWD = 1.0) (Table 2, Fig. 5B). Spot 1 gave the oldest concordant age of 1947 ± 17 Ma, whereas spot 3 gave the youngest age of 1891 ± 18 Ma. Another concordant age of 1924 ± 14 Ma is similar to the mean $^{207}Pb/^{206}Pb$ age (Table 2). All zircon grains from our study (Fig. 3B) are of metamorphic origin, or were completely recrystallized, obliterating old cores. Therefore, we suggest that the age of 1929 ± 8 Ma represents the time of granulite-facies metamorphism. However, a 1921 ± 1 Ma age reported by Guo et al. (2001a) was interpreted as a magmatic age.

Dike sample XWS01
(garnet two-pyroxene granulite)

Sample XWS01 was collected from a garnet two-pyroxene granulite dike (about 2 m wide and 2000 m long) in the SSZ south of Xiwangshan village, Xuanhua county (40°43' N, 115°3' E; Figs. 1B and 1C). The zircons are round, multifaceted, and commonly larger than 150 μm (Fig. 3C). Most grains contain several BSE domains constituting a pattern of inner core and outer rim. U contents of the core domains range from 352 to 1968 ppm, Th contents from 254 to 1929 ppm, and Th/U ratios of 0.23–1.14. The rim domains generally contain lower U from 342 to 1074 ppm, lower Th from 153 to 372 ppm, and have Th/U ratios of 0.32–0.49 (Table 2, Fig. 4). This phenomenon implies different origins for core and rim domains.

Twenty-three analysis spots on different zircon grains or domains give $^{207}Pb/^{206}Pb$ ages clustering in two groups around 1830 and 1970 Ma when plotted on a concordia diagram (Fig. 5C). Eleven analysis spots, measured on core domains, give older $^{207}Pb/^{206}Pb$ ages, but most of them are less concordant, except for concordant spots 7, 9, and 20 (Fig. 5F). Without spots 6, 13, and 22, the other eight analyses on core domains yield a weighted mean $^{207}Pb/^{206}Pb$ age of 1973 ± 4 Ma (MSWD = 1.4; Table 2, Figs. 3C and 5D). Both spots 13 and 22 give discordant ages. Spot 13 has the oldest $^{207}Pb/^{206}Pb$ age of 2001 ± 4 Ma, which may be due to inheritance. Spot 22 provided the youngest $^{207}Pb/^{206}Pb$ age of 1925 ± 8 Ma, which is similar to the metamorphic age of sample SX021 and probably represents that metamorphic event. The other 12 analysis spots, measured on the rim domains, provide more concordant U-Pb ages and yield a weighted mean $^{207}Pb/^{206}Pb$ age of 1834 ± 5 Ma (MSWD = 1.3; Table 2, Figs. 3C, 5E). Data for spot 6 probably represent a mixture of the core and rim domains. We interpret 1834 ± 5 Ma as the age of granulite-facies metamorphism and the 1973 ± 4 Ma age, as the crystallization age of the igneous precursor.

Dike sample SX020 (diabase)

Sample SX020 was collected from Fengzhen county (40°30' N, 113°2' E; Figs. 1B and 1D). Zircon grains are brownish and translucent, with some being prismatic, mostly less than 150 μm in length. BSE images show clear oscillatory zoning (Fig. 3D), implying magmatic origin (e.g., Vavra, 1994). U and Th contents are variable from 347 to 4698 ppm for U and from 61 to 12030 ppm for Th. Their Th/U ratios vary from 0.18 to 3.99 (Table 2, Fig. 4).

Fifteen analysis spots on different zircon grains were performed on this sample. Several analyses give reversely discordant U-Pb ages (Fig. 5D). Excluding spots 2, 9, and 13, the other 12 analyses yield a weighted mean $^{207}Pb/^{206}Pb$ age of 1778 ± 3

TABLE 2. SHRIMP U-Pb Zircon Data

Spot name	U, ppm	Th, ppm	Th/U	Rad. ^{206}Pb, ppm	^{204}Pb/^{206}Pb	^{207}Pb/^{206}Pb	^{208}Pb/^{206}Pb	^{206}Pb/^{238}U	^{208}Pb/^{232}Th	^{207}Pb/^{206}Pb age, Ma	Err. corr.
\multicolumn{12}{c}{SX009: Amphibolite, dike suite 1}
-1	399	466	1.17	92	9.5284E-05	0.1346 ± 0.0051	0.4507 ± 0.0040	0.8286 ± 0.0098	0.1121 ± 0.0202	2142 ± 10	0.97
-2	380	351	0.93	88	5.8623E-05	0.1340 ± 0.0050	0.3600 ± 0.0045	0.8344 ± 0.0078	0.1132 ± 0.0204	2141 ± 10	0.97
-3	300	246	0.82	72	1.3787E-04	0.1360 ± 0.0056	0.3191 ± 0.0143	0.9260 ± 0.0056	0.1173 ± 0.0253	2154 ± 11	0.96
-4	482	388	0.81	110	5.1426E-05	0.1347 ± 0.0044	0.3143 ± 0.0042	0.8072 ± 0.0079	0.1112 ± 0.0203	2151 ± 8	0.97
-5	728	793	1.09	83	2.0967E-04	0.1298 ± 0.0052	0.4146 ± 0.0044	0.5198 ± 0.0258	0.0546 ± 0.0206	2058 ± 12	0.97
-6	490	474	0.97	103	1.7835E-04	0.1342 ± 0.0081	0.3730 ± 0.0111	0.8386 ± 0.0071	0.1022 ± 0.0229	2123 ± 18	0.93
-7	438	388	0.89	100	7.0228E-05	0.1356 ± 0.0046	0.3442 ± 0.0043	0.8405 ± 0.0073	0.1114 ± 0.0204	2159 ± 9	0.97
-8	287	232	0.81	67	1.6319E-04	0.1366 ± 0.0056	0.3176 ± 0.0053	0.8222 ± 0.0065	0.1153 ± 0.0208	2156 ± 12	0.96
-9	359	417	1.16	83	1.4012E-04	0.1366 ± 0.0057	0.4483 ± 0.0043	0.8775 ± 0.0040	0.1121 ± 0.0204	2161 ± 11	0.96
-10	259	370	1.43	61	6.0025E-05	0.1335 ± 0.0060	0.5464 ± 0.0046	0.8388 ± 0.0065	0.1127 ± 0.0209	2134 ± 11	0.96
-11	637	636	1.00	146	5.6636E-05	0.1342 ± 0.0043	0.3840 ± 0.0040	0.8355 ± 0.0098	0.1100 ± 0.0200	2143 ± 8	0.98
-12	683	754	1.10	146	6.8347E-05	0.1344 ± 0.0039	0.4236 ± 0.0033	0.8232 ± 0.0029	0.1027 ± 0.0198	2144 ± 7	0.98
-13	275	226	0.82	65	5.2463E-05	0.1341 ± 0.0061	0.3196 ± 0.0057	0.8470 ± 0.0067	0.1158 ± 0.0211	2143 ± 12	0.96
-14	499	483	0.97	117	7.5098E-05	0.1348 ± 0.0043	0.3756 ± 0.0038	0.8368 ± 0.0098	0.1136 ± 0.0200	2148 ± 8	0.98
\multicolumn{12}{c}{SX021: Two-pyroxene granulite, dike suite 2}
-1	119	145	1.24	26	8.4950E-05	0.1206 ± 0.0093	0.4672 ± 0.0073	0.7247 ± 0.0065	0.1054 ± 0.0224	1947 ± 17	0.90
-2	50	33	0.68	11	1.0000E-32	0.1189 ± 0.0150	0.2532 ± 0.0165	0.7278 ± 0.0103	0.1001 ± 0.0299	1940 ± 27	0.84
-3	147	204	1.43	31	1.1568E-04	0.1173 ± 0.0088	0.5232 ± 0.0065	0.7544 ± 0.0061	0.0996 ± 0.0218	1891 ± 18	0.92
-4	141	206	1.49	31	1.4886E-04	0.1186 ± 0.0094	0.5629 ± 0.0069	0.7061 ± 0.0104	0.1042 ± 0.0235	1904 ± 24	0.90
-5	277	564	2.09	58	7.0605E-05	0.1185 ± 0.0066	0.7719 ± 0.0044	0.7353 ± 0.0046	0.0990 ± 0.0208	1920 ± 12	0.95
-6	109	152	1.42	24	8.8712E-05	0.1150 ± 0.0103	0.5266 ± 0.0076	0.7503 ± 0.0071	0.1043 ± 0.0228	1862 ± 19	0.89
-7	127	120	0.97	28	1.1517E-04	0.1203 ± 0.0095	0.3635 ± 0.0081	0.7314 ± 0.0074	0.1038 ± 0.0228	1937 ± 20	0.90
-8	106	114	1.10	23	1.7765E-04	0.1200 ± 0.0103	0.4196 ± 0.0083	0.7373 ± 0.0096	0.1060 ± 0.0231	1920 ± 22	0.89
-9	123	150	1.24	27	2.4113E-04	0.1213 ± 0.0092	0.4636 ± 0.0072	0.7255 ± 0.0082	0.1035 ± 0.0223	1927 ± 21	0.91
-10	180	195	1.11	39	6.7107E-05	0.1182 ± 0.0081	0.4170 ± 0.0066	0.7292 ± 0.0066	0.1027 ± 0.0219	1915 ± 15	0.93
-11	261	488	1.91	55	5.1838E-05	0.1201 ± 0.0067	0.7146 ± 0.0075	0.7216 ± 0.0046	0.0989 ± 0.0218	1948 ± 13	0.94
-12	119	158	1.37	25	9.9531E-05	0.1184 ± 0.0096	0.5195 ± 0.0072	0.7195 ± 0.0079	0.1019 ± 0.0223	1912 ± 19	0.90
-13	454	986	2.23	95	4.7198E-05	0.1197 ± 0.0050	0.8180 ± 0.0033	0.7394 ± 0.0087	0.0974 ± 0.0201	1942 ± 9	0.97
-14	144	168	1.20	32	6.3799E-05	0.1196 ± 0.0108	0.4523 ± 0.0078	0.7247 ± 0.0078	0.1054 ± 0.0232	1937 ± 20	0.88
-15	233	205	0.90	50	7.4695E-05	0.1188 ± 0.0069	0.3363 ± 0.0063	0.7251 ± 0.0047	0.1023 ± 0.0213	1924 ± 14	0.94
-16	99	98	1.02	21	1.3635E-04	0.1202 ± 0.0107	0.3854 ± 0.0154	0.7247 ± 0.0074	0.1033 ± 0.0266	1932 ± 22	0.88

XWS01: Garnet two-pyroxene granulite, dike suite 3									
−1	342	153	0.45	69	1.1366E-04	0.1117 ± 0.0060	0.6311 ± 0.0059	0.0946 ± 0.0211	1801 ± 13	0.96
−2	490	231	0.48	97	4.6154E-05	0.1131 ± 0.0049	0.6693 ± 0.0033	0.0934 ± 0.0204	1839 ± 9	0.97
−3	1074	338	0.32	210	2.1437E-05	0.1126 ± 0.0034	0.6519 ± 0.0078	0.0920 ± 0.0202	1838 ± 6	0.98
−4	431	182	0.42	89	4.3222E-05	0.1132 ± 0.0052	0.7030 ± 0.0039	0.0963 ± 0.0207	1842 ± 10	0.97
−5	608	290	0.49	122	2.4897E-05	0.1127 ± 0.0044	0.6963 ± 0.0030	0.0934 ± 0.0206	1838 ± 8	0.98
−6	352	270	0.78	77	1.0295E-04	0.1172 ± 0.0055	0.8811 ± 0.0040	0.1101 ± 0.0245	1893 ± 12	0.96
−7	1565	350	0.23	336	1.0516E-05	0.1213 ± 0.0026	0.8052 ± 0.0073	0.1047 ± 0.0214	1973 ± 5	0.99
−8	459	197	0.44	93	5.8235E-05	0.1116 ± 0.0052	0.6543 ± 0.0036	0.0960 ± 0.0207	1813 ± 10	0.97
−9	1771	1929	1.14	378	1.6930E-05	0.1212 ± 0.0024	0.7570 ± 0.0099	0.1036 ± 0.0195	1971 ± 4	0.99
−10	555	250	0.45	113	1.5248E-05	0.1124 ± 0.0047	0.6779 ± 0.0056	0.0969 ± 0.0209	1835 ± 9	0.97
−11	1847	1316	0.74	397	5.5835E-06	0.1214 ± 0.0024	0.8017 ± 0.0104	0.1036 ± 0.0248	1976 ± 4	0.99
−12	468	216	0.47	94	3.3550E-05	0.1121 ± 0.0051	0.6869 ± 0.0034	0.0952 ± 0.0207	1827 ± 10	0.97
−13	1968	1480	0.78	396	2.3602E-05	0.1234 ± 0.0023	0.7902 ± 0.0134	0.0948 ± 0.0244	2001 ± 4	0.99
−14	481	211	0.45	98	3.9238E-05	0.1131 ± 0.0067	0.6809 ± 0.0077	0.0948 ± 0.0205	1841 ± 13	0.94
−15	612	283	0.47	124	3.6267E-05	0.1131 ± 0.0044	0.6913 ± 0.0059	0.1084 ± 0.0202	1842 ± 8	0.98
−16	1703	482	0.29	349	3.2465E-05	0.1219 ± 0.0054	0.8156 ± 0.0138	0.1084 ± 0.0224	1978 ± 10	0.96
−17	585	254	0.45	121	4.0576E-05	0.1203 ± 0.0043	0.8106 ± 0.0069	0.1040 ± 0.0312	1953 ± 8	0.97
−18	1269	1315	1.08	269	2.2939E-05	0.1218 ± 0.0029	0.7819 ± 0.0116	0.1084 ± 0.0278	1978 ± 5	0.98
−19	1262	820	0.68	268	1.1805E-05	0.1211 ± 0.0030	0.7343 ± 0.0139	0.1042 ± 0.0210	1971 ± 5	0.99
−20	920	701	0.80	203	3.6410E-05	0.1222 ± 0.0037	0.9198 ± 0.0046	0.1062 ± 0.0209	1981 ± 7	0.98
−21	840	372	0.45	166	2.2834E-05	0.1122 ± 0.0039	0.6614 ± 0.0078	0.0929 ± 0.0211	1830 ± 7	0.98
−22	789	547	0.73	161	4.2253E-05	0.1185 ± 0.0041	0.7768 ± 0.0097	0.1045 ± 0.0303	1925 ± 8	0.97
−23	504	226	0.45	99	6.0905E-05	0.1132 ± 0.0051	0.6471 ± 0.0058	0.0927 ± 0.0206	1837 ± 10	0.97
SX020: Diabase (unmetamorphosed mafic dike swarm)									
−1	1106	565	0.52	223	3.5449E-05	0.1087 ± 0.0033	0.8106 ± 0.0064	0.0908 ± 0.0247	1769 ± 6	0.99
−2	1036	429	0.42	216	4.6128E-05	0.1082 ± 0.0032	0.9412 ± 0.0252	0.0886 ± 0.0250	1759 ± 6	0.99
−3	657	689	1.07	129	5.4715E-05	0.1099 ± 0.0044	0.7426 ± 0.0066	0.0912 ± 0.0207	1786 ± 9	0.97
−4	1007	1985	2.01	198	3.8755E-05	0.1099 ± 0.0036	0.7348 ± 0.0057	0.0902 ± 0.0211	1789 ± 7	0.98
−5	4228	10557	2.54	1009	7.6453E-06	0.1085 ± 0.0016	0.7719 ± 0.0170	0.1027 ± 0.0194	1773 ± 3	0.99
−6	1709	1946	1.17	366	4.4209E-05	0.1098 ± 0.0028	0.8732 ± 0.0025	0.0998 ± 0.0488	1786 ± 5	0.99
−7	347	61	0.18	68	1.8779E-04	0.1100 ± 0.0095	0.4318 ± 0.0294	0.0950 ± 0.0260	1757 ± 21	0.91
−8	2108	1185	0.57	493	2.1075E-05	0.1090 ± 0.0026	0.6744 ± 0.0175	0.1083 ± 0.0197	1778 ± 5	0.99
−9	991	3874	3.99	237	4.0710E-05	0.1108 ± 0.0037	1.1166 ± 0.0058	0.1066 ± 0.0258	1803 ± 7	0.98
−10	1384	2053	1.51	287	4.1443E-05	0.1096 ± 0.0031	1.0257 ± 0.0205	0.0976 ± 0.0200	1783 ± 6	0.98
−11	1506	1945	1.32	305	3.0426E-05	0.1092 ± 0.0032	1.4183 ± 0.0126	0.0893 ± 0.0196	1780 ± 6	0.99
−12	3133	7277	2.37	741	2.0872E-05	0.1077 ± 0.0048	0.5684 ± 0.0049	0.0982 ± 0.0217	1756 ± 9	0.97
−13	4698	12030	2.61	962	1.3388E-05	0.1084 ± 0.0019	0.4631 ± 0.0025	0.0894 ± 0.0196	1769 ± 3	0.99
−14	609	181	0.30	120	4.0578E-05	0.1100 ± 0.0048	0.7338 ± 0.0095	0.0940 ± 0.0219	1790 ± 9	0.97
−15	1648	1982	1.23	319	5.3530E-05	0.1095 ± 0.0030	0.7820 ± 0.0194	0.0923 ± 0.0213	1778 ± 6	0.98

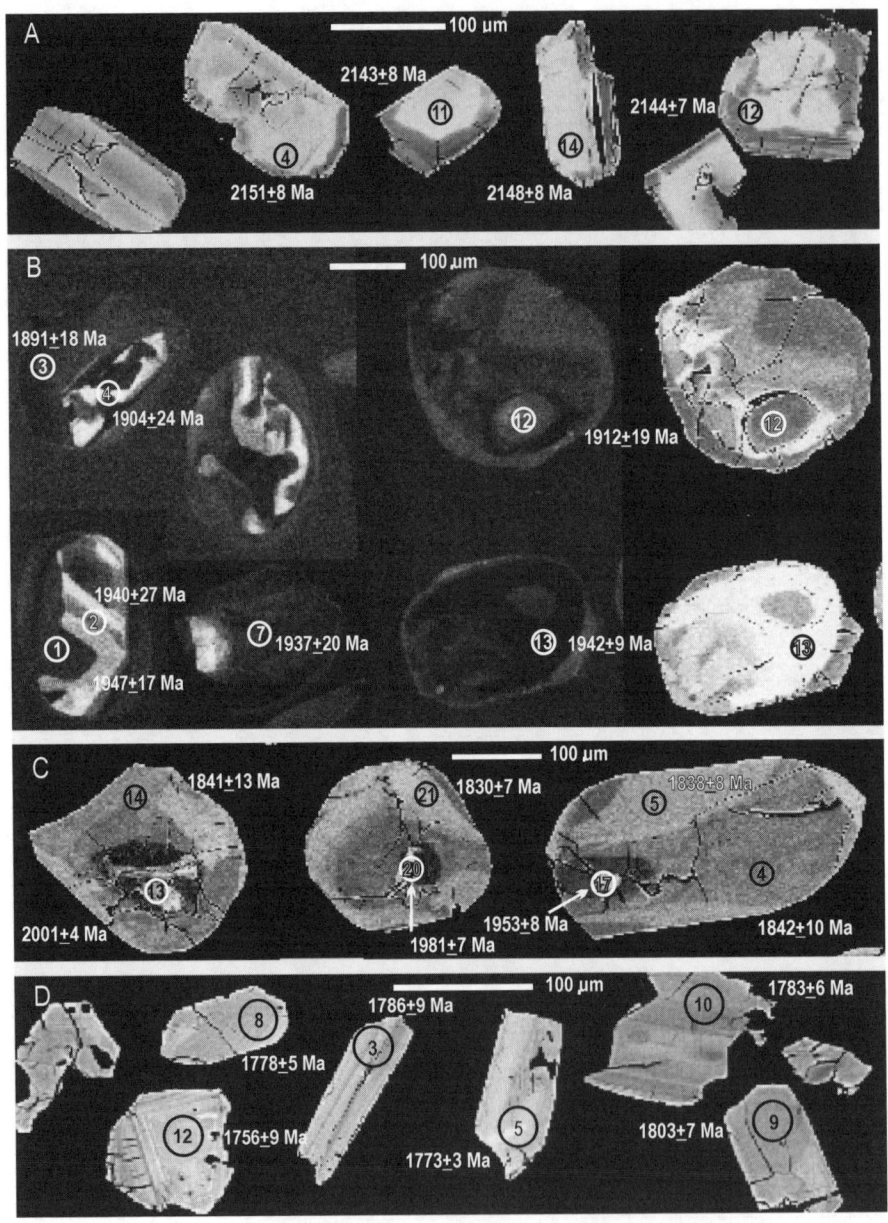

FIG. 3. Selected backscattered electron (BSE) and cathodoluminescence (CL) images of zircons. A. From sample SX009 (BSE). B. From SX021 (CL and BSE). C. From XWS01 (BSE). D. From SX020 (BSE).

Ma (MSWD = 1.7; Table 2, Figs. 3D and 5F). We take this to be the crystallization age.

Discussion and Tectonic Implications

There are numerous models regarding the Paleoproterozoic evolution of the NCC. Some geologists have suggested late Archaean collision models involving either uplift in an extensional tectonic regime at ~ 1.90–1.80 Ga (e.g., Li et al., 2000), or accretion of two orogenic belts in the north and east of the NCC respectively at ~2.20–1.85 Ga and subsequent rifting (Kusky and Li, 2003). For the Paleoproterozoic collision model, a major suturing event

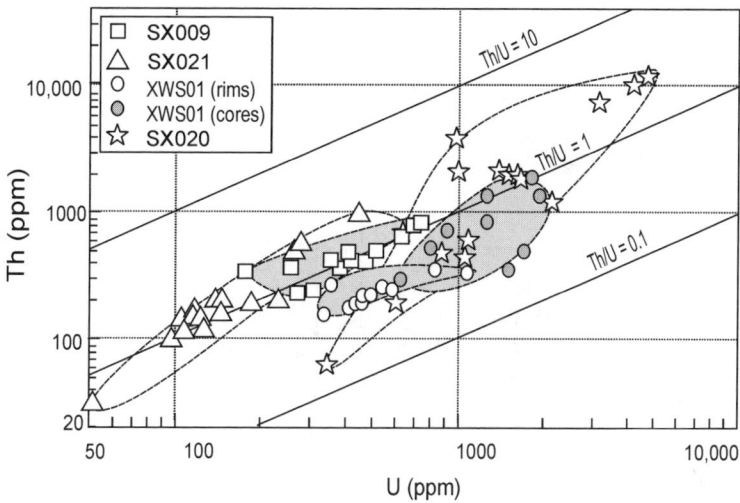

FIG. 4. Th and U contents of analyzed zircon grains from samples SX009, SX021, XWS01, and SX020.

at ~1.88–1.85 Ga constrained by the HP metamorphism ages (e.g. Zhao et al., 2001, 2002a; Guo et al., 2004; Kröner et al., 2004) is established; in a contrasting Paleoproterozoic model, a Fuping continental block and an arc terrane were postulated to have collided at ~2.05 Ga (Li et al., 1990). In the synthesis model, the NCC would have been cratonized at ~2.5 Ga, followed by its involvement in a global supercontinent at ~2.0 Ga and break-up between ~1.90 and 1.65 Ga (Zhai et al., 2003). In this model, mobile belts such as the Liaoji, Fengzhen, and Jinyü belts, are recognized (Fig. 1A). Mantle lithosphere peridotite with an isochron age of 1910 ± 220 Ma (MSWD = 35, initial γ_{Os} = 0.02 ± 0.79) sampled by Tertiary basalts in north-central China support an orogenic event around that time (Gao et al., 2002). In our opinion, a major collision possibly occurred between 2080 and 1980 Ma (~2000 Ma), constrained by: (1) the volcanism age of the Hutuo Group from Wilde et al. (2004) and the intrusive age of mafic dikes from this study, respectively; and (2) a mantle upwelling event at 1830–1750 Ma, resulting in the exhumation of the metamorphic dikes and the intrusion of mafic dikes throughout the NCC at 1780–1750 Ma (Figs. 6A, 6B, and 6C).

Zhao et al. (2001) summarized four distinct phases of deformation in the central NCC, represented as intrafolial, closed, rootless isoclinal folds (F_1); isoclinal folds accompanied by regional-scale ductile shear zones and thrusting (F_2); asymmetric upright folds (F_3); and a late phase as open folds. The main deformation leading to ductile fabrics is related to a major collision event (Kröner et al., 2002). Kusky and Li (2003) placed this suturing event at ~2.20–1.85 Ga on the northern margin of the NCC. There is no evidence that the dikes from Xiwangshan (Fig. 1C) experienced that deformation (F_2). Generally, dikes trend in the direction of maximum compressive stress and are oriented perpendicular to the minimum principle stress (e.g., Pollard, 1973; Féraud et al., 1987). So dikes should be perpendicular to the collision zone in pre-/syn-collisional environment, but parallel to the zone in a post-collisional extension environment. The mafic dike swarm (1973 Ma) in the orogenic zone is parallel to the SSZ and wall rock gneissosity. It seems to be intruded in an extensional setting (post-collisional) rather than before or contemporaneous to the collision (Fig. 6B). Hence the collision probably occurred before 1973 Ma. However, the 2087 ± 9 Ma age of Hutuo Group volcanism means that the deformation and metamorphism must have taken place after this time (Wilde et al., 2004); hence timing of collision was possibly between 2080 and 1980 Ma. ^{40}Ar-^{39}Ar spectra ages of 2017 ± 3 to 2025 ± 39 Ma (Zhao et al., 1993) and 1972 ± 3 Ma (Sang et al., 1996), and the ^{40}Ar-^{39}Ar isochron age of 1990 ± 22 Ma (Sang et al., 1996) reported from hornblende in late Archaean granulite in the northern NCC support a thermal event at ~2000 Ma, coincident with the collision age of 2080–1980 Ma.

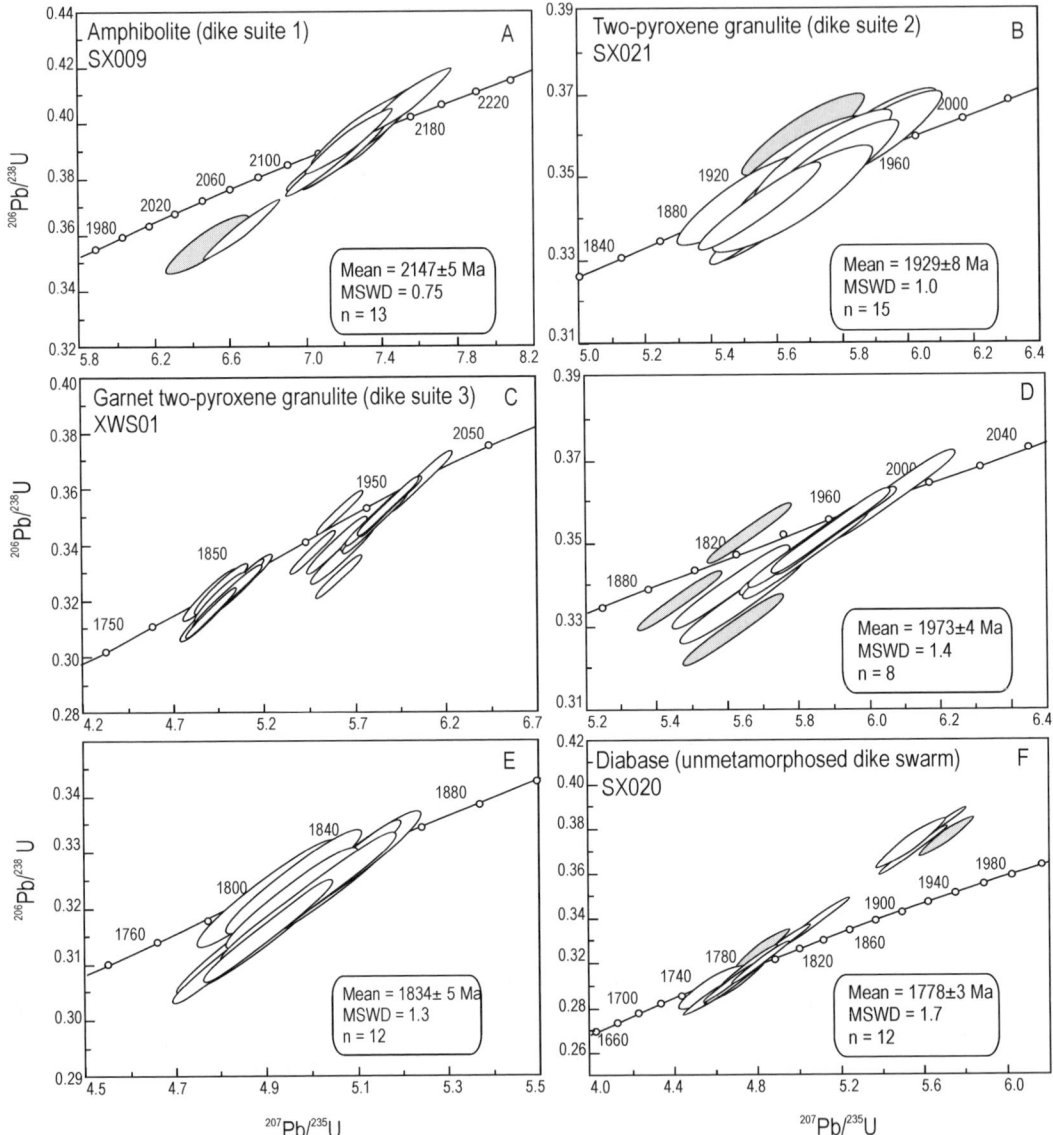

FIG. 5. U-Pb concordia diagrams. Error bars are 2-sigma. A. Thirteen spots without 1 spot (5) of sample SX009. B. Fifteen spots without 1 spot (6) of sample SX021. C. Eight spots without 3 spots (6, 13, 22) from the cores (D) and 12 spots from the rims of the zircon grains of sample XWS01 (E). F. Twelve spots without 3 spots (2, 9, 13) of sample SX020. The dark symbols refer to the spots excluded in calculating weighted mean ages.

The uplift event after 1830 Ma seems to be supported by the fact that high-grade metamorphism is dispersed in many parts of the NCC (Zhai, 1996; Zhai et al., 2004), and the typical structural style at 1900–1800 Ma is shallow-angle detachment associated with diapiric structures (Zhang et al., 1994; Zhang, 1997; Zhai et al., 2003). Kröner et al. (2002) also suggested that the mafic dikes experienced an extensive, penetrative ductile event under lower crustal conditions.

Unmetamorphosed mafic dike swarms developed across the NCC, and were emplaced after ~1800 Ma; these units are overlain by the Changcheng volcanosedimentary rock series, which marked the

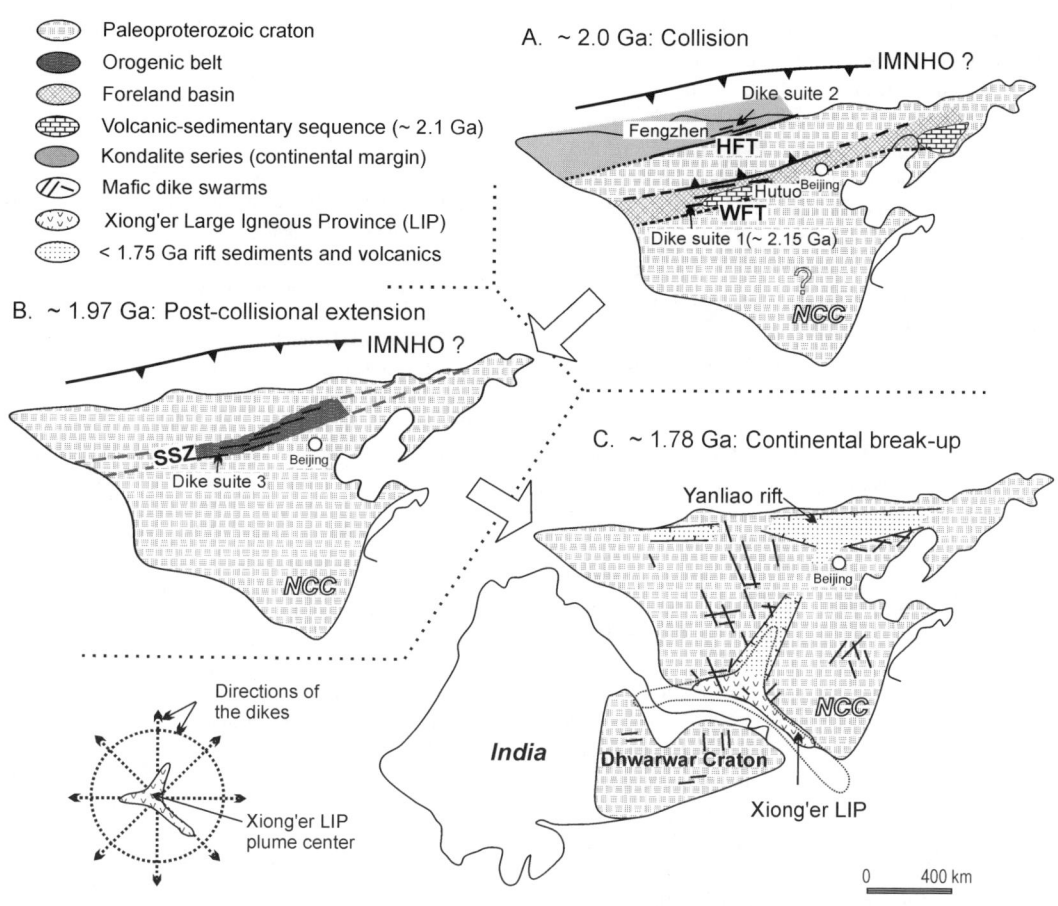

FIG. 6. Palinspastic maps showing the evolution of the NCC. A. At ~2.0 Ga. B. At ~1.97 Ga. C. At ~1.78 Ga. Abbreviations: IMNHO = Inner Mongolia–North Hebei Orogen, after Kusky and Li (2003). Configuration of the NCC with the Dhwarwar craton is inspired by Zhao et al. (2002a).

beginning of the Mesoproterozoic in the NCC, and whose lower age limit ranges from 2000 to 1640 Ma (Wang et al., 1995 and references therein). Nevertheless, Wang et al. (1995) favored ~1700 Ma as the low age limit. It has been suggested that this dike swarm is contemporary with and probably genetically related to the Xiong'er Group (Fig. 1A; Zhou et al., 2000; Zhao et al. 2002c), which is a large igneous province (LIP) that crops out over an area of 60,000 km² at the southern margin of the NCC. The Xiong'er Group is composed of mafic to felsic volcanic and minor sedimentary rocks, and its low age limit should be ~1750 Ma (Zhao et al., 2002c). Peng et al. (2004) interpreted these unmetamorphosed dikes as feeders of the Xiong'er LIP intruded at 1780–1750 Ma, and are probably plume related. It seems that a plume-driven upwelling beginning at 1830 Ma can explain the exhumation of the metamorphosed mafic dikes with the host rocks, subsequent triple-junction rifting and LIP at the south margin of the NCC, and radiating dike intrusion throughout the NCC and probably the Dhwarwar craton in the supercontinent Columbia (Fig. 6C).

The three metamorphic dike suites must have been intruded before 1930 Ma, and subsequently underwent multi-stages metamorphism at ~1930 to ~1830 Ma. Researchers (e.g. Zhao et al., 2001, Kröner et al., 2002) have suggested that the clockwise P/T path during this time period should be collision-related. This clockwise P/T path shows typical isothermal decompression and then cooling, but the preceding process is not well known. We think this kind of P/T trajectory reflects combined tectonic processes including crustal thickening and late extension, as suggested by Bohlen (1991); it probably was the result of thermal gradient changes due to collision in another belt, such as an Inner Mongolia–North Hebei Orogen (IMEHO, Fig. 6A) to the north, as suggested by Kusky and Li (2003). This idea is supported by the fact that the Fengzhen khondalite series and the dikes (dike suite 2) in it on the northern margin of the NCC also underwent granulite-facies metamorphism at the same time period from about 1.93 Ga (this study) to 1.87 Ga (Guo et al., 2001a).

Corresponding to dike suite 1 (2147 Ma), a large number of granitic and volcanic rocks formed around this time in the WFT (Kröner et al., 2002, 2004; Zhao et al., 2002b; Zhai et al., 2003; Wilde et al., 2004). The Dawaliang granite has been dated at 2176 ± 12 Ma (Fig. 1B); an anatectic granite gives an upper intercept SHRIMP U-Pb zircon age of 2113 ± 8 Ma and a $^{207}Pb/^{206}Pb$ evaporation age of 2112 ± 1 Ma (Kröner et al., 2002). In addition, a felsic tuffaceous rock in the Hutuo Group also shows an age population at 2180 ± 5 Ma (Wilde et al., 2004). The geochemistry of this suite shows within-plate basalt characteristics (Fig. 2D). So we think dike suite 1 in the northern WFT might have formed in the same tectonic environment as these volcanic and granitic rocks, most likely along a passive margin as suggested by Zhai et al. (2003), probably in a foreland basin (Figs. 1A and Fig. 6A). Although the crystallization age of dike suite 2 in the northern HFT has not been obtained, they probably formed at ca. 2.0 Ga. This suggestion is supported by the fact that many dikes from the study area have T_{DM}(Nd) of about 2200 Ma; however, some of them have younger T_{DM}(Nd) values, down to about 1850 Ma (Zhai et al. 2003). To clarify the reason for younger T_{DM} is beyond the scope of this study and awaits further research. The dikes are probably related to an arc or backarc basin event north of the HFT, as supposed by Kusky and Li (2003) (Fig. 6A), and supported by their geochemical affinities (Figs. 2C and 2D).

Acknowledgments

Financial support was provided by the China NSFC grant (40234050) and Chinese Academy of Sciences grant (KZCX1-07). We appreciate help from Y.-R. Shi, Y.-B. Zhang, H. Tao, B. Song, and D.-Y. Liu in SHRIMP zircon dating at the Beijing SHRIMP center. We thank J. G. Liou, Simon A. Wilde, Timothy M. Kusky, and F.-K. Chen for their incisive criticism and constructive reviews.

REFERENCES

Bohlen, S. R., 1991, On the formation of granulites: Journal of Metamorphic Geology, v. 9, p. 223–229.

Bai, J., Huang, X.-G., Dai, F.-Y., and Wu, C.-H., 1993, The Precambrian evolution of China: Beijing, China, Geological Publishing House, p. 65–79 (in Chinese with English abstract).

Chen, X.-D., and Shi, L.-B., 1983, Primary study on diabase dike swarms in the Wutai-Taihang area: Chinese Science Bulletin, v. 16, p. 1002–1005 (in Chinese).

Cheng, Y.-Q., and Zhang, S.-G., 1982, Notes on the metamorphic series and metamorphic belts of various metamorphic epochs of China and related problems: Regional Geology of China, v. 2, p. 1–14 (in Chinese with English abstract).

Féraud, G., Giannérini, G., and Campredon, R., 1987, Dyke swarms as paleostress indicators in areas adjacent to continental collision zones: Examples from the European and northwest Arabian plates, in Halls, H. C., and Fahrig, W. F., eds., Mafic dyke swarms: Geological Association of Canada Special Paper 34, p. 273–278.

Gao, S., Rudnick, R. L., Carlson, R. W., McDonough, W. F., and Liu, Y.-S., 2002, Re-Os evidence for replacement of ancient mantle lithosphere beneath the North China craton: Earth and Planetary Science Letters, v. 198, p. 307–322.

Guan, H., Sun, M., Wilde, S. A., Zhou, X. H., and Zhai, M.-G., 2002, SHRIMP U-Pb zircon geochronology of the Fuping complex: implication and assembly of the North China craton: Precambrian Research, v. 113, p. 1–18.

Guo, J.-H., Sun, M., Chen, F.-K., and Zhai, M.-G., 2004, Sm-Nd and SHRIMP U-Pb zircon geochronology of high-pressure granulites in the Sanggan area, North China Craton: timing of Paleoproterozoic continental collision: Journal of Asian Earth Sciences, in press.

Guo, J.-H., Zhai, M.-G., and Xu, R.-H., 2001a, Timing of the granulite facies metamorphism in the Sanggan area, North China craton: zircon U-Pb geochronology: Science in China (Series D), v. 44(11), p. 1010–1018.

Guo, J.-H., Wang, S.-S., Sang, H.-Q., and Zhai, M.-G., 2001b, ^{40}Ar-^{39}Ar age spectra of garnet porphyroblast: implications for metamorphic age of high-pressure granulite in the North China craton: Acta Petrologica Sinica, v. 17(3), p. 436–442. (in Chinese with English Abstract).

Guo, J.-H., and Zhai, M.-G., 2001, Sm-Nd age dating of high-pressure granulites and amphibolite from Sanggan area, North China craton: Chinese Science Bulletin, v. 46, p. 106–110.

Guo, J.-H., Zhai, M.-G., Li, J.-H., and Li, Y.-G.. 1996, Nature of the early Precambrian Sanggan structure zone in North China craton: evidence from rock association: Acta Petrologica Sinica, v. 12, p. 193–207 (in Chinese with English Abstract).

Guo, J.-H., Zhai, M.-G., Zhang, Y.-G., Li, Y.-G., Yan, Y.-H., and Zhang, W.-H., 1993, Early Precambrian Manjinggou high-pressure granulite mélange belt on the south edge of the Huai'an complex, North China craton: geological features, petrology and isotopic geochronology: Acta Petrologica Sinica, v. 9(4), p. 329–341 (in Chinese with English Abstract).

Halls, H. C., Li, J.-H., Davis, D., Hou, G.-T., Zhang, B.-X., and Qian, X.-L., 2000, A precisely dated Proterozoic paleomagnetic pole from the North China Craton, and its relevance to paleocontinental construction: Geophysical Journal International, v. 143, p. 185–203.

Hou, G.-T., Li, J.-H., and Qian, X.-L., 2001, Geochemical characteristics and tectonic setting of Mesoproterozoic dyke swarms in northern Shanxi: Acta Petrologica Sinica, v. 17 (3), p. 352–357 (in Chinese with English Abstract).

Hou, G.-T., Zhang, C., and Qian, X-L., 1998, The formation mechanism and tectonic stress field of the Mesoproterozoic mafic dyke swarms in the north China craton: Geological Review, v. 44 (3), p. 309–314 (in Chinese with English Abstract).

Huang, X.-L., Xu, Y.-G., Chu, X.-L., Zhang, H.-X., and Liu, C.-Q., 2001, Geochemical comparative studies of some granulite terranes and granulite xenoliths from north China craton: Acta Petrologica et Mineralogica, v. 20(3), p. 318–328.

Kretz, R., 1983, Symbols for rock-forming minerals: American Mineralogist, v. 68, p. 277–279.

Kröner, A., Wilde, S. A., Li, J.-H., and Wang, K.-Y., 2004, Age and evolution of a late Archaean to Paleoproterozoic upper to lower crustal section in the Wutaishan/Hengshan/Fuping terrain of north China: Journal of Asian Earth Sciences, in press.

Kröner, A., Zhao, G.-C., Wilde, S. A., Zhai, M.-G., Passchier, C. W., Sun, M., Guo, J.-H., O'Brien, P. J., and Walte, N., 2002, A late Archaean to early Proterozoic lower to upper crustal section in the Hengshan-Wutaishan area of northern China, Guidebook for Penrose Conference Field Trip, Sept. Penrose Conference, Beijing, p. 32–39.

Kusky, T. M., and Li, J.-H., 2003, Paleoproterozoic tectonic evolution of the North China Craton: Journal of Asian Earth Sciences, v. 22, p. 383–397.

Kusky, T. M., Li, J.-H, and Tucker, R. T., 2001, The Archaean Dongwazi ophiolite complex, North China craton: 2.505 billion year old oceanic crust and mantle: Science, v. 292, p. 1142–1145.

Li, J.-H., Kusky, T. M., and Huang, X.-N., 2002, Neoarchaean podiform chromitites and mantle tectonites in ophiolitic mélange, North China Craton: A record of early oceanic mantle processes: GSA Today, v. 12, p. 4–11.

Li, J.-H., Hou, G.-T., Qian, X.-L., Halls, H. C., and Don, D., 2001, Single-zircon U-Pb age of the initial Mesoproterozoic basic dyke swarms in Hengshan mountain and its implication for the tectonic evolution of the North China Craton: Geological Review, v. 47 (3), p. 234–238 (in Chinese with English Abstract).

Li, J.-H., Qian, X.-L., Huang, X.-N., and Liu, S.-W., 2000, The tectonic framework of the basement of north China craton and its implication for the early Precambrian cratonization: Acta Geologica Sinica, v. 16, p. 1–10 (in Chinese with English abstract).

Li, J.-H., Qian, X.-L., Zhai, M.-G., and Guo, J.-H., 1996, Tectonic division of a high-grade metamorphic terrain and late Archaean tectonic evolution in north-central part of the North China craton: Acta Petrologica Sinica, v. 12, p. 179–192 (in Chinese with English abstract).

Li, J.-L., Wang, K.-Y., Liu, X.-H., and Zhao, Z.-Y., 1990, Early Proterozoic collision orogenic belt in Wutaishan area, China: Scientia Geologica Sinica, v. 25(1), p. 1–11 (in Chinese with English abstract).

Li, T.-S., 1999, Taipingzai-Zunhua Neo-Archaean island arc terrain and continental growth in Eastern Hebei, North China: Doctoral paper, Institute of Geology and Geophysics, Chinese Academy of Sciences (CAS), Beijing, p. 73–88 (in Chinese with English abstract).

Lu, S.-N., Yang, C.-L., Li, H.-K., and Li, H.-M., 2002, A group of rifting events in the terminal Paleoproterozoic in the North China craton: Gondwana Research, v. 5, p. 123–131.

Ludwig, K. R., 1999, Users Manual for Isoplot/Ex, version 2.34, A geochronological toolkit for Microsoft Excel: Berkeley Geochronology Center, p. 1–43.

Mao, E.-B., Zhong, C.-T., Chen, Z.-H., Lin, Y.-X., Li, H.-M., and Hu, X.-D., 1999, The isotope ages and their geological implications of high-pressure basic granulites in north region to Chengde, Hebei province, China: Acta Petrologica Sinica, v. 15(4), p. 524–531 (in Chinese with English Abstract).

Mullen, E. D., 1983, MnO-TiO_2-P_2O_5: A minor element discriminant for basaltic rocks of oceanic environments and its implications for petrogenesis: Earth Planetary and Science Letters, v. 62, p. 53–62.

Pearce, J. A., and Cann, J. R., 1973, Tectonic setting of basic volcanic rocks determined using trace element analyses. Earth Planetary and Science Letters, v. 19(2), p. 290–300.

Peng, P., Zhai, M.-G., Zhang, H.-F., Zhao, T. P., and Ni, Z.-Y., 2004, Geochemistry and geological significance of the 1.8 Ga mafic dike swarms in the North China craton: An example from the juncture of Shanxi, Hebei, and Inner Mongolia: Acta Petrologica Sinica, v. 20, no. 3, p. 439–456 (in Chinese with English abstract).

Pidgeon, R. T., 1992, Recrystallization of oscillatory zoned zircon: some geochronological and petrological implications: Contribution to Mineralogy and Petrology, v. 110, p. 463–472.

Pollard, D. D., 1973, Derivation and evaluation of a mechanical model for sheet intrusions: Tectonophysics, v. 19, p. 233–269.

Qian, X.-L., and Chen Y.-P., 1987, Late Precambrian mafic dyke swarms of the North China craton. in Halls, H. C., and Fahrig, W. F., eds., Mafic Dyke Swarms: Geology Association of Canada Special Paper 34, p. 385–391.

Rogers, J. J. W., and Santosh, M., 2002, Configuration of Columbia, a Mesoproterozoic supercontinent: Gondwana Research, v. 5, p. 123–132.

Sang, H.-Q., Wang, S.-S., and Qiu, J., 1996, The ^{40}Ar-^{39}Ar ages of pyroxene, hornblende, and plagioclase in Taipingzhai granulites in Qianxi county, Hebei province, and their geological implications: Acta Petrologica Sinica. v. 12, p. 390–400 (in Chinese with English Abstract).

Shen, Q.-H., Zhang, Z.-Q., Geng, Y.-S., and Tang, S.-H., 1994, Petrology geochemistry and isotopic age of the garnet-bearing basic metamorphic rocks from Dadonggou, north-western Hebei province, in Qian, X.-L., and Wang, R.-M., eds., Geological evolution of the North China craton: Seismological Press, Beijing p. 120–129 (in Chinese with English Abstract).

Vavra, G., 1994, Systematics of internal zircon morphology in major Variscan granitoid types: Contribution to Mineralogy and Petrology, v. 117, p. 331–344.

Wang, S.-S, Sang, H.-Q, Qiu, J., Cheng, M.-E., and Li, M.-R., 1995, The metamorphic age of pre-Changcheng system in Beijing-Tianjin area and a discussion about the lower limit age of Changcheng system: Scientia Geologica Sinica, v. 3(4), p. 348–354.

Wilde, S. A., Zhao, G.-C., Wang, K.-Y., and Sun, M., 2004, First SHRIMP zircon U-Pb ages for Hutuo Group in Wutaishan: further evidence for Paleoproterozoic amalgamation of North China Craton: Chinese Science Bulletin, v. 19 (1), p. 83–90.

Wilde, S. A., Cawood, P. A., and Wang, K-Y., 1997, SHRIMP U-Pb data of granites and gneisses in the Taihangshan-Wutaishan area: implications for the timing of crustal growth in the North China craton: Chinese Science Bulletin, v. 43(sup), p. 144.

Wilde, S. A., Zhao, G.-C., and Sun, M., 2002, Development of the North China Craton during the Late Archaean and its final amalgamation at 1.8 Ga: some speculation on its position within a global Paleoproterozoic Supercontinent: Gondwana Research, v. 5, p. 85–94.

Williams, I. S., 1998, U-Th-Pb geochronology by ion microprobe, in McKibben, M. A., Shanks III, W. C., and Ridley, W. I., eds., Applications of microanalytical techniques to understanding mineralizing processes: Reviews in Economic Geology, v. 7, p. 1–35.

Wu, C.-H., and Zhang, C.-T., 1998, The Paleoproterozoic SW-NE collision model for the central north China Craton: Progress of Precambrian Research, v. 21, p. 28–50.

Wu, J.-S., Geng, Y.-S., and Shen, Q.-H., 1998, Archaean geology characteristics and tectonic evolution of Sino-Korea Paleo-continent: Geological Publication House, Beijing, p. 1–104 (in Chinese with English abstract).

Zhai, M.-G., and Liu, W.-J., 2003, Paleoproterozoic tectonic history of the North China craton: a review: Precambrian Research, v. 122, p. 183–199.

Zhai, M.-G., Guo, J.-H., Li, Y.-G., Liu, W.-J., Peng, P., and Shi, X., 2003, Two linear granite belts in the central-western North China Craton and their implication for late Neoarchaean-Paleoproterozoic continental evolution: Precambrian Research, v. 127, p. 267–283.

Zhai, M.-G., Guo, J.-H., and Liu W.-J., 2001, An exposed cross-section of early Precambrian continental lower crust in north China craton: Physics and Chemistry of the Earth, v. 26(9–10), p. 781–792.

Zhai, M.-G., Bian A.-G. and Zhao, T.-P., 2000, Amalgamation of the supercontinental of the North China craton and its break up during late-middle Proterozoic: Science in China (D), v. 43, p. 219–232.

Zhai, M.-G. (ed.), 1996, Granulite and Lower Continental Crust in North China Archaean Craton: Seismological Press, Beijing, p. 1–20.

Zhai, M.-G., Guo, J.-H., Yan, Y.-H., Li, Y.-G., and Zhang, W.-H., 1992, Preliminary study and discovery of high-pressure granulites in North China: China Science (B), v. 12, p. 1325–1330 (in Chinese).

Zhang, C., Hou, G.-T., and Qian, X.-L., 1994, Magnetic gabbroic evidence of the style of emplacement of late Precambrian mafic dyke swarms in the Lvliang-northern Shanxi region, north China: Geological Review, v. 40(2), p. 245–251 (in Chinese with English Abstract).

Zhang, J.-S., 1997, Extension and uplift of the Datong–Huaian granulite terrain: Geological Review, v. 43, p. 503–514 (in Chinese with English abstract).

Zhang, J.-S., Driks, P. H. G. M., and Passchier, C. W., 1994, Extensional collapse and uplift of a polymetamorphic granulite terrain in the Archaean of North China: Precambrian Research, v. 67, p. 37–57.

Zhao, G.-C., Sun, M., and Wilde, S. A., 2002a, Review of global 2.1-1.8 Ga orogens: implications for a pre-Rodinia supercontinent: Earth-Science Reviews, v. 59, p. 125–162.

Zhao, G.-C., Wilde, S. A., Cawood, P. A., and Sun, M., 2002b, SHRIMP U-Pb zircon ages of the Fuping complex: implications for late Archaean to Paleoproterozoic accretion and assembly of the North China Craton: American Journal of Sciences, v. 302, p. 191–226.

Zhao, G.-C., 2001, Paleoproterozoic assembly of the North China craton: Geological Magazine, v. 138 (1), p. 89–91.

Zhao, G.-C., Wilde, S. A., Cawood, P. A., and Sun, M., 2001, Archaean blocks and their boundaries in the

North China craton: lithological, geochemical, structural and *P-T* path constraints and tectonic evolution: Precambrian Research, v. 107, p. 45–73.

Zhao, T.-P., Zhou, M.-F., Zhai M.-G., and Xia, B., 2002c, Paleoproterozoic rift-related volcanism of the Xiong'er group, North China Craton: implications for the breakup of Columbia: International Geology Review, v. 44, p. 336–351.

Zhao, Z.-P., Zhai, M.-G., Wang, K.-Y., Yan, Y.-H., Guo, J.-H., and Liu, Y.-G., 1993, Precambrian crustal evolution of the Sino-Korean paraplatform: Science Press, Beijing, p. 333–399 (in Chinese).

Zhong, C.-T., 1999. The geological features and origin of two-stage high-pressure basic granulite from the Shanxi-Hebei-Inner Mongolia terrain in North China craton: Progress in Precambrian Research, v. 22(2), p. 53–58 (in Chinese with English Abstract).

Zhou, D.-W., Zhang, C.-L., Liu, L., Wang, J.-L., Wang, Y., and Liu, J.-P., 2000, Synthetic study on Proterozoic basic dike swarms in the Qinling orogenic belt and its adjacent block as well as a discussion about some questions related to them: Acta Petrologica Sinica, v. 16, p. 22–28 (in Chinese with English abstract).

Zhuang, Y.-X., Wang, X.-S., Xu, H.-L., Ren, Z.-K., Zhang, F.-Z., and Zhang, X.-M., 1997, Main geological events and crustal evolution in early Precambrian of Taishan region: Acta Petrologica Sinica, v. 13, p. 313–330 (in Chinese with English abstract).

Fluid-Rock Interaction in UHP Phengite-Kyanite-Epidote Eclogite from the Sulu Orogen, Eastern China

S. Ferrando,[1]

Dipartimento di Scienze Mineralogiche e Petrologiche, Via Valperga Caluso 35, I-10125 Torino, Italy and Dipartimento di Scienze della Terra, Via Laterina 8, I-53100 Siena, Italy

M. L. Frezzotti,

Dipartimento di Scienze della Terra, Via Laterina 8, I-53100 Siena, Italy

L. Dallai,

Geoscienze e Georisorse, Via G. Moruzzi 1, 56124 Pisa, Italy

and R. Compagnoni

Dipartimento di Scienze Mineralogiche e Petrologiche, Via Valperga Caluso 35, I-10125 Torino, Italy

Abstract

Combined petrographic, minerochemical, fluid inclusion, and stable isotope studies have been carried out on phengite-kyanite-epidote (Phe-Ky-Ep) eclogites from Hushan and Qinglongshan (Donghai area, UHP Sulu orogen, China) to unravel their metamorphic evolution and fluid-rock interaction. A complex metamorphic evolution, from coesite-eclogite (P = 3.5–4.0 GPa and T = 840 ± 50°C) to greenschist facies conditions (P ≅ 0.2 GPa and T ≅ 350°C), through quartz-eclogite and HT amphibolite-facies conditions, was recognized. The associated fluids are aqueous and show a progressive change in salinity and composition, reflecting different metamorphic stages. Stable isotope data confirm that, prior to subduction, the Sulu eclogite protolith experienced a meteoric-hydrothermal alteration, and indicate that during metamorphic evolution the rocks recrystallized without pervasive fluid infiltration.

Introduction

Recent petrological studies on high pressure (HP) and ultrahigh-pressure (UHP) rocks from the Dabie Sulu orogen have revealed the presence of (1) hydrous HP and UHP mineral assemblages (e.g., Zhang et al., 1995); and (2) eclogitic veins (e.g., Castelli et al., 1998). In contrast to older beliefs, these findings point to the presence of free fluid phases at great depths, as also suggested by fluid inclusions studies (e.g., Touret and Frezzotti, 2003). Moreover, stable isotope data indicate the persistence of a closed system during metamorphic evolution (e.g., Yui et al., 1995), and several authors have debated the origin of the fluid phase involved in the metamorphism (e.g., Ernst, 2000).

In this paper, petrological, fluid inclusion, and stable isotope data on UHP rocks are combined to constrain the P-T path, to determine composition and origin of the fluids, and to reconstruct the fluid-rock interaction. The rocks from the UHP Sulu orogen were selected because they preserve the peak mineral association, contain zoned UHP and HP minerals, show some fluid inclusions within UHP and HP minerals, and preserve very low $\delta^{18}O$ values.

Geological Setting and Sample Location

The Sulu belt (Fig. 1) is the easternmost extension of the Qinling-Dabie orogen, formed during Triassic collision between Yangtze and Sino-Korean plates. The Sulu orogen consists of UHP, HP (Zhang et al., 1995), and lower-grade metamorphic rocks intruded by post-orogenic Cretaceous granitic plutons. Mesozoic and younger sedimentary rocks are also exposed (Wallis et al., 2000). In the southern Sulu terrane, the UHP unit consists of amphibolite-facies granitoid gneisses, minor amphibolites, and ultramafics that locally include eclogite layers and boudins (e.g., Zhang et al., 1995). The age of the UHP metamorphism was determined at about 218–

---

[1]Corresponding author; email: simona.ferrando@unito.it

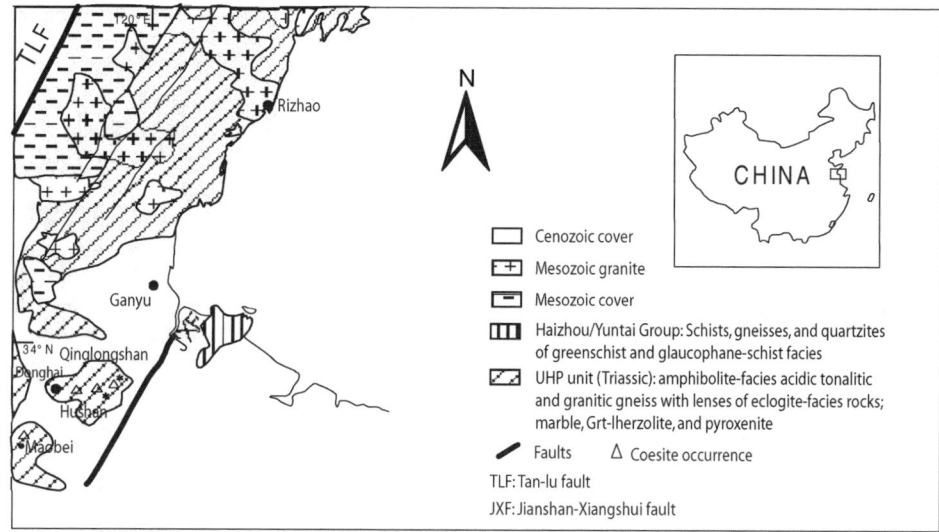

FIG. 1. Simplified tectonic sketch map of the southern Sulu region, showing major tectonic units, the distribution of coesite, and the locations of Hushan and Qinglongshan, Donghai area.

230 Ma (Ames et al., 1996; Yang et al., 2003), whereas the age of the gneiss protolith was estimated at about 700–800 Ma. In the UHP unit, the evidence of prograde metamorphism is represented by feldspars, amphiboles, paragonite, quartz, zoisite, chlorite, staurolite, and margarite as inclusions within peak garnet and pyroxene (Wallis et al., 2000 and references therein). Peak metamorphic conditions of Donghai UHP eclogites were estimated at T = 700–890°C and P > 2.8 GPa by Zhang et al. (1995), at 720–880°C and 3.2–4.0 GPa by Zhang et al. (2000), and at 750–880°C and 2.9–3.6 GPa by Hirajima and Nakamura (2003). As a whole, the retrograde P-T path is characterized by an initial strong decompression (P = 1.2–2.4 GPa) coupled with a moderate cooling (T = 600–750°C), followed by a cooling (T = 450–600°C) and moderate decompression (P = 0.6–1.9 GPa; Zhang et al., 1995).

Previous fluid inclusion studies on the Sulu rocks suggested that the fluids in equilibrium at metamorphic peak were Ca-rich brines ± CO_2 and/or N_2 (Fan et al., 2003; Fu et al., 2003a, 2003b). Rare CH_4-rich fluid inclusions, coexisting with high-salinity brine inclusions, were also found by Fu et al. (2003a). Mixed CO_2-H_2O fluids (Shen et al., 2002), N_2/CO_2/CH_4-H_2O fluids (Fu et al., 2003a), or CO_2 fluids (Fu et al., 2003b) were present during the initial stages of exhumation. CO_2 seems to be characteristic of HP-UHP rocks re-equilibrated under granulite-facies conditions (Fu et al., 2003a). All studies indicate the presence of low-salinity fluids during late stages of retrogression (Shen et al., 2002; Fan et al., 2003; Fu et al., 2003a).

Stable isotope studies of the Dabie-Sulu UHP rocks reveal low δD values of hydroxyl-bearing minerals (e.g., Phe from –127 to –83 ‰ and Zo from –6 to –49 ‰; Zheng et al., 2003) from eclogite, quartz-schist, paragneiss, and orthogneiss, whereas the eclogites from Qinglongshan show the lowest $\delta^{18}O$ values ever observed in eclogite-facies rocks (e.g., Qtz from –7.7 to +4.5 ‰ and Rt from –14.8 to –5.3‰; e.g., Rumble and Yui, 1998). Such values are interpreted as due to interaction of the protolith with meteoric-hydrothermal water under cold climate conditions. The preservation of the original isotopic values suggests that during prograde metamorphism, fluid flow occurred only at a small scale (e.g., Zheng et al., 2003).

The rocks analyzed in this study were sampled at Hushan and Qinglongshan, in the Donghai area. These localities are part of the southern UHP unit of Sulu (Fig. 1). The Qinglongshan eclogites (samples RPC541, RPC619, RPC624, RPC634, RPC742) were collected from a road cut exposed during work for the construction of a motorway. The Hushan eclogites (samples RPC778, RPC779) were collected near the village, southwest of a ridge consisting of granitoid gneiss. Despite the poor exposure, we have concluded that both eclogites occur as meter- to decameter-wide pods or layers within orthogneiss.

Analytical Methods

Chemical composition and X-ray elemental maps of minerals were obtained by a SEM Cambridge Instruments Stereoscan 360 equipped with an EDS Oxford Instruments Energy 200 at Dipartimento di Scienze Mineralogiche e Petrologiche, University of Torino, Italy. Operating conditions were 15 kV accelerating voltage, 1.35 nA beam current, and 50 s counting time for spot analyses, and 15 kV accelerating voltage, 4.52 nA beam current, acquisition for 12 hours, and 512 × 512 pixel size for X-ray elemental maps. Natural minerals and pure oxides were used as standards. Structural formulae of minerals were processed using the software of Ulmer (1986). The chemical composition of Cpx I was obtained using the module Point & ID of the INCA software on quantitative X-ray elemental maps (area: 66 × 36 μm) of clinopyroxene including SiO_2 rods perpendicular to the thin section. For amphiboles, the nomenclature of Leake et al. (2004) was followed. Mineral abbreviations are after Kretz (1983).

The garnet-omphacite geothermometer (Ravna, 2000; accuracy ±100°C), the garnet-omphacite-phengite geobarometer (Waters and Martin, 1996; accuracy ±0.1 GPa), the hornblende-plagioclase geothermometer (Holland and Blundy, 1994; accuracy ±35°C), and the Al-in-hornblende geobarometer (Schmidt, 1992; accuracy ±0.06 GPa) were used to estimate P-T conditions.

Petrogenetic grids were calculated in the CaO-Na_2O-MgO-Al_2O_3-SiO_2-H_2O-CO_2 (CNMASHC) system using the thermodynamic approach of Connolly (1990) and the database of Holland and Powell (1998). The isopleths modeling the jadeite content in omphacite, with composition Jd_{18-24} and Jd_{37-50}, were calculated using the diopside-jadeite solution model of Gasparik (1985) and the pseudocompound approximation of Connolly (1990). All equilibrium curves were calculated for X_{H2O} = 1, except for the equilibrium curve involving talc, where X_{H2O} = 0.90.

Microthermometry of fluid inclusions within doubly polished, 100 μm thick, sections was performed using a Linkam THMSG600 heating-freezing stage coupled with an Olimpus polarizing microscope (100× objective) at Dipartimento di Scienze Mineralogiche e Petrologiche of the University of Torino, Italy. The accuracy, estimated using synthetic fluid inclusion standards, is about 0.3°C at the eutectic temperature of CO_2. Freezing temperature (Tf), eutectic temperature (Te), final melting temperature (Tm_{ice} and Tm_{Hhl}), and homogenization temperature, always into the liquid (ThL_{H2O}), were measured. Heating rates were always of 0.1°C/min near Te, Tm_{ice}, and Tm_{Hhl}, and 0.5°C/min near ThL_{H2O}. Fluid inclusion compositions, densities, and isochores were determined using the software packages FLUID 1 (Bakker, 2003) and FLINCOR.97 (Brown, 1989).

Laser Raman analyses were made with a Labram microspectrometer (Jobin Yvon, Ltd.) at Dipartimento di Scienze della Terra, University of Siena, Italy. A polarized Ar^+-ion laser operating at 514.5 nm wavelength and 200–550 mW incident power, was used as the excitation source. The laser spot size was focussed to 1–2 μm. Accumulation times varied between 20 and 90 seconds. Calibration was performed by using the 1332 cm^{-1} diamond band.

Oxygen isotope compositions of mineral separates were measured at the CNR-Istituto di Geologia Ambientale e Geoingegneria (IGAG), Rome, Italy, using the laser fluorination technique of Sharp (1990). Mineral separates were obtained through handpicking from crushed, sieved (0.3 and 0.5 mm fraction), and ultrasonically cleaned samples. Analyses were duplicated on about 1–1.5 mg of material, and then averaged. A 15W Merchantek CO_2 laser was used for heating the sample in a F_2 atmosphere. The $\delta^{18}O$ values were measured on a Finnigan MAT Delta Plus mass spectrometer. The analytical precision was monitored by using laboratory standards QMS and Laus1 ($\delta^{18}O$ values = 14.05 and 18.10‰, respectively), and the results were always better than 0.12‰ (1σ). During the time of this study, the average composition of NBS 28 at IGAG was 9.54 ± 0.17‰ (n = 9). All isotope values are reported in the conventional $\delta^{18}O$ notation relative to SMOW.

Petrological Study

Petrography

The granoblastic eclogites are medium grained with zoned poikilitic porphyroblasts of epidote and amphibole, and consist of garnet (~ 30 vol%), clinopyroxene (~25 vol%), epidote (~15 vol%), blue-green amphibole (~15 vol%), kyanite (~7 vol%), phengite and paragonite (~ 5 vol%), quartz/(coesite)

(~ 3 vol%), and accessory rutile, apatite, zircon, and opaque ores. Petrography, combined with mineral chemistry and fluid inclusion studies, allowed us to recognize four garnet (Grt 0, Grt I, Grt II, Grt III), four clinopyroxene (Cpx I, Cpx II, Cpx III, Cpx IV), three kyanite (Ky I, Ky II, Ky III), and two phengite, epidote, and amphibole generations.

The earliest garnet generation (Grt 0), recognized only in RPC779, occurs as a relict portion in a Grt I included in the core of an epidote porphyroblast (Fig 2A). Grt 0 includes rutile and apatite. The second garnet generation (Grt I) consists of coarse-grained garnet including rutile, apatite, and locally multiphase solid inclusions (Fig. 2B). The third garnet generation (Grt II) is found in two structural positions: as overgrowth around Grt I (Fig. 2B) and in the core of medium-grained zoned idioblasts lacking solid inclusions (Figs. 2B and 2C). The fourth garnet generation (Grt III) overgrows Grt II (Fig. 2C).

The earliest clinopyroxene generation (Cpx I) consists of crystals that contain small needles (~ 3 × 50 µm) of SiO_2 (Fig. 3), suggesting a former "supersilicic" clinopyroxene composition (Chopin and Ferraris, 2003 and references therein). The second clinopyroxene generation (Cpx II) occurs both around Cpx I as coarse-grained zoned crystals (Fig. 3), and as core of medium-grained zoned crystals lacking SiO_2 needles (Fig. 2C). The third clinopyroxene generation (Cpx III) overgrows Cpx II (Figs. 2C and 3) and the fourth (Cpx IV) occurs together with oligoclase in the symplectite, developed from breakdown of Cpx II and Cpx III (Fig. 2D). Typically, Cpx II, III, and IV are partly replaced by fine- to very fine grained symplectites consisting of edenite (or pargasite) + calcic oligoclase, Mg-hornblende + sodic oligoclase, or actinolite + albite (Fig. 2D). In a vew occurrences, magnetite is present in such symplectites.

Zoned epidote is idioblastic and porphyroblastic (Fig. 2E), with different mineral associations included within the different zones: cores include polycrystalline quartz after coesite (Fig. 2F), Grt 0, Grt I, Grt II, and Cpx II, whereas rims include single quartz grains, Grt I, Grt II, Grt III, Cpx II, and Cpx III. In the most external part of the epidote rims, Ky II is also present.

Blue-green amphibole occurs as zoned porphyroblastic to interstitial crystals: the core is barroisite or winchite and the rim is Mg-taramite or Mg-katophorite + vermicular quartz. The outer part consists of a fine-grained symplectite of edenitic hornblende + oligoclase (Fig. 4A). Locally, the amphibole core includes Grt I, Grt II, Grt III, single quartz crystals, relics of clinopyroxene, and rare epidote; Ky II has never been found in clinoamphiboles.

The older kyanite generation (Ky I) consists of coarse-grained idioblasts (Fig. 4B) including zircon, rutile, and, locally, multiphase solid inclusions. The second generation (Ky II) occurs as small crystals, devoid of inclusions and associated with quartz (Fig. 4C). The third generation (Ky III) occurs together with oligoclase and hercynite in the symplectites after paragonite at the contact with kyanite (Ky I or Ky II), and quartz (Fig. 4D).

Phengite occurs as medium-grained zoned flakes (Fig. 2C) associated with medium-grained garnet (Grt II and Grt III) and clinopyroxene (Cpx II and Cpx III). Locally, it is partly replaced by symplectite consisting of biotite + andesine/oligoclase (Fig. 4E). Paragonite occurs either as bundles, locally associated with quartz, from retrogression of kyanite (Ky I and Ky II) and characteristically overgrowing porphyroblastic blue-green amphibole (Fig. 4F), or as fine-grained fracture-filling aggregates.

Minor quartz consists of medium- to fine-grained interstitial grains in the matrix rocks. Preserved coesite is absent, but rare polycrystalline quartz aggregates after former coesite are found only within Grt I, Cpx I, epidote core, and Ky I.

Mineral chemistry

Representative analyses of minerals are listed in Table 1.

Garnet. Different garnet generations are mainly characterized by a variation in the grossular content.

Grt 0 consists of equal amounts of almandine and grossular with minor pyrope (Fig. 5A). Grt I is homogeneous and lower in grossular and richer in almandine than Grt 0, whereas the pyrope content is unchanged (mean composition: $Alm_{45}Prp_{26}Grs_{27}Sps_{02}$; Fig. 5A). Around multiphase solid inclusions, Grt I is typically re-equilibrated to a garnet (Grt re-eq) slightly richer in almandine and grossular, and poorer in pyrope (Fig. 5A). Grt II (mean composition: $Alm_{44}Prp_{30}Grs_{25}Sps_{01}$) is generally grossular-poorer and pyrope-richer than Grt I (Fig. 5A), whereas Grt III (mean composition: $Alm_{43}Prp_{35}Grs_{20}Sps_{02}$) is characterized by increasing pyrope and decreasing grossular content. The heterogeneous compositions of both Grt II and Grt III probably are due to primary features (i.e., different microstructural sites).

Pyroxene. The average composition of the early "supersilicic" Cpx I is $Jd_{40}Di_{43}Aeg_{17}$. This composi-

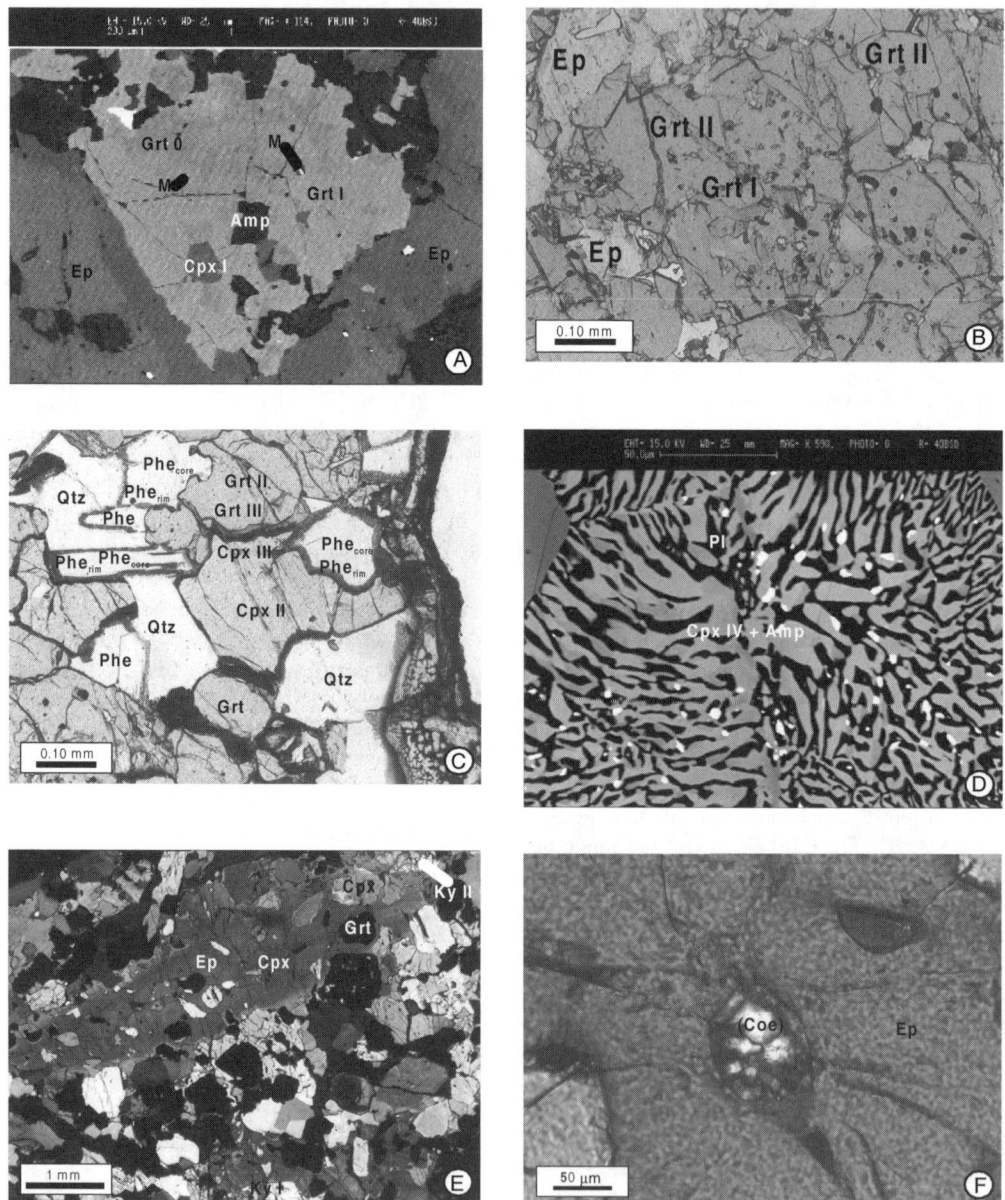

FIG. 2. Mineral associations of Phe-Ky-Ep eclogites from Hushan and Qinglongshan. A. Backscattered electron (BSE) image of Grt 0 relics within Grt I included in an epidote core. Multiphase solid inclusions (M) are present only within Grt I. Sample RPC779. B. Photomicrograph of the coarse-grained garnet (Grt I) that includes multiphase solid inclusions. Idioblastic Grt II is also evident. Sample RPC778, plane-polarized light (PPL). C. Photomicrograph of the medium-grained mineral association consisting of garnet (Grt II in the core, and Grt III in the rim), clinopyroxene (Cpx II in the core, and Cpx III in the rim), zoned phengite, and quartz. Sample RPC619, PPL. D. BSE image of Cpx IV + amphibole (Amp) + oligoclase (Pl) retrograde symplectite after clinopyroxene. Sample RPC619. E. Porphyroblastic and poikilitic zoned epidote. Sample RPC619, crossed polars (CP). F. Photomicrograph of an epidote (Ep) core, which includes polycrystalline quartz aggregates after former coesite [(Coe)]. Sample RPC779, CP.

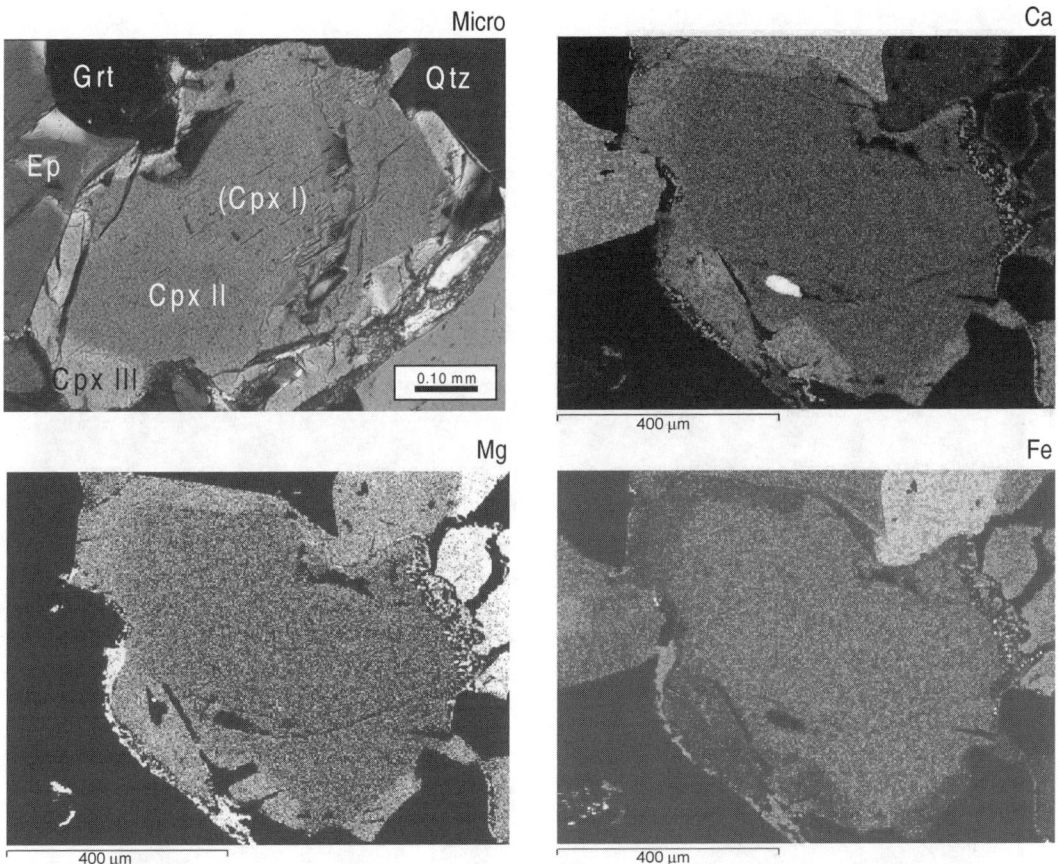

FIG. 3. Photomicrograph (CP) and X-ray elemental maps for Ca, Mg, and Fe of a coarse-grained zoned clinopyroxene (Cpx I, Cpx II and Cpx III). Cpx I is evident only in the photomicrograph for the presence of the SiO$_2$ rods. Sample RPC619. Scale bar for Ca, Mg, and Fe is 400 μm.

tion, plotted in Figure 5B, is very similar to that of Cpx II. However, inasmuch as the Si content is not systematically higher than that of other Cpx generations, the presence of SiO$_2$ needles is the only evidence that Cpx I was "supersilicic." Cpx II is a ferrian-omphacite ($X_{Aeg} \cong 0.15$), and Cpx III is an omphacite at the boundary with the ferrian-omphacite ($X_{Aeg} \cong 0.05$). The jadeite content (X_{Jd}) ranges between 0.47 and 0.40 for both Cpx II and Cpx III (Fig. 5B). The X-ray elemental maps of Figure 3 show a coarse-grained clinopyroxene containing SiO$_2$ rods in the core, whose zoning is consistent with compositions of Cpx I, Cpx II, and Cpx III, respectively: the core and the intermediate part (Cpx I and Cpx II) are Fe-richer, whereas the rim (Cpx III) is Ca- and Mg-richer.

The Cpx IV from retrogression symplectite after Na-clinopyroxenes is more calcic and less sodic than the other Cpx generations (Fig. 5B), varying in composition from a Jd-poor ferrian omphacite (mean composition: Jd$_{21}$Di$_{72}$Aeg$_7$) to an aluminian aegirin-augite (mean composition: Jd$_{10}$Di$_{79}$Aeg$_{11}$).

Epidote. Strong optical zoning of epidote is related to a Fe^{3+} decrease from 0.76 to 0.46 atoms per formula unit (a.p.f.u.), and an Al^{3+} increase (from 2.18 to 2.48 a.p.f.u.), from core to rim, respectively. Locally, the growth zoning is complicated by a partial retrograde re-equilibration around mineral inclusions.

Amphibole. The porphyroblastic blue-green amphibole is zoned from barroisite or winchite in the core to Mg-taramite or Mg-katophorite in the

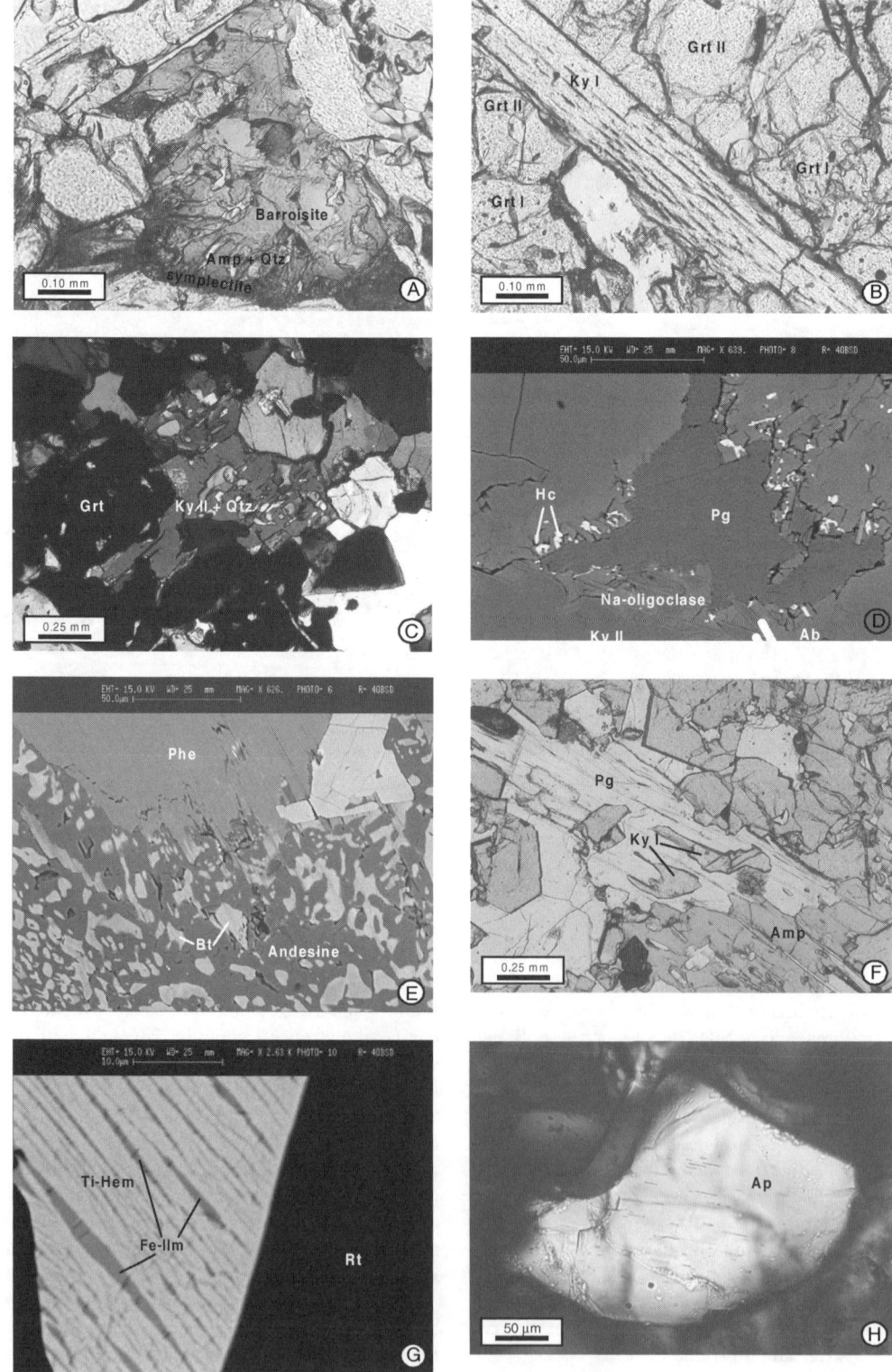

FIG. 4 (facing page). Mineral associations within Phe-Ky-Ep eclogites from Hushan and Qinglongshan. A. Photomicrograph of a zoned blue-green amphibole porphyroblast. The core is a barroisite and the rim is a Mg-taramite/Mg-katophorite + vermicular quartz. Porphyroblastic amphibole is partly replaced by a fine-grained symplectite consisting of edenitic hornblende + oligoclase. Sample RPC778, PPL. B. Photomicrograph of an idioblastic coarse-grained kyanite (Ky I). Sample RPC541, PPL. C. Photomicrograph of a medium-grained kyanite (Ky II) with quartz (Qtz). Sample RPC619, CP. D. BSE image of a retrograde symplectite, developed at the Ky II/paragonite (Pg) contact, which consists of Ky III + oligoclase (Na-oligoclase) + hercynite (Hc). Locally, albite (Ab) rims Ky II, Ky III, and paragonite. Sample RPC778. E. BSE image of the retrograde symplectite after phengite (Phe), consisting of biotite (Bt) + andesine/oligoclase (Andesine). Sample RPC619. F. Photomicrograph of paragonite bundles (Pg) after Ky I, that overgrow porphyroblastic blue-green amphibole (Amp). Sample RPC778, PPL. G. BSE image of rutile (Rt) that includes a relatively thick lamella of ilmeno-hematite, i.e., titano-hematite (Ti-Hem) with lamellae of ferrian-ilmenite (Fe-Ilm). Sample RPC779. H. Photomicrograph of an apatite crystal, showing oriented needles of an unknown non-Raman-active phase. Sample RPC619, PPL.

rim, Al^{IV} varying from 0.49 a.p.f.u. in the core to 1.86 a.p.f.u. in the rim, with Al^{VI} ranging from 1.06 to 1.41 a.p.f.u. (Fig. 5C). In symplectites after clinopyroxene and porphyroblastic amphibole (Fig. 5C), the amphibole ranges in composition from edenite through hornblendic-edenite or pargasite (Al^{IV} = 1.09–1.96 a.p.f.u.; Al^{VI} = 0.50–1.31 a.p.f.u.) to Mg-hornblende or actinolite (Al^{IV} = 0.88–0.11 a.p.f.u.; Al^{VI} = 0.52–0.38 a.p.f.u.).

Micas. Phengite is zoned with the Si-content decreasing from core (3.51 a.p.f.u.) to rim (3.23 a.p.f.u.). The maximum Ti content is 0.05 a.p.f.u. Paragonite has an Si content around 3.01 a.p.f.u., an average Ca content of 0.04 a.p.f.u., and a very low K content (average 0.02 a.p.f.u.). Symplectitic biotite after phengite has Al_{tot} = 1.49–1.75 a.p.f.u. and Al^{VI} = 0.34–0.56 a.p.f.u. The maximum Ti content is 0.23 a.p.f.u. and X_{Mg} varies between 0.67 and 0.70.

Plagioclase. The plagioclase composition ranges from andesine to pure albite (Fig. 5D). The most calcic plagioclase (X_{An} = 0.33–0.25) occurs together with biotite in the symplectite after phengite. Less calcic plagioclase (X_{An} = 0.25–0.21) occurs together with Ky III and hercynite in the symplectite after paragonite. Plagioclase in equilibrium with the Ca-rich Cpx IV from symplectite after Na-clinopyroxenes has X_{An} = 0.18. In the amphibole + plagioclase symplectite after all clinopyroxene generations and after porphyroblastic amphibole, Ca-rich oligoclase (X_{An} = 0.24–0.17) occurs with edenite, hornblendic-edenite or pargasite, Ca-poor oligoclase (X_{An} = 0.13–0.14) with Mg-horneblende, and albite (X_{An} = 0.07–0.01) with actinolite.

Rutile. Ubiquitous rutile commonly includes a relatively thick lamella (tens of μm) of ilmeno-hematite (Fig. 4G), which consists of titan-hematite with exsolution lamellae of ferrian-ilmenite (Rumble, 1981). This feature, also present in rutile from kimberlite, suggests that it may derive from a former UHP Ti-Fe-rich mineral (Haggerty, 1981).

Apatite. Apatite locally includes oriented rods or lamellae (Fig. 4H) of a mineral phase, that we were unable to identify because of its very small dimensions (< 20 × 1 μm). Some authors have reported the presence of oriented rods of monazite within apatite (Zhang and Liou, 2000): however, in our case, this mineral can be excluded inasmuch as it is not Raman active, as is monazite.

Metamorphic evolution

The metamorphic evolution of the Phe-Ky-Ep eclogite is summarized in Figure 6. Excluding the Grt 0—that is the only evidence of prograde metamorphism—seven stages have been distinguished that define the P-T path reported in Figure 7.

Stage 1. The metamorphic peak was characterized by the growth of a coarse-grained mineral assemblage consisting of Grt I + Cpx I + Ky I + Coe + Ap + Rt (Fig. 6). The presence of polycrystalline quartz within peak minerals, and of "supersilicic" Cpx I indicate UHP conditions (e.g., Chopin and Ferraris, 2003). Nevertheless, the lack of both microdiamond and significant K_2O contents in Cpx I suggest pressures lower than 4.0 GPa (e.g., Schmidt and Poli, 1998). Because the calculated pressure of the early decompression stage (stage 2) is 3.0 ± 0.1 GPa, a pressure of around 3.5 GPa for stage 1 is assumed. At this pressure, peak temperatures obtained from the garnet-omphacite geothermometer (Ravna, 2000) are 840 ± 50°C (Fig. 7).

Stage 2. The mineral assemblage stable during stage 2 is: Cpx II + Grt II + Phe core + Ep core + coesite/quartz + rutile + apatite (Fig. 6). During this stage the "supersilicic" Cpx I re-equilibrated to a ferrian-omphacite (Cpx II) by segregating SiO_2 needles. The titan-hematite lamellae within rutile and

TABLE 1. Representative Analyses of Minerals from the Phe-Ky-Ep Eclogite[1]

Analysis:	Grt52	Grt10	Grt19	Grt4	Grt73	Cpx med	Cpx 1	Cpx 34	Cpx 24	Ep38	Ep22	Amp42	Amp43	Amp 117	Amp 116	Amp 107	Wm47	Wm25	Pg120	Pl45	Pl110
Mineral:	Grt 0	Grt I	Grt I re	Grt II	Grt III	Cpx I	Cpx II	Cpx III	Cpx IV	Ep c	Ep r	Bar	Mg-Ktp	Ed	Mg-Hbl	Act	Phe c	Phe r	Pg	Pl	Pl
SiO_2	41.28	40.02	39.32	40.11	42.57	55.34	55.82	56.21	53.62	38.54	38.98	52.32	48.42	48.03	49.91	56.41	53.65	48.16	47.43	60.02	68.90
TiO_2	0.00	0.00	0.00	0.00	0.00	0.00	0.00	0.00	0.00	0.00	0.00	0.00	0.00	0.00	0.00	0.00	0.00	0.89	0.00	0.00	0.00
Al_2O_3	21.75	20.81	19.92	20.88	22.82	9.44	9.70	10.34	7.58	23.59	27.09	10.67	11.11	10.16	8.35	3.01	23.68	28.81	39.34	25.16	19.17
Fe_2O_3	0.00	0.16	2.06	0.00	0.00	6.10	7.70	1.87	2.74	12.86	7.94	3.38	3.98	0.00	1.36	0.00	0.51	3.28	0.59	0.00	0.00
FeO	17.34	21.54	21.02	20.72	18.40	3.39	2.52	3.53	5.41	0.00	0.00	6.63	9.17	11.05	9.91	7.33	2.07	0.00	0.00	0.00	0.00
MnO	0.83	0.91	0.84	0.83	0.91	0.00	0.00	0.00	0.00	0.00	0.00	0.00	0.00	0.00	0.00	0.00	0.00	0.00	0.00	0.00	0.00
MgO	6.16	6.94	6.28	8.43	9.29	6.34	6.23	8.00	9.19	0.30	0.00	13.50	12.44	14.20	14.80	18.03	4.29	2.91	0.00	0.00	0.00
CaO	12.64	10.17	10.89	8.10	7.00	11.32	10.44	13.42	18.42	23.78	24.13	6.17	8.29	11.97	11.36	10.93	0.00	0.17	0.35	6.97	0.32
Na_2O	0.00	0.00	0.00	0.00	0.00	7.99	8.57	6.95	4.04	0.00	0.00	5.13	4.57	2.79	2.66	1.84	0.51	0.50	7.30	7.76	12.39
K_2O	0.00	0.00	0.00	0.00	0.00	0.00	0.00	0.00	0.00	0.00	0.00	0.00	0.00	0.00	0.00	0.00	10.62	10.81	0.00	0.00	0.00
H_2O	0.00	0.00	0.00	0.00	0.00	0.00	0.00	0.00	0.00	1.91	1.93	2.15	2.09	2.09	2.10	2.14	4.50	4.48	4.71	0.00	0.00
Total	100.00	100.55	100.33	99.08	100.99	99.93	100.99	100.32	100.99	100.99	100.07	100.17	100.06	100.29	100.45	99.69	99.83	100.00	99.73	99.91	100.78
Si	3.150	3.058	3.034	3.086	3.184	1.994	1.989	1.998	1.943	3.026	3.033	7.306	6.939	6.906	7.121	7.889	3.574	3.226	3.020	2.674	2.974
Ti	0.000	0.000	0.000	0.000	0.000	0.000	0.000	0.000	0.000	0.000	0.000	0.000	0.000	0.000	0.000	0.000	0.000	0.045	0.000	0.000	0.000
Al	1.956	1.875	1.812	1.893	2.012	0.401	0.407	0.433	0.324	2.183	2.484	1.756	1.877	1.722	1.404	0.496	1.859	2.274	2.952	1.321	0.975
Fe^{3+}	0.000	0.009	0.120	0.000	0.000	0.165	0.207	0.050	0.075	0.760	0.465	0.355	0.429	0.000	0.146	0.000	0.025	0.165	0.028	0.000	0.000
Fe^{2+}	1.107	1.376	1.357	1.333	1.151	0.102	0.075	0.105	0.164	0.000	0.000	0.774	1.099	1.329	1.182	0.857	0.116	0.000	0.000	0.000	0.000
Mn	0.054	0.059	0.055	0.054	0.058	0.000	0.000	0.000	0.000	0.035	0.000	0.000	0.000	0.000	0.000	0.000	0.000	0.000	0.000	0.000	0.000
Mg	0.701	0.791	0.722	0.966	1.036	0.340	0.331	0.424	0.496	0.000	0.000	2.809	2.657	3.043	3.147	3.758	0.426	0.291	0.000	0.000	0.000
Ca	1.033	0.833	0.901	0.667	0.561	0.437	0.399	0.511	0.715	2.000	2.012	0.923	1.273	1.844	1.737	1.638	0.000	0.012	0.024	0.333	0.015
Na	0.000	0.000	0.000	0.000	0.000	0.558	0.592	0.479	0.284	0.000	0.000	1.390	1.271	0.778	0.736	0.499	0.066	0.065	0.902	0.670	1.037
K	0.000	0.000	0.000	0.000	0.000	0.000	0.000	0.000	0.000	0.000	0.000	0.000	0.000	0.000	0.000	0.000	0.902	0.924	0.000	0.000	0.000
OH	0.000	0.000	0.000	0.000	0.000	0.000	0.000	0.000	0.000	1.000	1.000	2.000	2.000	2.000	2.000	2.000	2.000	2.000	2.000	0.000	0.000

[1]Abbreviations: Grt I re = Grt I re-equilibrated; Brs = barroisite; Ep c = epidote core; Ep r = epidote rim; Phe c = phengite core; Phe r = phengite rim.

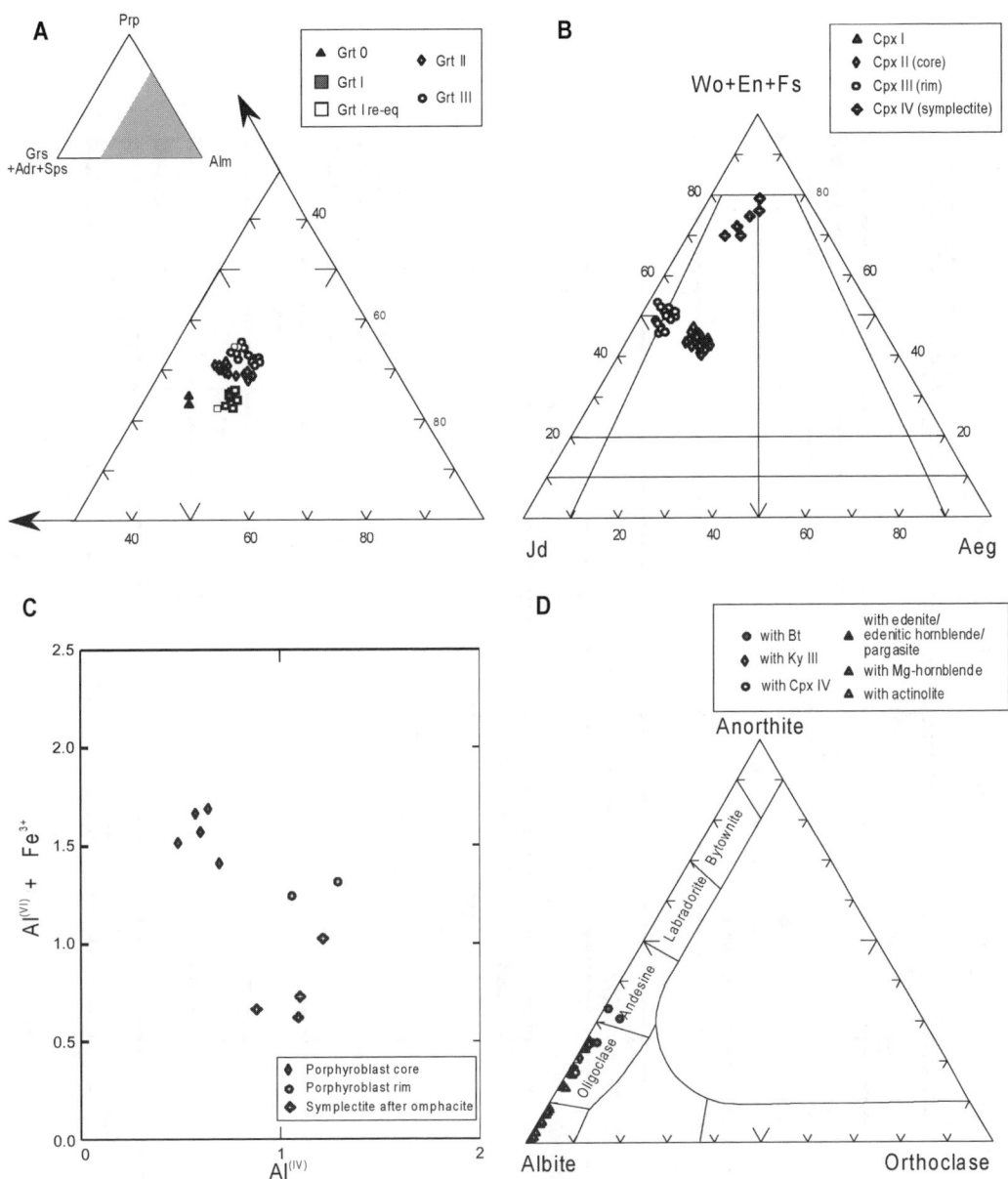

FIG. 5. A. Garnets in the (grossular + andradite + spessartine)-almandine-pyrope triangle. B. Clinopyroxenes in the jadeite-aegirine-(wollastonite + enstatite + ferrosilite) diagram (Rock, 1990). C. Amphiboles in the $Al^{VI} + Fe^{3+}$ vs. Al^{IV} diagram (Laird and Albee, 1981). D. Symplectitic plagioclase in the albite-anorthite-orthoclase diagram.

the undeterminable phase within apatite most probably exsolved during this stage.

The presence of polycrystalline quartz within the epidote core suggests that the early stage of growth was within the coesite stability field. For a nominal pressure of 3.0 GPa, temperatures of 790 ± 50°C were obtained using the garnet-omphacite geothermometer (Ravna, 2000). From the garnet-clinopyroxene-phengite geobarometer (Waters and Martin, 1996), pressures in the range 2.9 GPa and

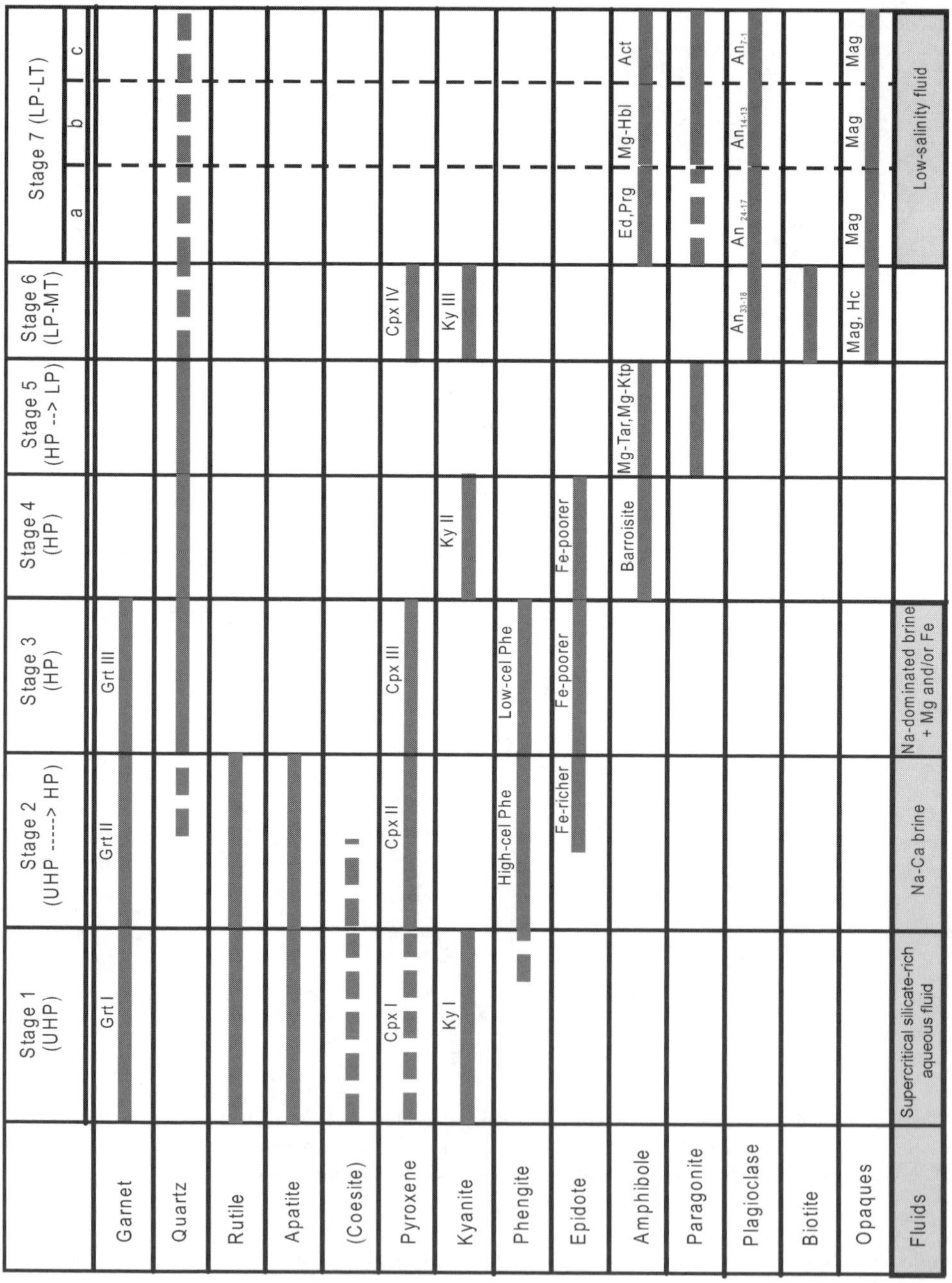

FIG. 6. Metamorphic evolution of eclogites and associated fluids. See text for discussion.

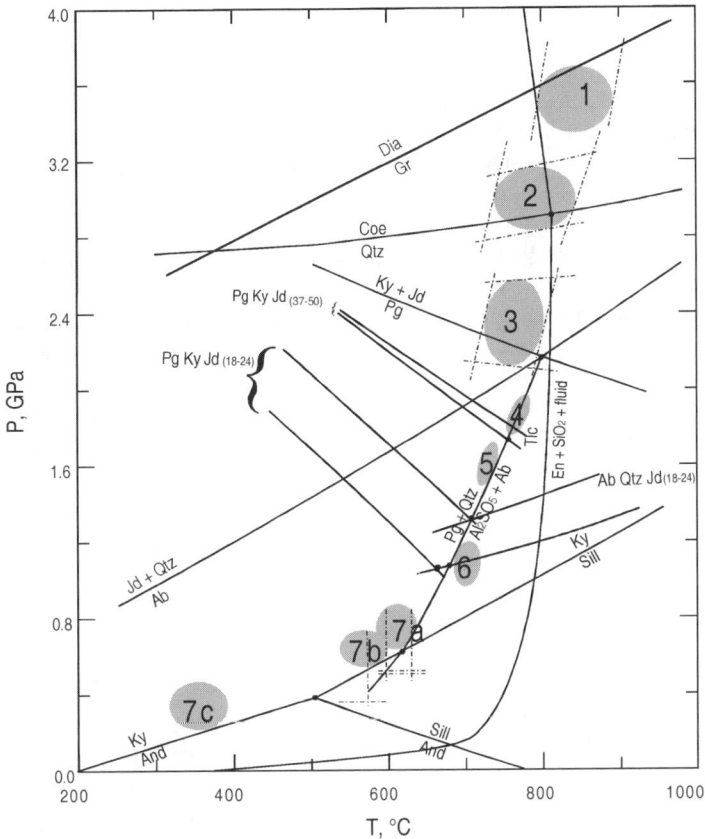

FIG. 7. Selected phase relationships in the CNMASHC system, calculated with the thermodynamic approach of Connolly (1990) with the database of Holland and Powell (1998). The equilibrium curve Dia-Gr is from Bundy (1980). The dots are invariant points. Fields labeled from 1 to 7c refer to the main metamorphic stages shown in Figure 6. Dashed-dotted lines bracket temperatures and pressures estimated from the geothermometers of Ravna (2000) and Holland and Blundy (1994), and from the geobarometers of Waters and Martin (1996) and Schmidt (1992). See text for discussion.

3.1 GPa were calculated, in agreement with the transition from UHP to HP conditions (Fig. 7).

Stage 3. Grt III, Cpx III, phengite rim, epidote rim, and quartz grew during this stage (Fig. 6).

The presence of quartz within the epidote rim suggests that stage 3 occurred within the quartz stability field. For a nominal pressure of 2.5 GPa, temperatures of 775 ± 40°C are obtained using the garnet-omphacite geothermometer (Ravna, 2000). From the garnet-clinopyroxene-phengite geobarometer (Waters and Martin, 1996), pressures in the range 2.1–2.6 GPa were obtained, which are in agreement with the presence of quartz instead of coesite in the mineral assemblage (Fig. 7).

Stage 4. This stage is characterized by the growth of winchite or barroisite, the most external epidote rims, and Ky II in association with quartz (Fig. 6). Garnet and clinopyroxene are partly replaced by blue-green amphibole, but are not yet breaking down to produce plagioclase-bearing symplectites. The isopleths of the reaction Ky + Jd = Ab for clinopyroxene with compositions X_{Jd} = 0.37 and 0.50 (similar to Cpx III) suggest pressures higher than 1.8 GPa, whereas the T-constraints given by previous and next stages and the reaction Pg + Qtz = Ky + Ab suggest temperatures around 750°C (Fig. 7).

Stage 5. Stage 5 is characterized by the growth of paragonite, in equilibrium with quartz, at the expense of kyanite (Ky I and Ky II) by the reaction Ky + Jd = Pg. Most likely Mg-taramite/Mg-katophorite (+ quartz), which are rimming winchite or barroisite, grew during this stage (Fig. 6); this is suggested by the relationships of Mg-taramite/Mg-katophorite rim with the minerals of stage 4, the absence of plagioclase, and the lack of this amphiboles in symplectites of stage 7. Pressures lower than 1.7 GPa are suggested by the calculated isopleths of the reaction Ky + Jd = Pg for clinopyroxenes with X_{Jd} = 0.37 and 0.50—i.e., compositions similar to Cpx III. The reaction Pg + Qtz = Ky + Ab implies temperatures lower than 750°C (Fig. 7).

Stage 6. This stage is characterized by the symplectitic breakdown of eclogite-facies minerals involving production of andesine or Ca-rich oligoclase: pyroxene is replaced by a Jd-poor ferrian-omphacite/aluminian aegirin-augite (Cpx IV) + oligoclase, phengite by biotite + andesine/oligoclase, and paragonite + quartz of stage 5 by Ky III + oligoclase ± hercynite symplectite according to the reaction Pg + Qtz = Ky + Ab (Fig. 6).

Isopleths of the reactions Ky + Jd = Pg and Jd + Qtz = Ab calculated for clinopyroxenes with X_{Jd} = 0.18 and 0.24 (i.e., compositions similar to Cpx IV) suggest pressures lower than 1.2 GPa. The reaction Pg + Qtz = Ky + Ab and the stability of the Ky III + oligoclase ± hercynite symplectite suggest minimum temperatures of 700°C. Hercynite + plagioclase symplectites after kyanite have also been described by Nakamura and Hirajima (2000) in granulitized eclogites from the northern Sulu orogen, which contain orthopyroxene, a mineral lacking in the rocks from the southern Sulu orogen. This is in agreement with the petrographic observation showing that quartz and hercynite are not in equilibrium because they were never in contact. Therefore, the estimated pressures (<1.2 GPa) and temperatures (>700°C) indicate HT amphibolite-facies conditions, near the HP granulite-facies (Fig. 7).

Stage 7. This stage is characterized by the growth of amphibole + plagioclase ± magnetite symplectites, and fine-grained paragonite. Three successive retrogression symplectitic assemblages, forming at the expense of clinopyroxene and porphyroblastic blue-green amphibole, have been recognized: edenite/edenitic hornblende/pargasite + Ca-richer oligoclase (stage 7a); Mg-hornblende + Ca-poorer oligoclase (stage 7b); actinolite + albite (stage 7c; Fig. 6). From the Al-in-hornblende barometer of Schmidt (1992), pressures of 0.55 ± 0.02 and 0.37 ± 0.02 GPa were obtained for the associations 7a and 7b, respectively. Probably, these pressures are underestimated because of the low Al content in amphiboles. The lack of sillimanite or andalusite suggests pressures between 0.7 and 0.2 GPa. For an assumed pressure of 0.5 GPa, equation B of the amphibole-plagioclase geothermometer of Holland and Blundy (1994) gives temperatures of 610°C and 570°C for stages 7a and 7b, respectively, corresponding to amphibolite-facies conditions (Fig. 7). As to stage 7c, the actinolitic composition of the amphibole prevents the use of the amphibole-plagioclase geothermometer; however, since actinolite + albite is a typical greenschist-facies association, temperatures around 350°C are inferred (Fig. 7).

In conclusion, the P-T path described above for the studied eclogites is in good agreement with that reported by Zhang at al. (1995) only from our stages 2 to 7b. In fact, a further higher pressure and temperature stage has been identified in the present study, which is characterized by the growth of coarser-grained anhydrous minerals. Most likely it is the real metamorphic peak.

Choice of the garnet-omphacite geothermometer

Table 2 shows temperatures of stages 1, 2, and 3, which have been estimated from the garnet-omphacite geothermometers, calibrated by Ravna (2000), Krogh (1988), Powell (1985), Ellis and Green (1979), Ganguly (1979), and Raheim and Green (1974), respectively. For all stages, the calibrations of Ravna (2000), Krogh (1988), Powell (1985), and Raheim and Green (1974) give identical or very similar temperatures (averaging around 840, 800, and 790°C, respectively), always lower than those estimated from the Ellis and Green (1979) and Ganguly (1979) calibrations.

As evident from Figure 7, a maximum temperature of 810°C for stage 2 is constrained by the upper stability of talc in a system with X_{H2O} = 0.90, i.e. the fluid composition close to that found within *type II* fluid inclusions (see below). Furthermore, petrographic study of the talc-bearing eclogite from Qinglongshan (RPC742, RPC787) has shown that the metamorphic peak (stage 1) is characterized by anhydrous minerals, talc growing only during an early decompression (stage 2), in agreement with the evidence reported in the present study, that hydrous minerals started to grow at the UHP-HP transition. For such reasons, Ravna's (2000) calibration has been selected, which also gives more consistent T values for stages 1, 2, and 3.

TABLE 2. Summary of Geothermometric Estimates Calculated from the Indicated Calibrations, °C

Geothermometer	Metamorphic stage		
	Stage 1	Stage 2	Stage 3
Ravna (2000)	840 ± 50	790 ± 50	775 ± 40
Krogh (1988)	840 ± 50	790 ± 60	790 ± 50
Powell (1985)	845 ± 90	800 ± 50	815 ± 40
Ellis & Green (1979)	860 ± 50	820 ± 50	830 ± 20
Ganguly (1979)	910 ± 30	890 ± 40	900 ± 30
Raheim & Green (1974)	845 ± 40	810 ± 60	820 ± 30

Fluid Inclusion Study

Fluid inclusion types and their occurrence

Rare fluid inclusions are observed in the studied samples. Aqueous inclusions and no gas has always been detected. A careful examination of the textural relationships between fluid inclusions and host minerals make it possible to recognize four different generations of fluid inclusions.

Type I. Grt I and Ky I locally contain multiphase solid inclusions that consist of mineral aggregates and cavities without visible fluid. Type I inclusions occur both in Grt I with random orientation (Fig. 8A) and in Ky I with the long dimension parallel to the mineral elongation. The inclusions have dimensions between 10 and 100 µm, and show a tendency toward negative crystal shape or a shape similar to that known from decrepitated fluid inclusions. Each multiphase solid inclusion contains various crystals (about 10), corresponding to 2–4 different mineral phases. Rutile (and/or an opaque mineral) and a colorless mineral are always present. A green pleochroic mineral with high relief and low birefringence is also recognized. One or more cavities are invariant within single inclusions, suggesting the former presence of a fluid phase. Locally, small trails of tiny fluid inclusions occur around single multiphase solid inclusions, which appear to have originated by decrepitation.

Type II. Type II fluid inclusions occur within the epidote core as isolated oriented inclusions (from 5 to 20 µm; Fig. 8B), and, locally, as secondary intragranular trails (van den Kerkhof and Hein, 2001). More rarely, intragranular trails, originating at grain boundary, are present within Ky I. Evidence for partial decrepitation are common (Roedder, 1984). Type II inclusions are biphase aqueous inclusions (L+V) with a degree of filling (d_f = L/L+V) of 80–90. Locally, the biggest inclusions within epidote core contain a birefringent solid phase.

Type III. Type III fluid inclusions occur mainly as isolated inclusions, from 3 to 20 µm in length, within epidote rims (Fig. 8C) and, locally, they form clusters within the undeformed matrix quartz. Rarely, intragranular trails are observed in both epidote and quartz. Generally, the inclusions show evidences of "post-trapping" changes (Roedder, 1984). Type III are three phases aqueous inclusions (L + V + S; d_f = 70–80). One to three anisotropic cubic phases, probably salts, are characteristically present within the inclusions.

Type IV. Type IV fluid inclusions, from 5 to 15 µm long, occur as isolated inclusions within matrix quartz, and as intragranular trails both within matrix quartz (Fig. 8D) and epidote. In a few of them, evidence of "post-trapping" changes is observed. Type IV are biphase (L + V) aqueous inclusions with d_f = 70–90.

Fluid inclusion analyses and compositions

EDS analysis indicates that the typical mineral association observed within multiphase solid inclusions (type I) is: paragonite + rutile (with exsolution lamella of ilmeno-hematite) + apatite. Ferroan pargasite is typically present in the inclusions within Grt I, whereas Zn-staurolite occurs within Ky I. Magnetite ± plagioclase ± zircon ± pyrite ± an "alunite-type" sulfate ± Zn-Mg-Fe-Al-Ti spinel are locally present. Trapped fluids are aqueous and contain very high amounts of Si, Ti, Al, Ca, Fe, Na, P, and Mg as dissolved ions, which allow us to identify type I fluid as supercritical silicate-rich aqueous fluid (Ferrando et al., in press).

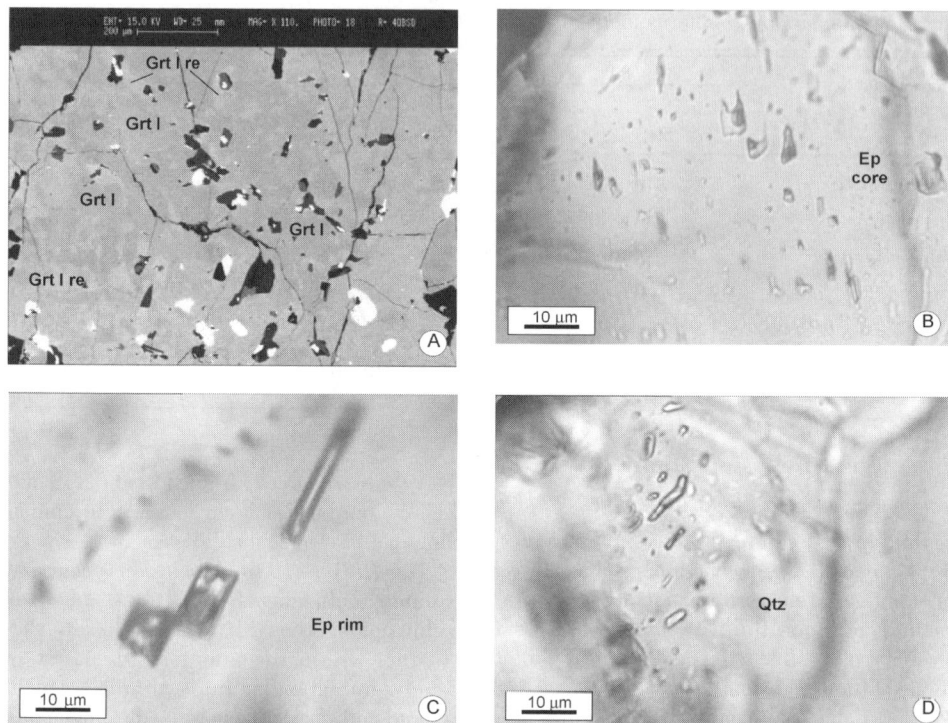

FIG. 8. Types of fluid inclusions within the Phe-Ky-Ep eclogites from Hushan and Qinglongshan. A. Photomicrograph of type I primary multiphase solid inclusions within Grt I. Note that garnet is re-equilibrated (Grt I re) close to the decrepitated multiphase solid inclusion. Sample RPC778, PPL. B. Photomicrograph of early type II aqueous fluid inclusions in the epidote core. RPC778, PPL. C. Photomicrograph of early type III aqueous fluid inclusions in the epidote rim. Note that some inclusions contain also two solids. RPC778, PPL. D. Photomicrograph of type IV aqueous fluid inclusions in the matrix quartz. Sample RPC778, PPL.

Figure 9A shows the range of freezing and melting temperatures for type II fluid inclusions. The Tf is between –62.2 and –58.9°C, and the first melting is recognized between –52.4 and –48.0°C. The Tm_{Hhl} are in the range from –30.1 to –22.0°C (with majority at –22.2°C) and the Tm_{ice} varies from –16.7 to –6.4°C (with the majority at –12.8°C). The ThL_{H2O} are very scattered between 139.8 and 372.1°C (Fig. 9B). Although the inclusions were heated up to 400°C, the solid crystals did not dissolve; Raman spectroscopy reveals that they are Mg-calcite (Fig. 9C). Microthermometric data (Table 3) indicate that fluid II is a brine with 13–14 wt% NaCl and 7–5 wt% $CaCl_2$. The total fluid density, calculated using the minimum value for ThL_{H2O}, is 1.04 g/cm³.

Tf of type III fluid inclusions is between –56.6 and –45.4°C, Te varies between –39.4 and –30.8°C, Tm_{ice} is between –27.5 and –20.7°C (with the majority at –21.6°C), and Tm_{Hhl} varies between –1.9 and –0.4°C (with the majority at –0.6°C; Fig. 9A). The ThL_{H2O} varies from 300.6 to 363.4°C with peak at around 340°C (Fig. 9B). The dissolution of the salts, present within the inclusions, is difficult to determine, but in one case it was observed at 204°C (Tm_{dm}). These data (Table 3) indicate that the fluid is a Na-dominated brine (up to 32 wt% NaCl eq.), and that Mg^{2+} and/or Fe^{2+} may be also present in the fluid, whereas Ca appears to be absent. The total fluid density, calculated using the minimum value for ThL_{H2O}, is 1.12 g/cm³.

Type IV fluid inclusions freeze (Tf) between –36.6 and –32.2°C, Te ranges between –22.3 and –19.0°C, and Tm_{ice} varies from –8.6°C to –4.5°C (with the majority at –4.5°C; Fig. 9A). ThL_{H2O} ranges from 330.0 to 367.0°C (Fig. 9B). These data (Table 3)

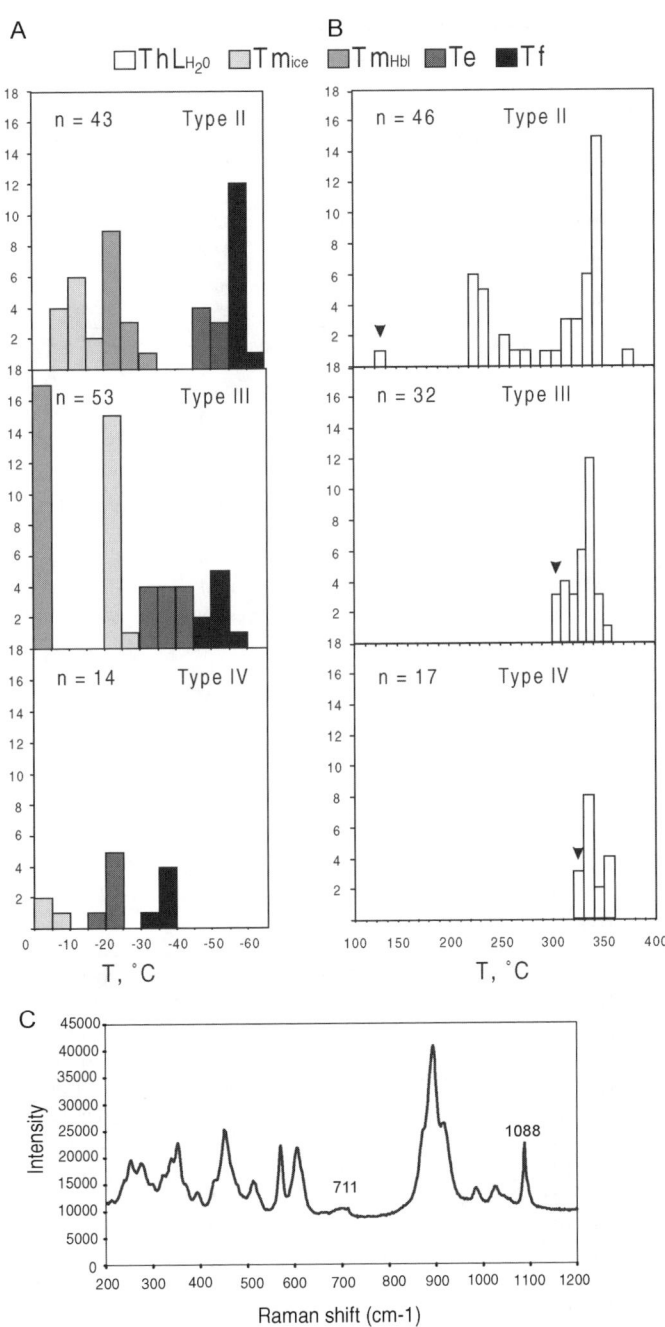

FIG. 9. Fluid inclusion data. A. Histograms showing freezing (T_f) and melting (T_m) temperatures for type II, III, and IV fluid inclusions, respectively. B. Histograms showing homogenization temperatures (ThL_{H2O}) for type II, III, and IV fluid inclusions, respectively. The arrow shows the value selected to calculate the isochore reported in Figure 10. C. Representative Raman spectrum of a Mg-calcite crystal. The numbered peaks (711 and 1088 cm^{-1}) are Mg-calcite vibrations. The spectrum also contains the peaks of the host epidote (unlabeled).

TABLE 3. Summary of Fluid Inclusion Data for the Phe-Ky-Ep Eclogite

Sample	Host mineral	Fluid inclusion type	Tf_{H2O}	Te	Tm_{Hhl}	Tm_{ice}	Tm_{dm}	ThL_{H2O}
778-9dA	Ep core	Type II	−62.2	−50.5	−22.0	−12.8		343.5
778-9dA	Ep core	Type II	58.3					342.9
778-9dA	Ep core	Type II		−48.5	−22.2	−7.1		347.2
778-9dA	Ep core	Type II	−59.6		−24.2	−7.9		338.4
778-11fA1	Ep core	Type II	−59.0			−11.5		294.7
778-11fA1	Ep core	Type II				−12.0		139.8
778-11dA	Ep core	Type II		−48.5	−22.2	−7.1		328.6
779-11A	Ep core	Type II			−30.1	−6.4		372.1
779-3aB	Ep core	Type II			−23.2			305.8
779-11dA	Ep core	Type II			−25.4	−11.5		326.2
778-9eC	Ep core	Type II	−59.4	−48.5	−25.4	−11.5		
778-9eC	Ep core	Type II						
778-9aA	Ky I	Type II		−51.5	−27.3	−11.1		
778-9aA	Ky I	Type II						
778-7cA	Ky I	Type II	−59.9	−48.0	−28.5	−12.2		272.0
778-7cA	Ky I	Type II	−59.9	−52.4	−22.4	−16.7		234.4
778-7cA	Ky I	Type II			−23.9	−15.5		227.0
778-3A	Ep rim	Type III		−36.3				331.2
778-3A	Ep rim	Type III	−45.4					331.7
779-4A	Ep rim	Type III			−1.9			339.5
779-4A	Ep rim	Type III		−30.8	−1.6			341.4
779-4A	Ep rim	Type III			−1.4			340.8
779-4D	Ep rim	Type III	−50.6	−36.6	−0.6			342.4
779-4B	Ep rim	Type III			−0.8			344.7
779-4B	Ep rim	Type III			−0.8			346.9
779-4C	Ep rim	Type III		−31.9	−1.2			363.4
779-4C	Ep rim	Type III				−21.4		347.1
779-4C	Ep rim	Type III				−21.2		350.8
779-4C	Ep rim	Type III			−0.5	−23.1		353.1
778-11aA	Ep rim	Type III	−49.3	−30.8	−0.5	−20.7		310.9
778-11aA	Ep rim	Type III	−50.2		−0.5			321.5
778-11aA	Ep rim	Type III	−50.2	−36.1	−0.4	−21.6		318.0
778-11aA	Ep rim	Type III	−50.2	−39.4	−0.4	−21.6	204.0	300.6
778-11aA	Ep rim	Type III			−0.6			313.8
778-11aB	Ep rim	Type III				−20.8		326.2
778-11aB	Ep rim	Type III			0.6	−21.3		330.7
778-11gC	Ep rim	Type III		−34.4		−23.5		343.7
778-11cA	Ep rim	Type III	−61.8		−0.4			326.2
778-11eD	Qtz	Type III	−56.6		−0.9			341.7
778-11eD	Qtz	Type III	−52.5					346.0
778-11fA2	Qtz	Type IV		−22.2				340.5
778-11fA2	Qtz	Type IV		−22.3				343.1
778-9eB	Qtz	Type IV	−36.3	−20.1		−4.5		364.2
778-9eB	Qtz	Type IV						324.6
778-9cB	Ep	Type IV	−36.6	−19.0				330.0
778-9cB	Ep	Type IV	−36.0	−21.3		−3.7		365.1
778-9cB	Ep	Type IV	−32.2	−20.0				330.0
778-9cB	Ep	Type IV	−35.0					353.4

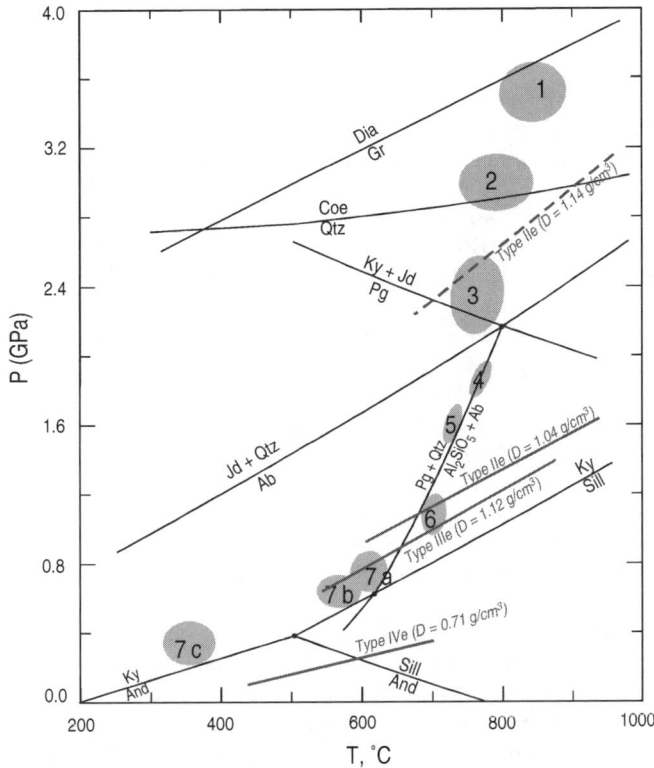

FIG. 10. Simplified petrogenetic grid and P-T path of Figure 7 in which are reported the isochores (grey lines) determined from microthermometric data of type II, III, and IV fluid inclusions. The dashed isochore is calculated for type II fluid inclusions with the maximum density. See text for discussion.

suggest that the fluid has low salinity and contains 2–6.5 wt% of NaCl. The total fluid density value is 0.74 g/cm^3.

Evolution of the fluid phase

The evolution of the fluid phase within the Phe-Ky-Ep eclogite is summarized in Figure 6. The calculated isochores for all fluid types are reported on the P-T path of Figure 10 and are all contained within the amphibolite-facies P-T space. This indicates that the density or the molar volume of all fluid inclusions re-equilibrated during exhumation, in agreement with the decrepitation structures and the scatter of ThL_{H2O}. It is possible, however, to reconstruct the evolution of the fluids on the basis of their textural relationships, because the well-constrained P-T conditions of mineral growth allow us to define the metamorphic stages at which the different kinds of fluid are in equilibrium.

The early type I fluids are recorded by the presence of the multiphase solid inclusions within peak Grt I and Ky I. The occurrence and composition of the inclusions suggest that the fluid, in equilibrium with the anhydrous mineral assemblage stable at peak conditions, was an aqueous fluid containing elevated amounts and kinds of dissolved ions (Fig. 6).

Type II fluid inclusions, occurring as early inclusions within epidote cores and as secondary inclusions within peak kyanite (Ky I), indicate that a Na-Ca brine was in equilibrium at the transition from UHP to HP conditions (stage 2; Fig. 6). Similar compositions of fluids included within both epidote cores and Ky I exclude the possibility that the Ca^{2+} content of the fluid results from "post-trapping" interaction between fluid inclusions and host mineral. Conversely, interaction at HP within the host epidote may have caused precipitation of Mg-calcite in the fluid inclusions. Re-equilibrated type II fluid inclusions (d = 1.04 g/cm³) give a pressure of 1.4 GPa (for T = 800°C), considerably lower than values calculated for stage 2 (2.9–3.1 GPa). However, to calculate the highest density attainable by a fluid with this composition, the cavity has been supposed to be filled only by liquid. The resulting density is 1.14 g/cm³, which for a nominal temperature of 800°C gives a pressure of 2.7 GPa, in good agreement with P-T conditions calculated for stage 2 (Fig. 10).

Type III fluid inclusions within epidote rim and matrix quartz were trapped during stage 3 (Fig. 6), in the range 2.1–2.6 GPa (Fig. 10). At this stage, the fluids are still brines, but show important differences from earlier type II fluids: higher salinity, presence of significant Mg and/or Fe, and lack of Ca^{2+}. Also in this case, original compositions were preserved inasmuch as similar type III brines are present in minerals with different compositions (epidote and quartz).

The latest fluid recognized is represented by early and secondary type IV fluid inclusions within matrix quartz (Fig. 6). Textural relationships indicate that the Na-bearing low-salinity and low-density fluid was probably trapped during late stages of exhumation (stage 7), under amphibolite- to greenschist-facies conditions (Fig. 10).

Oxygen Isotope Study

Isotope analysis

Mineral separates were obtained as described in section on Analytical Methods, selecting the most clear and coarser-grained crystals, and removing clinopyroxene and amphibole grains showing evidence of symplectitic retrogression. In the final separation, clear rutile, clear garnet, dark green clinopyroxene, dark yellow epidote, and dark blue amphiboles grains were selected, which assures that: (1) most separated clinopyroxenes were Cpx I or Cpx II, representative of stages 1 and 2, respectively; (2) most separated epidotes were the cores of the crystals, belonging to stage 2, although the presence of zoned crystals (stages 2 and 3) cannot be excluded; (3) separated amphiboles belong to stage 4; (4) separated clear garnet crystals were probably Grt II and Grt III, representative of stages 2 and 3, respectively, because Grt I has a dusty appearance due to the presence of the multiphase solid inclusions; and (5) separated rutiles, lacking the ilmenohematite lamella, do not represent stage 1 rutile.

Table 4 shows the $\delta^{18}O$ values of eclogites: they vary from –5.71‰ to –6.09‰ for quartz, from –9.04‰ to –9.66‰ for garnet, from –8.69‰ to –9.77‰ for epidote, from –7.66‰ to –9.36‰ for clinopyroxene, from –7.30‰ to –8.86‰ for amphibole, and from –12.10‰ to –14.16‰ for rutile. The $\delta^{18}O$ values are homogeneous with the exception of rutile and quartz, which show substantial variations. The $\delta^{18}O$ enrichment is quartz > garnet ≥ epidote ≥ pyroxene ≥ amphibole > rutile, which indicates equilibrium fractionation (Zheng, 1993a, 1993b).

In comparison with the $\delta^{18}O$ values (from –5.3 ‰ to +0.8 ‰) previously obtained by Rumble and Yui (1998) on Ep-eclogite from Qinglongshan, the present study gives systematically lower $\delta^{18}O$. On the contrary, these data are similar to those (from –14.6‰ to –4.9‰) reported by Rumble and Yui (1998) for eclogites s.s from Qinglongshan, to those (from –14.5‰ to +0.8‰) for Zo-eclogites (Zheng et al., 1998), to those (from –10.3‰ to –6.8‰) for an Ep-eclogite (Zheng et al., 1996), and to those (from –10.4‰ to –9.0‰) obtained by Yui et al. (1995) for Ep-eclogite from the same locality.

The $\delta^{18}O$ values of minerals increase, as expected from O-isotope fractionation under equilibrium conditions (Taylor and Epstein, 1962). Preservation of the pre-metamorphic isotopic signature demonstrates the absence of a pervasive fluid infiltration during subduction, metamorphic peak, and exhumation. This indicates that: (1) the eclogites recrystallized in a closed system; (2) fluids at UHP peak conditions had limited mobility; (3) hydration reactions, occurring during exhumation, were internally buffered. Moreover, the extremely negative $\delta^{18}O$ values measured in the mineral separates may be explained, in agreement with Rumble and Yui (1998), Yui et al. (1995), and Zheng et al. (1996, 1998, 2003), as the result of a meteoric-hydrothermal alteration of the eclogite protoliths under a cold climate.

Oxygen isotope geothermometry

The O-isotope fractionation between two coexisting phases A and B is defined as: $\Delta^{18}O_{(A-B)} = \delta^{18}O_A - \delta^{18}O_B \cong 1000 \ln \alpha_{A-B}$, where α is the fractionation factor related to the temperature according to the equation: $1000 \ln \alpha_{A-B} = a * 10^6/T^2 \pm b$. The constants a and b are determined empirically and/or experimentally, and T is the temperature in K. In this study, the coefficients of fractionation used are: $a_{(Qtz-Rt)} = 5.02$ (Matthews, 1994; temperature uncertainty ±40°C); $a_{(Qtz-Grt)} = 3.1 \pm 0.2$ (Sharp, 1995;

TABLE 4. Isotopic Compositions of Oxygen in the Phe-Ky-Ep Eclogite and Temperature Estimates

Sample	$\delta^{18}O$av. (‰)	ΔGrt-Rt	ΔGrt-Amp	ΔGrt-Ep	ΔGrt-Omp	ΔQtz-Grt	ΔQtz-Rt	ΔQtz-Amp	ΔQtz-Omp	ΔQtz-Ep	ΔAmp-Omp	ΔAmp-Ep	ΔAmp-Rt	ΔOmp-Ep	ΔOmp-Rt	ΔEp-Rt	$T_{Qtz-Grt}$, °C	T_{Qtz-Rt}, °C	$T_{Qtz-Omp}$, °C
RPC778 Grt	−9.08 ± 0.09	3.9	−1.0	−0.4	−0.7														
RPC778 Rt	−12.94																		
RPC778 Amp	−8.12 ± 0.08												4.8						
RPC778 Ep	−8.69 ± 0.11										0.3	0.6				4.3			
RPC778 Omp	−8.41 ± 0.04													0.3	4.5				
RPC619 Grt	−9.66	4.5	−0.8	−0.2	−0.3														
RPC619 Amp	−8.86 ± 0.18										0.5	0.6	5.3						
RPC619 Qtz	−6.09 ± 0.2					3.6	8.1	2.8	3.3	3.3							710 ± 25	565 ± 10	615 ± 25
RPC619 Omp	−9.36													0.1	4.8				
RPC619 Ep	−9.43 ± 0.12															4.7			
RPC779 Rt	−12.10 ± 0.12																		
RPC779 Amp	−7.30 ± 0.03										0.4		4.8						
RPC779 Ep	−9.77																		
RPC779 Omp	−7.66														4.4				
RPC779 Qtz	−5.71 ± 0.05					3.3	6.4	1.6	2.0								740 ± 25	665 ± 5	860 ± 15

TABLE 5. Comparison between Temperatures Obtained by Different Authors
Using Isotopic Geothermometry, °C

Author	Qtz-Grt	Qtz-Omp	Qtz-Rt
Rumble and Yui (1998)	725–782	600–610	576–630
Zheng et al. (1998)	650–765	655–760	390–510
This work	710–740	615–860	565–665

temperature uncertainty ±30°C); $a_{(Qtz-Omp)}$ = 2.75–1.06 X_{Jd} (Matthews, 1994; temperature uncertainty ±40°C), where X_{Jd} = molar fraction of jadeite in omphacite. Because coesite was the stable polymorph of silica at the inferred peak metamorphic conditions, a positive correction of about 50°C was made to the geothermometric estimates yielded by the Qtz-mineral pairs (Sharp et al., 1992). The $\delta^{18}O$ values for the coexisting minerals in the Sulu eclogites are reported in Table 4. Mineral pairs with large O-isotope fractionation generally yield more accurate thermometric estimates than minerals with similar $\delta^{18}O$ values, the latter resulting in $\delta^{18}O$ values close to the analytical precision (± 0.2‰). In this study, temperatures were calculated using Qtz-Grt, Qtz-Rt, and Qtz-Omp pairs of only two samples, due to the insufficient amount of quartz in sample RPC778. The calculated temperatures (corrected for coesite) are: 615–860°C for Qtz-Omp pairs (X_{Jd} = 0.43, i.e. the mean value of Cpx I and Cpx II), 710–740°C for Qtz-Grt pairs, and 565–665°C for Qtz-Rt pairs (Table 4).

The Qtz-Omp pair from sample RPC779 yielded a temperature very similar to that calculated using the Grt-Omp cationic geothermometry for stage 1 (840 ± 50°C), whereas the Qtz-Grt pair gave lower temperatures, corresponding to the values estimated by Grt-Omp cationic geothermometry for stage 2 (790 ± 50°C) or 3 (775 ± 40°C). Temperatures calculated using the Qtz-Rt pair are invariably lower than those obtained for the metamorphic peak or the earliest retrogression stage. A comparison with previous studies, reported in Table 5, indicates that the temperatures calculated from Qtz-Omp and Qtz-Grt pairs are similar to those obtained by Rumble and Yui (1998) with the Qtz-Grt pairs. Present data indicate that the minerals of the studied eclogite attained high-temperature oxygen-isotope equilibrium during the metamorphic peak under UHP conditions, followed by incomplete re-equilibration during decompression (Rumble and Yui, 1998).

Discussion and Conclusions: Fluid-Rock Interaction

The present multidisciplinary study on Phe-Ky-Ep eclogite from southern Sulu reveals that: (1) the rocks experienced a complex metamorphic evolution, which is recorded by different generations of zoned minerals; (2) the associated fluids were aqueous, and progressively changed composition and salinity; and (3) during metamorphic evolution, the fluid-rock system remained closed. This implies that fluid-rock interaction during prograde, peak, and retrograde metamorphism occurred without involvement of external fluids. The prograde evolution cannot be constrained, because of the lack of evidence, the only mineral being a very rare relict garnet (Grt 0). In contrast, the earliest well-recorded metamorphic event, which is the metamorphic peak, is characterized by an anhydrous (or nominally anhydrous) mineral assemblage (Grt I + Cpx I + Ky I + Coe) that developed at P-T conditions close to the graphite-diamond inversion curve. The fluid in equilibrium during stage 1 was a supercritical silicate-rich aqueous fluid, probably generated during prograde dehydration of the rocks, as confirmed by its composition consistent with that of the host eclogite. It is probable that the stability of anhydrous minerals during stage 1 was favored by low water activity of this kind of fluid (Aranovich and Newton, 1997). Textural evidence shows that the hydrous minerals, such as epidote and phengite, started to grow during stage 2, i.e., at the UHP to HP transition. During this stage, the fluid in equilibrium with the rock was a Na-Ca brine, in agreement with the coeval growth of Ca-bearing hydrated phases. Type II fluid might have been generated from type I fluid at high temperatures during decompression, which favored the precipitation of low-soluble elements, such as Ti, Si, and Al. The later almost isothermal decompression under HP conditions (stage 3) was characterized by decreasing Fe^{3+} and increasing Mg in the minerals. It is reasonable to suppose that the fluid in equilibrium during stage 2 underwent progressive Ca and H_2O depletion during mineral growth. The type III fluid, indeed, was a high-salinity, Na-dominated brine containing Mg and/or Fe but no Ca. It is probable that such a fluid promoted mineral zoning, but lack of clear textural evidence prevents us from determining if it was also responsible for the growth of blue-green amphibole of stage 4 and paragonite of stage 5. The LP evolution is initially characterized by "almost anhydrous" symplectites. The stability of andesine, or Ca-rich oligoclase, within the symplectites with Cpx IV, Ky III, biotite, and hercynite suggests that HT amphibolite-facies conditions at the boundary with HP granulite-facies conditions were reached (stage 6). Finally, stage 7, from amphibolite- to greenschist-facies conditions, was characterized by the occurrence of subsequent generations of "hydrous" symplectites, consisting of amphibole + plagioclase, probably promoted by low-salinity aqueous fluid in equilibrium at these P-T conditions. The origin of this late fluid is not clear: either it may have been internally derived from previous fluids or it may have had an external origin.

The overall fluid evolution in the studied rocks was dominated by aqueous fluids, with different salinity and cations in solution, and bears no evidence of CO_2 or other gases during the exhumation stages. Similar retrograde aqueous fluids have been observed in other HP-UHP rocks from China (e.g., Fu et al., 2001, 2002, 2003a, 2003b; Shen et al., 2002; Xiao et al., 2002; Fan et al., 2003), West Sudetes (Klemd and Bröcker, 1999), and Western Alps (e.g., Philippot et al., 1995; Scambelluri et al., 1998). According to many authors (e.g., Touret and Frezzotti, 2003) external CO_2 fluids in UHP rocks are related to granulite-facies retrogression and/or to the degree of deformation of the rocks. In this hypothesis, the lack of CO_2 fluids in the studied rocks is not surprising inasmuch as eclogites are granoblastic and underwent amphibolite-facies retrogression.

Acknowledgments

This work is part of the first author's Ph.D. thesis. The authors are grateful to D. Castelli and R. Cossio for help in calculating the petrogenetic grids and the quantitative analyses of Cpx I, respectively; T. Hirajima for constructive criticism and suggestions; and S. Xu, Y. Liu, W. Wu, and F. Rolfo for help and assistance during field work and for constructive comments. This work was supported by MURST, Finanziamento Convenzioni Interuniversitarie, Fondi Scambi Culturali Università di Torino, National Research Council of Italy, Institute of Geosciences and Earth Resources, Section of Torino, and "Programmi Ricerca di Ateneo 2002" of the University of Siena. Raman analytical facilities were provided by the Italian organization for research in Antarctica (P.N.R.A.).

REFERENCES

Ames, L., Zhou, G., and Xiong, B., 1996, Geochronology and isotopic character of ultrahigh-pressure metamorphism with implications for collision of the Sino-Korean and Yangtze cratons, central China: Tectonics, v. 15, p. 472–489.

Aranovic, L.Y., and Newton, R.C., 1997, H_2O activity in concentrated KCl and KCl-NaCl solutions at high temperatures and pressures measured by the brucite-periclase equilibrium: Contributions to Mineralogy and Petrology, v. 127, p. 261–271.

Bakker, R.J., 2003, Package FLUIDS 1. Computer programs for analysis of fluid inclusion data and for modelling bulk fluid properties: Chemical Geology, v. 194, p. 3–23.

Brown, P.E., 1989, FLINCOR: A microcomputer program for the reduction and investigation of fluid-inclusion data: American Mineralogist, v. 74, p. 1390–1393.

Bundy, F. P., 1980, The P, T phase and reaction diagram for elemental carbon: Journal of Geophysical Research, v. 85, p. 6930–6936.

Castelli, D., Rolfo, F., Compagnoni, R., and Xu, S., 1998, Metamorphic veins with kyanite, zoisite, and quartz in the Zhu-Jia-Chong eclogite, Dabie Shan, China: The Island Arc, v. 7, p. 159–173.

Connolly, J. A. D., 1990, Multivariable phase diagrams: an algorithm based on generalized thermodynamics: American Journal of Science, v. 290, p. 666–718.

Chopin, C., and Ferraris, G., 2003, Mineral chemistry and mineral reactions in UHPM rocks, in Carswell, D. A., and Compagnoni, R., eds., Ultrahigh-pressure metamorphism: E.M.U. Notes in Mineralogy, v. 5, p. 191–227.

Ellis, D. J., and Green, D. H., 1979, An experimental study of the effect of Ca upon garnet-clinopyroxene Fe-Mg exchange equilibria: Contributions to Mineralogy and Petrology, v. 71, p. 13–22.

Ernst, W. G., 2000, H_2O and ultrahigh-pressure subsolidus phase relations for mafic and ultramafic systems, *in* Ernst, W. G., and Liou, J. G., eds, Ultrahigh-pressure metamorphism and geodynamics in collision-type orogenic belts: Final report of the Task Group III-6 (1994-1998) of the International Lithosphere Project. Columbia, MD: Bellwether Publishing for Geological Society of America, International Book Series, v. 4., p. 121–129.

Fan, H. R., Guo, J. H., Chen, F. K., Jin, C. W., and Shen, K., 2003, Fluid evolution and exhumation history of ultra-high-pressure rocks at Lanshantou, Sulu terrane, Eastern China: Journal of Geochemical Exploration, v. 78-79, p. 51–54.

Ferrando, S., Frezzotti, M. L., Dallai, L., and Compagnoni, R., Multiphase solid inclusions in UHP (Su-Lu, China): Remnants of supercritical silicate-rich aqueous fluids released during continental subduction: Chemical Geology, in press.

Fu, B., Touret, J. L. R., and Zheng, Y.-F., 2001, Fluid inclusions in coesite-bearing eclogites and jadeite quartzite at Shuanghe, Dabie Shan (China): Journal of Metamorphic Geology, v. 19, p. 529–545.

Fu, B., Touret, J. L. R., and Zheng, Y.-F., 2003a, Remnants of pre-metamorphic fluid and oxygen-isotopic signatures in eclogites and garnet clinopyroxenite from the Dabie-Sulu terranes, eastern China: Journal of Metamorphic Geology, v. 21, p. 561–578.

Fu, B., Touret, J. L. R., Zheng, Y.-F., and Jahn, B.-M., 2003b, Fluid inclusions in granulites, granulitized eclogites, and garnet clinopyroxenites from the Dabie-Sulu terranes, eastern China: Lithos, v. 70, p. 293–319.

Fu, B., Zheng, Y.-F., and Touret, J. L. R., 2002, Petrological, isotopic and fluid inclusion studies of eclogites from Sujiahe, NW Dabie Shan (China): Chemical Geology, v. 187, p. 107–128.

Ganguly, J., 1979, Garnet and clinopyroxene solid solutions, and geothermometry based on Fe-Mg distribution coefficient: Geochimica et Cosmochimica Acta, v. 43, p. 1021–1029.

Gasparik, T., 1985, Experimentally determined compositions of diopside-jadeite pyroxene in equilibrium with albite and quartz at 1200-1350°C and 15-34 kb: Geochimica et Cosmochimica Acta, v. 49, p. 865–870.

Haggerty, S. E., 1981, Opaque mineral oxides in terrestrial igneous rocks, *in* Rumble, D., ed., Oxide minerals: Mineralogical Society of America, Reviews in Mineralogy, v. 3, p. Hg101–Hg300.

Hirajima, T., and Nakamura, D., 2003, The Dabie Shan–Sulu orogen, *in* Carswell, D.A., and Compagnoni, R., eds., Ultrahigh-pressure metamorphism: E.M.U. Notes in Mineralogy, v. 5, p. 105–144.

Holland, T. J. B., and Blundy, J. D., 1994, Non-ideal interactions in calcic amphiboles and their bearing on amphibole-plagioclase thermometry: Contributions to Mineralogy and Petrology, v. 116, p. 433–447.

Holland, T. J. B., and Powell, R., 1998, An internally consistent thermodynamic dataset for phases of petrological interest: Journal of Metamorphic Geology, v. 16, p. 309–343.

Klemd, R., and Bröcker, M., 1999, Fluid influence on mineral reactions in ultrahigh-pressure granulites: A case study in the Sniezznik Mts. (West Sudetes, Poland): Contributions to Mineralogy and Petrology, v. 136, p. 358–373.

Kretz, R., 1983, Symbols for rock-forming minerals: American Mineralogist, v. 68, p. 277–279.

Krogh, E. J., 1988, The garnet-clinopyroxene geothermometer: A reinterpretation of existing experimental data: Contributions to Mineralogy and Petrology, v. 99, p. 44–48.

Laird, J., and Albee, A. L., 1981, Pressure, temperature, and time indicators in mafic schist: Their application to reconstructing the polymetamorphic history of Vermont: American Journal of Science, v. 281, p. 127–175.

Leake, B. E., Woolley, A. R., Birch, W. D., Burke, E. A. J., Ferraris, G., Grice, J. D., Hawthorne, F. C., Kisch, H. J., Krivovichev, V. G., Schumacher, J. C., Stephenson, N. C. N., and Whittaker, E. J. W., 2004, Nomenclature of amphiboles: Additions and revisions to the International Mineralogical Association's amphibole nomenclature: European Journal of Mineralogy, v. 16, p. 191–196.

Matthews, A., 1994, Oxygen-isotope geothermometers for metamorphic rocks: Journal of Metamorphic Geology, v. 12, p. 211–219.

Nakamura, D., and Hirajima, T., 2000, Granulite-facies overprinting of ultrahigh-pressure metamorphic rocks, Northeastern Sulu Region, Eastern China: Journal of Petrology, v. 41, p. 563–582.

Philippot, P., Chevallier, P., Chopin, C., and Dubessy, J., 1995, Fluid composition and evolution in coesite-bearing rocks (Dora-Maira massif, Western Alps): Implications for element recycling during subduction: Contributions to Mineralogy and Petrology, v. 121, p. 29–44.

Powell, R., 1985, Regression diagnostics and robust regression in geothermometer/geobarometer calibration: The garnet-clinopyroxene geothermometer revisited: Journal of Metamorphic Geology, v. 3, p. 231–243.

Raheim, A., and Green, D. H., 1974, Experimental determination of the temperature and pressure dependence of the Fe-Mg partition coefficient for coexisting garnet and clinopyroxene: Contributions to Mineralogy and Petrology, v. 48, p. 178–302.

Ravna, E. J. K., 2000, The garnet-clinopyroxene Fe^{2+}-Mg geothermometer: an updated calibration: Journal of Metamorphic Geology, v. 18, p. 211–219.

Rock, N. M. S., 1990, The International Mineralogical Association (IMA/CNMMN) pyroxene nomenclature scheme: Computerization and its consequences: Mineralogy and Petrology, v. 43, p. 99–109.

Roedder, E., 1984, Fluid inclusions: Mineralogical Society of America, Reviews in Mineralogy, v. 12, 678 p.

Rumble, D., 1981, Oxide minerals in metamorphic rocks, *in* Rumble, D., ed., Oxide minerals, Mineralogical Society of America, Reviews in Mineralogy, v. 3, p. R1–R24.

Rumble, D., and Yui, T.-F., 1998, The Qinglongshan oxygen and hydrogen isotope anomaly near Donghai in Jiangsu Province, China: Geochimica et Cosmochimica Acta, v. 62, p. 3307–3321.

Scambelluri, M., Pennacchioni, G., and Philippot, P., 1998, Salt-rich aqueous fluids formed during eclogitization of metabasites in the Alpine continental crust (Austroalpine Mt. Emilius unit, Italian western Alps): Lithos, v. 43, p. 151–167.

Schmidt, M. W., 1992, Amphibole composition in tonalite as a function of pressure: An experimental calibration of the Al-in-hornblende barometer: Contributions to Mineralogy and Petrology, v. 110, p. 304–310.

Schmidt, M. W., and Poli, S., 1998, Experimentally based water budgets for dehydrating slabs and consequences for the arc magma generation: Earth and Planetary Science Letters, v. 163, p. 361–379.

Sharp, Z. D., 1990, A laser-based microanalytical method for the in situ determination of oxygen isotope ratios of silicates and oxides: Geochimica et Cosmochimica Acta, v. 54, p. 1353–1357.

Sharp, Z. D., 1995, Oxygen isotope geochemistry of the Al_2SiO_5 polymorphs: American Journal of Science, v. 295, p. 1058–1076.

Sharp, Z. D., Essene, E. J., and Smyth, J. R., 1992, Ultrahigh temperatures from oxygen isotope thermometry of a coesite-sanidine grospydite: Contributions to Mineralogy and Petrology, v. 112, p. 358–370.

Shen, K., Zhang, Z., Van den Kerkhof, A. M., Xiao, Y., Xu, Z., and Hoefs, J., 2002, High-density and saline aqueous inclusion in the ultra-high pressure metamorphic rocks in the southern Sulu region, eastern China [ext. abs.], *in* International workshop on Geophysiscs and Structural Geology of UHPM terranes, Beijing, Abs. vol., p. 121–124.

Taylor, H. P., Jr., and Epstein, S., 1962, Relationship between $^{18}O/^{16}O$ ratios in coexisting minerals of igneous and metamorphic rocks; Part 1: Principles and experimental results: Geological Society of America Bulletin, v. 73, p. 461–480.

Touret, J. L. R., and Frezzotti, M. L., 2003, Fluid inclusions in high pressure and ultrahigh pressure metamorphic rocks., *in* Carswell, D.A., and Compagnoni, R., eds., Ultrahigh-pressure metamorphism: E.M.U. Notes in Mineralogy, v. 5, p. 467–487.

Ulmer, P., 1986, NORM-Program for cation and oxygen mineral norms: Computer Library, Institute für Mineralogie und Petrographie, ETH-Zentrum, Zürich, Switzerland.

van den Kerkhof, A. M., and Hein, U. F., 2001, Fluid inclusion petrography: Lithos, v. 55, p. 27–47.

Wallis, S., Enami, M., and Banno, S., 2000, The Sulu UHP terrane: A review of the petrology and structural geology, *in* Ernst, W. G., and Liou, J. G., eds, Ultrahigh-pressure metamorphism and geodynamics in collision-type orogenic belts: Final report of the Task Group III-6 (1994-1998) of the International Lithosphere Project, International Book Series, v. 4., p. 190–204.

Waters, D. J., and Martin, N. H., 1996: The garnet-omphacite-phengite geobarometer. Recommended calibration and calculation method [http://www.earth.ox.ac.uk/~davewa/research/ecbarcal.html], updated 1 March 1996.

Xiao, Y. L., Hoefs, J., Van den Kerkhof, A. M., Simon, K., Fiebig, J., and Zheng, Y., 2002, Fluid evolution during HP and UHP metamorphism in Dabie Shan, China: Constraints from mineral chemistry, fluid inclusions and stable isotopes: Journal of Petrology, v. 43, p. 1505–1527.

Yang, J. S., Wooden, J. L., Wu, C. L., Liu, F. L., Xu, Z. Q., Shi, R. D., Katayama, I., Liou, J. G., and Maruyama, S., 2003, SHRIMP U-Pb dating of coesite-bearing zircon from the ultrahigh-pressure metamorphic rocks, Sulu terrane, east China: Journal of Metamorphic Geology, v. 21, p. 551–560.

Yui, T.-F., Rumble, D., and Lo, C.-H., 1995, Unusually low $\delta^{18}O$ ultra-high-pressure metamorphic rocks from the Sulu Terrain, eastern China: Geochimica et Cosmochimica Acta, v. 59, p. 2859–2864.

Zhang, R. Y., Hirajima, T., Banno, S., Cong, B., and Liou, J. G., 1995, Petrology of ultrahigh-pressure rocks from the southern Sulu region, eastern China: Journal of Metamorphic Geology, v. 13, p. 659–675.

Zhang, R. Y., and Liou, J. G., 2000, Exsolution lamellae in minerals from ultrahigh-pressure rocks, *in* Ernst, W. G., and Liou, J. G., eds, Ultrahigh-pressure metamorphism and geodynamics in collision-type orogenic belts: Final report of the Task Group III-6 (1994–1998) of the International Lithosphere Project, International Book Series, v. 4, p. 216–228.

Zhang, Z., Xu, Z., and Xu, H., 2000, Petrology of ultrahigh-pressure eclogites from the ZK703 drillhole in the Donghai, eastern China: Lithos, v. 52, p. 35–50.

Zheng, Y. F., 1993a, Calculation of oxygen isotope fractionation in anhydrous silicate minerals: Geochimica et Cosmochimica Acta, v. 57, p. 1079–1091.

Zheng, Y. F., 1993b, Calculation of oxygen isotope fractionation in hydroxyl-bearing silicates: Earth and Planetary Science Letters, v. 120, p. 247–263.

Zheng, Y. F., Fu, B., Gong, B., and Li, S., 1996, Extreme ^{18}O depletion in eclogite from the Sulu terrane in East China: European Journal of Mineralogy, v. 8, p. 317–323.

Zheng, Y. F., Fu, B., Gong, B., and Li, S., 2003, Stable isotope geochemistry of ultrahigh pressure metamorphic rocks from the Dabie-Sulu orogen in China: implications for geodynamics and fluid regime: Earth Science Reviews, v. 1276, p. 1–57.

Zheng, Y. F., Fu, B., Li, Y., Xiao, Y., and Li, S., 1998, Oxygen and hydrogen isotope geochemistry of ultrahigh-pressure eclogites from the Dabie Mountains and the Sulu terrane: Earth and Planetary Science Letters, v. 155, p. 113–129.

^{40}Ar-^{39}Ar Thermochronological Constraints on the Exhumation of Ultrahigh-Pressure Metamorphic Rocks in the Sulu Terrane of Eastern China

LI-HUNG LIN,[1]

Institute of Geosciences, National Taiwan University, Taipei, Taiwan

PEI-LING WANG,

Institute of Geosciences, National Taiwan University, Taipei, Taiwan and Institute of Oceanography, National Taiwan University, Taipei, Taiwan

CHING-HUA LO, CHIN-HO TSAI,

Institute of Geosciences, National Taiwan University, Taipei, Taiwan

AND BOR-MING JAHN

Institute of Geosciences, National Taiwan University, Taipei, Taiwan and Institute of Earth Sciences, Academia Sinica, Taipei, Taiwan

Abstract

In order to better understand the uplift history of the Sulu ultrahigh-pressure metamorphic terrane, ^{40}Ar-^{39}Ar thermochronological analysis was conducted on minerals extracted from both metamorphic (gneiss and retrograded eclogite) and igneous (pegmatite, granite, and mafic intrusion) rocks from this terrane. Samples of the major lithotectonic units were collected from key localities, including Rongcheng, Laoshan, Yangkou, Taohang, and Rizhao. The analyses of metamorphic rocks yielded ages for hornblende ranging from 222 to 186 Ma, biotite from 130 to 119 Ma, and K-feldspar from 112 to 108 Ma. Hornblende, biotite, and K-feldspar from igneous rocks yielded exclusively late Mesozoic ages from 137 to 97 Ma. A two-stage cooling path for the metamorphic rocks were revealed. A slow cooling (1.9 to 3.8°C/m.y.) during the period of ~220 to ~130 Ma was followed by fast cooling (6.8 to 16.7°C/m.y.) during ~130 to ~100 Ma. The time span and cooling rate of the second stage coincide with those of the analyzed post-collisional igneous rocks as well as with the formation of rift basins in the region. The study strongly indicates that extensional tectonics played an important role in the final stage of rapid exhumation of the ultrahigh-pressure rocks in the Sulu terrane.

Introduction

THE PRESENCE OF diamond and coesite inclusions within eclogites (Wang and Liou, 1989; Sobolev and Shatsky, 1990) and FeTiO$_3$ and chromite exsolutions within peridotite massifs (Dobrzhinetskaya et al., 1996) indicates subduction of continental plates to depths greater than 100 km. Many studies have focused on the process that facilitates the preservation of these ultrahigh-pressure (UHP) minerals during ascent to the surface (see references in Liou et al., 1996). The mechanism, however, remains controversial partly due to the lack of appropriate age constraints on the exhumation history.

The Sulu-Dabie-Hong'an orogen was formed by collision between the North China Block (NCB) and the South China Block (SCB) (Okay and Sengör, 1992; Yin and Nie, 1993). The easternmost part of this orogen, the Sulu terrane, was sinistrally offset by 500 to 700 km along the Tan-Lu fault (Fig. 1; Okay et al., 1993; Yin and Nie, 1993). Previous work focused on reconstruction of the pressure-temperature (P-T) evolution using various thermo-barometers and mineral assemblages. Zhang et al. (1994) reported a clockwise P-T path for the Sulu terrane with peak metamorphism occurring at 800–900°C and >30 kbar. Peak metamorphic P-T

---

[1]Corresponding author; email: lhlin@ntu.edu.tw

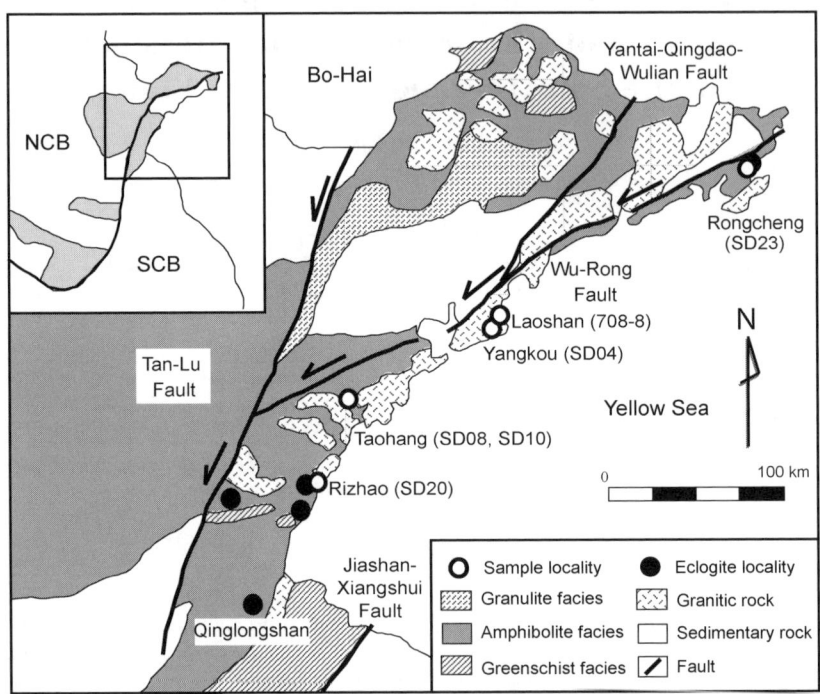

FIG. 1. Geological map of the Su-Lu terrane. Sample localities are shown as open circles, whereas solid circles represent the eclogite localities.

conditions progressively decrease westward in the Dabie and Hong'an terranes (Zhang and Liou, 1994; Zhang et al., 1994). In addition, the ages of UHP metamorphism have been well constrained at 220 to 240 Ma by various radiometric dating methods using different minerals (Li et al., 1993; Ames et al., 1993, 1996; Chavagnac and Jahn, 1996; Hacker et al., 1998). Extensive magmatic intrusions resulted from lithosphere melting, and cross-cut UHP metamorphic rocks during the late Mesozoic (Jahn et al., 1999). It has been suggested that igneous activity may have played a significant role in accelerating the final exhumation of UHP rocks because it coincided with the final cooling of UHP rocks from the Dabie and Hong'an terranes (Hacker et al., 2000). Whether the exhumation of UHP rocks from the Sulu terrane was driven by the same mechanism remains unclear and warrants further investigation.

Geochronological analyses of minerals provide critical constraints to temperature-time (T-t) paths for UHP rocks. The uplift history of an orogen can be better understood if geochronological data are combined with those from petrologic and structural analyses (McDougall and Harrison, 1999). ^{40}Ar-^{39}Ar systematics of metamorphic and igneous minerals is particularly useful because different minerals are closed to argon loss at specific blocking temperatures, thereby recording distinctive stages of metamorphic or magmatic cooling histories. In this paper, we report new ^{40}Ar-^{39}Ar thermochronological constraints on a retrograded UHP eclogite and gneisses, as well as post-UHP magmatic rocks from the Sulu terrane. A quantitative basis for geodynamic correlation and comparison with the well-studied Dabie and Hong'an terranes is outlined below.

Geological Setting and Previous Work

The Sulu terrane is composed of two fault-bounded metamorphic complexes, one a UHP and the other a high-pressure (HP) complex (Enami et al., 1993b; Hiramatsu and Hirajima, 1995). These two complexes are bounded by the Yantai-Qingdao-Wulian fault to the northwest and by the Jiashan-Xiangshui fault to the south (Fig. 1). Both complexes contain Mesozoic granites and mafic intrusions. The HP complex consists mainly of quartz-mica schist, chloritoid-kyanite-mica-quartz schist, marble, and rare blueschist, whereas the UHP complex is mainly composed of amphibolite, gneiss, and schist, with minor eclogite, marble, and peridotite (Zhang et al., 1995; Kato et al., 1997; Wallis et al., 1997). Eclogites occur as pods/boudins, discontinuous layers, or blocks within gneisses, marbles, and peridotites. Most eclogites show retrograde overprints of amphibolite- to greenschist-facies metamorphism; local granulite-facies overprinting occurs in the northeastern Sulu terrane (Nakamura and Hirajima, 2000). Peak metamorphic P-T conditions of the Sulu terrane are >30 kbar and 800–900°C (Zhang et al., 1994).

Compared with abundant age data for the Dabie and Hong'an terranes (Ames et al., 1993, 1996; Okay et al., 1993; Eide et al., 1994; Rowley et al., 1997; Hacker et al., 1998, 2000; Webb et al., 1999), the Sulu terrane is far less well geochronologically constrained. Most of the previous studies, however, center on the ages of protoliths and the peak metamorphism. Enami et al. (1993a) obtained ages of 1500–1720 Ma by Th-U-total Pb isochrons for monazite and zircon from gneisses in Laiyang. Their data represent protolith ages of basement rocks. Jahn et al. (1996) analyzed the Weihei eclogites and their data provided a constraint on the protolith age at about 1.7 Ga. Two eclogite samples from Rizhao yielded Sm-Nd garnet-omphacite isochron ages of 209 ± 31 and 221 ± 6 Ma (Li et al., 1993). Two eclogite samples from Qinglongshan and Zhubian analyzed by Sm-Nd and Rb-Sr methods yielded four isochron ages ranging from 220 to 228 Ma (Li et al., 1994). U-Pb analyses on zircon yielded a lower intercept age of 217 ± 9 Ma for an eclogite sample from the southern Sulu area (Ames et al., 1996). UHP gneisses and eclogites, retrograded gneisses, and amphibolized peridotites from the southwestern and northern Sulu yielded zircon U-Pb ages ranging from 200 to 250 Ma (Yang et al., 2003; Liu et al., 2004). In conclusion, the peak metamorphism of the Sulu UHP rocks is thought to occur in the Triassic (~220–235 Ma) (Li et al., 1993, 1994; Ames, et al., 1996; Jahn et al., 1996; Yang et al., 2003; Liu et al., 2004).

Sample Description and Analytical Procedures

Twelve samples were collected from five key localities (Rongcheng, Laoshan, Yangkou, Taohang, and Rizhao). They represent the major lithotectonic units (Table 1): one retrograded eclogite (SD20A), four gneisses (SD04, SD08A, SD20C, and SD20D), one pegmatite (SD08B), four granites (SD23, 708-8, SD10A, and SD20B2) and two mafic dikes (SD10B and SD20B1). Mineral separates were obtained by crushing, sieving, gravity/magnetic separation, and finally hand-picking to remove impurities. Samples were wrapped in aluminum foil, stacked with LP-6 biotite standards (Odin et al., 1982), and irradiated for 16 hours in the VT-C position of the THOR Reactor at the National Tsing-Hua University, Taiwan. The gradient of fast neutron flux across irradiation canister was less than 1%. After irradiation, samples were heated in vacuum, and degassed in steps from 550°C to 1200°C with a 30 minute–per step heating schedule. Gas was purified and analyzed with a VG3600 mass spectrometer at the National Taiwan University, Taiwan. Concentrations of ^{36}Ar, ^{37}Ar, ^{38}Ar, ^{39}Ar, and ^{40}Ar were corrected for system blanks, radiogenic decay of nucleogenic isotopes, and minor interference reactions involving Ca, K, and Cl, following the procedures by Lo and Lee (1994).

Quantitative electron microprobe analyses were performed on biotites to obtain Fe/(Fe+Mg) values for estimates of closure temperatures. A JEOL Superprobe 8900 at the Institute of Earth Sciences, Academia Sinica, Taiwan, was employed under operating conditions of 15 kV acceleration voltage and 10 µA beam current.

Results

The results of ^{40}Ar-^{39}Ar analyses are summarized in Table 1. Most samples yielded flat age spectra, except for a few showing disturbances. The adopted dates shown in Table 1 are either plateau or intercept dates obtained from isotope correlation plots.

K-feldspar

Seven K-feldspar separates (708-8f, SD04Af, SD08Af, SD08Bf, SD10Af, SD20B2f, and SD20Df) were obtained from granites, gneisses, and a pegmatite in Laoshan, Yangkou, Taohang, and Rizhao. All analyses yielded U-shaped spectra characterized by an old apparent date in the first step, followed by a relatively flat profile with 50 to 90% of $^{39}Ar_K$ released, and finally by a gradual increase to old apparent dates (Fig. 2). All age spectra yielded plateau dates ranging from 101.8 ± 0.7 to 121.9 ± 1.8 Ma. Except for 708-8f, SD04Af, and SD20Df (Figs.

TABLE 1. Summary of Results

Sample no.	Mineral type[1]	Rock type	Adopted date, Ma	Plateau date, Ma	Intercept date, Ma
		Rongcheng			
SD23b	Bt	Granite	106.3 ± 0.7	106.3 ± 0.7	104.4 ± 2.7
		Laoshan			
708-8h	Hbl	Granite	119.5 ± 2.1	119.5 ± 2.1	116.3 ± 4.4
708-8f	K-fsp		97.6 ± 1.4	101.8 ± 0.7	97.6 ± 1.4
		Yangkou			
SD04Ah	Hbl	Gneiss	186.7 ± 20.4	–	186.7 ± 20.4
SD04Ab	Bt		129.2 ± 2.0	–	–
SD04Af	K-fsp		111.5 ± 1.7	121.6 ± 1.8	111.5 ± 1.7
		Taohang			
SD08Af	K-fsp	Gneiss	119.3 ± 1.7	119.3 ± 1.7	116.9 ± 1.9
SD08Bm	Ms	Pegmatite	136.9 ± 1.9	136.9 ± 1.9	137.1 ± 2.0
SD08Bf	K-fsp		121.9 ± 1.8	121.9 ± 1.8	119.8 ± 1.9
SD10Ab	Bt	Granite	117.5 ± 1.7	117.5 ± 1.7	117.5 ± 1.7
SD10Af	K-fsp		118.9 ± 1.8	118.9 ± 1.8	119.9 ± 2.0
SD10B	WR	Mafic dike	108.9 ± 1.6	108.9 ± 1.6	105.8 ± 1.6
		Rizhao			
SD20Ah	Hbl	Eclogite	221.4 ± 3.9	–	221.4 ± 3.9
SD20B1	WR	Mafic dike	105.5 ± 1.5	105.5 ± 1.5	104.8 ± 1.6
SD20B2b	Bt	Granite	125.1 ± 1.8	125.1 ± 1.8	124.8 ± 1.8
SD20B2f	K-fsp		114.7 ± 1.7	114.7 ± 1.7	113.4 ± 1.9
SD20Cb	Bt	Gneiss	119.6 ± 1.7	119.6 ± 1.7	121.0 ± 2.2
SD20Df	K-fsp	Gneiss	108.9 ± 1.7	115.3 ± 1.7	108.9 ± 1.7

[1]Abbreviations of mineral types: Hbl = hornblende; Bt = biotite; K-fsp = K-feldspar; Ms = muscovite; WR = whole-rock.

2A, 2B, and 2G), the plateau dates were indistinguishable from their corresponding intercept dates (not shown). The initial $^{40}Ar/^{36}Ar$ intercept values, $(^{40}Ar/^{36}Ar)_i$, were consistent with the atmospheric value, indicating that excess argon contributes trivially to the observed ^{40}Ar. Contrarily, the intercept dates for 708-8f, SD04Af, and SD20Df were slightly younger than their respective plateau dates (Fig. 3). The $(^{40}Ar/^{36}Ar)_i$ values, which ranged from 350 to 425, were greater than the atmospheric value, suggesting a significant contribution of excess argon to the released ^{40}Ar (Fig. 3). For these samples, intercept dates are adopted for later discussion.

Micas

A muscovite separate (SD08Bm) from a pegmatite in Taohang yielded a flat profile with more than 95% of the cumulative $^{39}Ar_K$ released at 136.9 ± 1.9 Ma (Fig. 4A). The isotope correlation plot yielded an intercept date at 137.1 ± 2.0 Ma and a $(^{40}Ar/^{36}Ar)_i$ value of 302 (not shown), which are consistent with the plateau date and atmospheric value, respectively.

Four biotites from granites and gneisses (SD23b, SD10Ab, SD20B2b, and SD20Cb) in Rongcheng, Taohang, and Rizhao also yielded plateau dates ranging from 126 to 106 Ma (Fig. 4). These plateau dates are in agreement with their corresponding intercept dates (not shown). The age spectrum of SD04Ab is characterized by a lower apparent date at around 60 Ma in the first step, followed by a relatively "flat" region between 120 and 150 Ma, and ending with higher apparent dates between 160 and 260 Ma (Fig. 4C). No plateau date can be defined in

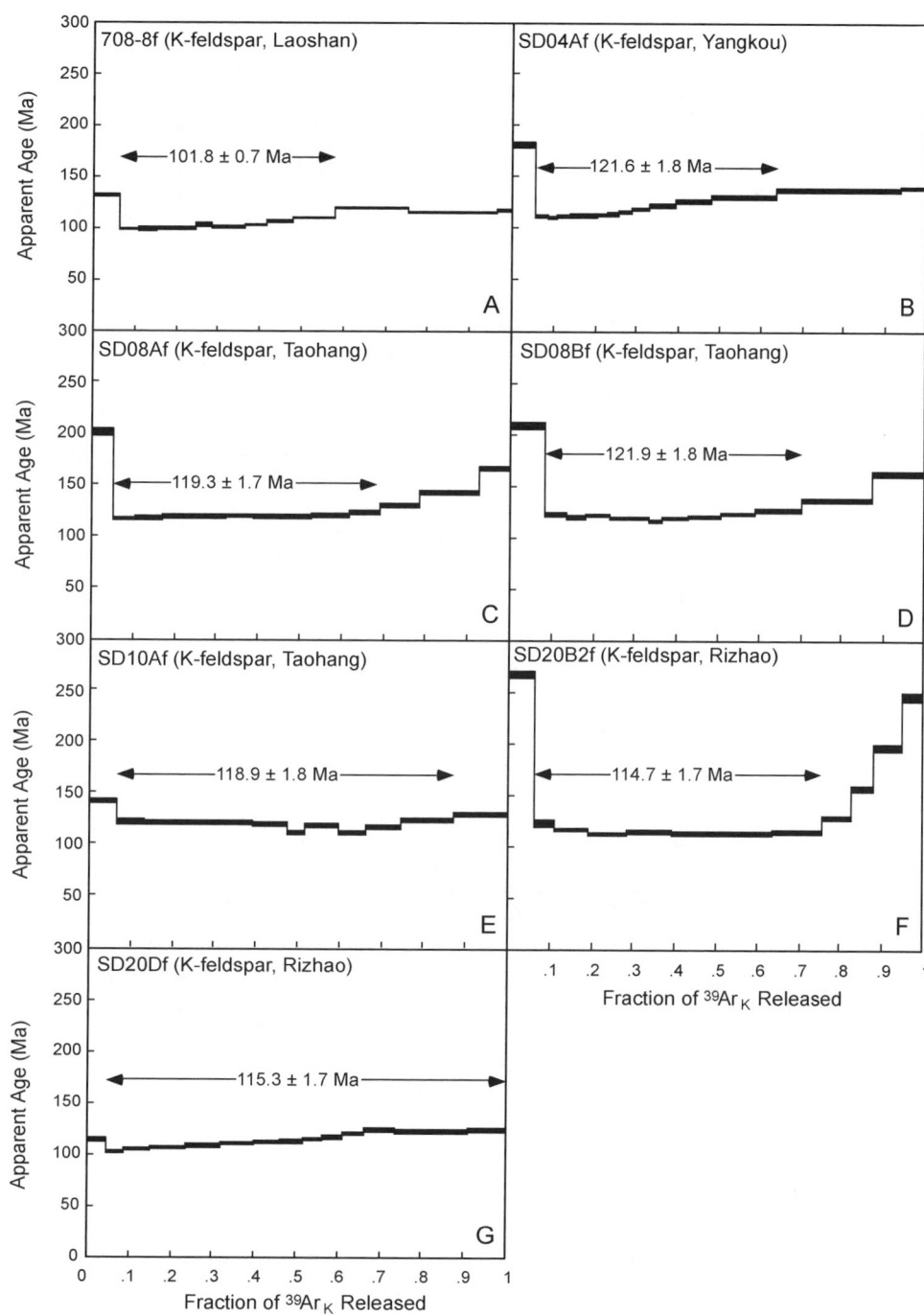

FIG. 2. Age spectra for K-feldspar. The arrow in the age spectrum indicates the area used to define the plateau date.

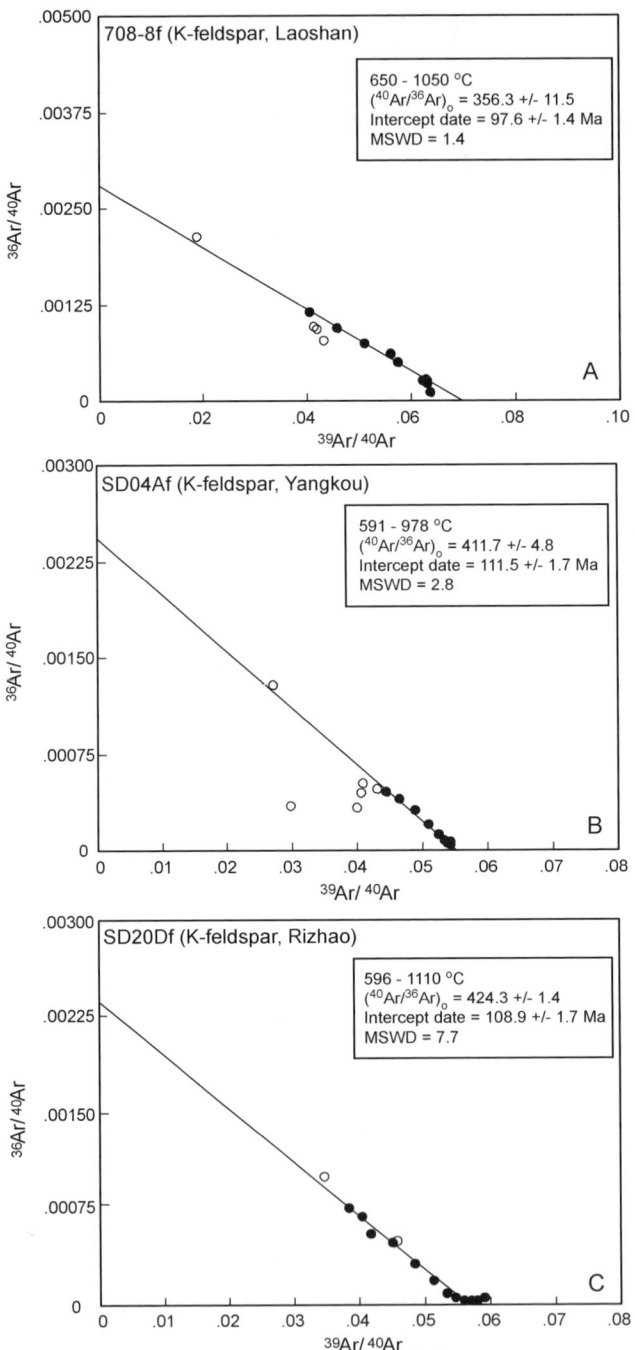

FIG. 3. $^{36}Ar/^{40}Ar$ vs. $^{39}Ar/^{40}Ar$ isotope correlation plots for K-feldspar with disturbed age spectra. Solid circles represent data used for the regression.

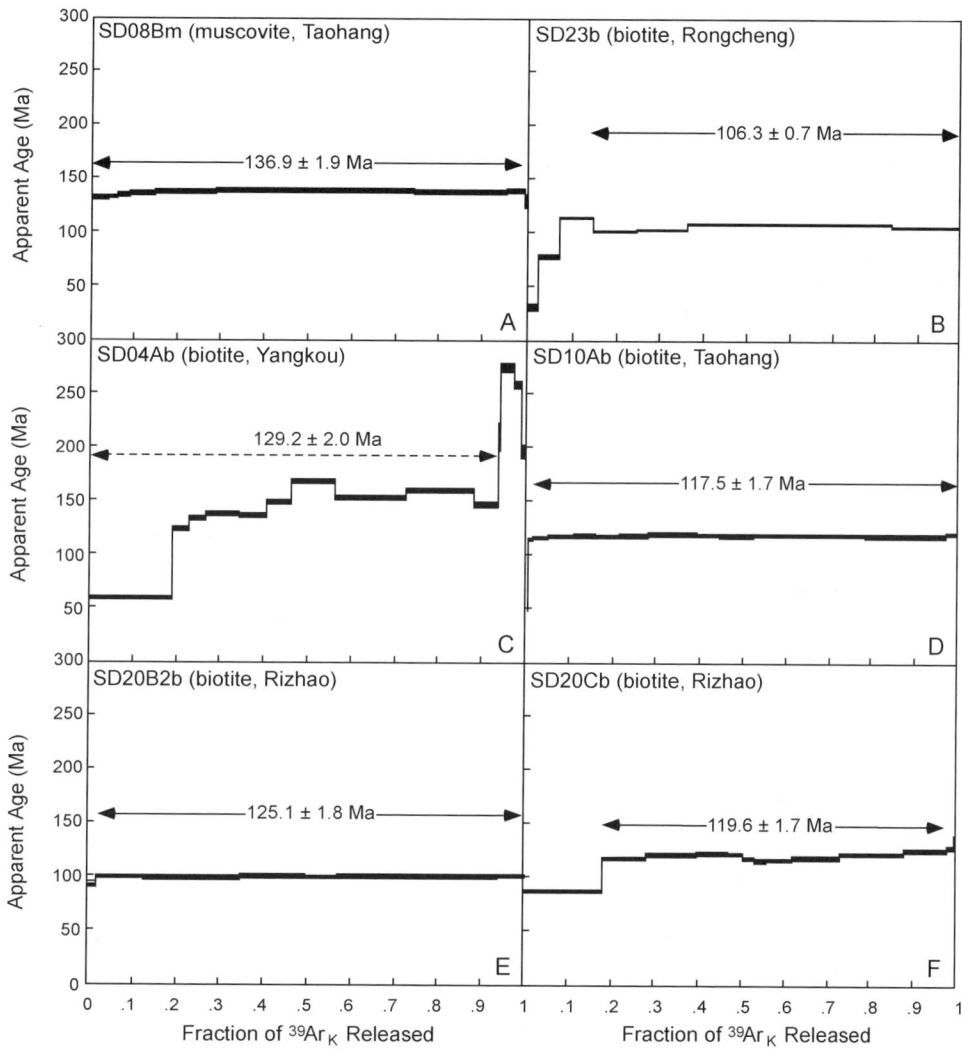

FIG. 4. Age spectra for muscovite and biotite. The arrow in the age spectrum indicates the area used to define the plateau date. Because no well-defined plateau was obtained for SD04Ab in (C), the dashed line with arrows indicates the integrated date that could represent the age for chloritized biotite.

the age spectrum (Fig. 4C), nor is a linear trend found in the isotope correlation plot (not shown). X-ray diffraction analysis, however, indicated the presence of chlorite, suggesting that some $^{39}Ar_K$ in biotite may be preferentially lost either to the vacuum or to the adjacent chlorite during neutron irradiation (Lo and Onstott, 1989). This recoil effect is further supported by the covariation of apparent dates with Ca/K ($^{37}Ar/^{39}Ar$) values (not shown) and multiple-humped age spectrum. We, therefore, used the integrated date (129.2 Ma) derived from the first 10 heating steps to represent the age for this sample.

Hornblende

Hornblende from a gneiss and a retrograded eclogite (SD04Ah and SD20Ah) in Yangkou and Rizhao yielded disturbed age spectra with apparent dates varying from 150 to 1000 Ma (Figs. 5A and 5B). In the isotope correlation plots, the data scattered widely and represented a three-component mixing, including radiogenic, excess, and atmospheric argon reservoirs (Fig. 6). The ^{40}Ar released during step-heating changed from the atmospheric component at low-temperature steps, through the radiogenic component at mid-temperature steps, and the excess argon component at high-temperature steps, and finally to the atmospheric component at the highest-temperature steps. The radiogenic component indicated intercept dates of ~187 Ma for SD04Ah (Fig. 6A) and ~221 Ma for SD20Ah (Fig. 6B). One hornblende separate from a granite (708-8h) in Laoshan yielded a flat profile with more than 80% of ^{39}Ar$_K$ released at 119.5 ± 2.1 Ma (Fig. 5C). The isotope correlation plot shows a (^{40}Ar/^{36}Ar)$_i$ value of 301 and an intercept date at 116.3 ± 4.4 Ma (not shown) that is in accordance with its plateau date.

Mafic Dikes

Two mafic dike samples yielded flat profiles with more than 80% of the ^{39}Ar$_K$ released, exhibiting plateau dates of 108.3 ± 1.6 Ma and 105.5 ± 1.5 Ma (Fig. 7). Both dates are consistent with their respective intercept dates (not shown).

In summary, the ^{40}Ar-^{39}Ar ages display a systematic younging trend from hornblende (222–186 Ma), muscovite (137 Ma), biotite (130–106 Ma), to K-feldspar (122–97 Ma). Such a pattern might be expected for a simple cooling model. Despite the field evidence that igneous rocks intruded to metamorphic rocks, all mica and K-feldspar ages for pegmatite and granite are compatible with those for their metamorphic equivalents at the same outcrops. They imply that both metamorphic and granitic rocks shared the same cooling history during the Jurassic to Cretaceous. The mafic dikes yielded ages younger than most of the samples dated (except 708-8f), consistent with our field observation that mafic intrusions postdated granites and pegmatites.

Thermochronology

^{40}Ar-^{39}Ar systematics in metamorphic and plutonic rocks record the times at which minerals cool through their closure temperatures. Thus, a cooling path can be constructed if closure temperatures of specific minerals are known. Inasmuch as the closure temperature for a specific mineral depends on the transport property of argon within the crystal structure, quantitative determination of a closure temperature requires diffusion parameters, such as the activation energy, pre-exponential factor, and diffusion size (Dodson, 1973). A number of studies have successfully estimated the diffusion parameters with step-heating and hydrothermal experiments (McDougall and Harrison, 1999, and references therein). Diffusion parameters for K-feldspar can be estimated by theoretical calculations based on the fractional loss of ^{39}Ar$_K$ measured in step-heating experiments, with the assumption that ^{39}Ar$_K$ diffusion obeys a slab or spherical geometry (e.g., Foland and Xu, 1990; Lovera et al., 1993). This method, however, cannot be directly applied to hydrous silicates, such as micas and hornblende, because elevated temperatures in step-heating experiments may cause decomposition of hydrous minerals (Gaber et al., 1988; Lee et al., 1991). The diffusion parameters derived from hydrothermal experiments (Robin, 1972; Gilletti, 1974; Norwood,

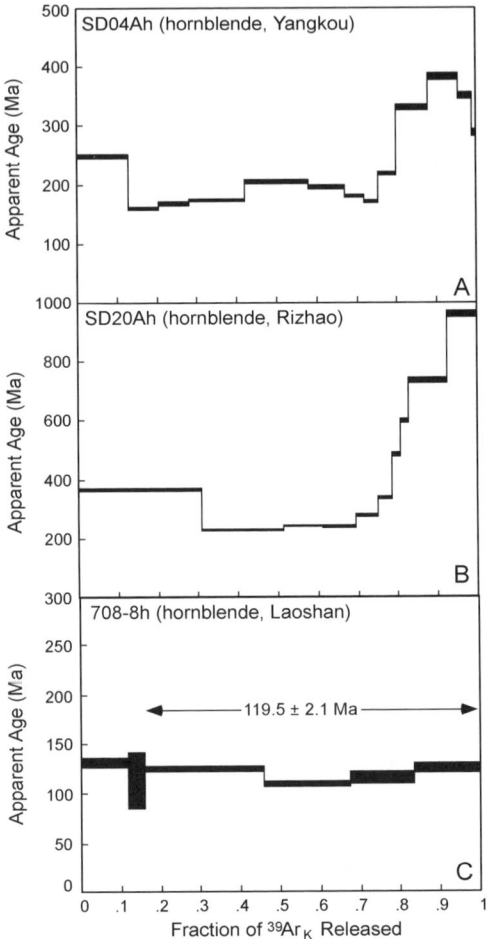

FIG. 5. Age spectra for hornblende. The arrow in the age spectrum indicates the area used to define the plateau date.

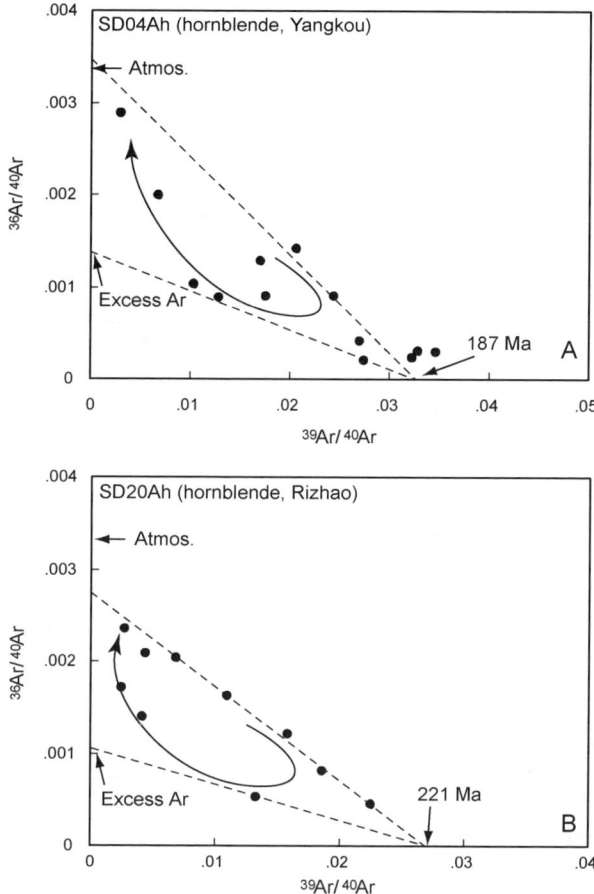

FIG. 6. $^{36}Ar/^{40}Ar$ vs. $^{39}Ar/^{40}Ar$ isotope correlation plots for hornblende with disturbed age spectra. The dashed lines outline three potential reservoirs (radiogenic, excess argon, and atmospheric components indicated by the straight lines with arrows) for the data points. The clockwise solid arrow indicates the evolution of the released argon.

1974; Harrison and McDougall, 1981; Harrison et al., 1985), which mimic the diffusion within hydrous minerals more closely, were adopted for the closure temperature calculation. The half mean size of mineral grains was estimated under a microscope and assumed as the size of diffusion (Table 2).

Based on Dodson's (1973) equation, closure temperatures for the analyzed samples were estimated as 490 to 520°C for hornblendes, 482°C for a muscovite, 290 to 410°C for biotites and 120 to 350°C for K-feldspars (Table 2). The T-t evolution for the Sulu terrane was compiled with incorporation of published U-Pb and Sm-Nd ages and their corresponding closure temperatures (Li et al., 1993; Ames et al., 1996; Jahn et al., 1996) (Fig. 8).

As shown in Figure 8, the Yangkou gneiss cooled slowly at a rate of 3.8°C/m.y. between 187 and 130 Ma. The cooling rate increased to 6.8°C/m.y. during Cretaceous time (130 to 111 Ma). The cooling path for the Rizhao gneiss and retrograded eclogite, however, is relatively complicated due to the lack of ages for multiple minerals from a single rock sample. We synthesized cooling paths for the gneiss and retrograded eclogite together, based on the assumption that eclogites and gneisses underwent the same metamorphism (Zhang et al., 1994). The Rizhao gneiss and eclogite cooled at a slow rate of 1.9°C/m.y. between 222 and 120 Ma. The cooling rate was enhanced significantly at 16.7°C/m.y. between 120 and 109 Ma (Fig. 8).

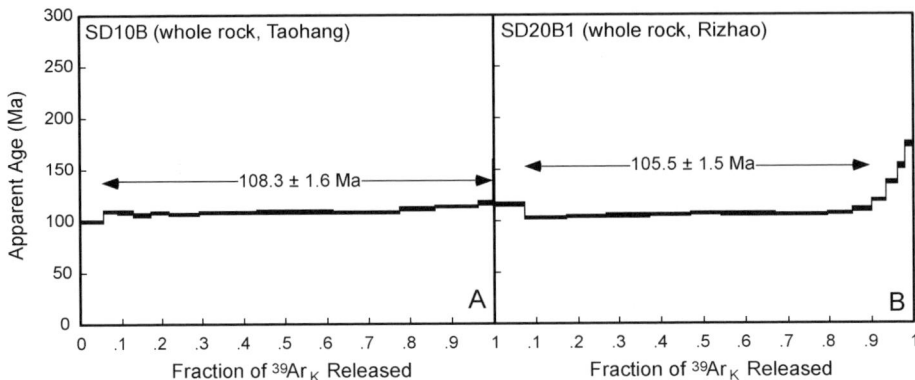

FIG. 7. Age spectra for whole-rock samples. The arrow in the age spectrum indicates the area used to define the plateau date.

The plutonic rocks exhibit cooling paths similar to the later cooling stage of the metamorphic rocks (Fig. 8). Hornblende and K-feldspar from the granite in Laoshan (708-8) yielded a cooling rate of 7.5°C/m.y. between 120 and 97 Ma. The rate is close to that of the metamorphic rock in the nearby Yangkou area during the same time period. The granites and pegmatite in Taohang and Rizhao cooled at rates greater than 18°C/m.y. between 137 and 114 Ma. The thermal relaxation of intrusive bodies may account for this rapid cooling during the Cretaceous. The intrusion of mafic dikes in Taohang and Rizhao appears to postdate the rapid cooling of the metamorphic rocks and granitic plutons during Cretaceous time.

Overall, the gneisses and retrograded eclogite underwent a slow cooling at rates of 3.8°C/m.y. between 187 and 130 Ma in the north (Yangkou), and 1.9°C/m.y. between 222 and 120 Ma in the south (Rizhao). The cooling rate increased to 6.8°C/m.y. between 130 and 111 Ma in Yangkou and a rapid cooling took place at a rate of 16.7°C/m.y. between 120 and 108 Ma in Rizhao (Fig. 8). The granites and pegmatite in Laoshan, Taohang and Rizhao also cooled rapidly at rates greater than 7°C/m.y. between 137 and 97 Ma (Fig. 8). The metamorphic and granitic rocks in the Sulu terrane share a similar cooling history at temperatures below ~300°C (Fig. 8). In the temperature range between 300 and 500°C, the metamorphic rocks cooled at a very slow rate during the Triassic/Jurassic. The onset of igneous intrusions, however, appears to be coeval with the rapid cooling/exhumation of metamorphic complexes and granitic plutons during the Cretaceous.

Despite of the general pattern of slow cooling of the Sulu terrane during the Triassic/Jurassic, followed by a relatively rapid cooling during the Cretaceous, the T-t paths varied slightly from locality to locality. The gneiss in the northern Sulu (Yangkou) cooled to a temperature of ~300°C at ~130 Ma, which was earlier than that (120 Ma) in the southern Sulu (Rizhao). The cooling rate of the northern Sulu was also much slower than that of the southern Sulu for a temperature range below 300°C (Fig. 8). This may reflect differential exhumation in the different areas. Alternatively, it could be explained by local thermal disturbance as a result of the widespread Mesozoic magmatic activity. If the activity perturbs the thermal structure of lithosphere locally, this transient heat would retard the cooling of the nearby or overlying crustal materials, and consequently result in different cooling paths. The analyzed metamorphic rocks, however, were collected far from large plutons in order to minimize the possible effect of igneous activity. The T-t path should reflect the cooling of the metamorphic rocks during exhumation. If the present geothermal gradient of 40°C/km (Tian et al., 1992) is employed, the uplift rate of metamorphic rocks in the Sulu region could have been about 0.5 mm/yr during the Cretaceous.

TABLE 2. Argon Diffusion Parameters for Closure Temperature Calculations

Sample	D_o (cm²/sec)	a (μm)[1]	$log(D_o/a^2)$ (1/sec)[2]	A. E. (kcal/mole)	Diffusion geometry	Fe/(Fe+Mg)[3]	T_c (°C)[4]	Reference
				Hornblende				
708-8h	0.024	100		64.1	Sphere		510	Harrison and
SD04Ah	0.024	135		64.1	Sphere		512	McDougall, 1981
SD20Ah	0.024	90		64.1	Sphere		493	
				Muscovite				
SD08Bb	6.03E-7	250		40	Slab		482	Robin, 1972
				Biotite				
SD23b	0.0376	410		50.6	Cylinder	0.36	385	Gilletti, 1974;
SD04Ab	0.0010	90		42.1	Cylinder	0.79	296	Norwood, 1974;
SD10Ab	0.0526	310		51.4	Cylinder	0.32	404	Harrision et al., 1985
SD20B2b	0.0209	450		49.3	Cylinder	0.43	403	
SD20Cb	0.0028	75		44.5	Cylinder	0.67	306	
				K-feldspar				
708-8f			4.83	54.5	Slab		349	
SD04Af			0.63	31.8	Slab		176	
SD08Af			1.65	28.3	Slab		189	
SD08Bf			0.83	30.9	Slab		189	
SD10Af			2.50	41.3	Slab		265	
SD20B2f			1.68	35.5	Slab		213	
SD20Df			1.69	29.1	Slab		128	

[1] a (diffusion size) is the mean half grain size measured under the microscope (at least 30 grains). The measured error is within 10% of the measured values.
[2] $log(D_o/a^2)$ is derived from the Arrhenius plot of K-feldspar by regressing the linear distribution of analytical results.
[3] Fe/(Fe+Mg) is used to correct the compositional dependence of biotite on the closure temperature.
[4] The uncertainties of closure temperatures are all within 20°C.

Tectonic Implications

Previous studies suggested that UHP metamorphism in the Dabie and Sulu terranes occurred at approximately the same time interval between 210 and 240 Ma (Ames et al., 1993, 1996; Li et al., 1993; Jahn et al., 1996; Hacker et al., 1998; Jahn et al., 2003; Yang et al., 2003; Liu et al., 2004). Because the temperature condition of peak metamorphism is estimated to be slightly higher than or equal to the closure temperature of zircon for U-Pb systematics (Lee et al., 1997), U-Pb dates could be regarded as the closest estimate for the timing of peak metamorphism. The consistency between U-Pb zircon ages of the Sulu and Dabie terranes effectively reduces the possibility of a hypothesis that the indentation of the SCB into the NCB occurred during the Triassic and gradually propagated westward from the Sulu through the Dabie to the Hong'an terrane, as proposed by Yin and Nie (1993) and Zhang et al. (1994). Instead, the entire Sulu-Dabie-Hong'an orogen may have been exhumed coevally since the Triassic. More geochronological data should be gathered to verify this assertion.

A three-stage exhumation for UHP rocks in the Sulu terrane is proposed as shown in Figure 8. During the Permo-Triassic, continuous northward subduction of the SCB beneath the NCB to depths of more than 100 km resulted in continent-continent

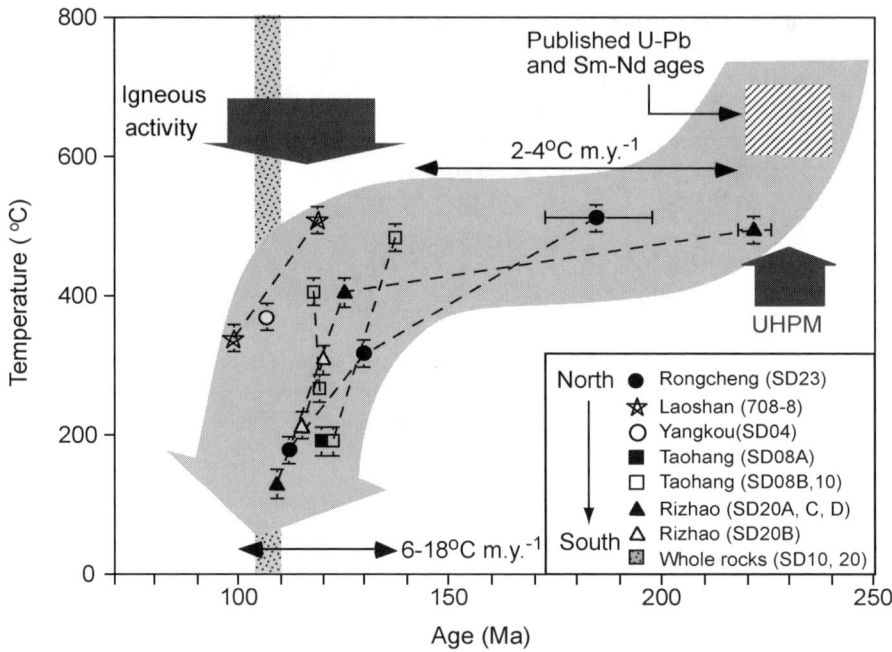

FIG. 8. Cooling history of the Sulu terrane. The solid symbol represents the T-t pair for metamorphic rocks, whereas the open symbol denotes igneous rocks. The dotted area around 105–108 Ma was derived from whole rock analyses of the mafic dike. Published Sm-Nd and U-Pb ages (Li et al., 1993; Ames et al., 1996; Jahn et al., 1996) are shown as the area of diagonal stripes in the upper right corner of the diagram. The grey region that encompasses all data points indicates the proposed cooling history of the Sulu terrane.

collision and triggered the UHP metamorphism. Perhaps due to lithosphere delamination (Platt, 1993) or slab break-off (Davies and von Blanckenburg, 1995; Jahn et al., 1999), subsequent isostatic adjustment rapidly extruded UHP rocks from the upper mantle to the lower crust (from ~800 to ~500°C). During the Late Triassic through the Jurassic, the Sulu terrane was almost stagnant, and slowly cooled with exhumation from the lower to the middle crust (from ~500 to ~300°C). The onset of Cretaceous magmatism may have induced much more rapid exhumation again during the Early Cretaceous (from ~300°C to surface temperature).

The exhumation of the Sulu terrane may be, in part, a result of lithospheric extension. The rift system in eastern Asia extended over 600 km during the Mesozoic and Cenozoic (Gilder et al., 1991; Tian et al., 1992). Extensional basins—including the Songliao, Bohai, Subei–South Yellow Sea, Meongyang, Pingyi-Feixian, and Jiaoli basins—started to develop during the Jurassic to Cretaceous (Hu et al., 1989; Ma et al., 1989; Zhang et al., 1989; Gilder et al., 1991; Ratschbacher et al., 2000).

Bimodal igneous intrusion (granitic and mafic) in the Sulu terrane was also coincident with the development of major faulting as indicated from sedimentary records of the Bohai and Subei–South Yellow Sea basins (Qiu et al., 2003). The exhumation of UHP rocks changed from a quiescent period to a faster stage at the same time.

Various tectonic models have advocated an intimate association of the development of sedimentary basins and invasion of bimodal magmatism, with rapid uplift of UHP rocks during the Jurassic to Cretaceous in the Sulu and Dabei terranes. Gilder et al. (1991) and Tian et al. (1992) proposed that these basins were genetically related to back-arc spreading resulting from the subduction of the Kula plate beneath the Eurasian plate since the Jurassic. Associated crustal thinning increased the thermal gradient and induced a large amount of intrusion/extrusion in eastern Eurasia. Alternately, the collision-extrusion model proposed by Kimura et al. (1990), an analogue of the Himalayan orogeny resulting from the collision between India and Eurasia, has also been used to explain the origin of these basins. Pre-existing structural features formed during the earlier collision between Indochina and Eurasia were re-activated to accommodate the stress propagation. Subsequently, many NNE-SSW transcurrent faults developed. The strike-slip movement of these faults may have triggered the formation of pull-apart basins distributed in similar directions.

Lithosphere thinning may have also played an essential role during coeval Jurassic to Cretaceous tectonism in the Sulu terrane. As demonstrated by theoretical considerations and case studies (e.g., King and Ellis, 1990; Platt, 1993), abnormal gravity produced during the accretion of two plates should be reduced toward isostatic balance. Continuous stretching of the accretionary wedge or even detachment of the root of an orogen reduces the thickness of continental lithosphere and initiates normal faulting and intrusion/extrusion of igneous rocks (Jahn et al., 1999). Eventually, the UHP rocks in the deep crust or mantle are exhumed to the surface along a normal fault.

Conclusions

^{40}Ar-^{39}Ar analyses yielded ages of 222–186 Ma for hornblende from a retrograded eclogite and gneisses, 130–119 Ma for biotite from gneisses, and 112–108 Ma for K-feldspar from gneisses. Ages for hornblende, biotite, and K-feldspar from granites, muscovite from a pegmatite, and whole-rock samples of mafic dikes fall between 137 and 97 Ma. The cooling history of the Sulu terrane was constructed using the mineral cooling ages with their closure temperatures. From the Late Triassic to Jurassic, the metamorphic rocks underwent a very slow cooling at a rate of ~2 to ~4°C/m.y. The cooling rate increased slightly to ~7°C/m.y. in the north (Yangkou) but increased significantly to ~17°C/m.y. in the south (Rizhao) during the Early Cretaceous. The cooling rate and time span of the later cooling stage for the metamorphic rocks coincide with those for the igneous intrusions and the development of rift basins near the Sulu terrane. This transition implies a change from compressional to extensional tectonics, which may have played an important role in the final exhumation of the Sulu UHP rocks.

Acknowledgments

We thank B. Cong, K. Ye, and Q. Wang for kindly providing logistical support during the field work, and C.-H. Chen for his kind help with microprobe analysis. We are indebted to J. G. Liou for advice and comments during the manuscript preparation. The work was supported by grants from the National Science Council of Taiwan. We dedicate this paper to W. G. Ernst for his contributions to the study of tectonics and metamorphism in Taiwan and in China.

REFERENCES

Ames, L., Tilton, G. R., and Zhou, G., 1993, Timing of collision of the Sino-Korean and Yangtze cratons: U-Pb zircon dating of coesite-bearing eclogites: Geology, v. 21, p. 339–342.

Ames, L., Zhou, G., and Xiong, B., 1996, Geochronology and isotopic character of ultrahigh-pressure metamorphism with implications for collision of the Sino-Korean and Yangtze cratons, central China: Tectonics, v. 15, p. 472–489.

Chavagnac, V., and Jahn, B. M., 1996, Coesite-bearing eclogites from the Bixiling Complex, Dabie Mountains, China: Sm-Nd ages, geochemical characteristics and tectonic implications: Chemical Geology, v. 133, p. 29–51.

Davies, J. H., and von Blanckenburg, F., 1995, Slab breakoff: A model of lithosphere detachment and its test in the magmatism and deformation of collision orogens: Earth and Planetary Science Letters, v. 129, p. 85–102.

Dobrzhinetskaya, L., Green, H. W., II, and Wang, S., 1996, Alpe Arami: A peridotite massif from depths of more than 300 kilometers: Science, v. 271, p. 1841–1845.

Dodson, M. H., 1973, Closure temperature in cooling geochronological and petrological systems: Contributions to Mineralogy and Petrology, v. 40, p. 259–274.

Eide, E. A., McWilliams, M. O., and Liou, J. G., 1994, ^{40}Ar/^{39}Ar geochronology and exhumation of high-pressure to ultrahigh-pressure metamorphic rocks in east-central China: Geology, v. 22, p. 601–604.

Enami, M., Suzuki, K., Zhai, M. and Zheng, X., 1993a, The chemical Th–U–total Pb isochron ages of Jiaodong and Jiaonan metamorphic rocks in the Shandong Peninsula, eastern China: The Island Arc, v. 2, p. 104–113.

Enami, M., Zang, Q., and Yin, Y., 1993b, High-pressure eclogites in northern Jiangsu–southern Shandong province, eastern China: Journal of Metamorphic Geology, v. 11, p. 589–604.

Foland, K. A., and Xu, Y., 1990, Diffusion of ^{40}Ar and ^{39}Ar in irradiated orthoclase: Geochimica et Cosmochimica Acta, v. 54, p. 3147–3158.

Gaber, L. J., Foland, K. A., and Corbato, C. E., 1988, On the significance of argon release from biotite and amphibole during ^{40}Ar/^{39}Ar vacuum heating:

Geochimica et Cosmochimica Acta, v. 52, p. 2457–2465.

Gilder, S. A., Keller, G. R., Luo, M., and Goodell, P. C., 1991, Timing and spatial distribution of rifting in China: Tectonophysics, v. 197, p. 225–243.

Gilleti, B. J., 1974, Studies in diffusion I: Ar in phlogopite mica, in Hofmann, A. W., Gilleti, B. J., Yorder, H. S., and Yund, R. A., eds., Geochemical transport and kinetics: Washington, DC, Carnegie Institute Publication, v. 634, p. 107–115.

Hacker, B. R., Ratschbacher, L., Webb, L., Ireland, T., Walker, D., and Dong, S., 1998, U/Pb zircon ages constrain the architecture of the ultrahigh-pressure Qinling-Dabie Orogen, China: Earth and Planetary Science Letters, v. 161, p. 215–230.

Hacker, B. R., Ratschbacher, L., Webb, L. E., McWilliams, M., Calvert, A., Dong, S., Wenk, H.-R., and Chateigner, D., 2000, Exhumation of ultrahigh-pressure rocks in the Dabie-Hong'an area: Late Triassic–Early Jurassic tectonic unroofing: Journal of Geophysical Research, v. 105, p. 13,339–13,364.

Harrison, T. M., Duncan, I., and McDougall, I., 1985, Diffusion of ^{40}Ar in biotite: Temperature, pressure, and compositional effects: Geochimica et Cosmochimica Acta, v. 49, p. 2461–2468.

Harrison, T. M., and McDougall, I., 1981, Diffusion of Ar in hornblende: Contributions to Mineralogy and Petrology, v. 78, p. 324–331.

Hiramatsu, N., and Hirajima, T., 1995, Petrology of the Hujialin garnet clinopyroxenite in the Su-Lu ultrahigh-pressure province, eastern China: The Island Arc, v. 4, p. 310–323.

Hu, J., Xu, S., Tong, X., and Wu, H., 1989, The Bohai Bay basin, in Zhu, X., ed., Chinese sedimentary basins: Amsterdam, The Netherlands, Elsevier, p. 89–106.

Jahn, B. M., Cornichet, J., Cong, B. L., and Yui, T. F., 1996, Ultrahigh-εNd eclogites from an ultrahigh pressure metamorphic terrane of China: Chemical Geology, v. 127, p. 61–79.

Jahn, B. M., Fan, Q., Yang, J. J., and Henin, O., 2003, Petrogenesis of the Maowu pyroxenite-eclogite body from the UHP metamorphic terrane of Dabieshan: Chemical and isotopic constraints: Lithos, v. 70, p. 243–267.

Jahn, B. M., Wu, F., Lo, C. H., and Tsai, C. H., 1999, Crust-mantle interaction induced by deep subduction of the continental crust: Geochemical and Sr-Nd isotopic evidence from post-collisional mafic-ultramafic intrusions of the northern Dabie Complex, central China: Chemical Geology, v. 157, p. 119–146.

Kato, T., Enami, M., and Zhai, M., 1997, Ultrahigh-pressure marble and eclogite in the Su-Lu-UHP terrane, eastern China: Journal of Metamorphic Geology, v. 15, p. 169–182.

Kimura, G., Takahashi, M., and Kono, M., 1990, Mesozoic collision-extrusion tectonics in eastern Asia: Tectonophysics, v. 181, p. 15–23.

King, G., and Ellis, M., 1990, The origin of large local uplift in extensional regions: Nature, v. 348, p. 689–693.

Lee, J. K. W., Onstott, T. C., Cashman, K. V., Cumbest, R. J., and Johnson, D., 1991, Incremental heating of hornblende in vacuo: Implications for ^{40}Ar/^{39}Ar geochronology and the interpretation of thermal history: Geology, v. 19, p. 872–876.

Lee, J. K. W., Williams, I. S., and Ellis, D. J., 1997, Pb, U and Th diffusion in natural zircon: Nature, v. 390, p. 159–162.

Li, S., Wang, S., Chen, Y., Liu, D., Qiu, J., Zhou, H., and Zhang, Z., 1994, Excess argon in phengite from eclogite: Evidence from dating of eclogite minerals by Sm-Nd, Rb-Sr and ^{40}Ar/^{39}Ar methods: Chemical Geology, v. 112, p. 343–350.

Li, S., Xiao, Y., Liou, D., Chen, Y., Ge, N., Zhang, Z., Sun, S. S., Cong, B., Zhang, R., Hart, S. R., and Wang, S., 1993, Collision of the North China and Yangtze blocks and formation of coesite-bearing eclogites: Timing and processes: Chemical Geology, v. 109, p. 89–111.

Liou, J. G., Zhang, R. Y., Wang, X., Eide, E. A., Ernst, W. G., and Maruyama, S., 1996, Metamorphism and tectonics of high-pressure belts in the Dabie-Sulu region, China, in Yin, A., and Harrison, T. M., eds., The tectonic evolution of Asia: Cambridge, UK, Cambridge University Press, p. 300–344.

Liu, F., Xu, Z., Liou, J. G., and Song, B., 2004, SHRIMP U-Pb ages of ultrahigh-pressure and retrograde metamorphism of gneisses, south-western Sulu, terrane, eastern China: Journal of Metamorphic Geology, v. 22, p. 315–326.

Lo, C. H., and Lee, C. Y., 1994, ^{40}Ar/^{39}Ar method of K-Ar age determination of geological samples using Tsing-Hua Open-Pool (THOPR) reactor: Journal of Geological Society of China, v. 37, p. 1–22.

Lo, C. H., and Onstott, T. C., 1989, ^{39}Ar recoil artifacts in chloritized biotite: Geochimica et Cosmochimica Acta, v. 53, p. 2697–2711.

Lovera, O. M., Heialer, M. T., and Harrison, T. M., 1993, Argon diffusion domains in K-feldspar II: Kinetic properties of MH-10: Contributions to Mineralogy and Petrology, v. 113, p. 381–393.

Ma, L., Yang, J., and Ding, Z., 1989, Songliao basin: An intracratonic continental sedimentary basin of combination type, in Zhu, X., ed., Chinese sedimentary basins: Amsterdam, The Netherlands, Elsevier, p. 77–88.

McDougall, I., and Harrison, T. M., 1999, Geochronology and thermochronology by ^{40}Ar/^{39}Ar method: New York, NY, Oxford University Press, 320 p.

Nakamura, D., and Hirajima, T., 2000, Granulite-facies overprinting of ultrahigh-pressure metamorphic rocks, northeastern Su-Lu region, eastern China: Journal of Petrology, v. 41, p. 563–582.

Norwood, C. B., 1974, Radiogenic argon diffusion in the biotite micas: Unpubl. M.S. thesis, Brown University, 58 p.

Odin, G. S. and 35 collaborators, 1982, Interlaboratory standards for dating purposes, *in* Odin, G. S., ed., Numerical dating in stratigraphy: New York, NY, John and Wiley Sons, p. 1040.

Okay, A. I., and Sengör, A. M. C., 1992, Evidence for intracontinental thrust-related exhumation of the ultra-high-pressure rocks in China: Geology, v. 20, p. 411–414.

Okay, A. I., Sengör, A. M. C., and Satir, M., 1993, Tectonics of an ultrahigh-pressure metamorphic terrane: The Dabie Shan/Tongbai Shan orogen, China: Tectonics, v. 12, p. 1320–1334.

Platt, J. P., 1993, Convective removal of lithosphere beneath mountain belt: Thermal and mechanical consequence: American Journal of Science, v. 293, p. 307–336.

Qiu, J. S., Jiang, S. Y., Xu, X. S., and Lo, C. H. 2003, Geochemistry of K-rich volcanic rocks along the middle-south parts of Tancheng-Lujiang deep fault zone, eastern China: Constraints on mantle source and petrogenesis: Geochimica et Cosmochimica Acta, v. 67, A386.

Ratschbacher, L., Hacker, B. R., Webb, L. E., McWilliams, M., Ireland, T., Dong, S., Calvert, A., Chateigner, D., and Wenk, H.-R., 2000, Exhumation of the ultrahigh-pressure continental crust in east-central China: Cretaceous and Cenozoic unroofing and the Tan-Lu fault: Journal of Geophysical Research, v. 105, p. 13,303–13,338.

Robin, G. A., 1972, Radiogenic argon diffusion in muscovite under hydrothermal conditions: Unpubl. M.S. thesis, Brown University, 47 p.

Rowley, D. B., Xue, F., Tucker, R. D., Peng, Z. X., Baker, J., and Davis, A., 1997, Ages of ultrahigh pressure metamorphism and protolith orthogneisses from the eastern Dabie Shan: U/Pb zircon geochronology: Earth and Planetary Science Letters, v. 151, p. 191–203.

Sobolev, N. V., and Shatsky, V. S., 1990, Diamond pseudomorph after coesite in garnets from metamorphic rocks: A new environment for diamond formation: Nature, v. 343, p. 742–746.

Tian, Z.-Y., Ping, H., and Xu, K.-D., 1992, The Mesozoic–Cenozoic East China rift system: Tectonophysics, v. 208, p. 341–363.

Wallis, S. R., Ishiwatari, A., Hirajima, T., Ye, K., Guo, J., Nakamura, D., Kato, T., Zhai, M., Enami, M., Cong, B., and Banno, S., 1997, Occurrence and field relationships of ultrahigh-pressure metagranitoid and coesite eclogite in the Su-Lu Terrane, eastern China: Journal of the Geological Society of London, v. 154, p. 45–54.

Wang, X., and Liou, J. G., 1989, Coesite-bearing eclogite from the Dabie Mountains in central China: Geology, v. 17, p. 1085–1088.

Webb, L. E., Hacker, B. R., Ratschbacher, L., McWilliams, M. O., and Dong, S., 1999, Thermochronologic constraints on deformation and cooling history of high- and ultrahigh-pressure rocks in the Qinling-Dabie orogen, eastern China: Tectonics, v. 18, p. 621–638.

Yang, J. S., Wooden, J. L., Wu, C. L., Liu, F. L., Xu, Z. Q., Shi, R. D., Katayama, I., Liou, J. G., and Maruyama, S., 2003, SHRIMP U-Pb dating of coesite-bearing zircon from the ultrahigh-pressure metamorphic rocks, Sulu terrane, east China: Journal of Metamorphic Geology, v. 21, p. 551–560.

Yin, A. and Nie, S. Y., 1993, An indentation model for the North and South China collision and the development of the Tan-Lu and Honam fault systems, eastern Asia: Tectonics, v. 12, p. 801–823.

Zhang, R. Y., Hirajima, T., Banno, S., Cong, B., and Liou, J. G., 1995, Petrology of ultrahigh-pressure rocks from the southern Su-Lu region, eastern China: Journal of Metamorphic Geology, v. 13, p. 659–675.

Zhang, R. Y., and Liou, J. G., 1994, Significance of magnesite paragenesis in ultrahigh-P metamorphic rocks: American Mineralogist, v. 79, p. 397–400.

Zhang, R. Y., Liou, J. G., and Cong, B., 1994, Petrogenesis of garnet-bearing ultramafic rocks and associated eclogites in the Su-Lu ultrahigh-P metamorphic terrane, eastern China: Journal of Metamorphic Geology, v. 12, p. 169–186.

Zhang, Y., Wei, Z., Xu, W., Tao, R., and Chen, R., 1989, The North Jiangsu–South Yellow Sea basin, *in* Zhu, X., ed., Chinese sedimentary basins: Amsterdam, The Netherlands, Elsevier, p. 89–106.

Limiting Effects of UHP Metamorphism on Length Scales of Oxygen, Hydrogen, and Argon Isotope Exchange: An Example from the Qinglongshan UHP Eclogites, Sulu Terrain, China

M. A. COSCA,[1] D. GIORGIS,

Institute of Mineralogy and Geochemistry, University of Lausanne, Lausanne 1015, Switzerland

D. RUMBLE,

Geophysical Laboratory, 5251 Broad Branch Road, NW, Washington, District of Columbia 20015-1305

AND J. G. LIOU

Department of Geological and Environmental Sciences, Stanford University, Stanford, California 94305

Abstract

New and previously published stable isotope (oxygen and hydrogen) and $^{40}Ar/^{39}Ar$ data are compared from selected outcrops of the Sulu UHP terrain near Qinglongshan (Jiangsu Province, China). These rocks exhibit unusually low, heterogeneous $\delta^{18}O$ and δD values acquired in a Neoproterozoic geothermal system despite undergoing Triassic (220–240 Ma) UHP metamorphism. Incremental heating $^{40}Ar/^{39}Ar$ analyses of muscovite, biotite, and K-feldspar from metagranite, quartzite, and gneiss (all metamorphosed to the coesite-eclogite facies) yield cooling ages between 190 and 204 Ma. In contrast, phengite $^{40}Ar/^{39}Ar$ data from eclogite, quartzite, and gneiss contain variable amounts of extraneous argon, consistent with inheritance from a Neoproterozoic protolith. Plots comparing phengite $\delta^{18}O$, δD, and $^{40}Ar/^{39}Ar$ total gas ages from different lithologies within individual Qinglongshan outcrops only meters apart highlight significant inter-outcrop isotopic heterogeneities ($\Delta D_{phengite}$ = 26‰; $\Delta^{18}O_{phengite}$ = 8.8‰; $\Delta^{40}Ar/^{39}Ar_{phengite}$ = 664 Ma); however maximum intra-outcrop isotopic variations between lithologies are limited ($\Delta D_{phengite}$ = 5‰; $\Delta^{18}O_{phengite}$ = 1.8‰). The oxygen and hydrogen isotopic variations are interpreted to reflect primary isotopic heterogeneities acquired during Neoproterozoic hydrothermal fluid circulation with cold-climate meteoric waters. The limited intra-outcrop isotopic variations suggest that extensive isotopic exchange occurred during UHP metamorphism within discrete outcrops, irrespective of lithology. Likewise, extraneous argon within the phengites reflects inheritance from a Neoproterozoic protolith, and the age variations between outcrops is probably due to differential argon loss during thermal and baric equilibration accompanying differential exhumation following UHP metamorphism. The retention of extraneous argon in phengite is partially controlled by the host rock lithology, inasmuch as as the armoring effects of basaltic eclogite are greater than quartzite or gneiss. Collectively, these data indicate that the length-scale of isotopic exchange is defined by contiguous blocks of coherent rock. Isotopic exchange during UHP metamorphism occurred between different lithologies restricted to discrete blocks within an accretionary prism, but did not communicate with other blocks. This study underscores the closed system (outcrop scale) behavior of isotopic exchange that can occur during continental subduction, collision, and uplift under UHP conditions.

Introduction

UNUSUALLY LOW VALUES of both $\delta^{18}O$ and δD reported from metamorphic minerals from UHP rocks near Donghai in Jiangsu Province, China are collectively referred to as the Qinglongshan isotope anomaly (e.g., Rumble and Yui, 1998). Garnets from coesite-bearing eclogites have $\delta^{18}O$ values (VSMOW) as low as –11‰, rutiles as low as –15‰, and phengites have δD values as low as –127‰. A number of investigations have shown that these isotopic signatures were acquired at or near the Earth's surface by alteration with meteoric water in a geothermal system prior to subduction and metamorphism at UHP conditions (e.g., Yui et al., 1995; 1997; Baker et al., 1997; Zheng et al., 1996, 1998a; 1998b; Rumble and Yui, 1998). Geochronological

---

[1]Corresponding author; email: Michael.Cosca@unil.ch

data yield early Proterozoic Nd model ages for protoliths of the Qinglongshan eclogites (Jahn, 1998), and U-Pb dating of zircon by SHRIMP yield Neoproterozoic ages for the Qinglongshan metagranite (Rumble et al., 2002). Additional observations including the exceedingly low $\delta^{18}O$ values (−15 to −1‰) measured in the UHP minerals support the hypothesis that this interaction occurred during a period of Snowball Earth (Rumble et al., 2002).

The preservation of isotopic values acquired near the surface despite undergoing high-grade metamorphism indicates limited metamorphic fluid involvement, and has been widely reported from a variety of geological settings (e.g., Valley and O'Neil, 1984; Eiler and Valley, 1994; Getty and Selverstone, 1994; Baker and Matthews, 1995; Cartwright and Barnicoat, 1999; Miller et., al, 2001; Putlitz et al., 2000, 2001). The Qinglongshan example is unique in that there appears to be a structural coherence to the exposed lithologies, consisting of a metagranite surrounded by volumetrically smaller amounts of basaltic eclogite, pelitic schist, orthogneiss, and quartzite all metamorphosed at UHP conditions (e.g., Rumble and Yui, 1998). Indeed, this field association, together with the stable isotope data, provides compelling evidence that much of the fossilized geothermal system is intact and presently exposed, including the granitic intrusion driving fluid circulation (e.g., Rumble and Yui, 1998). This relationship implies that the stable isotope signature acquired in rocks near the surface during Neoproterozoic time retained those values despite undergoing Triassic UHP metamorphism (Rumble and Yui, 1998; Rumble et al., 2002). Thus, any oxygen (and hydrogen) isotope exchange during the subduction, collision, and uplift history was limited to localized internal exchange within the protolith, perhaps restricted to a few centimeters (Rumble and Yui, 1998).

Phengite from Qinglongshan within the Sulu UHP terrain, eastern China, has been reported with extraneous argon (argon originating by processes other than *in situ* radioactive decay), representing more than 70% of the total ^{40}Ar budget (Li et al., 1994). Further investigation of this sample using *in situ* UV-laser ablation $^{40}Ar/^{39}Ar$ showed that the phengites were strongly zoned in terms of composition and extraneous argon (Giorgis et al., 2000). Together with textural and compositional observations, Giorgis et al. (2000) concluded that the extraneous argon originated from the eclogitic protolith, a conclusion also reached in phengites of the Dora Maira UHP rocks (Scaillet, 1996), and was preferentially concentrated into the phengite, and to a lesser extent epidote, during UHP metamorphism followed by limited intragrain argon isotopic redistribution and partial argon loss linked to local adjustments in major-element distributions during exhumation.

On the basis of these previous isotopic results from the Qinglongshan UHP rocks and their apparent ability to preserve protolith oxygen, hydrogen, and argon isotopic signatures, it was decided to compare these values from a suite of samples in order to examine internal isotopic variations and potential length scales of isotopic exchange. In this investigation, K-bearing minerals (phengite, muscovite, paragonite, biotite, K-feldspar) from a variety of lithologies were analyzed for their oxygen and argon isotope compositions and combined with additional data previously obtained by Rumble and Yui (1998). The combination of oxygen isotope, electron microprobe, and $^{40}Ar/^{39}Ar$ data not only supports the interpretation that such isotopic signatures can survive UHP metamorphism but the correlations observed between the argon, hydrogen, and oxygen isotope data suggest that isotopic exchange under UHP conditions occurs extensively within, and is limited to, contiguous outcrops.

Geological Framework

This investigation is centered in and around Qinglonshan, China (Fig. 1). Qinglonshan is located within the Sulu UHP metamorphic terrain, which together with the Dabieshan UHP terrain, formed during early Mesozoic collision of the Sino-Korean and Yangtze Cratons (Li et al., 1993, 1994; Eide et al., 1994; Hacker and Wang, 1995; Ames et al., 1996; Chavagnac and Jahn, 1996; Rowley et al., 1997; Hacker et al., 1998; Zhang and Liou, 1998). The minimum P-T conditions of UHP metamorphism resulting from this continental collision were 700°C and 30 kbar (Li et al., 1993; Zhang et al., 1995; Rumble and Yui, 1998), and the timing of peak metamorphism is estimated at 220–240 Ma (Li et al., 1993; Ames et al., 1996; Hacker et al., 1998).

Sample Descriptions

Rocks from this investigation, including gneiss, basaltic eclogite, quartzite, and schist, were collected from Qinglongshan and surrounding areas (Fig. 1). In general, outcrops are restricted to small hills above rice fields, and most occur as isolated

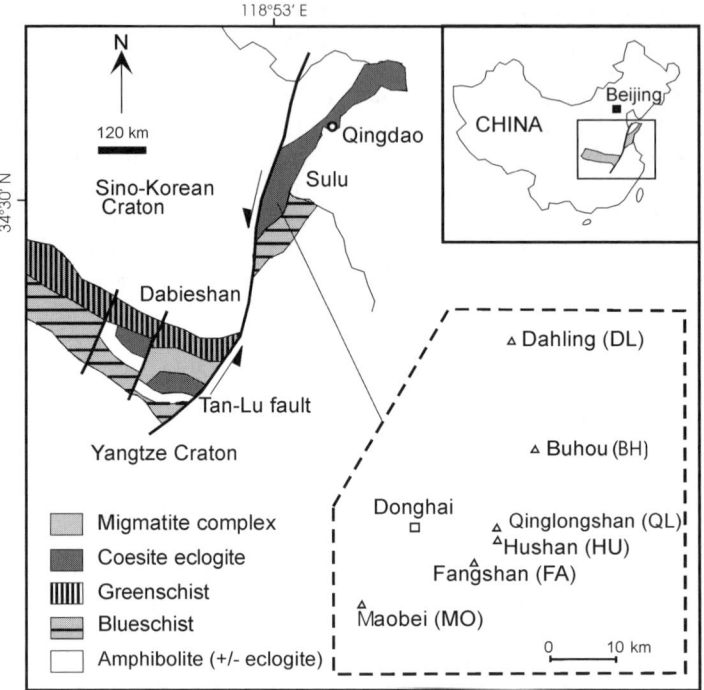

FIG. 1. Index map of the Dabieshan-Sulu UHP metamorphic terrains. Inset map shows the location of the Qinglongshan area and place names described in the text (after Rumble and Yui, 1998).

eclogitic blocks encased in quartzite, gneiss, and schist with poorly exposed contacts. The regional foliation trends ESE. In addition to fresh samples collected along a new highway roadcut (Fig. 2), some samples for which $^{40}Ar/^{39}Ar$ are reported include those previously analyzed for oxygen and/or hydrogen by Yui et al. (1995) and Rumble and Yui (1998).

Following the designations in Rumble and Yui (1998), the samples are referred to as eclogite, epidote eclogite (containing significant epidote), biotite-phengite gneiss, metagranite, orthogneiss, and quartzite. The main paragenesis in eclogite is garnet, omphacite, phengite, and epidote. Accessory minerals are coesite, kyanite, barroisite, rutile, apatite, and quartz. Inclusions of coesite and polycrystalline quartz (pseudomorphs after coesite) occur in garnet, omphacite, and epidote. Phengitic white micas in eclogite samples have variable compositions, with Si contents ranging from 3.3 to 3.6 p.f.u. (Table 1) and most display significant compositional zoning, reflecting Tschermaks exchanges (Fig. 3A) resulting from chemical reequilibration during decompression (Giorgis et al., 2000). Most of the primary UHP assemblages are well-preserved in the eclogites; however, rare cases of retrograde metamorphism may occur.

For example, sample (QL-99-11C) located in the rim of an eclogitic lens, contains amphibole and plagioclase with only relict garnet, omphacite, phengite, and titanite replacing rutile (Fig. 3C). The main assemblage in the biotite-phengite gneiss is phengite-quartz-plagioclase-biotite-microcline-epidote-garnet. The white mica is more muscovitic (Si varying from 3.1 to 3.4 p.f.u.), probably indicating significant chemical re-equilibration during exhumation (Fig. 3A; Table 1). The biotite from these samples contains interlayered chlorite (Fig. 3D). Quartzite contains quartz-phengite-epidote-kyanite-garnet. Phengites are slightly zoned with Si^{4+} contents varying from 3.3 to 3.6 p.f.u. The primary assemblage in the orthogneiss is plagioclase-microcline-quartz-hornblende-titanite. Metagranite samples were collected in active quarries from Hushan and Fangshan (see Fig. 2). The metagranite has a paragenesis of plagioclase-microcline-quartz-epidote-biotite±garnet (skeletal). Mafic segregations occur as lenses in the metagranite. Significant amounts of titanite, hornblende (with

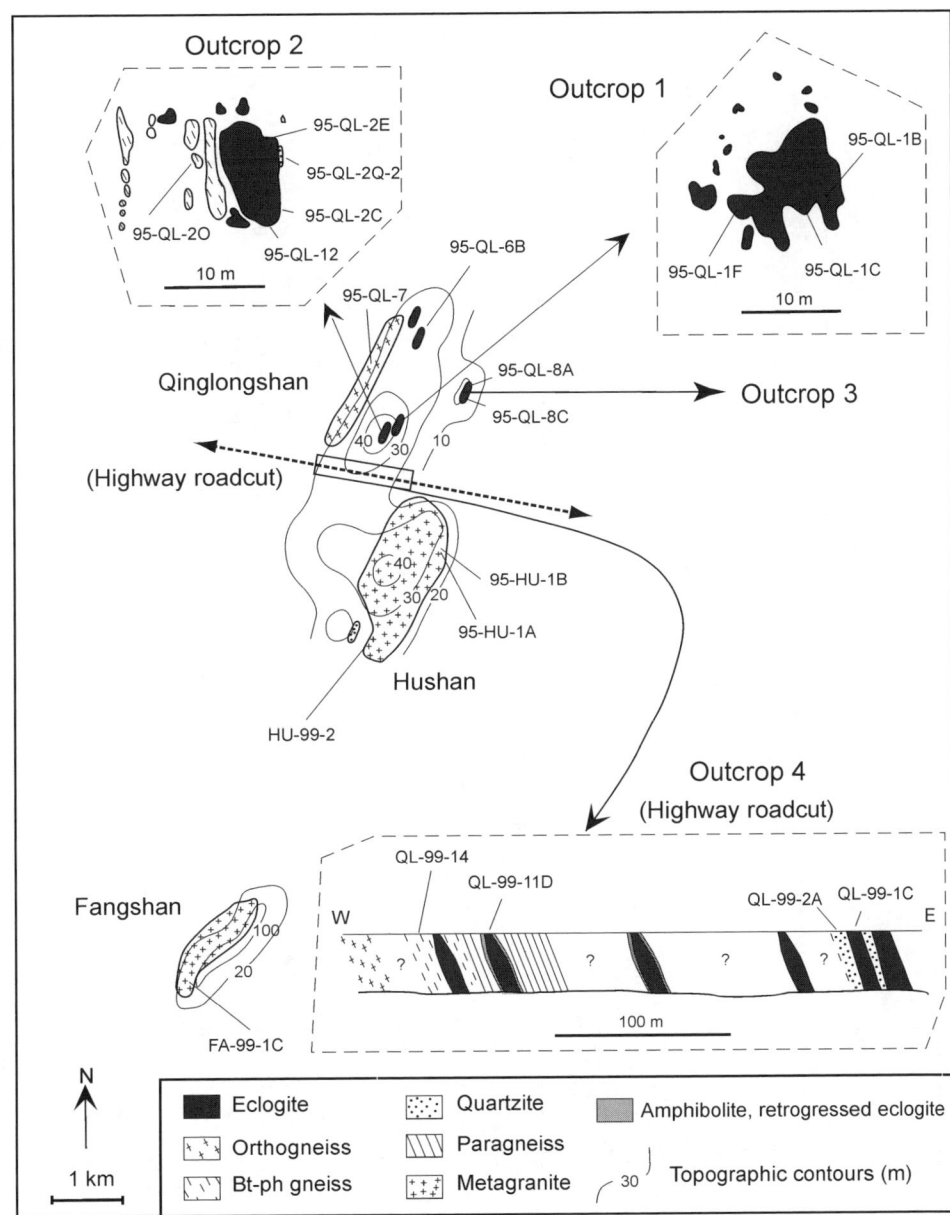

FIG. 2. Outcrop map and sample locations of the Qinglongshan area (after Rumble and Yui, 1998).

biotite inclusions), and epidote (cored by allanite) characterize these lenses. A sample of quartz-phengite-paragonite-kyanite white schist was found in the Qinglongshan metagranite (Fig. 2) and in this sample the phengite (Si = 3.25 to 3.45 p.f.u.) is rimmed by paragonite (Fig. 3E).

Analytical Procedures

Oxygen isotopic extraction was done using a CO_2 laser fluorination system at the Geophysical Laboratory, similar to the setup of Sharp (1990). Mineral fractions of 1.5–2.5 mg were used for each

TABLE 1. Representative Electron Microprobe Analyses and Mineral Formulae of White Mica from Different Lithologies Analyzed by $^{40}Ar/^{39}Ar$[1]

	Quartzite				Bt-Phn gneiss								Eclogite								
	QL-99-2A Ms		95-QL-2Q-2 Phn		QL-99-14 Ms		95-QL-20 Phn		95-QL-12 Phn		95-QL-8C Phn		95-QL-8A Phn		QL-99-11D Phn		95-QL-2C Phn		MO-99-1 Phn	
Oxide wt%:	Min	Max	Min	Max	Min	Max	Min	Max	Min	Max	Min	Max	Min	Max	Min	Max	Min	Max	Min	Max
SiO_2	46.23	48.20	50.77	51.21	45.87	46.30	45.80	50.61	48.91	52.16	50.04	52.33	51.39	54.48	52.05	53.28	50.96	51.59	53.41	52.70
TiO_2	0.68	0.61	0.26	0.28	0.77	0.77	0.70	0.33	0.25	0.26	0.33	0.32	0.30	0.30	0.29	0.29	0.23	0.25	0.31	0.27
Al_2O_3	28.40	31.12	24.61	24.10	26.34	26.61	31.90	26.69	27.87	24.68	26.62	23.64	26.30	22.45	24.72	22.98	23.45	23.31	24.43	25.63
Cr_2O_3	0.01	0.00	0.03	0.04	0.04	0.00	0.00	0.05	0.05	0.05	0.04	0.09	0.06	0.06	0.04	0.06	0.00	0.04	0.03	0.01
FeO	6.54	3.53	2.92	2.64	8.32	7.54	3.85	3.52	3.09	2.84	2.34	2.32	1.43	1.42	2.07	1.52	2.88	2.91	1.07	0.91
MnO	0.06	0.11	0.01	0.00	0.07	0.11	0.00	0.00	0.05	0.06	0.00	0.03	0.01	0.00	0.01	0.00	0.01	0.02	0.00	0.00
MgO	1.96	1.64	4.40	4.64	1.44	1.39	1.46	3.24	3.17	4.42	3.65	4.73	4.37	5.66	4.53	5.34	4.69	4.71	5.18	4.72
CaO	0.08	0.00	0.01	0.00	0.00	0.00	0.04	0.09	0.00	0.00	0.02	0.00	0.00	0.02	0.02	0.00	0.01	0.00	0.01	0.05
Na_2O	0.37	0.31	0.19	0.15	0.29	0.24	0.88	0.51	0.80	0.43	0.55	0.14	0.41	0.12	0.31	0.14	0.31	0.23	0.20	0.28
K_2O	10.11	10.10	10.73	10.76	11.13	11.11	10.03	10.03	9.61	10.09	10.18	11.44	11.09	11.16	10.05	10.29	10.85	10.95	11.16	10.88
F=O	–	–	–	–	–	–	–	–	0.43	0.58	–	–	–	–	–	–	–	–	–	–
Cl=O	–	–	–	–	–	–	–	–	–	–	–	–	–	–	–	–	–	–	–	–
Total	94.46	95.61	93.92	93.81	94.25	94.07	94.67	95.07	93.82	94.98	93.78	95.05	95.36	95.67	94.08	93.91	93.38	94.00	95.81	95.45
Si	6.26	6.37	6.85	6.91	6.35	6.41	6.15	6.74	6.57	6.92	6.74	7.01	6.82	7.20	6.95	7.11	6.93	6.97	7.04	6.96
Al^{IV}	1.74	1.63	1.15	1.09	1.65	1.59	1.85	1.26	1.43	1.08	1.26	0.99	1.18	0.80	1.05	0.89	1.07	1.03	0.96	1.04
Al^{VI}	2.79	3.21	2.76	2.74	2.65	2.75	3.20	2.93	2.98	2.77	2.97	2.75	2.94	2.69	2.83	2.73	2.70	2.69	2.83	2.94
Ti	0.07	0.06	0.03	0.03	0.08	0.08	0.07	0.03	0.02	0.03	0.03	0.03	0.03	0.03	0.03	0.03	0.02	0.03	0.03	0.03
Cr	0.00	0.00	0.00	0.00	0.00	0.00	0.00	0.00	0.01	0.01	0.00	0.01	0.01	0.01	0.00	0.01	0.00	0.00	0.00	0.00
Fe^{2+}	0.10	0.05	0.03	0.03	0.43	0.43	0.07	0.12	0.00	0.02	0.05	0.15	0.03	0.07	0.00	0.00	0.09	0.11	0.03	0.01
Fe^{3+}	0.64	0.34	0.30	0.27	0.53	0.44	0.36	0.27	0.36	0.29	0.22	0.11	0.13	0.09	0.23	0.17	0.23	0.22	0.09	0.09
Mg	0.01	0.01	0.00	0.00	0.01	0.01	0.00	0.00	0.01	0.01	0.00	0.00	0.00	0.00	0.00	0.00	0.00	0.00	0.00	0.00
Mn	0.40	0.32	0.88	0.93	0.30	0.29	0.29	0.64	0.64	0.87	0.73	0.95	0.87	1.11	0.90	1.06	0.95	0.95	1.02	0.93
Ca	0.01	0.00	0.00	0.00	0.00	0.00	0.01	0.01	0.00	0.00	0.00	0.00	0.00	0.00	0.00	0.00	0.00	0.00	0.00	0.01
Na	0.10	0.08	0.05	0.04	0.08	0.06	0.23	0.13	0.21	0.11	0.14	0.04	0.10	0.03	0.08	0.04	0.08	0.06	0.05	0.07
K	1.74	1.70	1.84	1.85	1.96	1.96	1.72	1.70	1.65	1.71	1.75	1.96	1.88	1.88	1.71	1.75	1.88	1.89	1.88	1.83
F	–	–	–	–	–	–	–	–	0.18	0.24	–	–	–	–	–	–	–	–	–	–
Cl	–	–	–	–	–	–	–	–	0.00	0.00	–	–	–	–	–	–	–	–	–	–

↑ *Table continues* ↑

[1]White mica formula calculated on the basis of 12 small cations. Phn = phengitic white mica; ms = muscovitic white mica.

analysis. An interlaboratory reference material, UWG-2 (Valley et al., 1995), was analyzed with each series of unknowns. Analyses were corrected to 5.8 ‰$_{VSMOW}$, the value recommended for standard UWG-2 (Valley et al., 1995). The reported $\delta^{18}O$ values in Table 1 are averages of duplicate or triplicate determinations with a precision of 0.1–0.2‰.

The samples and minerals of known age were irradiated for 20 and 40 MWh in separate experiments, in the USGS TRIGA reactor in Denver (CO), USA (Dalrymple et al., 1981). After irradiation, the samples were incrementally heated using a low-blank resistance furnace on-line to a MAP 215-50 mass spectrometer with an electron multiplier. Before introduction into the mass spectrometer, the gas was purified using activated Zr/Ti/Al getters and a metal cold finger maintained at liquid nitrogen temperatures. Time zero regressions were fit to data collected from eight scans over the mass range ^{40}Ar–^{36}Ar. Correction for the neutron flux (J value) was determined using the standard MMHB-1, assuming an age of 523.1 Ma (Renne et al., 1998). Moreover, analyses were corrected for backgrounds, blanks, mass discrimination (determined by online measurements of air), radioactive decay, and interfering isotopic reactions (Ca, K, and Cl derived isotopes of Ar). Isotopic production ratios for the Triga reactor were determined from analysis of irradiated salts (CaF_2 and KCl). Typical ^{40}Ar blank values were 4×10^{-15} moles at temperature below 1000°C, rising to 9×10^{-15} moles at 1600°C. All analytical uncertainties are given at the 2σ confidence level and all ages include a J error of 0.5%. Data are presented in Table 2 and Appendix Table 1 and $^{40}Ar/^{39}Ar$ age spectra are given in Figures 4 and 5.

Results

Oxygen isotopes

The $\delta^{18}O$ values obtained from phengite are consistent with data previously obtained by Rumble and Yui (1998) for samples from Qinglongshan (Table 2). For example, data reported here from sample 95-QL-12 comes from the same outcrop as samples 95-QL-2C and 95-QL-2E (Rumble and Yui, 1998), and the $\delta^{18}O$ values for phengite are very similar, ranging from –9.2 to –10‰. These data are more remarkable because the phengite $\delta^{18}O$ values exhibit little variation within an outcrop, and show no modal effects imposed by the host lithology. In addition, the new data reported here confirm that significant heterogeneities in phengite $\delta^{18}O$ values occur between outcrops with values ranging from

TABLE 1. (continued)

Oxide wt%	Epidote eclogite 95-QL-1C Phn min	max	95-QL-1F Phn min	max	QL-99-1C Phn min	max	White schist HU-99-2 Phn + Pg min	max
SiO_2	51.45	51.81	51.04	51.82	49.53	52.26	48.98	52.18
TiO_2	0.25	0.24	0.24	0.25	0.25	0.21	0.35	0.32
Al_2O_3	24.14	23.90	25.05	23.83	26.82	24.28	31.38	29.49
Cr_2O_3	0.04	0.06	0.09	0.01	0.08	0.03	0.00	0.00
FeO	2.67	2.89	3.07	3.12	2.95	2.63	1.34	1.34
MnO	0.01	0.01	0.05	0.01	0.00	0.03	0.00	0.00
MgO	4.60	4.75	4.30	4.75	3.74	4.71	1.64	1.78
CaO	0.00	0.01	0.04	0.02	0.01	0.00	0.00	0.00
Na_2O	0.36	0.31	0.34	0.18	0.75	0.31	1.13	1.03
K_2O	10.54	10.37	10.29	10.80	9.61	10.34	8.63	8.55
F	—	—	—	—	—	—	—	—
Cl	—	—	—	—	—	—	—	—
Total	94.05	94.34	94.51	94.79	93.74	94.80	93.46	94.71
Si	6.92	6.93	6.82	6.92	6.65	6.94	6.54	6.88
Al^{IV}	1.08	1.07	1.18	1.08	1.35	1.06	1.46	1.12
Al^{VI}	2.75	2.70	2.76	2.68	2.89	2.75	3.49	3.47
Ti	0.03	0.02	0.02	0.03	0.03	0.02	0.04	0.03
Cr	0.00	0.01	0.01	0.00	0.01	0.00	0.00	0.00
Fe^{2+}	0.05	0.01	0.00	0.05	0.00	0.01	0.06	0.15
Fe^{3+}	0.25	0.31	0.34	0.30	0.33	0.29	0.09	0.00
Mg	0.00	0.00	0.01	0.00	0.00	0.00	0.00	0.00
Mn	0.92	0.95	0.86	0.95	0.75	0.93	0.33	0.35
Ca	0.00	0.00	0.01	0.00	0.00	0.00	0.00	0.00
Na	0.09	0.08	0.09	0.05	0.20	0.08	0.29	0.26
K	1.81	1.77	1.75	1.84	1.65	1.75	1.47	1.44
F	—	—	—	—	—	—	—	—
Cl	—	—	—	—	—	—	—	—

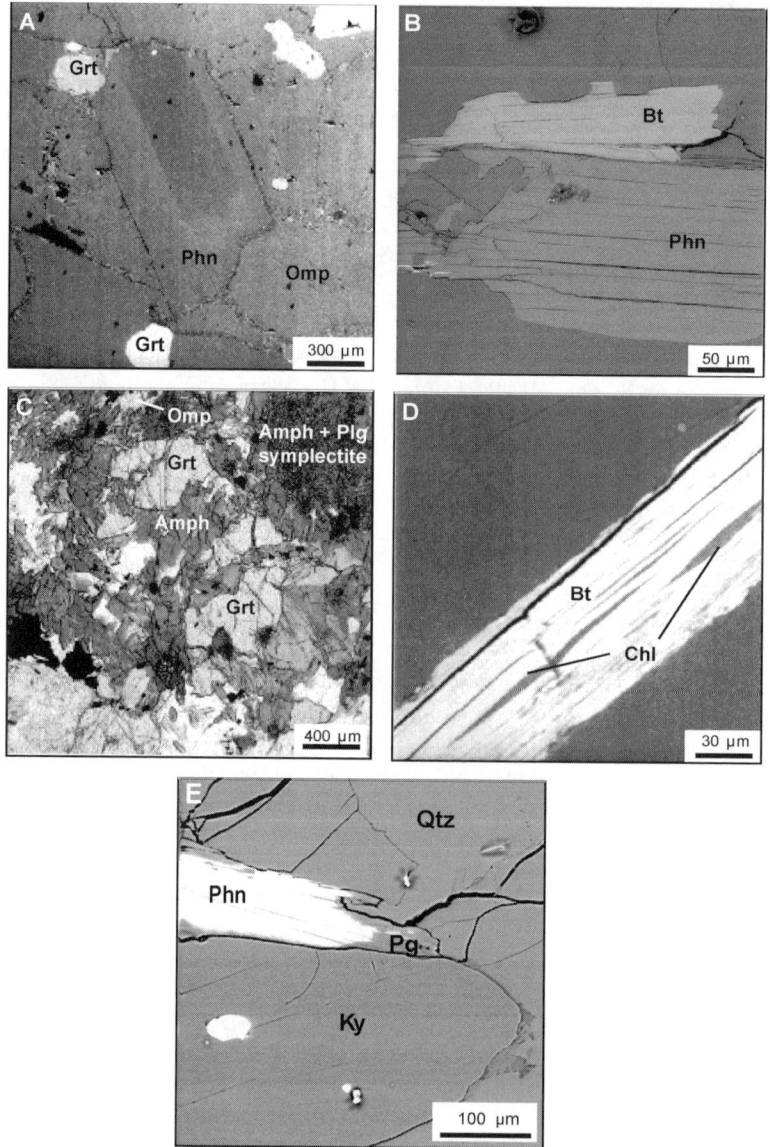

FIG. 3. Backscattered electron (BSE) and photomicrograph images of textures in the investigated samples. A. Eclogite sample QL-99-1C containing garnet (Grt), omphacite (Omp), and phengite (Phn) illustrating compositional variations within phengite developed during exhumation. B. BSE image from gneiss sample QL-99-14 illustrating the association of biotite (Bt) and phengite (Phn). C. Photomicrograph of amphibolite sample QL-99-11C occurring around the margin of an eclogitic lens, showing relict eclogitic garnets (Grt) and omphacite (Omp) together with amphibole and symplectite of amphibole and plagioclase (Plg). D. BSE image from sample QL-99-14 showing the occurrence of interlayered chlorite (chl) lenses in biotite cleavage. E. BSE image of white schist sample HU-99-2 showing the occurrence of kyanite (Ky), quartz (Qz), and phengite (Phn) rimmed by late paragonite. Mineral abbreviations from Kretz (1982).

–9.9 to –1.3‰. The results confirm that surface weathering had no significant effect on the oxygen isotopic signature in these rocks.

K-feldspar and biotite sampled from the Fangshan-Hushan metagranite yield $\delta^{18}O$ values from –3.7 to –3.8‰ and –6.1 to –6.7‰, respectively. The

TABLE 2. Total Fusion Ar-Ar ages and δD and δ^{18}O Values of Minerals from Qinglongshan Separated by Outcrops[1]

Sample	Rock	TF age (Ma)	Phm δD	Phm δ^{18}O	Ky δ^{18}O	Grt δ^{18}O	Ep δ^{18}O	Omp δ^{18}O	Rt δ^{18}O	Amp δ^{18}O	Qz δ^{18}O	Pl δ^{18}O	Bt δ^{18}O	Kfs δ^{18}O	Ms δ^{18}O
							Outcrop 1								
95-QL-1B*	Epidote eclogite	552	−124	−1.2	−0.5	−1.6	−2.4	−2.2	−5.3						
95-QL-1C	Epidote eclogite	449	−121	−1.5			−2.1								
95-QL-1F	Epidote eclogite	568	−124	−1.3*											
95-QL-1A	Epidote eclogite						4								
							Outcrop 2								
95-QL-2C	Amp- eclogite	901	−104	−10		−11.1	−11.1	−11.1	−14.8	−11					
95-QL-2E	Eclogite	925	−101	−9.4		−10.3	−10.8	−11.2	−14.2						
95-QL-2O	Bt-phm gneiss	590	−105	−9		−9.4					−6.5	−2.4			
95-QL-2Q-2	Quartzite	693	−106	−9.7			−10.1				−7.4				
95-QL-12	Eclogite	894		−9.2*											
95-QL-2D	Eclogite					−10.1	−9.9	−10.1	−14.1						
95-QL-2H	Eclogite					−10.3		−10.4	−14.1						
95-QL-2M	Eclogite			−10.7		−10.1	−10.4	−10.7	−14.6		−7.7				
95-QL-2N	Eclogite			−9.3		−10.3	−10.1	−10							
95-QL-2P	Eclogite			−8.9		−9.3	−9	−8.6	−13.4						
							Outcrop 3								
95-QL-8A	Eclogite	307	−123	−4.3		−4.9		−5.6	−8.2						
95-QL-8C	Eclogite	261	−123	−3.4*											
95-QL-8B	Eclogite		−118	−4.6		−5.4		−5.5	−8.7						
							Outcrop 4 (highway trace)								
QL-99-1C	Epidote eclogite	321		−9.9*											
QL-99-11D	Eclogite	314		−6.7*											
							Other samples								
HU-99-2	White schist	246		−8.9*											
QL-99-14	Bt-ms gneiss	204													
QL-99-2A	Quartzite	204													
95-HU-1B	Meta-granite	202(bt), 197 (kfs)									−2.4	−3.8	−6.5	−3.8	−1.3*
95-HU-1A	Mafic segregation	202(bt), 194 (kfs)					−5.6				−2.5		−6.1	−3.7	−9.2
95-QL-7	Gneiss	190								−1.1	2			−0.1*	
FA-99-1C	Meta-granite	202											−6.7*		

[1]Minerals: Phm = phengite, Ky = kyanite, Grt = garnet, Ep = epidote, Omp = omphacite, Rt = rutile, Amp = amphibole, Qz = quartz, Pl = plagioclase, Bt = biotite, Kfs = K-feldspar, Ms = Muscovite. Abbreviations from Kretz, 1983. Stable isotope data indicated by a (*) are from this study; otherwise stable isotope data from Rumble and Yui (1998). All Ar-Ar data are from this study.

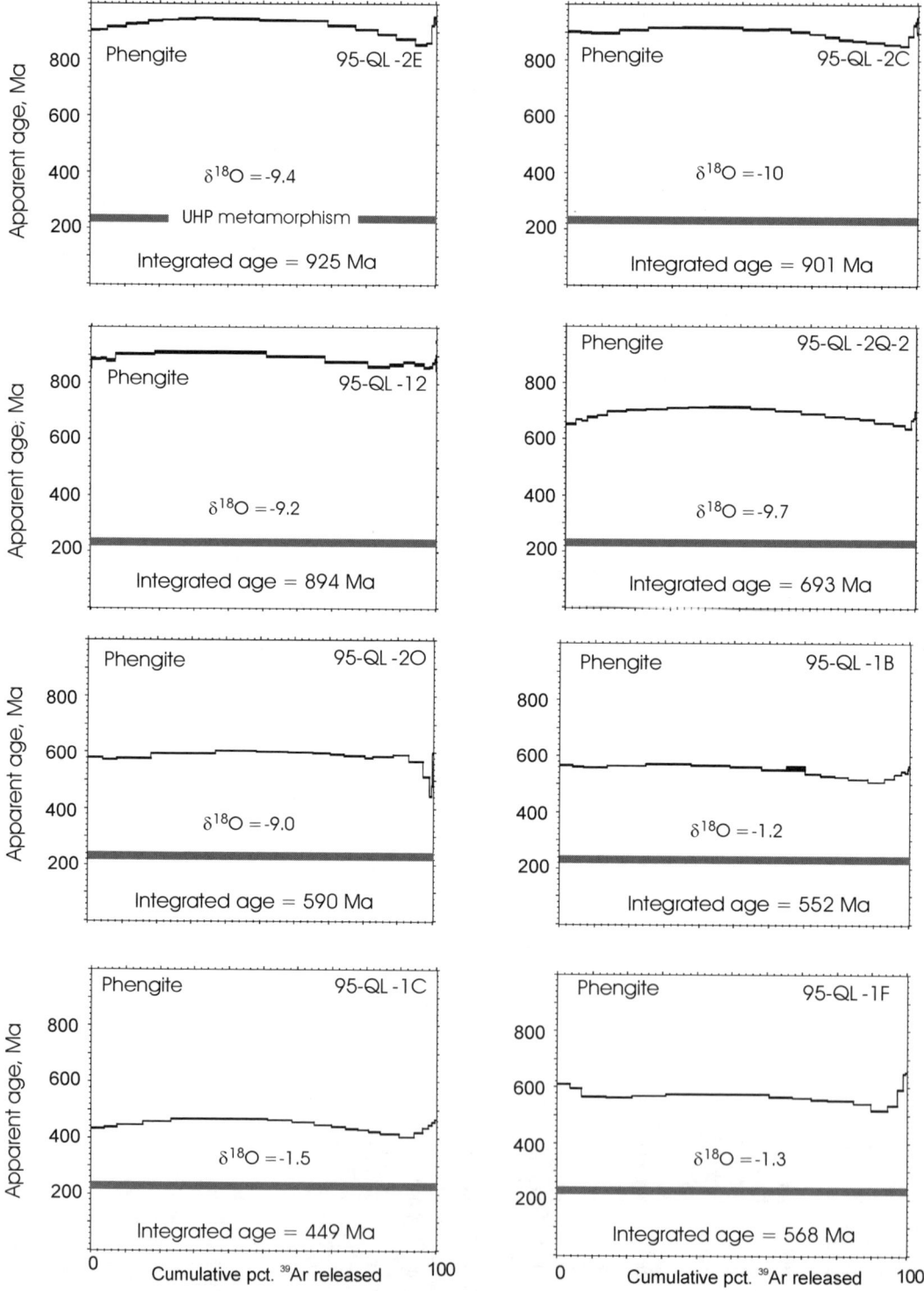

FIG. 4. For caption, see following page.

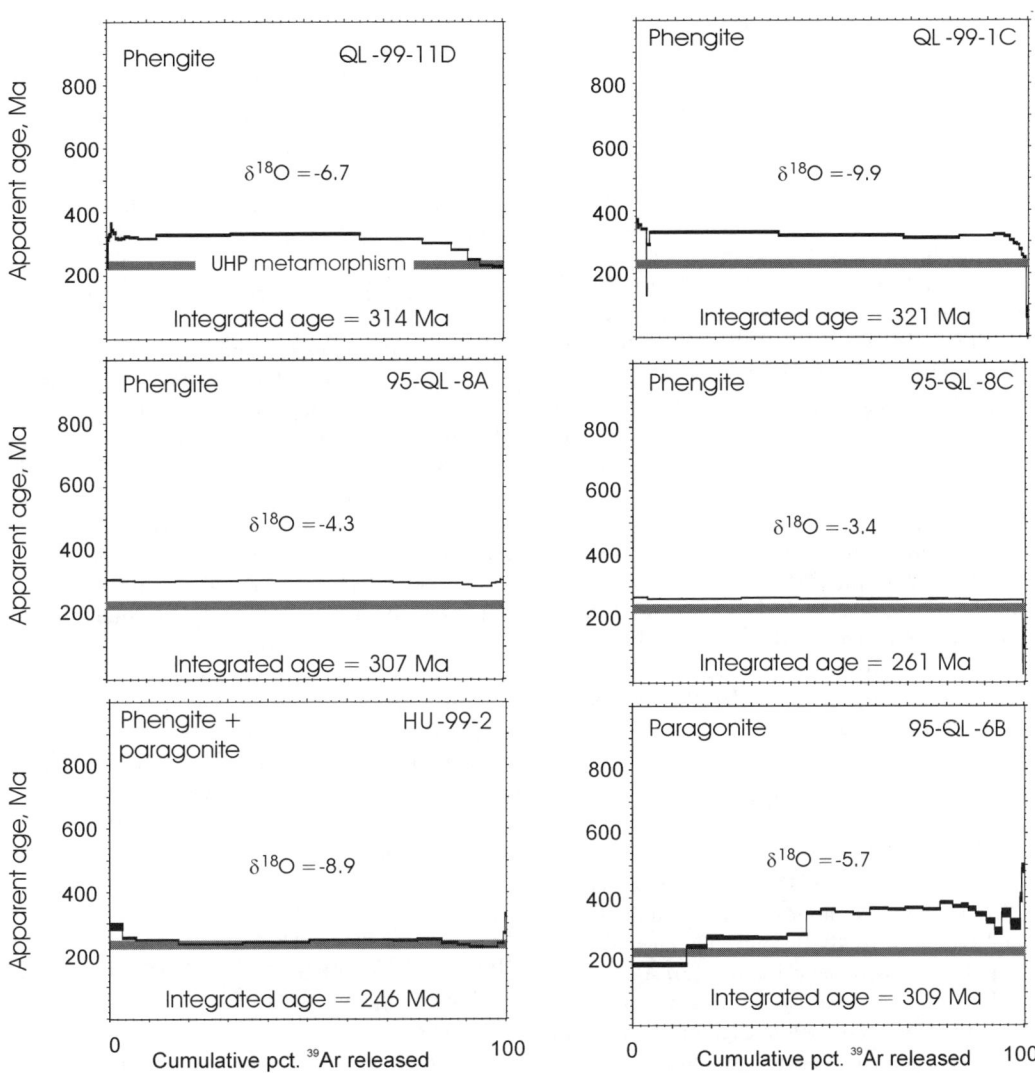

FIG. 4. $^{40}Ar/^{39}Ar$ incremental release spectra of phengite and paragonite from Qinglongshan UHP samples. The grey band at 225 Ma indicates the timing of UHP metamorphism. Note the presence of extraneous argon in all samples, which is inherited from the UHP protoliths. The amount of extraneous argon is a function of the protolith and its corresponding UHP lithology and amount of argon loss during exhumation.

K-feldspar from gneiss sample 95-QL-7 with a $\delta^{18}O$ value of –0.1‰ is in apparent isotopic equilibrium with the associated quartz with a $\delta^{18}O$ value of +2.0‰ (value of Rumble and Yui, 1998). In contrast, biotite from a bt-phn gneiss (sample QL-99-14) yielded a $\delta^{18}O$ value of –1.4‰, similar to the $\delta^{18}O$ value of associated phengites (–1.3‰), suggesting possible oxygen isotopic disequilibrium due to the alteration of biotite (Fig. 3D).

$^{40}Ar/^{39}Ar$ geochronology

Furnace step heating $^{40}Ar/^{39}Ar$ geochronology of 22 purified mineral separates of phengite, muscovite, paragonite, biotite, and K-feldspar from different lithologies around Qinglongshan and surrounding localities are presented in Figures 4 and 5. Without exception, the phengite $^{40}Ar/^{39}Ar$ age spectra exhibit variable amounts of extraneous argon yielding total gas ages varying between 925 Ma to 260 Ma (Fig. 4). In contrast, muscovitic white micas from samples QL-99-14 and QL-99-2A yield concordant plateau and isochron ages of 203 ± 1 Ma and 204 ± 1 Ma, and 202 ± 2 Ma and 204 ± 2 Ma, respectively. A phengite-paragonite intergrowth from white schist sample HU-99-2 yielded a ^{40}Ar/^{39}Ar total gas age of 246 Ma. Three samples of biotite from the Hushan-Fangshan metagranite (95-HU-1A, 95-HU-1B, FA-99-1C) and one sample of biotite from a bt-phn gneiss from Qinglongshan (QL-99-14) yield concordant ^{40}Ar/^{39}Ar plateau and isochron ages of 202–204 Ma (Fig. 5).

Mineral separates of K-feldspar from three samples of Hushan meta-granite (95-HU-1A, 95-HU-1B) and from the Qinglongshan gneiss (95-QL-7) yielded similar ^{40}Ar/^{39}Ar age results and argon degasing behavior during step heating (Fig. 5). The data do not allow a strict calculation of an ^{40}Ar/^{39}Ar age plateau, but the ^{40}Ar/^{39}Ar total gas ages obtained are between 197 and 190 Ma for all three samples. Isochron plots yield no additional information but a comparison between ages obtained on biotites and K-feldspar suggest that the presence of excess argon is unlikely.

Discussion

The Qinglongshan isotopic anomaly is manifested by extremely low δ^{18}O and δD values coupled with locally large variations within a limited area (Rumble and Yui, 1998). Because the unusually low δ^{18}O and δD values most likely represent stable isotopic signatures acquired during near-surface hydrothermal alteration before UHP metamorphism, their preservation provides insight into length scales of isotopic exchange throughout subduction and exhumation. Rumble and Yui (1998) concluded that stable isotopic exchange occurred over length scales not greater than a few centimeters. In their model, acquisition of low δ^{18}O and δD values took place in a shallow-level hydrothermal system driven by a near-surface granitic pluton (Hushan metagranite). Groundwater flow was driven through fractures by a lateral temperature gradient set up by the intrusion. In fractures, joints, and across bedding planes the water/rock ratios were very high, but a few centimeters into solid rock the water/rock ratios are low. Differences in water/rock ratios lead to local differences in ^{18}O/^{16}O as dictated by the geometry of the hydrothermal system, which locally can be significant. During Triassic subduction, collision, and UHP metamorphism, the ^{18}O/^{16}O ratios acquired during the Neoproterozoic hydrothermal alteration were retained. This is clear from the stable isotope data (Table 2), which show large variations in δ^{18}O and δD values for a given mineral in different outcrops.

Data from this study provide evidence that the apparent length scale of diffusional isotopic exchange for oxygen and hydrogen under UHP conditions was confined to the scale of contiguous rock within an outcrop. This limited isotopic exchange was not limited to oxygen and hydrogen, but is also recognized in the argon isotopic budget. This is best visualized by plotting the δ^{18}O and δD values of phengite against their ^{40}Ar/^{39}Ar total fusion ages (Figs. 6 and 7). These plots show a clear separation of δ^{18}O, δD, and ^{40}Ar/^{39}Ar total fusion ages by outcrop. Outcrops QL-1, QL-2, and QL-8 have minor internal variations in δ^{18}O for a given mineral, on the order of 1‰, and δD values are analytically indistinguishable. Significantly, the limited variations in the δ^{18}O and δD values of a given mineral within the same outcrop (Figs. 5 and 6, Table 1) despite large variations in host lithology suggests that isotopic exchange was extensive within each outcrop. This conclusion is based on the fact that large differences in lithology and modal mineralogy (e.g., quartzite versus eclogite) have a marked modal effect on stable isotope ratios and would normally result in differences in δ^{18}O greater than those observed for a given mineral (e.g., phengite) between lithologies within the same outcrop (e.g., Rumble, 1982; Eiler et al., 1993). Thus isotopic exchange appears to have been extensive within outcrops.

The indistinguishable δD values of hydrous minerals from different lithologies suggests that the length scale of isotopic exchange is probably longer for hydrogen than for oxygen (Rumble and Yui, 1998). Only outcrop QL-99, which is the highway transect, shows significant variation in the phengite δ^{18}O values, but given that the two samples are separated by several tens of meters and cannot be contiguously mapped in the field, this strongly suggests that these samples are not from the same contiguous and isotopically homogeneous outcrop. Thus, the variations observed in the δ^{18}O, δD, and ^{40}Ar/^{39}Ar values are consistent with preserved isotopic heterogeneities acquired during hydrothermal alteration, followed by complete isotopic exchange within contiguous outcrops irrespective of lithology during UHP metamorphism. This conclusion is in

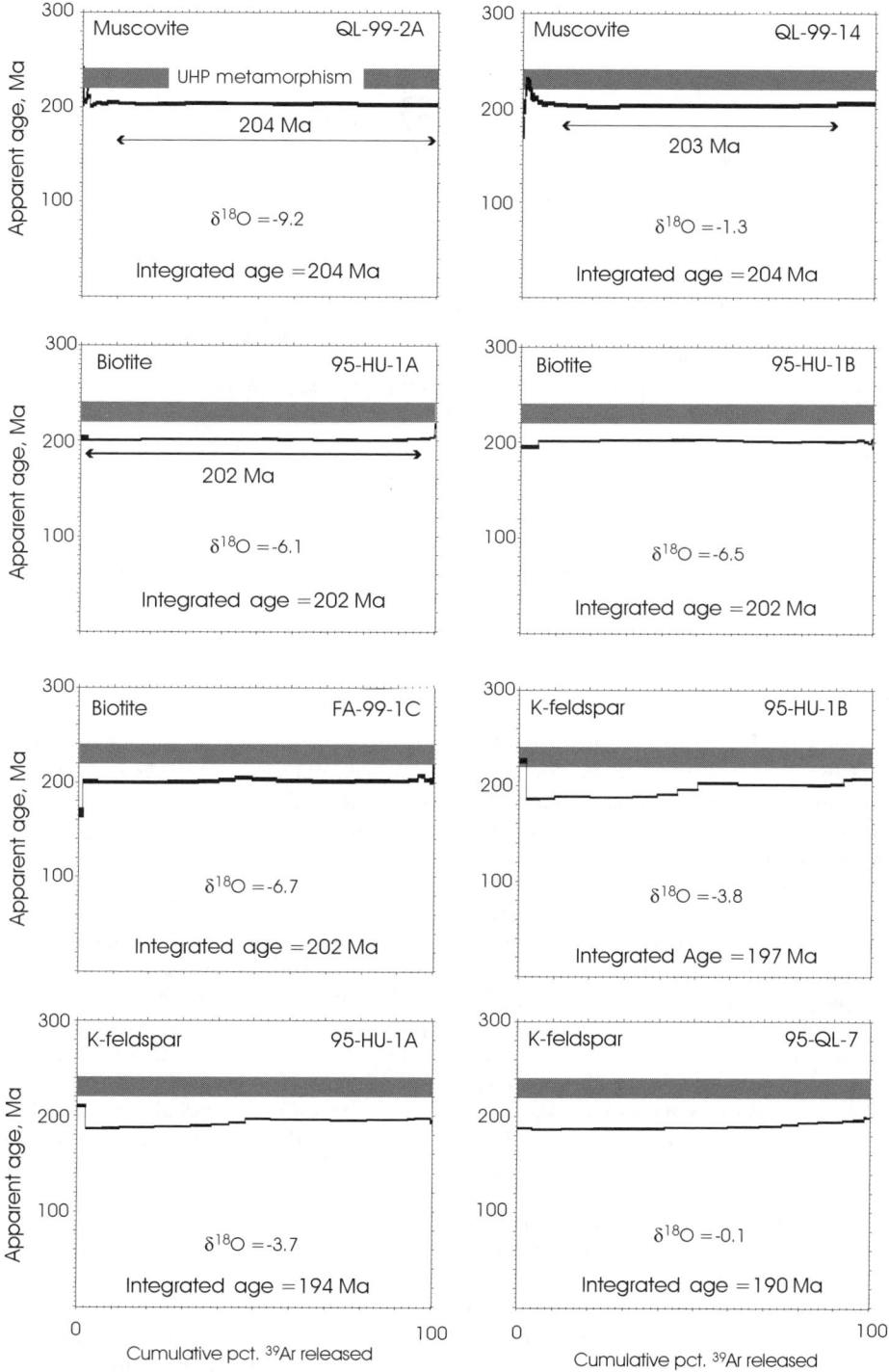

FIG. 5. $^{40}Ar/^{39}Ar$ incremental release spectra of muscovite, biotite, and K-feldspar. The grey band at 225 Ma indicates the timing of UHP metamorphism. Note that all samples show no evidence of extraneous argon and are interpreted as regional metamorphic cooling ages.

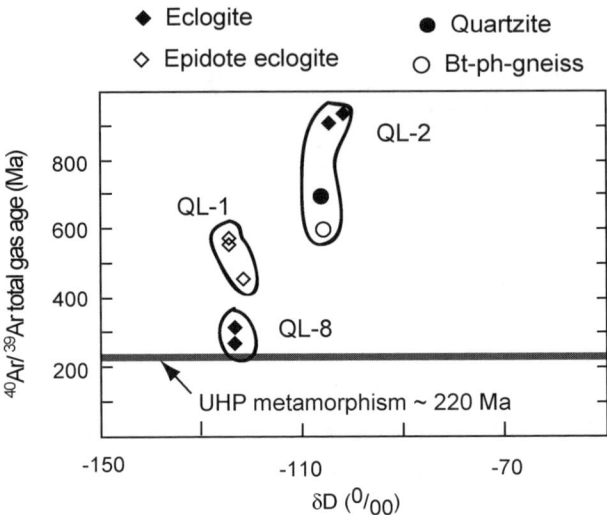

FIG. 6. Plot of integrated ^{40}Ar/^{39}Ar total fusion age versus δD in phengite from three outcrops of the Qinglonshan area (Fig. 2). Note the limited variation in δD values from each outcrop, despite large differences in lithology, suggesting isotopic exchange on the scale of the outcrop. The variations in ^{40}Ar/^{39}Ar age indicate a lithological control on the retention of argon in phengite.

agreement with Rumble and Yui (1998), who first recognized the persistent premetamorphic variations in outcrops separated by several meters. The ^{40}Ar/^{39}Ar apparent ages indicate the combined effects of variable amounts of extraneous (inherited) argon together with variable argon loss during exhumation. These results suggest that lithologically independent, contiguous blocks within an outcrop probably define the maximum length scale of oxygen, hydrogen, and argon isotopic exchange during UHP metamorphism. That is, subducted blocks now exposed as outcrops each behaved as a closed isotopic system, and did not communicate isotopically with other blocks now exposed only meters apart.

Empirical and theoretical studies have shown that argon retention in white mica is controlled to some extent by composition (e.g., Scaillet et al., 1992; Scaillet, 1998; Bosse et al., 2000), and that under UHP conditions argon can be inherited from its protolith (Scaillet, 1996; Giorgis et al., 2000). This study further demonstrates that rock bulk composition and paragenesis have an effect on the ^{40}Ar/^{39}Ar data. Note that phengites from basaltic eclogites collected from outcrops QL-2 and QL-8 yield well-defined groups of ages within each outcrop (Fig. 7). Similarly, the epidote eclogites from outcrop QL-1 have phengite ages that group together. However, phengite ^{40}Ar/^{39}Ar ages from different lithologies within a given outcrop behave much differently. Phengites from outcrop QL-2 yield much younger ^{40}Ar/^{39}Ar ages where collected from quartzites or gniesses, clearly illustrating that, during exhumation, the armoring effect of basaltic rocks on argon retention in phengite is measurably greater than in quartzites or gneisses (Fig. 7).

In contrast to ^{40}Ar/^{39}Ar data collected from phengitic white mica that reflect variable amounts of inherited argon, the musovitic white mica from the metagranitic rocks yield ^{40}Ar/^{39}Ar data that are consistent with metamorphic cooling following the Triassic UHP metamorphism (Fig. 8). Likewise, unaltered biotite and K-feldspar yield metamorphic cooling ages, consistent with other geochronological data from the Sulu and Dabieshan UHP terrains (e.g., Li et al., 1994, 2000). The muscovite, biotite, and K-feldspar were either not part of the UHP assemblages, or have been completely recrystallized during exhumation. This can explain the fact that the ^{40}Ar/^{39}Ar data yield geologically reasonable cooling ages. However, in contrast to the inert argon, the structurally bonded hydrogen and oxygen preserve their pre-UHP metamorphic signatures.

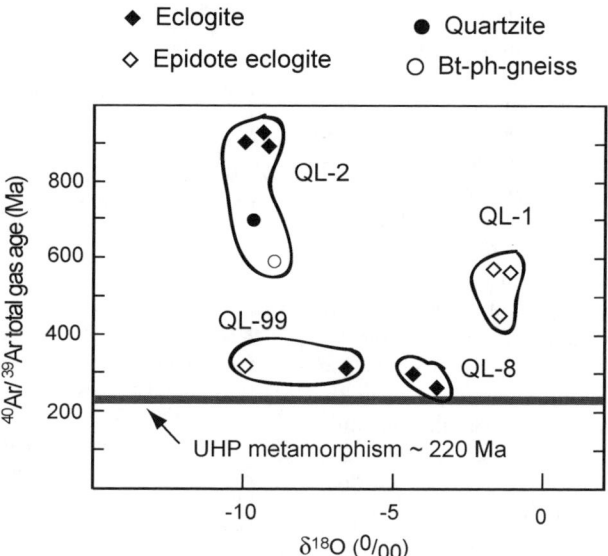

FIG. 7. Plot of integrated $^{40}Ar/^{39}Ar$ total fusion age versus $\delta^{18}O$ in phengite from four outcrops of the Qinglonshan area (Fig. 2). Note the limited variation in $\delta^{18}O$ values from each outcrop, despite large differences in lithology, suggesting isotopic exchange on the scale of the outcrop. Samples from outcrop QL-99 were collected several tens of meters apart along a roadcut for a new highway, and probably represent separate, isotopically distinct outcrops. The variations in $^{40}Ar/^{39}Ar$ age indicate a lithological control on the retention of argon in phengite.

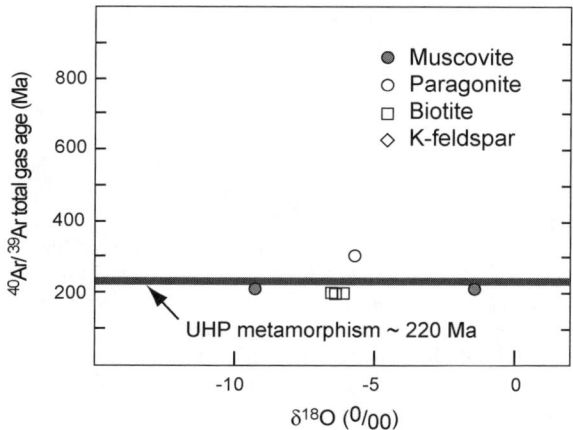

FIG. 8. Plot of $^{40}Ar/^{39}Ar$ total gas age versus $\delta^{18}O$ in muscovite, paragonite, biotite, and K-feldspar. The large variation in $\delta^{18}O$ values is consistent with preserved isotopic signatures obtained during pre-UHP near-surface hydrothermal alteration. The $^{40}Ar/^{39}Ar$ ages younger than 220 Ma are consistent with regional metamorphic cooling.

Conclusions

The combined $\delta^{18}O$, δD, and $^{40}Ar/^{39}Ar$ data from this study clearly illustrate that isotopic exchange for each system was limited to the scale of outcrops. The quasi-uniform $\delta^{18}O$ and δD values for a given mineral within a given outcrop, despite major differences in lithology, suggest nearly complete oxygen and hydrogen exchange within a given outcrop during UHP metamorphism. This may indicate that blocks of coherent rock within an accretionary prism behave as isolated isotopic reservoirs, permitting isotopic exchange only within its physical boundaries, once subducted to UHP depths. The variations in $^{40}Ar/^{39}Ar$ age reflect a complex effect of argon present in the protolith together with variable argon loss during different stages of exhumation. Phengites clearly retain argon under UHP conditions, and the age and composition of the protolith will, to some extent, determine whether geologically significant $^{40}Ar/^{39}Ar$ ages are obtainable from such UHP rocks. In the Qinglongshan example, it does not appear that geologically reliable ages can be obtained from UHP phengite. However, as an isotopic tracer, the argon data from this investigation clearly indicate that argon reaches upper mantle depths during subduction and the variable amounts of argon retained in phengite suggests that argon not only can be, but probably is, recycled into the mantle during UHP metamorphism.

Acknowledgments

We thank Dr. Ruth Zhang (Stanford University) and Z. M. Zhang and F. L. Liu (Chinese Academy of Geological Sciences) for field work assistance, fruitful discussions, and very pleasant moments in China. We also thank L. Baumgartner, B. Putlitz, and T. Vennemann for helpful discussions and comments. This study was supported by the Swiss and American National Science Foundations (FNRS grant 2000-65025.01 and NSF EAR grant 0003276).

REFERENCES

Ames, L., Zhou, G., and Xiong, B., 1996, Geochronology and isotopic character of ultrahigh-pressure metamorphism with implications for collision of the Sino-Korean and Yangtze cratons, central China: Tectonics, v. 15, p. 472–489.

Baker, J., and Matthews, A., 1995, The stable isotopic evolution of a metamorphic complex, Naxos, Greece: Contributions to Mineralogy and Petrology, v. 120, p. 391–403.

Baker, J., Matthews, A., Mattey, D., Rowley, D. B., and Xue, F., 1997, Fluid-rock interaction during ultrahigh-pressure metamorphism, Dabie Shan, China: Geochimica et Cosmochimica Acta, v. 61, p. 1685–1696.

Bosse, V., Feraud, G., Ruffet, G., Ballevre, M., Peucat, J. J., and De Jong, K., 2000, Late Devonian subduction and early-orogenic exhumation of eclogite-facies rocks from the Champtoceaux Complex (Variscan belt, France): Geological Journal, v. 35, p. 297–325.

Cartwright, I. and Barnicoat, A. C., 1999, Stable isotope geochemistry of Alpine ophiolites: A window to ocean-floor hydrothermal alteration and constraints on fluid-rock interaction during high-pressure metamorphism: International Journal of Earth Science, v. 88, p. 219–235.

Chavagnac, V., and Jahn, B. M., 1996, Coesite-bearing eclogites from the Bixiling Complex, Dabie Mountains, China: Sm-Nd ages, geochemical characteristics and tectonic implications: Chemical Geology, v. 133, p. 29–51.

Dalrymple, G. B., Alexander, E. C., Lanphere, M. A., and Kraker, G. P., 1981, Irradiation of samples for $^{40}Ar/^{39}Ar$ dating using the Geological Survey TRIGA reactor: United States Geological Survey Professional Paper 1176, 55 p.

Eide, E., McWilliams, M., and Liou, J., 1994, $^{40}Ar/^{39}Ar$ geohronology and exhumation of high-pressure to ultrahigh-pressure metamorphic rocks in east-central China: Geology, v. 22, p. 601-604.

Eiler, J. M., Valley, J. W., and Baumgartner, L. P., 1993, A new look at stable-isotope thermometry: Geochimica et Cosmochimica Acta, v. 57, p. 2571–2583.

Eiler, J. M, and Valley, J. W., 1994, Preservation of premetamorphic oxygen-isotope ratios in granitic orthogneiss from the Adirondack Mountains, New York, USA: Geochimica et Cosmochimica Acta, v. 58, 5525–5535.

Getty, S. R., and Selverstone, J., 1994, Stable isotope and trace element evidence for restricted fluid migration in 2 Gpa eclogites: Journal of Metamorphic Geology, v. 12, p. 747–760.

Giorgis, D., Cosca, M., and Li, S., 2000, Distribution and significance of extraneous argon in UHP eclogite (Sulu terrain, China): Insight from in situ $^{40}Ar/^{39}Ar$ UV-laser analysis: Earth and Planetary Science Letters, v. 181, p. 605–615.

Hacker, B. R., Ratschbacher, L., Webb, L., Ireland, T., Walker, D., and Dong, S., 1998, U/Pb zircon ages constrain the architecture of the ultrahigh-pressure Qinling-Dabie Orogen, China: Earth and Planetary Science Letters, v. 161, p. 215–230.

Hacker, B. R., and Wang, Q., 1995, Ar/Ar geochronology of ultrahigh-pressure metamorphism in central China: Tectonics, v. 14, p. 994–1006.

Jahn, B. M., 1998, Geochemical and isotopic characteristics of UHP eclogites and ultramafic rocks of the Dabie Orogen: Implications for continental subduction and collisional tectonics, *in* Hacker, B. R., and Liou, J. G., eds. When continents collide: Geodynamics and geochemistry of ultrahigh-pressure rocks: Dordrecht, Netherlands, Kluwer Academic Publishers, p. 203–239.

Kretz, R., 1983, Symbols for rock-forming minerals: American Mineralogist, v. 68, p. 277–279.

Li, S., Jagoutz, E., Chen, Y., and Li, Q., 2000, Sm-Nd and Rb-Sr isotopic chronology and cooling history of ultrahigh pressure metamorphic rocks and their country rocks at Shuanghe in the Dabie Mountains, Central China: Geochimica et Cosmochimica Acta, v. 64, p. 1077–1093.

Li, S., Wang, S., Chen, Y., Liu, D., Qui, J., Zhou, H., and Zhang, Z., 1994, Excess argon in phengite from eclogite: Evidence from dating eclogite minerals by Sm-Nd, Rb-Sr and $^{40}Ar/^{39}Ar$ methods: Chemical Geology, v. 112, p. 343–350.

Li, S., Xiao, Y., Zhang, Z., Chen, Y., Sun, S.-S., Cong, B., Liu, D., Ge, N., Hart, S. R., and Zhang, R., 1993, Collision of the North China and Yangtze blocks and formation of coesite-bearing eclogites: Timing and processes: Chemical Geology, v. 109, p. 70–89.

Miller, J. A., Cartwright, I., Buick, I.S., and Barnicoat, A.C., 2001, An O-isotope profile through the HP-LT Corsican ophiolite, France and its implications for fluid flow during subduction: Chemical Geology, v. 178, p. 43–69.

Putlitz, B., Katzir, Y., Matthews, A., and Valley, J. W., 2001, Oceanic and orogenic fluid-rock interaction in $^{18}O/^{16}O$-enriched metagabbros of an ophiolite (Tinos, Cyclades): Earth and Planetary Science Letters, v. 193, 99–113.

Putlitz, B., Mattthews, A., and Valley, J. W., 2000, Oxygen and hydrogen isotope study of high-pressure metagabbros and metabasalts (Cyclades, Greece): Implications for the subduction of oceanic crust: Contributions to Mineralogy and Petrology, v. 138, p. 114–126.

Renne, P. R., Swisher, C. C., Deino, A. L., Karner, D. B., Owens, T. L., and DePaolo, D. J., 1998, Intercalibration of standards, absolute ages and uncertainties in $^{40}Ar/^{39}Ar$ dating: Chemical Geology, v. 145, p. 117–152.

Rowley, D. B., Xue, F., Tucker, R. D., Peng, Z. X., Baker, J., and Davis, A., 1997, Ages of ultrahigh pressure metamorphism and protolith orthogneisses from the eastern Dabie Shan: U/Pb zircon geochronology: Earth and Planetary Science Letters, v. 151, p. 191–203.

Rumble, D., 1982, Stable isotope fractionation during metamorphic devolatilization reactions, in Ferry, J. M., ed., Characterization of metamorphism through mineral equilibria: Mineralogical Society of America, Reviews in Mineralogy, v. 10, p. 327–352.

Rumble, D., Giorgis, D., Ireland, T., Zhang, Z., Xu, H., Yui, T. F., Yang, J., Xu, Z., and Liou, J. G., 2002, Low $\delta^{18}O$ zircons, U-Pb dating, and the age of the Qinglongshan oxygen and hydrogen isotope anomaly near Donghai in Jiangsu Province, China: Geochimica et Cosmochimica Acta, v. 66, p. 2299–2306.

Rumble, D., and Yui, T.-F., 1998, The Qinglongshan oxygen and hydrogen anomaly near Donghai in Jiangsu Province, China: Geochimica et Cosmochimica Acta, v. 62, p. 3307–3322.

Scaillet, S., 1996, Excess ^{40}Ar transport scale and mechanism in high-pressure phengites: Case study from an eclogitized metabasite of the Dora Maira nappe, Western Alps: Geochimica et Cosmochimica Acta, v. 60, p. 1075–1090.

Scaillet, S., 1998, K-Ar ($^{40}Ar/^{39}Ar$) geochronology of ultrahigh pressure rocks, in Hacker, B. R., and Liou, J. G., eds., When continents collide: Geodynamics and geochemistry of ultrahigh-pressure rocks: Dordrecht, Netherlands, Kluwer Academic Publishers, p. 161–201.

Scaillet, S., Féraud, G., Ballèvre, M., and Amouric, M., 1992, Mg/Fe and [(Mg,Fe)Si-Al$_2$] compositional control on argon behaviour in high-pressure with micas: A $^{40}Ar/^{39}Ar$ continuous laser-probe study from the Dora-Maira nappe of the Internal Western Alps, Italy: Geochimica et Cosmochimica Acta, v. 56, p. 2851–2872.

Sharp, Z., 1990, A laser-based microanalytical method for the in situ determination of oxygen isotope ratios of silicates and oxides: Geochimica et Cosmochimica Acta, v. 54, p. 1353–1357.

Valley, J. W., Kitchen, N., Kohn, M. J., Niendorf, C. R., and Spicuzza, M. J., 1995, UWG-2, a garnet standard for oxygen isotope ratios: Strategies for high precision and accuracy with laser heating: Geochimica et Cosmochimica Acta, v. 59, p. 5223–5231.

Valley, J. W., and O'Neil, J. R., 1984, Fluid heterogeneity during granulite facies metamorphism in the Andirondacks: stable isotope evidence: Contributions to Mineralogy and Petrology, v. 85, p. 158–173.

Yui, T-F., Rumble, D., Chen, C-H., and Lo, C-H., 1997, Stable isotope characteristics of eclogites from the ultrahigh-pressure metamorphic terrain, east-central China: Chemical Geology, v. 137, p. 135–147.

Yui, T-F., Rumble, D., and Lo, C-H., 1995, Unusually low $\delta^{18}O$ ultra-high-pressure metamorphic rocks from the Sulu Terrain, eastern China: Geochimica et Cosmochimica Acta, v. 59, p. 2859–2864.

Zhang, R. Y., Hirajima, T., Banno, S., Cong, B., and Liou, J. G., 1995, Petrology of ultrahigh-pressure rocks from the southern Su-Lu region, eastern China: Journal of Metamorphic Geology, v. 13, p. 659–675.

Zhang, R. Y., and Liou, J. G., 1998, Ultrahigh-pressure metamorphism of the Sulu terrane, eastern China: Prospective View: Continental Dynamics, v. 3, p. 32–53.

Zheng, Y. F., Fu, B., Gong, B., and Li, S., 1996, Extreme ^{18}O depletion in eclogite from the Su-Lu terrane in east China: European Journal of Mineralogy, v. 8, p. 317–323.

Zheng, Y.F., Fu, B., Cong, Y., Xiao, Y., Wei, C., and Li, S., 1998a, Oxygen isotope constraints on fluid flow during eclogitization in the Sulu terrane: Progress in Natural Science, v. 8, p. 98–105.

Zheng, Y. F., Fu, B., Li, Y., Xiao, Y., and Li, S., 1998b, Oxygen and hydrogen isotope geochemistry of ultrahigh-pressure eclogites from the Dabie Mountains and the Sulu terrane: Earth and Planetary Science Letters, v. 155, p. 113–129.

APPENDIX TABLE 1. ^{40}Ar/^{39}Ar Analytical Data from Furnace Incremental Heating Experiment on Separated Grains[1]

T(°C)	^{40}Ar (10−13 moles)	^{39}Ar (10−14 moles)	^{38}Ar (10−16 moles)	^{37}Ar (10−16 moles)	^{36}Ar (10−17 moles)	Pct. ^{39}Ar of total	Pct. ^{40*}Ar	Date (Ma)	2σ
			Sample 95-QL-2E (Qinglongshan), phengitic white mica						
700	13.8052 ± 0.0099	0.9524 ± 0.0012	2.4015 ± 0.0829	1.3711 ± 0.0318	39.0571 ± 0.1039	2.5	91.6	904	± 8
750	10.8709 ± 0.0044	0.7867 ± 0.0009	1.6041 ± 0.0360	0.3985 ± 0.0299	13.2372 ± 0.0768	2.1	96.4	906	± 8
760	8.5320 ± 0.0061	0.6115 ± 0.0007	1.1841 ± 0.0165	0.2521 ± 0.0275	8.6678 ± 0.0701	1.6	97.0	918	± 8
775	10.6992 ± 0.0066	0.7613 ± 0.0010	1.4586 ± 0.0516	0.2600 ± 0.0278	13.8271 ± 0.0747	2.0	96.2	917	± 8
780	10.5030 ± 0.0101	0.7449 ± 0.0015	1.5640 ± 0.0361	0.2773 ± 0.0334	14.2137 ± 0.0776	2.0	96.0	918	± 8
790	15.7558 ± 0.0146	1.1007 ± 0.0013	2.3105 ± 0.0223	0.3851 ± 0.0256	22.0276 ± 0.0923	2.9	95.9	928	± 8
795	18.6155 ± 0.0128	1.3118 ± 0.0014	2.5775 ± 0.1110	0.2727 ± 0.0276	19.3754 ± 0.0935	3.4	96.9	929	± 8
800	22.4832 ± 0.0101	1.5666 ± 0.0015	3.1106 ± 0.0207	0.2205 ± 0.0342	21.3753 ± 0.0917	4.1	97.2	940	± 8
805	24.6421 ± 0.0119	1.7136 ± 0.0019	3.1693 ± 0.0216	0.2088 ± 0.0334	21.3877 ± 0.0908	4.5	97.4	943	± 8
810	26.6784 ± 0.0128	1.8480 ± 0.0017	3.4356 ± 0.0217	0.1908 ± 0.0424	22.4113 ± 0.0895	4.9	97.5	946	± 8
815	29.6838 ± 0.0174	2.0447 ± 0.0017	3.8241 ± 0.0324	0.1667 ± 0.0483	24.5315 ± 0.0891	5.4	97.6	951	± 8
820	31.5506 ± 0.0174	2.1832 ± 0.0024	4.1305 ± 0.0518	0.3150 ± 0.0380	26.4578 ± 0.1144	5.7	97.5	947	± 8
830	41.7749 ± 0.0295	2.8899 ± 0.0028	5.5166 ± 0.0324	0.2603 ± 0.0609	34.3387 ± 0.1217	7.6	97.6	948	± 8
840	57.5650 ± 0.0578	4.0096 ± 0.0040	7.6120 ± 0.0456	0.4745 ± 0.0415	47.4970 ± 0.1225	10.5	97.6	943	± 8
850	52.2444 ± 0.0578	3.6471 ± 0.0037	6.9585 ± 0.0418	0.2882 ± 0.0582	43.5203 ± 0.1137	9.6	97.5	941	± 8
860	42.8548 ± 0.0286	3.0679 ± 0.0023	5.7035 ± 0.0279	0.2219 ± 0.0410	35.6603 ± 0.1132	8.1	97.5	923	± 8
870	32.7090 ± 0.0211	2.3870 ± 0.0024	4.2670 ± 0.0324	0.1647 ± 0.0308	27.7989 ± 0.1011	6.3	97.5	908	± 8
890	28.4907 ± 0.0155	2.1295 ± 0.0024	3.8042 ± 0.0278	0.2101 ± 0.0390	24.5383 ± 0.0840	5.6	97.5	891	± 8
920	27.8717 ± 0.0192	2.1316 ± 0.0023	3.7070 ± 0.0244	0.2708 ± 0.0477	23.9428 ± 0.0906	5.6	97.5	875	± 7
950	14.9103 ± 0.0080	1.1750 ± 0.0012	2.1833 ± 0.0182	0.3089 ± 0.0314	12.2545 ± 0.0781	3.1	97.6	855	± 7
1000	6.9346 ± 0.0039	0.5414 ± 0.0009	0.9853 ± 0.0331	0.7112 ± 0.0218	5.6167 ± 0.0633	1.4	97.6	862	± 7
1050	3.8278 ± 0.0035	0.2725 ± 0.0005	0.5488 ± 0.0163	4.2359 ± 0.0382	3.3636 ± 0.0649	0.7	97.4	926	± 8
1150	2.2152 ± 0.0038	0.1520 ± 0.0004	0.3205 ± 0.0107	2.2818 ± 0.0303	1.7874 ± 0.0610	0.4	97.6	954	± 8
1400	0.8339 ± 0.0036	0.0627 ± 0.0003	0.1356 ± 0.0106	2.2333 ± 0.0286	0.6748 ± 0.0608	0.2	97.8	932	± 8
1501	0.0637 ± 0.0079	0.0081 ± 0.0002	0.0440 ± 0.0242	0.0681 ± 0.2243	0.5376 ± 0.1385	0.0	75.1	461	± 18

Weight = 17.39 mg; J value = 0.0049; total gas age = 925 Ma

Temp (°C)								Age (Ma)
Sample 95-QL-2C, phengitic white mica								
700	13.2762 ± 0.0084	0.9151 ± 0.0013	2.3244 ± 0.0208	1.2171 ± 0.0302	41.8210 ± 0.1034	3.4	90.7	903 ± 8
750	11.8682 ± 0.0069	0.8562 ± 0.0010	1.8238 ± 0.0717	0.4658 ± 0.0328	21.9194 ± 0.0855	3.2	94.5	900 ± 8
775	29.1368 ± 0.0193	2.1488 ± 0.0020	4.0397 ± 0.0209	0.4512 ± 0.0460	35.6696 ± 0.1061	7.9	96.4	898 ± 8
780	31.0367 ± 0.0201	2.2849 ± 0.0023	3.8843 ± 0.0207	0.4067 ± 0.0402	23.3057 ± 0.0880	8.4	97.8	909 ± 8
790	36.5777 ± 0.0249	2.6624 ± 0.0024	4.6593 ± 0.0279	0.4952 ± 0.0495	26.3313 ± 0.1062	9.8	97.9	918 ± 8
795	33.6426 ± 0.0156	2.4424 ± 0.0024	4.2253 ± 0.0343	0.4347 ± 0.0452	23.8866 ± 0.0978	9.0	97.9	920 ± 8
800	31.4102 ± 0.0137	2.2833 ± 0.0023	3.9501 ± 0.0308	0.3745 ± 0.0517	21.9194 ± 0.0925	8.4	97.9	920 ± 8
805	27.1306 ± 0.0129	1.9921 ± 0.0022	3.3666 ± 0.0235	0.3108 ± 0.0453	19.0244 ± 0.0818	7.3	97.9	912 ± 8
810	23.5138 ± 0.0147	1.7202 ± 0.0019	2.9759 ± 0.0250	0.3131 ± 0.0313	16.3247 ± 0.0863	6.3	97.9	915 ± 8
815	19.5671 ± 0.0090	1.4523 ± 0.0015	2.5995 ± 0.0208	0.2934 ± 0.0354	13.8546 ± 0.0830	5.3	97.9	905 ± 8
820	15.9303 ± 0.0070	1.1962 ± 0.0011	2.1532 ± 0.1090	0.2416 ± 0.0332	11.2446 ± 0.0712	4.4	97.9	896 ± 8
830	14.6948 ± 0.0063	1.1197 ± 0.0012	2.0267 ± 0.1004	0.1601 ± 0.0360	10.4917 ± 0.0710	4.1	97.9	886 ± 7
840	13.5511 ± 0.0060	1.0471 ± 0.0011	1.8743 ± 0.1001	0.1801 ± 0.0386	9.8634 ± 0.0742	3.9	97.8	876 ± 7
860	13.4572 ± 0.0097	1.0452 ± 0.0012	1.8709 ± 0.0208	0.1975 ± 0.0308	10.1591 ± 0.0774	3.9	97.8	872 ± 7
890	17.8092 ± 0.0094	1.3938 ± 0.0013	2.4113 ± 0.0243	0.2216 ± 0.0357	13.2136 ± 0.0799	5.1	97.8	867 ± 7
920	16.1826 ± 0.0073	1.2806 ± 0.0015	2.2410 ± 0.0181	0.2766 ± 0.0298	11.4994 ± 0.0776	4.7	97.9	860 ± 7
950	8.0848 ± 0.0041	0.6454 ± 0.0009	1.1036 ± 0.0147	0.2614 ± 0.0278	5.6276 ± 0.0660	2.4	97.9	854 ± 7
1000	5.0083 ± 0.0038	0.3795 ± 0.0006	0.6944 ± 0.0099	0.6033 ± 0.0260	3.3416 ± 0.0604	1.4	98.0	890 ± 8
1050	2.5379 ± 0.0035	0.1821 ± 0.0005	0.3514 ± 0.0193	2.0197 ± 0.0303	1.7624 ± 0.0603	0.7	98.0	929 ± 8
1150	1.3300 ± 0.0036	0.0937 ± 0.0003	0.1818 ± 0.0086	0.9285 ± 0.0242	1.0577 ± 0.0584	0.3	97.7	940 ± 8
1250	0.2228 ± 0.0035	0.0166 ± 0.0001	0.0281 ± 0.0089	0.4367 ± 0.0244	0.1757 ± 0.0622	0.1	97.7	901 ± 9
1400	0.5111 ± 0.0036	0.0358 ± 0.0002	0.0784 ± 0.0105	0.4003 ± 0.0313	0.1984 ± 0.0606	0.1	98.9	953 ± 9
Weight = 12.65 mg; J value = 0.004935; total gas age = 901 Ma								
Sample 95-QL-12, phengitic white mica								
700	2.8991 ± 0.0093	0.1909 ± 0.0004	1.5601 ± 0.0139	6.9613 ± 2.4093	58.6557 ± 0.3068	0.3	40.2	868 ± 15
750	1.1816 ± 0.0089	0.1766 ± 0.0004	0.3461 ± 0.0122	2.9865 ± 2.4473	2.4792 ± 0.2886	0.3	93.8	887 ± 5
800	3.0596 ± 0.0089	0.4648 ± 0.0006	0.8439 ± 0.0131	2.1273 ± 2.5186	4.8542 ± 0.2800	0.7	95.3	887 ± 5
850	6.7865 ± 0.0093	1.0372 ± 0.0006	1.8478 ± 0.0466	3.0172 ± 2.3190	9.4622 ± 0.2776	1.6	95.9	887 ± 5
875	6.2220 ± 0.0089	0.9519 ± 0.0009	1.6695 ± 0.0142	2.6054 ± 2.6175	7.9177 ± 0.2733	1.5	96.2	889 ± 5
900	10.9480 ± 0.0100	1.6574 ± 0.0015	3.0220 ± 0.0192	2.2793 ± 2.3585	20.9896 ± 0.2697	2.6	94.3	882 ± 5

Table continues

APPENDIX TABLE 1. (Continued)

T(°C)	^{40}Ar (10^{-13} moles)	^{39}Ar (10^{-14} moles)	^{38}Ar (10^{-16} moles)	^{37}Ar (10^{-16} moles)	^{36}Ar (10^{-17} moles)	Pct. ^{39}Ar of total	Pct. ^{40*}Ar	Date (Ma)	2σ
\multicolumn{10}{l}{Sample 95-QL-12, phengitic white mica (continued)}
925	48.6271 ± 0.0451	7.3130 ± 0.0060	12.5072 ± 0.0364	11.9041 ± 2.3160	52.1338 ± 0.2716	11.3	96.8	905	± 5
940	61.8488 ± 0.0511	9.3213 ± 0.0046	15.5436 ± 0.0464	14.6658 ± 2.3024	43.5475 ± 0.2708	14.5	97.9	912	± 5
955	75.2411 ± 0.0993	11.3465 ± 0.0139	18.9068 ± 0.0350	16.5581 ± 2.2327	50.9665 ± 0.2692	17.6	98.0	912	± 5
970	72.3198 ± 0.0628	11.1209 ± 0.0059	18.5731 ± 0.0309	17.5944 ± 2.2380	49.5200 ± 0.2702	17.2	98.0	898	± 5
985	51.0590 ± 0.0512	8.0720 ± 0.0053	13.4148 ± 0.0339	15.0147 ± 2.2793	36.0816 ± 0.2637	12.5	97.9	878	± 5
1000	25.5335 ± 0.0209	4.1318 ± 0.0025	6.8473 ± 0.0217	10.9518 ± 2.2310	19.5221 ± 0.2612	6.4	97.7	861	± 4
1015	16.5515 ± 0.0135	2.6458 ± 0.0012	4.4441 ± 0.0298	10.5245 ± 2.2001	13.1443 ± 0.2615	4.1	97.7	868	± 4
1030	13.1501 ± 0.0107	2.0732 ± 0.0013	3.4604 ± 0.0270	8.3873 ± 2.2797	9.9922 ± 0.2586	3.2	97.8	879	± 4
1050	10.2045 ± 0.0085	1.6308 ± 0.0010	2.7277 ± 0.1411	7.1248 ± 2.1573	7.3812 ± 0.2574	2.5	97.9	870	± 4
1075	6.7251 ± 0.0083	1.0935 ± 0.0009	1.7656 ± 0.0159	11.1169 ± 2.1852	4.7173 ± 0.2550	1.7	97.9	859	± 4
1100	3.6654 ± 0.0080	0.5926 ± 0.0006	0.9730 ± 0.0167	18.5577 ± 2.2024	2.6654 ± 0.2569	0.9	97.9	862	± 4
1126	2.0385 ± 0.0080	0.3231 ± 0.0005	0.5518 ± 0.0134	42.6571 ± 2.2454	1.5892 ± 0.2564	0.5	97.7	875	± 5
1150	1.0375 ± 0.0079	0.1615 ± 0.0003	0.2830 ± 0.0132	42.4749 ± 2.2426	0.9598 ± 0.2603	0.3	97.3	884	± 5
1200	0.6948 ± 0.0080	0.1079 ± 0.0003	0.1932 ± 0.0116	23.9840 ± 2.2861	0.5516 ± 0.2638	0.2	97.7	889	± 5
1400	0.3507 ± 0.0090	0.0568 ± 0.0004	0.0930 ± 0.0129	12.6525 ± 2.5275	0.3793 ± 0.3237	0.1	96.8	853	± 5
1601	0.4412 ± 0.0102	0.0635 ± 0.0004	0.1171 ± 0.0140	15.8365 ± 2.8334	1.5876 ± 0.3387	0.1	89.4	880	± 5
\multicolumn{10}{l}{Weight = 10.57 mg; J value = 0.01012; total gas age = 894 Ma}
\multicolumn{10}{l}{Sample 95-QL-2Q-2, phengitic white mica}
700	7.2611 ± 0.0074	0.6869 ± 0.0009	1.7859 ± 0.0548	0.6869 ± 0.0267	38.4664 ± 0.1019	2.7	84.3	654	± 6
750	4.3728 ± 0.0043	0.4452 ± 0.0007	0.8503 ± 0.0332	0.0890 ± 0.0292	9.5712 ± 0.0710	1.8	93.5	671	± 6
760	3.8124 ± 0.0043	0.3871 ± 0.0007	0.7625 ± 0.0156	0.1289 ± 0.0266	9.6768 ± 0.0681	1.5	92.5	666	± 6
775	7.1220 ± 0.0070	0.6961 ± 0.0009	1.5036 ± 0.0433	0.1371 ± 0.0263	21.3704 ± 0.0795	2.8	91.1	679	± 6
780	7.1893 ± 0.0045	0.7355 ± 0.0008	1.3092 ± 0.0198	0.0599 ± 0.0214	8.3851 ± 0.0737	2.9	96.6	686	± 6
790	13.3542 ± 0.0055	1.3439 ± 0.0012	2.2980 ± 0.0190	0.0706 ± 0.0234	12.5788 ± 0.0752	5.4	97.2	699	± 6
795	14.3585 ± 0.0059	1.4419 ± 0.0017	2.4512 ± 0.0147	0.1363 ± 0.0373	10.3238 ± 0.0741	5.7	97.9	704	± 6
800	15.0544 ± 0.0076	1.5057 ± 0.0015	2.5296 ± 0.0172	0.0000 ± 0.0000	10.1032 ± 0.0731	6.0	98.0	707	± 6

UHP METAMORPHISM

Sample 95-QL-2Q-2, phengitic white mica (*continued*)

T (°C)								Age (Ma)
805	15.0748 ± 0.0096	1.4952 ± 0.0016	2.4522 ± 0.0814	0.0867 ± 0.0380	9.7938 ± 0.0722	5.9	98.1	713 ± 6
810	15.2308 ± 0.0078	1.5077 ± 0.0016	2.5329 ± 0.0199	0.1553 ± 0.0288	9.9959 ± 0.0795	6.0	98.1	714 ± 6
815	15.3007 ± 0.0128	1.5109 ± 0.0017	2.5081 ± 0.0199	0.1407 ± 0.0330	9.8511 ± 0.0750	6.0	98.1	715 ± 6
820	15.3154 ± 0.0101	1.5119 ± 0.0014	2.5097 ± 0.0234	0.1330 ± 0.0388	9.9631 ± 0.0739	6.0	98.1	716 ± 6
830	17.4540 ± 0.0100	1.7388 ± 0.0019	2.7126 ± 0.0189	0.0935 ± 0.0505	11.5632 ± 0.0793	6.9	98.0	710 ± 6
840	18.5462 ± 0.0101	1.8700 ± 0.0016	2.8737 ± 0.0243	0.1214 ± 0.0369	12.5101 ± 0.0808	7.4	98.0	703 ± 6
850	16.8115 ± 0.0128	1.7264 ± 0.0022	2.6586 ± 0.0197	0.1298 ± 0.0439	11.7910 ± 0.0747	6.8	97.9	692 ± 6
860	13.5872 ± 0.0053	1.4156 ± 0.0014	2.3357 ± 0.0197	0.0869 ± 0.0231	9.9658 ± 0.0746	5.6	97.8	683 ± 6
870	10.4093 ± 0.0045	1.0937 ± 0.0013	1.8703 ± 0.0773	0.0649 ± 0.0288	7.8531 ± 0.0714	4.3	97.8	678 ± 6
890	10.0796 ± 0.0080	1.0688 ± 0.0016	1.7742 ± 0.1016	0.0308 ± 0.0284	7.9519 ± 0.0685	4.2	97.7	672 ± 6
920	12.5708 ± 0.0070	1.3628 ± 0.0017	2.2351 ± 0.0234	0.1853 ± 0.0260	9.3218 ± 0.0730	5.4	97.8	660 ± 6
950	7.4202 ± 0.0043	0.8180 ± 0.0009	1.3243 ± 0.0683	0.0811 ± 0.0213	5.0144 ± 0.0662	3.2	98.0	652 ± 6
1000	4.0374 ± 0.0039	0.4559 ± 0.0006	0.7157 ± 0.0418	0.1172 ± 0.0243	2.6897 ± 0.0581	1.8	98.0	639 ± 6
1050	1.6058 ± 0.0036	0.1712 ± 0.0004	0.2824 ± 0.0130	0.1349 ± 0.0229	1.2017 ± 0.0589	0.7	97.8	670 ± 6
1150	0.7905 ± 0.0035	0.0837 ± 0.0003	0.1415 ± 0.0129	0.0811 ± 0.0261	0.4336 ± 0.0617	0.3	98.4	677 ± 6
1400	0.6446 ± 0.0036	0.0655 ± 0.0003	0.0969 ± 0.0128	0.1303 ± 0.0232	0.3084 ± 0.0538	0.3	98.6	702 ± 7

Weight = 12.21 mg; J value = 0.0049; total gas age = 693 Ma

Sample 95-QL-2O, phengitic white mica

T (°C)								Age (Ma)
294	8.8191 ± 0.0118	1.0219 ± 0.0012	2.2946 ± 0.0204	0.3635 ± 0.0291	32.6431 ± 0.1843	4.3	89.1	580 ± 5
750	7.3828 ± 0.0128	0.9152 ± 0.0015	1.7114 ± 0.0818	0.2645 ± 0.0316	14.5518 ± 0.0811	3.8	94.2	574 ± 5
775	18.5822 ± 0.0314	2.3575 ± 0.0033	3.7012 ± 0.0406	0.2546 ± 0.0327	18.1526 ± 0.0837	9.9	97.1	578 ± 5
800	36.3142 ± 0.0493	4.4998 ± 0.0103	7.2897 ± 0.0513	0.5935 ± 0.0804	20.8791 ± 0.1154	18.9	98.3	596 ± 6
805	22.8447 ± 0.0483	2.7770 ± 0.0041	4.3599 ± 0.0304	0.4999 ± 0.0507	12.1911 ± 0.0896	11.6	98.4	606 ± 6
810	18.2020 ± 0.0333	2.2222 ± 0.0057	3.4667 ± 0.0252	0.3400 ± 0.0374	9.5778 ± 0.0778	9.3	98.4	604 ± 6
815	13.9157 ± 0.0164	1.7061 ± 0.0024	2.8150 ± 0.0223	0.3242 ± 0.0321	7.5920 ± 0.0693	7.2	98.4	602 ± 6
820	9.8689 ± 0.0051	1.2150 ± 0.0016	1.9925 ± 0.0147	0.2102 ± 0.0352	5.5281 ± 0.0661	5.1	98.3	600 ± 5
825	7.7140 ± 0.0076	0.9604 ± 0.0013	1.5785 ± 0.0173	0.1671 ± 0.0269	4.5236 ± 0.0696	4.0	98.3	593 ± 5
830	6.0464 ± 0.0053	0.7600 ± 0.0012	1.1856 ± 0.0665	0.0851 ± 0.0243	3.7012 ± 0.0648	3.2	98.2	588 ± 5
840	5.2542 ± 0.0037	0.6601 ± 0.0010	1.0628 ± 0.0172	0.0746 ± 0.0237	3.2149 ± 0.0661	2.8	98.2	588 ± 5
850	4.5031 ± 0.0039	0.5719 ± 0.0009	0.9093 ± 0.0524	0.0686 ± 0.0272	3.0196 ± 0.0649	2.4	98.0	582 ± 5

(*Table continues*)

APPENDIX TABLE 1. (Continued)

T(°C)	^{40}Ar (10^{-13} moles)	^{39}Ar (10^{-14} moles)	^{38}Ar (10^{-16} moles)	^{37}Ar (10^{-16} moles)	^{36}Ar (10^{-17} moles)	Pct. ^{39}Ar of total	Pct. $^{40*}Ar$	Date (Ma)	2σ
\multicolumn{10}{l}{Sample 95-QL-2O, phengitic white mica (continued)}
860	3.6612 ± 0.0039	0.4614 ± 0.0009	0.7659 ± 0.0439	0.0687 ± 0.0247	2.5376 ± 0.0625	1.9	98.0	586	± 5
870	3.2981 ± 0.0038	0.4152 ± 0.0008	0.6892 ± 0.0343	0.0229 ± 0.1862	2.3955 ± 0.0611	1.7	97.9	586	± 5
890	4.3453 ± 0.0037	0.5459 ± 0.0008	0.9117 ± 0.0138	0.1485 ± 0.0240	3.0517 ± 0.0606	2.3	97.9	587	± 5
920	8.5700 ± 0.0042	1.0701 ± 0.0013	1.7656 ± 0.0844	0.1166 ± 0.0332	4.7512 ± 0.0651	4.5	98.4	592	± 5
950	7.1587 ± 0.0042	0.9391 ± 0.0011	1.5213 ± 0.0163	0.1249 ± 0.0319	3.7468 ± 0.0653	3.9	98.5	568	± 5
1000	2.9434 ± 0.0035	0.4321 ± 0.0008	0.6698 ± 0.0213	0.1132 ± 0.0197	1.6767 ± 0.0589	1.8	98.3	515	± 5
1050	0.9082 ± 0.0034	0.1566 ± 0.0005	0.2661 ± 0.0112	0.1817 ± 0.0206	0.6622 ± 0.0566	0.7	97.8	445	± 4
1150	0.3407 ± 0.0034	0.0528 ± 0.0002	0.0691 ± 0.0097	0.1423 ± 0.0165	0.4206 ± 0.0545	0.2	96.4	483	± 5
1250	0.1190 ± 0.0034	0.0142 ± 0.0001	0.0181 ± 0.0065	0.0153 ± 0.1445	0.1998 ± 0.0517	0.1	95.1	599	± 7
1400	0.1187 ± 0.0038	0.0149 ± 0.0001	0.0372 ± 0.0108	0.2438 ± 0.0212	0.2701 ± 0.0654	0.1	93.3	564	± 8
1551	0.0748 ± 0.0123	0.0126 ± 0.0003	0.0867 ± 0.0240	1.1197 ± 0.0569	0.9800 ± 0.1923	0.1	61.4	298	± 16
\multicolumn{10}{l}{Weight = 11.08 mg; J value = 0.004935; total gas age = 590 Ma}
\multicolumn{10}{l}{Sample 95-QL-1B, phengitic white mica}
700	6.6246 ± 0.0052	0.7702 ± 0.0011	1.6405 ± 0.0132	1.0167 ± 0.0256	27.9580 ± 0.0898	3.6	87.5	567	± 5
750	5.2537 ± 0.0050	0.6458 ± 0.0011	1.1833 ± 0.0163	0.2596 ± 0.0265	14.4667 ± 0.0786	3.0	91.8	563	± 5
775	11.7561 ± 0.0076	1.4785 ± 0.0015	2.5874 ± 0.0260	0.2646 ± 0.0310	26.7607 ± 0.1008	6.9	93.3	559	± 5
790	18.9091 ± 0.0119	2.4237 ± 0.0040	3.7083 ± 0.0208	0.1893 ± 0.0305	23.7526 ± 0.0978	11.2	96.3	565	± 5
800	21.9386 ± 0.0183	2.7975 ± 0.0024	4.3362 ± 0.0277	0.2137 ± 0.0401	23.6952 ± 0.0928	12.9	96.8	571	± 5
805	18.7258 ± 0.0183	2.4052 ± 0.0032	3.6318 ± 0.0289	0.2328 ± 0.0391	20.0353 ± 0.0906	11.1	96.8	567	± 5
810	15.1083 ± 0.0086	1.9634 ± 0.0023	2.8273 ± 0.0172	0.1983 ± 0.0301	15.8641 ± 0.0825	9.1	96.9	562	± 5
815	11.5267 ± 0.0073	1.5284 ± 0.0015	2.4913 ± 0.1273	0.1359 ± 0.0306	12.3800 ± 0.0801	7.1	96.8	552	± 5
820	9.0133 ± 0.0841	1.1786 ± 0.0132	1.9683 ± 0.0628	0.0753 ± 0.0097	9.8652 ± 0.0931	5.5	96.8	558	± 10
830	7.2758 ± 0.0045	0.9930 ± 0.0011	1.5689 ± 0.0172	0.1072 ± 0.0246	8.2518 ± 0.0725	4.6	96.6	537	± 5
840	6.2748 ± 0.0050	0.8666 ± 0.0011	1.3779 ± 0.0148	0.1239 ± 0.0240	7.7475 ± 0.0721	4.0	96.4	530	± 5
860	5.8668 ± 0.0078	0.8162 ± 0.0011	1.2896 ± 0.0639	0.2065 ± 0.0220	7.6396 ± 0.0711	3.8	96.2	526	± 5

Sample 95-QL-1B, phengitic white mica (continued)							
890	8.5647 ± 0.0043	1.2140 ± 0.0013	1.8817 ± 0.0138	0.2258 ± 0.0248	10.4403 ± 0.0710	5.6	96.4	519 ± 5
920	7.1177 ± 0.0051	1.0346 ± 0.0011	1.6140 ± 0.0428	0.2949 ± 0.0269	8.1735 ± 0.0639	4.8	96.6	509 ± 5
950	4.3249 ± 0.0040	0.6136 ± 0.0008	0.9389 ± 0.0236	0.3498 ± 0.0220	4.6023 ± 0.0617	2.8	96.9	521 ± 5
1000	3.0795 ± 0.0035	0.4231 ± 0.0007	0.6444 ± 0.0198	1.2904 ± 0.0244	3.2620 ± 0.0593	2.0	96.9	535 ± 5
1050	1.7549 ± 0.0035	0.2348 ± 0.0005	0.3757 ± 0.0135	3.8537 ± 0.0307	2.0828 ± 0.0612	1.1	96.5	546 ± 5
1150	1.1208 ± 0.0035	0.1510 ± 0.0004	0.2371 ± 0.0115	4.0466 ± 0.0297	1.4692 ± 0.0579	0.7	96.2	541 ± 5
1250	0.3340 ± 0.0035	0.0437 ± 0.0002	0.0760 ± 0.0110	7.7711 ± 0.0248	0.4273 ± 0.0574	0.2	96.2	555 ± 6
1400	0.4454 ± 0.0036	0.0577 ± 0.0002	0.0882 ± 0.0093	6.6655 ± 0.0279	0.3754 ± 0.0645	0.3	97.5	567 ± 6
Weight = 11.60 mg; J value = 0.0049; total gas age = 552 Ma								
Sample 95-QL-1C, phengitic white mica								
700	6.1456 ± 0.0044	0.9367 ± 0.0009	1.9859 ± 0.0817	1.3583 ± 0.0243	33.4412 ± 0.0894	4.1	83.9	434 ± 4
750	5.3805 ± 0.0047	0.8826 ± 0.0010	1.5357 ± 0.0676	0.4607 ± 0.0253	15.7979 ± 0.0755	3.9	91.3	438 ± 4
775	11.0648 ± 0.0052	1.8041 ± 0.0017	2.8325 ± 0.0224	0.3825 ± 0.0316	26.3403 ± 0.0975	7.9	93.0	447 ± 4
780	12.2067 ± 0.0050	1.9994 ± 0.0021	2.9403 ± 0.0234	0.2479 ± 0.0236	15.9354 ± 0.0811	8.8	96.1	459 ± 4
790	15.1900 ± 0.0078	2.4460 ± 0.0017	3.6445 ± 0.0181	0.2429 ± 0.0310	17.8068 ± 0.0871	10.7	96.5	468 ± 4
795	14.4009 ± 0.0071	2.3194 ± 0.0019	3.3631 ± 0.0271	0.1329 ± 0.0340	15.7252 ± 0.0765	10.2	96.8	468 ± 4
800	13.2043 ± 0.0101	2.1311 ± 0.0024	3.1327 ± 0.0198	0.2685 ± 0.0309	14.5552 ± 0.0802	9.3	96.7	467 ± 4
805	11.1425 ± 0.0046	1.8139 ± 0.0016	2.5939 ± 0.0190	0.1959 ± 0.0268	12.0080 ± 0.0717	7.9	96.8	464 ± 4
810	9.3596 ± 0.0052	1.5478 ± 0.0017	2.3836 ± 0.0181	0.1402 ± 0.0210	10.4475 ± 0.0735	6.8	96.7	457 ± 4
815	7.4229 ± 0.0043	1.2569 ± 0.0014	1.8979 ± 0.0206	0.0719 ± 0.0325	8.4087 ± 0.0722	5.5	96.7	448 ± 4
820	5.6984 ± 0.0050	0.9821 ± 0.0009	1.4915 ± 0.0839	0.0980 ± 0.0227	6.5703 ± 0.0666	4.3	96.6	440 ± 4
830	4.9847 ± 0.0049	0.8720 ± 0.0013	1.3603 ± 0.0802	0.0942 ± 0.0174	6.1650 ± 0.0686	3.8	96.3	434 ± 4
840	4.3982 ± 0.0046	0.7748 ± 0.0012	1.1777 ± 0.0583	0.0968 ± 0.0213	5.7411 ± 0.0647	3.4	96.1	430 ± 4
860	4.8987 ± 0.0038	0.8773 ± 0.0011	1.3334 ± 0.0171	0.1202 ± 0.0285	6.6935 ± 0.0682	3.8	96.0	423 ± 4
890	7.0564 ± 0.0057	1.2888 ± 0.0018	1.9814 ± 0.1081	0.1353 ± 0.0233	9.1119 ± 0.0720	5.7	96.2	417 ± 4
920	5.5134 ± 0.0065	1.0417 ± 0.0013	1.5834 ± 0.0900	0.2167 ± 0.0217	6.8025 ± 0.0692	4.6	96.4	405 ± 4
950	3.2037 ± 0.0040	0.5810 ± 0.0008	0.8308 ± 0.0421	0.3323 ± 0.0266	3.7704 ± 0.0607	2.5	96.5	421 ± 4
1000	2.3902 ± 0.0037	0.4153 ± 0.0007	0.6230 ± 0.0409	1.3582 ± 0.0286	2.6706 ± 0.0616	1.8	96.7	438 ± 4
1050	1.4556 ± 0.0035	0.2466 ± 0.0006	0.4019 ± 0.0237	4.6131 ± 0.0363	1.6174 ± 0.0571	1.1	96.7	448 ± 4
1150	0.9799 ± 0.0036	0.1622 ± 0.0005	0.2189 ± 0.0128	4.5258 ± 0.0313	0.9747 ± 0.0590	0.7	97.1	459 ± 4

(Table continues)

APPENDIX TABLE 1. (Continued)

T(°C)	^{40}Ar (10^{-13} moles)	^{39}Ar (10^{-14} moles)	^{38}Ar (10^{-16} moles)	^{37}Ar (10^{-16} moles)	^{36}Ar (10^{-17} moles)	Pct. ^{39}Ar of total	Pct. $^{40*}Ar$	Date (Ma)	2σ
			Sample 95-QL-1C, phengitic white mica (continued)						
1250	0.2758 ± 0.0035	0.0455 ± 0.0002	0.0601 ±0.0104	1.4781 ± 0.0282	0.3312 ± 0.0596	0.2	96.5	458	± 5
1400	0.5162 ± 0.0036	0.0841 ± 0.0003	0.1379 ±0.0126	1.3104 ± 0.0231	0.4430 ± 0.0616	0.4	97.5	467	± 5
			Weight = 13.12 mg; J value = 0.004935; total gas age = 449 Ma						
			Sample 95-QL-1F, phengitic white mica						
700	12.0804 ± 0.0052	1.3427 ± 0.0011	3.3664 ±0.0253	1.3199 ± 0.0386	37.3282 ± 0.1055	3.7	90.9	612	± 6
750	9.6322 ± 0.0049	1.1415 ± 0.0011	2.0319 ±0.0232	0.3584 ± 0.0224	19.9770 ± 0.0953	3.2	93.9	595	± 5
775	20.0751 ± 0.0083	2.4971 ± 0.0020	4.3450 ±0.0207	0.4370 ± 0.0374	47.9451 ± 0.1135	6.9	93.0	566	± 5
780	21.0540 ± 0.0101	2.7240 ± 0.0020	4.3040 ±0.0235	0.3133 ± 0.0397	26.3413 ± 0.0913	7.6	96.3	564	± 5
790	26.8478 ± 0.0137	3.4527 ± 0.0019	5.3171 ±0.0279	0.2983 ± 0.0336	28.9677 ± 0.1051	9.5	96.8	570	± 5
800	32.0005 ± 0.0220	4.0675 ± 0.0029	6.2232 ±0.0253	0.3465 ± 0.0482	31.2789 ± 0.0981	11.3	97.1	577	± 5
805	28.1972 ± 0.0184	3.5973 ± 0.0033	5.4678 ±0.0399	0.3086 ± 0.0477	27.1593 ± 0.1009	10.0	97.2	576	± 5
810	23.8306 ± 0.0099	3.0402 ± 0.0018	4.5602 ±0.0261	0.2435 ± 0.0332	22.7708 ± 0.0912	8.4	97.2	576	± 5
815	18.8546 ± 0.0085	2.4513 ± 0.0019	3.6770 ±0.0270	0.1775 ± 0.0291	18.3849 ± 0.0875	6.8	97.1	566	± 5
820	14.8862 ± 0.0082	1.9471 ± 0.0015	2.8233 ±0.0181	0.1437 ± 0.0269	14.6810 ± 0.0722	5.4	97.1	563	± 5
830	12.4737 ± 0.0062	1.6495 ± 0.0013	2.5072 ±0.0224	0.1491 ± 0.0302	12.9815 ± 0.0743	4.6	96.9	557	± 5
840	10.2796 ± 0.0051	1.3623 ± 0.0014	2.1116 ±0.1183	0.1287 ± 0.0227	11.4708 ± 0.0786	3.8	96.7	555	± 5
860	10.1793 ± 0.0046	1.3480 ± 0.0016	2.0354 ±0.0231	0.1725 ± 0.0244	11.6867 ± 0.0751	3.7	96.6	555	± 5
890	13.2956 ± 0.0064	1.8046 ± 0.0014	2.6167 ±0.0207	0.2003 ± 0.0258	15.0868 ± 0.0762	5.0	96.6	543	± 5
920	12.2234 ± 0.0094	1.7520 ± 0.0021	2.5228 ±0.0216	0.0773 ± 0.0396	12.7192 ± 0.0779	4.8	96.9	519	± 5
950	7.1191 ± 0.0052	0.9839 ± 0.0010	1.5250 ±0.0197	0.2873 ± 0.0200	6.9559 ± 0.0638	2.7	97.1	537	± 5
1000	4.7250 ± 0.0046	0.5824 ± 0.0009	0.9143 ±0.0556	0.5882 ± 0.0217	4.4144 ± 0.0632	1.6	97.2	593	± 5
1050	2.1423 ± 0.0040	0.2359 ± 0.0006	0.3987 ±0.0184	2.0053 ± 0.0273	2.1846 ± 0.0584	0.7	97.0	651	± 6
1150	0.9482 ± 0.0035	0.1031 ± 0.0004	0.1753 ±0.0130	2.0952 ± 0.0236	1.1344 ± 0.0585	0.3	96.5	655	± 6
1250	0.1424 ± 0.0034	0.0155 ± 0.0001	0.0085 ±0.0074	0.2439 ± 0.0231	0.1215 ± 0.0533	0.0	97.5	662	± 8
1400	0.1724 ± 0.0037	0.0192 ± 0.0001	0.0515 ±0.0106	0.7783 ± 0.0238	0.3245 ± 0.0648	0.1	94.5	631	± 7
			Weight = 17.40 mg; J value = 0.004935; total gas age = 568 Ma						

Sample QL-99-11D, phengitic white mica

Temp							Age (Ma)
700	0.0630 ± 0.0000	0.0052 ± 0.0000	0.0405 ± 0.0007	0.2283 ± 0.0096	1.4979 ± 0.0063	0.1	29.7	286 ± 12
750	0.0029 ± 0.0000	0.0002 ± 0.0000	0.0018 ± 0.0001	0.0240 ± 0.0009	0.0759 ± 0.0003	0.0	23.9	248 ± 22
775	0.0424 ± 0.0000	0.0051 ± 0.0000	0.0208 ± 0.0009	0.2433 ± 0.0103	0.7868 ± 0.0037	0.1	45.2	294 ± 8
780	0.0444 ± 0.0000	0.0052 ± 0.0000	0.0188 ± 0.0007	0.1953 ± 0.0079	0.7953 ± 0.0033	0.1	47.0	317 ± 7
790	0.0649 ± 0.0000	0.0085 ± 0.0000	0.0278 ± 0.0007	0.1861 ± 0.0076	0.9967 ± 0.0046	0.1	54.6	326 ± 7
795	0.0881 ± 0.0000	0.0119 ± 0.0001	0.0333 ± 0.0027	0.2203 ± 0.0081	1.2819 ± 0.0059	0.1	57.0	330 ± 6
800	0.1493 ± 0.0000	0.0182 ± 0.0001	0.0998 ± 0.0015	0.2573 ± 0.0096	2.4585 ± 0.0078	0.2	51.3	330 ± 7
805	0.2733 ± 0.0001	0.0257 ± 0.0001	0.1801 ± 0.0026	0.0185 ± 0.0100	5.2244 ± 0.0128	0.3	43.5	359 ± 9
810	0.2724 ± 0.0001	0.0468 ± 0.0001	0.0918 ± 0.0017	0.0000 ± 0.0000	2.2384 ± 0.0106	0.5	75.7	343 ± 5
815	0.2625 ± 0.0001	0.0482 ± 0.0001	0.0843 ± 0.0046	0.0000 ± 0.0000	1.8644 ± 0.0089	0.5	79.0	336 ± 5
820	0.3034 ± 0.0001	0.0588 ± 0.0001	0.1193 ± 0.0024	0.4830 ± 0.0100	2.1798 ± 0.0082	0.6	78.8	319 ± 5
830	0.3201 ± 0.0002	0.0644 ± 0.0001	0.1178 ± 0.0024	0.4133 ± 0.0112	2.0793 ± 0.0094	0.6	80.8	315 ± 4
840	0.3848 ± 0.0001	0.0787 ± 0.0001	0.1415 ± 0.0017	0.3863 ± 0.0127	2.1874 ± 0.0091	0.8	83.2	319 ± 4
860	0.5641 ± 0.0001	0.1197 ± 0.0002	0.1915 ± 0.0101	0.4142 ± 0.0133	2.3821 ± 0.0095	1.2	87.5	323 ± 4
890	1.0558 ± 0.0002	0.2353 ± 0.0002	0.3670 ± 0.0040	0.4917 ± 0.0201	3.1290 ± 0.0109	2.3	91.2	321 ± 4
920	2.2496 ± 0.0010	0.5270 ± 0.0003	0.7642 ± 0.0092	0.5639 ± 0.0255	3.9684 ± 0.0143	5.2	94.8	317 ± 4
950	9.3318 ± 0.0021	2.1681 ± 0.0009	2.9703 ± 0.0210	0.6157 ± 0.0210	7.0246 ± 0.0207	21.3	97.8	329 ± 4
1000	16.4875 ± 0.0073	3.8138 ± 0.0011	4.8053 ± 0.0227	0.7361 ± 0.0190	11.3650 ± 0.0275	37.4	98.0	331 ± 4
1050	7.4896 ± 0.0011	1.8160 ± 0.0009	2.4698 ± 0.0171	0.7119 ± 0.0273	6.8464 ± 0.0235	17.8	97.3	315 ± 4
1150	2.9868 ± 0.0010	0.7309 ± 0.0004	1.0526 ± 0.0624	1.3303 ± 0.0243	6.2130 ± 0.0179	7.2	93.8	302 ± 4
1250	1.2637 ± 0.0004	0.3024 ± 0.0001	0.5382 ± 0.0049	2.3796 ± 0.0174	6.4403 ± 0.0140	3.0	85.0	281 ± 4
1400	0.3436 ± 0.0001	0.0927 ± 0.0001	0.1650 ± 0.0039	1.3887 ± 0.0088	1.7978 ± 0.0051	0.9	84.5	250 ± 4
1551	0.1742 ± 0.0001	0.0509 ± 0.0000	0.0861 ± 0.0028	0.4217 ± 0.0038	0.9116 ± 0.0023	0.5	84.5	232 ± 3

Weight = 11.18 mg; J value = 0.004935; total gas age = 314 Ma

Sample QL-99-1C, phengitic white mica

Temp							Age (Ma)
700	1.0346 ± 0.0017	0.1297 ± 0.0002	0.0206 ± 0.0108	2.0441 ± 0.0774	14.0526 ± 0.0864	0.1	59.8	368 ± 6
750	0.4532 ± 0.0017	0.0799 ± 0.0002	0.0259 ± 0.0076	0.4641 ± 0.0500	3.2029 ± 0.0574	0.1	79.1	348 ± 5
800	1.0586 ± 0.0016	0.2104 ± 0.0002	0.0163 ± 0.0078	0.7049 ± 0.0542	3.1944 ± 0.0601	0.2	91.1	354 ± 5
825	2.1990 ± 0.0018	0.4747 ± 0.0003	0.0159 ± 0.0140	0.5887 ± 0.0538	3.5701 ± 0.0624	0.4	95.2	342 ± 5
850	4.2586 ± 0.0062	0.9636 ± 0.0014	1.5495 ± 0.0478	100.1886 ± 1.1858	23.5116 ± 0.2096	0.8	83.9	292 ± 4

(Table continues)

APPENDIX TABLE 1. (Continued)

T(°C)	^{40}Ar (10^{-13} moles)	^{39}Ar (10^{-14} moles)	^{38}Ar (10^{-16} moles)	^{37}Ar (10^{-16} moles)	^{36}Ar (10^{-17} moles)	Pct. ^{39}Ar of total	Pct. ^{40*}Ar	Date (Ma)	2σ
				Sample QL-99-1C, phengitic white mica (continued)					
875	170.7566 ± 0.1998	38.7704 ± 0.0291	51.5647 ± 0.5278	144.6137 ± 1.7435	169.8145 ± 0.2766	33.2	97.1	332	± 5
900	161.4701 ± 0.1641	38.0375 ± 0.0291	49.6443 ± 0.3255	143.4014 ± 1.9701	166.6043 ± 0.2806	32.4	97.0	321	± 4
925	72.3326 ± 0.0566	17.1314 ± 0.0142	21.9282 ± 0.2404	132.0830 ± 1.4645	124.5452 ± 0.2422	14.6	94.9	313	± 4
975	48.3508 ± 0.0927	10.7189 ± 0.0156	15.3280 ± 0.7595	121.9864 ± 1.7466	150.0645 ± 0.2828	9.2	90.8	320	± 4
1000	15.6760 ± 0.0147	3.1741 ± 0.0038	5.0151 ± 0.1549	109.3602 ± 1.2756	84.7490 ± 0.2205	2.7	84.1	324	± 5
1025	8.7570 ± 0.0078	1.7633 ± 0.0014	3.0153 ± 0.1341	107.1778 ± 1.5038	53.4291 ± 0.2185	1.5	82.0	319	± 4
1050	5.5919 ± 0.0076	1.1745 ± 0.0014	2.0319 ± 0.0921	95.0632 ± 1.4636	35.3536 ± 0.2132	1.0	81.4	305	± 4
1075	4.1802 ± 0.0064	0.9102 ± 0.0008	1.6202 ± 0.0884	80.4043 ± 1.2115	26.2807 ± 0.2118	0.8	81.5	295	± 4
1100	3.5046 ± 0.0068	0.7693 ± 0.0008	1.2232 ± 0.0349	81.9662 ± 1.0486	22.6168 ± 0.2117	0.7	81.1	291	± 4
1126	3.0180 ± 0.0066	0.6771 ± 0.0009	1.1131 ± 0.0639	78.1640 ± 1.1470	22.1409 ± 0.2180	0.6	78.5	277	± 4
1150	2.5366 ± 0.0062	0.5734 ± 0.0006	1.1238 ± 0.0395	73.5446 ± 1.2939	22.7634 ± 0.2179	0.5	73.7	260	± 4
1200	2.4320 ± 0.0066	0.5403 ± 0.0008	1.1886 ± 0.0532	81.1326 ± 1.2735	24.9611 ± 0.2279	0.5	69.9	251	± 4
1250	1.8034 ± 0.0065	0.3969 ± 0.0007	0.9486 ± 0.0308	78.1186 ± 1.0392	23.7738 ± 0.2335	0.3	61.3	225	± 4
1300	0.6899 ± 0.0067	0.1412 ± 0.0006	0.5480 ± 0.0414	78.6542 ± 1.2696	19.9218 ± 0.2391	0.1	15.4	64	± 5
1400	0.6934 ± 0.0070	0.1325 ± 0.0006	0.7205 ± 0.0451	90.4270 ± 1.3992	23.6550 ± 0.2500	0.1	0.1	0	± 7
1601	1.5888 ± 0.0154	0.2050 ± 0.0008	1.2463 ± 0.0430	93.5576 ± 1.8124	46.8773 ± 0.4877	0.2	13.2	86	± 9
				Weight = 11.89 mg; J value = 0.00474; total gas age = 321 Ma					
				Sample 95-QL-8A, phengitic white mica					
700	6.1111 ± 0.0047	1.3567 ± 0.0016	2.4013 ± 0.0225	1.8179 ± 0.0327	29.1935 ± 0.0866	3.6	85.9	313	± 3
750	5.3664 ± 0.0076	1.2984 ± 0.0017	1.9476 ± 0.0163	0.6947 ± 0.0254	13.8930 ± 0.0852	3.5	92.3	309	± 3
760	5.3610 ± 0.0043	1.3183 ± 0.0014	1.9775 ± 0.0180	0.3863 ± 0.0259	12.1811 ± 0.0784	3.5	93.3	308	± 3
775	10.0615 ± 0.0053	2.4949 ± 0.0021	3.5178 ± 0.0180	0.3618 ± 0.0267	20.3333 ± 0.0860	6.7	94.0	307	± 3
780	10.2235 ± 0.0074	2.5876 ± 0.0020	3.4932 ± 0.0181	0.3441 ± 0.0221	12.8084 ± 0.0705	6.9	96.3	308	± 3
790	13.5604 ± 0.0064	3.4414 ± 0.0027	4.6803 ± 0.0181	0.2726 ± 0.0330	13.8687 ± 0.0854	9.2	97.0	310	± 3
795	13.7224 ± 0.0065	3.4856 ± 0.0023	4.6359 ± 0.0340	0.2140 ± 0.0335	12.3739 ± 0.0749	9.3	97.3	310	± 3
800	13.4862 ± 0.0092	3.4374 ± 0.0030	4.6061 ± 0.0278	0.1739 ± 0.0302	11.9965 ± 0.0775	9.2	97.4	310	± 3

Sample 95-QL-8A, phengitic white mica (continued)							
805	11.8655 ± 0.0052	3.0233 ± 0.0023	4.1419 ± 0.0173	0.1234 ± 0.0212	10.2490 ± 0.0754	8.1	97.4	310 ± 3
810	10.1706 ± 0.0063	2.6027 ± 0.0022	3.4356 ± 0.0214	0.1208 ± 0.0230	9.0834 ± 0.0725	7.0	97.4	308 ± 3
815	7.8823 ± 0.0054	2.0301 ± 0.0023	2.6798 ± 0.0179	0.1313 ± 0.0236	7.1663 ± 0.0657	5.4	97.3	306 ± 3
820	6.0043 ± 0.0048	1.5560 ± 0.0018	2.1384 ± 0.1353	0.0263 ± 0.0146	5.4617 ± 0.0679	4.1	97.3	305 ± 3
830	5.0736 ± 0.0059	1.3211 ± 0.0010	1.8759 ± 0.0164	0.0822 ± 0.0236	4.8880 ± 0.0663	3.5	97.2	303 ± 3
840	4.5462 ± 0.0045	1.1877 ± 0.0012	1.6271 ± 0.1085	0.0789 ± 0.0175	4.4893 ± 0.0632	3.2	97.1	302 ± 3
850	3.6397 ± 0.0043	0.9464 ± 0.0014	1.3154 ± 0.0155	0.1024 ± 0.0285	3.9368 ± 0.0637	2.5	96.8	302 ± 3
860	3.1390 ± 0.0043	0.8155 ± 0.0011	1.1336 ± 0.0624	0.1215 ± 0.0211	3.3845 ± 0.0607	2.2	96.8	303 ± 3
870	2.9158 ± 0.0047	0.7584 ± 0.0010	1.0846 ± 0.0658	0.1092 ± 0.0223	3.1930 ± 0.0620	2.0	96.8	302 ± 3
890	3.5062 ± 0.0038	0.9361 ± 0.0010	1.3910 ± 0.0703	0.1133 ± 0.0170	3.4544 ± 0.0634	2.5	97.1	296 ± 3
920	4.0777 ± 0.0042	1.1080 ± 0.0013	1.5733 ± 0.0983	0.1252 ± 0.0207	3.7782 ± 0.0668	3.0	97.3	292 ± 3
950	2.6839 ± 0.0041	0.7288 ± 0.0009	0.9693 ± 0.0567	0.2456 ± 0.0179	2.4779 ± 0.0603	1.9	97.3	292 ± 3
1000	1.5354 ± 0.0034	0.4045 ± 0.0007	0.5501 ± 0.0388	0.3964 ± 0.0233	1.2902 ± 0.0567	1.1	97.5	301 ± 3
1050	0.8055 ± 0.0034	0.2095 ± 0.0005	0.2891 ± 0.0151	1.2624 ± 0.0235	0.7416 ± 0.0538	0.6	97.3	304 ± 3
1150	0.7564 ± 0.0035	0.1966 ± 0.0005	0.3008 ± 0.0127	3.2659 ± 0.0274	0.5741 ± 0.0604	0.5	97.8	305 ± 3
1400	0.8798 ± 0.0036	0.2250 ± 0.0004	0.2966 ± 0.0185	0.3127 ± 0.0210	0.4657 ± 0.0618	0.6	98.4	312 ± 3
1501	0.1328 ± 0.0053	0.0351 ± 0.0002	0.0790 ± 0.0117	0.0020 ± 0.0045	0.3968 ± 0.0759	0.1	91.2	282 ± 4
Weight = 17.58 mg; J value = 0.0049; total gas age = 307 Ma								
Sample 95-QL-8C, phengitic white mica							
700	5.0245 ± 0.0066	1.2219 ± 0.0018	2.3339 ± 0.0188	2.5782 ± 0.0327	36.0464 ± 0.0942	3.7	78.8	268 ± 3
750	3.7618 ± 0.0040	1.0644 ± 0.0011	1.6605 ± 0.0147	0.9293 ± 0.0266	12.4539 ± 0.0757	3.2	90.2	264 ± 3
775	6.4705 ± 0.0048	1.8591 ± 0.0017	2.5655 ± 0.0232	0.9407 ± 0.0267	19.7062 ± 0.0886	5.6	91.0	262 ± 3
800	16.1388 ± 0.0193	4.8457 ± 0.0059	6.6954 ± 0.0235	0.7220 ± 0.0266	25.4401 ± 0.0910	14.6	95.3	263 ± 3
810	16.2368 ± 0.0183	4.9675 ± 0.0055	6.6401 ± 0.0382	0.3467 ± 0.0384	13.2633 ± 0.0799	14.9	97.6	264 ± 3
820	15.8224 ± 0.0239	4.8539 ± 0.0058	6.3730 ± 0.0369	0.3534 ± 0.0243	12.4261 ± 0.0827	14.6	97.7	263 ± 3
830	11.5075 ± 0.0258	3.5550 ± 0.0055	4.6214 ± 0.0252	0.2940 ± 0.0347	9.7761 ± 0.0833	10.7	97.5	261 ± 3
840	8.5358 ± 0.0239	2.6446 ± 0.0048	3.4644 ± 0.0190	0.3253 ± 0.0391	8.4626 ± 0.0732	8.0	97.1	259 ± 3
850	5.2530 ± 0.0164	1.5973 ± 0.0045	2.2202 ± 0.0212	0.3051 ± 0.0239	7.1718 ± 0.0660	4.8	96.0	261 ± 3
860	3.3975 ± 0.0091	1.0407 ± 0.0023	1.4362 ± 0.0187	0.2477 ± 0.0194	4.5583 ± 0.0657	3.1	96.0	259 ± 3
870	2.8154 ± 0.0051	0.8612 ± 0.0014	1.2070 ± 0.0147	0.2489 ± 0.0224	3.7374 ± 0.0648	2.6	96.1	260 ± 3

(Table continues)

APPENDIX TABLE 1. (Continued)

T(°C)	^{40}Ar (10^{-13} moles)	^{39}Ar (10^{-14} moles)	^{38}Ar (10^{-16} moles)	^{37}Ar (10^{-16} moles)	^{36}Ar (10^{-17} moles)	Pct. ^{39}Ar of total	Pct. $^{40*}Ar$	Date (Ma)	2σ
			Sample 95-QL-8C, phengitic white mica (continued)						
892	15.4592 ± 0.0078	4.5314 ± 0.0034	6.7064 ± 0.0359	2.8638 ± 0.0405	48.4212 ± 0.1211	13.7	90.7	256	3
1050	0.0524 ± 0.0024	0.0124 ± 0.0001	0.0239 ± 0.0064	1.7635 ± 0.0244	0.8406 ± 0.0461	0.0	52.9	189	9
1150	0.1024 ± 0.0030	0.0268 ± 0.0001	0.0746 ± 0.0068	2.7809 ± 0.0262	1.4687 ± 0.0506	0.1	57.8	186	4
1250	0.1122 ± 0.0033	0.0279 ± 0.0001	0.0948 ± 0.0083	2.3448 ± 0.0310	1.8986 ± 0.0583	0.1	50.2	171	5
1400	0.1431 ± 0.0038	0.0358 ± 0.0002	0.1486 ± 0.0099	2.9965 ± 0.0327	2.7471 ± 0.0682	0.1	43.4	148	5
1551	0.2076 ± 0.0063	0.0481 ± 0.0003	0.2530 ± 0.0125	4.4804 ± 0.0482	4.9734 ± 0.1089	0.1	29.4	110	6

Weight = 16.07 mg; J value = 0.004935; total gas age = 261 Ma

			Sample HU-99-2, phengitic white mica + paragonite						
700	1.0692 ± 0.0049	0.1199 ± 0.0005	0.0379 ± 0.0124	3.1064 ± 0.1161	21.1921 ± 0.3480	3.2	41.4	293	10
750	0.5787 ± 0.0046	0.1269 ± 0.0004	0.0325 ± 0.0165	1.7086 ± 0.0839	5.8739 ± 0.1064	3.3	70.0	255	4
800	1.3781 ± 0.0048	0.4084 ± 0.0005	0.0306 ± 0.0055	4.1123 ± 0.0870	3.5288 ± 0.1102	10.8	92.4	250	3
825	1.8929 ± 0.0048	0.6230 ± 0.0006	0.0593 ± 0.0520	7.7387 ± 0.0872	1.9875 ± 0.1121	16.4	96.9	237	3
850	2.0360 ± 0.0048	0.6454 ± 0.0005	0.0117 ± 0.0204	4.4856 ± 0.0947	2.9237 ± 0.1101	17.1	95.8	242	3
875	1.5764 ± 0.0048	0.4683 ± 0.0004	0.0571 ± 0.0270	1.0961 ± 0.0791	3.9570 ± 0.1148	12.4	92.6	249	3
900	1.2037 ± 0.0047	0.3387 ± 0.0004	0.0376 ± 0.0165	0.7892 ± 0.0862	4.9794 ± 0.1122	8.9	87.7	250	3
925	0.8531 ± 0.0047	0.2275 ± 0.0004	0.0325 ± 0.0058	0.6189 ± 0.0771	4.8690 ± 0.1541	6.0	83.1	250	4
975	0.8457 ± 0.0046	0.2327 ± 0.0003	0.0363 ± 0.0280	0.9246 ± 0.0809	3.8861 ± 0.1122	6.1	86.4	251	4
1000	0.5035 ± 0.0045	0.1477 ± 0.0003	0.0316 ± 0.0047	0.9952 ± 0.0859	2.0377 ± 0.1061	3.9	88.0	241	3
1025	0.3778 ± 0.0045	0.1170 ± 0.0003	0.0358 ± 0.0140	0.9614 ± 0.0786	1.2163 ± 0.1079	3.1	90.5	235	3
1050	0.3814 ± 0.0045	0.1268 ± 0.0003	0.0397 ± 0.0160	1.6085 ± 0.0839	0.7430 ± 0.1032	3.3	94.3	228	3
1075	0.3826 ± 0.0046	0.1329 ± 0.0003	0.0371 ± 0.0132	2.2560 ± 0.0791	0.3483 ± 0.1115	3.5	97.3	226	3
1100	0.1499 ± 0.0043	0.0495 ± 0.0003	0.0406 ± 0.0097	0.0000 ± 0.0000	0.0609 ± 0.0957	1.3	98.8	240	3
1126	0.0519 ± 0.0037	0.0144 ± 0.0002	0.0350 ± 0.0056	0.0000 ± 0.0000	0.0829 ± 0.0798	0.4	95.3	273	4
1150	0.0158 ± 0.0025	0.0030 ± 0.0001	0.0199 ± 0.0021	0.0702 ± 0.0386	0.1107 ± 0.0581	0.1	79.2	329	7
1200	0.0187 ± 0.0033	0.0021 ± 0.0002	0.0264 ± 0.0022	0.1254 ± 0.0508	0.2203 ± 0.0663	0.1	65.1	448	20
1250	0.0098 ± 0.0036	0.0008 ± 0.0016	0.0270 ± 0.0022	0.0452 ± 0.0617	0.0084 ± 0.0497	0.0	97.5	841	42

Weight = 10.36 mg; J value = 0.00477; total gas age = 246 Ma

				Sample 95-QL-6B, paragonite				
700	0.6910 ± 0.0035	0.0917 ± 0.0003	0.4895 ± 0.0120	4.6788 ± 0.0331	16.3595 ± 0.0773	14.9	30.1	190 ± 7
750	0.2758 ± 0.0034	0.0330 ± 0.0001	0.1727 ± 0.0092	3.6626 ± 0.0293	5.9907 ± 0.0648	5.4	35.9	248 ± 7
760	0.2014 ± 0.0033	0.0253 ± 0.0001	0.1128 ± 0.0091	3.4612 ± 0.0338	3.9360 ± 0.0580	4.1	42.4	276 ± 7
775	0.5068 ± 0.0035	0.0587 ± 0.0002	0.3034 ± 0.0093	12.6231 ± 0.0561	10.5379 ± 0.0701	9.5	38.7	274 ± 7
780	0.3018 ± 0.0034	0.0470 ± 0.0002	0.1500 ± 0.0098	10.0408 ± 0.0442	4.9337 ± 0.0626	7.6	51.9	273 ± 5
790	0.1638 ± 0.0032	0.0303 ± 0.0001	0.0678 ± 0.0090	4.1891 ± 0.0422	1.9888 ± 0.0546	4.9	64.3	284 ± 5
795	0.1165 ± 0.0031	0.0212 ± 0.0001	0.0534 ± 0.0070	1.0831 ± 0.0250	0.7854 ± 0.0507	3.4	80.1	353 ± 6
800	0.1166 ± 0.0031	0.0222 ± 0.0001	0.0258 ± 0.0072	0.4432 ± 0.0230	0.5161 ± 0.0467	3.6	87.0	364 ± 5
805	0.1319 ± 0.0031	0.0265 ± 0.0001	0.0392 ± 0.0066	0.3437 ± 0.0231	0.4904 ± 0.0495	4.3	89.1	355 ± 5
810	0.1341 ± 0.0031	0.0272 ± 0.0002	0.0381 ± 0.0070	0.3383 ± 0.0166	0.5197 ± 0.0512	4.4	88.6	350 ± 5
815	0.1366 ± 0.0031	0.0275 ± 0.0001	0.0406 ± 0.0077	0.3962 ± 0.0221	0.3541 ± 0.0479	4.5	92.4	367 ± 5
820	0.1280 ± 0.0031	0.0254 ± 0.0002	0.0345 ± 0.0079	0.3115 ± 0.0195	0.4165 ± 0.0485	4.1	90.4	363 ± 5
830	0.1366 ± 0.0031	0.0268 ± 0.0002	0.0386 ± 0.0081	0.3515 ± 0.0211	0.4290 ± 0.0472	4.4	90.7	368 ± 5
840	0.1394 ± 0.0031	0.0269 ± 0.0001	0.0358 ± 0.0078	0.5142 ± 0.0239	0.5712 ± 0.0484	4.4	87.9	363 ± 5
850	0.0941 ± 0.0029	0.0173 ± 0.0001	0.0199 ± 0.0067	0.3914 ± 0.0172	0.3416 ± 0.0437	2.8	89.3	385 ± 6
860	0.0662 ± 0.0027	0.0116 ± 0.0001	0.0278 ± 0.0058	0.4158 ± 0.0147	0.4099 ± 0.0426	1.9	81.7	372 ± 7
870	0.0513 ± 0.0025	0.0089 ± 0.0001	0.0118 ± 0.0054	0.3954 ± 0.0165	0.3267 ± 0.0402	1.4	81.2	375 ± 8
890	0.0620 ± 0.0026	0.0103 ± 0.0001	0.0146 ± 0.0059	0.8080 ± 0.0182	0.5092 ± 0.0453	1.7	75.8	365 ± 8
920	0.0982 ± 0.0028	0.0165 ± 0.0001	0.0349 ± 0.0070	1.6470 ± 0.0262	0.9109 ± 0.0491	2.7	72.7	348 ± 7
950	0.0732 ± 0.0026	0.0112 ± 0.0001	0.0366 ± 0.0065	1.2356 ± 0.0226	0.9524 ± 0.0442	1.8	61.7	326 ± 9
1000	0.0892 ± 0.0027	0.0105 ± 0.0001	0.0427 ± 0.0071	2.6513 ± 0.0327	1.7404 ± 0.0474	1.7	42.6	295 ± 12
1050	0.1380 ± 0.0030	0.0128 ± 0.0001	0.0655 ± 0.0063	5.6345 ± 0.0403	2.7903 ± 0.0524	2.1	40.6	351 ± 13
1150	0.0880 ± 0.0036	0.0125 ± 0.0003	0.0465 ± 0.0063	5.2880 ± 0.0568	1.3542 ± 0.0501	2.0	55.0	315 ± 18
1400	0.0213 ± 0.0064	0.0056 ± 0.0002	0.0156 ± 0.0186	1.6108 ± 0.0426	0.2351 ± 0.0995	0.9	133.1	399 ± 16
1501	0.0557 ± 0.0092	0.0102 ± 0.0002	0.0235 ± 0.0202	0.1501 ± 0.0517	0.3249 ± 0.1479	1.7	117.3	490 ± 15
			Weight = 16.74 mg; J value = 0.0049; total gas age = 309 Ma					
				Sample QL-99-2A, muscovitic white mica				
850	0.1021 ± 0.0831	0.0350 ± 0.0278	0.0720 ± 0.0472	0.0127 ± 0.1045	0.4183 ± 0.4443	0.3	87.9	207 ± 4
875	0.1433 ± 0.0746	0.0303 ± 0.0248	0.1236 ± 0.0421	0.0424 ± 0.0650	2.1448 ± 0.4082	0.3	55.7	212 ± 5
900	0.1419 ± 0.0697	0.0326 ± 0.0231	0.1108 ± 0.0391	0.0253 ± 0.0482	1.7592 ± 0.3812	0.3	63.3	221 ± 4

(Table continues)

APPENDIX TABLE 1. (Continued)

T(°C)	^{40}Ar (10^{-13} moles)	^{39}Ar (10^{-14} moles)	^{38}Ar (10^{-16} moles)	^{37}Ar (10^{-16} moles)	^{36}Ar (10^{-17} moles)	Pct. ^{39}Ar of total	Pct. $^{40*}Ar$	Date (Ma)	2σ
\multicolumn{10}{l}{Sample QL-99-2A, muscovitic white mica (continued)}
925	0.1240 ± 0.0647	0.0440 ± 0.0216	0.0791 ± 0.0362	0.0251 ± 0.0666	0.2901 ± 0.3658	0.4	93.1	211	3
975	0.2104 ± 0.0634	0.0787 ± 0.0210	0.1157 ± 0.0357	0.2825 ± 0.0598	0.4588 ± 0.3574	0.7	93.5	202	3
1000	0.2411 ± 0.0633	0.0917 ± 0.0211	0.1504 ± 0.0359	0.2343 ± 0.0610	0.3578 ± 0.3413	0.8	95.6	203	3
1025	0.3485 ± 0.0655	0.1314 ± 0.0218	0.2010 ± 0.0373	0.2194 ± 0.0635	0.5032 ± 0.3754	1.1	95.7	204	3
1050	0.5434 ± 0.0681	0.2037 ± 0.0227	0.2939 ± 0.0409	0.2465 ± 0.0633	0.9350 ± 0.3786	1.7	94.9	204	3
1075	0.9008 ± 0.0705	0.3446 ± 0.0235	0.4583 ± 0.0470	0.2457 ± 0.0674	0.7477 ± 0.3898	2.9	97.5	205	3
1100	1.3383 ± 0.0723	0.5191 ± 0.0240	0.6644 ± 0.0563	0.3488 ± 0.0690	0.7735 ± 0.3978	4.4	98.3	204	3
1126	1.9214 ± 0.0738	0.7503 ± 0.0245	0.9454 ± 0.0565	0.3534 ± 0.0718	0.9604 ± 0.4069	6.3	98.5	203	3
1150	2.1213 ± 0.0751	0.8307 ± 0.0250	1.0799 ± 0.0439	0.3439 ± 0.0714	0.6596 ± 0.4094	7.0	99.1	204	3
1200	3.3976 ± 0.0773	1.3305 ± 0.0257	1.7031 ± 0.0445	0.3606 ± 0.0746	0.7398 ± 0.4260	11.2	99.4	204	3
1250	3.9104 ± 0.0791	1.5346 ± 0.0262	1.9029 ± 0.1099	0.2854 ± 0.0762	0.8087 ± 0.4279	12.9	99.4	204	3
1300	3.4579 ± 0.0808	1.3549 ± 0.0268	1.7448 ± 0.0474	0.3022 ± 0.0768	0.9932 ± 0.4428	11.4	99.1	204	3
1400	4.2608 ± 0.0825	1.5520 ± 0.0276	2.1573 ± 0.0498	0.3337 ± 0.0784	10.9262 ± 0.4551	13.0	92.4	204	3
1601	8.6715 ± 0.1579	2.9923 ± 0.0524	4.2790 ± 0.3119	0.4788 ± 0.1502	38.0026 ± 0.8652	25.1	87.0	203	3
\multicolumn{10}{l}{Weight = 11.62 mg; J value = 0.00474; total gas age = 204 Ma}
\multicolumn{10}{l}{Sample QL-99-14, muscovitic white mica}
700	0.0905 ± 0.0001	0.0079 ± 0.0000	0.0630 ± 0.0011	0.0000 ± 0.0000	2.4768 ± 0.0057	0.1	19.1	179	11
750	0.0680 ± 0.0000	0.0080 ± 0.0000	0.0398 ± 0.0013	0.0092 ± 0.0036	1.6307 ± 0.0051	0.1	29.1	202	8
800	0.0854 ± 0.0000	0.0116 ± 0.0000	0.0467 ± 0.0014	0.0424 ± 0.0067	1.8324 ± 0.0051	0.1	36.5	219	6
825	0.0732 ± 0.0000	0.0094 ± 0.0000	0.0383 ± 0.0009	0.0244 ± 0.0051	1.5844 ± 0.0045	0.1	36.0	226	7
850	0.0897 ± 0.0000	0.0125 ± 0.0000	0.0503 ± 0.0020	0.0410 ± 0.0075	1.8563 ± 0.0048	0.1	38.8	226	6
875	0.1015 ± 0.0000	0.0146 ± 0.0000	0.0595 ± 0.0009	0.0440 ± 0.0057	2.0711 ± 0.0051	0.1	39.7	224	6
900	0.2390 ± 0.0001	0.0289 ± 0.0001	0.2005 ± 0.0027	0.1661 ± 0.0082	5.4920 ± 0.0122	0.3	32.0	215	7
925	0.2113 ± 0.0001	0.0345 ± 0.0001	0.1344 ± 0.0016	0.1430 ± 0.0063	4.1106 ± 0.0098	0.3	42.5	212	5
975	0.3257 ± 0.0002	0.0772 ± 0.0001	0.1729 ± 0.0056	0.1789 ± 0.0055	4.2444 ± 0.0095	0.7	61.4	211	3
1000	0.3550 ± 0.0001	0.0898 ± 0.0001	0.1957 ± 0.0098	0.2020 ± 0.0082	4.3087 ± 0.0114	0.8	64.1	206	3

Temp								Age (Ma)
1025	0.4164 ± 0.0002	0.1116 ± 0.0002	0.2287 ± 0.0091	0.1707 ± 0.0092	4.6076 ± 0.0110	1.0	67.2	204 ± 3
1050	0.5914 ± 0.0002	0.1756 ± 0.0001	0.3232 ± 0.0135	0.1950 ± 0.0081	5.0315 ± 0.0135	1.6	74.8	205 ± 3
1075	1.0734 ± 0.0004	0.3600 ± 0.0003	0.5615 ± 0.0320	0.2714 ± 0.0120	5.8313 ± 0.0141	3.3	83.9	204 ± 3
1100	1.7091 ± 0.0007	0.6110 ± 0.0004	0.9105 ± 0.0533	0.2322 ± 0.0110	6.2989 ± 0.0181	5.6	89.1	203 ± 3
1126	2.7683 ± 0.0005	1.0380 ± 0.0005	1.4325 ± 0.0862	0.3021 ± 0.0168	6.8405 ± 0.0221	9.4	92.7	201 ± 3
1150	3.3322 ± 0.0013	1.2570 ± 0.0008	1.6592 ± 0.0683	0.2841 ± 0.0148	6.9135 ± 0.0176	11.4	93.9	203 ± 3
1200	4.3582 ± 0.0011	1.6604 ± 0.0006	2.3080 ± 0.1392	0.2723 ± 0.0156	7.5548 ± 0.0195	15.1	94.9	203 ± 3
1250	4.4033 ± 0.0023	1.6709 ± 0.0009	2.2390 ± 0.1369	0.2924 ± 0.0174	7.9701 ± 0.0161	15.2	94.6	203 ± 3
1300	3.5404 ± 0.0018	1.3181 ± 0.0011	1.8190 ± 0.1136	0.2267 ± 0.0147	8.4885 ± 0.0224	12.0	92.9	203 ± 3
1400	3.5420 ± 0.0010	1.2215 ± 0.0011	1.8466 ± 0.0506	0.3213 ± 0.0138	16.6123 ± 0.0274	11.1	86.1	203 ± 3
1601	4.4710 ± 0.0032	1.2634 ± 0.0009	2.4257 ± 0.0836	0.3942 ± 0.0177	43.2072 ± 0.0504	11.5	71.4	206 ± 3

Weight = 11.64 mg; J value = 0.00479; total gas age = 204 Ma

Sample 95-HU-1A, biotite

Temp								Age (Ma)
600	2.5261 ± 0.0037	0.5753 ± 0.0009	1.5270 ± 0.0164	0.6616 ± 0.0233	38.8929 ± 0.1043	2.9	54.5	203 ± 3
700	11.4761 ± 0.0065	4.4561 ± 0.0025	6.1493 ± 0.0270	0.3792 ± 0.0285	30.7913 ± 0.0950	22.7	92.1	201 ± 2
800	19.4501 ± 0.0147	8.0627 ± 0.0039	10.3202 ± 0.0370	0.6587 ± 0.0378	8.3852 ± 0.0672	41.0	98.7	202 ± 2
825	4.9590 ± 0.0051	2.0603 ± 0.0016	2.5706 ± 0.0147	0.2003 ± 0.0233	1.6338 ± 0.0603	10.5	99.0	202 ± 2
850	3.1731 ± 0.0046	1.3149 ± 0.0017	1.7357 ± 0.1085	0.2551 ± 0.0242	1.2978 ± 0.0598	6.7	98.8	202 ± 2
875	3.1091 ± 0.0055	1.2907 ± 0.0020	1.7037 ± 0.0155	0.1575 ± 0.0197	1.4843 ± 0.0586	6.6	98.6	201 ± 2
900	4.0654 ± 0.0040	1.6870 ± 0.0015	2.2099 ± 0.0233	0.2379 ± 0.0217	1.8388 ± 0.0576	8.6	98.7	201 ± 2
925	4.6516 ± 0.0052	1.9370 ± 0.0024	2.3050 ± 0.0212	0.1048 ± 0.0218	1.8362 ± 0.0562	9.8	98.8	201 ± 2
950	4.3175 ± 0.0052	1.7956 ± 0.0018	2.1188 ± 0.0154	0.1252 ± 0.0233	1.6484 ± 0.0580	9.1	98.9	201 ± 2
975	3.8001 ± 0.0042	1.5834 ± 0.0022	2.1059 ± 0.0232	0.1207 ± 0.0253	1.6626 ± 0.0575	8.0	98.7	201 ± 2
1000	3.1434 ± 0.0041	1.3065 ± 0.0014	1.7246 ± 0.1083	0.1282 ± 0.0243	1.4763 ± 0.0604	6.6	98.6	201 ± 2
1025	2.8605 ± 0.0036	1.1845 ± 0.0013	1.5753 ± 0.0153	0.1694 ± 0.0195	1.5398 ± 0.0590	6.0	98.4	201 ± 2
1050	2.5943 ± 0.0037	1.0709 ± 0.0011	1.3815 ± 0.0168	0.0734 ± 0.0255	1.2316 ± 0.0577	5.4	98.6	202 ± 2
1075	2.0482 ± 0.0035	0.8465 ± 0.0009	1.0919 ± 0.0648	0.2142 ± 0.0279	0.7119 ± 0.0541	4.3	99.0	203 ± 2
1100	0.8062 ± 0.0035	0.3332 ± 0.0006	0.4199 ± 0.0157	0.3366 ± 0.0216	0.0986 ± 0.0465	1.7	99.6	204 ± 2
1150	0.2348 ± 0.0033	0.0973 ± 0.0002	0.1362 ± 0.0112	0.8765 ± 0.0228	0.0572 ± 0.0595	0.5	100.8	206 ± 2
1300	0.0699 ± 0.0034	0.0302 ± 0.0002	0.0338 ± 0.0075	0.2433 ± 0.0230	0.2524 ± 0.0491	0.2	110.7	216 ± 3

Weight = 14.87 mg; J value = 0.00497; total gas age = 202 Ma

(Table continues)

APPENDIX TABLE 1. (*Continued*)

T(°C)	^{40}Ar (10-13 moles)	^{39}Ar (10-14 moles)	^{38}Ar (10-16 moles)	^{37}Ar (10-16 moles)	^{36}Ar (10-17 moles)	Pct. ^{39}Ar of total	Pct. ^{40*}Ar	Date (Ma)	2σ
\multicolumn{10}{l}{Sample 95-HU-1B, biotite}									
600	4.4765 ± 0.0040	1.2632 ± 0.0013	2.9054 ± 0.1015	1.6187 ± 0.0221	52.4236 ± 0.1048	4.8	65.4	195	± 2
700	11.4721 ± 0.0061	4.6580 ± 0.0032	6.2883 ± 0.0287	0.5869 ± 0.0275	10.5271 ± 0.0736	17.6	97.3	202	± 2
800	13.2053 ± 0.0068	5.4341 ± 0.0035	7.2274 ± 0.0270	0.8097 ± 0.0293	3.8854 ± 0.0674	20.5	99.1	203	± 2
825	4.2784 ± 0.0051	1.7606 ± 0.0024	2.2183 ± 0.0147	0.3504 ± 0.0210	1.0352 ± 0.0583	6.6	99.3	203	± 2
850	4.0982 ± 0.0043	1.6318 ± 0.0012	2.2536 ± 0.0206	0.3246 ± 0.0251	1.0040 ± 0.0583	6.3	99.3	203	± 2
875	4.8950 ± 0.0052	2.0232 ± 0.0018	2.4885 ± 0.0197	0.3338 ± 0.0290	0.9408 ± 0.0555	7.6	99.4	202	± 2
900	5.3281 ± 0.0039	2.2044 ± 0.0014	2.7115 ± 0.0179	0.4299 ± 0.0231	1.1595 ± 0.0580	8.3	99.4	202	± 2
925	5.4232 ± 0.0055	2.2548 ± 0.0016	2.7960 ± 0.0170	0.4126 ± 0.0249	1.1432 ± 0.0550	8.5	99.4	201	± 2
950	4.5038 ± 0.0039	1.8663 ± 0.0020	2.2955 ± 0.0216	0.3005 ± 0.0262	0.8697 ± 0.0574	7.0	99.4	202	± 2
975	3.3421 ± 0.0036	1.3871 ± 0.0014	1.8449 ± 0.0154	0.3704 ± 0.0206	0.7477 ± 0.0575	5.2	99.3	201	± 2
1000	2.2177 ± 0.0036	0.9203 ± 0.0010	1.2240 ± 0.0121	0.3976 ± 0.0233	0.4381 ± 0.0545	3.5	99.4	202	± 2
1025	1.3566 ± 0.0035	0.5604 ± 0.0008	0.7565 ± 0.0319	0.4265 ± 0.0229	0.1732 ± 0.0617	2.1	99.6	203	± 2
1050	0.7579 ± 0.0034	0.3165 ± 0.0007	0.4716 ± 0.0274	0.6615 ± 0.0227	0.0299 ± 0.0419	1.2	99.9	201	± 2
1075	0.4181 ± 0.0033	0.1761 ± 0.0004	0.2660 ± 0.0180	0.7239 ± 0.0222	0.0175 ± 0.0632	0.7	99.9	200	± 2

Weight = 15.49 mg; J value = 0.004935; total gas age = 202 Ma

			Sample FA-99-1C, biotite						
600	0.6234 ± 0.0019	0.1068 ± 0.0003	0.4174 ± 0.0105	2.1063 ± 0.1165	13.6749 ± 0.0530	1.1	35.1	168	± 5
700	1.3007 ± 0.0015	0.3828 ± 0.0003	0.8000 ± 0.0322	2.0860 ± 0.1214	11.8453 ± 0.0433	4.0	73.0	201	± 3
800	5.1063 ± 0.0029	1.9273 ± 0.0010	2.6340 ± 0.1472	0.9521 ± 0.0937	11.5447 ± 0.0502	20.2	93.3	200	± 3
825	2.0517 ± 0.0021	0.8031 ± 0.0006	1.0842 ± 0.0658	0.1911 ± 0.0818	2.0398 ± 0.0387	8.4	97.1	201	± 3
850	1.4821 ± 0.0016	0.5639 ± 0.0004	0.7725 ± 0.0112	0.1771 ± 0.0918	2.6164 ± 0.0391	5.9	94.8	202	± 3
875	1.2490 ± 0.0014	0.4350 ± 0.0004	0.6655 ± 0.0357	0.1444 ± 0.0834	5.3499 ± 0.0409	4.6	87.3	203	± 3
900	1.3956 ± 0.0016	0.4488 ± 0.0004	0.7405 ± 0.0064	0.1221 ± 0.0760	8.7059 ± 0.0451	4.7	81.5	205	± 3
925	1.9512 ± 0.0018	0.6588 ± 0.0006	1.0277 ± 0.0580	0.4559 ± 0.0866	9.8318 ± 0.0443	6.9	85.0	204	± 3
950	2.6527 ± 0.0018	0.9649 ± 0.0008	1.3991 ± 0.0118	0.6108 ± 0.0905	8.2115 ± 0.0412	10.1	90.8	202	± 3
975	3.0706 ± 0.0024	1.1649 ± 0.0010	1.5930 ± 0.0088	0.4135 ± 0.1026	5.7893 ± 0.0410	12.2	94.4	201	± 3

			Sample FA-99-1C, biotite *(continued)*				
1000	2.3303 ± 0.0019	0.9142 ± 0.0007	1.1977 ± 0.0388	0.4964 ± 0.0826	3.2181 ± 0.0390	9.6	96.0	202 ± 3
1025	1.3658 ± 0.0016	0.5218 ± 0.0006	0.7097 ± 0.0087	0.5010 ± 0.0840	2.2126 ± 0.0386	5.5	95.2	202 ± 3
1050	0.6483 ± 0.0014	0.2398 ± 0.0003	0.3381 ± 0.0041	0.1300 ± 0.0951	1.6136 ± 0.0377	2.5	92.6	203 ± 3
1075	0.2557 ± 0.0013	0.0867 ± 0.0002	0.1240 ± 0.0051	0.0000 ± 0.0000	1.1103 ± 0.0369	0.9	87.2	208 ± 3
1100	0.2203 ± 0.0013	0.0744 ± 0.0002	0.1116 ± 0.0032	0.0000 ± 0.0000	0.9969 ± 0.0364	0.8	86.6	207 ± 3
1150	0.4375 ± 0.0014	0.1554 ± 0.0002	0.2285 ± 0.0131	0.1472 ± 0.0954	1.6163 ± 0.0387	1.6	89.0	203 ± 3
1300	0.3587 ± 0.0019	0.0812 ± 0.0004	0.2005 ± 0.0075	0.1152 ± 0.1089	5.2833 ± 0.0432	0.9	56.4	202 ± 4
1549	0.1670 ± 0.0030	0.0176 ± 0.0007	0.0966 ± 0.0067	0.1043 ± 1.3394	4.0386 ± 0.0713	0.2	28.5	219 ± 17
			Weight = 9.59 mg; J value = 0.00476; total gas age = 202 Ma					
			Sample 95-HU-1B, K-feldspar					
600	1.3909 ± 0.0036	0.3130 ± 0.0006	0.7668 ± 0.0172	0.9171 ± 0.0234	18.5292 ± 0.0825	2.0	60.6	227 ± 3
700	1.2579 ± 0.0038	0.5569 ± 0.0007	0.7295 ± 0.0486	0.7518 ± 0.0225	1.2474 ± 0.0588	3.6	97.1	187 ± 2
750	1.3734 ± 0.0037	0.6189 ± 0.0009	0.8017 ± 0.0123	0.8107 ± 0.0228	0.4784 ± 0.0575	4.0	99.0	187 ± 2
850	3.5303 ± 0.0042	1.5733 ± 0.0017	2.0768 ± 0.0206	2.2499 ± 0.0249	0.9692 ± 0.0579	10.2	99.2	189 ± 2
950	4.1567 ± 0.0048	1.8665 ± 0.0022	2.2957 ± 0.0233	1.9971 ± 0.0271	0.7596 ± 0.0550	12.1	99.5	188 ± 2
1000	2.3242 ± 0.0037	1.0352 ± 0.0011	1.3665 ± 0.0154	0.8065 ± 0.0224	0.5839 ± 0.0569	6.7	99.3	189 ± 2
1050	2.0488 ± 0.0037	0.8855 ± 0.0011	1.1777 ± 0.0785	0.7969 ± 0.0338	1.8064 ± 0.0587	5.8	97.4	191 ± 2
1100	2.2096 ± 0.0038	0.9006 ± 0.0009	1.2609 ± 0.0593	1.3690 ± 0.0323	4.2240 ± 0.0629	5.8	94.4	196 ± 2
1150	4.5363 ± 0.0040	1.7650 ± 0.0014	2.2062 ± 0.0232	4.4060 ± 0.0352	9.9190 ± 0.0725	11.5	93.5	204 ± 2
1200	7.6228 ± 0.0047	3.0849 ± 0.0023	4.0721 ± 0.0243	3.3317 ± 0.0375	9.4399 ± 0.0729	20.0	96.3	202 ± 2
1188	2.3885 ± 0.0038	0.9777 ± 0.0007	1.3199 ± 0.0816	0.4869 ± 0.0248	2.2487 ± 0.0590	6.4	97.2	201 ± 2
1300	1.4996 ± 0.0037	0.6025 ± 0.0008	0.8134 ± 0.0353	0.5935 ± 0.0255	2.1268 ± 0.0607	3.9	95.8	202 ± 2
1351	0.8759 ± 0.0037	0.3390 ± 0.0005	0.4576 ± 0.0208	0.2939 ± 0.0209	1.5119 ± 0.0622	2.2	94.9	207 ± 2
1450	2.2886 ± 0.0040	0.8783 ± 0.0011	1.2471 ± 0.0150	1.2120 ± 0.0266	4.4088 ± 0.0656	5.7	94.3	208 ± 2
1649	0.3064 ± 0.0111	0.0117 ± 0.0002	0.2606 ± 0.0149	0.0936 ± 0.0333	10.2462 ± 0.1750	0.1	1.2	28 ± 44
			Weight = 13.63 mg; J value = 0.00497; total gas age = 197 Ma					
			Sample 95-HU-1A, K-feldspar					
600	2.9967 ± 0.0038	0.9719 ± 0.0018	1.6425 ± 0.0672	1.1177 ± 0.0384	19.4374 ± 0.0836	2.4	80.8	211 ± 2
700	4.1129 ± 0.0041	1.8580 ± 0.0017	2.3225 ± 0.0182	1.1761 ± 0.0274	1.0015 ± 0.0612	4.6	99.3	187 ± 2
750	4.1967 ± 0.0041	1.9017 ± 0.0016	2.3961 ± 0.0241	1.0897 ± 0.0307	0.3043 ± 0.0549	4.7	99.8	187 ± 2

(Table continues)

APPENDIX TABLE 1. (Continued)

T(°C)	^{40}Ar (10^{-13} moles)	^{39}Ar (10^{-14} moles)	^{38}Ar (10^{-16} moles)	^{37}Ar (10^{-16} moles)	^{36}Ar (10^{-17} moles)	Pct. ^{39}Ar of total	Pct. $^{40*}Ar$	Date (Ma)	2σ
\multicolumn{10}{c}{Sample 95-HU-1A, K-feldspar(continued)}
850	8.7738 ± 0.0063	3.9521 ± 0.0044	4.9797 ± 0.0252	1.8575 ± 0.0366	0.8418 ± 0.0563	9.9	99.7	188	± 2
950	9.2782 ± 0.0055	4.1572 ± 0.0036	5.2796 ± 0.0242	2.7146 ± 0.0390	0.9354 ± 0.0588	10.4	99.7	189	± 2
1000	5.2188 ± 0.0057	2.3215 ± 0.0022	2.8322 ± 0.0231	0.8798 ± 0.0256	0.7243 ± 0.0543	5.8	99.6	190	± 2
1050	4.4354 ± 0.0038	1.9499 ± 0.0018	2.4373 ± 0.0215	1.0841 ± 0.0314	1.4449 ± 0.0590	4.9	99.0	191	± 2
1100	4.3906 ± 0.0044	1.8820 ± 0.0019	2.3902 ± 0.0172	1.6261 ± 0.0273	3.5759 ± 0.0614	4.7	97.6	193	± 2
1150	6.7339 ± 0.0068	2.8262 ± 0.0026	3.7024 ± 0.0225	3.0523 ± 0.0343	5.7373 ± 0.0663	7.0	97.5	197	± 2
1200	15.9148 ± 0.0119	6.7742 ± 0.0047	8.6709 ± 0.0262	3.1026 ± 0.0384	6.9774 ± 0.0689	16.9	98.7	197	± 2
1188	9.5755 ± 0.0110	4.1112 ± 0.0039	5.3034 ± 0.0290	1.3444 ± 0.0317	2.9518 ± 0.0623	10.3	99.1	196	± 2
1300	10.7470 ± 0.0051	4.5984 ± 0.0036	5.8859 ± 0.0313	18.4855 ± 0.0585	3.2695 ± 0.0618	11.4	99.1	197	± 2
1351	1.9564 ± 0.0038	0.8236 ± 0.0012	1.0872 ± 0.0607	10.7902 ± 0.0511	1.3590 ± 0.0593	2.1	98.0	197	± 2
1450	4.2457 ± 0.0039	1.7837 ± 0.0015	2.1761 ± 0.0219	2.6042 ± 0.0325	2.6934 ± 0.0652	4.5	98.1	198	± 2
1649	0.6426 ± 0.0094	0.1801 ± 0.0003	0.4322 ± 0.0127	0.0587 ± 0.0374	7.7608 ± 0.1481	0.4	64.3	195	± 3
\multicolumn{10}{c}{Weight = 14.87 mg; J value = 0.00497; total gas age = 194 Ma}
\multicolumn{10}{c}{Sample 95-QL-7, K-feldspar}
600	3.1955 ± 0.0037	1.2029 ± 0.0012	1.9525 ± 0.0173	0.8999 ± 0.0287	17.3013 ± 0.0803	4.0	84.0	188	± 2
700	5.6209 ± 0.0043	2.5358 ± 0.0022	3.2426 ± 0.0260	0.3271 ± 0.0267	0.5678 ± 0.0596	8.4	99.7	187	± 2
750	6.2322 ± 0.0046	2.7969 ± 0.0027	3.4719 ± 0.0225	0.2437 ± 0.0262	0.2915 ± 0.0558	9.2	99.9	188	± 2
850	13.5282 ± 0.0069	6.0676 ± 0.0040	7.5827 ± 0.0470	0.3419 ± 0.0277	0.5588 ± 0.0593	20.0	99.9	188	± 2
950	12.3197 ± 0.0052	5.5008 ± 0.0024	6.9090 ± 0.0342	0.2921 ± 0.0276	0.3615 ± 0.0532	18.2	99.9	189	± 2
1000	6.1072 ± 0.0044	2.7050 ± 0.0023	3.3386 ± 0.0214	0.0642 ± 0.0225	0.3661 ± 0.0531	8.9	99.8	190	± 2
1050	4.2828 ± 0.0043	1.8874 ± 0.0014	2.2617 ± 0.0189	0.1133 ± 0.0255	0.4623 ± 0.0540	6.2	99.7	191	± 2
1100	3.3897 ± 0.0037	1.4770 ± 0.0014	1.9069 ± 0.0600	0.0337 ± 0.0230	0.6746 ± 0.0528	4.9	99.4	192	± 2
1150	3.9953 ± 0.0043	1.7202 ± 0.0017	2.2072 ± 0.0719	0.0287 ± 0.0155	1.0866 ± 0.0578	5.7	99.2	194	± 2
1200	4.9306 ± 0.0041	2.1207 ± 0.0017	2.5601 ± 0.0180	0.0177 ± 0.0188	0.9360 ± 0.0607	7.0	99.4	195	± 2
1188	2.1523 ± 0.0038	0.9182 ± 0.0010	1.2001 ± 0.0161	0.0000 ± 0.0000	0.2407 ± 0.0664	3.0	99.7	197	± 2

			Sample 95-QL-7, K-feldspar (continued)				
1300	2.1621 ± 0.0068	0.9142 ± 0.0026	1.2011 ± 0.0706	0.0550 ± 0.0163	0.7543 ± 0.0580	3.0	99.0	197 ± 2
1351	0.4584 ± 0.0036	0.1891 ± 0.0007	0.2654 ± 0.0205	0.0510 ± 0.0277	0.3336 ± 0.0605	0.6	97.8	200 ± 2
1450	0.7151 ± 0.0038	0.2912 ± 0.0005	0.3753 ± 0.0186	0.0000 ± 0.0000	0.8446 ± 0.0601	1.0	96.5	199 ± 2
1649	0.2665 ± 0.0116	0.0222 ± 0.0002	0.2360 ± 0.0140	0.0000 ± 0.0000	8.5797 ± 0.1798	0.1	4.8	51 ± 18

Weight = 12.95 mg; J value = 0.004935; total gas age = 190 Ma

[1]Data corrected for system blanks, radioactive decay subsequent to irradiation, mass discrimination, and interfering K-, Ca-, and Cl-derived isotopes of argon.

SHRIMP U-Pb Dating of Zircon from the Xugou UHP Eclogite, Sulu Terrane, Eastern China

RUIXUAN ZHAO,[1] JUHN G. LIOU, RU Y. ZHANG,

Department of Geological and Environmental Sciences, Stanford University, Stanford, California 94305-2115

AND JOSEPH L. WOODEN

U.S. Geological Survey, Menlo Park, California 94025-3591

Abstract

Eclogites, together with garnet clinopyroxenites, occur as lenses within the Xugou garnet peridotite body in the southern Sulu ultrahigh-pressure (UHP) terrane. Combined cathodoluminescence (CL) imaging and SHRIMP U-Pb dating of zircon from two Xugou mafic eclogites provide added constraints on the timing of UHP metamorphism in this area. Zircons from both samples show sub-rounded to rounded shapes and patchy CL patterns without inherited igneous cores, indicating that they are metamorphic zircons. SHRIMP U-Pb analyses of these zircons yielded apparent U-Pb ages of 214–280 Ma, with a weighted mean age of 237 ± 8 Ma, which is consistent with previous reported UHP metamorphic ages from eclogite pods and country-rock gneisses. The Xugou mafic lenses may have formed by partial melting of the enclosing peridotites in the mantle before subduction (Zhang et al., 2003); then these eclogites, together with the host peridotites, were tectonically emplaced into the subduction zone and subjected to UHP metamorphism at 237 ± 8 Ma.

Introduction

DATING ULTRAHIGH-PRESSURE (UHP) metamorphism is a challenge for isotope geologists using different geochronological techniques. Most methods yield only minimum ages for the peak UHP metamorphism, except for the U-Pb zircon method. Excess Ar in phengite and hornblende in some UHP rocks could overestimate the metamorphic age when applying various Ar techniques (e.g., Mattauer et al., 1991; Kelley et al., 1994; Li et al., 1994; Hacker and Wang, 1995). In the case of Rb-Sr and Sm-Nd mineral isochrons, meaningless ages may be obtained because of isotopic disequilibrium or open-system behavior during UHP metamorphism (Okay and Sengor, 1993; Jagoutz, 1994; Li et al., 1994; Tilton et al., 1995).

On the other hand, the U-Pb zircon dating method is a very powerful tool because: (1) zircon is chemically stable and highly resistant to complete resetting; and (2) a combination of cathodoluminescence (CL) imaging and analyses on 20–30 μm diameter spots using sensitive high-resolution ion microprobe (SHRIMP) techniques generally can differentiate metamorphic from magmatic events (e.g., Gebauer, 1996; Vavra et al., 1996; Williams et al., 1996; Rubatto and Gebauer, 2000), resulting in ages that correspond to distinct geological processes. Therefore, SHRIMP dating linked with CL study has been used extensively to unravel the complex histories of UHP rocks.

The Dabie-Sulu terrane, one of the largest UHP metamorphic terranes in the world, has long been the focus of geochronological studies. Eclogites have received extensive attention because they preserve the most complete record of both prograde and retrograde events, and contain index minerals such as coesite and microdiamond (e.g., Liou et al., 1994; Coleman and Wang, 1995). Dabie-Sulu eclogites are classified into three types: mafic lenses in gneiss, ultramafic rocks, and marble based on country-rock lithologies. Most previous U-Pb zircon geochronological studies of the Dabie-Sulu terrane have focused on eclogites within country-rock gneisses, suggesting Triassic (202–228 Ma) UHP metamorphism of these rocks (Ames et al., 1993, 1996; Li et al., 1993a, 1993b; Rowley et al., 1997; Yang et al., 2003). Some of these eclogites also preserve Neoproterozoic protolith ages of 682–800 Ma (Ames et al., 1993, 1996; Li et al., 1993a, 1993b; Liu et al., 2004). However, eclogites enclosed within ultramafic rocks have not received much attention because of their limited occurrence. Only the Sm-Nd dating

---

[1]Corresponding author; email: rxzhao@stanford.edu

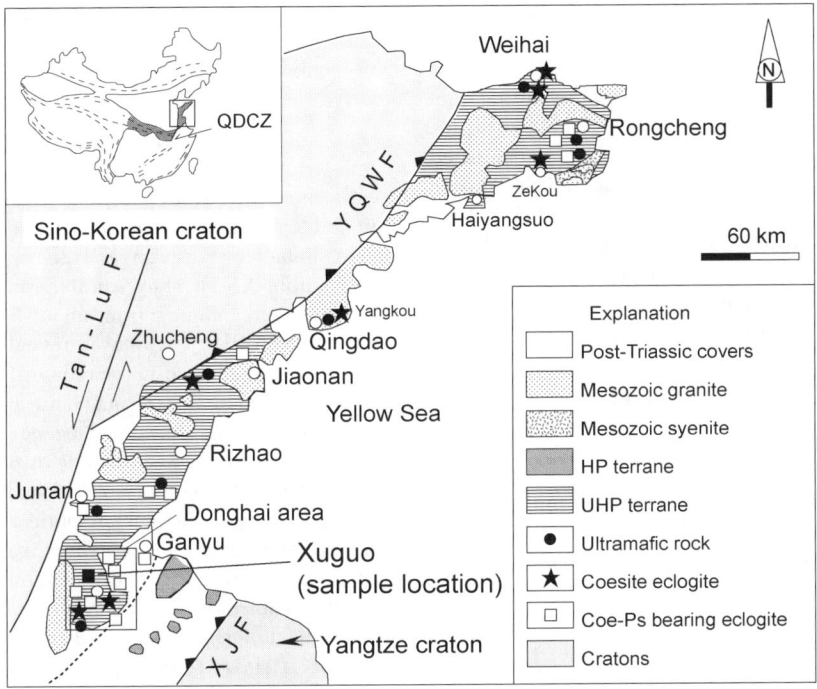

FIG. 1. Simplified geological map of the Sulu terrane, east-central China, and location of samples (modified after Zhang et al., 2003).

technique has been applied to eclogites in ultramafic rocks and associated garnet pyroxenites from the Dabie-Sulu terrane, yielding Triassic (209–224 Ma) metamorphic ages (Li et al. 1989a, 1989b, 1992a, 1993a, 1994). Study of these eclogites would be even more intriguing inasmuch as, together with ultramafic rocks, they may have come from the mantle wedge above the subducting slab, and could provide new information regarding mantle wedge processes. Therefore, this study aims to present new SHRIMP U-Pb zircon dating results on eclogites within ultramafic rocks and provide constraints on the timing of UHP metamorphism.

Geological Outline

The collision zone between the Sino-Korean and Yangtze cratons has received extensive investigation in recent years. The Sulu terrane is the eastern extension of the Qinling-Dabie belt, and is bounded by the Wulian-Qingdao-Yantai fault (WQYF) and the Jiashan-Xiangshui fault (JXF) (Fig. 1). It consists of fault-bounded UHP and HP belts, unconformably overlain by Jurassic clastic strata and Cretaceous volcanoclastic cover, and is intruded by post-collisional Mesozoic granites. Abundant coesite-bearing eclogites occur within gneiss, marble, and garnet peridotite (e.g., Liu et al., 2005). Eclogites in country rock gneiss are thought to have Proterozoic protolith ages (e.g. Jahn et al., 1996) and were subjected to Triassic (240–230 Ma) UHP metamorphism (e.g., Ames et al., 1996; Li et al., 1996; Hacker et al., 2000), and granulite- to amphibolite-facies overprinting.

The Xugou garnet peridotite body in the Donghai area, southern Sulu terrane, is well exposed in three large serpentinite quarries, and consists of garnet peridotite, garnet clinopyroxenite, and eclogite. It is located ~30 km northwest of the Chinese Continental Scientific Drilling (CCSD) site. The fault-bounded garnet peridotite body is about 1500 × 500 m and is enclosed in felsic gneiss. Harzburgite is dominant, whereas lherzolite is very minor. Small lenses of garnet clinopyroxenite and eclogite are enclosed in the peridotites. Most eclogite lenses range from 0.5 to 15 m in thickness, are enclosed concordantly in garnet peridotite, and consist mainly of garnet, omphacite, and rutile; inclusions of coesite pseudomorphs occur in the omphacite. Zhang et al. (2003) suggested that these mafic pods within the Xugou peridotite formed by high-P crystal accumulation + variable trapped melt resulting from partial melting of a mantle source. These eclogites together with garnet peridotites were later emplaced in the subduction zone and subjected to Triassic UHP metamorphism at 780–870°C, 5–6.7 GPa in the P-T "forbidden" zone (Zhang et al., 2003); multistage retrograde metamorphism took place under granulite-amphibolite and greenschist facies conditions during exhumation.

Sample Description (XJ-1E and XG-3C)

Two eclogite samples from the Xugou peridotite body were selected for SHRIMP age dating. Both samples are bimineralic with equigranular granoblastic textures, and consist mainly of garnet (0.8–1.2 mm), omphacite (1–1.5 mm), and minor rutile. Light brown apatite occurs as either inclusions in garnet and omphacite or as a matrix phase. Inclusions of quartz pseudomorphs after coesite are present within omphacite from sample XG-3C. Eclogite sample XJ-1E is crosscut by numerous prehnite veins; their occurrence is attributed to later low-T metasomatic alteration of the exhumed UHP terrane at shallow crustal depths. Each sample of fist size contains about 300–500 zircon grains. Most are anhedral to subhedral crystals, with diameters up to 250 μm.

Analytical Techniques

Zircon separates were mounted in epoxy, polished, and coated with gold before analysis. Cathodoluminescence examination was carried out using a scanning electron microscope; U-Pb analyses employed the SHRIMP-RG (sensitive high-resolution ion microprobe–reverse geometry) ion microprobe at Stanford University. Analytical spots of ~30 mm in diameter were sputtered using an ~5 nA O_2^- primary beam. Six scans through a 9 mass spectra were made for data collection. The measured $^{206}Pb/^{238}U$ ratios were calibrated against standard R-33 with an age of 419 Ma (John Aleinikoff, pers. commun., 2002), whereas the U contents were calibrated against CZ3 (550 ppm U). Data reduction and processing were conducted using the computer program SQUID and ISOPLOT provided by K. R. Ludwig at the Berkeley Geochemistry Center of the University of California (Ludwig, 2001a, 2001b). Isotope ratios and single ages are reported with 1σ errors, whereas those calculated for concordia intercept ages and weighted mean ages are at the 95% confidence level (2σ errors).

Geochronological Results

Zircon grains from sample XG-3C show subrounded to rounded shapes, ranging from 150 to 250 μm in diameter (Figs. 2A and 2B), whereas those from sample XJ-1E show similar rounded shapes with diameters ranging from about 50 to 100 μm (Figs. 2C and 2D). CL images reveal that zircons from both samples are not oscillatorily zoned, and have no obvious cores of magmatic or xenocrystic origin. Most zircons show a homogeneous CL image, in a few cases with a weak cloudy or patchy pattern (Fig. 2). Several zircon crystals also show irregular sector zoning with a nonsystematic alternation of brighter and darker domains. These features suggest that these zircons are mainly metamorphic and formed by subsolidus metamorphic crystal growth or recrystallization (see below for discussion).

The SHRIMP U-Pb data are summarized in Table 1. Twenty analyses were performed on sample XG-3C, and 10 were performed on sample XJ-1E. The zircon grains from both samples have variable but low U and Th contents (U: 4–423 ppm; Th: 0.2–67 ppm); Th/U ratios are also variable, ranging from 0.01 to 1.78 for sample XG-3C and 0.06 to 0.50 for sample XJ-1E (Table 1). There is a weak trend between Th/U ratio and $^{206}Pb/^{238}U$ age, as well as U vs. Th contents on the Th-U diagrams, although data from sample XG-3C are more scattered than that from XJ-1E (Fig. 3). Analyses of both samples fall into a single population on a Tera-Wasserburg (TW) diagram and a histogram (Fig. 4). U-Pb data of both samples yield apparent $^{206}Pb/^{238}U$ ages ranging from 214 to 280 Ma, and an error-weighted regression ($n = 30$, MSWD = 22) gave a lower concordia intercept at 236 ± 9 Ma, with an imprecise upper intercept age of 3356 ± 810 Ma on the Tera-Wasserburg diagram (Fig. 4A); the probability distribution of apparent $^{206}Pb/^{238}U$ ages indicates a single age peak with a weighted mean age of 237 ± 8 Ma ($n = 30$) (Fig. 4B). The weighted mean age (237 ± 8 Ma) and the lower intercept age (236 ± 9 Ma) on the TW diagram are identical within error, indicating the peak metamorphic age. The upper intercept age of 3356 ± 810 Ma on the TW diagram might suggest an older component in some of the studied zircons, but there is no independent evidence for this explanation.

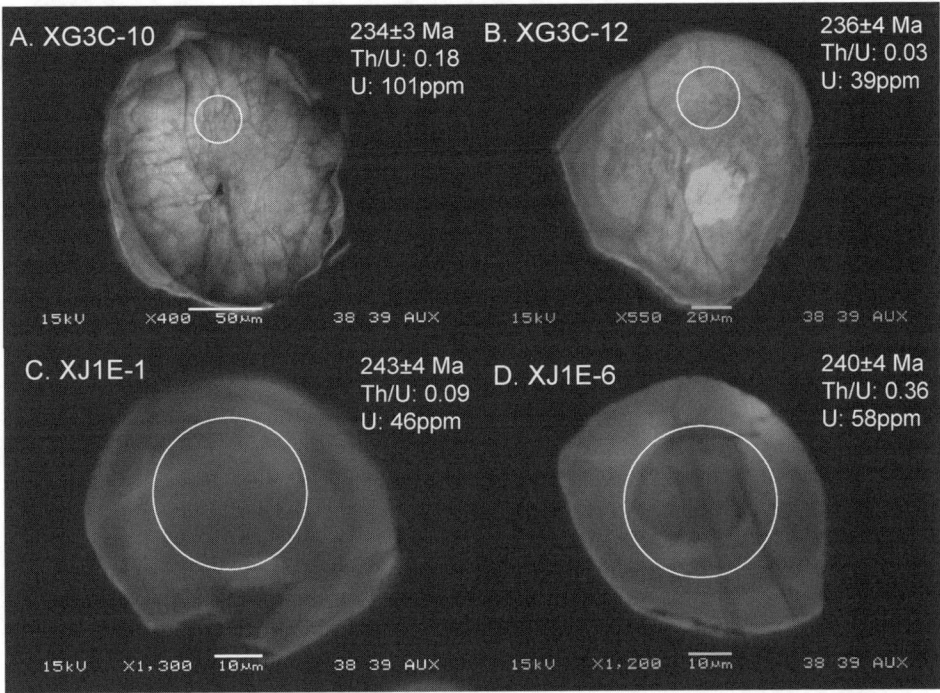

FIG. 2. Cathodoluminescence images of selected zircon grains from eclogite samples XG-3C and XJ-1E, showing the subrounded to rounded shapes and patchy patterns typical of metamorphic zircons. Spot ages, Th/U ratios, and U contents are also shown here.

Discussion

Most zircons from the Xugou eclogite show subrounded to rounded shapes, and exhibit cloudy or patchy CL patterns without magmatic or inherited cores, which is different from zircons from adjacent mafic eclogites in felsic gneisses (e.g., Liu et al., 2004). Th-U chemistry of studied zircons is characterized by low Th/U ratios ranging from 0.01 to 0.5 except for two data at 0.72 and 1.78 (Table 1). They also have reduced U and Th contents (from a few hundreds to less than 10 ppm). Morphology, CL patterns, and low Th/U ratios of the zircons suggest that they are metamorphic zircons. Vavra et al. (1999) suggested that metamorphic grade is an important factor controlling the morphology of zircon overgrowths. They considered that the morphology is the expression of the stability and the relative growth rates of crystal faces. The range of crystal habits of zircons (prismatic, stubby, equant/rounded) corresponds to a decreasing stability of planar crystal faces, therefore increasing metamorphic grade. Rounded or irregular morphology of studied zircons indicates that they formed when planar crystal faces were unstable, thus at high metamorphic grade. The investigated zircons contain eclogitic mineral inclusions such as garnet and omphacite, suggesting that the zircons grew during UHP metamorphism and did not recrystallize from pre-existing crystals.

Th/U ratios of zircons have also been suggested to be useful as indicators of magmatic or metamorphic growth (e.g., Williams and Claesson 1987; Maas et al., 1992; Williams et al., 1996; Carson et al., 2002). In general, zoned magmatic zircons have Th/U ratios larger than 0.5, whereas homogeneous or featureless metamorphic zircons give a wider range of Th/U ratios, most of them less than 0.5, with U contents up to 5000 ppm (Hokada et al., 2004). In the present study, we measured Th/U ratios of zircons in the range 0.01 to 1.78, with most less than 0.5 with U contents ranging from 4 to 423 ppm. This is consistent with other metamorphic features of our zircons. There is also a weak linear relation between Th/U ratios and $^{206}Pb/^{238}U$ age, as well as U and Th contents (Figs. 3A and 3D), indicating that these

TABLE 1. U-Th-Pb SHRIMP Zircon Data from Eclogite Samples (XG3C and XJ1E) from Xugou, Sulu Terrane[1]

Spot name	U, ppm	Th, ppm	Th/U	Common ^{206}Pb, %	Uncorrected ^{238}U / ^{206}Pb	Uncorrected ^{207}Pb / ^{206}Pb	^{206}Pb* / ^{238}U	^{206}Pb*/^{238}U age, Ma
				Eclogite XG3C				
XG3C-1	206	19	0.09	0.27	28.435 ± 1.0	0.053 ± 2.6	0.035 ± 0.00037	222 ± 2
XG3C-2	173	40	0.24	0.01	23.520 ± 1.0	0.052 ± 2.5	0.043 ± 0.00043	268 ± 3
XG3C-3	107	7	0.07	0.41	27.131 ± 1.2	0.054 ± 3.2	0.037 ± 0.00045	232 ± 3
XG3C-4	423	42	0.10	0.37	27.382 ± 0.8	0.054 ± 2.9	0.036 ± 0.00030	230 ± 2
XG3C-5	95	22	0.24	0.35	25.345 ± 1.2	0.054 ± 3.3	0.039 ± 0.00049	249 ± 3
XG3C-6	39	67	1.78	1.69	18.469 ± 1.6	0.067 ± 4.3	0.053 ± 0.00091	334 ± 6
XG3C-7	170	1	0.01	0.26	28.057 ± 1.0	0.053 ± 2.9	0.036 ± 0.00037	225 ± 2
XG3C-8	32	3	0.10	1.61	22.154 ± 1.9	0.065 ± 5.2	0.044 ± 0.00086	280 ± 5
XG3C-9	142	20	0.14	0.37	26.671 ± 1.1	0.054 ± 3.9	0.037 ± 0.00042	236 ± 3
XG3C-10	101	18	0.18	1.46	26.688 ± 1.2	0.062 ± 3.1	0.037 ± 0.00047	234 ± 3
XG3C-11	169	1	0.01	0.47	29.447 ± 1.0	0.054 ± 2.6	0.034 ± 0.00035	214 ± 2
XG3C-12	39	1	0.03	0.68	26.655 ± 1.8	0.056 ± 5.1	0.037 ± 0.00067	236 ± 4
XG3C-13	71	3	0.04	0.19	26.787 ± 1.3	0.052 ± 3.9	0.037 ± 0.00051	236 ± 3
XG3C-14	169	42	0.26	0.12	27.140 ± 1.0	0.052 ± 2.7	0.037 ± 0.00039	233 ± 2
XG3C-16	392	26	0.07	1.25	26.865 ± 0.8	0.061 ± 3.3	0.037 ± 0.00032	233 ± 2
XG3C-17	58	40	0.72	0.91	25.933 ± 1.5	0.058 ± 5.1	0.038 ± 0.00058	242 ± 4
XG3C-18	53	2	0.05	1.91	26.576 ± 1.5	0.066 ± 10.9	0.037 ± 0.00065	234 ± 4
XG3C-19	71	13	0.19	0.32	24.873 ± 1.3	0.054 ± 3.6	0.040 ± 0.00053	253 ± 3
XG3C-20	171	28	0.17	0.14	26.041 ± 1.0	0.052 ± 2.5	0.038 ± 0.00039	243 ± 2
XG3C-22	63	24	0.39	0.85	24.867 ± 1.5	0.058 ± 4.1	0.040 ± 0.00061	252 ± 4
				Eclogite XJ1E				
XJ1E-1	46	4	0.09	2.49	25.412 ± 1.6	0.071 ± 7.4	0.038 ± 0.00067	243 ± 4
XJ1E-2	66	27	0.41	2.11	25.138 ± 1.7	0.068 ± 5.8	0.039 ± 0.00070	246 ± 4
XJ1E-3	51	10	0.19	3.17	24.154 ± 1.6	0.077 ± 4.2	0.040 ± 0.00067	253 ± 4
XJ1E-4	86	42	0.50	0.63	25.701 ± 1.3	0.056 ± 3.4	0.039 ± 0.00050	245 ± 3
XJ1E-5	32	6	0.20	1.51	28.062 ± 1.9	0.063 ± 5.4	0.035 ± 0.00068	222 ± 4
XJ1E-6	58	20	0.36	2.18	25.750 ± 1.5	0.068 ± 6.4	0.038 ± 0.00061	240 ± 4
XJ1E-7	4	0.2	0.06	5.84	22.217 ± 4.8	0.098 ± 13.9	0.042 ± 0.00222	268 ± 14
XJ1E-8	44	6	0.15	1.13	26.380 ± 1.7	0.060 ± 4.7	0.037 ± 0.00065	237 ± 4
XJ1E-9	62	20	0.34	0.74	26.312 ± 1.5	0.057 ± 3.9	0.038 ± 0.00060	239 ± 4
XJ1E-10	45	12	0.28	2.7	25.044 ± 1.6	0.073 ± 10.1	0.039 ± 0.00073	246 ± 5

[1]Pb* corrected for common Pb using ^{207}Pb. All errors are 1 sigma of standard deviation.

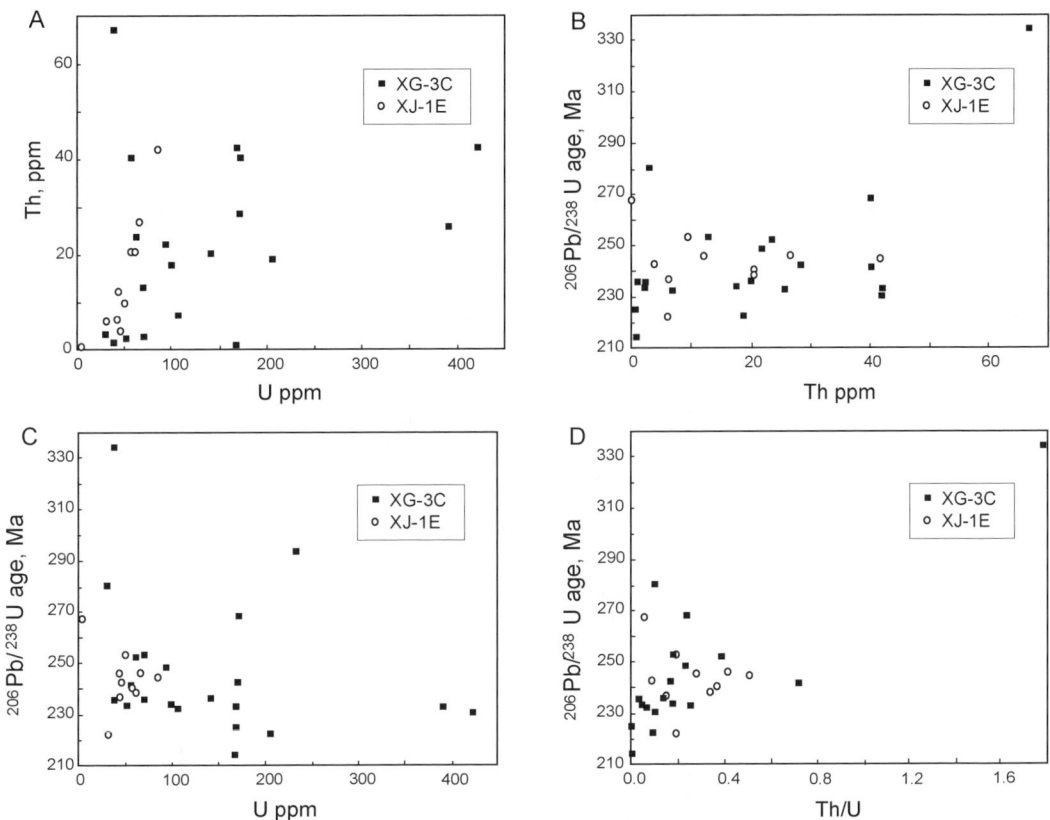

FIG. 3. Th-U diagrams for samples XG-3C and XJ-1E. Sample XG-3C shows more scattered plots than XJ-1E. A weak trend can be found on Th vs. U (A) and ^{206}Pb/^{238}U age vs. Th/U diagrams (D).

eclogites might have undergone systematic variation during UHP metamorphism. However, sample XG-3C has a more complicated history than XJ-1E, with more variable Th/U ratios (XG-3C: 0.01–1.78, XJ-1E: 0.06–0.50) and U contents (XG-3C: 32–423 ppm, XJ-1E: 4–86 ppm). On the U-Th diagrams, sample XG-3C also yields more scattered plots than XJ-1E (Fig. 3). But these are mainly secondary differences; hence we conclude that these are basically metamorphic zircons.

Zircon domains with similar features and reset U-Th-Pb systems are also described in other HP and UHP terranes (Gebauer, 1996; Gebauer et al., 1997; Rubatto et al., 1998). U-Pb ages obtained from these zircon domains in eclogite-facies rocks have been interpreted to represent UHP metamorphic ages for these terranes. Rubatto and Gebauer (2000) suggested that the formation of a metamorphic zircon domain could occur either by new growth from a metamorphic fluid or as subsolidus recrystallization of an already existing primary magmatic or detrital crystal. Most zircons from the eclogites within country rock gneiss from the Sulu UHP terrane recrystallized from pre-existing igneous zircons, inasmuch as they show igneous cores in CL images (Yang et al., 2003; Liu et al., 2002, 2004). These zircons yield UHP metamorphic (rim) ages of about 202 to 228 Ma as well as Proterozoic protolith (core) ages of 682–800 Ma, which indicate the time of original crystallization of the mafic rocks (Ames et al., 1993, 1996; Li et al., 1993a, 1993b). In contrast, most zircons from Xugou eclogites within the garnet peridotite do not show relict igneous cores. They may have only grown during UHP metamorphism close to peak metamorphic temperatures. U-Pb ages of 237 ± 8 Ma for our zircons are interpreted to be peak UHP metamorphic ages. This value is in good agreement with reported peak metamorphic ages from other eclogites and country-rock gneisses (for summary, see Table 2). Garnet peridotites from the

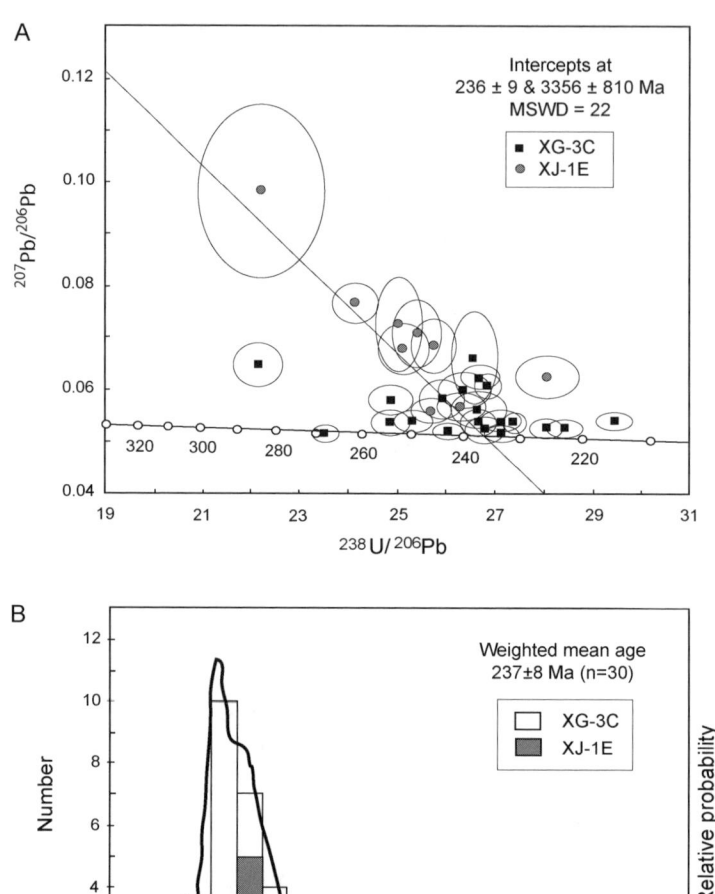

FIG. 4. A. Tera-Wasserburg diagram of SHRIMP analyses of zircons from eclogite samples XG-3C and XJ-1E. B. Histogram and relative probability age distribution curve for U/Pb zircon ages from both samples. The error ellipse of each analysis is 1-sigma confidence level, whereas intercept ages and weighted average ages are for 2-sigma uncertainties.

Zhimafang body in the same area were reported to have experienced UHP metamorphism at 226 ± 1 Ma (Zhang et al., 2005), consistent with our results. In essence, protoliths of the Xugou eclogites may have formed by partial melting of mantle peridotites before subduction (Zhang et al., 2003). These mafic lenses, together with host peridotites, were tectonically emplaced in the subduction zone, then were subjected to UHP metamorphism. The studied zircons thus recorded the timing of the UHP metamorphic event at 237 ± 8 Ma.

Conclusions

1. CL imaging reveals the morphology and internal structure of zircon grains from two Xugou UHP

TABLE 2. Summary of Previous U-Pb age Data for Different Rock Types in the Dabie-Sulu UHP Terrane

Locality	Rock type	U-Pb peak age, Ma	U-Pb protolith age, Ma	Reference
Sulu	Eclogite in garnet peridotite	237		This study
Sulu	Garnet peridotite	216–234	581	Rumble et al., 2002; Yang et al., 2003; Zhang et al., in review
Sulu	Eclogite in gneiss	202–228	682–800, 1821	Li et al., 1993a, 1993b; Ames et al., 1996; Yang et al., 2003
Dabie	Eclogite in gneiss	209–226	447–800	Ames et al., 1993, 1996; Rowley et al., 1997
Sulu	Gneiss	221–231	658–840	Rumble et al., 2002; Liu et al., 2004; Liu et al., 2005
Dabie	Gneiss	219–228	724–773	Rowley et al., 1997; Zheng et al., 2003

eclogites. Subrounded to rounded shapes and patchy CL patterns imply that these are metamorphic zircons.

2. SHRIMP U-Pb analyses of the studied zircons yield apparent ages of 214–280 Ma, with a weighted mean age of 237 ± 8 Ma, which is interpreted as the time of peak UHP metamorphism of the Xugou eclogites. This value is in good agreement with previously reported UHP metamorphic ages for the Sulu terrane.

Acknowledgments

This manuscript honors the retirement of Prof. W. G. Ernst, who has helped students and made contributions in the study of HP and UHP metamorphism. This study was supported in part by NSF EAR 0003355 to J. G. Liou, and in part by a GSA research grant and Stanford University McGee research grant to R. X. Zhao. We greatly appreciate critical reviews of the manuscript by W. G. Ernst and T. Tsujimori. We also thank J. S. Yang, T. F. Li, and S. Z. Chen of Chinese Academy of Geological Sciences for logistical support and assistance in our field work, and F. Mazdab for assistance with cathodoluminescence imaging and SHRIMP U-Pb zircon dating.

REFERENCES

Ames, L., Tilton, G. R., and Zhou, G., 1993, Timing of collision of the Sino-Korean and Yangtze cratons: U-Pb zircon dating of coesite-bearing eclogites: Geology, v. 21, p. 239–242.

Ames, L., Zhou, G. Z., and Xiong, B., 1996, Geochronology and geochemistry of ultrahigh-pressure metamorphism with implications for collision of the Sino-Korean and Yangtze cratons, central China: Tectonics, v. 15, p. 472–489.

Carson, C. J., Ague, J. J. and Coath, C. D., 2002, U-Pb geochronology from Tonagh Island, east Antarctica: Implication for the timing of ultra-high temperature metamorphism of the Napier Complex: Precambrian Research, v. 116, p. 237–263.

Coleman, R. G., and Wang, X., eds., 1995, Ultrahigh-pressure metamorphism: New York, NY, Cambridge University Press, 528 p.

Gebauer, D., 1996, A P-T-t path for a (ultra?-) high pressure ultramafic/mafic rock-associations and their felsic country-rock based on SHRIMP-dating of magmatic and metamorphic zircon domains. Example: Alpe Arami (Central Swiss Alps), in Basu, A., and Hart, S., eds., Earth processes: Reading the isotopic code: American Geophysical Union, Geophysical Monographs, p. 307–329.

Gebauer, D., Schertl, H.-P., Brix, M., and Schreyer, W., 1997, 35 Ma old ultrahigh-pressure metamorphism and evidence for very rapid exhumation in the Dora Maria Massif, Western Alps: Lithos, v. 41, p. 5–24.

Hacker, B. R., Ratschbacher, L., Webb, L., et al., 2000, Exhumation of ultrahigh-pressure continental crust in east central China: Journal of Geophysical Research, v. 105, p. 13,339–13,364.

Hacker, B. R. and Wang, Q., 1995, Ar/Ar geochronology of ultrahigh-pressure metamorphism in central China: Tectonics, v. 14, p. 994–1006.

Hokada, T., Misawa, K., Yokoyama, K., Shiraishi, K. and Yamaguchi, A., 2004, SHRIMP and electron microprobe chronology of UHT metamorphism in the Napier Complex, East Antarctica: Implications for zircon growth at >1,000°C: Contributions to Mineralogy and Petrology, v. 147, p. 1–20.

Jagoutz, E., 1994, Isotopic systematics of metamorphic rocks [abs.] in Abstracts of the Eighth International Conference on Geochronology, Cosmochronology, and Isotope Geology: ICOG 8, Berkeley, CA, 5–11 June 1994, US Geological Survey, Circular 1107, p. 156.

Jahn, B.-M., Cornichet, J., Cong, B., and Yui, T.-F., 1996, Ultrahigh-Nd eclogites from an ultrahigh-pressure metamorphic terrane of China: Chemical Geology, v. 127, p. 61–79.

Kelley, S. P., Arnaud, N. O., and Okay, A. I., 1994, Anomalously old Ar-Ar ages in high pressure metamorphic terraines: Mineralogical Magazine, v. 58A, p. 468–469.

Li, S., Chen, Y., Ge, N., Liou, D., Zhang, Z., Zhang, Q., and Zhou, D., 1993a, Zircon U-Pb ages of eclogite and gneiss from Qingdao area: Evidence for Jinning magma episode in Jiaonan Group: Chinese Science Bulletin, v. 38, p. 1773–1777 (in Chinese).

Li, S., Ge, N., Liou, D., Zhang, Z., Ye, X., Zheng, S., and Peng, C., 1989a, The Sm-Nd isotopic age of C-type eclogite from the Dabie Group in the Northern Dabie Mountains and its tectonic implication: Chinese Science Bulletin, v. 34, p. 1623–1628.

Li, S., Hart, S. R., Zheng, S., Liou, D., Zhang, G., and Guo, A., 1989b, Timing of collision between the North and South China blocks—Sm-Nd isotopic age evidence: Science in China (Ser. B), v. 32, p. 1391–1400.

Li, S., Liu, D., Chen, Y., and Ge, N., 1992, The Sm-Nd isotopic age of coesite-bearing eclogite from the southern Dabie Mountains: Chinese Science Bulletin, v. 37, p. 1638–1641.

Li, S., Wang, S., Chen, Y., Zhou, H., Zhang, Z., Liu, D., and Qiu, J., 1994, Excess argon in phengite of eclogite: Evidence from comparing dating of eclogite by Sm-Nd, Rb-Sr and $^{40}Ar/^{39}Ar$ isotope methods: Chemical Geology (Isotope Geoscience Section), v. 112, p. 343–350.

Li, S., Xiao, Y., Ge, N., and Chen, Y., 1996, Chronology of ultrahigh-pressure metamorphism in the Dabie Mountains and Su-Lu terrane. I. Sm-Nd isotope system: Science in China (Series D), v. 39, p. 597–609.

Li, S., Xiao, Y., Liou, D., Chen, Y., Ge, N., Zhang, Z., Sun, S.-S., Cong, B., Zhang, R., Hart, S. R. and Wang, S., 1993b, Collision of the North China and Yangtze Block and formation of coesite-bearing eclogites: Timing and processes: Chemical Geology, v. 109, p. 89–111.

Liou, J. G., Zhang, R. Y., and Ernst, W. G., 1994, An introduction to ultrahigh-P metamorphism: The Island Arc, v. 3, p. 1–24.

Liu, F. L., Xu, Z. Q., Liou, J. G., Katayama, I., Masago, H., Maruyama, S., and Yang, J. S., 2002, Ultrahigh-pressure mineral inclusions in zircons from gneissic core samples of the Chinese Continental Scientific Drilling Site in eastern China: European Journal of Mineralogy, v. 14, p. 499–512.

Liu, F., Liou, J. G., and Xu, Z., 2005, U-Pb SHRIMP ages recorded in the coesite-bearing zircon domains of paragneisses in the southwestern Sulu terrane, eastern China: New interpretation: American Mineralogist (in press).

Liu, F., Xu, Z., Liou, J. G., and Song, B., 2004, SHRIMP U-Pb ages of ultrahigh-pressure and retrograde metamorphism of gneisses, south-western Sulu terrane, eastern China: Journal of Metamorphic Geology, v. 22, p. 315–326.

Ludwig, K. R., 2001a, Squid 1.00: A user's manual: Berkeley, CA, Berkeley Geochronology Center, Special Publication, No. 2, 17 p.

Ludwig, K. R., 2001b, Isoplot/Ex rev. 2.49: A geochronological tool kit for Microsoft Excel: Berkeley, CA, Berkeley Geochronology Center, Special Publication, No. 1a, 55 p.

Maas, R., Kinny, P. D., Williams, I. S., Froude, D. O., and Compston, W., 1992, The earth's oldest known crust: A geochronological and geochemical study of 3900–4200 Ma old detrital zircons from Mt. Narryer and Jack Hills, Western Australia: Geochimica et Cosmochimica Acta, v. 56, p. 1281–1300.

Mattauer, M., Matte, P., Maluski, H., Xu, Z., Zhang, Q. W., and Wang, Y. M., 1991, Paleozoic and Triassic plate boundary between North and South China; new structural and radiometric data on the Dabie-Shan (eastern China): Comptes Rendus de L'Academie des Sciences, ser. 2, v. 312, p. 1227–1233.

Okay, A. I., and Sengor, A. M. C., 1993, Tectonics of an ultrahigh-pressure metamorphic terrane: The Dabie Shan-Tongbai Shan orogen, China: Tectonics, v. 12, p. 1320–1334.

Rowley, D. B., Xue, F., Tucker, R. D., Peng, Z. X., Baker, J., and Davis, A., 1997, Ages of ultrahigh pressure metamorphism and protolith orthogneisses from the eastern Dabie Shan: U/Pb zircon geochronology: Earth and Planetary Science Letters, v. 151, p. 191–203.

Rubatto, D., and Gebauer, D., 2000, Use of cathodoluminescence for U-Pb zircon dating by ion microprobe: Some examples from the Western Alps, in Pagel, M., Barbin, V., Blanc, P., and Ohnenstetter, D., eds., Cathodoluminescence in Geosciences: Berlin, Germany, Springer, p. 373–400.

Rubatto, D., Gebauer, D., and Fanning, M., 1998, Jurassic formation and Eocene subduction of Zermatt-Saas-Fee ophiolites: Implications for the geodynamic evolution of the Central and Western Alps: Contributions to Mineralogy and Petrology, v. 132, p. 269–287.

Rumble, D., Giorgis, D., Ireland, T., Zhang, Z. M., Xu, H. F., Yui, T. F., Yang, J. S., Xu, Z. Q., and Liou, J. G., 2002, $\delta 18O$ zircons, U-Pb dating, and the age of the Qinglingshan oxygen and hydrogen isotope anomaly near Donghai in Jiangsu province, China: Geochimica et Cosmochimica Acta, v. 66, p. 2299–2306.

Tilton, G. R., Ames, L., Schreyer, W., and Schertl, H.-P., 1995, Age determinations on rocks of an undeformed granite contact within the coesite-bearing unit of the Dora Maira Massif: Bochumer Geologische und Geotechnische Arbbeiten, v. 44, p, 245–247.

Vavra, G., Gebauer, D. and Schmidt, R., 1996, Multiple zircon growth and recrystallization during polyphase Late Carboniferous to Triassic metamorphism in granulites of the Ivrea Zone (Southern Alps): An ion microprobe (SHRIMP) study: Contributions to Mineralogy and Petrology, v. 122, p. 337–358.

Vavra, G., Schmid, R. and Gebauer, D., 1999, Internal morphology, habit, and U-Th-Pb microanalysis of amphibolite- to granulite-facies zircons: Geochronology of the Ivrea Zone (Southern Alps): Contributions to Mineralogy and Petrology, v. 134, p. 380–404.

Williams, I. S., Buick, S., and Cartwright, I., 1996, An extended episode of early Mesoproterozoic metamorphic fluid flow in the Reynolds Range, central Australia: Journal of Metamorphic Geology, v. 14, p. 29–47.

Williams, I. S., and Claesson, S., 1987, Isotopic evidence for the Precambrian provenance and Caledonian metamorphism of high-grade paragneisses from the Seve nappes, Scandanavia Caledonides: II. Ion microprobe zircon U-Th-Pb: Contributions to Mineralogy and Petrology, v. 97, p. 205–217.

Yang, J. S., Wooden, J. L., Wu, C. L., Liu, F. L., Xu, Z. Q., Shi, R. D., Katayama, I., Liou, J. G., and Maruyama, S., 2003, SHRIMP U-Pb dating of coesite-bearing zircon from the ultrahigh-pressure metamorphic rocks, Sulu terrane, east China: Journal of Metamorphic Geology, v. 21, p. 551–560.

Zhang, R. Y., Liou, J. G., Yang, J. S., and Ye, K., 2003, Ultrahigh-pressure metamorphism in the forbidden zone: The Xugou garnet peridotite, Sulu terrane, eastern China: Journal of Metamorphic Geology, v. 21, p. 539–550.

Zhang, R. Y., Yang, J. S., Wooden, J. L., Liou, J. G., and Li, T. F., 2005, Zircon SHRIMP U-Pb dating of garnet peridotite from the Chinese CCSD-PP1 hole: Implications for mantle metasomatism and subduction-zone UHP metamorphism: Earth and Planetary Science Letters (submitted).

Zheng, Y. -F., Gong, B., Zhao, Z.-F., Fu, B. and Li, Y.-L., 2003, Two types of gneisses associated with eclogite at Shuanghe in the Dabie terrane: carbon isotope, zircon U-Pb dating and oxygen isotope: Lithos, v. 70, p. 321–343.

Tectonic Division of the Sulu Ultrahigh-Pressure Region and the Nature of Its Boundary with the North China Block

MINGGUO ZHAI[1] AND WENJUN LIU

Key Laboratory of Mineral Resources, Institute of Geology and Geophysics, Chinese Academy of Sciences Beijing, 100029, China

Abstract

Detailed geological, geochemical, and geochronological data suggest that the boundary between the North China Block and the Sulu UHP region in eastern Shandong Peninsula is not a single fault as previously considered. Instead, it is a complicated magmatic-metamorphic zone about 40–60 km wide bounded by the Muping (MP) and Mishan (MS) faults, and herein termed the Kuyushan boundary zone. The complicated zone consists of a granitoid complex containing abundant lenses and pods of metamorphic rock of different protoliths and sizes that are tectonically juxtaposed. The Sulu UHP region is subdivided into two tectonic units: the Haiyangsuo eclogitized granulite unit, and the Rongcheng-Weihai UHP unit bounded by a large-scale ductile shear zone. The Haiyangsuo unit is considered to be a tectonic slab of the lower crust of the Yangtze Block, in which granulites underwent an eclogite-facies overprint. The Rongcheng-Weihai unit represents an ancient subduction relict of the Yangtze crust, and consists of two subunits. The Rongcheng subunit is a typical UHP unit, with a P-T path recording rapid uplifting from mantle (coesite-eclogite facies) to upper lower-middle crustal (amphibolite-granulite facies) depths. The Weihai subunit is a tectonic slab that was detached from the UHP zone during exhumation, and was tectonically emplaced into the lower crust of the North China Block above the subduction zone. UHP rocks of this unit have been strongly modified by subsequent granulite-facies metamorphism. The North China Block, the Kunyushan boundary zone, and the Haiyangsuo and Rongcheng-Weihai units are imbricated slabs. The Rongcheng subunit may have been subducted northwestward under the North China Block, but a detached UHP slab of the Weihai subunit was overthrust onto the North China Block. This inference is supported by a recent geophysical study.

Introduction

THE DABIE-SULU ULTRAHIGH-PRESSURE (UHP) and high-pressure (HP) belts between the North China and Yangtze blocks contain metamorphic rocks typically formed in a continent-continent collisional zone. It is generally considered that the Sulu UHP region is equivalent to Dabieshan and was translated northward for ~500–600 km by the left-lateral Tan-Lu fault. However, the boundary between the North China and the Yangtze blocks in East Shandong is not well constrained, and whether the UHP belt extends eastward into the Korea Peninsula is unknown. Answers to these questions are important for our understanding of the UHP collisional process and the tectonic framework of East Asia.

The northern Shandong Peninsula was traditionally divided into three parts: the northern Jiaodong (eastern Jiaodong) area, the southern Jiaodong (northern Jiangsu) area, and the Mesozoic Jiaoliao basin located between them. All metamorphic rocks of the Jiaodong and Jiaonan areas are composed of orthogneisses with lenses and pods of granulite-amphibolite facies supracrustal rocks. These rocks were previously regarded as part of the basement of the North China block (Zhang, 1984). The discovery of UHP metamorphic rocks leads most geologists to divide the eastern Shandong Peninsula into two parts: the Precambrian terrane and Sulu UHP region (Cao, 1990; Bai, 1993; Yin and Nie, 1993).

In addition to the Tan-Lu fault, a series of NNE-NE and ENE trending faults in East Shandong appears to have controlled the distribution of metamorphic rocks. However, geophysical data reveal that distribution of metamorphic rocks at depths is controlled by ENE-EW oriented structures. Therefore some geologists suggest that the EW-ENE–trending Wulian-Rongchen fault is a geotectonic boundary between the North China and Yangtze blocks (Tianjing Institute of Metallurgic Geology, 1996). However, the discovery of coesite and omphaite as inclusions in garnet of granulite-over-

[1]Corresponding author; email: mgzhai@mail.igcas.ac.cn.

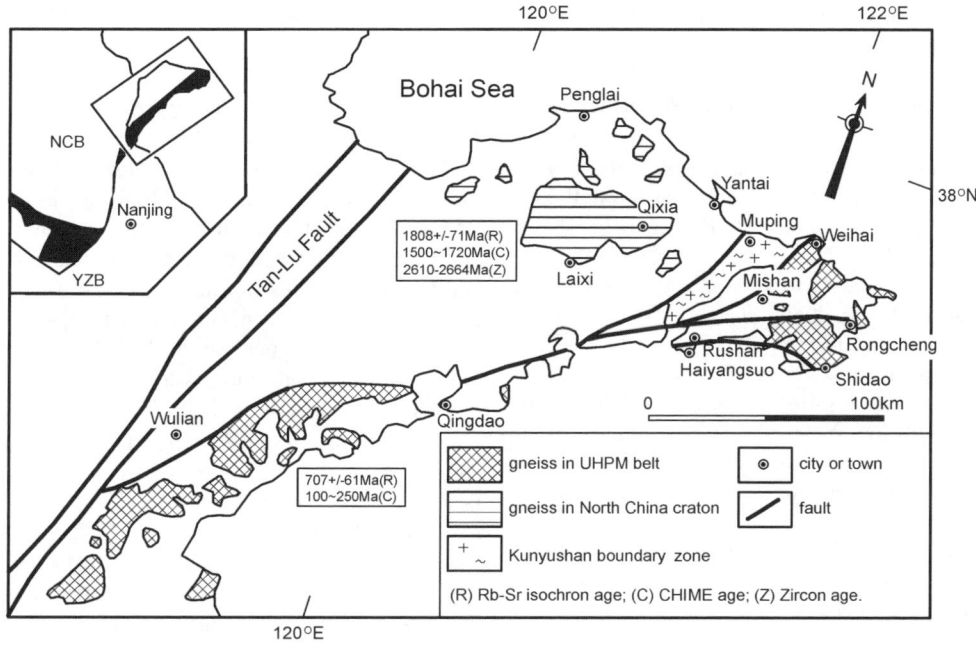

FIG. 1. Geological sketch map of Shandong Peninsula (enclosed in rectangle) (after Enami et al., 1993a; Zhai and Liu, 1998). Some isotopic ages are also shown. Abbreviations: NCB = North China Block; YZB = Yangtze Block.

printed eclogites from Weihai (Wang et al., 1993; Zhang et al., 1995) indicates that the granulites are retrogreaded from UHP eclogites, and the boundary should be located farther northwest of the Rongcheng fault. In this decade, some geologists (Enami et al., 1993a; Ishizaka et al., 1994; Zhai and Liu, 1998) have summarized characteristics of granitic gneisses in East Shandong, reporting obvious differences between the eastern and western areas bounded by the Muping (Yantai) fault. These data suggest that the Muping (Yantai) fault should be the boundary between the North China Block and the Sulu region (Fig. 1). Faure et al. (2001), from structural analyses, reported that all units in East Shandong—including the UHP unit, problematic Proterozoic (Sinian)-Paleozoic unit, and other metamorphic units—show a stretching lineation trending N130°–160°. Hacker et al. (1995, 2000) considered that the UHP metamorphism in East Shandong and the Dabieshan was overprinted by migmatization and ductile deformation, and is a top-to-the northwest shearing coeval with amphibolite facies overprint. Therefore they concluded that the boundary between the North China Block and the Sulu UHP region must be placed to the north of Shandong Peninsula.

Our studies (Zhai et al., 1998, 2001), however, suggest that the western boundary of the Sulu UHP region is located between the Muping and the Mishan faults, and consists of a complicated fault zone, termed as the Kunyushan boundary zone. The three high-grade metamorphic units containing granulites or eclogites in Sulu are distinctly different and are bounded by strike-slip faults. This paper deals with the tectonic division of East Shandong, especially the boundary between the North China Block and the Sulu UHP region.

Geological Setting

Regional geology

The study region is cut by the NE- trending Tan-Lu fault, to the west of which is the Neoarchaean Taishan gneissic complex (Fig.1, after Enami et al., 1993a; Zhai and Liu, 1998). The early Precambrian rocks metamorphosed to amphibolite-granulite facies occur to the north of the Muping (Yantai)–Wulian fault, and the metamorphic rocks on the south are dominated by quartz-feldspar orthogneisses with lenses of eclogites, ultramafic rocks, marbles, and other types of UHP metamorphosed rocks. The unmetamorphosed Cretaceous Qingshan Formation unconformably overlies units on both sides of the Muping (Yantai) fault. Voluminous granitoids of Jurrasic–Cretaceous ages intrude these metamorphic rocks.

Metamorphic conditions and peak metamorphic ages of UHP rocks in the region have been well documented (Hirajima et al., 1990; Enami et al., 1993b; Yang et al., 1993; Zhang et al., 1993; Liou et al., 1995; Ye et al., 1996; Kato et al., 1997; Nakamura and Banno, 1997; Zhang et al., 1997). The estimated metamorphic temperatures for the most of coesite-bearing eclogites are 700–750°C; the presence of coesite implies metamorphic depths in excess of 90 km. Ye et al. (2000a) reported that garnets in garnet-peridotite near Qingdao contain exsolution rods of clinopyroxene, rutile, and apatite along crystallographically controlled planes, indicating a metamorphic depth greater than 200 km. Typical reaction textures are symplectitic coronas surrounding garnet and breakdown of clinopyroxene. Estimated temperature and pressure of the retrograde metamorphism are 500–550°C at <10 kbar. However, some UHP eclogites, for example from Weihai, show higher retrograde metamorphic temperatures of granulite facies or a transitional facies of granulite-eclogite. All eclogites and other UHP rocks show decompression P-T paths. Hirajima et al. (1993) first reported eclogitized metagranitoid, indicating country gneisses also underwent UHP metamorphism. Recent study of granitic gneiss in the main drill hole of the Chinese Continental Scientific Drilling Project to a depth of 5000 m reveals that zircons in various metamorphic rocks commonly contain coesite (Liu et al., 2004). A variety of methods have been used to determine the ages of formation and metamorphism. U-Pb dating of zircons and whole-rock Rb-Sr dating suggest protolith ages of 800–700 Ma for the eclogites and orthogneisses (Ames et al., 1996; Ishizaka et al., 1994; Tang et al., 2003; Liu et al., 2003). The zircon U-Pb dating and Sm-Nd dating of eclogites gave ages of 240–218 Ma for eclogite-facies metamorphism (Li et al., 1994; Ames et al., 1996; Liu et al., 2003).

Main faults

A series of lithospheric faults, roughly trending NE-ENE, have been recognized east of the Tan-Lu fault (Cao, 1990; Jia et al., 1993; Tianjing Institute of Metallurgic Geology, 1996). The main ones from northwest to southeast are the Muping (Yantai), Mishan, Rongcheng, and Shidao (Qingdao) faults (Fig. 2). They converge southwestward near Rushan and diverge to the east, demonstrating a right-lateral rotation. All these faults show several stages of activity (Zhang and Gu, 1996).

Rongcheng fault. The Rongcheng fault is located south of Wulian and extends to Rongcheng with an ENE-EW trend (Fig. 2). It dips 60–70° to WNW-N and has a multi-stage history. An early-stage thrust and late-stage strike-slip movement can be distinguished. Geophysical data show that it penetrates the crust into the upper mantle. The Moho discontinuity is vertically offset by 3 km along the north side of the fault (Cao, 1990; Lai, 1997).

Muping (Yantai) fault zone. The Muping fault zone is composed of a series of parallel branch faults. It starts from the north of Muping, and extends southwestward to Wulian (Fig. 2). The fault zone is 20–30 km wide, and dips NW at 40–50°. Jia et al. (1993) suggested that this fault was sinistral in the Mesozoic and dextral in the Tertiary. Along the southeastern side of the fault zone, various metamorphic lenses or slabs of different sizes occur. The regional gneissosities of orthogneisses and other tectonic elements on either side of the fault are different. They strike NW-EW on the west side and NE-NS on the east side of the Muping fault. The regional granitic orthogneisses on both sides are also geochemically and geochronologically different (Enami et al., 1993a; Ishizaka et al., 1994; Li et al, 2000b), similar to the North China geochemical province and Dabieshan/Yangtze geochemical province, respectively (Wang et al., 1995; Zhang, 1995; Li et al., 1997; Zhai and Liu, 1997).

Mishan fault. The Mishan fault starts from Xiaoshidao, and extends southwestward via Mishan to the south of Rushan (Wang et al., 1995; Zhang and Wang, 1996). The fault roughly controls the southern boundary of the northern Kunyushan granitoid complex. The fault shows top to the southwest movement (Jia et al., 1993). An early stage of middle- to deep-crustal shearing and a late stage of shallow shearing (<20 km) can be distinguished (Tianjing Institute of Metallurgic Geology, 1996; Wallis et al., 1999). A series of lenses of graphite-bearing granulites and other granulite facies rocks occur along the western side of the fault, and are considered part of the North China Block (Zhai and Liu, 1997). Some geologists have insisted that the Mishan fault is a boundary fault between the North China and South China blocks, although they

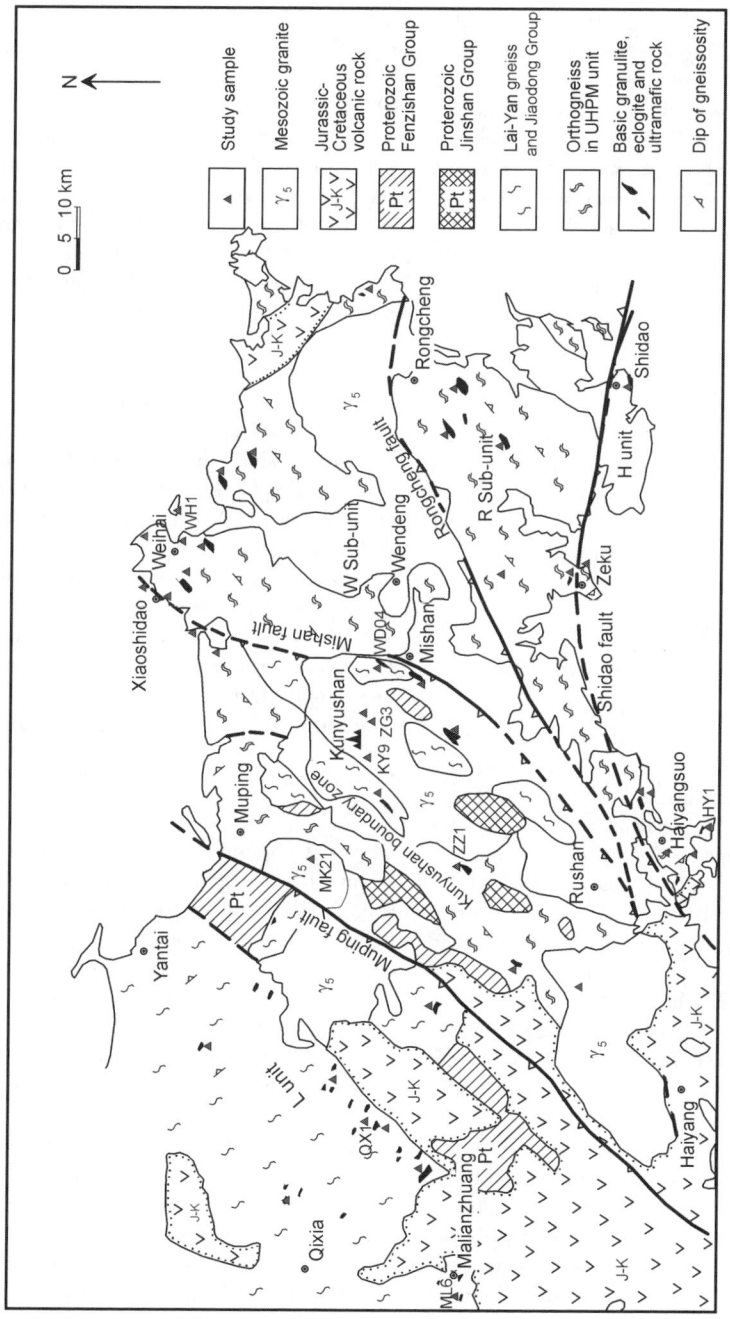

FIG. 2. Main lithotectonic units and major faults in the northern part of Shandong Peninsula. Abbreviations: H = Haiyangsuo unit; W = Weihai subunit; R = Rongcheng subunit; L = Lai-Yan unit.

proposed different tectonic models (Wang, 1994; Wang et al., 1995; Zhang and Gu, 1996; Lai, 1997).

Shidao fault. The Shidao fault extends E-W from Qingdao via Haiyangsuo and northward to Shidao (Cao, 1990; Kurahashi et al., 2001) and into the Yellow Sea (Fig. 2). Wang (1994) and Zhang and Wang (1996) claimed that the rocks to the south of the fault constitute a granulite-amphibolite facies zone, different from the UHP metamorphic rocks to the north. Ye et al. (1999) and Lai (1997) reported petrological evidence that metamorphic rocks of granulite and amphibolite facies on the south side underwent later eclogitization. Guo et al. (2001) suggested that these eclogitized rocks represent basement rocks of the Yangtze Block.

Petrotectonic Units

East Shandong can be divided into four petrotectonic units (Haiyangsuo, Weihai-Rongcheng, Kunyushan, and Lai-Yan) from southeast to northwest (Fig. 2), which are bounded by the above-mentioned four important faults, and each unit has its particular metamorphic history.

Haiyangsuo metamorphic unit (eclogitized lower crust unit of the Yangtze Block)

The Haiyangsuo metamorphic unit is bounded by the Shidao fault in the north. Its southern boundary lises under the Yellow Sea (Ji et al., 1988). The main rocks are at granulite facies and include two pyroxene mafic granulites, garnet gabbro, kyanite-bearing granulite, garnet amphibolite, serpentinized peridotite, pyroxene peridotite, and quartzite. They occur as lenses of different sizes in granitic gneisses. All the rocks, including metamorphic lenses and their hosts, underwent strong ductile deformation and mylonitization. The lenses of granulite facies are commonly overprinted by eclogite-facies metamorphism (Zhai and Cong, 1996a; Ye et al., 1999). Typical mineral assemblages in mafic granulites (sample HY1 in Fig. 2) is hypersthene (Hy) + clinopyroxene (Cpx) + garnet$_1$ (Grt1) + plagioclase (plg) + quartz (Qtz). Symplectite of Grt is composed of fine-grained mineral aggregate of garnet$_2$ (Grt2) + omphacite (Omp) and represents late eclogitization (Fig. 3A). Hy and Cpx are surrounded by later Grt2 and Omp (Fig. 3B). The Grt1 is also surrounded by fine-grained later Grt2. Grt2 contains higher pyrope and grossular components than Grt1. The estimated metamorphic temperature and pressure of corona minerals is 550–600°C and 13–15 kbar (Lai, 1997), which indicates significantly lower temperature and pressure conditions of eclogitization than metamorphic condition of coesite-bearing eclogite in the Rongcheng UHP metamorphic unit (Fig. 4; Ye, 1993). The eclogitized granulites record Sm-Nd and U-Pb zircon ages of 1.3–1.2 Ga, 0.8 Ga, and 0.2–0.25 Ga (Li, 1994; Lai, 1997). The ages of 1.3–1.2 Ga and 0.8 Ga correspond to the Sibao event and Jingning event in the South China Block, respectively (Li et al., 2001). The trace element geochemical characteristics and Nd-Sr isotopic composition of mafic granulites are also similar to those of metamorphic rocks in the Yangtze Block (Zhang, 1995; Guo et al., 2001). A proposed tectonic model that caused eclogitization of granulite is deep- crustal detachment. The continent-continent collision between the North China and Yangtze blocks and the accompanying UHP metamorphism probably began in the Early Triassic. Continental shortening at the northern margin of the Yangtze Block was achieved by northward crustal thrusting in the Yangtze Block (Okay and Sengör, 1992). Some lower crust was detached along a deep shear zone. Granulites of the lower crust were eclogitized (Guo et al., 2001). As a result, plagioclase and pyroxene were transformed to omphacite and garnet. This metamorphic process is similar to the granulite-eclogite transformation in Proterozoic orogens in central Australia (Ellis and Maboko, 1992).

Rongcheng-Weihai metamorphic unit (UHP metamorphic zone)

The Rongcheng-Weihai metamorphic unit is bounded by the Mishan fault and the Shidao fault (Fig. 2). UHP lenses and slabs ranging from meters to a kilometer in size are widely distributed, and the directions of their long axes are consistent with those of gneissosities of regional orthogneisses. The main types of UHP rocks are eclogites, ultramafic rocks, and metamorphosed sedimentary rocks. The UHP ultramafic rocks include garnet-bearing peridotite and pyroxenite. Protoliths were layered ultramafic-mafic bodies or a minor amount of upper mantle pods (Ye, 1993; Zhang et al., 1993). Eclogites are enclosed in three kinds of country granitic gneisses: ultramafic rock, marble, and gneiss. Most UHP rocks in the northwest area near the Rongcheng fault have been transformed to mafic granulites; however, those in the southeast partly underwent retrograde metamorphism, with some transformed to amphibolites. Regional granitic gneisses in the northwest area also underwent

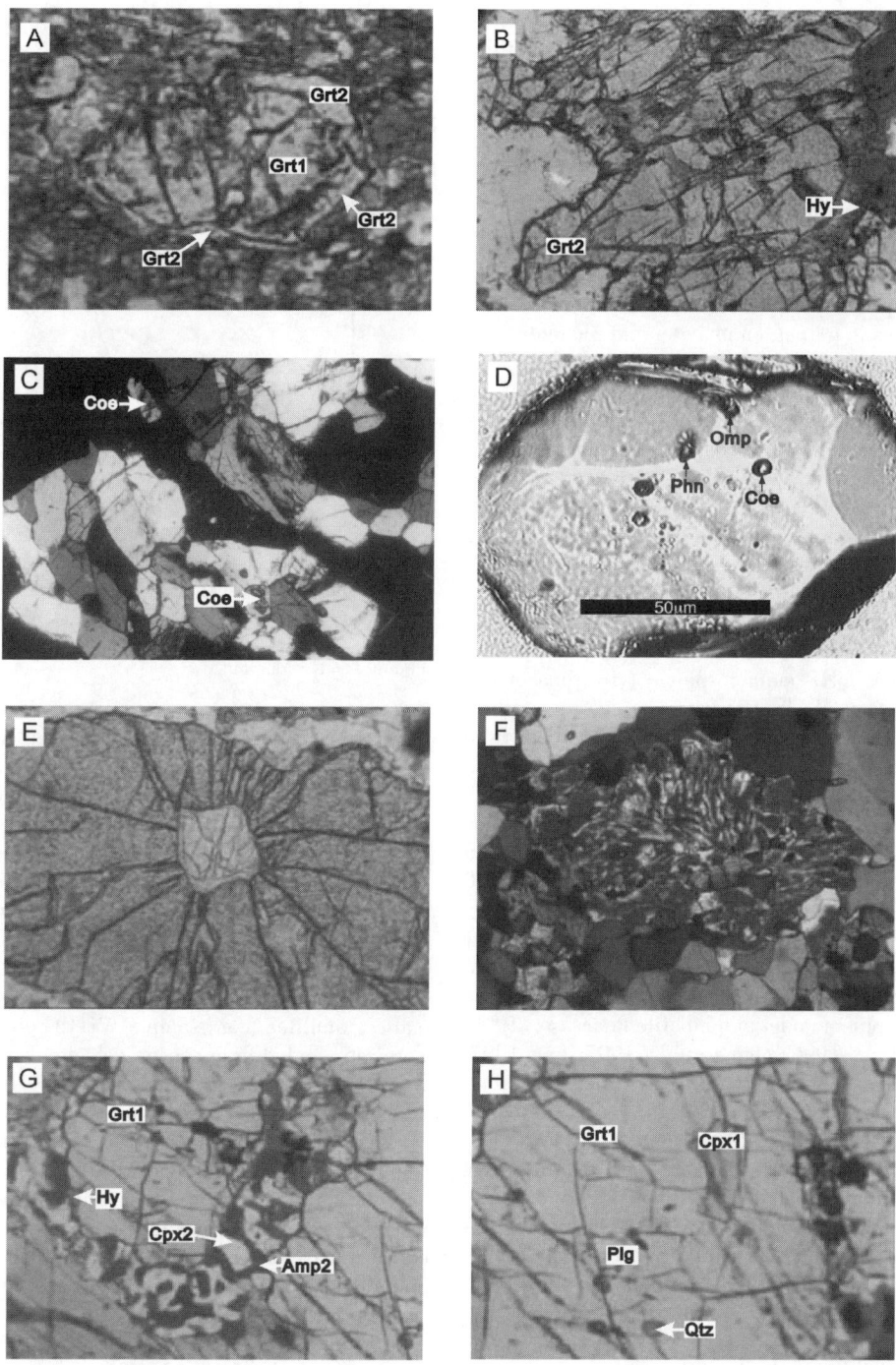

FIG. 3. Photomicrographs of mafic metamorphic rocks. A. Grt2 around Grt1 in Haiyangsuo unit. B. Grt2 around Hy in Haiyangsuo unit. A and B are from the granulite-eclogite transition zone. C. Coesite between matrix minerals in the Rongcheng subunit. D. Coesite (coe) in zircon in the Rongcheng subunit, showing evidence of UHPM regional metamorphism. E. Coesite in Grt in the Weihai subunit. F. Breakdown of Cpx in the Weihai subunit. E and F show evidence of the early UHPM stage in granulites (after Wang et al., 1993. G. Symplectitic Grt in the Lai-Yan unit. H. Cpx in Grt in the Lai-Yan unit.

migmatization. Therefore, the Rongcheng-Weihai unit can be further divided into two subunits: the Rongcheng and Weihai subunits, separated by the Rongcheng fault (Fig. 2).

Rongcheng metamorphic subunit (UHP metamorphic slab)

The Rongcheng metamorphic subunit is one of the representative UHP slabs in the Sulu region (Cong et al., 1996; Zhai and Cong., 1996a). Coesite and its pseudomorph are commonly preserved as inclusions in garnet, omphacite, and phengite, and rarely as a metamorphic matrix mineral along with garnet and omphacite in eclogite (Fig. 3C; Ye et al., 1996). Nearly without exception, mafic eclogites show major- and trace-element geochemical characteristics similar to island-arc basalts or cumulate gabbros (Zhai and Cong, 1996b). The UHP metasedimentary rocks include garnet-bearing mica schist, kyanite-bearing paragneiss, marble, and quartzite (Cong et al., 1996; Kato et al., 1997; Nakamura and Hirajima, 2000). Accumulating evidence indicates that country granitic gneisses also underwent UHP metamorphism (Hirajima et al., 1993; Liu et al., 2004): for example, coesite and omphacite are present as inclusions in zircons from orthogneiss (Fig. 3D; Ye et al., 2000b; Liu et al, 2001), indicating that the regional country gneisses, at least a part of them, were UHP rocks as well. Sm-Nd isotopic ages of UHP rocks fall within two groups: 240–218 Ma (Fig. 5A) and 800–700 Ma. The first group is proposed as the time of eclogite facies metamorphism. The second is generally considered to represent the residual ages of protolith rocks. The UHP rocks underwent retrograde metamorphism under amphibolite facies (Ye, 1993; Cong et al., 1996; Zhang et al., 1997; Lai, 1997) and granulite facies (Nakamura and Hirajima, 2000), with estimated temperature and pressure of 680–760°C and 8–12 kbar. Cong et al. (1994) and Wang and Cong (1999) suggested a multistage exhumation model. Firstly, the UHP rocks were uplifted rapidly from the mantle (>120 km) to mid-crustal levels at ~220–200 Ma, then to the upper crust at ~180–170 Ma under greenschist facies (Li et al., 2000a).

Weihai metamorphic subunit (granulitized UHP metamorphic slab)

The Weihai metamorphic subunit is situated between the Rongcheng and Mishan faults (Fig. 2). The main rock types are mafic granulites, ultramafic

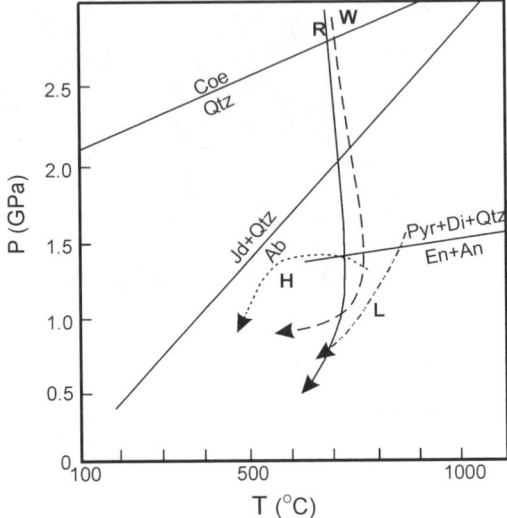

FIG. 4. PT diagram of different metamorphic units, showing their different metamorphic histories (after Wang et al., 1993; Cong et al., 1996; Zhai and Liu, 1997). Abbreviations: H = Haiyangsuo unit; W = Weihai subunit; R = Rongcheng subunit; L = Lai-Yan unit.

rocks, and some metamorphosed sedimentary rocks, including marble, schist, and quartzite (Cao, 1990; Wang, 1994; Wang et al., 1995; Zhang and Wang, 1996). The metamorphic rocks occur as lenses in granitic gneisses, which are characterized by migmatization and anatexis. However, they show similarities with those in the Rongcheng subunit in their geochemistry (Li et al., 2000b) and geochronology (Enami et al., 1993a; Ishizaka et al., 1994). Coesite was discovered as inclusions in garnets in mafic granulites (e.g., sample WH1; Fig. 3E) that were retrograded from coesite-eclogite facies (600–700°C/> 30 kbar) to granulite facies (720–850°C/9–12 kbar) with a sharp decompression and a slight increase of temperature P-T path (Wang et al., 1993; Cong et al., 1996; Banno et al., 2000). It is significant that this superimposed metamorphism of granulite facies is so strong that most of the coesite eclogites have retrograded to granulites. Jahn (1994) and Zhai et al. (2000) reported that four Weihai samples have rare εNd values ranging from + 129 to + 260 (Fig. 5B) with extreme Sm/Nd fraction (fSm / Nd = + 6.9 and + 4.63). The T_{DM} ages are 1.66 Ga and 1.61 Ga. These constrain the minimum age of the protolith as Mesoproterozoic and indicate a complicated metamorphic history with strong fluid-rock interaction. Considering that the country gneisses

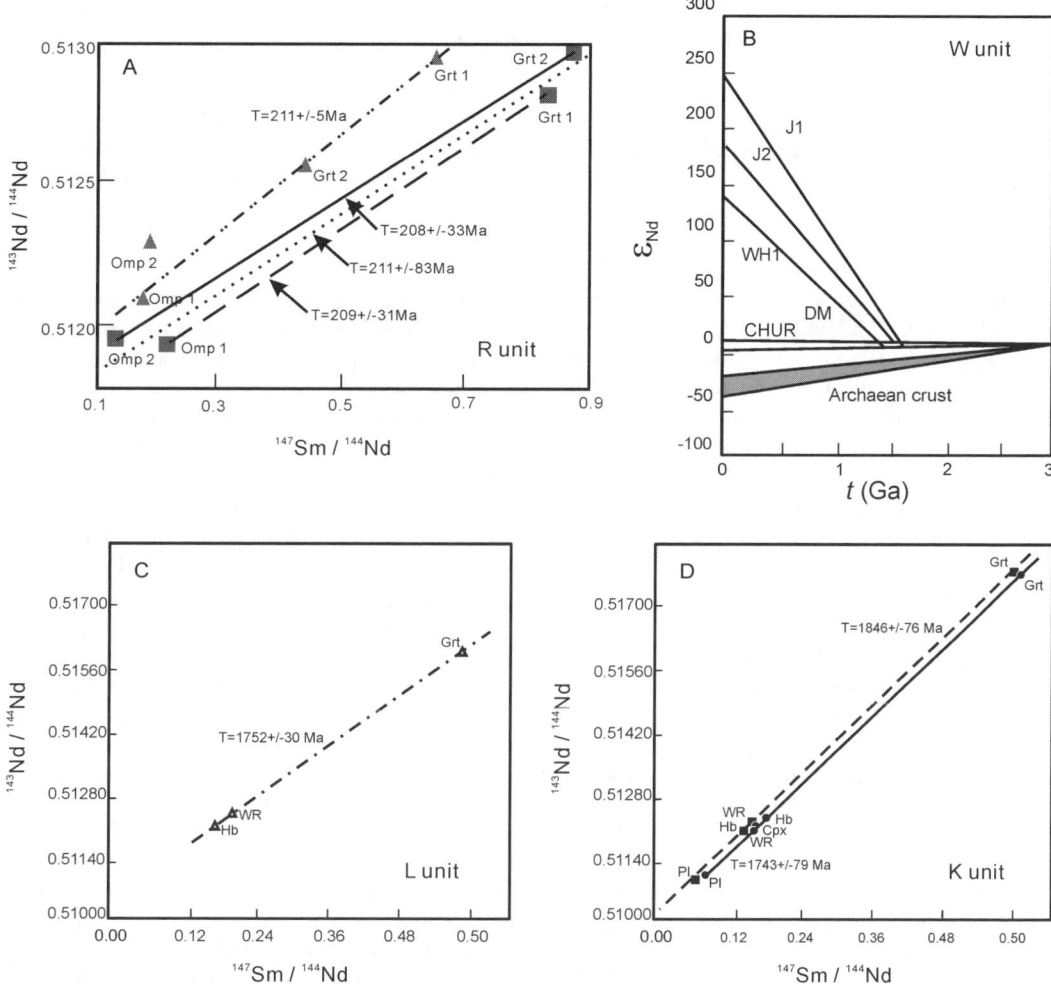

FIG. 5. Sm-Nd isochrons (A, C, D) and ε_{Nd}-t (age) (B) diagram for mafic metamorphic rocks. A. Isochrons for R unit. B. ε_{Nd}-t (age) for W unit. C. Isochrons for L unit. D. Isochron for K unit.

underwent strong migmatization (Li et al., 2000b), we suggest that the Weihai UHP rocks were not directly uplifted from the mantle to the middle crust; instead they were tectonically emplaced into a high heat flow area during exhumation (Zhai and Liu, 1998). The Weihai subunit can be interpreted as a high-temperature superimposed metamorphic slab separated from the UHP metamorphic zone during exhumation. Late-stage overprinting of amphibolite facies is also recorded, showing the further decrease of temperature and pressure with uplift (Fig. 4).

Lai-Yan metamorphic unit (basement of the North China Block)

The Lai-Yan (Laixi-Yantai) metamorphic unit refers to rocks in the area west of the Muping (Yantai) fault and is composed dominantly of an orthogneiss complex and metamorphosed supracrustal rocks at granulite and amphibolite facies. The supracrustal rocks include the Fenzishan and Jinshan groups, and are considered to be Neoarchean–Paleoproterozoic (Zhang, 1984). They consist of graphite gneisses, marbles, banded iron formations, mafic granulites, and amphibolites, and overlie the orthogneiss complex. The orthogneiss complex is tonalitic-trondhjemitic-granodioritic (TTG) metamorphosed to granulite facies (Wang and Yan, 1992). It has zircon U-Pb ages of 2.9–2.6 Ga (Wang and Jiang, 1985). Some lenses of granulite and banded iron formation–bearing metasediments are enclosed in the gneisses. A series of lenses of garnet-bearing mafic granulites or amphibolites, pyroxenites, and serpentinites constitute an uncontinuous belt in the Malianzhuang-Qixia-Yantai area (e.g., samples ML6 and Qx1 in Fig. 2), along the eastern margin of the Lai-Yan unit. Most amphibolite lenses were transformed from garnet granulites by retrogressive metamorphism (Li et al., 1997).

Three generations of mineral assemblages representing three main metamorphic episodes can be identified (Figs. 3G and 3H). The early assemblage is preserved only as inclusions in garnets, which are Cpx1 + Plg1 + Qtz + Grt1 (core). Cpx1 is a diopside-augite with 4.71 wt% Al_2O_3 and 0.98–2.12 wt% Na_2O. The second mineral assemblage is Cpx2 + Hy + Plg2 + Grt2 (rim) + Qtz ± brown $amphibole_2$ ($Amp2$), represented by symplectite around garnet and equilibrated with the matrix minerals. The matrix minerals are mainly Hy, Cpx2, Plg2, and quartz. The Cpx2 contains Al_2O_3 of 0.9–1.31 wt% and Na_2O of 0.27–0.42 wt%, lower than those in the Cpx1. Microprobe analyses show that garnets are compositionally zoned, with Mg content decreasing from core to rim (prp = 0.19–0.13), and Fe content increasing from core to rim (alm = 0.62–0.47) (Liu et al., 1998). The third mineral assemblage is Amp3 (green) + Plg3, partially replacing the second mineral assemblage. Metamorphic temperature and pressure conditions have been estimated, by using mineral thermobarometry and the TWQ program of Berman (1991), at 840–860°C and >1.4 GPa for the first episode, 720–780°C and 0.8–1.1 GPa for the second episode, and 600–700°C and 0.5–0.7 GPa for the third episode (Fig. 4; Liu et al., 1998). A mineral-WR Sm-Nd isochron age for the garnet granulite (ML6) from Malianzhuan near Laixi is 1752 ± 30 Ma (Fig. 5C). The T_{DM} age is 2.8 Ga, ε_{Nd} (0) value is –1.0 (Zhai et al., 2000). T_{DM} ages effectively constrain the minimum age of granulite protolith formation to the Archean. The Sm-Nd mineral-WR isochron ages are interpreted to date metamorphic closure temperatures of the second metamorphic episode (medium-pressure granulite facies), recorded during exhumation of earlier high-pressure granulites from deep in the continental crust. It is important to note that Sm-Nd characteristics are consistent with those of the Proterozoic medium-pressure granulites in the North China Block (Guo et al., 1996), but quite different from those of the eclogites in the Dabieshan-Sulu UHP rocks.

Kunyushan boundary metamorphic-magmatic complex zone

The Kunyushan boundary zone is bounded by the Mishan and the Muping faults. It can be subdivided into three parts. The first is composed of metamorphic rocks, traditionally included in the Jiaodong Group (Jia et al., 1993). The various rocks in thre complex are tectonically juxtaposed and intruded by late Mesozoic granotoids. Felsic gneisses, graphite gneisses, garnet mica schists, marbles, granulites, amphibolites, and other metamorphic rocks include the Jiaodong-Fengzishan-Jinshan complexes of Precambrian age.

The second part is the Kunyushan granitoid batholith (Yu, 1989; Jia et al., 1993; Lai, 1997). It can be subdivided into different phases with or without foliation, and is cut by small granite sills and veins of different ages. The isotopic ages range from Proterozoic to Cenozoic, with four populations at 2000–1900 Ma, 710–610 Ma (e.g., sample MK21a in Fig. 2), 180–160 Ma (e.g., sample MK21b), and 124–110 Ma (e.g., sample ZG3) (Lin et al., 1992; Xu et al., 1997; Guo et al., 2001). The batholith has no clear boundary with the country rocks, in some places showing a transitional contact relationship. Garnet and sillimanite commonly occur at the batholith margin. The composition of garnet is similar to that of garnet in the country rocks (Jia et al., 1993). Initial $^{87}Sr/^{86}Sr$ values of the batholith are concentrated at ca 0.708, but a few are up to 0.7105. Their Nd model ages (T_{DM}) range from 2015 to 752 and zircon SHRIMP U-Pb ages range from 162 to 142 Ma (Guo et al., 2002; Fan Hongrui, pers. commun., 2004). The Kunyushan granitoid batholith is considered to be an autochthonous and para-autochthonous S-type granite formed by partial melting of buried continental crust of the North China Block (Zhang et al., 1991; Lai, 1997) and mixed with some mantle materials (Guo et al., 2001).

The third component part is a series of mafic-ultramafic lenses in orthogneiss (e.g., samples WD04 and ZZ1 in Fig. 2). They include mafic granulite, garnet pyroxenite, and serpentinized peridotite. The metamorphic mineral assemblage of mafic granulite is Grt + Hy + Cpx + Plg + Qtz, and accessory minerals are magnetite and/or graphite. The abundance of Cpx1 is 25–30 vol% in the rocks, and it is an augite with 0.6–1.1 wt% of Na_2O and 1.3–2.4 wt% of Al_2O_3. Cpx1, Qtz, and Plg1 are present as inclusions in garnets. No minerals of eclogite facies have been found. Estimated metamorphic temperatures and pressures for inclusion and matrix minerals are 840–870°C and 13–14 kbar, and 820–850°C and 9–11 kbar, respectively (Zhai et al., 2000). The mafic granulites also underwent weak amphibolite-facies metamorphism, resulting in Plg3 + green Amp3. Two samples (KY9, ZZ1) of mafic granulites yielded Sm-Nd mineral-WR isochron ages of 1846 ± 76 Ma and 1743 ± 79 Ma (Fig. 5D), which are interpreted to date metamorphic closure temperatures of the second metamorphic episode (Li et al., 1997). The characteristics of the mafic granulites are similar to the mafic rocks of the Lai-Yan unit, indicating metamorphic rocks in the Kunyushan complex zone were derived from the North China Block. Moreover, some rocks possibly come from the UHP region. For example, lenses of garnet-bearing mafic granulites in Xiaoshidao near Weihai (Fig. 2) are very similar to the garnet granulites retrograded from UHP eclogites in the Weihai unit. However, the multistage overprinting of granulite facies and amphibolite facies is too strong to distinguish the P-T conditions.

Evidence for the Boundary between the North China Block and the Sulu UHP Metamorphic Region

Reflecting the discovery of coesite-eclogite at Sulu, this UHPM region has been separated from the North China Block, although their boundary in East Shandong has not been clearly defined. Zhai and Liu (1998) and Zhai et al. (2000) suggested that the boundary may be a complicated magmatic-metamorphic zone. This zone clearly divides the northern part of East Shandong into two parts, southeastern and northwestern, which belong to the Sulu UHP region and the North China Block, respectively. Differences between the two parts are obvious based on regional structure, rock association, metamorphism, geochemistry, and geochronology. The evidence to define the Kunyushan zone as the boundary between the North China Block and the Sulu region is summarized below.

Geochemistry and geochronology of regional orthogneisses

Orthogneisses dominate both sides of the Kunyushan boundary zone. Zhai and Liu (1998) termed them the Lai-Yan gneisses in the western part (the Lai-Yan unit) and the Sulu gneisses in the eastern part, respectively. The Lai-Yan gneisses are TTG rocks (Li et al., 2000b) with low total REE abundances and exhibit depleted HREE patterns. They have conventional zircon U-Pb ages of 2900–2600 Ma (Wang and Yan, 1992), zircon SHRIMP U-Pb ages of 2700–2600 Ma and 2000–1800 Ma (Tang et al., 2003), and Rb-Sr ages and U-Th-Pb chemical ages of 1600–2020 Ma (Enami et al., 1993a; Ishizaka et al., 1994). The HP granulites have yielded SHRIMP zircon ages of 2375 ± 9 Ma and 1794 ± 41 Ma and Sm-Nd mineralis isochron ages of 1800–1700 Ma (Li et al., 2000a; Tang et al., 2003). The Sulu gneisses are mainly trondhjemitic and granitic according to petrology and geochemistry. They have variable total REE abundances from depleted HREE patterns to undepleted HREE patterns with negative Eu anomalies (Li et al., 2000b). Their Rb-Sr and U-TH-Pb chemical ages range from 800 to 700 Ma and from 320 to 110 Ma, concentrating at 230–180 Ma (Enami et al., 1993a; Ishizaka et al., 1994). The Sulu gneisses and lenses within them underwent UHP metamorphism and were overprinted by granulite-facies or amphibolite-facies metamorphism (Liou and Zhang, 1996; Liu et al., 2004).

Mafic, ultramafic, and other rocks

Most of the BIF-bearing volcanic and sedimentary rocks in the Lai-Yan gneisses are granulite facies. Some garnet mafic granulites along the eastern margin of the Lai-Yan unit are high-pressure granulites, with 9–11 kbar pressures and ~1800 Ma isotopic ages (Zhai et al., 2000). Their metamorphic and petrological characteristics are the same as the high-P and high-T metamorphic rocks in the Hengshan, central North China Block (Li et al., 1997; Zhao et al., 1999). In contrast, the lenses in the Sulu gneisses are eclogites, garnet pyroxenites, garnet peridotites, UHP marbles, and other metasediments, in which garnets typically contain inclusions of coesite or its pseudomorphs. Isotopic data for the eclogite lenses, including Sm-Nd and zircon ages, range from 240 Ma to 208 Ma, and amphibole Ar-Ar ages are ~180 Ma (Li, 1994; Ames et al, 1993, 1996), corresponding to the results from eclogites in Dabieshan.

Metallogenesis

Gold deposits are well developed in East Shandong and are hosted by metamorphic rocks, orthogneisses, and Mesozoic granites. The Mishan fault is the eastern boundary of the East Shandong gold region, and abundant deposits are concentrated in the Lai-Yan unit and Kunyushan boundary zone, extending westward to eastern Hebei Province. There is no economic gold mineralization in the Weihai, Rongcheng, and the Haiyangsuo units of the Sulu region. Zhai et al. (2001) concluded that gold mineralization is strongly controlled by basement rocks of the North China Block. The time of the gold mineralization was from 130 to 110 Ma, and was related to large-scale Mesozoic granite intrusion (Yang et al., 2003; Li et al., 2003). The Mesozoic gold mineralization indicates a gold metallogenic specialization controlled by basement and/or mantle of the North China Block (Pei et al., 1999; Zhai et al., 2001, 2002).

Discussion

Several tectonic models have been proposed to explain UHP metamorphism in the Sulu region. Yin and Nie (1993) suggested an indentation model, whereby the Yangtze Block collided with the smooth southern edge of the North China Block. As a result, the initial position of the suture was displaced northward in the Sulu region by sinistral movement along the Tan-Lu fault. Li (1994) and Li et al. (1996) proposed a crustal-detachment model and suggested that the upper crust of the Yangtze Block was detached from the lower crust in the northern Jiangsu Province–Yellow Sea area, whereas the lower part of the lithosphere was subducted under the North China Block. These two models imply that Sulu and Dabieshan have differences in terms of their UHP metamorphic-tectonic processes.

Recent geophysical research suggested a crocodile mouth–like structural model (Xu et al., 2001). Figure 6A is a 3-D velocity image of the crust-mantle in northeastern Shandong. The UHPM crust of the Yangtze Block diverged into two branches; one (the Weihai subunit) overlies the North China crust; another (the Rongcheng subunit) represents an ancient subduction relic of the Yangtze crust (Zhai et al., 2000).

Zhai et al. (2001, 2002) considered that this imaged velocity structure possibly is not directly related to the collision between the North China and Yangtze blocks, but was strongly affected by tec-

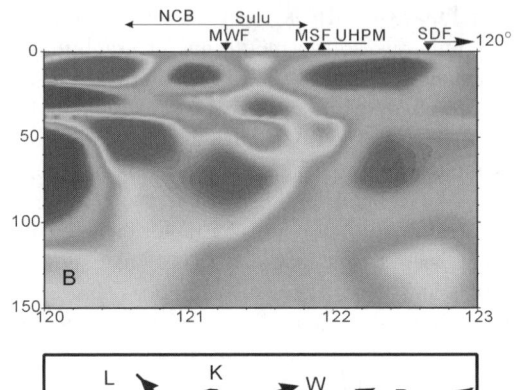
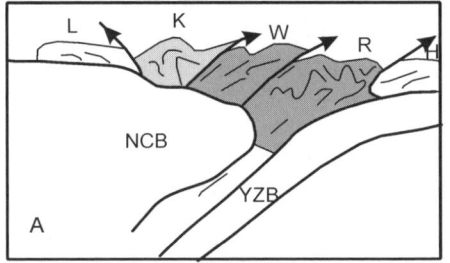

FIG. 6. A. Imaged velocity structure of the northern part of eastern Shandong. B. Proposed tectonic model (after Zhai et al., 1998; Xu et al., 2001). Abbreviations: NCB = North China Block; YB = Yangtze Block; MWF = Muping (Yantai)–Wulian fault; MSF = Mishan fault; SDF = Shidao fault.

tonic imbrication of rock slices postdating the collision. The present boundary between the North China Block and the Sulu region (subducted and metamorphosed Yangtze Block) is defined on the surface by the Muping and Mishan faults (Fig. 6B). The UHPM region represents the subducted Yangtze Block under the North China Block, bounded by the Mishan and Shidao faults. The UHPM region can be subdivided into two branches. The northwestern branch (the Weihai subunit) is located on buried basement of the North China Block, whereas the Haiyangsuo unit possibly represents a lower crust slab that was detached from the Yangtze Block during or a slightly after collision (Zhai et al., 1998), similar to the model of Li (1994).

The key points of the tectonic division proposed in this paper are summarized as follows:

1. The Haiyangsuo metamorphic unit is a lower-crust slab detached from the Yangtze Block. The lower-crustal granulites underwent partial eclogitization with the decrease of temperature associated with the ductile deformation zone.

2. The Rongcheng-Weihai metamorphic unit represents the Sulu UHP metamorphic region that was

a part of a deep zone, resulting from subduction of the Yangtze Block under the North China Block. The Rongcheng subunit has an ITD (iso-thermal decompression)–type path recording rapid uplift from the mantle to the lower-middle crust.

3. The Weihai metamorphic subunit is a tectonic slab that was separated from the UHP zone during exhumation and tectonically emplaced into an area of high heat flow above the subduction zone. The UHP rocks in the Weihai subunit show superimposition of granulite facies metamorphism and migmatization.

4. The Kunyushan complex zone is the boundary between the North China Block and the Sulu UHP region.

5. The Lai-Yan metamorphic unit is part of the basement of the North China Block.

6. Each of these units has its own distinct metamorphic history and is bounded by ductile shear zones.

Therefore, we propose an oblique collision–crustal detachment–thrusting model (Zhai et al., 2001). Affected by the rectangle-shaped northern edge of the Yangtze Block, an oblique collision took place in the Sulu area with the North China Block. The Yangtze Block was subducted northward under the North China Block, with the subduction zone represented by the Rongcheng-Weihai unit. This oblique collision led to the detachment of lower crust from the Yangtze Block (the Haiyangsuo unit) along a large-scale shear zone. Later collision occurred at Dabieshan, and exerted a further affect on the Sulu region. As a result, the Weihai subunit was separated from the Rongcheng UHP metamorphic sub-unit during exhumation, and was tectonically emplaced as a tectonic wedge above the subduction zone. When the Haiyangsuo unit and Rongcheng and Weihai subunits were uplifted to mid-crustal levels, these metamorphic units, together with a slab of the North China Block (the Lai-Yan unit) were further thrust and imbricated as a series of crystalline nappes (Zhai et al., 1998), similar to thin-skin tectonics (Fig. 6B). The Kunyushan unit is a zone of tectonic mixture of the metamorphic rocks of the North China Block and UHP rocks with granitoids.

The present distribution and contact relationship of these metamorphic units reflects collision and exhumation of UHPM rocks. Mesozoic tectonic inversion was prominent in East China (Guo et al., 2001). It is mainly reflected by a coupled tectonic system related to intracontinental tectonics and large-scale granitic magma melted from the lower crust (Zhai et al., 2001, 2002). The Tan-Lu fault had a multi-stage history of activity (Zhu et al., 2003). A series of branch faults possibly formed at Mid- to Late Mesozoic time and strongly affected deformation of high-grade metamorphic rocks in East Shandong (Zhai et al., 2001). Corresponding to the Mesozoic inversion, two or more generations of deformation can be determined in a few rock outcrops in the Lai-Yan and the Rongcheng-Weihai units (Zhai, 2002; Faure, 2002). Many granitic bodies and sills with 130–110 Ma isotopic ages intruded the metamorphic units in East Shandong. The present imbricated structure is the consequence of Mesozoic tectonics.

Acknowledgments

The National Natural Science Foundation of China (project No.40234050) and the Chinese Academy of Sciences (project No. KZCX1-07) are acknowledged for financial support. Simon Wilde, Takao Hirajima, Akira Ishiwatari, Shoufa Lin, P. Cawood, Zhongyan Zhao, Kai Ye, Qingren Meng, Guochun Zhao, Hongfu Zhang, and other geologists are thanked for their comments and review of this paper.

REFERENCES

Ames, L., Tilton, G. R., and Zhou, G., 1993, Timing of collision of the Sino-Korea and Yangtze cratons: U-Pb zircon dating of coesite-bearing eclogites: Geology, v. 21, p. 309–324.

Ames, L., Zhou, G., and Xiong, B., 1996, Geochronology and isotopic character of ultrahigh-pressure metamorphism with implications for collision on the Sino-Korea and Yangtze cratons, central China: Tectonics, v. 15, p. 472–489.

Bai, W. J., 1993, Mafic-ultramafic complexes and their mineralization: Lithosphere geotectonic evolution of the North China Block: Beijing, China, Seismological Press, p. 85–105.

Banno, S., Enami, M., Hirajima, T., Ishiwatari, A., and Wang, Q. C., 2000, Decompression P-T path of coesite eclogite to granulite from Weihai, eastern China: Lithos, v. 52, p. 97–108.

Berman, R. G., 1991, Thermobarometry using multi-equilibrium calculation: A new technique with petrologic applications: Canadian Mineralogist, v. 29, p. 833–855.

Cao, G. Q., 1990, The Jiaonan terrain and its geotectonics: Geology in Shandong, v.6 no. 2, p. 1–10 (in Chinese).

Cong, B. L., Wang, Q. C., Zhai, M. G., Zhang, R. Y., Zhao, Z. Y., and Ye, K., 1994, Ultra-high pressure metamorphic rocks in the Dabie-Su-Lu region, China: Their formation and exhumation: The Island Arc, v. 3, p. 35–50.

Cong, B. L., Zhang, R. Y., Liou, J. G., Ye, K., and Wang, Q. C., 1996, Metamorphic evolution of UHPM rocks, in Cong, B. L., ed., Ultrahigh-pressure metamorphic rocks in the Dabieshan-Sulu region of China: Beijing, China, Science Press–Kluwer Academic Publishers, p. 128–160.

Ellis, D. J., and Maboko, M. A. H., 1992, Precambrian tectonics and the geochemical evolution of the continental crust, I. The gabbro-eclogite transition revisited: Precambrian Research, v. 55, no. 1, p. 491–506.

Enami, K., Suzuki, A., Zhai, M. G., and Zheng, X. S., 1993a, The chemical U-Th-total Pb isochron age of Jiaodong and Jiaonan metamorphic rocks in Shandong Province, Eastern China: The Island Arc, v. 2, p. 104–113.

Enami, M., Zhang, Q., and Yin, Y., 1993b, High-pressure eclogites in northern Jiangsu–southern Shandong region, eastern China. Journal of Metamorphic Geology, v. 11, p. 589–603.

Faure, M., Lin, W., and Breton, N. L., 2001. Where is the North China–South China Block boundary in eastern China?: Geology, v. 29, p. 119–122.

Faure, M., Treton, N. L., Lin, W., and Monié, P., 2002, Where is the North China–South China Block boundary in eastern China? Reply: Geology, v. 30, p. 668.

Guo, J. H., Chen, F., Zhang, X., Fan, H. R., and Cong, B. L., 2001, Origin of post-collisional shoshonitic syenites and strongly peraluminous rocks in Sulu UHP belt, eastern China: Zircon U-Pb and petrologic-chemical data, in Jang, B. A. and Dekyo Cheong, eds., Proceeding of the 8th Korea-China Joint Symposium on Crustal Evolution in Northeast Asia: Kyunju, South Korea, Kongwon National University, p. 126–129.

Guo, J. H., Zhai, M. G., Li, J. H., and Li, Y. G., 1996, Nature of the early Precambrian Sanggan structure zone in the North China craton: Evidence from rock association: Acta Petrologica Sinica, v. 12, no. 2, p. 191–207 (in Chinese).

Guo, J. H., Zhai, M. G., Ye, K., Liu, W. J., and Cong, B. L., 2002, Petrochemistry and geochemistry of HP metabasites from Haiyangsuo, Sulu UHP belt, east China: Science in China, v. 45, p. 21–33.

Hacker, B. R., Ratsbacher, L., Webb, L., and Dong, S. W., 1995, What brought them up? Exhumation of the Dabieshan ultrahigh-pressure rocks: Geology, v. 23, p. 743–746.

Hacker, B. R., Ratsbacher, L., Webb, L., Ireland, T., Calvert, A., Dong, S. W., Wenk, H. R., and Chateigner, D., 2000, Exhumation of the ultrahigh-pressure continental crust in east-central China: Late Triassic–Early Jurassic extension: Journal of Geophysical Research, v. 105, p. 13,339–13,364.

Hirajima, T., Ishiwatari, A., Cong, B. L., Zhang, R. Y., Banno, S., and Nozaka, T., 1990, Coesite from Mengzhong eclogite at Donghai county, northeastern Jiangsu province, China: Mineralogical Magazine, v. 54, p. 579–583.

Hirajima, T., Wallis, S. R., Zhai, M., and Ye, K., 1993, Eclogitized metagranitoid from the Su-Lu ultra-high pressure (UHP) province, eastern Jiangsu province, eastern China: Proceedings of the Japan Academy, v. 69B, p. 49–254.

Ishizaka, K., Hiragima, T., and Zheng, X. S., 1994, Rb-Sr dating for the Jiaodong gneiss of the Su-Lu ultra-high pressure province, eastern China: The Island Arc, v. 3, p. 232–241.

Jahn B.-M., 1994, Geochemical and isotopic study of UHP terrain in China [abs.], in First Workshop on UHP Metamorphism and Tectonics: Stanford, CA, Stanford University, p. 71–74.

Ji, Z. Y., Zhao, H. J., and Zhao, G. H., 1988, Discovery of eclogite in Qianliyan island: Geology in Shandong, v. 8, no. 2, p. 123 (in Chinese).

Jia, D., He, Y. M., Shi, Y. S., and Lu, H. X., 1993, Structure of Shandong terrain and its joining dynamics: Nanjing, China, Nanjing University Press, p. 5–59 (in Chinese).

Kato, T., Enami, A., and Zhai, M. G., 1997, Ultrahigh-pressure marble and eclogite in the Sulu UHP terrane, eastern China: Journal of Metamorphic Geology, v. 15, p. 169–182.

Kurahashi, E., Nakajima, Y., and Ogasawara, Y., 2001, Coesite inclusions and prograde compositional zonation of garnet in eclogite from Zeku in the Su-Lu ultrahigh-pressure terrane, eastern China: Journal of Mineralogical and Petrological Sciences, v. 196, p. 100–108.

Lai, X. Y., 1997, UHP metamorphism and joining of terrains in East Shandong area: Unpubl. Ph.D. thesis, Chinese University of Geoscience, Beijing, p. 35–47.

Li, J. W., Vasconcelos, P. M., Zhang, J., Zhou, M. F., Zhang, X. J., and Yang, F. H., 2003, ^{40}Ar-^{39}Ar constraints on a temporal link between gold mineralization, magmatism, and continental margin transtension in the Jiaodong gold province, eastern China: Journal of Geology, v. 111, p. 741–751.

Li, S. G., Chen, Y. Z., Song, M. C., Zhang, Z. M., Yang, C., and Zhao, D. M., 1994, Zircon U-Pb ages of amphibolite in Haiyangsuo, east Shandong: Multi-stage metamorphism affecting to zircon concordant age: Journal of Geosciences, v. 1-2, p. 37–42.

Li, S. G., Jagoutz, E., Chen, Y. Z., and Li, Q. L., 2000a, Sm-Nd and Rb-Sr isotopic chronology and cooling history of ultrahigh pressure metamorphic rocks and their country rocks at Shuanghe in the Dabie Mountains, Central China: Geochimica et Cosmochimica Acta, v. 64, p. 1077–1093.

Li, Y. G., Guo, J. H., Zhai, M. G., Liu, W. J., and Guan, H., 2000b, Geochemical differences of basement felsic gneisses in Shandong Peninsula and its significance to the boundary between North China plate and Sulu UHP belt: Acta Petrologica Sinica, v. 15, p. 557–563 (in Chinese).

Li, Y. G., Zhai, M. G., Liu, W. J., and Guo, J. H., 1997, Sm-Nd geochronology of high-pressure granulite in Laixi, east Shandong: Scientia Geologica Sinica, v. 32, no. 3, p. 283–290 (in Chinese).

Li, Z. X., 1994, Collision between the North and South China blocks: A crustal-detachment model for suturing in the region east of the Tanlu fault: Geol.ogy, v. 22, p. 739–742.

Li, Z. X., Li, X. H., Zhou, H., and Kinny, P. D., 2001, Grenville-age continental collision in South China: New SHRIMP age constraints and implications to Rodinia configuration, in Sircombe, K. N. and Li, Z. X., eds., From basin to mountains: Rodinia at the turn of the century [abs.], in Chris Powell Memorial Symposium: Perth, Australia, Fineline Print and Copy Service, p. 78–79.

Li, Z. X., Zhang, L., and Powell, C. McA., 1996, Position of the east Asian cratons in the Neoproterozoic supercontinent Rodinia: Australian Journal of Earth Sciences, v. 43, p. 593–604.

Lin, J. Q., Tan, D. J., Chi, X. G., Bi, L. J., Xie, C. G., and Xu, J. L., 1992, Mesozoic granites in Jiao-Liao Peninsula: Beijing, China, Science Press, p. 208 (in Chinese).

Liou, J. G., Maruyama, S., and Ernst, W. G., 1995, Ultrahigh-pressure metamorphism and tectonics: The Island Arc, v. 4, p. 233–239.

Liou, J. G., and Zhang, R. Y., 1996, Occurrence of intergranular coesite in Sulu ultrahigh-P rocks from China: Implications for fluid activity during exhumation: American Mineralogist, v. 81, p. 1217–1221.

Liu, F. L., Xu, Z. Q., and Song, B., 2003, Determination of UHP and retrograded metamorphic ages of the Sulu terrane: Evidence from SHRIMP U-Pb dating on zircons of gneissic rocks: Acta Geologica Sinica, v. 77, p. 229–237 (in Chinese with English abstract).

Liu, F. L., Xu, Z. Q., Yang, J. S., Zhang, Z. M., Xue, H. M., and Li, T. F., 2004, Geochemical characteristics and UHP metamorphism of granitic gneisses in the main drilling hole of Chinese Continental Scientific Drilling Project and its adjacent area: Acta Petrologica Sinica, v. 20, p. 9–26 (in Chinese with English abstract).

Liu, J. B, Ye, K., Maruyama, S., Cong, B. L., and Fan, H. R., 2001, Mineral inclusion in zircon from gneisses in the ultrahigh-pressure zone of the Dabie Mountains, China: Journal of Geology., v. 109, p. 523–535 (in Chinese).

Liu, W. J., Zhai, M. G., and Li, Y. G., 1998, Metamorphism of the high-pressure basic granulites in Laixi, eastern Shandong, China: Acta Petrologica Sinica, v. 14, p. 449–459.

Nakamura, D., and Banno, S., 1997, Thermodynamics modeling of sodic pyroxene solid solution and its application in a garnet-omphacite-kyanite-coesite geothermobarometer for UHP metamorphic rocks: Contributions to Mineralogy and Petrology, v. 130, p. 93–102.

Nakamura, D., and Hirajima, T., 2000, Granulite-facies overprinting of ultrahigh-pressure metamorphic rocks, northeastern Su-Lu region, eastern China: Journal of Petrology, v. 41, p. 563–582.

Okay, A. I., and Sengör, A. M. C., 1992, Evidence for continental thrust–related exhumation of the ultrahigh pressure rocks in China: Geology, v. 20, p. 411–414.

Pei, R. F., Qiu, X. P., and Yin, B. C., 1999, The explosive anomaly of ore-forming processes and superaccumulation of metals: Mineral Deposits, v. 18, no. 4, p. 333–340 (in Chinese).

Tang, J., Zheng, Y. F., Wu, Y. B., and Zhou, J. B., 2003, SHRIMP zircon U-Pb dating and O-isotope study of metamorphosed rocks in western East Shandong [abs.], in Zheng, Y. F., ed., Proceedings for Geodynamics in Subduction and Collision Symposium: Hefei, China, Chinese University of Science and Technology, p. 115–116 (in Chinese).

Tianjing Institute of Metallurgic Geology, 1996, Comparative study of greenstone belt–type gold deposits in east Shandong, China and West Australia and prospecting significance: Tianjing, China, Tianjing Institute of Metallurgic Geology, Scientific Report, p. 17–29.

Wallis, S., Enami, M., and Banno, S., 1999, The Sulu UHP terrane: A review of the petrology and structural geology: International Geology Review, v. 41, p. 906–920.

Wang, L. M., 1994, Primary study on collisional zone in east Shandong: Geology in Shandong, v. 10, no. 1, p. 102–107 (in Chinese).

Wang, L. M., and Yan, Y. M., 1992, Archaean tonalities in Qixia, Shandong: Geology in Shandong, v. 1, p. 80–87 (in Chinese).

Wang, Q. C., and Cong, B. L., 1999, Exhumation of UHP terranes: A case study from the Dabie Mountains, eastern China: International Geology Review, v. 41, p. 94–118.

Wang, Q. C., Ishiwatari, A., Zhao, Z. Y., Hirajima, T., Enami, M., Zhai, M. G., Li, J. J., and Cong, B. L., 1993, Coesite-bearing granulite retrograded from eclogite in Weihai, eastern China: European Journal of Mineralogy, v. 5, p. 141–152.

Wang, R. M., An, J. T., and Lai, X. Y., 1995, Discovery of ophiolite suite in East Shandong and its geotectonic significance: Acta Petrologica Sinica (special issue), p. 221–240 (in Chinese).

Wang, Z. B., and Jiang, H. W., 1985, Division of metamorphic strata in Jiaonan Uplift and its geotectonic evolution: Geology in Shandong, v. 1, no. 1, p. 66–78.

Xu, H. L., Zhang, D. J., and Sun, G. Y., 1997, Characteristics of Kunyushan granite in east Shandong: Journal of Petrology and Mineralogy, v. 16, no. 2, p. 131–143.

Xu, P. F., Liu, F. T., Wang, Q. C., Cong, B. L., and Chen, H., 2001, Slab-like high velocity anomaly in the uppermost mantle beneath the Dabie-Sulu orogen: Geophysical Research Letters, v. 28, p. 1847–1850.

Yang, J., Godard, G., Kienast, J. R., Lu, Y., and Sun, J., 1993, Ultrahigh-pressure 60 kbar magnesite-bearing garnet perodotites from northeastern Jiangsu: China Journal of Geology, v. 101, p. 541–554 (in Chinese).

Yang, J. H., Wu, F. H., and Wildes, S. A., 2003, A review of the geodynamic settings of large-scale Late Mesozoic gold mineralization in the North China craton: An association with lithospheric thinning: Ore Geology Reviews, v. 23, p. 125–152.

Ye, K., 1993, Metamorphism and geotectonics of eclogites and related rocks in East Shandong: Unpubl. Ph.D. thesis, Institute of Geology, Academia Sinica, Beijing, China, p. 35–46.

Ye, K., Cong, B. L., Hirajima, T., and Banno, S., 1999, Transformation from granulite to transitional eclogite at Haiyangsuo, Rushan county, eastern Shandong peninsula: The kinetic process and tectonic implications: Acta Petrologica Sinica, v.15, no. 1, p. 21–36 (in Chinese).

Ye, K., Cong, B. L., and Ye, D. N., 2000a, The possible subduction of continental material to depths greater than 200 km: Nature, v. 407, no. 12, p. 734–736.

Ye, K., Hirajima, T., Ishiwatari, A., Guo, J. H., and Zhai, M. G., 1996, Discovery of coesite between matrix minerals in Yangkou, Qingdao and its geotectonic implication: Chinese Science Bulletin, v. 41, no. 15, p. 1407–1408 (in Chinese).

Ye, K., Yao, Y. P., Katayama, I., Cong, B. L., Wang, Q. C., and Maruyama, S., 2000b, Large areal extent of ultrahigh-pressure metamorphism in the Sulu ultrahigh-pressure terrane of East China: New implications from coesite and omphacite inclusions in zircon of granitic gneiss: Lithos, v. 52, p. 157–164.

Yin, A., and Nie, S., 1993, An indentation model for the north and south China collision and the development of the Tan-Lu and Honam fault systems, eastern China: Tectonics, v. 12, p. 801–813.

Yu, J. H., 1989. Formation age and petrogenesis of Kunyushan migmatitic complex in eastern east Shandong: Geological Review, v. 35, no. 4, p. 285–296 (in Chinese).

Zhai, M. G., 2002, Where is the North China–South China Block boundary in eastern China? Comment: Geology, v. 30, p. 667.

Zhai, M. G., and Cong, B. L., 1996a, Petrol-geotectonics in Sulu-Dabieshan UHP metamorphic zone: Sciences in China, v. 26, no. 3, p. 258–264.

Zhai, M. G., and Cong, B. L., 1996b, Major and trace element geochemistry of eclogites and related rocks, in Cong, B. L., ed., Ultrahigh-pressure metamorphic rocks in the Dabieshan–Sulu region of China: Beijing, China, p. 69–89.

Zhai, M. G., Cong, B. L., Guo, J. H., Liu, W. J., Li, Y. G., and Wang, Q. C., 2000, Sm-Nd geochronology and petrography of garnet pyroxene granulites in the northern Sulu region of China and their geotectonic implication: Lithos, v. 52, p. 23–33.

Zhai, M. G., Guo, J. H., Wang, Q. C., Ye, K., and Cong, B. L., 1998, Division of petrological-tectonic units in the northern Sulu UHP zone: An example of thin-skin thrust of crystalline units: Scientia Geologica Sinica, v. 17, no. 4, p. 539–549.

Zhai, M. G., and Liu, W. J., 1997, The boundary between Sino-Korea craton and Yangtze craton and its extension to the Korea Peninsula, in Jang, B. A., and Cheong Daekyo, eds., Crustal evolution in northeast Asia: Chuncheon, Korea, Kangweon National University, p. 10–12.

Zhai, M. G., and Liu, W. J., 1998, The boundary between Sino-Korea craton and Yangtze craton and its extension to the Korea Peninsula: Journal of the Petrology Society of Korea, v. 7, p. 15–26.

Zhai, M. G., Yang, J. H., and Liu, W. J., 2001, Large clusters of gold deposits and large-scale metallogenesis in the Jiaodong Peninsula, eastern China: Science in China, v. 44, p. 758–768.

Zhai, M. G., Yang, J. H., Miao, L. C., Fan, H. R., and Li, Y. G., 2002, Large-scale cluster of gold deposits and metallogenesis in eastern North China craton: International Geology Review, v. 44, p. 458–476.

Zhang, C. H., and Gu, D. L., 1996, Micro structure and deformation tectonism of sinistral ductile shear zone in middle part of northern Jiaonan Uplift, in Zhang, C. H., and Gu, D. L., eds., Structure characteristics and evolution in northern Jiaonan Uplift: Beijing, China, Chinese University of Geosciences Press, p. 96–104 (in Chinese).

Zhang, L. G., 1995, Geology of South Asia lithosphere: Geochemistry and dynamics of upper mantle, basement, and granite: Beijing, China, Science Press, p. 52–157 (in Chinese).

Zhang, Q. C., Shen, K., Zhao, Z. G., An, J. T., and Yu, D. B., 1991, Accessory mineral characteristics and petrogenesis of granite in Muping-Rushan area, East Shandong: Geology in Shandong, v. 7, no. 2, p. 60–75.

Zhang, Q. S., 1984, Early Precambrian geology and metallogenesis in China: Changchun, China, Jilin People's Press, p. 147–151.

Zhang, R. Y., Cong, B. L., and Liou, J. G., 1993, Ultra high-pressure metamorphic terrain and its explanation of petrogenesis: Acta Petrologica Sinica, v. 9, p. 211–226 (in Chinese).

Zhang, R. Y., Liou, J. G., and Cong, B. L., 1997, Petrogenesis of garnet-bearing ultramafic rocks and associated eclogites in the Su-Lu ultrahigh-P metamorphic terrain, eastern China: Journal of Metamorphic Geology, v. 12, p. 169–186.

Zhang, R. Y., Liou, J. G., and Ernst, W. G., 1995, Ultrahigh-pressure metamorphism and decompressional P-T path of eclogites and country rocks from Weihai, eastern China: The Island Arc, v. 4, p. 293–309.

Zhang, X. D., and Wang, L. M., 1996, Primary petrological study of granulite facies rocks in Weihai-Rushan, east Shandong: Chinese Regional Geology, v. 58, no. 3, p. 211–221 (in Chinese).

Zhao, G. C., Cawood, P. A., Wilde, S. A., Sun, M., and Lu, L. Z., 1999, Thermal evolution of two textural types of mafic granulites in the North China craton: Evidence for both mantle plume and collisional tectonics: Geological Magazine, v. 136, p. 223–240.

Zhu, G., Wang, D. X., Liu, G. S., Niu, M. I., and Song, C. Z., 2003, Evolution of the Tan-Lu fault zone and its response to plate movements in West Pacific Basin: Chinese Journal of Geology, v. 29, p. 36–49 (in Chinese).

Petrogenesis of UHP Metamorphic Crustal and Mantle Rocks from the Chinese Continental Scientific Drilling Pre-pilot Hole 1, Sulu Belt, Eastern China

ZEMING ZHANG,[1]

Institute of Geology, Chinese Academy of Geological Sciences, 26 Baiwanzhuang Road, Beijing, 100037, China

YILIN XIAO, JOCHEN HOEFS,

Geoscience Centre, University of Goettingen, Goldschmidtstrasse 1, D-37077 Goettingen, Germany

ZHIQIN XU,

Institute of Geology, Chinese Academy of Geological Sciences, 26 Baiwanzhuang Road, Beijing, 100037, China

AND J. G. LIOU

Department of Geological and Environmental Sciences, Stanford University, Stanford, California 94305

Abstract

The Pre-pilot Hole No. 1 of the Chinese Continental Scientific Drilling Project (CCSD-PPH1) in the southern Sulu terrane recovered a continuous core of eclogite, garnet peridotite, orthogneiss, paragneiss, and minor schist and quartzite. Geochemical characteristics indicate that the garnet peridotite was derived from depleted mantle; all other rock types are metamorphosed supracrustal rocks with continental affinities. The eclogite consists of garnet, omphacite, phengite, quartz (coesite), amphibole, rutile, and zircon; P-T estimates of peak metamorphism are 785–820°C and >2.7 to 3.7 GPa. The gneisses show common amphibolite-facies mineral assemblages consisting of plagioclase, K-feldspar, muscovite, and quartz, with minor garnet, epidote (or zoisite), and biotite; coesite inclusions in zircon indicate that the gneisses together with eclogites were subjected to an early UHP metamorphism prior to amphibolite-facies retrogression. The Garnet peridotites show porphyroblastic textures, and consist mainly of garnet, clinopyroxene, orthopyroxene, and olivine with minor phlogopite. Garnet and clinopyroxene porphyroblasts show significant compositional zoning. Applying relevant geothermobarometers, core compositions of the minerals indicate P-T conditions of 6.0–7.0 GPa and 1100–1200°C. For the rim, similar or slightly higher P-T conditions compared to the eclogites were obtained. We suggest that the former represent crystallization conditions of garnet peridotite in the upper mantle, whereas the latter reflect reequilibrium conditions during incorporation of mantle rocks into the subducted slab. We conclude that the garnet peridotite may have been derived from the mantle wedge above the subduction zone. If so, these mantle-derived rocks were sandwiched between continental-derived country rocks.

Introduction

THE DABIE-SULU OROGENIC BELT formed during the Triassic collision between the Yangtze and North China plates. Petrological studies over the last 15 years have documented that coesite-bearing eclogite and other ultrahigh-pressure (UHP) rocks are ubiquitous in the eastern Dabie-Sulu terrane. In Sulu, the coesite-bearing UHP belt extends about 320 km from Weihai, northeast Shangdong, to Donghai, northern Jiangsu (inset in Fig. 1). Although the Donghai area has poor outcrops, typical UHP metamorphic rocks, such as coesite-bearing eclogite, jadeite-quartzite, kyanite quartzite, and garnet peridotite have been described from a few restricted localities (Enami et al., 1993; Zhang et al., 1995; Yang et al., 1993; Z. Zhang et al., 1996, 2003). The southern Sulu area consists of the UHP belt in the north, and the high-pressure (HP) belt in the south. In Donghai, the UHP belt consists of metamorphosed mafic-ultramafic and supracrustal rocks, and abundant granitic gneisses (Fig. 1). The supracrustal rocks include biotite gneiss, two-mica

---

[1]Corresponding author; email: zzm@ccsd.org.cn

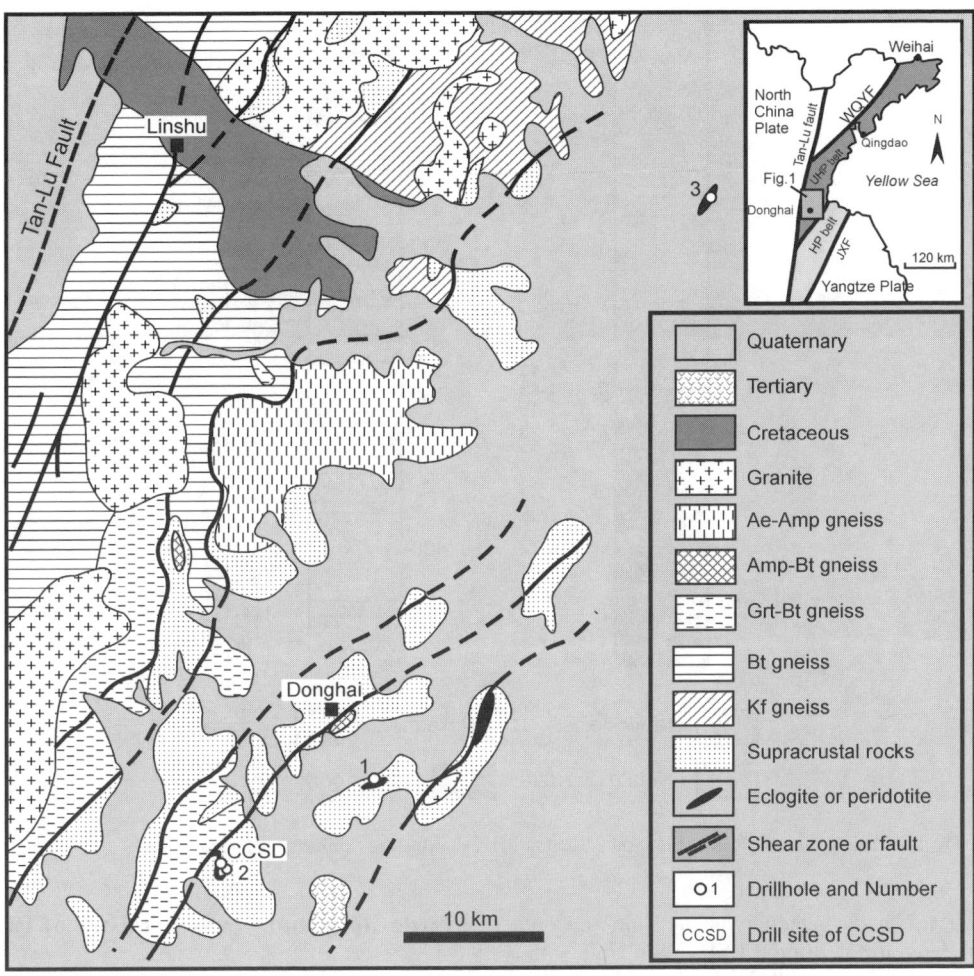

FIG. 1. Simplified geological map of Donghai, showing the locations of the three pre-pilot holes 1, 2, 3 and the main hole of CCSD. In the inset: WQYF = Wulian-Qingdao-Yantai fault, JXF = Jiashan-Xiangshui fault.

gneiss, quartzite, schist, marble, amphibolite, and eclogite; protoliths were probably volcanoclastic and sedimentary rocks (Zhang et al., 2002). The granitic gneisses include migmatitic K-feldspar gneiss, biotite-bearing gneiss, biotite- and garnet-bearing gneiss, aegirine- and sodic amphibole bearing gneiss as well as amphibole-biotite gneiss. These rocks have been intruded by a large volume of granites with radiometric ages from 200 to 115 Ma, and are overlain by Cretaceous sediments and Tertiary basalt (Zhang et al., 2002).

Recently, orogenic garnet peridotites have received much attention due to the finding of various exsolution lamellae in clinopyroxene and olivine, and supersilicic garnet (Dobrzhinetskaya et al., 1996, 1999; Harker et al., 1997; Liou and Zhang, 1998; Medaris, 1999; R. Zhang et al., 1999, 2003; Van Roermund et al., 2000; Bozhilov et al., 2003). This evidence has been interpreted to indicate that some garnet peridotites might have been derived from >180 km depth, or even from the mantle transition zone between 410 and 670 km (Dobrzhinetskaya et al., 1996), but what about the associated UHP supercrustal rocks? Were these supercrustal rocks recrystallized at similar or much shallower depths? This question lies at the heart of any tectonic model concerned with continental collision.

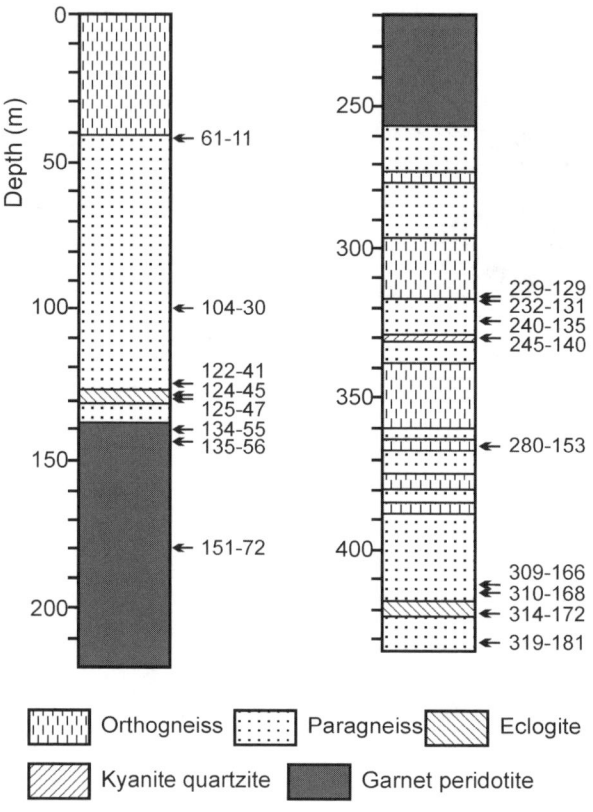

FIG. 2. Simplified lithological profile for the CCSD-PPH1, showing the main sample locations.

Prior to the site selection for the Chinese Continental Scientific Drilling (CCSD) Main Hole with a proposed depth of 5000 m, three pre-pilot holes (PPH) were drilled. The goal of CCSD-PPH1 was to recover garnet peridotite, and to determine its genetic relationship with eclogite and gneiss. Gneisses from the CCSD-PPH1 have been studied by Liu et al. (2001, 2002). Inclusions of coesite, jadeite, omphacite, phengite, and garnet were found in zircons separated from paragneiss, orthogneiss, quartzite, and amphibolite. This fact indicates that these rocks together with eclogite were subjected to UHP metamorphism. In this paper, petrography, petrochemistry, mineral chemistry, and metamorphism of various core samples from the CCSD-PPH1 are described, with the focus on peak metamorphic P-T conditions of garnet peridotite and its relationship to the UHP country rocks. Mineral abbreviations are those used by Kretz (1983), with these exceptions: Amp = amphibole, Coe = coesite, Cel = celadonite, and Phn = phengite.

Petrography and Petrochemistry of PPH1

The 432 m depth CCSD-PPH1 consists mainly of paragneisses, orthogneisses, garnet peridotites, and eclogites (Fig. 2). The paragneisses contain many layers of garnet-epidote amphibolite and epidote-biotite amphibolite, as well as minor kyanite quartzite and muscovite-quartz schist. Bulk rock compositions were analyzed in the National Geological Analytical Center of China in Beijing, major elements by X-ray fluorescence (XRF, Rigaku-3080) with RSD of < 0.5%, and trace elements by ICP-MS (TJA-PQ-ExCell). Analytical uncertainties are 1–5% when abundance >1 µg/g, and 5–10% when abundance <1 µg/g. Representative data for the various UHP rocks are listed in Table 1.

In the drill hole, ultramafic rocks at depths between 136 and 256 m, consist mainly of garnet peridotite and dunite, and are bordered by shear zones against the paragneisses. Most ultramafic

TABLE 1. Whole-Rock Chemical Compositions of Representative UHP rocks from the CCSD-PPH1[1]

Sample:	134-55	135-56	124-45	314-172	122-41	309-166	104-30	310-168	229-129	280-153	232-131	245-140
Rock:	Peridotite	Peridotite	Eclogite	Eclogite	Amphibolite	Amphibolite	Paragneiss	Paragneiss	Orthogneiss	Orthogneiss	Schist	Quartzite
SiO_2	42.37	44.01	49.16	45.55	49.37	52.56	70.66	68.70	78.57	78.51	72.36	82.65
TiO_2	0.02	0.02	0.76	1.61	2.02	0.81	0.41	0.75	0.04	0.07	0.65	0.63
Al_2O_3	1.43	2.73	14.62	16.31	14.23	15.83	13.57	14.25	11.51	11.90	14.74	14.63
Cr_2O_3	0.36	0.38	n.d.	n.d.	n.d.	n.d.	n.d.	n.d.	n.d.	n.d.	n.d.	n.d.
TFe_2O_3	8.38	8.14	12.81	12.93	12.94	9.46	3.93	4.42	1.68	1.44	3.18	0.83
MnO	0.11	0.12	0.21	0.21	0.23	0.15	0.11	0.14	0.04	0.06	0.03	0.01
MgO	45.99	43.19	7.21	8.82	6.36	6.20	1.22	1.20	0.17	0.10	0.76	0.05
CaO	1.08	1.25	11.25	10.41	8.41	8.58	2.36	2.43	0.07	0.30	0.20	0.11
Na_2O	0.10	0.07	2.85	2.79	3.98	3.08	4.45	4.39	4.00	4.02	0.36	0.00
K_2O	0.17	0.08	0.44	0.58	1.45	1.80	2.27	2.94	4.04	3.99	4.71	0.05
P_2O_5	<0.1	<0.1	<0.1	0.84	<0.1	0.36	<0.1	0.28	0.28	<0.1	0.11	<0.1
Total	100.21	100.61	99.98	100.43	99.80	99.98	99.37	99.85	100.40	100.39	100.06	99.75
La	12.0	2.5	7.4	38.7	28.4	31.7	27.8	53.1	24.3	54.7	61.7	42.5
Ce	24.4	6.6	14.9	91.3	43.5	50.4	49.8	86.3	44.8	86.2	99.5	72.3
Pr	2.3	0.7	2.1	10.5	4.8	5.7	5.8	9.2	4.6	8.7	10.2	6.5
Nd	8.5	2.9	9.2	50.1	24.4	23.6	25.9	43.9	19.6	34.1	46.0	28.1
Sm	1.7	0.7	3.1	10.5	5.9	4.8	6.2	9.0	5.1	6.8	8.3	4.7
Eu	0.4	2.2	1.2	3.3	1.9	1.4	1.6	2.3	0.4	0.3	2.1	1.0
Gd	1.7	0.6	10.9	10.1	13.5	4.4	7.7	8.7	4.9	5.5	7.7	2.0
Tb	0.3	0.1	0.6	1.5	1.7	0.4	1.3	1.2	0.8	0.8	1.3	0.6
Dy	1.3	0.5	5.1	8.0	10.1	5.0	6.4	7.7	5.9	7.0	6.5	2.4
Ho	0.3	0.1	1.0	1.7	2.2	1.0	1.5	1.5	1.3	1.5	1.5	0.4
Er	0.6	0.3	3.2	4.3	6.5	3.2	4.8	4.9	3.5	5.2	4.9	1.5
Tm	0.1	0.0	0.4	0.5	0.1	0.4	0.4	0.6	0.5	0.7	0.9	0.1
Yb	0.6	0.3	2.1	3.5	4.9	2.4	3.9	4.1	2.8	4.6	4.8	1.1
Lu	0.1	0.1	0.4	0.5	0.8	0.4	0.7	0.6	0.4	0.7	0.9	0.2
Nb	0.6	0.3	11.4	1.3	6.4	4.9	10.4	10.5	10.5	18.2	9.8	31.1
Rb	4.8	4.0	17.2	7.3	28.1	34.7	72.3	64.9	97.0	138.0	71.1	1.1
Sr	61.0	16.8	280.0	58.8	712.0	1015.0	256.0	183.0	30.7	21.6	98.2	1184.0
V	26.2	23.5	430.0	436.0	409.0	232.0	69.8	56.7	1.8	1.9	67.3	35.0
Ba	282	74	327	229	1075	1075	820	1343	567	344	1791	985
Zr	1.2	0.9	177.0	82.8	120.0	130.0	177.0	243.0	170.0	198.0	256.0	345.0
Cr	2189	2326	358	216	161	179	129	92	205	105	132	168
Ni	1807.0	1729.0	231.0	56.8	61.2	70.8	35.1	37.3	8.9	11.8	15.1	11.7
Y	6.7	2.7	36.2	25.1	57.7	27.5	37.6	43.9	31.4	41.9	37.9	9.6
Co	96.4	88.8	83.2	61.0	64.3	36.9	11.2	8.1	0.8	1.1	8.7	0.9
Th	0.8	0.2	0.8	0.5	1.5	3.7	3.6	6.0	3.5	3.6	5.2	3.0
Sc	7.7	17.5	35.7	51.0	32.0	28.2	12.6	11.5	4.3	5.6	15.0	10.7
Hf	0.9	0.1	2.5	2.1	2.7	3.6	4.0	5.9	4.8	3.8	5.8	4.4
Ta	0.1	0.1	2.0	0.3	<0.01	0.7	0.7	1.4	0.8	1.7	0.9	1.6
U	0.2	0.1	0.4	0.3	0.5	1.2	0.4	2.4	0.9	1.5	1.2	0.6

[1]Major elements in wt %; trace elements in ppm, n.d. = not determined.

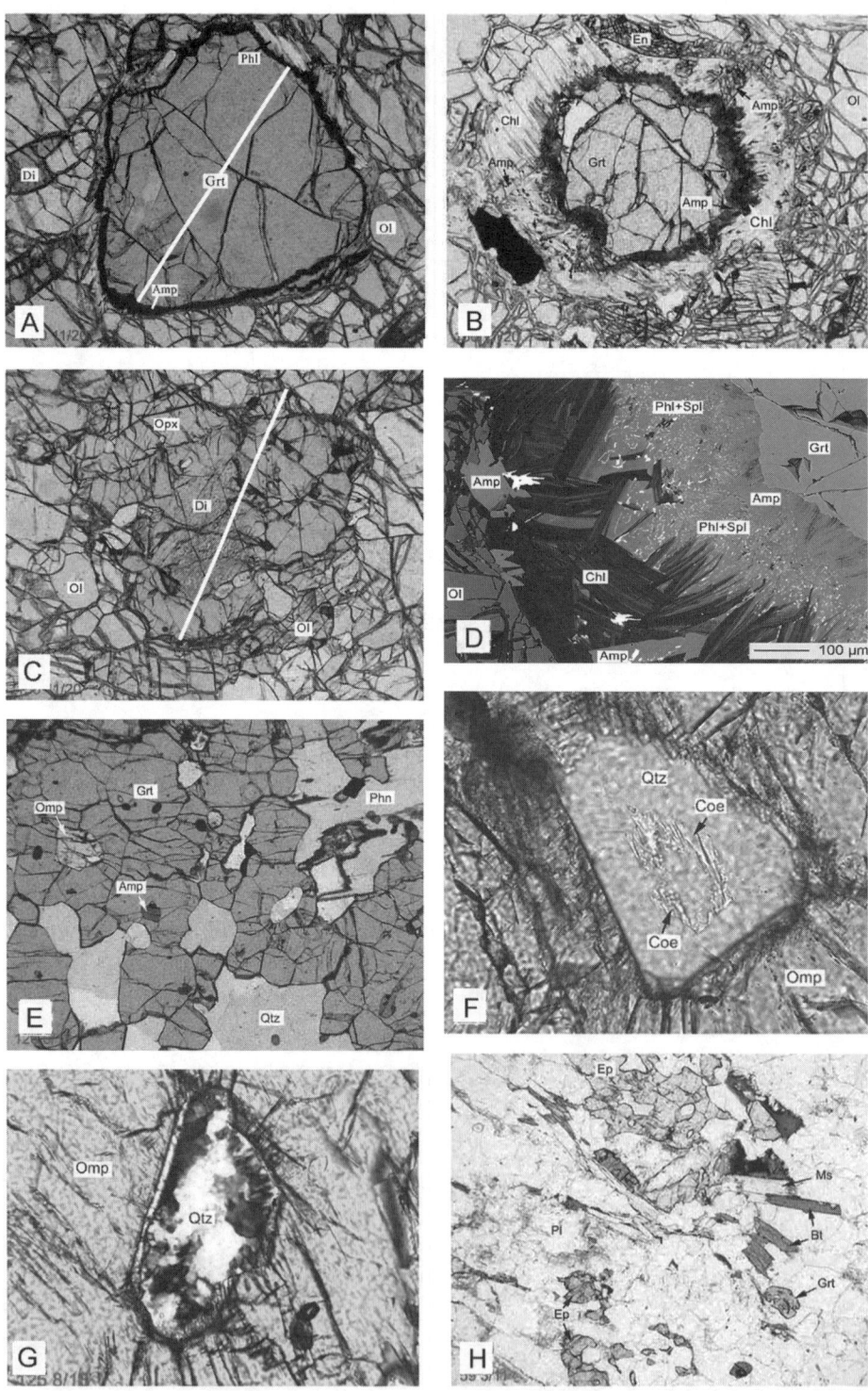

FIG. 3. Caption on facing page.

FIG. 3. *(facing page)* Photomicrographs and EBS image of UHP rocks from the CCSD-PPH1 hole. A. Garnet-peridotite (sample 134-55), consisting of diopside, olivine, phlogopite, and porphyroblast garnet in which the garnet is rimed by a continuous corona of amphibole (black) and a discontinuous phlogopite corona (top right corner and lower left corner of garnet). The white line is the location of microprobe analysis profile; plane light, width of view = 3.6 mm. B. Garnet peridotite (sample 135-56), the porphyroblastic garnet is rimmed by three layers of corona consisting of an amphibole layer, a phlogopite layer, and a chlorite + minor amphibole layer from the inner to the outer rim of garnet (for details, see Fig. 3D); plane light, width of view = 3.6 mm. C. Garnet peridotite (sample 134-55), consisting of diopside porphyroblasts with olivine and enstatite in matrix. The white line in the diopside is microprobe analysis profile; plane light, width of view = 3.6 mm. D. EBS image of the alteration corona of porphyroblastic garnet. Garnet is in the top right corner, and olivine is in the lower left corner. Three layers of alteration coronas rim the relict garnet, i.e., a radial amphiboles layer, a phlogopite + ferrospinel (white) symplectitic layer, and a chlorite (dark gray or black) + amphibole (light gray) + minor ferrospinel (white) symplectitic layer from the edge of garnet towards the left. E. Eclogite (sample 124-45), consisting of garnet, omphacite, phengite, and quartz; notably amphibole and polycrystalline quartz pseudomorphs after coesite occur as inclusions within garnet. Phengite is rimmed by a thin symplectitic corona of biotite and plagioclase; plane light, width of view = 3.6 mm. F. Coesite inclusion in omphacite of eclogite (sample124-45) in which coesite is replaced by polycrystalline quartz in the core and the rim, the host omphacite shows radial fractures; plane light, width of view = 0.1 mm. G. Polycrystalline quartz pseudomorph after coesite occurs as inclusion in omphacite of eclogite (sample 124-45), the host omphacite has radial fractures from the inclusion towards the edge; crossed polars, width of view = 0.1 mm. H. Gneiss (sample 61-11), containing epidote, plagioclase, quartz, biotite, garnet, and muscovite; plane light, width of view = 3.6 mm.

rocks exhibit variable degrees of hydration alteration; some have been completely serpentinized. Fresh garnet peridotite displays a porphyroblastic texture and contains garnet and clinopyroxene porphyroblasts (up to 3–10 mm size) within a finer grained matrix of garnet, clinopyroxene, olivine, and orthopyroxene (Figs. 3A–3C). Most ultramafic rocks contain variable amounts of phlogopite. In moderately altered peridotites, garnet is replaced by complex secondary phases. A typical texture rimming garnet consists of three successive layers of coronas outward from amphibole through phlogopite + ferrous spinel to magnesiochlorite + minor amphibole (Figs. 3B and 3D).

Garnet peridotites have a compositional range of SiO_2 = 42.0 to 44.5 wt%, MgO = 33.3–48.6 wt% and $K_2O + Na_2O$ < 1.5 wt% (Fig. 4). But, several extensively serpentinized garnet peridotites have higher SiO_2 (up to 49.6 wt%), K_2O (up to 3.2 wt%), and lower MgO (25.7 wt%) contents. This indicates that crustal materials have been added into the garnet peridotites by late hydration alteration. In terms of trace elements, the garnet peridotites have high Cr (up to 3,216 ppm) and Ni (2,436 ppm), and low Nb, Rb, Zr, Hf, Sr, and V contents. Chondrite-normalized REE patterns of fresh garnet peridotites are characterized by LREE enrichment. Primitive mantle normalized spidergrams show strongly negative Nb, Sr, Nd, Zr, and Ti anomalies (unpubl. data), which indicate that the garnet peridotites possibly were derived from a depleted mantle source. They are distinctly different from crustal garnet peridotites from Maowu and Bixiling of the Dabie belt, which were intruded into the crust prior to UHP metamorphism (Liou and Zhang, 1998; R. Zhang et al., 2000).

Two thick layers of eclogite occur at depths of 126 to 131 m and 418 to 422 m. Many thin layers of retrograded eclogites, including garnet-epidote amphibolite and epidote-biotite amphibolites, occur within paragneisses. The eclogites have similar mineral paragenesis of Grt + Omp + Phn + Qtz (or Coe) + Rt + Amp + Zrn (Fig. 3E). Relict coesite and polycrystalline quartz pseudomorphs after coesite occur as inclusions in garnet and omphacite (Figs. 3F and G). In the partly retrograded eclogites, garnets are replaced by a symplectitic corona of Amp + Pl + Ep, omphacite by Amp + Ab symplectite, phengite by a Pl + Bt symplectitic corona, and rutile by titanite or opaque minerals. Early formed subhedral amphiboles together with omphacites occur as inclusions in garnet (Fig. 3E), and have a different composition from those formed during amphibolite-facies retrogression. The eclogites and retrograde amphibolites show a wide range of SiO_2 from 43 to 57 wt.% (Fig. 4), have relatively high FeO and CaO contents, and the highest Sr (>400 ppm) and V (>150 ppm) concentrations among all recovered core samples. Eclogitic REE patterns show LREE enrichment with weak fractional HREE and weak negative Eu anomalies (unpubl. data), as commonly observed in Precambrian continental basalts and amphibolite (Jahn, 1998), rather than oceanic basalts. Trace element spidergrams of the eclogites show pronounced

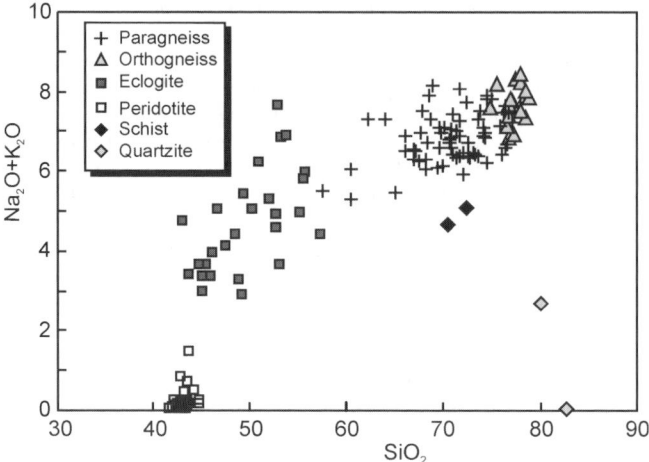

FIG. 4. SiO$_2$ vs. Na$_2$O+K$_2$O diagram for various UHP rocks.

negative Th, Nb, Sr, Nd, and Ti anomalies, which are somewhat different from the eclogites from the ZK703 drillhole of Maobie, without positive Th, negative K and Zr anomalies (Z. Zhang et al., 2000).

Gneisses contain the amphibolite-facies assemblage Pl + Kfs + Qtz ± Grt ± Bt ± Ms ± Ep ± Amp (Fig. 3H). However, symplectitic textures of biotite and plagioclase after phengite, and amphibole and plagioclase after omphacite and garnet occur in many gneiss cores. Moreover, coesite inclusions in zircons in most gneiss cores suggest that gneisses together with eclogites were subjected to early UHP metamorphism before amphibolite-facies retrogression (Liu et al., 2001). The paragneisses display a wide range of SiO$_2$ contents (58–78 wt%), and have relatively high Na$_2$O+K$_2$O (5.1–8.1wt%) (Fig. 4) and REE contents (20 to 900 times chondrite). Their REE distribution patterns are characterized by LREE enrichments and flat HREE with negative Eu anomalies. The granitic gneisses, however, have higher SiO$_2$ and Na$_2$O+K$_2$O contents than those of paragneisses, and show LREE enrichment patterns with pronounced negative Eu anomalies. These features indicate that protoliths of the paragneisses were sediments with relatively high maturity and continental affinity, whereas protoliths of orthogneiss were within-plate granitic intrusives.

Kyanite-muscovite quartzite and muscovite-quartz schist occur as thin layers (~10–30 cm) in the paragneiss, and consist mainly of quartz, white mica, kyanite, and minor rutile, topaz, zircon, monazite, and pyrite. The muscovite-quartz schists have similar SiO$_2$ contents as the average value of the gneisses, but have lower Na$_2$O + K$_2$O contents and higher FeO and MgO contents (Fig. 4), suggesting a metasedimentary protolith. The kyanite quartzite has the highest SiO$_2$ contents (up to 83 wt%) and lowest FeO contents (1.5 wt%), and lower Na$_2$O + K$_2$O contents (Fig. 4). Additionally, the quartzite has the highest Sr contents. Both schist and quartzite show LREE enrichment patterns and weak fractional or flat HREE. These facts indicate that the kyanite-muscovite quartzite may be metamorphosed Al$_2$O$_3$-bearing silicic rock, as suggested by Z. Zhang et al. (2003).

Mineral Chemistry

Mineral compositions were determined by electron-microprobe analysis using a JXA-8900RL Jeol Superprobe, equipped with WD/ED combined micro-analyzer at the Geoscience Center, University Gottingen. Analyses were performed at 15.0 kV accelerating voltage, 12 Na beam current, and 5µm probe diameter. Standards include silicates and pure oxides. Measurements were made on well-polished sections. The raw data were calculated by the CITZAF method of Armstrong (1991). Representative data of analyzed minerals from gneiss, eclogite, and peridotite are presented in Tables 2 to 7.

Garnets from garnet peridotite are characterized by high pyrope components, and low almandine and grossular components (Table 2 and Fig. 5). In contrast, eclogitic garnets contain higher almandine

TABLE 2. Representative Microprobe Analysis of Garnets in UHP Rocks from the CCSD-PPH1[1]

Sample:	61-11		124-45	134-55			135-56		151-72	240-135		314-172	319-181
				Rim	Core	Core							
SiO_2	37.36	36.98	38.17	40.87	41.16	41.37	41.60	41.33	41.71	37.52	37.66	39.43	37.83
TiO_2	0.04	0.19	0.052	0.04	0.05	0.04	0.03	0.021	0.00	0.04	0.03	0.02	0.08
Al_2O_3	20.13	19.18	21.37	21.42	21.72	21.62	21.62	21.65	21.86	20.89	20.78	22.21	20.89
Cr_2O_3	0.01	0.04	0.011	2.62	2.43	2.71	2.69	2.86	2.20	0.02	0.07	0.05	0.04
FeO	17.90	18.08	25.77	11.55	8.44	8.41	8.21	8.49	10.17	28.02	21.20	20.77	25.74
MnO	14.52	8.48	0.65	0.76	0.39	0.46	0.48	0.51	0.55	2.14	10.71	0.43	1.32
MgO	0.32	0.33	5.13	17.80	19.94	19.70	20.09	19.18	19.37	1.67	1.46	8.90	2.65
CaO	9.75	16.14	9.04	4.89	5.09	5.10	5.06	5.27	4.97	10.14	8.21	8.68	11.28
Na_2O	0.06	0.00	0.05	0.02	0.02	0.05	0.04	0.02	0.03	0.03	0.09	0.00	0.03
NiO	n.d.	n.d.	n.d.	n.d.	n.d.	0.011	n.d.	0.016	n.d.	n.d.	n.d.	n.d.	n.d.
Total	100.10	99.48	100.25	99.97	99.23	99.47	99.81	99.34	100.87	100.46	100.20	100.50	99.85
Si	3.006	2.964	2.967	2.964	2.958	2.972	2.973	2.981	2.967	2.977	3.008	2.977	2.989
Ti	0.002	0.012	0.003	0.002	0.003	0.002	0.001	0.001	0.000	0.002	0.002	0.001	0.004
Al	1.909	1.811	1.957	1.831	1.840	1.830	1.821	1.840	1.833	1.953	1.956	1.976	1.945
Cr	0.001	0.002	0.001	0.150	0.138	0.154	0.152	0.163	0.124	0.001	0.004	0.003	0.002
Fe^{3+}	0.082	0.235	0.110	0.089	0.102	0.073	0.082	0.033	0.113	0.090	0.034	0.063	0.070
Fe^{2+}	1.123	0.977	1.565	0.612	0.405	0.432	0.409	0.479	0.492	1.769	1.382	1.248	1.631
Mn	0.989	0.576	0.043	0.047	0.024	0.028	0.029	0.031	0.033	0.144	0.724	0.028	0.088
Mg	0.039	0.039	0.594	1.924	2.137	2.110	2.141	2.062	2.054	0.198	0.174	1.002	0.312
Ca	0.840	1.386	0.753	0.380	0.392	0.393	0.387	0.407	0.379	0.862	0.703	0.702	0.955
Na	0.009	0.000	0.008	0.002	0.003	0.006	0.005	0.002	0.004	0.004	0.014	0.000	0.004
Prp	0.013	0.013	0.201	0.650	0.723	0.712	0.722	0.692	0.694	0.066	0.058	0.336	0.105
Grs	0.281	0.466	0.255	0.128	0.133	0.132	0.131	0.137	0.128	0.290	0.235	0.236	0.320
Alm	0.375	0.328	0.530	0.206	0.137	0.146	0.138	0.161	0.166	0.595	0.463	0.419	0.546
Spe	0.331	0.193	0.015	0.016	0.008	0.010	0.010	0.010	0.011	0.048	0.243	0.009	0.030
And	0.041	0.115	0.053	0.046	0.053	0.038	0.043	0.018	0.058	0.044	0.017	0.031	0.035

[1] n.d. = not determined.

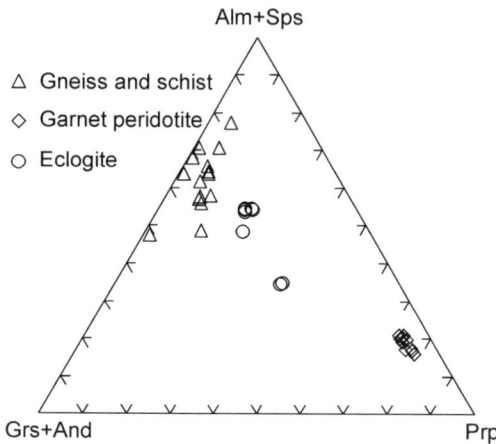

FIG. 5. Compositions of garnets from various UHP rocks.

and pyrope, but lower grossular components and are homogenous. Garnet from gneisses contains high almandine and grossular with variable spessartine components; they range even in a single grain with distinct variation in Mn (i.e., samples 61-11 and 240-135, Table 2). Garnets with lower spessartine contents may have formed during the UHP or HP metamorphic stage; those with higher spessartine contents, however, were formed during amphibolite-facies or even lower metamorphic grade.

Most garnet porphyroblasts from garnet peridotites show compositional zoning. For example, a garnet porphyroblast from sample 134-55 shows asymmetric compositional zoning (Fig. 6). Distinct compositional changes occur at the rims of garnet, in which Fe, Mn, and Cr contents increase, and Mg content decreases toward the margin. Previous investigators interpreted such garnet cores to reflect equilibrium at the peak stage, and rims to reflect subsequent retrograde effect (Brenker and Brey, 1997; Kadarusman and Parkinson, 2000; Paquin and Altherr, 2001). We suggest, however, that the garnet cores represent crystallization conditions in the mantle; the rims indicate the subsequent UHP metamorphism.

All eclogitic clinopyroxenes are omphacites with 45–50% jadeite component and less than 11% aegirine component (Table 3). Clinopyroxenes of garnet peridotites have higher CaO but lower Na_2O contents, and are diopsides; both jadeite and aegirine components are less than 10%. Compositional profiles across a porphyroblastic clinopyroxene from a garnet peridotite reveal a homogeneous core, surrounded by heterogeneous outer zones. In the outer zones, an increase in Fe and a decrease in Al, with variable Cr, Mg, Na, Ca and Si contents, are noted (Fig. 7). As in garnet porphyroblasts, we consider that the homogenous core compositions of clino-

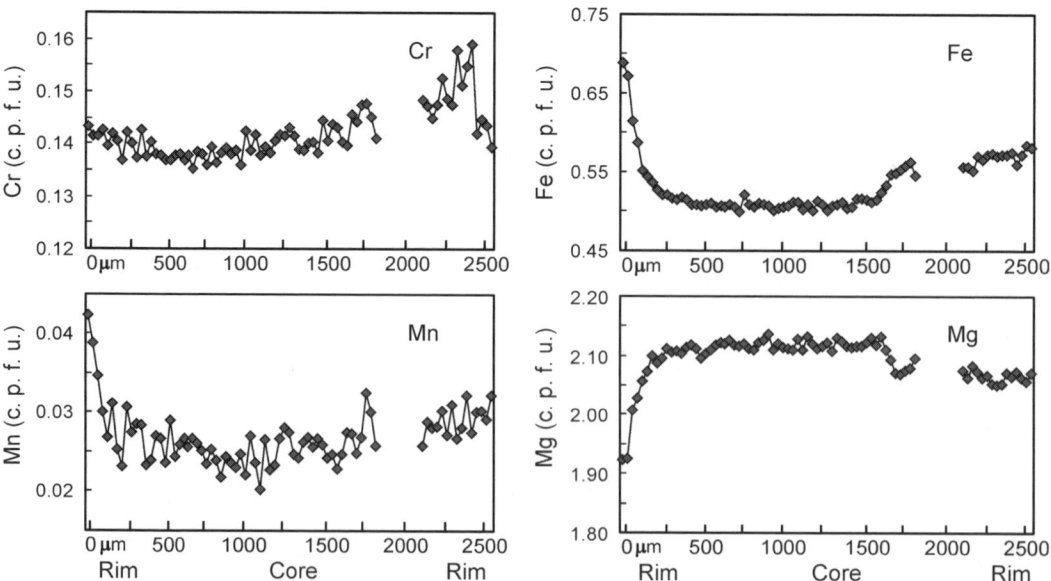

FIG. 6. Compositional profile across porphyroblastic garnet from garnet peridotite. The location of the analyzing line is referenced to Figure 3A.

TABLE 3. Representative Microprobe Analyses of Clinopyroxenes in UHP Rocks from the CCSD-PPH1

Sample:	124-45		134-55			314-172
			Single	Core	Rim	
SiO_2	55.88	55.95	54.56	54.27	54.69	55.58
TiO_2	0.10	0.09	0.02	0.04	0.01	0.04
Al_2O_3	10.83	10.67	1.51	2.10	1.57	10.52
Cr_2O_3	0.01	0.03	1.78	1.70	1.46	0.06
FeO	5.96	5.69	2.80	2.41	2.44	3.07
MnO	0.07	0.00	0.06	0.06	0.07	0.03
MgO	7.34	7.44	15.42	15.01	15.63	9.28
CaO	11.73	11.99	20.92	21.08	21.35	13.62
Na_2O	7.84	7.60	2.41	2.56	2.09	6.65
K_2O	0.00	0.02	0.00	0.02	0.02	0.01
Total	99.80	99.47	99.51	99.24	99.33	98.86
Si	1.987	1.997	1.980	1.971	1.988	1.985
Ti	0.003	0.002	0.001	0.001	0.000	0.001
Al	0.454	0.449	0.065	0.090	0.067	0.443
Cr	0.000	0.001	0.051	0.049	0.042	0.002
Fe^{2+}	0.177	0.170	0.085	0.073	0.074	0.092
Mn	0.002	0.000	0.002	0.002	0.002	0.001
Mg	0.389	0.396	0.834	0.813	0.847	0.494
Ca	0.447	0.459	0.813	0.820	0.831	0.521
Na	0.541	0.526	0.170	0.180	0.147	0.461
K	0.000	0.001	0.000	0.001	0.001	0.001
Cations	4.000	4.001	4.001	4.000	3.999	4.001
Jd	0.460	0.456	0.066	0.090	0.069	0.451
Ae	0.107	0.078	0.094	0.097	0.063	0.043
Aug	0.433	0.466	0.839	0.813	0.868	0.506

pyroxene represent crystallization of the mantle rocks, and the rims indicate the late re-equilibration during UHP metamorphism.

Orthopyroxene from garnet peridotite is characterized by a high enstatite component and very low Al_2O_3 content (Table 4), suggesting that it was formed under very high pressure. All analyzed olivine grains have very high MgO contents, and are forsterite (Fo = 91–93) with 0.38–0.44 wt% NiO (Table 5).

Phengitic mica of the eclogites is characterized by very high Si contents (3.43–3.55 p.f.u., O = 11) (Table 6) formed under UHP conditions. White mica in the gneisses can be divided into two groups: one with high Si (3.11–3.26), classified as muscovitic phengite, probably formed during the UHP or HP metamorphism; the other, with lower Si (<3.0), is muscovite formed during amphibolite-facies metamorphism.

In eclogites, amphiboles have distinct, variable compositions due to their formation at different metamorphic stages (Table 7). Amphiboles occurring as symplectite or symplectitic coronas after garnet and omphacite have higher CaO contents and are calcic amphiboles; matrix amphiboles with albite inclusions have higher Na_2O contents and are classified as sodic-calcic amphiboles, representing an early retrograde phase of the UHP eclogites. In a few samples, some sodic-calcic amphiboles together with omphacite occur as inclusions within garnet, and display lower SiO_2 and MgO, but higher Na_2O, Al_2O_3, and FeO contents than the early retrograde amphiboles. Therefore, these amphiboles may be a UHP phase coexisting with garnet and omphacite.

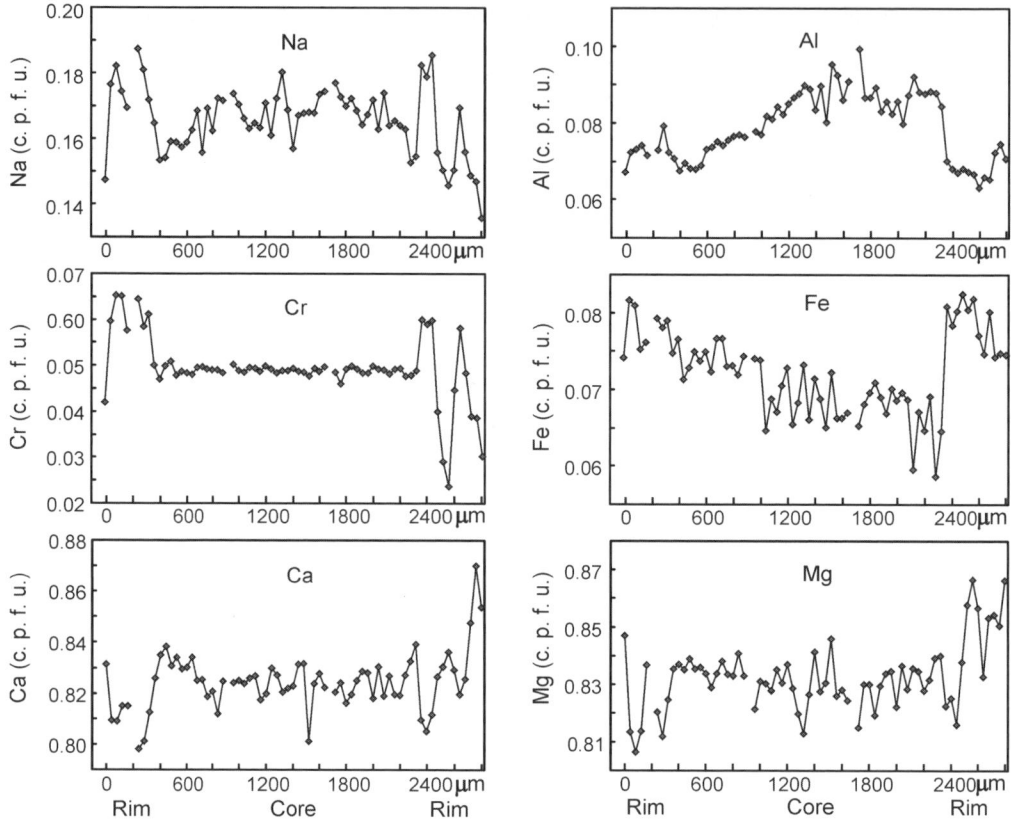

FIG. 7. Compositional profile across porphyroblastic clinopyroxene from garnet peridotite. The location of analyzing line is referenced to Figure 3C.

P-T Conditions of Metamorphism

The low-P limit of coesite-bearing eclogite is constrained by the coesite-quartz equilibrium curve (Bohlen and Boettcher, 1982) (Fig. 8). According to several studies (Carswell et al., 1997, 2000; Michael et al., 2000; Ravna and Milke, 2001), equilibrium pressures for the garnet-omphacite-phengite assemblage can be calculated through a KMASH-system reaction equilibrium, Prp + 2 Grs + 3 Cel = 6 Di + 3 Ms. The result for the analyzed phengite-bearing eclogite is located between the curves of the coesite-quartz and diamond-graphite transformation (Kennedy and Kennedy, 1976) (Fig. 8).

Most metamorphic temperatures of eclogite have been estimated from the Fe^{2+}-Mg exchange thermometer between coexisting garnet and clinopyroxene. Total Fe of garnet is taken as Fe^{2+}, whereas Fe^{2+} of omphacite is computed as Na minus Al, and using the geothermometers of Yang (1994), Krogh (1988), and Ravna (2000) at an assumed P of 3.0 GPa, temperature ranges from 813–840°C (sample 124-45) to 780–820°C (sample 314-172). Using the thermometer of Graham and Powell (1984) for amphibole inclusions in the adjacent garnet, a temperature range of 730–750°C is estimated; this value is somewhat lower than that derived from the coexisting garnet and omphacite, but still indicates that the amphibole inclusions may have formed under UHP conditions. Based on the above P-T estimates, we conclude that the UHP eclogites from this drill hole most likely formed at pressures >2.7 to 3.7 GPa and temperatures of 785 to 820°C (Fig. 8).

The same Fe^{2+}-Mg exchange thermometry was applied to garnet peridotites and yields temperature ranges of 1108–1179°C at 6.0 GPa and 1006–1076°C at 4.0 GPa for core compositions of porphyroblastic garnet (highest in MgO) and clinopyroxene. In contrast, rim compositions of

TABLE 4. Representative Microprobe Analyses
of Orthopyroxenes in UHP Rocks
from the CCSD-PPH1

Sample:	—— 134-55 ——		—— 135-56 ——	
SiO_2	57.65	57.50	58.63	58.63
TiO_2	0.03	0.00	0.01	0.01
Al_2O_3	0.17	0.18	0.13	0.10
FeO	5.22	5.26	4.81	4.81
Cr_2O_3	0.09	0.06	0.07	0.07
MnO	0.12	0.11	0.12	0.12
NiO	0.10	0.08	0.00	0.00
MgO	36.00	36.22	36.64	36.64
CaO	0.07	0.05	0.07	0.07
Na_2O	0.00	0.03	0.03	0.03
K_2O	0.00	0.00	0.00	0.00
Total	99.45	99.49	100.51	100.48
Si	1.984	1.976	1.993	1.993
Ti	0.001	0.000	0.000	0.000
Al	0.007	0.007	0.004	0.004
Fe^{3+}	0.021	0.042	0.010	0.010
Fe^{2+}	0.129	0.109	0.127	0.127
Cr	0.002	0.002	0.002	0.002
Ni	0.003	0.002	0.000	0.000
Mn	0.003	0.003	0.003	0.003
Mg	1.847	1.855	1.856	1.856
Ca	0.003	0.002	0.003	0.003
Na	0.000	0.002	0.002	0.002
K	0.000	0.000	0.000	0.000
Cations	4.000	4.000	4.000	4.000

porphyroblastic garnet (highest in FeO and MnO) and clinopyroxene (lowest in Na_2O), yield temperatures of 890–962°C at 6.0 GPa and 798–896°C at 4.0 GPa, about 200–300°C lower than the former. Apparently, the core compositions of garnet and clinopyroxene porphyroblasts represent the crystallization of garnet peridotite in the upper mantle, whereas the rim compositions were modified by late UHP metamorphism. Using the Fe-Mg exchange thermometer between garnet and olivine (O'Neill and Wood, 1979; O'Neill, 1980), a temperature range of 1067–1080°C at 6.0 GPa was calculated for the core composition of porphyroblast garnets and large olivine grains. In contrast, a lower temperature of 800°C was obtained for the rim of garnet and matrix olivine at the same pressure. A temperature difference of more than 200°C exists between the core and the rim of garnet and olivine. Formation pressure of garnet peridotite was calculated by applying the Al-in-Opx barometer (Brey and Kohler, 1990) in combination with thermometers based on the Fe-Mg exchange between garnet and clinopyroxene. For the core composition of porphyroblastic Grt and Cpx, 7.3–8.1 GPa at 1200°C and 6.5–7.4 GPa at 1100°C were calculated. On the other hand, the rim compositions of porphyroblastic Grt and Cpx yield 4.2–6.0 GPa at 800–900°C. This is additional evidence that the crystallization of garnet peridotite and its subsequent UHP metamorphism occurred at distinctly different P-T conditions.

As described above, the best P-T estimate for the original crystallization of garnet peridotite in the mantle from the CCSD-PPH1 appears to be 6.0–7.0°GPa, and 1100–1200°C (Fig. 8), and for the UHP metamorphic conditions may be 4.2–6.0 GPa at 800–900°C. The latter estimate is generally consistent with previous results for Sulu garnet peridotites (Yang and Jahn, 2000; R. Zhang et al., 2000). A very low geothermal gradient close to the P-T conditions of the forbidden zone of metamorphism in the Earth is obtained for both the crystallization and UHP metamorphism of the garnet peridotites.

Although minute coesite and phengite inclusions are common in zircon of gneisses from the PPH1, the composition and paragenesis of minerals presented above indicate that these rocks re-equilibrated at amphibolite-facies conditions after the UHP event. The P-T conditions are estimated by the TWQ program version 2.02 of Berman (1991) for the assemblages Grt + Bt + Ms + Pl + Kfs. Using garnets with low MnO contents, we obtain 540–710°C and 0.92–1.17 GPa (sample 61-11 and 240-135) (Fig. 8), within a P-T range of the upper amphibolite facies.

Discussion

P-T conditions and origins of garnet peridotite from continental orogenic belts have been the subject of a continuing controversy. Based on microstructural relationships between exsolved ilmenite lamellae and host olivine, Dobrzhinetskaya et al. (1996) proposed that the Alpe Arami garnet peridotite originally formed at very high P-T conditions (15 GPa and 1600°C) at a depth of >300 km. Recently, a pressure estimate of 9–12 GPa was given for the Alpe Arami garnet peridotite by a quantitative

TABLE 5. Representative Microprobe Analyses of Olivines
in UHP Rocks from the CCSD-PPH1[1]

Sample:	134-55			135-56			151-72
SiO_2	40.72	40.42	40.61	40.82	40.70	40.96	41.15
TiO_2	0.04	0.01	0.02	0.00	0.00	0.00	0.05
Al_2O_3	0.00	0.00	0.00	0.03	0.01	0.04	0.01
FeO	8.26	8.23	8.12	7.66	7.60	7.56	7.65
MnO	0.09	0.09	0.09	0.07	0.11	0.08	0.09
MgO	50.48	50.21	50.48	51.21	50.56	51.01	51.55
CaO	0.00	0.03	0.01	0.02	0.01	0.00	0.03
Na_2O	0.00	0.00	0.00	0.00	0.00	0.00	0.00
K_2O	0.00	0.00	0.00	0.00	0.00	0.00	0.00
NiO	0.38	0.39	0.38	0.44	0.39	n.d.	n.d.
Total	99.97	99.38	99.71	100.25	99.38	99.65	100.53
Si	0.993	0.992	0.992	0.990	0.995	0.996	0.993
Al	0.000	0.000	0.000	0.001	0.000	0.001	0.000
Ti	0.001	0.000	0.000	0.000	0.000	0.000	0.001
Fe^{2+}	0.168	0.169	0.166	0.155	0.155	0.154	0.154
Mn	0.002	0.002	0.002	0.001	0.002	0.002	0.002
Mg	1.835	1.837	1.839	1.852	1.843	1.850	1.855
Ca	0.000	0.001	0.000	0.001	0.000	0.000	0.001
Na	0.000	0.000	0.000	0.000	0.000	0.000	0.000
K	0.000	0.000	0.000	0.000	0.000	0.000	0.000
Ni	0.007	0.008	0.007	0.009	0.008	0.000	0.000
Cations	3.006	3.009	3.006	3.009	3.003	3.003	3.006

[1]n.d. = not determined.

estimate of ilmenite lamellae in olivine (Bozhilov et al., 2003). On the other hand, using a comprehensive set of new microanalytical data on the Alpe Arami garnet peridotite, Paquin and Altherr (2001) obtained much lower P-T estimates of about 1180°C and 5.9 GPa (Fig. 8) by applying a number of widely tested and up-to-date thermobarometers based on major and trace elements. No firm evidence was found for exhumation from depths >200 km as postulated by Dobrzhinetskaya et al. (1996). Paquin and Altherr (2001) recognized that mineral disequilibria appear to be the main cause for the lower P-T values suggested by some previous studies (e. g., Medaris, 1999).

Metamorphic conditions of 1000–1250°C and 6.0–6.5 GPa were originally estimated for garnet peridotites from southern Sulu based on the geothermobarometer by Yang et al. (1993). Later, Yang and Jahn (2000) reported a P-T estimate of around 1000°C and 5.1 GPa and a very complex metamorphic history. R Zhang et al. (2000) recognized that the garnet peridotites from the Dabie-Sulu UHP terrane have been subjected to recrystallization at 760–970°C and 4.0–6.5 GPa under an extremely low geothermal gradient of 5°C/km. Moreover, they differentiated two distinct types of garnet peridotite based on the mode of occurrence and petrochemical characteristics: Type A = mantle-derived peridotite and Type B = crustal peridotites. Garnet peridotites from southern Sulu, including those from this hole, have been recognized as Type A (R. Zhang et al., 2000).

Compositional zoning of porphyroblastic garnet and clinopyroxene has been revealed by our detailed microprobe analyses for the PPH1 garnet peridotite. Using equilibrium core compositions of the minerals for the relevant geothermobarometers, P-T conditions of 6.0–7.0 GPa and 1100–1200°C

TABLE 6. Representative Microprobe Analysis of Muscovites in UHP Rocks from the CCSD-PPH1

Sample:	61-11		124-45		240-135		314-172	
SiO_2	46.11	47.92	51.49	50.16	45.00	46.47	51.43	52.48
TiO_2	0.64	0.27	0.385	0.398	0.96	0.70	0.31	0.30
Al_2O_3	27.59	26.39	23.03	24.41	29.30	27.55	24.60	22.70
Cr_2O_3	0.06	0.06	0.053	0.022	0.08	0.04	0.06	0.07
FeO	7.03	6.44	2.67	2.76	4.17	4.52	1.31	1.33
MnO	0.10	0.16	0.023	0.00	0.05	0.10	0.01	0.00
MgO	1.64	2.32	4.73	4.05	2.00	2.50	4.88	5.38
CaO	0.04	0.04	0	0.01	0.01	0.01	0.00	0.00
Na_2O	0.21	0.22	0.158	0.366	0.54	0.44	0.33	0.17
K_2O	10.50	10.43	10.87	10.4	9.87	10.06	10.66	10.99
Total	94.21	94.54	93.41	92.58	93.75	93.77	94.63	93.85
Si	3.159	3.257	3.500	3.436	3.110	3.201	3.464	3.552
Ti	0.033	0.014	0.020	0.021	0.050	0.036	0.016	0.015
Al	2.227	2.114	1.845	1.971	2.386	2.237	1.953	1.811
Cr	0.003	0.003	0.003	0.001	0.004	0.002	0.003	0.004
Fe	0.403	0.366	0.152	0.158	0.241	0.260	0.074	0.075
Mn	0.006	0.009	0.001	0.000	0.003	0.006	0.001	0.000
Mg	0.167	0.235	0.479	0.414	0.206	0.257	0.490	0.543
Ca	0.003	0.003	0.000	0.001	0.001	0.001	0.000	0.000
Na	0.028	0.028	0.021	0.049	0.072	0.058	0.043	0.022
K	0.918	0.904	0.943	0.909	0.870	0.884	0.916	0.949
Cations	6.947	6.933	6.964	6.960	6.943	6.942	6.960	6.971

were obtained. This new P-T estimate is consistent with that derived from mineral exsolution textures (R. Zhang et al., 2003). Observed zonations within the garnet and clinopyroxene porphyroblasts provide a solution to the contradiction as to why the pressure calculated by geobarometers is always lower than that derived from mineral exsolution textures. We, therefore, conclude that temperature estimates using homogenous compositions of Grt + Cpx or their rims are too low; they possibly represent the UHP metamorphic condition of garnet peridotite transported into the subducted zone, rather than their original crystallization conditions. Our study also indicates that the Sulu garnet peridotite originally formed at great depth, possibly as deep as 180 km under a very low geothermal gradient. This thermal regime is restricted to a cold subduction zone and a mantle wedge close to a subduction zone. As demonstrated by thermal modeling (Peacock, 1991), rapid subduction of a cold plate will conduct heat from the mantle wedge directly above the subduction zone, resulting in a lower temperature than that of the normal mantle, and a reversed thermal gradient.

A mechanism explaining how fragments of the mantle wedge may have been incorporated in subducting supercrustal rocks has been proposed (Brueckner, 1998; Brueckner and Medaris, 2000; R. Zhang et al., 2000). Some garnet peridotites could have been transported to shallow mantle depths by mantle convection or upwelling, and later emplaced into subducted continental crust at about 100–120 km depth. Others may have been emplaced into the subducting slab at depths >120 km. In the latter case, the hosting supracrustal rocks would have been subducted to a similar depth as that postulated for type A garnet peridotites. The upper pressure limit of UHP eclogites has long been discussed. For eclogite from Qingdao of the northern Sulu belt, a very high pressure of >7 GPa has been claimed by Ye et al. (2000) based on the possible occurrence of majoritic garnet. However, no evidence indicates

TABLE 7. Representative Microprobe Analysis of Amphiboles
in UHP Rocks from the CCSD-PPH1

Sample:	125-47			124-45		134-55		135-56	
SiO_2	43.76	44.11	45.57	38.40	37.54	40.32	39.15	44.43
TiO_2	0.53	0.59	0.35	0.06	0.06	0.00	0.00	0.00
Al_2O_3	9.61	9.44	12.38	18.17	19.22	15.74	18.29	11.67
FeO	17.35	17.33	12.54	18.74	4.36	4.80	4.09	3.65
Cr_2O_3	0.05	0.03	0.02	0.01	2.37	1.56	2.58	1.81
MnO	0.39	0.46	0.07	0.17	0.04	0.03	0.10	0.06
MgO	10.53	10.41	12.21	6.92	18.65	18.58	17.85	19.00
CaO	11.23	11.11	8.44	9.56	10.09	11.73	10.95	11.79
Na_2O	1.78	1.89	4.44	4.16	3.40	2.82	2.86	2.53
K_2O	0.82	0.80	0.35	0.16	0.14	0.43	0.17	0.74
Total	96.05	96.17	96.37	96.35	95.87	96.01	96.04	95.68
TSi	6.61	6.66	6.64	5.81	5.22	5.71	5.48	6.35
TAl	1.39	1.34	1.36	2.19	2.79	2.29	2.52	1.66
CAl	0.32	0.34	0.77	1.04	0.36	0.33	0.50	0.31
CCr	0.01	0.00	0.00	0.00	0.26	0.17	0.29	0.20
CFe^{3+}	0.63	0.56	0.55	0.79	0.51	0.57	0.48	0.44
CTi	0.06	0.07	0.04	0.01	0.01	0.00	0.00	0.00
CMg	2.37	2.34	2.65	1.56	3.86	3.92	3.73	4.05
CFe^{2+}	1.56	1.63	0.98	1.58	0.00	0.00	0.00	0.00
CMn	0.05	0.06	0.01	0.02	0.01	0.00	0.01	0.01
BCa	1.82	1.80	1.32	1.55	1.50	1.78	1.64	1.80
BNa	0.18	0.20	0.68	0.45	0.50	0.22	0.36	0.20
ANa	0.34	0.35	0.57	0.77	0.42	0.55	0.42	0.51
AK	0.16	0.15	0.07	0.03	0.03	0.08	0.03	0.14
Cations	15.50	15.51	15.64	15.80	15.44	15.63	15.45	15.64

that eclogites from the southern Sulu area have been subducted to depths >120 km.

Based on the above discussion, a tectonic model of garnet peridotite sandwiched between UHP supracrustal rocks is suggested. We propose that the supercrustal rocks from this drill hole probably were recrystallized under much lower pressure than that of the garnet peridotite (Fig. 8). The garnet peridotite in the mantle wedge immediately above the subduction zone was emplaced tectonically into the subducted slab, and subjected to coeval UHP metamorphism together with deeply subducted supercrustal rocks. They were exhumed as a coherent slab toward the Earth's surface, and subjected to amphibolite-facies retrograde metamorphism. Therefore, the P-T path of mantle garnet peridotites is characterized by an early decrease in both pressure and temperature, and a late distinct decompression coupled with a slow temperature decrease (Fig. 8). Because of very short resident time and very rapid exhumation, many original characteristics of the garnet peridotites, such as higher metamorphic temperature and mantle compositions, have not been completely erased. This model is strongly supported by oxygen isotope data. All minerals in the eclogites, gneisses, schists, and quartzites from this drill hole have variable but negative oxygen isotopic values (Zhang et al., 2005). This indicates that their protoliths were subjected to widespread meteoric water-rock interactions prior to subduction. However, all fresh garnet, olivine, orthopyroxene, and clinopyroxene separates from the garnet peridotites have normal oxygen isotopic compositions as commonly obtained from mantle

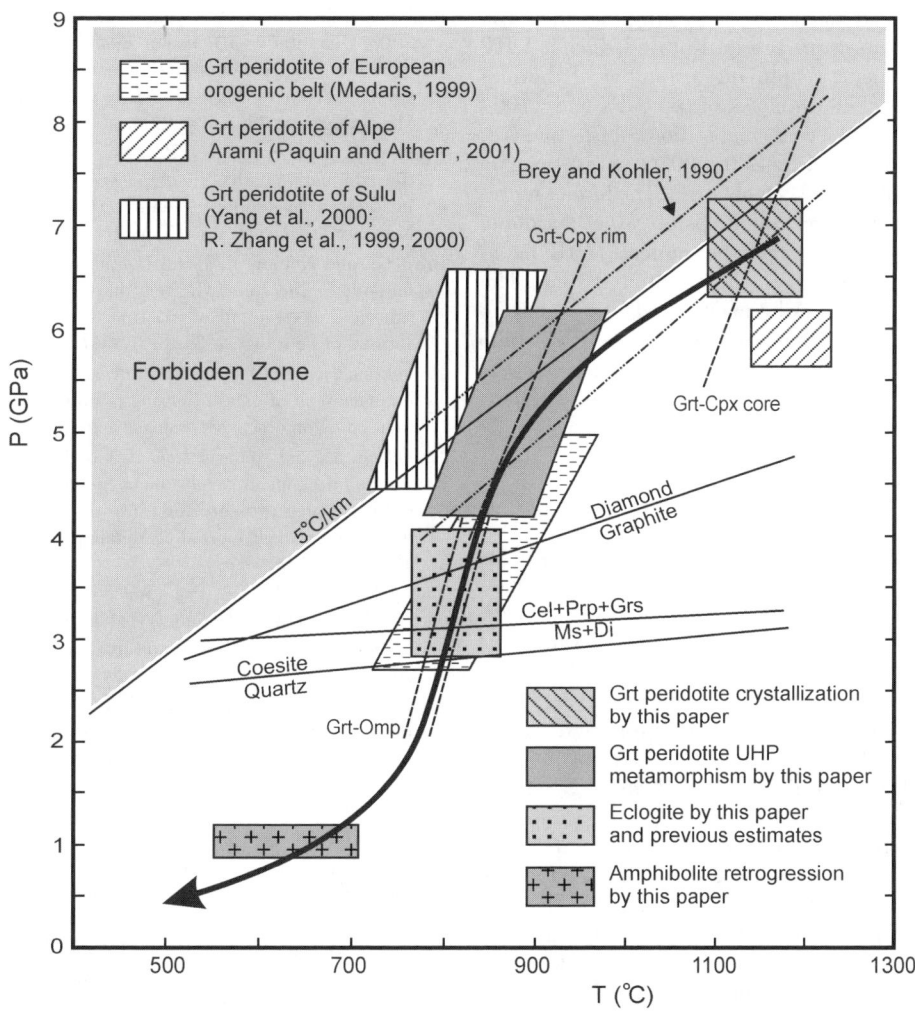

FIG. 8 Metamorphic conditions and P-T path of the eclogite and garnet peridotite, in comparison with other garnet peridotites from continental orogenic belts. Grt-Omp = the isothermal line calculated by the Fe^{2+}-Mg partitioning between coexisting garnet and omphacite in the eclogites; Grt-Cpx core and rim = the isothermal line by the Fe^{2+}-Mg partitioning between coexisting garnet and clinopyroxene core and rim compositions in the garnet peridotite, respectively; Brey and Kohler, 1990 = the isobaric line for the Al-in-orthopyroxene geobarometer. For others, see text.

rocks. This indicates that the UHP supercrustal rocks and the garnet peridotites had distinctly different histories prior to their merging in the subduction zone.

Conclusions

The CCSD-PPH1 recovered a continuous core of eclogite, garnet peridotite, orthogneiss, paragneiss, and minor schist and quartzite. Garnet peridotites were derived from a depleted mantle wedge above a subduction zone, whereas other rocks are metamorphosed supercrustal rocks with continental affinities. The gneiss, schist, and quartzite together with eclogite were subjected to an early UHP metamorphism and subsequent amphibolite-facies retrogression. Their P-T conditions are 780 to 850°C and >2.7 to 3.7 GPa, and 540–710°C and 1.0 GPa, respectively. Garnet and clinopyroxene porphyroblasts of garnet peridotite show distinct compositional zonations, suggesting that their rim compositions were re-equilibrated during UHP metamorphism. Employing a number of thermobarometers, P-T estimates of 6–7 GPa and 1100–1200°C using the core compositions of Cpx and Grt, and 4.2–6.0 GPa at 800–900°C using the rim compositions have been obtained. The former represents crystallization conditions of garnet peridotite in the upper mantle; the latter indicates UHP metamorphic conditions in the subducted zone. The garnet peridotite in the mantle wedge just above a subduction zone was emplaced tectonically into deeply subducted crustal rocks; both have common UHP and amphibolite-facies retrograde metamorphic overprints.

Acknowledgments

This paper is dedicated to W. G. Ernst for his contribution to the study of subduction zone metamorphism and his support for the Chinese Continental Scientific Drilling program. This work was funded by a Sino-German and a Sino-American project related to the Chinese Continental Scientific Drilling. These projects were supported by the Major State Basic Research Development Program (2003CB716501), the National Natural Scientific Foundation of China (40399142), the project of the Ministry of Land and Resources of China (2002207), the National Science Foundation of Germany (DFG, Ho 375/22), and the U.S. National Science Foundation (EAR 0003355). We sincerely thank Yang Jingsui, Liu Fulai, You Zhendong, Jin Zhenmin, Zhang Ruyuan, D. Rumble, and the scientists from the Geosciences Centre Goettingen for help during various stages of the research.

REFERENCES

Armstrong, J. T., 1991, Quantitative elemental analysis of individual microparticles with electron beam instruments, in Heinrich, K. F. J., and Newbury, D. E. eds., Electron probe quantization: New York, NY, Plenum Press, p. 261–285.

Berman, R. G., 1991, Thermobarometry using multi-equilibrium calculations: A new technique with petrological applications: Canadian Mineralogist, v. 29, p. 833–855.

Bohlen, S. R., and Boettcher, A. L., 1982, The quartz-coesite transformations: A pressure determination and the effects of other composition: Journal of Geophysical Research, v. 87, p. 7073–7078.

Bozhilov, K. N., Green, H. W., and Dobrzhinetsdaya, L. F., 2003, Quantitative 3D measurement of ilmenite abundance in Alpe Arami olivine by confocal microscopy: Confirmation of high-pressure origin: American Mineralogist, v. 88, p. 596–603.

Brenker, F. E., and Brey, G. P., 1997, Reconstruction of the exhumation path of the Alpe Arami garnet-peridotite body from depths exceeding 160 km: Journal of Metamorphic Geology, v. 15, p. 581–592.

Brey, G., and Kohler, T., 1990, Geothermobarometery in four-phase lherzolite II. New thermobaromters, and practical assessment of existing thermobarometers: Journal of Petrology, v. 31, p. 1353–1378.

Brueckner, H. H., 1998, Sinking intrusion model for the emplacement of garnet-bearing peridotites into continent collision orogens: Geology, v. 26, p. 478.

Brueckner, H. and Medaris, L. G., 2000, A general model for the intrusion and evolution of "mantle" garnet peridotites in high-pressure and ultra-high-pressure metamorphic terranes: Journal of Metamorphic Geology, v. 18, p. 123–133.

Carswell, D. A., O'Brien, P. J., Wilson, R. N., and Zhai, M., 1997, Thermobarometry of phengite- bearing eclogites in the Dabie Mountains of central China: Journal of Metamorphic Geology, v. 15, p. 239–252.

Carswell, D. A., Wilson, R. N., and Zhai, M., 2000, Metamorphic evolution, mineral chemistry, and thermobarometry of schists and orthogneisses hosting ultra-high pressure eclogites in the Dabieshan of central China: Lithos, v. 52, p. 121–155.

Dobrzhinetskaya, L., Bozhilov, K. N., and Green, H. W., 1999, The solubility of TiO_2 in olivine: implications for the mantle wedge environment: Chemical Geology, v. 160, p. 357–370.

Dobrzhinetskaya, L., Green, H., and Wang, S., 1996, Alpe Arami: A peridotite massif from depths of more than 300 kilometers: Science, v. 271, p. 1841–1845.

Enami, M., Zang, Q., and Yin, Y., 1993, High-pressure eclogites in northern Jiangsu–southern Sangdong province, eastern China: Journal of Metamorphic Geology, v. 11, p. 589–603.

Graham, C. M., and Powell, R., 1984, A garnet-hornblende geothermometers: Calibration, testing, and application to the Pelona Schist, Southern California: Journal of Metamorphic Geology, v. 2, p. 13–31.

Harker, B., Sharp, T., Zhang, R., Liou, J., and Hervig, R., 1997, Determining the origin of ultrahigh-pressure lherzolites: Science, v. 278, p. 702–704.

Jahn, B. M., 1998, Geochemical and isotopic characteristics of UHP eclogites and ultramafic rocks of the Dabie orogen, in Hacker, B. R., and Liou, J. G., eds., When continents collide: Geochemistry of ultrahigh-pressure rocks: Dordrecht, The Netherlands, Kluwer Academic Publishing, p. 203–239.

Kadarusman, A., and Parkinson, C. D., 2000, Petrology and P-T evolution of garnet peridotites from central

Sulawesi, Indonesia: Journal of Metamorphic Geology, v. 18, p. 193–209.

Kennedy, C. A., and Kennedy, G. C., 1976, The equilibrium boundary between graphite and diamond: Journal of Geophysical Research, v. 81, p. 2467–2470.

Kretz, R., 1983, Symbols for rock-forming minerals: American Mineralogist, v. 68, p. 277–279.

Krogh, E., 1988, The garnet-clinopyroxene Fe-Mg geothermometer—reinterpretation of existing experimental data: Contributions to Mineralogy and Petrology, v. 99, p. 44–48.

Liou, J., and Zhang, R., 1998, Petrogenesis of an ultrahigh-pressure garnet-bearing ultramafic body from Maowu, Dabie Mountains, east-central China: The Island Arc, v. 7, p. 115–134.

Liu, F., Xu, Z., Katayama, I., Yang, J., Maruyama, S., and Liou, J., 2001, Mineral inclusions in zircon of para- and orthogneiss from pre-pilot drillhole CCSD-PP1, Chinese continental scientific drilling project: Lithos, v. 59, p. 199–215.

Liu, F., Xu, Z., Liou, J. G., Katayama, I., Masago, H., Maruyama, S., and Yang, S., 2002, Ultrahigh-pressure mineral inclusions in zircons from gneissic core samples of the Chinese Continental Scientific Drilling Site in eastern China: European Journal of Mineralogy, v. 14, p. 499–512.

Medaris, G. L., 1999, Garnet peridotite in Eurasian HP and UHP terranes: A diversity of origins and thermal histories: International Geology Review, v. 41, p. 799–815.

Michael, P. T., Peter, B., and Ravna, E. J. K., 2000, Kyanite eclogite thermobarometry and evidence for thrusting of UHP over HP metamorphic rocks, Nordoyane, Western Gneiss Region, Norway: American Mineralogy, v. 85, p. 1637–1650.

O'Neill, H. St.C., 1980, An experimental study of Fe-Mg partitioning between garnet-olivine and its calibration as a geothermometer, corrections: Contributions to Mineralogy and Petrology, v. 72, p. 337.

O'Neill, H. StC., and Wood, B. J., 1979, An experimental study of Fe-Mg partitioning between garnet-olivine and its calibration as a geothermometer: Contributions to Mineralogy and Petrology, v. 70, p. 59–70.

Paquin, J., and Altherr, R., 2001, New constraints on the P-T evolution of the Alpe Arami Garnet peridotite body (Central Alps, Switzerland): Journal of Petrology, v. 42, p. 1191–1140.

Peacock, S. M., 1991, Numerical simulation of subduction zone pressure-temperature-time paths: constraints on fluid production and arc magmatism: Philosophical Transactions of the Royal Society of London, A, v. 335, p. 341–353.

Ravna, E. K., 2000, The garnet-clinopyroxene Fe^{2+}-Mg geothermometer: An updata calibration: Journal of Metamorphic Geology, v. 18, p. 211–219.

Ravna, K., and Milke, P. T., 2001, Geothermobarometry of phengite-kyanite-quartz/coesite eclogites: Eleventh Annual V. M. Goldschmidt Conference, Hot Springs, Virginia, p. 3145.

Van Roermund, H. L. M., Drury, M. R., Barnhoorn, A., and De Ronde, A. A., 2000, Super-silica garnet microstructures from an orogenic garnet peridotite: Evidence for an ultra-deep (>6 GPa) origin: Journal of Metamorphic Geology, v. 18, p. 135–147.

Yang, A., 1994, A revision of the garnet-clinopyroxene Fe^{2+}-Mg exchange geothermometer: Contributions to Mineralogy and Petrology, v. 115, p. 467–473.

Yang, J., Godard, G., Kienast, J., Lu, Y., and Sun, J., 1993, Ultrahigh-pressure (60 kbar) magnesite-bearing garnet peridotites from northeastern Jiangsu, China: Journal of Geology, v. 101, p. 541–554.

Yang, J., and Jahn, B., 2000, Deep subduction of mantle-derived garnet peridotites from the Su-Lu UHPM terrane in China: Journal of Metamorphic Geology, v. 18, p. 167–180.

Ye, K., Cong, B., and Ye, D., 2000, The possible subduction of continental material to depths greater than 200 km: Nature, v. 407, p. 734–736.

Zhang, R., Hirajima, T., Banno, S., Cong, B., and Liu, J. G., 1995, Petrology of ultrahigh-pressure rocks from the southern Sulu region, eastern China: Journal of Metamorphic Geology, v. 13, p. 659–675.

Zhang, R., Liou, J. G., Yang, J. S., and Ye, K., 2003, Ultrahigh-pressure metamorphism in the forbidden zone: The Xugou garnet peridotite, Sulu terrane, eastern China: Journal of Metamorphic Geology, v. 21, p. 539–550.

Zhang, R., Liou, J. G., Yang, J., and Yui, T., 2000, Petrochemical constraints for dual origin of garnet peridotites from the Dabie-Sulu UHP terrane, eastern-central China: Journal of Metamorphic Geology, v. 18, p. 149–166.

Zhang, R., Shu, J., Mao, H., and Liou, J. G., 1999, Magnetite lamellae in olivine and clinohumite from Dabie UHP ultramafic rocks, central China: American Mineralogist, v. 84, p. 564–569.

Zhang, Z., Xiao, Y., Liu, F., Liou, J. and Hoefs, J., 2005, Petrogenesis of UHP metamorphic rocks from Qinglongshan, southern Sulu, east-central China: Lithos, v. 73, p. 189–207.

Zhang, Z., Xu, Z., Li, F., Meng, F., Yang, T., Li, T., and Yin, X., 2002, Composition and metamorphism of the root of the southern Sulu orogen: Acta Geologica Sinica, v. 21, p. 609–616 (in Chinese with English abstract).

Zhang, Z., Xu, Z., and Xu, H., 2000, Petrology of ultrahigh-pressure eclogite from the ZK703 drillhole in the Donghai, eastern China: Lithos, v. 52, p. 35–50.

Zhang, Z., Xu, Z., and Xu, H., 2003, Petrology of the non-mafic UHP metamorphic rocks from a drillhole in the Southern Sulu orogenic belt, eastern-central China: Acta Geologica Sinica, v. 77, p. 173–186.

Zhang, Z., You, Z., Han, Y., and Sang, L., 1996, Petrology, metamorphic process and genesis of the Dabie-Sulu eclogite belt, Eastern-Central China: Acta Geologica Sinica, v. 9, p. 134–156.

Petrologic Study of Ultrahigh-Pressure Metamorphic Cores from 100 to 2000 m Depth in the Main Hole of the Chinese Continental Scientific Drilling Project, Eastern China

SHANGGUO SU,[1]

China University of Geosciences, Beijing, 100083, China

JUHN G. LIOU,

Department of Geological and Environmental Sciences, Stanford University, Stanford California 94305

ZHENDONG YOU,

China University of Geosciences, Wuhan, 430074, China

FENGHUA LIANG, AND ZEMING ZHANG

Institute of Geology, China Academy of Geosciences, Beijing, 100037, China

Abstract

The Chinese Continental Scientific Drilling (CCSD) Project in Donghai recovered more than 1000 m of eclogite and garnet peridotite cores for study. Examination of rocks from 100–2000 m of the main borehole has identified five major lithological types: (1) eclogite and garnet pyroxenite; (2) eclogitic gneiss; (3) garnet peridotite; (4) biotite (hornblende) two-feldspar gneiss; and (5) fault breccia and mylonite. The eclogite was further subdivided into two types: crustal eclogite and mantle-derived eclogite. Crustal eclogites are ubiquitous as layers of various thickness in gneissic rocks, and contain low-Prp (<40 mol%) garnet and omphacite. Mantle-derived eclogites are spatially associated with ultramafic cores composed mainly of garnet wehrlite, and have higher Prp-bearing (>40 mol%) garnet and low Jd-bearing clinopyroxene. Chemically, the crustal eclogites are relatively low in MgO and high in SiO_2, but have high, variable contents of Al_2O_3 and rare-earth elements. Most crustal eclogites range in SiO_2 content from 49 to 60 wt%, whereas mantle-derived eclogites are rich in MgO, and have SiO_2 content less than 49 wt%. Garnet peridotites consist of olivine (Fo = 85–91), enstatite, Mg-rich garnet, and diopsidic clinopyroxene; Ti-clinohumite is also widespread. Mineral paragenesis indicates that the garnet peridotites together with other lithologies underwent in situ ultrahigh-P metamorphism (UHPM).

Based on the differences in rock association, structural kinematics, and seismic characteristics, we have identified two different rock slices separated by a fault zone at 1600 m depth, where breccia and mylonite developed. Rutile eclogites are dominant in the upper slice, and phengite eclogites are layered with deformed tonalite and paragneiss in the lower slice. These UHPM rocks underwent variable retrograde metamorphism; eclogite is replaced by symplectite-bearing garnet amphibolite, and eclogitic gneiss is retrograded to biotite (hornblende) plagioclase gneiss. Late-stage crustal extension resulted in local cataclasis, forming tectonic breccia with the development of chlorite, calcite, hematite, and epidote under epidote amphibolite- to greenschist-facies conditions. Nearly 2000 m of recovered UHP core from the CCSD main hole reveals that voluminous crustal materials were subducted to mantle depths and rapidly returned to the surface. UHPM cores record subduction and exhumation processes of the continental crust and provide information for the study of continental subduction/collision and mantle dynamics.

Introduction

DABIE-SULU UHP metamorphic rocks have received both Chinese and international attention since the late 1980s, when coesite, diamond, and other UHP mineral assemblages were found in eclogites (Wang et al., 1989; Xu et al., 1992; Ye et al., 2000). Since the 1990s, regional geological mapping at a scale of 1:50,000 has been conducted in the Dabie-Sulu terrane by provincial geological bureaus; results reveal the two-dimensional structure of the collisional orogen. Several geophysical profiles, including seismic

---

[1]Corresponding author: email:susg@cugb.edu.cn

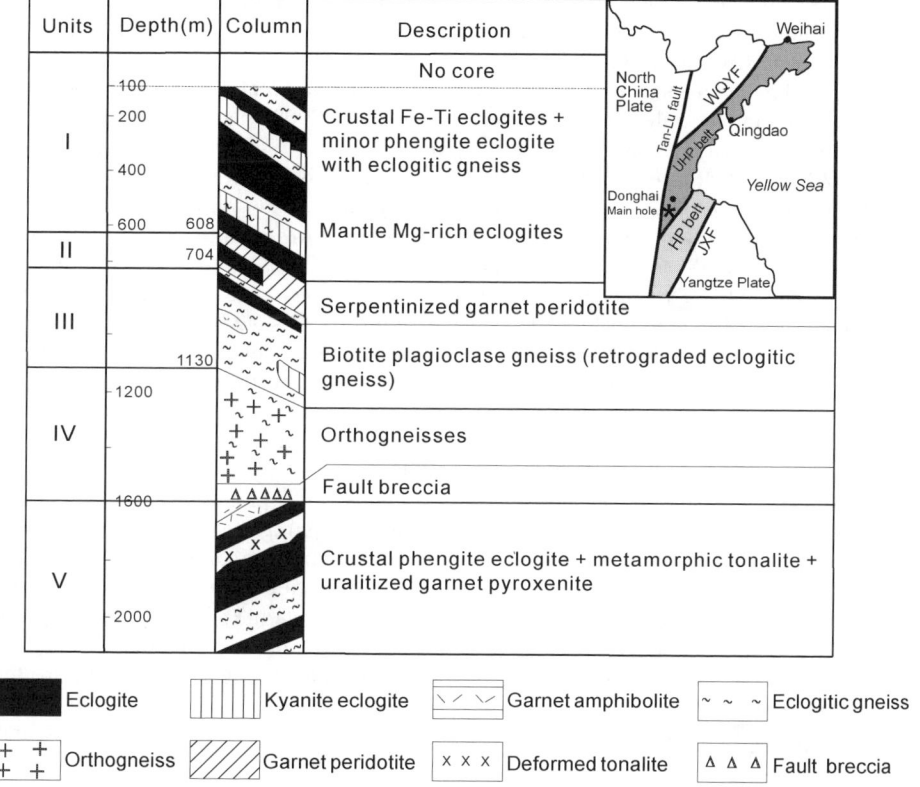

FIG. 1. A schematic lithological column from 100 to 2000 m depth in the main hole of the CCSD. Inset map shows the location of the main hole at Donghai, southeast of Beijing.

refraction studies, were conducted concurrently to interpret subsurface structures and the thickness of the UHP slices (Wang et al., 2000; Xu et al., 2001; Yang et al., 2004; Yuan et al., 2004). Implementation of the Chinese Continental Scientific Drilling (CCSD) Project provides a three-dimensional section of the collisional orogenic belt (Xu et al., 1998). By October 2002, the main drill hole had reached a depth of 2000 m. Reconnaissance descriptions of drill hole cores from 100–2000 m of the main hole have already been published in local Chinese journals (e.g., Liu et al., 2003a; Z. Zhang et al., 2003). On the basis of more than 1000 thin sections, we summarize rock associations and their petrological characteristics, identify two kinds of eclogites, and describe the metamorphic and deformational history of these UHPM rocks from the main borehole. This contribution provides the first description of the main CCSD drill hole cores for an international journal.

Geological Setting and Subdivision of Lithological Units

The main borehole is located near the village of Maobei, about 15 km southwest of Donghai County, Jiangsu Province, and lies at the southern end of the Sulu-UHPM belt (Fig. 1). Widespread eclogite and amphibolite lenses and blocks in orthogneisses and supracrustal rocks including biotite gneiss, kyanite quartzite, and marble occur in this region. Near Maobei, a large (3000 m × 250–300 m) N-S–trending rutile eclogite-ultramafic complex within gneissic rocks forms a NNE-trending overturned fold with a SE dipping axial plane; the main CCSD drill hole is located on the eastern limb of the fold (Fig. 1; Liu et al., 2001, 2002, 2003b).

No core sample was recovered in the uppermost 100 m of the main hole; only slushy fragments were obtained. The rock cores 100–2000 m and below are subdivided into 5 lithological units: (I) rutile eclogite and eclogitic gneiss with various degrees of retrogression (100–608 m); (II) garnet peridotite with layers of phengite-eclogite and Ti-clinohumite-bearing garnet pyroxenite (608–704 m); (III) epidote biotite (amphibole) gneiss intercalated with kyanite eclogite and amphibolite (704–1130 m); (IV) amphibole biotite plagioclase gneiss and felsic gneiss (1130–1600 m) with 40 m of tectonic breccia at its base; and (V) phengite-rutile-bearing eclogite with biotite gneiss and layers of highly deformed tonalitic gneiss (1600–2000 m) and uralitized pyroxenites (pyroxene entirely replaced by uralite with modal content 30–50%, up to 89% at 1998 m, and containing small amounts of phlogopite). These petrotectonic assemblages recovered from the main drill hole are well correlated with those exposed in the Sulu UHPM belt.

Two distinct structural breaks were recognized in this nearly continuous 2 km long core. One is a 40 m thick breccia zone at 1600 m depth separating the granitic gneiss of Unit IV from phengite eclogite + garnet amphibolite + dominant gneiss of Unit V (Z. Zhang et al., 2003); this breccia zone represents a lithological and tectonic boundary between two UHP slabs. The upper UHP slab consists of granitic gneiss and paragneiss with a variety of eclogitic rocks and garnet peridotite, whereas the lower slab consists of paragneiss, metamorphic tonalite and phengite eclogites. The foliation and lineation in the gneissic rocks also exhibit an abrupt change across the 1600 m plane. In the upper slice, the lineation trends ESE (100°) plunging 30–40°, whereas the lower slice trends SSE (160°) plunging >50°. This tectonic boundary is imaged on a seismic reflection profile at the depth of 1600 m (Yang et al., 2004). Pebbles in the brecciated zone are more-or-less rounded, suggesting that cataclasis must have occurred at an upper crustal level long after the exhumation of the UHP rocks.

At the 850–1200 m depth interval, a series of ductile shear zones is present as local mylonite zones (Xu et al., 2004). Petrographic study shows that the feldspars occur as porphyroclasts in mylonitic gneiss with a dominantly chlorite, albite and hematite matrix. The development of the ductile shear zones may be closely related to doming during later exhumation of the UHP terrane (e.g., Liou et al., 2004).

Major Rock Types

Chemical analyses of representative core samples are listed in Table 1. The compositions were determined by inductively coupled plasma spectrometry (ICP-MS) at the Institute of Rock and Mineral analysis, Chinese Academy of Geological Sciences. The analytical errors are discussed in Chen et al. (2002); major oxides are ± 3% and minor oxides are ± 5 % of the reported values.

Figure 2 is an A ($Na_2O + K_2O$) – F (FeOt) – M (MgO) diagram showing compositions of a few representative core samples. Most eclogites, either crustal- or mantle-derived, plot within the tholeiitic basalt field; eclogitic gneisses also plot within the tholeiite field, suggesting that they were derived from eclogites. Compositions of garnet peridotites are characterized by extremely low alkali contents; two samples lie on the FM join. Felsic gneiss plots near the AF join due to their high alkali and low MgO contents.

Figure 3 is a plot of SiO_2 vs. total alkaline after Irvine and Baragar (1971). All analyses are scattered along the boundary between the alkaline and subalkaline fields, but show distinct groupings: ultramafic for garnet peridotite, basaltic for eclogite, basaltic to andesitic for eclogitic gneisses, and tonalitic for felsic gneisses. Crustal eclogites vary in total alkali and SiO_2 contents as well as other major elements shown in Table 1, suggesting that they may have had different protoliths. Samples with high TiO_2 contain more rutile, whereas samples with high Al_2O_3 contain kyanite and zoisite. Again, analyses of eclogitic gneisses are higher in both alkali and SiO_2 contents than eclogites, consistent with the suggestion that they may have been derived from eclogitic rocks through metasomatic processes as suggested by R. Zhang et al. (2003b).[2]

Eclogites

Based on the bulk-rock compositions, eclogites are subdivided into five types, namely: high-silica, high-aluminum, high-titanium, high-magnesium and chemically normal basaltic types. However, on the basis of their tectonic origin, we subdivided the Sulu eclogites both from outcrops and drill hole cores into crustal and mantle-derived types. Crustal eclogites refer to those mafic to intermediate protoliths layered within supracrustal rocks. Mantle-derived eclogites refer to those enclosed within garnet peridotites, which were presumably incorporated in the complex during subduction, and have experienced UHP metamorphism. Criteria for

---

[2] Abbreviations of minerals used in this paper are after Kretz (1983).

TABLE 1. ICP-MS Chemical Analyses of Eclogitic and Other Rocks from 100 to 2000 m Depth, Main Hole of CCSD

Sample	depth	SiO_2	MgO	Al_2O_3	P_2O_5	Na_2O	K_2O	CaO	TiO_2	MnO	Fe_2O_3	FeO	H_2O+	CO_2	S	F	Total
								Garnet peridotite									
B308R253P4a	611.6	42.36	28.06	6.54	0.08	0.18	0.00	3.66	0.17	0.22	4.38	8.95	4.82	0.35	0.13	0.04	99.93
B315R259P6	625.2	35.99	33.65	2.99	0.01	0.00	0.00	0.88	0.16	0.19	7.94	5.62	12.08	0.35	0.04	0.02	99.91
							Mantle-derived eclogite										
B352R278P2k	684.1	43.78	15.78	18.44	0.18	0.62	0.01	7.72	0.13	0.27	3.52	8.95	0.22	0.44	0.01	0.02	100.08
							Crust-derived eclogite										
B132R114P1a	320.0	48.43	5.52	20.91	0.16	3.93	0.97	8.38	0.64	0.18	2.42	5.33	1.36	1.48	0.010	0.041	99.75
B118R104P3b	299.0	48.07	6.09	14.53	0.23	3.61	0.71	10.56	1.70	0.23	4.71	7.49	1.30	0.37	0.090	0.029	99.63
B166R146P1aL	376.2	48.72	3.64	16.29	0.79	3.25	0.52	8.24	3.36	0.42	2.88	11.53	0.46	0.05	0.005	0.068	100.22
B63R60P1a	212.5	49.62	5.38	15.17	0.31	3.06	0.79	9.44	1.78	0.27	2.93	9.73	0.88	0.21	0.017	0.14	99.73
B136R117P3a	326.4	45.72	5.22	14.89	2.07	2.52	0.02	10.08	3.69	0.30	2.22	12.20	0.38	0.22	0.021	0.11	99.66
B218R192P1c	463.1	45.11	4.93	15.15	0.87	3.38	0.16	9.11	3.79	0.35	3.77	12.50	0.32	0.05	0.23	0.064	99.55
							Eclogitic gneiss										
B192R171P1b	421.2	54.18	4.33	15.17	0.25	3.14	1.49	7.05	1.80	0.25	2.43	9.07	0.64	0.09	0.034	0.040	99.89
B87R82P2a	248.3	52.71	2.97	14.74	0.35	4.62	0.58	6.68	2.11	0.43	4.32	10.08	0.34	0.03	0.22	0.063	100.24
B91R86P1e	255.5	52.66	3.26	14.98	0.27	3.57	0.11	7.46	2.03	0.41	2.68	11.62	0.32	0.03	0.097	0.038	99.54
B7R11P1a	112.7	53.95	2.33	14.35	1.23	4.61	0.86	6.74	1.91	0.55	3.93	8.68	0.84	0.33	0.16	0.065	100.54
B10R15P1b	118.4	50.89	4.27	14.60	1.30	3.72	0.30	8.15	2.37	0.35	3.13	10.01	0.60	0.24	0.15	0.057	100.14
B225R196P1n	473.3	53.97	3.73	15.58	0.59	4.52	0.77	7.19	1.59	0.29	3.15	7.60	0.56	0.10	0.029	0.059	99.70
B242R208P1h	500.7	54.94	2.96	14.45	1.25	4.41	1.09	6.29	2.25	0.34	2.77	8.39	0.74	0.23	0.084	0.098	100.21
B242R208P1s	501.5	54.23	2.64	14.74	1.24	3.99	1.15	6.46	2.40	0.38	3.36	8.87	0.36	0.09	0.06	0.09	100.06
B6R10P2	112.3	52.02	2.86	14.24	1.00	4.67	0.67	7.15	2.03	0.56	5.19	8.64	0.90	0.41	0.18	0.053	100.57
							Felsic gneiss										
B33R30P2i	156.68	64.09	1.44	16.14	0.15	6.01	1.63	3.03	0.62	0.17	1.78	3.50	0.78	0.07	0.07	0.022	99.50
B100R92P1cA	269.6	56.84	3.54	15.89	0.25	4.03	2.84	5.27	1.13	0.25	3.79	5.16	1.12	0.09	0.037	0.11	100.35
B155R133P1	356.5	60.38	3.02	13.73	0.30	3.71	2.54	4.63	1.21	0.21	3.30	4.76	1.06	0.55	0.005	0.056	99.46
B105R97P1aL	278.6	58.46	3.26	15.00	0.38	4.76	1.57	5.43	1.22	0.24	2.77	5.80	0.74	0.03	0.028	0.082	99.77
B173R151P1d	387.1	58.65	3.93	14.65	0.58	3.94	1.80	6.45	1.26	0.21	2.26	6.38	0.50	0.03	0.047	0.067	100.64
B174R151P1j	389.1	59.49	2.95	14.13	0.74	4.09	1.96	5.45	1.34	0.28	2.14	6.25	0.48	0.05	0.005	0.082	99.44
B88R83P2aA	249.5	60.83	0.78	16.02	0.27	6.89	0.87	3.35	1.05	0.49	3.22	5.60	0.94	0.05	0.036	0.051	100.45

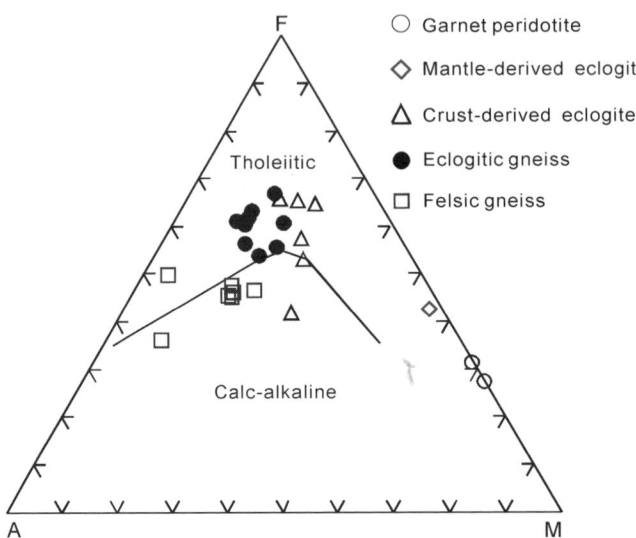

FIG. 2. AFM diagram for eclogitic rocks from the CCSD main hole.

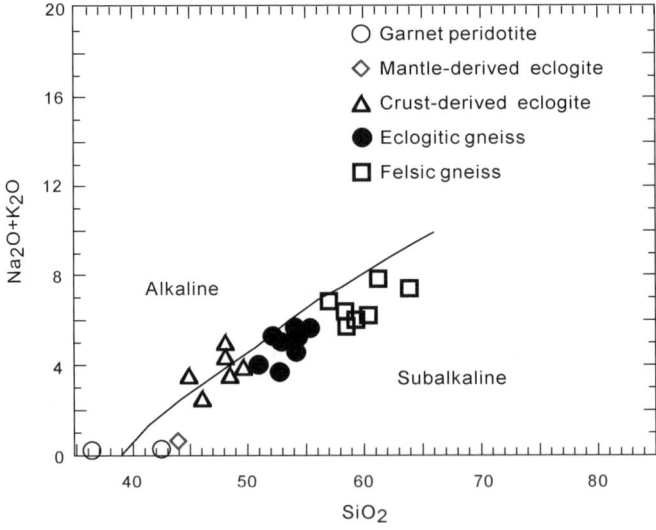

FIG. 3. SiO_2 vs total alkalis diagram for eclogitic rocks from the CCSD main hole.

distinguishing crustal- and mantle-derived eclogites are as follows.

1. Occurrence: Crustal eclogites are ubiquitous as layers of variable thickness in gneissic or other supracrustal rocks, including marble and quartzite; mantle-derived eclogites are restricted to ultramafic bodies, and are mainly distributed in the upper part of the bodies.

2. Mineralogy: Crustal eclogites are much more diverse, and commonly contain garnet and omphacite with additional phengite, epidote, kyanite, quartz, and minor rutile and apatite (Fig. 4A). Mantle-derived eclogites (mostly garnet clinopyroxenites) are mineralogically simple, containing garnet and omphacite or diopsidic clinopyroxene associated with minor rutile, ilmeno-magnetite, and

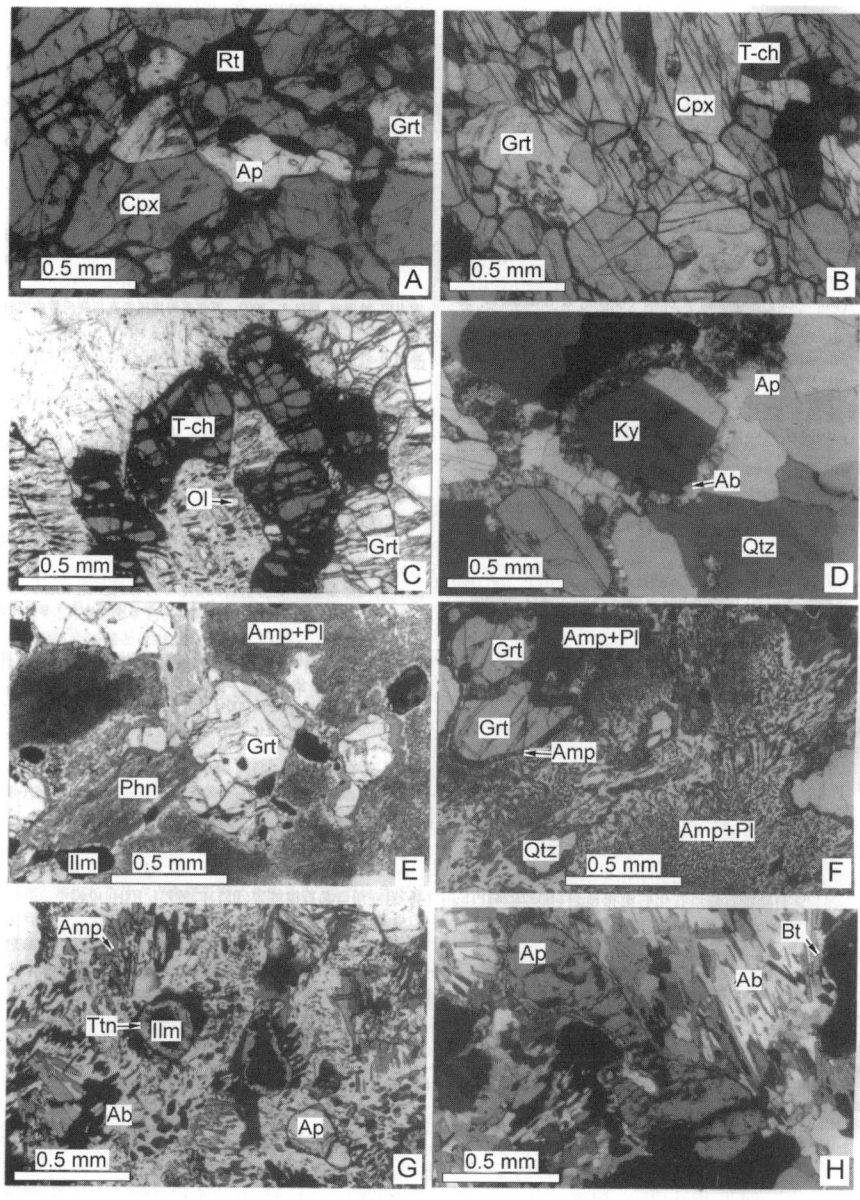

FIG. 4. Photomicrographs of representative core samples from the CCSD main hole. A. Fe-Ti eclogite (342.39 m) showing major phases of garnet, omphacite, and rutile with minor apatite. B. Ti-clinohumite–bearing garnet pyroxenite (611.90 m) showing brown Ti-clinohumite, clinopyroxene, and garnet. C. Serpentinized garnet peridotite (650.80 m) showing relict garnet (Grt), olivine (Ol), and Ti-clinohumite (T-Ch). D. Retrograde kyanite eclogite (1004.40 m) showing kyanite with albite corona. E. Retrograde phengite eclogite (524.80 m) showing phengite being replaced by biotite and Amp + Pl symplectites after omphacite. F. Retrograded quartz eclogite (112.70 m) showing Amp + Pl dendritic symplectite after omphacite and taramitic amphibole after garnet. G. Retrograde eclogitic gneiss (468.35 m) showing albite clouded with biotite and amphibole; ilmenite is rimmed by titanite. H. Biotite plagioclase gneiss (retrograded eclogitic gneiss, 489.30 m) showing relict coarse-grained apatite, matrix biotite + albite symplectite (cross-polarized light).

in a few samples accompanied by Ti-clinohumite (Fig. 4B).

3. Mineral chemistry: Garnets from crustal eclogites are low in Prp component (< 40 mol%) and most clinopyroxenes are omphacite. Garnets from mantle-derived eclogites are high in Prp component (> 40 mol%) and clinopyroxene contains little Jd.

4. Chemistry: Crustal eclogites are relatively low in MgO and CaO, but have higher, variable contents of SiO_2, Al_2O_3, and REEs. Most crustal eclogites range in SiO_2 content from 49 to 60 wt%, whereas the mantle-derived eclogites are high in MgO and have an SiO_2 content of less than 49 wt%.

Kyanite- or zoisite-eclogites are typically of crustal origin, and contain Al_2O_3 as high as 20.9 wt%; REE patterns are characterized by LREE enrichment and positive Eu anomalies. Spider diagrams of the trace element distribution are characterized by positive Th and Sr anomalies, consistent with the occurrence of zoisite and allanite in the rocks.

Eclogitic gneiss

Some UHP rocks with intermediate to silicic protoliths have similar mineral assemblages to those of eclogites, but with higher quartz content (>20%). Zhang et al. (1995) referred to these rocks as garnet-jadeite rocks; coesite pseudomorph inclusions occur in garnet. During retrograde recrystallization, jadeite reacted with quartz to form albite, and garnet was replaced by taramitic amphibole. Irregular strings of fine-grained aegirine augite occur at the boundary between omphacite and quartz aggregates. The rock is characterized by the predominance of quartz aggregates and saccharoidal albites (An = 0–7 mol%) with small amounts of sodic amphiboles and strings of aegirine augite. Phengite flakes of the rock are decomposed to fine-grained aggregates of biotite + plagioclase along margins; some are entirely replaced by biotite and sodic plagioclase. Some gneissic rocks contain zoned epidote minerals with allanite cores and epidote rims. The SrO content of a zoned epidote (at 111.9 m depth) ranges from 3.09→2.08→1.47 wt% from core to rim.

It is difficult to determine whether the eclogitic gneiss underwent UHP metamorphism by routine petrographic examination. Similar garnet-bearing jadeitite with inclusions of coesite pesudomorphs in garnet from the Dabie Mountains was described by Liou et al. (1997). Recently Liu et al. (2002, 2003b) successfully extracted 69 zircon grains from paragneiss and eclogitic gneiss core samples from the pilot hole, PP2, of the CCSD. Most zircon crystals contain micro-inclusions of coesite and other UHP minerals; these data undoubtedly indicate that the host rocks experienced UHP metamorphism (Liu et al., 2001, 2002, 2004a).

Garnet peridotite

Abundant garnet peridotite core samples were recovered from this main drill hole and most are intensively serpentinized; more than 58 thin sections were examined in this study. The major rock type is garnet wehrlite, containing Ol (75%), Cpx (15%), Grt (10%), and small amounts of Ti-clinohumite. Minor garnet lherzolite is found in the core samples. Analyzed olivines are fosterite (Fo = 85–91 mol%); most clinopyroxenes have a diopsidic composition, and garnets are pyrope rich. Intensive serpentinization fragmented olivine grains with segregation of magnetite. Garnet and clinopyroxene are chloritized. Orthopyroxene (En = 88 mol%) in lherzolite is serpentinized to bastite. Ti-clinohumite occurs as inclusions within garnet and olivine, and as trace phase in the matrix (Figs. 4B and 4C). Garnet wehrlite has a granoblastic texture and contains layers of mantle-derived eclogite.

Felsic gneiss

According to the relative abundance of plagioclase and K-feldspars, felsic gneisses are divided into plagioclase gneiss and two-feldspar gneiss; the latter has been described as orthogneiss (e.g., Liu et al., 2002, 2004b). Skeletal garnets with high MnO contents occur in a few gneissic cores. Epidotes are zoned with allanite cores, similar to those in the eclogitic gneiss mentioned above. K-feldspar of two-feldspar gneiss coexists with plagioclase in similar amounts, and is less euhedral than the plagioclase. The K-feldspar exhibits microcline twinning and contains fine-grained plagioclase inclusions; some grains show myrmekitic intergrowths. These textural relations suggest that the two-feldspar gneisses were originally granitic.

Many plagioclase gneisses are metasomatic products of retrograded eclogitic gneiss, such as those described by R. Zhang et al. (2003b). Alternatively, when eclogitic gneiss was uplifted to the mid- to lower crust, introduction of K-rich fluids from the crust at amphibolite-facies conditions may have caused partial melting. K-feldspar neoblasts crystallized; some eclogitic gneisses were transformed to migmatite and even to two-feldspar granitic gneiss. Therefore, some orthogneisses at depths in the 1130 to 1600 m interval may represent granitic rocks crystallized from partial melt at mid- to lower-crustal levels. Such granitic gneiss likely does not contain coesite inclusions in zircon, as documented by Liu et al. (2002).

Cataclasite and mylonite

Rocks are both brittlely and ductilely deformed throughout the main hole from 100 to 2000 m. Microscopic cataclasites including microbreccia are found in various rocks. Most of the fractured material is derived from their host rocks; the matrix consists primarily of carbonates (mainly calcite) with or without hematite. The predominant breccia layer (~40 m thick) occurs at 1600 m depth (Fig. 1) where felsic gneiss breccia forms an important boundary between the upper and lower slabs. The breccia consists of subrounded fragments in a fine-grained matrix of carbonates and chlorite. This fault breccia may have formed at a rather shallow depth during a late extensional event. Mylonitic rocks are less common, but this does not mean that ductile deformation is less prominent. Ductile deformation is clearly shown in penetrative foliations in both gneissic rocks and eclogites. In addition to a few mylonite layers at 850–1113 m depth in the upper slab, a layer of sheared tonalitic gneiss occurs at 1783–1810 m. Rotated plagioclase porphyroblasts occur at the base of this layer, whereas relict tonalite textures are preserved in the upper part of the protomylonites.

Representative Compositions of Minerals

More than 8 minerals from 36 core samples were analyzed using the EPMA lab at the China University of Geosciences in Beijing. More than 100 analyses were conducted by wavelength dispersive spectrometry with 15 kV accelerating voltage, 10 mm beam size, and 10 nA beam current; 76 analyses were obtained employing an energy dispersive system. Characteristic features of the main minerals are described below.

Garnets

Compositions of analyzed garnets are listed in Table 2 and plotted in Figure 5; compositional variations are directly related to the compositions of host rocks. Garnets from garnet peridotite are high in Pyp (> 60 mol%), whereas garnets of eclogitic gneiss contain mostly almandine, spessartine, and grossular components with less than 15 mol% Pyp.

The samples plotted as rhombs in Figure 5 are representative of mantle-derived eclogites that are high in Mg# and lower in SiO_2 content. Garnets from mantle-derived eclogites are Pyp rich, with values as high as 59 mol%. On the other hand, most garnets from crustal eclogites contain <40%Pyp and have high Alm with variable amounts of Grs.

Clinopyroxenes

Compositions of analyzed pyroxenes are listed in Table 3; 24 analyses plot on the Jd-Ae-WEF diagram of Figure 6. Clinopyroxenes from garnet peridotite plot in the WEF field, and contain 3.45 mol% Jd. Those from mantle-derived eclogites also plot into the WEF field, but have low Jd content of 13.53 mol%. Clinopyroxenes from crustal-derived eclogite have omphacite compositions. Most analyses of Cpx from eclogitic gneiss plot in the aegirine-augite field (Fig. 6); these clinopyroxenes occur as reaction rims around omphacite and quartz, and the relict omphacite cores contain 58.42 mol% Jd (Table 3; Fig. 6).

Amphiboles

Omphacites of most crustal eclogites are retrograded to amphibole and plagioclase symplectites. Analyzed amphiboles have both pargasitic and actinolitic compositions. The Si^{IV} atom of amphibole per formula unit has been suggested as an indicator of metamorphic grade (Spear, 1995). The Si^{IV} values of most analyzed amphiboles are less than 7.0, suggesting retrograde recrystallization occurred under amphibolite-facies conditions. A few analyzed samples with Si^{IV} values higher than 7.0 are actinolite formed during the late greenschist-facies overprint.

White Mica

Phengitic white mica is common in the crustal eclogites; most are 1.0–2.0 mm in size, with high Si values near 3.50 p.f.u, consistent with the UHP origin. Paragonite occurs as coronae around kyanite, and has high sodium (X_{na} = 0.9–1.29) content. Phengites are retrograded to biotite and plagioclase in association with K-feldspar.

Titanium minerals

The most common titanium mineral in the main hole is rutile, which is a major source of titanium of industrial interest in the Donghai region. Both rutile and ilmeno-magnetite are major accessory minerals in eclogites; rutile is common in the crustal eclogites whereas ilmeno-magnetite dominants over rutile in the mantle-derived eclogites and the garnet

TABLE 2. Electron Microprobe Analyses of Garnets from 100–2000 m Depth, Main Hole of CCSD

Sample	Depth (m)	SiO_2	TiO_2	Al_2O_3	Cr_2O_3	FeO	MnO	MgO	CaO	Na_2O	P_2O_5	K_2O	Total
					Garnet peridotite								
B4R7P2e	108.7m	37.73	0.06	21.36	1.33	17.39	0.64	14.57	5.20	0.00	0.00	0.01	98.29
B4R7P2e	108.7m	37.46	0.00	20.88	0.00	22.39	1.06	7.93	8.57	0.12	0.00	0.00	98.41
					Mantle-derived eclogite								
B308R253P7	611.90	40.53	0.03	22.12	0.00	15.50	0.52	15.50	4.11	0.00	0.09	0.01	98.41
					Crustal-derived eclogite								
B146 R126 P4a	342.30	37.76	0.06	21.45	0.00	25.78	0.69	6.70	6.23	0.02	0.01	0.00	98.70
B146 R126 P4a	342.30	38.51	0.03	21.53	0.00	26.10	0.71	6.60	5.94	0.00	0.00	0.00	99.42
B155R131 P1l	352.34	39.32	0.04	21.24	0.00	27.32	0.43	4.59	6.87	0.37	0.35	0.00	100.53
B161R140P1a	368.60	37.28	0.04	21.61	0.00	26.08	0.50	6.07	6.59	0.01	0.00	0.00	98.18
B161R140P1a	368.60	37.47	0.03	21.58	0.00	26.80	0.64	6.23	6.87	0.01	0.00	0.00	99.63
B173R151P1e	389.0	38.33	0.04	22.28	0.02	17.96	0.55	7.29	12.62	0.01	0.00	0.00	99.10
B208 R185 P2f	444.0	38.53	0.20	20.06	0.16	29.59	1.12	4.46	4.55	0.00	0.11	0.02	98.80
B289R236P2f	577.0	39.11	0.17	20.48	0.00	25.45	0.65	5.06	9.52	0.15	0.29	0.03	100.91
B289R236P2f	577.0	39.64	0.00	21.01	0.00	24.10	0.26	4.69	9.67	0.30	0.43	0.07	100.17
B285 R234 P1cf	570.0	39.16	0.04	20.09	0.00	23.57	0.13	5.15	9.91	0.25	0.08	0.08	98.46
B1049R640P17F	1992.3	39.11	0.01	21.21	0.00	22.05	0.60	6.87	8.45	0.02	0.32	0.04	98.68
B1363R124P13iF	2700.0	39.74	0.13	21.62	0.00	20.97	0.81	6.35	9.64	0.15	0.26	0.11	99.78
					Eclogitic gneiss								
B33R30P2i	156.68	38.55	0.09	20.76	0.00	20.35	2.08	1.32	18.03	0.06	0.11	0.01	101.36
B88R82P2g1	249.20	37.33	0.06	20.52	0.00	26.78	2.09	1.33	10.77	0.04	0.05	0.00	98.97
B88R82P2g1	249.20	37.93	0.06	21.47	0.00	26.70	1.98	1.32	10.89	0.04	0.08	0.00	100.47
B88R83P2a	249.50	36.78	0.11	20.96	0.00	26.87	2.60	0.81	11.24	0.03	0.00	0.00	99.40
B88R83P2a	249.50	37.64	0.10	20.90	0.00	26.00	2.45	0.77	11.16	0.05	0.05	0.00	99.12
B125 R110 P1b	309.09	39.14	0.06	21.83	0.00	22.37	0.49	5.59	11.18	0.02	0.03	0.01	100.72
B91R86p1c	354.35	37.25	0.07	21.17	0.00	23.63	1.23	2.33	12.21	0.13	0.00	0.03	98.05
B155R133P2f	357.50	37.92	0.07	21.04	0.00	24.46	1.06	3.10	11.78	0.04	0.00	0.00	99.47
B155R133P2f	357.50	37.73	0.03	21.44	0.00	25.01	1.05	3.07	11.03	0.01	0.00	0.01	99.38

(Table continues)

TABLE 2. (continued)

Sample	Si^{IV}	Al^{iv}	Al^{VI}	Fe^{3+}	Ti	Cr	Fe^{2+}	Mg	Mn	Ca	Na	Total cations	Alm	And	Grs	Pyp	Spe	Ura
							Garnet peridotite											
B4R7P2e	2.84	0.16	1.73	0.34	0.00	0.08	0.75	1.63	0.04	0.42	0.00	8.00	27.08	10.46	0.00	56.90	1.42	4.13
B4R7P2e	2.91	0.09	1.82	0.28	0.00	0.00	1.18	0.92	0.07	0.71	0.02	8.00	40.92	14.40	10.37	31.89	2.42	0.00
							Mantle-derived eclogite											
B308R253P7	3.02	0.00	1.94	0.02	0.00	0.00	0.94	1.72	0.03	0.33	0.00	8.00	29.31	1.07	10.07	58.44	1.11	0.00
							Crustal-derived eclogite											
B146 R126 P4a	2.96	0.04	1.95	0.08	0.00	0.00	1.61	0.78	0.05	0.52	0.00	8.00	46.17	4.70	16.13	31.18	1.82	0.00
B146 R126 P4a	3.01	0.00	1.98	0.00	0.00	0.00	1.70	0.77	0.05	0.50	0.00	8.00	55.90	0.09	16.61	25.82	1.58	0.00
B155R131 Pll	3.02	0.00	1.95	0.00	0.00	0.00	1.78	0.53	0.03	0.58	0.06	7.98	61.07	0.00	19.67	18.29	0.97	0.00
B161R140P1a	2.95	0.05	1.96	0.08	0.00	0.00	1.65	0.72	0.03	0.56	0.00	8.00	48.31	4.67	17.41	28.29	1.32	0.00
B161R140P1a	2.92	0.08	1.91	0.16	0.00	0.00	1.59	0.73	0.04	0.57	0.00	8.00	48.29	9.35	12.79	27.94	1.63	0.00
B173R151P1e	2.93	0.07	1.94	0.11	0.00	0.00	1.04	0.83	0.04	1.04	0.00	8.00	31.04	6.14	31.31	30.16	1.29	0.07
B208 R185 P2f	3.03	0.00	1.90	0.00	0.01	0.01	1.99	0.53	0.08	0.39	0.00	7.96	65.03	0.00	13.14	18.65	2.66	0.53
B289R236P2f	3.03	0.00	1.87	0.07	0.01	0.00	1.58	0.59	0.04	0.79	0.02	8.00	51.20	3.40	23.81	20.12	1.47	0.00
B289R236P2f	3.03	0.00	1.93	0.00	0.00	0.00	1.57	0.55	0.02	0.81	0.05	7.94	52.70	0.00	27.89	18.82	0.59	0.00
B285 R234 P1cf	3.03	0.00	1.87	0.00	0.00	0.00	1.55	0.61	0.01	0.84	0.04	7.95	48.19	0.00	29.89	21.61	0.31	0.00
B1049R640P17F	3.02	0.00	1.95	0.00	0.00	0.00	1.44	0.80	0.04	0.71	0.00	7.96	47.20	0.00	24.14	27.31	1.36	0.00
B1363R124P13iF	3.02	0.00	1.97	0.00	0.01	0.00	1.35	0.73	0.05	0.80	0.02	7.95	46.13	0.00	27.17	24.90	1.81	0.00
							Eclogitic gneiss											
B33R30P2i	2.99	0.01	1.88	0.13	0.01	0.00	1.19	0.15	0.14	1.50	0.01	8.00	40.07	6.27	43.96	5.12	4.58	0.00
B88R82P2g1	3.01	0.00	1.95	0.03	0.00	0.00	1.77	0.16	0.14	0.93	0.01	8.00	58.53	1.61	29.68	5.38	4.80	0.00
B88R82P2g1	3.01	0.00	2.00	0.00	0.00	0.00	1.77	0.16	0.13	0.92	0.01	8.00	59.32	0.00	31.00	5.23	4.46	0.00
B88R83P2a	2.96	0.04	1.94	0.09	0.01	0.00	1.72	0.10	0.18	0.97	0.01	8.00	50.09	5.35	33.55	3.90	7.11	0.00
B88R83P2a	3.03	0.00	1.98	0.00	0.01	0.00	1.75	0.09	0.17	0.96	0.01	8.00	58.89	0.00	32.38	3.11	5.62	0.00
B125 R110 P1b	3.00	0.00	1.97	0.02	0.00	0.00	1.42	0.64	0.03	0.92	0.00	8.00	46.82	0.96	29.77	21.38	1.07	0.00
B91R86p1c	2.99	0.02	1.98	0.04	0.00	0.00	1.54	0.28	0.08	1.05	0.02	8.00	52.28	1.98	33.50	9.42	2.83	0.00
B155R133P2f	2.99	0.01	1.95	0.06	0.00	0.00	1.56	0.37	0.07	1.00	0.01	8.00	52.13	2.79	30.52	12.20	2.37	0.00
B155R133P2f	2.98	0.02	1.98	0.03	0.00	0.00	1.62	0.36	0.07	0.93	0.00	8.00	45.04	1.96	35.63	14.55	2.83	0.00

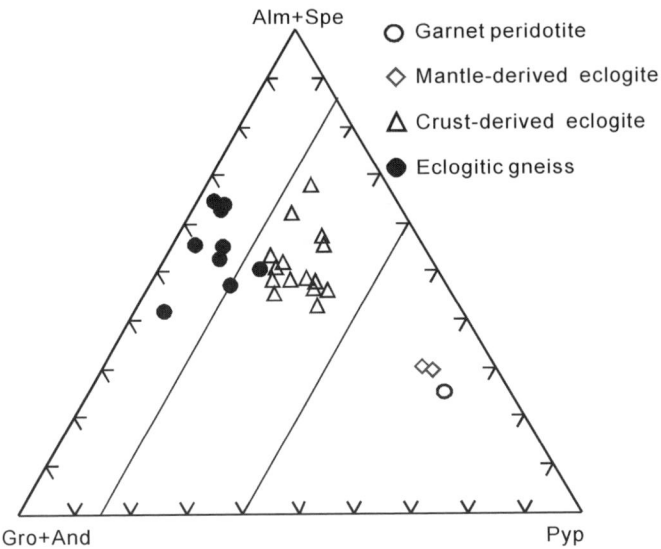

FIG. 5. Compositional plot of garnets from core samples at 100–2000 m depths of the CCSD main hole.

peridotites. Titanite is restricted to highly retrograded eclogitic gneiss and felsic gneiss.

Retrograde Metamorphism of Eclogite

Retrograde metamorphism proceeded from eclogite- through amphibolite- to greenschist-facies conditions (Z. Zhang et al., 2000). The metamorphic cores in 100–2000 m of the CCSD main hole display a spectrum of retrograde features. In the initial stage, very fine grained symplectitic aggregates formed after peak-stage minerals; with advanced recrystallization, symplectitic minerals increase in grain size, and eventually host minerals are replaced by mosaic-to-granular aggregates. In the eclogitic gneiss, for example, omphacite is first replaced by taramite + albite symplectite. Inasmuch as abundant quartz is present in the matrix, omphacite and quartz reacted according to the following reaction to form a moat of aegirine-augite at the boundary between omphacite and quartz:

Omphacite + garnet + quartz + H_2O =

taramite + albite + aegirine-augite.

In the products, taramite and albite occur as fine-grained symplectitic aggregates after omphacite and taramitic amphibole rims around garnet. As recrystallization proceeded, the albite became granular or saccharoidal aggregates with small amounts of needle-like taramite. The aegirine-augite retains a bead-like rim along the boundary between omphacite and quartz, and delineates the original border of the omphacite grains. Finally, the grain sizes of the retrograde phases become coarser, and the eclogitic gneiss transforms into garnet-bearing plagioclase (albite) gneiss (Figs. 4E–4H).

Rutile or ilmenite of eclogite-facies metamorphism is variably replaced by titanite. Most rutile grains are rimmed by a moat of titanite several tens of μm thick (Fig. 4E). In some intensively retrograded gneissic rocks, rutile is totally replaced by titanite, then decomposed to tiny anatase + calcite aggregates.

Deformational History

The main regional foliation and stretching lineation in the UHP rocks from Donghai dip gently and plunge to the southeast. Eclogite and ultramafic lenses of various sizes are clustered in paragneisses and granitic gneisses, with or without garnet relics. The long axis of the eclogite bodies coincide with the regional stretching lineation. Judging from mineral parageneses related to the deformation, the lineation formed during retrograde recrystallization under amphibolite-facies conditions.

The primary deformational fabrics formed during UHP metamorphism are preserved only in some larger eclogite bodies that record the overall early

TABLE 3. Electron Microprobe Analyses of Pyroxenes from 100–2000 m Depth, Main Hole of CCSD

Sample	depth (m)	SiO_2	TiO_2	Al_2O_3	FeO	Cr_2O_3	MnO	NiO	MgO	CaO	Na_2O	K_2O	P_2O_5	Total
						Garnet peridotite								
B4R7P2e	108.70	48.48	0.32	10.38	7.08	0.03	0.15	0.00	18.72	14.07	0.68	0.00	0.00	99.91
							Mantle-drived eclogite							
B352 R278 P2k	684.05	56.17	0.03	3.34	2.13	0.28	0.00	0.00	14.84	20.48	1.85	0.10	0.42	99.64
B308R253P7	611.90	57.78	0.01	0.11	8.39	0.00	0.11	0.00	34.03	0.10	0.00	0.00	0.00	100.53
							Crust-derived eclogite							
B132R114P1a	319.99	54.54	0.04	12.81	3.46	0.00	0.00	0.00	7.71	12.17	8.34	0.00	0.00	99.07
B136 R117 P3a	326.4	58.01	0.17	14.70	5.20	0.00	0.06	0.00	4.61	7.09	10.59	0.02	0.28	100.73
B146 R126 P4a	342.3	55.12	0.07	8.15	7.36	0.00	0.07	0.00	8.72	13.38	7.09	0.00	0.02	99.98
B146 R126 P4a	342.3	54.17	0.07	8.10	7.09	0.00	0.08	0.00	8.39	13.55	7.05	0.00	0.00	98.50
B146 R126 P4a	342.3	55.81	0.11	7.97	7.46	0.00	0.02	0.00	8.61	13.68	7.09	0.00	0.00	100.75
B155R131 P1l	352.34	55.91	0.13	7.57	8.93	0.00	0.07	0.00	8.19	12.35	6.94	0.08	0.38	100.55
B161R140P1a	368.6	54.49	0.07	7.38	7.99	0.00	0.00	0.00	8.86	13.94	6.71	0.01	0.02	99.47
B161R140P1a	368.6	54.75	0.06	6.47	8.51	0.00	0.04	0.00	9.33	15.33	6.00	0.01	0.04	100.54
B161R140P1a	368.6	54.61	0.04	7.13	7.68	0.00	0.03	0.00	9.18	14.40	6.46	0.01	0.00	99.54
B208 R185 P2f	444.0	54.94	0.28	6.45	10.41	0.10	0.00	0.00	7.25	12.16	6.64	0.04	0.40	98.67
B208 R185 P2f	444.0	55.22	0.13	6.71	11.07	0.10	0.00	0.00	7.50	12.05	6.07	0.13	0.39	99.37
B285 R234 P1cf	570	55.07	0.20	4.82	8.41	0.00	0.10	0.00	10.16	15.52	5.44	0.11	0.17	100.00
B289R236P2f	577	55.32	0.00	3.73	9.40	0.00	0.19	0.00	10.66	16.20	4.60	0.00	0.24	100.34
B290R237P1gf	578.5	55.08	0.18	3.51	9.25	0.00	0.11	0.00	10.41	16.44	4.59	0.05	0.42	100.04
B1049R640P17F	1992.3	55.92	0.17	7.02	5.68	0.00	0.16	0.00	9.99	15.22	5.25	0.04	0.10	99.55
							Eclogitic gneiss							
B88R82P2gl	249.20	54.00	0.10	0.89	24.58	0.00	0.42	0.00	3.04	7.57	10.44	0.00	0.00	101.04
B88R83P2a(c)	249.50	52.58	0.06	1.15	25.61	0.00	0.43	0.00	1.98	5.87	11.24	0.03	0.03	98.98
B88R83P2a(r)	249.50	53.54	0.06	1.32	24.24	0.00	0.70	0.00	2.03	5.92	10.93	0.02	0.09	98.85
B125 R110 P1b(c)	309.09	52.89	0.01	1.25	18.67	0.00	0.27	0.00	6.09	15.24	5.80	0.01	0.03	100.26
B125 R110 P1b(c)	309.09	52.24	0.02	0.94	22.61	0.00	0.34	0.00	4.03	12.14	6.93	0.02	0.00	99.27
B125 R110 P1b(r)	309.09	51.38	0.03	0.74	15.60	0.00	0.30	0.00	8.03	17.64	4.07	0.00	0.00	97.79
B155R133P2f	357.50	52.45	0.06	2.69	13.57	0.00	0.79	0.00	8.93	19.88	2.98	0.00	0.00	101.35

(Table continues)

TABLE 3. (continued)

Sample	Si^{IV}	Al^{IV}	Al^{VI}	Ti	Fe^{3+}	Mg	Fe^{2+}	Mn	Ca	Na	K	Total cations	WEF	Jd	Ae
						Garnet peridotite									
B4R7P2e	1.74	0.26	0.18	0.01	0.11	1.00	0.11	0.00	0.54	0.05	0.00	4.00	94.56	3.45	2.00
						Mantle-drived eclogite									
B352 R278 P2k	2.04	0.00	0.14	0.00	0.00	0.81	0.07	0.00	0.80	0.13	0.01	3.99	86.47	13.53	0.00
B308R253P7	2.00	0.00	0.00	0.00	0.01	1.75	0.24	0.00	0.00	0.00	0.00	4.00	100.00	0.00	0.00
						Crust-derived eclogite									
B132R114P1a	1.93	0.07	0.46	0.00	0.00	0.41	0.10	0.00	0.46	0.57	0.00	4.00	45.89	54.11	0.00
B136 R117 P3a	2.02	0.00	0.60	0.00	0.06	0.24	0.09	0.00	0.27	0.72	0.00	4.00	29.36	63.99	6.65
B146 R126 P4a	1.97	0.03	0.31	0.00	0.21	0.46	0.01	0.00	0.51	0.49	0.00	4.00	50.19	29.77	20.04
B146 R126 P4a	1.96	0.04	0.31	0.00	0.00	0.45	0.22	0.00	0.53	0.50	0.00	4.00	54.71	45.29	0.00
B146 R126 P4a	1.98	0.02	0.31	0.00	0.19	0.46	0.03	0.00	0.52	0.49	0.00	4.00	50.83	30.65	18.52
B155R131 P1l	2.01	0.00	0.32	0.00	0.15	0.44	0.12	0.00	0.48	0.48	0.00	4.00	51.79	33.14	15.06
B161R140P1a	1.96	0.04	0.28	0.00	0.23	0.48	0.01	0.00	0.54	0.47	0.00	4.00	52.29	26.14	21.57
B161R140P1a	1.96	0.04	0.24	0.00	0.21	0.50	0.04	0.00	0.59	0.42	0.00	4.00	57.53	22.31	20.16
B161R140P1a	1.97	0.04	0.27	0.00	0.22	0.49	0.02	0.00	0.56	0.45	0.00	4.00	54.15	25.43	20.42
B208 R185 P2f	2.03	0.00	0.28	0.01	0.12	0.40	0.20	0.00	0.48	0.48	0.00	4.00	53.24	32.83	13.92
B208 R185 P2f	2.03	0.00	0.29	0.00	0.07	0.41	0.27	0.00	0.48	0.43	0.01	4.00	57.21	34.57	8.22
B285 R234 P1ef	1.99	0.01	0.20	0.01	0.18	0.55	0.07	0.00	0.60	0.38	0.01	4.00	61.61	20.06	18.33
B289R236P2f	2.01	0.00	0.16	0.00	0.14	0.58	0.15	0.01	0.63	0.33	0.00	4.00	67.76	17.34	14.90
B290R237P1gf	2.02	0.00	0.15	0.01	0.13	0.57	0.15	0.00	0.65	0.33	0.00	4.00	67.70	17.13	15.18
B1049R640P17F	2.02	0.00	0.30	0.01	0.02	0.54	0.16	0.01	0.59	0.37	0.00	4.00	63.62	34.40	1.98
						Eclogitic gneiss									
B88R82P2gl	1.98	0.02	0.02	0.00	0.73	0.17	0.02	0.01	0.30	0.74	0.00	4.00	25.13	2.14	72.73
B88R83P2a(c)	1.97	0.03	0.02	0.00	0.00	0.11	0.80	0.00	0.01	0.24	0.82	0.00	41.58	58.42	0.00
B88R83P2a(r)	2.01	0.00	0.06	0.00	0.72	0.11	0.05	0.00	0.02	0.24	0.80	0.00	20.83	5.96	73.21
B125 R110 P1b(c)	1.98	0.02	0.04	0.00	0.41	0.34	0.18	0.01	0.61	0.42	0.00	4.00	57.46	3.33	39.21
B125 R110 P1b(c)	1.99	0.01	0.03	0.00	0.49	0.23	0.23	0.01	0.50	0.51	0.00	4.00	48.52	3.11	48.37
B125 R110 P1b(r)	1.97	0.03	0.00	0.00	0.33	0.46	0.17	0.01	0.73	0.30	0.00	4.00	69.30	0.30	30.41
B155R133P2f	1.94	0.06	0.06	0.00	0.21	0.49	0.21	0.03	0.79	0.21	0.00	4.00	77.99	4.79	17.22

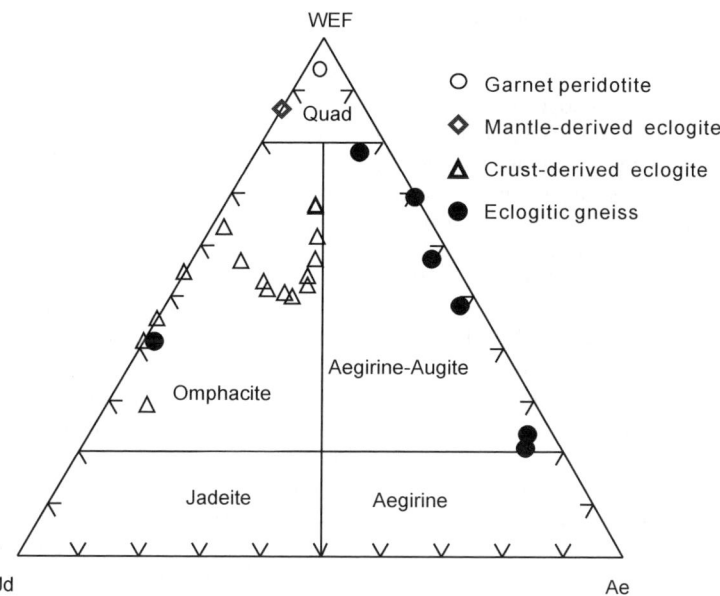

FIG. 6. Compositional plot of clinopyroxenes from core samples at 100–2000 m depths of the CCSD main hole.

compressional regime (Suo et al., 2001, 2002). Compositional layering of eclogites is common; garnet and omphacite are concentrated in alternative dark red and light green bands, respectively. The main foliation in the Donghai eclogite and ultramafic bodies is parallel to the compositional layering (Suo et al., 2002).

In the main hole of the CCSD, eclogites, especially the crustal eclogites, exhibit a prominent foliation of this earlier deformation stage. Phengite is aligned parallel to the foliation; garnet and omphacite crystals are differentially elongated. Some omphacite grains are intensively elongated with an aspect ratio ranging from 1:7 to as high as 1:23. Eclogite bodies contain both massive and foliated portions, indicating that the deformation partitioning occurred during the UHPM. Moreover, there are macroscopic fold closures and the indentation of less-competent omphacite-rich layers into eclogite boudins; these structures indicate that during the peak metamorphic stage, some UHP rocks were subjected to strong syntectonic compressional deformation.

Microtextural and petrological studies of the UHPM rocks indicate that the peak metamorphic stage took place in the UHP eclogite-facies. Maximum metamorphic T and P have been estimated at 700–850°C and about 3.0 GPa (Zhang et al., 1993, 2000). Many peak-stage minerals are retrograded; omphacite is decomposed to amphibole and plagioclase symplectites along its crystal boundaries, phengites are replaced by biotite + plagioclase symplectites (Fig. 4E), and garnet by pargasitic amphibole + plagioclase + magnetite along contacts with retrograded omphacite (Fig. 4F). All of these features indicate amphibolite-facies retrogression. In addition, kyanite in eclogite is rimmed with albite and/or paragonite coronae (Fig. 4D); the source of Na was from the decomposition of omphacite. It is evident that during retrogression, significant chemical exchanges with infiltrating fluids occurred. Most retrograde symplectites developed along foliation and formed during exhumation.

Almost all retrograde metamorphic minerals are not aligned with the primary foliation of the peak-stage metamorphism. From this, we infer that retrograde metamorphism and recrystallization occurred during exhumation of the UHP rocks. During later extension at shallow crustal levels, the complex underwent a greenschist-facies overprint.

Discussion and Conclusions

1. Eclogite is the dominant rock type in most lithological units except for unit IV of the CCSD main hole, and can be divided into mantle- and

crustal-derived eclogites. Mantle-derived eclogites are geochemically akin to ultramafic rocks with lower SiO_2, lower REE, and higher MgO contents than crustal analogues. Mineralogically, most mantle-derived eclogites are bimineralic, composed of Pyp-rich garnet and Ca-Mg-Fe clinopyroxene with a very low Jd content. Accessory minerals include mostly ilmeno-magnetite and minor rutile ± Ti-clinohumite. Crustal eclogites are geochemically akin to mafic rocks with higher SiO_2, higher REE, and lower MgO contents, and are mineralogically more diverse. In addition to the bimineralic varieties of eclogite with colorful garnet, omphacite, and accessory rutile, additional minor phengite, kyanite, zoisite, and epidote are present.

2. Ultramafic rocks of the Dabie-Sulu belt are of dual origin (Zhang and Liou, 1998; R. Zhang et al., 2003a). One type represents pre-existing mafic-ultramafic intrusives in the continental crust prior to subduction, such as the Bixiling peridotite-eclogite layered complex. Another type includes massive to layered ultramafics, interpreted as derived from the overlying mantle wedge and being incorporated into the subducting slab. Judging from geochemical characteristics, the garnet peridotites of the CCSD main hole possess characteristics of depleted mantle, and have P-T estimates of T = 1100–1200°C and P = 6.5–8.0 GPa, suggesting that they were derived from the mantle wedge above the subducting Yangtze plate (e.g., Z. Zhang et al., pers. commun., 2004).

3. Eclogitic gneiss is a special rock type that lies geochemically and mineralogically between eclogite and UHP quartzite, and occurs as layers in crustal eclogites. The gneiss contains similar compositions and mineral parageneses as in the eclogites, except for higher modal quartz (SiO_2 content is up to 65 wt%). They are intermediate to siliceous equivalents in the UHPM rock series, implying that their protoliths were mainly pre-existing volcanic piles of basalt and dacite.

4. Felsic gneiss is widespread from depths of 1130 to 1600 m of the main hole. This includes two-feldspar orthogneiss, amphibole schist, and epidote biotite gneiss. The occurrence of relict eclogitic minerals and textures implies that some felsic gneisses are the products of retrograde recrystallization, and fluid-induced K-metasomatism along foliation at mid- to lower-crustal depths.

5. From diverse rock types in the main drill hole, and the occurrences of coesite and other UHP mineral inclusions in zircons from many core samples from the CCSD, we conclude that voluminous supracrustal materials were subducted to mantle depths, experienced UHP metamorphism, and were rapidly exhumed to the upper crust. Therefore, the UHPM rocks preserve significant information about tectonics of continental subduction/collision and mantle dynamics (Xu, 2003a, 2003b).

Acknowledgments

We prepared this manuscript in honor of W. G. Ernst who has promoted several Sino-American projects, and made many contributions to the study of HP and UHP metamorphism and tectonics. This study represents one product of the Sino-American cooperative project and is financially supported by the key project of the National Science Foundation of China (40399142 and 40234048) and Major State Basic Research Development Program (2003CB716501) of China and the U.S. National Science Foundation EAR 0003355. We appreciate the detailed review by Mary Leech.

REFERENCES

Chen, B., Jahn, B. M., Ye, K., and Liu, J. B., 2002, Cogenetic relationship of the Yangkou gabbro-to-granite unit, Su-Lu terrane, eastern China, and implications for UHP metamorphism: Journal of the Geological Society, London, v. 159, p. 457–467.

Irvine, I. N., and Baragar, W. R. A., 1971, A guide to the chemical classification of the common volcanic rocks: Canadian Journal of Earth Science, v. 8, p. 523–548.

Kretz, R., 1983, Symbols for rock-forming minerals: American Mineralogist, v. 68, p. 277–279.

Liou, J. G., Tsujimori, T., and Zhang, R. Y., 2004, Global UHP metamorphism and continental subduction/collision: The Himalayan model: International Geology Review, v. 46, p. 1–27.

Liou, J. G., Zhang, R. Y., and Jahn, B. M., 1997, Petrogenesis of ultrahigh-pressure jadeite quartzite from the Dabie region, East- central China: Lithos, v. 41, p. 59–78.

Liu, F. L., Xu, Z. Q., Katayama, I., Yang, J. S., Maruyama, S., and Liou, J. G, 2001, Mineral inclusions in zircons of para- and orthogneiss from pre-pilot drillhole CCSD-PP1, Chinese Continental Scientific Drilling Project: Lithos, v. 59, p. 199–215.

Liu, F. L., Xu, Z., and Liou, J. G., 2004a, Tracing the boundary between UHP and HP metamorphic belt in the southwestern Sulu terrane, eastern China: Evidence from mineral inclusions in zircons from metamorphic rocks: International Geology Review, v. 46, p. 409–425.

Liu, F. L., Xu, Z. Q., Liou, J. G., Katayama, I., Masago, H., Maruyama, S., and Yang, J. S., 2002, Ultrahigh-pressure mineral inclusions in zircons from gneissic core samples of the Chinese Continental Scientific Drilling Site in eastern China: European Journal of Mineralogy, v. 14, p. 499–512.

Liu, F., Xu, Z., Liou, J. G., and Song, B., 2004b, SHRIMP U - Pb ages of ultrahigh-pressure and retrograde metamorphism of gneissic rocks, southwestern Sulu terrane, eastern China: Journal of Metamorphic Geology, 22, p. 315–326.

Liu, F. L., Xu, Z. Q., and Song, B., 2003a, Determination of UHP and retrograde metamorphic ages of the Sulu terrane: Evidence from SHRIMP U-Pb dating on zircons of gneissic rocks: Acta Geologica Sinica, v. 77, p. 229–237 (in Chinese with English abstract).

Liu, F. L., Zhang, Z. M., and Xu, Z. Q., 2003b, Three-dimensional distribution of ultrahigh minerals in Sulu terrane: Acta Geologica Sinica, v. 77, p. 69–84 (in Chinese with English abstract).

Spear, F. S., 1995, Metamorphic phase equilibrium and pressure-temperature-time paths: Mineralogical Society of America monograph (second printing), p. 425–427.

Suo, S. T., Zhong, Z. Q., and You, Z. D., 2001, Extensional tectonic framework of the Dabie-Sulu UHP-HP metamorphic belt, central China, and its geodynamical significance: Acta Geologica Sinica, v. 75, p. 14–24 (in Chinese with English abstract).

Suo, S. T., Zhong, Z. Q., and Zhou, H. W., 2002, Reclined folds within the UHP metamorphic belt in the Donghai area, northern Jiangsu, China: Geological Science and Technology Information, v. 21, p. 1–7 (in Chinese with English abstract).

Wang, C. Y., Zeng, R. S., Mooney, W. D., and Hacker, B. R., 2000, A crustal model of the ultrahigh-pressure Dabieshan orogenic belt, China, derived from deep seismic refraction profiling: Journal of Geophysical Research, v. 105, p. 10,857–10,869.

Wang, X., Liou, J. G., and Mao, H. K., 1989, Coesite bearing eclogite from the Dabie mountains in central China: Geology, v. 17, p. 1085–1088.

Xu, P. F., Liu, F. T., Wang, Q. C., Cong, B. L., and Chen, H., 2001, Slab-like high velocity, anomaly in the uppermost mantle beneath the Dabie-Sulu orogen: Geophysical Research Letters, v. 28, p. 1847–1850.

Xu, S., Okay, A. I., Ji, S., Sengor, A. M. C., Su, W., Liu, Y., and Jiang, L., 1992, Diamond from the Dabie Shan metamorphic rocks and its implication for tectonic setting: Science, v. 256, p. 80–82.

Xu, Z. Q., Yang, W. C., Zhang, Z. M., and Yang, T. N., 1998, Scientific significance and site selection researches of the first Chinese continental scientific deep drillhole: Continental Dynamics, v. 3, p. 1–13.

Xu, Z. Q., Zhang, Z. M., and Liu, F. L., 2003a, Exhumation structure and mechanism of the Sulu ultrahigh-pressure metamorphic belt, Central China: Acta Geologica Sinica, v. 77, p. 433–450 (in Chinese with English abstract).

Xu, Z. Q., Zhang, Z. M., and Liu, F. L., 2004, The structure profile of 0–1200 m in main borehole, Chinese Continental Scientific Drilling and its preliminary deformation analysis: Acta Petrologica Sinica, v. 20, p. 53–72.

Xu, Z. Q., Zhao, Z. X., Yang, J. S., Yuan, X. C., and Jiang, M., 2003b, Tectonics beneath plates and mantle dynamics: Geological Bulletin of China, v. 22, p. 149–159 (in Chinese with English abstract).

Yang, W. C., Yang, W. Y, and Chen, Z. Y., 2004, Interpretation of 3-D seismic reflection data in the Chinese Continental Scientific Drilling site: Acta Petrologica Sinica, v. 20, p. 127–137.

Ye, K., Cong, B., and Ye, D., 2000, The possible subduction of continental material to depths greater than 200 km: Nature, v. 407, p. 734–736.

Yuan, X. C., Klemperer, S. L., and Teng, W. B., 2004, Crustal structure and exhumation of the Dabieshan ultrahigh-pressure orogen eastern China, from seismic reflection profiling: Geology, v. 31, p. 435–438.

Zhang, R. Y., Cong, B. L., and Liou, J. G., 1993, Sulu ultrahigh pressure metamorphic terrane and explanation of its origin: Acta Petrologica Sinica, v. 9, p. 211–225.

Zhang, R. Y., Hirajimam, T., Banno, S., Cong, B., and Liou, J. G., 1995, Petrology of ultrahigh-pressure rocks from the southern Sulu region, eastern China: Journal of Metamorphic Geology, v. 13, p. 659–675.

Zhang, R. Y., and Liou, J. G., 1998, Dual origin of garnet peridotites of Dabie Sulu UHP terrane, eastern central China: Episodes, v. 21, p. 229–234.

Zhang, R. Y., Liou, J. G., Ye, K., and Yang, J., 2003a, Ultrahigh-pressure metamorphism in the forbidden zone: the Xugou garnet peridotite, Sulu terrane, eastern China: Journal of Metamorphic Geology, v. 21, p. 539–550.

Zhang, R. Y., Liou, J. G., Zheng, Y. F., and Fu, B., 2003b, Transition of UHP eclogites to gneissic rocks of low-amphibolite facies during exhumation: evidence from the Dabie terrane, central China: Lithos, v. 70, p. 269–291.

Zhang, Z. M., Xu, Z. Q., and Xu, H. F., 2000, Petrology of ultrahigh-pressure eclogites from the ZK703 drillhole in the Donghai, eastern China: Lithos, v. 52, p. 35–50.

Zhang, Z. M., Xu, Z. Q., and Xu, H. F., 2003, Petrology of the non-mafic UHP metamorphic rocks from the drillhole in the southern Sulu orogenic belt, eastern central China: Acta Geologica Sinica, v. 77, p. 173–186 (in Chinese with English abstract).

Low-Grade Metamorphic Rocks in the Dabie-Sulu Orogenic Belt: A Passive-Margin Accretionary Wedge Deformed during Continent Subduction

YONG-FEI ZHENG,[1] JIAN-BO ZHOU, YUAN-BAO WU, AND ZHI XIE

CAS Key Laboratory of Crust-Mantle Materials and Environments, School of Earth and Space Sciences, University of Science and Technology of China, Hefei 230026, China

Abstract

Greenschist-facies metasedimentary and meta-igneous rocks occur continuously along the northern margin, and sporadically in the interior, of the Dabie-Sulu orogenic belt in east-central China. An integrated study of geochronological, petrological, and paleontological observations demonstrates that precursors of flysch-facies metasedimentary rocks were deposited along the northern, passive continental margin of the Yangtze plate prior to the Triassic, and that protoliths of the meta-igneous rocks are a product of Middle Neoproterozoic bimodal magmatism along the northern margin of this plate. Except for the striking contrast in metamorphic grade, these low-grade rocks generally can be correlated in protolith origin and age with ultrahigh-pressure metamorphic rocks within the orogenic belt. Relationships in time and space between these rocks of contrasting grades can be reasonably interpreted through an accretionary wedge model that links their evolution with continent subduction. The low-grade metamorphic rocks of the subducting accretionary wedge consist of two parts: (1) large masses of metasedimentary rocks (including slates, schists, phyllites, metasandstones, and marble) along the northern margin of the Dabie-Sulu orogenic belt and deformed igneous rocks of Middle Neoproterozoic age; (2) sporadic outcrops in the interior of the belt of metavolcanics, metaclastics, phyllite, and marbles. During Triassic subduction of the Yangtze plate, the sedimentary cover and its underlying basement were partly scraped off by the overthrusted North China plate. The scraped-off materials accumulated in front of the overriding plate, forming an accretionary wedge that underwent deformation and metamorphism under greenschist-facies conditions. The present study also provides a constraint on the location of the Triassic suture zone between the North China and Yangtze plates. It is located below, or north of, the accretionary wedge (i.e., the Beihuaiyang zone in the Dabie region and the Wulian-Penglai zone in the Sulu region) rather than along the northern margin of the ultrahigh-pressure metamorphic zones.

Introduction

DURING SUBDUCTION of an oceanic plate, marine sediments and underlying rocks generally are scraped off from the subducting oceanic plate and accumulate as a wedge-shaped mass. It is usually called an active-margin accretionary wedge and uniquely develops along the boundary of the non-subducted plate (e.g., Karig and Sharman, 1975; Cloos, 1984; Sengor and Okurogullari, 1991; Tarbuck and Lutgens, 1994). This tectonic unit is commonly reported to occur in an orogenic belt formed by the subduction of an oceanic plate, but it is less often reported for an orogenic belt resulting from the subduction of a continental plate that may contain a passive-margin accretionary prism along its leading edge. The initial stage of the continental subduction process may resemble subduction of oceanic plates. Sediments covering the subducting continental margin and the continental crust may be scraped off during this process, forming an accretionary wedge similar to, but much larger than, oceanic accretionary prisms (Chemenda et al., 1995). The wedge includes scraped-off blocks of the upper layers of the continental crust, which may correspond to the low-grade metamorphic mélanges that are tectonically juxtaposed against high to ultrahigh-pressure (UHP) metamorphic zones in continental collisional belts.

The Dabie-Sulu terranes in east-central China represent a continent-continent collisional orogenic belt due to Triassic subduction of the Yangtze plate

---

[1]Corresponding author; email: yfzheng@ustc.edu.cn

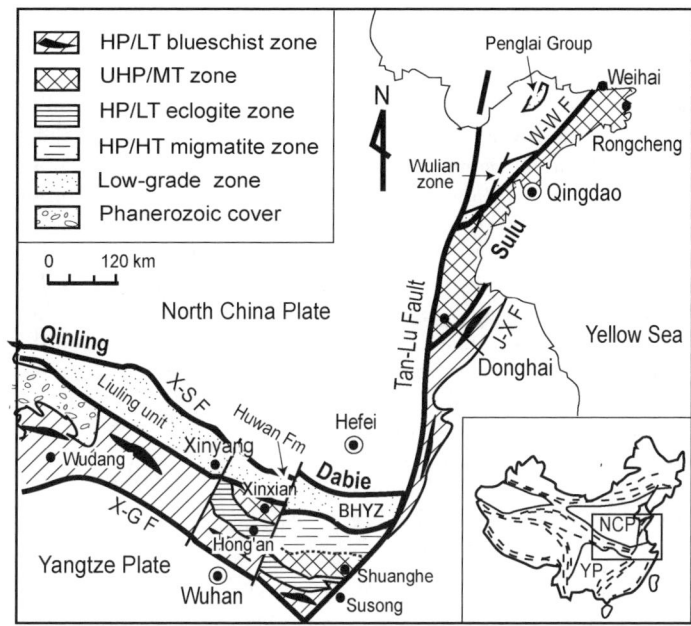

FIG. 1. Sketch map of the general geology in the Dabie-Sulu orogenic belt of central-east China. The surface exposure of low-grade units is exaggerated, and Mesozoic igneous rocks are omitted. Insert indicates the geographical position of the Dabie-Sulu orogenic belt in China, where NCP and YP denote the North China plate and the Yangtze plate, respectively. Abbreviations: J-X F = Jiashan-Xiangshui fault; W-W F = Wulian-Weihai fault; X-S F = Xinyang-Shucheng fault; X-G F = Xiangfan-Guangji fault; and BHYZ = Beihuanyang zone.

beneath the North China plate (e.g., Yin and Nie, 1993; Ernst and Liou, 1995; Hacker et al., 2000). UHP metamorphism has been recognized by the occurrence of coesite- and microdiamond-bearing eclogites, gneisses, and marbles within the belt (e.g., Wang et al., 1995; Cong, 1996; Liou et al., 1996; Xu et al., 2003). A great deal of low-grade metasedimentary and meta-igneous rock occurs not only along the northern margin of the UHP metamorphic belt, but also in its interior. All of the rocks experienced varying degrees of folding and associated metamorphism, mostly under greenschist-facies conditions. Considerable controversy exists concerning the origin and genesis of these low-grade metamorphic rocks. For example, molassic foredeep, clastic wedge, and forearc settings were proposed by Mattauer et al. (1985), Okay et al. (1993) and Xu et al. (1994), respectively, for the Foziling Group metasediments in the Beihuaiyang zone (BHYZ in Fig. 1), along the northern edge of the Dabie orogenic belt. The western extension of the Foziling Group to the Qinling orogenic belt corresponds to the Xinyang Group (the Liuling unit in Fig. 1), which was considered by Hsu et al. (1987) as the accretionary wedge on the active margin of the North China plate, where the Paleotethys ocean floor was being subducted during the Triassic.

Debate concerning location of the suture zone between the North China and Yangtze plates also has been associated with the low-grade metamorphic zone along the northern margin of the Dabie-Sulu orogenic belt. By considering the rocks in the Beihuaiyang zone as a forearc flysch formation, it was suggested that the boundary between the Beihuaiyang zone and the North Dabie HT/HP migmatitic unit (called North Dabie hereafter) denotes the plate suture (Liu and Hao, 1989; Xu et al., 1994; Hacker et al., 1995). Cong et al. (1994) postulated that the North Dabie could represent an island arc near the North China plate, and that the Beihuaiyang zone formed as a back-arc basin, and thus suggested that the suture would be located on the south boundary of North Dabie. Assuming the Beihuaiyang zone to be the active continental margin of the North China plate, Li et al. (2001) suggested that the suture was located along the

Xiaotian-Mozitan fault, the boundary between North Dabie and the Beihuaiyang zone.

On the other hand, Okay et al. (1993) proposed that the suture is located north of the Beihuaiyang zone that probably represents the passive continental margin of the Yangtze plate in the Dabie orogen. A similar conclusion was reached by Hacker et al. (1998) on the basis of U-Pb and Ar-Ar dates and eclogite distribution, placing the suture in the northern margin of the Huwan shear zone for the Hong'an terrane, in the western part of the Dabie orogen. Further westward in the Qinling orogen, Ratschbacher et al. (2003) proposed that the suture is located within the Liuling unit including the Xinyang Group (Fig. 1). Because of the similarity of petrologic and structural features in the north zone between the Dabie and Sulu regions and the lack of oceanic basin lithologies in the Sulu region, Faure et al. (2001, 2002) argued that the suture boundary between the Yangtze and North China plates lies north of the Penglai Group in the northeastern part of the Sulu region (Fig. 1). This differs from the common hypothesis that the Wulian-Weihai fault between the UHP metamorphic zone and the Penglai-Wulian zone represents the collision suture between the two plates east of the Tanlu fault (e.g., Wang and Cong, 1998; Cong and Wang, 1999; Zhai et al., 2000a; Yang, 2002; Zhai, 2002). Moreover, Li (1994) proposed a crustal-detachment model such that the surface suture occurs in the Sulu orogen, but a deep suture lies much to the south in Nanjing, parallel to the Lower Yangtze Valley fault zone.

Although there is agreement regarding the occurrence of low-grade metamorphic rocks in the interior of the western Sulu UHP belt (Dong et al., 1996; Song and Song, 1998; Zhou et al., 2001), debate has arisen concerning the nature and nomenclature of metavolcanoclastics, mylonite, and UHP metabasalt at Ganghe in the central Dabie (Tang et al., 1995; Dong et al., 1997, 2002; Zhai et al., 2000b; Zhou et al., 2001; Oberhaensli et al., 2002; Schmid et al., 2003). As documented in this paper, these debates can be reconciled by interpreting the low-grade metamorphic rocks as an accretionary wedge that formed by the Triassic subduction of the Yangtze continental margin beneath the North China plate.

This paper is devoted to an integrated study of all available observations from geochronology, petrology, and paleontology for low-grade meta-igneous and metasedimentary rocks in the northern margin and the interior of the orogenic belt. The results are used to demonstrate that the low-grade metamorphic rocks were scraped off from the subducting Yangtze plate during Triassic collision, and thus correspond to the accretionary wedge of the continent subduction. This model involving the formation of an accretionary wedge is strongly supported by documented temporal and spatial relationships between the low-grade and the UHP metamorphic rocks in the Dabie-Sulu region. Despite the striking contrast in metamorphic grades of the UHP versus low-grade metamorphic rocks, the similar lithologies and protolith ages indicate correlation. Our study also places a constraint on location of the Triassic suture zone between the North China and Yangtze plates.

Geological Setting

The Dabie-Sulu orogenic belt is separated into two terranes by approximately 500 km of left-lateral strike-slip along the Tan-Lu fault (Fig. 1). Three main tectonic zones are generally identified in both terranes (e.g., Wang et al., 1995; Cong, 1996; Liou et al., 1996; Hirajima and Nakamura, 2003), with an ultrahigh-pressure/high-temperature (UHP/HT) migmatitic zone in the north, an ultrahigh-pressure/mid-temperature (UHP/MT) eclogite-facies zone in the central, and high-pressure/low-temperature (HP/LT) eclogite-facies/blueschist-facies lithologies in the south. Early Cretaceous intrusive rocks of bimodal composition widely crop out in the three zones.

The central zone is composed of typical UHP/MT metamorphic rocks in a tract up to about 160 km wide, including eclogites, gneisses, and marbles. Occurrence of microdiamond and coesite as inclusions in such minerals as garnet and zircon in the rocks of the central zone reveals temperature-pressure conditions of 740 to 840°C and >2.8 GPa for the UHP/MT metamorphism (e.g., Wang et al., 1995; Cong, 1996). Triassic ages of 245 to 210 Ma were obtained by Sm-Nd and U-Pb isotope techniques for the UHP rocks (e.g., Ames et al., 1993, 1996; Li et al., 1993, 2000; Okay et al., 1993; Chavagnac and Jahn, 1996; Hacker et al., 1998; Zheng et al., 2003b), but the exact timing of peak UHP metamorphism was constrained to occur in the Middle Triassic rather than Late Triassic (e.g., Ernst and Liou, 1999; Zheng et al., 2002). This constraint is quantitatively confirmed by new SHRIMP U-Pb dates of 234 ± 4 to 227 ± 2 Ma for the coesite-bearing domain of metamorphic zircons from granitic gneisses in the Sulu terrane (Liu and Xu, 2004; Liu et al., 2004a, 2004b).

The south zone is bounded by a foreland fold-thrust zone to the south, by the Xiangfan-Guangji fault in the Dabie region and Jiashan-Xiangshui fault in the Sulu region (upper right insert in Fig. 2). It comprises quartz-rich amphibolite, biotite gneiss, minor eclogites, schist, marble, metasandstone, metavolcanics, and metaphosphorite, and shows Triassic ages of metamorphism (Sang et al., 1987; Liou et al., 1995; Faure et al., 1998; Chen et al., 2003b; Li et al., 2003). This unit is called the Susong Group in the Dabie region, and it shows a general decrease in metamorphic grade southward over tens of kilometers from low-T eclgite-facies to amphibolite-facies through blueschist-facies to greenschist-facies. The Susong Group contains a prominent phosphorite-bearing metamorphic sequence that allows it to be temporally correlated with the Hong'an Group in the western part of the Dabie region and the Zhangbaling (or Haizhou) Group in the southwestern margin of the Sulu region. Low-T eclogites occur in both the Hong'an Group of western Dabie and the northernmost part of the Susong Group (the Huangzhen area). Coesite pseudomorph has been found in the Huangzhen eclogite with Triassic metamorphic ages (Li et al., 2004), upgrading this HP unit into a UHP one.

The north zone contains synclinal structures and napes of different scales from hundreds of meters to tens of kilometers. In the Dabie region, it is referred to as north Dabie in this paper, but sometimes as the Northern Orthogneiss Unit in the literature (e.g., Hacker et al., 1998). A number of outcrops with relict eclogite paragenesis have recently been discovered in North Dabie (e.g., Wei et al., 1998; Faure et al., 1999; Tsai and Liou, 2000; Xu et al., 2000; Zhou et al., 2000; Liu et al., 2005). The mineralogical compositions of garnet and omphacite in some of the eclogites have been identified as corresponding to UHP conditions (Tsai and Liou, 2000; Su et al., 2001), and microdiamond recently was identified in eclogite at Baizhangyan and Huangweihe (Xu et al., 2003). Although the emplacement of Cretaceous granitoids has markedly reset the radiometric systems of the orthogenesis in North Dabie (e.g., Chen et al., 1995; Hacker et al., 1998, 2000; Ratschbacher et al., 2000; Bryant et al., 2004), a number of Triassic ages have been obtained for eclogites and orthogneisses from this unit. These include: (1) a mineral Sm-Nd isochron age of 244 ± 11 Ma for eclogite associated with intensely deformed ultramafic rocks at Raobazhai (Li et al., 1993); (2) concordant zircon TIMS U-Pb ages of 227 to 244 Ma for felsic granulite associated with mafic granulite at Huilanshan (Chen et al., 1996); (3) hornblende Ar-Ar plateau ages of 237 ± 5 and 230 ± 5 Ma for tonalitic orthogneiss (Liou et al., 1999); (4) discordant zircon U-Pb TIMS ages of 229 ± 18 Ma for granitic gneiss at Shizhuhe (Xie et al., 2001), 213 ± 4 Ma and 262 ± 28 Ma for tonalitic gneiss at Zhujiapu and granitic gneiss at Hongmiao, respectively (Zheng et al., 2004), and a weighted mean SIMS age of 212 ± 21 Ma for granitic gneiss at Baizhangyan (Xie et al., 2004a); (5) mineral Sm-Nd isochron ages of 219 ± 11 and 229 ± 13 Ma for amphibolitized eclogite at Baizhangyan (Xie et al., 2004b), and 212 ± 4 Ma for granulitized eclogite at Huangweihe (Liu et al., 2005). Therefore, the metamorphic rocks in north Dabie also underwent Triassic UHP metamorphism related to subduction of the Yangtze plate, but experienced granulite-facies overprinting due to a differential two-stage uplift relative to the central Dabie (Zheng et al., 2001).

A low-grade metamorphic zone 10 to 14 km wide occurs between the north zone and the North China craton (Anhui, 1987, 1997; Henan, 1989, 1997; Jiangsu, 1997; Shandong, 1987, 1997). It is composed of folded metasedimentary and meta-igneous rocks, mostly metamorphosed under greenschist-facies conditions. In the Dabie region, it is called the Beihuaiyang zone, and is approximately bounded by the Xiaotian-Mozitan fault on the south and the Xingyang-Shucheng fault in its north (Fig. 2). In the Sulu region, it is called the Wulian-Penglai zone, which is approximately bounded by the Wulian-Weihai fault on the southeast (Fig. 2) and the Tan-Lu fault on the northwest. Scattered low-grade metamorphic rocks also occur in the interior of UHP zones (Figs. 2 and 3). The genesis of low-grade metamorphic rocks and tectonic implications will be discussed in this paper.

A number of tectonic models have been advanced for the formation and exhumation of the HP and UHP metamorphic rocks in the Dabie-Sulu orogenic belt (e.g., Okay et al., 1993; Maruyama et al., 1994; Eide, 1995; Hacker et al., 1995, 1996, 2000; Liou et al., 1996; Dong et al., 1998; Faure et al., 1999; Ratschbacher et al., 2000; Wang and Cong, 1999; Zhong et al., 1999), and for the collision and suture between the Yangtze plate and the North China plate (e.g., Yin and Nie, 1993; Li, 1994; Zhang, 1997; Gilder et al., 1999; Faure et al.,

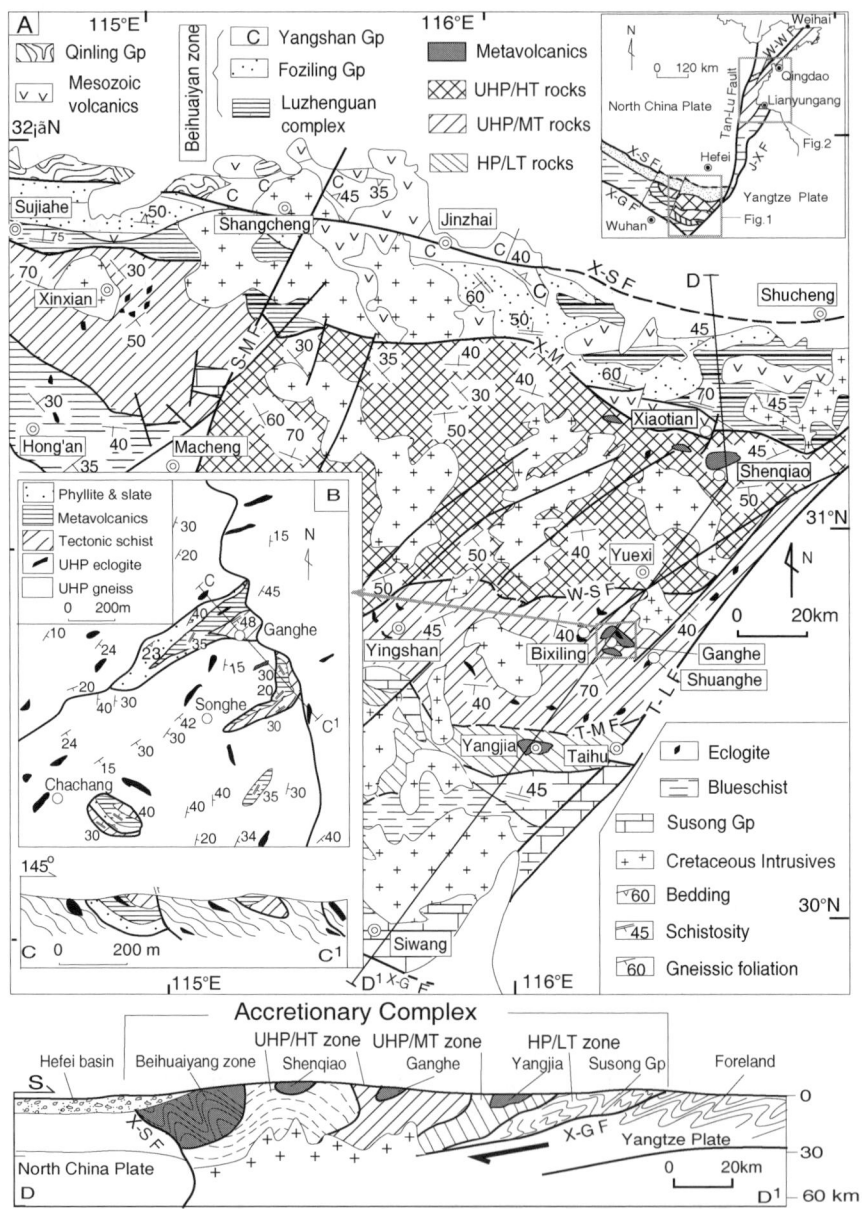

FIG. 2. Sketch map of the geology of the Dabie region. The line D–D' denotes the location of the cross-section in the eastern part of the Dabie belt. Abbreviations: J-X F = Jiashan-Xiangshui fault; W-W F = Wulian-Weihai fault; X-S F = Xinyang-Shucheng fault; X-G F = Xiangfan-Guangji fault; T-L F = Tan-Lu fault; X-M F = Xiaotian-Mozitan fault; S-M F = Shangcheng-Macheng fault; W-S F = Wuhe-Shuihou fault; T-M F = Taihu-Mamiao fault. Insert map B shows low-grade metamorphic rocks in the interior of the UHP metamorphic zone, located at Shenqiao in the north Dabie (31°07'25" N and 116°29'07" E), at Ganghe in the central Dabie (30°41'17" N and 116°14'54" E), and at Yangjia in the south Dabie (33°32'12" N and 116°08'21" E). C–C' denotes the Ganghe profile at Changpu in Yuexi. The surface exposure of low-grade units is somewhat exaggerated on the cross-section relative to the map, and several Mesozoic plutons on the line of the section on the map are arbitrarily omitted from the cross-section.

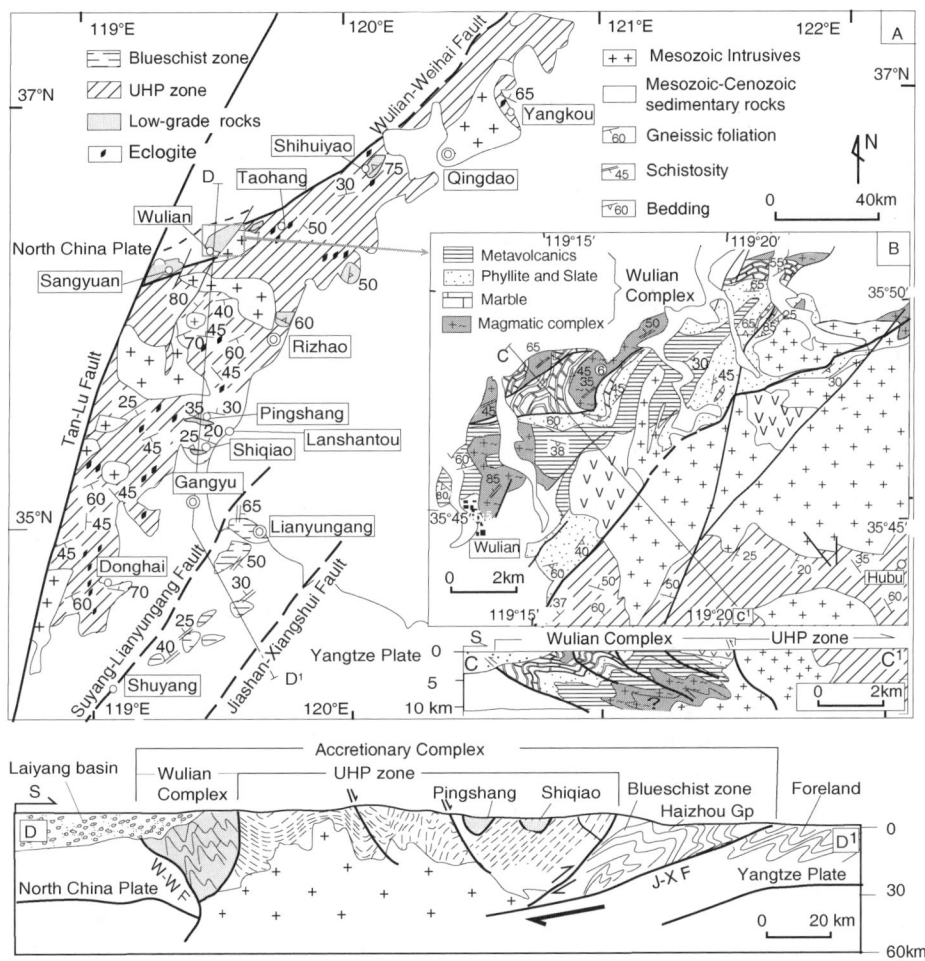

FIG. 3. Sketch map of the geology of the Sulu region, with the Penglai Group on the northeast and the Zhangbaling (or Haizhou) Group on the southwest. The line D–D' denotes the location of the cross-section in the western part of the Sulu belt. Abbreviations: J-X F = Jiashan-Xiangshui fault; W-W F = Wulian-Weihai fault; and T-L F = Tan-Lu fault. Insert map B shows the low-grade metamorphic rocks in the northern margin of the UHP metamorphic zone, located at Shihuiyao in the north (36°14'50" N and 119°40'47" E), at Pingshang in the central (35°07'33" N and 119°03'56" E) and at Shiqiao in the south (35°02'32" N and 119°10'32" E). C–C' denotes the Wulian profile in Shandong. The surface exposure of low-grade units is somewhat exaggerated on the cross-section relative to the map, and several Mesozoic plutons on the line of the section on the map are arbitrarily omitted from the cross-section.

2001). According to Hacker et al. (2000) and Ratschbacher et al. (2000), the exhumation was primarily accomplished by lithosphere-scale normal shear within and along the top of the upward and eastward-moving HP-UHP continental wedge. Because of the aggregate buoyancy of continental-crust-capped slab (Cloos, 1993; Ernst et al., 1997), it is possible that the wholesale entrace of a broad continental margin into the jaws of the subduction zone would probably have resulted in the shutdown of convergence, or stepout of the plate junction, and thus the exhumation of the deeply subducted continent. A comprehensive study of oxygen isotopes in minerals from the Dabie-Sulu UHP eclogites indicates that both subduction and exhumation of the Yangtze continental plate would proceed at rapid rates with a short residence time at mantle depths for UHP metamorphism (Zheng et al.,

1998, 1999), which can be modeled by a series of processes like frying ice cream (Zheng et al., 2003a).

Low-Grade Rocks of the Northern Margin

The low-grade metamorphic rocks along the northern margin of the Dabie region are composed of Upper Neoproterozoic (i.e., the Sinian system in China) to Lower Paleozoic flysch sediments of the Foziling Group, Upper Paleozoic metaclastic rocks of the Yangshan-Meishan Group, and the Luzhenguan magmatic complex of Middle Neoproterozoic age (Ma, 1991; Xu et al., 1994; Anhui, 1997; Li et al., 1997; Ma et al., 1997, 2001; Liu et al., 1998; Xie et al., 2002). Contemporaneous units in the Sulu region are the Wulian zone in the west (Zhao et al., 1995) and the Penglai Group in the east (Faure et al., 2001). They are composed of metaclastics, gneissic granite-diorite-gabbro, and associated metaclastics and marbles (Fig. 3B). In the Wulian zone, the metasedimentary unit is called the Wulian Group, and the meta-igneous unit is known as the Wulian complex. The southern side of the low-grade metamorphic rocks is regionally separated from the UHP zones by steeply dipping faults, and the northern side is covered by Mesozoic strata.

The Foziling Group is commonly regarded as a flysch sequence dominated by fined-grained, monotonous siltstones and shales, with metamorphic conglomerates, pebbles, graywackes, and distal turbidites. It is in steep fault contact with a sequence of medium-grained, banded, orthopyroxene-bearing felsic gneiss and augen-gneiss that contains minor discontinuous bands of mafic gneiss. A number of Yangtze-type microfossils were found in the Foziling flysch sediments, including *Protosphaeridium* sp., *Trematosphaeridium* sp., *Taeniatum* sp., and *Stenomarginata* sp., corresponding to the time span from Late Neoproterozoic to Early Paleozoic in the more southerly Yangtze strata (Zheng, 1964; Anhui, 1987, 1997). The Yangshan-Meishan Group contains the following fossil association of Upper Paleozoic marine strata: *Panderodus unicostatus* Brason et. Mehl, *Palaworda*, *Carwoodia* sp., and *Heliolites* cf. *anhuiensin* (Jin et al., 1987; Li et al., 1997). *Heliolites* cf. *anhuiensis* Deng (sp. nov), a kind of coral that occurs in the basal conglomerate of the Yangshan-Meishan Group, is a key fossil in the Lower Paleozoic strata on the Yangtze plate. Devonian microfossils were reported in the Xinyang Group of western Dabie (corresponding to the Foziling Group in the eastern Dabie) (Gao and Liu, 1990; Shan et al., 1992); some Carboniferous clastic rocks are even unmetamorphosed. These Paleozoic sediments represent continental shelf deposition in the northern margin of the Yangtze plate, but later underwent ductile shearing.

The flysch zone does not crop out in the Sulu region, but many Yangtze-type microfossils occur in marbles and quartzites of the Wulian Group (Fig. 3B), including *Leiopsopheara* sp., *Wulianensis* Yan sp., *Lophuminusxula* cf. *prima* Naum, *Wuliania Minor* Yan, *W. wulianesis* Yan, *Ceratophyton Vernicosum*, *Teophipolia Lacelata*, *Lophosphaeridium* sp., and *Preasolenopore* sp. (Shandong, 1987, 1997; Zhao et al., 1995), also corresponding to the period of Late Neoproterozoic to Early Paleozoic in the Yangtze strata. Fossils such as *Schubertella lata* Lee et Chen, *Palaeotextularia* sp., *Climacammina* sp., *Tetrataxis* sp., and *Eotuberitina* sp. occur in some limestone gravels within sandstone and conglomerate that were deposited during Late Jurassic time in the southern part of the Laiyang Basin (Guo and Sun, 1985). These fusulinids and foraminifera in the limestone gravels are Late Carboniferous fossils common in the northern part of the Yangtze plate. Therefore, the flysch sediments appear to have been deposited in the Sulu region during the Paleozoic, but were eroded later.

The magmatic complex at Luzhenguan of the Beihuaiyang zone in the Dabie region corresponds to that at Wulian in the Sulu region (Figs. 3B and 3C). The complexes are composed of gneissic or deformed granitoid plutons (granite, diorite, and gabbro) and associated metaclastics-marbles that underwent metamorphism at greenschist facies (locally at low amphibolite facies) with strong ductile deformation in some localities (Fig. 4C). They are commonly considered to represent part of the basement of the Foziling Group.

Our zircon U-Pb dating for deformed granitoids at Luzhenguan gave a SHRIMP $^{206}Pb/^{238}U$ mean-age of 783 ± 22 Ma and a TIMS discordia intercept age of 762 ± 16 Ma (Zheng et al., 2004). Our zircon LA-ICP-MS U-Pb dating for gabbro and gneissic granite at Haiyankou, Sanyuan, and Dajinyu in the Wulian complex resulted in concordant ages of 738 ± 10 to 758 ± 5 Ma (Wu et al., 2004). Our hornblende Ar-Ar dating on the gabbro at Haiyankou yielded consistent plateau and isochron ages of 719.1 ± 3.7 and 717.9 ± 7.2 Ma, respectively (*ibid.*). These radiometric ages provide direct dating for the timing of granitoid emplacement. The influence of Triassic UHP metamorphism on mineral U-Pb and

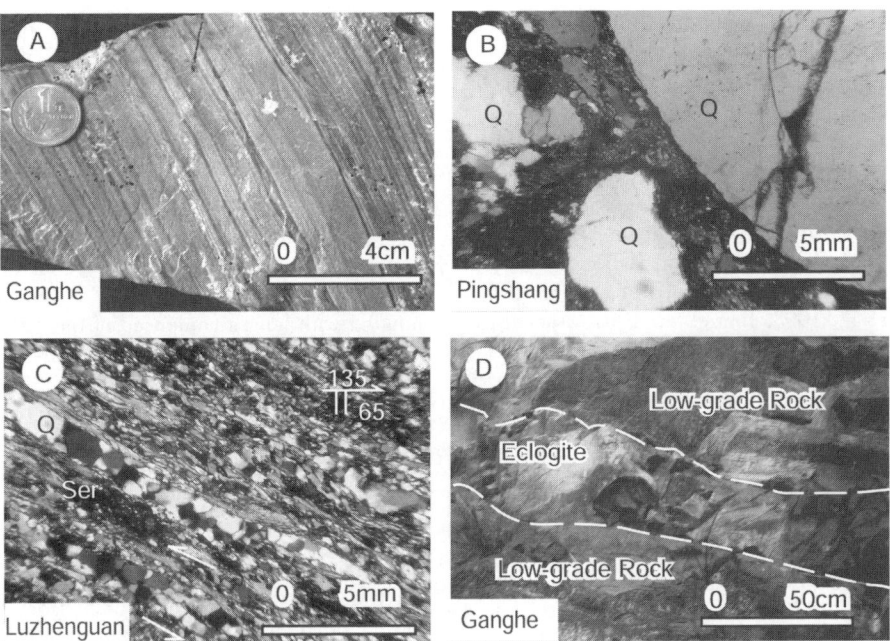

FIG. 4. Macroscale or microscale images of low-grade metamorphic rocks in the interior of the Dabie-Sulu UHP metamorphic belt. A. Laminated siliceous slate at Ganghe in the central Dabie. B. Gravel-bearing coarse metasandstone at Pingshang in southwest Sulu. C. Granitic mylonite at Luzhenguan in Beihuanyang north to the Dabie terrane (GPS: 31°11'30" N, 116°41'18" E). D. Tectonic lense of eclogite dike within metavolcanics at Ganghe in the central Dabie (field photo; the low-grade metamorphic rock was mylonitized to be metavolcanic breccia, and the pseudomorph of coesite is found in the eclogite dike).

Ar-Ar isotope systems is insignificant in these deformed intrusives, consistent with the field, petrographic, and cathodoluminescence observations that they only underwent low-grade metamorphism under greenschist-facies conditions.

Concordant U-Pb ages of 755 to 758 Ma were dated by Xue et al. (1997) for zircon from mylonitic granite at the 50.5 km kilometer marker along the main road west of Xiaotian, along the northern boundary of the north Dabie near the Xiaotian-Mozitan fault. Six discordia upper-intercept ages of 719 ± 2 to 766 ± 4 Ma were obtained by Chen et al. (2003a) for zircons from granitoids in the Luzhenguan complex. Hornblende Ar-Ar plateau ages of 742 to 770 Ma were obtained by Hacker et al. (2000) for deformed gabbro and gabbrodiorite of the Luzhenguan complex. These ages are interpreted in this study to represent the timing of rift magmatism during the Middle Neoproterozoic along the northern margin of the Yangtze craton, as previously concluded for Middle Neoproterozoic protoliths of UHP meta-igneous rocks in the Dabie region (e.g., Ames et al., 1996; Rowley et al., 1997; Hacker et al., 1998; Zheng et al., 2003a, 2003b, 2004). In particular, preservation of the hornblende Ar-Ar ages of 742 to 770 Ma for the deformed granitoids in the Luzhenguan complex (Hacker et al., 2000) and 717 to 719 Ma for the gabbro in the Wulian complex (Wu et al., 2004) demonstrate that the Middle Neoproterozoic plutons did not undergo metamorphism at temperatures higher than 500°C, because the closure temperature for Ar diffusion in hornblende is about 500 ± 20°C (e.g., Harrison, 1981; Dodson and McClelland-Brown, 1985).

A number of Sinian microfossils (e.g., *Leiopsopheara* sp., *Wulianensis* Yan sp., *Lophosphaeridium* sp., *Preasolenopore* sp.) occur in quartzite-marble relict strata at Luzhenguan (Anhui, 1987, 1997), allowing correlation with fossils from the Doushantuo Formation of the Sinian System on the Yangtze plate. The Neoproterozoic magmatic-sedimentary association may represent Precambrian basement of the Yangtze craton. It underwent metamorphism under greenschist-facies (locally low amphibolite-facies) conditions with strong ductile deformation and isoclined folding (Fig. 3C). K-Ar and Ar-Ar ages of 202 to 242 Ma were obtained for biotite and muscovite from the low-grade metamorphic rocks in the Beihuaiyang zone in the Dabie region (Chen et al., 1993; Hacker et al., 2000).

However, numerous Paleozoic ages of 480 to 300 Ma have been obtained by various radiometric methods from the metamorphic rocks of different grades in the northern part of the western Dabie (e.g., Ye et al., 1993; Jian et al., 1997, 2001; Li et al., 1998; Zhai et al., 1998; Xu B. et al., 2000; Sun et al., 2002). These dates lend support to the previous suggestion by Lerch et al. (1995) that an Early to Middle Paleozoic magmatic arc occurred along the southern margin of the North China plate in the Qinling-Dabie orogenic belt. Subduction of oceanic lithosphere beneath the North China plate and subsequent Late Carboniferous HP eclogite-facies metamorphism was also suggested by a combined study of petrology, stable and radiogenic isotope and fluid inclusions in eclogites from Sujiahe (Fu et al., 2002) as well as zircon U-Pb SHRIMP dating and REE partition (Sun et al., 2002). Both the Carboniferous and the Triassic events in the Dabie region apparently are consistent with the timings of the two suturing events by subduction-collision within the Qinling orogen between the Yangtze plate and North China plate—the Carboniferous suture at Shangdan in the north and the Triassic suture at Mianlu in the south (Meng and Zhang, 1999). Several muscovite Ar-Ar ages of ~205 to 263 Ma were obtained for eclogites from the Huwan shear zone in northwestern Dabie (Hacker et al., 1998; Xu B. et al., 2000), suggesting the influence of the Triassic subduction of the Yangtze plate on the shear zone. Greenschist-facies overprinting is also evident for the Carboniferous eclogites and their country rocks, which may have reset their mica Ar-Ar radiometric systems.

Low-Grade Rocks of the Interior

Low-grade schist in the interior of the UHP zones occur at Shenqiao within the UHP/HT zone in the North Dabie (Fig. 2A), and as metavolcanoclastics at Ganghe within the UHP/MT zone in the Central Dabie (Figs. 2B, 2C, and 4A) and at Yangjia in the HP/LT zone of South Dabie (Fig. 2A). Those within the UHP zone in western Sulu are metasedimentary rocks (Fig. 3A), occurring in outcrops at Shihuiyao in the north, Pingshang in the central (Fig. 4B), and Shiqiao in the south. The low-grade rocks occur as tectonic lenses with variable outcrop areas ranging from about 0.5 to 2.5 km^2 (Figs. 2B and 2C). Deformed metagranitoids at Bixiling in the Central Dabie (Liu et al., 1999, 2003) and at Zhimafang (near Donghai) in western Sulu (Liu et al., 2001), which occur in the interior of the UHP zone with fault contacts but contain neither UHP mineral paragenesis nor UHP index minerals, may also belong to this category. All of them have fault contacts with their country rocks, and in some cases are mingled with lenses composed of the UHP granitic gneiss or eclogites (Fig. 4D). A multi-grain zircon U-Pb dating gave a concordant age of 729 ± 4 Ma for a weakly gneissic metagranite at Bixiling in the Central Dabie (Liu et al., 2003), located about 200 m west of coesite-bearing eclogite.

Some of the low-grade metamorphic rocks have preserved the structures of their protoliths, including interlayered metaclastics (metasandstone, slate, phyllite, quartzite, and metavolcanoclastics) and marble; others underwent folding and ductile shearing, becoming schists and mylonites. Sinian algae fossils were found in the low-grade metamorphic rocks, for example, *Trematosphaerdium Mior* Sin et Lin, *Leiopsosphaera densa* Sin et Lin. at Shiqiao (Dong et al., 1996); *Tremafospaerium* sp., *Teophipolia lacelata*, and *Lophosphaeridium* sp. at Pingshang (Zhang et al., 1994); and *Leiosphaeridia* sp., *Trachysphaeridium* sp., *Prototrachites* sp., and *Retinaritse* sp. at Ganghe (Qian, 1996). These results are consistent with those from the Wulian Group, and thus indicative of a Late Neoproterozoic shallow-sea environment on the Yangtze plate.

Debate concerning the low-grade rocks at Ganghe in the Central Dabie mainly reflects observational contrasts reported by different workers at various localities in this area. In the field, however, the metavolcanoclastics, mylonite, and UHP metabasalt are associated with each other to differing degrees. Dong et al. (1997) obtained zircon U-Pb ages of 760 ± 9 to 802 ± 3 Ma and whole-rock Rb-Sr isochron ages of 232 ± 8 Ma for the metavolcanoclastics at Ganghe, which were interpreted to indicate that the volcanoclastics were erupted during Neoproterozoic time but underwent Triassic metamorphism. A U-Pb isochron age of 761 ± 33 Ma was also dated for zircon grains from felsic volcanoclasts at Ganghe (Schmid et al., 2003). No indication of UHP or HP metamorphism was observed in the felsic metavolcanics at this locality (Oberhaensli et al., 2002).

Protolith Correlation between Low-Grade and UHP Metamorphic Rocks

The low-grade metamorphic rocks in the Dabie-Sulu orogenic belt have following characteristics in common: (1) They were dynamically metamorphosed, generally at greenschist-facies conditions. Typical mineral parageneses are chlorite + sericite + quartz for metasandstone, chlorite + biotite + sericite + quartz for slate, chlorite + sericite + quartz + albite for phyllite, and epidote + sericite + quartz + albite for metavolcanoclastics. These are significantly different, not only from the mineral parageneses of the UHP rocks, but also from the mineral assemblages of sedimentary rocks from the foreland fold belt to the south of the HP-UHP zones. (2) They can be classified into two groups. One is an association with Precambrian complexes, including the Luzhenguan and Wulian magmatic intrusions as well as low-grade metavolcanic rocks in the interior of the Dabie-Sulu UHP zones, and corresponding to the Precambrian basement of the Yangtze plate. The other is the metaclastics along the northern margin of the orogenic belt, corresponding to the Paleozoic sedimentary cover along the northern margin of the Yangtze plate. (3) They underwent ductile deformation and low-grade metamorphism, forming the tectonic association of fold + fault + mélange, suggestive of compression settings. This is similar to the deformation-metamorphism in low-grade metamorphic rocks of the accretionary wedge that formed during subduction of oceanic plates. (4) Triassic metamorphic ages were obtained for the low-grade metamorphic rocks in the Beihuaiyang zone (Chen et al., 1993; Hacker et al., 2000) and the metavolcanoclastics at Ganghe in the central Dabie (Dong et al., 1997), being in agreement with the Triassic ages for the UHP metamorphism in the Dabie-Sulu orogenic belt (Li et al., 1999). Thus the formation of the low-grade metamorphic rocks can be related to geological processes accompanying Triassic subduction of the Yangtze plate beneath the North China plate.

Protoliths of the UHP metamorphic rocks are of two types. One is Neoproterozoic to Paleozoic sedimentary rocks and their underlying metasedimentary basement, including limestone (now coesite-bearing marbles) with positive carbon isotope anomalies ($\delta^{13}C$ of 2 to 6‰ after Zheng et al., 1998; Rumble et al., 2000). These positive $\delta^{13}C$ values may correspond to $\delta^{13}C$ values for limestone from the Doushantuo Formation and the Dengying Formation of the Sinian system above the Nantuo tillites in the Yangtze plate (Lambert et al., 1987; Yang et al., 1999). The other protolith is igneous rocks emplaced during the Middle Neoproterozoic into the metasedimentary basement, corresponding to bimodal magmatism at 700 to 800 Ma along the northern margins of the Yangtze plate (Ames et al., 1996; Rowley et al., 1997; Xue et al., 1997; Hacker et al., 1998, 2000; Zheng et al., 2003a, 2003b, 2004). Petrologic and geochemical studies suggested that the protoliths of most eclogites within granitic orthogneiss in the Dabie-Sulu terranes were continental basalts and gabbros (e.g., Jahn, 1998; Zhang and Liou, 1998; Zheng et al., 2002); eclogites associated with marbles must have been derived from marl, calcareous silicate layers, or basaltic rocks contaminated with calcic sediments. Because of the occurrence of granitic orthogneiss associated with the eclogites, it appears that a bimodal composition is responsible for protoliths of the UHP meta-igneous rocks, and thus the bimodal magmatism is assumed to have occurred in rift tectonic zones during the Middle Neoproterozoic along the northern margin of the Yangtze craton (Zheng et al., 2003a, 2004).

Protoliths of the low-grade metamorphic rocks are of three types: (1) sedimentary cover of Paleozoic age, i.e., the Foziling Group (including the Wulian and Penglai groups) and the Yangshan-Meishan Group in the Yangtze plate; (2) Neoproterozoic low-grad metasedimentary rocks overlying the crystalline basement, corresponding to UHP metasedimentray rocks; (3) Middle Neoproterozoic igneous rocks, including the Luzhenguan and Wulian complexes along the northern margin, as well as the deformed metagranite at Bixiling and metavolcanoclastics at Ganghe in the interior of the Dabie UHP belt. Most of the cover sediments and a few rocks from the underlying basement were scraped off by overthrusting of the North China plate during subduction of the Yangtze plate, whereas most of the underlying Precambrian metasedimentary and igneous rocks were carried into the deep mantle and subjected to UHP metamorphism. In this regard, the protoliths of the low-grade metamorphic rocks are generally comparable to those of the UHP rocks, and both of them are parts of rock associations now present along the northern margin of the Yangtze plate.

Geodynamic Model for Formation of Accretionary Wedge

Accretionary wedges resulting from oceanic plate subduction have been described in a number of orogenic belts that occur along the margins of convergent plates—for example, the Franciscan Complex in the western United States (Cloos, 1983), the Southern Uplands in Scotland (McKerrow et al., 1977; Leggett et al., 1979), and southern Mongolia (Sengor et al., 1993). Much study has been devoted to tectonic evolution, metamorphism, deformation, and fluid activity in the accretionary wedges (e.g., Davis et al., 1983; Davis and Hyndman, 1989; Moore, 1989; von Huene and Scholl, 1991). With respect to the type of accretion, the wedge would progressively grow by accumulation of off-scraped rocks from the obducting or subducting plate. In these settings, HP to UHP rocks are tectonically emplaced by channel flow as mélange, or as more coherent slabs at the base of the accretionary wedge due to buoyancy after plate breakoff, resulting in extension or erosion of the accretionary wedge and relatively rapid exhumation of the HP to UHP rocks (e.g., Platt, 1986; Ring and Brandob, 1999; Ernst, 2001).

Triassic subduction of the Yangtze continental plate beneath the North China plate has been demonstrated to have taken place in the Dabie-Sulu regions of east-central China (e.g., Ernst and Liou, 1995; Wang et al., 1995; Cong, 1996; Liou et al., 1996). Subhorizontal extrusion of thin, fault-bounded allochthons of HP and UHP rocks was proposed by Maruyama et al. (1994) and Maruyama (1997) to account for the present distribution of these rocks in the orogenic belt. Hacker et al. (1995) proposed a contrasting tectonic model involving the northward emplacement of an UHP nappe, on the order of 5 km (10 km maximum) thick, which arrived at mid-crustal levels by Early Jurassic time. The occurrence of the low-grade metamorphic rocks along the northern margin and in the interior of the Dabie-Sulu orogenic belt provides insight into the mechanism of continental plate subduction and exhumation. A geodynamic model, incorporating the observation for the low-grade metamorphic rocks, is proposed as follows (Fig. 5).

Continental collision usually involves the attempted subduction of the continental passive margin following the subducted oceanic plate into the trench. It is thus assumed that deep underflow of an ancient, cold continental slab constituting the leading edge of the Yangtze craton occurred during the Triassic, and that slab breakoff is a natural consequence of the attempted subduction of continental crust. Before the Triassic collision of continental margins, the pre-Triassic subduction of preexisting oceanic crust is required to bring about the continental subduction with deformation of continental shelf sediments during the Permian (Fig. 5A).

From a tectonic point of view, the accretionary complex can belong to the obducting or subducting plate; a decollement level between them separates the upper and lower plates. The conventional accretionary wedge model assumes that the deformed sediments were derived from the overriding plate and deep oceanic sediments (for example, the Franciscan Complex in the western U.S.). In our tectonic model, however, the deformed materials include not only the crystalline rocks derived from the Yangtze craton of Precambrian basement (consisting of Paleoproterozoic to Archean metamorphic complexes and Middle Neoproterozoic igneous rocks) but also the overlying sedimentary cover of Neoproterozoic to Paleozoic age. The bulldozing of the continental margin deposits forms a pre-collision complex. Because of both the buoyancy of continental crust relative to oceanic crust and the resistance of crystalline continental basement, subduction of the continental plate may have been initiated during the Permian at much slower rates than the final subduction to mantle depths in the Middle Triassic. Consequently, an initial collision may have occurred between the leading edge of the Yangtze craton and the southern margin of the North China craton. A vertical detachment of the continental margin may be triggered by the processes of protracted subduction-collision, resulting in the decollement of the upper layer but HP metamorphism of the lower layer. In this regard, the Late Paleozoic ages dated for some of the metamorphic rocks in the Dabie terrane may have been caused by pre-Triassic subduction.

During the Triassic subduction of the Yangtze plate beneath the North China plate, subsurface rocks and sediments, which have relatively low density, high porosity and high relief, were more easily scraped off than the underlying basement. Strong folding and faulting accompanied the synsubduction low-grade metamorphism under greenschist-facies conditions. Rocks of differing types and ages were tectonically juxtaposed to produce an accretionary wedge consisting of low-grade rocks (Fig. 5B). This process resulted in the low-grade metamorphic

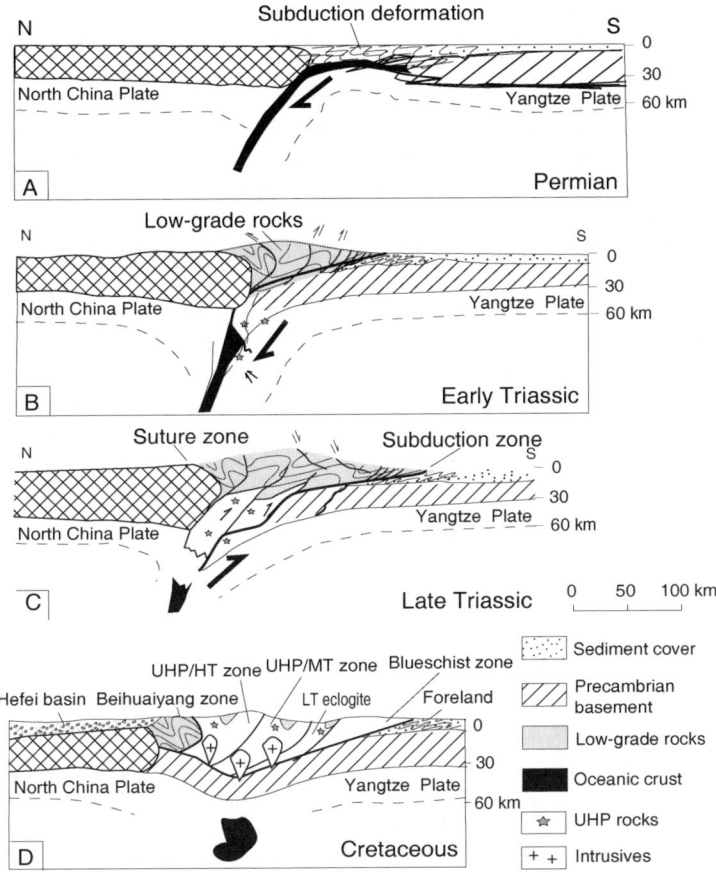

FIG. 5. Proposed geodynamic model for formation of a passive-margin accretionary wedge deformed during continental plate subduction in the Dabie-Sulu orogenic belt. A. Permian subduction of the leading edge of the Yangtze craton, which caused the deformation of continental shelf sediments along the passive margin of the subducting continent due to the obduction of the overriding plate. Subduction of the preexisting oceanic crust is assumed to have occurred before Triassic subduction of continental crust. B. Triassic subduction of the Yangtze plate beneath the North China plate generates the accretionary wedge of continent subduction, in which the off-scraped materials came from the subducting Yangtze plate. C. Tectonic extrusion of UHP slices upward into the overlying low-grade layer during the Late Triassic, resulting in a tectonic mélange within the UHP metamorphic zone. D. Metamorphic rocks of different grades occurring as a tectonic mélange in the Dabie-Sulu orogenic belt that is essentially the residual parts of a large-scale accretionary complex.

rocks now occurring in the Beihuaiyang and the Wulian zones, within which the Luzhenguan and Wulian meta-igneous complexes correspond to part of the crystalline basement underlying the Paleozoic sediments; both of them escaped the Triassic continental deep subduction, as did the Huwan shear zone in the northwestern Dabie that only experienced the greenschist-facies overprinting during the Triassic subduction of the Yangtze plate.

On the other hand, the subduction process would produce a layer of "flake tectonics" as envisaged by Oxburgh (1972) and Li (1998), resulting in the low-grade metamorphic rocks within the UHP zone that now crop out as relics after tectonic erosion. Li (1994) proposed a crustal-detachment model by envisaging northward obduction of the upper crust of the Yangtze plate east of the Tanlu fault over the North China plate but the northward subduction of the lower crust and upper mantle underneath the North China plate. According to this model, Li (1994) interpreted the linear magnetic low trending easterly from Nanjing as reflecting relics of the lower-lithospheric subduction zone beneath Nanjing, whereas a mid-crustal seismic reflector in the Subei region represents the thrust plane between the upper crust of the Yangtze plate and the North China plate. The obducted upper crust may correspond to the layer of low-grade metamorphic rocks discussed in this study.

With the progressive subduction of the Yangtze plate, an approximately 10 km high pile of the scraped-off materials would accumulate along the leading edge of this continent. As soon as the vast edge-load became available, appropriate tensile and compressive stresses would be generated to speed the deep subduction of the Yangtze continent. As a result, the Precambrian basement and some of the cover sediments on its northern margin were rapidly carried to deep-crustal and mantle depths and underwent UHP metamorphism, forming coesite- and microdiamond-bearing eclogites and gneisses in the Dabie-Sulu orogenic belt. Depending on depths of plate subduction, in fact, a series of different P-T assemblages was developed, from UHP hot eclogites to UHP cold eclogites to epidote amphibolites to blueschist-facies rocks to greenschist-facies rocks, all of which are characterized by subduction-related deformation and metamorphism. Inasmuch as the subduction of old, cold continental rocks cannot produce sufficient fluid to metasomatize the overlying mantle wedge for arc magmatism (Zheng et al., 2003a), synsubduction igneous rocks did not develop either in the hanging wall (the North China plate) or within the Dabie-Sulu orogenic belt. After breakoff of the subducted Yangtze plate at mantle depths due to buoyancy (Ernst and Liou, 1995; Dobretsov, 2000), the UHP slabs returned to the subsurface and were emplaced in the overlying layer of accretionary wedge. This process proceeded due to a change in tectonic regimes from contractional deformation during subduction to extension faulting during exhumation. Decompression exsolution of hydroxyl from nominally anhydrous minerals in the UHP metamorphic rocks may deliver aqueous fluid to catalyze the shear remelting of the uplifting slab itself to result in synexhumation magmatism (Zheng et al., 2003a). Differential return toward the surface of various subducted integral blocks as a consequence of decoupling from the sinking oceanic lithosphere may account for extremely high pressures of some of the recovered section and the contrasting metamorphic facies observed in several segments of the orogenic belt. As a result, tectonic juxtaposition of the UHP and the low-grade rocks occurred to form the accretionary complex including mélange materials during exhumation (Fig. 5C).

Major fault slip is required for the association of greenschist-facies rocks with the HP and UHP metamorphic rocks. Low-angle ductile normal faults (or extensional allochthons) were invoked (Faure et al., 1999), involving the exhumation of low-grade and UHP metamorphic rocks during Late Triassic to Early Jurassic time, to explain the great differences in estimated peak pressures (over 2 GPa). Our model emphasizes the tectonic extrusion of UHP slices upward into the overlying low-grade layer in this period to result in a type of tectonic mélange within the UHP zone (Fig. 5C). Underplating of the HP and UHP rocks caused further extension and erosion of mixed covers. Extensional collapse and/or erosional removal of the ascending crustal overburdon likely promoted the continued exhumation of the deeply subducted, but buoyant, continental materials (Fan and Shi, 2001).

Regional magma emplacement during the Early Cretaceous played contrasting roles in the different units of the Dabie-Sulu orogenic belt. On the one hand, the heat effect was significant in the western part of the north Dabie, resulting in coeval migmatization and gneissification. A mantle superplume event in this period has been suggested to trigger the large-scale coeval bimodal magmatism in these regions (Zhao et al., 2004, 2005). On the other hand, the associated tectonic uplift caused the low-grade metamorphic rocks to be eroded, resulting in further exposures of UHP metamorphic rocks. In this regard, the observable metamorphic rocks of different grades in these regions are residual parts of this large-scale accretionary complex (Fig. 5D). Preservation of the low-grade metamorphic rocks in the interior of the Dabie-Sulu orogenic belt, however, suggests that the depth of tectonic erosion was limited to several kilometers since the Jurassic. Moreover, field geometry of the main faults and foliation directions in the D–D' profile at the bottom of Figures 2 and 3 are different from the present ones due to significant modification by Early Cretaceous uplift after the exhumation of HP to UHP rocks during Late Triassic to Early Jurassic time (Faure et al., 1998, 1999).

Implications for the Relationship between Metavolcanics and Eclogites

The present study provides a resolution to the active debate concerning the relationship between metavolcanics and eclogites in Ganghe in the Dabie region—i.e., "foreign" versus "in-situ," based on the occurrence of low-grade metamorphic rocks in the interior of UHP metamorphic belt (Figs. 2B and 2C). The "foreign" relationship postulates that the low-grade metamorphic rocks were not subducted to mantle depths, and thus did not undergo UHP metamorphism, because they only have low metamorphic grade as indicated by retention of primary volcanic texture and stratigraphic sequence (Tang et al., 1995; Qian, 1996; Dong et al., 1997; Zhou et al., 2001). As shown in Figure 4D, however, a coesite-bearing eclogite occurs like a "dike" cutting the low-grade metavolcanic sequence (Gao et al., 1997). On the basis of this observation, it was concluded that both the low-grade metavolcanics and the UHP rocks were subducted to the mantle depths and experienced UHP metamorphism, but the low-grade metavolcanics were not changed into UHP rocks due to dynamic factors (Dong et al., 2002; Oberhaensli et al., 2002). This hypothesis may be true for the metagranite (Wallis et al., 1997) and metagabbro (Zhang and Liou, 1997) that occur at Yangkou in the Sulu orogen, and thus favors the "in situ" relationship.

However, if the low-grade metamorphic rocks could be subducted to mantle depths (>100 km), they would have been subjected to UHP metamorphism, hence should not retain their low metamorphic grade. Even if the subducted continental plate remained at depth for a very short time, as modeled by processes such as frying ice cream (Zheng et al., 2003a), the volcanic rocks, which are cryptocrystalline, should be much easier to recrystallize than the crystalline intrusion rocks. It thus seems impossible that the mafic dikes intercalated with the volcanic rocks and the surrounding intrusions all have been metamorphosed into the coesite-bearing eclogite, whereas the volcanic rocks only suffered the low-grade metamorphism at greenschist-facies under the UHP conditions of 700 to 900°C and >3.2 GPa.

We argue that all these volcanic rocks, as well as the metagranite at Bixiling nearby, the Luzhenguang complex in the northern part of the Dabie orogen, and the Wulian complex in the northwestern part of the Sulu orogen, were not brought to mantle depths during the Triassic continental subduction. They were scraped off from the subducted continental crust, and are similar to layers of "flake tectonics" during continental collision. Deformed and structurally recrystallized mylonites were found in close contact to the low-grade metavolcanics at Ganghe (Zhai et al., 2000b), which may correspond to friction-fracture zones formed during tectonic extrusion of UHP slabs. The coesite-bearing eclogite dikes within the low-grade metavolcanic rocks may have been extruded during the Late Triassic as plastic veins during the exhumation of UHP eclogites. This interpretation implies that the UHP slabs would move sideways to rise into the accretionary wedge and would be emplaced near the pre-Triassic trench, probably marking a basic difference in exhumation fashion between the subduction zones of oceanic and continental plates.

Constraints on the Location of the Suture Zone

Assuming that the suture zone represents the original movement interface between the two continental terranes, the present study places an important constraint on the location of the Triassic suture zone between the North China plate and the Yangtze plate. All of the available observations indicate that the metamorphic rocks of different grades in the Dabie-Sulu orogenic belt share some common features: (1) field relationships demonstrating structurally coherent deformation; (2) comparable protolith compositions and origins; and (3) coeval protolith ages, based on zircon U-Pb dating and rock assemblages. The occurrence of bimodal magmatism at 700 to 800 Ma is typical along the northern margin of the Yangtze plate (Zheng et al., 2003a, 2004; Wu et al., 2004), and the Sinian and Paleozoic fossils in the low-grade metasedimentary rocks are correlated with more southerly Yangtze strata. These provide important evidence that these rocks were all portions of the Yangtze plate.

On the other hand, different layers of the entire orogen, some containing coesite-bearing rocks on a scale of tens of kilometers, did not retain complete coherency during the Triassic subduction. In other words, the surface layer of the subducting Yangtze plate was decoupled from the underlying layers due to the progressive off-scraping of the overriding North China plate. As a result, the sedimentary cover and underlying basement of the Yangtze plate sheared off from the subducting lithosphere at relatively shallow depths, whereas the more coherent, old continental crust continued on to greater depths before breaking off from the descending plate. The subducted crust sank into the mantle to form the UHP metamorphic rocks, whereas the off-scraped materials accumulated as low-grade metamorphic rocks in front of the overriding plate and on the surface of the subducting plate.

During Triassic subduction of the Yangtze plate, the Neoproterozoic to Paleozoic sediments on the passive continental margin and their underlying basement rocks were evidently scraped off from the surface of the subducting continental plate to form a large accretionary wedge. Low-grade metamorphic rocks in the Dabie-Sulu orogenic belt represent residual slices of continental crust off-scraped during this subduction. Therefore, occurrence of the Triassic deformed rocks as the accretionary wedge indicates the location of the Triassic subduction zone. The northern boundary is the accretionary wedge itself, consisting of the low-grade metamorphic rocks in the northern part of the Dabie-Sulu orogenic belt (Fig. 1)—i.e., the Beihuaiyang zone (including the Huwan shear zone) in the Dabie region, the Wulian zone in western Sulu, and the Penglai Group in the eastern Sulu. Together with the present occurrences of the UHP metamorphic rocks in the two regions, constraints can be placed on the location of the Triassic suture zone between the Yangtze plate and the North China plate. We propose that this suture is located below, or north of, the accretionary wedge rather than along the northern margin of the UHP metamorphic zones (Fig. 1).

The proposed suture zone corresponds to the Beihuaiyang zone (close to the Xinyang-Shucheng fault) in the Dabie region and to the Wulian zone in the western Sulu. Its western extension corresponds to the boundary marked by Hacker et al. (1988) in the north to the Huwan Formation in the western Dabie, and its eastern extension corresponds to the boundary proposed by Faure et al. (2001) in the northern part of the Penglai Group in the eastern Sulu. Along with the previous interpretation concerning the suture location within the Liuling unit, including the Xinyang Group along the Qinling orogen (Ratschbacher et al., 2003), it appears that the boundary between the Yangtze plate and the North China plate is generally located below, or north of, the low-grade metamorphic zones rather than along the northern margin of HP to UHP metamorphic belt in the Qinling- Dabie-Sulu orogen.

Acknowledgments

This study was supported by funds from the Chinese Academy of lSciences (KZCX3-SW-141) and the Natural Science Foundation of China (40334036 and 40272100). Thanks are due to Drs. Gao Tianshan and Li Long for their assistance with field trips, and to Profs. Dong Shuwen, Jiang Laili, Li Shuguang, Liu Deliang, Wang Qingcheng, Xu Shutong, Yin An, and Zhang Guowei for stimulating discussions of the senior author's idea. Comments by Drs. G. E. Bebout, M. Cloos, W. G. Ernst, M. Faure, M. Leech, Z.-X. Li, and A. Yin on earlier versions of this manuscript in different phases greatly helped clarify some ambiguities as well as the English presentation.

REFERENCES

Ames, L., Tilton, G. R., and Zhou, G., 1993, Timing of collision of the North China and Yangtse cratons: U-Pb zircon dating of coesite-bearing eclogites: Geology, v. 21, p. 339–342.

Ames, L., Zhou, G., and Xiong, B., 1996, Geochronology and geochemistry of ultrahigh-pressure metamorphism with implications for collision of the North China and Yangtze cratons, central China: Tectonics, v. 15, p. 472–489.

Anhui (Anhui Bureau of Geology and Mineral Resources), 1987, Regional geology of Anhui Province: Beijing, China, Geological Publishing House, 672 p. (in Chinese).

Anhui (Anhui Bureau of Geology and Mineral Resources), 1997, Stratigraphy of Anhui Province. Wuhan, China, China University of Geosciences Press, 271 p. (in Chinese).

Bryant, D. L., Ayers, J. C., Gao, S., Miller, C. F., and Zhang, H., 2004, Geochemical, age, and isotopic constraints on the location of the Sino-Korean/Yangtze suture and evolution of the Northern Dabie Complex, east central China: Geological Society of American Bulletin, v. 116, p. 698–717.

Chavagnac, V. , and Jahn. B.-M., 1996, Coesite-bearing eclogites from the Bixiling Complex, Dabie Mountains, China: Sm-Nd ages, geochemical characteristics and tectonic implications: Chemical Geology, v. 133, p. 29–51.

Chemenda, A. I., Mattauer, M., Malavieille, J., and Bokum, A. N., 1995, A mechanism for syn-collisional rock exhumation and associated normal faulting: Results from physical modeling: Earth and Planetary Science Letters, v. 132, p. 225–232.

Chen, F.-K., Guo, J.-H., Jiang, L.-L., Siebel, W., Cong, B.-L., and Satir, M., 2003a, Provenance of the Beihuaiyang lower-grade metamorphic zone of the Dabie ultrahigh-pressure collisional orogen, China: Evidence from zircon ages: Journal of Asian Earth Sciences, v. 22, p. 343–352.

Chen, F.-K., Siebel, W., Guo, J.-H., Cong, B.-L., and Satir, M., 2003b, Late Proterozoic magmatism and metamorphism recorded in gneisses from the Dabie high-pressure metamorphic zone, eastern China: evidence from zircon U-Pb geochronology: Precambrian Research, v. 120, p. 131–148.

Chen, J.-F., Dong, S.-W., Deng, Y., and Chen, Y., 1993, Interpretation of K-Ar ages of the Dabie orogen—a differential uplifted block: Geological Review, v. 39, p. 15–22 (in Chinese with English abstract).

Chen, J.-F., Xie, Z., Liu, S.-S., Li, X.-M., and Foland, K. A., 1995, Cooling age of Dabie orogen, China, determined by ^{40}Ar-^{39}Ar and fission track techniques: Science in China (Series B), v. 38, p. 749–757.

Chen, N.-S., You, Z.-D., Suo, S.-T., Yang, Y., and Li, H.-M., 1996, Zircon U-Pb ages of intermediate granulite and deformed granites in the Dabie Mountains, central China: Chinese Science Bulletin, v. 41, p. 1886–1890.

Cloos, M., 1983, Comparative study of mélange matrix and metashales from the Franciscan subduction complex with the basal Great Valley sequence, California: The Journal of Geology, v. 91, p. 291–306.

Cloos, M., 1984, Flow mélanges and the structural evolution of accretionary wedges, in Mélanges—their nature, origin, and significance: Special Paper of the Geological Society of America, v. 198, p. 71–79.

Cloos, M., 1993, Lithospheric buoyancy and collisional orogenesis: Subduction of oceanic plateaus, continental margins, island arcs, spreading ridges, and seamounts: Geological Society of America Bulletin, v. 105, p. 715–737.

Cong, B. L., 1996, Ultrahigh-pressure metamorphic rocks in the Dabieshan-Sulu region of China: Beijing, China, Science Press, 224 p.

Cong, B.-L., and Wang, Q.-C., 1999, The Dabie-Sulu UHP rocks belt: Review and prospect: Chinese Science Bulletin, v. 44, p. 1074–1086.

Cong, B., Wang, Q., Zhai, M., Zhang, R., Zhao, Z., and Ye, K., 1994, Ultra-high pressure metamorphic rocks in the Dabie-Sulu region, China: Their formation and exhumation: The Island Arc, v. 3, p. 135–150.

Davis, D. M., Suppe, J., and Dahlen, F. A., 1983, Mechanics of fold and thrust belts and accretionary wedges: Journal of Geophysical Research, v. 88, p. 1153–1172.

Davis, E. E., and Hyndman, R. D., 1989, Accretion and recent deformation of sediments along the northern Cascadia subduction zone: Geological Society of American Bulletin, v. 101, p. 1465–1480.

Dobretsov, N. L., 2000, Collision processes in Paleozoic foldbelts of Asia and exhumation mechanisms: Petrology, v. 8, p. 403–427.

Dodson, M. H., and McClelland-Brown, E., 1985, Isotopic and palaeomagnetic evidence for rates of cooling, uplift, and erosion, in Snelling, N. J., ed., The chronology of the geological record: Geological Society of London, Memoir, no. 10, p. 315–325.

Dong, S.-W., Chen, J.-F. and Huang, D.-Z., 1998, Differential exhumation of tectonic units and ultrahigh-pressure metamorphic rocks in the Dabie Mountains, China: The Island Arc, v. 7, p. 174–183.

Dong, S.-W., Oberhaensli, R., Schmidt, R., Liu, X.-C., Tang, J.-F., and Xue, H.-M., 2002, Occurrence of metastable rocks in deeply subducted continental crust from the Dabie Mountains, central China: Episodes, v. 25, p. 84–89.

Dong, S.-W., Wang, X.-F., and Huang, D.-Z., 1997, Discovery of low-grade metavolcanic rock sheets within UHP belt in Dabieshan and its implications: Chinese Science Bulletin, v. 42, p. 1199–1203.

Dong, S.-W., Zhang, Y., and Huang, D.-Z., 1996, Geological feature of a tectonic window at Shiqiao within ultrahigh pressure metamorphic belt in North Jiangsu: Anhui Geology, v. 6, no. 1, p. 9–13 (in Chinese with English abstract).

Eide, E. A., 1995, A model for the tectonic history of HP and UHPM regions in east central China, in Coleman, R. G., and Wang, X.-M., eds., Ultrahigh pressure metamorphism: Cambridge, UK, Cambridge University Press, p. 391–426.

Ernst, W. G., 2001, Subduction, ultrahigh-pressure metamorphism, and regurgitation of buoyant crustal slices—implications for arcs and continental growth: Physics of Earth and Planetary Interiors, v. 127, p. 253–275.

Ernst, W. G., and Liou, J. G., 1995, Contrasting plate-tectonic styles of the Qinling-Dabie-Sulu and Franciscan metamorphic belts: Geology, v. 23, p. 253–256.

Ernst, W. G., and Liou, J. G., 1999, Overview of UHP metamorphism and tectonics in well-studied collisional orogens: International Geology Review, v. 41, p. 477–493.

Ernst, W. G., Maruyama, S., and Wallis, S., 1997, Buoyancy-driven rapid exhumation of ultrahigh-pressure metamorphosed continental crust: Proceedings of National Academy of Sciences, v. 94, p. 9532–9537.

Fan, T.-Y., and Shi, Y.-L., 2001, Thermodynamic modeling of P-T-t paths of Dabie-Sulu ultra-high pressure metamorphism: Chinese Journal of Geophysics, v. 44, no. 5, p. 627–635 (in Chinese with English abstract).

Faure, M., Breton, N. L., Lin, W., and Monie, P., 2002, Where Is the North China–South China block boundary in eastern China? Reply: Geology, v. 30, p. 668.

Faure, M., Lin, W., and Le Breton, N., 2001, Where Is the North China–South China block boundary in eastern China?: Geology, v. 29, p. 119–122.

Faure, M., Lin, W., Shu, L.-S., Sun, Y., and Schaerer, U., 1999, Tectonics of the Dabieshan (eastern China) and possible exhumation mechanism of ultrahigh-pressure rocks: Terra Nova, v. 11, p. 251–258.

Faure, M., Lin, W., and Sun, Y., 1998, Doming in the southern foreland of the Dabieshan (Yangtse block), China: Terra Nova, v. 10, p. 307–311.

Fu, B., Zheng, Y.-F., and Touret, J. L. R., 2002, Petrological, isotopic and fluid inclusion studies of eclogites from Sujiahe, NW Dabie Shan (China): Chemical Geology, v. 187, p. 107–128.

Gao, L., and Liu, Z., 1990, Microfossils and geological significance of the Nanwan Formation, Xinyang Group, Henan, China: Regional Geology of China, v. 5, p. 421–428 (in Chinese).

Gao, T.-S., Tan, J.-F., Zhou, C.-T., Hu, M.-J., and Qian, C.-C., 1997, The discovery of eclogite dikes in low-greenschist facies volcaniclastic rocks of the Dabie Mountains: Chinese Science Bulletin, v. 42, p. 1726–1729.

Gilder, S. A., Leloup, P. H., Courtillot, V. , Chen, Y., Coe, R. S., Zhao, X. X., Xiao, W. J., Halim, N., Cogne, J. P., and Zhu, R. X., 1999, Tectonic evolution of the Tancheng-Lujiang (Tan-Lu) fault via Maddle Triassic to Early Cenozoic paleomagnetic data: Journal of Geophysics Research, v. 104B, p. 15,365–15,390.

Guo, Z.-Y., and Sun, X.-Z., 1985, Discovery of oölitic limestone gravels and foraminifer and fusulinid fossils in the Upper Jurassic on the southern margin of the Jiao-Lai depression, eastern Shandong and their tectonic significance: Geological Review, v. 31, no. 2, p. 179–182 (in Chinese with English abstract).

Hacker, B. R., Ratschbacher, L., Webb, L., and Dong, S., 1995, What brought them up? Exhumation of the Dabie Shan ultrahigh-pressure rocks: Geology, v. 23, p. 743–746.

Hacker, B. R., Ratschbacher, L., Webb, L., Ireland, T., Walker, D., and Dong, S., 1998, U/Pb zircon ages constrain the architecture of the ultrahigh-pressure Qinling-Dabie Orogen, China: Earth and Planetary Science Letters, v. 161, p. 215–230.

Hacker, B. R., Ratschbacher, L., Webb, L., McWilliams, M. O., Ireland, T., Calvert, A., Dong, S., Wenk, H.-R., and Chateigner, D., 2000, Exhumation of ultrahigh-pressure continental crust in east central China: Late Triassic–Early Jurassic tectonic unroofing: Journal of Geophysics Research, v. 105B, p. 13,339–13,364.

Hacker, B. R., Wang, X., Eide, E. A. and Ratschbacher, L., 1996, Qinling-Dabie ultrahigh-pressure collisional orogen, in Yin, A., and Harrison, T. M., eds., The tectonic evolution of Asia: Cambridge, UK, Cambridge University Press, p. 345–370.

Harrison, T. M., 1981, Diffusion of ^{40}Ar in hornblende: Contributions to Mineralogy and Petrology, v. 78, p. 324–331.

Henan (Henan Bureau of Geology and Mineral Resources), 1989, Regional geology of Henan Province: Beijing, China, Geological Publishing House, 687 p. (in Chinese).

Henan (Henan Bureau of Geology and Mineral Resources), 1997, Stratigraphy of Henan Province: Wuhan, China, China University of Geosciences Press, 299 p. (in Chinese).

Hirajima, T., and Nakamura, D., 2003, The Dabie Shan-Sulu orogen, in Carswell, D. A., and Compagnoni, R., eds., Ultrahigh pressure metamorphism: European Mineralogical Union Notes in Mineralogy, v. 5, p. 105–144.

Hsu, K. J., Wang, Q.-C., Li, J.-L., Zhou, D., and Sun, S., 1987, Tectonic evolution of Qinling Mountains, China: Eclogae Geologicae Helvetiae, v. 80, p. 735–752.

Jahn, B.-m., 1998, Geochemical and isotpic characteristics of UHP eclogites and ultramafic rocks of the Dabie orogen: Implications for continental subduction and collisional tectonics, in Hacker, B., and Liou, J. G., eds., When continents collide: Geodynamics and geochemistry of ultrahigh-pressure rocks: Dordrecht, Netherlands, Kluwer Academic Publishers, p. 203–239.

Jian, P., Liu, D.-Y., Yang, W.-R., and Williams, I. S., 2001, SHRIMP dating of zircons from the Caledonian Xiongdian eclogites, western Dabie Mountains, China: Chinese Science Bulletin, v. 46, p. 77–79.

Jian, P., Yang, W.-R., Li, Z.-C., and Zhou, H.-F., 1997, Isotopic geochronological evidence for the Caledonian Xiongdian eclogite in the northwestern Dabie Mountains, China: Acta Geologica Sinica, v. 71, p. 455–465.

Jiangsu (Jiangsu Bureau of Geology and Mineral Resources), 1997, Stratigraphy of Jiangsu Province. Wuhan, China, China University of Geosciences Press, 288 p. (in Chinese).

Jin, F.-Q., Yan, H.-X., and Nue, P.-J., 1987, New progress in the study of stratigraphy in Beihuaiyang region: Bulletin of Hefei University of Technology, no. 9, p. 3–10 (in Chinese).

Karig, D. E., and Sharman, G. F., 1975, Subduction and accretion in trenches: Geological Society of American Bulletin, v. 86, p. 377–389.

Lambert, I. B., Walter, M. R., Zang, W., Lu, S., and Ma, G., 1987, Paleoenvironment and carbon isotope stratigraphy of Upper Proterozoic carbonates of the Yangtze Platform: Nature, v. 325, p. 140–142.

Leggett, J. K., McKerrow, W. S., and Eales, M. H., 1979, The Southern Uplands of Scotland: A Lower Paleozoic accretionary prism: Journal of the Geological Society, London, v. 136, p. 755–770.

Lerch, M. F., Xue, F., and Kroner, A., 1995, A Paleozoic magmatic arc in the Heihe area, Qinling orogenic belt, central China: The Journal of Geology, v. 103, p. 437–449.

Li, J.-Y., Yang, T.-N., Chen, W., and Zhang, S.-H., 2003, ^{40}Ar/^{39}Ar dating of deformation events and reconstruction of exhumation of ultrahigh-pressure metamorphic rocks in Donghai, East China: Acta Geologica Sinica, v. 77, p. 155–168.

Li, S.-G., Han, W.-L., Huang, F., Zheng, Y.-F., Zhang, S.-Q., and Zhang, Z.-H., 1998, Sm-Nd and Rb-Sr ages and geochemistry of metavolcanics from the Dingyuan Formation in Dabie Mountains, Central China: Evidence for the Paleozoic magmatic arc: Scientia Geologica Sinica, v. 7, p. 461–470.

Li, S.-G., Jagoutz, E., Chen, Y.-Z., and Li, Q.-L., 2000, Sm-Nd and Rb-Sr isotopic chronology and cooling history of ultrahigh pressure metamorphic rocks and their country rocks at Shuanghe in the Dabie Mountains, Central China: Geochimica et Cosmochimica Acta, v. 64, p. 1077–1093.

Li, S.-G., Jagoutz, E., Lo, C.-H., Chen, Y.-Z., Li, Q.-L., and Xiao, Y.-L., 1999, Sm/Nd, Rb/Sr, and $^{40}Ar/^{39}Ar$ isotopic systematics of the ultrahigh-pressure metamorphic rocks in the Dabie-Sulu belt, Central China: A retrospective view: International Geology Review, v. 41, p. 1114–1124.

Li, S.-G., Huang, F., Nie, Y.-H., Han, W.-L., Long, G., Li, H.-M., Zhang, S.-Q., and Zhang, Z.-H., 2001, Geochemical and geochronological constraints on the suture location between the North and South China Blocks in the Dabie orogen, Central China: Physics and Chemistry of the Earth (part A), v. 26, p. 655–672.

Li, S.-G., Xiao, Y.-L., Liou, D.-L., Chen, Y.-Z., Ge, N.-J., Zhang, Z.-Q., Sun, S.-s., Cong, B.-L., Zhang, R.-Y. Hart, S. R., and Wang, S.-S., 1993, Collision of the North China and Yangtse blocks and formation of coesite-bearing eclogites: Timing and processes: Chemical Geology, v. 109, p. 89–111.

Li, X.-P., Zheng, Y.-F., Wu, Y.-B., Chen, F.-K., Gong, B., and Li, Y.-L., 2004, Low-T eclogite in the Dabie terrane of China: Petrological and isotopic constraints on fluid activity and radiometric dating: Contributions to Mineralogy and Petrology, v. 148, p. 443–470.

Li, Y.-J., Hu, S.-L., and Jin, F.-Q., 1997, The genetic type of Yangshan upper Paleozoic basin and its relation to the Tongbai-Dabie orogenic belt: Scientia Geologica Sinica, v. 32 no. 1, p. 19–26 (in Chinese with English abstract).

Li, Z.-X., 1994, Collision between the North and South China blocks: A crustal-detachment model for suturing in the region east of the Tanlu fault: Geology, v. 22, p. 739–742.

Li, Z.-X., 1998, Tectonic history of the major East Asian lithospheric blocks since the mid-Proterozoic—a synthesis, in Martin, F. J. et al., eds., Mantle dynamics and plate interactions in East Asia: AGU Geodynamics Series, v. 27, p. 221–243.

Liou, J.-G., Wang, Q.-C., Zhai, M.-G., Zhang. R.-Y., and Cong, B.-L., 1995, Ultrahigh-P metamorphic rocks and their associated lithologies from the Dabie terrain, Central China: A field trip guide to the 3rd international eclogite field symposium: Chinese Science Bulletin, v. 40 (suppl.), p. 1–40.

Liou, J. G., Zhang, R.-Y., Eide, E. A., Wang, X. M., Ernst, W. G., and Maruyama, S., 1996, Metamorphism and tectonics of high-pressure and ultra-high-pressure belts in the Dabie-Sulu region, China, in Harrison, M. T., and Yin, A., eds., The tectonics of Asia: Cambridge, UK, Cambridge University Press, p. 300–344.

Liou, Y.-S., Lo, C.-H., Tsai, C.-H., Wang, P. -L., and Chen, C.-H., 1999, Thermochronological study of the Dabie Shan ultrahigh-pressure metamorphic terrane, east central China: Journal of the Geological Society of China, v. 42, p. 159–188.

Liu, F.-L., Xu, Z.-Q., Katayama, I., Yang, J.-S., Maruyama, S., and Liou, J.G., 2001, Mineral inclusions in zircons of para- and orthogneiss from pre-pilot drillhole CCSD-PP1, Chinese Continental Scientific Drilling Project: Lithos, v. 59, p. 199–215.

Liu, F.-L., and Xu, Z. Q., 2004, Fluid inclusions hidden in coesite-bearing zircons in ultrahigh-pressure metamorphic rocks from southwestern Sulu terrane in eastern China: Chinese Science bulletin, v. 49, p. 396–404.

Liu, F.-L., Xu, Z. Q., Liou, J. G., and Song, B., 2004a, SHRIMP U-Pb ages of ultrahigh-pressure and retrograde metamorphism of gneisses, southwest Sulu terrane, eastern China: Journal of Metamorphic Geology, v. 2, p. 315–326.

Liu, F.-L., Xu, Z. Q., and Xue, H. M., 2004b, Tracing the protolith, UHP metamorphism, and exhumation ages of orthogneiss from the SW Sulu terrane (eastern China): SHRIMP U-Pb dating of mineral inclusion–bearing zircons: Lithos, v. 78, p. 411–429.

Liu, X.-C., Dong, S.-W., Xue, H.-M., and Zhou, J.-X., 1999, Significance of allanite-(Ce) in granitic gneisses from the ultrahigh-pressure metamorphic terrane, Dabie Shan, central China: Mineralogical Magazine, v. 63, p. 579–586.

Liu, X.-C., Jahn, B.-M., Dong, S.-W., Li, H.-M., and Oberhaensli, R., 2003, Neoproterozoic granitoid did not record ultrahigh-pressure metamorphism from the Southern Dabieshan of China: The Journal of Geology, v. 111, p. 719–732.

Liu, X. H., and Hao, J., 1989, Structure and tectonic evolution of the Tongbai-Dabie range in the east Qinling collision belt, China: Tectonics, v. 8, p. 637–645.

Liu, Y.-C., Li, S.-G., Xu, S.-T., Jahn, B.-m., Zheng, Y.-F., Zhang, Z.-Q., Jiang, L.-L., Chen, G.-B., and Wu, W.-P. , 2005, Geochemistry and geochronology of eclogites from the north Dabie Mountains, central China: Journal of Asian Earth Sciences, v. 25, p. 431–443.

Liu, Y.-C., Xu, S.-T., and Jiang, L.-L., 1998, The meta-flysch nappe in the northern part of the Dabie mountains: Regional Geology of China, v. 17, no. 2, p. 156–162 (in Chinese with English abstract).

Ma, W.-P. , 1991, The carboniferous at the northern foot of the Dabie Mountains and its tectonic implications: Acta Geologica Sinica, v. 65, no. 1, p. 17–26 (in Chinese with English abstract).

Ma, W.-P. , Liu, W.-C., and Wang, G.-S., 1997, The Meishan Group in north Dabie and its tectonic implication: Geoscience, v. 11, no. 1, p. 95–101 (in Chinese with English abstract).

Ma, W.-P., Wang, G.-Y., and Wang, G.-S., 2001, Jinningian plutonic belt in the Foziling Group and its tectonic implication: Geological Review, v. 47, no. 5, p. 476–482 (in Chinese with English abstract).

Maruyama, S., 1997, Pacific-type orogeny revisited: Miyashiro-type orogeny proposed: The Island Arc, 6, 91–120.

Maruyama, S., Liou, J. G., and Zhang, R. Y., 1994, Tectonic evolution of the ultrahigh-pressure (UHP) and high-pressure (HP) metamorphic belts from central China: The Island Arc, v. 3, p. 112–121.

Mattauer, M., Matte, P., Malavieille, J., Tapponnier, P., Maluski, H., Qin, X. Z., Lun, L.Y., and Qin, T. Y., 1985, Tectonics of the Qinling Belt: Build-up and evolution of eastern Asia: Nature, v. 317, p. 496–500.

McKerrow, W. S., Leggett, J. K., and Eales, M. H., 1977, Imbricate thrust model of the Southern Uplands of Scotland: Nature, p. 267, v. 237–239.

Meng, Q.-R., and Zhang, G.-W., 1999, Timing of collision of the North and South China blocks: Controversy and reconciliation: Geology, v. 27, p. 123–126.

Moore, J. C., 1989, Tectonics and hydrogeology of accretionary prisms: Role of the decollement zone: Journal of Structural Geology, v. 11, p. 95–106.

Oberhaensli, R., Martinotti, G., Schmid, R., and Liu, X., 2002, Preservation of primary volcanic textures in the ultrahigh-pressure terrain of Dabie Shan: Geology, v. 30, p. 699–702.

Okay, A. I., Sengor, A. M. C., and Satir, M., 1993, Tectonics of an ultrahigh-pressure metamorphic terrane: The Dabie Shan/Tongbai Shan orogen, China: Tectonics, v. 12, p. 1320–1334.

Oxburgh, E. R., 1972, Flake tectonics and continental collision: Nature, v. 239, p. 202–204.

Platt, J. P., 1986, Dynamics of orogenic wedges and the uplift of high-pressure metamorphic rocks: Geological Society of American Bulletin, v. 97, p. 1037–1053.

Qian, C.-C., 1996, Geological features and tectonic settings of low-grade metamorphic volcanoclastics within eclogite belt in Dabieshan: Anhui Geology, v. 6, no. 2, p. 15–20 (in Chinese with English abstract).

Ratschbacher, L., Hacker, B. R., Calvert, A., Webb, L. E., Grimmer, J. C., McWilliams, M. O., Ireland, T., Dong, S.-W., and Hu, J.-M., 2003, Tectonics of the Qinling (Central China): Tectonostratigraphy, geochronology, and deformation history: Tectonophysics, v. 366, p. 1–53.

Ratschbacher, L., Hacker, B. R., Webb, L. E., McWilliams, M., Ireland, T., Dong, S., Calvert, A., Chateigner, D., and Wenk, H.-R., 2000, Exhumation of the ultrahigh-pressure continental crust in east central China: Cretaceous and Cenozoic unroofing and the Tan-Lu fault: Journal of Geophysical Research, v. 105B, p. 13,303–13,338.

Ring, U., and Brandob, M. T., 1999, Ductile deformation and mass loss in the Franciscan subduction complex; implications for exhumation processes in accretionary wedges, in Ring, U., Brandon, M. T., Lister, G. S., and Willett, S. D., eds., Exhumation processes: Normal faulting, ductile flow, and erosion: Geological Society of London Special Publications, v. 154, p. 55–86.

Rowley, D. B, Xue, F., Tucker, R. D., Peng, Z. X., Baker, J., and Davis, A., 1997, Ages of ultrahigh pressure metamorphism and protolith orthogneisses from the eastern Dabie Shan: U/Pb zircon geochronology: Earth and Planetary Science Letters, v. 151, p. 191–203.

Rumble, D., Wang, Q., and Zhang, R., 2000, Stable isotope geochemistry of marbles from the coesite UHP terrains of Dabieshan and Sulu, China: Lithos, v. 52, p. 79–95.

Sang, B., Chen, Y., and Shao, G., 1987, Rb-Sr ages of metamorphic series of the Susong Group at southeastern foot of the Dabie Mountains, Anhui Province, and their tectonic significance: Regional Geology of China, v. 4, p. 364–370 (in Chinese).

Schmid, R., Romer, R. L., Franz, L., Oberhaensli, R., and Martinotti, G., 2003, Basement-cover sequences within the UHP unit of the Dabie Shan: Journal of Metamorphic Geology, v. 21, p. 531–538.

Sengor, A. M. C, Natal'in, B. A., and Burtman, V. S., 1993, Evolution of the Altaid tectonic collage and Paleozoic crustal growth in Eurasia: Nature, v. 364, p. 299–304.

Sengor, A. M. C, and Okurogullari, A. H., 1991, The role of accretionary wedges in the growth of continents: Asiatic examples from Argand to plate tectonics: Eclogae Geologicae Helvetiae, v. 84, p. 535–597.

Shan, Q., Xue, S., and Cao, Z., 1992, The discovery of fossils remained in Guishan Formation of Xinyang Group: Henan Geology, no. 10, p. 40–46 (in Chinese).

Shandong (Shandong Bureau of Geology and Mineral Resources), 1987, Regional geology of Shandong Province: Beijing, China, Geological Publishing House, 595 p. (in Chinese).

Shandong (Shandong Bureau of Geology and Mineral Resources), 1997, Stratigraphy of Shandong Province: Wuhan, China, China University of Geosciences Press, 279 p.

Song, M.-C., and Song, Z.-Y., 1998, Some new viewpoints on Pengheshi Group in Jiaonan orogen: Shandong Geology, v. 14, no. 1, p. 25 (in Chinese).

Su, W., You, Z.-D., Wang, R.-C., and Liu, X.-W., 2001, Quartz and clinoenstatite exsolutions in clinopyroxene of garnet-pyroxenolite from the North Dabie Mountains, eastern China: Chinese Science Bulletin, v. 46, p. 1482–1485.

Sun W.-D., Williams, I. S., and Li, S.-G., 2002, Carboniferous and Triassic eclogites in the western Dabie Mountains, east-central China: Evidence for protracted convergence of the North and South China blocks: Journal of Metamorphic Geology, v. 20, p. 873–886.

Tang, J.-F., Qian, C.-C., and Gao, T.-S., 1995, Association of low-grade metamorphic volcanoclastics within eclogite belt in Dabieshan and its geological significance: Anhui Geology, v. 5, no. 2, p. 29–36 (in Chinese with English abstract).

Tarbuck, E. J., and Lutgens, F. K., 1994, Earth science, 7th ed.: New York, NY, Macmillan College Publishing Company, 659 p.

Tsai, C.-H., and Liou, J. G., 2000, Eclogite-facies relics and inferred ultrahigh-pressure metamorphism in the North Dabie Complex, central-eastern China: American Mineralogist, v. 85, p. 1–8.

von Huene, R., and Scholl, D. W., 1991, Observation at convergent margins concerning sediment subduction, subduction erosion, and the growth of continental crust: Reviews of Geophysics, v. 29, p. 279–316.

Wallis, S. R., Ishiwatari, A., Hirajima, T., Ye, K., Guo, J., Nakamura, D., Kato, T., Zhai, M., Enami, M., Cong, B., and Bano, S., 1997, Occurrence and field relationships of ultrahigh-pressure metagranitoid and coesite eclogite in the Su-Lu terrane, eastern China: Journal of Geological Society, London, v. 154, p. 45–54.

Wang, Q.-C., and Cong, B.-L., 1999, Exhumation of UHP terranes: A case study from the Dabie Mountains, eastern China: International Geology Review, v. 41, p. 994–1004.

Wang, Q.-C. and Cong, B.-L, 1998, Tectonic framework of ultrahigh pressure metamorphic belt in Dabieshan: Acta Petrologica Sinica, v. 14, p. 481–492 (in Chinese with English abstract).

Wang, X.-M., Zhang, R. Y., and Liou, J. G., 1995, UHPM terrane in east central China, in Coleman, R., and Wang, X.-M., eds., Ultrahigh pressure metamorphism: Cambridge, UK, Cambridge University Press, p. 356–390.

Wei, C.-J., Shan, Z.-G., Zhang, L.-F., Wang, S.-G., and Chang, Z.-G., 1998, Determination and geological significance of the eclogites from the northern Dabie Mountains, central China: Chinese Science Bulletin, v. 43, p. 253–256.

Wu, Y.-B., Zheng, Y.-.F., and Zhou, J.-B., 2004, Neoproterozoic granitoid in northwest Sulu and its bearing on the North China–South China Block boundary in east China: Geophysical Research Letters, v. 31 [L07616, doi:10.1029/2004GL019785].

Xie, Z., Chen, J.-F., Zhang, X., Gao, T.-S., Dai, Z.-Q., Zhou, T.-X., and Li, H.-M., 2001, Zircon U-Pb dating of gneiss from Shizhuhe in North Dabie and its geological implications: Acta Petrologica Sinica, v. 17, p. 139–144 (in Chinese with English abstract).

Xie, Z., Chen, J.-F., Zhang, X., Li, H.-M., Zhou, T.-X., and Yang, G., 2002, Geochronology of Neoproterozoic mafic intrusions in North Huaiyang region: Acta Geoscientia Sinica, v. 23, no. 6, p. 517–520 (in Chinese with English abstract).

Xie, Z., Gao, T. S., and Chen, J. F., 2004a, Multi-stage evolution of gneiss from North Dabie: Evidence from zircon U-Pb chronology: Chinese Science Bulletin, v. 49, p. 1963–1969.

Xie, Z., Zheng Y.-F., Jahn, B.-M., Ballevre, M., Chen, J.-F., Gautier, P., Gao, T. S., Gong, B., and Zhou, J. B., 2004b, Sm-Nd and Rb-Sr dating for pyroxene-garnetite from North Dabie in east-central China: Problem of isotope disequilibrium due to retrograde metamorphism: Chemical Geology, v. 206, p. 137–158.

Xu, B., Grove, M., Wang, C.-Q., Zhang, L.-F. and Liu, S.-W., 2000, $^{40}Ar/^{39}Ar$ thermochronology from the northwestern Dabie Shan: Constraints on the evolution of Qinling-Dabie orogen, east-central China: Tectonophysics, v. 322, p. 279–301.

Xu, S.-T., Liu, Y.-C., Jiang, L.-L., Su, W., and Ji, S.-Y., 1994, Tectonic regime and evolution of the Dabie Mountains: Beijing, China, Science Press, 175 p. (in Chinese with English abstract).

Xu, S.-T., Liu, Y.-C., Chen, G.-B., Compagnoni, R., Rolfo, F., He, M.-C., and Liu, H.-F., 2003, New finding of micro-diamonds in eclogites from the Dabie-Sulu region in east-central China: Chinese Science Bulletin, v. 48, p. 988–994.

Xu, S.-T., Liu, Y.-C., Su, W., Wang, R.-C., Jiang, L.-L., and Wu, W.-P., 2000, Discovery of the eclogite and its petrography in Northern Dabie Mountains: Chinese Science Bulletin, v. 45, p. 273–278.

Xue, F., Rowley, D. B., Tucker, R. D., and Peng, Z. X., 1997, U-Pb zircon ages of granitoid rocks in the North Dabie complex, eastern Dabie Shan, China: The Journal of Geology, v. 105, p. 744–753.

Yang, J.-D., Sun, W.-G., Wang, Z.-Z., Xue, Y.-S., and Tao, X.-C., 1999, Variations in Sr and C isotopes and Ce anomalies in successions from China: Evidence for the oxygenation of Neoproterozoic seawater?: Precambrian Research, v. 93, p. 215–233.

Yang, W.-C., 2002, Geophysical profiling across the Sulu ultra-high-pressure metamorphic belt, eastern China: Tectonophysics, v. 354, p. 277–288.

Ye, B.-D., Jian, P., Xu, J.-W., Cui, F., Li, Z.-C., and Zhang, Z.-H., 1993, The Sujiahe terrane collage belt and its constitution and evolution among the north hillslope of the Tongbai-Dabie orogen: Wuhan, China, China University of Geosciences Press, 81 p. (in Chinese with English abstract).

Yin, A., and Nie, S., 1993, An indentation model for the North and South China collision and the development of the Tan-Lu and Honam fault systems, eastern Asia: Tectonics, v. 12, p. 801–813.

Zhai, M.-G., 2002, Where in the North China–South China block boundary in eastern China? Comment: Geology, v. 30, p. 667.

Zhai, M.-G., Cong, B.-L., Guo, J.-H., Liu, W.-J., Li, Y.-G., and Wang, Q.-C., 2000a, Sm-Nd geochronology and petrology of garnet pyroxene granulites in the northern Sulu region of China and their geotectonic implication: Lithos, v. 52, p. 23–33.

Zhai, M.-G., Jiang, L.-L., Wang, Q.-C., and Cong, B.-L., 2000b, Call in question and discussion: Are there sandwiched low-grade metamorphic slabs within UHP metamorphic belt in the Dabie terrane?: Chinese Science Bulletin, v. 45, p. 181–189.

Zhai, X.-M., Day, H. W., Hacker, B. R. and You, Z. D., 1998, Paleozoic metamorphism in the Qinling orogen, Tongbai Mountains, central China: Geology, v. 26, p. 371–374.

Zhang, K.-J., 1997, North and South China collision along the eastern and southern North China margins: Tectonophysics, v. 270, p. 145–156.

Zhang, R. Y., and Liou, J. G., 1997, Partial transformation of gabbro to coesite-bearing eclogite from Yangkou, the Sulu terrane, eastern China: Journal of Metamorphic Geology, v. 15, p. 183–202.

Zhang, R. Y., and Liou, J. G., 1998, Ultrahigh-pressure metamorphism of the Sulu terrane, eastern China: A prospective view: Continental Dynamics, v. 3, p. 32–53.

Zhang, Z.-Q., Chi, S.-X., and Song, Z.-Y., 1994, Redetermination of Jiaonan Group in Shandong and establishment of Pengheshi Group: Regional Geology of China, v. 13, no. 4, p. 354–356 (in Chinese).

Zhao, D., Cheng, L.-R., and Liu, M.-X., 1995, The discovery of Solenopora in the Wulian group in the Jiaonan area, Shandong, and its significance: Regional Geology of China, v. 14, no. 4, p. 379–384 (in Chinese with English abstract).

Zhao, Z.-F., Zheng, Y.-F., Wei, C.-S., and Wu, Y.-B., 2004, Zircon isotope evidence for recycling of subducted continental crust in post-collisional granitoids from the Dabie terrane in China: Geophysical Research Letters, v. 31, L22602 [doi:10.1029/2004GL021061].

Zhao, Z.-F., Zheng, Y.-F., Wei, C.-S., Wu, Y.-B., Chen, F.-K., and Jahn, B.-m., 2005, Zircon U-Pb age, element, and C-O isotope geochemistry of post-collisional mafic-ultramafic rocks from the Dabie orogen in east-central China: Lithos, v. 82, p. 1–28.

Zheng, W.-W., 1964, Division of Foziling Group in eastern Dabie and its times: Geological Review, v. 22, no. 5, p. 338–347 (in Chinese).

Zheng, Y.-F., Fu, B., Gong, B., and Li, L., 2003a, Stable isotope geochemistry of ultrahigh pressure metamorphic rocks from the Dabie-Sulu orogen in China: Implications for geodynamics and fluid regime: Earth Science Review, v. 62, p. 105–161.

Zheng, Y.-F., Fu, B., Gong, B., and Wang, Z.-R., 1998, Carbon isotope anomaly in marbles associated with eclogites from the Dabie Mountains in China: Journal of Geology, v. 106, p. 97–104.

Zheng, Y.-F., Fu, B., Li, Y.-L., Wei, C.-S. and Zhou, J.-B., 2001, Oxygen isotope composition of granulites from Dabieshan in eastern China and its implications for geodynamics of Yangtze plate subduction: Physics and Chemistry of the Earth (part A), v. 26, p. 673–684.

Zheng, Y.-F., Fu, B., Xiao, Y.-L., Li, Y.-L., and Gong, B., 1999, Hydrogen and oxygen isotope evidence for fluid-rock interactions in the stages of pre- and post-UHP metamorphism in the Dabie Mountains: Lithos, v. 46, p. 677–693.

Zheng, Y.-F., Gong, B., Zhao, Z.-F., Fu, B. and Li, Y.-L., 2003b, Two types of gneisses associated with eclogite at Shuanghe in the Dabie terrane: carbon isotope, zircon U-Pb dating and oxygen isotope: Lithos, v. 70, p. 321–343.

Zheng, Y.-F., Wang, Z.-R., Li, S.-G., and Zhao, Z.-F., 2002, Oxygen isotope equilibrium between eclogite minerals and its constraints on mineral Sm-Nd chronometer: Geochimica et Cosmochimica Acta, v. 66, p. 625–634.

Zheng, Y.-F., Wu, Y.-B., Chen, F.-K., Gong, B., and Zhao, Z.-F., 2004, Zircon U-Pb and oxygen isotope evidence for a large-scale ^{18}O depletion event in igneous rocks during the Neoproterozoic: Geochimica et Cosmochimica Acta, v. 68, p. 4145–4165.

Zhong, Z.-Q., Suo, S.-T., and You, Z.-D., 1999, Regional-scale extensional tectonic pattern of ultrahigh-pressure and high-pressure metamorphic belts from the Dabie massif, China: International Geology Review, v. 41, p. 1033–1041.

Zhou, C.-T., Gao, T.-S., Tang, J.-F., Shen, H.-S., and Hu, Y.-Q., 2000, Distribution and main characteristics of eclogite in the northern Dabie Mountains, Anhui: Regional Geology of China, v. 19, p. 253–257 (in Chinese with English abstract).

Zhou, J.-B., Zheng, Y.-F., Li, L., and Xie, Z., 2001, On low-grade metamorphic rocks within Dabie-Sulu ultrahigh pressure metamorphic belt: Acta Petrologica Sinica, v. 17, no. 1, p. 39–48 (in Chinese with English abstract).

TAIWAN

Isotopic Composition of Carbonaceous Material in Metamorphic Rocks from the Mountain Belt of Taiwan

TZEN-FU YUI[1]

Institute of Earth Sciences, Academia Sinica, P.O. Box 1-55, Nankang, Taipei, Taiwan, R.O.C.

Abstract

Carbonaceous (organic) material was extracted from pelitic rocks and marbles collected along three highways transecting the mountain belt of Taiwan. $\delta^{13}C$ values of carbonaceous matter show a general increasing trend with elevated metamorphic grade. The first prominent, systematic change in $\delta^{13}C$ values occurs in samples of the middle greenschist facies, with $\delta^{13}C$ values higher than –24‰. Carbon isotope fractionation between coexisting calcite and carbonaceous material suggests that this marked change might have taken place at around 300–350°C. From systematic temperature variations, two thermal maxima, 430°C and 460–490°C, were delineated in the central part of the Hsuehshan Range and the Backbone Range, respectively, and are attributed to Cenozoic mountain building. High temperatures (>500°C) were detected in some marble samples adjacent to Mesozoic granite intrusions, indicating that pre-existing thermal records are preserved in these marbles. The carbon isotope fractionation preserved in rocks across the mountain belt therefore may not have been synchronous.

Introduction

CARBONACEOUS MATERIAL derived from organic debris is a common, minor constituent in metamorphosed sedimentary rocks. Systematic changes in its atomic order, crystal structure, and chemical composition with increasing metamorphic grade have been reported (French, 1964; Griffin, 1967; Izawa, 1968; Landis, 1971; Grew, 1974; Diessel et al., 1978; Itaya, 1981; Wopenka and Pasteris, 1993; Yui et al., 1996b; Beyssac et al., 2002). Structurally poorly organized kerogen- or coal-like organic material is transformed into well-crystallized graphite with concomitant chemical maturation during loss of H (in the form of H_2O, CH_4), O (in the form of CO_2, H_2O), and N. Such a transformation is referred to as graphitization. With progressive maturation, $\delta^{13}C$ of the carbonaceous material also changes as a result of the different binding energies of the various C-isotope species (Barker and Friedman, 1969; Hoefs and Frey, 1976; Arneth et al., 1985) as well as through C-isotope exchange with other C-bearing phases such as coexisting carbonates (Eichmann and Schidlowski, 1975; Wada and Oana, 1975; Hoefs and Frey, 1976; Pineau et al., 1976; Valley and O'Neil, 1981; Kreulen and van Beek, 1983; Wada and Suzuki, 1983; Morikiyo, 1984; Arneth et al., 1985; Dunn and Valley, 1992; Kitchen and Valley, 1995). Carbon isotope fractionation between coexisting calcite and graphite has therefore been proposed as a potential geothermometer (Bottinga, 1969; Valley and O'Neil, 1981; Wada and Suzuki, 1983; Morikiyo, 1984; Chacko et al., 1991; Dunn and Valley, 1992; Kitchen and Valley, 1995; Polyakov and Kharlashina, 1995). Due to the isotopic inertness of carbonaceous material, this thermometer was also suggested as being able to "see through" regional metamorphism to provide semi-quantitative temperature information in polymetamorphic settings (Dunn and Valley, 1992).

The Taiwan orogen, resulting from ongoing collision between Eurasia and the Philippine Sea plate since the Plio-Pleistocene, is situated along the Eurasian plate margin (Ho, 1986). The main part of the mountain belt consists of an Upper Paleozoic–Mesozoic basement complex, and an unconformably overlying Cenozoic cover series. The geohistory and tectonic evolution of the belt was discussed by Liou and Ernst (1984) and by Yui et al. (1990). Both papers suggested a complicated metamorphic history. Based on available geologic-mineralogic-petrologic data, Chen et al. (1983) presented a metamorphic facies map of Taiwan. Using K-mica–chlorite crystallinity, Chen (1984) further refined the lower greenschist-facies boundary in the mountain belt. All these studies, however, only gave

---

[1] email: tfyui@earth.sinica.edu.tw

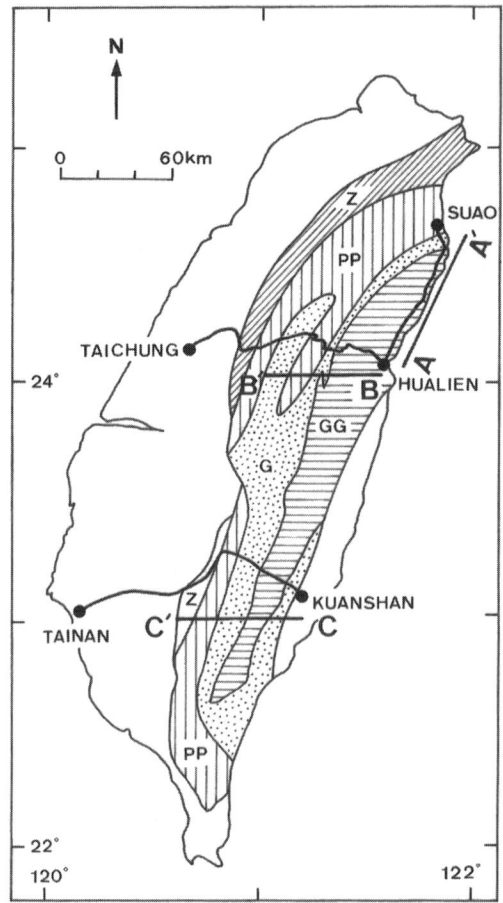

FIG. 1. Simplified metamorphic facies map of Taiwan (after Chen et al., 1983). Abbreviations: S = unmetamorphosed sedimentary rocks; Z = zeolite-facies rocks, PP = prehnite-pumpellyite–facies rocks; G = greenschist-facies rocks; GG = polymetamorphosed greenschist-facies rocks. Samples were collected along three highways. A–A', B–B', and C–C' are three lines projecting the sample positions shown in Figures 2, 3, and 4, respectively.

qualitative descriptions. Quantitative estimations of metamorphic temperatures in the Taiwan mountain belt are scarce (see Chen and Wang, 1995, and references therein) due to a shortage of appropriate geothermometers. Systematic, quantitative studies of metamorphic rocks in Taiwan therefore are still rare. To fully understand the mountain-building processes, the thermal structure and its evolution, which are important in controlling the rheology of deforming rocks within the mountain belt, are vitally needed data. Inasmuch as calcite and carbonaceous material are common constituents in metamorphosed sedimentary rocks in the mountain belt of Taiwan, this paper presents systematic carbon isotope data for these rocks, with the purpose of semi-quantitatively estimating metamorphic temperature variations.

General Geology and Sample Collection

The Cenozoic Taiwan orogenic belt consists of two parallel mountain ranges: the western Hsuehshan Range and the eastern Backbone Range. The Backbone Range extends throughout Taiwan from the northeast to the southwest, whereas the Hsuehshan Range dies out southward in central Taiwan (Ho, 1986). According to the metamorphic facies map (Chen et al., 1983; Chen and Wang, 1995), rocks in Taiwan can be divided into: (1) an amphibolite facies (A-) group of the pre-Tertiary basement complex in the Backbone Range; (2) a polymetamorphosed greenschist facies (GG-) group of the pre-Tertiary basement complex in the Backbone Range; (3) a greenschist facies (G-) group of the Tertiary formations in both the Backbone and the Hsuehshan ranges; (4) a prehnite-pumpellyite facies (PP-) group of the slate belt in both the Backbone and the Hsuehshan ranges; (5) a zeolite facies (Z-) group of the Western Foothills; and (6) unmetamorphosed sedimentary rocks (S) (Fig. 1). Lithologically, the A-group consists of scattered amphibolite and epidote amphibolite, as well as granitic gneisses in the pre-Tertiary basement (i.e., the Tananao Metamorphic Complex). The GG-group is composed of marble/schist/phyllite of the Tananao Metamorphic Complex. The G- and PP-group mainly comprise phyllite/slate/argillite/metasandstone of the Eo-Oligocene Tachien and Chiayang formations, the Eocene Pilushan Formation, and the Miocene Lushan Formation. The Z- and S-groups comprise shale/sandstone of Miocene and younger formations (Ho, 1986).

Pelitic rocks and marbles were collected along three highways crossing the mountain belt: the Suhua Highway (A–A'), the Central E-W Cross-Island Highway (B–B') and the Southern E-W Cross-Island Highway (C–C') (Fig. 1). Samples along the A–A' and C–C' sections are all collected from the Backbone Range, while samples along the B–B' are from both the Hsuehshan and the Backbone ranges. In the latter case, samples with a sample number less than C-43 are from the Hsuehshan Range (see the following sections). For all samples studied, marble of the GG-group mainly consists of calcite, with minor amounts of quartz, talc, phlogopite, and carbonaceous material. The pelitic schist of the GG-group contains quartz, albite, white mica, biotite, chlorite, calcite, titanite, epidote, and carbonaceous material, whereas the pelitic schist/phyllite of the G-group consists of quartz, albite, white mica, calcite, titanite, and carbonaceous material. The phyllite/slate of the PP-group comprises chlorite-illite, kaolinite, quartz, albite, calcite, and carbonaceous material. The argillite of the Z-group contains illite/smectite, kaolinite, quartz, feldspar, calcite, and carbonaceous material. The shale of the S-group consists of smectite, kaolinite, feldspar, quartz, calcite, mixed-layer clay, and carbonaceous material. Some pelitic samples also contain a very small amount of dolomite and ankerite, in addition to calcite. The former carbonate phases are considered not to be important in the following discussion. Minor opaque minerals are also commonly present in pelitic samples. Most of the carbonaceous materials in the present study belong to graphite-d_1, graphite-d_{1A}, graphite-d_2, or graphite-d_3 following Landis's scheme (1971). Fully ordered graphite is only present in a few high-temperature marble samples along A–A' and B–B' sections (Warneke and Ernst, 1984; author's unpublished data).

Analytical Methods

Carbonaceous material was extracted from rock samples, mainly following the procedures suggested by French (1964), Grew (1974), Itaya (1981) and Wedeking et al. (1983). Twenty to 30 grams of rock powders sieved through 200 mesh were first treated twice with concentrated HCl, and then with a 4:1 mixture of HF and HCl, in four repetitions, at 80°C. Carbonates, silicates, and most of the oxides were dissolved. Some insoluble precipitates, formed during acid digestion, were then removed by further treatment with an $AgCl_3$ solution. The residue obtained contains predominately carbonaceous material with minor amounts of acid-insoluble minerals such as zircon, rutile, and pyrite. The residue was washed and dried for further study.

For isotopic analyses, the carbonaceous material was heated at 900°C in an evacuated sealed Vycor tube with CuO as the oxidant and silver metal as the catalyst (Wedeking et al., 1983). The resultant CO_2 gas was then purified for mass-spectrometer measurement. The results are reported as per mil $\delta^{13}C$ values relative to PDB (Craig, 1957). The reproducibility is about ±0.2‰. During the period of this study, the NBS-21 was analyzed with an averaged $\delta^{13}C$-value of –28.2‰.

The isotopic compositions of calcite in the rock samples were also analyzed by the conventional method of reaction with 100% H_3PO_4 at 25°C to liberate CO_2 gas (McCrea, 1950). The acid fractionation factor used for oxygen between liberated CO_2 and calcite was 1.01025 (Sharma and Clayton, 1965). The results are reported as per mil $\delta^{18}O$ values relative to SMOW, and per mil $\delta^{13}C$ values relative to PDB (Craig, 1957). Reproducibility for both O- and C-isotope analyses is better than ±0.1‰. During the period of study, the resulting $\delta^{18}O$ and $\delta^{13}C$-value of NBS-20 were +26.69‰ and –1.03‰, respectively.

Results

The isotopic compositions for calcite and carbonaceous material from samples collected along the Suhua Highway, the Central E-W Cross-Island Highway, and the Southern E-W Cross-Island Highway are listed in Tables 1–3, respectively. All these data are plotted in Figures 2–4.

The calcite (cc) and the carbonaceous material (org) in samples from the Suhua Highway exhibit $\delta^{18}O(cc)$ = +11.2 to +24.1‰, $\delta^{13}C(cc)$ = –12.2 to +4.0‰ and $\delta^{13}C(org)$ = –25.1 to –4.7‰ (Table 1); those from the Central E-W Cross-Island Highway, $\delta^{18}O(cc)$ = +10.6 to +23.6‰, $\delta^{13}C(cc)$ = –8.1 to +5.3‰, and $\delta^{13}C(org)$ = –25.9 to –3.1‰ (Table 2); and those from the Southern E-W Cross-Island Highway, $\delta^{18}O(cc)$ = +8.9 to +24.4‰, $\delta^{13}C(cc)$ = –10.0 to +0.8‰, and $\delta^{13}C(org)$ = –26.8 to –14.1‰ (Table 3). Generally speaking, C-isotope compositions of these phases from marbles are, as expected, higher than the corresponding ones from pelitic samples. On the other hand, the O-isotope compositions of calcite in marbles cannot be distinguished from that in pelitic samples.

Discussion

Variation of isotope compositions

The $\delta^{13}C$ value of the carbonaceous material from pelitic samples ranges from –26.8 to –14.1‰, and from marbles, from –18.4 to –3.1‰ (Tables 1–3). These values are higher than the $\delta^{13}C$ range (i.e., –26 ± 7‰) of the common sedimentary carbonaceous material (mainly organic matter; Degens, 1969; Eichmann and Schidlowski, 1975; Hayes et al., 1983; Schidlowski et al., 1983). The $\delta^{13}C$ value of the carbonaceous material was shown to increase during diagenesis and metamorphism (McKirdy and Powell, 1974; Hayes et al., 1983). Such an increase in $\delta^{13}C$ value could be accounted for by the degassing of ^{12}C-rich methane during chemical maturation of carbonaceous material, such as dehydrogenation or oxidizing of organic matter, or by high-temperature C-isotope exchange with ^{13}C-rich coexisting sedimentary carbonates. Empirical observations and theoretical calculations show that graphitization alone could at most only cause a small C-isotope shift of the carbonaceous material, on the order of 3–5‰ (McKirdy and Powell, 1974; Chung and Sackett, 1979; Arneth et al., 1985). The isotopic exchange between the carbonaceous material and the coexisting calcite must have involved a larger C-isotope shift, especially for those in the marble samples.

Upon close inspections of the data along the Suhua Highway, the $\delta^{13}C$ value of the carbonaceous material ranges from –25.1 to –22.8‰ for the G-group samples and from –21.9 to –4.7‰ for the GG-group samples (Fig. 2). Along the Central E-W Cross-Island Highway, the $\delta^{13}C$ value of the carbonaceous material ranges from –25.9 to –25.0‰ for the PP-group samples, from –25.5 to –18.4‰ for the G-group samples, and from –23.1 to –3.1‰ for the GG-group samples (Fig. 3). And along the Southern E-W Cross-Island Highway, the $\delta^{13}C$ value of the carbonaceous material ranges from –25.9 to –25.3‰ for the S-group samples, from –25.3 to –21.7‰ for the Z-group samples, from –26.8 to –22.9‰ for the PP-group samples, from –24.7 to –19.0‰ for the G-group samples, and from –22.1 to –14.1‰ for the GG-group samples (Fig. 4). It is clear that carbonaceous matter from the S-, Z-, PP-, and the lower part of the G-group samples exhibit similar $\delta^{13}C$ values. This may be due to the presence of carbonaceous materials of different structural states within the low-grade samples in which graphitization was not significant. However, from the middle part of the G-group to the GG-group samples, the $\delta^{13}C$-value of the carbonaceous material, especially discernible in pelitic rocks, shows a systematic increasing trend (>–24‰) across the three sections (Figs. 2–4). This may result from higher degrees of graphitization, chemical maturation, and isotopic exchange with coexisting calcite. These results are quite similar to those reported for metamorphic rocks from the Swiss Alps by Hoefs and Frey (1976). They showed that the most rapid change in C-isotope composition of the carbonaceous material (i.e., from < –22‰ to –18/–16‰) probably took place within greenschist-facies samples at temperature conditions of around 350–400°C.

In contrast to the carbonaceous material, both $\delta^{18}O$ and $\delta^{13}C$ values of calcite, especially in pelitic samples, roughly decrease with increase in metamorphic grade. The trends are more prominent for the high-grade rocks (Figs. 2–4). Such trends can be readily explained as a result of isotopic exchange with coexisting silicates, carbonaceous material, and the ambient fluid phase during metamorphism (Taylor et al., 1963). The extent of this isotopic exchange depends on the relative amounts and proportions of the oxygen/carbon atoms, the metamorphic temperature, the

TABLE 1. Isotope Compositions of Calcite and Carbonaceous (organic) Matter,
and Calculated Isotopic Temperatures from Samples Collected along the Suhua Highway

Sample no.[1]	Grade[2]	$\delta^{18}O$ (‰) calcite	$\delta^{13}C$ (‰) calcite	$\delta^{13}C$ (‰) organic	$\Delta^{13}C$ cc.-org.	T(°C)[3]	T(°C)[4]
100	G	–	–	–25.1			
LB-14	G	+18.0	–4.4	–24.1	19.7	303	237
101	G	–	–	–24.6			
102	G	–	–	–22.8			
B-16	GG	+16.4	–5.5	–21.9	16.4	342	280
103	GG	–	–	–20.1			
LB-19	GG	+16.5	–5.1	–19.0	13.9	378	320
106	GG	–	–	–19.8			
NG-27	GG	–	–	–19.9			
LB-21	GG	–	–	–18.7			
L80-3	GG	+15.6	–7.6	–18.7	11.1	426	378
LB-22	GG	–	–	–19.4			
LB-25	GG	–	–	–20.1			
LB-26	GG	–	–	–21.6			
110	GG	+14.4	–11.3	–21.4	10.1	446	403
LB-27	GG	+15.7	–10.5	–19.8	9.3	464	425
LB-28	GG	+14.7	–12.7	–21.2	8.5	482	450
125m	GG	+11.2	–4.5	–6.3	1.8	727	875
LB-29	GG	+13.4	–8.1	–20.2	12.1	408	355
111m	GG	+19.7	+3.8	–6.7	10.5	438	393
124m	GG	+16.0	+3.3	–6.8	10.1	446	403
112m	GG	+16.4	+1.1	–7.0	8.1	492	463
113m	GG	+18.2	+2.5	–6.7	9.2	466	428
114m	GG	+23.0	+1.8	–10.1	11.9	411	360
115m	GG	+24.1	+2.0	–6.0	8.0	495	467
116m	GG	+16.5	+0.9	–10.5	11.4	421	371
117m	GG	+23.7	+2.9	–4.7	7.6	505	481
118m	GG	+22.9	+3.8	–7.6	11.4	421	371
119m	GG	+16.9	+3.8	–7.7	11.5	419	369
122m	GG	+19.7	+4.0	–7.6	11.6	417	366
120m	GG	+18.3	+2.5	–7.6	10.1	446	403

[1]Sample number with the suffix "m" denotes marble samples.
[2]G = greenschist facies; GG = polymetamorphosed greenschist facies.
[3]Calculated temperatures based on the calibration curve of Morikiyo (1984).
[4]Calculated temperatures based on the calibration curve of Dunn and Valley (1992).

isotopic contrasts among phases, and the degree of isotopic equilibrium. That calcite and carbonaceous material in marble samples generally exhibiting higher C-isotope compositions than those in pelitic rocks is a good example showing the mass balance effect.

TABLE 2. Isotope Compositions of Calcite and Carbonaceous (organic) Matter, and Calculated Isotopic Temperatures from samples collected along the Central E-W Cross-Island Highway

Sample no.[1]	Grade[2]	$\delta^{18}O$ (‰) calcite	$\delta^{13}C$ (‰) calcite	$\delta^{13}C$ (‰) organic	$\Delta^{13}C$ cc.-org.	T(°C)[3]	T(°C)[4]
C-64	PP	–	–	–25.0			
C-60	G	+10.6	–8.0	–24.5	16.5	341	278
C-59	G	–	–	–24.0			
C-55	G	–	–	–24.7			
C-52	G	+11.5	–8.4	–19.5	11.1	426	378
C-51	G	–	–	–20.2			
C-49	G	+15.7	–9.1	–21.2	12.1	408	355
C-48	G	+19.8	–8.6	–24.0	15.4	356	295
C-47	G	–	–	–24.0			
C-46	G	+17.9	–5.9	–23.6	17.7	329	262
C-45	G	+17.9	–6.4	–25.5	19.1	310	244
C-43	PP	+21.9	–1.8	–25.9	24.1	261	193
C-41	PP	+19.2	–3.2	–25.1	21.9	281	214
C-38	PP	+19.3	–2.1	–25.6	23.5	266	199
C-35	G	+20.8	–5.3	–25.4	20.1	299	232
C-33	G	+17.1	–5.9	–25.0	19.1	310	244
C-31	G	+16.2	–2.8	–21.5	18.7	314	249
C-29	G	+16.0	–4.4	–22.1	17.7	326	262
C-27	G	+17.1	–3.2	–24.1	20.9	291	224
C-7m	G	+19.7	+1.3	–18.4	19.7	303	237
C-26	G	–	–	–23.8			
C-25	GG	+15.1	–8.1	–22.7	14.6	367	308
C-24	GG	–	–	–23.0			
C-23	GG	–	–	–22.7			
C-22	GG	–	–	–23.1			
C-21	GG	–	–	–20.7			
C-20	GG	+10.6	–8.9	–21.6	12.7	397	343
C-19	GG	–	–	–19.0			
C-14	GG	+16.8	–6.6	–15.9	9.3	464	425
C-13m	GG	+14.6	+3.3	–7.5	10.8	432	385
C-11m	GG	+19.1	+3.4	–9.7	13.1	391	335
C-10m	GG	+15.7	+2.0	–3.1	5.1	581	595
C-7m	GG	+19.8	+1.9	–9.1	11.0	428	380
C-5	GG	+23.6	–4.7	–16.4	11.7	415	364
121m	GG	+13.7	+5.3	–5.1	10.4	440	395

[1]Sample number with suffix 'm' denotes marble samples.
[2]PP: prehnite–pumpellyite zone; G: greenschist facies; GG: polymetamorphosed greenschist facies.
[3]Calculated temperatures based on the calibration curve of Morikiyo (1984).
[4]Calculated temperatures based on the calibration curve of Dunn and Valley (1992).

TABLE 3. Isotope Compositions of Calcite and Carbonaceous (organic) Matter, and Calculated Isotopic Temperatures from Samples Collected along the Southern E-W Cross-Island Highway

Sample no.	Grade[1]	$\delta^{18}O$ (‰) calcite	$\delta^{13}C$ (‰) calcite	$\delta^{13}C$ (‰) organic	$\Delta^{13}C$ cc.-org.	$T(°C)^2$	$T(^{13}C)^3$
S-1	S	+23.9	−1.2	−25.3	24.1	261	193
S-3	S	−	−	−25.9			
S-5	Z	+19.2	−4.8	−24.2	19.4	306	241
S-6	Z	+22.4	−3.5	−			
S-8	Z	+20.2	−4.8	−23.9	19.1	310	244
S-9	Z	+22.1	−3.4	−24.6	21.2	288	221
S-15	Z	+21.2	−2.1	−25.3	23.2	269	201
S-17	Z	+19.6	0.8	−21.7	22.5	275	208
S-23	Z	+20.3	−4.3	−24.0	19.7	303	237
S-28	Z	+20.1	−4.6	−25.2	20.6	294	227
S-34	Z	+19.7	−4.4	−24.6	20.2	298	232
S-43	Z	+16.5	−6.2	−24.6	18.4	318	253
S-49	Z	+20.3	−5.3	−			
S-65	Z	+17.9	−2.6	−23.4	20.8	292	225
S-58	PP	+18.4	−0.3	−22.9	22.6	274	207
S-72	PP	+16.7	−4.3	−			
S-78	PP	+19.3	−3.0	−25.5	22.5	275	208
S-81	PP	+24.4	−7.2	−26.8	19.6	304	238
S-87	PP	+16.5	−4.1	−26.4	22.3	277	210
S-101	PP	+15.6	−2.1	−24.7	22.6	274	207
S-114	G	+14.9	−6.0	−24.4	18.4	318	253
S-120	G	−	−	−22.7			
S-137	G	+11.3	−6.7	−22.6	15.9	349	287
S-142	G	+8.9	−6.7	−21.9	15.2	359	298
S-146	G	−	−	−21.6			
S-152	G	+23.6	−10.0	−20.5	10.5	438	393
S-161	G	+17.9	−8.0	−19.0	11.0	428	380
S-184	GG	−	−	−22.1			
S-188	GG	−	−	−20.9			
S-190	GG	−	−	−20.6			
S-194	GG	+9.7	−4.2	−14.1	9.9	451	408
S-194.1	GG	+13.8	−4.7	−14.2	9.5	459	420
S-194.5	GG	−	−	−16.3			
S-196	GG	+12.3	−6.9	−19.1	12.2	406	353
S-200	GG	+14.5	−9.4	−21.4	12.0	410	358
S-201	GG	+11.3	−9.2	−20.1	10.9	430	383
S-203	GG	+14.7	−8.9	−21.2	12.3	404	351
S-204	G	+24.3	−6.5	−21.1	14.6	367	308
S-205	G	−	−	−23.1			

[1] S = unmetamorphosed zone; Z = zeolite facies; PP = prehnite-pumpellyite zone; G = greenschist facies; GG = polymetamorphosed greenschist facies.
[2] Calculated temperatures based on the calibration curve of Morikiyo (1984).
[3] Calculated temperatures based on the calibration curve of Dunn and Valley (1992).

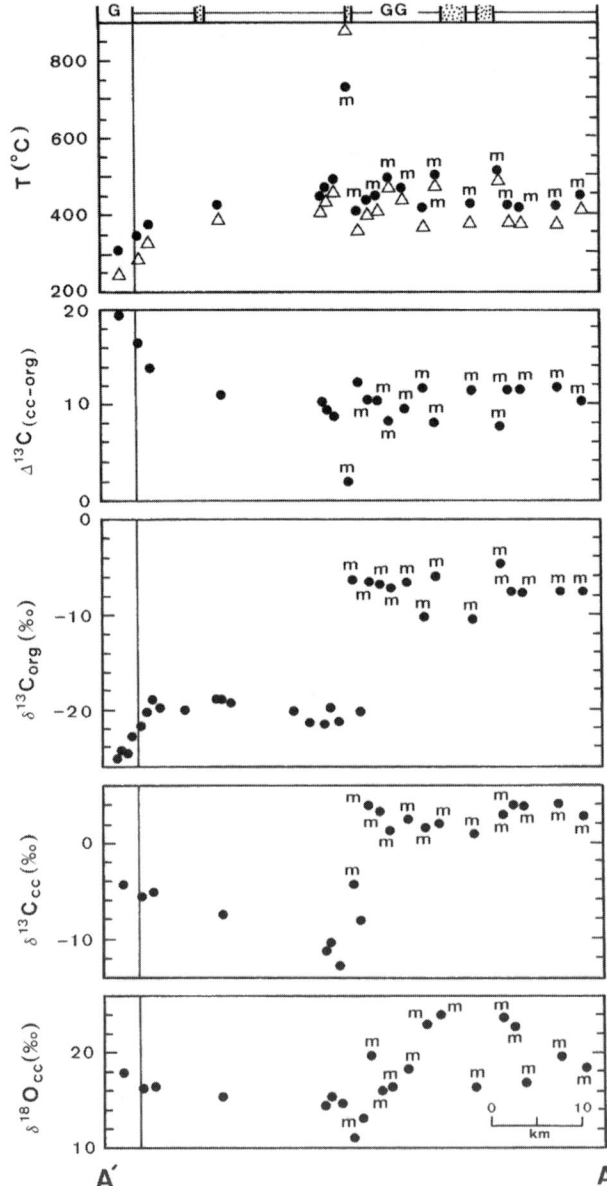

FIG. 2. Plot of $\delta^{18}O$, $\delta^{13}C$ of calcite (cc), $\delta^{13}C$ of carbonaceous matter (org), $\Delta^{13}C$(cc-org), and estimated temperatures for samples from the Suhua Highway versus the sample position as projected on the A–A' line shown in Figure 1. The metamorphic facies groups are after Chen et al. (1983; see Fig. 1). The stipple symbol on top of the figure marks the positions where Mesozoic granites occur. Legend for temperature estimates: circles = Morikiyo's (1984) calibration; triangles = Dunn and Valley's (1992) calibration. See text for details.

Calcite–carbonaceous material (graphite) isotope thermometer

C-isotope fractionation between calcite and graphite has been theoretically, experimentally, and empirically studied by various authors (Bottinga, 1969; Valley and O'Neil, 1981; Wada and Suzuki, 1983; Morikiyo, 1984; Chacko et al., 1991; Dunn and Valley, 1992; Scheele and Hoefs, 1992; Kitchen

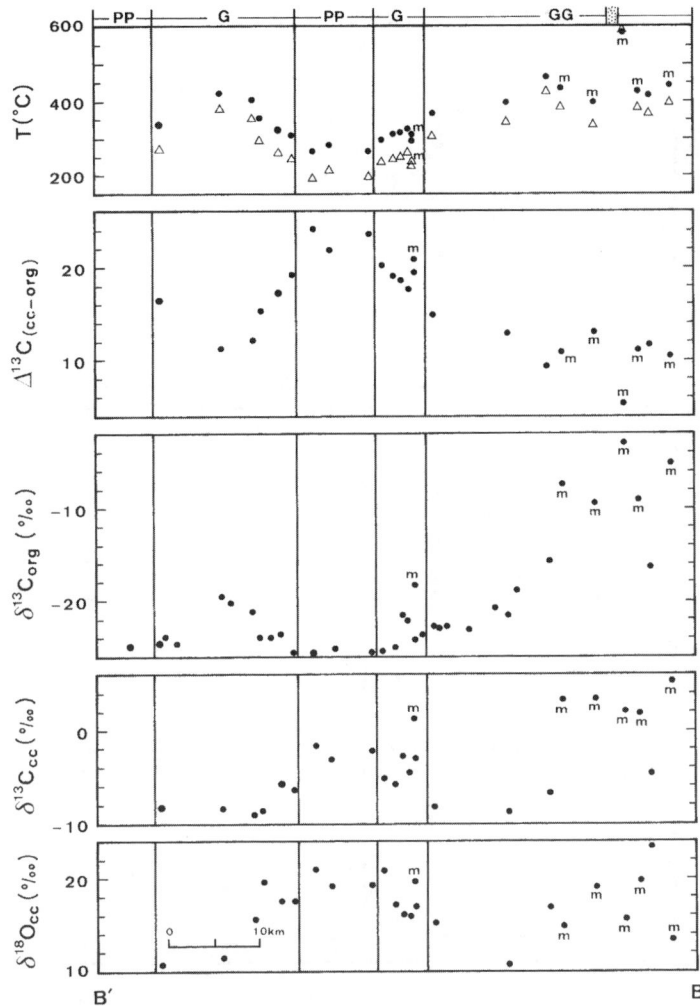

FIG. 3. Plot of $\delta^{18}O$, $\delta^{13}C$ of calcite (cc), $\delta^{13}C$ of carbonaceous matter (org), $\Delta^{13}C$(cc-org), and estimated temperatures for samples from the Central E-W Cross-Island Highway versus the sample position as projected on the B–B' line shown in Figure 1. The metamorphic-facies groups are after Chen et al. (1983) (see Fig. 1). The stipple symbol on top of the figure marks the positions where Mesozoic granites occur. Legend for temperature estimates: circles = Morikiyo's (1984) calibration; triangles = Dunn and Valley's (1992) calibration. See text for details.

and Valley, 1995; Polyakov and Kharlashina, 1995). The results differ significantly (Fig. 5), probably due to the different partition function ratios used (Chacko et al., 1991), as well as the slow rate of carbon diffusion in carbonaceous material (Feates, 1968). Based on available data, it is now generally agreed that isotopic equilibrium between calcite and carbonaceous material may not be attained until the temperature is above 650°C (Valley and O'Neil, 1981; Kreulen and van Beek, 1983; Arneth et al., 1985). Any empirical relation below this temperature is obviously invalid in terms of isotopic equilibrium, but might represent isotopic fractionations resulting from integrated kinetic effects. The metamorphic temperature in the Taiwan mountain belt is mostly equal to or lower than that of greenschist-facies conditions (Chen et al., 1983). Isotopic equilibrium between calcite and carbonaceous matter may not be expected. Therefore, it may be not suitable to apply theoretical or experimental carbon

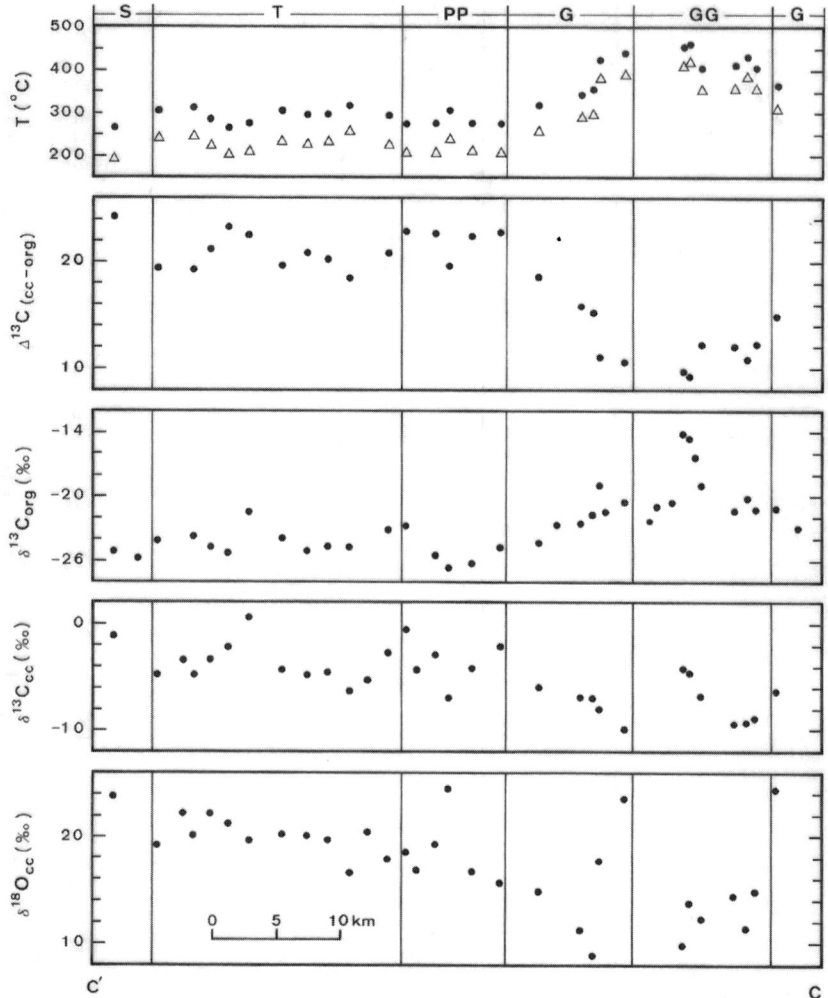

FIG. 4. Plot of $\delta^{18}O$, $\delta^{13}C$ of calcite (cc), $\delta^{13}C$ of carbonaceous matter (org), $\Delta^{13}C$(cc-org), and estimated temperatures for samples from the Southern E-W Cross-Island Highway versus the sample position as projected on the C–C' line shown in Figure 1. The metamorphic facies groups are after Chen et al. (1983) (see Fig. 1). Legend for temperature estimates: circles = Morikiyo's (1984) calibration; triangles = Dunn and Valley's (1992) calibration. See text for details.

isotope fractionation relations between calcite and carbonaceous material for samples in this study. Nevertheless, empirical relations reaching down to temperatures of greenschist-facies conditions might yield semi-quantitative estimates of the metamorphic temperature if the integrated kinetic isotope effect is approximately the same between studied areas. In the present study, the Morikiyo (1984) and the Dunn and Valley (1992) schemes are employed to give a temperature range estimate.

Pelitic rocks and marbles are the two rock types sampled in this study. Although calcite and carbonaceous material in marble might be theoretically suitable for metamorphic peak temperature estimates (Eiler et al., 1992), the two phases, being minor components, in pelitic rocks might be more easily subjected to retrograde reactions. Yui (1995) tested this problem from a 200 m thick marble-schist interlayer located close to sample LB-19 along the A–A' section. The results (Table 4) clearly show that: (1) isotopic equilibrium between calcite and carbonaceous material was not attained in either rock type; and (2) no significant difference in $\Delta^{13}C$ (calcite-carbonaceous material) can be

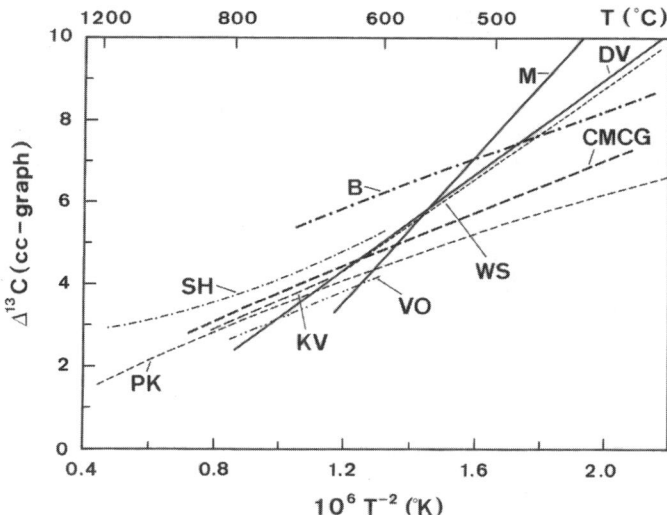

FIG. 5. Temperature dependence of carbon isotope fractionations between calcite and carbonaceous material (graphite) as determined theoretically by Bottinga (1969; B), Chacko et al. (1991; CMCG), and Polyakov and Kharlashina (1995; PK at 7 kbar); experimentally by Scheele and Hoefs (1992; SH); and empirically by Valley and O'Neil (1981; VO), Wada and Suzuki (1983; WS), Morikiyo (1984; M), Dunn and Valley (1992; DV), and Kitchen and Valley (1995; KV).

discerned between pelitic rocks and marbles. This may indicate that retrograde effects on the carbon isotope system were not prominent in the pelitic samples, probably because of the carbon isotopes were largely buffered by the rocks.

The calculated temperature estimates along the three sections are listed in Tables 1–3 and plotted in Figures 2–4. The resulting temperatures range from 261 to 727°C, following the relation of Morikiyo (1984) and from 193 to 875°C following the scheme of Dunn and Valley (1992). At approximately 300°C, the calculated temperatures from the former relation are higher by 60–70°C than those from the latter relation; around 400°C, the difference is about 40–50°C; around 500°C, 20–30°C; but at temperatures higher than 580°C, the difference is reversed.

Carbon isotope fractionation between calcite and carbonaceous material, as well as the deduced temperature estimates along three sections in this study show some common features. First, there is no systematic temperature variation among the S-, Z-, PP-, and lower part of the G-group samples. Systematic temperature increase is apparent for the middle part of the G- to the GG-group samples. The latter change is parallel with the $\delta^{13}C$ value increase of the carbonaceous material mentioned above, indicating that substantial isotopic exchange between calcite and carbonaceous matter took place only in the middle part of the G- and the GG-group samples, but not in the S-, Z-, PP-, and lower part of the G-group samples. The temperature for the beginning of this prominent carbon isotope exchange would be around 300–350°C (Morikiyo, 1984)/250–300°C (Dunn and Valley, 1992).

Second, along with systematic temperature variations, two thermal maxima are evident, in the central part of the Hsuehshan Range (sample C-52, 426/378°C), and in the Backbone Range (sample C-14, 464/425°C; and sample S-194.1, 459/420°C), respectively. This is prominent for the B–B' and C–C' sections. A vague thermal maximum along the A–A' section in the Backbone Range could also be discerned, probably near sample 112m (492/463°C). The temperature of the thermal maximum in the Backbone Range shows a slight decrease toward the south.

Third, crossing the Tananao Metamorphic Complex and the overlying Cenozoic sediments in the Backbone Range (i.e., crossing the G- to the GG-group samples in the three sections), there is no distinct break in temperature or isotopic variations. This suggests that Cenozoic collision tectonism was the major reason for the observed metamorphism across the mountain belt. Because the Tananao Metamorphic Complex has been subjected to at least two to three stages of metamorphism (Liou and

TABLE 4. Comparison of C-isotope Fractionation between Calcite and Carbonaceous (organic) Material between Pelitic and Marble Samples

Sample no.	$\delta^{13}C$ (‰) calcite	$\delta^{13}C$ (‰) organic	$\Delta^{13}C$ cc.-org.
WWL-4[1]	−5.1	−18.2	13.1
WWL-7[1]	−4.3	−16.6	12.3
WWWL-8-T[1]	−4.6	−17.4	12.8
WWL-22[1]	−5.3	−16.6	11.3
WWL-23[1]	−5.8	−17.5	11.7
WL-5	+2.1	−11.0	13.1
WWL-8-5	+1.4	−9.8	11.2
WWL-12-5	+2.3	−10.4	12.7
WL-20	+0.9	−10.2	11.1
WWL-35-2	+2.3	−8.8	11.1
WWL-35-4	+2.3	−9.5	11.8

[1]Represents pelitic samples; others are marbles. Average $\Delta^{13}C$(cc.-org) is 12.2 ± 0.6‰ for five pelitic samples and 11.8 ± 0.7‰ for six marbles.
Source: (after Yui, 1995)

Ernst, 1984; Yui et al., 1990), the postulated Mesozoic metamorphism(s) should have comparable or less intensity compared to the Cenozoic one.

Finally, superimposed on the features above is the presence of some high-temperature marbles in contact with granitic rocks, such as sample C-10m along the B–B' section and samples 125m, 115m and 117m along the A–A' section. The high-temperature records in these marble samples may most probably be inherited from contact thermal aureoles around granitic intrusions, although the present marble-granite contact may well be a tectonic rather than the original one. It should be noted that these granites were shown to have Mesozoic ages (Jahn et al., 1986; Yui et al., 1996a). It seems therefore that the earlier carbon isotope system of marble can survive a later, low-temperature thermal event as was pointed out by Dunn and Valley (1992). The temperature variation trends for marble samples along the southern part of section A–A' and eastern part of section B–B' are not very distinct. The possibility that some marbles, other than those in contact with granites, might have also been influenced by the Mesozoic thermal event can not be excluded.

To evaluate the temperature estimates derived from carbon isotope fractionation between calcite and carbonaceous material presented in this study, other independent geothermometers are needed.

Tsao et al. (1992, 1996) presented zircon fission track (FT) data and <2 μm white mica K-Ar data along the B–B' and C–C' sections. According to their results, samples that belong to the partial annealing zone (PAZ) of both thermochronologic systems are present in these two sections (Fig. 6). In principle, samples that belong to the PAZ of the zircon FT system/white mica K-Ar system should have been subjected to metamorphic temperatures not higher than the respective blocking temperature, i.e., 235 ± 50°C/350 ± 50°C (Jager, 1979; Liu, 1982). As shown in Figure 6, samples in the present study close to the base of the PAZ of FT/K-Ar thermochronologic system yield metamorphic temperature estimates of 250–300°C/300–350°C according to the scheme of Morikiyo (1984), and of 200–250°C/250–300°C following the relation given by Dunn and Valley (1992). Note that the blocking temperature of zircon FT (i.e., 235 ± 50°C) is beyond the temperature range for empirical relations of Morikiyo (1984; i.e., 270–650°C) and Dunn and Valley (1992; i.e., 400–800°C). In addition, calcite and carbonaceous material in samples within the zircon FT PAZ also do not show prominent carbon isotope exchange, as discussed above. The estimated temperature range for samples close to the base of zircon FT PAZ may therefore be unreliable. On the other hand, the scheme of Morikiyo (1984)

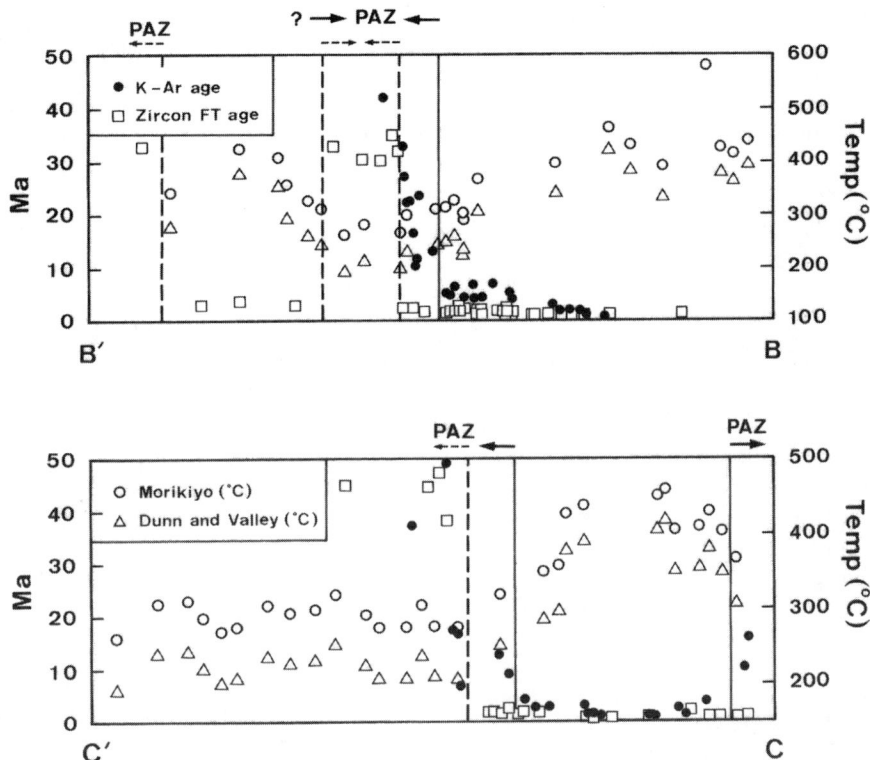

FIG. 6. Zircon fission-track data and <2 μm white mica K-Ar data across the B–B' and C–C' sections given by Tsao et al. (1992, 1996). Also shown are carbon-isotope temperature estimates of samples in the present study. Dashed and solid lines/arrows indicate the base of partial annealing zone of zircon fission track system and white mica K-Ar system, respectively.

gives a more reasonable temperature range of 300–350°C for samples close to the base of white mica PAZ than the relation of Dunn and Valley (1992) which gives an estimate of 250–300°C. The latter is also lower than its empirical temperature range. Note that Morikiyo (1984) derived his carbon isotope fractionation relation from Ryoke metamorphic rocks, whereas most samples studied by Dunn and Valley (1992) were from contact aureoles. Isotopic kinetic effects in these two sets of samples may be different. It is therefore conceivable that the scheme of Morikiyo (1984) gives more reasonable semi-quantitative temperature estimates for most of the samples from the Taiwan mountain belt, which has been subjected to Cenozoic subduction/collision processes. In this regard, the relation given by Dunn and Valley (1992) might give better temperature estimates for those marble samples in contact with granites along section A–A' and B–B'. Following the scheme of Morikiyo (1984) the metamorphic temperatures for the Tananao basement would be around 350–500°C along the A–A' section, 350–460°C along the B–B' section, and 400–460°C along the C–C' section. These results are also consistent with those (425 ± 75°C) estimated for the Tananao basement along the B–B' section by Ernst (1983), as well as other reports documented in Chen and Wang (1995).

Concluding Remarks

This paper presents carbon isotope data on pelitic rocks and marbles along three sections across the Cenozoic Taiwan mountain belt, which consists of the western Hsuehshan Range and the eastern Backbone Range. Although carbon isotope equilibrium between coexisting calcite and carbonaceous material may not have been attained for most samples, the carbon isotope fractionation relation given by Morikiyo (1984), tested independently using a white mica K-Ar thermochronologic data set, has shown itself to be a promising scheme to give reasonable semi-quantitative metamorphic temperature estimates for rocks in the Taiwan mountain belt. The major results of this study are as follows.

1. Prominent carbon isotope exchange between coexisting calcite and carbonaceous material begins at temperatures of ~300–350°C.

2. Systematic temperature variation is shown in the G- to GG-group samples. The temperature variations define two metamorphic thermal highs in the core of the Hsuehshan Range and the Backbone Range, respectively. The recorded metamorphic temperature maximum, attributed to Cenozoic tectonic processes, is about 430°C in the former, and 460–490°C in the latter.

3. Some marble samples in contact with Mesozoic granitic rocks record abnormally high temperatures, which may be explained as a result of contact metamorphism. The recorded temperature data across the mountain belt may therefore not have been synchronous.

Two thermal maxima in two parallel mountain ranges are a characteristic feature of the mountain belt of Taiwan. Teng et al. (1991) proposed that the Hsuehshan sediments were originally deposited within a half graben bounded by a normal fault in the east during the Paleogene. These Paleogene rocks were then deformed and exhumed as a large-scale pop-up structure during the Plio-Pleistocene continent-arc collision and the east-trending normal fault was reactivated as a thrust (Clark et al., 1993). Yui and Chu (2000) also postulated an upward extrusion process to account for the observed stratigraphic, structural, and metamorphic grade variations across the Backbone Range. The role of pre-existing structures, as well as the effect on stress propagation during collision tectonism, therefore, ought to be evaluated. The present results on thermal distribution across the Taiwan mountain belt must also be taken into consideration in future thermo-kinetic modeling of Taiwan mountain building.

Acknowledgments

The author would like to dedicate this paper to W. G. Ernst on the occasion of his retirement. He led a U.S.-Taiwan cooperation project studying the Taiwan mountain belt in the late 1970s and early 1980s, and introduced new ideas on metamorphic petrology and tectonism that significantly improved our understanding of this mountain belt. The author also wishes to thank J.G. Liou for his helpful review and editorial suggestions. This study was financially supported by the National Science Council of Taiwan.

REFERENCES

Arneth, J. D., Schidlowski, M., Sarbos, B., Goerg, U., and Amstutz, G. C., 1985, Graphite content and isotopic fractionation between calcite-graphite pairs in metasediments from the Mgama Hills, southern Kenya: Geochimica et Cosmochimica Acta, v. 49, p. 1553–1560.

Barker, F., and Friedman, I., 1969, Carbon isotopes in pelites of the Precambrian Uncompahgre Formation, Needle Mountains, Colorado: Geological Society of America Bulletin, v. 80, p. 1403–1408.

Beyssac, O., Goffe, B., Chopin, C., and Rouzaud, J. N., 2002, Raman spectra of carbonaceous material in metasediments: a new geothermometer: Journal of Metamorphic Geology, v. 20, p. 859–871.

Bottinga, Y., 1969, Calculated fractionation factors for carbon and hydrogen isotope exchange in the system calcite-carbon dioxide-graphite-methane-hydrogen-water vapor: Geochimica et Cosmochimica Acta, v. 33, p. 49–64.

Chacko, T., Mayeda, T. K., Clayton, R. N., and Goldsmith, J. R., 1991, Oxygen and carbon isotope fractionations between CO_2 and calcite: Geochimica et Cosmochimica Acta, v. 55, p. 2867–2882.

Chen, C. H., 1984, Determination of lower greenschist facies boundary by K-mica-chlorite crystallinity in the Central Range of Taiwan: Proceedings of Geological Society of China, v. 27, p. 41–44.

Chen, C. H., Chu, H. T., Liou, J. G., and Ernst, W. G., 1983, Explanatory notes for the Metamorphic Facies Map of Taiwan: Taipei, Taiwan, Central Geological Survey, Special Publication, v. 2, 32 p.

Chen, C. H., and Wang, C. H., 1995, Explanatory notes for the Metamorphic Facies Map of Taiwan: Taipei, Taiwan, Central Geological Survey, Special Publication, 2nd ed., 51 p.

Chung, H. M., and Sackett, W. M., 1979, Use of stable carbon isotope compositions of pyrolytically derived methane as maturity indices for carbonaceous materials: Geochimica et Cosmochimica Acta, v. 43, p. 1979–1988.

Clark, M. B., Fisher, D. M., Lu, C. Y., and Chen, C. H., 1993, Kinematic analyses of the Hsuehshan Range, Taiwan: A large-scale pop-up structure: Tectonics, v. 12, p. 205–217.

Craig, H., 1957, Isotopic standards for carbon and oxygen and correction factors for mass-spectrometric analysis of carbon dioxide: Geochimica et Cosmochimica Acta, v. 12, p. 133–149.

Degens, E. T., 1969, Biogeochemistry of stable isotopes, in Eglinton, G., and Murphy, M. T. J., eds., Organic geochemistry: Methods and results: New York, NY, Springer-Verlag, p. 304–329.

Diessel, C. F. K., Brothers, R. N., and Black, P. M., 1978, Coalification and graphitization in high-pressure schists in New Caledonia: Contributions to Mineralogy and Petrology, v. 68, p. 63–78.

Dunn, S. R., and Valley, J. W., 1992, Calcite-graphite isotope thermometry: A test for polymetamorphism in marble, Tudor gabbro aureole, Ontario, Canada: Journal of Metamorphic Geology, v. 10, p. 487–501.

Eichmann, R., and Schidlowski, M.. 1975, Isotopic fractionation between coexisting organic carbon-carbonate pairs in Pre-Cambrian sediments: Geochimica et Cosmochimica Acta, v. 39, p. 585–595.

Eiler, J. M., Baumgartner, L. P., and Valley, J. W., 1992, Intercrystalline stable isotope diffusion: A fast grain boundary model: Contributions to Mineralogy and Petrology, v. 112, p. 543–557.

Ernst, W. G., 1983, Mineral parageneses in metamorphic rocks exposed along Tailuko Gorge, Central Mountain Range, Taiwan: Journal of Metamorphic Geology, v. 1, p. 305–329.

Feates, F. S., 1968, The diffusion of carbon in single crystal graphite: Journal of Nuclear Materials, v. 27, p. 325–330.

French, B. M., 1964, Graphitization of organic material in a progressively metamorphosed iron formation: Science, v. 146, p. 917–918.

Grew, E. S., 1974, Carbonaceous material in some metamorphic rocks of New Zealand and other areas: Journal of Geology, v. 82, p. 50–73.

Griffin, G. M., 1967, X-ray diffraction technique applicable to studies of diagenesis and low rank metamorphism in humic sediments: Journal of Sedimentary Petrology, v. 37, p. 1006–1011.

Hayes, J. M., Kaplan, I. R., and Wedeking, K. W., 1983, Precambrian organic geochemistry, preservation of the record, in Schopf, J. W., ed., The origin and evolution of the Earth's earliest biosphere: Princeton, NJ, Princeton University Press, p. 93–134.

Ho, C. S., 1986, An introduction to the geology of Taiwan: Explanatory text of the Geologic Map of Taiwan, 2nd ed.: Taipei, Taiwan, Central Geological Survey, 163 p.

Hoefs, J., and Frey, M., 1976, The isotopic composition of carbonaceous matter in a metamorphic profile from the Swiss Alps: Geochimica et Cosmochimica Acta, v. 40, p. 945–951.

Itaya, T., 1981, Carbonaceous material in pelitic schists of the Sanbagawa metamorphic belt in central Shikoku, Japan: Lithos, v. 14, p. 215–224.

Izawa, E., 1968, Carbonaceous matter in some metamorphic rocks in Japan: Journal of Geological Society of Japan, v. 74, p. 427–432.

Jager, E., 1979, Introduction to geochronology, in Jager, E., and Hunziker, J. C., eds., Lectures in isotope geology: Berlin, Germany, Springer-Verlag, p. 1–16.

Jahn, B. M., Martineau, F., Peucat, J. J., and Cornichet, J., 1986, Geochronology of the Tananao schist complex, Taiwan, and its regional tectonic significance: Tectonophysics, v. 125, p. 103–124.

Kitchen, N. E., and Valley, J. W., 1995, Carbon isotope thermometry in marbles of the Adirondack Mountains, New York: Journal of Metamorphic Geology, v. 13, p. 577–594.

Kreulen, R., and van Beek, P. C. J. M., 1983, The calcite-graphite isotope thermometer; data on graphite-bearing marbles from Naxos, Greece: Geochimica et Cosmochimica Acta, v. 47, p. 1527–1530.

Landis, C. A., 1971, Graphitization of dispersed carbonaceous material in metamorphic rocks: Contributions to Mineralogy and Petrology, v. 30, p. 34–45.

Liou, J. G., and Ernst, W.,G., 1984. Summary of Phanerozoic metamorphism in Taiwan: Memoirs of Geological Society of China, v. 6, p. 133–152.

Liu, T. K., 1982, Tectonic implication of fission track ages from the Central Range, Taiwan: Proceedings of the Geological Society of China, v. 25, p. 22–37.

McCrea, J. M., 1950, On the isotope chemistry of carbonates and a paleotemperature scale: Journal of Chemistry and Physics, v. 18, p. 849–857.

McKirdy, D. M., and Powell, T. G., 1974. Metamorphic alteration of carbon isotopic composition in ancient sedimentary organic matter: New evidence from Australia and South Africa: Geology, v. 2, p. 591–595.

Morikiyo, T., 1984, Carbon isotopic study on coexisting calcite-graphite in the Ryoke metamorphic rocks, northern Kiso district, central Japan: Contributions to Mineralogy and Petrology, v. 87, p. 251–259.

Pineau, F., Latouche, L., and Javoy, M., 1976, L'origine du graphite et les fractionnements isotopiques du carbone dans les marbres metamorphiques des Gour Oumelalen (Ahaggar, Algerie), des Adirondacks (New York, U.S.A.) et du Damara (Namibie, Sud-Ouest-Africain): Bulletin de la Société Géologique de France, v. 7, p. 1713–1723.

Polyakov, V. B., and Kharlashina, N. N., 1995, The use of heat capacity data to calculate carbon isotope fractionation between graphite, diamond, and carbon dioxide: A new approach: Geochimica et Cosmochimica Acta, v. 59, p. 2561–2572.

Scheele, N., and Hoefs, J., 1992, Carbon isotope fractionation between calcite, graphite and CO_2: An experimental study: Contributions to Mineralogy and Petrology, v. 112, p. 35–45.

Schidlowski, M., Hayes, J. M., and Kaplan, I. R., 1983, Isotopic inferences of ancient biochemistries: Carbon, sulfur, hydrogen, and nitrogen, in Schopf, J. W., ed., The origin and evolution of the Earth's Earliest Biosphere: Princeton, NJ, Princeton University Press, p. 149–186.

Sharma, T., and Clayton, R. N., 1965, Measurement of $^{18}O/^{16}O$ ratios of total oxygen from carbonates: Geochimica et Cosmochimica Acta, v. 29, p. 1347–1354.

Taylor, H. P., Albee, A. L. and Epstien, S., 1963, $^{18}O/^{16}O$ ratios of coexisting minerals in three assemblages of kyanite zone pelitic schists: Journal of Geology, v. 71, p. 513–522.

Teng, L. S., Wang, Y., Tang, C. H., Huang, C. Y., Huang, C. T., Yu, M. S., and Ke, A., 1991, Tectonic aspects of the Oligocene depositional basin of northern Taiwan: Proceedings of Geological Society of China, v. 34, p. 313–336.

Tsao, S., Law, E., Ho, H. C., Lee, Y. H., Jiang, W. T., and Chen, C. H., 1996, The geological significances of K-Ar ages of metapelites from the Central Range, Taiwan: Bulletin, Central Geological Survey (Taiwan), v. 11, p. 37–84.

Tsao, S., Li, T. C., Tien, J. L., Chen, C. H., Liu, T. K., and Chen, C. H., 1992, Illite crystallinity and fission-track ages along the east Central Cross-Island Highway of Taiwan: Acta Geologica Taiwanica, v. 30, p. 65–94.

Valley, J. W., and O'Neil, J. R., 1981, $^{13}C/^{12}C$ exchange between calcite and graphite: A possible thermometer in Grenville marbles: Geochimica et Cosmochimica Acta, v. 45, p. 411–419.

Wada, H., and Oana, S., 1975, Carbon and oxygen isotope studies of graphite-bearing carbonates in the Kasuga area, Gifu Prefecture, central Japan: Geochemical Journal, v. 9, p. 149–160.

Wada, H., and Suzuki, K., 1983, Carbon isotope thermometry calibrated by dolomite-calcite solvus temperatures: Geochimica et Cosmochimica Acta, v. 47, p. 697–706.

Warneke, L. A., and Ernst, W. G., 1984, Progressive Cenozoic metamorphism of rocks cropping out along the Southern East-West Cross-Island Highway, Taiwan: Memoirs of the Geological Society of China, v. 6, p. 105–132.

Wedeking, K. W., Hayes, J. M., and Matzigkeit, U., 1983, Procedures of organic geochemical analysis, in Schopf, J. W., ed., The origin and evolution of the Earth's earliest biosphere: Princeton, NJ, Princeton University Press, p. 428–441.

Wopenka, B., and Pasteris, J. D., 1993, Structural characterization of kerogens to granulite-facies graphite: Applicability of Raman microprobe spectroscopy: American Mineralogist, v. 78, p. 533–557.

Yui, T. F., 1995, Stable isotope studies of marble/schist in the Tungao area, northeastern Taiwan: Characteristics of the metamorphic fluid: Journal of the Geological Society of China, v. 38, p. 1–14.

Yui, T. F., and Chu, H. T., 2000, "Overturned" marble layers: Evidence for upward extrusion of the Backbone Range of Taiwan: Earth and Planetary Science Letters, v. 179, 351–361.

Yui, T. F., Heaman, L., and Lan, C. Y., 1996a, U-Pb and Sr isotopic studies on granitoids from Taiwan and Chinmen-Lieyu and tectonic implications: Tectonophysics, v. 263, p. 61–76.

Yui, T. F., Huang, E., and Xu, J., 1996b, Raman spectrum of carbonaceous material: A possible metamorphic grade indicator for low-grade metamorphic rocks: Journal of Metamorphic Geology, v. 14, p. 115–124.

Yui, T. F., Lu, C. Y., and Lo, C. H., 1990, Tectonic evolution of the Tananao Schist Complex of Taiwan, in Aubouin, J., and Bourgois, J., eds., Tectonics of circum-Pacific continental margins: Zeist, Netherlands, VSP, p. 193–209.

SOUTH KOREA

Metamorphic Evolution of the Ogcheon Belt, Korea: A Review and New Age Constraints

MOONSUP CHO[1] AND HYEONCHEOL KIM

School of Earth and Environmental Sciences, Seoul National University, Seoul, 151-747, South Korea

Abstract

The Ogcheon metamorphic belt (OMB), which is often correlated with the Dabie-Sulu ultrahigh-pressure (UHP) belt in China, comprises Neoproterozoic to Paleozoic metasedimentary and -volcanic sequences representing a stack of synmetamorphic nappes. Regional metamorphic grade increases northwestward to produce the assemblage, biotite + garnet ± staurolite ± kyanite + plagioclase + quartz. Garnet porphyroblasts show chemical zoning typical for prograde metamorphism, with decreasing Mn and increasing Fe and Mg from core to rim. P–T conditions were estimated to be 4.2–9.4 kbar and 490–630°C, corresponding to the medium-pressure type. In addition, the GIBBS calculation suggests a clockwise P–T–t path, corroborating a crust-thickening event associated with regional peak metamorphism. Using various isotopic techniques, the timings for intracontinental rift volcanism and syntectonic metamorphism were determined. New SHRIMP U–Pb zircon ages from a felsic tuff reflect the Neoproterozoic rifting at ~750 Ma. Regional metamorphism has been recently dated at ~285 Ma, based upon several independent sets of radiometric ages: U–Pb ages of monazite inclusions in garnet, Pb–Pb whole-rock ages of black slates, chemical ages of uraninite, and SHRIMP U–Pb zircon ages from a granitic gneiss pebble in metadiamictite. Thus, the Ogcheon orogeny is newly defined as an earliest Permian event, preceding a high-pressure metamorphic event in the Imjingang belt, central Korea, by ~30 Ma. During the subsequent Songrim orogeny, at ~250–220 Ma, the OMB experienced a second regional-thermal metamorphism under greenschist- to amphibolite-facies conditions. This Triassic event is interpreted to correspond to a major collisional orogeny in the Dabie-Sulu belt. These multiple metamorphic events in the Ogcheon belt are readily correlated with those reported from not only the UHP belt in east-central China, but also the Hida and Renge belts in Japan. Thus, eastward extension of the Dabie-Sulu belt through the Korean Peninsula to the Hida belt (Ernst and Liou, 1995) remains a valid hypothesis.

Introduction

THE EURASIAN CONTINENT is a collage of at least six major cratons, amalgamated during a number of Phanerozoic orogens (Liou et al., 1989). In particular, the Triassic growth of the east Eurasian landmass is prominent: (1) continental collision between the North and South China blocks (e.g., Mattauer et al., 1985; Okay et al., 1993; Yin and Nie, 1993, 1996; Hacker et al., 1998); and (2) enormous addition of juvenile mantle materials through igneous activities in the Central Asian fold belt (Jahn et al., 2000; Wu et al., 2000; Fig. 1A). Triassic continental collision associated with the UHP metamorphism in the Qinling-Dabie-Sulu belt, China, has been well documented by many studies (e.g., Liou et al., 1996; Zheng et al., 2003). Interpretations of the complex and prolonged history of collisional processes, however, have led to a variety of tectonic models and some controversies. Among them, diverse opinions still exist regarding the location of the suture in the Sulu UHP belt (Zhai et al., 2000; Faure et al., 2001, 2003), and its eastward continuation into the Korean Peninsula and, possibly, the Japanese Islands (e.g., Yin and Nie, 1993; Li, 1994; Ernst and Liou, 1995; Ree et al., 1996, 2001; Chen and Jahn, 1998; Cho, 2001; Ishiwatari and Tsujimori, 2001; Lee et al., 2003).

The Ogcheon belt, situated in the southern part of the Korean Peninsula, consists of two tectonic provinces: the Taebaegsan basin and Ogcheon metamorphic belt (OMB; Fig. 1A). According to a recent tectonic model of Chough et al. (2000), these provinces were separated before Triassic time and belonged to the North and South China blocks, respectively. Thus, understanding the tectonometamorphic evolution of the Ogcheon belt is one of the key issues for establishing regional correlation in

---

[1]Corresponding author; email: moonsup@snu.ac.kr

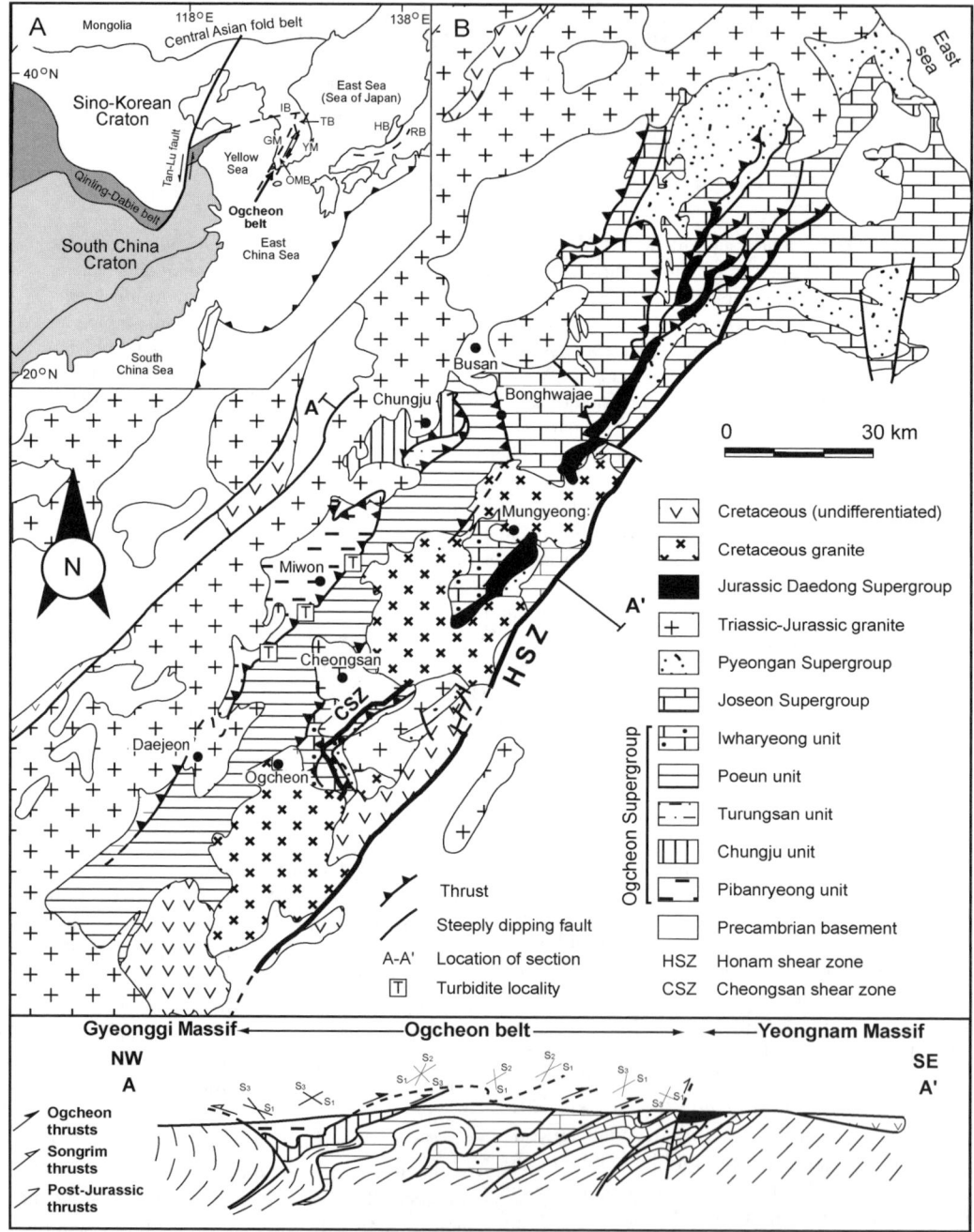

FIG. 1. A. A simplified tectonic province map of East Asia showing the present-day subduction zone along the Pacific margin (after Ernst et al., 1988). B. A geologic map showing various lithotectonic units in the Ogcheon metamorphic belt and the Taebaegsan basin (modified from Cluzel et al., 1990). Note that a north-striking Bonghwajae thrust between Busan and Mungyeong is the boundary between the Ogcheon metamorphic belt and the Taebaegsan basin (Chough et al., 2000). Thus, a lithotectonic unit to the east of this thrust, originally assigned to the Iwharyeong unit by Cluzel et al. (1990), belongs to the Joseon Supergroup. A simplified cross-section of the central Ogcheon metamorphic belt is adopted from Cluzel et al. (1990). Mesozoic granites are omitted for clarity. Abbreviations: TB = Taebaegsan basin; IB = Imjingang belt; GM = Gyeonggi Massif; YM = Yeongnam Massif; OMB = Ogcheon metamorphic belt; HB = Hida belt; RB = Renge belt.

East Asia including China and Japan. Together with the Imjingang belt, central Korea, this belt is often referred to as the possible continuation of the Qinling-Dabie-Sulu belt in east-central China (e.g., Ernst and Liou, 1995; Fig. 1A).

The tectonometamorphic evolution of the Ogcheon belt has been the subject of many investigations for the past three decades (e.g., Reedman and Um, 1975; Kim, 1990; Cluzel et al., 1990, 1991; Cho et al., 1994; Kim et al., 1995; Koh and Kim, 1995; Oh et al., 1995; Min and Cho, 1998; Chough et al., 2000; Cho and Kim, 2002). Although many aspects of the metamorphism, structures, and tectonics of the OMB are well understood (see reviews by Chough et al., 2000, and Cho and Kim, 2002), some problems are yet to be resolved. In particular, geochronological data are lacking for constraining not only depositional ages of metasedimentary and -volcaniclastic rocks but also metamorphic ages of polycyclic regional events. The paucity of reliable ages is primarily attributed to complex and prolonged histories of tectono-metamorphism in the OMB. Recently, however, isotope studies have revealed the Neoproterozoic age of bimodal volcanism (Lee et al., 1998) followed by the earliest Permian, medium-pressure type metamorphism (Kim et al., 2001; Cheong et al., 2003).

In this study, we review the petrological and geochronological data available on the OMB, and provide some new age constraints from the Sensitive High-Resolution Ion MicroProbe (SHRIMP) ages of zircon. Taken together, we suggest that two major tectonometamorphic events have occurred in the OMB, in the Permian and Triassic, respectively. We mainly discuss various aspects of tectono-metamorphism in the central and northern parts of the OMB, because these areas are critical for unraveling the tectonic relationship with the Taebaegsan basin. Further details of the metamorphic evolution in the southwestern part of the OMB are available in Oh et al. (1998) and Kim et al. (2002). Chough et al. (2000) recently reviewed some geological and tectonic features of the Ogcheon belt, but the regional metamorphism and P–T–t evolution were not discussed. Thus, our review is complementary to that of Chough et al. (2000).

Geological Outline and Polycyclic Deformation

The NE-trending Ogcheon belt, bounded by Precambrian gneisses of the Gyeonggi and Yeongnam massifs, comprises a fold-and-thrust belt (OMB) to the southwest, and the Paleozoic Taebaegsan basin to the northeast (Fig. 1A; Chough et al., 2000). The OMB comprises unfossiliferous metasedimentary and metavolcanic rocks whose protolith ages possibly range from Neoproterozoic to Paleozoic. The Taebaegsan basin consists of unmetamorphosed to weakly metamorphosed, fossiliferous sedimentary strata of Early Cambrian to Early Jurassic age. Along the boundary with Precambrian massifs, the OMB is commonly in contact with three distinct groups of Mesozoic granitoids that were intruded in: (1) the Late Triassic (~220 Ma, Cho et al., 2001; Ree et al., 2001); (2) the Middle Jurassic (~180 Ma, Ree et al., 2001; Cheong et al., 2003); and (3) the Late Cretaceous (~90 Ma, Cheong and Chang, 1997). Thermal aureoles developed around each Mesozoic pluton, and their widths may be up to a few kilometers. Moreover, the Triassic granitoids in southeastern OMB are typically truncated by dextral strike-slip shear zones such as the Cheongsan shear zone (Chough et al., 2000; Ree et al., 2001).

Different types of sedimentary and volcanic rocks were repeatedly deformed and metamorphosed in the OMB (Chough et al., 2000; Cho and Kim, 2002). Stratigraphic relationships among these protoliths are equivocal, mainly because of the lack of fossils and the obliteration of primary structures. In addition, the lack of reliable geochronological data does not permit us to establish the tectonostratigraphy of the various metasedimentary rocks. Nevertheless, primary sedimentary structures such as graded bedding or cross bedding are locally preserved in the OMB (Lee et al., 1980; Chough and Bahk, 1992). We have also found metaturbidites that are traceable for approximately 10 km along the strike in the Miwon area, central OMB (Fig. 2; see Fig. 1B for location). The turbidite sequences are interpreted to belong to Facies Class C2.3 (thin-bedded sand-mud couplets) and Facies Class D2.3 (thin, regular silt, and mud laminae), respectively, based on the classification scheme of Pickering et al. (1986). These turbidites show fining-upward units that indicate the sequence to be the right-way-up. This observation is consistent with the northwestward-younging direction reported by Choi and Kim (1981).

Polyphase deformation of the OMB has been described in detail by many structural investigations (e.g., Cluzel et al., 1990, 1991; Kang et al., 1993; Koh and Kim, 1995; Kang and Ryoo, 1997; Lee, 2000; Kang and Lee, 2002). Four major

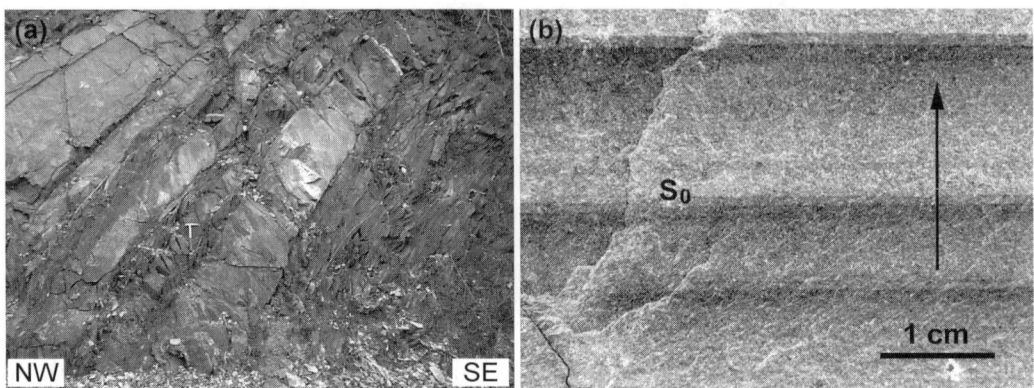

FIG. 2. A. Transposed bedding planes preserving the turbidite sequence, defined by alternating sand and shale layers in the Poeun unit. B. Rock slab photograph showing gradual change in grain size and color. Color changes are attributed to different compositions of each layer, in particular, the amount of biotite and carbonaceous materials. S_0 and arrow denote bedding plane and stratigraphically younging direction, respectively.

deformational events are generally defined, as summarized below. Among them, the D_1 to D_3 phases are synmetamorphic.

Non-coaxial strain features indicate that the D_1 and D_2 are characterized by the top-to-the-southeast ductile movement. Both events are interpreted to define a progressive deformation in a subhorizontal shear regime related to the southeastward stacking of nappes (Cluzel et al., 1990, 1991). Cluzel et al. (1990) also pointed out that the D_{1-2} deformation is absent from the Carboniferous—Permian Pyeongan Supergroup of the Taebaegsan basin.

S_1, the planar fabric produced during D_1, is rarely observed at outcrop, but is well defined under the microscope by inclusion trails of quartz in garnet and biotite porphyroblasts. S_2 is the predominant penetrative foliation, accompanying peak regional metamorphism of the medium-pressure type. Porphyroblasts of garnet, staurolite, and biotite are wrapped by the S_2 foliation, consisting of biotite, muscovite, chlorite, and rare ilmenite. It is thus apparent that these porphyroblasts grew prior to or during the D_2 deformation. Previous studies (Cluzel et al., 1991; Kim et al., 1995; Koh and Kim, 1995) have reported the syn-tectonic growth of helicitic garnet porphyroblasts. However, helicitic textures are absent from the psammitic, garnet-bearing schists (Cho and Kim, 2002).

The D_3 deformation reactivated D_{1-2} structural elements under brittle to semiductile conditions, producing chlorite and muscovite. During D_3, subhorizontal, asymmetric crenulation cleavages develop axial planar to mesoscopic chevron- to kink-style folds. Because of strain partitioning, D_3 structures are commonly observed in incompetent pelitic slates and phyllites of the Poeun unit, but are rare in psammitic layers. Finally, D_4 is characterized by map-scale, E-trending, open folds, and is not accompanied by distinct metamorphic minerals.

Regional structures in the metamorphic rocks of the OMB are generally governed by large-scale, SE-verging D_2 thrust faults. In particular, Cluzel et al. (1990) divided the OMB into five thrust-bounded lithotectonic units: the structurally upper units of Pibanryeong and Chungju and the lower ones of Turungsan, Poeun, and Iwharyeong (Fig. 1B). The Pibanryeong unit consists primarily of biotite and garnet-biotite schists containing rare kyanite and staurolite, whereas the Poeun unit contains apparently lower grade rocks such as biotite-bearing slates and phyllites (Kim and Cho, 1999; Cho and Kim, 2002). On the other hand, the Chungju and Turungsan units are characterized by bimodal metavolcanics represented by quartzofeldspathic schists, muscovite schists, amphibole schists, and amphibolites. Quartzofeldspathic schists of felsic volcanic origin are also present in the Iwharyeong unit near Ogcheon (see Fig. 1B for location), suggesting that the bimodal volcanism was pervasive throughout the OMB.

Mafic bodies represented by garnet-free amphibolites and epidote-amphibole schists occur primarily along the southeastern boundary, including the boundary between the OMB and the Taebaegsan basin (Bonghwajae thrust; Fig. 1B), and to a lesser extent inside the Chungju and Turungsan units. It is

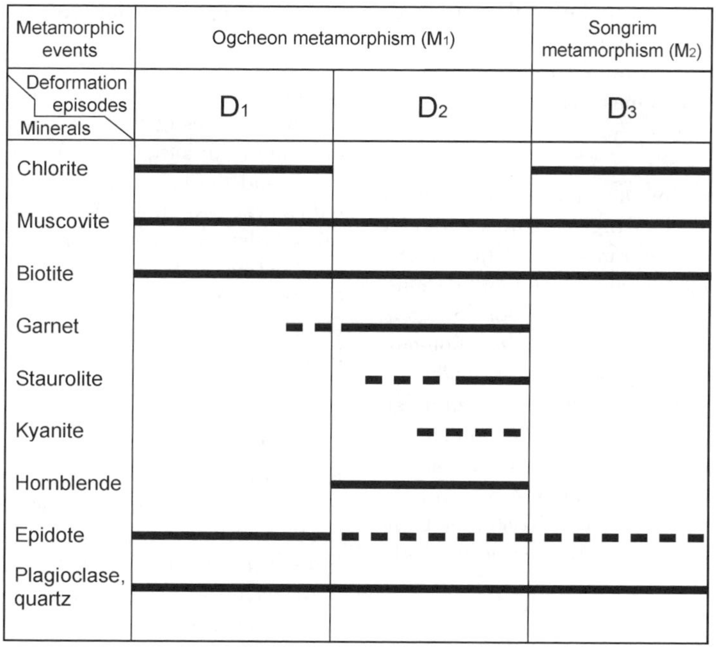

Metamorphic events	Ogcheon metamorphism (M₁)		Songrim metamorphism (M₂)
Deformation episodes / Minerals	D₁	D₂	D₃
Chlorite	▬▬▬▬▬▬		▬▬▬▬▬▬
Muscovite	▬▬▬▬▬▬▬▬▬▬▬▬▬▬▬▬▬▬		
Biotite	▬▬▬▬▬▬▬▬▬▬▬▬▬▬▬▬▬▬		
Garnet		─ ─ ▬▬▬▬▬	
Staurolite		─ ─ ─	
Kyanite		─ ─ ─	
Hornblende		▬▬▬▬▬▬	
Epidote	▬▬▬▬▬▬	─ ─ ─ ─	─ ─ ─ ─
Plagioclase, quartz	▬▬▬▬▬▬▬▬▬▬▬▬▬▬▬▬▬▬		

FIG. 3. Mineral parageneses of metasedimentary rocks in the Ogcheon metamorphic belt (modified from Koh, 1995).

noted that a lithotectonic unit to the east of the Bonghwajae thrust, originally assigned to the Iwharyeong unit by Cluzel et al. (1990), belongs to the Joseon Supergroup (Chough et al., 2000). Mafic schists and amphibolites in the Chungju and Turungsan units have major- and trace-element compositions of basanites and trachybasalts, whereas the felsic metavolcanics are trachytes and rhyolites (Cluzel, 1992a; Kwon and Lee, 1992; Lee and Chang, 1997). Petrological and geochemical studies also demonstrated that these bimodal compositions are attributed to intra-continental rift magmatism (Cluzel, 1992a; Kwon and Lee, 1992; Lee et al., 1998). Moreover, the large volume of felsic metavolcanics and the bimodal chemistry of metavolcanic rocks suggest that the OMB was a high-volcanicity rift, following the classification of Barberi et al. (1982).

Metamorphic Zones

Three metamorphic zones—the biotite, garnet, and Al-silicate zones—are defined on the basis of the first appearance of biotite, garnet, and andalusite in metapelites. The former two zones are the product of dynamothermal metamorphism of the medium-pressure type. On the other hand, the Al-silicate zone, mostly less than ~1 km in width, resulted from later thermal metamorphism, caused by the intrusions of Late Triassic to Cretaceous granites at higher crustal levels. The Al-silicate isograd is nearly parallel to the granite boundary and overprints the regional metamorphic assemblages. Towards the granite contact, andalusite coexists with fibrolitic sillimanite and cordierite in metapelites. Neoblasts of garnet and staurolite are also present in the contact aureole around the Jurassic granite. In calc-silicate rocks or marbles, thermal metamorphism produced various assemblages consisting mainly of diopside, calcic amphiboles, forsterite, andradite-grossular garnet, wollastonite, and epidote (e.g., Shin and Lee, 2003).

Parageneses of regional metamorphic minerals in metasedimentary rocks are summarized in Figure 3. Typical assemblages in phyllites and schists consist of biotite ± garnet ± muscovite ± chlorite ± epidote ± plagioclase + quartz. Kyanite and staurolite are rare as porphyroblasts in pelitic schists. Kyanite was commonly replaced by white mica during retrogression or by other Al-silicates during thermal metamorphism (Kim and Cho, 1999). Staurolite, garnet, and biotite were also altered to white mica, chlorite, and margarite during the retrograde metamorphism. In particular, margarite crosscuts the main foliation, S_2, defined by the preferred orientation of biotite and muscovite.

Biotite zone

Major lithologies of the biotite zone include metapelites, pebble-bearing phyllites, phyllitic schists, and recrystallized limestone. Pelitic assemblages are represented by chlorite + biotite + muscovite + quartz ± plagioclase ± epidote, whereas calcareous assemblages comprise calcite + dolomite + quartz + muscovite + plagioclase ± K-feldspar ± actinolite ± clinozoisite ± chlorite. Opaque minerals are common and consist of graphite, ilmenite, hematite, pyrite, and pyrrhotite. Most biotite crystals are aligned in S_2, but some of them are pre-S_2 porphyroblasts showing quartz inclusion trails oblique to the external foliation (Chang, 1988; Kim et al., 1995; Oh et al., 1995).

Garnet zone

Quartzose schists derived from psammitic or felsic volcanic protoliths predominate in the garnet zone, and commonly consist of high-variance assemblages such as biotite + muscovite ± epidote + albite + quartz. However, pelitic to semipelitic schists, although less abundant than quartzose schists, contain garnet-bearing assemblages typical of medium-pressure type metamorphism: i.e., garnet + biotite ± kyanite ± staurolite + muscovite ± plagioclase + quartz. On the other hand, chlorite schists probably originated from mafic volcaniclastics generally contain garnet together with chlorite, muscovite, epidote, plagioclase, quartz, magnetite, and ilmenite. Euhedral to subhedral porphyroblasts of garnet commonly show pressure shadows of quartz and chlorite, suggesting syn-kinematic growth (Min and Cho, 1998).

Garnet porphyroblasts are chemically zoned, and their compositions continuously change from core to rim (Kim et al., 1995; Min and Cho, 1998; Kim and Cho, 1999; Cho and Kim, 2002). The almandine [$=Fe/(Mg+Fe+Mn+Ca)$] content is inversely proportional to the spessartine content and increases toward the rim in most samples. The pyrope [$= Mg/(Mg + Fe + Mn + Ca)$] content slightly increases in the outer part by 1–2 mol%. Zoning and inclusion patterns of garnet suggest two different garnet growth stages in the Pibanryeong unit (Cho and Kim, 2002).

Metamorphic Conditions and P–T Paths

In order to estimate P–T conditions of garnet-free metapelites in the Poeun unit, central OMB, Kim et al. (2000) investigated the b_0 parameter (pressure indicator; Guidotti, 1984; Massone and Schreyer, 1987) and the crystallinity index (CI, temperature indicator) of K-white micas in slate, phyllite, and schist. Slate and phyllite have the b_0 parameters ranging from 8.995 Å to 9.031 Å, with an average of 9.010 ± 0.007 Å ($n = 63$), whereas those in the vicinity of the Jurassic Poeun granite have slightly lower values of 8.988–9.010 Å with an average of 8.998 ± 0.006 Å ($n = 31$). On the other hand, b_0 parameters in muscovite schists vary from 8.993 Å to 9.042 Å ($n = 57$), and show no apparent correlation with the granite intrusive. The CI values of muscovite in slates and phyllites range from 0.10° to 0.24° $\Delta 2\theta$, with an average of 0.12 ± 0.01° $\Delta 2\theta$, indicating an upper epizone grade (greenschist facies; Kim et al., 2000). These results suggest that garnet-free metapelites of the central OMB have experienced a medium-pressure type metamorphism.

Metamorphic conditions reached the upper greenschist facies (biotite zone) and partly lower epidote-amphibolite facies (garnet zone). Furthermore, Kim et al. (2000) suggested that the metamorphic grade of the central OMB increases northwestward across the boundary between the Pibanryeong and Poeun units without an apparent discontinuity. It is thus likely that metamorphism postdates the thrust movement along the unit boundary, or at least that vertical displacement after the peak metamorphism was insignificant.

In garnet-bearing metapelites, peak metamorphic conditions were estimated using various geothermobarometers, such as the garnet-biotite geothermometer of Ferry and Spear (1978), together with the garnet-plagioclase-biotite-muscovite or quartz geobarometer of Hoisch (1989, 1990). Pressures and temperatures estimated for regional metamorphism are in the range of 4.2–9.4 kbar and 490–630°C, respectively (Fig. 4; Cho et al., 1994; Kim et al., 1995; Oh et al., 1995; Min and Cho, 1998; Kim and Cho, 1999). Such P–T conditions are consistent with the occurrences of kyanite and staurolite, indicating the medium-pressure type metamorphism. In the thermal aureoles, in spite of the presence of contact metamorphic minerals such as sillimanite and andalusite, the estimated P–T ranges indicate conditions in the kyanite stability field (Kim and

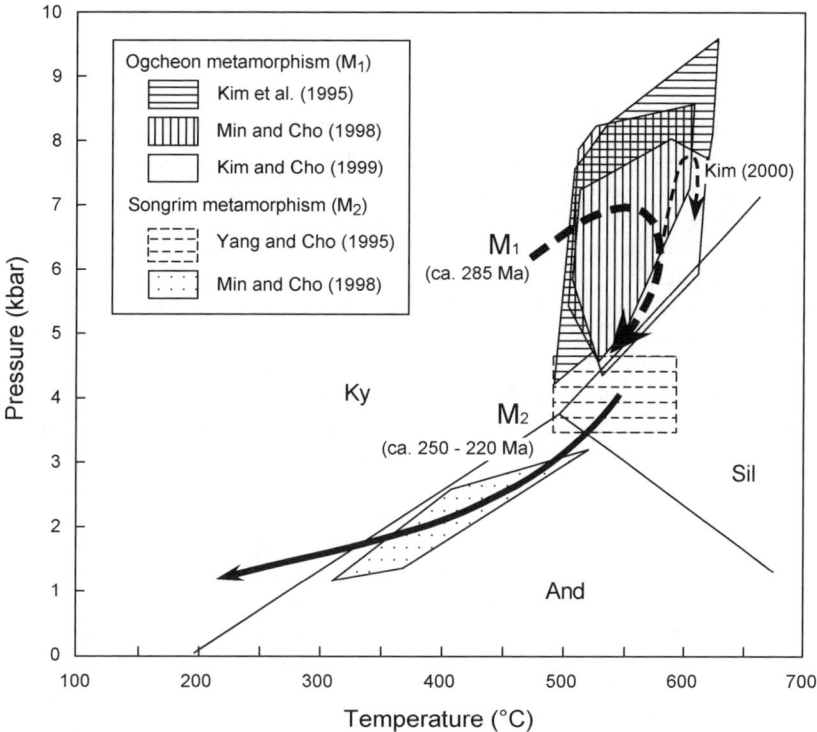

FIG. 4. A pressure-temperature diagram showing schematic P-T-t paths of regional metamorphic events, M_1 and M_2. The first metamorphism is attributed to the Ogcheon orogeny at ~285 Ma, whereas the second is assigned to the Songrim orogeny at ~250–220 Ma. The peak metamorphic condition for M_2 is not well defined, but is probably close to that determined from the metapelites of the Pyeongan Supergroup by Yang and Cho (1995).

Cho, 1999). It is thus apparent that thermal metamorphism was not sufficiently pervasive to re-equilibrate all of the regional metamorphic minerals.

The calculated P–T conditions increase northwestward across the strike of the OMB. For example, in the central OMB, Kim et al. (1995) reported a progressive increase in P–T conditions from 4.2–8.2 kbar and 490–540°C in the Poeun unit in the southeast, to 5.4–9.4 kbar and 520–630°C in the Pibanryeong unit in the northwest. No apparent discontinuity in P–T conditions was observed between upper and lower units, as noted from the results of b_0 parameters and crystallinity indices. Furthermore, the P–T conditions of the upper unit are higher than those of the lower unit. Such an inverted metamorphic grade is different from the P–T architecture commonly envisaged for thermally relaxed, thrust-bound regional metamorphic terranes. This apparent discrepancy may be partly attributed to the difference in original sedimentation thickness resulting from the asymmetric morphology of the Ogcheon basin. Although Koh and Kim (1995) suggested a wedge-shaped thrust stacking model (Coward and Butler, 1985) to account for the inverted metamorphic grade, further studies are necessary to elucidate temporal relationships between deformation and metamorphism of each lithotectonic unit.

The tectonometamorphic evolution of an orogenic belt can be better understood by quantitative calibration of the P–T path. Thus, many workers have attempted to elucidate P–T paths in the OMB (Cho et al., 1994; Kim et al., 1995; Oh et al., 1995, 1998; Min and Cho, 1998; Cho and Kim, 2002). All the studies except for that of Oh et al. (1995) suggested a clockwise P–T path, but neither prograde nor retrograde path was determined quantitatively.

Metamorphic P–T paths can be determined from the compositions of zoned minerals using computer programs such as GIBBS (Spear and Menard, 1989;

Spear et al., 1991) and THERMOCALC (Powell and Holland, 1998). Kim (2000) calculated the P–T path of a garnet-staurolite schist in the Pibanryeong unit, using the GIBBS program in the model system K_2O-Na_2O-CaO-MnO-FeO-MgO-Al_2O_3-SiO_2-H_2O. Staurolite, garnet, biotite, muscovite, chlorite, plagioclase, quartz, and H_2O were considered as phase components, but the mass balance of phase components at each step of the calculation was ignored. It was further assumed that all phase components are present and equilibrated with each other during the growth of garnet. The starting P–T conditions, at 7.2 kbar and 600°C, were estimated from the geothermobarometry, using the rim composition of garnet; ΔAlm, ΔSps, and ΔGrs of garnet were selected as monitor parameters. The result gave a clockwise P–T path with high dP/dT gradient and increases in pressure (~2 kbar) and temperature (~ 25°C) during the garnet growth (Fig. 4).

Subsequent to the peak metamorphism, the OMB experienced a second regional metamorphism at greenschist- to low-rank amphibolite-facies conditions, as suggested by Cho et al. (1995) and Min and Cho (1998). This regional-thermal metamorphism produced widespread retrograde phases such as chlorite, muscovite, margarite, biotite, and pyribole in pelitic schists (Kim et al., 1995; Ahn and Cho, 1996). On the other hand, hornblendes in amphibolites are commonly replaced by actinolite-actinolitic hornblende, epidote, and chlorite. These results are compatible with P–T conditions estimated from fluid inclusions in deformed quartz veins by Min and Cho (1998). According to their study, during exhumation, the OMB has passed through the P–T range 1–3 kbar and 350–500°C, following the isochore curves of the CO_2 inclusions (Fig. 4). However, the prograde P–T path and the peak condition of the second regional metamorphism are not known quantitatively.

Ages of Basin Formation and Metamorphism

Protolith and metamorphic ages of the OMB have long been controversial, primarily because of the complexity of the analyzed isotopic systems. Difficulties in interpreting various isotopic ages result from the overprinting effect of polymetamorphic episodes of similar metamorphic grade ranging from the greenschist to amphibolite facies. Recently, however, applications of modern techniques have provided some age constraints critical for understanding the P–T–t evolution of the OMB. These results together with those from the earlier workers are shown in Figure 5 and summarized below.

Neoproterozoic rifting

The protolith age of metamorphic rocks in the Ogcheon belt has not been well constrained, but commonly is thought to be Cambro-Ordovician by correlation with the Joseon Supergroup in the Taebaegsan basin. Studies of early Paleozoic fossils in the OMB (Lee et al., 1972, 1989) apparently reinforced this notion of the Paleozoic basin formation throughout the Ogcheon belt. The presence of fossils in amphibolite-facies marbles is, however, questionable in view of the likely destruction of any biogenic material during high-temperature metamorphism and deformation. Furthermore, failure to find any more of these so-called Paleozoic fossils casts doubt on their presence in the OMB (Chough et al., 2000). On the other hand, Oh and Trichet (1990) suggested that microfossils in uraniferous slates of the central Poeun unit are similar to those encountered in the Doushantuo Formation, China, dated at 650–700 Ma.

The historical view that the formation of the Ogcheon basin took place in the early Paleozoic was challenged as a result of a Sm–Nd isotopic hornblende age of 677 ± 91 Ma (Kwon and Lan, 1991; Fig. 5). Using K–Ar hornblende dating, a similar age of 675 ± 30 Ma was reported from four amphibolites by Min et al. (1995). Our subsequent study, however, revealed that these hornblende samples do not define $^{40}Ar/^{39}Ar$ plateau ages (Cho et al., unpubl. data). Thus, the $^{40}Ar/^{39}Ar$ dating of hornblende suggests that the K–Ar hornblende ages may be geologically meaningless. Nevertheless, using the isotope dilution technique, Lee et al. (1998) confirmed a Neoproterozoic age for a U–Pb zircon age of 756 ± 1 Ma from a felsic metavolcanic rock in the Turungsan unit.

Recently, Cho (unpubl. data) dated zircon grains of a felsic tuff in the Turungsan unit, using a SHRIMP ion microprobe (Fig. 6). The weighted mean $^{206}Pb/^{238}U$ zircon ages obtained from 11 spot analyses of 9 grains provide an essentially concordant age of 736 ± 10 Ma. If two spot analyses showing apparent Pb loss are excluded, the rest gives a slightly older age of 742 ± 10 Ma. This result corroborates the conventional U–Pb zircon age for the Neoproterozoic bimodal volcanism. Thus, the basins associated with intracontinental, high-volcanicity rift in the OMB are most likely to have formed at

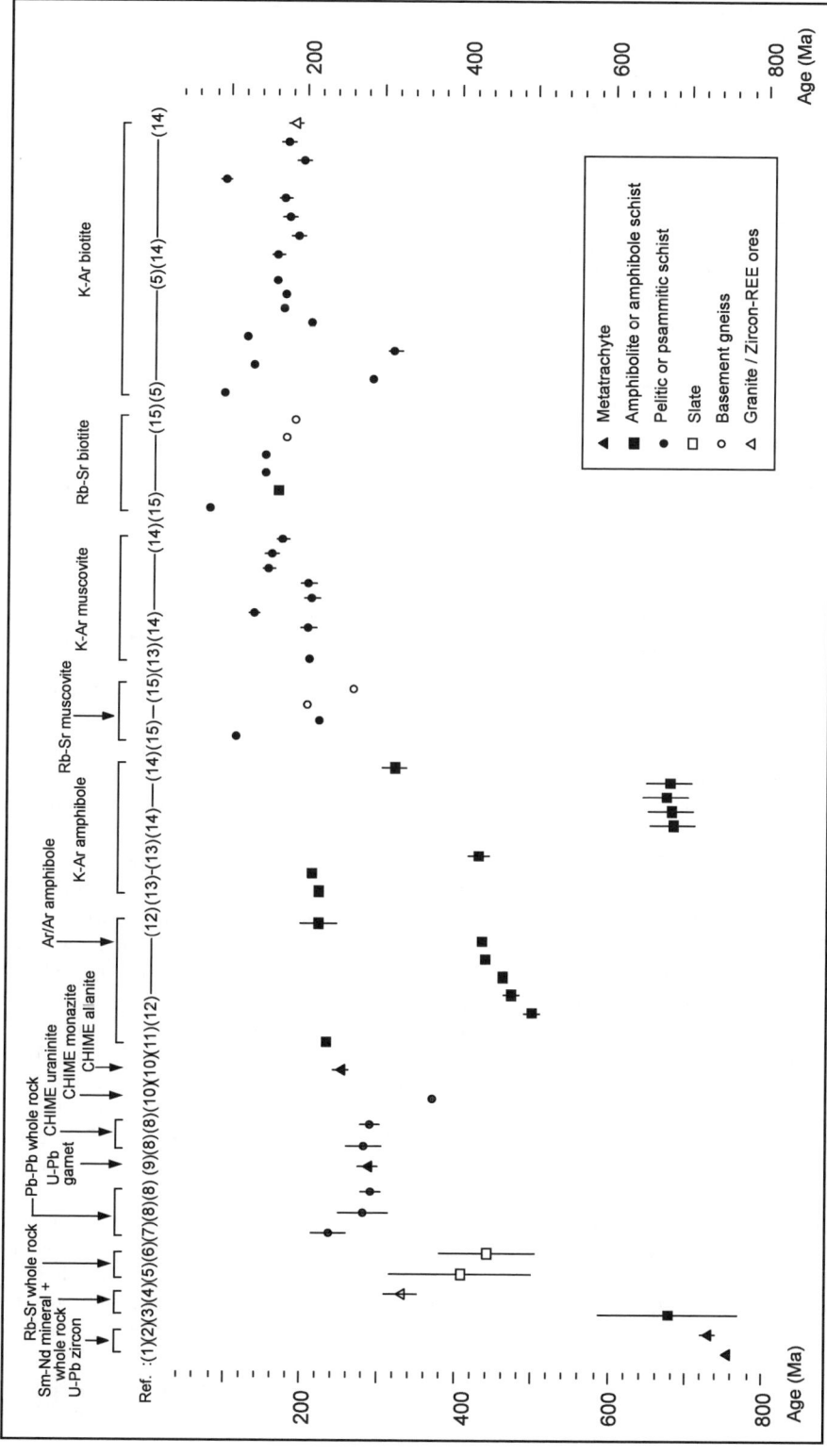

FIG. 5. Radiometric ages of metamorphic rocks from the Ogcheon metamorphic belt. Various isotopic techniques used for the age determination are given at the top of figure, and numbers in parentheses refer to the source of data: 1 = Lee et al. (1998); 2 = Cho (2001); 3 = Kwon and Lan (1991); 4 = Park and Kim (1995); 5 = Kim (1990; re-calculated by Park and Cheong, 1998); 6 = Park and Cheong (1998); 7 = Park (2001); 8 = Cheong et al. (2003); 9 = Kim et al. (2001); 10 = Adachi et al. (1996); 11 = Lee (1983); 12 = Cho et al. (1999); 13 = Oh et al. (1995); 14 = Min et al. (1995); 15 = Cliff et al. (1985).

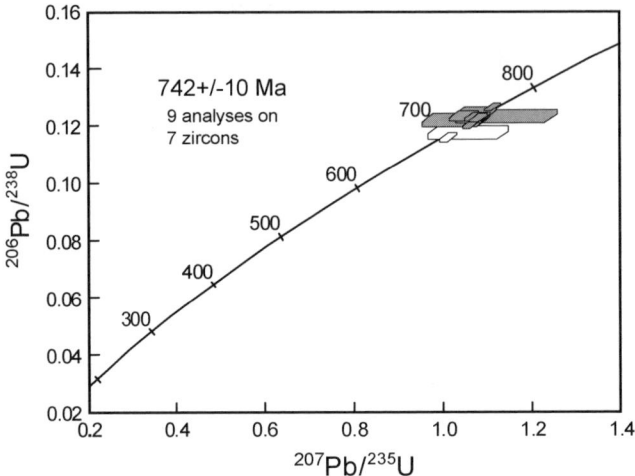

FIG. 6. Concordia diagram showing SHRIMP U–Pb ages of zircon grains from a felsic metavolcanic rock. Two spot analyses not included in the age calculation are shown as open boxes. Data-point error boxes and calculated ages are at the 95% confidence level.

~750 Ma. Furthermore, the occurrences of bimodal metavolcanics in both structurally upper and lower units, in conjunction with consistency in their geochemical characteristics (Cluzel, 1992a; Lee and Chang, 1997), allows us to speculate that the whole Ogcheon basin formed in Neoproterozoic time. The cessation of sedimentation before the Paleozoic is consistent with an apparent lack of fossils in both clastic and carbonate protoliths of the OMB, in marked contrast with those of the Taebaegsan basin.

Early Permian metamorphism and Triassic overprint

The timing of regional metamorphism in the OMB has been traditionally construed as late Paleozoic to early Mesozoic because: (1) the Permian strata are metamorphosed and deformed, whereas Early to Middle Jurassic granitoids are free of tectonometamorphism; and (2) radiometric dating of metamorphic muscovite and biotite has commonly yielded early Mesozoic ages, as summarized in Figure 5. For example, Cliff et al. (1985) suggested a Late Triassic metamorphism followed by slow cooling, based on whole-rock Rb–Sr ages of ~200 Ma, and younger 190–150 Ma biotite ages. The Triassic age of metamorphism is further supported by a $^{40}Ar/^{39}Ar$ hornblende age of 230 Ma (Lee, 1988). Some $^{40}Ar/^{39}Ar$ hornblende ages in the range of ~500–400 Ma, however, were interpreted to be geologically meaningless owing to the presence of excess argon (Cho et al., 1999). On the other hand, a Chemical Th–U-total Pb Isochron Method (CHIME) age obtained from allanite in muscovite-chlorite-quartz schist in the Turungsan unit yielded 251 ± 10 Ma (Adachi et al., 1996). This result, together with the Rb–Sr muscovite age ~219 ± 3 Ma in the Turungsan schist (Cliff et al., 1985), supports the interpretation of Permo-Triassic metamorphism followed by slow cooling.

Recent radiometric dating, however, suggests that the OMB experienced peak metamorphic conditions at ~285 Ma (Kim et al., 2001; Cheong et al., 2003; Cho et al., unpubl. data). Cheong et al. (2003) reported two $^{207}Pb/^{206}Pb$ whole-rock ages of 270 ± 28 Ma and 261 ± 17 Ma from the black slates of the Poeun unit. These ages, in conjunction with CHIME uraninite ages of 283 ± 26 Ma and 285 ± 45 Ma, were interpreted by Cheong et al. (2003) as the time of peak metamorphism. On the other hand, Kim et al. (2001) reported earliest Permian (~285 ± 5 Ma) U–Pb ages of garnet from several pelitic schists of the Pibanryeong unit. Although these ages probably represent those of monazite inclusions in garnet, they are consistent with each other among three analyzed samples. It is thus concluded that major tectonometamorphism in the OMB has occurred during Early Permian time. Finally, an Rb–Sr muscovite age of 266 ± 2 Ma was obtained from the highly deformed basement gneiss near Busan by Cliff et al. (1985; see Fig. 1B for location). This age could represent a cooling age after the peak metamorphism at ~285 Ma, inasmuch as the regional metamorphism also affected the Busan gneiss and L-tectonite underlying the OMB. This interpretation differs from that of Cliff et al. (1985), who attributed the Permian muscovite age to the inheritance of an old component.

The Early Permian age for the regional metamorphism is also supported by preliminary SHRIMP U–Pb zircon ages obtained from a granitic gneiss pebble in metadiamictite (the so-called Hwanggangri Formation), defining a discordia with lower intercept age of ~300 ± 40 Ma (Cho et al., unpubl. data). It is thus likely that a significant Pb loss occurred during the Permian peak metamorphism in the Ogcheon metamorphic belt.

Discussion

Tectonometamorphic evolution of the OMB

Various lines of evidence deduced from petrologic and geochronologic data suggest that the metamorphic evolution of the OMB is characterized by a composite clockwise P–T path (Fig. 4): (1) medium-pressure type regional metamorphism at 4.2–9.4 kbar and 490–630°C in association with the D_{1-2} deformation; and (2) subsequent regional-thermal metamorphism at 1–5 kbar and 350–500°C, probably associated with D_3. The first regional metamorphism, dated at ~285 Ma, was a product of the Ogcheon collisional orogeny (Cluzel et al., 1990). However, the timing of this orogeny should be redefined as earliest Permian, in contrast to Silurian-Devonian as postulated by Cluzel et al. (1990).

Many aspects of the second metamorphic event, including the areal extent, are uncertain, primarily because of the lack of detailed studies. Nevertheless, it probably occurred between 250 Ma and 220 Ma, because of the recurrence of Triassic ages in various isotopic systems and the lack of amphibolite-facies regional metamorphism in the Upper Triassic Daedong Supergroup (Fig. 5). The heat source for the second regional-thermal event is not well defined, but could be attributed to either syntectonic Triassic igneous activity or a crust-thickening event during the Songrim (or Indosinian) orogeny. Further studies are also necessary to determine what took place between the Ogcheon and Songrim orogenies.

Tectonometamorphic evolution of the OMB is summarized in Figure 7, which shows the relationship between pressure (= depth) and time. The Ogcheon orogeny at ~285 Ma was responsible for the peak metamorphism at ~4–9 kbar, whereas the Songrim orogeny was expressed by the regional-thermal metamorphic event at lower pressures. During the Songrim orogeny, the Ogcheon metamorphic belt probably amalgamated with the Taebaegsan basin to form the Ogcheon belt. As a consequence, the Carboniferous–Permian sequences of the Taebaegsan basin (Pyeongan Supergroup) were buried to a substantial depth, producing anchizone to epizone assemblages. Moreover, at least at a few localities, kyanite and staurolite porphyroblasts are reported from graphitic schists of the Pyeongan Supergroup, suggesting burial to ~15 km (Lee, 1992; Yang and Cho, 1995).

Subsequent to the regional metamorphism, the Ogcheon belt was intruded by ~220 Ma granitoids, which were affected by dextral shearing to produce ductile shear zones (e.g., Cheongsan shear zone; Ree et al., 2001). Because the ~180–170 Ma granitic batholiths along the OMB are generally free of deformation, the timing of dextral movement is best estimated as Late Triassic to Early Jurassic. Ree et al. (2001) inferred from the contrasting temperature regimes between ductile shearing and subsequent thrusting in the south-central OMB that the timing of the ductile shearing event was approximately Late Triassic. Based on this argument, Ree et al. (2001) suggested that the Cheongsan shear zone developed during the late stage of the Songrim orogeny, and is distinct from the Early Jurassic Honam shear zone (Yanai et al., 1985; Otoh and Yanai, 1996). In addition, Ree et al. (2001) proposed a tectonic model where the Cheongsan shear zone represents the major plate boundary between the South China block (Gyeonggi Massif and Ogcheon belt) and the North China block (Yeongnam Massif).

In the absence of precise geochronologic data, however, it is not yet clear whether the Cheongsan shear zone is the product of the Songrim orogeny. Furthermore, temporal relationships between the high-temperature Cheongsan shear zone and lower temperature thrusts in the Taebaegsan basin are too poorly known to support the tectonic model of Ree et al. (2001), concerning the boundary between the South and North China blocks. Nevertheless, their model is consistent with our contention, based on the P–T–t history of the Ogcheon belt, that the Taebaegsan basin and the OMB are separated by a suture boundary. These two terranes would be mutually allochthonous, because neither disturbance nor

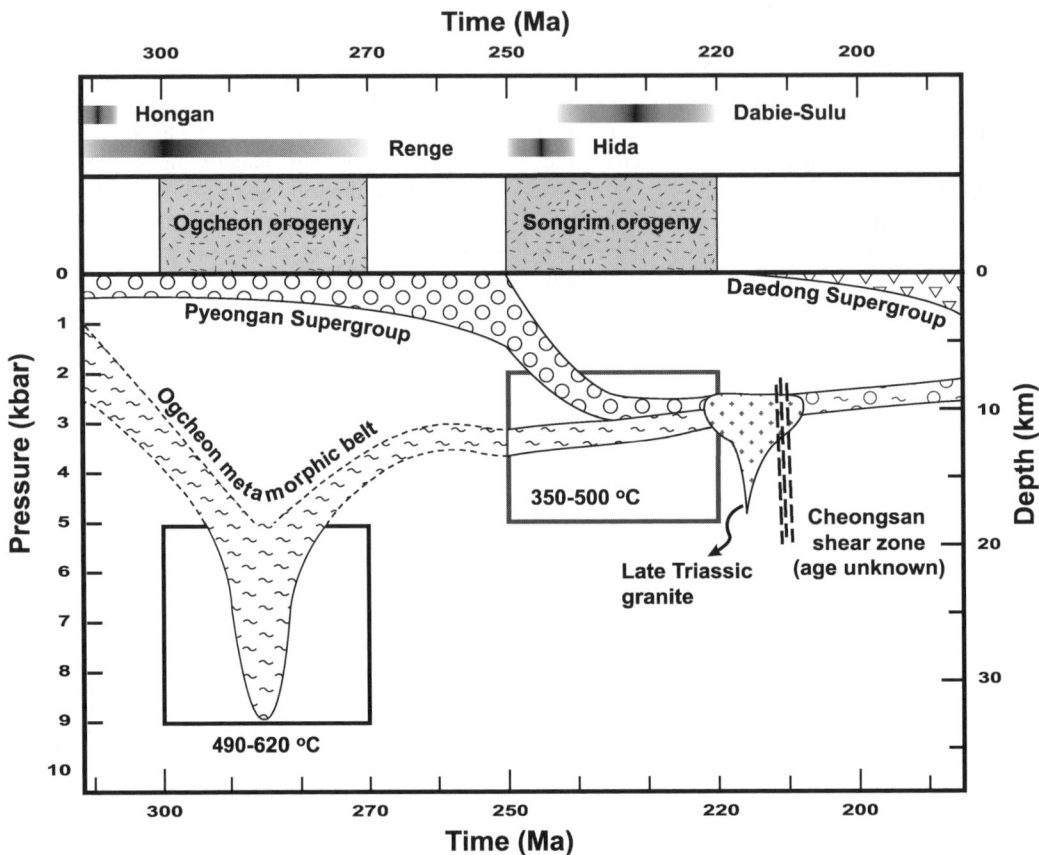

FIG. 7. A schematic diagram showing the pressure (or depth)–time evolution of the Ogcheon metamorphic belt. In this belt, the Ogcheon orogeny at ~285 Ma is responsible for the peak metamorphism at ~5–9 kbar and 490–630°C. Another regional metamorphic event at ~1–5 kbar and 350–500°C occurred during the Songrim orogeny at ~250–220 Ma. Then the Ogcheon belt, comprising both the Ogcheon metamorphic belt and the Taebaegsan basin, was intruded by ~220 Ma granitoids, which were subsequently affected by dextral ductile shearing (e.g., Cheongsan shear zone, Ree et al., 2001). Finally, nonmarine sedimentary rocks of the Daedong Supergroup were deposited during Late Triassic to Early Jurassic time, and were affected by the Daebo orogeny (e.g., Cluzel, 1992b). The Daebo metamorphism is probably limited in regional extent, and its P–T conditions are poorly constrained.

metamorphism in the Early Permian is recorded in the Pyeongan Supergroup. This conclusion is in marked contrast with the suggestion of Cluzel et al. (1990) that the protoliths of the Ogcheon basin are the deeper-facies equivalent of the shallow-marine Cambro-Ordovician sequences in the Taebaegsan basin.

Finally, nonmarine sedimentary rocks of the Daedong Supergroup were deposited during Late Triassic to Early Jurassic time, and were affected by the Daebo orogeny (e.g., Cluzel, 1992b; Fig. 7). The Daebo metamorphism is probably limited in areal extent, and its P–T conditions are poorly constrained. However, common occurrences of andalusite and cordierite around Jurassic plutons suggest low-pressure conditions less than ~3–4 kbar.

Tectonic and geochronologic correlations

The Ogcheon orogeny is a major tectonometamorphic event responsible for not only the SE-vergent, NE-trending structures but also the medium-pressure type of metamorphism in the OMB (Cluzel et al., 1990; Koh and Kim, 1995). As discussed above, this orogeny culminated during the earliest Permian (ca. 285 Ma), in contrast to previous suggestions of Silurian-Devonian (Cluzel et al., 1990; Kim, 1990) or Triassic-Jurassic (Kim, 1971; Reedman and Um, 1975; Cliff et al., 1985) ages. The Ogcheon orogeny was followed by another tectonometamorphic event, the Songrim orogeny, at ~250–220 Ma. This Triassic event tectonically juxtaposed the medium-pressure metamorphic rocks of the OMB over the Paleozoic sedimentary sequences of the Taebaegsan basin (Chough et al., 2000). On the basis of coeval radiometric ages, it is further inferred that the Songrim orogeny was associated with continental collision between the North and South China blocks (Ree et al., 2001; Cho and Kim, 2002).

In the absence of precise geochronologic data, it is uncertain whether the major orogenies in the Ogcheon belt represent two distinct tectonometamorphic events or a prolonged event spanning several tens of million years. Nevertheless, it is interesting to compare Permo-Triassic tectonometamorphic events in the Ogcheon belt with those reported from the Qinling-Dabie-Sulu belt of China and the Hida metamorphic belt in Japan, inasmuch as these belts are often correlated with each other (e.g., Ernst and Liou, 1995).

The timing for the UHP metamorphism in China has been well dated at ~245–220 Ma, using various isotopic systematics (e.g., Hacker et al., 1998). Recently, however, Sun et al. (2002) suggested that prior to the main Triassic collision, eclogite-facies metamorphism occurred as early as Late Carboniferous (309 ± 3 Ma) in the northwestern Dabie Mountains. This result in conjunction with middle Paleozoic collision in the Qinling belt (Kröner et al., 1993; Xue et al., 1996; Meng and Zhang, 1999) suggests that multiple collisions produced the complex and prolonged histories along the suture between the North and South China blocks. The same complexity may be applicable to the Ogcheon belt, but further studies are necessary to reveal its tectonic correlation with Chinese UHP belts. It is noted, however, that the prevalence of ~750 Ma magmatism is apparent in the South China block, probably in association with the Neoproterozoic breakup of the Rodinian supercontinent (Hacker et al., 1998; Chen et al., 2003; Li et al., 2003). It is thus likely that bimodal volcanism in the OMB is correlative with the rifting event in the South China block.

Permian and Triassic ages of two tectonometamorphic events in the Ogcheon belt are compatible with geochronologic data recently reported from the Hida and Renge (or Hida Marginal) belts in Japan (e.g., Nishimura, 1998; Sano et al., 2000; Kunugiza et al., 2001). Nishimura (1998) suggested that the Renge belt is characterized by high-pressure metamorphism that took place at 330–280 Ma. In the Hida belt, an early metamorphic stage is dated at ~270 Ma on the basis of the CHIME ages of monazite as well as SHRIMP zircon ages in metagranites and migmatites (Kunugiza et al., 2001). Sano et al. (2000) also reported a minor thermal event at ~285 Ma in the Hida terrane, based on a concordant SHRIMP age of inherited zircon in a paragneiss. The peak metamorphism in the Hida belt, however, is well constrained at ~250–240 Ma: CHIME monazite age of 250 ± 10 Ma (Suzuki and Adachi, 1994); SHRIMP zircon age of 245 ± 15 Ma (Sano et al., 2000); and CHIME ages of ~240 Ma estimated from zircon, monazite and uraninite (Kunugiza et al., 2001). Thus, the coincidence in geochronologic data suggests that the Hida and Ogcheon belts are correlative. This correlation is corroborated by the similarities in lithologies and medium-pressure type metamorphism dominant in Carboniferous metasedimentary rocks in both belts (Hiroi, 1981; Cluzel, 1991; Yang and Cho, 1995). Hence, as first suggested by Ernst and Liou (1995), it is still a valid hypothesis that the Qinling-Dabie-Sulu UHP belt in China extends eastward through the Imjingang and Ogcheon belts in Korea to the Hida and Renge belts in Japan.

Acknowledgments

We thank W. R. Fitches and J. G. Liou for helpful and constructive comments, and W. G. Ernst for his support and interest in tectonics of the Korean Peninsula. This research was sponsored by the KOSEF grant R14-2003-017-01002-0. Kim's postdoctoral research at SNU was supported through the BK21 program.

REFERENCES

Adachi, M., Suzuki, K., and Chwae, U. C., 1996, CHIME age determination of metamorphic rocks in the Okchon belt, Korea [abs.]: Annual Meeting of the Geological Society of Japan, p. 80.

Ahn, J. H., and Cho, M., 1996, High-resolution transmission electron microscopy of structural defects in hornblendes of Ogcheon amphibolites: Journal of the Geological Society of Korea, v. 32, p. 334–344.

Barberi, F., Santacroe, R., and Varet, J., 1982, Chemical aspects of rift magmatism, in Palmason, G., ed., Continental and oceanic rifts: American Geophysical Union, Washington, DC, p. 223–258.

Chang, T. W., 1988, Time-relationships between the growth of biotite porphyroblast and deformation, and the development of slaty cleavage in the metapelitic rocks of Ogcheon Group: Journal of the Geological Society of Korea, v. 24, p. 127–139 (in Korean with English abstract).

Chen, J., and Jahn, B., 1998, Crustal evolution of southeastern China: Nd and Sr isotopic evidence: Tectonophysics, v. 284, p. 101–133.

Chen, F., Guo, J.-H., Jiang, L.-L., Siebel, W., Cong, B., and Satir, M., 2003, Provenance of the Beihuaiyang lower-grade metamorphic zone of the Dabie ultrahigh-pressure collisional orogen, China: Evidence from zircon ages: Journal of Asian Earth Sciences, v. 22, p. 343–352.

Cheong, C.-S., and Chang, H. W., 1997, Sr, Nd, and Pb isotope systematics of granitic rocks in the central Ogcheon Belt, Korea: Geochemical Journal, v. 31, p. 17–36.

Cheong, C.-S., Jeong, G. Y., Kim, H., Lee, S.-H., Choi, M. S., and Cho, M., 2003, Early Permian peak metamorphism recorded in U–Pb system of black slates from the Ogcheon metamorphic belt, South Korea, and its tectonic implication: Chemical Geology, v. 193, p. 81–92.

Cho, D.-L., Kwon, S.-T., Sagong, H., Cheong, C.-S., and Armstrong, R., 2001, Precise cooling histories of three neighboring plutons in the central Okcheon belt: Implications for magma movement rate and tectonics [abs.], in 54th Annual Meeting of the Geological Society of Korea, p. 90.

Cho, M., 2001, A continuation of Chinese ultrahigh-pressure belt in Korea: Evidence from ion microprobe U–Pb zircon ages [abs.]: Gondwana Research, v. 4, p. 708.

Cho, M., and Kim, H., 2002, Metamorphic evolution of the Ogcheon metamorphic belt: Review of recent studies and remaining problems: Journal of the Petrological Society of Korea, v. 11, p. 121–137 (in Korean with English abstract).

Cho, M., Kim, H., Lo, C.-H., Min, K., and Ahn, J. H., 1999, Ordovician (or "Caledonian") Okchon orogeny: Evidence from $^{40}Ar/^{39}Ar$ hornblende ages [abs.], in 52th Annual Meeting of the Geological Society of Korea, p. 66.

Cho, M., Kim, I. J., Kim, H., Min, K., Ahn, J.-H., and Nagao, K., 1995, K-Ar biotite age of pelitic schists in the Jeungpyeong-Deokpyeong area, central Ogcheon metamorphic belt, Korea: Journal of the Petrological Society of Korea, v. 4, p. 178–185 (in Korean with English abstract).

Cho, M., Min, K., and Kim, H., 1994, Metamorphism in the central Ogcheon belt, in Ree, J.-H., Cho, M., Kwon, S.-T., and Kim, J. H., eds., Structure and metamorphism of the Ogcheon Belt—field trip guidebook:

Seoul, Korea, Hanlimwon Publishers, IGCP 321 4th International Symposium, p. 97–120.

Choi, W., and Kim, D. H., 1981, The study of the Ogcheon geosynclinal belt: Report of the Korean Institute of Geoscience and Mineral Resources, v. 11, p. 19–44.

Chough, S. K., and Bahk, K. S., 1992, The Hwangkangri Formation in the Okchon Basin, in Chough, S. K., ed., Sedimentary basins in the Korean Peninsula and adjacent seas: Seoul, Korea, Hanlimwon Publishers, Korean Sedimentology Research Group, Special Publication, p. 77–101.

Chough, S. K., Kwon, S.-T., Ree, J.-H., and Choi, D. K., 2000, Tectonic and sedimentary evolution of the Korean peninsula: A review and new view: Earth Science Reviews, v. 52, p. 175–232.

Cliff, R. A., Jones, G., Choi, W. C., and Lee, T. J., 1985, Strontium isotopic equilibration during metamorphism of tillites from the Ogcheon Belt, South Korea: Contributions to Mineralogy and Petrology, v. 90, p. 346–352.

Cluzel, D., 1991, Late Paleozoic to early Mesozoic geodynamic evolution of the Circum-Pacific orogenic belt in South Korea and Southwest Japan: Earth and Planetray Science Letters, v. 108, p. 289–305.

Cluzel, D., 1992a, Ordovician bimodal magmatism in the Ogcheon belt (South Korea): Intracontinental rift-related volcanic activity: Journal of Southeast Asian Earth Sciences, v. 7, p. 195–209.

Cluzel, D., 1992b, Formation and tectonic evolution of early Mesozoic intramontane basins in the Ogcheon belt (South Korea): A reappraisal of the Jurassic "Daebo orogeny": Journal of Southeast Asian Earth Sciences, v. 7, p. 223–235.

Cluzel, D., Cadet, J. P., and Lapierre, H., 1990, Geodynamics of the Ogcheon belt (South Korea): Tectonophysics, v. 183, p. 41–56.

Cluzel, D., Jolivet, L., and Cadet, J. P., 1991, Early Middle Paleozoic intraplate orogeny in the Ogcheon belt (South Korea): A new insight on the Paleozoic buildup of East Asia: Tectonics, v. 10, p. 1130–1151.

Coward, M. P., and Butler, R. H. W., 1985, Thrust tectonics and deep structure of the Pakistan Himalaya: Geology, v. 13, p. 417–420.

Ernst, W. G., Cao, R., and Jiang, J., 1988, Reconnaissance study of Precambrian metamorphic rocks, northeastern Sino-Korean shield, People's Republic of China: Geological Society of America Bulletin, v. 100, p. 692–701.

Ernst, W. G., and Liou, J. G., 1995, Contrasting plate-tectonic styles of the Qinling-Dabie-Sulu and Franciscan metamorphic belts: Geology, v. 23, p. 353–356.

Faure, M., Lin, W., and Le Breton, N., 2001, Where is the North China-South China block boundary in eastern China?: Geology, v. 29, p. 119–122.

Faure, M., Lin, W., Monie, P., Le Breton, N., Poussineau, S., Panis, D., and Deloule, E., 2003, Exhumation tectonics of the ultrahigh-pressure metamorphic rocks in the Qinling orogen in east China: New petrological-structural-radiometric insights from the Shandong Peninsula: Tectonics, v. 22, 1019 [doi:10.1029/2002TC001450].

Ferry, J. M., and Spear, F. S., 1978, Experimental calibration of the partitioning of Fe and Mg between biotite and garnet: Contributions to Mineralogy and Petrology, v. 66, p. 113–117.

Guidotti, C. V., 1984, Micas in metamorphic rocks, in Bailey, S. W., ed., Micas: Mineralogical Society of America, Reviews in Mineralogy, v. 13, p. 357–467.

Hacker, B. R., Ratschbacher, L., Webb, L., Ireland, T., Walker, D., and Dong, S., 1998, U/Pb zircon ages constrain the architecture of the ultrahigh pressure Qinling–Dabie orogen, China: Earth and Planetray Science Letters, v. 161, p. 215–230.

Hiroi, Y., 1981, Subdivision of the Hida metamorphic complex, central Japan, and its bearing on the geology of the Far East in pre-Sea of Japan time: Tectonophysics, v. 76, p. 317–333.

Hoisch, T. D., 1989, A muscovite-biotite geothermometer: American Mineralogist, v. 74, p. 565–572.

Hoisch, T. D., 1990, Empirical calibration of six geobarometers for the mineral assemblage quartz + muscovite + biotite + plagioclase + garnet: Contributions to Mineralogy and Petrology, v. 104, p. 225–234.

Ishiwatari, A., and Tsujimori, T., 2001, Late Paleozoic high-pressure metamorphic belts in Japan and Sikhote-Alin: Possible oceanic extension of the Chinese Dabie-Su-Lu suture detouring Korea [ext. abs.]: Gondwana Research, v. 4, p. 636–638.

Jahn, B.-M., Wu, F., and Chen, B., 2000, Massive granitoid generation in Central Asia: Nd isotopic evidence and implication for continental growth in the Phanerozoic: Episodes, v. 23, p. 82–92.

Kang, J. H., Hara, I., Hayasaka, Y., Sakurai, Y., Shiota, T., and Umemura, H., 1993, Time relationship between deformation and metamorphism of the Ogcheon zone in the Ogcheon district, South Korea: Memoirs of the Geological Society of Japan, v. 42, p. 63–90.

Kang, J. H., and Lee, J.-Y., 2002, Geological structure of Okcheon metamorphic zone in the Miwon-Boeun area, Korea: Journal of the Petrological Society of Korea, v. 11, p. 234–249 (in Korean with English abstract).

Kang, J. H., and Ryoo, C. R.,1997, Igneous activity and geological structure of the Ogcheon metamorphic zone in the Kyemyeongsan area, Chungju, Korea: Journal of the Petrological Society of Korea, v. 6, p. 151–165 (in Korean with English abstract).

Kim, H., 2000, Metamorphic evolution of low- to medium-pressure rocks in central Ogcheon metamorphic belt, Korea and on Barton Peninsula, King George Island, Antarctica: Ph.D. dissertation, Seoul National University, 172 p.

Kim, H., Cheong, C.-S., Cho, M., Jeong, G. Y., and Choi, M. S., 2001, Geochronologic evidence for Late Paleozoic orogney in the Ogcheon metamorphic belt, South Korea [abs.]: Geological Society of America, Abstract with Programs, v. 33, p. 380.

Kim, H., and Cho, M., 1999, Polymetamorphism of Ogcheon Supergroup in the Miwon area, central Ogcheon metamorphic belt, South Korea: Geosciences Journal, v. 3, p. 151–162.

Kim, H., Cho, M., and Ahn, J. H., 2000, A study on the b_0 parameter and crystallinity index of K-white micas from low-grade metapelites in Deokpyeong and Miwon areas, central Ogcheon metamorphic belt, Korea: Geosciences Journal, v. 4, p. 201–209.

Kim, H., Cho, M., and Koh, H. J., 1995, Tectonometamorphic evolution of the central Ogcheon belt in the Jeungpyeong-Deokpyeong area: Journal of the Geological Society of Korea, v. 31, p. 299–314 (in Korean with English abstract).

Kim, H. S., 1971, Metamorphic facies and regional metamorphism of Ogcheon metamorphic belt: Journal of the Geological Society of Korea, v. 7, p. 221–256.

Kim, J. H., 1990, Middle Paleozoic isotopic ages of the Ogcheon Group in Korea and their significance, in Aubouin, J., and Bourgois, J., eds., Tectonics of circum-Pacific continental margins: Utrecht, Netherlands, VSP, p. 181–191.

Kim, S. W., Itaya, T., Hyodo, H., and Matsuda, T., 2002, Metamorphic K-feldspar in low-grade metasediments from the Ogcheon metamorphic belt in South Korea: Gondwana Research, v. 5, p. 849–855.

Koh, H. J., 1995, Structural analysis and tectonic evolution of the Ogcheon Belt, Korea: Unpubl. Ph.D. dissertation, Seoul National University, 282 p.

Koh, H. J., and Kim, J. H., 1995, Deformation sequence and characteristics of the Ogcheon Supergroup in the Geosan area, central Ogcheon belt, Korea: Journal of the Geological Society of Korea, v. 31, p. 271–298.

Kröner, A., Zhang, G. W., and Sun, Y., 1993, Granulites in the Tongbai area, Qinling belt, China: Geochemistry, petrology, single zircon geochronology, and implications for the tectonic evolution of eastern Asia: Tectonics, v. 12, p. 245–255.

Kunugiza, K., Tsujimori, T., and Kano, T., 2001, Evolution of the Hida and Hida marginal belts, in Kano, T., ed., ISRGA Field Workshop guidebook for major geologic units of southwest Japan: Osaka, Japan, Field Science Publishers, p. 75–131.

Kwon, S. T., and Lan, C. Y., 1991, Sm–Nd isotopic study of the Ogcheon amphibolite, Korea: Preliminary report: Journal of the Korean Institute of Mining Geology, v. 24, p. 277–285 (in Korean with English abstract).

Kwon, S.-T., and Lee, D. H., 1992, Petrology and geochemistry of the Ogcheon metabasites in Poun, Korea: Journal of the Petrological Society of Korea, v. 1, p. 104–123 (in Korean with English abstract).

Lee, C. H., 1992, Metamorphism of the Pyeongan Supergroup in the Kangreung-Bukpyung area: Journal of the Geological Society of Korea, v. 28, p. 553–570 (in Korean with English abstract).

Lee, C. H., Lee, M. S., and Park, B. S., 1980, Explanatory notice of the 1:50,000 geologic map of Korea, Miweon sheet: Korea Research Institute of Geoscience and Mineral Resources, Seoul, 29 p. (in Korean with English summary).

Lee, D. S., Chang, K. H., and Lee, H. Y., 1972, Discovery of Archaeocyatha from Hyangsanri Dolomite Formation of the Ogcheon System and its significance: Journal of the Geological Society of Korea, v. 8, p. 191–197 (in Korean with English abstract).

Lee, H., 2000, Significance of systematic changes in crenulation asymmetries within metasediments across the Ogcheon Supergroup in the Goesan area, southern Korea: Geosciences Journal, v. 4, p. 115–134.

Lee, J.-H., Lee, H. Y., Yu, K.-M., and Lee, B.-S., 1989, Discovery of microfossils from limestone pebbles of the Hwanggangri formation and their stratigraphic significance: Journal of the Geological Society of Korea, v. 25, p. 1–15.

Lee, K. S., and Chang, H. W., 1997, Geochemistry and Sr–Nd–Pb isotopic systematics of the Ogcheon amphibolites from the central Ogcheon belt, Korea: Implication for the source heterogeneity: Geochemical Journal, v. 31, p. 223–243.

Lee, K. S., Chang, H. W., and Park, K.-H., 1998, Neoproterozoic bimodal volcanism in the central Ogcheon belt, Korea: Age and tectonic implication: Precambrian Research, v. 89, p. 47–57.

Lee, M. S., 1988, Geochemistry of amphibolites from the Ogcheon belt and its application to the tectonic setting of the Korean peninsula [abs.]: International Symposium on Geodynamic Evolution of the Eastern Eurasian Margin, Paris, France, p. 56.

Lee, S. R., Cho, M., Hwang, J. H., Lee, B., Kim, Y., and Kim, J. C., 2003, Crustal evolution of the Gyeonggi massif, South Korea: Nd isotopic evidence and implications for continental growths of East Asia: Precambrian Research, v. 121, p. 25–34.

Li, Z. X., 1994, Collision between the North and South China blocks: A crustal-detachment model for suturing in the region east of the Tanlu fault: Geology, v. 22, p. 739–742.

Li, Z. X., Li, X.-H., Kinny, P. D., Wang, J., Zhang, S., and Zhou, H., 2003, Geochronology of Neoproterozoic syn-rift magmatism in the Yangtze Craton, South China and correlations with other continents: Evidence for a mantle superplume that broke up Rodinia: Precambrian Research, v. 122, p. 85–109.

Liou, J. G., Maruyama, S., Zhang, Z. M., Coleman, R. G., and Wang, X. M., 1989, Blueschists in major suture zones of China: Tectonics, v. 8, p. 609–619.

Liou, J. G., Zhang, R., Wang, X., Eide, E. A., Ernst, W. G., and Maruyama, S., 1996, Metamorphism and tectonics of high-pressure and ultrahigh-pressure belts in Dabie-Sulu regions, eastern central China, in Yin, A., and Harrison, T. M., eds., The tectonic evolution of Asia: New York, NY, Cambridge University Press, p. 300–344.

Massone, H. V., and Schreyer, W., 1987, Phengite geobarometry based on the limited assemblage with K-feldspar, phlogopite, and quartz: Contributions to Mineralogy and Petrology, v. 96, p. 212–224.

Mattauer, M., Matte, P., Malavieille, J., Tapponnier, P., Maluski, H., Xu, Z., Lu, Y., and Tang, Y., 1985, Tectonics of the Qinling belt: Build-up and evolution of eastern Asia: Nature, v. 317, p. 496–500.

Meng, Q.-R., and Zhang, G., 1999, Timing of collision of the North and South China blocks: Controversy and reconciliation: Geology, v. 27, p. 123–126.

Min, K., and Cho, M., 1998, Metamorphic evolution of the northwestern Ogcheon metamorphic belt, South Korea: Lithos, v. 43, p. 31–51.

Min, K., Cho, M., Kwon, S.-T., Kim, I. J., Nagao, K., and Nakamura, E., 1995, K-Ar ages of metamorphic rocks in the Chungju area: Late Proterozoic (675 Ma) metamorphism of the Ogcheon belt: Journal of the Geological Society of Korea, v. 31, p. 315–327 (in Korean with English abstract).

Nishimura, Y., 1998, Geotectonic subdivision and areal extent of the Sangun belt, Inner zone of southwest Japan: Journal of Metamorphic Geology, v. 16, p. 129–140.

Oh, C. H., Kim, S. T., and Lee, J. H., 1995, The P-T condition and timing of the main metamorphism in the southwestern part of the Ogcheon metamorphic belt: Journal of the Geological Society of Korea, v. 31, p. 343–361.

Oh, C. H., Kim, S. T., and Lee, J. H., 1998, A study on the regional and contact metamorphism in the southwestern part of the Ogcheon metamorphic belt: Journal of the Geological Society of Korea, v. 34, p. 311–332 (in Korean with English abstract).

Oh, C.-H., and Trichet, J., 1990, Uranium in the black schists of Goesan (Ogcheon, Korea): Relationships between organic matter and uranium: Compte Rendus Academie des Sciences, Paris, t. 310, Series II, p. 241–245.

Okay, A., Sengör, A. M. C., and Satir, M., 1993, Tectonics of an ultrahigh-pressure metamorphic terrane: The Dabieshan/Tongbaishan orogen, China: Tectonics, v. 12, p. 1320–1334.

Otoh, S., and Yanai, S., 1996, Mesozoic inversive wrench tectonics in far east Asia: examples from Korea and Japan, in Yin, A., and Harrison, T. M., eds., The tectonic evolution of Asia: New York, NY, Cambridge University Press, p. 401–419.

Park, K.-H., 2001, The study of age determination using stepwise dissolution technique: Journal of Petrological Society of Korea, v. 10, p. 133–147 (in Korean with English abstract).

Park, K.-H., and Cheong, C.-S., 1998, $^{87}Sr/^{86}Sr$ ratios and ages of the Ogcheon metasedimentary rocks: Journal of

Geological Society of Korea, v. 34, p. 81–93 (in Korean with English abstract).

Park, M.-E., and Kim, G.-S., 1995, Genesis of the REE ore deposits, Chungju district, Korea: Occurrence features and geochemical characteristics: Economic and Environmental Geology, v. 28, p. 599–612 (in Korean with English abstract).

Pickering, K., Stow, D., Watson, M., and Hiscott, R., 1986, Deep-water facies, processes, and models: A review and classification scheme for modern and ancient sediments: Earth Science Reviews, v. 23, p. 75–174.

Powell, R., and Holland, T. J. B., 1998, An internally consistent thermodynamic dataset with uncertainties and correlations: 3: Application methods, worked examples, and a computer program: Journal of Metamorphic Geology, v. 6, p. 173–204.

Ree, J.-H., Cho, M., Kwon, S.-T., and Nakamura, E., 1996, Possible eastward extension of Chinese collision belt in South-Korea: The Imjingang belt: Geology, v. 24, p. 1071–1074.

Ree J.-H., Kwon, S.-H., Park, Y., Kwon, S.-T., and Park, S.-H., 2001, Petrotectonic and post-tectonic emplacements of the granitoids in the south central Okchon belt, South Korea: Implications for the timing of strike-slip shearing and thrusting: Tectonics, v. 20, p. 850–867.

Reedman, A. J., and Um, S. H., 1975, Geology of Korea: Seoul, Korea, Geological and Mineralogical Institute of Korea, 139 p.

Sano, Y., Hidaka, H., Terada, K., Shimizu, H., and Suzuki, M., 2000, Ion microprobe U-Pb zircon geochronology of the Hida gneiss: Finding of the oldest minerals in Japan: Geochemical Journal, v. 34, p. 135–153.

Shin, D., and Lee, I., 2003, Evaluation of the volatilization and infiltration effects on the stable isotopic and mineralogical variations in the carbonate rocks adjacent to the Cretaceous Muamsa Granite, South Korea: Journal of Asian Earth Sciences, v. 22, p. 227–243.

Spear, F. S., and Menard, T., 1989, Program GIBBS: A generalized Gibbs method algorithm: American Mineralogist, v. 74, p. 942–943.

Spear, F. S., Peacock, S. M., Kohn, M. J., Florence, F. P., and Menard, T., 1991, Computer programs for petrologic P–T–t path calculations: American Mineralogist, v. 76, p. 2009–2012.

Sun, W., Williams, I. S., and Li, S., 2002, Carboniferous and Triassic eclogites in the western Dabie mountains, east-central China: evidence for protracted convergence of the North and South China blocks: Journal of Metamorphic Geology, v. 20, p. 873–886.

Suzuki, K., and Adachi, M., 1994, Middle Precambrian detrital monazite and zircon from the Hida gneiss on Oki-Dogo Island, Japan: Their origin and implications for the correlation of basement gneiss of southwest Japan and Korea: Tectonophysics, v. 235, p. 277–292.

Wu, F.-Y., Jahn, B.-M., Wilde, S. A., and Sun, D.-Y., 2000, Phanerozoic continental crustal growth: U-Pb and Sr-Nd isotopic evidence from the granites in northeastern China: Tectonophysics, v. 328, p. 89–113.

Xue, F., Lerch, M. F., Kröner, A., and Reischmann, T., 1996, Tectonic evolution of the east Qinling Mountains, China, in the Paleozoic: A review and new tectonic model: Tectonophysics, v. 253, p. 271–284.

Yanai, S., Park, B. S., and Otoh, S., 1985, The Honam shear zone, South Korea: Deformation and tectonic implication in the Far East: Science Papers, College of Arts and Sciences, University of Tokyo, v. 35, p. 181–210.

Yang, P., and Cho, M., 1995, The staurolite-biotite-andalusite-garnet assemblage in the Dueumri Formation near the Chunyang granite: Algebraic analysis: Journal of the Petrological Society of Korea, v. 4, p. 49–60 (in Korean with English abstract).

Yin, A., and Nie, S., 1993, An indentation model for the North and South China collision and the development of the Tan-Lu and Honam fault systems, eastern Asia: Tectonics, v. 12, p. 801–813.

Yin, A., and Nie, S., 1996, A Phanerozoic palinspastic reconstruction of China and its neighboring regions, in Yin, A., and Harrison, T. M., eds., The tectonic evolution of Asia: New York, NY, Cambridge University Press, p. 442–485.

Zhai, M., Cong, B., Guo, J., Liu, W., Li, Y., and Wang, Q., 2000, Sm-Nd geochronology and petrography of garnet pyroxene granulites in the northern Sulu region of China and their geotectonic interpretation: Lithos, v. 52, p. 23–33.

Zheng, Y.-F., Fu, B., Gong, B., and Li, L., 2003, Stable isotope geochemistry of ultrahigh pressure metamorphic rocks from the Dabie-Sulu orogen in China: Implications for geodynamics and fluid regime: Earth Science Reviews, v. 55, 1–57.

Ridge Subduction–Related Jurassic Plutonism in and around the Okcheon Metamorphic Belt, South Korea, and Implications for Northeast Asian Tectonics

S. W. KIM,

Basic Science Research Institute, Chonbuk National University, Chonju 561-756, South Korea

C. W. OH,[1]

Department of Earth and Environmental Sciences, Chonbuk National University, Chonju 561-756, South Korea

S. G. CHOI,

Department of Earth and Environmental Sciences, Korea University, Seoul 136-701, South Korea

I.-C. RYU

Department of Geology, Kyungpook National University, Daegu, 702-701, South Korea

AND T. ITAYA

Research Institute of Natural Science, Okayama University of Science, Okayama 700-0005, Japan

Abstract

The Okcheon metamorphic belt (OMB) in central South Korea is surrounded by Middle Jurassic granitoid batholiths that intruded South Korea extensively; the granitic bodies form a complex about 200 km long and 150 km wide as part of a Mesozoic granite belt along the East Asian continental margin. Middle Jurassic magmatism was related to ridge subduction that occurred around 200 to 166 Ma, with the main magmatic period between 175 and 166 Ma; main cooling ages range from 168 to 152 Ma. The magmatism was divided into two stages: (1) a deeper, earlier stage, which resulted in emplacement of diorite, granodiorite and granite as shown in the northeast OMB; and (2) a shallower, younger stage, which resulted in emplacement of granite and two-mica granite as shown in the southwest OMB. Most granitoids are peraluminous to metaluminous I-type granitoids that originated in a volcanic arc; an exception is an S-type two-mica granite. Inherited cores of 998 and 262 Ma U-Pb SHRIMP II zircon ages from the two-mica granite indicate that two-mica granite is reworked crustal material formed by earlier magmatism before the Middle Jurassic event. Together with previous studies on the Middle Jurassic granitoids, the present result indicates that subduction of the Farallon-Izanagi ridge beneath Asia caused widespread igneous activity throughout South Korea, especially during Middle Jurassic ridge subduction.

Introduction

SOUTH KOREA is located on the East Asian continental margin and consists of two Precambrian basements (the Gyeonggi and Yeongnam massifs), two narrow Phanerozoic belts (the Okcheon and Imjingang belts), and the Cretaceous sedimentary Gyeongsang basin (Fig. 1). Mesozoic batholithic granitoids in South Korea are part of an extensive Mesozoic granitic belt that formed along the continental margin, extending from the southern region of Guangxi in South China to the far eastern part of Russia. Many researchers have suggested that eastern Eurasian Mesozoic plutonism might be ascribed to active ridge subduction (Isozaki and Maruyama, 1991; Kiminami et al., 1993; Kinoshita, 1995, 2002; Maruyama et al., 1997). Mesozoic granitoids in South Korea can be divided into three groups: Middle to Late Triassic granitoids, Middle Jurassic granitoids, and Cretaceous granitoids. The Middle Jurassic granitoids are the most prevalent, distributed throughout an extensive area approximately 350 km long and 200 km wide (Fig. 1). The areal extent of Middle Jurassic plutonism in South Korea is too large to be the result of simple subduction of

---

[1]Corresponding author; email: ocwhan@chonbuk.ac.kr

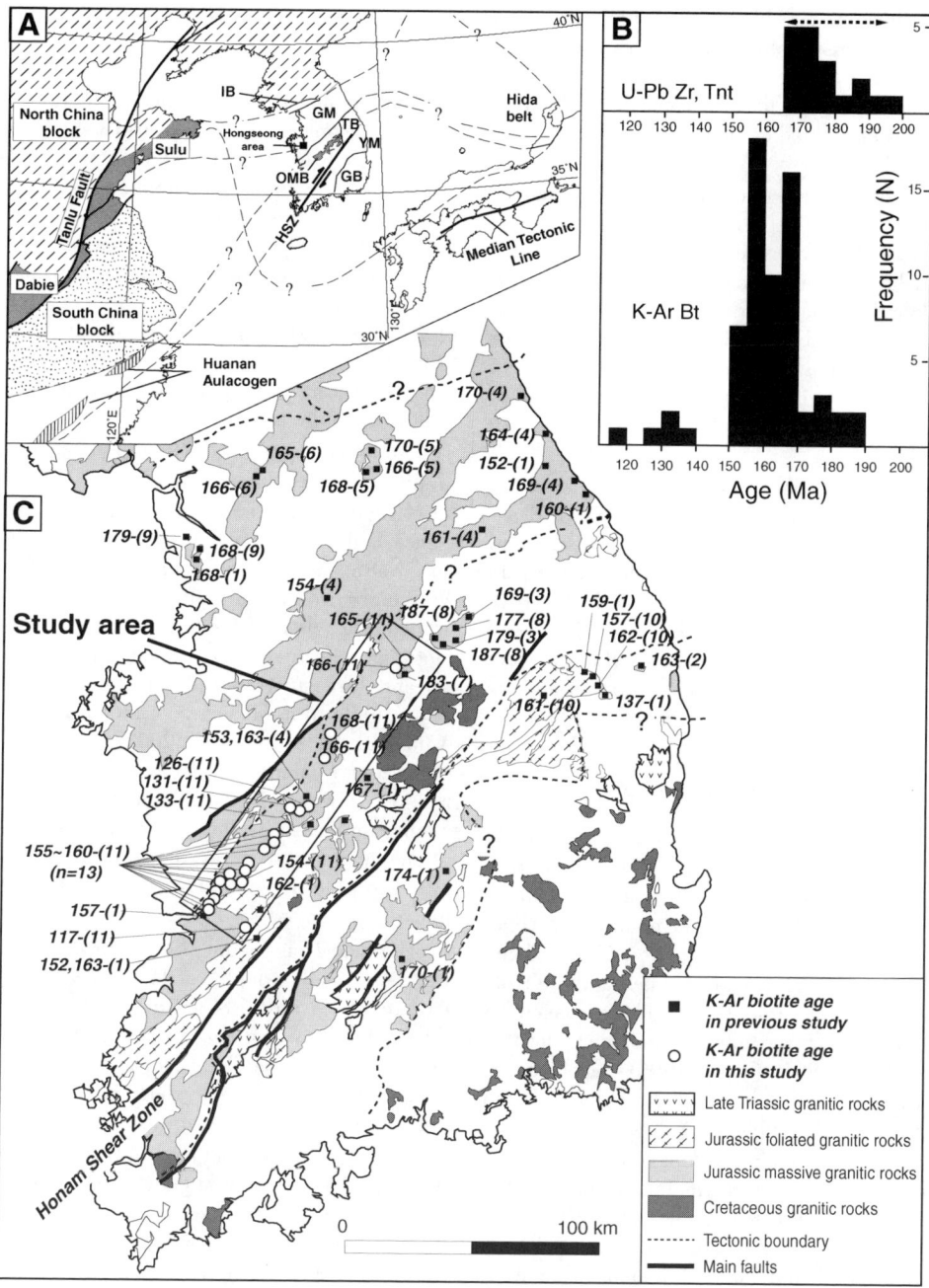

FIG. 1. A. Tectonic map of northeast Asia. B. Frequency distribution diagrams for U-Pb zircon and titanite and K-Ar biotite ages for the Jurassic granitoids. C. Distribution map of Mesozoic granitoids in South Korea (modified from Chough et al., 2000). A series of large-scale dextral strike-slip fault movements defines the Honam shear zone. Abbreviations: IB = Imjingang belt; GM = Gyeonggi massif; OMB = Okcheon metamorphic belt; TB = Taebacksan Basin; YM = Yeongnam massif; GB = Gyeongsang Basin. References of K-Ar biotite ages (in parentheses, Fig. 1C): 1 = Kim, 1971; 2 = Yun and Silberman, 1979; 3 = Shibata et al., 1983; 4 = Jin et al., 1984; 5 = Jin et al., 1992; 6 = Yun, 1995; 7 = Min et al., 1995; 8 = Jin et al., 1995; 9 = Kim et al., 1998; 10 = Jin and Jang, 1999; 11 = this study. References of conventional and SHIRIMP U-Pb zircon and titanite ages: Kim and Turek, 1996; Kim et al., 1999; Ree et al., 2001; Lee et al., 2003; Cheong et al., 2003b; C. B. Kim et al., 2003; Oh et al., 2004; this study.

oceanic lithosphere. Although regionally synchronous Middle Jurassic plutonism has been suspected to be related to the subduction of the Farallon-Izanagi ridge beneath the Asian plate, no detailed study has been carried out to constrain the origin of of this granitic belt.

Middle Jurassic granitic batholiths are present in and around the Okcheon metamorphic belt (OMB) in central South Korea; the map of the OMB resembles an island within a sea of granitoid bodies (Fig. 1). Middle Jurassic granitoids in and around the OMB have been subject to less attention than similar granitic batholiths elsewhere in South Korea. This lack of study is one of the obstacles to better understanding the origin of the Middle Jurassic granitoids.

This paper describes new whole-rock chemistry, trace element data, ion microprobe (SHRIMP II) U-Pb dating of zircons, and K-Ar dating of biotite and muscovite separates of Middle Jurassic granitoids in and around the Okcheon metamorphic belt. Petrochemical and geochronological data from the present and previous studies were used to interpret the magmatic evolution of these granites and its implication for northeast Asian tectonics.

Geologic Background

The OMB is a NE-trending fold-thrust belt about 150 km long and 20 km wide (Fig. 1), and consists mainly of metamorphosed volcanosedimentary rocks (e.g., Kim, 1971, 1996; Lee et al., 1998; Oh et al., 1998; Chough et al., 2000). The OMB has been correlated with the Hida marginal belt and the Hida metamorphic belt in southwest Japan, based on similar geochronology, lithology, and metamorphic histories (Hiroi, 1981). Chang (1996) suggested that the OMB is the continuation of the Huanan failed rift in the South China block.

Recent geochronological data have constrained the timing of the main M1 metamorphism in the OMB to 280–300 Ma (Kim et al., 2001; Cheong et al., 2003a). Oh et al. (2004) suggested that collision between the North and South China blocks occurred between 230 and 300 Ma in the Korean Peninsula, and that the OMB underwent intermediate-P/T type M1 metamorphism in the peripheral area of the collision zone. Sparse Middle to Late Triassic granites of uncertain origin occur mainly in the Precambrian Yeongnam massif (Fig. 1; Cho et al., 2001; C. B. Kim et al., 2003). In contrast, Middle Jurassic granites are quite abundant, and are distributed mainly in the Gyeonggi massif and OMB (Kim et al., 1999; Ree et al., 2001; Cheong et al., 2003b; C. B. Kim et al., 2003; Lee et al., 2003; Oh et al., 2004). This plutonism caused widespread Middle Jurassic M2 thermal metamorphism in the OMB (S. W. Kim et al., 2003; Oh et al., 2004). Oh et al. (2004) suggested that the Jurassic plutonism was caused by subduction of the Farallon-Izanagi plate. In the Early Cretaceous, regional plutonism and volcanism occurred in the Gyeongsang Basin and along the margin of the OMB due to subduction of the Izanagi plate (Lee, 1991; Maruyama et al., 1997; Okada, 2000). At the margin of the OMB, volcanism occurred within pull-apart basins along a sinistral strike-slip fault.

Late Triassic granitoids consist of coarse- to medium-grained hornblende-biotite granodiorite and K-feldspar megacryst-bearing biotite granite. In contrast, Middle Jurassic granitoids are mainly composed of coarse- to medium-grained biotite granite with minor two-mica granite and hornblende-biotite granodiorite, diorite, K-feldspar megacryst-bearing biotite granite, and leucocratic granite. Some Triassic and Jurassic granitoids are foliated, resulting from Middle Jurassic dextral strike-slip fault motion (Cho et al. 1999; Lee et al., 2003) along the Honam shear zone (Yanai et al., 1985; Cluzel et al., 1991; Fig. 1) between the OMB and the Precambrian Yeongnam massif. Cretaceous granitoids consisting of coarse- to medium-grained biotite granite and alkali granite are characterized by quartz and/or feldspar phenocrysts.

The Late Triassic and Middle Jurassic granitoids were emplaced at depths of 12–28 km, whereas the Cretaceous granitoids solidified at depths less than10 km (Cho and Kwon, 1994). The Cretaceous granitoids occur mainly as stocks with contemporaneous volcanic rocks; many have miarolitic pockets, indicating shallow levels of intrusion.

Based on a geochemical study of granitoids in the central and northwest OMB, Cheong and Chang (1996a, 1996b, 1997) suggested the following: (1) the Triassic granitoids formed as a result of subduction-related primitive arc magmatism, and reflect assimilation of crustal material; (2) the Middle Jurassic granitoids formed in a calc-alkaline continental arc environment, based on the comparatively high concentrations of large ion lithophile (LIL) element and lower Ta/Hf ratios. Lee (1991) suggested that the Cretaceous granitoids originated from the lower crust due to magmatism related to subduction of the Kula plate.

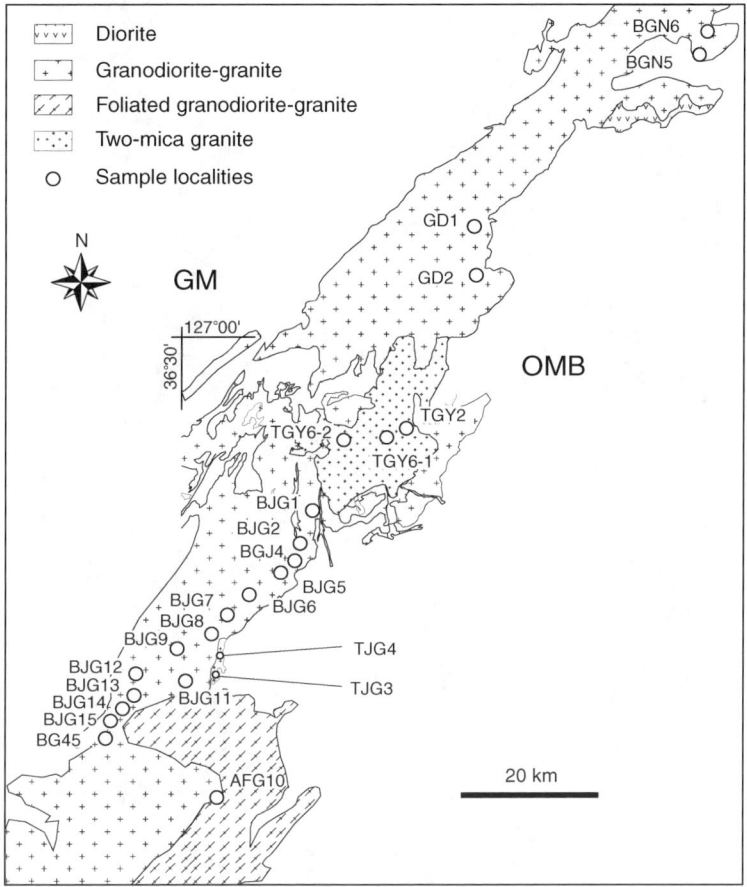

FIG. 2. Distribution map of Middle Jurassic granitoids and sample location map in the study area.

In the study area (Fig. 2), Jurassic granitoids have intruded the OMB and surrounding Gyeonggi massif, and consist of biotite granite, granodiorite, diorite, two-mica granite, and foliated biotite granite; biotite granite is most common, whereas granodiorite bodies are present in and around the central and northeastern OMB. Diorite is only present in and around the northeastern OMB. Two-mica granites intruded the southern part of the central OMB, and are massive or weakly foliated. A separate, small body of two-mica granite is also present in the southwestern OMB. Jurassic granitoids are strongly foliated along the Honam shear zone.

U-Pb Geochronology

SHRIMP II U-Pb analyses of zircons from a two-mica granite (TJG4, Fig. 2) in the southwest OMB were performed by Activation Laboratories Ltd., Canada. Zircon grains were hand-selected from heavy mineral concentrates and mounted in epoxy, together with chips of the FC1 (Duluth gabbro) and SL13 (Sri Lankan gem zircon) reference zircons. The zircons have slender, elongate, or lozenge shapes, most with pyramidal terminations (Fig. 3). Many of the grains contain opaque or dusty inclusions that define an inherited central domain. Each grain mount was cut in half and polished. Cathodoluminescence (CL) SEM images were prepared for all grains (Fig. 3).

Each U-Pb analysis of a zircon consisted of six scans through the mass range. The data were reduced according to the method of Williams (1998) using the SQUID Excel Macro of Ludwig (2000). The Pb/U ratios were normalized using 0.1859 for the $^{206}Pb/^{238}U$ ratio of the FC1 reference zircons,

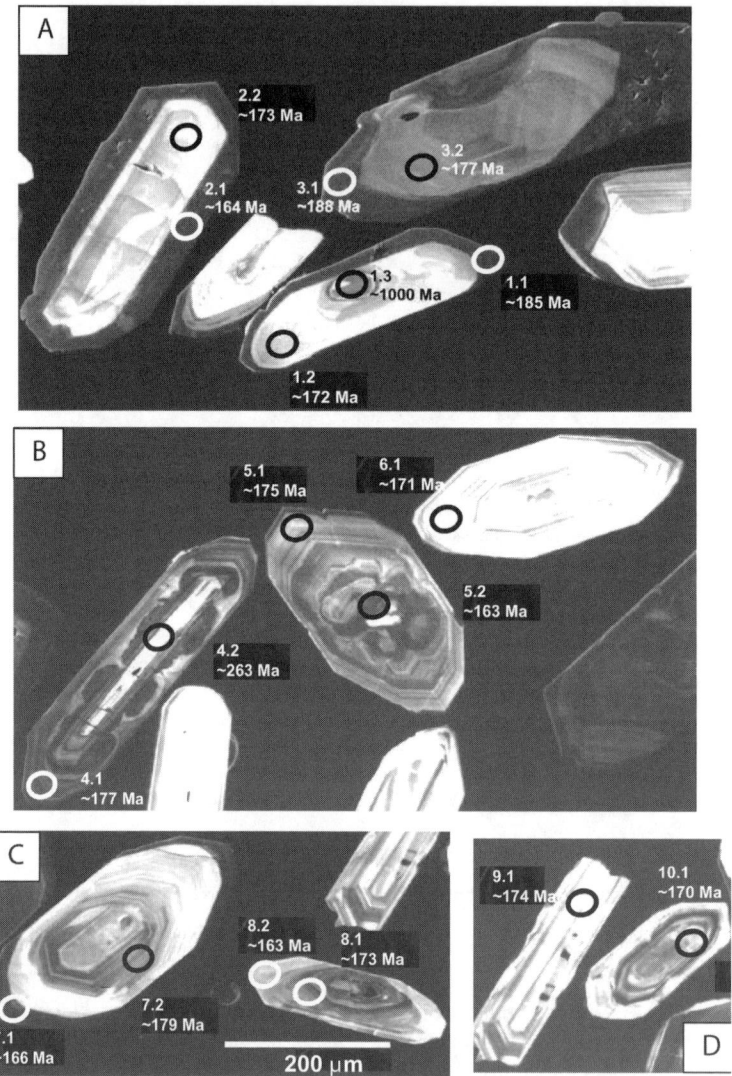

FIG. 3. Cathodoluminescence (CL) image of zircons analyzed from sample TJG4. Areas analyzed are shown with circle labeled as in Table 1.

whose age is 1099 Ma (Pearce and Miller, 1993). Uncertainties for individual analyses (ratio and ages) are at the 2σ level, and calculated weighted mean ages are within 95% confidence limits. The Tera and Wasserburg (1972) concordia plot, probability density plot, and weight mean age calculations were done using ISOPLOT/EX (Ludwig, 1999).

SHRIMP U-Pb results are presented in Table 1 and Figure 3. They record the occurrence of Grenville (998 Ma) and Permian (263 Ma) inherited components within Jurassic zoned igneous zircons that crystallized at 173 ± 3 Ma (Fig. 4). The significantly older $^{206}Pb/^{238}U$ ages in the inherited cores (analyses 1.3 and 4.2; Fig. 3 and Table 1) reflect the ages of protoliths in the source area. Two rims with high U content yield $^{206}Pb/^{238}U$ ages of 180 Ma (analysis 1.1 with 7700 ppm U) and 188 Ma (analysis 3.1 with 8600 ppm U). Therefore, these two ages are not considered to reflect the true times of crystallization (Fig. 3A and Table 1).

A dark CL rim with 7100 ppm U (analysis 2.1) gives a $^{206}Pb/^{238}U$ age of 164 Ma (Fig. 3A). Analysis

TABLE 1. Summary of SHRIMP U-Th-Pb Zircon Data for Sample TJG4

Grain spot	U, (ppm)	Th, (ppm)	$^{232}Th/^{238}U$	$^{206}Pb^*$ (ppm)	$^{204}Pb/^{206}Pb$	f_{206} %	Total $^{238}U/^{206}Pb$	±, %	Total $^{207}Pb^*/^{206}Pb^*$	±, %	Radiogenic $^{206}Pb^*/^{238}U$	±, %	Age, Ma $^{206}Pb/^{238}U$	±
1.1	7702	1070	0.14	194	0.000386	0.71	34.142	0.349	0.0550	0.0002	0.0291	0.0003	184.8	200
1.2	419	161	0.38	10	–	<0.01	37.016	0.468	0.0479	0.0010	0.0271	0.0003	172.2	53
1.3	1295	35	0.03	195	0.000010	4.64	5.693	0.123	0.1113	0.0012	0.1675	0.0039	998.4	53
2.1	7105	384	0.05	163	0.000950	3.85	37.367	0.428	0.0800	0.0088	0.0257	0.0004	163.8	35
2.2	549	179	0.33	13	0.000128	0.01	36.877	0.431	0.0496	0.0009	0.0271	0.0003	172.5	36
3.1	8566	332	0.04	220	0.000319	0.95	33.438	0.372	0.0574	0.0003	0.0296	0.0003	188.2	35
3.2	1233	89	0.07	29	0.000133	<0.01	36.038	0.426	0.0493	0.0006	0.0278	0.0003	176.5	73
4.1	1965	291	0.15	47	0.000016	0.01	35.862	0.392	0.0497	0.0005	0.0279	0.0003	177.3	34
4.2	1196	236	0.20	43	–	<0.01	24.028	0.355	0.0512	0.0005	0.0416	0.0006	262.9	45
5.1	768	115	0.15	18	–	0.02	36.428	0.406	0.0497	0.0007	0.0274	0.0003	174.5	39
5.2	1319	35	0.03	29	0.000213	0.74	39.011	0.424	0.0552	0.0006	0.0254	0.0003	162.0	49
6.1	358	122	0.34	8	–	0.34	37.007	0.457	0.0522	0.0011	0.0269	0.0003	171.3	39
7.1	1007	404	0.40	23	–	0.08	38.385	0.425	0.0500	0.0007	0.0260	0.0003	165.7	35
7.2	1078	191	0.18	26	0.000076	0.42	35.419	0.384	0.0530	0.0006	0.0281	0.0003	178.7	35
8.1	1898	230	0.12	44	–	<0.01	36.750	0.392	0.0492	0.0005	0.0272	0.0003	173.1	46
8.2	584	113	0.19	13	–	0.11	38.979	0.545	0.0502	0.0009	0.0256	0.0004	163.1	35
9.1	554	257	0.46	13	–	0.05	36.580	0.427	0.0500	0.0009	0.0273	0.0003	173.8	35
10.1	1355	169	0.12	31	–	<0.01	37.365	0.402	0.0487	0.0006	0.0268	0.0003	170.4	35

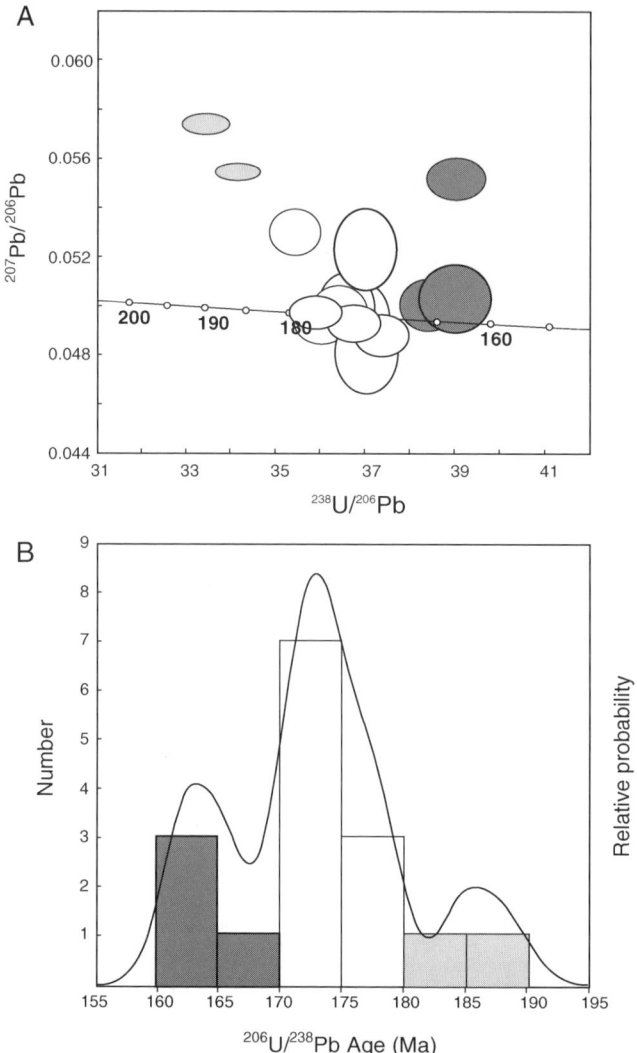

FIG. 4. A. Enlarged Tera-Wasserburg concordia plot of the SHRIMP U-Pb zircon analyses from sample TJG4. Analyses plotted as total ratios, uncorrected for common Pb and are shown as one σ error ellipses. B. Probability density plot, with stacked histogram, for the ^{206}Pb/^{238}U zircon ages from sample TJG4.

7.1 is also a dark rim under CL, but has 1010 ppm U with a ^{206}Pb/^{238}U age of 166 Ma (Fig. 3C). A bright-luminant rim with low U (analysis 8.2) records a similar ^{206}Pb/^{238}U age of 163 Ma (Fig. 3C). The weighted mean age for the three rims is 165 ± 3 Ma (Fig. 4B), which is interpreted as the crystallization age of the youngest component in this suite of zircons.

All other areas analyzed are bright-luminant under CL (relative to dark CL rims that have very high U), and have U concentrations that range from 420 to 2000 ppm. These areas are generally low in common Pb and form concordant clusters near the enlarged Tera-Wasserburg concordia (Fig. 4A). They give a dominant age group averaging 173 ± 3 Ma on the probability density histogram (Fig. 4B).

In South Korea, recently obtained U-Pb ages of Jurassic granitoids range from 165 to 179 Ma except for some 183–195 Ma for foliated granitoids at the boundary between the OMB and the Yeongnam massif (Fig. 1B). U-Pb ages obtained from Jurassic granitoids in South Korea are as follows: 166-179

Ma conventional and SHRIMP U-Pb zircon ages from biotite granite (Kim and Turek, 1996; Lee et al., 2003; C. B. Kim et al., 2003; Oh et al., 2004), 173–176 Ma conventional and SHRIMP U-Pb zircon ages from foliated granitoids (Lee et al., 2003; C. B. Kim et al., 2003) and 172–175 Ma conventional U-Pb titanite ages from biotite granite in the OMB (Ree et al., 2001; Cheong et al., 2003b), 165 Ma conventional U-Pb zircon age from granitoids in the Gyeonggi massif (Kim et al., 1999), and 177–179 Ma conventional U-Pb zircon age from granitoids in the Yeongnam massif (Kim and Turek, 1996; C. B. Kim et al., 2003). The 173 ± 3 Ma age for two-mica granite in the southwestern OMB, together with these U-Pb ages from Jurassic granitoids in South Korea, strongly point out a regional plutonism in South Korea during the Middle Jurassic.

K-Ar Geochronology

K-Ar age dating on biotite and muscovite from granitoids collected in the OMB was carried out using the method described by Itaya et al. (1991), at Okayama University of Science, Japan; the results are shown in Figure 2. Rock samples were crushed with a jaw-crusher and then sieved to obtain an appropriate size fraction. The sieved fraction was washed with deionized water and dried at 70°C. The residue was washed repeatedly with distilled water, employing both an ultrasonic bath and a centrifuge. Analyses of potassium and argon in the biotite and muscovite separates, and calculation of age and error, were carried out using the method described by Nagao et al. (1984) and Itaya et al. (1991). Potassium was analyzed by flame photometry using a 2000-ppm Cs buffer; the concentrations listed in Table 2 have an analytical error within 2% at a 2-6 confidence level. Argon was analyzed on a 150 mm radius sector-type mass spectrometer with a single collector system, using an isotopic dilution method and argon 38 spike (Itaya et al., 1991). Multiple runs of standard JG-1 (91 Ma) biotite indicated that the error of argon analysis is about 1% at the 2σ confidence level. The decay constants for ^{40}K to ^{40}Ar and for ^{40}K to ^{40}Ca, and the ^{40}K content in the potassium used in the age calculation, are 0.581×10^{-10}/y, 4.962×10^{-10}/y, and 0.0001167, respectively (Steiger and Jäger, 1977).

K-Ar age results are listed in Table 2. K-Ar ages for biotite from Jurassic granitoids (BGN5 and BGN6) in the northwest OMB are 165-166 Ma. In the middle part of the central OMB, biotite K-Ar ages from amphibole-biotite Jurassic granodiorite (GD1 and GD2) are 166–168 Ma, similar to biotite ^{40}Ar/^{39}Ar plateau ages of 160–168 Ma, reported from biotite granite and amphibole-biotite granodiorite in the same region (S. W. Kim et al., 2003). On the other hand, biotite separates from Jurassic two-mica granite in the central OMB (TGY2, TGY6-1, and TGY6-2) gave younger ages of 126–133 Ma. In general, the difference between the conventional and SHRIMP U-Pb zircon ages and K-Ar biotite ages for the granitoids in the OMB is around 10 Ma. Therefore, it is possible that the younger two-mica granites were emplaced during Late Jurassic time and are different from the widespread Middle Jurassic granitoids. In the southwest OMB, K-Ar biotite ages in the Middle Jurassic biotite granites (BJG) lie in the range 150–159 Ma and muscovite ages from a Jurassic two-mica granite (TJG3 and TJG4) are 156–162 Ma. The foliated granite (AFG10) gave a younger K-Ar biotite age of 117 Ma but has SHRIMP U-Pb zircon ages of 173 Ma (Lee et al., 2003), within the range of intrusion ages for Middle Jurassic granitoids (167–173 Ma). Younger biotite ages in the foliated granite may indicate the timing of final, strong ductile shearing along the Honam shear zone, instead of cooling ages.

K-Ar biotite ages of Middle Jurassic granitoids in the central and northwest OMB (165–168 Ma) are older than those in the southwest OMB (150–159 Ma). Compared to the intrusion age of the plutons (167–173 Ma), these data can be interpreted as the age of uplift of the Middle Jurassic granitoids; the age distribution indicates that uplifting becomes younger towards the southwest. The K-Ar biotite ages reported from Middle Jurassic granitoids in the Gyeonggi and Yeongnam massifs range from 187 to 154 Ma, concentrating between 170 and 152 Ma (Fig. 1B). K-Ar biotite ages indicate that the Middle Jurassic granitoids were uplifted mainly between 150 and 170 Ma in South Korea.

Geochemistry of Granitoids around the OMB

Whole-rock geochemical analyses were carried out for nine biotite granites, three two-mica granites, and two amphibole-biotite foliated granites around the southwest OMB area, and four two-mica granites and three granodiorites around the central OMB (Table 3 and Fig 2). Major-, trace-, and rare earth–element (REE) concentrations were determined by ICP-AES (Termo Jarrel-Ash ENVIRO II) and ICP-

TABLE 2. K-Ar Age Data of Granitic Rocks of the Northern Okcheon Metamorphic Belt

Area	Sample no.	Mineral	Fraction	Potassium, (wt%)	Rad. ^{40}Ar (10^{-8}ccSTP/g)	Age, Ma	Non-rad. ^{40}Ar, (%)
SW OMB				Biotite granite			
	BJG1	Biotite	104-147μm	6.25±0.13	3953±39	155.9±3.3	1.7
	BJG2	Biotite	104-147μm	7.33±0.15	4626±45	155.7±3.3	0.7
	BJG4	Biotite	104-147μm	7.27±0.15	4614±54	156.5±3.5	1.0
	BJG5	Biotite	104-147μm	6.93±0.14	4402±48	156.8±3.4	0.7
	BGJ6	Biotite	104-147μm	6.94±0.14	4381±54	155.8±3.5	1.7
	BJG7	Biotite	104-147μm	7.40±0.15	4671±64	155.7±3.6	1.2
	BJG8	Biotite	104-147μm	6.82±0.14	4361±54	157.7±3.5	1.5
	BJG9	Biotite	104-147μm	7.17±0.14	4562±62	156.9±3.6	0.7
	BJG11	Biotite	104-147μm	7.36±0.15	4760±56	159.4±3.5	1.1
	BJG12	Biotite	104-147μm	7.40±0.15	4712±58	157.1±3.5	0.7
	BJG13	Biotite	104-147μm	7.43±0.15	4677±55	155.3±3.5	1.1
	BJG14	Biotite	104-147μm	5.83±0.12	3638±43	154.1±3.4	1.3
	BJG15	Biotite	104-147μm	6.33±0.13	4012±47	156.4±3.5	1.7
	BG45	Biotite	104-147μm	7.30±0.15	4636±58	156.6±3.5	1.1
				Two-mica granite			
	TJG3	Muscovite	104-147μm	8.43±0.17	5396±53	155.7±3.4	0.8
	TJG4	Muscovite	104-147μm	8.44±0.17	5543±65	161.7±3.6	0.8
				Amphibole-biotite foliated-granite			
	AFG10	Biotite	104-147μm	4.86±0.10	2281±27	117.0±2.6	1.9
				Two-mica granite			
C OMB	TGY2	Biotite	104-147μm	7.43±0.15	3864±45	126.0±2.8	2.6
	TGY6-1	Biotite	104-147μm	6.49±0.13	3410±41	130.5±2.9	1.9
	TGY6-2	Biotite	104-147μm	6.61±0.13	3543±44	133.1±3.0	1.9
				Granodiorite			
	GD1	Biotite	104-147μm	6.81±0.13	4653±45	167.6±3.5	1.2
	GD2	Biotite	104-147μm	6.89±0.14	4637±32	166.3±3.5	1.1
				Biotite granite			
NW OMB	BGN5	Biotite	104-147μm	6.40±0.13	4278±50	164.5±3.6	0.9
	BGN6	Biotite	104-147μm	7.21±0.15	4863±49	166.1±3.5	1.0

MS (Perkin Elmer Optima 3000) at Activation Laboratories Ltd., Canada. Major-element, trace-element, and REE abundances of the analyzed Jurassic granitoids are listed in Table 3. To evaluate geochemical characteristics of Jurassic plutonism in and around the OMB, the geochemical data obtained in this study are interpreted together with geochemical data for Jurassic granitoids analyzed by Cheong and Chang (1996b; Figs. 5–10).

Biotite, two-mica, and foliated biotite granites in the southwest OMB have 70.4–72.1%, 73.8–74.2%, and 70.4–70.6% SiO_2, respectively (Fig. 5A). SiO_2

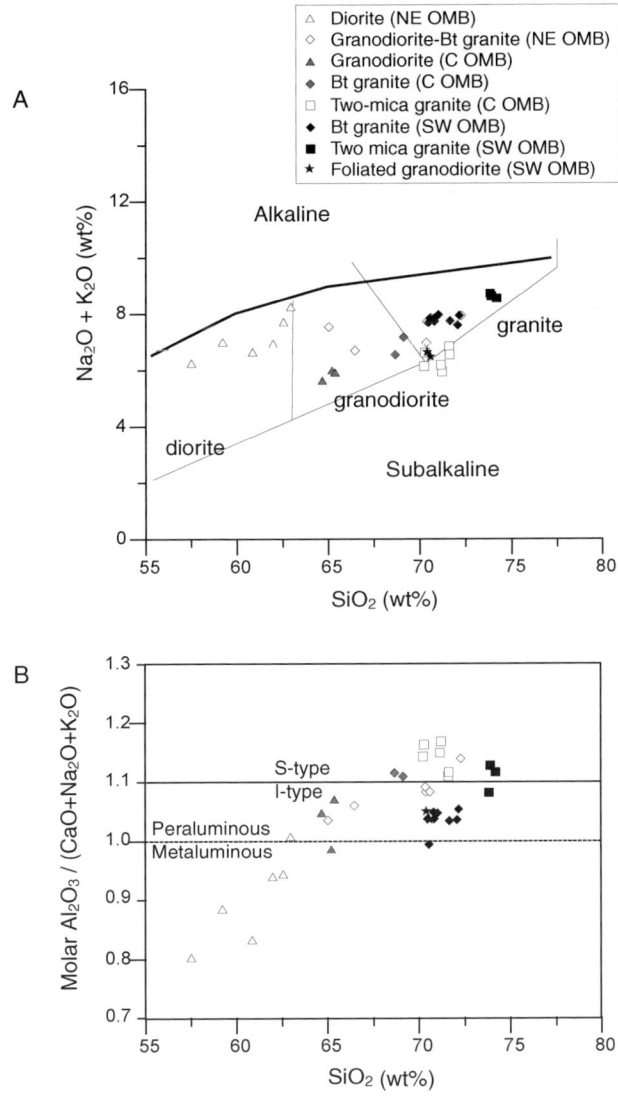

FIG. 5. A. Alkali vs. silica diagram for Middle Jurassic granitoids in and around the OMB. B. Molar Al_2O_3/$(CaO+Na_2O+K_2O)$ vs. SiO_2 diagram for Middle Jurassic granitoids.

contents of two-mica granite and granodiorite of the central OMB are 71.1–71.6% and 64.7–69.0%, respectively (Fig. 5A). Diorite, granodiorite, and granite from the northeastern OMB area have 57.5–63.0%, 65.0–66.5%, and 70.4–72.3% SiO_2, respectively. The range of SiO_2 content increases from northeast to southwest. All Jurassic granitoids in the OMB occupy the subalkaline field on the Na_2O + K_2O vs. SiO_2 diagram (Fig. 5A; Irvine and Baragar, 1971). SiO_2 content shows a negative correlation with P_2O_5, MgO, Fe_2O_3*, CaO, and TiO_2, a weak positive correlation with Na_2O, and no apparent correlation with K_2O and Al_2O_3 (Fig. 6). Diorites and most granodiorites and granites are I-type, whereas one granite, two granodiorites, and most two-mica granites are S-type (Fig. 5B). Diorites are metaluminous, but the other rock types are mostly peraluminous.

Trace elements for Jurassic granitoids are plotted against SiO_2 (Fig. 7). Y, Sr, Ba, and V show a

TABLE 3. Major and Trace Element Compositions of Granitoids in the Northern OMB

	SW OMB Biotite granite								SW OMB Amphibole-biotite foliated granite	
	BJG2	BJG4	BJG5	BJG6	BJG7	BJG8	BJG9	BJG11	AFG10	AFG13
SiO_2	70.54	70.48	70.79	72.16	70.83	71.65	72.07	71.01	70.39	70.57
TiO_2	0.336	0.431	0.403	0.347	0.406	0.37	0.346	0.406	0.385	0.342
Al_2O_3	14.74	15.11	15.21	14.87	15.43	14.6	14.26	15.38	14.83	14.58
FeO*	2.52	2.55	2.38	1.97	2.37	2.31	2.18	2.27	2.73	2.85
MnO	0.028	0.025	0.025	0.019	0.024	0.023	0.023	0.025	0.047	0.043
MgO	0.56	0.56	0.5	0.35	0.51	0.47	0.42	0.48	0.99	0.92
CaO	2.24	2.25	2.18	1.84	2.29	2.02	1.92	2.17	2.67	2.72
Na_2O	4.05	3.84	3.88	3.88	3.93	3.67	3.66	3.77	3.69	3.75
K_2O	3.78	3.84	3.84	4.05	3.92	4.07	3.93	4.19	2.95	2.73
P_2O_5	0.1	0.13	0.13	0.09	0.13	0.11	0.1	0.12	0.1	0.11
LOI	0.36	0.44	0.52	0.44	0.24	0.43	0.35	0.39	0.99	0.95
Total	99.27	99.66	99.85	100.01	100.09	99.72	99.28	100.21	99.78	99.565
Ba	817	1174	1104	1321	1284	1083	998	1146	532	636
Sr	452	533	515	511	614	486	435	448	377	438
Zr	138.87	184.38	183.74	173.29	185.68	181.29	175.62	184.35	144.61	137.90
Ni		5.14	5.83	3.15	2.39	3.16	1.86	1.96	8.44	8.78
Ga	23.32	22.99	24.05	21.68	22.77	21.93	22.18	23.04	21.33	20.79
V	25	27	20	16	23	18	19	12	23	21
Hf	3.93	5.12	5.37	5.22	5.05	5.42	5.12	5.45	4.36	4.31
Nb	9.18	8.79	8.66	9.20	8.89	8.88	9.37	11.04	13.08	13.28
Ta	0.68	0.93	1.16	0.89	0.61	0.72	0.89	0.92	1.63	1.73
Rb	154.77	162.30	174.43	157.32	140.76	154.99	166.31	166.69	150.49	167.57
Y	4	7	7	5	4	7	6	6	10	11
Cs	3.10	3.07	3.31	3.34	2.52	1.86	3.57	3.10	3.98	3.57
U	3.18	5.79	3.76	2.81	2.78	5.33	2.42	2.77	1.91	1.90
Th	11.56	20.52	20.55	17.74	15.43	21.72	21.36	18.21	16.72	17.09
Pb	21.02	23.03	25.20	32.81	25.35	27.78	29.65	26.45	27.35	29.56
La	24.95	58.40	59.30	52.96	47.65	55.33	51.29	56.47	38.31	31.35
Ce	44.61	103.82	107.12	95.14	87.13	100.35	92.28	100.72	65.73	63.13
Pr	5.24	10.07	10.28	9.28	8.65	9.71	8.86	9.71	6.15	5.87
Nd	19.30	35.36	36.29	33.11	31.14	34.19	31.38	34.20	21.05	20.13
Sm	3.55	5.71	6.10	5.63	5.01	5.40	5.27	5.79	3.48	3.40
Eu	0.84	1.25	1.20	1.18	1.25	1.16	1.08	1.24	0.85	0.76
Gd	2.32	3.77	4.08	3.49	3.24	3.58	3.51	3.78	2.85	2.75
Tb	0.24	0.39	0.47	0.33	0.31	0.35	0.35	0.39	0.35	0.32
Dy	1.06	1.49	1.92	1.19	1.13	1.28	1.29	1.49	1.74	1.55
Ho	0.14	0.21	0.28	0.16	0.15	0.18	0.18	0.20	0.31	0.28
Er	0.37	0.57	0.71	0.41	0.40	0.46	0.46	0.53	0.88	0.78
Tm	0.04	0.07	0.09	0.05	0.05	0.06	0.06	0.07	0.13	0.12
Yb	0.28	0.43	0.52	0.30	0.33	0.35	0.37	0.40	0.83	0.76
Lu	0.04	0.06	0.07	0.04	0.05	0.05	0.05	0.05	0.13	0.12
Eu/Eu*	0.90	0.82	0.73	0.81	0.95	0.81	0.77	0.81	0.83	0.76
$(La/Yb)_N$	59.53	90.72	77.33	117.49	98.36	105.59	94.46	96.25	31.04	27.85

Table continues

Table 3. *Continued*

	SW OMB Muscovite-biotite granite			Central OMB Muscovite-biotite granite				Central OMB Granodiorite		
	TJG3-1	TJG3-2	TJG4	TGY6-1	TGY6-2	TGY1	TGY2	GD1	GD1-1	GD2
SiO_2	73.91	73.84	74.19	71.6	71.62	71.2	71.14	64.68	65.21	65.37
TiO_2	0.059	0.118	0.085	0.196	0.196	0.221	0.226	0.697	0.506	0.592
Al_2O_3	14.67	14.32	14.41	15.79	15.84	16.17	16.21	16.58	15.81	16.37
FeO*	1.14	1.4	1.28	1.84	1.77	1.96	2.04	5.39	4.96	4.83
MnO	0.039	0.033	0.034	0.041	0.044	0.042	0.043	0.085	0.057	0.068
MgO	0.11	0.19	0.13	0.47	0.47	0.57	0.55	1.96	1.83	1.78
CaO	0.79	1	0.86	2.73	2.88	3.02	3	4.33	4.17	3.83
Na_2O	3.99	3.52	3.71	3.32	3.33	3.41	3.49	3.3	3.5	3.41
K_2O	4.63	5.2	4.84	3.52	3.21	2.53	2.69	2.33	2.49	2.51
P_2O_5	0.1	0.09	0.09	0.09	0.09	0.12	0.13	0.22	0.21	0.19
LOI	0.57	0.35	0.43	0.57	0.56	0.47	0.46	0.57	0.43	0.54
Total	100.00	100.05	100.059	100.17	100.01	99.713	99.97	100.15	99.173	99.49
Ba	177	445	100.059	960	823	812	768	799	783	946
Sr	49	137	118	475	479	485	486	463	461	547
Zr	49.20	99.46	59.24	120.71	112.98	124.35	127.46	239.41	219.26	235.88
Ni	1.50	1.13	0.21	3.17	2.68	3.21	3.93	7.69	8.34	7.06
Ga	24.26	22.88	23.12	21.03	20.55	21.42	21.57	22.77	23.17	21.77
V	−5	−5	−5	12	9	15	17	64	57	71
Hf	1.91	3.56	2.96	3.61	3.58	3.71	3.88	6.40	6.46	6.38
Nb	24.25	21.32	24.85	12.58	12.85	13.66	13.71	14.33	13.98	13.26
Ta	6.11	6.83	6.69	1.16	1.24	1.26	1.28	0.80	0.73	0.81
Rb	432.03	307.25	411.37	139.94	117.40	121.63	113.53	130.00	140.49	104.00
Y	11	8	10	11	10	11	11	15	16	15
Cs	24.08	16.41	17.19	5.12	5.35	3.69	3.48	6.97	5.76	6.34
U	4.11	4.22	4.20	2.20	1.29	1.48	1.32	1.76	1.59	1.68
Th	14.43	28.09	25.99	7.21	7.18	8.54	8.71	9.80	9.76	8.57
Pb	49.25	45.79	47.21	25.45	28.75	26.84	24.95	18.31	16.08	19.37
La	18.52	38.51	25.90	33.34	31.45	36.95	38.44	40.64	37.56	43.83
Ce	34.17	67.01	56.24	58.12	55.37	62.97	65.54	72.13	67.38	85.34
Pr	3.99	7.67	6.60	5.46	5.31	5.91	6.22	8.73	8.03	9.13
Nd	13.74	26.52	23.90	19.25	19.02	22.04	21.79	32.99	27.88	24.54
Sm	3.25	4.75	4.79	3.45	3.42	3.78	3.92	6.30	6.02	5.81
Eu	0.24	0.54	0.44	1.03	1.04	1.03	1.02	1.52	1.43	1.33
Gd	2.44	3.08	2.99	2.83	2.77	3.13	3.17	4.75	4.56	4.08
Tb	0.40	0.36	0.38	0.40	0.39	0.40	0.44	0.63	0.59	0.58
Dy	2.06	1.68	1.88	1.95	1.98	2.10	2.14	3.27	3.07	3.05
Ho	0.30	0.24	0.24	0.35	0.36	0.36	0.37	0.55	0.57	0.50
Er	0.78	0.62	0.69	0.92	1.04	1.04	1.04	1.49	1.39	1.27
Tm	0.10	0.08	0.08	0.13	0.14	0.14	0.14	0.20	0.19	0.17
Yb	0.60	0.49	0.52	0.84	0.88	0.87	0.86	1.27	1.21	1.17
Lu	0.08	0.07	0.07	0.13	0.13	0.12	0.12	0.18	0.17	0.16
Eu/Eu*	0.26	0.43	0.35	1.01	1.04	0.91	0.89	0.85	0.84	0.83
$(La/Yb)_N$	20.86	53.15	33.40	26.77	24.22	28.77	30.36	21.71	20.99	25.35

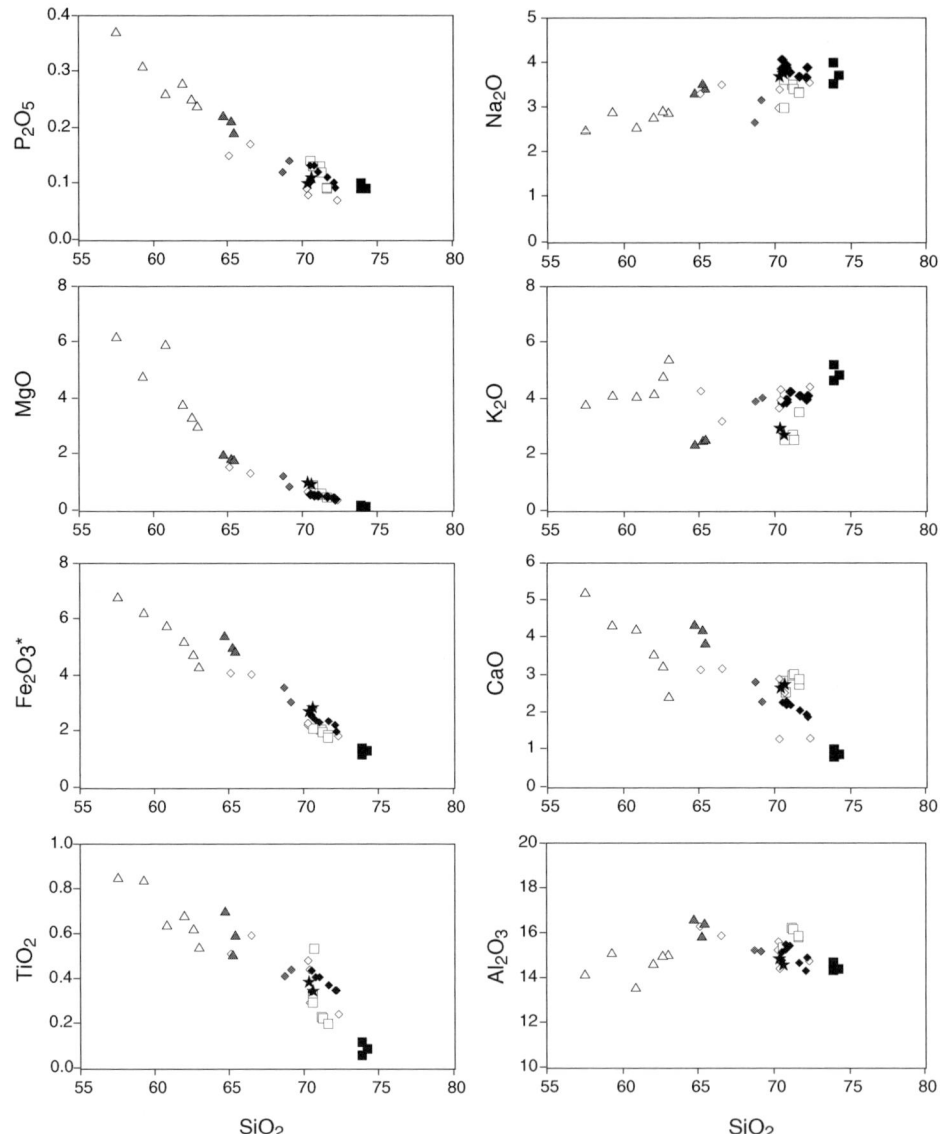

FIG. 6. Harker variation diagrams for Middle Jurassic granitoids in and around the OMB. Symbols are the same as those in Figure 5.

negative correlation with SiO_2, reflecting a progressive decrease from diorite through granodiorite to granite. In contrast, Nb shows a positive correlation, and increases from diorite to granite. Rb, Zr, and Ta do not display any correlation with SiO_2.

On a trace element abundance diagram normalized to MORB by Pearce (1983, Fig. 8), Jurassic granitoids all display similar characteristics, with the decoupled LIL/HFS (HFS = high-field-strength) elements pattern, Ta-Nb trough, and depletion in Y, Yb, and Ti as a characteristic of subduction zone magmas (except for two-mica granites around the southwestern OMB). Depletions in Y and Yb are stronger in granite and granodiorite than in diorite

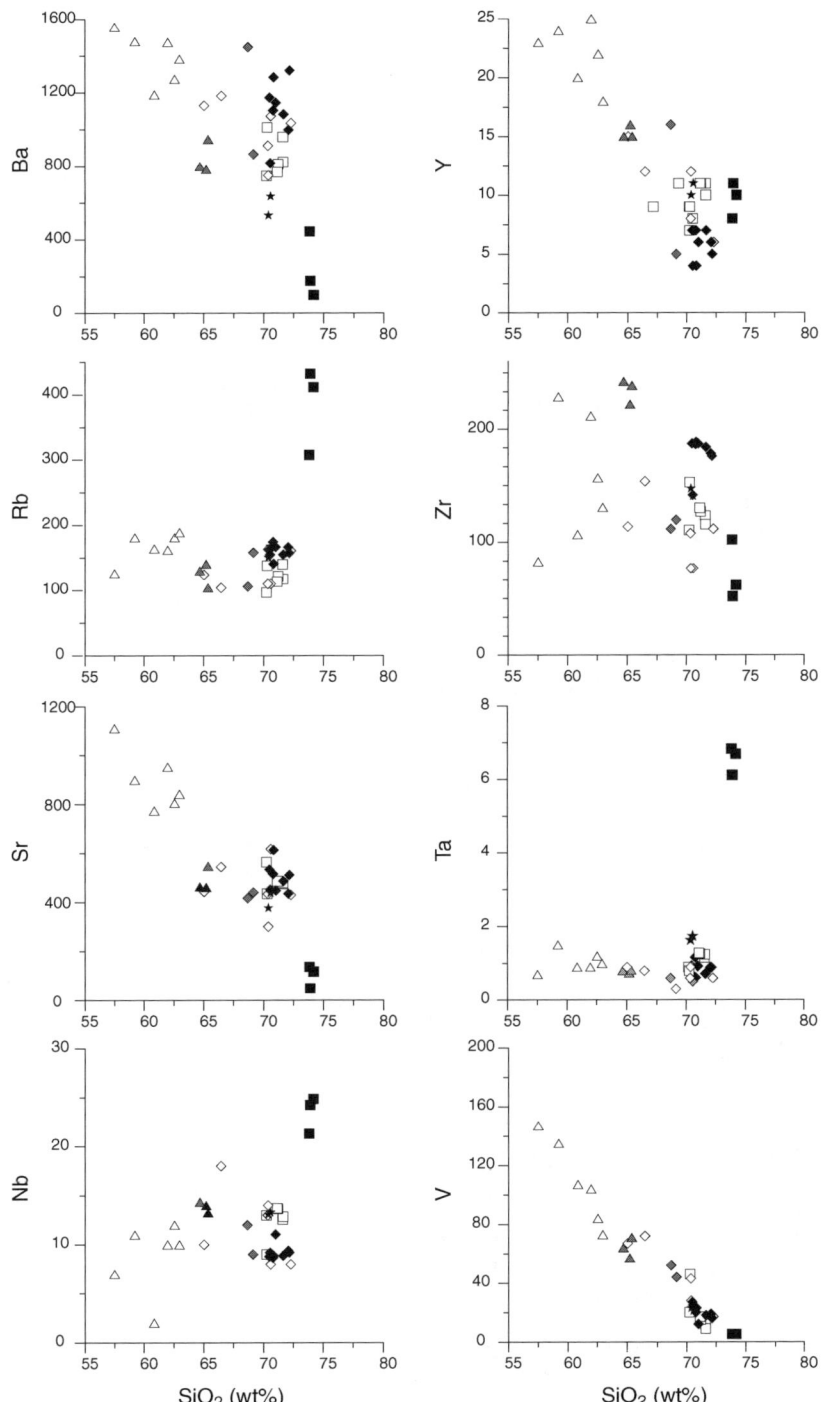

FIG. 7. Trace element variation diagrams for Middle Jurassic granites. Symbols are the same as those in Figure 5.

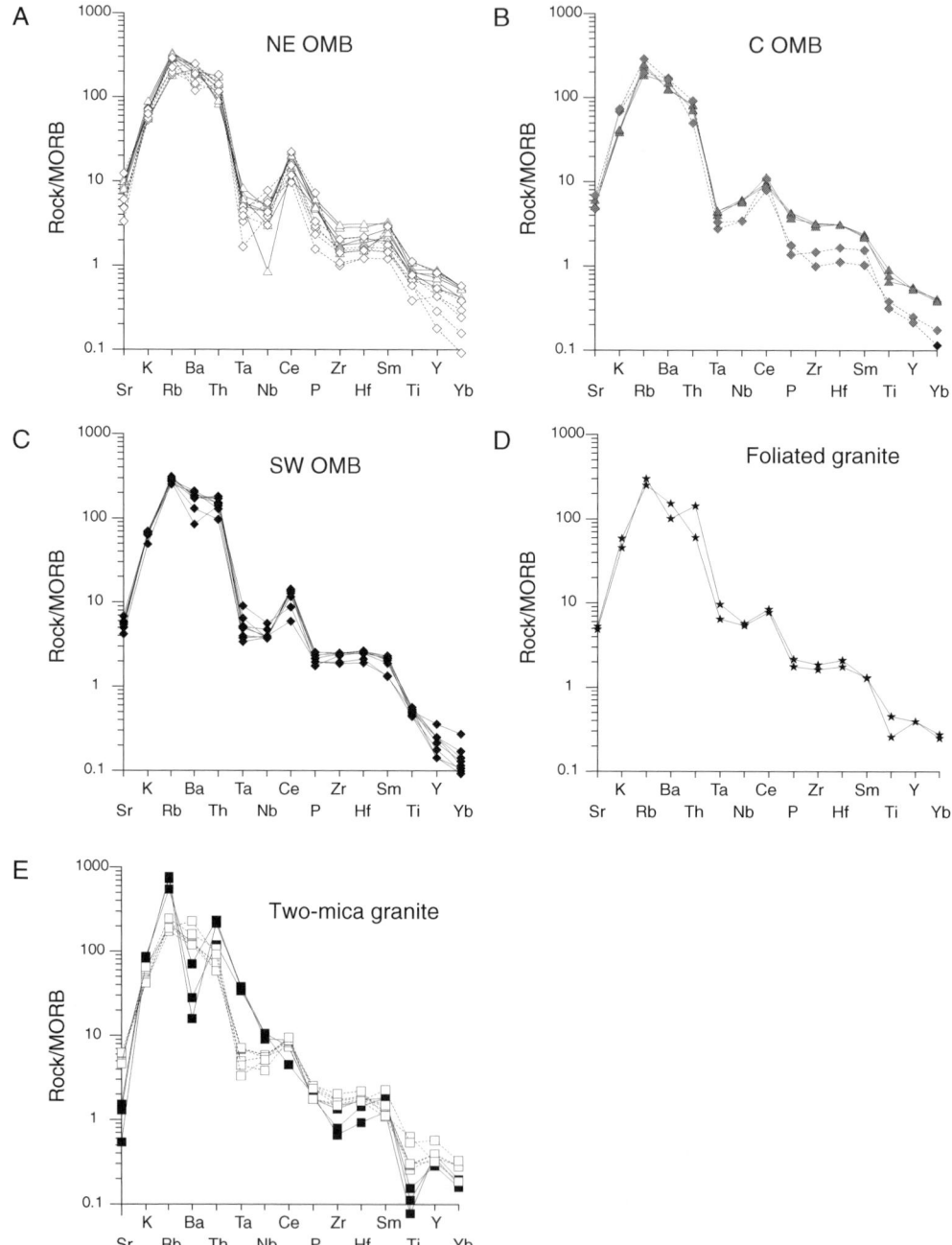

FIG. 8. Chondrite-normalized REE patterns of Middle Jurassic granitoids. Symbols are the same as those in Figure 5.

(Fig. 8). Two-mica granites in the southwestern OMB clearly exhibit high Rb, Ta, and Nb contents and low Zr and Ti contents (Fig. 8).

Jurassic biotite granites in and around the OMB show similar chrondrite-normalized REE patterns (Fig. 9). Biotite granites are depleted in heavy

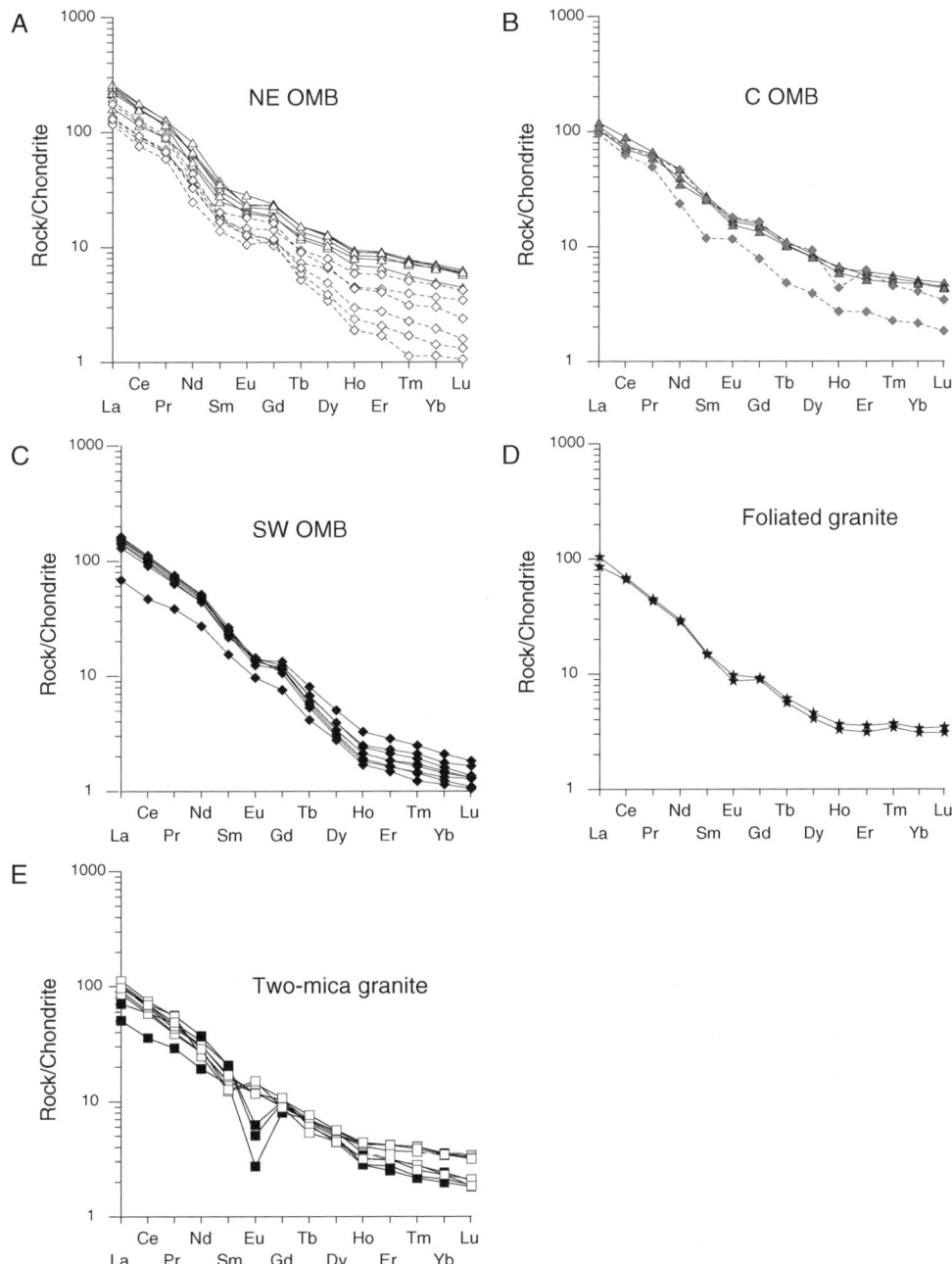

FIG. 9. MORB-normalized spider diagram (Pearce, 1983) for Middle Jurassic granitoids to the north of the OMB. Symbols are same as those in Figure 5.

REE (HREE), as compared with Jurassic diorite, granodiorite, and foliated granite (Fig. 9). Two-mica granites in the southwest OMB show strongly negative Eu anomalies (Eu/Eu* = 0.26-0.43, Table 3); this indicates fractionation of plagioclase from the melt or inheritance from the source material. In

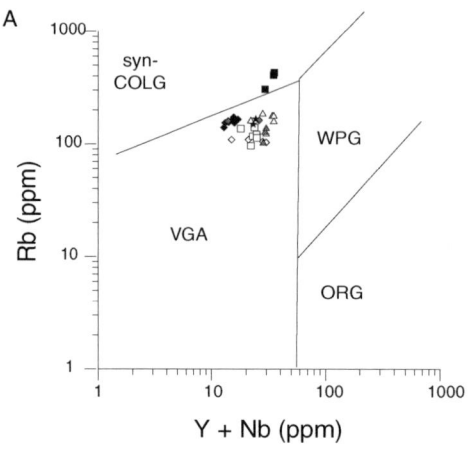

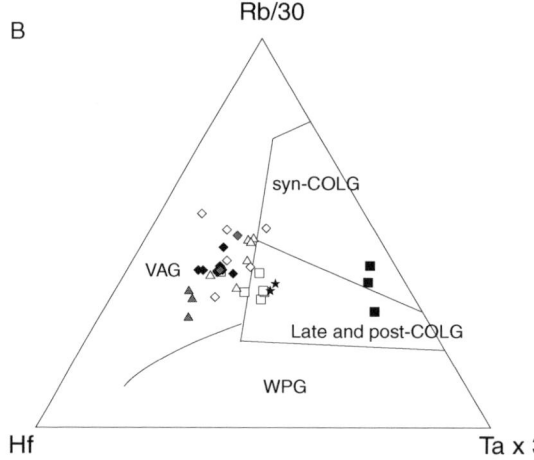

FIG. 10. A. Rb vs. Y + Nb (Pearce et al., 1984). B. Rb/30-Hf-Ta×3 (Harris et al., 1986) diagrams for Middle Jurassic granites. Abbreviations: VAG = volcanic-arc granite; syn-COLG = syn-collisional granite; ORG = oceanic-ridge granite; WPG = within-plate granite. Symbols are the same as those in Figure 5.

contrast, Late Jurassic two-mica granites in the central OMB have only weakly negative or positive Eu anomalies (Eu/Eu* = 0.89-1.41, Table 1).

Patterns of major and trace elements, together with age-dating results, indicate that most of the Jurassic granitoids (except for the two-mica granite) in and around the OMB originated from the same magma source, and evolved from diorite through granodiorite to granite. Middle Jurassic two-mica granite, however, seems to have been affected by crustal assimilation that resulted in higher Rb and Ta, which are abundant in upper crustal rocks.

According to the classification of Pearce et al. (1984), all of the Jurassic granitoids, with the exception of the two-mica granites in the southwestern OMB, occupy the VAG field of the Y + Nb-Rb diagram (Fig. 10). In the Hf-Rb/30-Ta×3 diagram (Harris et al., 1986), most of the Jurassic granitoids occupy the VAG field, with the exception of the foliated granites and two-mica granites (Fig. 10). As described above, two-mica granites are S-type and are enriched in Rb and Ta due to crustal assimilation. Such assimilation resulted in the two-mica granites plotting in the collision field. Plotting of foliated granite in the late and post-collisional field

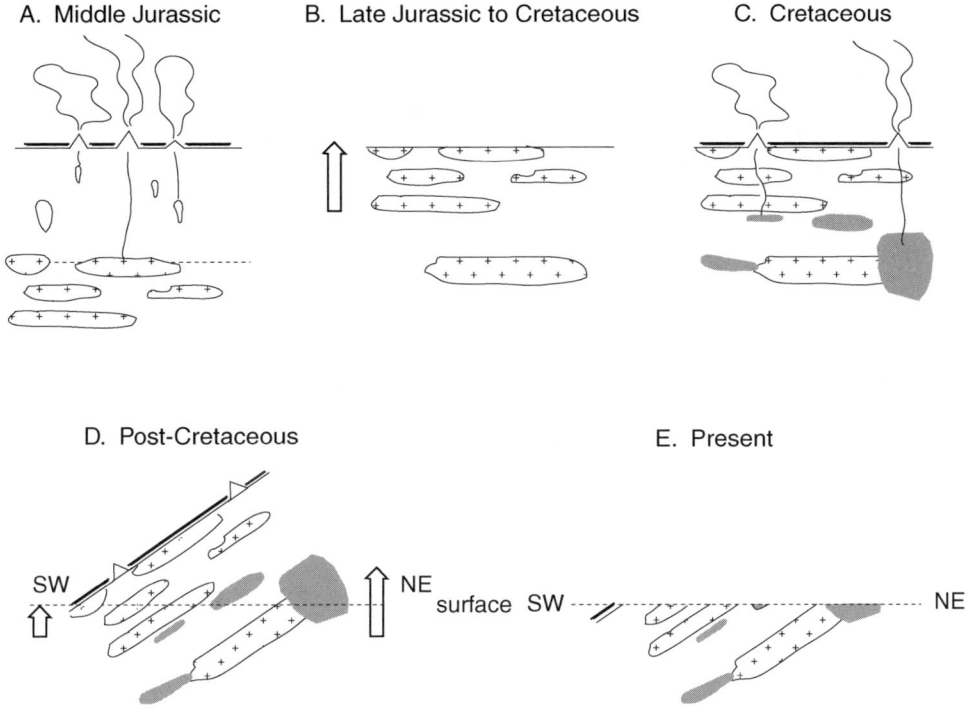

FIG. 11. Tectonic model for Mesozoic magmatism and subsequent uplifting and erosion along SW-NE cross-section of the study area from the Middle Jurassic to the present.

may be the result of contamination due to fluid infiltration during shearing, caused by strike-slip movement along the Honam shear zone. Consequently, most Jurassic granitoids of the OMB can be regarded as originating in a volcanic arc.

Tectonic Implications

Kinoshita and Itô (1988, 1990) classified granitoids along the northeast Asian continental margin into an older granitic belt on the continental side and a younger granitic belt on the oceanic side, based on geochronological data. The older granitic belt is arranged as follows: a Triassic granitic belt in South China (245–203 Ma), a Jurassic granitic belt in South Korea (180 Ma), the Cretaceous West Sikhote Alin belt (120 Ma), and the Okhotsk-Chukotka belt (from the Albian to the early Cenomanian). Based on these age distributions, Kinoshita (1995) suggested migration of igneous activity resulting from the migration of the Farallon-Izanagi ridge subduction toward the northeast. However, until now, no one has succeeded in confirming the possibility of ridge subduction–related igneous activity in the Korean Peninsula.

The present study concludes that most granitoids around the OMB formed during the Middle Jurassic from a common magmatic source beneath the volcanic arc. The exception to this is the two-mica granite. As shown in Figure 1, the major U-Pb zircon and titanite ages and K-Ar biotite ages of Jurassic granitoids in the Gyeonggi and Yeongnam massifs have similar ranges (165–179 Ma and 152–170 Ma) as those associated with the OMB (166–176 Ma and 150–168 Ma). This similarity indicates that the Middle Jurassic granite intruded not only the OMB but also the Gyeonggi and Yeongnam massifs, forming a complex about 200 km long and >150 km wide. Uyeda and Miyashiro (1974) suggested that such a spatial breadth for Middle Jurassic magmatism occurred not by simple subduction of oceanic lithosphere but by ridge subduction. Together with previous studies, our data indicate that Middle Jurassic magmatism was related to ridge subduction as suggested by Kinoshita (1995, 2002) during the period 200 to 166 Ma, with the principal magmatic period

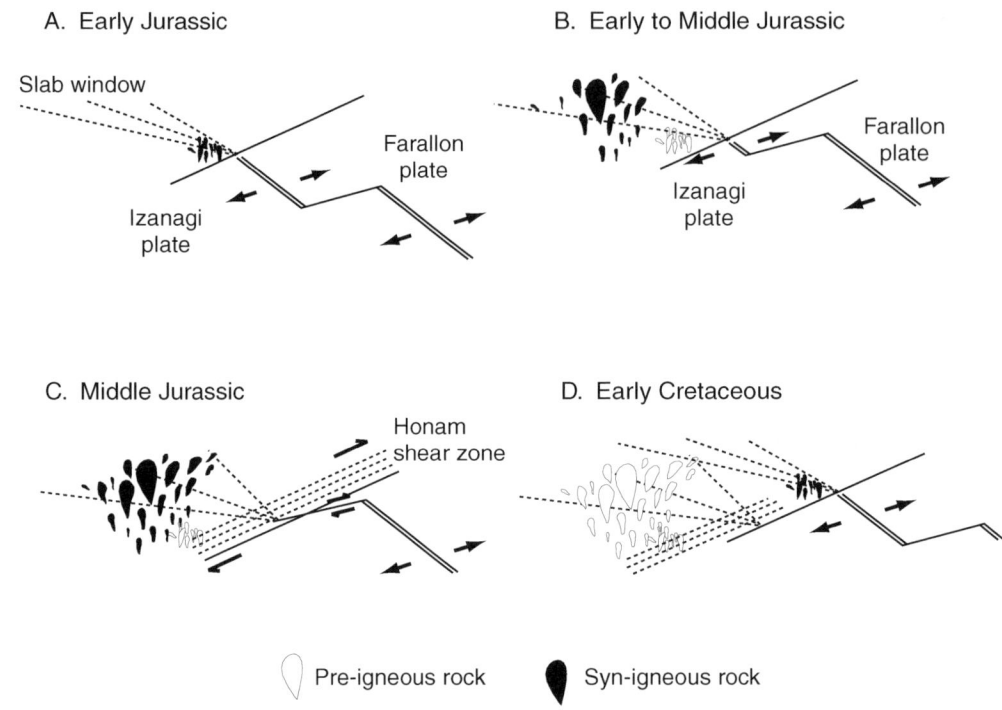

FIG. 12. Tectonic model associated with magmatism and transfault movement in South Korea due to the migration of ridge subduction of the Farallon-Izanagi plate during Mesozoic time.

between 175 and 166 Ma, and main cooling ranging from 168 to 152 Ma (Fig. 1B). For each pluton, the difference between the intrusion age, represented by U-Pb zircon and titanite dates, and the cooling age, determined by K-Ar biotite dating, is around 10 Ma, indicating a cooling rate of 45°C/m.y.

In the northeastern OMB, Cretaceous granite is now exposed on the surface together with Middle Jurassic granite, but in the southwestern OMB, Cretaceous volcanic rock is associated with Middle Jurassic granitoids. Based on hornblende geobarometric calibrations, Cho and Kwon (1994) suggested that the average emplacement depths of Jurassic and Cretaceous granites are 12–28 km and less than 10 km, respectively. This indicates that uplift after Middle Jurassic igneous activity elevated the Jurassic batholith from a depth of 12–28 km to 10 km, prior to the intrusion of Cretaceous granite (Fig. 11).

After the Cretaceous igneous activity ceased, a second phase of asymmetric uplift took place. Uplift was faster in the northeastern OMB, resulting in exposure of deeper parts of the crust. It is inferred, therefore, that Middle Jurassic granitoids now exposed in the northeastern part of the OMB represent deeper igneous activity, whereas those exposed in the southwest represent shallower igneous activity. The Jurassic granitoids have regionally variable magnetic susceptibility (MS, Jin et al., 2001). Granitoids in the southwest are magnetite-series, with high MS (more than 5×10^{-3} SI unit), whereas those in the central and northeastern areas are ilmenite-series, with relatively low MS (less than 3×10^{-3} SI unit). Given that the granitoids are inferred to have had the same source, this difference in redox state probably reflects different emplacement depths. The MS pattern supports the conclusion that the granitoids in the northern area formed at greater depths than those in the southwestern area.

Magmatism can be divided into a deeper, earlier stage, which resulted in emplacement of diorite, granodiorite, and granite as evident in the northeast OMB, and a shallower, younger stage, which resulted in emplacement of granite and two-mica granite in the southwest OMB (Fig. 11). At intermediate depths, represented by granitoids of the central OMB, granodiorite and granite formed. The inherited cores of 998 and 262 Ma SHRIMP II U-Pb zircon ages from two-mica granite indicate that two-mica granite is reworked crustal material formed by earlier magmatism before the Middle Jurassic event.

The Middle Jurassic granite in the Honam shear zone between the OMB and the Precambrian Yeongnam massif is strongly foliated. Recent SHRIMP U-Pb zircon and CHIME monazite dates for foliated granites (Cho et al., 1999; Lee at al., 2003) indicate that large-scale, dextral strike-slip movement along the Honam shear zone began around 180–173 Ma and continued until at least 160 Ma. Slip can be related to the subduction of a transform fault, as shown in Figure 12. Subduction of a transform fault between ridges took place during an interval in which the locus of ridge subduction jumped, completing subduction of the old ridge, and initiating new subduction of another ridge not yet begun. During subduction, dextral movement of the transform fault reactivated the pre-existing zones of weakness such as the Honam shear zone, one of the area's important tectonic boundaries.

Subduction of the Fallon-Izanagi ridge under the Korean Peninsula started around 200 Ma, initiating Jurassic magmatism and the so-called Daebo orogeny. Ridge subduction ended at around 180–173 Ma, and was followed by subduction of a transform fault, which caused dextral strike-slip along the Honam shear zone until at least 160 Ma. The subducted ridge continued to cause magmatism in the Korean Peninsula until around 165 Ma, with main magmatic period between 175 and 166 Ma.

Acknowledgments

This study was supported in part by Grants from the Post-Doctoral Program of Chonbuk National University (2002) and Grant-in-Aid, KRF-2003-070-C00046, from the Korea Science and Engineering Foundation.

REFERENCES

Chang, E. Z., 1996, Collisional orogen between north and south China and its eastern extension in the Korean Peninsula: Journal of Southeast Asian Earth Sciences, v. 13, p. 267–277.

Cheong, C. S., and Chang, H. W., 1996a, Tectono-magmatism, -metamorphism, and -mineralization of the central Ogcheon belt, Korea (1): Journal of the Geological Society of Korea, v. 32, p. 91–116.

Cheong, C. S., and Chang, H. W., 1996b, Geochemistry of the Daebo granitic batholith in the central Ogcheon belt, Korea: A preliminary report: Economic and Environmental Geology, v. 29, p. 483–493.

Cheong, C. S., and Chang, H. W., 1997, Sr, Nd, and Pb isotope systematics of granitic rocks in the central Ogcheon Belt, Korea: Geochemical Journal, v. 31, p. 17–36.

Cheong, C. S., Cheong, K. Y., Kim, H., Choi, M. S., Lee, S., and Cho, M., 2003a, Early Permian peak metamorphism recorded in U-Pb system of black slates from the Ogcheon metamorphic belt, South Korea, and its tectonic implication: Chemical Geology, v. 193, p. 81–92.

Cheong, C. S., Cheong, Y. C. Kil, Y. U., and Cheong, K. Y., 2003b, U-Pb sphene age in the Cheongju granite [abs.]: Abstract volume for proceedings of the Annual Joint Conference, Mineralogical Society of Korea and Petrological Society of Korea, p. 53 (in Korean).

Cho, D. L., and Kwon, S. T., 1994, Hornblende geobarometry of the Mesozoic granitoids and evolution of crustal thickness: Journal of the Geological Society of Korea, v. 30, p. 41–61.

Cho, D. L., Kwon, S. T., Sagong, H., Cheong, C. S., and Armstrong, R., 2001, Precise cooling histories of three neighboring plutons in the central Okcheon belt: Implication for magma movement rate and tectonics: Abstract volume for the conference of the Geological Society of Korea, p. 90.

Cho, K. H., Takagi, H., and Suzuki, K., 1999, CHIME monazite age of granitic rocks in the Sunchang shear zone, Korea: Timing of dextral ductile shear: Geoscience Journal, v. 3, p. 1–16.

Chough, S. K., Kwon, S. T., Ree, J. H., and Choi, D. K., 2000, Tectonic and sedimentary evolution of the Korean Peninsula: A review and new view: Earth-Science Reviews, v. 52, p. 175–235.

Cluzel, D., Lee, B. J., and Cadet, J.-P., 1991, Indosinian dextral ductile system and synkinematic plutonism in the southwest of the Ogcheon belt (South Korea): Tectophysics, v. 194, p. 131–151.

Harris, N. B. W., Pearce, J. A., and Tindle, A. G., 1986, Geochemical characteristics of collision-zone magmatism, in Coward, M. P., and Reis, A. C., eds., Collision tectonics: Geological Society Special Publication, v. 19, p. 67–81.

Hiroi, Y., 1981, Subdivision of the Hida metamorphic complex, central Japan, and its bearing on the geology of the far east in pre-sea of Japan time: Tectonophysics, v. 76, p. 317–333.

Irvine, T. N., and Baragar, W. R., 1971, A guide to the chemical classification of the common igneous rocks: Canadian Journal of Earth Science, v. 8, p. 523–548.

Isozaki, Y., and Maruyama, S., 1991, Studies on orogeny based on plate tectonics in Japan and new geotectonic subdivision of the Japanese Islands: Journal of Geography, v. 100, p. 697–761 (in Japanese with English abstract).

Itaya, T., Nagao, K., Inoue, K., Honjou, Y., Okada, T., and Ogata, A., 1991, Argon isotopic analysis by a newly developed mass spectrometric system for K-Ar dating: Mineralogical Journal, v. 15, p. 203–221.

Jin, M.-S., Gleadow, A. J. W., and Lovering, J. F., 1984, Fission track dating of apatite from the Jurassic and Cretaceous granites in South Korea: Journal of the Geological Society of Korea, v. 20, p. 257–265.

Jin, M.-S., and Jang, B.-A., 1999, Thermal history of the Late Triassic to Early Jurassic Yeongju-Chunyang Granitoid in the Sobaegsan Massif, South Korea, and its tectonic implication: Journal of the Geological Society of Korea, v. 35, p. 189–200.

Jin, M.-S., Kim, S. J., Shin, S. C., and Choi, S. J., 1992, Thermal history of the Jecheon granite pluton in the Ogcheon fold belt, South Korea: Journal of the Petrological Society of Korea, v. 1, p. 49–57.

Jin, M.-S., Kim, S. J., Shin, S. C., Choo, S. H., and Choi, S. J., 1995, Geochronology and cooling history of the Mesozoic granite pluton in the central part of the Ogcheon Fold Belt, South Korea: Journal of Petrological Society of Korea, v. 4, p. 153–167 (in Korean with English abstract).

Jin, M.-S., Lee, Y. S., and Ishihara, S., 2001, Granitoids and their magnetic susceptibility in South Korea: Resource Geology, v. 51, p. 189–203.

Kim, C.-B., Andrew, T., Chang, H.-W., Park, Y.-S., and Ahn, K.-S., 1999, U-Pb zircon ages for Precambrian and Mesozoic plutonic rocks in the Seoul-Chooncheon area, Gyeonggi massif, Korea: Geochemical Journal, v. 33, p. 379–397.

Kim, C.-B., Chang, H.-W., and Andrew, T., 2003, U-Pb zircon ages and Sr-Nd-Pb isotopic compositions for Permian–Jurassic plutons in the Ogcheon belt and Ryeongnam massif, Korea: Tectonic implications and correlation with the China Qinling-Dabie belt and the Japan Hida belt: The Island Arc, v. 12, p. 366–382.

Kim, C.-B., and Turek, A., 1996, Advances in U-Pb zircon geochronology of Mesozoic plutonism in the southwestern part of Ryeongnam massif, Korea: Geochemical Journal, v. 30, p. 323–338.

Kim, H., Cheong, C. S., Cho, M., Jeong, G. Y., and Choi, M. S., 2001, Geochronological evidence for late Paleozoic orogeny in the Ogcheon metamorphic belt, South Korea [abs.]: Abstract volume for Annual Meeting of the Geological Society of America, p. 33.

Kim, H. S., 1971, Metamorphic facies and regional metamorphism of Ogcheon metamorphic belt: Journal of the Geological Society of Korea, v. 7, p. 221–256.

Kim, J. H., 1996, Mesozoic tectonics in Korea: Journal of Southeast Asian Earth Sciences, v. 13, p. 251–265.

Kim, K. H., Tanaka, T., and Nagao, K., 1998, Nd and Sr Isotopes and K-Ar Ages of the granitic and rhyolitic rocks from the Bupyeong Silver Mine Area: Economic and Environmental Geology, v. 31, p. 149–158 (in Korean with English abstract).

Kim, S. W., Oh, C. W., Lee, D. S., and Lee, J. H., 2003, K-Ar and $^{40}Ar/^{39}Ar$ from metasediments in the Okcheon metamorphic belt and their tectonic implication: Journal of Petrological Society of Korea, v. 12, p. 79–99 (in Korean with English abstract).

Kiminami, K., Miyashita, S., and Kawabata, K., 1993, Active ridge-forearc collision and its geological consequence: An example from Late Cretaceous southwest Japan: Memoirs of the Geological Society of Japan, v. 42, p. 167–182 (in Japanese with English abstract).

Kinoshita, O., 1995, Migration of igneous activity related to ridge subduction in southwest Japan and the East Asian continental margin from the Mesozoic to the Paleogene: Tectonophysics, v. 245, p. 25–35.

Kinoshita, O., 2002, Possible manifestations of slab window magmatisms in Cretaceous southwest Japan: Tectonophysics, v. 344, p. 1–13.

Kinoshita, O., and Itô, H., 1988, Cretaceous magmatism in southwest and Northeast Japan related to two ridge subduction and Mesozic magmatism along East Asia continental margin: Journal of Geological Society of Japan, v. 94, p. 925–944 (in Japanese with English abstract).

Kinoshita, O., and Itô, H., 1990, Reconstruction of southwest Japan and northeast Japan based on trend of Mesozoic igneous rock ages: Journal of Geological Society of Japan, v. 96, p. 821–838 (in Japanese with English abstract).

Lee, J. I., 1991, Petrology, mineralogy, and isotopic study of the shallow-depth emplaced granitic rocks, southern part of the Kyoungsang basin, Korea: Unpubl. Ph.D. thesis, University of Tokyo, 197 p.

Lee, K. S., Chang, H. W., and Park, K. H., 1998, Neoproterozoic bimodal volcanism in the central Ogcheon belt, Korea: Age and tectonic implication: Precambrian Research, v. 89, p. 47–57.

Lee, S. R., Lee, B. J., Cho, D. L., Kee, W. S., Koh, H. J., Kim, B. C., Song, K. Y., Hang, J. H., and Choi, B. Y., 2003, SHRIMP U-Pb zircon age from granitic rocks in Jeonju shear zone: Implications for the age of the Honam shear zone [abs.]: Abstract volume for proceedings of the Annual Joint Conference, Mineralogical Society of Korea and Petrological Society of Korea, p. 55 (in Korean).

Ludwig, K. R., 1999, User's manual for Isoplot/Ex, Version 2.10, A geochronological toolkit for Microsoft Excel: Berkeley, CA, Berkeley Geochronology Center, Special Publication no. 1a, 46 p.

Ludwig, K. R., 2000, SQUID 1.00, A user's manual: Berkeley, CA, Berkeley Geochronology Center, Special Publication no. 2, 17 p.

Maruyama, S., Isozaki, Y., Kimura, G., and Terabayashi, M., 1997, Paleogeographic maps of the Japanese Islands: plate tectonic synthesis from 750 Ma to the present: The Island Arc, v. 6, p. 121–142.

Min, K., Cho, M., Kwon, S.-T., Kim, I. J., Nagao, K., and Nakamura, E., 1995, K-Ar ages of metamorphic rocks in the Chungju area: Late Precambrian (675 Ma) metamorphism of the Ogcheon Belt: Journal of Geological Society of Korea, v. 31, p. 315–327 (in Korean with English abstract).

Nagao, K., Nishido, H., Itaya, T., and Ogata, K., 1984, K-Ar age determination method: Bulletin of Research Institute of Natural Sciences Okayama University of Science, v. 9, p. 19–38.

Oh, C. W., Kim, S. W., and Lee, J. H., 1998, A study on the regional and contact metamorphism in the southwestern part of the Ogcheon Metamorphic Belt: Journal of Geological Society of Korea, v. 34, p. 311–332 (in Korean with English abstract).

Oh, C. W., Kim, S. W., Ryu, I.-C., Okada, T., Hyodo, H., and Itaya, T., 2004, Tectono-metamorphic evolution of the Okcheon Metamorphic Belt, South Korea: Tectonic implications in East Asia: The Island Arc, v. 13, p. 387–402.

Okada, H., 2000, Nature and development of Cretaceous sedimentary basins in East Asia: A review: Geoscience Journal, v. 4, p. 271–282.

Pearce, J. A., 1983, Role of sub-continental lithosphere in magma genesis at active continental margins, *in* Hawkesworth, C. J., and Norry, M. J. eds., Continental basalts and mantle xenoliths: Nantwich, UK, Shiva Publishing, p. 230–249.

Pearce, J. A., Harris, N. B. W., and Tindle, A. G., 1984, Trace element discrimination diagrams for the tectonic interpretation of granitic rocks: Journal of Petrology, v. 25, p. 956–983.

Pearce, J. B., and Miller, J. D., 1993, Precise U-Pb ages of Duluth Complex and related mafic intrusions, northeastern Minnesota: Geochronological insights to physical, petrogenetic, paleomagnetic, and tectonomagmatic process associated with the 1.1 Ga Midcontinent Rift System: Journal of Geophysical Research, v. 98, p. 13997–14013.

Ree, J. H., Kwon, S. H., Park, Y., Kwon, S. T., and Park, S. H., 2001, Pre- and post-tectonic emplacements of the granitoids in the central-southern Okchon belt, South Korea: Implications for the timing of the strike-slip shearing and thrusting: Tectonics, v. 20, p. 850–867.

Shibata, T., Park, N. Y., Uchiumi, S., and Ishihara, S., 1983, K-Ar ages of the Jeocheon granitic complex and related molybdenite deposits in South Korea: Mining Geology, v. 33, p. 193–197.

Steiger, R., and Jäger, E., 1977, Subcommission of geochronology: Convention on the use of decay constants in geo- and cosmo-chronology: Earth and Planetary Science Letters, v. 36, p. 359–362.

Tera, F., and Wasserburg, G., 1972, U-Th-Pb systematics in three Apollo 14 basalts and the problem of initial Pb in lunar rocks: Earth and Planetary Science Letters, v. 14, p. 281–304.

Uyeda, S., and Miyashiro, A., 1974, Plate tectonics and the Japanese islands: Geological Society of America Bulletin, v. 9, p. 19–38.

Williams, I. S., 1998, U-Th-Pb geochronology by ion microprobe, *in* McKibben, M. A., Shanks, W. C., III, and Ridley, W. I., eds., Applications of microanalytical techniques to understanding mineralizing processes: Reviews in Economic Geology, v. 7, p. 1–35.

Yanai, S., Park, B. S., and Otoh, S., 1985, The Honam shear zone (South Korea): Deformation and tectonic implication in the Far East: Scientific Paper of the College of Arts and Science, Tokyo University, v. 35, p. 181–210.

Yun, H. S., 1995, Occurrence and petrochemistry of the granites in the Pocheon-euijeongbu area: Journal of Petrological Society of Korea, v. 4, p. 91–103.

Yun, S., and Silberman, M. L., 1979, K-Ar Geochronology of igneous rocks in the Yeonhwa-Ulchin zinc-lead district and southern margin of the Taebaegsan Basin, Korea: Journal of Geological Society of Korea, v. 15, p. 89–100 (in Korean with English abstract).

Metamorphic Evolution of the Southwest Okcheon Metamorphic Belt in South Korea and Its Regional Tectonic Implications

S. W. KIM,[1]

Basic Science Research Institute, Chonbuk National University, Chonju 561-756, South Korea

C. W. OH,

Department of Earth and Environmental Sciences, Chonbuk National University, Chonju 561-756, South Korea

H. HYODO, T. ITAYA,

Research Institute of Natural Science, Okayama University of Science, Okayama 700-0005, Japan

AND J. G. LIOU

Department of Geological and Environmental Sciences, Stanford University, Stanford California 92304-2115

Abstract

The Hwasan area of the southwest Okcheon metamorphic belt in South Korea underwent intermediate-P/T *M1* metamorphism in the Late Carboniferous to Early Permian. Mica K-Ar and ^{40}Ar/^{39}Ar ages of the pelitic and psammitic rocks are concentrated at ~160 Ma. Granites in the northern and western parts of the Hwasan area have K-Ar mica ages of ~156 Ma. Carbonaceous material throughout the Hwasan metasediments exhibits a narrow range in d_{002} of 3.353 to 3.359 Å, corresponding to that of fully ordered graphite, consistent with amphibolite-facies conditions of ~500°C. Graphitic carbon is also distributed throughout lower-grade greenschist-facies rocks formed during the main *M1* metamorphic episode. These results, along with the occurrence of a thermal metamorphic aureole (>500°C) in the metasediments close to the mid-Jurassic granites, suggest that the study area was overprinted by low-P/T *M2* metamorphism due to the intrusions of the plutons. This latter event, however, failed to re-equilibrate mineral assemblages of the earlier metamorphism, except within a narrow aureole of 1–2 km around the Jurassic granites.

Introduction

THE PRINCIPAL STRUCTURAL units of South Korea are two Precambrian basement terranes (the Gyeonggi and Yeongnam massifs), two narrow Phanerozoic belts (the Okcheon and Imjingang belts), and the Cretaceous sedimentary Gyeongsang Basin (Fig. 1A). The Okcheon belt lies between the two Precambrian blocks and is subdivided into the Okcheon metamorphic belt (OMB) in the southwest and the Taebaeksan Basin (TB) in the northeast (Fig. 1B). The OMB consists of non-fossiliferous metasediments that underwent mainly an intermediate-P/T type metamorphic event (e.g., Kim, 1971; Kim et al., 1995; Oh et al., 1995a; Oh et al., 1998), whereas the Taebaeksan Basin consists of non- and weakly metamorphosed, fossiliferous sedimentary rocks.

The OMB has been correlated with the Hida marginal belt and the Hida metamorphic belt in southwest Japan, inasmuch as all three belts possess a similar geochronology, lithology, and metamorphic evolution (Hiroi, 1981; Kunugiza et al., 2001). In addition, Chang (1996) suggested that the OMB is the continuation of the Huanan failed rift within the South China block. Consequently, a proper understanding of the OMB is one of the keys to interpreting the tectonic evolution of northeast Asia.

During the last decade, two different tectonic models for the OMB have been proposed. The first model suggests that the OMB and the Imjingang belt in South Korea are the eastward extensions of the Dabie-Sulu collision belt in China (e.g., Liu, 1993; Ernst and Liou, 1995; Ree et al., 1996; Zhai and Liu, 1997). However, no evidence of a suture zone in the OMB has been found, such as the occurrence of high-pressure minerals and ophiolite suites. The second model proposes that the OMB represents an intracontinental aborted rift (Cluzel et al., 1990; Lee

---

[1]Corresponding author; E-mail: sungwon@chonbuk.ac.kr

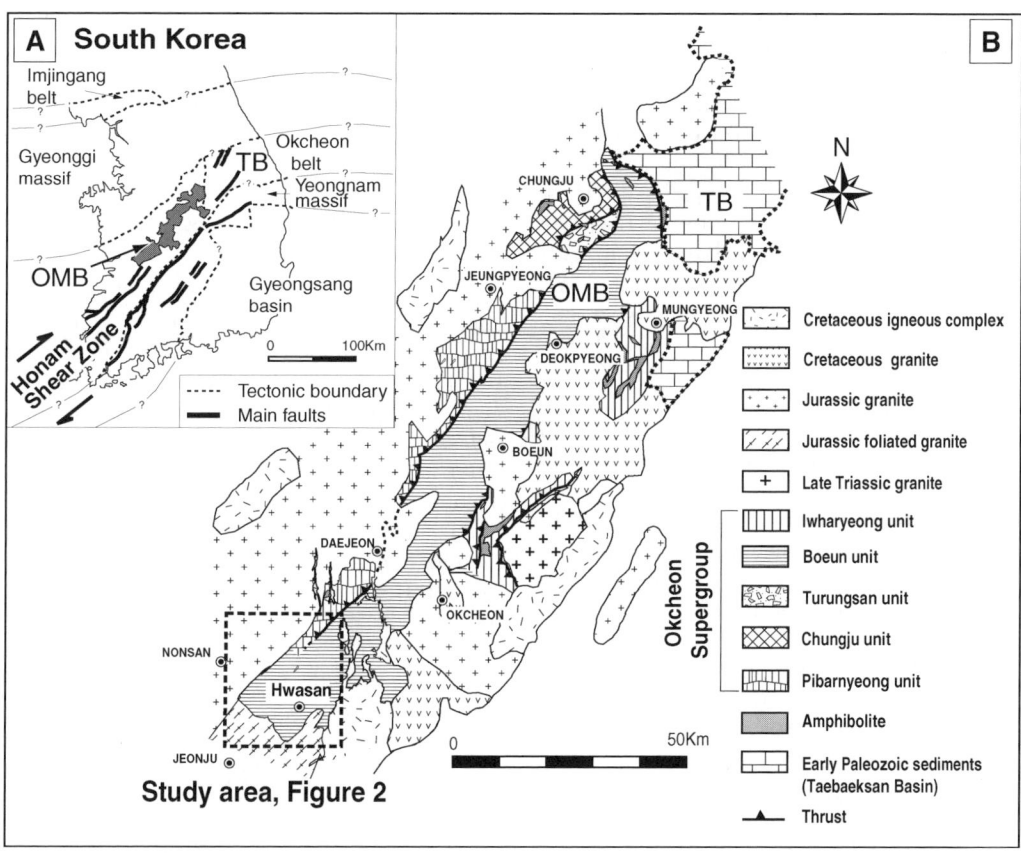

FIG. 1. A. Simplified geological sketch map of the main tectonic units, South Korea (adapted from Chough et al., 2000). A series of large-scale dextral strike-slip faults defines the Honam shear zone. B. Lithotectonic map of the OMB (adapted from Cluzel et al., 1990).

et al., 1998). The discrepancy between these two models arises in large part from a lack of petrological data that could define the stage of evolutionary development of this tectono-metamorphic terrane. Typically, insufficient data regarding the effect of Mesozoic magmatism is one of the most important reasons for discrepancy, because Mesozoic granitic activity was widespread in the OMB.

The present paper re-examines the metamorphic evolution of the OMB and its tectonic significance. The study is focused primarily on: (1) metamorphic evolution of the Hwasan area in the southwestern part of the OMB, as expressed in the composition and paragenesis of minerals; and (2) K-Ar dating of biotite and muscovite separates, and $^{40}Ar/^{39}Ar$ dating of single biotite and muscovite crystals. Graphitization of carbonaceous material in metasediments is also examined. A correlation of the results has allowed the nature of the late-stage thermal metamorphism caused by Mesozoic granite plutons to be ascertained, and various problems arising from previous geochronological studies to be clarified. Both of these areas hold important implications for the tectonic history of the belt. Abbreviations of minerals used in this paper are from Kretz (1983).

Outline of Geology

Geological history of the OMB

The OMB is a fold-and-thrust belt that trends NE-SW and extends over a distance of 150 km, with an average width of 20 km (Fig. 1). It developed initially within an intraplate rift environment during the Neoproterozoic (Cluzel et al., 1990; Lee et al. 1998). Although the stratigraphy of the OMB is still a matter of debate, Cluzel et al. divided it into five lithotectonic units: the Iwharyeong, Boeun, Turungsan, Chungju, and Pibanryeong (Fig. 1B). They considered the metasediments of the Iwharyeong, Boeun, and Turungsan units as deposited in a transitional domain located between an outer shelf and a deep basin, and those of the Chungju and Pibanryeong units deposited in a deeper basin. These units are bounded by thrust faults having a NE strike, i.e., parallel to the regional trend of the belt.

Cluzel et al. (1990, 1991a) considered the OMB to represent a stack of metamorphic, southeastward-verging nappes. They used chronological data to define four large-scale deformation events (D_1-D_4): (a) the Caledonian (or Okcheon) orogeny (Silurian-Devonian) for D_1 and D_2 deformation events; (b) the Indosinian (or Songrim) orogeny (late Permian–Triassic) for a D_3 deformation event; and (c) the Daebo orogeny (mid-Jurassic–Early Cretaceous) for a D_4 deformation event. They regarded D_1 to D_2 as synmetamorphic phases associated with the *M1* metamorphism responsible for formation of most metamorphic mineral assemblages in the OMB. Koh and Kim (1995) agreed that the geological structure of the belt was consistent with four deformational events, including two phases of thrusting.

However, recent Pb-Pb whole-rock ages (283–291 Ma) and concordant CHIME uraninite ages (281–283 Ma) in central OMB indicate that the Okcheon orogeny for the *M1* metamorphic event probably occurred in the late Carboniferous to early Permian (Cheong et al., 2003). Several studies have been undertaken in the last decade to establish the timing of the *M2* metamorphism and D_3 deformation (Songrim or Daebo orogeny) (Cho et al., 1995; Cho and Kim, 2002; Kim et al., 2003). Cho and Kim proposed that *M2* metamorphism occurred as a result of the Triassic Songrim orogeny, but most K-Ar and $^{40}Ar/^{39}Ar$ ages on muscovite and biotite obtained from metasediments throughout the belt are concentrated in the mid-Jurassic (Kim et al., 2003; Oh et al., 2004). Furthermore, recent U-Pb zircon and titanite ages indicate that most granites that surround the OMB, and give it the shape of an island within a surrounding granite sea, have mid-Jurassic intrusion ages (166–176 Ma; Cho et al., 2001; Ree et al., 2001; Cheong et al., 2003; Lee et al., 2003; Oh et al., 2004), and produced thermal metamorphism at contact aureoles (Oh et al., 1997, 1998, 1999; Min and Cho, 1998). Cluzel et al. (1991b) suggested that D_3 deformation caused the development of a dextral ductile fault system (Honam shear zone; Fig. 1A) during the Triassic Songrim orogenic event. However, new dating indicates that the dextral ductile faults formed during the mid-Jurassic Daebo orogenic event (173–176 Ma; Lee et al., 2003). The Daebo event was followed, in turn, by the Cretaceous Bulguksa event that attended igneous activity.

Geological setting and structural framework of the Hwasan area

The Hwasan area, the subject of the present study, is located in the southwestern part of the belt and divided into the Pibanryeong unit in the northwest and the Boeun unit in the southeast. The former consists of pelites, well-sorted fine-grained psammites, carbonaceous psammites, and quartzites. The latter is composed of pelites, psammites, carbonaceous pelites, limestones, and pebble-bearing quartzites. Kim et al. (2002) established three metamorphic zones for pelitic rocks of the Hwasan area, Zones I, II, and III grading from southeast to northwest (Fig. 2). Zones I and II are correlated with the Boeun unit, and Zone III with the Pibanryeong unit. Quartzites of both units have a pronounced NE trend, and provide marker beds for the regional structure and stratigraphy. Structural analysis southeast of the Hwasan area by Cluzel et al. (1991a) suggested that the main foliation (S_1) and the associated stretching lineation appear to be closely related to the ductile thrusts (D_1), with mainly isoclinal folds (F_1). In the present study, the dominant S_1 foliations in three zones generally strike NE-SW with a regional dip in Zones I and II to the northwest, and in Zone III to the southeast (Fig. 3). S_1 is commonly subparallel to the bedding plane (S_0). Locally, the disjunctive and crenulation cleavage (S_2) developed by D_2 displays a similar orientation to S_1 (Cluzel et al., 1991a: Fig. 3).

In the southwestern and northern parts of the Hwasan area, granites have either intrusive or fault contacts with the metasediments (Fig. 2). Mid-Jurassic granitic rocks to the north are massive, but are strongly foliated in the southwest. The foliated granitic rocks in this area have SHRIMP U-Pb zircon ages of 173–176 Ma (Lee et al., 2003). Their foliation is a result of ductile shearing within the Honam shear zone, and is recognized as the D_3 deformation by Cluzel et al. (1991a). Therefore, the D_3 deformation occurred after 173-178 Ma. In the southeast, the metasediments are covered by younger Cretaceous volcanic or pyroclastic rocks and are intruded by minor younger granites (Fig. 2).

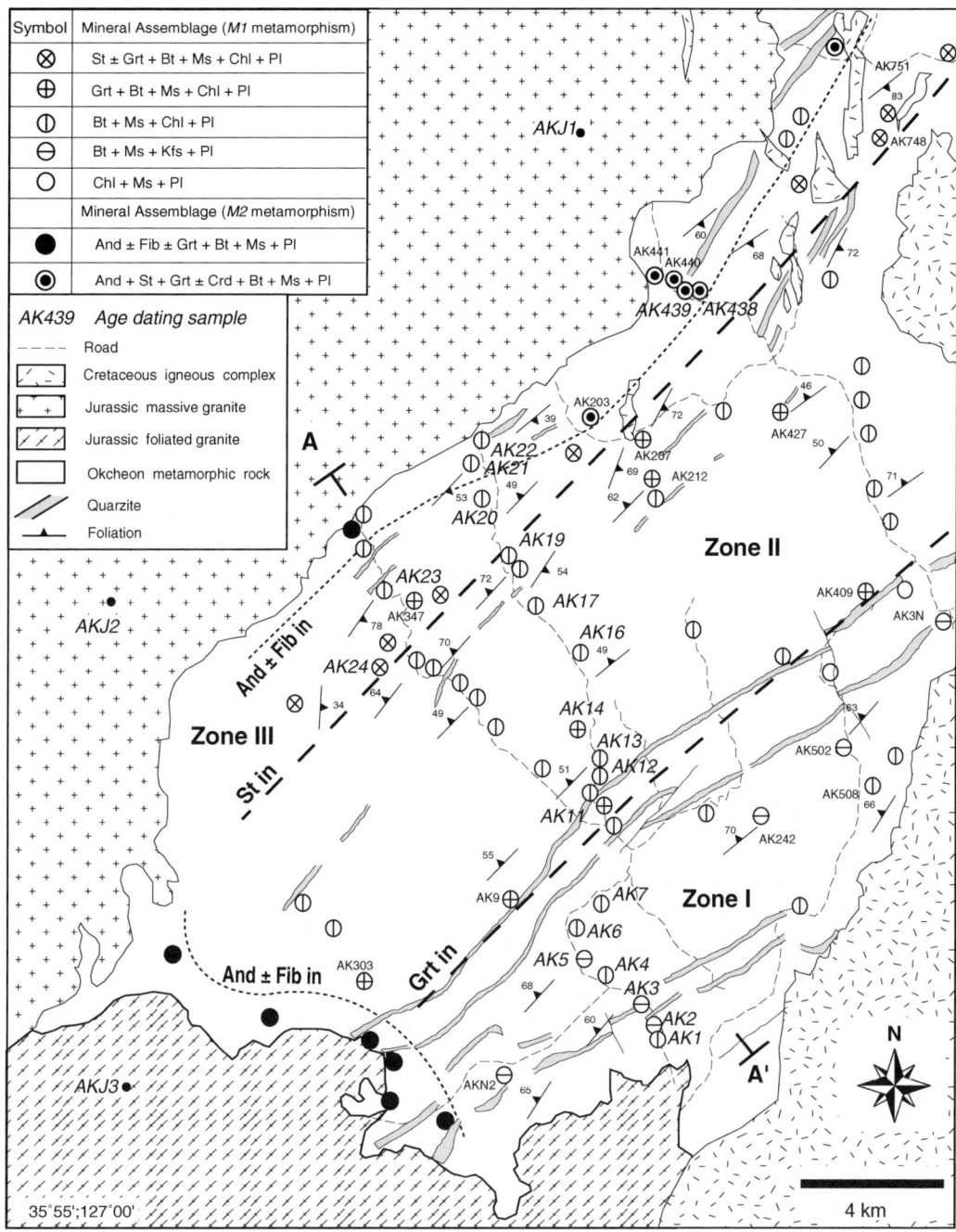

FIG. 2. Geological and metamorphic sketch map of the Hwasan area, southwest Okcheon Metamorphic Belt, showing sample localities and metamorphic zones, modified after Kim et al. (2002). Abbreviations: And = andalusite; Bt = biotite; Chl = chlorite; Crd = cordierite; Fib = fibrolitic sillimanite; Grt = garnet; Kfs = K-feldspar; Ms = muscovite; Pl = plagioclase; St = staurolite. Mineral abbreviations are adopted from Kretz (1983).

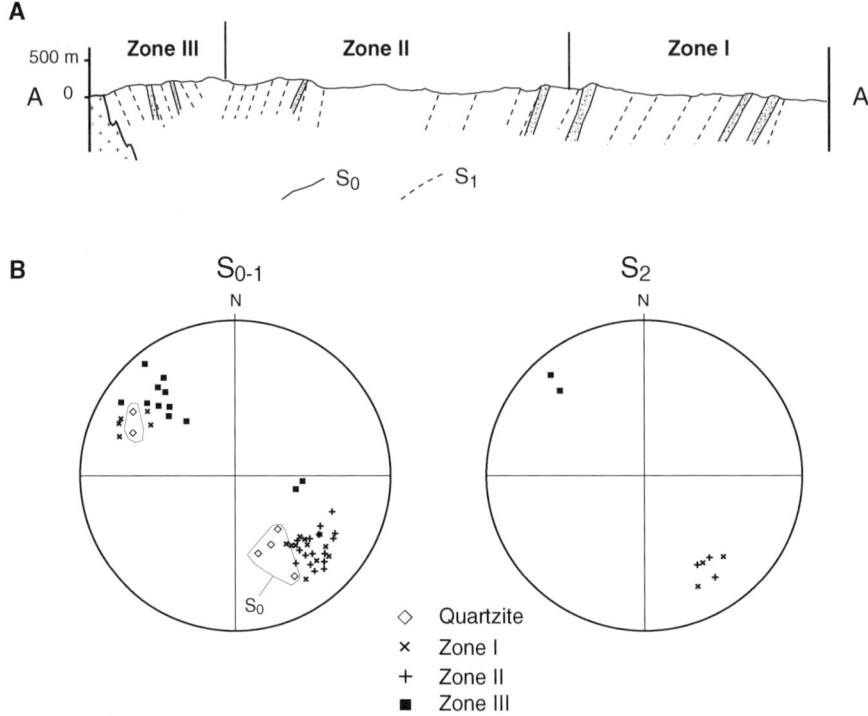

FIG. 3. Cross section (A) and stereoplots (B) of selected structural elements of the Hwasan area, Okcheon Metamorphic Belt.

Petrography and Paragenetic Assemblages

The Hwasan pelitic rocks are divided into Zones I, II, and III, based on mineral parageneses (Fig. 2). The Zone I assemblage consists of Bt + Ms + Pl + Qtz ± Chl; Zone I was subdivided into lower- and higher-grade sectors. Low-grade carbonaceous pelites contain extremely fine grained K-feldspar as anhedral to subhedral crystals coexisting with fine-grained muscovite, biotite, plagioclase, and quartz (Kim et al., 2002). Some rocks have biotite porphyroblast as single grains that have rectangular outlines and are 0.5–2 mm in diameter. They contain quartz inclusions, exhibit pressure shadows, and are associated with fine-grained biotite (30–60 µm). The cleavage defined by the biotite porphyroblasts is oblique to the main foliation (Fig. 4A). Some biotites are partly replaced by chlorite. Coarse K-feldspar grains in psammitic rocks are detrital. In the matrix of higher Zone I rocks, biotite and muscovite increase in grain size, and it is difficult to distinguish porphyroblastic biotite from matrix biotite.

Zones II and III units are characterized by the appearance of garnet and staurolite, respectively. Typical assemblages of Zone II pelitic rocks are Grt + Bt + Ms + Chl + Pl + Qtz. In Zone III, they contain additional staurolite with or without chlorite. Garnet occurs as euhedral to subhedral crystals, 0.5–3 mm in diameter, that exhibit pressure shadows. Many Zone II garnets contain inclusions of quartz as well as minor apatite and ilmenite (Fig. 4B). Inclusions in Zone III garnet are quartz, apatite, biotite, tourmaline, muscovite, plagioclase, ilmenite, and rutile; ilmenite inclusions are distributed from core to rim whereas rutile inclusions, 50–70 µm across are restricted to the rim only (Fig. 4C). Staurolite shows poikiloblastic textures (Fig. 4D) and contains garnet inclusions. Most staurolite grains are replaced by chlorite and fine-grained muscovite.

Matrix ilmenite and rutile occur as independent, elongated grains in pelitic rocks of Zones I and II, but in Zone III they typically form composite ilmenite-rutile grains (Fig. 4E). Kyanite has been reported from Zone III rocks (Oh et al., 1995a,

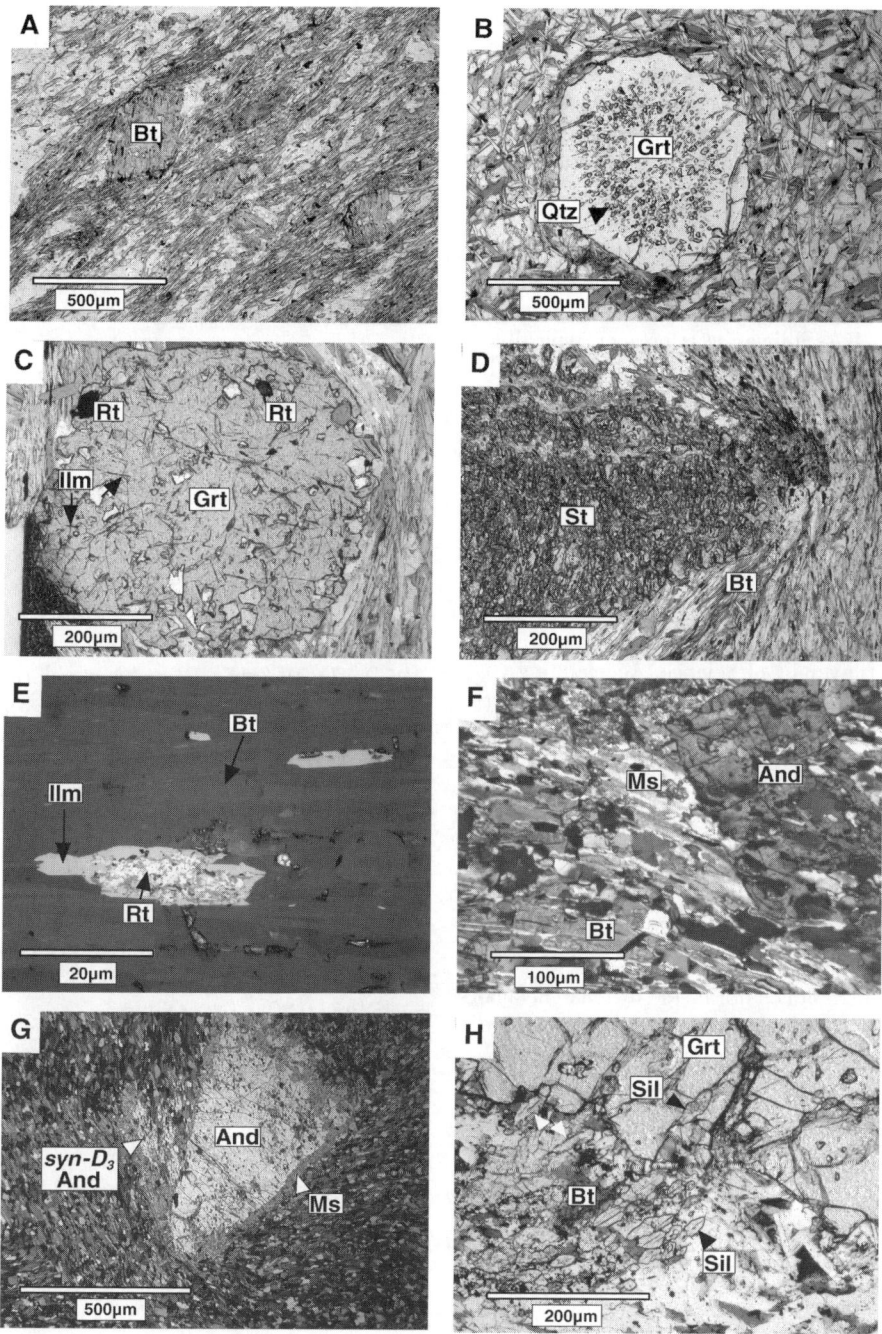

FIG. 4. Photomicrographs showing the modes of occurrence of biotite, garnet, staurolite, sillimanite, andalusite, and ilmenite-rutile composite in pelitic rocks. All photomicrographs are plane-polarized light except for E (crossed nichols). A. Biotite porphyroblast having a rectangular or diamond section. B. Idoblastic garnet occurring as a porphyroblast. C. Garnet with rutile and ilmenite inclusions. D. Staurolite showing poikiloblastic texture. E. Ilmenite occurring as a composite grain with rutile. F. Euhedral andalusite showing post-D_2 texture. G. Andalusite showing syn-D_3 texture. H. Sillimanite displaying a rhombic section. Abbreviations: Ilm = ilmenite; Qtz = quartz; Rt = rutile; Sil = sillimanite. Other abbreviations are the same as in Figure 2.

1998) but none was found in the Hwasan area. Instead, euhedral to anhedral andalusite, commonly associated with fibrous sillimanite, occurs in both pelitic and psammitic rocks close to Jurassic granite plutons, pointing to its formation during thermal metamorphism during emplacement of the granitoids. Andalusite crosscuts the foliation, D_{1-2} (Fig. 4F) and in a few occurrences replaces staurolite. Mid-Jurassic syn-D_3 andalusite occurs, indicating that ductile D_3 deformation and Jurassic granite intrusion occurred concurrently in some restricted areas (Fig. 4G). Sillimanite is present, typically as fibrous crystals and uncommonly in rhomboid-shaped grains in and around garnets in pelitic schists (Fig. 4H). Cordierite occurs with andalusite in staurolite-bearing pelitic rocks close to granite.

Mineral Chemistry

Chemical analyses of rock-forming minerals were carried out with a JEOL electron-probe microanalyzer (EPMA) JXA-8900R Superprobe at Okayama University of Science. Accelerating voltage, beam current, and probe diameter were, respectively, 15 kV, 12 nA, and 3–5 μm for spot analyses, and 15 kV, 200 nA, and 1–2 μm for X-ray maps. The ZAF method was employed for matrix corrections. Natural and synthetic minerals were used as standards. Tables 1, 2, 3, and 4 list representative compositions of garnet, biotite-muscovite, chlorite, staurolite, and plagioclase.

Garnet

Garnet that formed during the regional metamorphism shows a growth zoning in which its internal texture conforms to the external grain morphology, as shown in Figure 5. In general, X_{Fe} (almandine), X_{Mg} (pyrope), and X_{Ca} (grossular) all increase from core to rim, whereas X_{Mn} (spessartine) decreases (Table 1). However, in garnet from sample AK440 from a granite contact, only the inner portion of larger grains displays the general compositional gradient. In the outer rim of these crystals, X_{Fe} and X_{Mg} decrease whereas X_{Mn} and the Fe/(Fe + Mg) ratios increase outward (Fig. 6). The same outer rim pattern occurs in fine-grained garnets less than 300 μm in diameter (also in sample AK 440).

Muscovite, biotite, and chlorite

Muscovite, biotite, and chlorite are common in all zones. Si values of muscovite based on 22 oxygens for Zones I, II, and III are 6.20–6.58, 6.19–6.42, and 6.14–6.32, respectively. The Na/(Na + K) values are 0.02–0.11 in Zone I, 0.04–0.13 in Zone II, and 0.09–0.17 in Zone III. The average celadonite content decreases with increasing metamorphic grade, whereas the average Na/(Na + K) value increases slightly. In lower-grade Zone I rocks, muscovite in K-feldspar–bearing pelitic rocks exhibits higher Mg/(Mg + Fe) values (0.80–0.88) than that in K-feldspar–free pelitic rocks, ranging from 0.48 to 0.68. Biotites in K-feldspar–free pelitic rocks in Zone I have Al^{IV} values of 2.29–2.71, and Mg/(Mg + Fe) ranges from 0.46 to 0.58. Biotite in K-feldspar–bearing pelitic rocks has significantly higher Mg/(Mg + Fe) ratios (0.78–0.83) than that in K-feldspar–free pelitic rocks. Zone I porphyroblastic biotite and associated fine-grained biotite have similar composition. Matrix biotite in Zones II and III is relatively homogeneous compared to Zone I matrix biotite. Zone III biotite included in garnet rims is higher in Mg/(Mg + Fe) than is matrix biotite (Table 2). Chlorite in all zones is either ripidolite or pycnochlorite, according to the classification of Hey (1954). The Mg/(Mg + Fe) value of chlorite in these zones ranges from 0.41 to 0.72 (Table 3). Zone III chlorite, which commonly occurs as a secondary phase replacing biotite, garnet, staurolite, and cordierite, has Mg/(Mg + Fe) ratios ranging from 0.41 to 0.55.

Staurolite

The Mg/(Fe + Mg) values of some well-preserved staurolite range from 0.13 to 0.14, whereas that of staurolite relics armored by muscovite aggregates range from 0.20 to 0.23 (Table 3).

Plagioclase

Plagioclase occurs as anhedral to subhedral grains in all zones. In the lower grade part of Zone I, plagioclase occurs as extremely fine-grained crystals, 5–15 μm in width and 15–25 μm in length, and has X_{An} from 0.10 to 0.36. Zone I coarse-grained (~200 μm) plagioclase has a large range for X_{An} of 0.40 to 0.80 (Fig. 7 and Table 4), suggesting that this grain is detrital in origin. Plagioclase in the upper-grade part of Zones I and II ranges X_{An} from 0.30 to 0.52, whereas Zone III plagioclase has X_{An} of 0.34–0.78 (Fig. 7). Plagioclase inclusions, 10-50 μm in diameter, in Zone III garnet are not significantly zoned. Plagioclase included in garnet cores has a substantially higher X_{An}, in the range of 0.58–0.89, than that in garnet rims, ranging from 0.41–0.50.

TABLE 1. Representative Microprobe Analyses of Garnet (formula based on 12 oxygens)[1]

Sample no.:	AK9	AK9	AK14	AK14	AK751	AK751	AK24	AK24	AK347	AK347	AK203	AK203	AK438	AK438	AK439	AK439	AK440	AK440
Metamorphic zone:	Zone II										Zone III							
Position:	core	rim	core	rim	core	rim	core	rim	core	rim	core	rim	core	rim	core	rim	core	rim
SiO_2	37.95	37.60	37.83	38.17	38.31	38.32	37.14	37.23	37.45	37.69	38.56	38.52	38.37	39.00	38.71	38.35	38.34	38.13
TiO_2	0.06	0.01	0.05	0.12	0.10	0.14	0.00	0.00	0.00	0.00	0.00	0.04	0.05	0.00	0.05	0.09	0.04	0.04
Al_2O_3	20.94	20.82	21.15	21.22	21.38	21.24	20.51	20.47	20.13	20.60	20.51	20.83	20.64	20.70	20.72	21.01	20.74	20.79
Cr_2O_3	0.00	0.04	0.06	0.00	0.00	0.00	0.00	0.00	0.01	0.00	0.00	0.02	0.00	0.00	0.02	0.02	0.00	0.00
FeO*	31.20	37.32	28.28	26.51	27.02	27.20	31.05	32.26	28.06	32.55	28.22	28.47	30.53	30.87	27.01	28.17	32.59	31.28
MnO	5.57	1.44	6.42	4.19	5.60	5.27	6.91	4.62	8.08	2.05	7.08	5.47	4.89	1.33	6.23	4.28	3.07	5.87
MgO	1.12	1.53	2.04	1.95	1.82	1.78	2.40	2.55	2.14	3.13	3.73	3.18	2.38	2.57	3.13	3.46	3.11	2.26
CaO	3.08	1.24	4.55	7.18	6.24	6.64	2.56	2.85	2.58	3.01	1.95	3.81	2.87	5.71	4.15	4.97	2.51	2.08
Na_2O	0.07	0.01	0.03	0.03	0.00	0.03	0.03	0.03	0.03	0.04	0.02	0.00	0.05	0.01	0.00	0.08	0.00	0.02
K_2O	0.00	0.03	0.00	0.02	0.00	0.00	0.00	0.01	0.00	0.01	0.00	0.00	0.01	0.00	0.03	0.00	0.02	0.00
Total	99.99	100.04	100.41	99.39	100.47	100.62	100.60	100.02	98.49	99.08	100.08	100.34	99.78	100.19	100.06	100.44	100.42	100.47
Si	3.059	3.046	3.022	3.046	3.038	3.037	2.997	3.010	3.061	3.042	3.071	3.057	3.076	3.089	3.073	3.033	3.055	3.054
Ti	0.004	0.000	0.003	0.007	0.006	0.008	0.000	0.000	0.000	0.000	0.000	0.002	0.003	0.000	0.003	0.005	0.003	0.003
Al	1.990	1.988	1.991	1.996	1.998	1.984	1.951	1.950	1.939	1.959	1.926	1.949	1.951	1.932	1.939	1.958	1.948	1.963
Cr	0.000	0.003	0.004	0.000	0.000	0.000	0.000	0.000	0.001	0.000	0.000	0.001	0.000	0.000	0.001	0.001	0.000	0.000
Fe	2.104	2.528	1.889	1.769	1.792	1.803	2.095	2.181	1.918	2.197	1.880	1.890	2.048	2.045	1.793	1.863	2.172	2.095
Mn	0.381	0.099	0.434	0.283	0.376	0.354	0.472	0.316	0.560	0.140	0.478	0.367	0.332	0.089	0.419	0.287	0.207	0.398
Mg	0.134	0.185	0.243	0.232	0.215	0.210	0.289	0.307	0.261	0.377	0.443	0.376	0.284	0.303	0.370	0.408	0.369	0.270
Ca	0.266	0.108	0.389	0.614	0.530	0.564	0.221	0.247	0.226	0.260	0.167	0.324	0.246	0.485	0.353	0.421	0.214	0.178
Na	0.010	0.002	0.005	0.005	0.000	0.005	0.005	0.005	0.005	0.005	0.002	0.001	0.007	0.002	0.000	0.013	0.000	0.002
K	0.000	0.003	0.000	0.002	0.000	0.000	0.000	0.001	0.001	0.001	0.000	0.000	0.001	0.000	0.003	0.000	0.002	0.000
Total	7.948	7.962	7.980	7.954	7.955	7.965	8.030	8.017	7.971	7.982	7.967	7.967	7.949	7.946	7.954	7.989	7.970	7.963
X_{Fe}	0.729	0.866	0.639	0.610	0.615	0.615	0.681	0.715	0.647	0.739	0.633	0.639	0.704	0.700	0.611	0.625	0.733	0.712
X_{Mg}	0.046	0.063	0.082	0.080	0.074	0.072	0.094	0.101	0.088	0.127	0.149	0.127	0.098	0.104	0.126	0.137	0.125	0.092
X_{Mn}	0.132	0.034	0.147	0.098	0.129	0.121	0.153	0.104	0.189	0.047	0.161	0.124	0.114	0.030	0.143	0.096	0.070	0.135
X_{Ca}	0.092	0.037	0.132	0.212	0.182	0.192	0.072	0.081	0.076	0.087	0.056	0.110	0.085	0.166	0.120	0.141	0.072	0.061
Mg/Fe	0.064	0.073	0.129	0.131	0.120	0.116	0.138	0.141	0.136	0.172	0.236	0.199	0.139	0.148	0.206	0.219	0.170	0.129

[1]X_{Fe} = Fe/(Fe + Mn + Mg + Ca); X_{Mg} = Mg/(Fe + Mn + Mg + Ca); X_{Mn} = Mn/(Fe + Mn + Mg + Ca); X_{Ca} = Ca/(Fe + Mn + Mg + Ca); * = total iron as FeO.

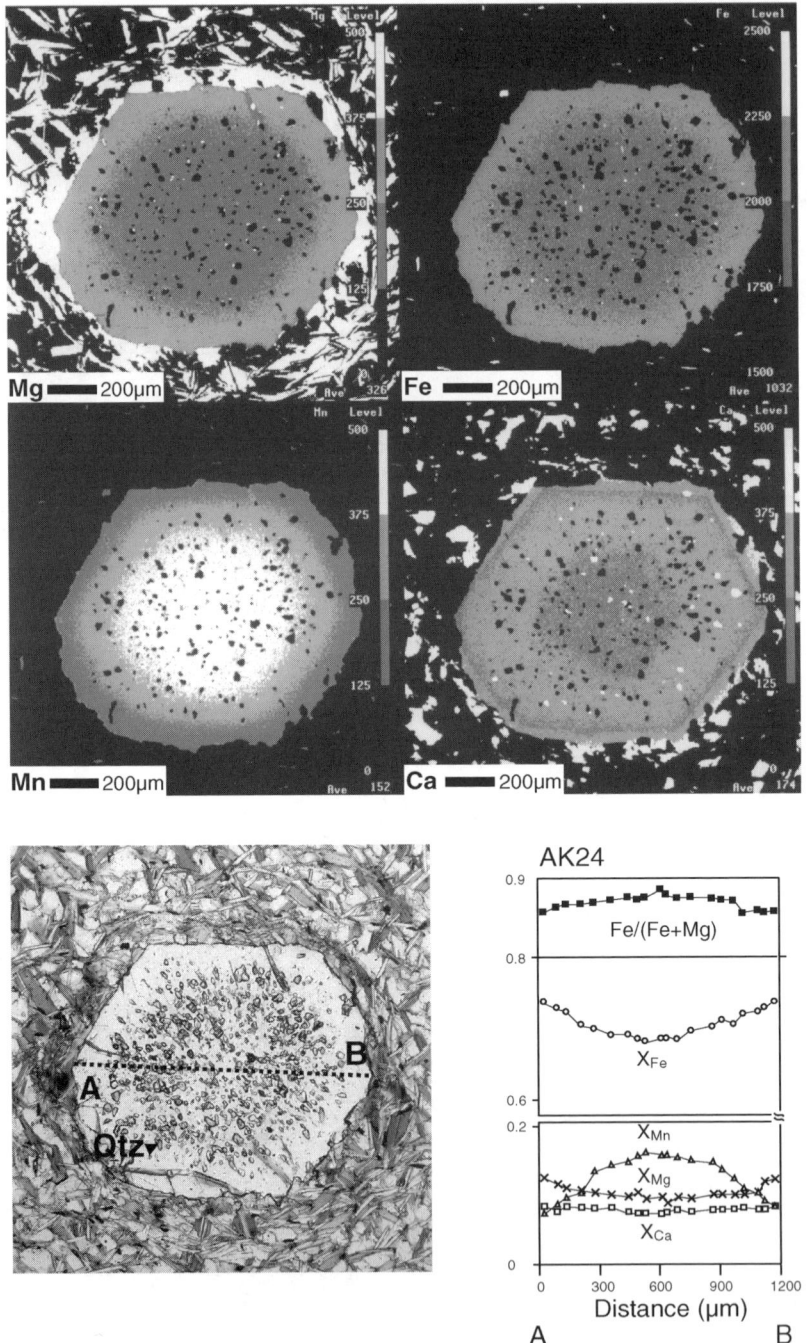

FIG. 5. X-ray mapping image, photomicrograph, and compositional profile of a garnet from sample AK24 in Zone III. Abbreviations: X_{Fe} = almandine; X_{Ca} = grossular; X_{Mg} = pyrope; X_{Mn} = spessartine.

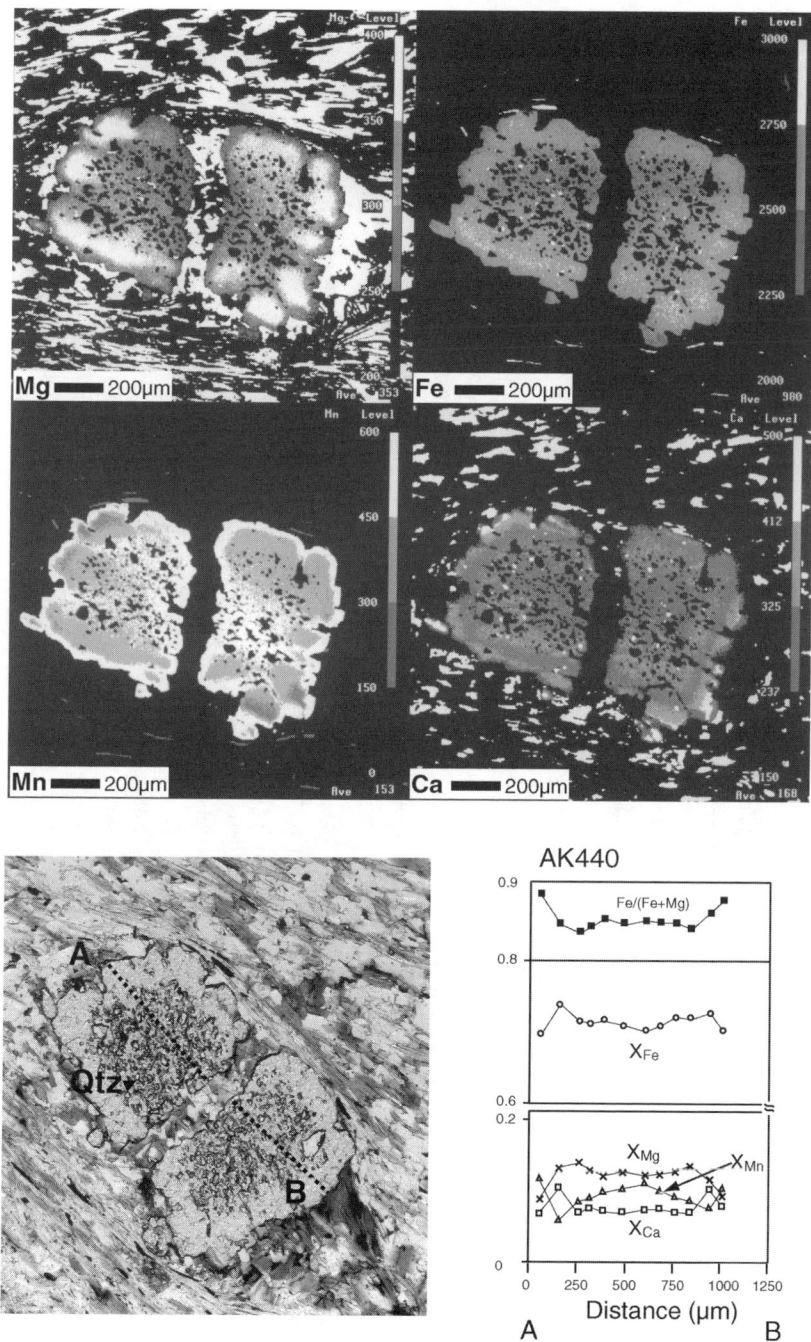

FIG. 6. X-ray mapping image, photomicrograph, and compositional profile of a garnet from sample AK440 in Zone III close to a granite. Abbreviations are the same as those for Figure 5.

TABLE 2. Representative Microprobe Analyses of Biotite and Muscovite (formula based on 22 oxygens)

Mineral:	Biotite													Muscovite		
Sample no.:	AK24	AK24	AK347	AK203	AK203	AK438	AK438	AK439	AK439	AK440	AK440	AK203	AK203	AK439	AK439	AK440
Metamorphic zone:						Zone III								Zone III		
Position:	grt incl.	matrix	grt incl.	grt incl.	matrix	grt incl.	grt incl.	grt incl.	grt incl.	grt incl.	matrix	grt incl.	matrix	grt incl.	matrix	matrix
SiO_2	36.20	36.6	36.90	36.62	36.03	36.98	37.94	36.63	37.00	36.25	36.40	45.85	46.46	46.58	46.22	47.24
TiO_2	1.46	1.74	1.15	1.65	1.42	1.55	1.58	1.11	1.64	1.30	1.43	0.46	0.46	0.38	0.23	0.52
Al_2O_3	18.09	18.32	19.15	19.07	19.12	17.67	18.28	19.70	19.90	18.72	19.45	0.01	34.34	35.75	35.61	34.87
Cr_2O_3	0.00	0.00	0.07	0.00	0.00	0.03	0.04	0.00	0.02	0.07	0.05	33.86	0.02	0.00	0.00	0.00
FeO*	18.32	18.21	17.70	15.27	16.22	16.07	17.60	11.97	15.76	17.65	18.63	2.58	1.17	1.46	1.22	0.90
MnO	0.10	0.11	0.06	0.16	0.17	0.00	0.08	0.19	0.15	0.16	0.00	0.18	0.01	0.07	0.00	0.02
MgO	10.87	10.65	10.56	11.97	12.12	12.23	10.51	15.07	11.69	11.38	10.37	1.70	0.75	0.53	0.62	0.62
CaO	0.03	0.00	0.06	0.00	0.01	0.01	0.01	0.03	0.00	0.00	0.03	0.01	0.02	0.01	0.00	0.02
Na_2O	0.31	0.09	0.18	0.35	0.33	0.11	0.24	0.37	0.30	0.24	0.20	0.90	0.82	0.81	0.98	1.01
K_2O	9.02	9.62	8.76	9.66	9.67	9.62	9.55	9.06	9.54	8.89	8.09	9.24	10.00	8.80	10.81	9.95
Total	94.39	95.34	94.60	94.75	95.11	94.27	95.83	94.12	95.99	94.66	94.67	94.78	94.05	94.40	95.68	95.16
Si	5.521	5.529	5.563	5.492	5.418	5.593	5.662	5.419	5.471	5.487	5.491	6.155	6.246	6.194	6.143	6.262
Ti	0.167	0.197	0.131	0.186	0.160	0.177	0.177	0.124	0.182	0.148	0.163	0.046	0.046	0.038	0.022	0.052
Al	3.251	3.262	3.404	3.369	3.389	3.149	3.216	3.438	3.465	3.332	3.458	5.358	5.437	5.604	5.573	5.448
Cr	0.000	0.000	0.009	0.000	0.000	0.003	0.005	0.000	0.003	0.008	0.006	0.001	0.002	0.000	0.000	0.000
Fe	2.337	2.301	2.232	1.915	2.040	2.032	2.197	1.486	1.949	2.238	2.350	0.290	0.132	0.162	0.136	0.100
Mn	0.012	0.015	0.008	0.021	0.022	0.000	0.010	0.024	0.019	0.020	0.000	0.020	0.001	0.008	0.000	0.003
Mg	2.470	2.399	2.374	2.677	2.718	2.757	2.337	3.333	2.577	2.569	2.331	0.340	0.150	0.105	0.123	0.122
Ca	0.005	0.000	0.011	0.000	0.002	0.001	0.001	0.005	0.001	0.000	0.005	0.002	0.002	0.001	0.000	0.004
Na	0.091	0.028	0.052	0.101	0.097	0.031	0.069	0.106	0.085	0.070	0.059	0.234	0.213	0.210	0.252	0.259
K	1.755	1.855	1.684	1.849	1.856	1.856	1.818	1.711	1.799	1.714	1.558	1.582	1.716	1.493	1.834	1.683
Total	15.610	15.586	15.468	15.610	15.703	15.599	15.493	15.646	15.551	15.586	15.422	14.028	13.945	13.817	14.083	13.933
Na/(Na +K)												0.129	0.110	0.123	0.121	0.133
Mg/(Mg + Fe)	0.514	0.510	0.515	0.583	0.571	0.576	0.515	0.692	0.569	0.534	0.498	0.540	0.532	0.393	0.475	0.550

TABLE 3. Representative Microprobe Analyses of Chlorite and Staurolite

Mineral:	Chlorite												Staurolite			
Sample no.:	AK508	AK2	AK7	AK10	AK16	AK218	AK24	AK751	AK203	AK438	AK439	AK440	AK24	AK203	AK438	AK439
Metamorphic zone:		Zone I			Zone II				Zone III					Zone III		
SiO$_2$	27.10	27.72	25.86	27.17	26.29	25.75	25.10	25.64	26.03	25.53	25.68	25.41	27.27	27.88	28.46	27.69
TiO$_2$	0.04	0.34	0.31	0.14	0.04	0.12	0.05	0.08	0.08	0.12	0.11	0.14	0.75	52.72	52.88	0.58
Al$_2$O$_3$	23.86	19.62	22.42	21.43	23.07	22.55	23.62	23.33	22.84	23.68	23.15	22.73	52.07	12.01	11.08	53.80
Cr$_2$O$_3$	0.01	0.05	0.01	0.01	0.00	0.00	0.01	0.00	0.06	0.01	0.05	0.01	0.01	2.06	1.70	0.04
FeO*	19.51	25.94	22.52	16.77	17.13	22.16	21.99	21.93	19.23	20.67	19.87	23.32	14.01	0.00	0.00	11.71
MnO	0.14	0.01	0.18	0.06	0.05	0.14	0.12	0.15	0.29	0.25	0.30	0.19	0.23	0.04	0.10	0.34
MgO	15.51	11.77	16.29	20.73	19.52	15.39	15.25	16.30	18.08	17.22	18.15	15.16	1.30	0.00	0.00	1.47
CaO	0.00	0.00	0.04	0.04	0.00	0.00	0.00	0.02	0.03	0.00	0.00	0.00	0.00	0.72	0.54	0.00
Na$_2$O	0.04	0.07	0.03	0.00	0.00	0.00	0.04	0.02	0.03	0.01	0.00	0.02	0.01	0.55	0.42	0.06
K$_2$O	0.52	0.84	0.73	0.04	0.05	0.35	0.04	0.01	0.02	0.05	0.02	0.17	0.02	0.00	0.00	0.00
Total	86.74	86.37	88.40	86.39	86.15	86.46	86.22	87.48	86.71	87.56	87.33	87.16	95.66	95.99	95.19	95.70
	Formula based on 28 oxygens												Formula based on 46 oxygens			
Si	5.563	5.945	5.339	5.533	5.376	5.406	5.271	5.300	5.359	5.244	5.269	5.324	7.762	7.836	8.009	7.779
Ti	0.007	0.056	0.049	0.022	0.006	0.019	0.007	0.013	0.013	0.019	0.017	0.022	0.161	17.463	17.537	0.123
Al	5.773	4.959	5.454	5.143	5.560	5.579	5.845	5.684	5.543	5.731	5.598	5.613	17.469	2.822	2.607	17.812
Cr	0.002	0.009	0.001	0.001	0.000	0.000	0.002	0.000	0.010	0.001	0.009	0.002	0.002	0.863	0.713	0.009
Fe	3.350	4.652	3.888	2.856	2.929	3.891	3.863	3.790	3.311	3.550	3.410	4.087	3.335	0.000	0.000	2.751
Mn	0.024	0.003	0.031	0.010	0.008	0.025	0.022	0.026	0.050	0.044	0.052	0.034	0.055	0.025	0.056	0.081
Mg	4.747	3.763	5.012	6.293	5.951	4.818	4.775	5.022	5.549	5.273	5.553	4.737	0.551	0.000	0.000	0.617
Ca	0.000	0.000	0.010	0.008	0.000	0.000	0.000	0.005	0.007	0.000	0.000	0.000	0.000	0.152	0.115	0.000
Na	0.018	0.031	0.013	0.002	0.001	0.000	0.018	0.008	0.012	0.005	0.001	0.008	0.004	0.131	0.100	0.033
K	0.135	0.230	0.192	0.012	0.013	0.093	0.009	0.003	0.006	0.013	0.004	0.044	0.007	0.000	0.000	0.000
Total	7.980	7.954	7.954	7.955	7.965	7.974	8.030	7.954	7.989	7.970	7.963	7.959	29.346	29.292	29.137	29.205
Mg/(Mg + Fe)	0.586	0.447	0.563	0.688	0.670	0.553	0.553	0.570	0.626	0.598	0.620	0.537	0.142	0.234	0.215	0.183

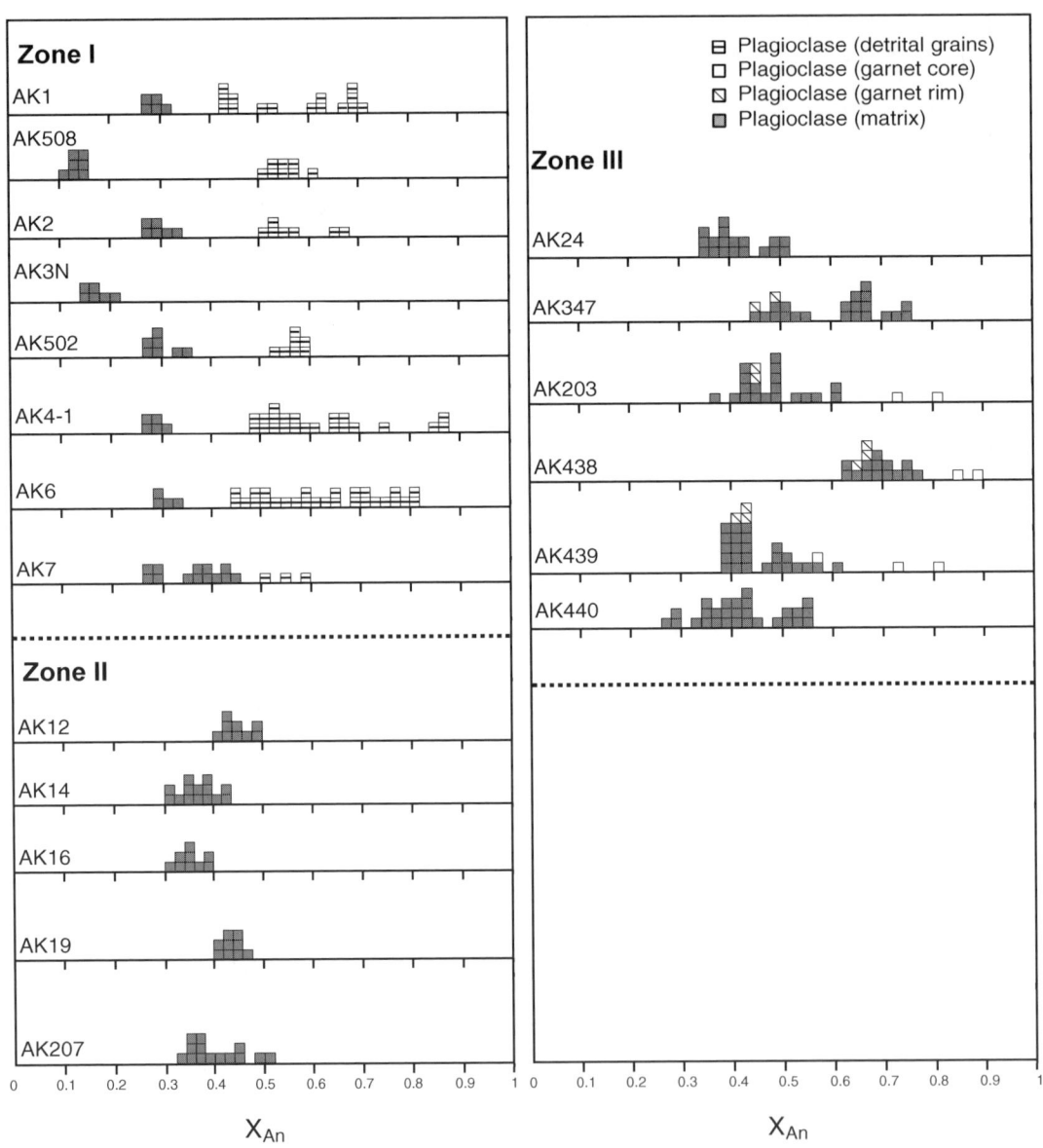

FIG. 7. Compositional diagram of plagioclases from each zone of the Hwasan area. $X_{An} = Ca/(Ca + Na + K)$.

Plagioclase in pelitic rock close to granite contacts ranges in X_{An} from 0.26 to 0.56.

P-T Conditions and P-T Paths

Qualitative prograde P-T paths for the OMB has been proposed by Oh et al. (1995b) and Kim et al. (1995). In this study, the prograde P-T path of the OMB is better constrained using P-T estimates and mineral parageneses. Retrogressive effects are common around garnets in high-grade pelitic and psammitic rocks from the Hwasan area. For example, the plagioclase composition is variable and/or chlorite commonly replaces biotite in contact with garnet. Such local disequilibria make it difficult to constrain the metamorphic P-T conditions. Fortunately, mineral inclusions in Zone III zoned garnets do not show important retrograde metamorphism. These

TABLE 4. Representative Microprobe Analyses of Plagioclase in Zone III (formula based on 8 oxygens)[1]

Sample no.:	AK1	AK2	AK502	AK7	AK12	AK14	AK207	AK19	AK24	AK347	AK203	AK23	AK438	AK439	AK440
Metamorphic zone:		Zone I				Zone II						Zone III			
SiO_2	61.76	60.59	60.72	63.03	57.37	59.55	60.31	57.05	58.60	56.90	56.19	46.51	57.05	59.37	57.22
TiO_2	0.01	0.00	0.00	0.04	0.01	0.02	0.03	0.01	0.00	0.02	0.03	0.01	0.01	0.02	0.02
Al_2O_3	24.14	24.23	24.84	22.97	26.77	25.60	25.25	26.95	26.47	26.26	27.68	34.62	26.95	25.65	26.90
Cr_2O_3	0.00	0.00	0.00	0.00	0.02	0.00	0.00	0.02	0.00	0.04	0.00	0.00	0.02	0.03	0.00
FeO*	0.10	0.20	0.07	0.21	0.08	0.06	0.11	0.07	0.05	0.69	0.07	0.11	0.07	0.20	0.11
MnO	0.06	0.03	0.00	0.00	0.00	0.03	0.00	0.03	0.00	0.05	0.00	0.03	0.03	0.00	0.02
MgO	0.00	0.00	0.00	0.02	0.01	0.01	0.00	0.00	0.02	0.01	0.03	0.00	0.00	0.00	0.00
CaO	5.96	6.02	6.76	4.96	9.10	7.78	7.02	9.49	8.39	9.30	10.35	18.06	9.49	8.03	9.02
Na_2O	8.46	8.21	7.71	8.67	6.37	7.32	7.55	6.22	6.82	6.11	5.88	1.18	6.22	7.21	6.52
K_2O	0.10	0.12	0.07	0.69	0.24	0.06	0.09	0.07	0.05	0.06	0.05	0.00	0.07	0.07	0.05
Total	100.59	99.40	100.17	100.59	99.97	100.43	100.36	99.91	100.40	99.44	100.28	100.52	99.91	100.58	99.86
Si	2.730	2.713	2.696	2.784	2.575	2.647	2.676	2.563	2.609	2.573	2.521	2.128	2.563	2.639	2.569
Ti	0.000	0.000	0.000	0.001	0.001	0.001	0.001	0.000	0.000	0.001	0.001	0.000	0.000	0.001	0.001
Al	1.257	1.278	1.300	1.196	1.416	1.342	1.321	1.427	1.389	1.400	1.464	1.867	1.427	1.344	1.424
Cr	0.000	0.000	0.000	0.000	0.001	0.000	0.000	0.001	0.000	0.002	0.000	0.000	0.001	0.001	0.000
Fe	0.004	0.007	0.003	0.008	0.003	0.002	0.004	0.003	0.002	0.026	0.003	0.004	0.003	0.007	0.004
Mn	0.002	0.001	0.000	0.000	0.000	0.001	0.000	0.001	0.000	0.002	0.000	0.001	0.001	0.000	0.001
Mg	0.000	0.000	0.000	0.001	0.001	0.001	0.000	0.000	0.001	0.001	0.002	0.000	0.000	0.000	0.000
Ca	0.282	0.289	0.322	0.235	0.437	0.371	0.334	0.457	0.400	0.451	0.498	0.885	0.457	0.383	0.434
Na	0.725	0.713	0.664	0.743	0.555	0.631	0.649	0.542	0.589	0.536	0.512	0.105	0.542	0.622	0.568
K	0.006	0.007	0.004	0.039	0.014	0.003	0.005	0.004	0.003	0.003	0.003	0.000	0.004	0.004	0.003
Total	5.007	5.008	4.988	5.007	5.001	4.998	4.990	4.996	4.993	4.995	5.004	4.990	4.996	5.001	5.004
X_{Ab}	0.716	0.707	0.671	0.731	0.551	0.628	0.657	0.540	0.594	0.541	0.505	0.106	0.540	0.616	0.565
X_{An}	0.278	0.286	0.325	0.231	0.435	0.369	0.338	0.456	0.403	0.455	0.491	0.894	0.456	0.379	0.432
X_{Or}	0.006	0.007	0.004	0.038	0.014	0.003	0.005	0.004	0.000	0.003	0.000	0.000	0.004	0.004	0.000

[1]X_{Ab} = Na/(Na + Ca + K); X_{An} = Ca/(Na + Ca + K); X_{Or} = K/(Na + Ca + K).

TABLE 5. Representative Garnet and Inclusion Compositional Data used in the Geothermobarometries

Sample:	AK24	AK347	AK203	AK203	AK438	AK438	AK439	AK439	AK440
Garnet: $Xi = i/(Fe + Mn + Mg + Ca)$									
	rim	rim	core	rim	core	rim	core	rim	rim
X_{Fe}	0.739	0.739	0.643	0.632	0.704	0.700	0.611	0.605	0.733
X_{Mg}	0.129	0.127	0.118	0.139	0.098	0.104	0.126	0.115	0.125
X_{Mn}	0.043	0.047	0.177	0.104	0.114	0.030	0.143	0.129	0.070
X_{Ca}	0.089	0.087	0.062	0.125	0.085	0.166	0.120	0.151	0.072
Mg/Fe	0.175	0.172	0.184	0.221	0.139	0.148	0.205	0.191	0.168
Biotite (in garnet): $Xi = i/\Sigma[VI]$									
	rim	rim	core	rim	core	rim	core	rim	rim
$X^{[VI]}_{Al}$	0.134	0.169	0.148	0.136	0.174	0.157	0.147	0.163	0.141
X_{Ti}	0.029	0.023	0.031	0.031	0.010	0.032	0.021	0.031	0.026
X_{ann}	0.390	0.372	0.327	0.335	0.337	0.393	0.255	0.391	0.386
X_{phl}	0.412	0.396	0.438	0.456	0.479	0.418	0.572	0.413	0.443
Mg/Fe	1.057	1.064	1.339	1.364	1.421	1.064	2.243	1.057	1.148
Plagioclase (in garnet): $Xi = i/(Ca + Na + K)$									
	rim	rim	core	rim	core	rim	core	rim	
X_{Ca}	0.349	0.457	0.815	0.464	0.900	0.650	0.683	0.414	
X_{Na}	0.651	0.543	0.185	0.536	0.100	0.350	0.315	0.583	
Muscovite (in garnet): $X^{[VI]}Al = X^{[VI]}Al/\Sigma[VI]$; $XK = K/2$									
				rim				rim	
$X^{[VI]}_{Al}$				0.834				0.924	
X_K				0.870				0.820	

inclusions allow for a better P-T estimate than those matrix crystals. The P-T conditions of Table 5 are estimated solely using analyses of inclusions within garnet rims.

Five geothermobarometers were employed: (1) the garnet-biotite geothermometer (Ferry and Spear, 1978: the activity model of Berman, 1990; Hodges and Spear, 1982); (2) the garnet-plagioclase-biotite-muscovite geobarometer (Hodges and Crowley, 1985; Hoisch, 1990); (3) the garnet-plagioclase-muscovite-quartz geobarometer (Hodges and Crowley, 1985; Hoisch, 1990); (4) the garnet-plagioclase-biotite-quartz geobarometer (Hoisch, 1990); and (5) the garnet-plagioclase-rutile-ilmenite-quartz geobarometer (Bohlen and Liotta, 1986). Results are presented in Figure 8.

Representative compositions of plagioclase and biotite inclusions and adjacent garnet grains are listed in Table 5. P-T estimates for samples AK24, AK347, and AK203 in Zone III, which use analyses of biotite and plagioclase inclusions in the rim and from the adjacent garnet, are 5.0–6.0, 5.5–6.7, and 6.8–8.8 kbar, and 560–610, 570–615, and 570–620°C, respectively. Correspondingly, P-T estimates for samples AK438 and AK439 in Zone III are 2.0–5.0 kbar and 420–460°C for inclusions in the core, and 6.8–9.0 kbar and 580–620°C for those in the rim (Fig. 8). These P-T estimates by inclusions within the garnet support the prograde P-T path.

Metamorphic pressures involved in the formation of the K-feldspar-biotite-muscovite-quartz assemblage in Zone I were deduced using the Massonne and Schreyer (1987) geobarometer. The highest Si value of muscovite in pelitic rocks gives a range of 3.6–4.4 kbar at 350–450°C (Kim et al., 2002; Fig. 8). The above results for Zones III and I indicate that the Hwasan area experienced an intermediate-P/T type metamorphic event (*M1* metamorphism), in which the peak P-T conditions failed to attain those necessary for the formation of kyanite.

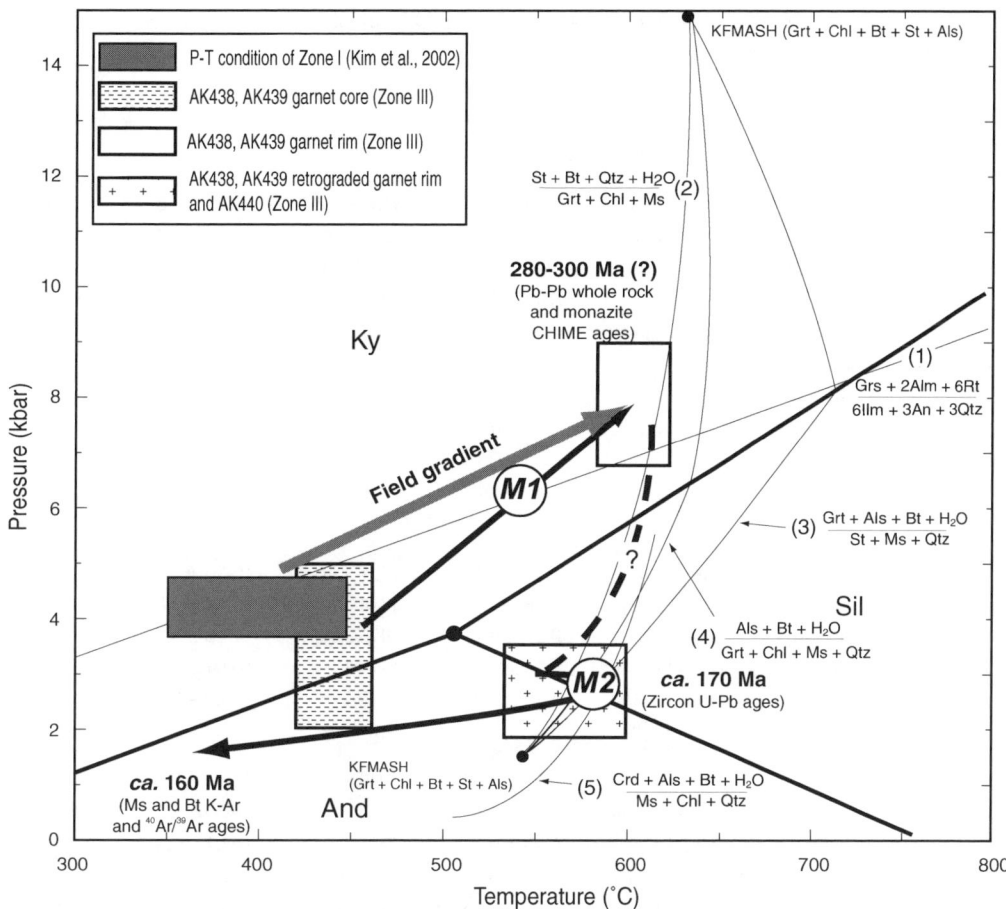

FIG. 8. *P-T* paths of the Hwasan area rocks, determined from reaction history and *P-T* estimates of Zone III made in this study. Thin line (1) denotes the reaction from Bohlen and Liotta (1986) and the thin curves (2)–(6) denote the KFMASH reactions adapted from Spear and Cheney (1989) and Spear et al. (1995). Reaction curves among Al-silicates are from Holdaway (1971). Geochronological data of Cheong et al. (2003a) are also shown.

However, computed P-T conditions using retrograded garnet rims, rim compositions of plagioclase, and matrix compositions of biotite and muscovite for samples AK438 and AK439 in Zone III gave 535–600°C and 2.9–3.5 kbar (Fig. 8). P-T estimates for samples AK 440 and AK441 sited near granite also yield values of 530–600°C and 1.7–3.5 kbar, based on analyses of biotite and plagioclase in contact with the garnet rim. These results are consistent with the occurrence of andalusite in these samples. Along with the prograde P-T path determined by inclusions in garnet, these data also indicate that metasediments close to the granites underwent a clockwise P-T path in which retrograde low-P/T M2 contact metamorphism arising from intrusion of the granites occurred after the peak regional metamorphism (Fig. 8).

Age Determinations

K-Ar and $^{40}Ar/^{39}Ar$ age determinations were carried out at Okayama University of Science, Japan, on muscovite and biotite in pelitic and psammitic rocks collected systematically from Zones I, II, and III (Fig. 2). Muscovite and biotite from both massive and foliated granites were also analyzed by the K-Ar method. Analytical methods are identical to those described by Itaya etal. (1991) and Hyodo et al. (1999).

TABLE 6. K-Ar Age Data of Metasediments and Granites in the Hwasan Area

Sample no.	Mineral	Fraction wt%	Potassium, 10^{-8}ccSTP/g	Rad. ^{40}Ar,	Age, Ma	Non-rad. ^{40}Ar, %
			Zone I			
AK1	Biotite	61–74 µm	5.29 ± 0.11	3188 ± 32	148.9 ± 3.2	2.9
AK2	Biotite	61–74 µm	6.65 ± 0.13	4203 ± 39	155.8 ± 3.3	1.4
AK3	Muscovite	2–4 µm	5.87±0.12	3714±35	156.1±3.2	3.3
AK3	Muscovite	4–8 µm	6.03 ± 0.12	3854 ± 40	157.5 ± 3.4	1.9
AK4	Biotite	61–74 µm	6.73 ± 0.14	4207 ± 41	150.9 ± 3.2	3.6
AK4	Muscovite	61–74 µm	5.60 ± 0.11	3725 ± 39	164.3 ± 3.5	3.6
AK5	Biotite	61–74 µm	6.54 ± 0.13	4124 ± 41	155.6 ± 3.3	0.9
AK6	Biotite	61–74 µm	7.22 ± 0.14	4550 ± 45	155.6 ± 3.3	1.8
AK6	Muscovite	61–74 µm	6.38 ± 0.13	4043 ± 48	156.3 ± 3.4	2.1
AK7	Muscovite	61–74 µm	6.41 ± 0.13	4108 ± 38	158.1 ± 3.4	2.0
			Zone II			
AK14	Biotite	61–74 µm	7.16 ± 0.14	4724 ± 46	162.6 ± 3.5	0.9
AK14	Muscovite	61–74 µm	5.78 ± 0.12	3853 ± 38	164.2 ± 3.5	1.9
AK17	Muscovite	61–74 µm	8.29 ± 0.17	6075 ± 60	179.6 ± 3.8	1.3
AK19	Biotite	104–147 µm	6.54 ± 0.13	4370 ± 44	164.4 ± 3.5	1.3
AK19	Muscovite	104–147 µm	4.42 ± 0.09	3212 ± 32	178.0 ± 3.8	2.1
			Zone III			
AK24	Muscovite	104–147 µm	6.19 ± 0.12	4043 ± 41	156.7 ± 3.4	2.2
AK20	Muscovite	104–147 µm	6.82 ± 0.14	4656 ± 47	167.9 ± 3.6	1.8
AK20	Biotite	104–147 µm	6.96 ± 0.14	4641 ± 32	164.1 ± 3.5	1.1
AK21	Biotite	61–74 µm	7.31 ± 0.15	4852 ± 47	163.4 ± 3.5	0.6
AK22	Biotite	61–74 µm	7.43 ± 0.15	4908 ± 48	162.7 ± 3.5	1.4
AK23	Muscovite	61–74 µm	7.94 ± 0.16	5134 ± 52	159.4 ± 3.4	2.3
AK438	Muscovite	61–74 µm	5.91 ± 0.12	3826 ± 37	159.7 ± 3.4	1.1
AK439	Biotite	61–74 µm	5.35 ± 0.11	3402 ± 34	156.8 ± 3.4	3.1
AK439	Muscovite	61–74 µm	5.97 ± 0.12	3852 ± 39	159.8 ± 3.4	3.5
			Granite			
AKJ1	Biotite	104–147 µm	6.25 ± 0.13	3953 ± 39	155.9 ± 3.3	1.7
AKJ2	Biotite	104–147 µm	7.33 ± 0.15	4626 ± 45	155.7 ± 3.3	0.7
AKJ3	Muscovite	104–147 µm	8.43 ± 0.17	5396 ± 53	155.7 ± 3.4	0.8

K-Ar age results are listed in Table 6. Muscovite and biotite K-Ar age ranges for Zone I are 156–158 and 149–156 Ma, respectively, although one muscovite yielded an older age of 164 Ma. Muscovite separates from two size fractions of 2–4 µm and 4–8 µm in AK3 yielded ages of 156.1 ± 3.2 and 157.5 ± 3.4 Ma, respectively. Biotite and muscovite ages in Zone II are 163–164 Ma, and 164–180 Ma, respectively. Muscovites of samples AK17 and AK19 gave a range of 178–180 Ma, significantly older than the biotite age in the same sample. In Zone III, biotite and muscovite ages are similar in the same sample with a range of 157–168 Ma. Some older muscovite ages in Zone II suggest that Zone II has been less affected by M2 metamorphism than Zones I and III, which are closer to the granites. The majority of biotite and muscovite ages from all zones are concentrated at ~160 Ma. Biotites in massive granites

AKJ1 and AKJ2 gave identical ages of 156 Ma, similar to the muscovite in foliated granite AKJ3. The intrusion ages of AKJ1 and AKJ2, determined using U-Pb zircon isotope ages by our recent work (Oh et al., 2004) are 167–169 Ma.

The age results of single-grain ^{40}Ar/^{39}Ar analysis are listed in Table 7. In Zone I, a coarse-grained biotite from sample AK3 yielded a well-defined plateau age of 158.2 ± 0.5 Ma with 96% of the total argon fraction, which is same as the K-Ar muscovite ages from the same sample (Fig. 9A). In Zone II, muscovite and biotite also display well-defined plateau ages with 50–95 % of the total argon fraction. Muscovites from samples AK11, AK12, AK13, and AK16 yielded 161.5 ± 0.5, 174.5 ± 1.8, 166.2 ± 1.2, and 166.1 ± 0.7 Ma, respectively, and biotites, 154.2 ± 1.4, 163.7 ± 2.5, 159.8 ± 0.6, and 160.2 ± 0.7 Ma, respectively (Fig. 9B). The muscovite of AK12 (175 Ma) in Zone II is 11 Ma older than the corresponding biotite. In Zone III, muscovite AK24 gave a plateau age of 158.3 ± 2.4 Ma with 75% of the total argon fraction. Biotites from AK24 and AK20 gave plateau ages of 159.7 ± 0.6 and 161.5 ± 0.9 Ma, respectively, with 50–60% of the total argon fraction (Fig. 9C). The ^{40}Ar/^{39}Ar ages from the three zones are in accord with the K-Ar ages: K-Ar and ^{40}Ar/^{39}Ar ages are identical within the error limits for samples AK24 and AK20 in Zone III and AK3 in Zone I. These data indicate that both metamorphic rocks and granites in the study area cooled through a temperature range of ~270–350°C in the mid-Jurassic.

d_{002} Values for Carbonaceous Material in Pelitic and Psammitic Rocks

The state of graphitization of carbonaceous material in the Hwasan area was investigated by X-ray diffraction analysis of 49 samples (Fig. 10A) using the method of Itaya (1981). Advancing graphitization is a well-accepted indicator of the thermal structure of a metamorphic sequence insofar as it depends mainly on the temperature of metamorphism (Itaya, 1981; Wang, 1989; Nishimura et al., 2000). Although metamorphic grade during regional *M1* metamorphism increases progressively from the greenschist to upper amphibolite facies toward the north, carbonaceous materials do not show a corresponding variation in d_{002} values. Instead, the present d_{002} values for carbonaceous materials fall within a narrow range, from 3.353 to 3.359 Å (3.354–3.359 Å in Zone I, 3.353–3.358 Å in Zone II, and 3.353–3.357 Å in Zone III), consistent with amphibolite-facies metamorphism having affected the whole area (Fig. 10B). Regional *M1* metamorphism for Zone I rocks is at greenschist-facies conditions. However, the d_{002} values for carbonaceous materials for these greenschist-facies rocks show fully ordered graphite (ca. 500°C), which is consistent with amphibolite-facies conditions, and point to the occurrence of a two-stage regional metamorphism.

The Effect of M2 Regional Thermal Metamorphism

Mesozoic granite plutons are widespread in and around the OMB, such that the shape of the OMB resembles an island within a sea of granite (Fig. 1B). Effects from an increase in the geothermal gradient as a result of the regional intrusion of the Mesozoic granite and a subsequent regional thermal effect are to be expected. Recent U-Pb age dating on Mesozoic granites reveals that most of their ages range from the late Triassic to mid-Jurassic (Cho et al., 2001; Ree et al., 2001, Lee et al., 2003; Cheong et al., 2003; Oh et al., 2004). However, most U-Pb ages of the Mesozoic granites are concentrated between 166 and 178 Ma, and the Late Triassic granites occur in a restricted area, separated from the main body of the OMB by a thrust fault (Fig. 1B).

Metapelites within 1–2 km of the boundary of the mid-Jurassic granite plutons experienced a M2 metamorphic overprint during which andalusite, sillimanite, and cordierite formed (Oh et al., 1997, 1998, 1999; Min and Cho, 1998; Kim and Cho, 1999). P-T conditions estimated from the metapelites are 2.1–5.0 kbar, 540–698°C and correspond to a low-*P/T* type metamorphism (Oh et al., 1997, 1998, 1999). According to Oh et al. (1997) the mid-Jurassic Boeun granite pluton, which has a roughly triangular outline with sides of about 20 km, heated the metamorphic rocks within 1 km of its boundary to temperatures in excess of 600°C, and those within 4 km to temperatures above 480 °C by about 0.5 m.y. after the intrusion. Many granite plutons larger than the Boeun pluton occur around the OMB, and might have heated areas lying further than 4 km from their boundaries. In addition, many granite plutons may exist beneath the OMB. Given that the width of the OMB is less than 20 km (Fig. 1), the mid-Jurassic intrusion of these vast granite plutons presumably heated the whole OMB to temperatures in excess of 480°C, and

TABLE 7. ^{40}Ar/^{39}Ar Analytical Data of Muscovite and Biotite from Metasediments in the Hwasan Area[1]

T(°C)	^{39}Ar Frac[1]	Non-rad. ^{40}Ar (%)	^{40}Ar*/^{39}Ar	Error (±)	^{37}Ar$_{Ca}$/^{39}Ar$_K$	Error (±)	Age (Ma)	Error (±)
AK3 biotite								
500	0.003	90.4	4.65	2.79	2.975	0.115	41.15	24.39
600	0.011	66.3	8.20	1.27	0.750	0.028	71.92	10.90
700	0.056	5.8	18.91	0.17	0.120	0.006	161.71	1.45
800	0.247	2.2	18.50	0.10	0.035	0.001	158.35	0.90
900	0.379	1.7	18.72	0.09	0.095	0.002	160.14	0.80
950	0.495	2.3	18.28	0.17	0.744	0.006	156.55	1.43
1000	0.639	1.3	18.74	0.09	0.294	0.002	160.36	0.78
1050	0.745	0.4	18.57	0.12	0.042	0.003	158.91	1.02
1100	0.875	3.5	18.11	0.13	0.081	0.002	155.20	1.14
1200	0.962	1.4	18.24	0.14	0.231	0.011	156.25	1.19
1500	1	5.4	18.27	0.21	0.410	0.016	156.46	1.78
							Integrated age = 157.2 ± 0.5	
AK11 biotite								
500	0.003	100.7	–	6.937	0.000	0.000	–	–
600	0.014	41.2	21.322	0.926	0.619	0.016	181.35	7.50
700	0.089	19.1	19.846	0.138	0.000	0.000	169.36	1.17
800	0.363	11.9	18.250	0.104	0.029	0.001	156.32	0.91
900	0.581	0.0	18.915	0.062	0.005	0.001	161.77	0.60
950	0.801	0.6	18.877	0.071	0.037	0.001	161.45	0.67
1000	0.892	0.3	19.421	0.127	0.009	0.003	165.90	1.09
1050	0.958	0.0	19.282	0.103	0.162	0.003	164.77	0.90
1100	1	0.0	19.375	0.201	0.122	0.005	165.52	1.67
							Integrated age = 161.1 ± 0.5	
AK11 muscovite								
500	0.000	39.0	514.770	3244.000	144.574	904.686	2287.21	8169.23
600	0.037	154.5	–	9.399	4.334	0.665	–	–
700	0.123	32.4	9.553	2.282	1.857	0.062	83.51	19.50
800	0.468	43.9	10.414	0.916	3.703	0.336	90.85	7.80
900	0.662	11.0	17.975	0.709	0.847	0.018	154.06	5.84
950	0.711	0.0	18.812	5.169	0.928	0.218	160.92	42.30
1000	0.806	0.0	18.601	0.892	0.400	0.025	159.15	7.69
1050	0.878	0.0	18.422	0.930	0.597	0.034	157.39	7.70
1200	1	1.2	19.401	0.968	0.354	0.015	165.74	7.91
							Integrated age = 149.3 ± 5.3	
AK11 muscovite								
500	0.010	84.0	1.994	5.029	3.399	0.146	17.75	44.56
600	0.031	34.9	12.030	2.305	1.731	0.044	104.55	19.47
700	0.075	21.1	14.854	0.601	0.782	0.015	128.24	5.02
800	0.228	13.7	17.175	0.565	0.778	0.022	147.48	4.67
900	0.551	8.9	17.676	0.280	0.200	0.002	151.60	2.32
950	0.683	3.9	18.398	0.162	0.298	0.006	157.53	1.37
1000	0.775	0.0	19.194	0.211	0.411	0.010	164.04	1.75
1050	0.902	0.7	18.493	0.240	0.070	0.010	158.31	1.99
1100	1	1.0	19.661	0.245	0.124	0.006	167.86	2.02
							Integrated age = 152.1 ± 1.3	
AK12 biotite								
600	0.020	0.0	33.193	0.897	0.306	0.393	244.95	6.34
700	0.102	0.0	23.609	0.460	0.187	0.063	177.57	3.45
800	0.278	0.0	23.585	0.257	0.105	0.028	177.41	2.11
850	0.579	7.3	21.297	0.259	0.022	0.024	160.94	2.09
900	0.827	5.1	21.757	0.899	0.100	0.020	164.94	6.56
950	0.898	4.8	22.482	1.036	0.428	0.075	168.16	7.82
1000	1	0.0	22.692	2.480	0.293	0.047	170.99	2.04
							Integrated age = 168.4 ± 2.5	

Table continues

TABLE 7. Continued

T(°C)	^{39}Ar Frac[1]	Non-rad. ^{40}Ar (%)	^{40}Ar*/^{39}Ar	Error (±)	^{37}Ar$_{Ca}$/^{39}Ar$_K$	Error (±)	Age (Ma)	Error (±)
AK12 muscovite								
600	0.013	1.3	22.429	2.863	0.739	0.142	169.11	20.63
700	0.041	0.3	22.942	1.357	0.116	0.061	172.80	9.80
800	0.138	3.5	24.036	0.450	0.070	0.017	180.63	3.38
900	0.496	0.9	22.883	0.244	0.035	0.007	172.37	2.01
950	0.640	0.0	22.923	0.213	0.082	0.013	172.65	1.83
1000	0.796	1.9	23.008	0.292	0.000	0.000	173.27	2.32
1050	0.885	0.1	23.395	0.422	0.172	0.019	176.04	3.19
1100	1	1.4	23.825	0.384	0.185	0.018	179.12	2.94
							Integrated age = 174.4 ± 1.8	
AK13 biotite								
500	0.007	31.0	4.319	0.819	0.787	0.009	38.24	7.17
600	0.018	1.2	11.657	0.405	0.410	0.007	101.40	3.43
700	0.034	2.7	16.837	0.141	0.337	0.004	144.68	1.20
800	0.073	0.8	17.954	0.165	0.138	0.002	153.89	1.39
900	0.435	2.3	18.738	0.127	0.192	0.004	160.31	1.09
950	0.571	1.0	18.600	0.157	0.108	0.006	159.18	1.32
1000	0.660	1.7	18.541	0.116	0.114	0.001	158.70	1.00
1050	0.838	0.5	18.685	0.089	0.066	0.000	159.88	0.80
1100	1	0.4	18.845	0.087	0.108	0.001	161.19	0.78
							Integrated age = 158.1 ± 0.6	
AK13 muscovte								
500	0.006	21.0	72.562	10.893	4.722	0.720	554.56	71.68
600	0.086	0.0	21.635	0.280	0.740	0.026	183.88	2.29
700	0.289	0.0	20.049	0.247	0.212	0.011	171.02	2.04
800	0.465	0.6	19.782	0.257	0.154	0.013	168.40	2.13
900	0.604	0.0	19.601	0.177	0.239	0.028	167.36	1.48
950	0.807	0.0	19.871	0.195	0.088	0.018	169.57	1.62
1000	0.911	0.0	19.298	0.246	0.257	0.019	164.90	2.03
1100	1	0.0	19.582	0.299	0.508	0.026	167.21	2.46
							Integrated age = 170.6 ± 1.2	
AK16 biotite								
500	0.005	98.8	0.870	1.474	2.924	0.153	7.76	13.13
600	0.017	90.8	2.212	1.342	1.112	0.208	19.69	11.88
700	0.068	22.3	18.937	0.288	0.221	0.032	161.94	2.38
800	0.222	21.0	17.502	0.326	0.103	0.003	150.17	2.70
900	0.377	6.2	18.299	0.182	0.000	0.000	156.72	1.53
950	0.617	1.4	19.476	0.095	0.052	0.002	166.35	0.84
1000	0.835	2.0	19.047	0.104	0.036	0.026	162.85	0.91
1050	0.963	3.5	18.434	0.142	0.045	0.002	157.83	1.20
1100	1	0.0	19.441	0.310	0.158	0.006	166.06	2.55
							Integrated age = 157.8 ± 0.7	
AK16 muscovite								
500	0.005	0.0	13.862	1.073	3.350	0.177	119.95	8.99
600	0.020	0.0	14.443	0.348	0.768	0.031	124.81	2.91
700	0.042	3.1	19.099	0.401	0.153	0.024	163.27	3.30
800	0.208	1.9	19.531	0.199	0.099	0.008	166.80	1.66
900	0.577	0.0	19.597	0.082	0.073	0.001	167.34	0.75
950	0.652	1.9	19.130	0.195	0.098	0.006	163.53	1.62
1000	0.757	0.0	19.539	0.091	0.173	0.007	166.86	0.81
1050	0.915	1.2	19.106	0.152	0.070	0.004	163.33	1.28
1100	1	0.0	20.619	0.127	0.112	0.005	175.65	1.08
							Integrated age = 166.1 ± 0.6	

[1] ^{40}Ar* = radiogenic ^{40}Ar; *= normalized ^{39}Ar cumulate.

Table continues

TABLE 7. Continued

T(°C)	^{39}Ar Frac[1]	Non-rad. ^{40}Ar (%)	^{40}Ar*/^{39}Ar	Error (±)	^{37}Ar$_{Ca}$/^{39}Ar$_K$	Error (±)	Age (Ma)	Error (±)
AK24 biotite								
500	0.006	107.4	2.061	2.399	1.343	0.053	18.53	21.68
600	0.014	52.1	7.797	0.888	1.406	0.034	68.44	7.65
700	0.096	10.0	17.482	0.150	0.067	0.003	150.01	1.27
800	0.512	2.5	18.331	0.183	0.063	0.001	156.98	1.54
900	0.750	1.2	18.541	0.148	0.032	0.001	158.70	1.26
950	0.842	3.2	18.913	0.103	0.062	0.002	161.75	0.90
1000	0.891	2.5	18.879	0.185	0.379	0.005	161.47	1.55
1050	0.924	0.0	19.081	0.138	0.378	0.008	163.12	1.17
1100	1	1.6	18.248	0.155	0.968	0.010	156.30	1.31
							Integrated age = 155.9 ± 0.8	
AK24 muscovite								
500	0.009	0.0	13.216	0.831	0.000	0.000	114.53	6.98
600	0.038	14.4	5.974	3.790	0.000	0.000	52.68	32.93
700	0.132	0.0	17.840	1.579	0.000	0.000	152.95	12.98
800	0.755	0.3	18.585	0.252	0.020	0.001	159.06	1.87
850	0.816	0.0	15.080	3.128	0.000	0.000	130.12	26.04
900	0.825	0.0	17.455	0.516	0.000	0.000	149.78	4.26
900	0.848	0.0	3.499	4.403	0.000	0.000	31.04	38.73
1000	1	2.9	14.124	2.824	0.183	0.025	122.14	23.61
							Integrated age = 144.7 ± 5.0	
AK20 muscovite								
500	0.01	48.1	14.428	1.191	1.326	0.046	124.69	9.95
600	0.037	18.2	16.786	0.238	0.609	0.017	144.26	1.98
700	0.149	3.2	18.783	0.258	0.154	0.004	160.68	2.14
800	0.341	1.3	19.773	0.354	0.091	0.003	168.77	2.90
900	0.603	1.4	18.738	0.158	0.048	0.002	160.31	1.33
950	0.834	4.2	18.304	0.188	0.089	0.002	156.76	1.57
1000	0.935	6.5	19.505	0.150	0.069	0.006	166.58	1.27
1050	0.974	11.2	17.708	0.221	0.314	0.015	151.87	1.84
1100	1	38.4	11.891	0.804	0.638	0.019	103.38	6.79
							Integrated age = 159.2 ± 0.9	

hence caused the entire Hwasan area to have experienced a low-P/T type regional thermal metamorphic episode.

In the Hwasan area, the conclusion of a regional contact *M2* thermal metamorphic event is supported by the following geochronological data: (1) Most biotite and muscovite K-Ar and ^{40}Ar/^{39}Ar ages of OMB metasediments are younger than the intrusion age of mid-Jurassic granites, but are similar to the biotite and muscovite K-Ar and ^{40}Ar/^{39}Ar ages of mid-Jurassic granites. (2) The d_{002} values of carbonaceous materials exhibit a limited range from 3.353 to 3.359Å, consistent with amphibolite-facies regional metamorphism. And (3) Jurassic massive and foliated granites have contact metamorphic aureoles of 1–2 km.

Despite the occurrence of a regional low-P/T thermal metamorphic event, low-P/T mineral assemblages have been recognized only in a few metasediments within 1–2 km of the mid-Jurassic granites (Kim et al., 1995; Oh et al., 1995a, 1998, 1999; Min and Cho, 1998; Kim and Cho, 1999; Cho and Kim, 2002). A similar pattern was found in the Hwasan area, as shown in Figure 2; several factors may contribute to the scarcity of low-P/T assemblages outside the contact aureole. Both Zones II and III underwent an earlier metamorphism at temperatures above 570°C during the *M1* intermediate-P/T type metamorphism. Consequently, the *M2* amphibolite facies thermal overprint may not have sufficient heat to generate distinctive low-P/T mineral assemblages in areas farther than 1–2 km

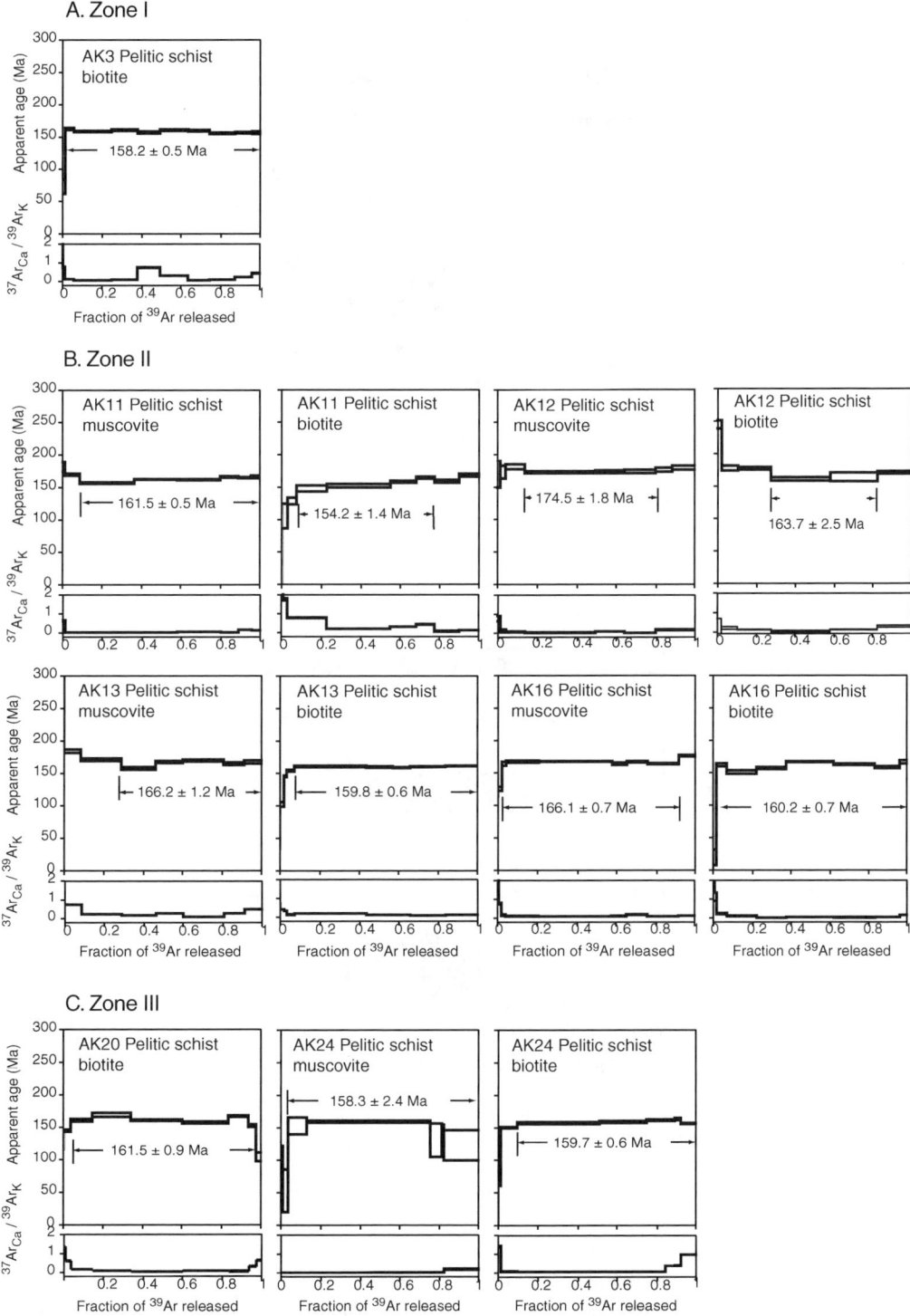

FIG. 9. $^{40}Ar/^{39}Ar$ age spectra and Ca/K ratios from single biotite and muscovite grains in pelitic rocks from Zone I (A), Zone II (B), and Zone III (C).

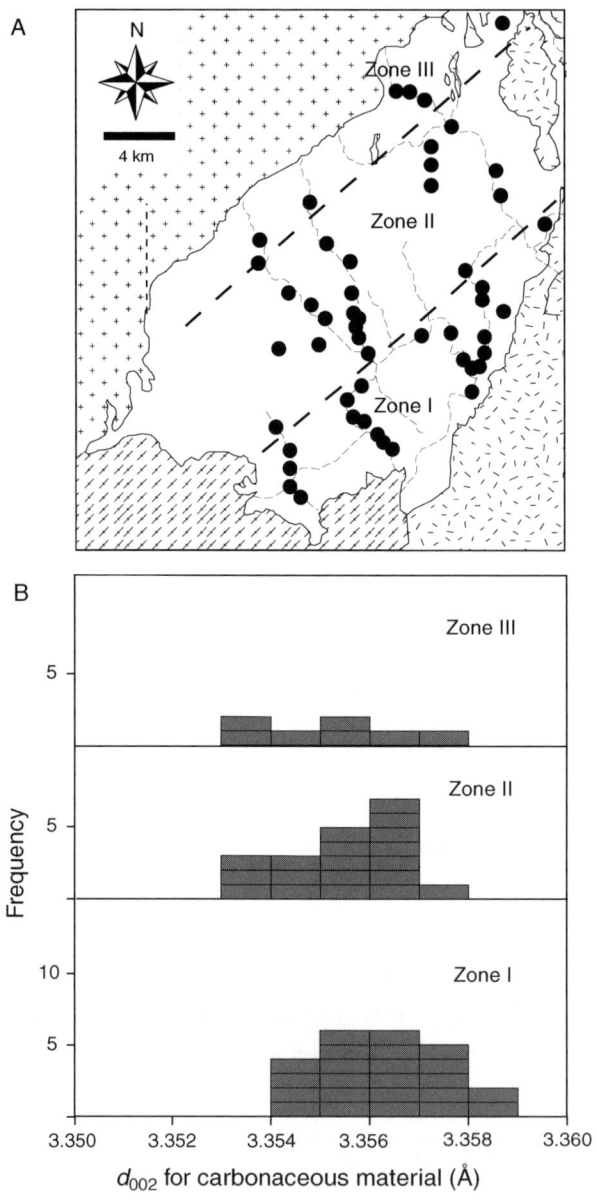

FIG. 10. A. Sample localities in the Hwasan area from which carbonaceous material was obtained. B. Frequency distribution of d_{002} values for carbonaceous materials from pelitic and psammitic rocks in Zones I, II, and III.

from the granite contacts. Furthermore, the similar ages of biotite, muscovite, and zircon in the granites suggest very fast cooling of these plutons. As a result, the duration of heating may have been insufficient to generate a distinctive low-P/T mineral assemblage in those areas beyond 1–2 km, but long enough to allow conversion of carbonaceous materials to graphite.

Tectonic Implications

Although recent geochronological data have constrained the main *M1* metamorphism at between 280 and 300 Ma (Kim et al. 2001; Cheong et al. 2003), the causal mechanism for this metamorphic event in the OMB remains unknown. No evidence has been found for a collision zone in the OMB, such as the occurrence of high-pressure minerals or ophiolite suites that would allow the OMB to be considered as an eastward extension of the Dabie-Sulu collisional belt in east central China.

P-T estimates and the discovery of high-pressure minerals by Oh et al. (2002) led them to propose a 225–257 Ma collisional event in the Hongseong area of the southwestern part of the Gyeonggi massif, located 100 km from the northern boundary of the OMB. They further suggested that, instead of the Imjingang belt, the Hongseong area is an extension of the Dabie-Sulu collision belt in China. If this suggestion is indeed valid, then the OMB could have undergone the *M1* intermediate-P/T metamorphic episode as a result of compression caused by the distal collisional event. Inasmuch as no clear evidence of collision in the Imjingang belt exists, the high-pressure amphibolite facies metamorphism in the Imjingang belt during the Late Permian and Early Triassic is possibly a byproduct of intermediate-P/T metamorphism in the area peripheral to the collision, as it is in the OMB.

Middle to Late Triassic granites are few and have a restricted distribution in the OMB. However, the mid-Jurassic granite intruded not only the OMB but also the Gyeonggi and Yeongnam massifs, forming a granitic complex about 200 km long and >150 km wide. Such a spatial breadth for Middle Jurassic magmatism is too large to be due solely to simple subduction of oceanic lithospere. Kinoshita and Itô (1988, 1990) classified granitoids along the northeast Asian continental margin into an older granitic belt on the continental side and a younger granitic belt on the oceanic side, based on available geochronological data. The older granitic belt is arranged as follows: a Triassic granitic belt in South China (245–203 Ma), a Jurassic granitic belt in South Korea (180 Ma), the Cretaceous West Sikhote Alin belt (120 Ma), and the Okhotsk-Chukotka belt (from the Albian to the early Cenomanian). The younger granitic belt forms a sequence from the Late Jurassic granitic belt in South China (160–130 Ma), through the Cretaceous granitic belts in Gyeongsang Basin, South Korea (100 Ma), southwestern Japan (80 Ma), the East Sikhote Alin belt (60 Ma), and southeast Hokkaido (60 Ma), to Cenozoic granitic belts in Kamchatka (45 Ma) and the Aleutian islands (30 Ma). Based on these age and spatial distributions, Kinoshita (1995) suggested migration of igneous activity resulting from subduction of the Farallon-Izanagi ridge (older granitic belt) and the Kula-Pacific ridge (younger granitic belt) toward the northeast. We conclude that, as Kinoshita (1995, 2002) suggested, subduction of the Farallon-Izangi ridge caused extensive mid-Jurassic magmatism in South Korea. The extensive regional plutonism was the direct cause of the *M2* regional thermal metamorphic event in the OMB.

Recent U-Pb zircon and CHIME monazite dates show that large-scale dextral strike-slip movement occurred along the Honam shear zone between the OMB and the Precambrian Yeongnam massif during the mid-Jurassic (Kim and Turek, 1996; Cho et al. 1999). Large-scale strike-slip during the Middle Jurassic may be related to dextral transform motion between ridges (Kim et al., in press). This feature indicates that plutonism and dextral strike-slip overlapped, resulting in the production of syn-D_3 andalusite during the mid-Jurassic in some local areas of the OMB.

Conclusions

1. *M1* metamorphism in the OMB is an intermediate-P/T type regional metamorphism, and probably occurred in the Late Carboniferous to Early Permian. During *M1* metamorphism, metamorphic grade increased northwestward to yield three metamorphic zones. The P-T conditions of garnet cores and rims in Zone III are, respectively, 2.0–5.0 kbar and 410–540°C and 6.8–9.0 kbar and 580–620°C. Conditions in the garnet rim were not sufficient to allow kyanite to form.

2. The massive, foliated granites that encircle the Hwasan area generated thermal metamorphic zones in the metasediments only adjacent to the plutons. The P-T conditions of thermal metamorphism, estimated at the contact aureole, are 530–600°C and 1.7–3.5 kbar.

3. Most K-Ar and $^{40}Ar/^{39}Ar$ ages for muscovite and biotite from metasediments in the Hwasan area are concentrated near 160 Ma. K-Ar ages of muscovite and biotite from the mid-Jurassic granites are 156 Ma.

4. The limited range in d_{002} values of 3.353 to 3.359 Å of carbonaceous material of all analyzed

metasediments in the Hwasan area is consistent with amphibolite-facies regional metamorphism. In Zone I of the Hwasan area, regional metamorphic grade, indicated by intermediate-P/T metamorphic mineral assemblages, is at greenschist-facies conditions. However, d_{002} values of carbonaceous materials in this same sector are those of the fully ordered graphite (ca. 500°C), typical of the amphibolite facies. This result, along with the concentration of mid-Jurassic mica ages of metasediments, the presence of a contact metamorphic zone close to the Jurassic granites, and the mid-Jurassic intrusion and cooling ages of these plutons, point to a mid-Jurassic low-P/T *M2* regional contact thermal overprint. This process failed to reset *M1* mineral assemblages, except within a narrow (1–2 km) aureole around the Jurassic granites, largely because of a relatively short duration of the thermal event as a result of rapid cooling of the Jurassic granites.

5. Both the large-scale dextral strike-slip movements along the Honam shear zone and regional plutonism in and around the OMB during the mid-Jurassic appear to be related to northwestward ridge subduction of the Farallon-Izanagi plate. The regional plutonism was a direct cause of the *M2* metamorphic overprint.

Acknowledgments

This paper is dedicated to W. G. Ernst, who was one of the first geologists to propose an eastward extension of the Dabie-Sulu HP/UHP belt to the Korean Peninsula. We would like to thank T. Okada and T. Tsujimori for their help with analyses of potassium and argon and with formulae calculations of EPMA data. The critical comments of Aley El-Shazley, Y. Osanai, and M. Cho helped to improve the paper, and are greatly appreciated. This study was supported, in part, by a grant from the Post-Doctoral Program of Chonbuk National University (2002) and Grant-in-Aid, KRF-2003-070-C00046, from the Korea Research Foundation.

REFERENCES

Berman, R. G., 1990, Thermobarometry using multi-equilibrium calculations: A new technique, with petrological applications: Canadian Mineralogist, v. 29, p. 833–855.

Bohlen S. R., and Liotta, J. J., 1986, A barometer for garnet amphibolites and garnet granulites: Journal of Petrology, v. 27, p. 1025–1034.

Chang, E. Z., 1996, Collisional orogen between north and south China and its eastern extension in the Korean Peninsula: Journal of Southeast Asian Earth Sciences, v. 13, p. 267–277.

Cheong, C. S., Cheong, K. Y., Kim, H., Choi, M. S., Lee, S., and Cho, M., 2003, Early Permian peak metamorphism recorded in U-Pb system of black slates from the Ogcheon metamorphic belt, South Korea, and its tectonic implication: Chemical Geology, v. 193, p. 81–92.

Cho, D. L., Kwon, S. T., Sagong, H., Cheong, C. S., and Armstrong, R., 2001, Precise cooling histories of three neighboring pluton in the central Okcheon belt: Implication for magma movement rate and tectonics [abs.], *in* Abstract volume for the conference of the Geological Society of Korea, p. 90.

Cho, K. H., Takagi, H., and Suzuki, K., 1999, CHIME monazite age of granitic rocks in the Sunchang shear zone, Korea: Timing of dextral ductile shear: Geoscience Journal, v. 3, p. 1–16.

Cho, M., and Kim, H., 2002, Metamorphic evolution of the Ogcheon metamorphic belt: Review of recent studies and remaining problems: Journal of Petrological Society of Korea, v. 11, p. 121–137 (in Korean with English abstract).

Cho, M., Kim, I. J., Kim, H., Min, K., Ahn, J.-H., and Nagao, K., 1995, K-Ar biotite ages of pelitic schists in the Jeungpyeong-Deokpyeong area, central Ogcheon metamorphic belt, Korea: Journal of Petrological Society of Korea, v. 4, p. 178–185 (in Korean with English abstract).

Chough, S. K., Kwon, S. T., Ree, J. H., and Choi, D. K., 2000, Tectonic and sedimentary evolution of the Korean Peninsula: A review and new view: Earth-Science Reviews v. 52, p. 175–235.

Cluzel, D., Cadet, J. P., and Lapierre, H., 1990, Geodynamics of the Ogcheon belt (South Korea): Tectonophysics, v. 183, p. 41–56.

Cluzel, D., Jolivet, L., and Cadet, J.-P., 1991a, Early Middle Paleozoic intraplate orogeny in the Ogcheon Belt (South Korea): A new insight on the Paleozoic buildup of East Asia: Tectonics, v. 10, p. 1130–1151.

Cluzel, D., Lee, B. J., and Cadet, J.-P., 1991b, Indosinian dextral ductile system and synkinematic plutonism in the southwest of the Ogcheon belt (South Korea): Tectonophysics, v. 194, p. 131–151.

Ernst, W. G., and Liou, J. G., 1995, Contrasting plate-tectonic styles of the Qinling-Dabie-Sulu and Franciscan metamorphic belts: Geology, v. 23, p. 353–356.

Ferry, J. M., and Spear, F. S., 1978, Experimental calibration of the partitioning of Fe and Mg between biotite and garnet: Contributions to Mineralogy and Petrology, v. 66, p. 113–117.

Hey, M. H., 1954, A new review of the chlorites: Mineralogical Magazine, v. 25, p. 277–292.

Hiroi, Y., 1981, Subdivision of the Hida metamorphic complex, central Japan, and its bearing on the geology of the far east in pre-sea of Japan time: Tectonophysics, v. 76, p. 317–333.

Hodges, K. V., and Crowley, P. D., 1985, Error estimation and empirical geothermobarometry for pelitic system: American Mineralogist, v. 70, p. 702–709.

Hodges, K. V., and Spear, F. S., 1982, Geothermometry, geobarometry and Al_2SiO_5 triple point at Mt. Moosilauke, New Hampshire: American Mineralogist, v. 67, p. 1118–1134.

Hoisch, T. D., 1990, Empirical calibration of six geobarometers for the mineral assemblage quartz + muscovite + biotite + plagioclase + garnet: Contributions to Mineralogy and Petrology, v. 104, p. 225–234.

Holdaway, M. J., 1971, Stability of andalusite and the aluminum silicate phase diagrams: American Journal of Science, v. 271, p. 97–131.

Hyodo, H., Kim, S. W., Itaya, T., and Matsuda, T., 1999, Homogeneity of neutron flux during irradiation for $^{40}Ar/^{39}Ar$ age dating in the research at Kyoto University: Journal of Minerlogists, Petrologists, and Economic Geologists, v. 94, p. 329–337.

Itaya, T., 1981, Carbonaceous material in pelitic schists of the Sanbagawa metamorphic belt in central Shikoku, Japan: Lithos, v. 14, p. 215–224.

Itaya, T., Nagao, K., Inoue, K., Honjou, Y., Okada, T., and Ogata, A., 1991, Argon isotopic analysis by a newly developed mass spectrometric system for K-Ar dating: Mineralogical Journal, v. 15, p. 203–221.

Kretz, R., 1983, Symbols for rock-forming minerals: American Mineralogist, v. 68, p. 277–279.

Kim, H., Cheong, C. S., Cho, M., Jeong, G. Y., and Choi, M. S., 2001, Geochronological evidence for late Paleozoic orogeny in the Ogcheon metamorphic belt, South Korea [abs.], in Abstract volume for Annual Meeting of the Geological Society of America, p. 33.

Kim, H., and Cho, M., 1999, Polymetamorphism of Ogcheon Supergroup in the Miwon area, central Ogcheon metamorphic belt, South Korea: Geoscience Journal, v. 3, p. 151–162.

Kim, H., Cho, M., and Koh, H. J., 1995, Tectonometamorphic evolution of the central Ogcheon belt in the Jeungpyeong-Deokpyeong area: Journal of Geological Society of Korea, v. 31, p. 299–314 (in Korean with English abstract).

Kim, H. S., 1971, Metamorphic facies and regional metamorphism of Ogcheon metamorphic belt: Journal of Geological Society of Korea, v. 7, p. 221–256.

Kim, J. B., and Turek, A., 1996, Advances in U-Pb zircon geochronology of Mesozoic plutonism in the southwestern part of Ryeongnam massif, Korea: Geochemical Journal, v. 30, p. 323–338.

Kim, S. W., Itaya, T., Hyodo, H., and Matsuda, T., 2002, Metamorphic K-feldspar in low-grade meta-sediments from the Ogcheon metamorphic belt in South Korea: Gondwana Research, v. 5, p. 849–855.

Kim, S. W., Oh, C. W., Choi, S. G., and Itaya, T., in press, Ridge subduction–related Jurassic plutonism in and around the Okcheon Metamorphic Belt, South Korea, and implications for Northeast Asian tectonics: International Geology Review.

Kim, S. W., Oh, C. W., Lee, D. S., and Lee, J. H., 2003, K-Ar and $^{40}Ar/^{39}Ar$ from metasediments in the Okcheon metamorphic belt and their tectonic implication: Journal of Petrological Society of Korea, v. 12, p. 79–99 (in Korean with English abstract).

Kinoshita, O., 1995, Migration of igneous activity related to ridge subduction in southwest Japan and the East Asian continental margin from the Mesozoic to the Paleogene: Tectonophysics, v. 245, p. 25–35.

Kinoshita, O., 2002, Possible manifestations of slab window magmatisms in Cretaceous southwest Japan: Tectonophysics, v. 344, p. 1–13.

Kinoshita, O., and Itô, H., 1988, Cretaceous magmatism in southwest and Northeast Japan related to two-ridge subduction and Mesozic magmatism along East Asia continental margin: Journal of the Geological Society of Japan, v. 94, p. 925–944 (in Japanese with English abstract).

Kinoshita, O., and Itô, H., 1990, Reconstruction of southwest Japan and northeast Japan based on trend of Mesozoic igneous rock ages: Journal of Geological Society of Japan, v. 96, p. 821–838 (in Japanese with English abstract).

Koh, H. J., and Kim, J. H., 1995, Deformation sequence and characteristics of the Ogcheon Supergroup in the Goesan Area, Central Ogcheon Belt: Journal of the Geological Society of Korea, v. 31, p. 271–298 (in Korean with English abstract).

Kretz, R., 1983, Symbols for rock-forming minerals: American Mineralogist, v. 68, p. 277–279.

Kunugiza, K., Tsujimori, T., and Kano, T., 2001, Evolution of the Hida and Hida marginal belts, in ISRGA Field Workshop Guidebook, Osaka, Japan, p. 75–131.

Lee, K. S., Chang, H. W., and Park, K. H., 1998, Neoproterozoic bimodal volcanism in the central Ogcheon belt, Korea: Age and tectonic implication: Precambrian Research, v. 89, p. 47–57.

Lee, S. R., Lee, B. J., Cho, D. L., Kee, W. S., Koh, H. J., Kim, B. C., Song, K. Y., Hang, J. H., and Choi, B. Y., 2003, SHRIMP U-Pb zircon age from granitic rocks in Jeonju shear zone: Implications for the age of the Honam shear zone [abs.], in Abstract volume for proceedings of the Annual Joint Conference, Mineralogical Society of Korea and Petrological Society of Korea, p. 55 (in Korean).

Liu, X., 1993, High-P metamorphic belt in central China and its possible eastward extension to Korea: Journal of the Petrological Society of Korea, v. 2, p. 9–18.

Liou, J. G., Zhang, R. Y., and Ernst, W. G., 1994, The Triassic Qinling-Dabie collision and ultrahigh-P metamorphism in East-Central China: Implication for a similar collision in the Korean Peninsula [abs.], in 4th IGCP 321 Abstract Volume, p. 76–78.

Massonne, H. J., and Schreyer, W., 1987, Phengite geobarometry based on the limiting assemblage with K-feldspar, phlogophite, and quartz: Contributions to Mineralogy and Petrology, v. 96, p. 212–224.

Min, K., and Cho, M., 1998, Metamorphic evolution of the northwestern Ogcheon metamorphic belt, South Korea: Lithos, v. 43, p. 31–51.

Nishimura, Y., Coombs, D. S., Landis, C. A., and Itaya, T., 2000, Continuous metamorphic gradient documented by graphitization and K-Ar age, southeast Otago, New Zealand: American Mineralogist, v. 85, p. 1625–1636.

Oh, C. W., Kim, C. S., and Park, Y. D., 1997, The contact metamorphism due to the intrusion of the Ogcheon and Boeun granites: Journal of the Petrological Society of Korea, v. 6, p. 133–149 (in Korean with English abstract).

Oh, C. W., Kim S. T., and Lee, J. H., 1995a, The metamorphic evolution in the southwestern part of the Ogcheon metamorphic belt: Journal of the Geological Society of Korea, v. 31, p. 21–31 (in Korean with English abstract).

Oh, C. W., Kim, S. T., and Lee, J. H., 1995b, The P-T condition and timing of the main metamorphism in the southern part of the Ogcheon metamorphic belt: Journal of the Geological Society of Korea, v. 31, p. 343–361.

Oh, C. W., Kim, S. W., and Lee, J. H., 1998, A study on the regional and contact metamorphism in the southwestern part of the Ogcheon metamorphic belt: Journal of the Geological Society of Korea, v. 34, p. 311–332 (in Korean with English abstract).

Oh, C. W., Kim, S. W., Ryu, I.-C., Okada, T., Hyodo, H., and Itaya, T., 2004, Tectono-metamorphic evolution of the Okcheon metamorphic belt, South Korea: Tectonic implications in East Asia: The Island Arc, v. 13, p. 387–402.

Oh, C. W., Kwon, Y. W., and Kim, S. W., 1999, Metamorphic evolution of the central Ogcheon metamorphic belt in the Cheongju-Miwon area, Korea: Journal of the Petrological Society of Korea, v. 8, p. 106–124 (in Korean with English abstract).

Ree, J. H., Cho, M., Kwon, S. T., and Nakamura, E., 1996, Possible eastward extension of Chinese collision belt in the South Korea: the Imjingang belt: Geology, v. 24, p. 1071–1074.

Ree, J. H., Kwon, S. H., Park, Y., Kwon, S. T., and Park, S. H., 2001, Pre- and post-tectonic emplacements of the granitoids in the central-southern Okchon belt, South Korea: Implications for the timing of the strike-slip shearing and thrusting: Tectonics, v. 20, p. 850–867.

Spear, F. S., and Cheney, J. T., 1989, A petrogenetic grid for pelitic schists in the system SiO_2-Al_2O_3-FeO-MgO-K_2O-H_2O: Contributions to Mineralogy and Petrology, v. 101, p. 149–164.

Spear, F. S., Kohn, M. J., and Paetzold, S., 1995, Petrology of the regional sillimanite zone, west-central New Hampshire, U.S.A., with implications for the development of inverted isograds: American Mineralogist, v. 80, p. 361–376.

Wang, G., 1989, Carbonaceous material in the Ryoke metamorphic rocks, Kinki district, Japan: Lithos, v. 22, p. 305–316.

Zhai, M., and Liu, W., 1997, The boundary between Sino-Korea and Yangtze Craton and its extension to the Korea Peninsula, in Proceedings of the 4th Sino-Korean Joint Symposium on crustal evolution in Northeast Asia, p. 21–28.

JAPAN

Eclogite-Facies Mineral Inclusions in Clinozoisite from Paleozoic Blueschist, Central Chugoku Mountains, Southwest Japan: Evidence of Regional Eclogite-Facies Metamorphism

T. TSUJIMORI[1] AND J. G. LIOU

Department of Geological and Environmental Sciences, Stanford University, Stanford, California 94305

Abstract

An eclogite-facies assemblage garnet + omphacite + rutile + glaucophane + quartz occurs as inclusions in clinozoisite porphyroblasts in a high-grade blueschist block in the Paleozoic Osayama serpentinite mélange of the Chugoku Mountains, southwest Japan. Textual relations, and parageneses and compositions of minerals, indicate that the high-grade block experienced peak eclogite-facies metamorphism (M_1) and subsequent blueschist-facies overprinting (M_2). Moreover, low-P amphibolite-facies minerals (M_0) that may represent remnants of pre-subduction conditions are locally preserved in the protolith. Omphacite + adjacent garnet inclusion pairs in clinozoisite give a minimum P = 1.1–1.3 GPa at T = ~480–550°C. Many petrologic similarities between the Osayama eclogite and the Hida Mountains eclogite provide evidence for a regional eclogite-facies metamorphism in Paleozoic terranes of southwest Japan. Considering available petrotectonic information regarding eastern Asia, we suggest two alternative models: (1) Japanese Paleozoic eclogites may represent "Pacific-type" subduction prior to collision between the Sino-Korean and Yangtze blocks, which occurred on both western and eastern sides of the Qinling-Dabie-Sulu belt; or (2) Japanese Paleozoic eclogites may represent part of a Paleozoic "Pacific-type" subduction chain, which is continuous from the China-Russia border through Japan southward to North Korea.

Introduction

BLUESCHIST AND LOW-T eclogite of the "Franciscan-type" (Ernst, 1988) = "Pacific-type" (Maruyama et al., 1996) orogenic belt reflect the metamorphism of oceanic materials at an inter-oceanic subduction zone; such high-pressure (HP) metamorphic rocks in Pacific-type orogens generally occur intercalated among accretionary complexes and ophiolites. In contrast, ultrahigh-pressure (UHP) metamorphic rocks of the "Alpine-type" (Ernst, 1988) = "collisional-type" (Maruyama et al., 1996) orogenic belt display essential differences in protoliths and convergent tectonics; they involve deeply subducted continental basement complexes and overlying supracrustal rocks, and record almost isothermal decompression. The Japanese Islands are a typical example of the "Pacific-type" orogenic belt; this belt comprises mainly accretionary complexes with arc-related granitic plutons and volcanic rocks of Mesozoic and Cenozoic age. However, minor fragments of Paleozoic terranes are sporadically distributed. These include Paleozoic blueschist-eclogite terranes that represent the key to understanding the original geologic continuity between the Japanese orogen and suture zones between the Sino-Korean–Yangtze and/or Sino-Korea–Khanka cratons before the Miocene opening of the Japan Sea (e.g., Ernst and Liou, 1995; Ishiwatari and Tsujimori, 2003).

Recently, regional eclogite-facies metamorphism has been locally confirmed in the Paleozoic terrane of the Hida Mountains, central Honshu (Tsujimori et al., 2000; Tsujimori, 2002), where the glaucophane-eclogite with Grt + Omp ($Jd_{~52}$) + Gln + Rt + Qtz assemblage (mineral abbreviations after Kretz, 1983) occurs as mafic layers in pelitic schist characterized by Grt + Phe + Pg ± Czo ± Phe + Rt + Qtz. Moreover, Tsujimori (1998) has also identified relict eclogite-facies assemblages in Paleozoic blueschists of the Chugoku Mountains, western Honshu. In this paper, we present new petrologic data for eclogite-facies metamorphism in the Chugoku Mountains. These, together with available geochronological data, are used to evaluate the petrotectonic evolution of a Paleozoic suture. Then we discuss the regional tectonic problem between the Japanese Paleozoic terrane and suture zones in eastern Asia.

---

[1]Corresponding author; email: tatsukix@pangea.stanford.edu

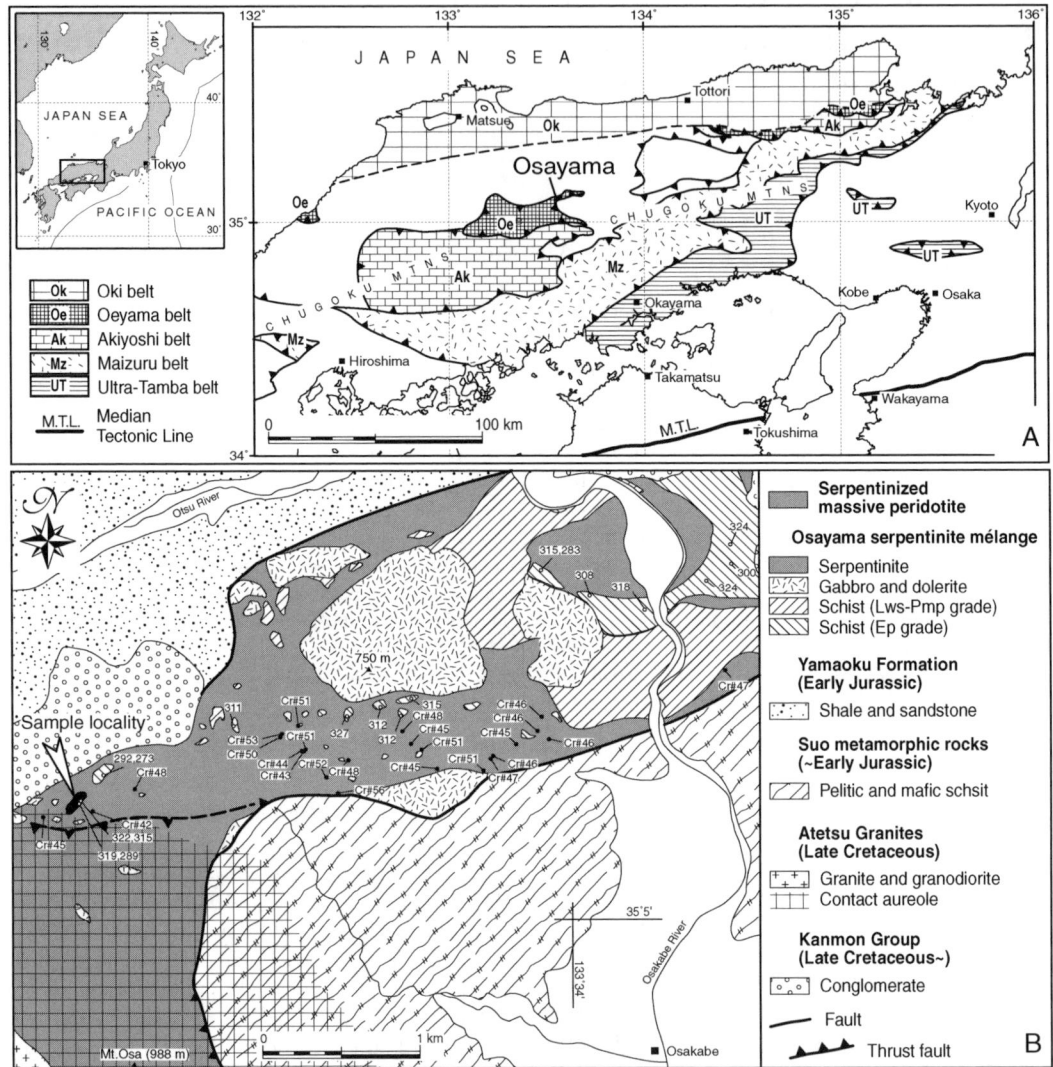

FIG. 1. A. simplified map of the Chugoku Mountains, showing various pre-Triassic petrotectonic units and the Osayama area. B. Geologic map of the Osayama serpentinite mélange, showing sample locality of the investigated blueschist block (after Tsujimori, 1998; Tsujimori and Itaya, 1999). Triple-digit numbers on the map represent phengite K-Ar ages from blueschist-facies blocks. Cr# with double-digit numbers on the map mean $100 \times Cr/(Cr + Al)$ for relict chromian spinel in serpentinites.

Geologic Setting

Paleozoic rocks in the Chugoku Mountains

In the Chugoku Mountains of southwest Japan, tectonic superposition of almost E-W–trending Paleozoic ophiolites and accretionary complexes occupies the highest structural positions in Phanerozoic Pacific-type orogens of southwest Japan (e.g., Ishiwatari and Tsujimori, 2003; Fig. 1). Pre-Triassic rocks occur as five petrotectonic units: (1) Oki belt; (2) Oeyama belt; (3) Akiyoshi belt; (4) Maizuru belt; and (5) Ultra-Tamba belt (Fig. 1A).

The Oki belt consists of low-P pelitic and psammitic gneisses with minor marble and amphibolite. Microprobe Th-U-total Pb chemical ages of metamorphic monazite and zircon in gneisses yielded

~250 Ma for amphibolite-facies metamorphism (Suzuki and Adachi, 1994), whereas conventional zircon U-Pb age shows ~1.9 Ga (Yamashita and Yanagi, 1994). The early Paleozoic Oeyama belt consists mainly of serpentinized harzburgite with minor gabbroic rocks (e.g., Arai, 1980); Sm-Nd study yields ages for gabbroic intrusions of ~560 Ma, suggesting a Cambrian ophiolite sequence (Hayasaka et al., 1995). Two different HP metamorphic rocks are associated with the Oeyama belt as tectonic blocks: epidote-amphibolite–facies gabbroic rocks with 470–400 Ma hornblende K-Ar age (e.g., Kuroda et al., 1976; Nishimura and Shibata, 1989; Tsujimori and Liou, 2004a), and blueschist-facies pelitic and mafic schists with phengite K-Ar ages of 330-280 Ma (Nishimura, 1998; Tsujimori and Itaya, 1999). In particular, the younger HP rocks have been regarded as fragments of the late Paleozoic Renge blueschist belt, tectonically underlying the Oeyama belt (e.g., Ishiwatari, 1991; Tsujimori, 1998). The Akiyoshi belt is a Permian accretionary complex consisting mainly of a thick limestone-greenstone complex with pelagic to semi-pelagic sediments; limestone contains Early Carboniferous to Middle Permian fossils (Kanmera et al., 1990). The Maizuru belt is a Late Permian ophiolitic unit capped by sedimentary cover; ophiolitic metabasalt-metagabbro-metaperidotite (rare) complexes in this belt are distinguished as the "Yakuno" ophiolite (e.g., Ishiwatari, 1985; Ichiyama and Ishiwatari, 2004). Metagabbroic hornblende yields K-Ar ages of ~280-240 Ma (Shibata et al., 1977), and zircon of plagiogranite yields U-Pb ages of ~280 Ma (Herzig et al., 1997). The Ultra-Tamba belt is a Late Permian accretionary complex structurally underlying the Maizuru belt; it consists mainly of pelagic to semi-pelagic sediments with minor greenstones (Ishiga, 1990). All these Paleozoic rocks occur as a nappe pile, with the older units in structurally higher positions, except for the Oki belt. The structural relationship between the Oki belt and other petrotectonic units is not well known.

Geology of blueschist-bearing serpentinite mélange in Chugoku Mountains

A blueschist-bearing serpentinite mélange is developed in the Osayama area (Tsujimori, 1998; Tsujimori and Itaya, 1999; Fig. 1B); called the "Osayama serpentinite mélange"; it is tectonically underlain by Early Jurassic pumpellyite-actinolite to greenschist-facies schists of the Suo metamorphic belt, and is overlain by a massive serpentinized peridotite unit. The matrix of the serpentinite mélange consists of schistose, friable, fine-grained serpentinite with pebble- to boulder-size fragments of serpentinized peridotite. Chrysotile and lizardite are the dominant serpentine minerals; rare winchitic amphibole occurs in the serpentinite. The protolith of the serpentinite was moderately depleted harzburgite and minor cumulus dunite; relict olivine ($Fo_{90.5-91.5}$), orthopyroxene (2.4–3.0 wt% Al_2O_3), clinopyroxene, and chromian spinel posessing a Cr/(Cr+Al) ratio = 0.40–0.57 occur in the serpentinized harzburgite. Several jadeitites and omphacitites have been described (Kobayashi et al., 1987; Tsujimori, 1997; Tsujimori and Liou, 2004b). Late Cretaceous granitic intrusions caused thermal metamorphism in the western part of the area.

Blueschist blocks in the Osayama serpentinite mélange consist mainly of metasediments (pelitic, psammitic, and siliceous schists), metabasites, and minor marble. The low-grade blueschists were subdivided into lawsonite-pumpellyite and epidote grades, based on mineral assemblages of the mafic schist; blocks of the lawsonite-pumpellyite grade are the most abundant rock type. Gabbro and dolerite blocks within serpentinite mélange also contain blueschist-facies assemblages of lawsonite-pumpellyite grade. Some pelitic schists of epidote grade contain garnet porphyroblast ($Prp_{1-2}Alm_{23-33}Sps_{41-57}Grs_{19-25}$). Phengite K-Ar ages of the blueschist blocks are in the range 273–327 Ma (Tsujimori and Itaya, 1999).

Eclogite-facies mineral assemblages occur in a blueschist block (~150 × 50 m). This blueschist is exceptionally high grade, and contains retrograde mineral assemblage comparable to that of the lawsonite-pumpellyite grade described later. This block occurs near the thermal aureole of Cretaceous granitic intrusion, but phengite K-Ar ages show no rejuvenation, hence the thermal effect was insignificant.

Analytical Methods

Concentrations of major (Si, Ti, Al, Fe, Mn, Mg, Ca, Na, K and P) and trace (Ni, Cu, Zn, Pb, Y and V) elements were analyzed employing a Rigaku System 3270 X-ray fluorescence spectrometer with Rh tube at Kanazawa University. Operating conditions for both major and trace elements were 50 kV accelerating voltage and 20 mA beam current. Other trace elements (Sc, Cr, Co, Rb, Sr, Zr, Nb, Ba, La, Ce, Pb, Sm, Eu, Yb, Lu, Hf, Ta, Th and U) were determined by instrumental neutron activation analysis (INAA method). INAA samples were activated at the Kyoto University Reactor, and gamma-ray spectroscopic analyses were done at the Radioisotope Laboratory of Kanazawa University.

Electron microprobe analysis and X-ray element mapping were carried out with a JEOL JXA-8800R at Kanazawa University and JEOL JXA-8900R at Okayama University of Science. Quantitative analyses of rock-forming minerals were performed with 15 kV accelerating voltage, 12 nA beam current, and 3-5 μm beam size. Natural and synthetic silicates and oxides were used for calibration. The ZAF method (oxide basis) was employed for matrix corrections. X-ray element maps were collected at 15 kV accelerating voltage and 100 nA beam current.

Petrography of a Blueschist Block with Eclogite-Facies Minerals

A blueschist block preserving eclogite-facies metamorphic phase assemblage is a well-foliated schist; three different petrographic features were identified: clinozoisite-rich blueschist (type-I); blueschist with relict Hbl (type-II); and phengite-rich blueschist (type-III). Eclogite-facies minerals are present in the type-I rock that occurs as layers within type-II and type-III rocks. The investigated blueschist block is lithologically heterogeneous and severely weathered, but more than 30 samples of various rock types were collected for this study.

Type-I: Clinozoisite-rich blueschist

This rock type is characterized by a primary Gln + Czo + Grt + Rt ± Phe + Qtz assemblage; the matrix is heterogeneously replaced by the secondary assemblage Fgl + Pmp + Chl + Ttn + Ab. Garnet occurs as porphyroblast (1–5 mm; Fig. 2A), and is commonly replaced by secondary Fgl + Chl ± Ab (Fig. 2B). Most mineral inclusions in garnet are chloritized, but tiny rutile and rare quartz granules are well preserved (Fig 2A); inclusion trails are roughly parallel to the foliation of the matrix. Clinozoisite occurs as porphyroblastic anhedral crystals (2–4 mm; Fig. 2C), and contains inclusions of Grt + Omp + Rt + Gln + Qtz (Figs. 2D and 2E); pumpellyite locally replaces the rims and extension veins of boudinaged clinozoisite. Glaucophane in the matrix is generally rimed by secondary ferroglaucophane, but the core contains inclusions of Grt + Rt + Qtz. Secondary ferroglaucophane in some samples accompanies aggregates of prismatic pumpellyite (Fig. 2F). Rutile in the matrix is commonly replaced by secondary titanite.

Type-II: Blueschist with relict hornblende

This rock type contains the same primary mineral assemblages as type-I rocks, but is characterized by the presence of relict green hornblende and less abundant clinozoisite. K-feldspar and allanite also occur as relict phases. Hornblende-rich amphibolitic lenticular clots (2 to 30 cm in length) are rare. Relict hornblende occurs as cores of zoned glaucophane (Fig. 2G), and rarely contains tiny ilmenite inclusions. Amphibolitic clots consist mainly of hornblende with small amounts of K-feldspar and rare garnet (Fig. 2H); hornblende contains tiny ilmenite grains, whereas K-feldspar contains tiny quartz and rare plagioclase (An_{29}) inclusion. Garnet-rich bands (1 to 6 mm wide) parallel to foliation occur in some samples; the band consists of highly fractured garnet of various size (0.01–1.0 mm) and secondary chlorite (Fig. 3A). However, some garnet fragments are overgrown by neoblastic garnet (Figs. 3B and 3C); other euhedral garnets display two discontinuous stages of garnet growth, with earlier garnet containing ilmenite instead of rutile.

Type-III: Phengite-rich blueschist

This rock type is well-foliated micaceous blueschists. It is characterized by a primary mineral assemblage of Gln + Phe + Qtz + Rt, and Ttn. Chl and Ab replace the primary minerals.

Bulk-Rock Chemistry

One specimen each of type I and II were selected for geochemical study. Analyzed bulk compositions are listed in Table 1. Both type I and II are mafic in composition (48.6–51.4 wt% SiO_2), and they are characterized by significantly high Sr (783–929 ppm) and Ba (937–3895 ppm) contents in comparison with Sambagawa mafic schists (Fig. 4A). Nearly flat REE patterns (Fig. 4B) and relationships of Zr (103 ppm)–Zr/Yi (2) and Nb/Y (0.1)–Ti/Y (213) suggest MORB-like affinity for type-I (Pearce, 1982, 1983). In contrast, type II is characterized by high K_2O (5.6 wt%) and Al_2O_3 (14.9 wt%); it shows high LREE enrichments in La, Ce, and Sm concentrations. This may suggest a protolith as mixture of mafic rocks with some sediment component. High concentrations of Th (10.5 ppm) and U (2.8 ppm) of type II may be reflected in modal allanite.

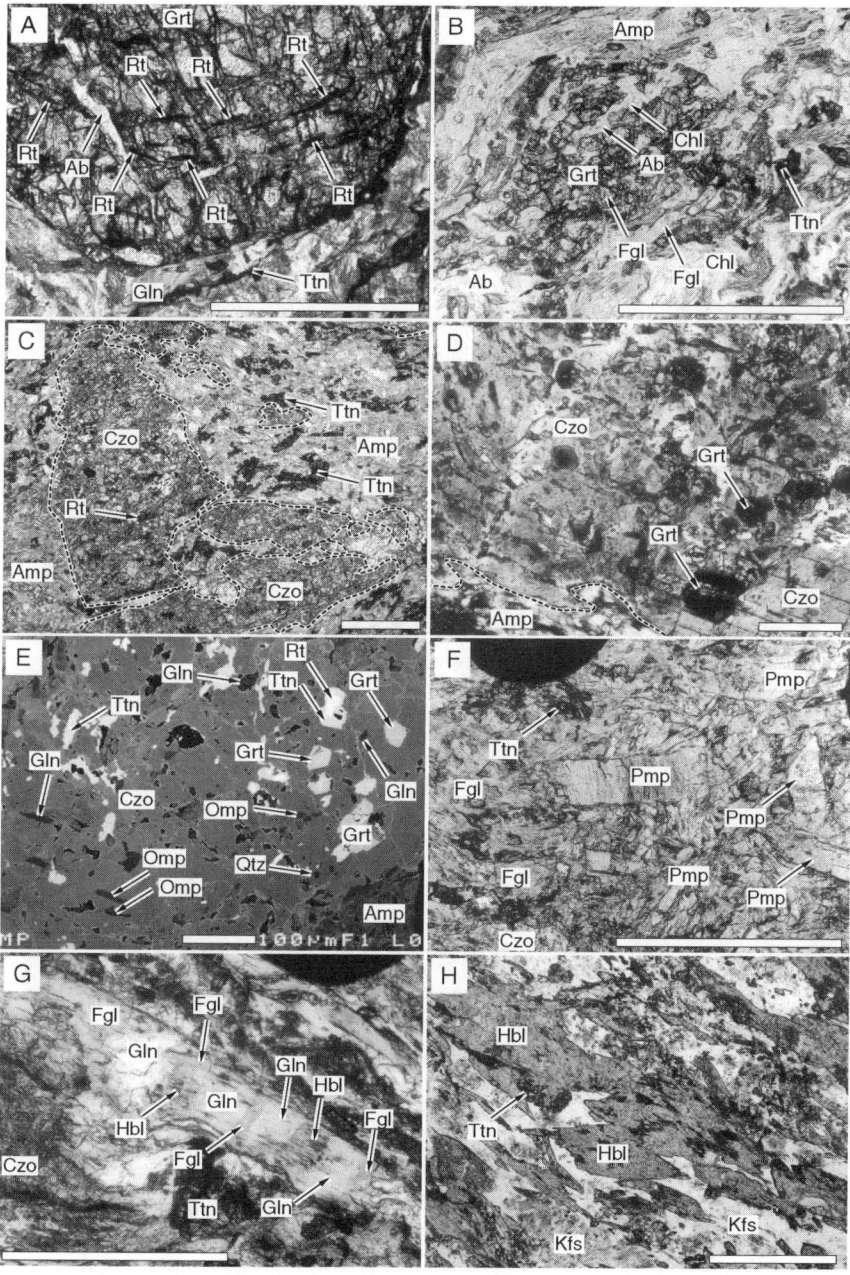

FIG. 2. Microtextures of the investigated samples. A. Garnet porphyroblast of type-I rock (plane-polarized light = PPL); garnet contains rutile as inclusions, and secondary albite fills fractures of garnet. B. Garnet pseudomorph of type-I rock (PPL); garnet is replaced by chlorite, ferroglaucophane, and albite. C. Clinozoisite porphyroblasts of type-I rock (PPL). Broken lines indicate optical grain boundaries. D. Cross-polarizer view of garnet inclusions within clinozoisite porphyroblast of type-I rock. E. Backscattered electron image of eclogitic inclusions within clinozoisite porphyroblast of the Type-I rock; eclogitic assemblage, garnet + rutile (partly replaced by titanite) + omphacite + quartz + glaucophane, is shown. F. Retrograde pumpellyite coexisting ferroglaucophane and titanite of type-I rock. G. Multiple growth of amphibole in type-II rock; extension vein of boudinaged glaucophane with detrital hornblende core filled by late-stage ferroglaucophane. H. Hornblende- and K-feldspar–rich part in amphibolitic clot of type-II rock.

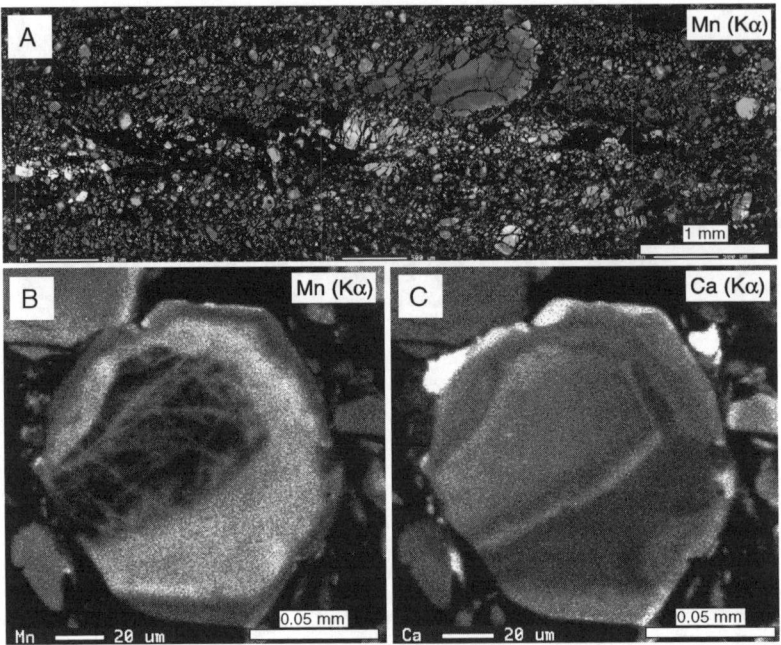

FIG. 3. X-ray image of garnets of type-II rock. A. Mn ($K\alpha$) of garnet-rich band with cataclastic garnets. Concentric compositional zoning is preserved in a fragmented garnet. The chemical composition is not modified along the fractures. B. Mn ($K\alpha$) of a garnet in garnet-rich layer. C. Ca ($K\alpha$) of same view as Figure 3B.

Mineral Chemistry

Representative electron microprobe analyses of rock-forming minerals are presented in Table 2.

Amphiboles

The Fe^{2+}/Fe^{3+} ratios of amphiboles were estimated assuming total cations = 13 (O = 23) excluding Ca, Na, and K. Na-amphiboles from both type I and II show two growth stages marked by a distinct compositional anomaly. Glaucophane is rimmed by late-stage ferroglaucophane (Fig. 4A); cores are characterized by X_{Mg} (= $Mg/(Mg+Fe^{2+})$) = 0.54–0.72 and $Fe^{3+}/(Fe^{3+}+Al)$ < 0.15, and rims by X_{Mg} = 0.39–0.52 and $Fe^{3+}/(Fe^{3+}+Al)$ < 0.08-0.21 (Fig. 5A). Some ferroglaucophane rims are enriched in MnO relative to the cores. Glaucophane inclusions of clinozoisite in type-I rocks show the same composition range as the cores of zoned Na-amphiboles in the matrix (Fig. 5A). Na-amphibole of type-III rocks is glaucophanitic with X_{Mg} = 0.50–0.60 and $Fe^{3+}/(Fe^{3+}+Al)$ < 0.11. Glaucophanes of type-I and -II rocks are characterized by higher glaucophane component than most Na-amphiboles in low-grade blueschists of the Osayama area (Fig. 5B); they also have similar compositional trends as those in the Hida Mountains eclogite (Fig. 5A). In type-II rocks, relict hornblendes are magnesiohornblendes to actinolitic hornblendes, with relatively high Na + K in the A-site (up to 0.48 $(Na + K)_A$ p.f.u.) and low Na in the A-site (Na_B p.f.u. < 0.18; Figs. 6A and 6B); X_{Mg} ranges from 0.52 to 0.68.

Garnet

Analyzed garnet compositions are plotted on Mn-Fe-Mg and Ca-(Fe+Mn)-Mg ternary diagrams (Fig. 7). Garnets in type-I rocks are characterized by high Alm components with moderate Grs and low Prp and Sps (Alm = 65–71%, Grs = 16–27%, Prp = 3–18%, Sps = 0–18%; Figs. 7A and 7B); thin rims with higher Grs (34–35%) are recognized in one garnet porphyroblast. Garnet inclusions in clinozoisite have slightly Mg-richer compositions with Prp component up to 18%, and are similar to rim compositions of the matrix garnet. In contrast, garnets of type-II rocks show wide compositional ranges (Figs. 7C and 7D); their chemical trends are different in each grain, even in a single thin-section.

Omphacite

The Fe^{2+}/Fe^{3+} ratios of clinopyroxene were estimated on the assumption that total cations sum to 4; the end-member components were calculated as $Jd = 100 \times Al^{VI}/(Ca+Na)$, $Ae = 100 \times Fe^{3+}/(Ca+Na)$ and $Aug = 100 \times Ca/(Ca+Na)$. Omphacite inclusions in clinozoisite contain $Jd = 34$–48%, $Ae = 0$–9% and $Aug = 52$–57% (Fig. 8); X_{Mg} values range from 0.70 to 0.81. The Fe^{2+}-Mg distribution coefficient, K_D ($= (Fe^{2+}/Mg)_{Grt}/(Fe^{2+}/Mg)_{Omp}$), between omphacite inclusions and adjacent garnet inclusion ranges from 11.5 to 18.3.

Other minerals

Clinozoisite with eclogite-facies mineral inclusions from type-I rocks is characterized by X_{Fe3+} ($= Fe^{3+}/(Fe^{3+}+Al)$) values ranging from 0.12 to 0.20; both core and mantle are homogeneous, respectively at $X_{Fe3+} = 0.12$–0.15. Clinozoisite of type-II rocks has a homogeneous composition with $X_{Fe3+} = 0.11$–0.16. Pumpellyite in type-I rocks shows similar compositional trends to those in low-grade blueschist of the Osayama area: Al/(Al+Fe+Mg) ratios range from 0.77 to 0.82. Phengite in type-III rocks has 3.4–3.6 Si p.f.u. (O = 11), and the X_{Mg} values range from 0.62 to 0.73.

P-T Conditions of Metamorphism

Based on observed petrographic features, and compositions and parageneses of minerals, at least three different metamorphic stages were identified. A summary of mineral paragenesis for each stage is given in Figure 9. Characteristic features for each stage are described below:

Precursor stage (M_0)

The M_0 stage reflects assemblages formed prior to the production of eclogite. Relict M_0 minerals identified in type-II rocks include Hbl + Kfs (with Qtz + Pl inclusions) + Grt + Ilm. These relics are fragmented and corroded due to later recrystallization and reaction. The low Na_B (< 0.18) and moderate Al contents of hornblendes suggest low-P type recrystallization of metabasites (e.g., Terabayashi, 1993). Moreover, the presence of ilmenite instead of rutile also suggests a lower-P and moderate- to high-T condition (e.g., Liu et al., 1996). The mineral paragenesis and compositional characteristics of hornblende indicate a low-P, moderate-T amphibolite-facies condition (Fig. 10).

TABLE 1. Bulk-Rock Compositions of Type-I and -II Rocks of the Osayama High-Grade Blueschist

Rock type	Type I	Type II
Major-element compositions, wt%		
SiO_2	48.56	51.37
TiO_2	1.78	1.05
Al_2O_3	12.71	14.93
FeO^1	11.96	7.76
MnO	0.17	0.17
MgO	9.38	8.32
CaO	9.62	6.51
Na_2O	1.90	0.87
K_2O	1.31	5.57
P_2O_5	0.14	0.21
Total	97.39	96.55
Trace-element compositions, ppm		
Sc	50.3	24.7
Cr	127	136
Co	66	30
Ni	132	117
Cu	23	13
Zn	131	96
Rb	23	112
Sr	929	783
Y	50	32
V	471	165
Zr	103	159
Nb	5	18
Pb	6	12
Ba	937	3895
La	4.9	26.9
Ce	11.2	58.3
Sm	4.3	6.2
Eu	1.5	1.3
Yb	4.6	4.0
Lu	0.7	0.4
Hf	2.4	4.1
Th	n.d.	10.5
U	n.d.	2.8

[1]Total Fe as FeO.

Peak eclogite-facies stage (M_1)

The M_1 stage represents the eclogite-facies stage, represented by the Grt + Omp + Rt + Gln + Qtz + Czo assemblage. These phases occur as mineral inclusions in clinozoisite. As described below,

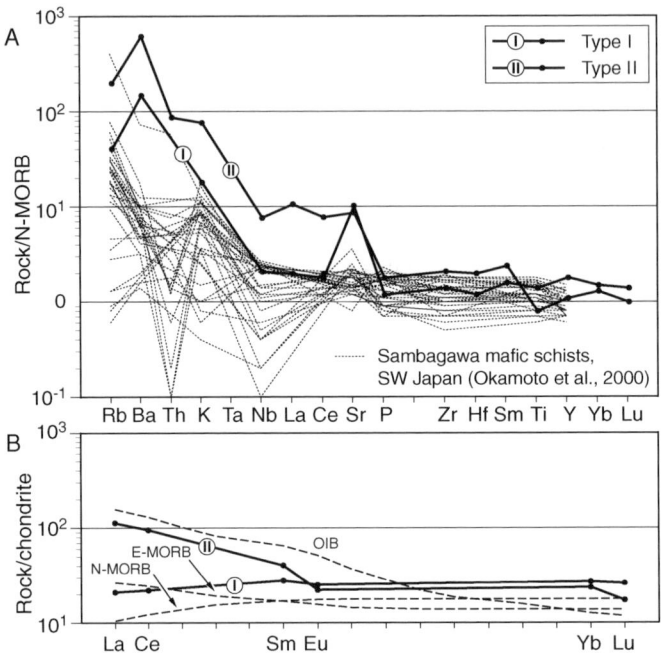

FIG. 4. Bulk-rock chemistry of the investigated samples. A. N-MORB–normalized incompatible element patterns for analyzed samples of type-I and -II rocks. For comparison, the element abundance patterns of the mafic schist of the Sambagawa metamorphic belt, southwest Japan (Okamoto et al., 2000) are shown. B. Chondrite-normalized rare-earth element (REE) patterns for type-I and -II rocks. Normalizing values are from Sun and McDonough (1989).

matrix garnet has rim compositions similar to garnet inclusions in clinozoisite. This may indicate that matrix garnet was formed progressively during the M_1 stage, and preserved its compositions. Although omphacite was recognized only as inclusions within clinozoisite in type-I rocks, eclogitic M_1 minerals including Grt + Rt + Gln are widespread in the matrix of type-I and -II rocks.

A low-T eclogitic mineral assemblage is common in HP terranes throughout the world (e.g., Schliestedt, 1986; Clarke et al., 1997), similar to Paleozoic glaucophane-eclogites in the Hida Mountains (Tsujimori, 2002). The Jd + Qtz sliding equilibrium (Holland, 1983), and Grt-Cpx thermometry (Ravna, 2000) for omphacite + adjacent garnet inclusion pairs in clinozoisite give a minimum P = 1.1–1.3 GPa at T = ~480–550°C (Fig. 10). Because many petrologic similarities exist between the investigated eclogite and the Hida Mountains eclogites, the M_1 stage may record eclogite-facies metamorphism and cooling histories similar to those of Paleozoic eclogites throughout southwest Japan (P> 1.1-1.3 GPa at T= ~450-520 °C).

Blueschist-facies overprinting (M_2)

M_2 stage minerals occur as replacements of pre-existing minerals, and consist of Fgl + Pmp + Chl + Ab + Qtz + Ttn. Some of those phases were recrystallized during later ductile deformation, and filled the boudin-necks and fractures of pre-existing minerals. This stage represents blueschist-facies overprinting after the peak M_1 stage. According to the petrogenetic grid for low-grade mafic rocks erected by Liou et al. (1985) and Frey et al. (1991), the M_2 assemblage is stable in a field transitional between blueschist and pumpellyite-actinolite facies. Moreover, the Na-amp + Pmp assemblage is also stable at lower P than the Na-amp + Lws field, and at a lower T than the Na-amp + Czo field (Katzir et al., 2000). The mineral paragenesis indicates approximate P-T conditions as P = ~0.7–0.8 GPa and T = ~250–300°C (Fig. 10).

P-T path

Detailed petrographic/petrologic examination of various rock types within a high-grade block in the

TABLE 2. Representative Electron-Microprobe Analyses of Rock-Forming Minerals in the Osayama High-Grade Blueschist

	M_0 Precursor stage (type-II rocks)					M_1 Peak eclogite-facies stage (type-I rocks)								M_2 stage (type-I rocks)		
	Hbl	Grt (matrix)	Grt (Grt-band)	Kfs	Pl in kfs	Grt (in Czo)	Grt (matrix) rim	Grt (matrix) core	Omp (in Czo)	Omp	Gln (in Czo)	Gln	Gln (matrix)	Czo	Fgl	Pmp	Phe
SiO_2	43.81	37.85	37.10	65.31	60.11	38.75	38.34	38.74	37.75	56.59	56.44	58.18	58.19	58.85	58.14	38.89	57.35	37.84	52.12
TiO_2	0.03	0.23	0.13	0.01	0.04	0.06	0.21	0.12	0.07	0.09	0.01	0.02	0.02	0.03	0.01	0.12	0.06	0.08	0.07
Al_2O_3	12.18	20.52	20.69	18.60	24.97	22.35	21.66	22.12	21.04	10.78	10.33	11.59	11.87	11.83	11.33	27.52	10.41	25.76	23.79
Cr_2O_3	0.01	0.01	0.03	0.00	0.04	0.00	0.09	0.00	0.01	0.01	0.01	0.01	0.01	0.10	0.01	0.00	0.00	0.02	0.01
Fe_2O_3*	–	–	–	–	0.51	–	–	–	–	–	–	–	–	–	–	8.07	–	–	–
FeO**	16.70	15.32	25.80	–	–	25.16	28.28	26.16	30.84	4.19	4.50	10.90	10.50	9.60	11.00	–	17.06	3.61	3.14
MnO	0.36	0.49	12.23	0.01	0.00	0.38	0.65	0.38	1.48	0.01	0.07	0.00	0.07	0.03	0.02	0.03	0.22	0.72	0.00
MgO	9.48	11.03	0.33	0.00	0.17	4.49	3.82	4.33	1.40	8.40	8.77	9.71	9.62	10.23	9.45	0.17	5.66	2.81	4.69
CaO	11.74	12.33	4.47	0.00	6.07	9.44	7.66	8.88	8.09	13.58	13.56	0.93	0.72	0.60	1.02	23.57	0.51	23.00	0.02
Na_2O	1.35	0.78	0.01	0.32	7.72	0.02	0.00	0.00	0.00	6.79	6.49	7.00	7.05	7.16	6.97	0.01	6.80	0.20	0.15
K_2O	0.68	0.36	0.01	16.08	0.60	0.01	0.00	0.00	0.00	0.01	0.00	0.02	0.01	0.01	0.01	0.00	0.01	0.00	10.02
Total	96.34	96.97	100.86	100.34	100.23	100.66	100.71	100.73	100.68	100.44	100.18	98.35	98.06	98.44	97.96	98.38	98.08	94.04	94.01
O=	23	23	12	8	8	12	12	12	12	6	6	23	23	23	23	25	23	24.5	22
Si	6.577	3.025	3.002	3.001	2.676	2.994	2.997	3.001	3.008	2.001	2.004	7.924	7.931	7.953	7.972	6.029	8.055	6.019	7.052
Ti	0.003	0.014	0.008	0.000	0.001	0.003	0.012	0.007	0.004	0.002	0.000	0.002	0.002	0.003	0.001	0.014	0.006	0.010	0.007
Al	2.155	1.933	1.973	1.007	1.310	2.035	1.996	2.019	1.976	0.449	0.432	1.860	1.907	1.884	1.831	5.028	1.723	4.829	3.794
Cr	0.001	0.001	0.002	0.000	0.001	0.000	0.005	0.000	0.001	0.000	0.000	0.000	0.001	0.011	0.001	0.000	0.000	0.003	0.001
Fe^{3+}	0.384	0.048	–	–	0.017	–	–	–	–	0.010	0.005	0.166	0.152	0.142	0.069	0.941	0.147	0.480	0.355
Fe^{2+}	1.712	1.849	1.746	0.000	–	1.626	1.849	1.695	2.055	0.114	0.129	1.076	1.045	0.943	1.193	–	1.857	0.097	0.000
Mn	0.046	0.061	0.842	0.000	0.000	0.025	0.043	0.025	0.100	0.000	0.002	0.000	0.008	0.003	0.003	0.004	0.026	0.666	0.946
Mg	2.122	2.434	0.039	0.000	0.011	0.517	0.445	0.500	0.166	0.443	0.464	1.972	1.955	2.061	1.931	0.039	1.185	3.920	0.003
Ca	1.888	1.955	0.383	0.000	0.290	0.782	0.642	0.737	0.690	0.514	0.516	0.136	0.105	0.087	0.149	3.915	0.077	0.062	0.039
Na	0.393	0.224	0.002	0.029	0.666	0.003	0.000	0.000	0.000	0.465	0.447	1.848	1.863	1.876	1.853	0.003	1.852	0.000	1.730
K	0.130	0.068	0.002	0.943	0.034	0.001	0.000	0.000	0.000	0.003	0.000	0.003	0.002	0.002	0.002	0.000	0.002	–	–
Total	15.411	15.247	8.004	4.980	5.008	7.987	7.990	7.983	8.000	4.000	4.000	14.987	14.970	14.965	15.004	15.974	14.930	16.086	13.927

*Total Fe as Fe_2O_3.
**Total Fe as FeO.

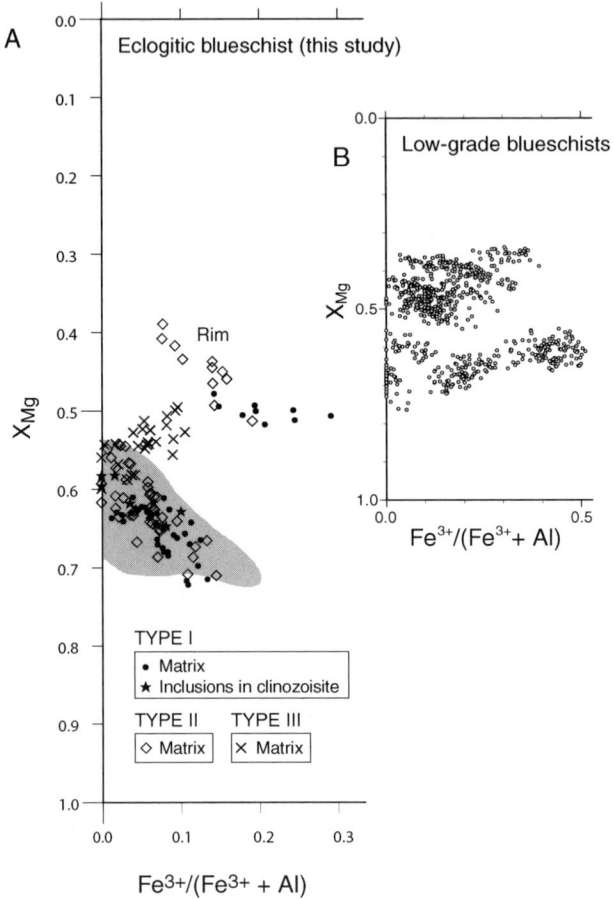

FIG. 5. Chemical compositions of Na-amphiboles in an X_{Mg} versus $Fe^{3+}/(Fe^{3+} + Al)$ diagram. A. Na-amphibole in the investigated blueschist. Grey area represents compositional range of Na-amphiboles from eclogites in the Hida Mountains, southwest Japan. B. Na-amphibole in the low-grade blueschist blocks in the Osayama serpentinite mélange.

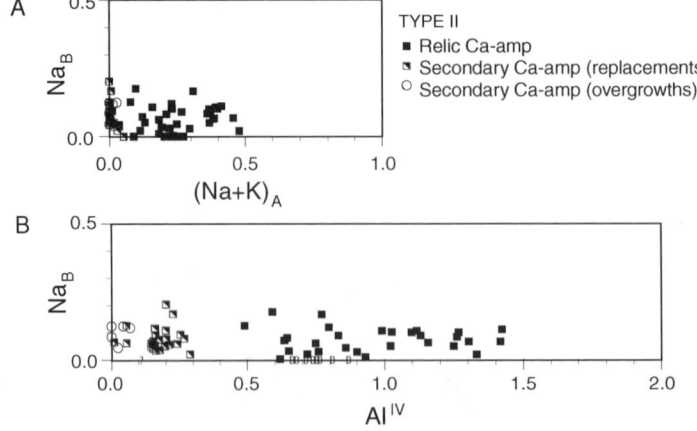

FIG. 6. Chemical compositions of relict and secondary Ca-amphiboles in type-II rock. A. Na_B versus Al^{IV} diagram. B. Na_B versus $(Na + K)_A$ diagram.

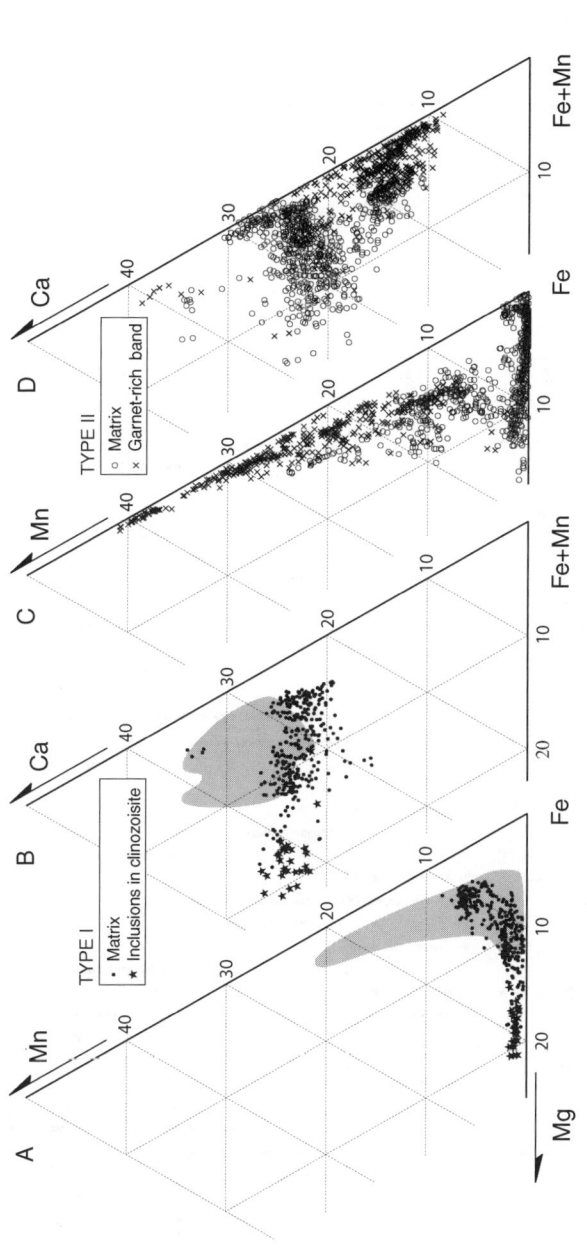

FIG. 7. Chemical compositions of garnets in the Mn-Fe-Mg and Ca-(Fe+Mn)-Mg ternary diagrams. A–B. Type-I rock. Grey area represents compositional range of Na-amphiboles from eclogites in the Hida Mountains, southwest Japan. C–D. Type-II rock.

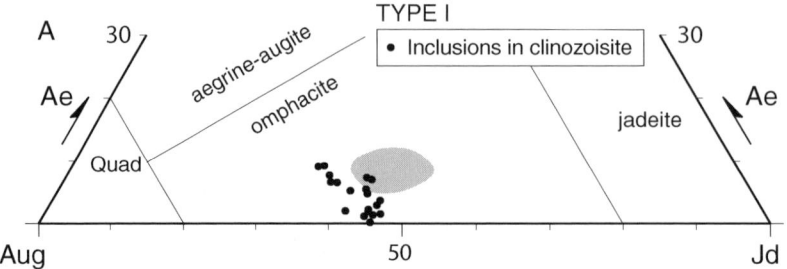

FIG. 8. Chemical compositions of omphacite in the Jd-Ae-Aug ternary system. Grey area represents compositional range of Na-amphiboles from the eclogites in the Hida Mountains, southwest Japan.

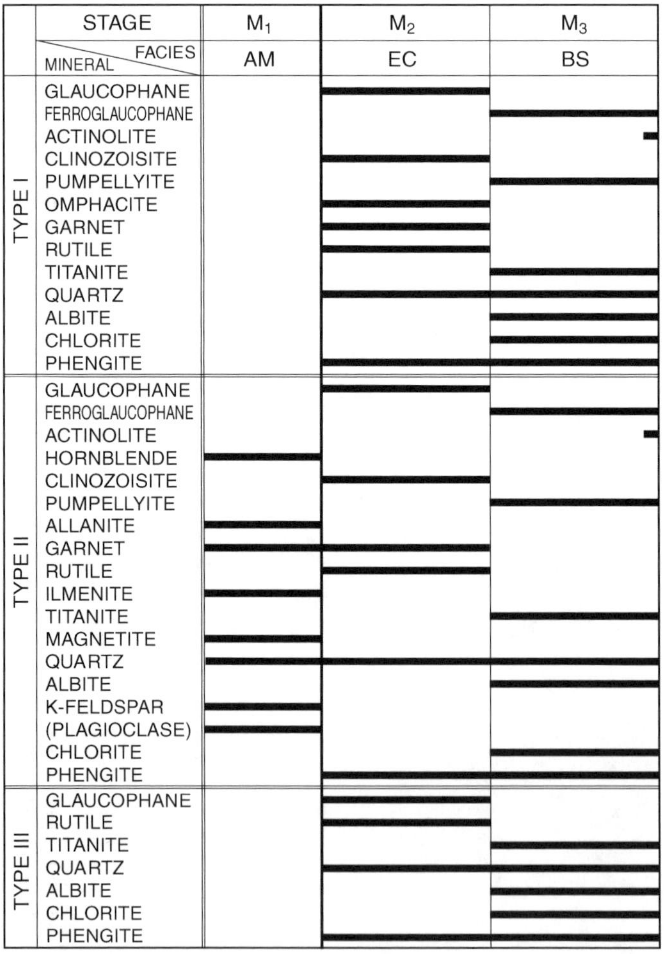

	STAGE	M_1	M_2	M_3
	MINERAL \ FACIES	AM	EC	BS
TYPE I	GLAUCOPHANE		▬	
	FERROGLAUCOPHANE			▬
	ACTINOLITE			▪
	CLINOZOISITE		▬	▬
	PUMPELLYITE			▬
	OMPHACITE		▬	
	GARNET		▬	
	RUTILE		▬	
	TITANITE			▬
	QUARTZ		▬	▬
	ALBITE			▬
	CHLORITE			▬
	PHENGITE		▬	▬
TYPE II	GLAUCOPHANE		▬	
	FERROGLAUCOPHANE			▬
	ACTINOLITE			▪
	HORNBLENDE	▬		
	CLINOZOISITE	▬	▬	▬
	PUMPELLYITE			▬
	ALLANITE	▬		
	GARNET	▬	▬	
	RUTILE	▬	▬	
	ILMENITE	▬		
	TITANITE			▬
	MAGNETITE	▬		
	QUARTZ	▬	▬	▬
	ALBITE			▬
	K-FELDSPAR	▬		
	(PLAGIOCLASE)	▬		
	CHLORITE			▬
	PHENGITE		▬	▬
TYPE III	GLAUCOPHANE		▬	
	RUTILE		▬	
	TITANITE			▬
	QUARTZ		▬	▬
	ALBITE			▬
	CHLORITE			▬
	PHENGITE		▬	▬

FIG. 9. Mineral parageneses for the different stages of metamorphic evolution.

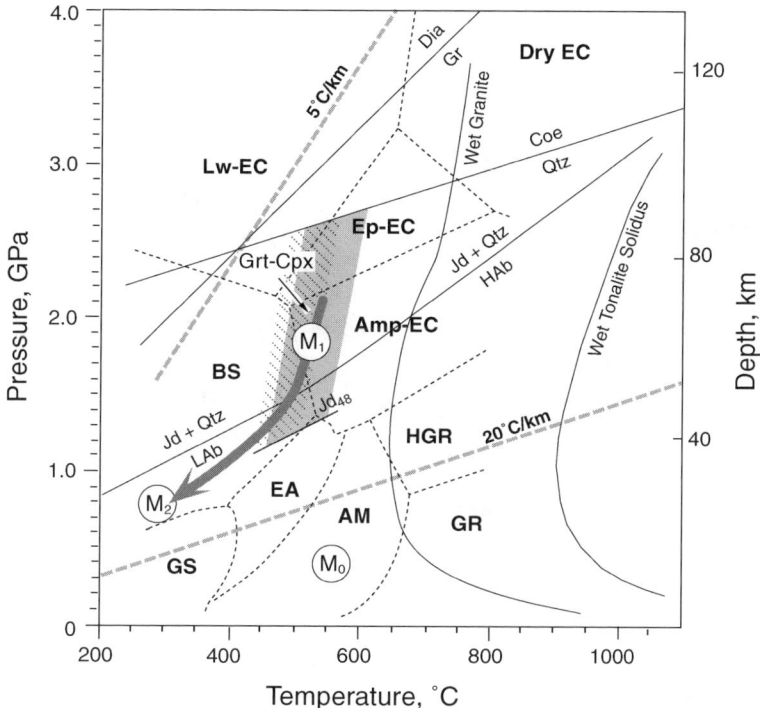

FIG. 10. P-T diagrams showing a qualitative retrograde P-T path, from the M_1 to M_2, of the investigated blueschist. Grey area represents a P-T field constrained by the Jd + Qtz sliding equilibrium (Holland, 1983), and Grt-Cpx thermometry (Ravna, 2000). Hatched area represents the P-T condition of peak eclogite-facies metamorphism in the Hida Mountains eclogites (Tsujimori, 2002). Metamorphic facies, their abbreviations, and phase equilibria are after Liou et al. (2004).

Osayama serpentinite mélange reveals peak eclogite-facies metamorphism and partial preservation of relict amphibolite-facies assemblages (Fig. 10). The observed petrographic signatures imply that the investigated high-grade block underwent eclogite-facies metamorphism (M_1) and later, blueschist-facies overprinting (M_1). Although the M_1 prograde evolution is unclear, the inferred retrograde P-T path after M_1 is similar to the "Franciscan-type" (Ernst, 1988) retrograde path of "Pacific-type" subduction zones. The M_2 blueschist-facies overprint over the low-T eclogite assemblage is interpreted as cooling during decompression representing tectonic exhumation involving significant refrigeration. The remarkable compositional zoning from Fe^{2+}-poor core (Gln) to Fe^{2+}-rich rim (Fgl) in Na-amphibole may relate to a discontinuous reaction, which consumed garnet and omphacite during retrogression, such as 5Alm + 4Grs + 10Jd + 11Qtz + 8H$_2$O = 5Fgl + 6Czo. The breakdown of garnet is also supported by the enrichment of MnO in retrograde ferroglaucophane. In the Osayama serpentinite mélange, low-grade blueschists record P-T conditions comparable to the M_2 stage. This is confirmed by similar cooling ages: phengite K-Ar ages of the investigated high-grade block (289–322 Ma) overlap those from low-grade blueschists (273–327 Ma), in particular those from the lawsonite-pumpellyite grade (311–315 Ma; Tsujimori and Itaya, 1999). On the other hand, there is no evidence to tie the P-T path from M_0 to M_1. M_0 minerals may represent remnants of the pre-subduction conditions of the protolith. Limited fluid infiltration and continuous subduction of colder materials at the bottom of the exhuming blueschist-bearing slab is suggested to explain both the persistence of relict amphibolitic minerals and the blueschist-facies overprinting.

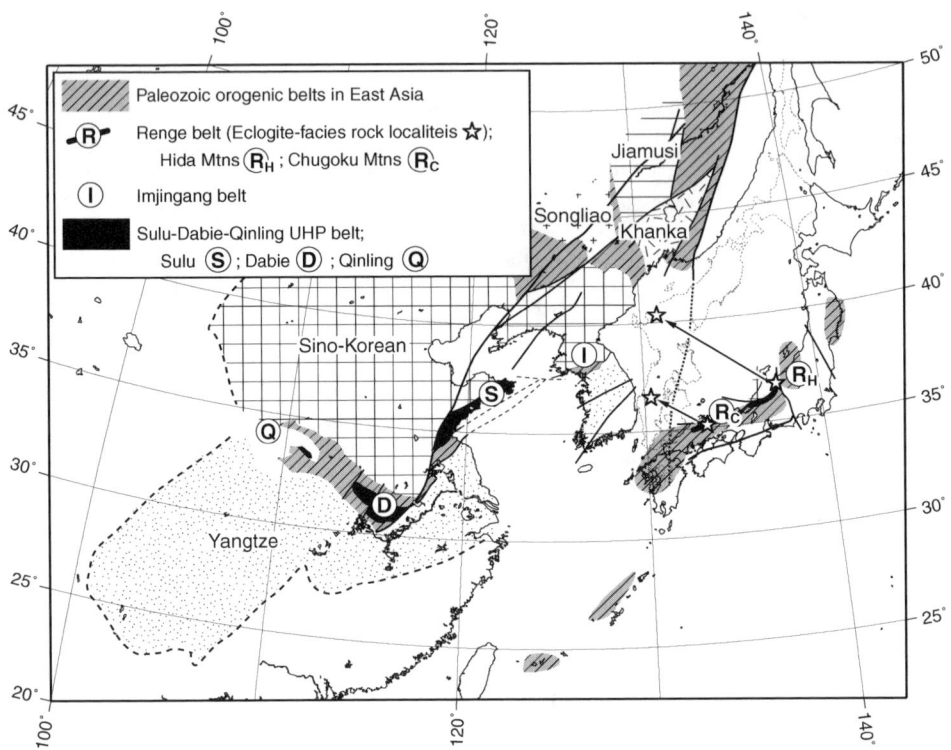

FIG. 11. Simplified geologic map of eastern Asia, showing tectonic correlation between the Qinling-Dabie-Sulu belt and the Renge belt (modified after Maruyama et al., 1994; Khanchuk, 2001).

Petrotectonic Link to Suture Zones in Eastern Asia

Significance of Paleozoic eclogite-facies metamorphism in Japan

How does the occurrence of a Paleozoic eclogite belt in southwest Japan fit into the regional tectonic framework of eastern Asia? The huge Qinling-Dabie-Sulu (QDS) HP-UHP belt is recognized as the east-central Chinese collisional suture between the Sino-Korean and Yangtze blocks (e.g., Ernst et al., 1991; Liou et al., 1996; Hacker et al., 2000; Fig. 11). It has been postulated that the eastern limb of the Qinling-Dabie-Sulu belt extends across the Korean Peninsula (Ernst and Liou, 1995; Ree et al. 1996). A possible geotectonic correlation between the Korean Barrovian-type terranes and the Japanese Hida Mountains has also been pointed out (Hiroi, 1983; Isozaki, 1996). However, uncertainty remains because of the absence of critical evidence of HP-UHP metamorphism in the Korean Peninsula, despite minor eclogite relicts (Grt + Cpx with up to 25 mol% Jd was reported by Oh et al., 2004). On the other hand, recent studies of Paleozoic terranes in southwest Japan show the existence of regional eclogite-facies metamorphism in Renge metamorphic rocks of the Hida Mountains (Tsujimori et al., 2000; Tsujimori, 2002; Fig. 11). Although the Osayama area of the Chugoku Mountains is about 400 km away from the eclogite locality of the Hida Mountains, eclogitic mineral inclusions described in this paper provide additional evidence for an early Paleozoic eclogite-facies metamorphism in southwest Japan.

The "Renge HP eclogite" in Japan may be a representative of the eastern extension of the QDS suture, as hypothesized by Ernst and Liou (1995). However, metamorphic characteristics, protoliths, and ages are different. For instances, protoliths of "Renge HP eclogites" include trench-fill sediments and fragments of oceanic crust, and they are generally associated with ophiolitic serpentinites; these are typical "Pacific-type (Type-B)" protoliths

(Maruyama et al., 1996). In contrast, coesite- and diamond-bearing UHP eclogites in the QDS belt indicate subduction of continental crust to >20 km (Liou et al., 1996, 2004), and are characterized as "collisional-type (Type-A)" protoliths (Maruyama et al., 1996). Moreover, radiometric ages from the QDS belt are >100 Ma younger than K-Ar and ^{40}Ar-^{39}Ar ages of the Renge eclogites (340–280 Ma). Numerous QDS belt radiometric ages have been obtained, and have confirmed the Triassic UHP metamorphic event. Recent zircon U-Pb SHRIMP age dating combined with studies of micro-inclusions in zircon has further documented the 230 Ma UHP and 210 Ma retrograde metamorphism in east-central China (e.g., Liu et al., 2004).

On the other hand, two UHP metamorphic events have been recognized in the central orogenic belt of China, and a 4000 km long UHP belt extends from the Altun–North Qadiam through Qinling to the Dabie-Sulu. This belt was delineated based on numerous findings of coesite and microdiamond inclusions in UHP rocks and systematic SHRIMP age dating of zircon separates from eclogites and associated gneisses. Where does the Japanese Paleozoic eclogitic terrane fit? Based on available petrotectonic information around Eastern Asia, the following two models should be considered.

Model 1: Pre-collisional "Pacific-type" subduction on both side of the QDS belt

Early Paleozoic terranes consisting of accretionary complex, ophiolite, and island arc–type plutons and HP rocks of typical "Pacific type" occur in both the North Qilian (for summary see Yang et al., 2001) and the northwest Qinling-Dabie Mountains (e.g., Kröner et al., 1993; Maruyama et al., 1994; Lerch et al., 1995). Recently, Ratschbacher et al. (2003) comprehensively summarized pre-existing data together with their new data, and pointed out that a Silurian-Permian, active subduction-accretion system developed in the Qinling orogen. Moreover, ~320 Ma zircon SHRIMP U-Pb ages have been reported from HP eclogites from the Xiongdian and Hujiawan areas (Sun et al., 2002). We postulate continuous subduction of oceanic crust and subsequent exhumation of HP rocks prior to collision between the Sino-Korean and Yangtze blocks. Such a geotectonic configuration is also observed in the northwest Himalaya; "Pacific-type" subduction formed a Cretaceous accretionary complex with 80 Ma blueschists along the southern margin of the Kohistan-Ladakh arc (e.g., Anczkiewicz et al., 2000), whereas the Indian block collided with Asia to form the UHP rocks at 46 Ma (Kaneko et al., 2003). This model suggests that "Pacific-type" subduction generated early to late Paleozoic HP eclogite-facies metamorphism on both western and eastern sides of the QDS belt.

Model 2: Paleozoic "Pacific-type" subduction chain from further north to southwest Japan

Several stages of collision and amalgamation of microcontinental blocks are recognized in the Russia–China–North Korean border region; Paleozoic orogenic belts are developed between these micro-continental blocks (Fig. 11; e.g., Maruyama and Liou, 1998; Khanchuk, 2001; Jia et al., 2004). In particular, a N-S–trending ophiolitic suture zone along the western margin of the Jiamusi block contains 450–410 Ma blueschists (Ye and Zhang, 1994; Maruyama and Liou, 1998). This trend is almost perpendicular to the general E-W trend of the current southwest Japanese terranes, as shown on Figure 11. However, if we restore the paleogeography of southwest Japan prior to the opening of the Japan Sea at about 15 Ma (Otofuji and Matsuda, 1984; Ernst and Liou, 1995), southwest Japan was located directly south of the Russian Primorye, and to the east of the Korean Peninsula (e.g., Kojima, 1989; Isozaki, 1997; Khanchuk, 2001). In fact, Permian–Jurassic accretionary complexes in the Russian Primor'ye and southwest Japan have very similar age-lithology relationships (Kojima, 1989; Ishiwatari and Tsujimori, 2003). Khanchuk (2001) also suggested that the N-S–trending ophiolite belts in the Russia–China–North Korean border area are possibly extensions of the Paleozoic ophiolite belts in southwest Japan. If we adopt this model, late Paleozoic HP eclogite-facies metamorphism in southwest Japan does not need to be correlated with the QDS belt. Instead it may represent part of the Paleozoic "Pacific-type" subduction chain that extended from the present Russia–China–North Korea border. In this case, the QDS belt terminates somewhere east of the Sulu region.

Acknowledgments

We dedicate this paper to W. G. Ernst who was the first to propose an eastern extension of the Chinese UHP belt to southwest Japan. This research was supported financially in part by a JSPS Research Fellowship for Young Scientists (1998–1999, 2001–2003), and a JSPS Research

Fellowship for Research Abroad (2004–) of the first author. Preparation of this manuscript was supported by NSF EAR-0003355. The first author thanks to A. Ishiwatari for discussions during his stay at Kanazawa University (1994–2000).

REFERENCES

Ankiewicz, R., Burg, J. P., Villa, I. M., and Meier, M., 2000, Late Cretaceous blueschist metamorphism in the Indus suture zone, Shangla region, Pakistan Himalaya: Tectonophysics, v. 324, p. 111–134.

Arai, S., 1980, Dunite-harzburgite-chromitite complexes as refractory residues in the Sangun-Yamaguchi zone, western Japan: Journal of Petrology, v. 21, p. 141–165.

Clarke, G. L., Aitchison, J. C., and Cluzel, D., 1997, Eclogites and blueschists of the Pam Peninsula, NE New Caledonia: A reappraisal: Journal of Petrology, v. 38, p. 843–876.

Frey, M., de Capitani, C., and Liou, J. G., 1991, A new petrogenetic grid for low-grade metabasites: Journal of Metamorphic Geology, v. 9, p. 497–509.

Ernst, W. G., 1988, Tectonic history of subduction zones inferred from retrograde blueschist P-T paths: Geology, v. 16, p. 1081–1084.

Ernst, W. G., and Liou, J. G., 1995, Contrasting plate-tectonic styles of the Qinling-Dabie-Sulu and Franciscan metamorphic belts: Geology, v. 23, p. 353–356.

Ernst, W. G., Zhou, G., Liou, J. G., Eide, E., and Wang, X., 1991, High-pressure and superhigh-pressure metamorphic terranes in the Qinling-Dabie mountain belt, central China: Early- to mid-Phanerozoic accretion of the western Paleo-pacific rim: Pacific Science Association Information Bulletin, v. 43, p. 6–15.

Hacker, B. R., Ratschbacher, L., Webb, L. E., McWilliams, M., Calvert, A., Dong, S., Wenk, H.-R., and Chateigner, D., 2000, Exhumation of ultrahigh-pressure rocks in east-central China: Late Triassic–Early Jurassic tectonic unroofing: Journal of Geophysical Research, v. 105, p. 13339–13364.

Hayasaka, Y., Sugimoto, T., and Kano, T., 1995, Ophiolitic complex and metamorphic rocks in the Niimi-Katsuyama area, Okayama Prefecture: Excursion Guidebook of 102nd Annual Meeting of the Geological Society of Japan, p. 71–87 (in Japanese).

Herzig, C. T., Kimborough, D. L., and Hayasaka, Y., 1997, Early Permian zircon uranium-lead ages for plagiogranites in the Yakuno ophiolite, Asago district, southwest Japan: The Island Arc, v. 6, p. 396–403.

Hiroi, Y., 1981, Subdivision of the Hida metamorphic complex, central Japan, and its bearing on the geology of the Far East in pre–Sea of Japan time: Tectonophysics, v. 76, p. 317–333.

Holland, T. J. B., 1983, The experimental determination of activities in disorderd and short-range ordered jadeitic pyroxenes: Contributions to Mineralogy and Petrology, v. 82, p. 214–220.

Ichiyama, Y., and Ishiwatari, A., 2004, Petrochemical evidence for off-ridge magmatism in a back-arc setting from the Yakuno ohhiolite, Japan: The Island Arc, v. 13, p. 157–177.

Ishiga, H., 1990, Ultra-Tamba terrane, in Ichikawa, K., Mizutani, S., Hara, I., Hada, S., and Yao, A., eds., Pre-Cretaceous terranes of Japan: Osaka, Japan, Osaka City University, Publication of IGCP Project 224, p. 97–107.

Ishiwatari, A., 1985, Igneous petrogenesis of the Yakuno ophiolite (Japan) in the context of the diversity of ophiolites: Contributions to Mineralogy and Petrology, v. 89, p. 155–167.

Ishiwatari, A., 1991, Ophiolites in the Japanese Islands: Typical segment of the circum-Pacific multiple ophiolite belts: Episodes, v. 14, p. 274–279.

Ishiwatari, A., and Tsujimori, T., 2003, Paleozoic ophiolites and blueschists in Japan and Russian Primorye in the tectonic framework of East Asia: A synthesis: The Island Arc, v. 12, p. 190–206.

Isozaki, Y., 1996, Anatomy and genesis of a subduction-related orogen: A new view of geotectonic subdivision and evolution of the Japanese Islands: The Island Arc, v. 5, p. 289–320.

Isozaki, Y., 1997, Contrasting two types of orogen in Permo-Triassic Japan: Accretionary versus collisional: The Island Arc, v. 6, p. 2–24.

Jia, D. C., Hu, R. H., Lu, Y., and Qiu, X. L., 2004, Collision belt between the Khanka block and the North China block in the Yanbian Region, Northeast China: Journal of Asian Earth Sciences, v. 23, p. 211–219.

Kaneko, Y., Katayama, I., Yamamoto, H., Misawa, K., Ishikawa, M., Rehman, H. U., Kausar, A. B., and Shiraishi, K., 2003, Timing of Himalayan ultrahigh-pressure metamorphism: Sinking rate and subduction angle of the Indian continental crust beneath Asia: Journal of Metamorphic Geology, v. 21, p. 589–599.

Kanmera, K., Sano, H., and Isozaki, Y., 1990, Akiyoshi terrane, in Ichikawa, K., Mizutani, S., Hara, I., Hada, S., and Yao, A., eds., Pre-Cretacious terranes of Japan: Osaka, Japan, Osaka City University, Publication of IGCP Project 224, p. 49–62.

Katzir, Y., Avigad, D., Matthewas, A., Garfunkel, Z., and Evans, B. W., 2000, Origin, HP/LT metamorphism, and cooling of ophiolitic mélanges in southern Evia (NW Cyclades), Greece: Journal of Metamorphic Geology, v. 18, p. 699–718.

Khanchuk, A.I., 2001, Pre-Neogene tectonics of the Sea of Japan region: A view from the Russian side: Earth Science (Chikyu Kagaku), v. 55, p. 275–291.

Kretz, R., 1983, Symbols for rock-forming minerals: American Mineralogist, v. 68, p. 277–279.

Kobayashi, S., Miyake, H., and Shoji, T., 1987, A jadeite rock from Oosa-cho, Okayama Prefecture, southwestern Japan: Mineralogical Journal, v. 13, p. 314–327.

Kojima, S., 1989, Mesozoic terrane accretion in northeast China, Sikhote-Alin, and Japan regions: Palaegeography, Palaeoclimatology, Palaeoecology, v. 69, p. 213–232.

Kröner, A., Zhang, G. W., and Sun, Y., 1993, Granulites in the Tongbai area, Qinling belt, China: Geochemistry, petrology, single zircon geochronology, and implications for the tectonic evolution of eastern Asia: Tectonics, v. 12, p. 245–255.

Kuroda, Y., Kurokawa, K., Uruno, K., Kinugawa, T., Kano, H., and Yamada, T., 1976, Staurolite and kyanite from epidote-hornblende rocks in the Oeyama (Komori) ultramafic mass, Kyoto Prefecture, Japan: Earth Sciences (Chikyu Kagaku), v. 30, p. 331–333.

Lerch, M. F., Xue, F., Kröner, A., Zhang, G. W., and Todt, W., 1995, A Middle Silurian–Early Devonian magmatic arc in the Qinling Mountains of central China: Journal of Geology, v. 103, p. 437–449.

Liou, J. G., Maruyama, S., and Cho, M., 1985, Phase equilibria and mineral parageneses of metabasites in low-grade metamorphism: Mineralogical Magazine, v. 49, p. 321–333.

Liou, J. G., Tsujimori, T., Zhang, R. Y., Katayama, I., and Maruyama, S., 2004, Global UHP metamorphism and continental subduction/collision: The Himalayan model: International Geology Review, v. 46, p. 1–27.

Liou, J. G., Zhang, R. Y., Eide, E. A., Maruyama, S., Wang, X., and Ernst, W. G., 1996, Metamorphism and tectonics of high-P and ultrahigh-P belts in Dabie-Sulu regions, eastern central China, in Yin, A., and Harrison, T. M., eds., The tectonic evolution of Asia: New York, NY: Cambridge University Press, p. 330–343.

Liu, F., Xu, Z., Liou, J. G., and Song, B., 2004, SHRIMP U-Pb ages of ultrahigh-pressure and retrograde metamorphism of gneisses, south-western Sulu terrane, eastern China: Journal of Metamorphic Geology, v. 22, p. 315–326.

Liu, J., Bohlen, S. R., and Ernst, W. G., 1996, Stability of hydrous phases in subducting oceanic crust: Earth and Planetary Science Letters, v. 143, p. 161–171.

Maruyama, S., and Liou, J. G., 1998, Initiation of ultrahigh-pressure metamorphism and its significance on the Proterozoic–Phanerozoic boundary: The Island Arc, v. 7, p. 6–35.

Maruyama, S., Liou, J. G., and Terabayashi, M., 1996, Blueschists and eclogites of the world and their exhumation: International Geology Review, v. 38, p. 485–594.

Maruyama, S., Liou, J. G., and Zhang, R. Y., 1994, Tectonic evolution of the ultrahigh-pressure (UHP) and high-pressure (HP) metamorphic belts from central China: The Island Arc, v. 3, p. 112–121.

Nishimura, Y., 1998, Geotectonic subdivision and areal extent of the Sangun belt, inner zone of southwest Japan: Journal of Metamorphic Geology, v. 16, p. 129–140.

Nishimura, Y., and Shibata, K., 1989, Modes of occurrence and K-Ar ages of metagabbroic rocks in the "Sangun metamorphic belt," southwest Japan: Memoirs of Geological Society of Japan, no. 33, p. 343–357. (in Japanese with English abstract)

Oh, C. W., Choi, S. G., Song, S. H., and Kim, S. W., 2004, Metamorphic evolution of the Baekdong metabasite in the Hongseong area, South Korea and its relationship with the Sulu Collision belt of China: Gondwana Research, v. 7, p. 809–816.

Okamoto, K., Maruyama, S., and Isozaki, Y., 2000, Accretionary complex origin of the high-P/T Sambagawa metamorphic rocks in central Shikoku, Japan: Journal of Geological Society of Japan, v. 106, p. 70–86.

Otofuji, Y., and Matsuda, T., 1984, Timing of rotational motion of southwest Japan inferred from paleomagnetism: Earth and Planetary Science Letters, v. 70, p. 373–382.

Pearce, J. A., 1982, Trace element characteristics of lavas from destructive plate bounders, in Thorpe, R. S., ed., Andesites: Orogenic andesites and related rocks: New York, NY, John Wiley and Sons, Inc., p. 525–548.

Pearce, J. A., 1983, The role of sub-continental lithosphere in magma genesis at disructive plate margins, in Hawkesworth, C. J., and Norry, M. J., eds., Continental basalts and mantle xhenoliths: Nantwich, UK Shiva, p. 230–249.

Ratschbacher, L., Hacker, B. R., Calvert, A., Webb, L. E., Grimmer, J. C., McWilliams, M., Ireland, T., Dong, S., and Hu, J., 2003, Tectonics of the Qinling (Central China): Tectonostratigraphy, geochronology, and deformation history: Tectonophysics, v. 366, p. 1–53.

Ravna, E. K., 2000, The garnet–clinopyroxene Fe^{2+}–Mg geothermometer: An updated calibration: Journal of Metamorphic Geology, v. 18, p. 211–219.

Ree, J.-H., Cho, M., Kwon, S. T., and Nakamura, E., 1996, Possible eastward extension of Chinese collision belt in South Korea: The Imjingang belt: Geology, v. 24, p. 1071–1074.

Schliestedt, M., 1986. Eclogite-blueschist relationships as evidenced by mineral equilibria in the high-pressure metabasic rocks of Sifnos (Cycladic Islands), Greece: Journal of Petrology, v. 27, p. 1437–1459.

Shibata, K., Igi, S., and Uchiumi, S., 1977, K-Ar ages of hornblendes from gabbroic rocks in Southwest Japan: Geochemical Journal, v. 11, p. 57–64.

Sun, S. S., and McDonough, W. E., 1989, Chemical and isotopic systematics of ocean basalts: Implications for mantle composition and processes, in Saunders, A. D., and Norry, M. J., eds., Magmatism in the ocean basins, in Saunders, A. D. and Norry, M. J., eds., Magmatism in the ocean basins: Oxford, UK, Blackwell Scientific, Geological Society Special Publication, p. 313–345.

Sun, W. D., Williams, I. S., and Li, S. G., 2002, Carboniferous and Triassic eclogites in the western Dabie Mountains, east-central China: Evidence for protracted convergence of the North and South China blocks: Journal of Metamorphic Geology, v. 20, p. 873–886.

Suzuki, K., and Adachi, M., 1994, Middle Precambrian detrital monazite and zircon from the Hida Gneiss on Oki-Dogo Island, Japan: Their origin and implications for the correlation of basement gneiss of southwest Japan and Korea: Tectonophysics, v. 235, p. 277–292.

Terabayashi, M., 1993, Compositional evolution in Ca-amphibole in the Karmutsen metabasites, Vancouver Island, British Columbia, Canada: Journal of Metamorphic Geology, v. 11, p. 677–690.

Tsujimori, T., 1997, Omphacite-diopside vein in an omphacitite block from the Osayama serpentinite melange, Sangun-Renge metamorphic belt, southwestern Japan: Mineralogical Magazine, v. 61, p. 845–852.

Tsujimori, T., 1998, Geology of the Osayama serpentinite melange in the central Chugoku Mountains, southwestern Japan: 320 Ma blueschist-bearing serpentinite melange beneath the Oeyama ophiolite: Journal of Geological Society of Japan, v. 104, p. 213–231 (in Japanese with English abstract).

Tsujimori, T., 2002, Prograde and retrograde P-T paths of the late Paleozoic glaucophane eclogite from the Renge metamorphic belt, Hida Mountains, southwest Japan: International Geology Review, v. 44, p. 797–818.

Tsujimori, T., Ishiwatari, A., and Banno, S., 2000, Eclogitic glaucophane schist from the Yunotani valley in Omi Town, the Renge metamorphic belt, the Inner Zone of southwestern Japan: Journal of Geological Society of Japan, v. 106, p. 353–362 (in Japanese with English abstract).

Tsujimori, T., and Itaya, T., 1999, Blueschist-facies metamorphism during Paleozoic orogeny in southwestern Japan: Phengite K-Ar ages of blueschist-facies tectonic blocks in a serpentinite mélange beneath Early Paleozoic Oeyama ophiolite: The Island Arc, v. 8, p. 190–205.

Tsujimori, T., and Liou, J. G., 2004a, Metamorphic evolution of kyanite-staurolite-bearing epidote-amphibolitic rocks from the Early Paleozoic Oeyama belt, SW Japan: Journal of Metamorphic Geology, v. 22, p. 301–314.

Tsujimori, T., and Liou, J. G., 2004b, Coexisting chromian omphacite and diopside in tremolite schist from the Chugoku Mountains, SW Japan: The effect of Cr on the omphacite-diopside immiscibility gap: American Mineralogist, v. 89, p. 7–14.

Yamashita, K., and Yanagi, T., 1994, U-Pb and Rb-Sr dating of Oki metamorphic rocks, the Oki Island, Southwest Japan: Geochemical Journal, v. 28, p. 333–339.

Yang, J., Xu, Z., Zhang, J., Chu, C.-Y., Zhang, R., and Liou, J. G., 2001, Tectonic significance of early Paleozoic high-pressure rocks in Altun-Qaidam-Qilian Mountains, northwest China, in Hendrix, M. S., and Davis, G. A., eds., Paleozoic and Mesozoic tectonic evolution of central Asia: From continental assembly to intracontinental deformation: Geological Society of America Memoir 194, p. 151–170.

Ye, H. W., and Zhang, X. Z., 1994, The ^{40}Ar-^{39}Ar age of the vein crossite in blueschist in Mudanjiang area, NE China and its geological implication: Journal of Changchun University of Earth Sciences, no. 24, p. 369–372 (in Chinese with English abstract).

Accretionary Complex Origin of the Mafic-Ultramafic Bodies of the Sanbagawa Belt, Central Shikoku, Japan

MASARU TERABAYASHI,[1]

Department of Safety Systems Construction Engineering, Kagawa University, Kagawa 761-0396, Japan

KAZUAKI OKAMOTO,[2]

Riken, the Institute of Physical and Chemical Research, Saitama 351-0198, Japan

HIROSHI YAMAMOTO,

Department of Earth and Environmental Sciences, Kagoshima University, Kagoshima 890-0065, Japan

YOSHIYUKI KANEKO,[3]

Geoscience and Technology, Geological Survey of Japan, National Institute of Advanced Industrial Science and Technology, Tsukuba 305-8567, Japan

TSUTOMU OTA,[4] SHIGENORI MARUYAMA, IKUO KATAYAMA,[5] TSUYOSHI KOMIYA, AKIRA ISHIKAWA,[6]

Department of Earth and Planetary Sciences, Tokyo Institute of Technology, Tokyo 152-8551, Japan

RYO ANMA,

Institute of Geoscience, University of Tsukuba, Ibaraki 305-8571, Japan

HIROAKI OZAWA,

Department of Geology, Naruto University of Education, Tokushima 772-0051, Japan

BRIAN F. WINDLEY,

Department of Geology, Leicester University, Leicester LE1 7RH, United Kingdom

AND J. G. LIOU

Department of Geological and Environmental Sciences, Stanford University, Stanford, California 94305

Abstract

In the high-grade Cretaceous Sanbagawa high-pressure (HP) metamorphic belt, our new 1:5000 scale mapping of eclogitic mafic-ultramafic bodies and their surrounding epidote-amphibolite–facies schists has revealed a duplex structure formed by the subduction of the Izanagi-Pacific oceanic plate. Lithologies of the two largest mafic-ultramafic bodies in the Sanbagawa belt, the Iratsu eclogite and the Higashi-Akaishi peridotite, strike WNW-ESE and dip N; the upper boundary with the surrounding schist is a normal fault, whereas the lower boundary is a thrust. The Iratsu body is subdivided into at least two tectonic units; the unit boundary is subparallel to a lithological boundary. Protoliths of the upper unit are gabbro, basalt, minor quartz rock, and pelite, and those of the lower unit are pyroxenite, gabbro, basalt, chert, and marble, in ascending order. The lower unit is characterized by layers of alternating eclogitic metagabbro and pyroxenite. The layers are exten-

---

[1]Corresponding author; email: tera@eng.kagawa-u.ac.jp
[2]Present address, Institute of Earth Sciences, Academia Sinica, P.O. Box 1-55 Nankang, Taipei 115, Taiwan, R.O.C.
[3]Present address: Graduate School of Environment and Information Sciences, Yokohama National University, Kanagawa 240-8501, Japan.
[4]Present address: The Pheasant Memorial Laboratory, Institute for Study of the Earth's Interior, Okayama University, Tottori 682-0193, Japan.
[5]Present address: Department of Geology and Geophysics, Yale University, New Haven, CT 06520.
[6]Present address: The Pheasant Memorial Laboratory, Institute for Study of the Earth's Interior, Okayama University, Tottori 682-0193, Japan.

sive at the bottom of the Iratsu eclogite, and transient toward the Higashi-Akaishi body. Eclogite-facies metapsammite is intercalated between the Iratsu and Higashi-Akaishi bodies. Our mapping has revealed the following: (1) a duplex structure of the mafic-ultramafic bodies indicating their accretionary complex origin; (2) reconstructed oceanic plate stratigraphy in ascending order of peridotite, gabbro, basalt, limestone, minor chert, and pelite, suggesting that different parts of the protolith were derived from a mid-oceanic topographic high, an oceanic island or plateau, and an overlying trench turbidite; and (3) a change in the convergent motion of the oceanic plate from NW to NE during the accretion of the large oceanic island or plateau.

Introduction

THE MECHANISM OF emplacement of eclogite-ultramafic rocks in ultrahigh-pressure (UHP) and high-pressure (HP) metamorphic belts has long been debated. In HP belts, related to oceanic plate subduction (Pacific or B-type orogens: Matsuda and Uyeda, 1971; Maruyama et al., 1996), occurrences of eclogitic high-grade rocks and ultramafic rocks have been considered as tectonic blocks with complex thermal histories and different origins from surrounding schists (e.g., Franciscan: Cloos, 1986; Maruyama and Liou, 1988; Oh and Liou, 1990; Sanbagawa: Kunugiza et al., 1986; Takasu, 1989; Kamuikotan of Japan: Takayama, 1986). In UHP belts, eclogite-facies rocks and garnet-bearing ultramafic rocks occur as blocks or lenses in host gneisses related to continent-continent collision zones (A-type blueschist belts: Maruyama et al., 1996). Alpine-type peridotites, or simply orogenic peridotites, were thought to be incorporated into regional rocks by plate convergence (e.g., Alpe Arami: Ernst, 1978; Western Gneiss Region of Norway: Carswell and Harvey, 1985; Bohemian body: Medaris et al., 1995). Recent studies of UHP belts identified coesite and micro-diamond inclusions in zircons and garnets from host gneiss and marble, and suggest that UHP eclogites were subducted with supracrustal rocks, experienced UHP metamorphism, and were exhumed as a coherent body (e.g., Alps: Chopin, 1984; Dabie Mountains: Tabata et al., 1998; Ye et al., 2000; Liu et al., 2001; Kokchetav body: Sobolev and Shatsky, 1990; Katayama et al., 2000). Finally, deeply subducted buoyant continental materials have incorporated fragments of garnet peridotite, and returned to the surface.

The Sanbagawa HP metamorphic belt in the orogenic core of the Cretaceous orogen, southwestern Japan, is a key area to understand Pacific-type orogenic processes, because it has the best exposure from low to high grades of any region in the Pacific Rim. The Sanbagawa belt extends over 800 km from east to west, and has a maximum width of 30 km (Fig. 1A). It is a tectonic slice less than 2 km thick, and is separated by subhorizontal faults from weakly metamorphosed accretionary complexes (Isozaki and Maruyama, 1991; Maruayama et al., 1996).

In the last 30 years, the Sanbagawa protoliths have been interpreted to be derived from passive or active continental margins. Modes of occurrence of metachert or quartz schist and greenstone or mafic schist have long been believed to be coherent. Kawachi et al. (1982) considered that the Sanbagawa protoliths originated from a marginal basin in which mafic volcanic rocks of mainly alkaline basalt composition erupted during sedimentation of continent-derived terrigenous clastics. Ishizuka et al. (1989) supported this interpretation based on geochemical signatures of the Sanbagawa greenstones. Faure (1985) proposed that the protoliths were deposited or erupted in a passive continental margin that flanked a microcontinent, and collided against the Asian continental margin. On the other hand, Ernst (1975) speculated that the Sanbagawa belt formed as a Cordilleran-type orogenic belt based on the occurrence of an ophiolitic unit known as the Mikabu greenstones. Later, detailed mapping of metachert and geochemistry of greenstones indicated that they formed in the Pacific domain, and accreted later as a subduction complex at a trench (Okamoto et al., 2000). Furthermore, Okamoto et al. (2000) and Sakakibara et al. (2001) showed from petrochemical data of basaltic rocks that the protoliths were more compatible with mid-ocean ridge basalt (MORB) or tholeiitic basalt of an oceanic plateau than with basalts from a passive continental margin or a back-arc basin. Thus these recent studies concluded that the Sanbagawa belt was part of an accretionary complex, and that the protoliths were derived from oceanic crust with pelagic sediments from a mid-oceanic region and turbidites from a trench.

However, the geology, petrology, and chronology of the highest P-T rocks including eclogite and Alpine-type peridotite in orogenic core of the Sanbagawa belt are not yet fully understood because of

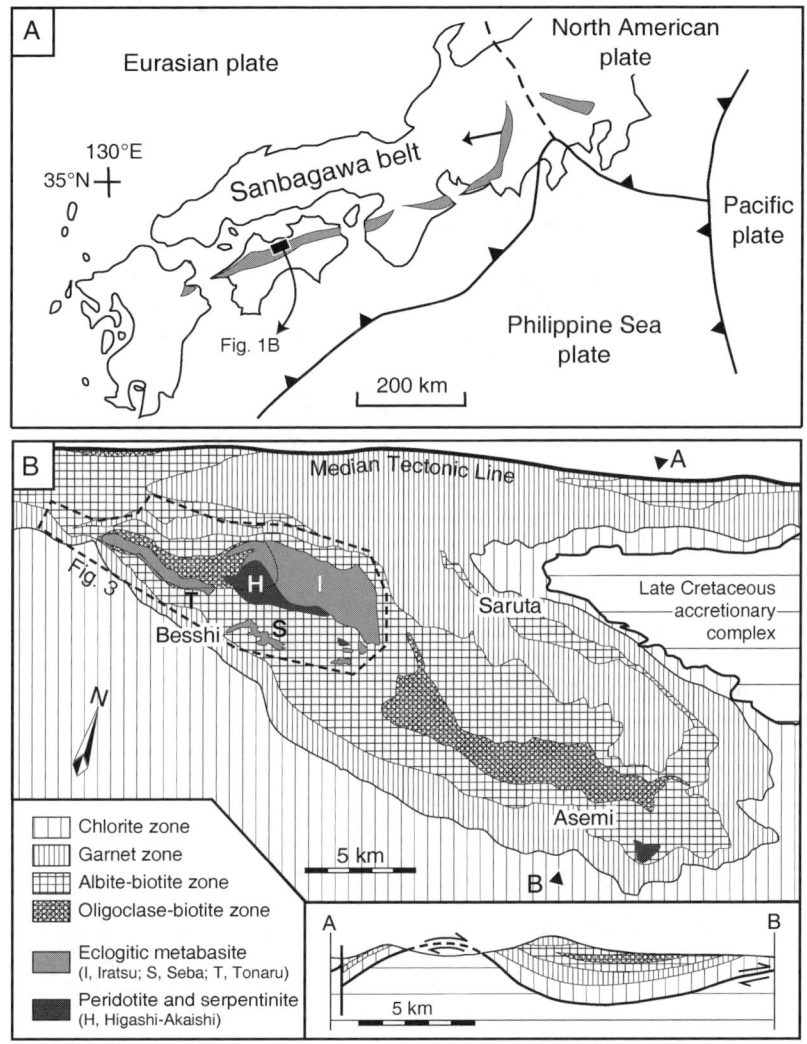

FIG. 1. A. Index map showing the location of the Sanbagawa belt. B. Index map of central Shikoku, Japan, showing metamorphic zones and cross-section of the Sanbagawa belt (modified from Higashino, 1990). The mapped area in Figure 3 is also shown.

the inaccessibility of outcrops in this mountainous region. Protoliths of the eclogitic mafic-ultramafic bodies in the Sanbagawa belt were considered to be mafic-ultramafic bodies derived from layered gabbros or cumulates of the lower crust and underlying upper mantle of an island arc or a continent (Mori and Banno, 1973; Kunugiza et al., 1986). Rare occurrences of a carbonate horizon and a quartz-bearing eclogite (Banno et al., 1976) included in the mafic-ultramafic bodies contradicted this interpretation. Nevertheless, Banno and Yokoyama (1977) modeled a fractional crystallization path to explain the derivation of peridotite-gabbro-quartz–bearing eclogite from a parental picritic magma, suggesting that the eclogite formed from a silica-rich residual liquid in the lower continental crust. From field relationships Takasu (1989) argued that the protolith of the quartz-bearing eclogite was continental crust with deep-sea sediment or volcanoclastic rocks; Kugimiya and Takasu (2002) proposed that the carbonate rock was derived from reef limestone on a seamount, oceanic plateau, or island arc. Okamoto et al. (2004) reported detrital zircons with inherited cores yielding Early Proterozoic ages from the quartz-bearing eclogite, supporting its sedimentary origin. However, the origin of other metagabbros and ultramafic rocks is still controversial.

To clarify the evolution of the HP metamorphic belts, it is essential to constrain the protoliths of the eclogitic rocks and ultramafic rocks that are widely regarded as tectonic blocks. In order to obtain these constraints, geological and structural evidence based on field mapping of the lithological succession is required. We have recently completed a new lithotectonic map at 1:5000 scale for the highest grade areas in central Shikoku, with special attention paid to the duplex structure and protolith occurrences. In addition, we collected about 1400 samples in the mapped area during a comprehensive investigation from 2000 to 2002, which included geology (Maruyama et al., 2001), metamorphic petrology (Terabayashi et al., 2001; Ota et al., 2004), structural geology (Yamamoto et al., 2001, 2004), and geochronology (Okamoto et al., 2004). In this paper we report a reconstructed duplex structure of the eclogitic metagabbro-peridotite and surrounding schists, and discuss the tectonic implications, particularly for the origin of the metagabbro-peridotite bodies.

Geological Overview of the Sanbagawa Belt

The Sanbagawa belt in southwestern Japan extends for over 800 km from the Kanto Mountains near Tokyo to the eastern tip of Kyushu (Fig. 1A). The belt is less than 30 km wide, less than 2 km thick, and occurs as a subhorizontal tectonic slice bounded by a normal fault on the top and a thrust fault on the bottom (Isozaki and Maruyama, 1991; Maruyama et al., 1996). The structurally underlying unit is the Late Cretaceous (ca. 80 Ma) Shimanto accretionary complex, weakly metamorphosed under prehnite-actinolite to zeolite facies conditions (Yamato-Omine Research Group, 1981; Sasaki and Isozaki, 1992; Ishihama and Kiminami, 1999; Masago et al., 2005). The overlying unit is a Jurassic subduction complex, metamorphosed under prehnite-pumpellyite facies conditions at ca. 140 Ma (Isozaki et al., 1990; Kawato et al., 1991; Masago et al., 2005).

Protoliths of the Sanbagawa schists include basaltic pillow lava and related clastic rocks, and nonclastic rocks. The predominant metasedimentary rocks are pelitic to psammitic schists and phyllites, with rare metachert (quartz schist) and rare calcareous schists (Ernst et al., 1970). Fossil ages of the protoliths range from Carboniferous to Jurassic (Fig. 2: Isozaki and Itaya, 1990). To examine the timing of the Sanbagawa metamorphic event, Itaya and Takasugi (1988) reported mid- to Late Cretaceous ages based on about 70 K-Ar dates for the schists in a N-S cross-section in central Shikoku. These data were later confirmed by Ar-Ar dates (Takasu and Dallmeyer, 1990; Takasu et al., 1994; Dallmeyer et al., 1995) that were interpreted as cooling ages during exhumation (Isozaki and Itaya, 1990). The peak Sanbagawa metamorphism has a late Early Cretaceous Rb-Sr isochron age of 116 ± 10 Ma (Minamishin, 1979). Zircon SHRIMP U-Pb ages of 110–120 Ma from quartz-bearing eclogites (Okamoto et al., 2004) define the time of eclogite-facies metamorphism (Fig. 2).

A kinematic analysis of the Sanbagawa schists indicated that the belt underwent high-strain ductile flow during or shortly after the peak metamorphism, resulting in an orogen-parallel stretching lineation and a penetrative foliation (D1 foliation) (Toriumi, 1982; Faure, 1983; Hara et al., 1992; Wallis, 1992, 1998). Two subsequent phases of deformation are characterized by recumbent folding (D2) and upright folding (D3) (e.g., Hara et al., 1977; Faure, 1985; Wallis, 1990).

The Sanbagawa metamorphism has been studied in detail from pumpellyite-actinolite, through the blueschist-greenschist transition, up to epidote-amphibolite facies (for summary, see Banno and Sakai, 1989). In addition, four mineral zones of chlorite, garnet, albite-biotite, and oligoclase-biotite, with a progressive increase of metamorphic temperature, are based on assemblages and chemistry of diagnostic minerals in pelitic schists (Fig. 1B: Kurata and Banno, 1974; Higashino, 1975, 1990; Enami, 1982). In central Shikoku the distribution of individual lithologies has revealed the principal geological and thermal structure; the highest-grade rocks are situated in an intermediate structural level within several eclogitic metabasite and peridotite bodies (Kojima, 1951; Hide, 1961; Banno, 1964; Kawachi, 1968; Higashino, 1975; Banno et al., 1978). As noted above, these bodies were considered to be tectonic blocks recording complex thermal histories with eclogite-facies metamorphism prior to the epidote-amphibolite-facies Sanbagawa metamorphism (e.g., Kunugiza et al., 1986; Takasu, 1989). However, minor occurrences of eclogitic

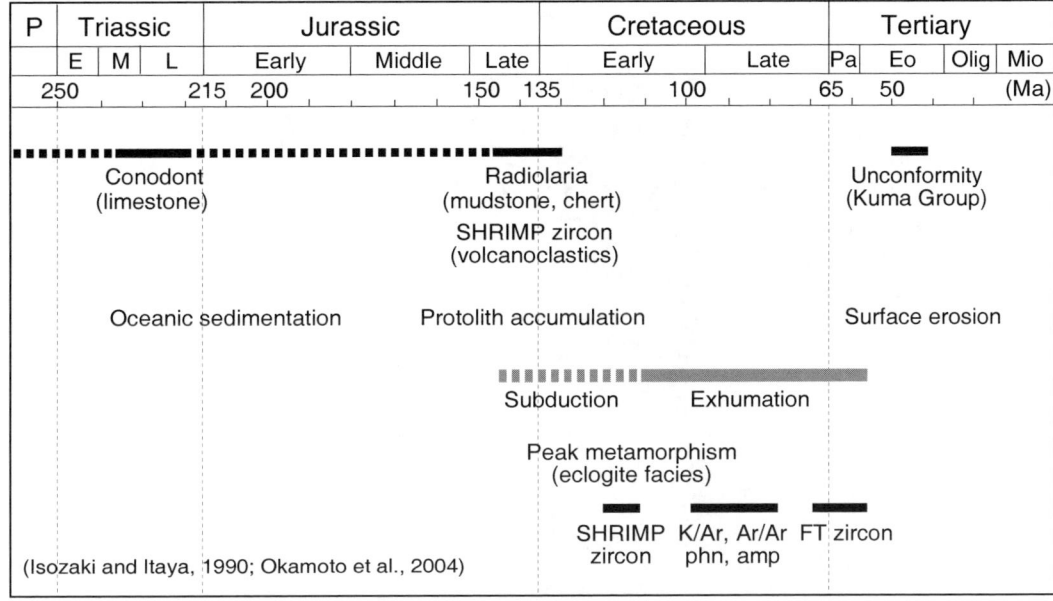

FIG. 2. Chronological summary of the Sanbagawa metamorphism modified after Isozaki and Itaya (1990), with recent age data.

schists, formed by prograde metamorphism, were discovered by Takasu (1984). Wallis and Aoya (2000) and Aoya (2001) proposed an eclogite nappe based on the regional distribution of eclogitic schists in the coherent schists surrounding an eclogitic metagabbro body. These workers emphasized the apparent gaps in P-T conditions between the eclogite nappe and the coherent schists. Ota et al. (2004) defined a progressive thermobaric structure around the largest eclogitic mafic-ultramafic bodies, and argued that they represented the highest-grade part of the Sanbagawa belt, which formed together with the surrounding schists under a specific thermal gradient at depth in the subduction zone.

Geology of the Iratsu and Higashi-Akaishi Bodies and Their Surroundings

In order to clarify the distribution of eclogites and peridotites, and their structural relationships with surrounding rocks, we mapped the lithologies of the Iratsu and Higashi-Akaishi bodies and their surroundings (Fig. 3). Data by Hara et al. (1992), Okamoto (1998), MMEAJ (1999), and Okamoto et al. (2000) were compiled from their northwestern parts. There are numerous faults subparallel to lithological boundaries; boundaries between pelitic and mafic schists, and between pelitic and quartz schists, typically are faulted. In Figure 3, only mappable faults are shown.

The Sanbagawa belt is mainly composed of mafic, quartzose, and pelitic-psammitic schists, and some eclogitic metagabbro and ultramafic bodies. Most schists belong to the albite-biotite zone, some to the oligoclase-biotite zone and the garnet zone (Fig. 1B). The eclogitic metagabbro bodies, Iratsu, Tonaru, and Seba, were overprinted by epidote-amphibolite-facies metamorphism; they are included in the thermal core of the Sanbagawa schists (Banno et al., 1976; Kunugiza et al., 1986; Takasu, 1984, 1989). The Higashi-Akaishi peridotite body underlyies the southwestern side of the Iratsu body; these two comprise a composite body (Fig. 4A). The northern (top) boundary between the Iratsu body and the overlying schists is a layer-parallel normal fault, whereas the southern (bottom) boundaries of the Tonaru body and the Iratsu and Higashi-Akaishi composite body are thrusts that caused mylonitization. The bottom boundary of the composite body trends E-W, and cuts the boundary between the Iratsu and Higashi-Akaishi bodies, striking northwest and dipping northeast at about 40°.

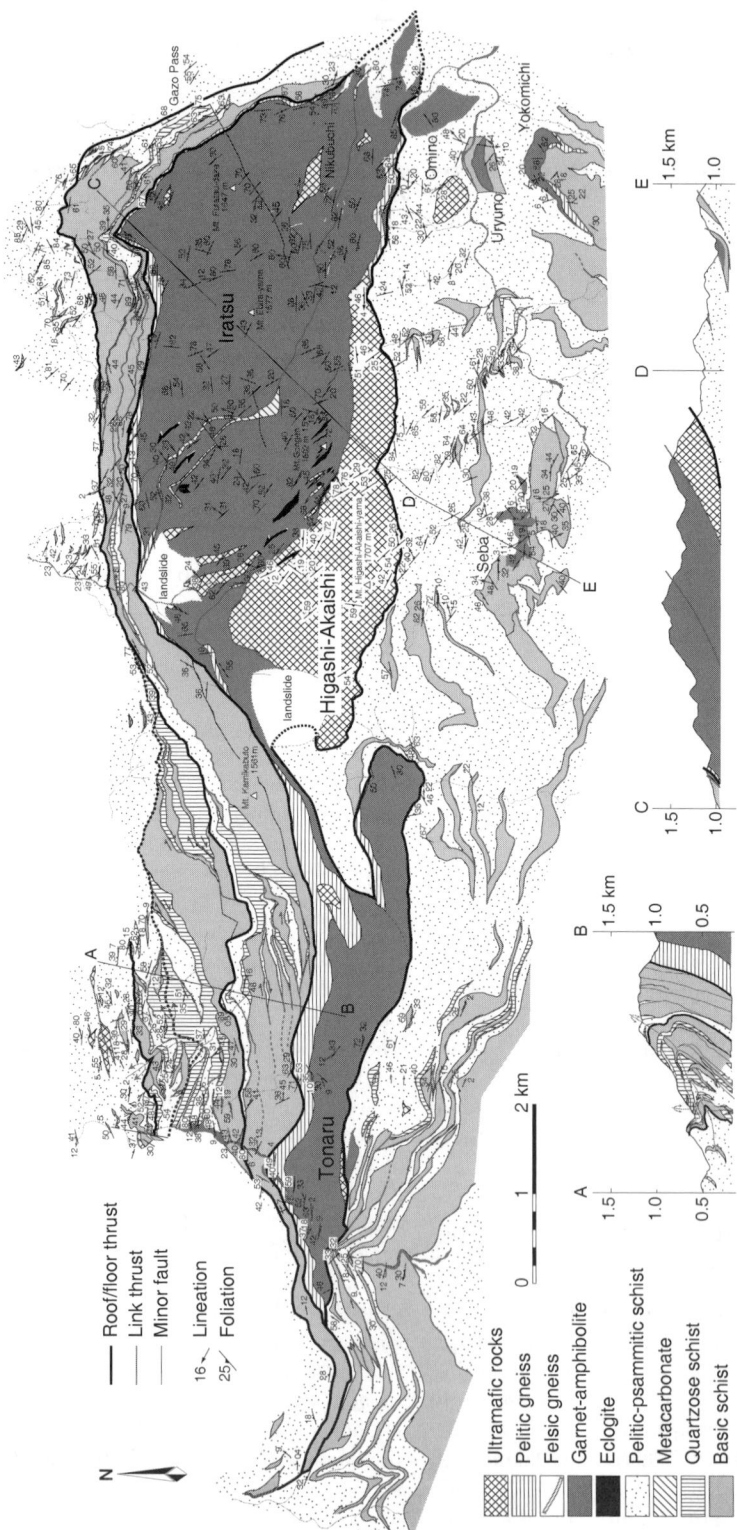

Fig. 3. Geologic map of the Sanbagawa belt around the Iratsu, Higashi-Akaishi, and Tonaru bodies in the Besshi area, central Shikoku. Cross-sections A–B and C–D–E are from Okamoto et al. (2000) and this study, respectively.

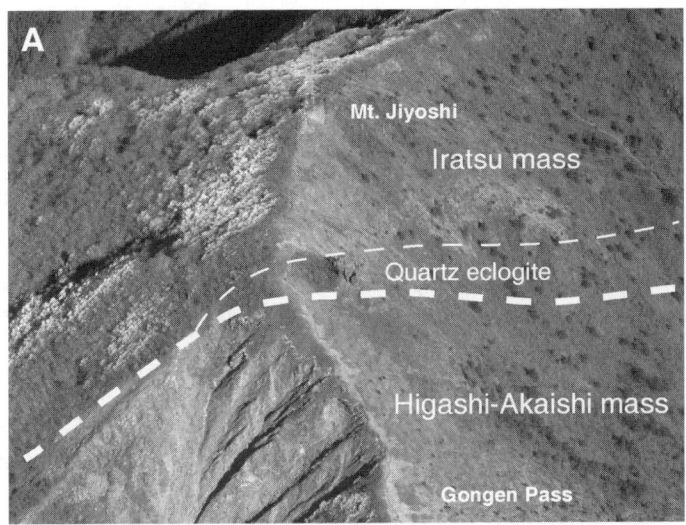

FIG. 4. A. Birds-eye view of the boundary between the Iratsu and Higashi-Akaishi bodies by Asia Air Survey CO., Ltd., Japan. The Iratsu metagabbro overlies the Higashi-Akaishi peridotite along a shear zone. B. Duplex structure of garnet clinopyroxenite-peridotite. Shear sense indicates a top-to-north movement.

A metapelite extends from the northern side of the Tonaru and Iratsu bodies to the southeastern side of the Iratsu unit, and is inserted within the most eastern and western parts of the latter. This metapelite is characterized by larger mica flakes with or without garnet porphyroblasts, and is coarser grained than the surrounding pelitic schists. In the northeastern side of the Iratsu body, some lenses of eclogite alternate with the metapelite (Gazo body: Sakurai, 2000). Thus, we regard this metapelite as a constituent lithology of the Iratsu body, and hereafter refer to it as pelitic gneiss in order to distinguish it from the surrounding pelitic schists.

On the northern side of the Tonaru and Iratsu bodies, mafic schists predominate with associated quartz schists and pelitic schists. Although planar structures of the surrounding schists to the north of the Tonaru body have undergone two phases of large-scale folding (Okamoto et al., 2000), most of the quartz schists are associated with subjacent or superjacent mafic schists. Such a mafic schist–quartz schist association is interpreted as derived from a basalt-chert layer in an oceanic plate stratigraphy (Okamoto et al., 2000). In contrast, pelitic-psammitic schist is widespread on the southern side, where some metagabbro or peridotite bodies are present, including those at Omino, Uryuno, Yokomichi, and Seba. Among these, only the Seba body and the adjacent mafic schists exhibit eclogitic assemblages (Takasu, 1984; Naohara and Aoya, 1997; Aoya, 2001).

On its southeastern margin, the Iratsu body is divided into thin layers, and separated into the isolated bodies of Omino, Uryuno, and Yokomichi. The Omino amphibolite body lies on the same structural horizon as the Iratsu body, whereas the Omino peridotite and Uryuno bodies occur in a lower structural horizon than the Iratsu body. The Yokomichi body with a wedge shape terminating to the south is considered to be the structural equivalent to the northern (upper) part of the Iratsu body; it is wrapped around by the pelitic gneiss and further by mafic schists. The boundary with the schists is a fault that cuts the boundary between the Yokomichi body and the pelitic gneiss, as well as the top boundary fault of the Iratsu body.

Lithology of the Iratsu Body

Our 1:5000-scale mapping based mainly on the protolith distribution has determined the internal structure of the Iratsu body. No prior researchers have visualized an internal structure of the entire body based on the lithological succession. Rocks in the body include garnet amphibolite, eclogite (or eclogitic metagabbro), felsic gneiss, pelitic gneiss, metapsammite, metacarbonate, peridotite, and quartz schist (Fig. 3).

As shown by the dip and strike of bedding and lithologies in Figure 3, the marginal structure of the Iratsu body is strongly affected by the surrounding schists. However, the internal lithology generally trends SE-NW, and dips north at moderate to high angles. Therefore, the structural succession is from SW to NE. Based on differences of the protoliths, the Iratsu body can be separated into three tectonic units—upper, middle, and lower. On the northeastern side of the Iratsu body (the upper unit), an alternation of garnet amphibolite and felsic gneiss is obvious at the outcrop scale. In less amphibolitized domains, eclogite-facies mineral assemblages are preserved (Ota et al., 2004). The Nikubuchi peridotite (Yokoyama, 1980) is included in the lower part of the upper unit. The bottom of this unit is characterized by thin but continuous horizons of felsic gneiss.

The middle unit only occurs in the northwestern part of the Iratsu body, and is characterized by quartz schist and metacarbonate. The quartz schist layer is several 10 cm to a meter thick and is associated with micaceous garnet amphibolite. The metacarbonate layers alternate with or include garnet amphibolites.

In the lower unit, less amphibolitized eclogites or eclogitic metagabbros become dominant toward the bottom. In the eastern top of the lower unit, it is characterized by the presence of metacarbonate blocks ranging in size from 2 to 3 m, which include eclogite and pyroxenite pods 10 to 20 cm in diameter (Fig. 5). Peridotite layers, partly serpentinized, also increase toward the bottom and alternate with eclogites or eclogitic metagabbros. On the actual boundary and in the uppermost part of the Higashi-Akaishi peridotite body, eclogitic metagabbros (or garnet pyroxenites) alternate with peridotites on a length-scale of tens of centimeters (Fig. 4B).

In the bottom of the lower unit, a quartz-bearing eclogite extends along the boundary between the Iratsu and Higashi-Akaishi bodies. It occurs as a lens with compositional banding at a centimeter scale, and is composed of a quartz layer and a mafic (eclogitic) layer; in places, it is intercalated in serpentinized peridotite. As mentioned above, this rock type is regarded as sedimentary in origin (Takasu, 1989; Kugimiya and Takasu, 2002); the compositional banding reflects an original sedimentary structure. In addition, its bulk composition indicates that the protolith is most likely sandstone quite different from basalt (58–79 wt% SiO_2: Banno et al., 1976; Okamoto et al., 2004). Therefore, we hereafter refer to this as an eclogitic metapsammite.

Duplex Structure in and around the Iratsu and Higashi-Akaishi Bodies

Our new lithotectonic map (Fig. 3) indicates that the internal lithology in the Iratsu body generally trends NW-SE or WNW-ESE and dips to the NE or NNE at moderate to high angles. Foliations of the internal lithology are slightly oblique to those of the surrounding schists, which trend WNW-ESE and dip gently or steeply to the north or south (Fig. 3). Such obliquity is characteristic of layers of the supracrustal rocks, metacarbonate and pelitic gneiss that extend from the northwestern to the southeastern parts of the Iratsu body. Assuming link

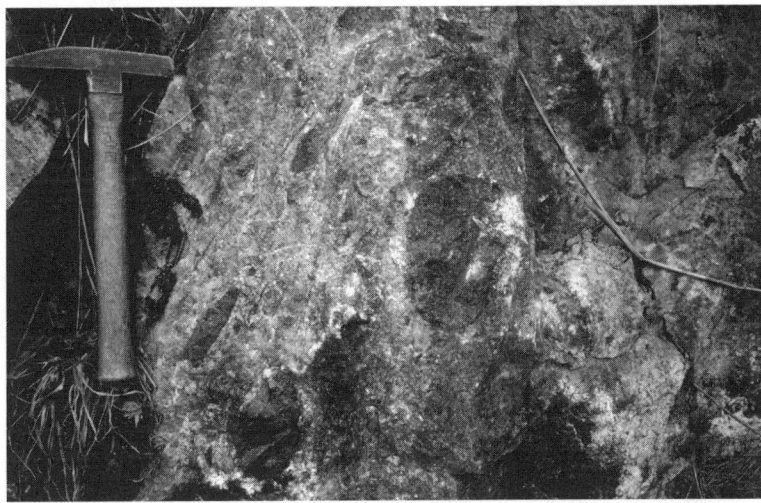

FIG. 5. Photograph showing mode of occurrence of metacarbonate, including abundant eclogite and peridotite pods, at the southeastern end of the Iratsu body.

thrusts along the extension of these lithologies, we consider that the Iratsu and Higashi-Akaishi bodies form a duplex structure with an eastward convergence (preliminarily reported by Maruyama et al., 2001; Ota et al., 2001).

A simplified structural map of the relevant area is shown in Figure 6. Each duplex, bounded by layer-parallel faults or link thrusts, consists of several horses. In the upper duplex I and middle duplex II, the layer-parallel boundary thrusts among horses, inter- and intra-duplex thrusts show right-lateral displacement in their present attitude. In the upper duplex I, horse A was accreted (underthrust), then horses B to E were successively underthrust beneath horse A. Horses F to M belong to the middle duplex. In the middle duplex II, horse F was accreted first beneath duplex I, and horses G to M were successively underthrust. In duplexes I and II, each horse converges to the west or north. Westward underthrusting and duplexing of the original sequence of basalt, chert, and turbidite have created an imbricate stack. In lower duplex III, the Iratsu, Higashi-Akaishi, and Tonaru bodies are composed of four horses bounded by link thrusts dipping to the NE or NNE. Horse N was accreted, and then horses O, P, Q, and R were underthrust in this order. The link thrusts bounding these horses show left-lateral displacement, converging to the east or south. Duplex III is separated on its top by the roof thrust from duplex II and on the bottom by a floor thrust from the underlying schists; both thrusts generally trend E-W and dip to the north.

Through our field mapping, we have reconstructed lithostratigraphic columnar sections for horse G in duplex II, and for horses N, O, P, Q, and R in duplex III (Fig. 7). In a columnar section from horse G (section G), mafic schist is overlain by quartz schist, and then by pelitic schist. Section N is composed of lower gabbroic and upper basaltic layers with peridotite and gabbro intercalations that are overlain by pelitic gneiss of probable hemipelagic origin. In the lower gabbroic layers, corresponding to a horizon of the Nikubuchi peridotite body, some metagabbros contain granulite-facies relicts that formed prior to the subduction-related (Sanbagawa) metamorphism (Yokoyama and Mori, 1975; Yokomaya, 1976, 1980). Section O, covering the middle unit of the Iratsu body, consists of garnet amphibolite, quartz schist, and pelitic gneiss in ascending order, and is comparable with section G. Section P covers the lower unit of the Iratsu body. In the middle and lower parts, several eclogite (or eclogitic metagabbro) layers were resistant to retrograde amphibolitization. Metacarbonate, approximately 10 m thick, overlies basalt and cumulate sections. Section Q corresponds to the Higashi-Akaishi peridotite body and an overlying eclogitic metapsammite. Section R includes the Tonaru body, and is mainly composed of a basaltic layer with underlying gabbro and peridotite and overlying pelitic gneiss.

Discussion

Accretionary complex origin of the Iratsu and Higashi-Akaishi bodies

The Iratsu and Higashi-Akaishi bodies represent a pile of imbricated thrust sheets derived from a primary rock succession that consists in ascending order of: (1) a gabbro-peridotite unit; (2) a basalt unit; and (3) a limestone-clastic unit. Minor quartz schist occurs immediately above the basalt unit. Such a succession is comparable with oceanic plate stratigraphy from a mid-oceanic topographic high, seamount, oceanic island, or plateau capped by pelagic limestone and minor chert on their flanks formed before the arrival at a trench (Fig. 8). Our reconstructed duplex structure at a mappable scale and the oceanic plate stratigraphy suggest that the Iratsu, Higashi-Akaishi, and Tonaru bodies and adjacent schists were parts of an accretionary complex that incorporated oceanic crustal materials in a trench with arc-derived turbidites during progressive plate subduction.

Okamoto et al. (2000) confirmed a duplex structure and an oceanic plate stratigraphy in the western part of duplex I from their detailed field mapping, although the duplex structures were multiply folded and overturned by later recumbent folding related to exhumation. Kimura et al. (1996) investigated the Franciscan complex in the Pacheco Pass area, Diablo Range, California, and demonstrated the presence of duplex structures. The metamorphic grade in these areas does not exceed blueschist or epidote-amphibolite facies, but in the study area it correspond to the eclogite facies, and the highest-grade part it reached UHP conditions (Enami et al., 2004; Ota et al., 2004).

Isozaki et al. (1990) and Kimura and Ludden (1995) documented the volume ratio or thickness of accreted oceanic crustal materials such as mid-oceanic ridge basalts and deep-sea chert in Mesozoic accretionary complexes of the Pacific Rim. They suggested that the volume of oceanic materials accreted to the overriding plate during subduction was a function of the depth at which they accreted, i.e., their metamorphic grade. In fact, the volume ratios of metabasalt and metachert in the duplex of the blueschist-facies Franciscan complex (Kimura et al., 1996) are clearly less than those in the

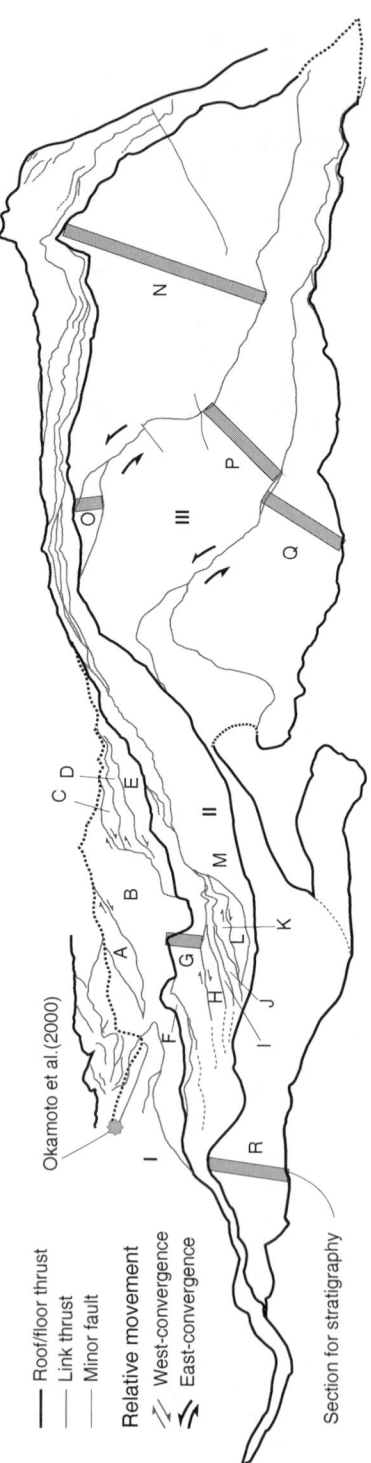

FIG. 6. Simplified structural map of the Besshi area, showing duplexes bounded by layer-parallel faults. Three sets of duplexes consist of several horses. In the top and middle units (I and II), the layer-parallel boundary thrusts among horses, both interduplexes and intraduplex thrusts, are right-lateral, and relative direction of fault displacement is shown by small arrows. They converge to the west or north. In the lower unit (III), the layer-parallel boundary thrusts are left-lateral, and they converge to the east or south.

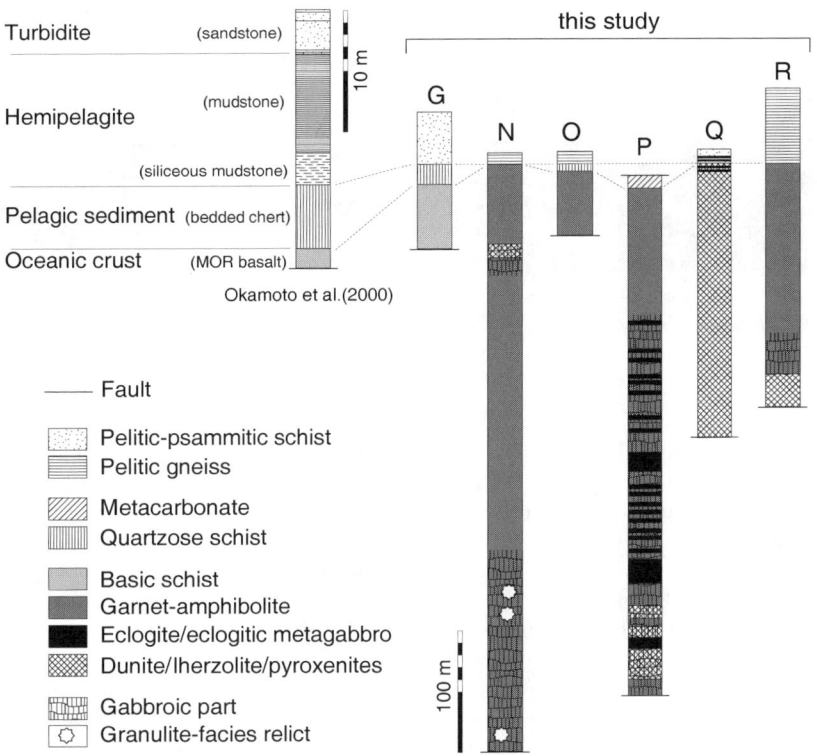

FIG. 7. Columnar sections representing the lithostratigraphy of the Sanbagawa metagabbro-peridotite bodies in the Besshi area (alphabetical symbols correspond to those in Figure 6), with that of the Sanbagawa schists along the Kokuryo River, by Okamoto et al. (2000). Note that the section from Okamoto et al. (2000) differs in scale from the others.

epidote-amphibolite–facies Sanbagawa schists (Okamoto et al., 2000). Most oceanic materials make up our duplex III, composed of the Iratsu, Higashi-Akaishi, and Tonaru bodies that include gabbro and ultramafic rocks from a deep section of the oceanic crust. This explains why their metamorphic grade reached eclogite facies.

Origin of the Sanbagawa mafic-ultramafic bodies

The origin of the Sanbagawa mafic-ultramafic bodies has long been discussed from various viewpoints. Petrological characteristics of the bodies, with the granulite-facies relics in the Nikubuchi- and Higashi-Akaishi bodies, require unusual HP conditions, because of the presence of spinel (Nikubuchi: Yokoyama, 1980) and garnet (Higashi-Akaishi: Kunugiza et al., 1986). These metagabbro and ultramafic rocks have been regarded as cumulates formed by igneous processes in the lower continental crust, and trapped in the mantle wedge of a subduction zone prior to juxtaposition with the Sanbagawa schists at further depths in the subducting plate (e.g., Kunugiza et al., 1986; Takasu, 1989; Enami et al., 1994).

Ota et al. (2004) investigated the spatial variations of peak P-T conditions in the entire Iratsu body, in order to establish the thermobaric structure around the body, integrated with previously published P-T estimates for various rock types in the Sanbagawa belt. The thermobaric structure suggests that the highest P-T area was situated in the upper part of the Higashi-Akaishi body near the boundary with the Iratsu body, and metamorphic conditions systematically decreased toward the upper and lower structural levels within those bodies, and also in the surrounding lower grade schists. This implies that the Iratsu and Higashi-Akaishi composite body was metamorphosed together with the surrounding schists under a specific thermal gradient at depth in the subduction zone. In addition, Enami et al. (2004) re-investigated garnet-bearing ultramafic rocks in the Higashi-Akaishi

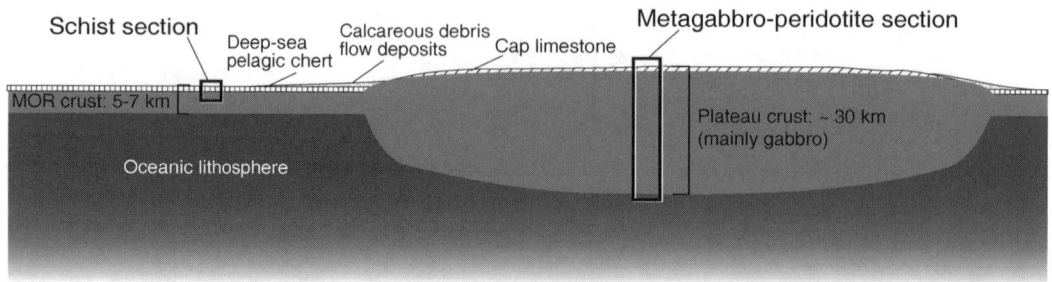

FIG. 8. A schematic diagram showing the primary sites, deduced from the oceanic plate stratigraphy of the schists and the metagabbro-peridotite body in the Sanbagawa belt.

body, and revealed a prograde P-T path with a steep P/T gradient, including a record of low P-T conditions at an early stage of their HP-UHP metamorphism. These studies questioned whether the Sanbagawa metagabbros and ultramafic rocks were "directly" derived from a mantle wedge adjacent to a subduction zone (Enami, 1996).

Furthermore, the Iratsu and Higashi-Akaishi bodies, which make up the duplex structure, together with supracrustal metacarbonate, quartz schist, and pelitic gneiss, formed prior to their subduction, suggesting that these bodies were produced in the lower crust and underlying upper mantle beneath an oceanic topographic high. This indicates that the Iratsu and Higashi-Akaishi bodies were not derived from the hanging wall of a mantle wedge above a subducting plate, but formed as part of a seamount, oceanic island, or plateau in a mid-oceanic region, and were subsequently subducted together with the protoliths of the Sanbagawa schists, from a trench down to considerable depth in the subduction zone. In other words, geological, metamorphic, and petrological evidence indicates that the Iratsu and Higashi-Akaishi bodies, as the largest mafic-ultramafic bodies in the Sanbagawa belt, have an ocean-floor origin

Considering the fact that the oceanic plate stratigraphy of the Iratsu, Higashi-Akaishi, and Tonaru bodies is comparable with the crustal succession of mid-oceanic topographic highs, the occurrence of granulite-facies relics (ca. 750°C at 5–10 kbar: Yokoyama, 1980) in the Iratsu body (section N: Fig. 7) indicates an original crustal thickness of 15–30 km. We calculate that the present thickness must be less than a hundredth of the original due to tectonic thinning during exhumation. Oceanic lower crust can be recrystallized under amphibolite- to granulite-facies conditions by ocean-floor metamorphism (hydrothermal alteration around mid-oceanic ridges), but the granulite-facies relics in the Iratsu body exhibit unusually high pressure conditions, considering the average thickness of the Phanerozoic oceanic crust (5–7 km). However, the crustal structure of large igneous provinces (LIP) in oceanic regions raises the possibility that such a granulite-facies crystallization may occur in the lower part of an oceanic island or plateau, because their crustal thickness typically reaches 20–40 km (e.g., Cretaceous Ontong Java Plateau in the South Pacific; Furumoto et al., 1976; Neal et al., 1997). Not only olistoliths from fracture zones associated with oceanic ridges and with collapse of oceanic topographic highs approach a trench, but also unroofed oceanic lower crust (gabbro and peridotite) exposed on the ocean floor (Atlantis Bank of the SW Indian Ridge: Dick et al., 2000). Therefore, it seems likely that deep sections of oceanic crust would mix with sedimentary rocks in the mid-oceans or near trenches.

It is well known that the Pacific superplume has been episodically active to form a number of oceanic LIPs: the Ontong-Java Plateau, Caribbean Plateau, Mid-Pacific seamount chains, etc. that formed during the mid-Cretaceous (e.g., Larson, 1991). Also, several fragments of these oceanic LIPs have been recognized as accreted fragments in accretionary complexes around the Pacific Rim (e.g., Kimura et al., 1994). The Cretaceous paleogeography of the Pacific Ocean based on paleo-plate reconstructions (Engebretson et al., 1984) and compilation of accreted oceanic crusts and plateaus in the Circum-Pacific orogenic belts, have led to the idea of a huge composite volcanic continent formed from the South Pacific superplume in the earliest Cretaceous (Suzuki et al., 2000). Our reconstructed oceanic topographic high might have been derived from this volcanic continent. As mentioned above, our duplex structure and reconstructed oceanic plate stratigraphy suggest the subduction polarity was always northward, but the relative convergent plate motion was estimated to have changed from NW to NE. Perhaps this was caused by accretion of the huge oceanic topographic high.

Acknowledgments

We dedicate this manuscript to Prof. W. G. Ernst, who introduced the plate tectonic model to the Sanbagawa belt in the 1970s, promoted several successful U.S.-Japan projects, and hosted many Japanese scientists at UCLA and Stanford. We are grateful to Phil Skemer for his encouragement in improving the manuscript. The aerial photograph in Figure 4A was taken and supplied by Asia Air Survey Co., Ltd., Japan, which we appreciate.

REFERENCES

Aoya, M., 2001, P-T-D path of eclogite from the Sambagawa belt deduced from combination of petrological and microstructural analysis: Journal of Petrology, v. 42, p. 1225–1248.

Banno, S., 1964, Petrologic studies of the Sanbagawa crystalline schists, in the Bessi-Ino district, central Shikoku, Japan: Journal of the Faculty of Science, University of Tokyo, Sec. II, v. 15, p. 203–319.

Banno, S., Higashino, T., Otsuki, M., Itaya, T., and Nakajima, T., 1978, Thermal structure of the Sanbagawa belt in central Shikoku: Journal of the Physics of the Earth (suppl.), p. 345–356.

Banno, S., and Sakai, C., 1989, Geology and metamorphic evolution of the Sanbagawa metamorphic belt, Japan, in Daly, J. S., Cliff, R. A., and Yardley, B. W. D., eds., Evolution of metamorph;ic belts: Geological Society of London Special Publication, v. 43, p. 519–532.

Banno, S., and Yokoyama, K., 1977, Peridotite-metagabbro complex in central Shikoku, in Hide, K., ed., The Sanbagawa belt: Hiroshima, Japan, Hiroshima University Press, p. 57–68 (in Japanese with English abstract).

Banno, S., Yokoyama, K., Enami, M., Iwata, O., Nakamura, K., and Kasashima, S., 1976, Petrology of the peridotite-metagabbro complex in the vicinity of Mt. Higashi-akaishi, central Shikoku. Part I. Megascopic textures of the Iratsu and Tonaru epidote amphibolite bodies: Science Report of Kanazawa University, v. 21, p. 139–159.

Carswell, D. A., and Harvey, M. A., 1985, The intrusive history and tectono-metamorphic evolution of the Basal Gneiss Complex in the Moldefjord area, west Norway, in Gee, D. G., and Sturt, B. A., eds., The Caledonide Orogen–Scandinavia and Related Areas: Chichester, UK, Wiley, p. 843–857.

Chopin, C., 1984, Coesite and pure pyrope in high-grade blueschists of the western Alps: A first record and some consequences: Contributions to Mineralogy and Petrology, v. 86, p. 107–118.

Cloos, M., 1986, Blueschists in the Franciscan Complex of California: Petrotectonic constraints on uplift mechanism, in Evans, B. W., and Brown, E. H., eds., Blueschists and eclogites: Geological Society of America Memoir, v. 164, p. 77–94.

Dallmeyer, R. D., Takasu, A., and Yamaguchi, K., 1995, Mesozoic tectonothermal development of the Sanbagawa, Mikabu, and Chichibu belts, south-west Japan: Evidence from $^{40}Ar/^{39}Ar$ whole-rock phyllite ages: Journal of Metamorphic Geology, v. 13, p. 271–286.

Dick, H. J. B., Natland, J. H., Alt, J. C., Bach, W., Bideau, D., Gee, J. S., Haggas, S., Hertogen, J. G. H., Hirth, G., and Holm, P. M., 2000, A long in situ section of the lower ocean crust: Results of ODP Leg 176 drilling at the Southwest Indian Ridge: Earth and Planetary Science Letters, v. 179, p. 31–51.

Enami, M., 1982, Oligoclase-biotite zone of the Sanbagawa metamorphic terrain in the Besshi district, central Shikoku, Japan: Journal of the Geological Society of Japan, v. 88, p. 887–900 (in Japanese with English abstract).

Enami, M., 1996, Petrology of kyanite-bearing tectonic blocks in the Sanbagawa metamorphic belt of the Bessi area, central Shikoku, Japan, in Shimamoto, T., Hayasaka, Y., Shiota, T., Oda, M., Takeshita, T., Yokoyama, S., and Ohtomo, Y., eds., Tectonics and metamorphism (Memorial volume for Prof. Hara, I.): Hiroshima, Japan, Hiroshima University Press, p. 47–55 (in Japanese with English abstract).

Enami, M., Mizukami, T., and Yokoyama, K., 2004, Metamorphic evolution of garnet-bearing ultramafic rocks from the Gongen area, Sanbagawa belt, Japan: Journal of Metamorphic Geology, v. 22, p. 1–15.

Enami, M., Wallis, S. R, and Banno, Y., 1994, Paragenesis of sodic pyroxene-bearing quartz schists: Implications for the P-T history of the Sanbagawa belt: Contributions to Mineralogy and Petrology, v. 116, p. 182–198.

Engebretson, D. C., Cox, A., and Gordon, R. G., 1984, Relative motions between oceanic plates in the Pacific basin: Journal of Geophysical Research, v. 89, p. 10,291–10,310.

Ernst, W. G., 1975, Systematics of large-scale tectonics and age progressions an Alpine and circum-Pacific blueschist belts: Tectonophysics, v. 26, p. 229–246.

Ernst, W. G., 1978, Petrochemical study of lherzolitic rock from the Western Alps: Journal of Petrology, v. 28, p. 341–392.

Ernst, W. G., Onuki, H., Seki, Y., and Gilbert, M. C., 1970, Comparative study of low-grade metamorphism in the

California Coast Ranges and Outer Metamorphic Belt of Japan: Geological Society of America Memoir 124, 276 p.

Faure, M., 1983, Eastward ductile shear during the early tectonic phase in the Sanbagawa belt: Journal of the Geological Society of Japan, v. 89, p. 319–329.

Faure, M., 1985, Microtectonic evidence for eastward ductile shear in the Jurassic orogen of SW Japan: Journal of Structural Geology, v. 7, p. 175–186.

Furumoto, A. S., Webb, J. P., Odegard, M. E., and Hussong, D. M., 1976, Seismic studies on the Ontong Java Plateau: Tectonophysics, v. 34, p. 71–90.

Hara I., Hide, K., Takeda K., Tsukuda, E., Tokuda M., and Shiota T., 1977, Tectonic movement in the Sambagawa belt, in Hide, K., ed., The Sambagawa Belt: Hiroshima, Japan, Hiroshima University Press, p. 309–390.

Hara, I., Shiota, T., Hide, K., Kanai, K., Goto, M., Seki, S., Kaikiri, K., Takeda, K., Hayasaka, Y., Miyamoto, T., Sakurai, Y., and Ohtomo, Y., 1992, Tectonic evolution of the Sambagawa schists and its implication in convergent margin processes: Journal of Science, Hiroshima University, Series C 9, p. 495–595.

Hide, K., 1961, Geological structure and metamorphism of the Sanbagawa crystalline schists of the Besshi-Shirataki mining district, Kochi Prefecture: Geological Report, Hiroshima University, v. 9, p. 1–87 (in Japanese with English abstract).

Higashino, T., 1975, Biotite zone of Sanbagawa metamorphic terrain in the Siragayama area, central Sikoku, Japan: Journal of the Geological Society of Japan, v. 81, p. 653–670 (in Japanese with English abstract).

Higashino, T., 1990, The higher grade metamorphic zonation of the Sambagawa metamorphic belt in central Shikoku, Japan: Journal of Metamorphic Geology, v. 8, p. 413–423.

Ishihama, S., and Kiminami, K., 1999, Geochemical relationships between psammitic/pelitic schist of the Sanbagawa Belt and sandstone/shale of the Southern Chichibu–Northern Shimanto belts [abs.]: Abstracts of the 106th Annual Meeting of the Geological Society of Japan, p. 121 (in Japanese).

Ishizuka, H., Masaoka, M., and Katsuya, M., 1989, Chemical compositions of metabasites from the Sanbagawa belt in Shikoku [abs.]: Abstracts of the 96th annual meeting of the Geological Society of Japan, p. 577 (in Japanese).

Isozaki, Y., and Itaya, T., 1990, Chronology of Sanbagawa metamorphism: Journal of Metamorphic Geology, v. 8, p. 401–411.

Isozaki, Y., and Maruyama, S., 1991, Studies on orogeny based on plate tectonics in Japan and new geotectonic subdivision of the Japanese Islands: Journal of Geography, v. 100, p. 697–761 (in Japanese with English abstract).

Isozaki, Y., Maruyama, S., and Furuoka, F., 1990, Accreted oceanic materials in Japan: Tectonophysics, v. 181, p. 179–206.

Itaya, T., and Takasugi, H., 1988, Muscovite K-Ar ages of the Sanbagawa schists, Japan and argon depletion during cooling and deformation: Contributions to Mineralogy and Petrology, v. 100, p. 281–290.

Katayama, I., Zayachikovsky, A. A., and Maruyama, S., 2000, Prograde P-T records from inclusions in zircons from UHP-HP rocks of the Kokchetav massif, northern Kazakhstan: The Island Arc, v. 9, p. 417–427.

Kawachi, Y., 1968, Large-scale overturned structure in the Sambagawa metamorphic zone in central Shikoku, Japan: Journal of the Geological Society of Japan, v. 74, p. 607–616.

Kawachi, Y., Watanabe, T., and Landis, A., 1982, Origin of mafic volcanogenic schists and related rocks in the Sanbagawa belt, Japan: Journal of the Geological Society of Japan, v. 88, p. 797–817.

Kawato, K., Isozaki, Y., and Itaya, T., 1991, Geotectonic boundary between the Sanbagawa and Chichibu belts in central Southwest Japan: Journal of Geological Society of Japan, v. 97, p. 959–975 (in Japanese with English abstract).

Kimura, G., and Ludden, J., 1995, Peeling oceanic crust in subduction zones: Geology, v. 23, p. 217–220.

Kimura, G., Maruyama, S., Isozaki, Y., and Terabayashi, M., 1996, Well-preserved underplating structure of the jadeitized Franciscan complex, Pacheco Pass, California: Geology, v. 24, p. 75–78.

Kimura, G., Sakakibara, M., and Okamura, M., 1994, Plumes in central Panthalassa? Deductions from accreted oceanic fragments in Japan: Tectonics, v. 13, p. 905–916.

Kojima, G., 1951, Stratigraphy and geological structure of the crystalline schist region in central Shikoku: Journal of the Geological Society of Japan, v. 57, p. 177–190 (in Japanese with English abstract).

Kugimiya, Y., and Takasu, A., 2002, Geology of the Western Iratsu mass within the tectonic melange zone in the Sambagawa metamorphic belt, Besshi district, central Shikoku, Japan: Journal of the Geological Society of Japan, v. 108, p. 644–662 (in Japanese with English abstract).

Kunugiza, K., Takasu, A., and Banno, S., 1986, The origin and metamorphic history of the ultramafic and metagabbro bodies in the Sanbagawa metamorphic belt: Geological Society of America, Memoir 164, p. 375–385.

Kurata, H., and Banno, S., 1974, Low-grade progressive metamorphism of pelitic schists of the Sazare area, Sambagawa metamorphic terrain in central Shikoku: Journal of Petrology, v. 15, p. 361–382.

Larson, R. L., 1991, Latest pulse of Earth: Evidence for a mid-Cretaceous superplume: Geology, v. 19, p. 547–550.

Liu, F. L., Xu, Z. Q., Katayama, I., Yang, J. S., Maruyama, S., and Liou, J. G., 2001, Mineral inclusions in zircons of para- and orthogneiss from pre-pilot drillhole CCSD-PP1, Chinese Continental Scientific Drilling Project: Lithos, v. 59, p. 199–215.

Maruyama, S., and Liou, J. G., 1988, Petrology of Franciscan metabasites along the jadeite-glaucophane type facies series, Cazadero, California: Journal of Petrology, v. 29, p. 1–37.

Maruyama, S., Liou, J. G., and Terabayashi, M., 1996, Blueschists and eclogites of the world and their exhumation: International Geology Review, v. 38, p. 485–594.

Maruyama, S., Kaneko, Y., Ota, T., Terabayashi, M., Yamamoto, H., Katayama, I., Okamoto, K., Komiya, T., Ozawa, H., Anma, R., Ishikawa, A., Shinjoe, H., Liou, J. G., and Windley, B. F., 2001, Mode of occurrence, protolith, and tectonic history of the Iratsu eclogite body, Sanbagawa belt, Shikoku, with special emphasis on late-stage hydration [abs.]: Abstracts, 6th International Eclogite Conference, Niihama, Japan, p. 86.

Masago, H., Okamoto, K., and Terabayashi, M., 2005, Exhumation tectonics of the Sanbagawa high-pressure metamorphic belt, SW Japan—constraints from the upper and lower boundary faults: International Geology Review, in press.

Matsuda, T., and Uyeda, S., 1971, On the Pacific-type orogeny and its model: Extension of the paired belts concept and possible origin of marginal seas: Tectonophysics, v. 11, p. 5–27.

Medaris, G. J., Jelinek, E., and Misar, Z., 1995, Czech eclogites: Terrane setting and implications for Variscan tectonic evolution of the Bohemian Massif: European Journal of Mineralogy, v. 7, p. 7–28.

Minamishin, M., 1979, Whole rock Rb-Sr age of the Sanbagawa schists in the Asemi area, central Shikoku (abstract): Journal of Japanese Association of Mineralogy, Petrology, and Economic Geology, v. 74, p. 153.

MMEAJ (Metallic Minerals Exploration Agency of Japan), 1999, Geological map (1:20,000) in the Mt. Shiraga district: Tokyo, Japan, Metallic Minerals Exploration Agency of Japan.

Mori, T., and Banno, S., 1973, Petrology of peridotite and garnet clinopyroxenite of the Mt. Higashi-Akaishi mass, central Sikoku, Japan—subsolidus relation of anhydrous phases: Contributions to Mineralogy and Petrology, v. 41, p. 301–323.

Naohara, R., and Aoya, M., 1997, Prograde eclogites from Sanbagawa mafic schists in the Sebadani area, central Shikoku, Japan: Memoirs, Faculty of Science and Engineering, Shimane University (Series A), v. 30, p. 63–73.

Neal, C. R., Mahoney, J. J., Kroenke, L. W., Duncan, R. A., and Petterson, M. G., 1997, The Ontong Java Plateau, in Mahoney, J. J., and Coffin, M. F., eds., Large igneous provinces: Continental, oceanic, and planetary flood volcanism: Geophysical Monographs, v. 100: Washington, DC, American Geophysical Union, p. 183–216.

Oh, C. W., and Liou, J. G., 1990, Metamorphic evolution of two different eclogites in the Franciscan Complex, California, U.S.A: Lithos, v. 25, p. 41–53.

Okamoto, K., 1998, Inclusion-trail geometry of albite porphyroblasts in a fold structure in the Sambagawa belt, Central Shikoku, Japan: The Island Arc, v. 7, p. 283–294.

Okamoto, K., Maruyama, S., and Isozaki, Y., 2000, Accretionary complex origin of the Sanbagawa, high P/T metamorphic rocks, Central Shikoku, Japan -Layer-parallel shortening structure and greenstone geochemistry: Journal of the Geological Society of Japan, v. 106, p. 70–86.

Okamoto, K., Shinjoe, H., Katayama, I., Terada, K., Sano, Y., and Johnson, S., 2004, SHRIMP U-Pb zircon dating of quartz-bearing eclogite from the Sanbagawa Belt, south-west Japan; implications for metamorphic evolution of subducted protolith: Terra Nova, v. 16, p. 81–89.

Ota, T., Terabayashi, M., Kaneko, Y., Yamamoto, H., Okamoto, K., Katayama, I., and Komiya, T., 2001, A newly found fragment of Cretaceous oceanic LIP derived from Pacific superplume; an example from the Sanbagawa eclogite-peridotite mass in Shikoku, Japan [abs.]: EOS (Transactions of the American Geophysical Union), v. 82(47), Fall Meeting, abstract T42A-0919.

Ota, T., Terabayashi, M,. and Katayama, I., 2004, Thermobaric structure and metamorphic evolution of the Iratsu eclogite body in the Sanbagawa belt, central Shikoku, Japan: Lithos, v. 73, p. 95–126.

Sakakibara, M., Satake, A., Okamoto, K., and Moriyama, Y., 2001, Well preserved oceanic plate stratigraphy of the Sanbagawa metamorphic rocks in western and central Shikoku, Japan [abs.], in Abstracts 6th International Eclogite Conference, Niihama, Japan, p. 137.

Sakurai, T., 2000, Chemical compositions of the constituent minerals of the Gazo mass, a tectonic block in the Sambagawa metamorphic belt, Besshi district, central Shikoku, Japan: Geoscience Report, Shimane University, v. 19, p. 167–185.

Sasaki, H., and Isozaki, Y., 1992, Low-angle thrust between the Sanbagawa and Shimanto belts in central Kii Peninsula, Southwest Japan: Journal of the Geological Society of Japan, v. 98, p. 57–60 (in Japanese).

Sobolev, N. V., and Shatsky, V. S., 1990, Diamond inclusions in garnets from metamorphic rocks: Nature, v. 343, p. 742–746.

Suzuki, N., Kadarusman, A., and Maruyama, S., 2000, Gossira continent: A new name of composite volcanoes formed at South Pacific superplume during the Cretaceous [abs.]: EOS (Transactions of the American Geophysical Union), v. 81(48), Fall Meeting, abstract V62B-33.

Tabata, H., Yamauchi, K., Maruyama, S., and Liou, J. G., 1998, Tracing the extent of a UHP metamorphic terrane: Mineral-inclusion study of zircons in gneisses from the Dabie Shan, in Hacker, B. R., and Liou, J. G., eds., When continents collide: Geodynamics and geochemistry of ultrahigh-pressure rocks: Dordrecht, Netherlands, Kluwer Academic Publishers, p. 261–273.

Takasu, A., 1984, Prograde and retrograde eclogites in the Sambagawa metamorphic belt, Besshi district, Japan: Journal of Metamorphic Geology, v. 8, p. 413–423.

Takasu, A., 1989, P-T histories of peridotite and amphibolite tectonic blocks in the Sambagawa metamorphic belt, Japan, in Daly, J. S., Cliff, R. A., and Yardley, B. W. D., eds., Evolution of metamorphic belts: Geological Society of London, Special Publication, v. 43, p. 533–538.

Takasu, A., and Dallmeyer, R. D., 1990, $^{40}Ar/^{39}Ar$ mineral age constraints for the tectonothermal evolution of the Sanbagawa metamorphic belt, central Shikoku, Japan: A Cretaceous accretionary prism: Tectonophysics, v. 185, p. 111–139.

Takasu, A., Wallis, S. R., Banno, S., and Dallmeyer, R. D., 1994, Evolution of the Sanbagawa metamorphic belt, Japan: Lithos, v. 33, p. 119–133.

Takayama, M., 1986, Mode of occurrence and significance of jadeite in the Kamuikotan metamorphic rocks, Hokkaido, Japan: Journal of Metamorphic Geology, v. 4, p. 445–454.

Terabayashi, M., Ota, T., and Katayama, I., 2001, Retrogressive hydration of eclogites from the Iratsu mass: Constraints for the metamorphic evolution of the Sanbagawa belt, central Shikoku, Japan, in UHPM Workshop 2001 "Fluid/Slab/Mantle Interactions and Ultrahigh-P Minerals," p. 266–269.

Toriumi, M., 1982, Stress, strain and uplift: Tectonics, v. 1, p. 57–72.

Wallis, S. R., 1990, The timing of folding and stretching in the Sambagawa belt: The Asemigawa region, central Shikoku: Journal of the Geological Society of Japan, v. 96, p. 345–352.

Wallis, S. R., 1992, The timing of folding and stretching in the Sanbagawa Belt: The Asemigawa region, central Shikoku: Journal of the Geological Society of Japan, v. 96, p. 345–352.

Wallis, S. R., 1998, Exhuming the Sanbagawa metamorphic belt; the importance of tectonic discontinuities: Journal of Metamorphic Geology, v. 16, p. 83–95.

Wallis, R., and Aoya, M., 2000, A re-evaluation of eclogite facies metamorphism in SW Japan: Proposal for an eclogite nappe: Journal of Metamorphic Geology, v. 18, p. 653–664.

Yamamoto, H., Okamoto, K., Kaneko, Y., and Terabayashi, M., 2001, Wedge-extrusion exhumation of eclogite-bearing metagabbro-peridotite bodies in the Sanbagawa belt, central Shikoku, Japan [abs.]: Abstracts, 6th International Eclogite Conference, Niihama, Japan, p. 175.

Yamamoto, H., Okamoto, K., Kaneko, Y., and Terabayashi, M., 2004, Southward extrusion of eclogite-bearing mafic-ultramafic bodies in the Sanbagawa belt, central Shikoku, Japan: Tectonophysics, v. 387, p. 151–168.

Yamato-Omine Research Group, 1981, Paleozoic and Mesozoic rocks in central Kii mountains: Excursion Guidebook, 35th Annual Meeting of the Chidanken, 88 p. (in Japanese).

Ye, K., Yao, Y., Katayama, I., Cong, B. L., Wang, Q. C., and Maruyama, S., 2000, Large areal extent of ultrahigh-pressure metamorphism in the Sulu ultrahigh-pressure terrane of east China: New implications from coesite and omphacite inclusions in zircon of granitic gneiss: Lithos, v. 52, p. 157–164.

Yokoyama, K., 1976, Finding of plagioclase-bearing granulite from the Iratsu epidote amphibolite mass in central Shikoku: Journal of Geological Society of Japan, v. 82, p. 549–551.

Yokoyama, K., 1980, Nikubuchi peridotite body in the Sanbagawa metamorphic belt; thermal history of the "Al-pyroxene-rich suite" peridotite body in high pressure metamorphic terrain: Contributions to Mineralogy and Petrology, v. 73, p. 1–13.

Yokoyama, K., and Mori, T., 1975, Spinel-garnet-two pyroxene rock from the Iratsu epidote amphibolite mass, central Japan: Journal of the Geological Society of Japan, v. 81, p. 29–37.

Exhumation Tectonics of the Sanbagawa High-Pressure Metamorphic Belt, Southwest Japan— Constraints from the Upper and Lower Boundary Faults

HIDEKI MASAGO,[1]

Center for Deep Earth Exploration, Japan Marine Science and Technology Center, 3173-25 Showa-machi, Yokohama 236–0001, Japan

KAZUAKI OKAMOTO,

Institute of Earth Sciences, Academia Sinica, P.O. Box 1–55 Nankang, Taipei, 115, Taiwan

AND MASARU TERABAYASHI

Department of Safety Systems Construction Engineering, Kagawa University, 2217-20 Hayashi-cho, Takamatsu, Kagawa 761-0396, Japan

Abstract

We have determined the exhumation process of the Sanbagawa Belt based on kinematic analyses of the upper and the lower boundary faults. The Sanbagawa Belt is tectonically intercalated as a thin subhorizontal sheet between overlying, weakly metamorphosed Jurassic and underlying Cretaceous accretionary complexes (e.g., Maruyama et al., 1996). On the lower boundary in Kii Peninsula, pumpellyite–actinolite facies metabasites have undergone semibrittle deformation, indicating a top-to-the south sense of thrusting. The upper boundary in central Shikoku is north-vergent, and indicates a top-to-the north sense of shear; this suggests that the original normal fault on the boundary was warped by later doming. These results support a model of selective exhumation of the Sanbagawa Belt from a depth of 30 km, and its juxtaposition against the over- and underlying accretionary complexes by orogen-orthogonal movements.

Introduction

EXHUMATION TECTONICS of high-pressure (high-P) rocks is a key to the understanding of orogenic processes at oceanic plate subduction zones and continental collision zones. A number of models have been proposed to explain the exhumation of high-P metamorphic belts: buoyancy uplift (e.g., Ernst, 1971); a "two-way-street" model combining underplating of newly accreted materials and surface erosion (e.g., Suppe, 1972); a mélange or diapir model, uplift by net buoyancy owing to the relatively low density matrix (e.g., Cloos, 1982); extensional tectonics (e.g., Lister et al., 1984), and their combinations. Maruyama et al. (1996) reviewed 250 blueschist belts of the world in terms of their geological and thermobaric structure and chronology, and proposed a "wedge extrusion" model to explain their exhumation. The essence of the model is a selective denudation of a thin slice of the high-P belt with normal and reverse faults on its upper and lower boundaries, respectively. All other models involve different combinations of the fault kinematics; hence determination of the kinematic sense of the boundary faults is of fundamental importance.

The Cretaceous high-P Sanbagawa metamorphic belt in southwest Japan is regarded as a product of ancient subduction along the continental margin of Asia. Numerous studies have been performed on the internal geologic and thermobaric structure of this belt. However, only a few were concerned with the upper and lower boundaries of this high-P unit; exceptions include Kawato et al. (1991), Sasaki and Isozaki (1992), Isozaki and Maruyama (1991) and Maruyama et al. (1996). The former two determined the structural boundaries between the Sanbagawa Belt and the overlying and underlying accretionary complexes, but did not report fault kinematics. The latter two outlined possible tectonic models for exhumation of the fault-bounded high-P belt, but did not provide structural evidence for fault kinematics to verify their models.

---

[1]Corresponding author; email: masagoh@jamstec.go.jp

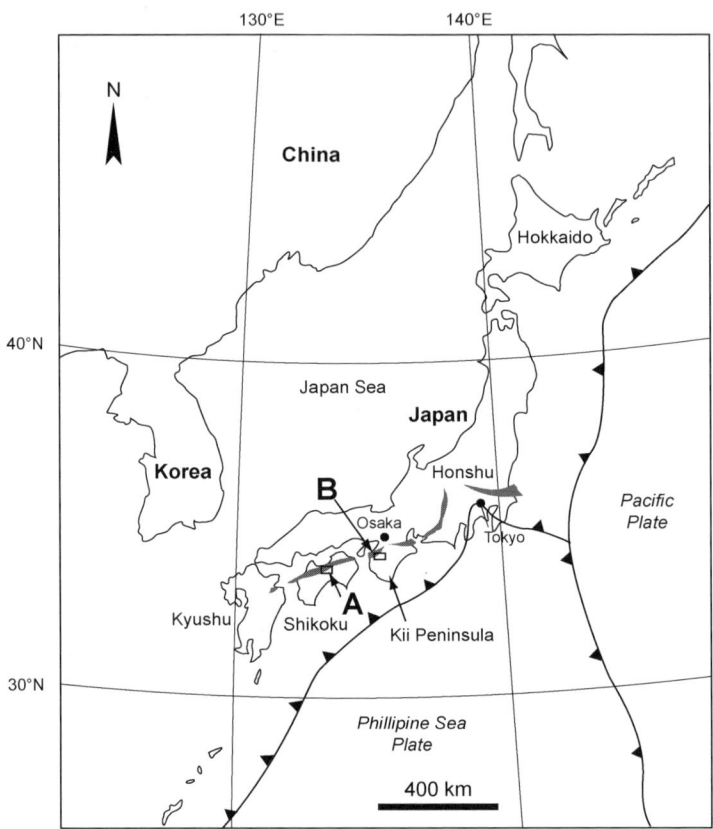

FIG. 1. Index map showing the distribution of the Sanbagawa Belt (shaded) in Japan. Study areas of the upper and lower boundaries are indicated as A and B, respectively.

In general, subduction zone-related large tectonic boundary faults are commonly reactivated numerous times during subduction and subsequent exhumation. Hence, it is particularly difficult to distinguish kinematic features related to the individual events. Nevertheless, we have discovered outcrops that preserve evidence for shear deformation during exhumation of the Sanbagawa Belt. This paper presents new structural data on the fault kinematics along its upper and lower boundaries. Mineral abbreviations used in the text and figures are after Kretz (1983).

Geologic Outline

The Sanbagawa Belt is a major Cretaceous metamorphic belt in Japan, and runs from the Kanto Mountains near Tokyo, through Kii Peninsula, across Shikoku Island to Kyushu Island for over 1000 km, almost parallel to the trench of the southwest Japan arc (Fig. 1). Protoliths are dominated by sandstone and shale with a small amount of basalt, chert, and carbonates. Recent detailed mapping in central Shikoku confirmed a duplex structure and an oceanic plate stratigraphy (Okamoto et al., 2000; Terabayashi et al., 2005). Metamorphic grade ranges from prehnite–pumpellyite facies through blueschist/greenschist facies to amphibolite facies and to eclogite facies. Metamorphic facies series show some variation along strike, and the highest-grade part occurs in central Shikoku (e.g. Maruyama et al., 1996). Metamorphic zonation is quite well studied in this region, where the highest-grade rocks crop out in the structurally intermediate part (Banno et al., 1978; Banno and Sakai, 1989; Ota et al., 2004).

A number of K–Ar (e.g., Banno and Miller, 1965; Itaya and Takasugi, 1988), and $^{40}Ar-^{39}Ar$ (Takasu

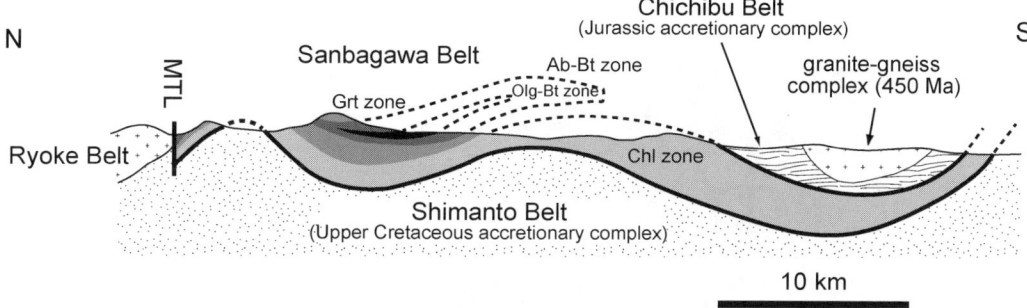

FIG. 2. Schematic cross-section showing the relationship between the Sanbagawa Belt and the overlying and underlying accretionary complexes in southwest Japan. The internal thermobaric structure of the Sanbagawa Belt is also shown. Abbreviation: MTL = Median Tectonic Line.

and Dallmeyer, 1990) age data have been reported from the Sanbagawa rocks. However, most from higher grade zones are cooling ages ranging from 65 to 90 Ma. The peak stage of the Sanbagawa metamorphism is considered to have been late Early Cretaceous, based on an Rb–Sr isochron age (116 ± 10 Ma; Minamishin et al., 1979), and zircon SHRIMP ages (110–120 Ma; Okamoto et al., 2004).

Compared to its length along strike, the width of the Sanbagawa Belt is less than 30 km, and its thickness is approximately 2 km (Isozaki and Maruyama, 1991), although other workers have estimated it as more than 5 km (S. Wallis, pers. commun., 2001). The deformation in the Sanbagawa metamorphic belt is characterized by high-strain ductile shear (e.g. Toriumi, 1985; Faure, 1985), and means that the Sanbagawa Belt is a thin, fault-bound, subhorizontal tectonic slice (Fig. 2), although the belt was flattened. The belt is intercalated between the overlying and underlying Jurassic Chichibu and Cretaceous Shimanto accretionary complexes, respectively. The upper and the lower boundaries of the Sanbagawa Belt are well exposed in central Shikoku and the Kii Peninsula, respectively.

Geology of the Tosayama area, central Shikoku (upper boundary)

The upper boundary fault of the Sanbagawa Belt is exposed in the Tosayama area in central Shikoku. In this area, the Sanbagawa and overlying Jurassic accretionary complex (Chichibu Belt) are bounded by the E-W–striking, southward-dipping Sasagatani fault (Fig. 3). The Sanbagawa Belt in this area is composed of basic and pelitic phyllites with allochthonous blocks of chert, limestone, and dolomite, and possesses metamorphic mineral parageneses of blueschist facies overprinted by pumpellyite–actinolite facies (Kawato et al., 1991). The Chichibu Belt is composed of pebbly mudstone with allochthonous blocks of chert, greenstone, sandstone and limestone, and is characterized by pumpellyite + chlorite metamorphic mineral parageneses.

The protolith ages of the Sanbagawa and the Chichibu belts in this area determined by microfossils are Early Cretaceous and Middle Jurassic, respectively. K-Ar ages of metamorphic white micas concentrate around 115 Ma for the Sanbagawa rocks, and about 140 Ma for the Chichibu rocks (Kawato et al., 1991).

Geology of the Higashi-Yoshino area, central Kii Peninsula (lower boundary)

In southwest Japan, the lower boundary of the Sanbagawa Belt is generally obscured, and the overlying Jurassic accretionary belt (Chichibu Belt) is in direct contact with the underlying Shimanto Belt. In the central Kii Peninsula (Fig. 4), however, structural relationships among these three geologic units are clear, and the lower boundary of the Sanbagawa Belt is well exposed in the Higashi-Yoshino area (Yamato-Omine Research Group, 1981; Takeuchi and Yamato-Omine Research Group, 1984). The Higashi-Yoshino area is located in the middle of the Kii Peninsula, just south of the Median Tectonic Line (MTL), where the Sanbagawa Belt is cut off by faults toward the east, and the Shimanto Belt is in direct contact with the MTL (Fig. 4). The boundary between the Sanbagawa and Shimanto belts is an E-W–striking fault reported by Sasaki and Isozaki (1992).

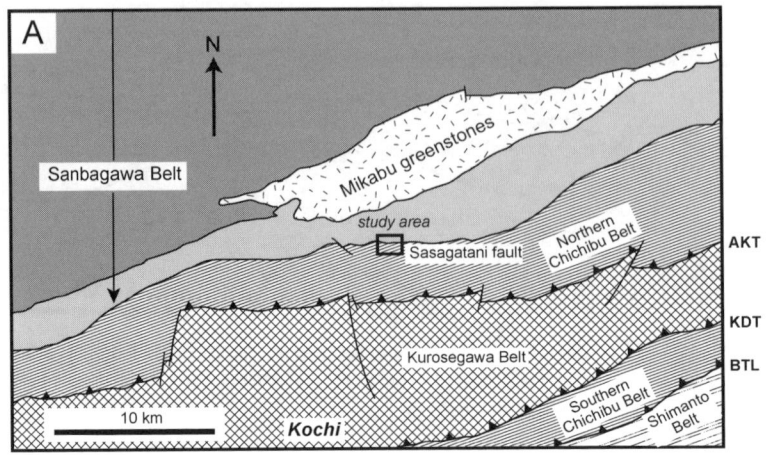

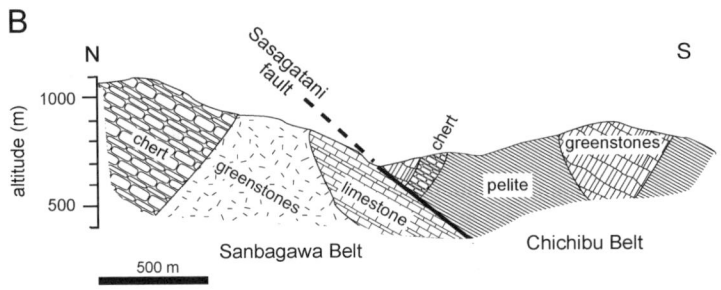

FIG. 3. Simplified geologic map (A) and profile (B) of the Tosayama area, central Shikoku. The Jurassic accretionary complex of the Chichibu Belt overlies the Sanbagawa Belt across the south-dipping Sasagatani fault in this area (modified after Kawato et al. 1991). Abbreviations: BTL = Butsuzo Tectonic Line; KDT = Kanbaradani thrust; AKT = Agekura thrust.

The Sanbagawa Belt situated on the northern side of the boundary fault is composed predominantly of pelitic schist with well-developed planar and deformation structures involving sheets and lenses of basic and psammitic schists at the meter to kilometer scale. Epidote, chlorite, actinolite, and white micas are common metamorphic minerals, with alkali amphibole occurring in some samples.

The Shimanto Belt in the south mainly consists of pelitic phyllite and lensoidal blocks of basic rocks and/or sandstones. The blocks range from a centimeter to a kilometer in length. Planar structures, which are well-developed in the northern part of the unit, decrease in intensity southward. The degree of recrystallization of the pelitic phyllite is generally low, and therefore many detrital quartz and plagioclase grains are preserved. The basic schist contains epidote, chlorite, actinolite, and white micas.

Fault Zone Descriptions

The upper boundary fault

Kawato et al. (1991) identified three exposures of the boundary in the Tosayama area based on the distribution of fault breccia and fault gouge. We re-examined each exposed boundary and considered Loc. 7 of Kawato et al. (1991) as the best representative exposure where a contact between massive Sanbagawa chert and overlying altered metabasites of the Chichibu Belt is clearly observed. A dominant 10 m wide fault gouge zone is evident. The boundary fault strikes N83°E and dips 46° to the south; the fault zone is comprised of alternating green and black clay layers representing altered and highly sheared metabasites.

A 50 cm wide intensely sheared zone is present toward the center of the fault where fragmented

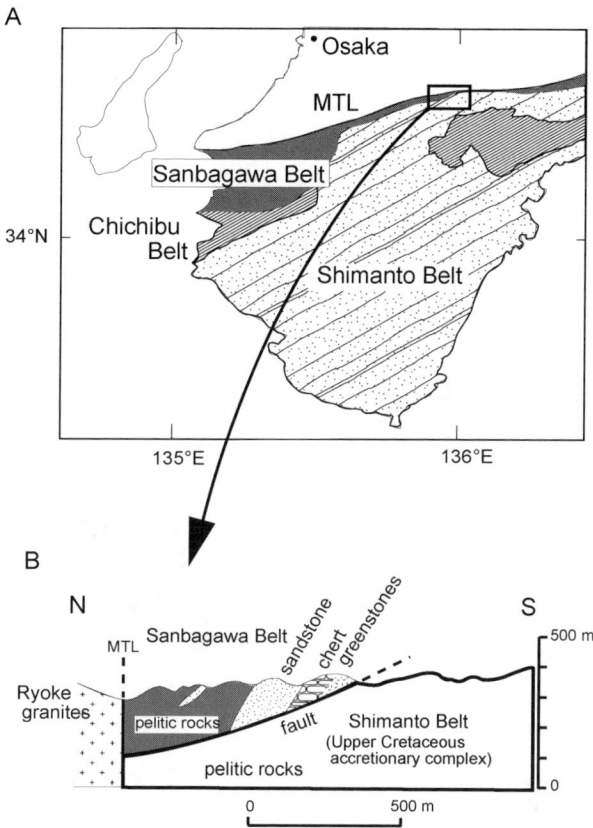

FIG. 4. Simplified geological map (A) and profile (B) in Kii Peninsula (modified after Sasaki and Isozaki, 1992). The distribution of the Sanbagawa, Shimanto, and Chichibu belts are indicated. Abbreviation: MTL = Median Tectonic Line.

black material is dominant. Kink-like recumbent folds are evident just above the shear zone. Well-developed axial plane cleavages are parallel to the fault zone and another set of cleavages intersects them (Fig. 5). From their geometric relationship, these cleavages are interpreted as Riedel shear planes R_1 and P. The sense of shear deduced from the asymmetry of the kink fold and the Riedel shear planes is top-to-the N35°E with a thrust sense.

The lower boundary fault

Sasaki and Isozaki (1992) reported 11 outcrops of the possible lower boundary of the Sanbagawa Belt in the Higashi-Yoshino area. The attitudes of the fault measured at these outcrops are rather uniform, striking N70–90° E and dipping 20–40° N. This is consistent with the regional extent of the fault. Some 1 m wide subhorizontal fracture zones propagate from the main fault into the underlying unit. An excellent outcrop of the lower boundary exposed in a new roadcut was found during this study. Foliations defined by mica minerals in the Shimanto pelitic schists are subparallel to the boundary fault, whereas those in the Sanbagawa metabasites are slightly oblique to the fault plane. Mineral lineation in the matrix metabasites generally has an E-W azimuth and a gentle plunge to the west, but it has different attitudes only in the vicinity of the boundary fault. Figure 6A shows the change in attitude of the mineral lineation approaching the boundary fault. The azimuth of the lineation swings from E-W to SE–NW in a narrow zone less than 30 cm from the boundary fault. The plunge of the lineation also changes to the northwest.

As shown in Figure 6B, the fault has a rough surface, suggesting that faulting occurred in a semi-brittle manner. Figure 7A is a close-up view

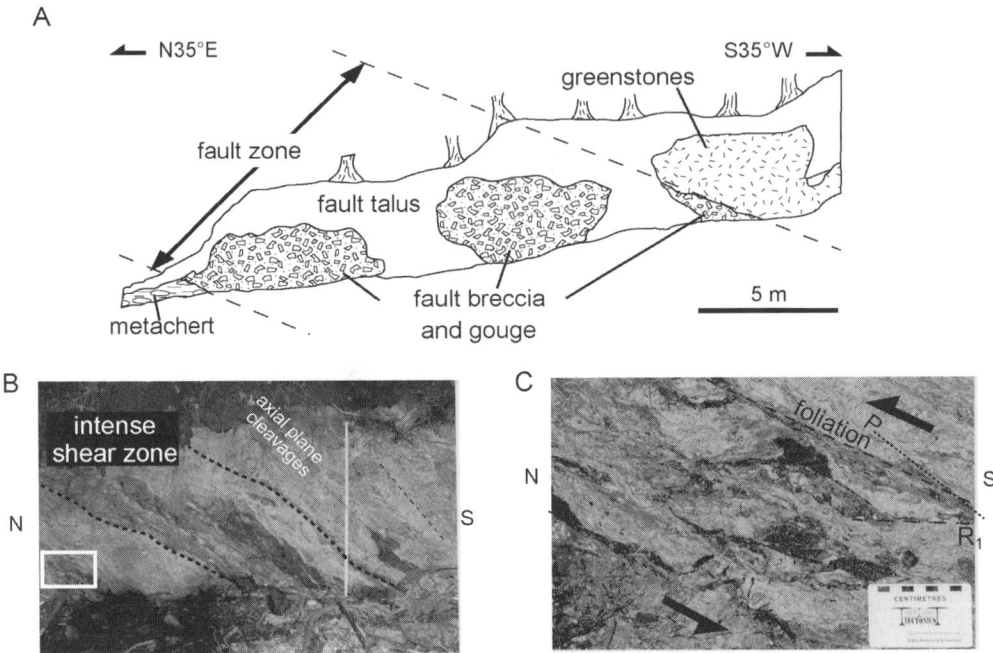

FIG. 5. Upper boundary of the Sanbagawa Belt in the Tosayama area, central Shikoku. A. Sketch of the outcrop of the fault zone. B. Field photograph of the fault zone. Metabasites have undergone alteration related to the faulting. A 50 cm wide intensely sheared zone is in the center of the field of view. Kink folds with well-developed axial plane cleavages are formed on both sides of the shear zone. The scale bar indicates 1 m. A focused view of the open box on the left side is shown in (C). C. Close-up view of the outcrop-scale deformation of the footwall. Riedel shear planes indicating top-to-the north movement are prominent in the fault gouge.

showing anastomosing, sigmoidal shear planes, and the boundary itself is a completely deformed zone. A metabasite layer has been dislocated, and some fragments are included in the footwall metapelite (Fig. 7B). The host metapelite and metabasite blocks are thoroughly recrystallized, and contain a foliation that is parallel to the main fault plane. A sheared metabasite sample contains the following mineral assemblage: Chl + Ab + Qtz + Cal + Hem ± Pmp ± Act ± Ep ± Ttn ± white mica. The three-phase mineral paragenesis Pmp + Act + Ep indicates pumpellyite–actinolite facies metamorphic conditions. Epidote and albite occur as porphyroclasts set within a chloritic matrix. A relict igneous augite porphyroclast is surrounded by actinolite pressure fringes. Epidote porphyroclasts have asymmetric tails, mostly of quartz and/or albite, and albite porphyroclasts have asymmetric tails of chlorite. In albite porphyroclasts, abundant inclusion trails mostly consist of epidote, and are generally oblique to the external foliation defined by chlorite (Fig. 7C). The thrust sense of shear was deduced from these porphyroclasts as top-to-the south.

Discussion

Exhumation tectonics of the Sanbagawa Belt

Our study has revealed the shear sense of both representative boundary faults. The lower boundary fault in the Higashi-Yoshino area has a thrust sense indicating top-to-the south movement of the Sanbagawa Belt over the Shimanto Belt. This is consistent with the fact that deeply subducted high-P rocks overlie feebly metamorphosed accretionary material.

The observed sense of shear on the upper boundary in the Tosayama area is top-to-the N35°E. At this locality, the boundary fault plane dips moderately to the south where it forms the southern limb of a regional-scale antiform. This sense apparently implies northward thrusting of the overlying Chichibu Belt onto the Sanbagawa Belt. However,

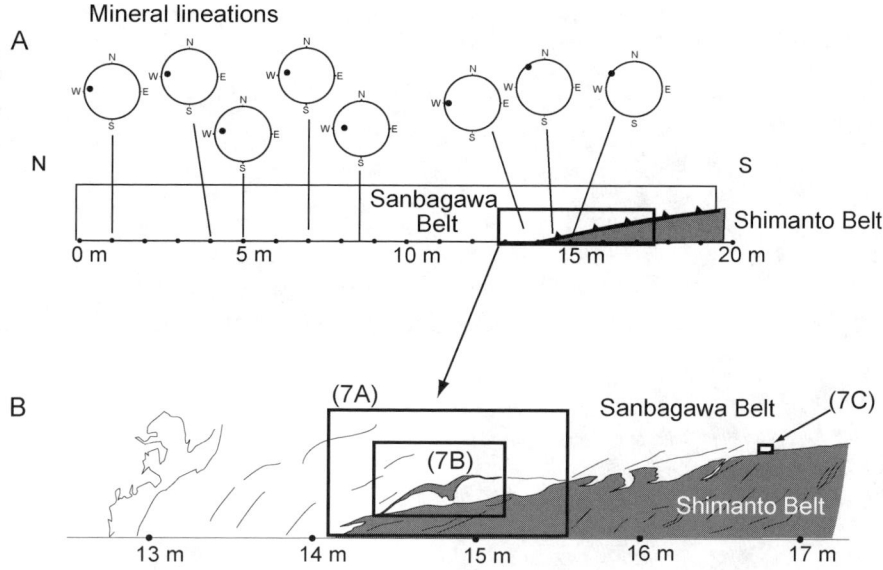

FIG. 6. A. Stereoplot of the mineral stretching lineations in the vicinity of the lower boundary fault of the Sanbagawa Belt. E-W–trending mineral lineation in the Sanbagawa Belt abruptly changes to a NW-SE trend just above and below the boundary fault. B. Sketch of the fault zone. The lower part of the Sanbagawa Belt is dislocated, and pelitic schist of the Shimanto Belt is included in the overlying Sanbagawa schist as a lensoidal block. The photos of open boxes (Figs. 7A–7C) are shown in Figure 7.

regionally this fault plane is north-dipping, with local re-orientation occurring during later antiformal doming, which changed the present attitude of the fault plane. The movement sense of this upper boundary is thus considered as normal and not reverse. Thus our results support the model that the fault-bound Sanbagawa Belt was selectively denuded as a wedge between the Chichibu and the Shimanto accretionary complexes.

We emphasize that the mineral lineation changes its azimuth from E-W to NW-SE just above the boundary fault. This implies that syndeformational recrystallization of the lower boundary occurred under pumpellyite–actinolite facies conditions when the Sanbagawa Belt was thrust over the Shimanto Belt. Internal ductile deformation in the Sanbagawa Belt was characterized by high-strain ductile flow associated with the formation of an orogen-parallel (E-W) stretching lineation and a penetrative foliation (Faure, 1985; Toriumi, 1985; Wallis, 1992, 1998). The general sense of shear is complex: top-to-the east (Faure, 1985), top-to-the west (Wallis, 1990; Wallis et al., 1992), inversion of top-to-the east and top-to-the west in the outcrop to map scales (Hara et al., 1992), and top-to-the east in the Kanto Mountains (Guidi et al., 1984; Otoh and Yanai, 1989; Abe et al., 2001). The E-W–directed shear deformation of low non-coaxiality is interpreted to have taken place during formation of the foliation during extensional flow (e.g., Wallis et al., 1992; Wallis, 1995; Yagi and Takeshita, 2002), and shear strain is variable ranging from constrictional, plane strain to uni-axial flattening (e.g., Shimizu and Yoshida, 2004) although oblique plate subduction along an E-W trench axis was once considered as the principal cause of the constriction-dominant stress regime (Faure, 1985; Toriumi, 1985; Abe et al., 2001). However, the N-S–trending mineral lineation developed only in the vicinity of the lower boundary fault, thus indicating that exhumation of the Sanbagawa Belt was orthogonal to the orogen.

The deduced kinematics of the boundary faults are also consistent with the exhumation tectonics of the eclogite-peridotite masses in the structural and thermal core of the Sanbagawa Belt. In the Iratsu and Higashi-Akaishi areas, central Shikoku, selective southward exhumation of the eclogite-peridotite masses is reported (Yamamoto et al., 2004). In the highest-grade parts, orogen-oblique or vertical lineations, predating formation of the penetrative

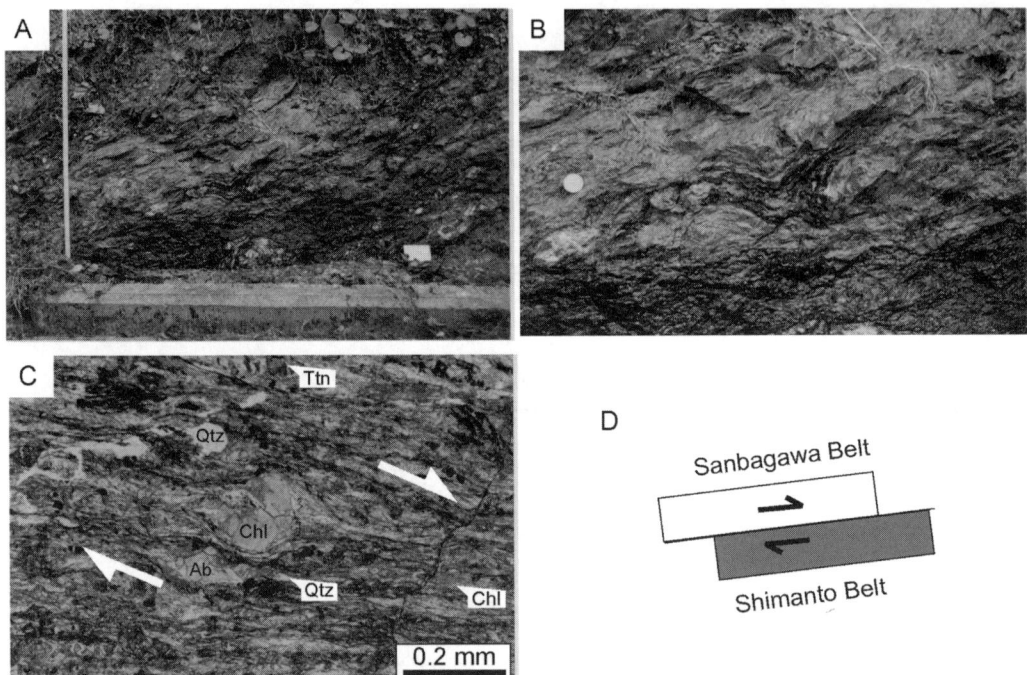

FIG. 7. Field and microscopic photos of the boundary fault of the Sanbagawa Belt. A. Sanbagawa metabasites overlie Shimanto metapelites. Foliations in the Sanbagawa Belt dip moderately to the north, and those in the Shimanto Belt dip gently to the north. The boundary fault is parallel to the foliations in the Shimanto schists. The vertical scale bar is 1 m. B. Sheared lensoidal block of the Shimanto metapelite incorporated in the overlying Sanbagawa schist. C. Photomicrograph of Sanbagawa metabasite. Porphyroclast systems indicate a dextral (top-to-the-south) sense of shear. D. Schematic diagram showing low-angle overthrusting of the Sanbagawa Belt onto the Shimanto Belt.

orogen-parallel fabrics, were also recognized (Kohsaka, 1986; Yamamoto et al., 2004).

Many models have been proposed to explain the exhumation of regional metamorphic belts. They are classified into the following five groups. The first model is buoyancy uplift of a thin slice (Ernst, 1971; Ernst and Liou, 1995). This model simply explains that the exhumation force is related to the buoyancy of the relatively less dense metamorphic rocks themselves. The second is a mélange or diapir model (England and Holland, 1979; Carlson and Rosenfeld, 1981; Cloos, 1982; Moore, 1984), and is essentially similar to the first model, buoyancy-driven uplift. The difference is that buoyancy is driven not by the metamorphic rocks themselves but by significantly less dense matrix materials. This model assumes a corner flow in a serpentinized mantle wedge, and metamorphic rocks are incorporated as tectonic blocks and transported to the surface (Fig. 8A). The third is the "two-way-street" model (Suppe, 1972; Platt, 1975). This model is a combination of jack-up by underplating of newly accreted materials and surface erosion and collapse of the overburden materials (Fig. 8B). The fourth is extension tectonics (Lister et al., 1984; Lister and Davis, 1989; Blake and Jayko, 1990; Jolivet et al., 1994). The continental crust is thinned in an extensional stress field forming a listric normal fault system, and deep metamorphic rocks are exposed at the surface (Fig. 8C). The fifth is the wedge extrusion model (Maruyama, 1990; Maruyama et al. 1994, 1996). The essence of this model is selective uplift of the fault-bound metamorphic belt to the mid-crust by compression of the mantle wedge due to change in subduction angle (Fig. 8D). Subsequent doming unroofs the overlying accretionary complex.

Both the first (buoyancy) and the fifth (wedge extrusion) models were suggested on the basis of physical modeling designed under the regime of continental collision tectonics and more compre-

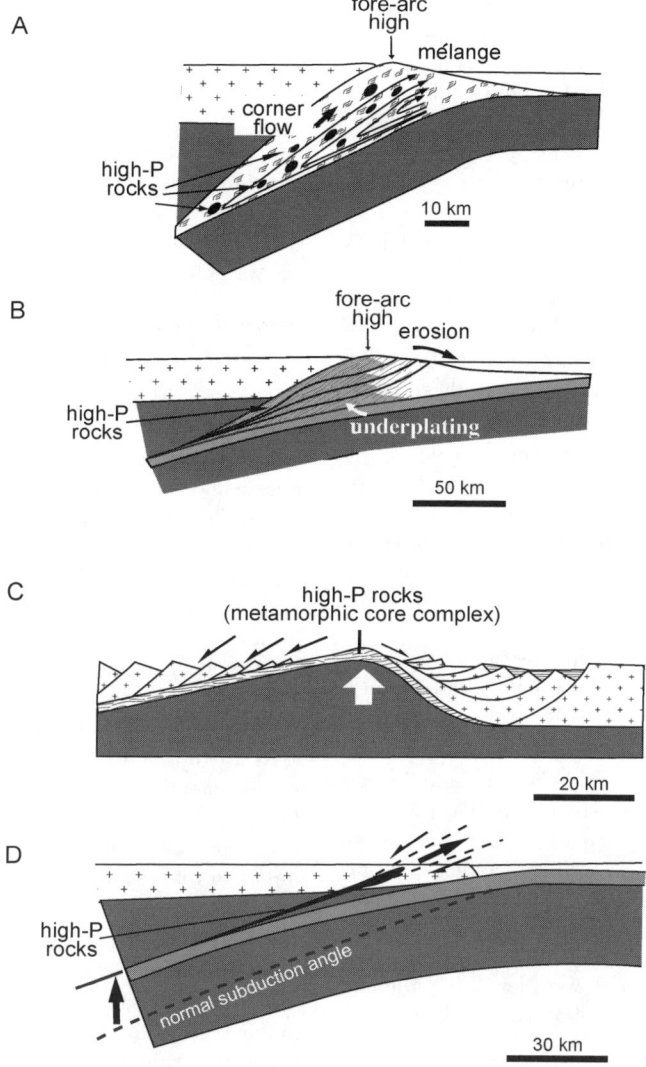

FIG. 8. Representative exhumation tectonic models of the high-P metamorphic belt. A. Mélange model, simplified after Cloos (1982). B. Two-way-street model, simplified after Suppe (1972) and Platt (1975). C. Extension model, simplified after Lister and Davis (1989). D. Wedge extrusion model, simplified after Maruyama et al. (1996). See text for details.

hensive covering of the exhumation of metamorphic rocks in collisional orogens and subduction zones of oceanic plates (Chemenda et al., 1995, 1996). However, the first model is possible only with erosion of the relief, which provides an unloading effect, allowing the subducted crustal slice to rise, and the fifth model is appropriate for high-compression regimes characterized by high pressure between the overriding and subducting plates. In general (low-compression regimes), a rapid spontaneous crustal uplift (intrusion into the interplate zone) brings the deeply subducted crust to the mid-crust, and further exhumation is by change of subduction angle (due to slab breakoff or ridge subduction) and subsequent doming. That is, the fifth model is applicable in the more general case.

A major criticism of the first three models, is a lack of a large sedimentary basin to account for the voluminous eroded materials caused by uplift of the metamorphic belt. Because these models require erosion of the overburden of the metamorphic belt, voluminous amounts of eroded materials should have been supplied to the foreland basin of the exhumed metamorphic belt. However, the amount of sediments in the foreland basins of high-P belts is too small to account for erosion of the overlying materials of > 30 km thick.

A further criticism of the first three models is that these models cannot explain the episodic uplift of high-P/T metamorphic rocks that occurs every ~100 m.y. documented detail in southwest Japan (Isozaki and Itaya, 1990; Maruyama et al., 1996). Therefore, the exhumation process of these metamorphic belts requires selective extrusion.

The fourth model is more accommodating on these points; however, both upper and lower boundary faults are normal, i.e., extensional (Fig. 8C). This is inconsistent with our observation in the Sanbagawa Belt where the lower boundary fault is a thrust. In addition, mantle materials should be present below the metamorphic belt in this model. However, at present no drilling research has identified mantle materials beneath the high-P metamorphic belts.

The wedge extrusion model is currently applied to many regional metamorphic belts of both the collision type, such as the Himalayas (Grujic et al., 1996; Kaneko, 1997) and the Kokchetav Massif (Kaneko et al., 2000), as well as Pacific-type (Cordilleran-type) plate-convergent margins such as the Franciscan complex (Terabayashi et al., 1996; Terabayashi and Maruyama, 1998). Crocodile-type extrusion (Meissner, 1989; Wheeler, 1991; Michard et al., 1993; Henry et al., 1993; Caby, 1994) and tectonic wedging (Wakabayashi and Unruh, 1995) are essentially similar. The following three features are essential to the application of the wedge extrusion model: (1) a subhorizontal structure of the metamorphic belt; (2) normal and reverse fault at the upper and lower boundaries of the metamorphic belt, respectively; and (3) a gap of metamorphic grade (chiefly pressure) across the boundary faults. These features are relatively well documented in collision-type orogens. Compared to the collision type, applications of the model to Pacific-type orogens are inadequately confirmed. Although the subhorizontal, sandwiched structure of metamorphic belts is recognized in many orogens (see review in Maruyama et al., 1996), the kinematics of the boundary faults and the gap in metamorphic grade are poorly recognized in these orogens. In the case of the Sanbagawa Belt, a subhorizontal structure has long been recognized (e.g., Higashino, 1990), and gaps of metamorphic grade between the overlying Jurassic Chichibu and the underlying Upper Cretaceous Shimanto accretionary belts have also been demonstrated (Isozaki and Maruyama, 1991). Therefore, our kinematic results combined with the above-mentioned features clearly suggest that the wedge extrusion model can be applied to the Sanbagawa Belt.

Contrasting natures of faulting conditions of the upper and lower boundaries

The conditions and nature of faulting at the upper and lower boundaries are significantly different. The upper fault zone is composed of fault breccia and gouge, without metamorphic recrystallization. This suggests that faulting took place under a brittle regime. However, a 50 cm wide fracture zone around the boundary underwent extensive argillization, suggesting localized fluid flow during the faulting. Kink folds are developed just above this boundary. In general, kink folds are formed in relatively competent layers under dry and high interlayer friction conditions (Honea and Johnson, 1976; Reches and Johnson, 1976). Considering the features mentioned above, the faulting condition on the upper boundary was generally dry and brittle at low temperature, with localized fluid flow playing an important role as a hydrosoftening agent.

In contrast, the lower boundary is coherent and well recrystallized under pumpellyite–actinolite facies conditions. The absence of fault gouge suggests that faulting occurred at higher temperature and/or under fluid-rich conditions. However, considering that thickness of the metamorphic belt is less than 2 km, large differences in temperature are not expected. This may imply abundant fluid supply from the descending slab beneath the ascending Sanbagawa Belt. For northern Kazakhstan, Terabayashi et al. (2002) demonstrated that underlying unit was thermally metamorphosed by the tectonic juxtaposition of a hot Kokchetav ultrahigh-P metamorphic slab, and substantial amounts of fluid were transported to the overlying high-P–ultrahigh-P unit. Metamorphic belts are hydrated during exhumation, which enhances retrograde mineral recrystallization. The contrasting nature of the upper and lower boundaries in term of the extent of hydration suggests that the main source of the retrograde fluid was the underlying Shimanto Belt.

Acknowledgments

We thank Prof. B. F. Windley and Dr. S. Johnson for their constructive comments on the early draft and for grammatical improvements. We also thank Prof. W. G. Ernst, Prof. M. Cloos, and Prof. S. Maruyama for critical advice and discussions, and Dr. I. Katayama for his assistance during the field work. Prof. Y. Isozaki and Mr. H. Sasaki kindly provided detailed information about the geology of the Higashi-Yoshino area, central Kii Peninsula. A part of the field investigation was supported by a research grant from the Japan Society for the Promotion of Science for Young Scientists (Masago, No. H13-05232).

REFERENCES

Abe, T., Shimada, K., Takagi, H., Kimura, S., Ikeyama, S., and Miyashita, A., 2001, Ductile shear deformation of the Sanbagawa metamorphic rocks in the Kanto Mountains: Journal of Geological Society of Japan, v. 107, p. 337–353 (in Japanese with English abstract).

Banno, S., Higashino, T., Otsuki, M., Itaya, T., and Nakajima, T., 1978, Thermal structure of the Sanbagawa metamorphic belt in central Shikoku: Journal of Physics of the Earth (supplement), p. 245–356.

Banno, S., and Miller, J. A., 1965, Additional data on the age of metamorphism of the Ryoke-Abukuma and Sanbagawa metamorphic belt, Japan: Japanese Journal of Geology and Geography, v. 36, p. 17–22.

Banno, S., and Sakai, C., 1989, Geology and metamorphic evolution of the Sanbagawa metamorphic belt, Japan, in Daly, J. S., Cliff, R. A., and Yardley, B. W. D., eds., Evolution of metamorphic belts: Geological Society of London, Special Publication, v. 43, p. 519–532.

Blake, M. C., Jr., and Jayko, A. S., 1990, Uplift of the very high pressure rocks in the western Alps: Evidence for structural attenuation along low-angle faults: Mémoire de Societé de Géology de France, N. S., v. 156, p. 237–246.

Caby, R., 1994, Precambrian coesite from northern Mali: First record and implications for plate tectonics in the trans-Saharan segment of Pan-African belt: European Journal of Mineralogy, v. 6, p. 235–244.

Carlson, W. D., and Rosenfeld, J. L., 1981, Optical determination of topotactic aragonite–calcite growth kinetics: Metamorphic implications: Journal of Geology, v. 89, p. 615–638.

Chemenda, A., Mattauer, M., Malavieille, J., and Bokun, A. N., 1995, A mechanism for syn-collisional rock exhumation and associated normal faulting: Results from physical modelling: Earth and Planetary Science Letters, v. 132, p. 225–232.

Chemenda, A., Mattauer, M., and Bokun, A. N., 1996, Continental subduction and a mechanism for exhumation of high-pressure metamorphic rocks: New modelling and field data from Oman: Earth and Planetary Science Letters, v. 143, p. 173–182.

Cloos, M., 1982, Flow mélange: Numerical modeling and geologic constraints on their origin in the Franciscan subduction complex, California: Geological Society of America Bulletin, v. 93, p. 330–345.

England, P. C., and Holland, T. J. B., 1979, Archimedes and the Tauern eclogites: The role of buoyancy in the preservation of exotic eclogite blocks: Earth and Planetary Science Letters, v. 53, p. 63–68.

Ernst, W. G., 1971, Do mineral parageneses reflect unusually high pressure conditions of Franciscan metamorphism?: American Journal of Science, v. 270, p. 81–108.

Ernst, W. G., and Liou, J. G., 1995, Contrasting plate-tectonic styles of Qinling-Dabie-Sulu (Alpine-type) and Franciscan (Pacific-type) metamorphic belts: Geology, v. 23, p. 353–356.

Faure, M., 1985, Eastward ductile shear during the early tectonic phase in the Sanbagawa Belt: Journal of the Geological Society of Japan, v. 89, p. 319–329.

Guidi, A., Charvet, J., and Sato, T., 1984, Finding of granitic olistoliths and pre-Cretaceous radiolarians in the northwestern Kanto Mountains, Gunma Prefecture, Central Japan: Journal of the Geological Society of Japan, v. 90, p. 853–856.

Grujic, D., Casey, M., Davidson, C., Hollister, L. S., Kündig, R., Pavlis, T., and Schmid, S., 1996, Ductile extrusion of the Higher Himalayan Crystalline in Bhutan: Evidence from quartz microfabrics: Tectonophysics, v. 260, p. 21–43.

Hara, I., Shiota, T., Hide, K., Kanai, K., Goto, M., Seki, S., Kaikiri, K., Takeda, K., Hayasaka, Y., Miyamoto, T., Sakurai, Y., and Ohtomo, Y., 1992, Tectonic evolution of the Sambagawa schists and its implications in convergent margin processes: Journal of Science of Hiroshima University, Series C, v. 9, p. 495–595.

Henry, C., Michard, A., and Chopin, C., 1993, Geometry and structural evolution of the high-pressure rocks from the Dora-Maira massif, western Alps, Italy: Journal of Structural Geology, v. 15, p. 965–981.

Higashino, T., 1990, Metamorphic zones of the Sambagawa metamorphic belt in central Shikoku, Japan: Journal of the Geological Society of Japan, v. 96, p. 703–718 (in Japanese with English abstract).

Honea, E., and Johnson, A. M., 1976, A theory of concentric, kink, and sinusoidal folding and of monoclinical flexuring of compressible, elastic multilayers. Part IV: Development of sinusoidal and kink folds in multilayers confined by rigid boundaries: Tectonophysics, v. 30, p. 197–239.

Isozaki, Y., and Itaya, T., 1990, Chronology of the Sambagawa metamorphism: Journal of Metamorphic Geology, v. 8, p. 401–411.

Isozaki, Y., and Maruyama, S., 1991, Studies on orogeny based on plate tectonics in Japan and new geotectonic subdivision of the Japanese islands: Journal of Geology of Japan, v. 100, p. 697–761 (in Japanese with English abstract).

Itaya, T., and Takasugi, H., 1988, Muscovite K-Ar ages of the Sanbagawa schists, Japan and argon depletion during cooling and deformation: Contributions to Mineralogy and Petrology, v. 100, p. 281–290.

Jolivet, L., Daniel, J. M., Truffert, C., and Goffé, B., 1994, Exhumation of the deep crustal metamorphic rocks and crustal extension in arc and back-arc regions: Lithos, v. 33, p. 3–30.

Kaneko, Y., 1997, Two-step exhumation model of the Himalayan Metamorphic Belt, central Nepal: Journal of the Geological Society of Japan, v. 103, p. 203–226.

Kaneko, Y. Maruyama, S., Terabayashi, M., Yamamoto, H., Ishikawa, M., Anma, R., Parkinson, C. D., Ota, T., Nakajima Y., Katayama, I., and Yamauchi, K., 2000, Geology of the Kokchetav UHP–HP metamorphic belt, northern Kazakhstan: The Island Arc, v. 9, 264–283.

Kawato, K., Isozaki, Y., and Itaya, T., 1991, Geotectonic boundary between the Sanbagawa and Chichibu Belts in central Shikoku, Southwest Japan: Journal of the Geological Society of Japan, v. 97, p. 959–975 (in Japanese with English abstract).

Kohsaka, Y., 1986, Structure and recrystallization of amphibolite masses from Tonaru and Iratsu regions: Unpubl. Master's thesis, Ehime University, 107 p.

Kretz, R., 1983, Symbols for rock-forming minerals: American Mineralogist, v. 68, p. 277–279.

Lister, G. S., Banga, G., and Feenstra, A., 1984, Metamorphic core complexes of Cordilleran type in the Cyclades, Aegean Sea, Greece: Geology, v. 12, p. 221–225.

Lister, G. S., and Davis, G. A., 1989, The origin of metamorphic core complexes and detachment faults forming during Tertiary continental extension in the northern Colorado River region, U. S. A: Journal of Structural Geology, v. 11, p. 65–94.

Maruyama, S., 1990, Denudation process of high-pressure metamorphic belt [abs.]: Geological Society of Japan, 97th Annual Meeting Abstracts, p. 484.

Maruyama, S., Liou, J. G., and Zhang, R. Y., 1994, Tectonic evolution of the ultrahigh-pressure (UHP) and high-pressure (HP) metamorphic belts from central China: The Island Arc, v. 3, p. 112–121.

Maruyama, S., Liou, J. G., and Terabayashi, M., 1996, Blueschists and eclogites of the world and their exhumation: International Geology Review, v. 38, p. 485–594.

Meissner, R., 1989, Rupture, creep, lamellae, and crocodiles: Happenings in the continental crust: Terra Nova, v. 1, p. 17–28.

Michard, A., Chopin, C., and Henry, C., 1993, Compression versus extension in the exhumation of the Dora-Maira coesite-bearing unit, Western Alps, Italy: Tectonophysics, v. 221, p. 173–193.

Minamishin, M., Yanagi, T., and Yamaguchi, M., 1979, Rb-Sr whole-rock age of the Sanbagawa metamorphic rocks in central Shikoku, Japan: Isotope geosciences of Japanese Islands: Geological Survey of Japan, p. 68–71.

Moore, D. E., 1984, Metamorphic history of a high-grade blueschist exotic block from the Franciscan Complex, California: Journal of Petrology, v. 25, p. 126–150.

Okamoto, K., Maruyama, S. and Isozaki, Y., 2000, Accretionary complex origin of the Sanbagawa, high-P/T metamorphic rocks, central Shikoku, Japan—layer-parallel shortening structure and greenstone geochemistry: Journal of the Geological Society of Japan, v. 106, p. 70-86.

Okamoto, K., Shinjoe, K., Katayama, I., Terada, K., and Sano, Y., 2004, SHRIMP U-Pb zircon dating of quartz-bearing eclogite from the Sanbagawa Belt, south-west Japan: Implications for metamorphic evolution of subducted protolith: Terra Nova, v. 16, p. 81–89.

Ota, T., Terabayashi, M., and Katayama, I., 2004, Thermobaric structure and metamorphic evolution of the Iratsu eclogite body in the Sanbagawa belt, central Shikoku, Japan: Lithos, v. 73, p. 95–126.

Otoh, H., and Yanai, S., 1989, Evidence of eastward ductile overthrust shear in the Mikabu and Chichibu Belts, Kanto Mountains: Journal of Geological Society of Japan, v. 95, p. 711–714.

Platt, J. P., 1975, Metamorphic and deformational processes in the Franciscan Complex, California: Some insights from the Catalina Schist terrane: Geological Society of America Bulletin, v. 86, p. 1337–1347.

Reches, Z., and Johnson, A. M., 1976, A theory of concentric, kink, and sinusoidal folding and of monoclinical flexuring of compressible, elastic multilayers. Part VI: Asymmetric folding and monoclinical kinking: Tectonophysics, v. 35, p. 295–334.

Sasaki, H., and Isozaki, Y., 1992, Low-angle thrust between the Sanbagawa and Shimanto Belts in central Kii Peninsula, Southwest Japan: Journal of the Geological Society of Japan, v. 98, p. 57–60 (in Japanese with English abstract).

Shimizu, I., and Yoshida, S., 2004, Strain geometries in the Sanbagawa Metamorphic Belt inferred from deformation structures in metabasite: The Island Arc, v. 13, p. 95–109.

Suppe, J., 1972, Interrelationships of high-pressure metamorphism, deformation, and sedimentation in Franciscan tectonics, U.S.A: 24th International Geological Congress, v. 3, p. 552–559.

Takasu, A., and Dallmeyer, R. D., 1990, $^{40}Ar/^{39}Ar$ mineral age constraints for the tectonothermal evolution of the Sanbagawa metamorphic belt, central Shikoku, Japan: A Cretaceous accretionary prism: Tectonophysics, v. 185, p. 111–139.

Takeuchi, Y. and Yamato-Omine Research Group, 1984, Chichibu and Shimanto belts in the Central Kii Mountains 11, Otaki region [abs.]: Geological Society of Japan, Annual Meeting Abstracts, v. 91, p. 172 (in Japanese).

Terabayashi, M., and Maruyama, S., 1998, Large pressure-gap between the coastal and central Franciscan belts, northern and central California: Tectonophysics, v. 285, p. 87-101.

Terabayashi, M., Maruyama, S., and Liou, J. G., 1996, Thermobaric structure of the Franciscan Complex in the Pacheco Pass region, Diablo Range, California: Journal of Geology, v. 104, p. 617-636.

Terabayashi, M., Okamoto, K., Yamamoto, H., Kaneko, Y., Ota, T., Maruyama, S., Katayama, I., Komiya, T., Ishikawa, A., Anma, R., Ozawa, H., Windley, B. F., and Liou, J. G., 2005, Accretionary complex origin of the mafic-ultramafic bodies of the Sanbagawa belt, central Shikoku, Japan: International Geology Review, v. 47, 2005, p. 1058–1073.

Terabayashi, M., Ota, T., Yamamoto, H., and Kaneko, Y., 2002, Contact metamorphism of the Daulet Suite by Solid-State Emplacement of the Kokchetav UHP–HP metamorphic slab: International Geology Review, v. 44, p. 819–830.

Toriumi, M., 1985, Two types of ductile deformation/regional metamorphic belt: Tectonophysics, v. 113, p. 1307–1326.

Wakabayashi, J., and Unruh, J. R., 1995, Tectonic wedging, blueschist metamorphism, and exposure of blueschists: Are they compatible?: Geology, v. 100, p. 19–40.

Wallis, S. R., 1990, The timing of folding and stretching in the Sanbagawa Belt, the Asemigawa region, Central Shikoku: Journal of the Geological Society of Japan, v. 96, p. 345–552.

Wallis, S. R., 1992, Vorticity analysis in a metachert from the Sanbagawa Belt, SW Japan: Journal of Structural Geology, v. 14, p. 2271–2280.

Wallis, S. R., 1995, Vorticity analysis and recognition of ductile extension in the Sanbagawa metamorphic belt, SW Japan: Journal of Structural Geology, v. 17, p. 1077–1093.

Wallis, S. R., 1998, Exhuming the Sanbagawa metamorphic belt: The importance of tectonic discontinuities: Journal of Metamorphic Geology, v. 16, p. 83–95.

Wallis, S. R., Banno, S., and Radvanec, M., 1992, Kinematics, structure and relationship to metamorphism of the east-west flow in the Sanbagawa Belt, southwest Japan: The Island Arc, v. 1, p. 176–185.

Wheeler, J., 1991, Structural evolution of a subducted continental sliver: The northern Dora Maira massif, Italian Alps: Journal of Geological Society of London, v. 148, p. 1101–1113.

Yagi, K., and Takeshita, T., 2002, Regional variation in exhumation and strain rate of the high-pressure Sambagawa metamorphic belt, in central Shikoku, Southwest Japan: Journal of Metamorphic Geology, v. 20, p. 633–647.

Yamamoto, H., Okamoto, K., Kaneko, Y., and Terabayashi, M., 2004, Southward extrusion of the eclogite-bearing mafic–ultramafic bodies in the Sanbagawa belt, central Shikoku, Japan: Tectonophysics, v. 387, p. 151–168.

Yamato-Omine Research Group, 1981, Paleozoic and Mesozoic units in central Kii Mountains, in Association of Geological Collaboration, Excursion Guidebook, v. 35, p. 88 (in Japanese).

U-Pb Dating of Large Zircons in Low-Temperature Jadeitite from the Osayama Serpentinite Mélange, Southwest Japan: Insights into the Timing of Serpentinization

T. TSUJIMORI,[1] J. G. LIOU,

Department of Geological and Environmental Sciences, Stanford University, Stanford, California 94305

J. WOODEN,

United States Geological Survey, 345 Middlefield Road, Menlo Park, California 94025

AND T. MIYAMOTO

Department of Earth and Planetary Sciences, Faculty of Sciences, Kyushu University, Fukuoka, 812-8581, Japan

Abstract

Crystals of zircon up to 3 mm in length occur in jadeitite veins in the Osayama serpentinite mélange, Southwest Japan. The zircon porphyroblasts show pronounced zoning, and are characterized by both low Th/U ratios (0.2–0.8) and low Th and U abundances (Th = 1–81 ppm; U = 6–149 ppm). They contain inclusions of high-pressure minerals, including jadeite and rutile; such an occurrence indicates that the zircon crystallized during subduction-zone metamorphism. Phase equilibria and the existing fluid-inclusion data constrain P-T conditions to P > 1.2 GPa at T < 350°C for formation of the jadeitite. Most U/Pb ages obtained by SHRIMP-RG are concordant, with a weighted mean $^{206}Pb/^{238}U$ age of 472 ± 8.5 Ma (MSWD = 2.7, n = 25). Because zircon porphyroblasts contain inclusions of high-pressure minerals, the SHRIMP U-Pb age represents the timing of jadeitite formation, i.e., the timing of interaction between alkaline fluid and ultramafic rocks in a subduction zone. Although this dating does not provide a direct time constraint for serpentinization, U-Pb ages of zircon in jadeitite associated with serpentinite result in new insights into the timing of fluid-rock interaction of ultramafic rocks at a subduction zone and the minimum age for serpentinization.

Introduction

ZIRCON OCCURS as an accessory phase in igneous and metamorphic rocks and is the best mineral for U-Pb age dating, with a closure temperature around 900°C (Cherniak and Watson, 2001). On the other hand, in low-temperature (LT) environments such as blueschist- and greenschist-facies conditions, new zircon growth and overgrowth are very rare. Hence U-Pb dating of zircon in subduction-zone metamorphic rocks has been very limited. However, crystallization of zircon from hydrothermal fluid has been documented (e.g., Rubin et al., 1989; Rubatto and Hermann, 2003). In such LT environments, hydrothermal fluid may cause partial dissolution of pre-existing zircon and precipitation of neoblastic zircon (e.g., Hansen and Friderichsen, 1989; Sinha et al., 1992; Vavra et al., 1999). These processes accompany Pb or U loss and cause isotopic discordance at LT conditions.

Serpentinite is a unique environment for precipitation of hydrothermal zircon. Zircon has been reported from serpentinite-related LT metasomatic rocks such as rodingite blackwall and jadeitite (Kobayashi et al., 1987; Harlow, 1994; Bröcker and Enders, 1999; Miyajima et al. 2001; Dubińska et al. 2004); reconnaissance U-Pb dating for such LT zircon has just began (e.g., Bröcker and Enders, 1999; Kunugiza et al., 2002; Dubinska et al., 2004). Isotopic and chemical information on these zircon crystals provide a key not only to understand the fluid history concerning serpentinization but also to evaluate the hydrothermal effects on both growth and dissolution of zircon at LT conditions.

In this contribution, we present SHRIMP (sensitive, high-resolution ion microprobe) U-Pb isotopic data for large hydrothermal zircon crystals, up to 3 mm in size, in jadeitite (nearly monomineralic

---

[1]Corresponding author; email: tatsukix@pangea.stanford.edu

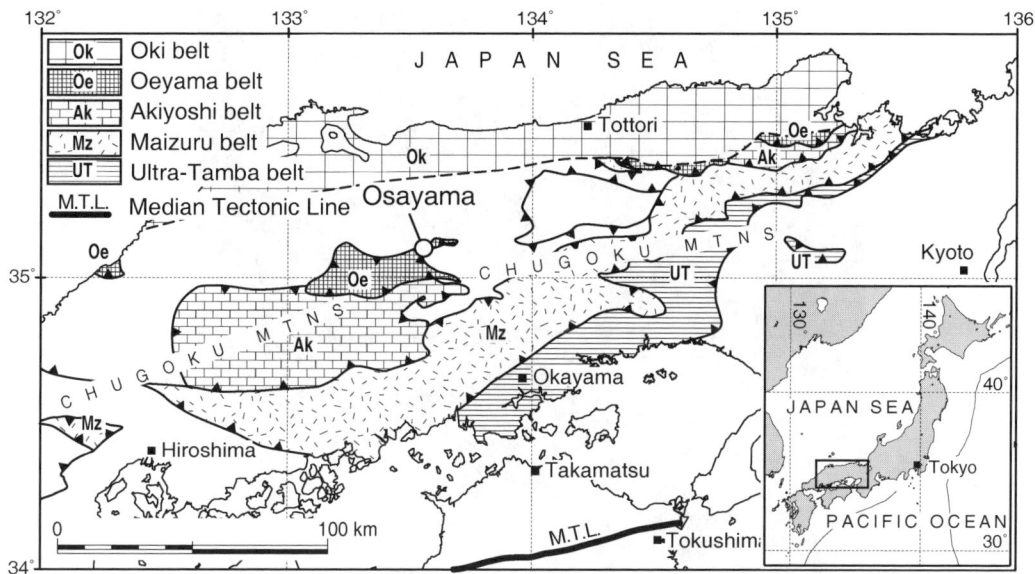

FIG. 1. Simplified map of the Chugoku Mountains of Southwest Japan, showing various pre-Triassic geotectonic units and the sample locality.

jadeite-rich rock) from a serpentinite-matrix mélange in the Osayama Mountain, Southwest Japan. Although zircon porphyroblasts from this locality have been previously considered as a relict igneous phase (Miyamoto and Yanagi, 1998), we have recently found a few jadeite inclusions in zircon. We also describe the internal texture and U-Th chemistry of hydrothermal zircon, discuss the implications for the zircon crystallization in jadeitite, and comment on the U-Pb age dating for serpentinization at a subduction zone.

Geologic Outline

Jadeitite, omphacitite, albitite, rodingite, and blueschist-facies schist have been identified as blocks in the Osayama serpentinite mélange of the central Chugoku Mountains (Tsujimori, 1997; Tsujimori and Itaya, 1999; Tsujimori and Liou, 2004a, 2005a, 2005b) (Fig. 1). Serpentinized peridotite bodies and related mélanges occur as an Early Paleozoic ophiolitic nappe that occupies the highest structural position of the Oeyama belt (Ishiwatari and Tsujimori, 2003). Chrysotile and lizardite are major serpentine-group minerals in the Osayama mélange (Tsujimori, 1998). Protoliths of the serpentinites consist mainly of harzburgite with minor dunite, podiform chromitite, and gabbroic dikes (e.g., Arai, 1980); the gabbroic intrusives with MORB-like geochemical signatures yield Sm-Nd isochron ages of ca. ~560 Ma (Hayasaka et al., 1995). Phengites from blueschist-facies schists at Osayama yield K-Ar ages of 327–273 Ma (Tsujimori and Itaya, 1999).

Jadeitites in the Osayama serpentinite mélange occur as veins up to 4 m wide. They are composed mainly of jadeite (generally 75% by volume) with minor amounts of grossular (nearly pure), and trace rutile and zircon. The jadeite-rich matrix is heterogeneous and shows variable degrees of retrogression. Omphacite, analcime, pectolite, stronalsite, thomsonite, natrolite, vesuvianite, prehnite, phlogopite, serpentinite, and titanite are retrograde phases (Kobayashi et al., 1987; Tsujimori, 1998). Shoji and Kobayashi (1988) described primary fluid inclusions in jadeite. Preliminary U-Pb ages of large zircon porphyroblasts in jadeitite using convectional thermal ionization mass spectrometry yield ca. 500–450 Ma (Miyamoto and Yanagi, 1998); the data was interpreted as the crystallization age of an igneous protolith for the jadeitite.

Sample Description and Condition of Zircon Crystallization

Electron microprobe analysis for rock-forming minerals was carried out with a JEOL JXA-8900R at

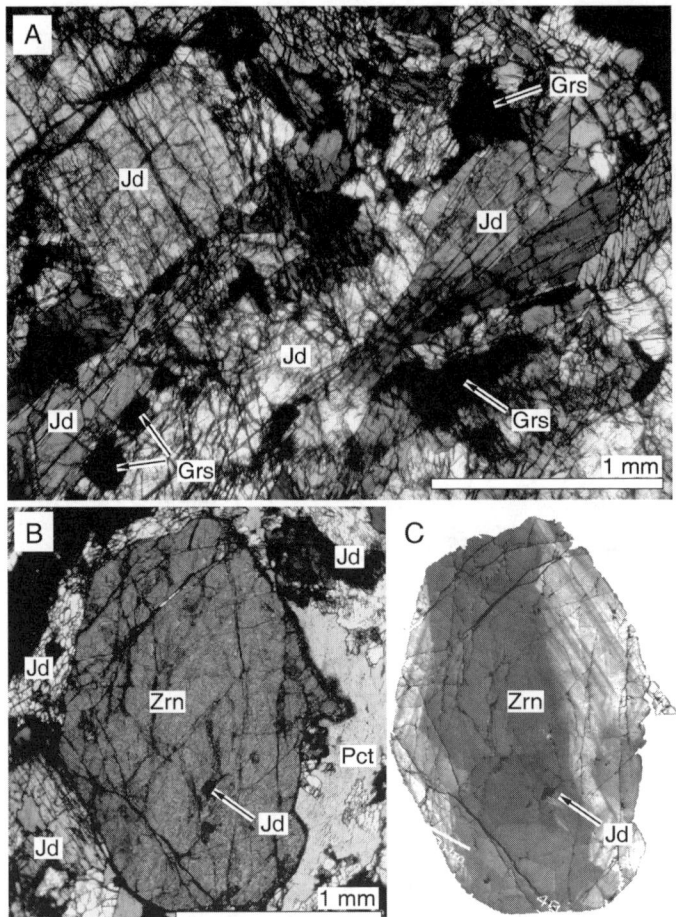

FIG. 2. Microtexture of the investigated jadeitite. A. Plane polarizer view of the occurrence of jadeite (Jd) with grossular (Grs). B. Plane polarizer view of occurrences of zircon (Zrn); matrix minerals including jadeite (Jd) and secondary pectolite (Pct) are also shown. C. Cathodoluminescence image of zircon of (B). Texture shows zircon consisting of at least two crystals.

Okayama University of Science. Quantitative analyses were performed with 15 kV accelerating voltage, 12 nA beam current, and a 3 μm beam size. The ZAF (oxide basis) method was employed for matrix corrections. Cathodoluminescence (CL) imaging was obtained using a JEOL 5600LV Scanning Electron Microscope (SEM) equipped with a HAMAMATSU photo multiplier tube at Stanford University.

In the investigated sample, jadeite with 91–99 mol% $NaAl_2Si_2O_6$ (Jd) component occurs as either subhedral to anhedral prismatic crystals or radial aggregates (Figs. 2A and 3); most are less than 0.5 mm in length, but some are up to 5 mm. Jadeite contains rare grossular, zircon, and rutile as tiny inclusions (< 0.1 mm). Some secondary omphacites (39–58 mol% Jd) epitaxially overgrow jadeite crystals and fill fractures together with analcime (Fig. 3). In the altered sections of the sample, jadeite was extensively replaced by pectolite. Coarse-grained rutile prisms, up to 4 mm long, occur in the jadeite-dominant matrix and are rimed by secondary titanite that is in contact with secondary pectolite.

Zircon occurs as discrete euhedral to subhedral crystals up to 3 mm in length or as twinned crystals or aggregates in the matrix (Fig. 2B and 2C); zircon contains tiny inclusions of rutile and rare jadeite (97 mol% Jd; Fig. 3). Fine-grained zircon also occurs as

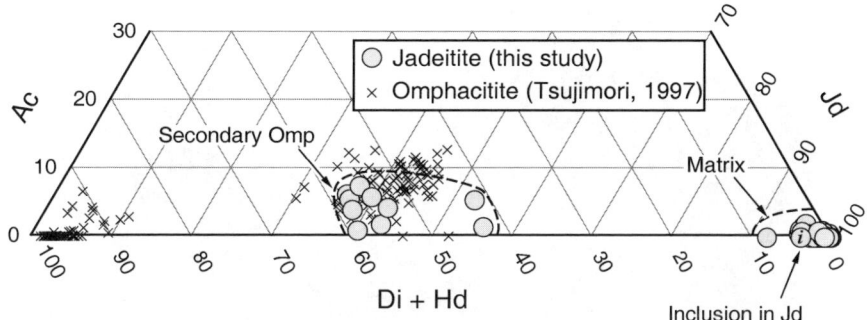

FIG. 3. Compositions of sodic pyroxenes from the investigated jadeitite on the jadeite (Jd)–acmite (Ac)–diopside + hedenburgite (Di + Hd) ternary diagram. The Fe^{2+}/Fe^{3+} ratio for clinopyroxene was estimated assuming a cation total of 4.00.

inclusions in a single jadeite crystal (Kobayashi et al., 1987). This evidence indicates synchronous growth of zircon and jadeite.

Jadeitite formation requires HP condition (P > 0.5–0.6 GPa at T = 200–400°C) and infiltration of Na- and Al-rich alkaline fluid (Harlow, 1994; Shi et al., 2003). It is thought that zircon crystals directly crystallized from a hydrothermal fluid during the jadeitite formation in serpentinites. In the Osayama jadeitite, the homogenization T of fluid inclusions in jadeite examined by Shoji and Kobayashi (1988) requires a minimum of ~345°C for crystallization of the primary jadeitite minerals. Thermodynamic calculation employing the THERMOCALC program (ver. 3.21) (Powell et al., 1998) indicates P > 1.2 GPa at a nominal T = ~350°C for the primary jadeite + rutile + grossular assemblage, and P = 0.4–0.6 GPa at 200–300°C for the secondary omphacite + analcime assemblage. The presence of grossular and absence of carbonate minerals suggest low X_{CO2} of the hydrothermal fluid.

U-Th Chemistry and U-Pb Isotopic Results

U-Th-Pb analyses were performed with the SHRIMP-RG (sensitive high-resolution ion microprobe-reverse geometry) in the Stanford-USGS cooperative ion microprobe facility at Stanford University. Instrumental conditions and data acquisition are similar to the procedures described by Williams (1998). Analytical spots ~ 40 µm in diameter were sputtered using an ~5 nA O_2^- primary beam. The data were collected in sets of six scans through nine mass spectra. The primary beam was rastered across the analytical spot for 120 s before analysis to reduce surficial common Pb resulting from sample preparation and Au-coating. Concentration data were calibrated against CZ3 zircon (550 ppm U), and isotope ratios were calibrated against R33 (419 Ma, John Aleinikoff, pers. commun., 2002). Data reduction follows Williams (1998), and utilized Squid (Ludwig, 2001). Isoplot 3 (Ludwig, 2003) was used to calculate all ages, which are reported here at the 95% confidence level.

Representative CL images of three zircon grains are shown in Figure 3; the Th and U abundance from 28 spots of 8 zircon grains are presented in Figure 4. The dating results are shown in Table 1. As shown in the images of Figure 4, zircon porphyroblasts in the Osayama jadeitite have preserved remarkable internal zoning textures including fine, euhedral growth layers with sharp contrasts in CL brightness; these zircons are characterized by low Th and U abundances. Based on textural and geochemical features, analyzed zircons are subdivided into two different zones. Type I is a normal growth zone that is commonly characterized by a broad, stubby core with concentric oscillatory-zoned rims. This texture, shown in Figure 4A, is similar to the growth pattern of hydrothermal zircon crystallized from fluid in serpentinite (e.g., Fig. 7 of Dubińska et al., 2004), and provides evidence of crystallization from fluid. Type I zones in zircon crystals show a wide geochemical variation; Th and U exhibit a positive correlation—Th increases from 3 to 37 ppm as U increases from 11 to 82 ppm. However, concentrations fluctuate greatly from one layer to another (Figs. 4A and 5A). Type II zones in zircon crystals, with dark CL, represent texturally inherited cores rarely preserved in Type I zones (Fig. 4B) and have

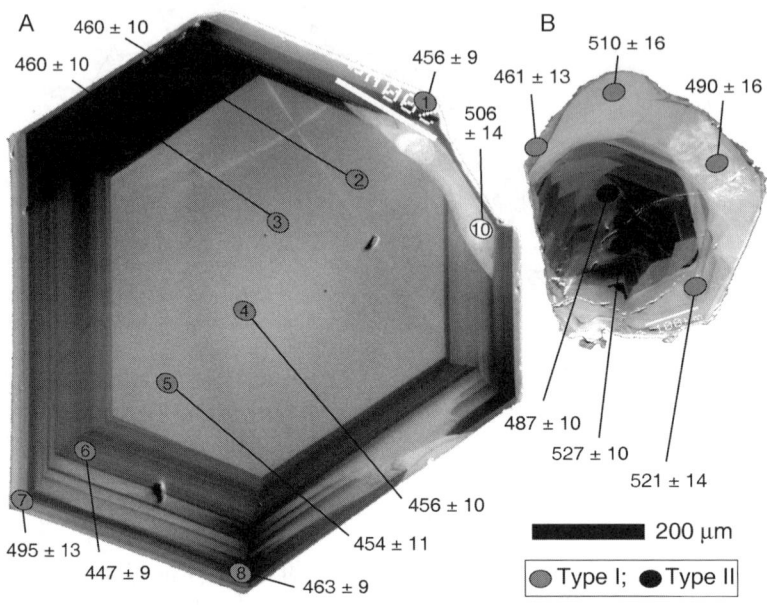

FIG. 4. Cathodoluminescence images of representative zircon separates from the Osayama jadeitite showing growth zoning layers and three zircon zones described in the text. Ellipses within crystal represent primary beam pits; SHRIMP U-Pb results are also shown as apparent $^{206}Pb/^{238}U$ ages (Ma) with 1σ errors. For explanation see text. Scale bar represents 200 μm.

very irregular zoning patterns; Th/U ratios of Type II zones are significantly higher (0.70–0.80) than that of the others (0.21–0.48) (Fig. 5B). Textures and different Th/U ratios indicate partial dissolution of early-formed zircon crystals.

The analyzed U-Pb isotopic ratios are plotted on a Tera-Wasserburg diagram with 1σ errors (Fig. 6). Most of the data points are concordant within error. Twenty-five analyses of the Type I zone yields $^{206}Pb/^{238}U$ ages arranging 452 to 521 Ma, whereas three analyses of the Type II zone yields 488 to 523 Ma (Table 1). Although the $^{206}Pb/^{238}U$ ages are slightly scattered within the same zones in same zircon crystals (Fig. 4), the lack of noticeable correlation between apparent $^{206}Pb/^{238}U$ age and either U abundance or Th/U ratio (Fig. 5) indicates that the observed chemical variations may not be related to zircon alteration and the U-Pb system may not be significantly disturbed by later events. When the Type II zone is excluded, the type I zone yields a weighted mean $^{206}Pb/^{238}U$ age of 472 ± 8.5 Ma (MSWD = 2.7), which is in agreement with the conventional data (Fig. 6).

Zircon Growth in Jadeitite from Serpentinite

Most ultramafic rocks contain less than 20 ppm Zr (Erlank et al., 1978); this value is less than 2% of the Zr content of highly evolved granite (>1000 ppm). In fact, zircon in ultramafic rocks is itself very rare except for a few metasomatized peridotites from the Kokchetav Massif, Kazakhstan (e.g., Katayama et al., 2003). Despite such Zr-depleted environments, however, many zircon occurrences in jadeitite and rodingite in serpentinized mantle peridotites have been commonly documented, as mentioned above. On the other hand, formation of nearly monomineralic jadeitite is rare but is almost always restricted to serpentinites from blueschist or eclogite terranes, clearly indicating a subduction-zone environment with a low geotherm (e.g., Coleman, 1961; Harlow, 1994). It is well known that the hydration of ultramafic rocks results in high-pH alkaline fluid (pH ~10) in serpentinites from both orogenic belts and modern seafloor environments (e.g., Coleman, 1971; Kelley et al., 2001). The fluid-inclusion studies of Guatemalan jadeitites (Johnson

TABLE 1. Ion-Probe U-Th-Pb Isotope Data of Zircon in Jadeitite
from the Osayama Serpentinite Mélange, Southwest Japan[1]

Label	Zone	U (ppm)	Th (ppm)	P^{brad}* (ppm)	*f (%)	Th/U	^{238}U/^{206}Pb	Error	^{207}Pb/^{206}Pb	Error	^{207}Pb/^{206}Pb age** (Ma)	Error
					Grain 01							
#01-1	Type I	11.1	2.7	0.8	2.23	0.25	12.00	4.2	0.0686	10.1	509.0	21.3
#01-2	Type I	11.7	2.4	0.7	–1.48	0.21	13.62	4.0	0.0609	10.8	454.2	18.2
					Grain 05							
#05-11	Type I	15.4	3.7	1.1	–0.96	0.25	12.21	3.4	0.0660	8.5	502.2	16.9
#05-13	Type I	12.5	3.4	0.8	4.79	0.28	13.56	4.1	0.0668	12.4	452.8	18.8
					Grain 07							
#07-34	Type I	21.8	6.4	1.4	–0.78	0.30	13.13	3.1	0.0603	8.1	471.1	14.3
#07-35	Type I	18.2	4.9	1.3	–0.76	0.28	12.37	3.1	0.0614	9.6	498.7	15.8
					Grain 09 (Fig. 3A)							
#09-15 (3A-3)	Type I	60.4	20.5	3.9	1.26	0.35	13.31	2.2	0.0690	4.9	460.0	10.3
#09-17 (3A-7)	Type I	46.3	14.2	3.2	3.15	0.32	12.26	2.6	0.0748	8.2	494.7	13.0
#09-18 (3A-6)	Type I	78.8	26.6	4.9	0.99	0.35	13.87	2.1	0.0601	4.6	446.5	9.3
#09-19 (3A-5)	Type I	53.2	17.4	3.4	2.04	0.34	13.64	2.3	0.0601	5.6	454.0	10.6
#09-20 (3A-4)	Type I	58.5	19.9	3.7	1.12	0.35	13.58	2.2	0.0602	5.1	455.7	10.1
#09-21 (3A-1)	Type I	94.1	37.3	6.0	0.89	0.41	13.58	2.0	0.0600	4.1	455.9	9.0
#09-22 (3A-2)	Type I	58.8	20.1	3.8	–0.34	0.35	13.48	2.3	0.0583	5.4	460.2	10.4
#09-23 (3A-8)	Type I	81.8	29.5	5.3	0.69	0.37	13.36	2.0	0.0593	4.1	463.6	9.2
					Grain 10							
#10-24	Type I	13.1	3.5	0.9	4.41	0.27	12.72	3.97	0.0682	10.7	481.1	19.2
#10-27	Type I	11.0	2.9	0.8	–1.22	0.27	12.21	3.69	0.0692	9.3	500.0	18.5
					Grain 11 (Fig. 3B)							
#11-3	Type II	85.2	68.0	6.3	–0.149	0.82	11.70	1.89	0.0605	3.5	526.9	9.8
#11-4	Type I	17.7	4.7	1.2	–0.804	0.27	12.18	3.14	0.0550	8.5	510.3	15.9
#11-5	Type I	17.3	3.6	1.2	–0.897	0.21	12.52	3.20	0.0659	8.5	490.0	15.7
#11-6	Type I	25.4	7.5	1.6	–0.586	0.30	13.39	2.86	0.0611	6.8	461.5	13.1
#11-7	Type II	63.5	45.0	4.3	–0.256	0.73	12.69	2.12	0.0588	4.7	487.7	10.2
#11-36	Type I	26.7	7.9	1.9	–0.536	0.30	11.77	2.73	0.0651	6.5	521.0	14.2
					Grain 14							
#14-28	Type I	10.4	2.2	0.7	4.56	0.22	12.669	3.94	0.0575	10.3	489.4	19.2
#14-30	Type I	60.1	24.9	4.0	–0.28	0.43	12.994	2.16	0.0572	4.9	477.6	10.2
#14-31	Type I	46.4	21.7	3.1	–0.37	0.48	12.853	2.34	0.0633	5.5	479.2	11.2
#14-32	Type I	23.1	6.5	1.5	–0.70	0.29	13.217	2.93	0.0541	8.0	471.5	13.7
					Grain 15							
#15-37	Type I	26.1	7.0	1.8	–0.52	0.28	12.204	2.70	0.0613	6.5	505.3	13.6
#15-38	Type II	83.7	61.8	5.9	0.39	0.76	12.210	1.93	0.0596	3.8	506.1	9.6

[1]All errors are 1 σ of standard deviation. * = percentage of common ^{206}Pb in total measured ^{206}Pb. A negative value indicates that the number of ion counts was indistinguishable from background. In such case no common Pb correction is made; ** = commom Pb correction based on ^{207}Pb.

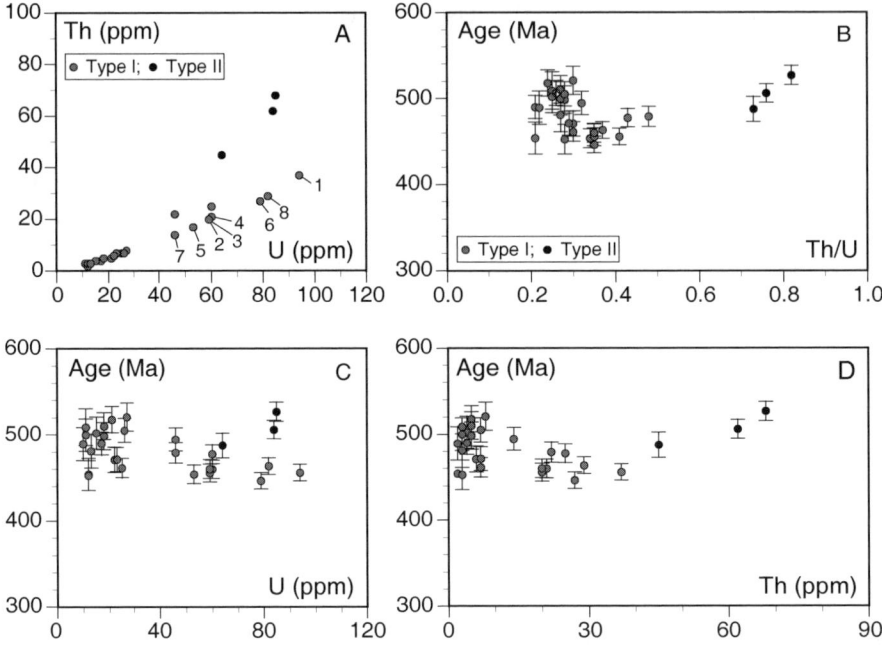

FIG. 5. Plots of apparent $^{206}Pb/^{238}U$ ages (Ma) and Th and U abundance (ppm). Labeled data within plot 5A relate to analyses of zircon shown in Figure 3A. For explanation see text.

and Harlow, 1999) revealed that the infiltration of seawater-like, high-salinity fluid (up to 8 wt% NaCl equivalent) into serpentinites might have played a significant role in jadeitite formation. It is thus possible that serpentinization and subsequent migration of high-pH, high-salinity fluid in serpentinites were almost contemporaneous. Moreover, some geologic and experimental evidence for Zr mobility in hydrothermal environments indicates that Si-undersaturated fluid containing F, Cl, Na, and Ca can behave as a corrosive solution that promotes high concentrations of high-field-strength elements (HFSEs) (e.g., Hansen and Friderichsen, 1989; Sinha et al., 1992; Vard and Williams-Jones, 1993; Rubatto and Hermann, 2003).

Considering these synthetically, fluid-rock interaction associated with the serpentinization of ultramafic rocks appears to be the only reasonable process that may account for the Zr concentration and for the precipitation of zircon. Although the source of Zr remains unclear, Zr saturation was most likely supplied by infiltration of an HFSE-enriched subduction-zone fluid or internal source such as clinopyroxene breakdown during serpentinization, inasmuch as mantle peridotite clinopyroxene contains 10–30 ppm Zr (e.g., Batanova and Sobolev, 2000). In any case, such hydrothermal fluid may have contributed to zircon crystallization during serpentinization in the subduction-zone environment.

Time Constraints on Fluid-Rock Interaction

The investigated zircon porphyroblasts have been previously considered as a relict igneous phase (Miyamoto and Yanagi, 1998). Our work has confirmed the result of earlier study that a zircon U-Pb age dated by conventional technique is Ordovician. However, we have also revealed syn-metamorphic, simultaneous growth of zircon with jadeite from the serpentinite-related hydrothermal fluid at a condition of P > 1.2 GPa at T < 350°C. Inasmuch as the P-T condition of zircon precipitation is well below the closure temperature of the zircon U-Pb system, we interpret the SHRIMP U-Pb ages as the timing of zircon crystallization from alkaline fluid during the formation of the jadeitite. As mentioned above, the fluid circulation in ultramafic rocks and the subsequent fluid-rock interaction under the LT-HP condition are required to form nearly monomineralic

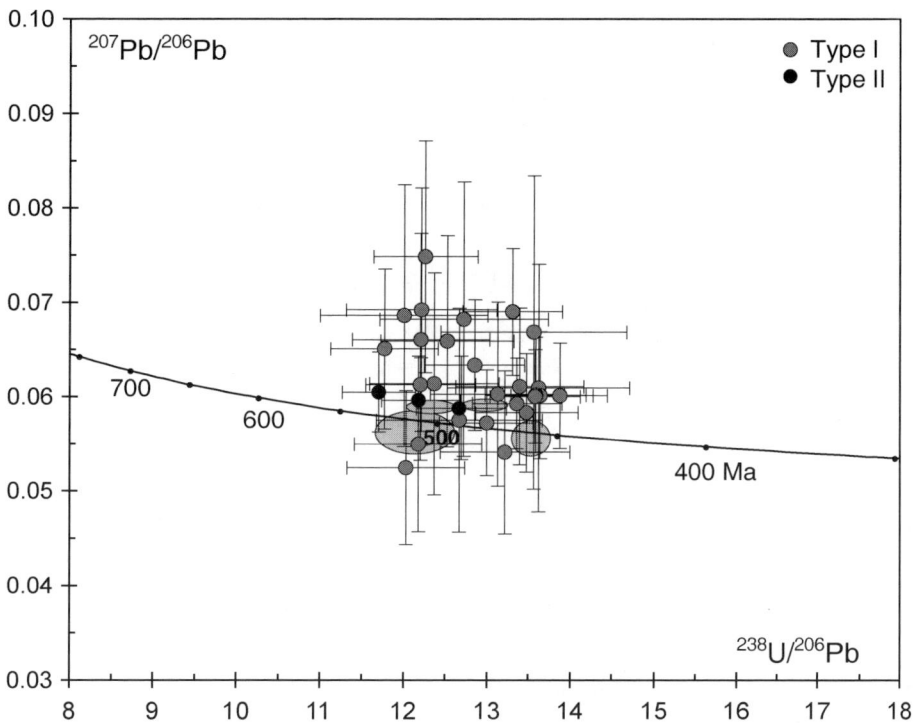

FIG. 6. Tera-Wasserburg diagram for SHRIMP analyses of zircons from the Osayama jadeitite. The grey ellipses represent conventional data by Miyamoto and Yanagi (1998).

jadeitite. Thus, an age of 472 Ma may represent the timing of interaction between hydrothermal fluid and ultramafic rocks in a subduction zone. Fluid circulation may have resulted in varying degrees of metasomatism leading to mineralization of jadeite and zircon during serpentinization. Despite its geologic or tectonic importance, however, radiometric dating of serpentinization is a difficult problem. Although this dating does not provide a direct time constraint for the serpentinization, U-Pb ages of zircon in jadeitite can result in new insights into the minimum age of serpentinization by the fluid-rock interaction of ultramafic rocks at a subduction zone.

The hydrothermal mineralization of 472 Ma is significantly older than the K-Ar ages, 327–273 Ma, of blueschist-facies blocks from the Osayama serpentinite mélange. This suggests the jadeitite formation is not related to regional Late Paleozoic blueschist-facies metamorphism. In the eastern Chugoku Mountains, Early Paleozoic HP epidote-amphibolites with hornblende K-Ar age of 443–403 Ma occur in a serpentinized peridotite body located about 180 km to the east of the Osayama serpentinite mélange (Tsujimori and Liou, 2004b). The jadeitite formation may be related to this Early Paleozoic HP metamorphism.

Acknowledgments

This research was supported financially in part by JSPS Research Fellowship for Research Abroad of the first author. Preparation of this manuscript and SHRIMP dating was supported by NSF EAR-0003355. We thank R. E. Jones, A. Meibom, J. Hourigan, and R. Y. Zhang for their help in the laboratory, and W. G. Ernst, I. Katayama, G. E. Harlow, and C. G. Mattinson for critical review of an early version of this manuscript.

REFERENCES

Arai, S., 1980, Dunite-harzburgite-chromitite complexes as refractory residue in the Sangun-Yamaguchi Zone, western Japan: Journal of Petrology, v. 21, p. 141–165.

Batanova, V. G., and Sobolev, A. V., 2000, Compositional heterogeneity in subduction-related mantle peridotites, Troodos massif, Cyprus: Geology, v. 28, p. 55–58.

Bröcker, M., and Enders, M., 1999, U-Pb zircon geochronology of unusual eclogite-facies rocks from Syros and Tinos (Cyclades, Greece): Geological Magazine, v. 136, p. 111–118.

Cherniak, D. J., and Watson, E. B., 2001, Pb diffusion of zircon: Chemical Geology, v. 172, p. 5–24.

Coleman, R.G., 1961, Jadeite deposits of the Clear Creek area, New Idria district, San Benito County, California: Journal of Petrology, v. 2, p. 209–247.

Coleman, R. G., 1971, Petrologic and geophysical nature of serpentinites: Geological Society of America, Bulletin, v. 82, p. 897–918.

Dubińska, E., Bylina, P., Kozlowski, A., Dörr, W., Nejbert, K., and Schastok, J., 2004, U-Pb dating of serpentinization: Hydrothermal zircon from a metasomatic rodingite shell (Sudetic ophiolite, SW Poland): Chemical Geology, v. 203, p. 183–203.

Erlank, A. J., Smith, A. S., Marchant, J. W., Cardoso, M. P., and Ahrens, L. H., 1978, Zr, in Wedepohl, K. H., ed., Handbook of geochemistry, v. II-4, section 40: New York, NY, Springer.

Harlow, G. E., 1994, Jadeitites, albitites and related rocks from the Motagua Fault Zone, Guatemala: Journal of Metamorphic Geology, v. 12, p. 49–68.

Hansen, B. T., and Friderichsen, J. D., 1989, The influence of recent Pb-loss on the interpretation of disturbed U-Pb systems in zircons from igneous rocks in East Greenland: Lithos, v. 23, p. 209–223.

Hayasaka, Y., Sugimoto, T., and Kano, T., 1995, Ophiolitic complex and metamorphic rocks in the Niimi-Katsuyama area, Okayama Prefecture: Excursion Guidebook of 102nd Annual Meeting of the Geological Society of Japan, p. 71–87 (in Japanese).

Hoskin, P. W. O., 2000, Patterns of chaos: Fractal statistics and the oscillatory chemistry of zircon: Geochimica et Cosmochimica Acta, v. 64, p. 1905–1923.

Ishiwatari, A., and Tsujimori, T., 2003, Paleozoic ophiolites and blueschists in Japan and Russian Primorye in the tectonic framework of East Asia: A synthesis: The Island Arc, v. 12, p. 190–206.

Isozaki, Y., 1996, Anatomy and genesis of a subduction-related orogen: A new view of geotectonic subdivision and evolution of the Japanese Island: The Island Arc, v. 5, p. 289–320.

Johnson, C. A., and Harlow, G. E., 1999, Guatemala jadeitites and albitites were formed by deuterium-rich serpentinizing fluids deep within a subduction-channel: Geology, v. 27, p. 629–632.

Kelley, D. S., Karson, J. A., Blackman, D. K., Fruuh-Green, G. L., Butterfeld, D. A., Lilley, M. D., Olson, E. J., Schrenk, M. O., Roe, K. V., Lebonk, G. T., Rivizzigno, P., and the AT3-60 Shipboard Party, 2001, An off-axis hydrothermal vent field near the Mid-Atlantic Ridge at 30°N: Nature, v. 412, p. 145–149.

Katayama, I., Mukou, A., Iizuka, T., Maruyam, S., Terada, K., Tsutsumi, Y., Sano, Y., Zhang, R. Y., and Liou, J. G., 2003, Dating of zircon from Ti-clinohumite–bearing garnet peridotite: Implication for timing of mantle metasomatism: Geology, v. 31, p. 713–716.

Kobayashi, S., Miyake, H., and Shoji, T., 1987, A jadeite rock from Oosa-cho, Okayama Prefecture, southwestern Japan: Mineralogical Journal, v. 13, p. 314–327.

Kunugiza, K., Nakamura, E., Miyajima, H., Goto, A., and Kobayashi, K., 2002, Formation of jadeite-natrolite rocks in the Itoigawa-Ohmi area of the Hida marginal belt inferred from U-Pb zircon dating: 2002 Joint Annual Meeting of The Japanese Association of Mineralogists, Petrologists, and Economic Geologists, and The Mineral Society of Japan, Abstracts (in Japanese).

Ludwig, K. R., 2001, SQUID 1.02: Berkeley, CA, Berkeley Geochronology Center, Special Publication no. 2.

Ludwig, K. R., 2003, Isoplot 3: Berkeley, CA, Berkeley Geochronology Center, Special Publication, no. 4.

Miyajima, H., Matsubara, S., Miyawaki, R., and Hirokawa, K., 2001, Rengeite, $Sr_4ZrTi_4Si_4O_{22}$, a new mineral, the Sr-Zr analogue of perrierite from the Itoigawa-Ohmi district, Niigata Prefecture, central Japan: Mineralogical Magazine, v. 65, p. 111–120.

Miyamoto, T., and Yanagi, T., 1998, U-Pb dating of zircon in jadeite rock in Oosayama serpentinite melange from Okayama Pref., Southwest Japan: 1998 Joint Annual Meeting of the Japanese Association of Mineralogists, Petrologists, and Economic Geologists, and The Mineral Society of Japan, Abstract, p. 190 (in Japanese).

Powell, R., Holland, T. J. B., and Worley, B., 1998, Calculating phase diagrams involving solid solutions via non-linear equations, with examples using THERMOCALC: Journal of Metamorphic Geology, v. 16, p. 577–588.

Rubatto, D., and Hermann, J., 2003, Zircon formation during fluid circulation in eclogites (Monviso, Western Alps): Implications for Zr and Hf budget in subduction zones: Geochimica et Cosmochimica Acta, v. 67, p. 2173–2187.

Rubin, J. N., Henry, C. D., and Prince, J. G., 1989, Hydrothermal zircons and zircon overgrowths, Sierra Blance Peals, Texas: American Mineralogist, v. 74, p. 865–869.

Sinha, A. K., Wayne, D. M., and Hewitt, D. A., 1992, The hydrothermal stability of zircon: Preliminary experimental and isotopic studies: Geochimica et Cosmochimica Acta, v. 56, p. 3551–3560.

Shi, G. H., Cui, W. Y., Tropper, P., Wang, C. Q., Shu, G. M., and Yu, H., 2003, The petrology of a complex sodic and sodic-calcic amphibole association and its implications for the metasomatic processes in the jadeitite area in northwestern Myanmar, formerly Burma: Contributions to Mineralogy and Petrology, v. 145, p. 355–376.

Shoji, T., and Kobayashi, S., 1988, Fluid inclusions found in jadeite and stronalsite, and a comment on the jadeite-analcime boundary: Journal of Mineralogy, Petrology, and Economic Geology, v. 83, p. 1–8.

Tsujimori, T., 1997, Omphacite-diopside vein in an omphacitite block from the Osayama serpentinite melange, Sangun-Renge metamorphic belt, southwestern Japan: Mineralogical Magazine, v. 61, p. 845–852.

Tsujimori, T., 1998, Geology of the Osayama serpentinite melange in the central Chugoku Mountains, southwestern Japan: 320 Ma blueschist-bearing serpentinite melange beneath the Oeyama ophiolite: Journal of the Geological Society of Japan, v. 104, p. 213–231 (in Japanese with English abstract).

Tsujimori, T., and Itaya, T., 1999, Blueschist-facies metamorphism during Paleozoic orogeny in southwestern Japan: Phengite K-Ar ages of blueschist-facies tectonic blocks in a serpentinite melange beneath Early Paleozoic Oeyama ophiolite: The Island Arc, v. 8, p. 190–205.

Tsujimori, T. and Liou, J. G., 2004a, Coexisting chromian omphacite and diopside in tremolite schist from the Chugoku Mountains, SW Japan: The effect of Cr on the omphacite-diopside immiscibility gap: American Mineralogist, v. 89, p. 7–14.

Tsujimori, T., and Liou, J. G., 2004b, Metamorphic evolution of kyanite-staurolite-bearing epidote-amphibolitic rocks from the Early Paleozoic Oeyama belt, SW Japan: Journal of Metamorphic Geology, v. 22, p. 301–313.

Tsujimori, T., and Liou, J. G., 2005a, Eclogite-facies mineral inclusions in clinozoisite from Paleozoic blueschist, central Chugoku Mountains, SW Japan: Evidence of regional eclogite-facies metamorphism: International Geology Review, v. 47, p. 215–232.

Tsujimori, T., and Liou, J. G., 2005b, Low-pressure and -temperature K-bearing kosmochloric diopside from the Osayama serpentinite mélange, SW Japan: American Mineralogist, v. 90, in press.

Vard, E., and William-Jones, A. E., 1993, A fluid inclusion study of vug minerals in dawsonite phonolite sillis, Montreal, Quebec: Implications for HFSE mobility: Contributions to Mineralogy and Petrology, v. 113, p. 410–423.

Vavra, G., Schmid, R., and Gebauer, D., 1999, Internal morphology, habit and U-Th-Pb microanalysis of amphibolite-to-granulite facies zircons: Geochronology of the Ivrea zone (Southern Alps): Contributions to Mineralogy and Petrology, v. 134, p. 380–404.

Williams, I. S., 1998, U-Th-Pb geochronology by ion microprobe, *in* Mickibbeen, M. A., Shanks, W. C., III, and Ridley, W. I., eds., Application of microanalytical techniques to understanding mineralizing processes: Society of Economic Geology, Reviews in Economic Geology, v. 7, p. 1–35.

WESTERN NORTH AMERICA

Fast Cooling and Exhumation of the Valhalla Metamorphic Core Complex, Southeastern British Columbia

FRANK S. SPEAR

Department of Earth and Environmental Sciences, Rensselaer Polytechnic Institute, Troy, New York 12180

Abstract

High-grade, migmatitic paragneisses (820°C, 8 kbar peak conditions) of the Valhalla metamorphic core complex experienced the retrograde net transfer reaction (ReNTR):

garnet + melt + K-feldspar = sillimanite + biotite + plagioclase

during cooling and crystallization of the melt, resulting in an increase in the Fe/(Fe + Mg) of both garnet and biotite at their margins. Diffusion profiles generated by this increase in Fe/(Fe + Mg) have been modeled numerically. Optimal fits were obtained with an initial, short period of rapid cooling (> 200°C/m.y. for < 0.5 m.y.) followed by slower (but still rapid) cooling at 20–30°C/m.y. for 10–20 m.y.

Four generations of monazite have been observed. Monazite inclusions within garnet contain only generations 1 and 2, and generations 3–4 are found only in matrix monazites. Generations 1–2 record ages of 75–85 Ma, which is believed to represent the prograde metamorphism of the complex. A single age of ~105 Ma may represent an early contact metamorphic episode. Third-generation monazite is 60 ±2 Ma, and reflects near-peak conditions. Generation 4 monazite, which occurs only on the outer rims of monazite, has not been dated, but is believed to represent monazite produced during melt crystallization. A hornblende K-Ar age of 58±2 Ma suggests a cooling rate of several tens to several hundreds of degrees/m.y. immediately following the metamorphic peak.

The extremely rapid cooling (hundreds of degrees/m.y.) at ~60 Ma must have been tectonically induced and it is suggested that transport of the hot Valhalla complex onto cooler basement up a thrust ramp along the underlying Gwillim Creek shear zone was the cause. Two-dimensional thermal modeling indicates that thrust transport of several centimeters/year up a 20° ramp will produce the observed cooling rates. The subsequent slower cooling rate (tens of degrees/m.y.) is believed to have been the result of extensional unroofing by low-angle normal faulting along the Slocan Lake fault.

Introduction

THIS CONTRIBUTION examines the cooling history of the Valhalla complex based on a new interpretation of the origin of chemical zoning observed on the rims of garnet in paragneisses from the central part of the complex. In a previous contribution (Spear and Parrish, 1996), the petrologic cooling rate was constrained by models in which diffusion was driven only by retrograde Fe-Mg exchange reaction (ReER) between biotite and garnet; the exchange reaction mechanism was ensured by examining only inclusions of biotite within garnet. In this contribution, the implications of retrograde net transfer reaction (ReNTR) on the garnet rim composition and resultant diffusional zoning is modeled. The cooling rate required to match the garnet rim zoning pattern is considerably faster than previously believed (over 100°C/Ma), and a 2-D thermal model involving significant tectonic shortening along a thrust ramp prior to the onset of extension is presented as a probable explanation. The proposed rapid cooling is supported by new, in situ dates on monazite, which constrain the shortening event to have occurred over a very short time period (less than 1 Ma) at ~60 Ma. These results are entirely consistent with the conclusions of Spear and Parrish (1996) and Schaubs et al. (2002), who suggest cooling of the complex was controlled by thrusting onto a cold footwall.

Geologic Setting and Petrology

The Valhalla complex (Fig. 1) is one of several metamorphic core complexes in the Shuswap terrane of eastern British Columbia and vicinity. The complex (Fig. 1) is comprised of Cretaceous orthogneisses, Paleocene–Eocene granitoids, and paragneisses of unknown depositional age. The complex

554 FRANK S. SPEAR

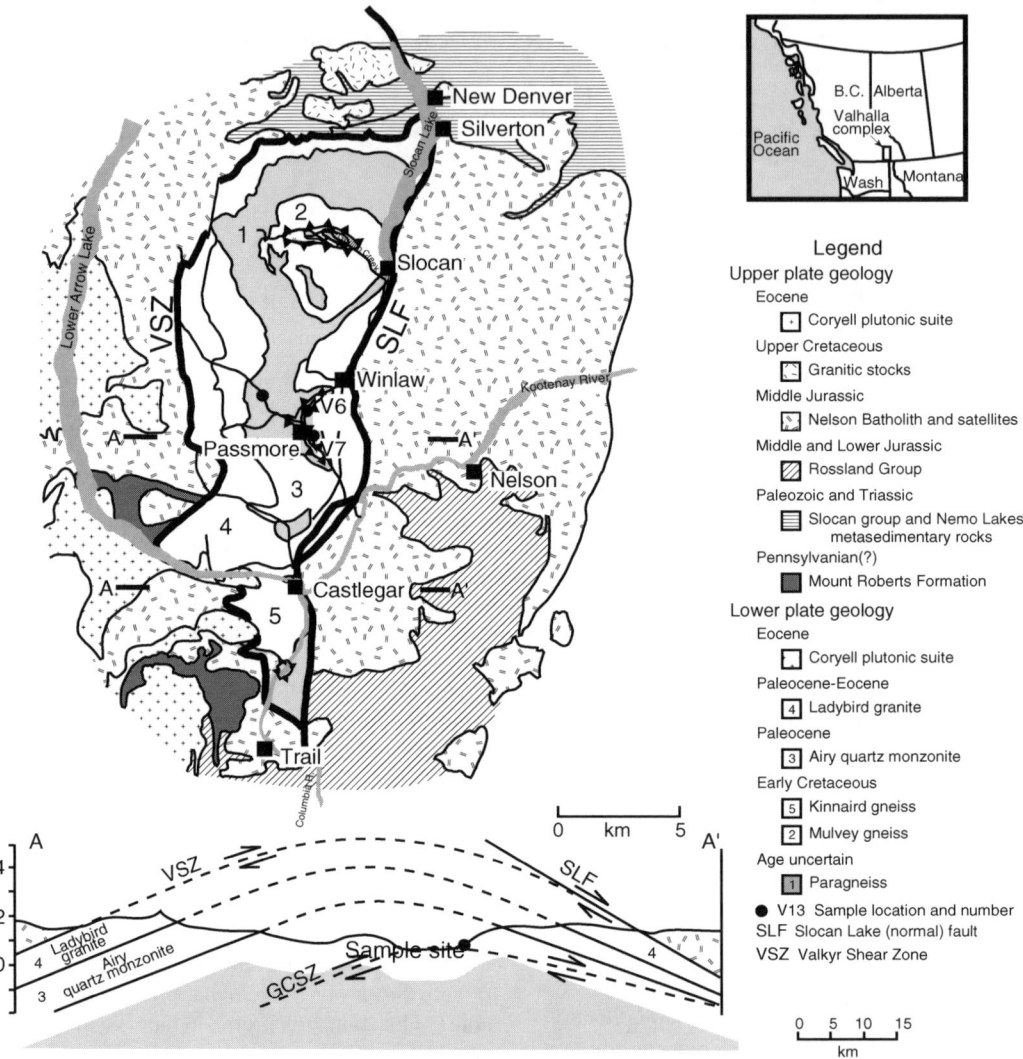

FIG. 1. Geological map of the Valhalla complex showing the location of samples described in this study (V6 and V7). Inset shows location of map in southeastern British Columbia. The lower panel shows a cross section along line A-A'. Note the proximity of the samples to the overhanging Slocan Lake fault and underlying Gwillim Creek Shear Zone (GCSZ).

is bounded on the east by the Slocan Lake low-angle normal fault and on the north, west, and south by the Valkyr shear zone, both with shear indicators indicating top down to the east. The complex is floored by the Gwillim Creek shear zone, which occurs at the deepest exposed structural levels throughout the complex. A Lithoprobe traverse across the complex identifies the Gwillim Creek shear zone as a major crustal reflector (Cook et al., 1988).

Paragneisses of the Valhalla complex experienced peak metamorphic conditions of approximately 820°C, 0.8 GPa at 65–75 Ma (Fig. 2; see also Spear and Parrish, 1996; Schaubs et al., 2002). Exhumation of the complex commenced at ~57 Ma (Carr et al., 1987) by low-angle normal faulting along the Valkyr shear zone and the Slocan Lake fault. The average cooling rate during this exhumation is well constrained by several geochronologic

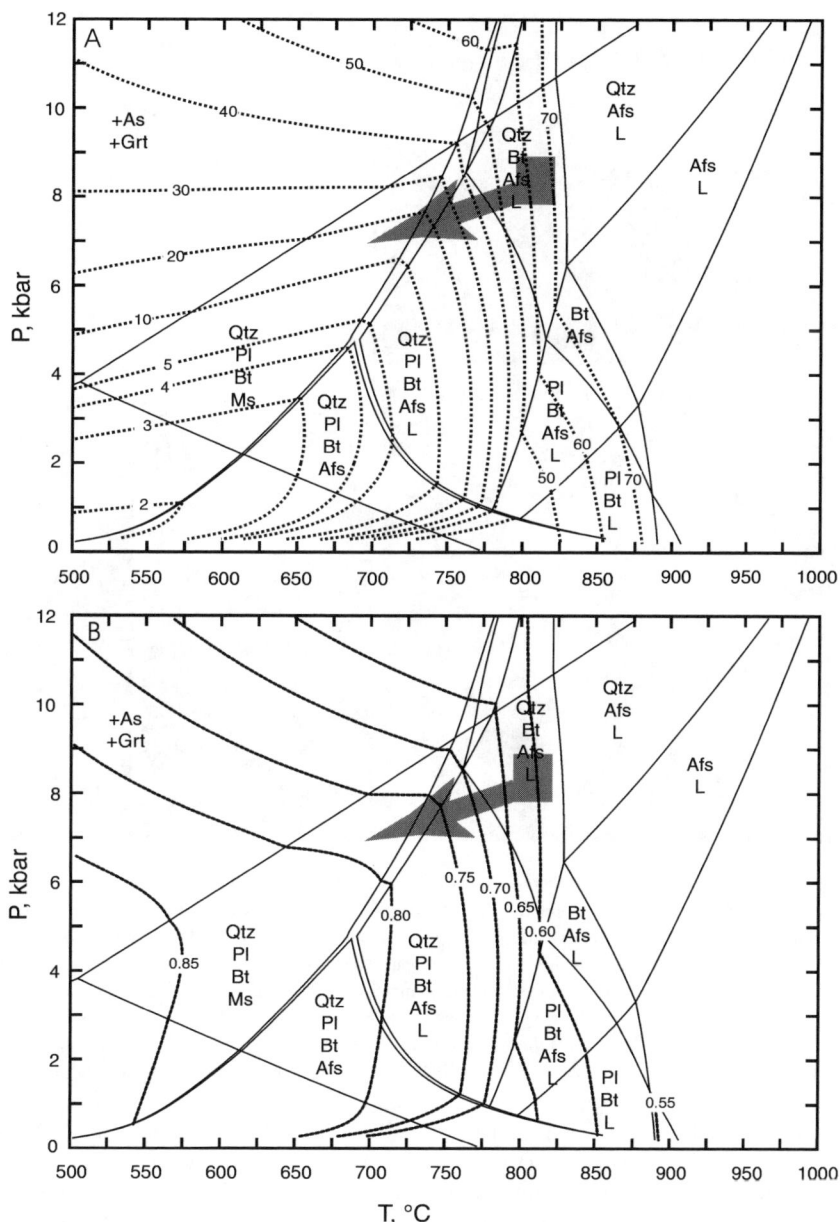

FIG. 2. P-T pseudosection (stability diagram) for a typical paragneiss bulk composition. Fields are labeled with stable mineral assemblages (all assemblages contain Al_2SiO_5 + garnet). The peak P-T conditions and retrograde P-T path are shown in gray. A. Diagram contoured for garnet abundance (millimoles garnet/100 cm³ of rock). B. Diagram contoured for Fe/(Fe+Mg) in garnet. Note that along the retrograde P-T path (gray arrow), garnet is consumed and the Fe/(Fe+Mg) of the remaining garnet increases.

systems to around 25°C/m.y. (Carr et al., 1987; Parrish et al., 1988, Parrish, 1995; Spear and Parrish, 1996).

The samples selected for detailed study (V6B and V7C) are from the Passmore gneiss. The location of the samples with respect to major faults and

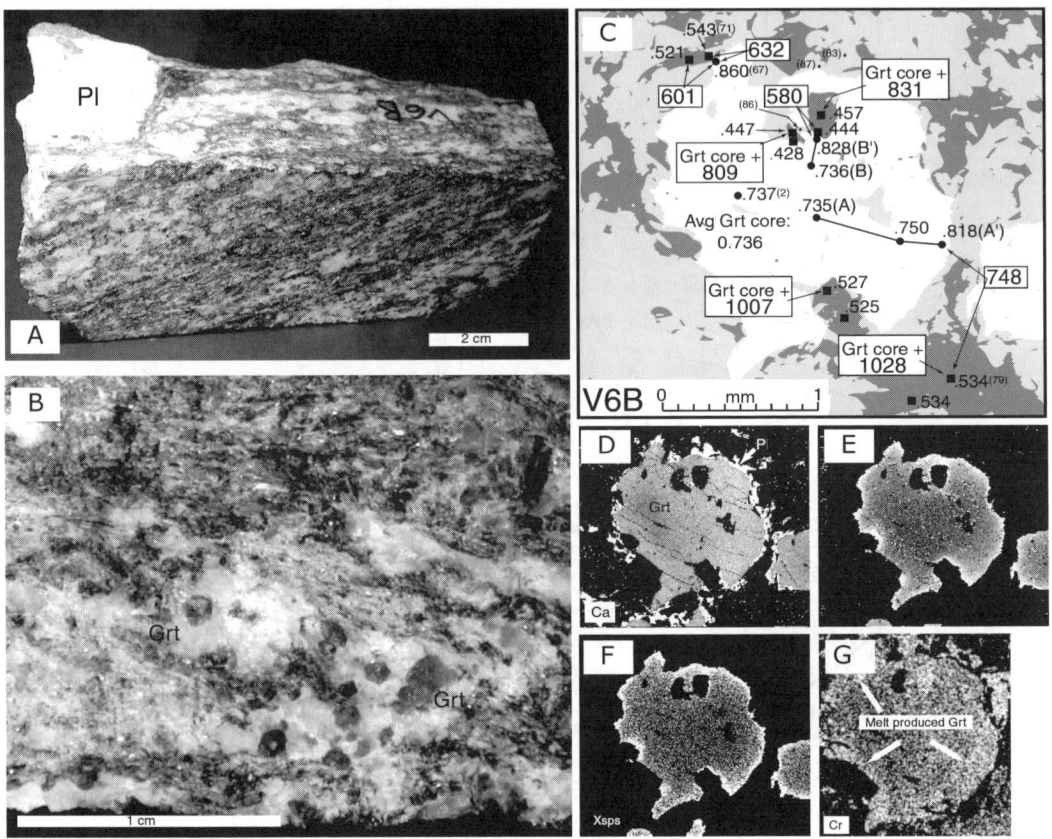

FIG. 3. A and B. Hand sample photographs of migmatite from sample V6B. Note garnet-bearing pods of leucosomes in B. C. Sketch of part of a thin section from sample V6B showing Fe/(Fe + Mg) of garnet and biotite (dark grey). Numbers in boxes are temperatures calculated from Fe-Mg exchange thermometry based on analyses from the garnet and biotite spots indicated. Note that the temperature obtained from the garnet core + matrix biotite is in excess of 1000°C (peak T ~820°C) and the temperature obtained from the garnet rim + matrix biotite is 748°C (the T at which reaction 1r ceased). D–G. X-ray maps of garnet shown in (C) showing zoning in Ca (D), Fe/(Fe + Mg) (E), Mn (F), and Cr (G). Ca and Cr zoning are growth zoning; Cr zoning shows the part of the garnet produced from prograde reaction 1. Fe/(Fe + Mg) and Mn zoning are diffusion zoning; the Mn zoning is evidence that the ReNTR (1r) has consumed garnet during cooling.

shear zones is important for the thermal modeling presented later and warrants discussion. The samples are located within domain P-I of Schaubs et al. (2002), within and immediately above highly sheared migmatites that are interpreted as splays of the Gwillim Creek shear zone. The samples are also located approximately four kilometers beneath the Slocan Lake fault (Fig. 1 cross section; see also Schaubs et al., 2002).

Reaction History

A pseudosection depicting the stable mineral assemblages for a paragneiss similar to those of the Valhalla complex is shown in Figure 2, contoured for moles of garnet (Fig. 2A) and Fe/(Fe + Mg) in garnet (Fig. 2B). The calculated P-T conditions are well within the field for dehydration melting, and paragneisses from the Valhalla complex contain abundant leucosomes (Fig. 3A). Garnet is found in both melanosomes and leucosomes (Fig. 3B), indicating that at least some garnet grew by the reaction

$$\text{sillimanite + biotite + plagioclase + quartz =}$$
$$\text{garnet + melt + K-feldspar,} \quad (1)$$

consistent with the garnet molar isopleths (Fig. 2A). Spear and Parrish (1996) identified the parts of garnet that had grown by reaction 1 as having higher Ca than the garnet cores (Figs. 3B and 5B). The Ca zoning in Figure 3D is a bit ambiguous, but is interpreted as reflecting growth zoning based on other samples. The Cr zoning (Fig. 3G) clearly shows a rim of higher Cr, which is interpreted as garnet produced by reaction 1.[1] Fe, Mg, and Mn display only diffusional zoning (Figs. 3E and 3F).

There is strong evidence in these paragneisses for progress of both the retrograde net transfer reactions (ReNTR)

garnet + melt + K-feldspar =
sillimanite + biotite + plagioclase + quartz   (1r)

and the retrograde exchange reaction (ReER)

pyrope + annite = almandine + phlogopite.   (2r)

ReER (2r) is especially obvious around inclusions of biotite within garnet (e.g., Fig. 3E), where the Fe/(Fe+Mg) increases toward the inclusion. Evidence for reaction (1r) operating in a retrograde sense comes from several sources, as indicated below.

1. Garnets are embayed on their margins (e.g., Figs. 3D, 3E, and 4), consistent with crossing the garnet molar isopleths along the retrograde path shown in Figure 2A.

2. Zoning of elements that display growth zonation (e.g., Ca and Cr) is truncated (e.g., Figs. 3D and 3E).

3. Fe, Mg, and Mn zoning is symmetric around garnet (c.f. Tracy et al., 1976).

4. Mn is zoned at the garnet rim, indicating garnet resorption.

5. ReNTR products (sillimanite + biotite + plagioclase + quartz) are concentrated at the garnet rim. This can be seen in Figure 3D where plagioclase decorates the upper right-hand side of the garnet crystal. Figure 4 displays ReNTR products, including quartz + plagioclase symplectite at the garnet rim.

6. Thermometry using garnet core and matrix biotite results in temperatures significantly higher than the peak temperature (i.e., over 1000°C; Fig. 3C). This occurs because matrix biotite is driven toward more Fe-rich compositions as a result of the ReNTR.

---

[1]See Spear and Kohn, 1996, for discussion of the partitioning arguments for this zoning behavior.

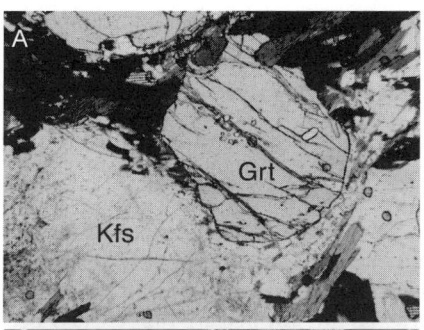
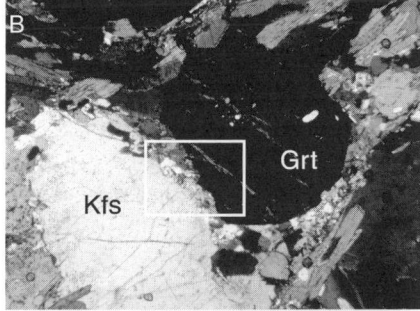
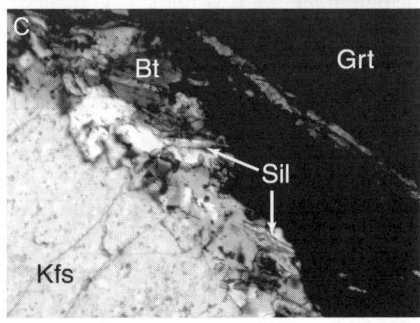

FIG. 4. Sample V6B; photomicrographs of textures produced by ReNTR 1r. A. Garnet rim is resorbed everywhere adjacent to matrix. B. Same view as (A) in crossed polars. Box shows location of (C). C. Close-up view (crossed polars) showing sillimanite + biotite + quartz + plagioclase symplectite produced along the garnet rim by operation of ReNTR 1r.

7. Garnet Fe/(Fe + Mg) increases towards the rim, consistent with the isopleths in Figure 2B, and biotite also becomes more Fe-rich as the garnet rim is approached. This zoning is consistent with a ReNTR but not a ReER.

In summary, biotite inclusions within garnet experienced only retrograde Fe-Mg exchange with enclosing garnet. In contrast, the rims of most garnets in sillimanite-bearing paragneisses are zoned as the result of diffusion driven by compositional changes on the garnet rim caused by the retrograde net transfer reaction (1r).

A comparison of the zoning profiles produced by ReNTR (1r) and ReER (2r) is shown in Figure 5. Although broadly similar, zoning produced by ReER (2r) has a slightly shorter penetration distance for Fe/(Fe+Mg), and does not show significant Mn zoning at the boundary. Another significant difference is that locally biotite modified by ReER (2r) becomes more Mg-rich towards the adjacent garnet, whereas biotite affected by ReNTR (1r) becomes more Fe-rich toward garnet. It should be pointed out that on the garnet rim where ReNTR (1r) was operative, equilibrium described by ReER (2r) was also maintained, the significant point being that the boundary composition was controlled by both reactions 1r and 2r acting together. The significance of identifying the correct retrograde reaction history for diffusion modeling will be discussed in more detail below.

Gwillim Creek shear zone. The timing of the deformation along the Gwillim Creek shear zone is important for the tectonic history discussed below, so evidence that relates the movement along the shear zone to the rock history is critical. A photomicrograph showing the texture of a sample from locality 6 (sample V6E) was presented by Spear and Parrish (1996, Fig. 8). The onset of shearing relative to onset of ReNTR (1r) is constrained by the presence of biotite produced by this reaction in the pressure shadows of deformed garnets (Fig. 8C in Spear and Parrish, 1996). Therefore, some deformation must have occurred before ReNTR (1r) went to completion.

The cessation of shearing relative to the ReNTR is constrained by the following observations. Sillimanite and biotite produced by the ReNTR (1r) are aligned in the shear fabric, some leucosomes produced by the ReNTR are aligned in the foliation, and garnet is broken and has a sigmoid shape. These observations indicate that deformation continued after some progress on the ReNTR had occurred. Cessation of major shearing in sample V6E must have occurred before or soon after the final completion of reaction (1r) because the garnet diffusion zoning produced in response to reaction (1r) is not disrupted. Biotite adjacent to garnet (Figs. 8B and 8C in Spear and Parrish, 1996) is zoned with higher Fe/Mg closest to the garnet, indicating that little or no retrograde Fe-Mg exchange occurred after cessation of reaction (1r). Garnet-biotite exchange thermometry on the garnet-biotite pair in closest proximity (adjoining rim analyses) yields 713°C (Hodges and Spear, 1982 calibration), indicating that this is the temperature of cessation of reaction (1r). Not coincidentally, this is also the approximate temperature of the solidus in this sample, suggesting that reaction (1) ceased when the last melt crystallized. Consequently, shearing must have ceased at a temperature slightly below the solidus. In summary, the available evidence suggests that shearing on the Gwillim Creek shear zone was more or less coincident with progress of ReNTR (1), and that shearing ceased at conditions near the solidus.

Diffusion Modeling

Since the pioneering study of Lasaga et al. (1977), a number of papers have used diffusion modeling of zoning in garnet to constrain the cooling history of metamorphic rocks. To the author's knowledge, all of these studies with the exception of Spear and Florence (1992) assumed Fe-Mg exchange to be the reaction responsible for the changing boundary concentration on the garnet rim that drove diffusion. As discussed above, ReER (2r) was the only reaction that operated between biotite inclusions and enclosing garnet, but ReNTR (1r) (with ReER 2r) operated between the garnet rim and the matrix, and is responsible for the diffusive zoning on the garnet rim.

The significance of this difference for diffusion modeling can be seen in the T-X section of Figure 6A. Garnet rim compositions change along two different T-X paths, depending on the reaction(s) that are operative. Specifically, the Fe/(Fe + Mg) of garnet rim changes much more gradually (dX/dT is smaller) if ReER (2r) is the only reaction operating. For example, consider the T-X curves in Figure 6A starting at a peak temperature of 820°C and a garnet with a core Fe/(Fe + Mg) of 0.645 and a hypothetical rim Fe/(Fe + Mg) of 0.78. If only ReER (2r) has operated on the garnet rim, then the rim composition would not be realized until the temperature reaches < 500°C. In contrast, if ReNTR (1r) has operated, the garnet rim composition is realized when the temperature reaches ca. 760°C.

The effect on the diffusion modeling can be seen in Figure 6B. A cooling rate of 100°C/m.y. generates a diffusion penetration distance of ~400 μm with the ReNTR as the boundary condition, whereas the same cooling rate with the ReER as the boundary condition only generates a penetration distance of 100 μm or so. Consequently, attempts to model a

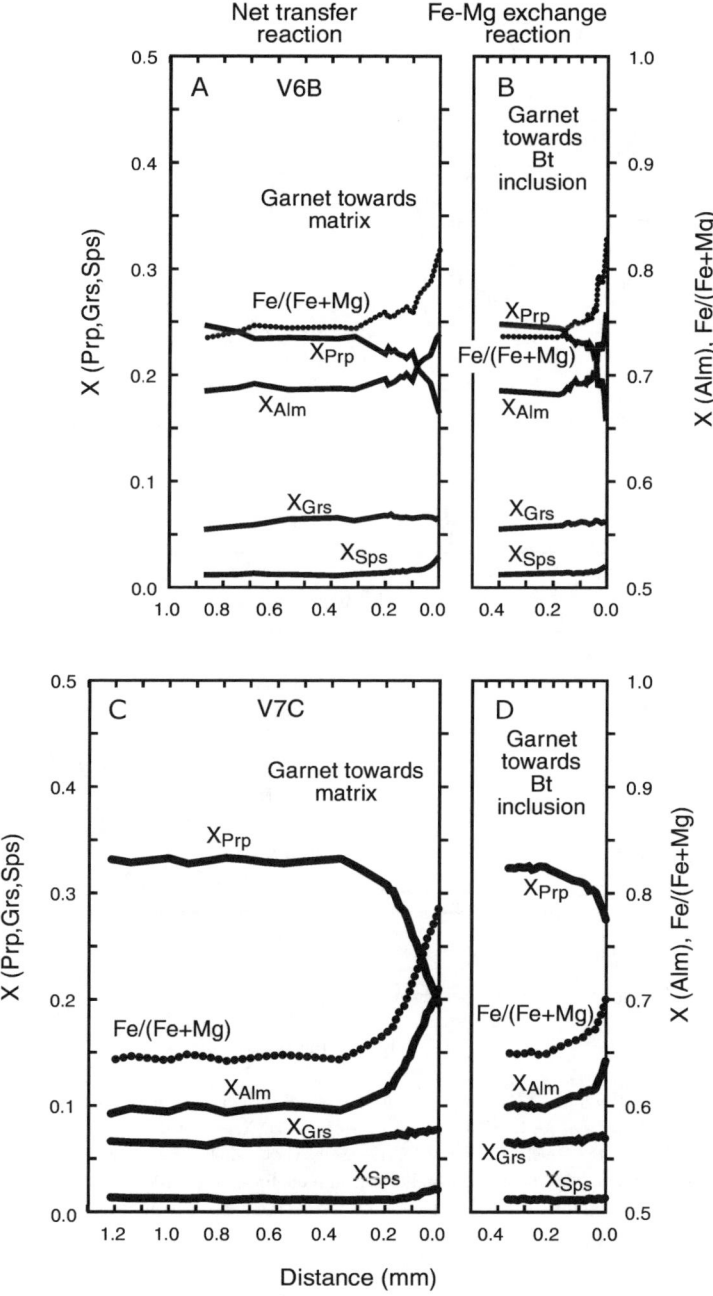

FIG. 5. Line traverses showing garnet zoning from samples V6B (A, B) and V7C (C, D). A and C. Core to rim traverses. Note upturn of Mn on the garnet rim toward the matrix, indicating that ReNTR 1r has operated at this interface. B and D. Traverse toward a biotite inclusion within garnet. Note the absence of significant increase in Mn toward the biotite inclusion, indicating that only ReER 2r has been operative at this interface.

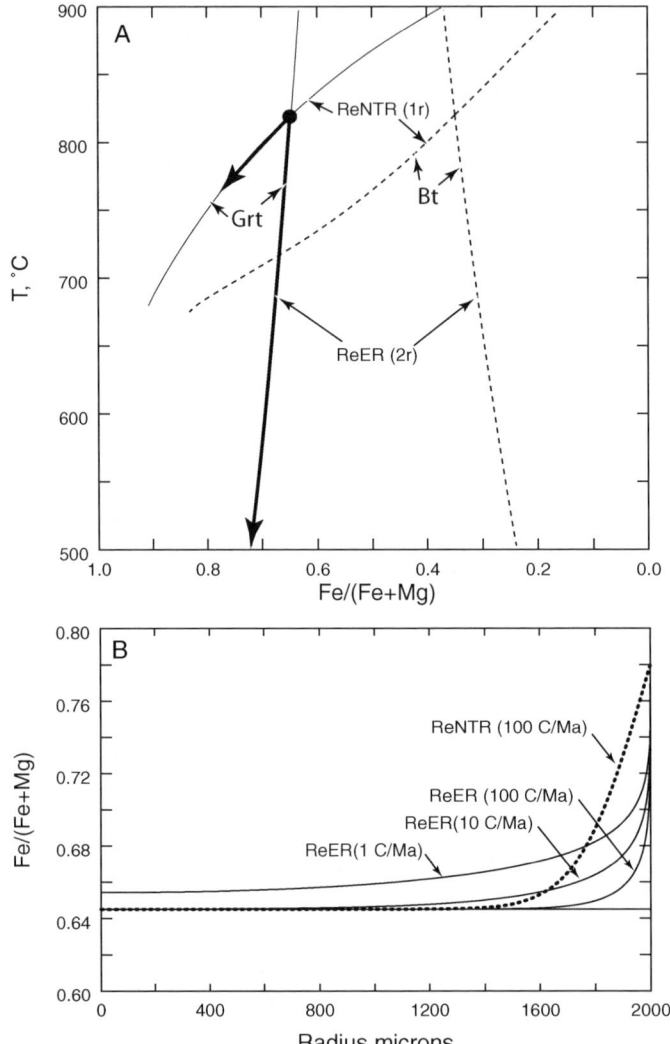

FIG. 6. A. T-X diagram (P = 8 kbar) showing how the Fe/(Fe + Mg) in garnet (solid lines) and biotite (dashed lines) change with temperature based on the ReNTR 1r and the ReER (2r). Note that dX[Fe/(Fe + Mg)]/dT is significantly larger for the ReNTR than for the ReER. B. Plot of Fe/(Fe + Mg) versus distance from garnet rim, showing calculated diffusion profiles assuming a ReNTR boundary condition with a cooling rate of 100°C/m.y. (dotted line) and a ReER boundary condition at three different cooling rates (solid lines). Note the difference in the shapes of the profiles, and the marked difference in diffusion penetration distance.

particular zoning profile using a ReER as the boundary condition will always result in a significantly slower cooling rate than modeling the same profile using a ReNTR as the boundary condition.

Based on the above discussion, it is clear that correct assignment of the boundary condition governing reaction (ReNTR versus ReER) is one of the most critical factors in modeling garnet zoning profiles. The problem can be abstracted somewhat by realizing that the key factor is the temperature at which the boundary condition is set—i.e., the T-X trajectory of the garnet rim.

Constraints on the T-X path of a garnet rim that is governed by a ReER are relatively straightforward because the exchange reaction can be used to calculate the garnet rim composition based on the ΔH of reaction and mass balance assumptions (i.e., how much biotite is present in the sample) (Lasaga et al., 1977; Lasaga, 1983). Figure 6A shows the exchange reaction T-X path for sample V7C calculated for the garnet-biotite exchange thermometer. The path is nearly linear, and suggests that the garnet rim composition around biotite inclusions (where only ReERs are operative) was set at less than 600°C.

Constraints on the T-X path of a garnet rim that is governed by a ReNTR can be obtained by: (a) thermodynamic modeling of the T-X behavior of the inferred governing reaction (e.g., Figs. 2A and 6A); (b) direct measurement of the temperature at which the rim composition was set using thermometry; and (c) inferences based on phase equilibria considerations. Thermodynamic modeling of sample V7C (Fig. 6A) indicates that the rim Fe/(Fe + Mg) was reached at a temperature of approximately 760°C. The garnet rim–matrix biotite temperature yielded by this sample is 709°C (see Fig. 5D of Spear and Parrish, 1996) and the solidus temperature is ~710°C.[2] Thermodynamic modeling of sample V6B (not shown) suggests the rim Fe/(Fe + Mg) was established at ~785°C; the garnet-biotite temperature for the garnet rim/matrix biotite is 748°C (see Fig. 4D of Spear and Parrish, 1996) and the solidus temperature is again ~710°C. The approach taken here will be to model the garnet rim zoning using the extremes of the boundary T-X constraints, which will yield a maximum and minimum cooling rate for each sample.

Results

An explicit finite difference diffusion model (written in FORTRAN) was used for all calculations. The program is designed so that the T-X path of a garnet rim composition can be specified as starting and ending T-X points, with intermediate compositions calculated by linear interpolation. Linear cooling rates are used. Several attempts at modeling the zoning profiles using a single cooling rate resulted in poor fits; therefore, a strategy of breaking the T-t history into a number of linear segments with different cooling rates was adopted, with significantly improved fits. Diffusivities of Chakraborty and Ganguly (1992) were used for all calculations.

Results of the diffusion models are shown in Figures 7 and 8 and Table 1. As discussed above, the cooling rates required to match the diffusion profiles are particularly sensitive to the T-X relation for the garnet rim. To explore the range of possible cooling rates, two models of the ReNTR were run for each sample. The first uses the highest temperature estimate for the establishment of the rim boundary condition (i.e., 785°C and 762°C for V6B and V7C, respectively), which comes from the calculated T-X diagram (e.g., Fig. 6). The second ReNTR model assumes that the final rim condition was not set in until the solidus was reached at approximately 710°C. The first ReER models used the temperature inferred from the calculated T-X diagram (e.g. 440°C and 550°C, Fig. 6) and the second models were based on calculated garnet-biotite temperatures along mutual contacts in the sample (e.g. 550°C and 582°C).

Because the garnet rim composition controlled by the ReNTR "freezes in" at high temperature, it is the high-temperature cooling history that is most sensitive to the ReNTR diffusion models. Similarly, the ReER freezes in at lower temperature, and it is this part of the T-t history that is most sensitive to the ReER diffusion models.

The high-temperature T-t histories determined for each of the four ReNTR models are relatively similar (Fig. 8A). Each shows an initial pulse of rapid cooling followed by more gradual cooling to ca. 40 Ma. The cooling rates required to match the zoning profiles vary somewhat (Fig. 8B), but are all in excess of 200°C/m.y. One model (V6B ReNTR 785) requires a cooling rate of 1000°C/m.y. for approximately 0.1 Ma to achieve a good fit.

The lower temperature T-t histories as constrained by the ReERs are similar to each other (dashed lines in Fig. 8A), but reflect somewhat slower cooling than suggested by models of the ReNTRs. Indeed, the cooling rates inferred from the ReERs are inconsistent with the available geochronology (dots in Fig. 8A), and with the rocks being at ambient temperature at the surface today. The reason for this is unknown, but the only way to increase the cooling rate over this interval using the same diffusivities is to increase the temperature at which the garnet rim temperature was "frozen." Values of approximately 650°C yield cooling rates of ~25°C/m.y., consistent with the geochronology.

Discussion

Spear and Parrish (1996) presented results of a diffusion model in which only biotite inclusions within garnet were modeled, restricting the type of retrograde reaction to the garnet-biotite Fe-Mg

---

[2]Note that the ReNTR requires the presence of melt so the solidus is the lower T limit.

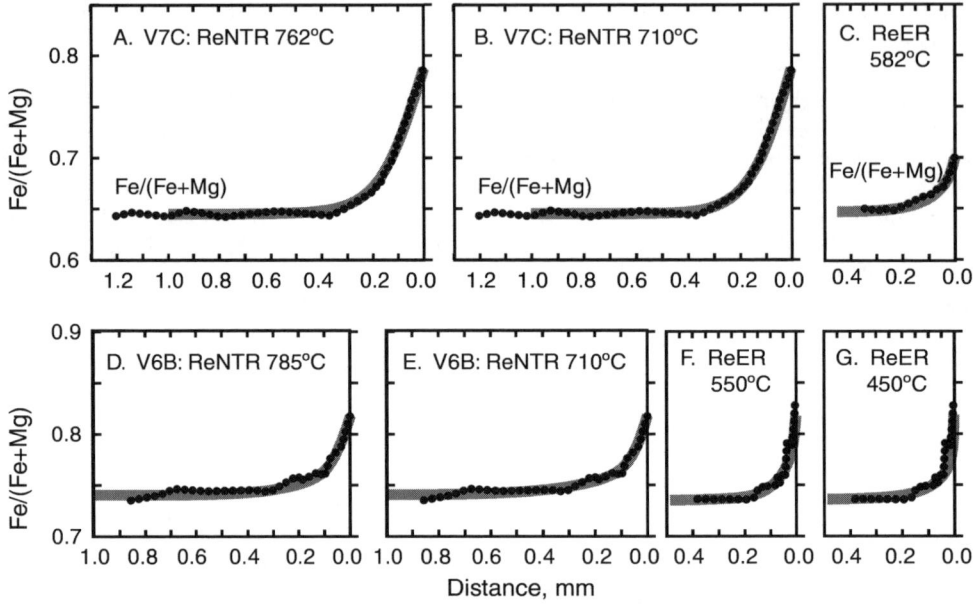

FIG. 7. Plots of Fe/(Fe + Mg) versus garnet radius (distance in mm) showing measured Fe/(Fe + Mg) zoning profiles (dotted lines) and diffusion model fits (solid lines). A–C. Sample V7C. D–G. Sample V6B. Each model begins at 820°C with a homogeneous garnet and assumes a different T-X (or X-t) path for the boundary condition (Fe/(Fe + Mg) in garnet). Models (A), (B), (D), and (E) assume the ReNTR boundary condition and models (C), (F), and (G) assume the ReER boundary condition. In all cases, the rim garnet Fe/(Fe + Mg) is attained at the temperature indicated in the panel.

ReER (2r). Best-fit cooling rates based on these calculations were ~25°C/m.y. (range of 3–80°C/m.y.) for rocks from the same locality studied here and using the same diffusivities (Chakraborty and Ganguly, 1992). These values are completely consistent with the cooling rates obtained for ReERs in the present study.

Ducea et al. (2003) also modeled the zoning on the rim of garnet from sample V7C and obtained a cooling rate of 2–13°C/m.y. based on the assumption that the garnet rim composition was controlled by ReER (2r). This result is consistent with the modeling presented here, but our present interpretation of the garnet rim is that it was controlled by ReNTR (1r), and the initial cooling rate should be 1–2 orders of magnitude faster.

Geochronology

Age dating

Monazite occurs in a number of textural settings in migmatites from the Valhalla complex. Figure 9A shows a BSE image of a paragneiss containing two large garnet crystals. Monazite is located both inside garnet (grain 4) and in the matrix (grains 3 and 5).

Chemical zoning (especially Y) is different in each of the texturally distinct monazites (Figs. 9C–9E), and at least four distinct generations of monazite have been identified based on the Y zoning. Monazite inside of garnet (grain 4; Fig. 9B) is comprised of high- and low-Y zones (zones 1 and 2, respectively). The same high- and low-Y patches (zones 1 and 2) can be seen in the monazites in the matrix (grains 5 and 3; Figs. 9C–9D). Matrix monazites have an additional zone outboard from zones 1 and 2 that contains the lowest Y values (zone 3), and locally tips on grains that are intermediate Y (zone 4). Pyle and Spear (2003) documented four generations of monazite growth in samples of similar metamorphic grade from central New Hampshire.

In situ ages were obtained using the IMS 1270 ion microprobe at Woods Hole Oceanographic Institute and UCLA. Procedures were the same as those discussed by Spear et al. (2004). The Pb/Th ages are reported here.

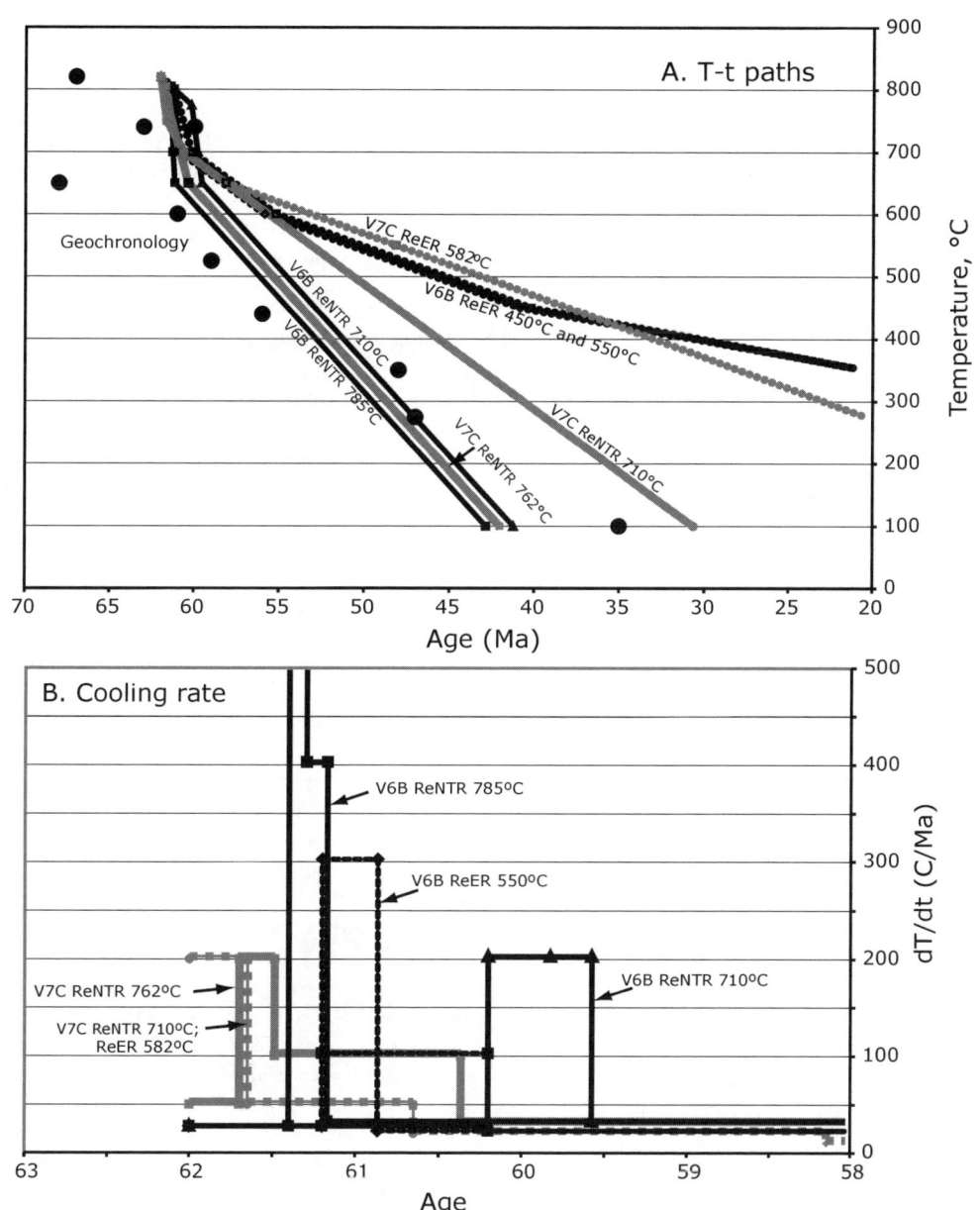

FIG. 8. Summary of results of diffusion models. A. T versus t plot. B. dT/dt versus t plot. Each curve is one of the diffusion models shown in Figure 7. Solid lines are models with ReNTR boundary conditions and dotted lines are models with ReER boundary conditions. Dots in (A) are T-t constraints from geochronology. Note that the ReER models give cooling rates that are slower than those assuming ReNTR boundary condition, and are not consistent with cooling rates from geochronology.

Zone 1–2 monazite ages span the range 75–85 Ma (Fig. 9). Although the statistical uncertainty in the ages (ca. ± 1 Ma) is smaller than the spread, the relative uncertainties must be larger owing to effects other than counting statistics (e.g., sample charging), so it is concluded that zone 1–2 ages are distinguishable.

The age of zone 3 monazite is significantly younger (~60 Ma) and, unfortunately, no zone 4

TABLE 1. T-t Histories and Cooling Rates
Used in Diffusion Models

	T	t, m.y.	DT/dt
	V6B		
ReNTR 785	820	0.00	
	805	0.60	25
	700	0.11	1000
	650	0.13	400
	100	18.33	30
ReNTR 710	820	0.00	
	775	1.80	25
	700	0.38	200
	650	0.25	200
	100	18.33	30
ReER 550	820	0.00	
	800	0.80	25
	700	0.33	300
	600	5.00	20
	450	15.00	10
	100	70.00	5
ReER 450	820	0.00	
	800	0.80	25
	700	1.00	100
	600	5.00	20
	450	15.00	10
	100	70.00	5
	V7C		
ReNTR 762	820	0.00	
	805	0.30	50
	762	0.22	200
	650	1.12	100
	100	18.33	30
ReNTR 710	820	0.00	
	750	0.35	200
	700	1.00	50
	650	2.50	20
	100	27.50	20
ReER 582	820	0.00	
	750	0.35	200
	700	1.00	50
	650	2.50	20
	550	10.00	10
	100	45.00	10

monazite was dated. Based on comparison with the study of Pyle and Spear (2003), in which monazites from similar rocks are well characterized in terms of their paragenesis, the zone 4 monazite is interpreted to have grown during melt crystallization. Zone 3 monazite must, therefore, represent monazite that was produced either during or prior to melting reactions. It also requires that melt crystallization occurred later than approximately 60 Ma.

Discussion of age data

An extensive amount of geochronology has been done on the Valhalla complex (e.g., Parrish 1995; Parrish et al, 1988; summarized in Spear and Parrish, 1996 and Schaubs et al., 2002), much of which is multigrain TIMS analyses of monazite. The complex zoning displayed by the monazites examined in this study suggest that many of the previous analyses may represent mixed domains. There is no evidence in the Y zoning that diffusion has homogenized Y contents during peak metamorphism, and similar studies have revealed no Pb diffusional zoning (e.g., Crowley and Ghent, 1999). Therefore, it is inferred that the previous multi-grain TIMS ages on monazite are mixed, and do not reflect Pb loss (e.g., Harrison et al., 2002).

Spear and Parrish (1996) argued from these earlier data that the metamorphic peak in the Valhalla complex occurred at 67–80 Ma. The present results, on the other hand, indicate that cooling of the leucosomes could not have occurred until after ~60 Ma, and that anatexis may have been progressing as late as ~60 Ma.

A recent Sm-Nd age on the core and rim of garnet from sample V7C (Ducea et al., 2003) yielded 67.3 ± 2.3 for the core and 59.8 ± 2 Ma for the rim. Ducea et al. (2003) assumed that the peak metamorphic age was close to the core 67.3 Ma determination and that the rim age of 59.8 reflected Nd loss during relatively slow cooling of a few degrees C/m.y. Slow cooling from the metamorphic peak is inconsistent with the diffusion calculations presented here. However, it is entirely possible that slow cooling did occur between ca 67 and 60 Ma followed by a pulse of rapid cooling.

Thermal/Tectonic Model for Rapid Cooling of the Valhalla Complex

The diffusion modeling discussed above indicates that the zoning observed on garnet rims required a very rapid episode of cooling from the

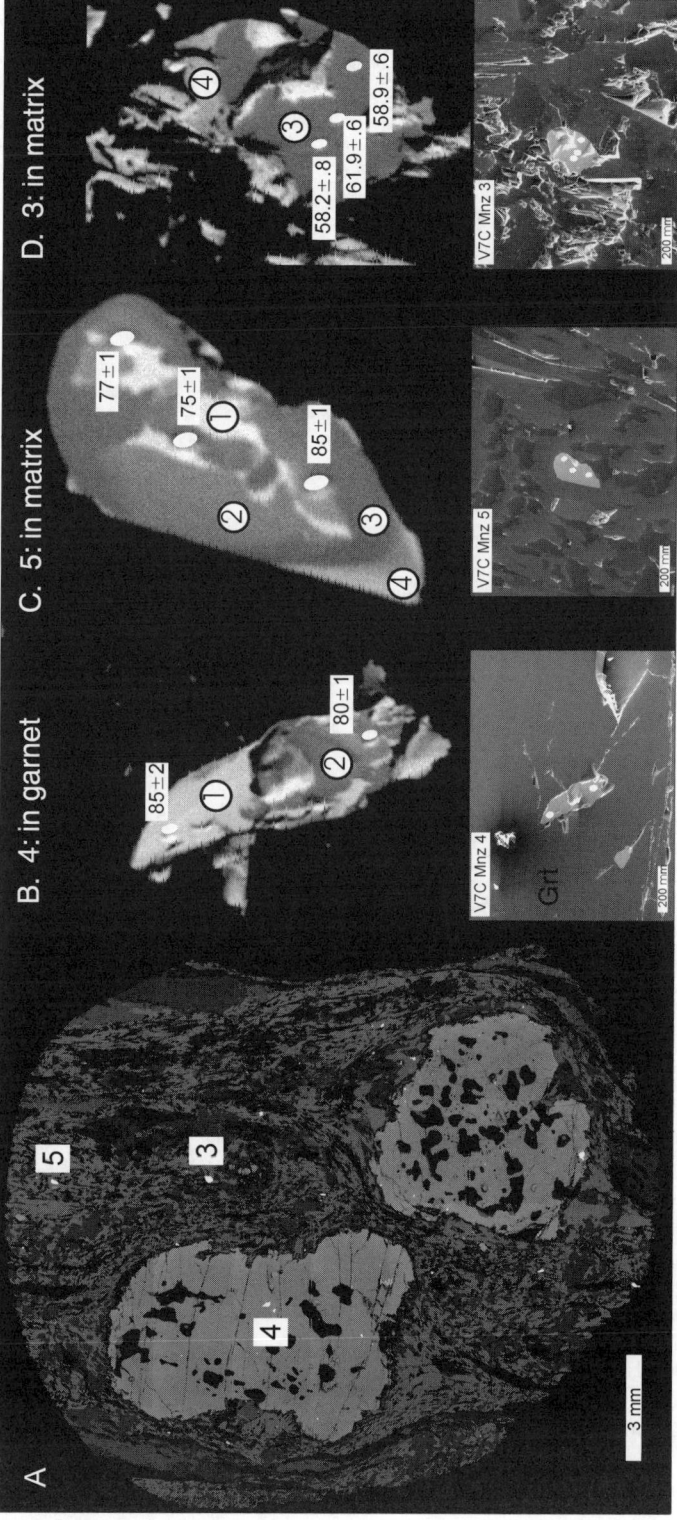

FIG. 9. Sample V7C. A. BSE image showing location of analyzed monazite crystals in sample. B–D. Upper panels are yttrium X-ray maps depicting the distribution of Y in analyzed monazites. Numbers in circles are monazite growth zones (1–4) based on Y zoning. Lower panels are SE images showing location of SIMS analysis spots (white ellipses). Ages are shown in boxes in the upper panels.

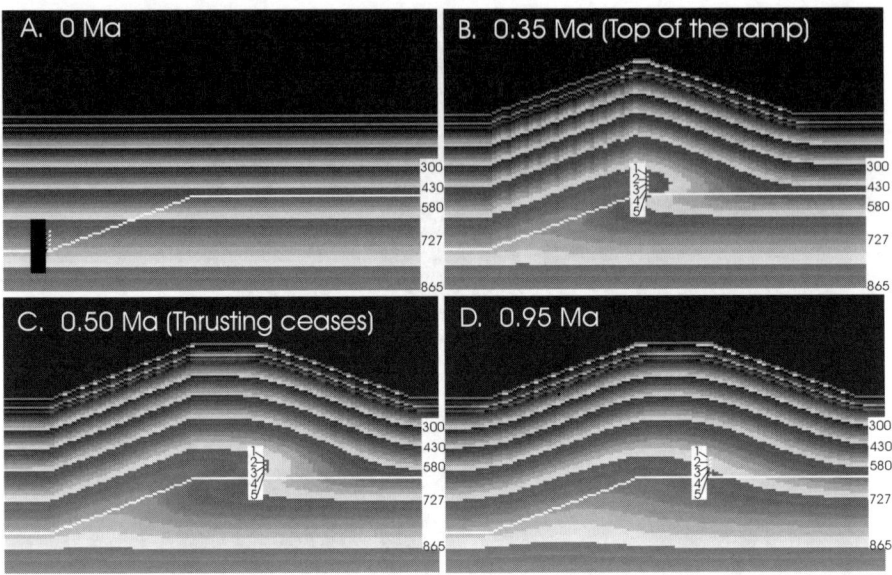

FIG. 10. Cross sections showing the evolution of two-dimensional thermal model with a 20° thrust ramp. Transport velocity is 10 cm/year for 0.5 m.y. Note the thermal overhang in (C). Temperatures of isotherms are given on the right side of each panel. Numbered dots (1–5) refer to rocks for which paths are plotted in Figure 11.

metamorphic peak during melt crystallization in order to preserve the zoning. Furthermore, in situ geochronology indicates that the crystallization of leucosomes did not begin until after ~60 Ma, thereby greatly compressing the time available to achieve cooling of the complex.

Rapid cooling at rates constrained by this study is only possible by mechanisms that involve tectonic juxtaposition of rocks of different temperature. Erosion as a means of cooling rocks (e.g., England and Richardson, 1977; England and Thompson, 1984) is limited by the thermal diffusivity of rocks and length scales involved (tens of kilometers of crust) and a maximum of a few tens of degrees/Ma is possible by this mechanism.

Tectonic denudation by normal faulting provides a mechanism for rapid exhumation and cooling of a core complex, and the Slocan Lake low-angle normal fault was certainly a major factor in the exposure of the Valhalla complex during the early Tertiary. However, the Slocan Lake fault cannot be responsible for the initial, very rapid cooling of the paragneisses for two reasons. First, rapid cooling of footwall rocks will only occur in rocks immediately below the fault. Rocks farther from the fault will cool more slowly as thermal conduction permits. The samples discussed here are 4 km structurally below the Slocan Lake fault, and would not, therefore, have cooled at the rates observed. Second, and most convincingly, exhumation by faulting results in P-T paths of footwall rocks that would first show nearly isobaric unloading, followed by cooling. The rocks of the Valhalla complex display no evidence whatsoever for decompression during cooling; rather the post-peak P-T path is nearly isobaric cooling (e.g., Fig. 2), thus ruling out the possibility that the Slocan Lake fault was responsible for the initial rapid cooling observed.

One possible explanation of the rapid cooling is that it was the result of transport of hot, high-grade rocks of the Valhalla complex up a thrust ramp onto a cooler footwall, which provided the heat sink for cooling the complex, as suggested by Spear and Parrish (1996) and Schaubs et al. (2002). Such a mechanism would result in nearly isobaric cooling paths, as observed. The rate of cooling would be determined by the rate of transport, the thrust ramp angle, the lateral thermal gradient (if any), and the proximity of the rocks in the hanging wall to the heat sink below the fault. Indeed, Schaubs et al. (2002) provide evidence that rocks closer to the Gwillim Creek shear zone cooled more rapidly than those farther away, consistent with this "footwall quenching" model.

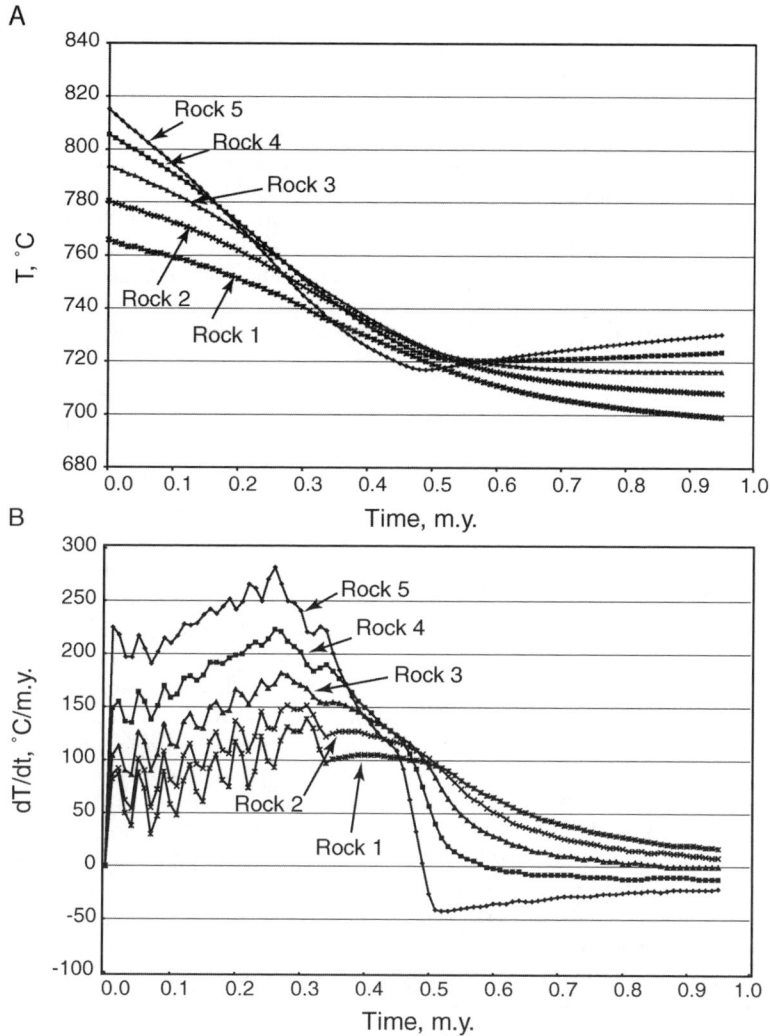

FIG. 11. Plots of (A) temperature versus time and (B) cooling rate versus time for the five rocks shown in the 2-D thermal model in Figure 10 at distances of 4.5, 3.5, 2.5, 1.5, and 0.5 km from the fault (rocks 1–5, respectively). Note that cooling rates of all rocks within 4 km of the fault are in excess of 100°C/m.y.

In order to determine whether this mechanism could produce the calculated cooling rates given reasonable input parameters, a two-dimensional thermal model was invoked (Fig. 10). The most important model parameters for determining the cooling rate are the thrust velocity and the ramp angle: the steeper the ramp and the faster the thrust rate, the faster the rocks will cool. Several models were examined in detail and a representative one with a ramp angle of 20° and a thrust velocity of 10 cm/year are shown in Figures 10 and 11. Thrusting at 10 cm/year up a 20° ramp results in a thermal overhang, as can be seen in Figures 10B and 10C, which then decays with time (Fig. 10D). Slower thrusting rates (e.g., 5 cm/year) or gentler ramps (e.g. 10° dip) do not produce thermal overhangs, but still produce rapid cooling.

Results are also shown in T versus t and dT/dt versus t plots (Fig. 11) for five rock positions located as shown in Figure 10 (4.5, 3.5, 2.5, 1.5, and 0.5 km above the thrust). Cooling is most rapid in rock 5 (located 0.5 km from the thrust) and for this model

(10 cm/year; 20° ramp) rock 5 cools in excess of 200°C/m.y. Even rocks several km from the thrust ramp (e.g., rock 1) cool in excess of 100°C/m.y. Note that there is a major inflection in the temperature-time plots (Fig. 11A) at the time in which thrusting ceases. Indeed, the rock closest to the fault (Rock 5) cools initially but then heats up for the duration of the model (to 1 m.y.) after thrusting ceases. This is a consequence of thermal relaxation following the formation of the thermal overhang (Figs. 10C and 10D). Late heating following thrusting is not suspected to have happened in the Valhalla complex because unroofing due to normal faulting and erosion will have mitigated this effect. However, such a phenomenon might be manifest in some terranes above thrust faults, resulting in differing P-T paths over short distances with no intervening structures.

The cooling-rate plots (Fig. 11B) show two inflections. The major inflection occurs at 0.5 m.y. when thrusting ceased (note that the cooling rate for Rock 5 becomes negative as the rock heats up). A second inflection occurs at ~0.35 m.y. when the rocks reach the top of the thrust ramp (Fig. 11B).

Additional models have been run to examine the range of possible thrust ramp angles and thrust velocities. For example, a thrust velocity of 2 cm/year up a 20° ramp results in cooling rates of ~70–80°C/m.y.; a thrust velocity of 10 cm/year up a 10° ramp results in cooling rates of 60–100°C/m.y. Because of the wide range of calculated cooling rates for the Valhalla complex (~100–1000°C/Ma), it was not deemed appropriate to attempt to fit the thermal history of the complex any more accurately than has been presented. The goal of this exercise was to demonstrate that with reasonable geological parameters, cooling rates of several hundreds of degrees/m.y. are expected. Based on the above thermal models, a thrust rate of several cm/year or faster up a thrust ramp of 10 degrees or more will produce the observed cooling rates greater than 100°C/m.y.

Discussion

Rapid cooling of the Valhalla complex through quenching by emplacement on a cool footwall ramp was argued by both Spear and Parrish (1996) and Schaubs et al. (2002); post-metamorphic juxtaposition of units of different metamorphic histories is also reported from the Monashee complex (Crowley and Parrish, 1999). This paper presents the first direct evidence that the cooling rate was in excess of 100°C/m.y., virtually requiring a thrust-ramp mechanism.

This is the period of Late Cretaceous shortening in the Omenica belt. The reason for such rapid thrust emplacement may be directly tied to the presence of melt in the Valhalla paragneisses. Hollister and Crawford (1986) suggested that major deformations might be significantly enhanced by the presence of melt (anatexis), and the petrology and geochronology of the Valhalla complex is consistent with this model. It is also probably no coincidence that deformation on the Gwillim Creek shear zone apparently ceased when the rocks cooled through the solidus. Indeed, it is entirely possible that the timing and amount of shortening in the Late Cretaceous was controlled by the existence of melt in the middle crust. Once the melt crystallized, further shortening was not possible and strain was partitioned elsewhere.

Extremely rapid cooling as has been documented here is not believed to be unique to the Valhalla complex. Other examples of high-grade garnets with similar diffusion zoning have been described in the literature and, based on the discussion here, are probably the result of net transfer reaction rather than exchange reaction as was assumed. If so, then based simply on the similarity of diffusion penetration distance of Fe/(Fe + Mg) in garnet from these samples, the cooling rates of these other samples must have been quite rapid. It is possible that rapid thrusting and subsequent quenching of migmatite terranes is a common phenomenon.

Acknowledgments

The author wishes to thank J. T. Cheney for invaluable insights during many late night discussions on the interpretation of pelitic schist phase equilibria and monazite geochronology. T. Mark Harrison and Graham Layne provided necessary access to and assistance with the SIMS analyses reported here. X-ray maps of monazite were collected by J. Pyle. Insightful and constructive reviews were provided by S. Carr and J. Ganguly.

REFERENCES

Carr, S. D., Parrish, R. R., and Brown, R. L., 1987, Eocene structural development of the Valhalla complex, southeastern British Columbia: Tectonics, v. 6, p. 175–196.

Chakraborty, S., and Ganguly, J., 1992, Cation diffusion in aluminosilicate garnets—experimental determination in spessartine-almandine diffusion couples, evaluation of effective binary diffusion coefficients, and applications: Contributions to Mineralogy and Petrology, v. 111, p. 74–86.

Cook, F. A., Green, A. G., Simony, P. S., Price, R. A., Parrish, R. R., Milkereit, B., Gordy, P. L., Brown, R. L., Coflin, K. C., and Patenaude, C., 1988, Lithoprobe seismic reflection structure of the southeastern Canadian Cordillera: Initial results: Tectonics, v. 7, p. 157–180.

Crowley, J. L., and Ghent, E. D., 1999, An electron microprobe study of the U-Pb-Th systematics of metamorphosed monazite: The role of Pb diffusion versus overgrowth and recrystallization: Chemical Geology, v. 157, p. 285–302.

Crowley, J. L., and Parrish, R., 1999, U-Pb isotopic constraints on diachronous metamorphism in the northern Monashee complex, southern Canadian Cordillera: Journal of Metamorphic Geology, v. 17, p. 483–502.

Ducea, M. N., Ganguly, J., Rosenberg, E. J., Patchett, P. J., Cheng, W., and Isachsen, C., 2003, Sm-Nd dating of spatially controlled domains of garnet single crystals: A new method of high temperature thermochronology: Earth and Planetary Science Letters, v. 213, p. 31–42.

England, P. C., and Richardson, S. W., 1977, The influence of erosion upon the mineral facies of rocks from different metamorphic environments: Journal of the Geological Society of London, v. 134, p. 201–213.

England, P. C., and Thompson, A. B., 1984, Pressure-temperature-time paths of regional metamorphism, Part I: Heat transfer during the evolution of regions of thickened continental crust: Journal of Petrology, v. 25, p. 894–928.

Harrison, T. M., Catlos, E. J., and Montel, J.-M., 2002, U-Th-Pb dating of phosphate minerals, in Kohn, M. J., Rakovan, J., and Hughes, J. M., eds., Phosphates: Geochemical, geobiological, and materials importance: Mineralogical Society of America, Reviews in Mineralogy and Geochemistry, no. 48, p. 523–558.

Hodges, K. V., and Spear, F. S., 1982, Geothermometry, geobarometry and the Al_2SiO_5 triple point at Mt. Moosilauke, New Hampshire: American Mineralogist, v. 67, p. 1118–1134.

Hollister, L. S., and Crawford, M. L., 1986, Melt-enhanced deformation—a major tectonic process: Geology, v. 14, p. 558–561.

Lasaga, A. C., 1983, Geospeedometry: An extension of geothermometry, in Saxena, S. K., ed., Kinetics and equilibrium in mineral reactions: New York, NY, Springer-Verlag, p. 81–114.

Lasaga, A. C., Richardson, S. M., and Holland, H. D., 1977, The mathematics of cation diffusion and exchange between silicate minerals during retrograde metamorphism, in Saxena, S. K., and Bhattacharji, S., eds., Energetics of geological processes: New York, NY, Springer-Verlag, p. 354–387.

Parrish, R., 1995, Thermal evolution of the southeastern Canadian Cordillera: Canadian Journal of Earth Sciences, v. 18, p. 944–958.

Parrish, R. R., Carr, S. D., and Parkinson, D. L., 1988, Eocene extensional tectonics and geochronology of the southern Omineca belt, British Columbia and Washington: Tectonics, v. 7, p. 181–212.

Pyle, J. M., and Spear, F. S., 2003, Four generations of accessory phase growth in low-pressure migmatites from SW New Hampshire: American Mineralogist, v. 88, p. 338–351.

Schuabs, P. M., Carr, S. D., and Berman, R. G., 2002, Structural and metamorphic constraints on ~70 Ma deformation of the northern Valhalla complex, British Columbia: Implications for the tectonic evolution of the southern Omineca belt: Journal of Structural Geology, v. 24, p. 1195–1214.

Spear, F. S., Cheney, J. T., Pyle, J. M., Harrison, M., and Layne, G., 2004, Monazite geochronology on a transect across central New England: Journal of Petrology, in review.

Spear, F. S., and Florence, F. P., 1992, Thermobarometry in granulites: Pitfalls and new approaches: Journal of Precambrian Research, v. 55, p. 209–241.

Spear, F. S., and Kohn, M. J., 1996, Trace element zoning in garnet as a monitor of crustal melting: Geology, v. 24, p. 1099–1102.

Spear, F. S., and Parrish, R., 1996, Petrology and petrologic cooling rates of the Valhalla Complex, British Columbia, Canada: Journal of Petrology, v. 37, p. 733–765.

Tracy, R. J., Robinson, P., and Thompson, A. B., 1976, Garnet composition and zoning in the determination of temperature and pressure of metamorphism, central Massachusetts: American Mineralogist, v. 61, p. 762–775.

Multi-stage Origin of the Coast Range Ophiolite, California: Implications for the Life Cycle of Supra-subduction Zone Ophiolites

JOHN W. SHERVAIS,[1]

Department of Geology, Utah State University, 4505 Old Main Hill, Logan, Utah 84322-4505

DAVID L. KIMBROUGH,

Department of Geological Sciences, San Diego State University, San Diego California 92182-1020

PAUL RENNE,

Berkeley Geochronology Center, 2455 Ridge Road, Berkeley, California 94709 and Department of Earth and Planetary Science, University of California, Berkeley, California 94720

BARRY B. HANAN,

Department of Geological Sciences, San Diego State University, San Diego California 92182-1020

BENITA MURCHEY,

United States Geological Survey, 345 Middlefield Road, Menlo Park California 94025

CAMERON A. SNOW,

Department of Geology, Utah State University, 4505 Old Main Hill, Logan, Utah 84322-4505 and Department of Geological and Environmental Sciences, Stanford University, Stanford, California 94305

MARCHELL M. ZOGLMAN SCHUMAN,[2] AND JOE BEAMAN[3]

Department of Geological Sciences, University of South Carolina, Columbia, South Carolina 29208

Abstract

The Coast Range ophiolite of California is one of the most extensive ophiolite terranes in North America, extending over 700 km from the northernmost Sacramento Valley to the southern Transverse Ranges in central California. This ophiolite, and other ophiolite remnants with similar mid-Jurassic ages, represent a major but short-lived episode of oceanic crust formation that affected much of western North America. The history of this ophiolite is important for models of the tectonic evolution of western North America during the Mesozoic, and a range of conflicting interpretations have arisen. Current petrologic, geochemical, stratigraphic, and radiometric age data all favor the interpretation that the Coast Range ophiolite formed to a large extent by rapid extension in the fore-arc region of a nascent subduction zone. Closer inspection of these data, however, along with detailed studies of field relationships at several locales, show that formation of the ophiolite was more complex, and requires several stages of formation.

Our work shows that exposures of the Coast Range ophiolite preserve evidence for four stages of magmatic development. The first three stages represent formation of the ophiolite above a nascent subduction zone. Rocks associated with the first stage include ophiolite layered gabbros, a sheeted complex, and volcanic rocks with arc tholeiitic or (more rarely) low-K calc-alkaline affinities. The second stage is characterized by intrusive wehrlite-clinopyroxenite complexes, intrusive gabbros, Cr-rich diorites, and volcanic rocks with high-Ca boninitic or tholeiitic ankaramite affinities. The third stage includes diorite and quartz diorite plutons, felsic dike and sill complexes, and

---

[1] Corresponding author; email: shervais@cc.usu.edu
[2] Present address: 2623 Withington Peak Drive North, Rio Rancho, NM 87144.
[3] Present address: 9405 Greenfield Drive, Raleigh, NC 27615.

calc-alkaline volcanic rocks. The first three stages of ophiolite formation were terminated by the intrusion of mid-ocean ridge basalt dikes, and the eruption of mid-ocean ridge basalt or ocean-island basalt volcanic suites. We interpret this final magmatic event (MORB dikes) to represent the collision of an active spreading ridge. Subsequent reorganization of relative plate motions led to sinistral transpression, along with renewed subduction and accretion of the Franciscan Complex. The latter event resulted in uplift and exhumation of the ophiolite by the process of accretionary uplift.

Introduction

THE WESTERN MARGIN of North America is characterized by extensive tracts of ophiolitic basement with radiometric ages of 173 to 160 Ma or younger (Hopson et al., 1981; Saleeby et al., 1982; Wright and Sharp, 1982; Saleeby, 1983; Coleman, 2000). The Coast Range ophiolite of California is the most extensive of these tracts, which also include the Fidalgo ophiolite complex in Washington state (Brown et al., 1979), the Smartville ophiolite in the Sierra Foothills (Xenophontos and Bond, 1978), the Josephine ophiolite of southern Oregon and northern California (Harper, 1984; Harper et al., 1985), the Preston Peak ophiolite in the Klamath Mountains (Snoke, 1977), and the Cedros ophiolite in Baja California (Kimbrough, 1982). The regional extent of these ophiolite belts, and the narrow range in their ages of formation, make their petrogenesis one of the more important (and contentious) tectonic problems in the Cordillera (e.g., Dickinson et al., 1996).

The Coast Range ophiolite (CRO) was first recognized by Bailey et al. (1970) and later studied in detail by Hopson et al. (1981). Early studies of the CRO emphasized its similarity to oceanic crust formed at mid-ocean ridge spreading centers, and tectonic interpretations focused on the obduction of intact oceanic lithosphere (e.g., Bailey et al., 1970; Page, 1972; Hopson and Frano, 1977; Hopson et al., 1981, 1992, 1997). These studies provided the fundamental petrologic and structural framework for later investigations, and established the overall "oceanic" nature of the ophiolite.

Later petrologic and geochemical studies demonstrated that the Coast Range Ophiolite formed in a supra-subduction zone environment, possibly by fore-arc or intra-arc rifting (Evarts, 1977; Blake and Jones, 1981; Saleeby, 1982; Shervais and Kimbrough, 1985a, 1985b; Lagabrielle et al., 1986; Shervais, 1990, 2001; Stern and Bloomer, 1992). Petrologic evidence for this interpretation includes the presence of both keratophyre and quartz keratophyre (metamorphosed andesite and rhyolite) in the hypabyssal and volcanic sections, magmatic hornblende in the plutonic section, and the crystallization of clinopyroxene before plagioclase in the cumulates. Geochemical evidence includes low Ti, Na, and Cr concentrations in relict clinopyroxene, relative depletion of high-field-strength trace elements in extrusive rocks, and a trend toward silica-saturation with increasing FeO/MgO (Bailey and Blake, 1974; Evarts, 1977; Blake and Jones, 1981; Shervais and Kimbrough, 1985a; Shervais, 1990, 2001; Giaramita et al., 1998).

More recent work on the CRO suggests that none of the simple, single-stage interpretations advanced earlier are entirely correct, and that the CRO had a brief but complex history. In particular, the occurrence of MORB-like volcanic and hypabyssal rocks within stratigraphic or tectonic settings characteristic of the CRO suggests that models that propose a supra-subduction zone origin are too simple. New data gathered over the last 10 years suggest that although a supra-subduction zone setting was the dominant focus of CRO origin and development, the interaction between tectonic and igneous processes that form mid-ocean ridges and convergent plate boundaries was important, especially during the later stages of ophiolite development (Shervais, 1993, 2001; Murchey and Blake, 1993). These observations suggest that our current concepts on the development of supra-subduction zone ophiolites are too simple and need to be revised.

In this contribution we synthesize the results of recent and ongoing studies of the CRO and reexamine previous ideas about its development in light of this synthesis. We then generalize these observations to place the CRO within the context of other supra-subduction zone ophiolites, and assess the implications of our results for the tectonic evolution of the western Cordillera.

Geology and Setting of the Coast Range Ophiolite

Field setting

The Coast Range ophiolite is represented by a series of dismembered ophiolitic fragments scattered along the Coast Range fault (the tectonic

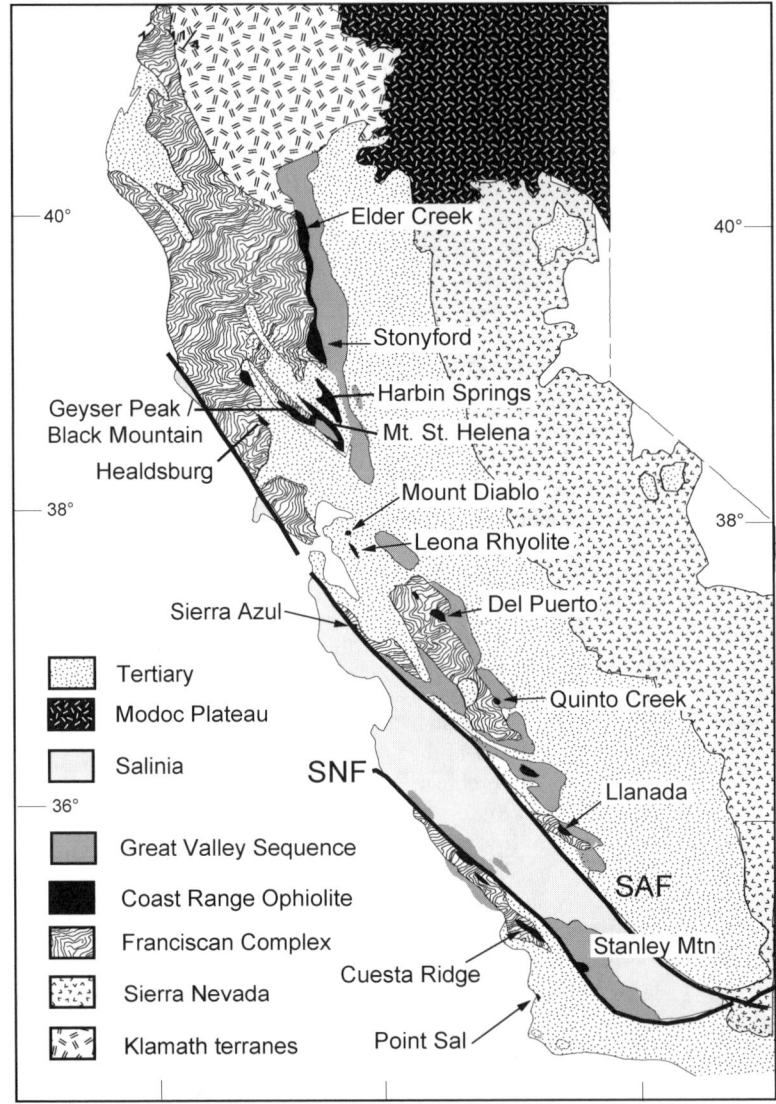

FIG. 1. Simplified geologic map of California showing Coast Range ophiolite localities discussed in this paper (black), along with the main geologic subdivisions of the state: batholiths and metamorphic rocks of the Sierra Nevada, Klamaths, and Salinia, sediments of the Jurassic–Cretaceous Great Valley Sequence, the Jurassic–Paleogene Franciscan Complex, Tertiary rocks of the Modoc Plateau, and Cenozoic rocks unrelated to the other units. Abbreviations: SAF = San Andreas fault system; SNF = Sur-Nacimiento fault system.

contact between the CRO and the underlying Franciscan Complex; see Jayko et al., 1987) from Elder Creek in the north to Point Sal in the south (Fig. 1). When Neogene wrench faulting along the San Andreas and Sur-Nacimiento fault systems is restored, the ophiolite belt defines an original extent of over 1300 km (Hopson et al., 1981). The ophiolite is overlain depositionally by Upper Jurassic strata of the basal Great Valley Series (Knoxville Formation in northern California, Toro and Espada formations in central California, Riddle Formation in southwest Oregon), a flysch deposit derived from emergent arc terranes to the north and east (Bailey et al., 1970; Ingersoll and Dickinson, 1981). Basal Great Valley sediments in northern California contain *Buchia rugosa* of late Kimmeridgian to early Tithonian age (~152–148 Ma; Palfey et al., 2000), which places an upper limit on the time of ophiolite formation (Jones, 1975; Imlay, 1980).

Exposures of CRO vary greatly in their character from place to place (Table 1). Some CRO remnants feature more or less complete ophiolite stratigraphy (although all are complexly faulted and dismembered), volcanic rocks with arc-like geochemistry, and depositional contacts with overlying Upper Jurassic sediments of the Great Valley Sequence. At other locales, plutonic rocks are scarce, the volcanic rocks are more like "oceanic" basalts (that is, basalts erupted at oceanic spreading centers or within oceanic plates), and contacts with sediments of the Great Valley Sequence may be faulted. The differences between these occurrences suggest that the CRO represents a complex composite terrane formed through the interplay of different tectonic processes, as originally proposed by Shervais and Kimbrough (1985a). In order to determine what processes were involved, and their influence on ophiolite formation, we need to review in some detail the results of work carried out over the last decade at various CRO localities. Much of this work has not been published except in abstracts or theses, so this summary represents the first description of geologic relations in several of these locales.

The CRO in California is divided geographically into three distinct belts: the Northern or Sacramento Valley belt, the Central or Diablo Range belt, and the Southern or Coastal belt, which is separated from the other belts by two major wrench fault systems (the San Andreas and Sur-Nacimiento fault systems) and by the Salinia terrane. Ophiolites in the Sacramento Valley belt are typically (but not everywhere) overlain by an ophiolite-derived sedimentary breccia, whereas ophiolites in the other belts are typically overlain by tuffaceous chert or cherty tuff (Hopson et al., 1981). Lesser known remnants of Jurassic ophiolite in southwest Oregon that are correlated with the CRO of California crop out west of the Rogue-Chetco arc complex (Blake et al., 1985; Kosanke et al., 1999).

Northern (Sacramento Valley) belt

The Northern or Sacramento Valley belt of CRO crops out along the western margin of the Sacramento Valley and in small remnants that have been offset by late Neogene wrench faulting (Fig. 1). The CRO and overlying Great Valley Sequence in this belt are separated from blueschist-facies schists and semi-schists of the Franciscan Complex by serpentinite-matrix mélange that crops out continuously along the western edge of the valley (Bailey et al., 1964; Jayko and Blake, 1987). This serpentinite mélange is commonly considered to be part of the CRO, but recent data suggest that a mixed CRO/Franciscan provenance is more likely (Jayko et al., 1987; MacPherson et al., 1990; Huot and Maury, 2002). Pillow lava knockers in this serpentinite matrix mélange resemble backarc basin basalts, and are overlain conformably by banded radiolarian cherts resembling those in the Franciscan Complex (Huot and Maury, 2002).

There are three main ophiolite locales in this belt—the Elder Creek ophiolite, the Stonyford Volcanic Complex, and the Black Mountain/Geyser Peak ophiolite—plus three smaller remnants: Mount Saint Helena, Harbin Springs, and Healdsburg. These ophiolites collectively exhibit the entire range of lithologic and tectonic diversity observed in the CRO. Additional fragments not discussed here include the Fir Creek, Wilbur Springs, and Bennett Creek ophiolite remnants (e.g., Hopson et al., 1981). Ophiolite remnants in the Clear Lake–Sonoma County area (Black Mountain/Geyser Peak, Mt. St. Helena, Harbin Springs, Healdsburg) have all been displaced northwestward during the Neogene along faults related to the San Andreas system (Fig. 1).

Elder Creek. The Elder Creek ophiolite remnant is the northernmost exposure of CRO in California, and also one of the best preserved (Fig. 2). It comprises a nearly complete ophiolite sequence, with cumulate mafic and ultramafic rocks, non-cumulate gabbro and diorite, dike complex, and volcanic rocks (Bailey et al., 1970; Hopson et al., 1981; Shervais, 2001).

Igneous rocks of the ophiolite are overlain by the Crowfoot Point Breccia, a coarse, unsorted fault-scarp talus breccia which varies from <10 m to over 1000 m in thickness (Lagabrielle et al., 1986; Robertson, 1990). This unit was deposited on an eroded surface that crosscuts all other units of the ophiolite (from cumulate ultramafics through dike complex). The Crowfoot Point Breccia contains clasts that range in composition depending on the level of erosion, such that volcanic clasts are most common where the breccia sits on eroded dike complex, and plutonic clasts are most common where the breccia sits on eroded plutonic complex. The Crowfoot Point

TABLE 1. Comparison of Major Occurrences of Coast Range Ophiolite in California
and Their Correspondence with the Proposed Life Cycle for Suprasubduction Zone (SSZ) Ophiolites

Life cycle:	Stage 1: Birth	Stage 2: Youth	Stage 3: Maturity	Stage 4: Death	Stage 5: Resurrection	Primary references
Events:	Initial spreading hinge rollback?	Refractory melts, second-stage melting	Stable arc (calc-alkaline or arc tholeiite)	Ridge subduction	Accretionary uplift or obduction	For all: Hopson et al., 1981; Shervais and Kimbrough, 1985a
Ophiolite	Arc tholeiite					
Elder Creek volcanics	Dike complex, pillows, clasts in Crowfoot Point Breccia	Clasts in Crowfoot Point Breccia (ol-px phyric)	Clasts in Crowfoot Point Breccia (andesite)	MORB dikes cut gabbro-diorite	Accretionary uplift by Franciscan Complex	Hopson et al, 1981; Shervais and Kimbrough, 1985a; Lagabrielle et al, 1986; Shervais and Beaman, 1989, 1991, 1994
Elder Creek plutonics:	Layered ± foliated gabbro, dunite	Gabbro pegmatoid–wehrlite–clinopyroxenite high-Cr diorites	Hb diorite, qtz diorite plutons	Franciscan high-grade blocks (U/Pb, Ar/Ar ≈162–159 Ma)	Accretionary uplift by Franciscan Complex	
Stonyford Complex volcanics	Melange blocks of pillow lava	Melange blocks of pillow lava		Stonyford volcanics	Accretionary uplift by Franciscan Complex	Hopson et al, 1981; Shervais and Kimbrough, 1985a; Shervais and Hanan, 1989; Zoglman, 1991; Zoglman and Shervais, 1991
Stonyford Complex plutonics	Layered gabbro blocks in mélange	Wehrlite-clinopyroxenite blocks in mélange	Qtz diorite blocks in mélange below volcanics	Franciscan high-grade blocks (U/Pb, Ar/Ar ≈162–159 Ma)	Accretionary uplift by Franciscan Complex	
Black Mountain volcanics	Diabase sill complex	?	?	MORB pillow lava, diabase breccia	Accretionary uplift by Franciscan Complex	McLaughlin and Pessagno, 1978; McLaughlin and Ohlin, 1984; McLaughlin, 1978; Giaramita et al., 1998
Black Mountain plutonics	Layered gabbro	?	?		Accretionary uplift by Franciscan Complex	
Mount Diablo volcanics	Arc basalt pillow lava, diabase sill complex		Keratophyre (andesite)	MORB dikes cut gabbro-diorite	Accretionary uplift by Franciscan Complex	Williams, 1983; Hagstrum and Jones, 1998
Mount Diablo plutonics						
Leona Rhyolite volcanics			Rhyolite vitrophyre	Mid-Ocean Ridge Basalt glass inclusions	Accretionary uplift by Franciscan Complex	Jones and Curtis, 1991; Shervais, 1993
Leona Rhyolite plutonics	Layered gabbros underlie rhyolite					

COAST RANGE OPHIOLITE 575

Location	Unit	Column A	Column B	Column C	Late features	Emplacement	References
Del Puerto Canyon	Volcanics:	Pillow lava, massive flows, sills (cpx-plg phyric)	Pillow lava, massive flows (ol-cpx phyric), boninites	Volcanics: Lotta Creek tuff, Leona rhyolite; qtz keratophyre lavas, sills	late MORB dikes	Accretionary Uplift by Franciscan Complex	Evarts, 1977; Evarts and Schiffman, 1983; Hopson et al, 1981; Evarts et al, 1992, 1999; Robertson, 1989
	Plutonics:	Plutonics: dunite, layered gabbro	Plutonics: wehrlite, clinopyroxenite, dunite	Plutonics: Hb qtz diorite plutons, 200m+ thick	Franciscan High grade blocks (U/Pb, Ar/Ar ~ 162–159)		
Quinto Creek		Pillow Lava Layered Gabbros				Accretionary Uplift by Franciscan Complex	Bailey et al, 1970, Hopson et al, 1981
Llanada	Volcanics:	Pillow lava (cpx-plg phyric)	Pillow lava (ol-cpx phyric)	Lotta Creek tuff; qtz keratophyre lavas; andesite breccia, lahars	Not seen	Accretionary Uplift by Franciscan Complex	Giaramita et al, 1998; Lagabrielle et al, 1986; Robertson, 1989; Hopson et al, 1981; Emerson, 1979; Bailey et al, 1970
	Plutonics:	Layered gabbro	Feldpathic wehrlite, dunite	Keratophyre sills			
Sierra Azul	Volcanics:	Pillow basalt	?	Keratophyre, andsite breccias and tuffs	MORB dikes cut gabbro-diorite	Accretionary Uplift by Franciscan Complex	McLaughlin et al., 1991, 1992
	Plutonics:	Layered gabbro	Ultramafics?	Diorite, Qtz diorite,			
Cuesta Ridge	Volcanics:	??	Boninite lavas	Basaltic andesite to dacite lavas	MORB dikes cut gabbro-diorite	Accretionary Uplift by Franciscan Complex	Page 1972; Pike, 1974
	Plutonics:	Rare gabbro	Wehrlite, lherzolite, clinopyroxenite	Qtz diorite, keratophyre Qtz gabbro, diabase			
Point Sal	Volcanics:	Lower pillow lavas (plag-cpx phyric)	Upper pillow lavas (olivine-phyric)	Tuffaceous chert	Not seen	Accretionary Uplift by Franciscan Complex	Hopson and Frano, 1977; Hopson et al, 1981; Mattinson and Hopson, 1992
	Plutonics:	Layered, foliated gabbro, dunite	Wehrlite-clinopyroxenite	Keratophyre sill complex, plagiogranite			
Snow Camp Mtn. (Oregon)	Volcanics:			Hornblende quartz diorite dikes		Accretionary Uplift by Dothan Formation Franciscan Complex	Blake et al., 1985; Harper, personal communication, 1999
	Plutonics:	Layered gabbro					
Wild Rogue Wilderness (Oregon)	Volcanics:	Pillow lava Dike Complex Group 1	basalt, basaltic andesite: (lower volcaniclastics) Dike Complex Group 2	Hornblende andesite, dacite dikes, lavas (upper volcaniclastics)	Rare MORB dikes	Accretionary Uplift by Dothan Formation	Blake et al., 1985; Harper, personal comm., 1999; Kosanke, 2000
	Plutonics:	Layered gabbro	High Cr diorites	Hornblende quartz diorite stocks, dikes			

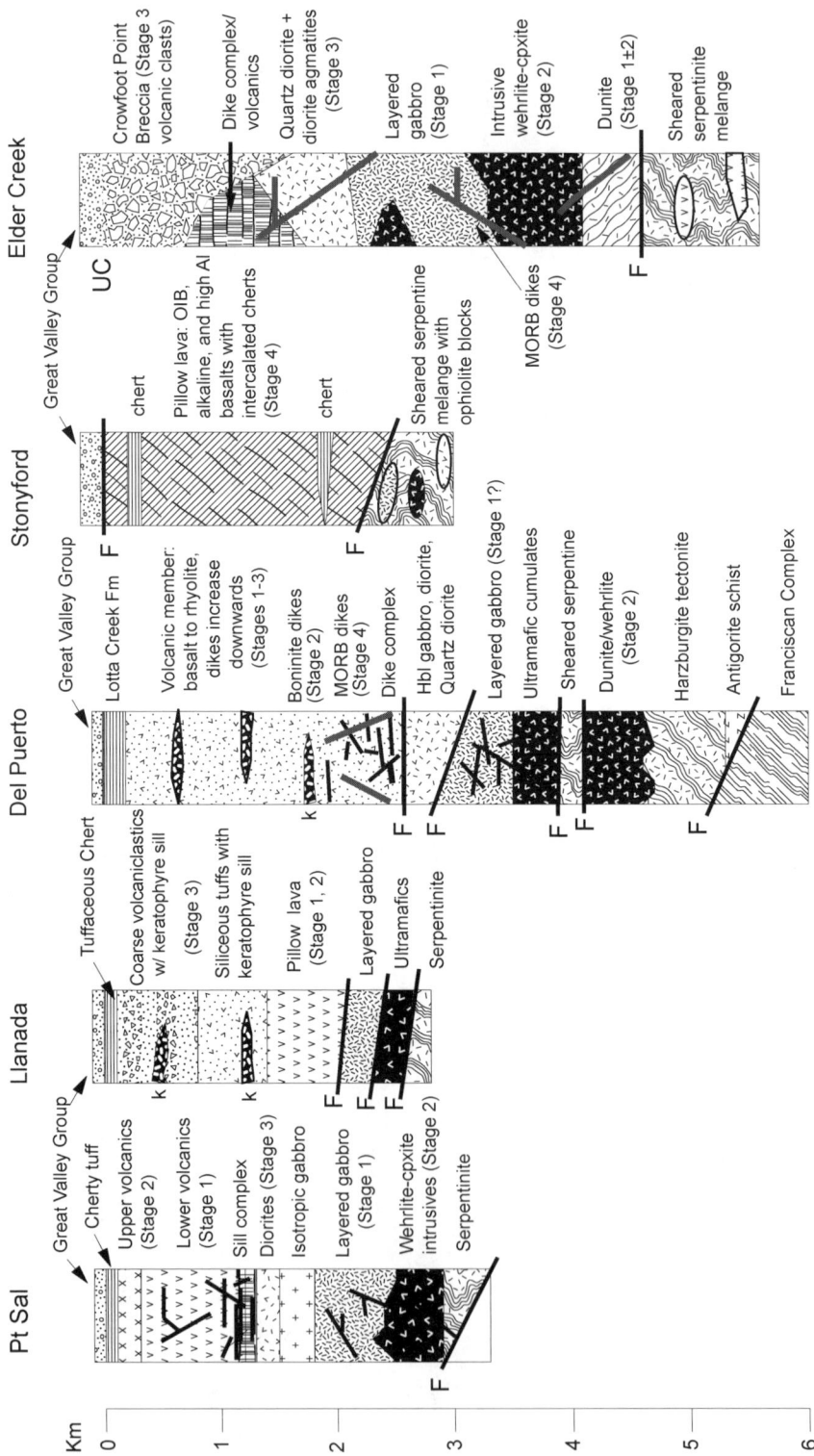

FIG. 2. Simplified stratigraphic columns of five representative Coast Range ophiolite localities: Point Sal, Llanada, Del Puerto Canyon, Stonyford, and Elder Creek. Thick lines crossing sections and labelled "F" are faults; crosscutting lines within sections are dikes of various generations and crosshatched wedges labelled "k" are keratophyre sills or dikes. Point Sal section after Hopson et al. (1981); Llanada section after Emerson (1979) and Hopson et al. (1981); Del Puerto section after Evarts et al. (1999); Stonyford Volcanic Complex and Elder Creek from unpublished data of Shervais and master's theses by Zoglman (1991) and Beaman (1989).

Breccia is itself overlain depositionally by the Upper Jurassic Great Valley Sequence, which here contains fossils of *Buchia rugosa* of late Kimmeridgian age (Bailey et al., 1970; Jones, 1975; Blake et al., 1987).

Field relations and geochemistry of the Elder Creek remnant indicate that four magmatic episodes are required for its formation (Beaman, 1989; Shervais 2001). The first magma series comprises cumulate dunite and gabbro, isotropic gabbro, dike complex, and by some of the massive volcanics. The cumulate gabbros may exhibit foliation formed by extension, or be cut by ductile faults. The second magmatic series comprises clinopyroxenite-wehrlite intrusions with less common gabbro and gabbro pegmatoid. Intrusive relations are shown by xenoliths of layered gabbro or dunite within pyroxenite and gabbro pegmatoid, and by truncation of layering along intrusive contacts between the two series. The third magmatic episode is represented by isotropic gabbro, agmatite composed of xenoliths of cumulate or foliated gabbro, volcanic rock, and dike complex in a diorite or quartz diorite matrix; diorite and quartz diorite stocks and dikes that intrude all of the older lithologies; and felsite dikes that are marginal to the quartz diorite plutons. This episode is also represented by volcanic clasts in the Crowfoot Point breccia that are calc-alkaline in composition. The fourth magma series is represented by basaltic dikes that crosscut rocks of the older episodes, and basaltic pillow lavas that form volcanic lenses within the Crowfoot Point Breccia.

Geochemical data are consistent with formation of the first three series in a supra-subduction zone (arc) environment, but all require distinct parent magmas. Rocks of the first magmatic series represent primitive arc tholeiite magmas, with a narrow range of silica contents (53–66% SiO_2), low TiO_2 and Cr concentrations (Fig. 3), low Ti/Zr and Ti/V ratios (Fig. 4), and LREE-depleted REE patterns. In contrast, parent magmas for the second magmatic series were more primitive in their major-element characteristics, similar to boninites and tholeiitic ankaramites found in many western Pacific arcs. These rocks have high Cr concentrations over a wide range of silica contents (Fig. 3) but also have low Ti/Zr and Ti/V ratios (Fig. 4). Rocks of the third magmatic series (diorites, quartz diorites, basaltic andesites, andesites, dacites) form a classic low-K calc-alkaline magma series (Fig. 3). These rocks do not represent a later, superimposed volcanic arc, but are an integral part of the ophiolite stratigraphy.

Dikes and pillows of the final magma series are characterized by MORB-like major and trace element compositions: low silica, high TiO_2 that increases with silica and Zr, and high Ti/V ratios (Figs. 3 and 4). The dikes crosscut all of the older suites, and the pillows are intercalated within the Crowfoot Point Breccia, showing that this magma series represents the final magmatic event in the life cycle of the Elder Creek ophiolite.

Stonyford. The Stonyford Volcanic complex (SFVC), which lies just 35 miles south of Elder Creek, is unique within the CRO (Fig. 1). This seamount complex consists almost entirely of volcanic flows (pillow lava) with subordinate diabase, hyaloclastite breccia, and minor sedimentary intercalations of chert and limestone (Fig. 2; Shervais et al., 2003). The rocks are exceptionally fresh for the CRO, as shown by the preservation of primary igneous plagioclase, clinopyroxene, and basaltic glass in most of the volcanic rocks, and (rarely) olivine phenocrysts within volcanic glass (Shervais and Hanan, 1989). The preservation of unaltered basaltic glass shows that these rocks could not have been subducted, and therefore must have formed in the upper plate of the subduction zone.

The SFVC crops out as four large tectonic blocks (up to 5 × 3 km in areal extent) within or overlying a sheared serpentinite-matrix mélange. Structurally below the largest blocks of SFVC are dismembered remnants of CRO plutonic and volcanic rock, including dunite, wehrlite, clinopyroxenite, gabbro, diorite, quartz diorite, and keratophyre pillow lava. Quartz diorite also occurs as dikes within mélange blocks of isotropic gabbro. At other locations in the mélange, tectonic blocks include unmetamorphosed volcanogenic sandstones (correlative with the Crowfoot Point breccia near Elder Creek), foliated metasediments (possibly correlative with the Galice Formation), and pale green metavolcanic rocks (Zoglman and Shervais, 1991; Shervais et al., 2003). The occurrence of foliated Galice-like metasediments suggests tectonic transport of the CRO across Sierran basement (e.g., Jayko and Blake, 1987).

Volcanic rocks of the SFVC comprise three distinct petrologic groups: (1) oceanic tholeiite basalts; (2) transitional alkali basalts and basaltic glasses; and (3) high-alumina, low-Ti tholeiites (Zoglman, 1991). These compositionally distinct magma series are intercalated at all stratigraphic levels of the seamount. The oceanic tholeiite suite has high concentrations of TiO_2 (Fig. 3) with MORB-like Ti/Zr and Ti/V ratios (Fig. 4). The alkali-suite lavas have lower

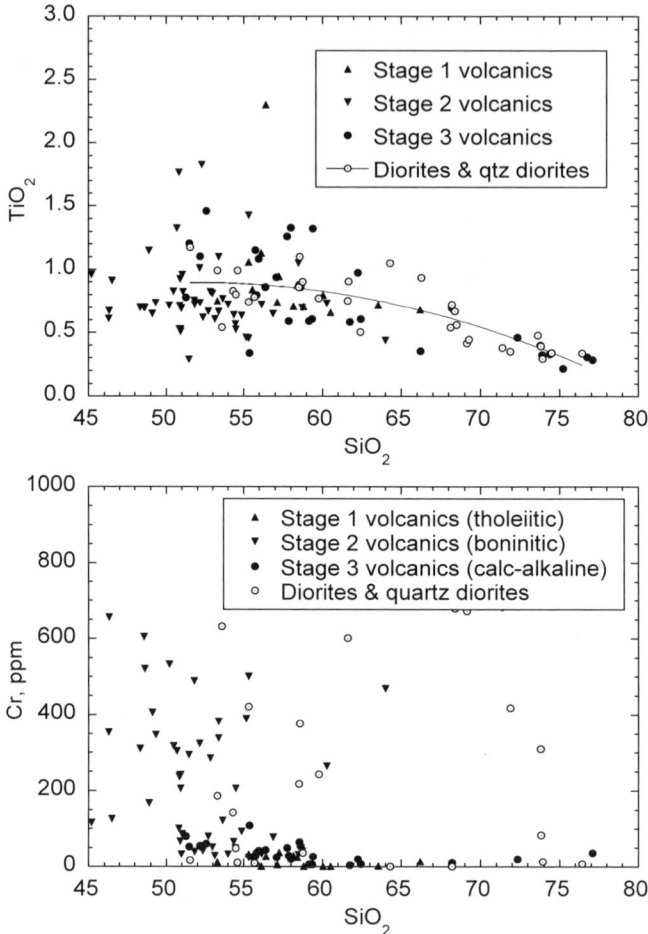

FIG. 3. Harker diagrams for TiO_2 and Cr in volcanic rocks of the Coast Range ophiolite. Upright triangles = stage 1 arc tholeiite lavas (Elder Creek, Healdsburg, Del Puerto, Cuesta Ridge, Stanley Mountain, Point Sal); inverted triangles = stage 2 refractory lavas (tholeiitic ankaramites or boninites: Elder Creek, Del Peurto, Llanada, Quinto Creek, Cuesta Ridge, Point Sal); filled circles = stage 3 calc-alkaline lavas (Elder Creek, Llanada, Cuesta Ridge); open circles = stage 4 mid-ocean ridge basalt/ocean island basalt (Elder Creek, Stonyford, Black Mountain). Data sources: Shervais and Kimbrough (1985a), Lagabrielle et al. (1986), Beaman (1989), Shervais (1990), Zoglman (1991), and Giaramita et al. (1998).

TiO_2 but higher Ti/V ratios (Fig. 4). The high-Al, low-Ti lavas have low Ti/V ratios that are typical of arc-volcanic rocks (Shervais, 1982). Initial Pb isotopic ratios (at 164 Ma) for the volcanic glasses show a wide range in $^{206}Pb/^{204}Pb$ (Hanan et al., 1991, 1992), similar to Pacific oceanic basalts currently found along the East Pacific Rise. The trace-element and Pb isotopic data indicate that the oceanic tholeiites and alkalic lavas of the Stonyford complex were derived from a heterogeneous mantle source comprising two components: depleted MORB-asthenosphere and an enriched OIB-like component (Hanan et al., 1991, 1992). The high-Al, low-Ti lavas resemble high-Al island-arc basalts; second-stage melts of MORB asthenosphere that form by melting of plagioclase lherzolite at low pressures have similar compositions (Shervais et al., 2003). The occurrence of the SFVC structurally overlying dismembered ophiolite lithologies that resemble the Elder Creek ophiolite remnant suggests that the

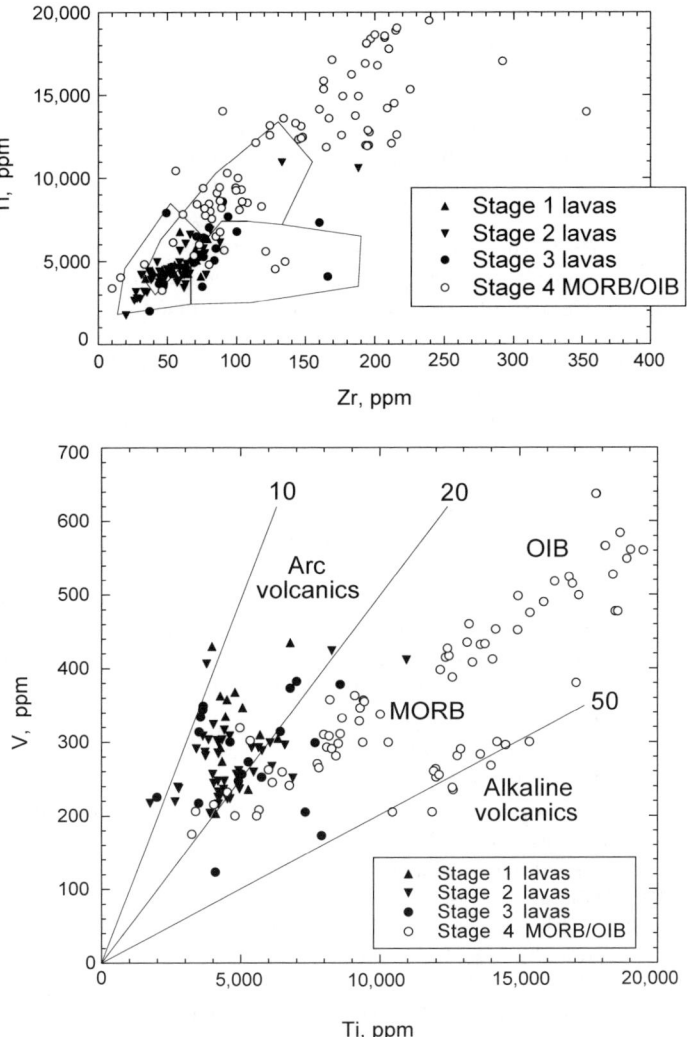

FIG. 4. Ti-Zr plot (Pearce and Cann, 1973) and Ti/V plot (Shervais, 1982) for volcanic rocks of the Coast Range ophiolite. Symbols and data sources are the same as Figure 3.

volcanic complex was built on older, subduction-related ocean crust prior to or during deformation of the older crust.

Shervais and Kimbrough (1985a, 1987) did not distinguish between the Stonyford Volcanic Complex and the Snow Mountain Volcanic Complex of MacPherson (1983). However, as noted by MacPherson and Phipps (1985), these units are distinct and unrelated. The Snow Mountain Volcanic Complex contains incipient blueschist metamorphic phases and lies west of the serpentinite belt that marks the Coast Range fault (MacPherson, 1983, 1986; MacPherson and Phipps, 1985). In contrast, the SFVC contains unaltered volcanic glass, lies entirely within the Coast Range fault serpentinite belt, and is closely associated with (but is not directly overlain by) sediments of the Jurassic basal Great Valley Series (Shervais and Hanan, 1989; Zoglman, 1991; Zoglman and Shervais, 1991). The SFVC also contains radiolarian chert intercalations with faunal assemblages that are similar to those found at other CRO locations, but distinct from faunal assemblages derived from Franciscan cherts. Thus, the Snow Mountain Volcanic Complex is clearly part of the Franciscan assemblage, whereas the SFVC is best regarded as part of the CRO (contrary to our former correlation of both units with the Franciscan assemblage, e.g., Shervais and Kimbrough, 1987; Shervais and Hanan, 1989; Shervais, 1990). This distinction is important because it has significant tectonic implications for the origin and evolution of the CRO.

Black Mountain/Geyser Peak. The Black Mountain/Geyser Peak ophiolite remnant consists largely of pillow lava and diabase dikes and sills (McLaughlin, 1978; McLaughlin and Pessagno, 1978; McLaughlin & Ohlin, 1984; McLaughlin et al., 1988). The western edge of this remnant comprises a thin strip of serpentinized peridotite and an adjacent strip of uralitic gabbro, in contact with a diabase sill complex that underlies much of Geyser Peak (McLaughlin, 1978; McLaughlin and Pessagno, 1978). A steep fault juxtaposes this sill complex against pillow lava and pillow breccia that underlie much of Black Mountain to the east. The pillow lava is overlain in turn by ophiolitic breccias containing clasts of microgabbro, diabase, diorite, and basalt, in a fine, sandy matrix (McLaughlin and Pessagno, 1978; Lagabrielle et al., 1986). These breccias are similar to the Crowfoot Point Breccia and occupy the same stratigraphic horizon.

Geochemical investigations of the Black Mountain pillow lavas show that they are MORB-like basalts that contain a minor subduction component (Lagabrielle et al., 1986; Giaramita et al., 1998). The basalts are characterized by increasing TiO_2 relative to increasing silica (Fig. 3), high Ti relative to Zr, and Ti/V ratios of ~22–30 (Fig. 4; Giaramita et al., 1998). These characteristics are nearly identical to the "oceanic tholeiite" suite in the SFVC, but without the accompanying alkali basalt and low-Ti/high-Al basalt suites. The only hint of a subduction component in these rocks are the elevated Th concentrations relative to MORB.

Mount Saint Helena, Harbin Springs. The Harbin Springs ophiolite remnant crops out about 10 miles north of Mount Saint Helena and the Mount Saint Helena ophiolite remnant (Fig. 1). Both ophiolites have the same fault-bounded assemblages: serpentinized harzburgite ± dunite or sheared serpentinite, cumulate gabbro, and gabbro/diabase breccia (Bezore, 1969; McLaughlin and Pessagno, 1978; Hopson et al., 1981; Robertson, 1990). Contacts between units are faults, and the breccia is overlain by tuffaceous sediments of the lower Great Valley Sequence. The cumulate gabbros include both ol + cpx gabbros and cpx gabbros with well-developed planar laminations and adcumulus textures, and may be cut by dikes. The breccia is dominantly clast supported, crudely stratified, with intercalations of finer-grained sand and silt of similar composition (Robertson, 1990). The breccia at these locales correlates with the Crowfoot Point breccia at Elder Creek. No geochemistry has been reported from either location.

Healdsburg. The Healdsburg ophiolite remnant comprises sheared serpentinite (mostly harzburgite), cumulate gabbro and olivine gabbro, hornblende diorite and hornblende quartz diorite, diabase dike complex, and up to 300 m of volcanic rocks (Hopson et al., 1981). Contacts are either faults or covered. The volcanic section includes both spilite (basalt) and keratophyre (andesite-dacite). Geochemical data for two samples from Healdsburg in Shervais (1990) show that these rocks range from 56% to 71% SiO_2, with low TiO_2 and Ti/V ratios of 10–15. These values are characteristic of arc volcanics, but it is not clear how they relate to one another or how well they reflect the true proportions of volcanic rocks present.

Central (Diablo Range) belt

The Central or Diablo Range belt crops out in the Diablo Range, along the western margin of the San Joaquin Valley (Fig. 1). Major CRO locales in this belt include Mount Diablo, the Leona Rhyolite, Del Puerto Canyon, Quinto Creek, and Llanada. Also included in this belt here is the recently described Sierra Azul ophiolite in the Santa Cruz Mountains, which is separated from the Diablo Range locales by the Hayward-Calaveras fault system. Ophiolites of the Diablo Range belt are typically overlain conformably by silicic tuff, tuffaceous chert, and sediments of the Great Valley Sequence (Hopson et al., 1981; Evarts et al., 1999).

Mount Diablo. The Mount Diablo ophiolite remnant consists of serpentinite, diabase sill complex, and pillow lava, juxtaposed against Franciscan mélange (Williams, 1983). Cumulate plutonic rocks are absent, and all contacts between adjacent units are faults (Williams, 1983; Hagstrum and Jones, 1998). Zircons from small keratophyre dikes yield a U-Pb age of 165 Ma (J. M. Mattinson, reported in Mankinen et al., 1991), essentially identical to U-Pb zircon dates from other CRO locales (Mattinson and Hopson, 1992).

Two geochemically distinct suites of igneous rocks can be distinguished in the Mount Diablo ophiolite remnant: an older suite of island arc basalts and andesites (keratophyres), and a younger suite of pillow lava and diabase with geochemical affinities to mid-ocean ridge basalts (Williams, 1983). The older island-arc suite volcanic rocks are cross-cut by dikes and sills of MORB-like basalt; similar cross-cutting relations are observed at Elder Creek and Del Puerto Canyon.

Leona Rhyolite. The Leona Rhyolite crops out extensively in the Berkeley Hills, forming prominent local landmarks such as Indian Rock and the Cragmont Crags. Originally mapped as a Miocene felsic volcanic unit, the Leona Rhyolite is now interpreted to correlate with the Lotta Creek Formation in Del Puerto Canyon because it rests depositionally on mafic rocks of the CRO, and is overlain depositionally by Jurassic sediments of the Great Valley Sequence (Jones and Curtis, 1991). A more robust correlation may be with felsic volcanic rocks in Del Puerto Canyon that underlie the Lotta Creek Formation, based on the absence of hornblende phenocrysts in the Leona Rhyolite (Evarts, pers. commun., 1999). Our data show that rhyolite samples from outcrops near Leona Regional Park in east Oakland consists of clear volcanic glass with small phenocrysts of plagioclase ($\approx An_{55-75}$), augite ($\approx Wo_{38}En_{33}$), hypersthene ($\approx Wo_{3.6}\ En_{48}$), and ferropigeonite ($\approx Wo_{14}En_{13}$). In thin section, the glass contains domains that resemble collapsed pumice, suggesting that it may have formed as a submarine pumice-obsidian flow.

Another interesting feature of the Leona Rhyolite is the occurrence of immiscible blobs (= 3 mm in diameter) of brown basaltic glass within the clear rhyolite glass (Shervais, 1993). Electron microprobe and ion microprobe analyses of these brown glass inclusions show that they have major-element chemistry similar to mid-ocean ridge basalts of the Pacific basin, LREE-depleted trace element patterns, and no dissolved water (microprobe totals = 99.5%; Shervais, unpubl.). The brown glass inclusions may simply represent the product of liquid immiscibility; alternatively, they may reflect the mixing of MORB magmas into the rhyolite prior to eruption. If so, this implies the intersection of an active ridge segment with the CRO forearc and intrusion of mid-ocean ridge basalt magmas into magma chambers of the overlying plate (Shervais, 1993).

Del Puerto Canyon. The Del Puerto Canyon ophiolite remnant is the largest and best preserved ophiolite fragment in the Diablo Range (Evarts, 1977; Evarts and Schiffman, 1983; Evarts et al., 1999). It crops out as four large, fault-bounded blocks that together preserve a relatively complete ophiolite stratigraphy and are overlain depositionally by sediments of the Great Valley Sequence, which contain *Buchia piochii* of Tithonian age (Evarts et al., 1999). These blocks are underlain tectonically by blueschist-facies rocks of the Franciscan assemblage. Despite its relatively complete stratigraphy (Fig. 2), much of the plutonic section appears to be missing, along with portions of the volcanic section (Evarts et al., 1999).

Four magmatic series are evident, especially in the plutonic section of the ophiolite. The oldest is represented by cumulate dunites and gabbros; associated volcanic rocks may include low-K, LREE-depleted massive flows in the volcanic section, but there is no clear distinction between "early" lavas and "late" lavas as there is at Point Sal. The second magmatic series is represented by pyroxenite, wehrlite, feldspathic peridotite, and some gabbronorites, and by sparse high-Mg andesites with "boninitic" affinities (high Mg, Cr, and Ni relative to Si; high Cr spinels; Shervais, 1990; Evarts et al., 1992, 1999). The third magmatic series comprises a calc-alkaline suite of pyroxene and hornblende gabbro, hornblende diorite, hornblende quartz diorite, keratophyre dikes and sills, and calc-alkaline volcanic rocks ranging from basalt to rhyolite in composition. The fourth magmatic suite observed here comprises rare basaltic dikes with MORB geochemistry (Ti/V $\approx$ 30, $La_n/Yb_n \approx 0.5$, Nb $\approx$ 8 ppm; Evarts et al., 1992). Tuffaceous radiolarian cherts, tuffs, and volcanic breccias of the Lotta Creek Formation seem to postdate the MORB dikes, and may represent volcanic-arc activity unrelated to the ophiolite.

Age relations in the Del Puerto ophiolite remnant are based on a range of techniques and rock types. Hornblende gabbro has been dated at $\approx$157 Ma (^{40}Ar-^{39}Ar hornblende; Evarts et al., 1992), and a plagiogranite dike that crosscuts quartz diorite has the same age ($\approx$157 Ma, U-Pb zircon; Hopson et al., 1981). Felsic dikes that cut cumulates, rhyolite lava, and andesite boulders in a Lotta Creek Formation conglomerate bed have all been dated at $\approx$150 Ma (zircon fission track and ^{40}Ar-^{39}Ar hornblende; Evarts et al., 1992). These are the youngest ages yet reported from the CRO and overlap in age with basal Knoxville strata of the Great Valley Sequence; they are clearly much younger than other CRO locales (e.g., Point Sal, $\approx$165–173 Ma; Mattinson and Hopson, 1992).

The Del Puerto ophiolite remnant bears many similarities to the Elder Creek remnant in that: (1) the first magma series is represented primarily within the plutonic section; (2) the third magmatic series dominates the volcanic section and is well represented by dikes, sills, and sill-like plutons in the plutonic section; and (3) the fourth magmatic event, which may have overlapped with stage 3 volcanism, is represented by rare basaltic dikes with MORB-affinity. But Del Puerto differs from other CRO locales in several important ways as well: (1) the first magmatic suite is low-K calc-alkaline rather than arc tholeiite; (2) Del Puerto extends to younger ages than other well-dated CRO remnants; and (3) calc-alkaline volcanism seems to have continued (or restarted) after the late MORB dikes, as shown by the hornblende-phyric Lotta Creek tuff (Evarts et al., 1999). The extended magmatic history of this ophiolite remnant may be related to late Jurassic arc activity in the western Sierran foothills, or may reflect variations in plate margin geometry.

Quinto Creek. This small ophiolite remnant comprises serpentinized harzburgite, cumulate and non-cumulate gabbro; quartz gabbro/diorite; plagiogranite dikes;, a sheeted complex of diabase, basalt, andesite, and microdiorite; and a volcanic complex of pillowed and massive basalts (Bailey et al., 1970; Hopson et al., 1981). Volcanic breccia horizons are intercalated with the pillows, similar to the volcanic breccias at Cuesta Ridge (Robertson, 1989). Limited geochemical data suggest that the volcanic rocks here are borderline between island-arc lavas and MORB (Shervais and Kimbrough, 1985a; Shervais, 1990), but the data are insufficient to identify distinct suites or relationships.

Llanada. The Llanada ophiolite remnant was originally mapped by Enos (1963) and was later mapped in detail by Emerson (1979). It has been studied in reconnaissance by many investigators, including Bailey et al. (1970), Hopson et al. (1981), Robertson (1989), Lagabrielle et al. (1986), and Giaramita et al. (1998). The Llanada remnant contains the following members: (1) cumulate plutonic rocks, including gabbro, olivine gabbro and melagabbro, feldspathic wehrlite, and possibly dunite or harzburgite; (2) a lower volcanic member consisting of massive and pillowed lava flows; and (3) an upper volcanic member consisting of siliceous tuffs, basaltic breccias, and massive, coarsening upwards laharic breccias and tuff breccias composed mainly of andesite (Fig. 2; Emerson, 1979; Hopson et al., 1981; Lagabrielle et al., 1986; Robertson, 1989).

The upper volcanic section at Llanada corresponds to a calc-alkaline volcanic arc sequence, grading upwards from distal to proximal in depositional facies; this is similar to the third magmatic series observed at other CRO localities (e.g., Elder Creek, Del Puerto). The lower volcanic series of massive and pillowed basalt has the geochemical characteristics of supra-subduction zone volcanism: low TiO_2; low Ti/V ratios; high Th, Rb, and Ba; and REE abundances that range from slightly LREE depleted to slightly LREE enriched (Shervais and Kimbrough, 1985a; Lagabrielle et al., 1986; Shervais, 1990; Giaramita et al., 1998). Although not true boninites, these lavas are also enriched in Cr and Ni relative to normal arc tholeiites, and contain Cr-spinels that have Cr/Al ratios higher than typical MORB spinels (Giaramita et al., 1998). Thus these lavas, along with the feldspathic wehrlites of the plutonic section, may correspond to the second magmatic series observed at other CRO localities (e.g., the upper pillow lavas and wehrlites at Point Sal (see below), and the wehrlite intrusions at Elder Creek). There is no clear evidence for a typical "arc tholeiite" series that might correspond to the lower pillow lavas or the early cumulate gabbros at Point Sal, although the Llanada cumulate gabbros may represent this magma series. There is also no indication of a later MORB-like magma series.

Sierra Azul. This recently discovered ophiolite locality in the Santa Cruz Mountains near Los Gatos contains a nearly complete ophiolite stratigraphy, with cumulate ultramafic rocks, gabbros, diorites, quartz diorites, andesites, massive basalt flows, and overlying andesite/quartz keratophyre tuffs and breccias (McLaughlin et al., 1991, 1992). Intrusive relations of the plutonic rocks have not been described in detail, so it is not possible to tell how many magmatic series are present. Almost all of the volcanic rocks and dike rocks have arc-like geochemical features (low TiO_2, high SiO_2) but a few are apparently MORB-like in character (McLaughlin et al., 1991, 1992). Overall, this ophiolite remnant appears similar to the occurrences found at Elder Creek and Del Puerto Canyon.

Southern (Coastal) belt

The Southern or Coastal belt of the CRO is separated from the other belts by the San Andreas and Sur-Nacimiento fault systems (Fig. 1). There are three main ophiolite locales (Cuesta Ridge, Stanley Mountain, and Point Sal) plus several smaller remnants not described here (Marmalejo Creek, San

Simeon, and Santa Cruz Island; Hopson et al., 1981). All of these ophiolites were likely derived from locations much farther to the south, based on the aggregate offset along these fault systems.

Cuesta Ridge. The Cuesta Ridge ophiolite (Page, 1972; Pike, 1974; Snow, 2002) is exposed in a large, open syncline along Cuesta Ridge, just north of San Luis Obispo. It consists of harzburgite ± dunite tectonite, cumulate mafic and ultramafic rocks, and up to 1.2 km of pillow lava and breccias. Layered gabbros are rare; olivine clinopyroxenites, wehrlites, and lherzolites with interstitial hornblende and plagioclase are more common. The dominant plutonic rock is quartz diorite, which forms the entire sheeted sill complex mapped at the east end of Cuesta Ridge and a 400 m thick high-level intrusive complex (quartz gabbro, quartz diorite, keratophyre, and quartz diabase sills 5–10 m thick) near Cerro Alto Peak (Snow, 2002).

Arc tholeiite volcanics are present low in the volcanic sequence but are less common than other lavas. More common are volcanic rocks with boninitic affinities (high Mg, Cr, Ni, and silica contents) that are intercalated with calc-alkaline–series volcanics up-section. These include massive volcanic flows, pillow lavas, and breccias with basaltic andesite to dacite compositions (Snow, 2002). Late-stage dikes of olivine tholeiite that crosscut the quartz diorite sill complex have major and trace element composition similar to MORB (e.g., $\approx$ 1.1 to 2.2 wt% TiO_2); late pillow lavas that crop out just below the chert in some places have similar MORB-like compositions (Snow, 2002).

Stanley Mountain. The Stanley Mountain ophiolite remnant crops out along Alamo Creek in southern San Luis Obispo County (Fig. 1). This ophiolite comprises a thick section of pillow lava and volcanic breccias overlain by more than 100 m of chert, tuffaceous chert, and mudstone, in turn overlain conformably by basal sediments of Great Valley Series (Hopson et al, 1981; Robertson, 1989). The radiolaria assemblages from the cherts and mudstones have been described in great detail by Hull (1995, 1997), and by Hull and Pessagno (1994, 1995). A paleomagnetic study Hagstrum and Murchey (1996) showed that these cherts did not form south of the equator, but relatively close to their present latitude (see also Murchey and Hagstrum, 1997). Little information is available on the chemistry of the underlying pillow lava; Shervais (1990) published one whole-rock analysis: it is an andesite with low Mg, Cr, and Ni, and low Ti/V $\approx$ 12, i.e., a fairly typical low-K arc tholeiite.

Point Sal. The Point Sal ophiolite remnant is one of the best exposed and most thoroughly studied of all the CRO fragments (Hopson and Frano, 1977; Hopson et al., 1981; Mattinson and Hopson, 1992; Fig. 1). Coastal exposures of the plutonic section reveal mantle harzburgite, dunite, layered and in some cases foliated cumulate gabbros, intrusive wehrlite-clinopyroxenite bodies that crosscut layering in the gabbros, isotropic high-level gabbros, diorites and quartz diorites with miarolytic cavities that form arrays of vertically oriented pipes, thin sheets of plagiogranite, and a sheeted sill complex with basaltic, diabase, microdiorite, and epidosite sills and dikes that penetrate the base of the volcanic section (Fig. 2). The volcanic section consists of a lower volcanic zone of augite-plagioclase microphyric basaltic-andesite to andesite, and an upper volcanic zone of olivine-augite microphyric basalt. The upper volcanic zone is overlain by a condensed sequence of grey-green to brown tuffaeous radiolarian cherts.

All of the volcanic rocks at Point Sal have trace element systematics characteristic of supra-subduction zone basalts (Figs. 3 and 4; Shervais and Kimbrough, 1985a; Shervais, 1990). The lower volcanic zone lavas are arc tholeiites, whereas the upper volcanic zone lavas have high-Ca boninitic affinities. Based on phenocryst assemblages in the volcanic rocks and cumulate phase assemblages in the plutonic rocks, the lower pillow lavas correlate with the early cumulate gabbros, whereas the upper pillow lavas correlate with the intrusive wehrlite-clinopyroxenites. These magma groups correspond to the first and second magma series observed at Elder Creek. High-level diorites, microdiorites, and plagiogranites are rare, and there is no evidence for a final, MORB-like lava series.

Southwest Oregon

Ophiolite remnants in the Wild Rogue Wilderness area have been correlated with the CRO of California (Gray et al., 1982; Saleeby et al., 1984; Blake et al., 1985; Kosanke et al., 1999; Kosanke, 2000). These rocks crop out west of the Rogue-Chetco arc complex and its back-arc basin, the Josephine ophiolite complex (Harper, 1984; Harper et al., 1985, 1994), and structurally overlie rocks of the Franciscan Complex. The structural position of these remnants overlying Franciscan units, their stratigraphic position below Great Valley equivalent rocks (Riddle Formation, Days Creek Formation), and their lack of Nevadan deformation or metamorphism all support their correlation with the Coast Range ophiolite in California (Saleeby, 1984; Blake et al., 1985; Kosanke et al., 1999). A similar ophiolite remnant, the Snow Camp Mountain ophiolite, appears to represent an outlier of the Josephine ophiolite (Schoonmaker et al., 2003).

The Wild Rogue Wilderness ophiolite remnant has been studied in detail by Kosanke, from whom this description has been abstracted (Kosanke et al., 1999; Kosanke, 2000). Cumulate plutonic rocks are found only as screens in the sheeted complex, but isotropic gabbros, diorites, tonalites, and volcanic rocks are well represented. The volcanic rocks include pillow lava and the overlying Mule Mountain metavolcanics, which consists of basalt/basaltic andesite flows in its lower part and andesite/dacite flows in its upper part. The pillow lavas are LREE-depleted basaltic andesites with island-arc tholeiite affinities, similar to the first magma series at Elder Creek, Point Sal, and other CRO locales in California. Basalts in the lower part of the Mule Mountain metavolcanics include pyroxene porphyries with high Cr contents similar to high-Ca boninites (e.g., Crawford, 1989); these rocks correspond to the second magma series at Elder Creek and Point Sal. Andesites and dacites in the upper part of the Mule Mountain metavolcanics are calc-alkaline lavas and volcaniclastics similar to those found a Llanada and Del Puerto; these are inferred to correspond to the third magma series at Elder Creek and other California CRO locales.

Similar groupings are found in the hypabyssal rocks: isotropic gabbros and diabase dikes have transitional MORB/IAT compositions that correspond to the earliest magma series, whereas felsic diorites include both high-Cr varieties that correspond to the refractory magma series and hornblende gabbros and diorites that appear to correspond to later calc-alkaline magmas (Kosanke, 2000). Where cross-cutting relations can be observed, the diabase dikes are always older than the dioritic dikes. Two dikes with MORB-like compositions were described by Kosanke, but their timing is not clear. The latest dikes are garnet-muscovite tonalites with "S-type" granite affinities that are similar to late biotite granites in the Semail ophiolite (Lippard et al., 1986).

Age of the Coast Range ophiolite

Age relations in the CRO are critical for understanding the tectonic evolution of the Western Cordillera, especially Jurassic orogeny in the Sierra Nevada and Klamath mountains. Evolution of the CRO is closely tied in space and time to arc volcanism, deformation, and fabric development in the western Sierra foothills terranes. A number of models for Jurassic orogeny in the foothills metamorphic belt are specifically linked to the igneous age of the CRO and subsequent biostratigraphic ages of overlying sediments (e.g., Hopson et al, 1981; Dickinson et al., 1996).

Published ages for the Coast Range ophiolite are summarized in Table 2. These dates can be evaluated in three principal groups: (1) K-Ar dates on igneous hornblendes from gabbros and diorites (Lanphere, 1971; McDowell et al., 1984); (2) older U-Pb zircon dates, which predate high-precision, multi-collector mass spectrometers (Hopson et al., 1981); and (3) newer U-Pb zircon ages, which are generally significantly older than both the K-Ar ages and the older U-Pb zircon ages (J. M. Mattinson reported in Mankinen et al., 1991 and in Pessagno et al., 1993; Evarts et al., 1992; Mattinson and Hopson, 1992). Unfortunately, almost all of the more recent ages are reported either in abstracts (Mattinson and Hopson, 1992; Evarts et al., 1992) or as personal communications in other papers (J. M. Mattinson, reported in Mankinen et al., 1991, and Pessagno et al., 1993), with no supporting data and in some cases, no assignment to specific locations within the ophiolite. As a result, these ages are of limited value at this time.

The data suggest that CRO formation began circa 172 Ma and ended in most areas by ~160–164 Ma. Volcanism persisted until ~150 Ma in the Del Puerto ophiolite, possibly in response to progradation of arc volcanics from the Sierra foothills. Early K-Ar ages are now considered unreliable in light of pervasive low-temperature metamorphism, whereas the older U-Pb zircon ages (Hopson et al., 1981) are considered to be too young by 5–10 million years (Mattinson and Hopson, 1992).

Discussion

Multi-stage, supra-subduction zone origin for the Coast Range ophiolite

It now seems relatively well established that the CRO of California formed in a fore-arc setting,

TABLE 2. Summary of Published Age Dates for the Coast Range Ophiolite[1]

Locale	Rock type	Age	System	Reference
Elder Creek	Hb gabbro	154 ± 5[2]	K-Ar Hb	Lanphere, 1971
	Hb gabbro	166 ± 3, 143–144 ± 3	K-Ar Hb	McDowell et al., 1984
Wilbur Springs	Hb gabbro	143–144 ± 3	K-Ar Hb	McDowell et al., 1984
Harbin Springs	Plagiogranite	169[3]	U-Pb zircon	*J. M. Mattinson, reported in McLaughlin and Ohlin, 1984*
Healdsburg	Plagiogranite	163 ± 2	U-Pb zircon	Hopson et al., 1981
Mt Diablo	Plagiogranite	**165**[3]	U-Pb zircon	*J. M. Mattinson, reported in Mankinen et al., 1991*
Del Puerto (Red Mtn.)	Hb gabbro	162 ± 5[2]	K-Ar Hb	Lanphere, 1971
	Hb peridotite	164 ± 5[2]	K-Ar Hb	Lanphere, 1971
	Plagiogranite	154–157 ± 2	U-Pb zircon	Hopson et al., 1981
	Hb gabbro	157[3]	^{40}Ar-^{39}Ar Hb	*Evarts et al., 1992*
	Rhyolite, andesite	**150**[3]	^{40}Ar-^{39}Ar, fission track	*Evarts et al., 1992*
Llanada	Hb albitite	163–165 ± 2		Hopson et al., 1981
Cuesta Ridge	Plagiogranite	152–153 ± 3	U-Pb zircon	Hopson et al., 1981
Stanley Mtn	Plagiogranite	**166**[3]	U-Pb zircon	*J. M. Mattinson, reported in Pessagno et al., 1993*
Point Sal	Plagiogranite	160–162 ± 2	U-Pb zircon	Hopson et al., 1981
	Plagiogranite	**165-173**[3]	U-Pb zircon	*Mattinson and Hopson, 1992*
Santa Cruz Island	Plagiogranite	161–167 ± 2	U-Pb zircon	Hopson et al., 1981
	Plagiogranite	144–148 ± 2	U-Pb zircon	Hopson et al., 1981

[1]Older K-Ar dates recalculated by McDowell et al., 1984. Most of the recent (post-1990) dates are reported in abstracts or as personal communications, with no supporting data or uncertainties (references shown in italics). Dates in bold are considered to be the most reliable at this time (dates based on modern U-Pb zircon work [since 1990].
[2]Ages recalculated by McDowell et al., 1984, using new decay constants.
[3]Ages reported in abstracts or as personal communication in unrelated paper, with no supporting data.

probably above the east-dipping proto-Franciscan subduction zone (e.g., Evarts, 1977; Shervais and Kimbrough, 1985a, 1985b; Lagabrielle et al., 1986; Robertson, 1989, 1990; Shervais, 1990, 2001; Shervais and Beaman, 1989, 1991, 1994; Stern and Bloomer, 1992; Giaramita et al., 1998). Three distinct magmatic suites can be found in many remnants of the Coast Range ophiolite, all of which formed in this supra-subduction zone setting; these suites and their occurrence are summarized in Table 1.

The first magmatic suite includes dunites, layered and foliated cumulate gabbros, some isotropic gabbros, portions of the sheeted complex, and a lower or older volcanic series (arc tholeiite or, less commonly, low-K calc-alkaline), where these are preserved. Volcanic rocks of this suite are characterized by low Ti and Cr over a range of silica contents (dominantly basaltic andesite to andesite in composition; Fig. 3), and low Ti/Zr and Ti/V ratios (Fig. 4). The second magmatic suite includes intrusive wehrlites, lherzolites, clinopyroxenites, primitive gabbros, and gabbronorites in the plutonic sections, portions of the sheeted complex, and an upper volcanic series with high-Ca boninitic affinities. The volcanic rocks of this suite are

A. Stage 1: Birth

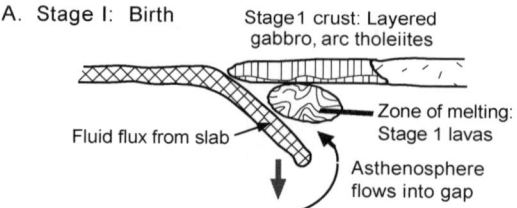

B. Stage 2: Youth

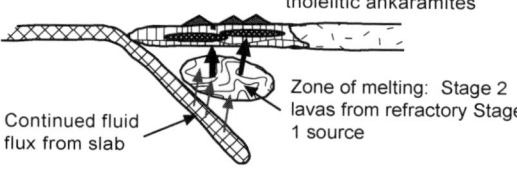

C. Stage 3: Maturity

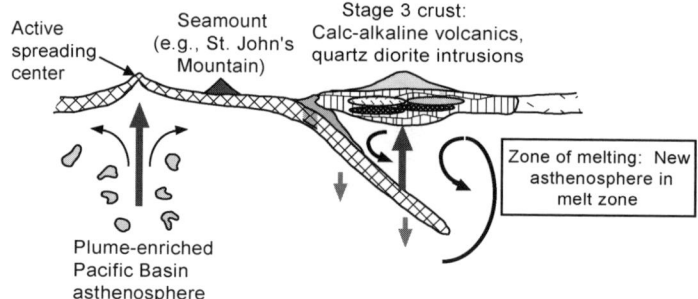

D. Stage 4: Death

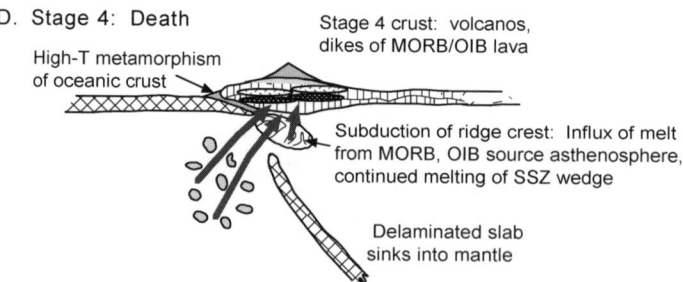

E. Stage 5: Resurrection

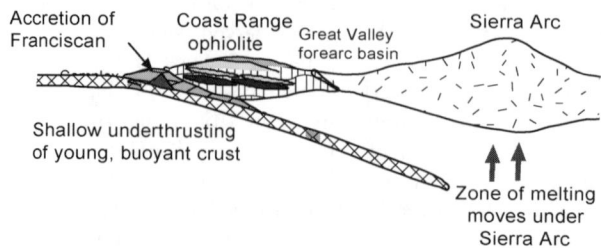

See caption on facing page.

FIG. 5. Schematic diagrams showing model for the development of the Coast Range ophiolite by a five-stage process, as recorded in the magmatic and tectonic record (after Shervais, 2001). These stages include the following. A. Stage 1, "Birth" = formation of the oldest magma series (arc tholeiite) in a nascent subduction zone during hinge rollback of the sinking slab. B. Stage 2, "Youth" = formation of a refractory magma series by remelting the mantle asthenosphere depleted during stage 1 melting, in response to fluid flux from the sinking slab. C. Stage 3, "Maturity" = onset of stable calc-alkaline volcanism as the subduction zone matures. D. Stage 4, "Death" = end of ophiolite formation caused by ridge subduction, and seen as oceanic magmas overlying the older ophiolite lithologies. E. Stage 5, "Resurrection" = uplift and emplacement of the ophiolite in response to accretionary uplift.

characterized by high Cr over a range of silica contents (Fig. 3), and by low TiO_2, Ti/Zr, and Ti/V ratios (Fig. 4). The third magmatic suite includes hornblende diorites, quartz diorites, plagiogranites, diorite-matrix agmatites, portions of the sheeted complex, and the volcaniclastic cover ("non-ophiolite volcanics") of andesite to rhyolite composition (Fig. 3) that rests on the underlying massive pillow lavas. The third magmatic suite is calc-alkaline in character and is often not included with the "true" ophiolite magma series, even though all three lie beneath the radiolarian cherts and clastic sediments that define the upper surface of the ophiolite.

The three magmatic episodes described above correspond to the three progressive stages in ophiolite formation as defined by Shervais (2001). These include: (1) birth—formation of incipient island-arc crust above a nascent subduction zone (e.g., Casey and Dewey, 1984; Hawkins et al., 1984; Stern and Bloomer, 1992); (2) youth—rifting and deformation of this crust, which allowed the intrusion of primitive boninitic magmas; and (3) maturity—transition to more normal calc-alkaline magmatism. Application of this model to the CRO is depicted schematically in Figure 5.

These three stages and their associated magma series form the classic ophiolite stratigraphy and comprise the bulk of all supra-subduction zone ophiolites. These stages and their associated magma series always form in the same relative order, and if evidence for one magma series is missing at any given locale, the other series still exhibit the same order of formation. They also seem to occur in almost all supra-subduction zone ophiolites. Shervais (2001) interpreted this sequence in terms of a model of progressive ophiolite formation, starting with a nascent subduction zone and progressing toward a stable calc-alkaline arc. Ophiolite formation may be terminated by collision with a spreading center, but this is not always apparent. Subsequent emplacement of the ophiolite may occur by obduction onto a passive continental margin (Tethyan ophiolites) or by accretionary uplift (Cordilleran ophiolites; Shervais, 2001).

Ridge collision: Death of an ophiolite

The fourth magma series, found in only a few CRO locales, represents the influx of true oceanic basalt magmas to form dikes and pillows that crosscut or overlie all of the older igneous rock series (Fig. 2; Table 1). Evidence for this late, oceanic basalt event includes late dikes of MORB composition that intrude older calc-alkaline intrusives or volcanics (Elder Creek, Mount Diablo, Del Puerto Canyon, Sierra Azul, Cuesta Ridge), lavas of MORB or oceanic basalt composition that overlie older subduction-related igneous suites (Stonyford volcanic complex, Geyser Peak/Black Mountain, Mount Diablo, Cuesta Ridge), intercalation of oceanic basalts with more arc-like lavas (Stonyford volcanic complex), and inclusions of MORB glass in rhyolite glass (Leona Rhyolite). There are three possible interpretations for these data: (1) the change to MORB-like chemistry represents continued rifting of the arc to form a backarc basin in which lavas have more oceanic compositions (e.g., Shervais and Kimbrough, 1985a, for the Mount Diablo ophiolite); (2) the propagation of a pre-existing backarc basin spreading center into the forearc (e.g., Harper, 2003); or (3) collision of the subduction zone with an active spreading center, which leaks oceanic magmas through the overlying plate (Shervais, 1993).

We favor the third explanation here (Fig. 5). Evidence for ridge collision includes the abrupt nature of the change from calc-alkaline to oceanic lavas, the intercalation of plume-enriched arc lavas with oceanic tholeiites and alkali basalts at Stonyford, the forearc location of the ophiolite (which would place the ophiolite directly over the mid-ocean ridge when it collides), and the presence of mafic crust under the Great Valley Sequence that is too thick to represent normal backarc basin crust (e.g., Godfrey and Klemperer, 1998). Further evidence comes from accreted terranes in the Klamath Mountains, the

Sierra Nevada, and the Franciscan Complex, where formation ages and oceanic residence times show a reversal in the early Late Jurassic, indicating that a spreading center was consumed at that time (Murchey and Blake, 1993).

Ophiolite locales such as Black Mountain may represent "captured" segments of the subducting ridge axis, where the subtle geochemical overprint of subduction enrichment (e.g., elevated Th concentrations) could result from the backflow of subduction-zone fluids through a slab window (Casey and Dewey, 1984). Similar scenarios may apply to other ophiolite fragments in the CRO, where no clear arc-related basement can be demonstrated. In other cases (e.g., Mount Diablo, Stonyford) the demonstrated pre-existence of arc-related crust implies that the ophiolite remnant formed in the upper plate of the subduction zone, but was penetrated by magmas from the ridge axis. This seems to be the case for most CRO locales, where oceanic basalts (MORB, alkali) intrude or overlie older arc-related basement. These locales include Elder Creek, Stonyford, Mount Diablo, the Berkelely Hills (Leona Rhyolite), Del Puerto Canyon, Sierra Azul, and Cuesta Ridge—that is, most of the known CRO remnants.

Collision with an active spreading ridge would likely end ophiolite formation in most locations because of the resulting change in relative plate motions, and because the young lithosphere adjacent to the spreading center is too bouyant to sink into the mantle (Cloos, 1993). As a result, continued subduction would take place at an angle too low to support magma formation in the forearc, and magmatic activity would shift inland, toward the volcanic arc, after a hiatus to establish normal subduction. The change from "hinge-rollback"–driven spreading (e.g., Stern and Bloomer, 1992) to shallow underthrusting of hot oceanic asthenosphere would initiate high-temperature metamorphism of the upper oceanic crust in the thrust zone, and compressional deformation of the forearc region.

As noted by Cloos (1993, p. 733), "The subduction of spreading ridges will cause vertical isostatic uplift and subsidence of as much as 2 to 3 km in the forearc region compared to when 80 m.y.-old oceanic lithosphere is subducted. The subduction of an active spreading center causes such a major perturbation in the margins thermal structure that evidence of the event is likely to be recorded widely in the geology of the forearc block." Thus, it may be no accident that high-grade metamorphic blocks in the Franciscan Complex in northern California (garnet amphibolite, eclogite) have maximum ages ($\approx$160–162 Ma; Coleman and Lanphere, 1971; Mattinson, 1986, 1988; Ross and Sharp, 1986) that correspond to a ridge-subduction event ($\approx$164 Ma) in this area, and not to initial ophiolite formation (~172 Ma). Models that link formation of the high-grade blocks to subduction initiation are not supported by the age data, which show that initiation of CRO formation began ~173 Ma and ended about 164–162 Ma.

Eclogite-facies metamorphism appears to overprint an older amphibolite-facies assemblage in many high-grade blocks (Moore and Blake, 1989); this would result from the cooling and continued subduction of amphibolites formed at hotter, shallower conditions. In addition, many CRO localities (especially those in the Northern/Sacramento Valley Belt) show evidence for uplift, high-angle faulting, and erosion prior to deposition of the overlying Great Valley Sequence (Hopson et al., 1981; Phipps and MacPherson, 1992).

The presence of oceanic magmas as the final magmatic event in ophiolite formation is not unique to the CRO. The Semail ophiolite (Oman) is underlain by garnet amphibolites with oceanic provenance that are only slightly younger than the third magmatic suite (diorites, plagiogranites), and is overlain by oceanic alkali basalts (Salahi volcanics) that reflect leakage of the oceanic spreading center through the overlying plate (Shervais, 2001).

Accretionary uplift as an alternative to obduction

One consequence of ophiolite formation in the upper plate of a subduction zone is that emplacement of the ophiolite onto a continental margin may occur by obduction (Gealey, 1977; Coleman, 1981). Obduction occurs when the ophiolite subduction zone attempts to subduct a passive continental margin (Gealey, 1977; Coleman, 1981; Pearce et al., 1981; Searle and Cox, 1999). Thrusting of the ophiolite over continental crust is terminated by the buoyant rise of the continental lithosphere, which lifts the ophiolite and exposes it subaerially (Gealey, 1977).

An alternative to obduction that may expose ophiolites is the process that has been defined previously as "accretionary uplift" (Shervais, 2001). Accretionary uplift occurs in response to continued subduction, regardless of whether or not a ridge collision event occurs. If a passive continental margin is not encountered, a thick accretionary complex will form and the ophiolite will be exposed by progressive uplift as the accretionary complex thickens. When the accretionary complex becomes overthickened and gravitationally unstable, unroofing and exhumation may occur by extensional denudation (e.g., Platt, 1986; Jayko et al., 1987), by rapid erosion (Ring and Brandon, 1999), or both.

This is the process that we believe is responsible for uplift and emplacement of the CRO. Formation of the Franciscan complex throughout the late Mesozoic and early Paleogene underplated and thickened crust beneath the CRO, causing uplift and extensional deformation of the ophiolite (Jayko et al., 1987; Harms et al., 1992). Relatively early exhumation of the ophiolite is documented by the occurrence of sedimentary serpentinite debris flows in the lower Great Valley sequence (Moiseyev, 1970; Phipps, 1984). After subduction ceased in the Miocene, deformation was dominated by wrench faulting, compression, and tectonic wedging of Franciscan rocks under the CRO/Great Valley section (Wentworth et al., 1984; Glen, 1990; Blake et al., 1992; Unruh et al., 1995).

Implications for Jurassic tectonics of the Western Cordillera

It has been suggested that the CRO formed in a backarc basin behind Middle Jurassic arc rocks of the western Sierra Foothills belt and was emplaced during an arc-arc collision (the Late Jurassic Nevadan orogeny, ≈155 Ma) that resulted in formation of a new, proto-Franciscan subduction zone to the west (e.g., Schweikert and Cowan, 1975; Dickinson et al., 1996; Godfrey and Klemperer, 1998). These models are based on a view of Jurassic orogeny in the Cordillera that is no longer valid (e.g., Wright and Fahan, 1988). It is now known that orogenic activity during the Jurassic involved a complex series of events that began in the Early to Middle Jurassic and continued into the Late Jurassic, ending largely by ≈145 Ma (Harper and Wright, 1984; Wright and Fahan, 1988; Saleeby, 1982, 1983, 1990; Tobisch et al., 1989; Wolf and Saleeby, 1995). Many of the structures and fabrics attributed to Nevadan deformation in the Sierra Foothills are now known to be early Middle Jurassic in age (pre-168 Ma, post-190 Ma) and record deformation that occurred prior to formation of the CRO (Edelman et al., 1989; Edelman and Sharp, 1989; Saleeby, 1990). Girty et al. (1995) have shown that Nevadan age deformation took place within the arc, and that the forearc region lay to the west of this arc. They also suggest that a new subduction zone may have formed in the early Middle Jurassic (≈174 Ma) in response to an arc collision or collapse of a fringing arc against the continental margin (Girty et al., 1995). This is slightly older than the oldest ophiolitic rocks that have been dated in the CRO, and slightly younger than the initial opening of central Atlantic at about 175–180 Ma (Klitgord and Schouten, 1986).

Geochemical data show that the stage 2 volcanics found at many CRO locales are related to high-Ca boninites (Figs. 3 and 4), which are characteristic of forearc volcanism during brief episodes of rapid spreading. The occurrence of wehrlite intrusives (which represent ultramafic cumulates of high-Ca boninite magmas) into the older layered gabbros (arc tholeiite cumulates), and deformational fabrics within the older layered gabbros, imply that the CRO formed initially by rapid spreading in the forearc, as proposed by Stern and Bloomer (1992). In this model, extensions deformation of the older arc gabbros is common prior to intrusion of the later ultramafic cumulates (Shervais, 2001).

Ward (1995) has shown that subduction was not continuous during the Jurassic, as assumed previously, but varied in response to changes in plate motions and convergence. He has shown that both relative and absolute plate motions changed significantly at the beginning of the Late Jurassic—coincident with the ridge subduction event postulated here, and just prior to the Nevadan orogeny. Plate motion studies (Ward, 1995) and structural analysis of dike swarms in the Foothills terrane (Wolf and Saleeby, 1995) both show a change from relative convergence of North America and plates of the Pacific basin in the middle Jurassic to left-lateral transtension (in the upper plate) during the Late Jurassic. Ward (1995) suggested that this change must coincide with a ridge "collision" event.

In summary, the model presented here is consistent with our current understanding of tectonic evolution in the Western Cordillera during the mid- to Late Jurassic. Models that call for arc collision during the late Jurassic and formation of the CRO in a backarc basin are not compatible with these data, or with the chemical and petrologic characteristics of the CRO.

Implications for the origin and evolution of ophiolites

The data presented here show that the Coast Range ophiolite of California followed a consistent progression during its formation over a primary linear extent of at least 1300 km, from Elder Creek in the north to Point Sal in the south (when Neogene motion on the San Andreas fault system is restored), or 1600 km if the southwest Oregon remnants are included. This progression in magmatic evolution, from primitive arc tholeiite to boninitic to calc-alkaline, is the same progression followed by other ophiolites worldwide (Shervais, 2001). The global occurrence of this consistent progression implies that ophiolite formation is not a stochastic event, but is a natural consequence of the tectonic setting in which ophiolites form. The first three stages all form in a supra-subduction zone setting, that is, in the upper plate of a subduction zone (Shervais, 2001). The most likely setting appears to be subduction initiation and hinge rollback (Stern and Bloomer, 1992), but other settings are possible. In any event, models of ophiolite formation that call on unusual combinations of circumstances, or on non-uniformitarian interpretations of ocean crust formation, are not compatible with this observed progression.

The termination of ophiolite formation by ridge collision is a natural consequence of ophiolite formation above a subduction zone whenever the plate being consumed is actively generated at a spreading center and the convergence rate exceeds the spreading rate (Shervais, 2001). This termination will be diachronous unless the ridge is parallel to the trench. Evidence for ridge collision may include the intrusion of oceanic lavas into the ophiolite complex, the eruption of oceanic lavas on top of the ophiolite complex, or the occurrence of high-grade metamorphic soles with oceanic affinities. The occurrence of high-grade metamorphic soles in which the protoliths are oceanic basalts is common to many ophiolites, suggesting that ridge collision is more common than the current literature suggests.

Finally, the variations observed among different ophiolite remnants in the CRO shows that concentrating our collective efforts on studying a few occurrences within any given ophiolite, and ignoring those that are less well exposed, obscures the insights that can be gained by considering all ophiolite occurrences in sufficient detail to understand their fundamental characteristics and evolution. Indeed, focusing on one or two classic occurrences would lead to significant misconceptions regarding how the CRO formed and evolved. The CRO of California shows how detailed investigations of many ophiolite occurrences are needed to understand better the origin and evolution of all ophiolites.

Acknowledgments

This paper would not have been possible without the pioneering work and insights of Cliff Hopson, who introduced us (Shervais, Kimbrough, Hanan) to the Coast Range ophiolite and who has provided the inspiration for our continued work there. This research was supported by NSF grants EAR8816398 and EAR9018721 (Shervais) and EAR9018275 (Kimbrough and Hanan), and by the U.S. Geological Survey, National Cooperative Geologic Mapping Program (EDMAP grant 01HQAG0152) to support mapping in the Cuesta Ridge area (Shervais and Snow). The views and conclusions contained in this document are those of the authors and should not be interpreted as necessarily representing the official policies, either expressed or implied, of the U.S. Government. Many thanks to Thomas Moore for a thoughtful review.

REFERENCES

Bailey, E. H., and Blake, M. C., Jr., 1974, Major chemical characteristics of Mesozoic Coast Range ophiolite in California: U.S. Geological Survey Journal of Research, v. 2, p. 637–656.

Bailey, E. H., Blake, M. C., Jr., and Jones, D. L., 1970, Onland Mesozoic oceanic crust in California Coast Ranges: U.S. Geological Survey Professional Paper 700-C, p. C70–80.

Bailey, E. H., Irwin, W. P., and Jones, D. L., 1964, Franciscan and related rocks: California Division of Mines and Geology, Bulletin 183, 177 p.

Beaman, B. J., 1989, Petrology and geochemistry of the Elder Creek ophiolite, California: Unpubl. M. S. thesis, University of South Carolina, 155 p.

Bezore, S. P., 1969, The Mount Saint Helena ultramafic-mafic complex of the northern Coast Ranges [abs.], in Geological Society of America, Abstracts with Programs, v. 1, p. 5.

Blake, M. C., Engebretson, D. C., Jayko, A. S., and Jones, D. L., 1985, Tectonostratigrapic terranes in southwest Oregon, in Howell, D. G., ed., Tectonostratigraphic terranes of the Circum Pacific region: Circum Pacific Council for Energy and Mineral Resources, Earth Science Series, v. 1, p. 147–157.

Blake, M. C., Jayko, A. S., Jones, D. L., and Rogers, B. W., 1987, Unconformity between Coast Range ophiolite and part of the lower Great Valley Sequence, South Fork of Elder Creek, Tehama County, California: Geological Society of America, Centennial Field Guide–Cordilleran Section, 1987, p. 279–282.

Blake, M. C., Jr., Jayko, A. S., Murchey, B. L., and Jones, D. L., 1992, Formation and deformation of the Coast Range Ophiolite and related rocks near Paskenta,

California: Bulletin of the American Association of Petroleum Geologists, v. 76, p. 417.

Blake, M. C., Jr., and Jones, D. L., 1981, The Franciscan assemblage and related rocks in northern California: A re-interpretation, *in* Ernst, W. G., ed., The geotectonic development of California, Rubey Volume I: Englewood Cliffs, NJ, Prentice-Hall, p. 306–328.

Brown, E. H., Bradshaw, J. Y., et al., 1979, Plagiogranite and keratophyre in ophiolite on Fidalgo Island, Washington: Geological Society of America Bulletin, v. 90, p. 493–507.

Casey, J. F., and Dewey, J. F., 1984, Initiation of subduction zones along transform and accreting plate boundaries, triple-junction evolution, and forearc spreading centres-implications for ophiolitic geology and obduction, *in* Gass, I. G., Lippard, S. J., and Shelton, A. W., eds., Ophiolites and oceanic lithosphere: Geological Society of London Special Publication 13, p. 269–290.

Cloos, Mark, 1993, Lithospheric bouyancy and collisional orogenesis: Subduction of oceanic plateaus, continental margins, island arcs, spreading ridges, and seamounts: Geological Society of America Bulletin, v. 105, p. 715–737.

Coleman, R. G., 1981, Tectonic setting for ophiolite obduction in Oman: Journal of Geophysical Research, v. 86, p. 2497–2508.

———, 2000, Prospecting for ophiolites along the California continental margin, *in* Dilek, Y., Moores, E. M., Elthon, D., and Nicolas, A. eds., Ophiolites and oceanic crust: New insights from field studies and the Ocean Drilling Program: Geological Society of America, Special Paper 349, p. 351–364.

Coleman, R. G., and Lanphere, M. A., 1971, Distribution and age of high-grade blueschist, associated eclogites, and amphibolites from Oregon and California: Geological Society of America Bulletin, v. 82, p. 2397–2412.

Crawford, A. J., Falloon, T. J., and Green, D. H., 1989, Classification, petrogenesis, and tectonic setting of boninites, *in* Crawford, A. J., ed., Boninites and related rocks: London, UK, Unwin Hyman, p. 2–49.

Dickinson, W. R., Hopson, C. A., Saleeby, J. B., Schweickert, R. A., Ingersoll, R. V., Pessagno, E. A., Jr., Mattinson, J. M., Luyendyk, B. P., Beebe, W., Hull, D. M., Munoz, I. M., and Blome, C. D., 1996, Alternate origins of the Coast Range Ophiolite (California): Introduction and implications: GSA Today, v. 6, no. 2, p. 1–10.

Edelman, S. H., Day, H. W., and Bickford, M. E., 1989, Implications of U-Pb zircon ages for the Smartville and Slate Creek complexes, northern Sierra Nevada, California: Geology, v. 17, p. 1032–1035.

Edelman, S. H., and Sharp, W. D., 1989, Terranes, early faults, and pre-late Jurassic amalgamation of the western Sierra Nevada metamorphic belt, California: Geological Society of America Bulletin, v. 101, p. 1420–1433.

Emerson, N. L., 1979, Lower Tithonian volcaniclastic rocks above the Llanada ophiolite, California: Unpubl. M.S. thesis, University of California, Santa Barbara, 65 p.

Enos, P., 1963, Jurassic age of Franciscan Formation south of Panoche Pass, California: Bulletin of the American Association of Petroleum Geologists, v. 47, p. 158–163.

Evarts, R. C., 1977, The geology and petrology of the Del Puerto ophiolite, Diablo Range, central California Coast Ranges, *in* Coleman, R. G., and Irwin, W. P., eds., North American ophiolites: Oregon Department of Geology and Mineral Industries Bulletin 95, p. 121–140.

Evarts, R. C., Coleman, R. G., and Schiffman, P., 1999, The Del Puerto ophiolite: Petrolgy and tectonic setting, *in* Wagner, D. L., and Graham, S. A., eds., Geologic field trips in Northern California: California Division of Mines and Geology, Special Publication 119, p. 136–149.

Evarts, R. C., and Schiffman, P., 1983, Submarine hydrothermal metamorphism of the Del Puerto ophiolite, California: American Journal of Science, v. 283, p. 289–340.

Evarts, R. C., Sharp, W. D., and Phelps, D. W., 1992, The Del Puerto Canyon remnant of the Great Valley ophiolite: Geochemical and age constraints on its formation and evolution: Bulletin of the American Association of Petroleum Geologists, v. 76, p. 418.

Gealey, W. K., 1977, Ophiolite obduction and geologic evolution of the Oman Mountains and adjacent areas: Geological Society of America Bulletin, August, v. 88, p. 1183–1191.

Giaramita, M. I., MacPherson, G. J., and Phipps, S. P., 1998, Petrologically diverse basalts from a fossil oceanic forearc in California: The Llanada and Black Mountain remnants of the Coast Range ophiolite: Geological Society of America Bulletin, v. 110, p. 553–571.

Girty, G. H., Hanson, R. E., Girty, M. S., Schweickert, R. A., Harwood, D. S., Yoshinobu, A. S., Bryan, K. A., Skinner, J. E., and Hill, C. A., 1995, Timing of emplacement of the Haypress Creek and Emigrant Gap plutons, *in* Miller, D. M., and Busby, C., eds., Jurassic magmatism and tectonics of the North American Cordillera: Geological Society of America, Special Paper 299, p. 191–202.

Glen, R. A., 1990, Formation and thrusting in some Great Valley rocks near the Franciscan complex, California, and implications for the tectonic wedging hypothesis: Tectonics, v. 9, p. 1451–1477.

Godfrey, N. J., and Klemperer, S. L., 1998, Ophiolitic basement to a forearc basin and implications for continental growth; the Coast Range/Great Valley Ophiolite, California: Tectonics, v. 17, p. 558–570.

Gray, F., Ramp, L., Moring, B., Douglas, I., and Donahoe, J., 1982, Geologic map of the Wild Rogue Wilderness,

Coos, Curry, and Douglas Counties, Oregon: U.S. Geological Survey Miscellaneous Field Studies Map MF-1381-A.

Hagstrum, J. T., and Jones, D. L., 1998, Paleomagnetism, paleogeographic origins, and uplift history of the Coast Range ophiolite at Mount Diablo, California: Journal of Geophysical Research, v. 103, B1, p. 597–603.

Hagstrum, J. T., and Murchey, B. L., 1996, Paleomagnetism of Jurassic radiolarian chert above the Coast Range ophiolite at Stanley Mountain, California, and implications for its paleogeographic origins: Geological Society of America Bulletin, v. 108, p. 643–652.

Hanan, B. B., Kimbrough, D. L., and Renne, P. R., 1992, The Stonyford Volcanic Complex: A Jurassic seamount in the northern California Coast Ranges: Bulletin of the American Association of Petroleum Geologists, v. 76, p. 421.

Hanan, B. B., Kimbrough, D. L., Renne, P. R., and Shervais, J. W., 1991, The Stonyford Volcanic Complex: Pb isotopes and $^{40}Ar/^{39}Ar$ ages of volcanic glass from a Jurassic seamount in the northern California Coast Ranges [abs.]: Geological Society of America, Abstracts with Programs, v. 23, no. 5, p. A395.

Harms, T. A., Jayko, A. S., and Blake, M. C., Jr., 1992, Kinematic evidence for extensional unroofing of the Franciscan Complex along the Coast Range Fault, northern Diablo Range, California: Tectonics, v. 11, no. 2, p. 228–241.

Harper, G. D., 1984, The Josephine ophiolite, northwestern California: Geological Society of America Bulletin, v. 95, p. 1009–1026.

———, 2003, Fe-Ti basalts and propagating-rift tectonics in the Josephine ophiolite: Geological Society of America Bulletin, v. 115, p. 771–787.

Harper, G. D., Saleeby, J. B., and Heizler, M., 1994, Formation and emplacement of the Josephine Ophiolite and the Nevadan Orogeny in the Klamath Mountains, California-Oregon; U/Pb zircon and $^{40}Ar/^{39}Ar$ geochronology: Journal of Geophysical Research, v. B99, no. 3, p. 4293–4321.

Harper, G. D., Saleeby, J. B., and Norman, E. A. S., 1985, Geometry and tectonic setting of sea floor spreading for the Josephine ophiolite, and implications for Jurassic accretionary events along the California margin, in Howell, D. G., ed., Tectonostratigraphic terranes of the Circum-Pacific region: Circum-Pacific Council for Energy and Mineral Resources Series, no. 1, p. 239–257.

Harper, G. D., and Wright, J. E., 1984, Middle to Late Jurassic tectonic evolution of the Klamath Mountains, California-Oregon: Tectonics, v. 3, p. 759–772.

Hawkins, J. W., Bloomer, S. H., Evans, C. A., and Melchior, J. T., 1984, Evolution of intra-oceanic arc-trench systems: Tectonophysics, v. 102, p. 175–205.

Hopson, C. A., and Frano, C. J., 1977, Igneous history of the Point Sal ophiolite, southern California, in Coleman, R. G., and Irwin, W. P., eds., North American ophiolites: Oregon Department of Geology and Mineral Industries Bulletin 95, p. 161–183.

Hopson, C. A., Mattinson, J. M., and Pessagno, E. A., 1981, Coast Range Ophiolite, western California, in Ernst, W. G., ed., The geotectonic development of California, Rubey Volume I: Englewood Cliffs, NJ, Prentice-Hall, p. 418–510.

Hopson, C. A., Mattinson, J. M., Luyendyk, B. P., Beebe, W., Pessagno, E. A., Jr., Hull, D. M., and Blome, C. D., 1992, California Coast Range ophiolite: Jurassic tectonic history: Bulletin of the American Association of Petroleum Geologists, v. 76, p. 422.

Hopson, C. A., Mattinson, J. M., Luyendyk, B. P., Beebe, W. J., Pessagno, E. A., Jr., Hull, D. M., Munoz, I. M., and Blome, C. D., 1997, Coast Range ophiolite: Paleoequatorial ocean-ridge lithosphere: Bulletin of the American Association of Petroleum Geologists, v. 81, p. 687.

Hull, D. M., 1995, Morphologic diversity and paleogeographic significance of the Family Parvicingulidae (Radiolaria): Micropaleontology, v. 41, p. 1–48.

———, 1997, Upper Jurassic Tethyan and southern Boreal radiolarians from western North America: Micropaleontology, v. 43, Suppl. no. 2, p. 202.

Hull, D. M., and Pessagno, E. A., Jr., 1994, Upper Jurassic radiolarian biostratigraphy of Stanley Mountain, Southern California Coast Range, in Cariou, E., and Hantzpergue, P., eds., 3eme Symposium international de stratigraphie du Jurassique: Geobios, Memoire Special: Lyon, France, Universite Claude Bernard, Departement de Geologie, p. 309–315.

———, 1995, Radiolarian stratigraphic study of Stanley Mountain, California, in Baumgartner, P. O., O'Dogherty, L., Gorican, S., Urquhart, E., Pillevuit, A., and De Wever, P., eds., Middle Jurassic to Lower Cretaceous radiolaria of Tethys: Occurrences, systematics, biochronology: Lausanne, Switzerland, Memoires de Geologie, p. 985–996.

Huot, F., and Maury, R. C., 2002, The Round Mountain serpentinite melange, northern Coast Ranges of California: An association of backarc and arc-related tectonic units: Geological Society of America Bulletin, v. 114, p. 109–123.

Imlay, R. W., 1980, Jurassic paleobiogeography of the conterminous United States in its continental setting: U.S. Geological Survey Professional Paper, v. 1062, p. 1–134.

Ingersoll, R. V., and Dickinson, W. R., 1981, Great Valley Group (sequence), Sacramento Valley, California, in Frizzel, V., ed., Upper Mesozoic Franciscan rock and Great Valley Sequence, Central California Coast Ranges, California: Los Angeles, CA, Pacific Section, Society of Economic Paleontologists and Mineralogists, 33 p.

Jayko, A. S. and Blake, M. C., Jr., 1987, Significance of Klamath rocks between the Franciscan complex and

Coast Range ophiolite, northern California: Tectonics, v. 5, p. 1055–1071.

Jayko, A. S., Blake, M. C., Jr., and Harms, T., 1987, Attenuation of the Coast Range ophiolite by extensional faulting and the nature of the Coast Range thrust, California: Tectonics, v. 6, p. 475–488.

Jones, D. L., 1973, Structural significance of upper Mesozoic biostratigraphic units in northern California and southwest Oregon [abs.]: Geological Society of America, Abstracts with Programs, v. 5, no. 7, p. 684–685.

———, 1975, Discovery of *Buchia rugosa* of Kimmeridgian age from the base of the Great Valley Sequence [abs.]: Geological Society of America, Abstracts with Programs, v. 7, p. 330.

Jones, D. L., and Curtis, G., 1991, Guide to the geology of the Berkeley Hills, Central Coast Ranges, California, *in* Sloan, D., and Wagner, D. L., eds., Geologic excursions in Northern California: San Francisco to the Sierra Nevada: California Department of Conservation, Division of Mines and Geology, Special Publication 109, p. 63–73.

Kimbrough, D. L., 1982, Structure, petrology, and geochronology of Mesozoic paleo-oceanic terranes of Cedros Island and the Vizcaino Peninsula: Unpubl. Ph.D. dissertation, University of California, Santa Barbara, 395 p.

Klitgord, K. D., and Schouten, H., 1986, Plate kinematics of the central Atlantic, *in* Vogt, P. R., and Tucholke, B. E., eds., Decade of North American geology, the geology of North America: The western North Atlantic region: Boulder, CO, Geological Society America, 1986, p. 351–378.

Kosanke, S. B., 2000, The geology, geochronology, structure, and geochemistry of the Wild Rogue Wilderness remnant of the Coast Range ophiolite, SW Oregon: Implications for the tectonic evolution of the Coast Range ophiolite: Unpubl. Ph.D. dissertation, State University of New York, Albany, 572 p.

Kosanke, S. B., and Harper, G. D., 1997, Geochemical variations in volcanic rocks from a fragment of the Jurassic Coast Range Ophiolite, Oregon: Possible indication of fore-arc rifting [abs.]: Geological Society of America, Abstracts with Programs, v. 29, no. 6, p. 159.

Kosanke, S. B., Harper, G. D., and Heizler, M., 1999, Younger extrusive and intrusive volcanic arc rocks and Nevadan-age ductile deformation in the 164 Ma Coast Range Ophiolite, SW Oregon: Geological Society of America, Abstracts with Programs, v. 31, no. 6, p. 71.

Lagabrielle, Y., Roure, F., Coutelle, A.., Maury, R., Joron, J. L., and Thonon, P., 1986, The Coast Range ophiolite (northern California) possible arc and marginal basin remnants, their relations with the Nevadan orogeny: Bulletin de la Société Géologique de France, v. 8, p. 981–999.

Lanphere, M. A., 1971, Age of the Mesozoic oceanic crust in the California Coast Ranges: Geological Society of America Bulletin, v. 82, p. 3209–3211.

Lippard, S. J., Shelton, A. W., and Gass, I. G., 1986, The ophiolite of northern Oman: Oxford, UK, The Geological Society, Memoir no. 11, Blackwell Scientific Publications, 178 p.

MacPherson, G. J., 1983, The Snow Mountain Complex: An on-land seamount in the Franciscan terrain, California: Journal of Geology, v. 91, p. 73–92.

———, 1986, Incipient blueschist facies mineral assemblges from metavolcanic rocks of the Snow Mountain Complex [abs.]: Geological Society of America, Abstracts with Programs, Cordilleran section.

MacPherson, G. J., and Phipps, S. P., 1985, Geochemical evidence for the origin of the Coast Range ophiolite: Comment: Geology, v. 13, p. 827–828.

MacPherson, G. J., Phipps, S. P., and Grossman, J. N., 1990, Diverse sources for igneous blocks in Franciscan melanges, California Coast Ranges: Journal of Geology, v. 98, p. 845–862.

Mankinen, E. A., Gromme, C. S., and Williams, K. M., 1991, Concordant paleolatitudes from ophiolite sequences in the northern California Coast Ranges, U.S.A.: Tectonophysics, v. 198, p. 1–21.

Mattinson, J. M., 1986, Geochronology of high-pressure–low-temperature Franciscan metabasite: A new approach using the U-Pb system, *in* Evans, B. W., and Brown, E. H., eds., Blueschists and eclogites: Geological Society of America, Memoir 164, p. 95–105.

———, 1988, Constraints on the timing of Franciscan metamorphism; geochronological approaches and their limitations, *in* Ernst, W. G., ed., Rubey colloquium on metamorphism and crustal evolution of the Western United States, Rubey Volume 7: Englewood Cliffs, NJ, Prentice-Hall, p. 1023–1034.

Mattinson, J. M., and Hopson, C. A., 1992, U/Pb ages of the Coast Range Ophiolite: A critical reevaluation based on new high-precision Pb/Pb ages: Bulletin of the American Association of Petroleum Geologists, v. 76, p. 425.

McDowell, F. W., Lehman, D. H., Gucwa, P. R., Fritz, D., and Maxwell, J. C., 1984, Glaucophane schists and ophiolites of the Northern California Coast Ranges: Isotopic ages and their tectonic implications: Geological Society of America Bulletin, v. 95, p. 1373–1382.

McLaughlin, R. J., 1978, Preliminary geologic map and structural sections of the central Mayacamas Mountains and the Geysers steam field, Sonoma, Lake, and Mendecino Counties, California: U.S. Geological Survey Open-file Map 78-389, scale 1:24,000.

McLaughlin, R. J., Blake, M. C., Griscom, A., Blome, C. D., and Murchey, B., 1988, Tectonics of formation, translation, and dispersal of the Coast Range ophiolite of California: Tectonics, v. 7, p. 1033–1056.

McLaughlin, R. J., Elder, W. P., and McDougall, K., 1991, Tectonic framework of the Loma Prieta area, *in* Sloan,

D., and Wagner, D. L., eds., Geologic excursions in Northern California: San Francisco to the Sierra Nevada: California Department of Conservation, Division of Mines and Geology, Special Publication 109, p. 45–54.

McLaughlin, R. J., Kistler, R. W., Wooden, J. L., and Franck, C. R., 1992, Coast Range ophiolite of the Sierra Azul block southwest of Los Gatos, California: Bulletin of the American Association of Petroleum Geologists, v. 76, p. 425.

McLaughlin, R. J., and Ohlin, H. N., 1984, Tectonostratigraphic framework of the Geysers–Clear Lake region, California, in Blake, M. C., ed., Franciscan geology of Northern California: Los Angeles, CA, Pacific Section, Society of Economic Paleontologists and Mineralogists, no. 43, p. 221–254.

McLaughlin, R. J., and Pessagno, E., 1978, Significance of age relations above and below upper Jurassic ophiolite in The Geysers–Clear Lake region, California: U.S. Geological Survey Journal of Research, v. 6, p. 715–726.

Moiseyev, A. N., 1970, Late serpentinite movements in the California Coast Ranges: New evidence and its implications: Geological Society of America Bulletin, v. 81, p. 1721–1732.

Moore, D. E., 1984, Metamorphic history of a high-grade blueschist exotic block from the Franciscan complex: Journal of Petrology, v. 25, p. 126–150.

Moore, D. E., and Blake, M. C., Jr., 1989, New evidence for polyphase metamorphism of glaucophane schist and eclogite exotic blocks in the Franciscan complex, California and Oregon: Journal of Metamorphic Petrology, v. 7, p. 211–228.

Murchey, B. L., and Blake, M. C., Jr., 1993, Evidence for subduction of a major ocean plate along the California margin during the middle to early late Jurassic, in Dunne, G. C., and McDougall, K., eds., Mesozoic paleogeography of the western United States II: Los Angeles, CA, Society of Economic Paleontologists and Mineralogists, Pacific Section, v. 71, p. 1–17.

Murchey, B. L., and Hagstrum, J. T., 1997, Paleomagnetism of Jurassic radiolarian chert above the Coast Range Ophiolite at Stanley Mountain, California, and implications for its paleogeographic origins: Reply: Geological Society of America Bulletin, v. 109, p. 1633–1639.

Page, B. M., 1972, Oceanic crust and mantle fragment in subduction zone omplex near San Luis Obispo, California: Geological Society of America Bulletin, v. 83, p. 957–972.

Palfy, J., Smith, P. L., and Mortensen, J. K., 2000, A U-Pb and $^{40}Ar/^{39}Ar$ time scale for the Jurassic: Canadian Journal of Earth Sciences/Revue Canadienne des Sciences de la Terre, v. 37, p. 923–944.

Pearce, J. A., Alabaster, T., Shelton, A. W., and Searle, M. P., 1981, The Oman ophiolite as a Cretaceous arc-basin complex: Evidence and implications: Philosophical Transactions of the Royal Society of London, v. A300, p. 299–317.

Pearce, J. A., and Cann, J. R., 1973, Tectonic setting of basic volcanic rocks determined using trace element analyses: Earth and Planetary Science Letters, v. 19, p. 290–300.

Pessagno, E. A., Blome, C. D., Hull, D. M., and Six, W. M., 1993, Jurassic Radiolaria from the Josephine ophiolite and overlying strata, Smith River subterrane (Klamath Mountains), northwestern California and southwestern Oregon: Micropaleontology, v. 39, p. 93–166.

Phipps, S. P., 1984, Ophiolitic olistostromes in the basal Great Valley Sequence, Napa County, northern California Coast Ranges, in Raymond, L. A., ed., Melanges: Their nature, origin and significance: Boulder, CO, Geological Society of America, Special Paper 198, p. 103–125.

Phipps, S. P., and MacPherson, G. J., 1992, The Coast Range Ophiolite: Polygenetic crust in a late Mesozoic oceanic fore arc: Bulletin of the American Association of Petroleum Geologists, 76, p. 428.

Pike, J. E. N., 1974, Intrusions and intrusive complexes in the San Luis Obispo ophiolite: A chemical and petrologic study: Unpubl. Ph.D. dissertation, Stanford University, 212 p.

Platt, J. P., 1986, Dynamics of orogenic wedges and uplift of high-pressure metamorphic rocks: Geological Society of America Bulletin, v. 97, p. 1037–1053.

Ring, U., and Brandon, M. T., 1999, Ductile deformation and mass loss in the Franciscan subduction complex; implications for exhumation processes in accretionary wedges, in Ring, U., Brandon, M. T., Lister, G. S., and Willett, S., eds., Exhumation processes: Normal faulting, ductile flow, and erosion: London, UK, Geological Society of London, Special Publication 154, p. 55–86.

Robertson, A. H. F., 1989, Paleoceanography and tectonic setting of the Jurassic Coast Range ophiolite, central California: Evidence from the extrusive rocks and the volcaniclastic sediment cover: Marine and Petroleum Geology, v. 6, p. 194–220.

_____, 1990, Sedimentology and tectonic implications of ophiolite-derived clastics overlying the Jurassic Coast Range ophiolite, northern California: American Journal of Science, v. 290, p. 109–163.

Ross, J. A., and Sharp, W. D., 1986, $^{40}Ar/^{39}Ar$ and Sm/Nd dating of garnet amphibolite in the Coast Ranges, California: EOS (Transactions of the American Geophysical Union), v. 67, no. 44, p. 1249.

Saleeby, J. B., 1982, Polygenetic ophiolite belt of the California Sierra Nevada: Geochronological and tectonostratigraphic development: Journal of Geophysical Research, v. B87, p. 1803–1824.

_____, 1983, Accretionary tectonics of the North American Cordillera: Annual Reviews in Earth Science, v. 15, p. 45–73.

_____, 1990, Geochronological and tectonostratigraphic framework of Sierran-Klamath ophiolitic assemblages, *in* Harwood, D. S., and Miller, M. M., eds., Paleozoic and early Mesozoic paleogeographic relations: Sierra Nevada, Klamath Mountains, and related terranes: Geological Society of America, Special Paper 255, p. 93–114.

Saleeby, J. B., Harper, G. D., Snoke, A. W., and Sharp, W. D., 1982, Time relations and structural-stratigraphic patterns in ophiolite accretion, west-central Klamath Mountains, California: Journal of Geophysical Research, v. 87, p. 3831–3848.

Saleeby, J. B., Blake, M. C., Jr., and Coleman, R. G., 1984, Pb/U zircon ages on thrust plates of west central Klamath Mountains and Coast Ranges, northern California and southern Oregon: EOS (Transactions of the American Geophysical Union), v. 65, p. 1147.

Schoonmaker, A., Harper, G., and Heizler, M., 2003, Tectonic history of the Snowcamp remnant of the Coast Range ophiolite, Game Lake Peak area, SW Oregon—outlier of the Josephine ophiolite? [abs.]: Geological Society of America, Abstracts with Programs, paper no. 104-9.

Schweickert, R. A., and Cowan, D. S., 1975, Early Mesozoic tectonic evolution of the western Sierra Nevada, California: Geological Society of America Bulletin, v. 86, p. 1329–1336.

Searle, M., and Cox, J., 1999, Tectonic setting, origin, and obduction of the Oman ophiolite: Geological Society of America Bulletin, v. 111, p. 104–122.

Shervais, J. W., 1982, Ti-V plots and the petrogenesis of modern and ophiolitic lavas: Earth and Planetary Science Letters, v. 59, p. 101–118.

_____, 1990, Island arc and ocean crust ophiolites: Contrasts in the petrology, geochemistry, and tectonic style of ophiolite assemblages in the California Coast Ranges, *in* Malpas, J., Moores, E. M., Panayiotou, A., and Xenophontos, C., eds., Ophiolites: Oceanic crustal analogues, Proceedings of the Symposium Troodos 1987: Nicosia, Cyprus, Geological Survey Department, p. 507–520.

_____, 1993, Tectonic implications of oceanic basalts in the Coast Range Ophiolite, California: Evidence for ridge subduction as final ophiolite-forming event [abs.]: Geological Society of America, Abstracts with Programs, v. 25, no. 5, p. 445..

_____, 2001, Birth, death, and resurrection: The life cycle of supra-subduction zone ophiolites: Geochemistry, Geophysics, Geosciences, v. 2 [2000GC000080].

Shervais, J. W., and Beaman, B. J., 1989, Evidence for multi-stage magmatic history in the Elder Creek ophiolite remnant, Northern California Coast Ranges: EOS (Transactions of the American Geophysical Union), v. 70, p. 1399.

_____, 1991, The Elder Creek Ophiolite: Multi-stage magmatic history in a fore-arc ophiolite, northern California Coast Ranges [abs.]: Geological Society of America, Abstracts with Programs, v. 23, no. 5, A387.

_____, 1994, Major and trace element geochemistry of the Elder Creek ophiolite, California Coast Ranges [abs.]: Geological Society of America, Abstracts with Programs, v. 26, no.7, p. A39.

Shervais, J. W., and Hanan, B. B., 1989, Jurassic volcanic glass from the Stonyford volcanic complex, Franciscan assemblage, northern California coast ranges: Geology, v. 17, p. 510–514.

Shervais, J. W., and Kimbrough, D. L., 1985a, Geochemical evidence for the origin of the Coast Range ophiolite: A composite island arc–oceanic crust terrane in western California: Geology, v. 13, p. 35–38.

_____, 1985b, Geochemical evidence for the origin of the Coast Range ophiolite: Reply: Geology, v. 13, p. 828–829.

_____, 1987, Alkaline and transitional subalkaline metabasalts in the Franciscan complex melange: Geological Society of America, Special Paper 215, p. 167–182.

Snoke, A. W., 1977, A thrust plate of ophiolitic rocks in the Preston Peak area, Klamath Mountains, California: Geological Society of America Bulletin, v. 88, p. 1641–1659.

Snow, C. A., 2002, Geology of the Cuesta Ridge ophiolite remnant near San Luis Obispo, California: Evidence for the tectonic setting and origin of the Coast Range ophiolite: Unpubl. M.S. thesis, Utah State University, 150 p.

Stern, R. J., and Bloomer, S. H., 1992, Subduction zone infancy: Examples from the Eocene Izu-Bonin-Mariana and Jurassic California arcs: Geological Society of America Bulletin, v. 104, p. 1621–1636.

Tobisch, O. T., Paterson, S. R., Saleeby, J. B., and Geary, E. E., 1989, Nature and timing of deformation in the Foothills Terrane, central Sierra Nevada, California: Its bearing on orogenesis: Geological Society of America Bulletin, v. 101, p. 401–413.

Unruh, J. R., Loewen, B. A., and Moores, E. M., 1995, Progressive arcward contraction of a Mesozoic Tertiary forearc basin, southwestern Sacramento Valley, California: Geological Society of America Bulletin, v. 107, p. 38–53.

Ward, P. L., 1995, Subduction cycles under western North America during the Mesozoic and Cenozoic eras, *in* Miller D. M., and Busby, C., eds., Jurassic magmatism and tectonics of the North American Cordillera: Geological Society of America, Special Paper 299, p. 1–46.

Wentworth, C. M., Blake, M. C., Jones, D. L., Walther, A. W., and Zoback, M. D., 1984, Tectonic wedging associated with emplacement of the Franciscn assemblage, California Coast Ranges, *in* Blake, M. C., ed., Franciscan geology of Northern California: Los Angeles, CA, Pacific Section, Society of Economic Paleontologists and Mineralogists, no. 43, p. 163–174.

Williams, K. M., 1983, The Mount Diablo ophiolite, Contra Costa County, California: Unpubl. M.S. thesis, San Jose State University, 156 p.

Wolf, M. B., and Saleeby, J. B., 1995, Late Jurassic dike swarms in the southwestern Sierra Nevada Foothills terrane, California, in Miller D. M., and Busby, C., eds., Jurassic magmatism and tectonics of the North American Cordillera: Geological Society of America, Special Paper 299, p. 203–228.

Wright, J. E., and Fahan, M. R., 1988, An expanded view of Jurassic orogenesis in the western United States Cordillera; Middle Jurassic (pre-Nevadan) regional metamorphism and thrust faulting within an active arc environment, Klamath Mountains, California: Geological Society of America Bulletin, v. 100, p. 859–876.

Wright, J. E. and Sharp, W. D., 1982, Mafic-ultramafic intrusive complexes of the Sierra-Klamath region, California: Remnants of a middle Jurassic arc complex [abs.]: Geological Society of America, Abstracts with Programs, v. 14, p. 245–246.

Xenophontos, C., and Bond, G. C., 1978, Petrology, sedimentation, and paleogeography of the Smartville terrane (Jurassic)—bearing on the origin of the Smartville ophiolite, in Howell, D. G., and McDougall, K. A., eds., Mesozoic paleogeography of the western United States: Los Angeles, CA, Pacific Section, Society of Economic Paleontologists and Mineralogists, p. 291–302.

Zoglman, M. M., 1991, Petrology and structure of the Stonyford volcanic complex, California: Unpubl. M.S. thesis, University of South Carolina, 136 p.

Zoglman, M. M., and Shervais, J. W., 1991, The Stonyford Volcanic Complex: Petrology and structure of a Jurassic seamount in the Northern California Coast Ranges [abs.]: Geological Society of America, Abstracts with Programs, v. 23, no. 5, p. A395.

Field and Isotopic Evidence for Fluid Mobility in the Franciscan Complex: Forearc Paleohydrogeology to Depths of 30 Kilometers

SETH J. SADOFSKY[1] AND GRAY E. BEBOUT[2]

Department of Earth and Environmental Sciences, 31 Williams Drive, Lehigh University, Bethlehem, Pennsylvania 18015

Abstract

In the Franciscan Complex exposed in the California Coast and Diablo ranges (representing depths of up to 30 km in a Mesozoic to Early Tertiary accretionary prism), differences in porosity, permeability, and rheology impacted geometries and scales of fluid mobility and the deep entrainment of pore water (chemically evolved seawater). Carbon and O isotope compositions of abundant, texturally diverse $CaCO_3$ veins (most veins with $\delta^{13}C_{VPDB}$ = –11.0 to –3.0‰, $\delta^{18}O_{VSMOW}$ = +12.0 to +18.5‰) are in part consistent with control of fluid isotopic compositions by relatively local-scale exchange with large volumes of metaclastic host-rocks. Although this apparent local-scale equilibration, observed for some veins, complicates assessment of external sources for the fluids that produced these veins, many veins with elevated $\delta^{18}O$ (relative to calculated rock-buffered values) could reflect up-dip flow of H_2O released at greater depths and previously equilibrated with similar lithologies at higher temperatures (and sluggish reequilibration of the rocks with these externally derived fluids). In the Coastal Belt, differences in vein $\delta^{13}C$ in adjacent coherent graywacke and shaley mélange zones may be due to preferential infiltration of the more permeable mélange zones by deeply derived CH_4-bearing fluids, perhaps in part during exhumation and cooling. A number of variations in vein isotopic composition in individual exposures can be attributed to vein formation over a range of increasing T during underthrusting, then decreasing T during exhumation, and related to varying rheology (affecting permeability) within intercalated highly disrupted shales, sandstone-shale sequences (with interbedding at mm to cm scales), and more massive metagraywacke.

Calculated fluid $\delta^{18}O$ for veins in the Franciscan units peak metamorphosed at lower temperatures (Coastal Belt) spans the range of fluids venting in active accretionary prisms and producing forearc serpentinite seamounts (mostly 0 ± 3‰). In general, higher flux of aqueous fluid from depth in accretionary prisms, preventing reequilibration with rocks along its path, should favor expulsion of fluid with $\delta^{18}O$ higher than that of seawater that could be traced in fluids along fault structures at shallower levels. Salinities of fluid inclusions in vein quartz are mostly lower than that of seawater, consistent with delivery of "fresher" aqueous fluids toward the surface along structural heterogeneities. Recent work on volatile and trace element contents of the Coast and Diablo ranges exposures of the Franciscan Complex indicates significant loss of structurally bound H_2O (i.e., not pore water) due to clay-to-mica transitions, but at shallower levels than those for peak metamorphism of all units studied here (i.e., at <5 km). Much of the fluid flux through these rocks involved fluids liberated at greater depths, precipitating calcite and quartz along down-T, down-P paths, and mobilizing trace elements (e.g., B, Cs, perhaps also K, Rb, As, and Sb) liberated at higher temperatures. Calcite cement in the Coastal Belt (representing very shallow offscraping, likely to <5 km) is absent in higher-grade rocks representing deeper subduction of similar rocks, consistent with significant loss of carbonate cement during decarbonation reactions at 5–10 km depths (evidenced by significant depletions in Ca and Sr in the higher-grade rocks; Sadofsky and Bebout, 2003).

Introduction

CYCLING THROUGH subduction zones represents a major variable in the global mantle-ocean-atmosphere mass balance for some volatile components (Berner and Lasaga; 1989; Zhang and Zindler, 1993; Bebout, 1995, 1996; Javoy, 1997, 1998; Deming, 1999; Franck et al., 1999; Jarrard, 2003). Much effort has recently been aimed at constraining the quantities and geochemistry of subducting sediment (von Huene and Scholl, 1991; Rea and Ruff, 1996; Plank and Langmuir, 1998; Jarrard, 2003), and evaluating the magnitude and mechanisms of the movement of volatiles (emphasizing H_2O and C)

---

[1]Present address: SFB 574, University of Kiel, Wischhofstraße 1-3, D-24148 Kiel, Germany [ssadofsky@geomar.de]
[2]Corresponding author; email: geb0@lehigh.edu

being subducted in sediments in active convergent margins. Some subducted H_2O likely reaches the mantle to participate in arc magma genesis, but pore fluid expulsion and, at greater depth, metamorphic devolatilization affect the proportion of the initially subducted H_2O budget that is more deeply subducted (see Moore and Vrolijk, 1992; Jarrard, 2003). Fluid released during mechanical expulsion and metamorphic reactions may be transferred toward the surface along fracture networks in evolving accretionary prisms (Fisher, 1996; Carson and Screaton, 1998; Saffer et al., 2000; Saffer, 2003) or may contribute to hydration of the forearc mantle wedge (Peacock, 1993). Additions from deep devolatilization to shallower fluids migrating along accretionary prism fracture systems have been difficult to trace (possible examples discussed by Elderfield et al., 1990; Kastner et al., 1991; Moore and Vrolijk, 1992; Sample et al., 1993; Sample, 1996; Sample and Reid, 1998; Mottl et al., 2003, in press).

Forearc, high-P/T metamorphic suites in the Circum-Pacific (e.g., Franciscan Complex, Western Baja Terrane, Catalina Schist; Kodiak Complex; Sanbagawa Complex) are dominantly composed of terrigenous sedimentary lithologies thought to dominate the subducted sedimentary inventory of graywacke and shale at modern convergent margins (see Sadofsky and Bebout, 2003; cf. Plank and Langmuir, 1998). As an example, Bailey et al. (1964) estimated that the Franciscan Complex consists of 80% graywacke, 10% micrograywacke and shales, 10% mafic igneous rocks, and < 0.5% chert and associated shale and limestone (the proportions of these lithologies vary among the Coastal, Central, and Eastern belts of the Franciscan Complex; Bailey et al., 1964). The Catalina Schist, California, is also dominantly metasedimentary (~70% metagraywacke and metashale), with smaller amounts of metabasalt/metagabbro, metachert, and mélange (Bebout and Barton, 1993). Incorporating the massive trench sedimentation and the additions through subduction erosion, von Huene and Scholl (1991, 1993) estimated that the subducting sediments are ~95% terrigenous, ~3% biogenic carbonate, and ~2% biogenic siliceous rocks. However, the more carbonate-rich pelagic section (~15% biogenic carbonate on a global basis; Rea and Ruff, 1996), known to be subducting largely in just a few margins (e.g., Central America), may commonly escape offscraping in accretionary prisms (von Huene and Scholl, 1991, 1993), resulting in its preferential deeper subduction (Bebout, 1995).

Previous study of the Catalina Schist suggests that at high-P/T, low-T conditions, such as those of a thermally mature subduction zone (see thermal models of Peacock, 1996), a large fraction of the structurally bound H_2O (in hydrous mineral phases) in sediments is retained to great depth (up to at least 45 km; Bebout, 1995). This surprisingly efficient retention of H_2O to such depths was recently confirmed by a detailed study of devolatilization on Franciscan Complex rocks in the Coast and Diablo ranges, California and on similar paleoaccretionary rocks in Baja California (Western Baja Terrane; Sadofsky and Bebout, 2003). The minimal loss of structural H_2O in Franciscan rocks implies that any H_2O-rich fluids mobilized at these levels are likely dominantly fluids released by devolatilization reactions at greater depths or pore fluids entrained to the levels represented by these rocks. In this study, we consider the role that fluid-rock interactions in sediment-rich accretionary complexes play in convergent-margin cycling of H_2O through examination of fluid-rock interactions preserved in low-grade, extremely high P/T metamorphic rocks of the Franciscan Complex (California Coast and Diablo ranges; early results were presented by Sadofsky and Bebout, 2001). Our work extends the earlier stable isotope investigation of Magaritz and Taylor (1976), on some of the same exposures, by more thoroughly relating variations in vein C- and O-isotopic compositions to considerations of larger-scale fluid mobility in the context of texture, mineralogy, and devolatilization history.

The Franciscan Complex contains well-preserved metasedimentary rocks that experienced a range of P-T conditions similar to those at depth in many active accretionary complexes (Peacock, 1992, 1996). In the Coast Ranges, peak P-T conditions range from <3 kbar and ~125 °C (Coastal Belt) to ~300°C and 8–9 kbar (Eastern Belt; Blake et al., 1988; Fig. 1A; also see Terabayashi and Maruyama, 1998). In the Pacheco Pass area of the Diablo Range, peak conditions reached 7–8 kbar, ~200 ± 50°C (Ernst, 1971, 1993; Maruyama et al., 1985; Terabayashi et al., 1996; Dalla Torre et al., 1996). The Franciscan Complex, and the lowest-grade units of the Catalina Schist (Grove and Bebout, 1995), record conditions similar to those expected in a relatively cool and mature subduction zone at P <10 kbar and T <350°C (Peacock, 1992). Appropriate P-T paths, including relatively cool retrograde histories (Tagami and Dumitru, 1996) and lithological similarities to many other ancient and active

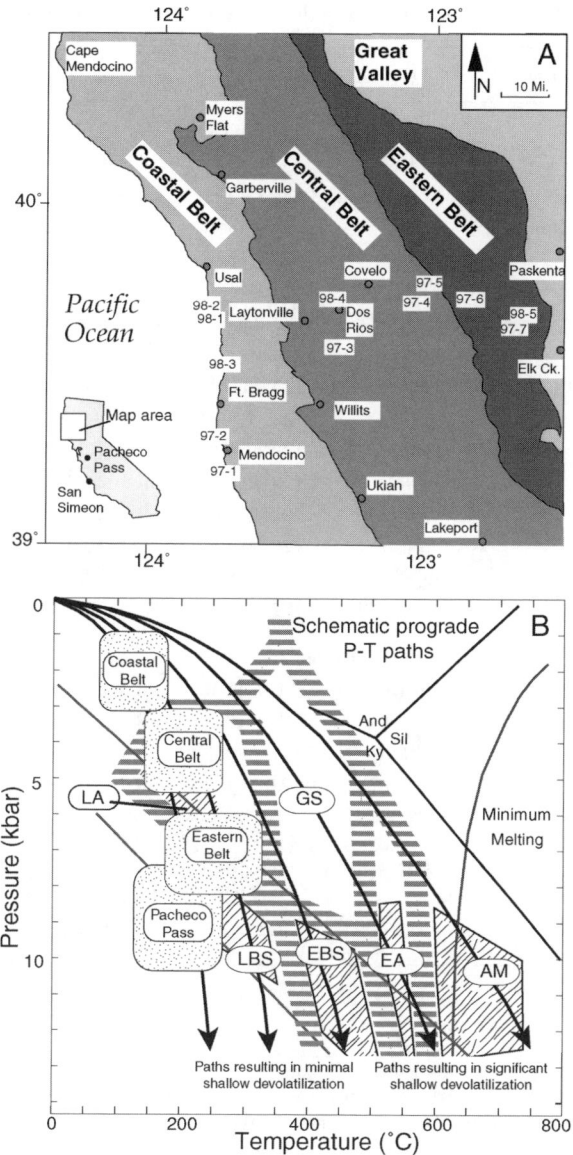

FIG. 1. A. Map of the Coast Ranges showing the locations of analyzed samples and their relations to the main lithotectonic subdivisions of the Franciscan Complex, after Blake et al. (1988). Numbers represent sample sites from the northern Coast Ranges. B. Pressure-temperature diagram showing estimates of peak metamorphic recrystallization of each of the Franciscan units studied here and the subterranes of the Catalina Schist (exposed on Santa Catalina Island, California). P-T estimates are from Blake et al. (1988), Jayko and Blake (1989), Ernst (1993), Grove and Bebout (1995); for the units of the Catalina Schist, Dalla Torre et al. (1996), Tagami and Dumitru (1996), and Terabayashi et al. (1996). Generalized phase equilibria for relevant volcanic and volcaniclastic rocks are from Liou et al. (1987), Evans (1980), and Frey et al. (1991). Abbreviations: LA = lawsonite-albite; LBS = lawsonite-blueschist; EBS = epidote blueschist, EA = epidote amphibolite; A = amphibolite; GS = greenschist.

TABLE 1. Localities of Samples Collected from the Franciscan Complex

	Also known as	Approx. Lat.	Approx. Long.	Nearby landmark
	Coast Range outcrops			
97-1	Van Damme or VD	39°16'N	123°46'W	Cliffs on Beach near Van Damme State Park
97-2	Mendo or MH	39°18'N	123°46'W	Cliffs on Beach near Mendocin Headlands
98-1	KW98-1-28, 52-68	39°42'N	123°45'W	Cliffs near Westprot-Union State Beach
98-2	KW98-33-50	39°14'N	123°45'W	Navarro Point
98-3	KW98-85-94	39°34'N	123°44'W	
97-3	Eel	39°40'N	123°23'30"W	Bank of Eel Riv. SR 162, mile 10.42
98-4	P&S	39°45'N	123°28'W	Presley and Smith Gravel Co. Quarry
98-5	Dos Rios or DR	39°50'N	123°23'30"W	Dos Rios
97-4	JC	39°49'N	122°58'30"W	Jumpoff Creek crosses SR 162
97-5	AP	39°52'N	122°57'30"W	Anthony Peak
97-6	CR	39°46'N	122°40'15"W	SR 162 ~ 1.5 mi. E. of Mendocino Pass
97-7	Alder or AS	39°40'N	122°40'15"W	SR 162 near turnoff for Alder Springs
98-6	AS2	39°39'30"N	122°39'30"W	SR 162 1 mi. E of 97-7
97-8	T	39°58'N	122°45'30"W	Toomes Rd crossed by Griffin Ck.
NR	Nicasio	38°05'30"N	122°44'30"W	Metabasites near Nicasio Reservoir
	Diablo Range[1]			
EP-3	East Pass	37°4'15"N	121°11'W	SR 152 2.0 mi. E. of Pacheco Pass
EP-4	East Pass	37°4'20"N	121°10'W	SR 152 2.45 mi. E. of Pacheco Pass
PP	Pacheco Mafic	37°3'N	121°14'45"W	SR 152 near BM 1065, W. of Pacheco Pass
WP	West Pass	37°3'15"N	121°14'15"W	SR 152 0.5 mi. W. of Dinosaur Pt. Turnoff
DPC-4	Del Puerto Canyon	37°23'30"N	121°29'W	San Antonio Valley Rd. @ Mines Rd.
	Other sample collection areas			
CCK		38°30'30"N	122°22'30"W	Chiles and Pope Valley Rd. @ Chiles Valley Rd
8/3/98-4		39°0'30"N	122°30'W	SR 20 1 mi. E of Lake/Colusa County line

[1]Including Pacheco Pass area.

accretionary systems, make the Franciscan Complex an excellent representative of deep accretionary processes.

Geological Setting

Sampling locations from the Coast Ranges for which data are presented in this paper are shown on Figure 1A (coordinates of the localities are presented in Table 1). Several outcrops in the Pacheco Pass area of the Diablo Range, central California (see inset map of Fig. 1A) were also investigated.

Field locations differ significantly in lithology among the major belts of the Coast Ranges (Coastal, Central, and Eastern belts) and the rocks in the Diablo Range. Rocks in the Coastal Belt appear only mildly metamorphosed, and are composed primarily of graywacke varying in scale of interbedding from several centimeters to several tens of meters (see Figs. 2A–2C). The graywacke generally occurs as lenticular blocks surrounded by a more deformed shale matrix. The graywackes preserve brittle deformation features such as veins, in some cases with complex cross-cutting relations, whereas in more shaley units, many veins are highly deformed (occurring now as highly disaggregated vein material) and only later-stage veins are intact. Graywackes of the Coastal Belt are dominated by clastic quartz, feldspars, clays, and opaques, with other minor detrital grains and lithic fragments.

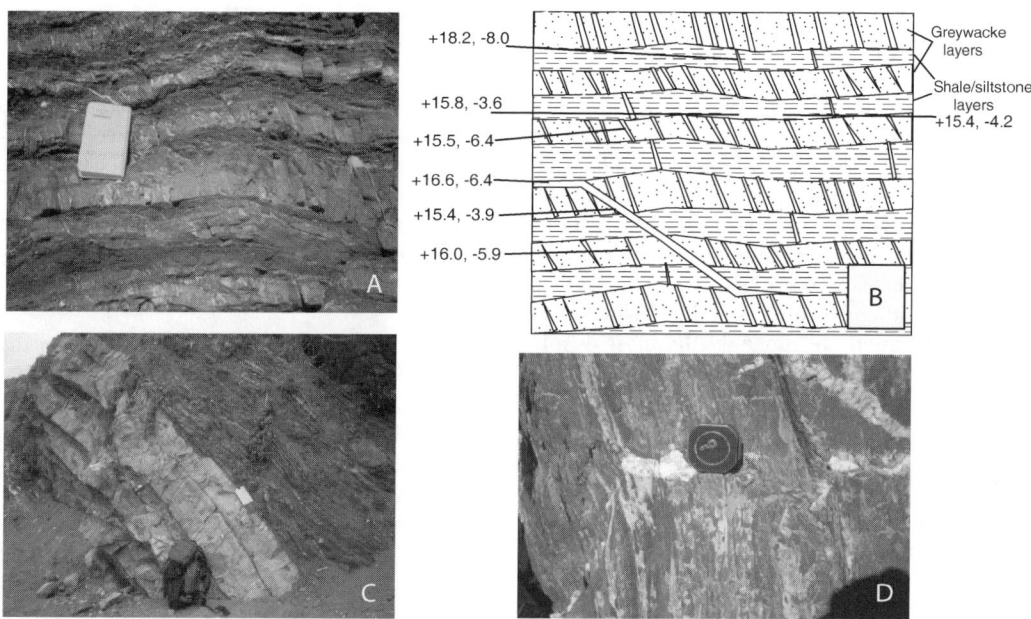

FIG. 2. Representative examples of lithologies, deformation styles, veining textures, and isotopic sampling approach (see corresponding vein textural information in Tables 2 and 3). A. Veining relationships in interlayered sandstone-shale sequences exposed on the beach in the Coastal Belt. Note the extensional veins in sandy layers. In some cases, particularly in the Coastal and Central belts where shale:sandstone ratios are higher, the sandstone layers are more "broken," producing boudin. B. C and O isotope data for one exposure of Coastal Belt sandstone-shale sequence similar to that in Figure 2A, illustrating typical variation in isotopic composition within a single outcrop. Non-vein data are for finely disseminated carbonate in whole-rock samples of sandy and shaley lithologies. C. Example of interlayering on larger scale of sandstone (light-colored layer dipping to lower right) with sandstone-shale sequences (darker lithology above light-colored layer), the latter appearing more disrupted by deformation. D. Typical Eastern Belt metasedimentary exposure of sandstone with some finely interlayered shaley lithologies. Note the more strongly developed deformation fabric in these rocks relative to Coastal Belt exposures in (A–C), and also the transposed veins parallel to the cleavage and cross-cutting, less deformed veins.

Prograde minerals such as white mica (mainly as sericitization of feldspar), laumontite (mainly in veins), chlorite, and calcite are also present.

The Central Belt has a macroscopic texture characterized as blocks in mélange (see Cloos, 1986) and, like the Coastal Belt, coarser graywackes preserve more diverse and abundant veins than the shale matrix (see Figs. 2A and 2C). Metasedimentary rocks in the Central Belt retain some detrital grains, like those of the Coastal Belt, but there is a greater degree of recrystallization in the Central Belt. Fewer lithic clasts remain, and those that do remain are more altered than in the Coastal Belt. These rocks differ from the Coastal Belt in several ways: there are fewer clays present, and much more white mica, lawsonite, and pumpellyite; much of the calcite has been transformed to aragonite (Terabayashi and Maruyama, 1998); most detrital K-feldspar has been replaced by albite (Blake et al., 1988).

Eastern Belt rocks show a much stronger metamorphic deformation fabric (Blake et al., 1988; Fig. 2D; also see discussion of deformation history by Bolhar and Ring, 2001) and contain more high-P indicator minerals, such as lawsonite, sodic amphibole, and sodic pyroxene, but with no detrital K-feldspar remaining (see textural zones of Blake et al., 1988; Jayko and Blake, 1989). Cleavage is well developed in pelitic rocks, and some outcrops show development of segregation textures. Samples analyzed in this study are composed of quartz, albite, white-mica (phengite and paragonite; Jayko et al., 1986), lawsonite, stilpnomelane, and pumpellyite, with relatively minor epidote, titanite, and zircon. Other phases include sodic pyroxene, sodic amphibole, chlorite, tourmaline, and apatite (Blake et al., 1988; Tagami and Dumitru, 1996).

Samples analyzed from the Pacheco Pass area, mostly relatively massive metagraywackes, are composed primarily of quartz, albite, sodic pyroxene, white-mica, and lawsonite, with more abundant but minor chlorite, carbonate, titanite, sodic amphibole, stilpnomelane, zircon, and prehnite. Other phases present include pumpellyite and calcite or aragonite. A wealth of mineral-chemical studies has provided a solid understanding of the P-T history of these rocks, which are probably the most deeply subducted intact tract of metasedimentary rocks in the Franciscan Complex (Ernst, 1993).

Analytical Methods

For stable-isotope analyses, CO_2 from carbonate veins and cements was prepared by dissolution in 100% phosphoric acid at 25°C overnight, following McCrea's techniques (1950), and the liberated CO_2 gas was analyzed on the Finnigan MAT 252 at Lehigh University. Oxygen and C isotopic values are reported relative to the Vienna standard mean ocean water (V-SMOW) and PeeDee Belemnite (V-PDB), respectively, and reported in the standard notation of ‰ (per mil) according to the following:

$$\delta^{18}O = 1000[((^{18}O/^{16}O) \text{ sample} / (^{18}O/^{16}O) \text{ standard}) - 1], \quad (1)$$

$$\delta^{13}C = 1000[((^{13}C/^{12}C) \text{ sample} / (^{13}C/^{12}C) \text{ standard}) - 1]. \quad (2)$$

Proper standardization for the O- and C-isotope analyses was verified by analysis of various laboratory and international carbonate standards including NBS-19 (calcite). For ~145 analyses of one internal calcite standard (sample 8-3-7v, calcite vein) over a nine-year period, uncertainties (expressed as 1σ) are ~0.15‰ and ~0.10‰ for $\delta^{18}O$ and $\delta^{13}C$, respectively.

Fluid-inclusion microthermometry measurements were made on an USGS-type heating-freezing stage at the University of Pittsburgh. Most of the inclusions observed contain two phases present at room temperature (liquid + vapor), with freezing points near 0°C, and are therefore presumed to be H_2O-rich fluids. Only inclusions with textures that indicate a "primary" origin were analyzed. These inclusions are round or oval-shaped in appearance and occur as isolated inclusions, irregular clusters, and rare planar clusters. Excessively oblong inclusions, inclusions that appear to have been stretched, and inclusions that appear to occur along cracks or in linear clusters were avoided. Measurements were made of the inclusion size, relative volumes of liquid and vapor, melting temperature, homogenization temperature, and the phase to which the inclusions homogenize. The data were reduced using the computer program MacFlinCor (Brown and Hagemann, 1995).

Results

Samples of each host-rock lithology, and multiple samples of each texturally or compositionally distinct vein type (see Fig. 2), were collected at each outcrop, and a subset of these $CaCO_3$ veins was analyzed for C-O-isotopic compositions. Most veins are composed of $CaCO_3$ (mainly calcite, some aragonite), quartz, or a combination of the two phases. Veins containing phases such as laumontite and jadeitic pyroxene were also collected.

Table 2 lists vein types, vein-host rock relations, crosscutting relationships, types of analyses, and representative samples from outcrops studied in the Franciscan Complex. Table 3 presents host-rock type, mineralogy, and some textural and structural information for the $CaCO_3$ veins that were analyzed isotopically. Phases present in the veins were determined in a variety of ways; most were analyzed petrographically, and a subset was analyzed by x-ray diffraction (XRD) to identify minerals that are difficult to distinguish optically, such as calcite and aragonite. The XRD results were then extrapolated to other veins, from the same outcrop, that are texturally and optically similar.

Figures 2A–2C demonstrate some typical veining styles in an outcrop from the Coastal Belt of the Franciscan. Figures 2A and 2C show typical sections of massive graywacke and more finely interlayered graywacke-shale. The massive graywacke contains variably deformed veins at a variety of scales, with two generations of small veins (1–2 mm in aperture), one crosscutting the other. A set of larger veins is also present in these coarsely bedded graywackes; this set appears to be composed of more interconnected veins (<5 mm in aperture), in some cases fed by a second set of fine veins (see sketch in Fig. 2B). In Figure 2C, thinner graywacke layers (finely interbedded with darker shale) are less densely veined than thicker layers, and it is not possible to identify two generations of small veins;

TABLE 2. Characterization of Veins from the Franciscan Complex[1]

Outcrop	Host	Series	Mineralogy	Texture relative to host + other series	Rel. age	Analyses	Typical samples
Coast Ranges-Coastal Belt							
97-1 (VD)	GW-massive	V1	cc	Small early stage veins	1	^{13}C-^{18}O	VD5b, VD8b
97-1 (VD)	GW	V2	cc	Small x-cutting veins, cut by minor faults	2	^{13}C-^{18}O, XRD	VD5a. VD9b
97-1 (VD)	GW	V3	cc	Large, possibly fed by V2	3	^{13}C-^{18}O	VD5d, VD8a
97-1 (VD)	GW	V4	cc	Large vein, bedding parallel		^{13}C-^{18}O, XRD	VD9a
97-1 (VD)	Conglom	V5	cc	Small veins cut through cobbles			VD6a
97-1 (VD)	Conglom	V6	cc	Large veins flow around cobbles			VD6b
97-1 (VD)	Siltstone	V7	cc	Small veins x-cut bedding			VD7b
97-2 (MH)	GW/SH	V1	cc ± Q	Small veins coarse layers, xcut beds	1	^{13}C-^{18}O, XRD	MH2, MH4, MH8
97-2 (MH)	GW/SH	V2	cc ± Q	Small veins fine layers, xcut beds	1	^{13}C-^{18}O, XRD	MH3, MH6, MH7, MH9
97-2 (MH)	GW/SH	V3	cc	Larger veins, x-cut beds @ lowñ	2	^{13}C-^{18}O	MH1, MH5, MH10
97-2 (MH)	GM-massive	V1	Q ± lm	Very small, not oriented	1		MH13c
97-2 (MH)	GM-massive	V2	Q ± lm	Anastamosing	2		MH13b
97-2 (MH)	GM-massive	V3	Q ± lm	~1cm orthogonal sets	3		MH13a, d, e
Coast Ranges-Central Belt							
97-3 (Eel)	Chert/Mélange	V1	Q + cc	Veins in chert block	1	^{13}C-^{18}O	Eel2c
97-3 (Eel)	GW	V1	cc ± ar	Small wispy veins, some defm'd	1	^{13}C-^{18}O	Eel4a
97-3 (Eel)	GW	V2	cc ± Q	Larger, less defm'd veins, xcut 1	2	^{13}C-^{18}O, XRD	Eel4b
97-3 (Eel)	GW	V3	cc ± Q	Large veins @ GW/SH interface	3	^{13}C-^{18}O	Eel 43
98-4 (P&S)	GW	V1	Q ± cc ± albite	Small veins	1		P&S3
98-4 (P&S)	GW	V2	Q ± cc ± albite	Large bedding parallelveins	2		P&S1
98-4 (P&S)	GW	V3	Q ± cc ± albite	Large bedding normal veins, xcut 2	3		P&S2
98-5 (DR)	Eclogite	V1	CaCO3	5 mm thick defm'd + faulted veins			DR3b
98-5 (DR)	Eclogite	V2	CaCO3	1 mm thick defm'd veins			DR3b
98-5 (DR)	GW	V3	cc ± Q	Small deformed veins			DR1
98-5 (DR)	Blueschist	V4	Q ± cc ± pyx	Small defm'd veins in mafic blocks			DR2, DR4
97-4 (JC)	Chert	V1	Q ± cc ± ar	Small veins in chert	1	^{13}C-^{18}O, XRD	JC3a, JC3b
97-4 (JC)	Chert Mélange	V2	Q ± cc ± ar	Larger veins at interface (xcutting)	2	^{13}C-^{18}O	JC3c
97-4 (JC)	SH	V3	cc ± Q ± ar	Cleavage parallel veins	1	^{13}C-^{18}O, XRD	JC-6
97-4 (JC)	GW	V4	cc ± Q	~Cleavage parallel anastamosingveins	1	^{13}C-^{18}O	JC-9
97-4 (JC)	GW/SH	V5	cc ± Q	Large veins xcut WG/SH layers	2	^{13}C-^{18}O	JC-8
97-5 (AP)	SH/Silty GW	V1	cc	cc pods in foliation, ~1x5 cm	1	^{13}C-^{18}O	AP4a
97-5 (AP)	SH/Silty GW	V2	Q	Small veins xcut pods	2	^{13}C-^{18}O	AP4b
97-5 (AP)	Sandy GW	V3	Q ± cc	Small defm'd veins in wacke	1	^{13}C-^{18}O	AP3c
97-5 (AP)	Sandy GW	V4	Q + cc	Large polygenerational veins, xcut 1	2	^{13}C-^{18}O	AP5a, AP8a
97-5 (AP)	Sandy GW	V5	Q ± cc	Orthogonal late-stage veins	3		
8398-4a	Mafic block	V1	Q	Quartz veins in mafic block	1	FL.Inc.	8398-4a
Coast Ranges-Eastern Belt							
97-6 (CR)	Greenstone	V1	cc ± Q ± ar	Small veins parallel to rock fabric	1	^{13}C-^{18}O	CR1c
97-6 (CR)	Greenstone	V2	cc ± Q ± ar	Large meandering veins, xcut 1	2	^{13}C-^{18}O, XRD	CR1a
97-6 (CR)	Greenstone	V3	cc ± Q ± ar	Small veins, appear to fill fractures, xcut 2	3	^{13}C-^{18}O	CR1b
97-6 (CR)	Shaley schist	V1	Q	Large veins cross foliation		Fl. Inc	CR2a, b
97-7 (AS)	Shale/silt	V1	Q ± cc	Cleavage parallel crenulated veins	1	^{13}C-^{18}O	AS 1b
97-7 (AS)	Shale/silt	V2	Q ± cc	Broken, xcutting veins	2	^{13}C-^{18}O	AS4b, AS1a
97-7 (AS)	Shale/silt	V3	Q ± cc	Larger straight veins	3	^{13}C-^{18}O	AS2b
98-6 (AS2)	GW	V1	Q ± cc ± ar	Small highly defm'd veins	1		AS2-1
98-6 (AS2)	GW	V2	Q + cc ± ar	Larger veins, somewhat defm's, xcut 1	2	Fl. Inc	AS2-2
97-8	Schistose	V1	Q ± cc	Early veins, caught in fabric & defm'd	1		T3a, T4a
97-8	Schistose	V2	Q ± cc	Straighter veins, xcut foliation	2		T3b, T4b
97-8	Schistose	V3	Q ± cc	Xcuts all, fault related?	3		T6c
Diablo Range							
EP	Silty GW	V1	Q + cc	q-cc boudins, roughly in fabric,	1	^{13}C-^{18}O	Ep4a
EP	Silty GW	V2	Q + cc	Small veins xcut fabric and 1	2	^{13}C-^{18}O, XRD	EP4b
EP	Sandy GW	V1	Q + cc	Small veins throughout GW	1	^{13}C-^{18}O	EP3a
EP	GW	V3	cc ± Q	Larger veins xcut several layers	3	^{13}C-^{18}O	Ep3c
WP	Coarse GW	V1	Q ± cc	Small veins, mildly defm'd	1		WP2b
WP	Coarse GW	V2	Q ± cc	Large (3–5 cm thick) xcut 1	2	Fl. Inc	WP2a
PP	Metabasalt	V1	± Q ± cc ± ar ± jd	Small, defm'r veins	1	^{13}C-^{18}O	PP2
PP	Metabasalt	V2	Q ± cc	Small, undefm'd veins, xcut 1	2	^{13}C-^{18}O	PP4
PP	Metabasalt	V3	Q ± cc ± ab	Large planar veins	3	^{13}C-^{18}O	PP1
DPC	GW	V1	Q	Early, complex vein caught in fabric	1	Fl. Inc	DPC4c
DPC	GW	V2	Q	Small defm'd veins	2		
DPC	GW	V3	Q	Large xcutting veins, xcut 1 and 2	2		

[1]Information about outcrops 98-1, 98-2, and 98-3 is presented in Wroblewski, 1999. Abbreviations used above: GW = graywacke; GM = greywacke/mélange interlayers; SH = shale; Vn = specific group of veins; Q= quartz; cc = calcite; ar = aragonite; lm = laumontite;^{13}C–, ^{18}O designates isotopic analyses; Fl. inc. = fluid inclusion microthermometry.

TABLE 3. Mineralogy, Isotopic Composition, and Texture of Analyzed Veins

Site	Sample ID	$\delta^{18}O_{VSMOW}$	$\delta^{13}C_{PDB}$	Host rock	Mineralogy	Structure and texture
			Coast Ranges—Coastal Belt			
97-1	VD 5a (x)	16.7	−4.6	GW-Lm-grade	cc – CXT	Small vein, crosscuts 5b
97-2	VD 5b	15.1	−4.5	GW-Lm-grade	cc – CXT	Small, early vein
97-3	VD 5d	16.1	−4.9	GW-Lm-grade	cc – CXT	Larger vein, may be latestage
97-4	VD 9a1 (x)	16.0	−11.0	GW-Lm-grade	cc – CXT	Large, bedding normal vein
97-5	VD 9a2	14.8	−12.6	GW-Lm-grade	cc – CXT	Large, bedding normal vein
97-6	VD 9b1	16.4	−11.6	GW-Lm-grade	cc – CXT	Small fine vein in greywacke
97-2	MH 1	15.4	−3.9	GW-Lm-grade	cc – CXT	Vein crosscuts sandy and silty layers at a low angle
97-3	MH 2	15.4	−4.2	GW-Lm-grade	cc – CXT	Bedding normal vein in silty layer
97-4	MH 3	15.7	−4.0	GW-Lm-grade	cc + Q – CXT	Bedding normal vein in sandy layer
97-5	MH 4	18.2	−8.0	GW-Lm-grade	cc – CXT	Bedding normal vein in silty layer
97-6	MH 5 (x)	15.8	−3.6	GW-Lm-grade	cc – CXT	Bedding normal vein in silty layer
97-7	MH 6 (x)	16.0	−5.9	GW-Lm-grade	cc + Q – CXT	Bedding normal vein in sandy layer
97-8	MH 7	15.5	−6.4	GW-Lm-grade	cc + Q – CXT	Bedding normal vein in sandy layer
97-9	MH 11	16.6	−6.4	GW-Lm-grade	cc – CXT	Bedding parallel vein in sandy layer
98-1[2]	KW98-1a	15.7	−4.7	GW-Lm-grade	cc – MXT	Vein on fault plane in greywacke
98-1[2]	KW98-1b	14.5	−4.6	GW-Lm-grade	cc – CXT	Vein on fault plane in greywacke
98-1[2]	KW98-3a	14.2	−4.6	GW-Lm-grade	cc – lm CEG	Small veins in greywacke
98-1[2]	KW98-3b	15.8	−4.6	GW-Lm-grade	cc – lm Granular	3b xcuts 3a
98-1[2]	KW98-6a	14.6	−4.5	GW-Lm-grade	cc – lm – Q	Fine crystalline vein
98-1[2]	KW98-7a	17.5	−6.1	GW-Lm-grade	cc + Q	cc is CEG, Q is FEG
98-1[2]	KW98-21a	17.0	−4.2	GW-Lm-grade	coarse cc + finer lm	Crosscutting veins (21a->21b) on
98-1[2]	KW98-21b	14.0	−5.3	GW-Lm-grade	coarse cc + finer lm	Fracture in greywacke
98-1[2]	KW98-52a	17.5	−5.4	GW-Lm-grade	cc + lm	Granular, parallel to fabric
98-1[2]	KW98-52b	16.0	−4.8	GW-Lm-grade	cc – fibrous	Normal to fabric
98-1[2]	KW98-52c	17.2	−5.5	GW-Lm-grade	cc – granular	Normal to fabric
98-1[2]	KW98-68a	15.0	−5.2	GW-Lm-grade	cc + Q – granular	
98-2[2]	KW98-33a	13.6	−3.4	GW-Lm-grade	cc – CEG	Crystalline vein in metagreywacke
98-2[2]	KW98-45a	17.9	−0.4	SH-Lm-grade	cc + Q	Anastamosing vein of granular crystals
98-2[2]	KW98-46a	15.6	−1.0	SH-Lm-grade	cc + Q CEG	Normal to rock fabric
98-3[2]	KW98-85a (x)	14.4	−1.4	Mélange	cc + Q	Broken vein
98-3[2]	KW98-91a (x)	13.2	0.3	GW-Lm-grade	cc + lm CEG	Normal to fabric
98-3[2]	KW98-94a	12.2	−1.9	Mélange	cc granular	Crosscuts fabric
Nicasio	NR 1b	15.3	−0.8	Mafic	cc (ar?)	Samll vein in metabasalt
Nicasio	NR 1c	16.1	0.8	Mafic	cc (ar?)	Samll vein in metabasalt
			Coast Ranges—Central Belt			
97-3	Eel 2c	13.8	−6.7	Chert/mélange	Q + cc	Vein at border of chert block and mélange
97-3	Eel 4a1	14.5	−10.7	GW-LA grade	cc – CEG	Small wispy vein
97-3	Eel 4b1	13.2	−10.0	GW-LA grade	cc	Larger and only mildly deformed
97-3	Eel 4b2 (x)	13.6	−11.4	GW-LA grade	cc (CEG) + Q (finer)	Larger and only mildly deformed
97-3	Eel 4b3 (x)	14.2	−13.6	GW-LA grade	cc	Larger and only mildly deformed
97-3	Eel 4c1	14.6	−10.5	GW-LA grade	cc ± ar?	Large vein at and parallel to greywacke-shale interface
97-3	Eel 4c2	15.0	−11.6	GW-LA grade	cc ± ar?	Large vein at and parallel to greywacke-shale interface
97-3	Eel 4c3	14.7	−11.0	GW-LA grade	cc ± ar?	Large vein at and parallel to greywacke-shale interface
97-4	JC 3a1 (x)	19.4	−16.9	Chert	Q + cc (blocky) + ar	Small vein @ chert/GW interface
97-4	JC 3b1	18.6	−11.3	Chert	Q + cc (blocky)	Small vein @ chert/GW interface
97-4	JC 3c-1	18.5	−12.7	Mélange	Q + cc (blocky)	Large vein @ chert/mélange interface
97-4	JC 6-1	15.8	−8.7	SH-LA grade	Q + cc	Cleavage parallel segregation
97-4	JC 6-3 (x)	16.1	−9.1	SH-LA grade	cc + Q	Cleavage parallel segregation
97-4	JC 6a1 (x)	16.5	−8.2	SH-LA grade	cc + ar	Cleavage parallel segregation
97-4	JC 8-1 (x)	17.3	−9.9	GW-LA grade	cc + Q	Vein crosscuts GW/SH layers
97-4	JC 8-2	16.9	−9.3	GW-LA grade	cc + Q	Vein crosscuts GW/SH layers
97-4	JC 9a1	16.4	−9.3	GW-LA grade	cc (CXT) + Q	Large meandering, within GW layer
97-4	JC 9b1	15.8	−9.1	GW-LA grade	cc (CXT) + Q	Large meandering, within GW layer
97-4	JC 9b2	16.9	−9.3	GW-LA grade	cc (CXT) + Q	Large meandering, within GW layer

Table continues

TABLE 3. Continued

Site	Sample ID	$\delta^{18}O_{VSMOW}$	$\delta^{13}C_{PDB}$	Host rock	Mineralogy	Structure and texture
			Coast Ranges—Central Belt (continued)			
97-4	P&S 1e	17.9	−10.2	GW-LA grade	Q − cc − ab	Mainly coarse, xtalline quartz, large vein
97-5	AP 3c1	15.4	−8.4	GW-LA grade	Q + cc (CXT)	Slightly deformed vein ~1cm x 2-3m
97-5	AP 3c2	15.7	−8.5	GW-LA grade	Q + cc (CXT)	Slightly deformed vein ~1cm x 2-3m
97-5	AP 4a1	15.0	−8.7	silty GW	Q + cc (CXT)	CC pod
97-5	AP 4a2 (x)	15.0	−8.7	silty GW	Q + cc (CXT)	CC pod
97-5	AP 5a1	15.7	−11.0	GW-LA grade	Q + cc (CXT)	Large, polygenerational vein
97-5	AP 5a2 (x)	15.3	−11.7	GW-LA grade	Q + cc (CXT)	Large, polygenerational vein
97-5	AP 5a3	15.0	−11.8	GW-LA grade	Q + cc (CXT)	Large, polygenerational vein
97-5	AP 8a1	15.3	−10.5	GW-LA grade	cc (CXT)	Large, polygenerational vein
97-5	AP 8a2	14.1	−11.7	GW-LA grade	cc (CXT)	Large, polygenerational vein
97-5	AP 8b2	14.0	−12.4	GW-LA grade	cc (CXT)	Small vein appears to feed larger vein
97-5	AP 10	14.8	−8.9	GW-LA grade	cc (CXT)	Slightly deformed vein ~1cm x 2-3m
97-5	AP 98-3	14.5	−4.2	GW-LA grade	Q + cc (CXT)	Large, polygenerational vein
			Coast Ranges—Eastern Belt			
97-6	CR 1a-1	15.8	−12.0	Mafic	Q − cc − ar (CXT)	Large (2–3 cm aperture) deformed vein
97-6	CR 1a2	16.3	−9.9	Mafic	Q − cc − ar (CXT)	Large (2–3 cm aperture) deformed vein
97-6	CR 1a3 (x)	17.1	−10.8	Mafic	Q − cc − ar (CXT)	Large (2–3 cm aperture) deformed vein
97-6	CR 1b1	17.1	−9.2	Mafic	Q − cc − ar (CXT)	Small (<1mm aperture) deformed vein
97-6	CR 1c2	16.7	−12.5	Mafic	Q − cc − ar (CXT)	Parallel to rock fabric, appears crosscut by above
97-7	AS 1a1	19.4	−2.4	SH	Q + cc (minor)	Small cleavage normal vein
97-7	AS 1a2	19.8	−2.9	SH	Q + cc (minor)	Small cleavage normal vein
97-7	AS 1b1	18.3	−2.9	SH	Q + cc (minor)	Small cleavage parallel vein
97-7	AS 1b2	17.5	−5.0	SH	Q + cc (minor)	Small cleavage parallel vein
97-7	AS 2a (x)	18.4	−4.7	GW	cc + Q	Small planar vein in sandier rock
97-7	AS 2b	20.1	−3.5	GW	Q (FG) + minor cc	Small planar vein-crosscuts AS 2a
97-7	AS 3a1	18.8	−3.8	fine GW	Q (CU) + cc (finer)	Small planar vein-crosscut by AS 2samples
97-7	AS 3a2	18.0	−2.7	fine GW	Q (CU) + cc (finer)	Small planar vein-crosscut by AS 2samples
97-7	AS 4a1	18.0	−3.4	fine GW	Q (CU) + cc (finer)	Large (2–3 cm aperture) cleavage parallel vein
97-7	AS 4a2 (x)	16.4	−3.2	fine GW	Coarse dfm'd cc + Q	Large (2–3 cm aperture) cleavage parallel vein
97-7	AS 4b1	20.0	−4.7	fine GW	Q (dfm'd) + minor cc	Smaller (<1 cm aperture) xcuts fabric and 4a
97-7	AS 4b2	21.0	−3.4	fine GW	Q (dfm'd) + minor cc	Smaller (<1 cm aperture) xcuts fabric and 4a
			Pacheco Pass Area			
Clastic rocks	EP 3a1	13.1	−7.4	GW	cc ± ar	Small (<2mm wide) Bedding parallel
	EP 3a2	18.9	−10.4	GW	Q + cc ± ar	Small (<2mm wide) Bedding parallel
	EP 3c1	19.6	−7.0	GW	Q + cc ± ar	Larger, crosscuts bedding
	EP 3c2	22.1	−11.7	GW	cc ± ar	Larger, crosscuts bedding
	EP 3c3	14.6	−8.2	GW	Q + cc ± ar	Larger, crosscuts bedding
	EP 4a1	12.9	−10.4	silty GW	Coarse cc ± ar minor Q	Pods of Q/CC
	EP 4a2	16.1	−9.8	silty GW	Coarse cc ± ar minor Q	Pods of Q/CC
	EP 4b1	17.9	−9.6	silty GW	Coarse cc ± ar minor Q	Small, deformed, crosscuts foliation
	EP 4b2 (x)	20.0	−10.2	silty GW	Very coarse cc/ar + Q	Small, deformed, crosscuts foliation
	EP 4b3 (x)	20.9	−9.7	silty GW	Very coarse cc/ar + Q	Small, deformed, crosscuts foliation
	EP 4c1	14.0	−11.2	silty GW	Very coarse cc/ar + Q	Larger, less deformed, foliation parallel
	EP 4c2	20.5	−6.3	silty GW	Very coarse cc/ar + Q	Larger, less deformed, foliation parallel
			Pacheco Pass Area			
Clastic rocks	WP2a1	14.6	−8.5	GW	Very coarse cc/ar + Q	
	WP2a2	13.6	−8.3	GW	Very coarse cc/ar + Q	
Mafic rocks	PP 1b (x)	13.5	−7.5	Mafic	Coarse cc+ar + Q	Large planar-late stage
	PP 1d	13.8	−5.7	Mafic	Coarse cc/ar + Q	Large planar-late stage
	PP 1e (x)	13.9	−4.7	Mafic	Coarse cc + Q	Large planar-late stage
	PP 2a	13.9	−4.2	Mafic	Coarse cc/ar + Q	Mildly deformed, 1mm to 1cm wide
	PP 2e	15.0	0.7	Mafic	Coarse cc/ar + Q	Mildly deformed, 1mm to 1cm wide
	PP 2f	13.8	−7.3	Mafic	Coarse cc/ar + Q	Mildly deformed, 1mm to 1cm wide
	PP 4a	13.5	−2.7	Mafic	Coarse cc/ar + Q	Small, straight, crosscut type 2
	PP 4b	13.4	−3.8	Mafic	Coarse cc/ar + Q	Small, straight, crosscut type 2

[1]Abbreviations: Q = quartz; cc = calcite; ar = aragonite; ab = albite; lm = laumontite; CEG = coarse equant granular; CXT = coarsely crystalline; MEG = medium crystallize; CU = coarse undulose; SH = shale; GW = graywacke; x = mineralogy verified by X-ray diffraction.
[2]See Wroblewski, 1999.

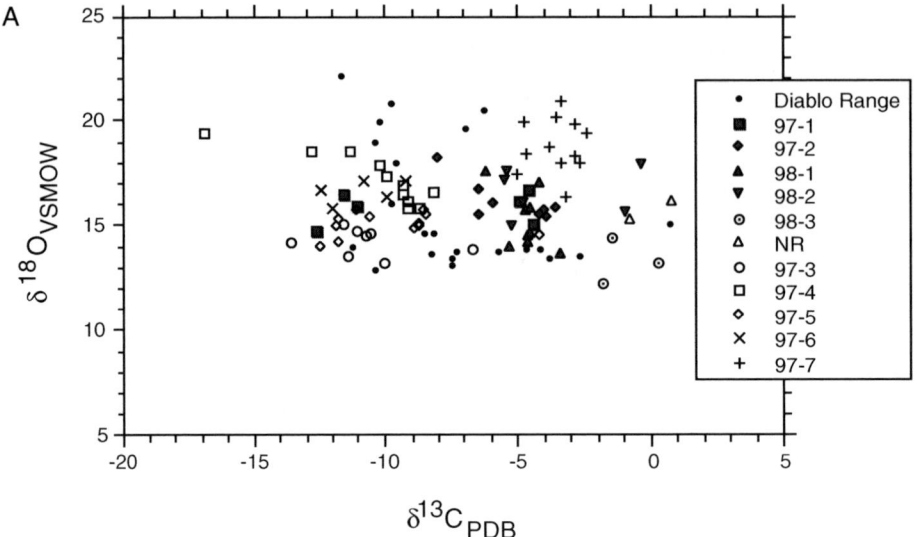

FIG. 3. A. Carbon and O isotopic compositions of all $CaCO_3$ veins analyzed as a part of this study. Legend corresponds to the listing of outcrops in Table 1. B (facing page). Histograms showing $\delta^{18}O$ and $\delta^{13}C$ of $CaCO_3$ from veins in the Franciscan Complex. The individual plots show data for each of the subterranes (Coastal, Central, and Eastern belts of the Coast Ranges; Pacheco Pass, Diablo Range; see Fig. 1A).

however, the veining is otherwise very similar to that shown in Figures 2A and 2B. Some larger, undeformed, nearly vertically dipping (striking NNW) veins in this outcrop were avoided because they appear to be very late stage, and conceivably were associated with late-stage uplift and the migration of the Mendocino Triple Junction (see Mashburn, 1985). Figure 2D shows typical veining styles in a silty/shaley outcrop from the Eastern Belt of the Coast Ranges. Early-formed small veins are deformed and transposed into the foliation (to varying degrees), later-stage veins crosscut the foliation, and less-deformed, foliation-parallel veins appear to crosscut the second group.

Isotopic results are summarized in Figures 3A and 3B and Table 3. In the Coastal Belt, $\delta^{13}C$ values range from −12.6 to +0.8‰ (mean = −4.7‰), and $\delta^{18}O$ ranges from +12.2 to +18.2‰ (mean = 15.6‰). Central Belt veins have $\delta^{13}C$ of −16.9 to −6.7‰ (mean = −10.4‰) and $\delta^{18}O$ of +13.2 to +19.4‰ (mean = 15.6‰). Eastern Belt samples have $\delta^{13}C$ ranging from −12.5 to −2.4‰ (mean = −5.7‰) and $\delta^{18}O$ ranging from +15.8 to +21.0‰ (mean = +18.2‰). Pacheco Pass vein-$\delta^{13}C$ values range from −11.7 to +0.7‰ and tend to be higher in mafic rocks; $\delta^{18}O$ ranges from +12.9 to +22.1‰, and is typically higher in the metasedimentary rocks. Metagraywackes from the Coastal Belt and the Central Belt contain up to 5 wt% of finely disseminated calcite (Bebout, 1991a, 1991b); this calcite generally has $\delta^{13}C$ and $\delta^{18}O$ similar to that of the veins in the same exposure. Finely disseminated calcite is not observed in the higher-grade units.

Salinities of the H_2O-rich fluid inclusions (expressed as wt% NaCl equivalent) are shown in Figure 4. Note that, whereas there is significant variability in salinity within subunits, the fluid inclusions of the Eastern Belt and Diablo Range are, on average, less saline than those of the lower-grade rocks. Homogenization temperatures of all analyzed veins are shown on Figure 5, and isochores calculated from the average T_h of each studied vein are shown on Figure 6 (see Table 4). One sample from the lowest-grade part of the Coastal Belt contains inclusions that appear to be rich in CH_4. These fluid inclusions contain only one phase at room temperature, and generally homogenize at temperatures in the range of −90 to -82°C. Using the phase diagram for CH_4 (see Mullis, 1987), it is possible to estimate the bulk density of the fluid inclusions and, therefore, the likely isochoric paths that they may have experienced. Homogenization temperatures and density estimates for these inclusions are presented in Table 5.

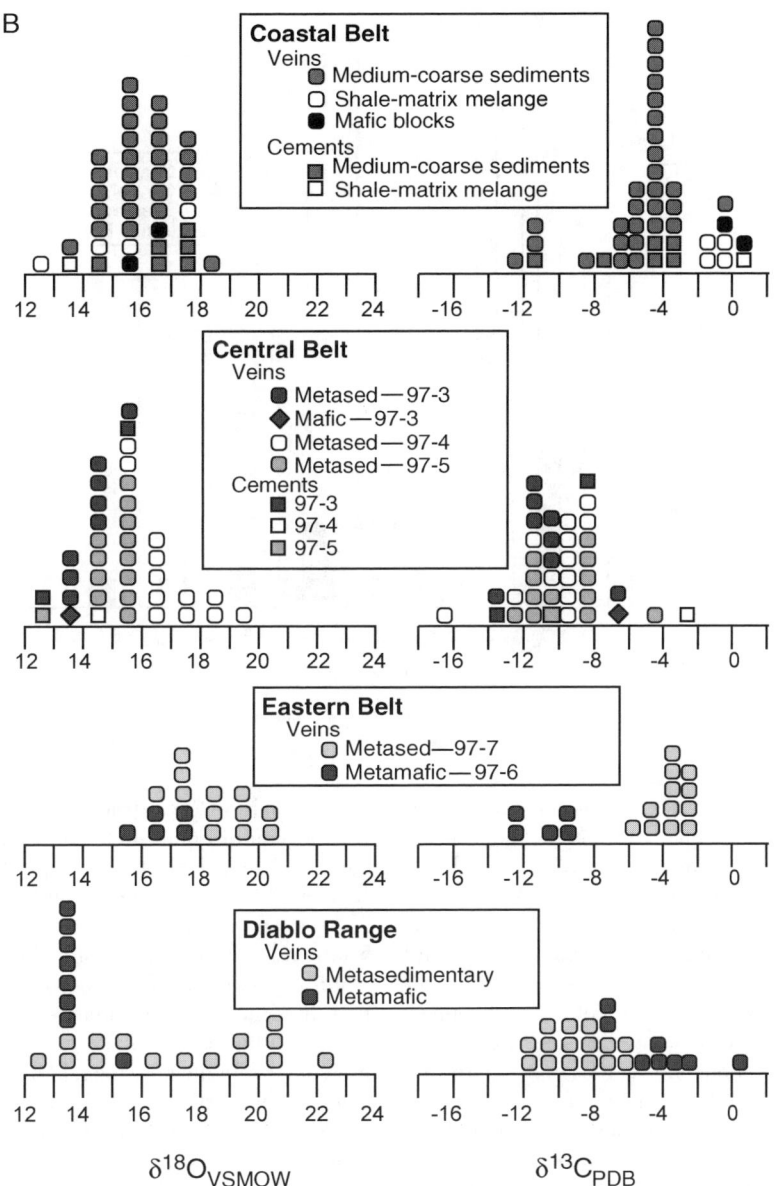

Fig. 3B

Discussion

Models of fluid sources and vein formation based on isotope compositions

Toward interpretation of the isotopic compositions observed in the veins from the Franciscan Complex, it is necessary to explore some simple models of isotopic composition for the fluids and veins. End members for the possible sources of the fluids that produced these veins are seawater (subducted pore water), the host rocks themselves as they devolatilize, or externally derived fluids that may have interacted with the rocks. Seawater $\delta^{18}O$ is by definition 0‰, and most pore waters recovered from shallow sediments are within ~3‰ of 0 (i.e., Vrolijk et al., 1990; Kastner et al., 1991). The

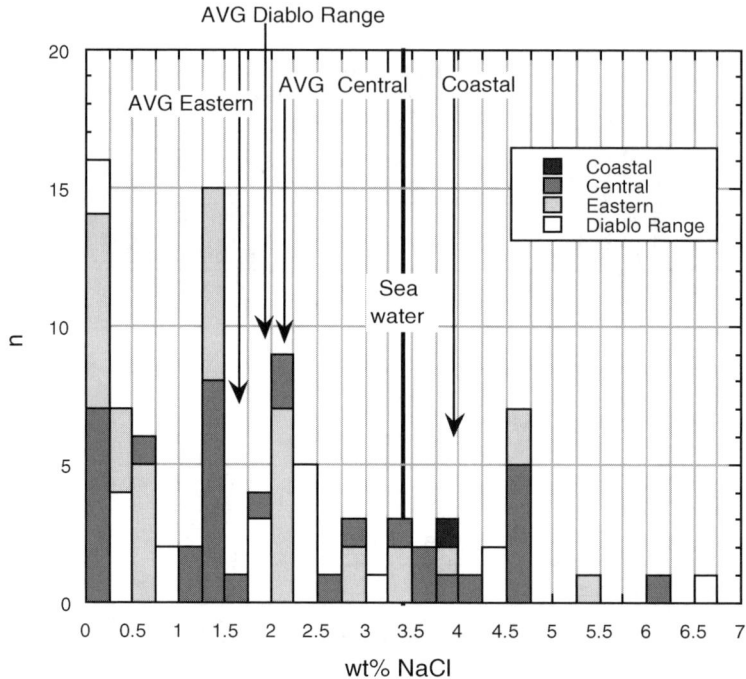

FIG. 4. Salinity of H_2O-rich fluid inclusions from the Coast and Diablo Ranges (expressed as wt% NaCl equivalent).

host-rock $\delta^{18}O$ is variable, but the graywackes contain abundant clastic quartz with $\delta^{18}O_{VSMOW}$ of +12 to +19‰ (Magaritz and Taylor, 1976; Bebout, 1991a, 1991b). Figures 7A and 7B show the expected O-isotope composition of $CaCO_3$ veins formed in equilibrium with a seawater-like fluid (with $\delta^{18}O = 0‰$) and likely high and low values for the effects of rock buffering as a function of temperature, taking into account the +12 to +19‰ range of quartz $\delta^{18}O$ (fractionation factors from Sharp and Kirschner, 1994). Whereas it is unlikely that a pore fluid would remain to the depths of metamorphism without any exchange with the surrounding rocks, a fluid with an isotopic composition similar to seawater could remain if a relatively large amount of pore water was retained in these rocks and relatively little exchange took place at the relatively low metamorphic temperatures between the pore water and surrounding rocks until veins were precipitated.

The other possible extreme conceptual end member is that a fluid is entirely rock buffered in its isotopic composition. Such a condition does not allow determination of fluid source(s), but only indicates that the fluid equilibrated with the host rocks (or lithologies similar to host rocks), and that any infiltration by externally derived fluid initially out of isotopic equilibrium with the host rocks was insufficient to change the isotopic composition of the rock. It is possible to create a rock-buffered fluid by mineral dehydration, in which case fluid released from a mineral initially would probably be in isotopic equilibrium with it, by retention of a small amount of pore fluid (must be less than ~2.5% by volume to stay rock buffered if the fluid-$\delta^{18}O = 0‰$) that exchanges with the host-rock, or by infiltration of an external fluid if that fluid exchanges with the local rocks or travels over a long distance through similar lithologies (see related discussion by Gray et al., 1991). However, if large volumes of seawater are retained in such rocks during metamorphism, these fluids would be likely to buffer the isotopic composition of the veins and, depending on permeability, interact with the rock and change the bulk isotopic composition of the rock. Note also that very high vein-calcite $\delta^{18}O$ values can be produced by low-T equilibration with seawater, and that there is a range of temperatures at which seawater buffering and rock buffering would produce the same calcite-$\delta^{18}O$ (see Fig. 7A). Figure 7B shows the $\delta^{13}C$ values of $CaCO_3$ veins that would be expected to form in

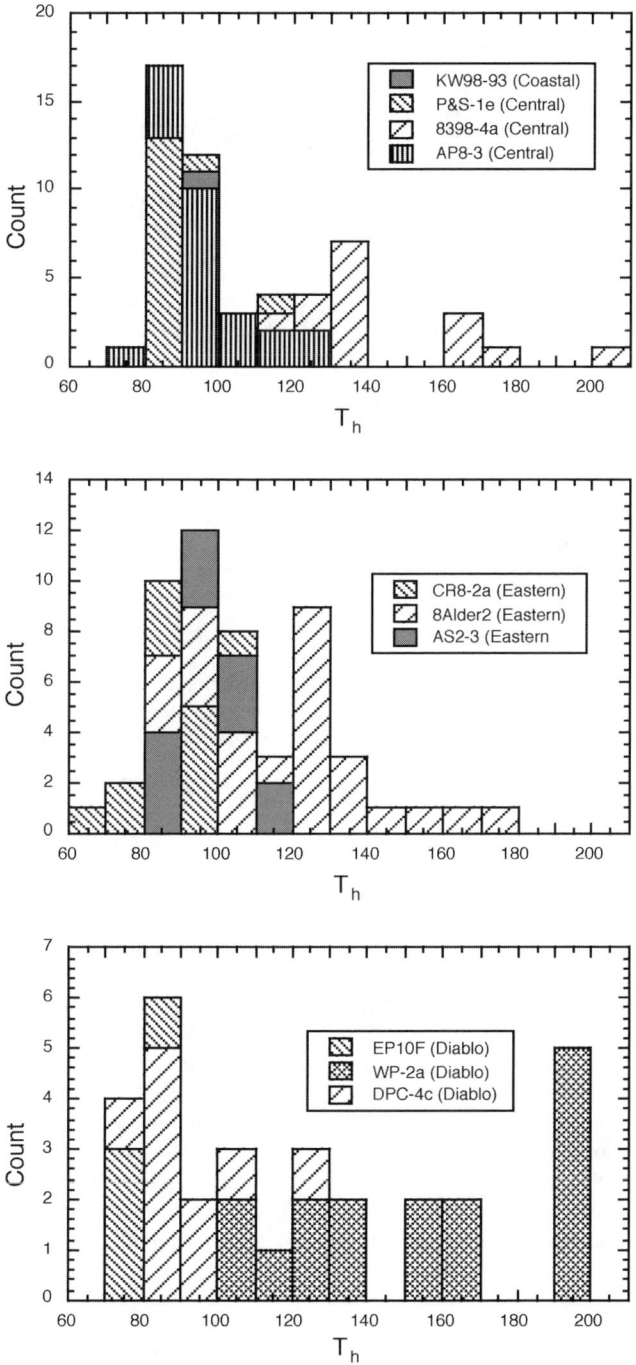

FIG. 5. Homogenization temperatures of H_2O-rich fluid inclusions from the Franciscan Complex.

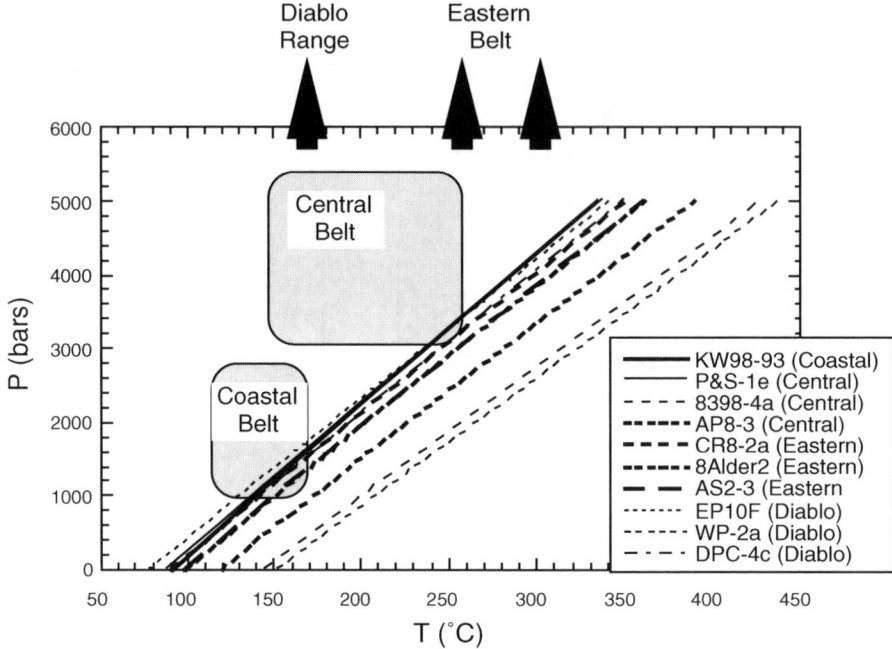

FIG. 6. Isochores for the H$_2$O-rich fluid inclusions. Calculated isochores are based on average T$_h$ calculated for each vein. Shaded fields show inferred peak P-T conditions for the Coastal and Central belts, whereas peak conditions for the Diablo Range and Eastern Belt are higher in P (at similar T) than conditions shown on this diagram.

equilibrium with reduced C in the host rocks ($\delta^{13}C_{VPDB} \approx -25‰$; Sadofsky and Bebout, 2003). Note the trend of lower $\delta^{13}C$ as temperature increases and the fractionation factor decreases (Friedman and O'Neil, 1977). No attempt was made to model the effect of C dissolved in the original pore water on the isotopic compositions of the veins because the isotopic composition of dissolved C is extremely variable in sedimentary pore fluids (i.e., Gamo et al., 1993).

Seawater buffering end member. In order to evaluate the end-member models of vein-forming fluids described above, it is necessary to compare the model calculations with the data collected from the veins themselves. Beginning with the model of seawater buffering of O-isotope composition (Fig. 8), it is apparent that there are several possible ways to explain the observed data. There is some significant scatter in both $\delta^{18}O$ and $\delta^{13}C$ in vein samples from the Coastal Belt. However, most veins have $\delta^{18}O$ values that may have been produced in equilibrium with pore fluids with seawater-like $\delta^{18}O$ at temperatures in the range of 100 to 150°C (see Fig. 8A). This is consistent with the possibility of vein production by original sedimentary pore fluid that may have been released at or near peak metamorphic temperatures, estimated to be ≤ 150°C (Blake et al., 1988).

Examination of the $\delta^{18}O$ values of the Central Belt veins, outcrop by outcrop, shows several interesting relationships (Fig. 8B). Veins from outcrop 97-3 have $\delta^{18}O$ of +13 to +15‰, suggesting that they could have been formed in equilibrium with a seawater-like fluid at temperatures of ~120 to ~140°C. Many of the veins from outcrops 97-4 and 97-5 have $\delta^{18}O$ values significantly higher than those of samples from outcrop 97-3, suggesting that, if they were formed by rock-buffered fluids, they would have to have formed at temperatures <120°C. This is unlikely because these rocks have experienced significantly higher temperatures, and these isotopic compositions are high throughout several texturally distinguished generations of crosscutting veins. Therefore, it appears that most of the veins from outcrops 97-4 and 97-5 probably were not produced by equilibration with a fluid with seawater-like $\delta^{18}O$. Veins from Eastern Belt metasedimentary rocks (Fig. 8C) trend to higher $\delta^{18}O$ that could only have been buffered by a seawater-like fluid at temperatures of <100°C. This possibility of low-T

TABLE 4. Results of Fluid-Inclusion Microthermometry of H_2O-Rich Fluids[1]

System H_2O-NaCl-[KCl] Sample	Inclusion	Size (μm) L	W	Vol. frac. vap.	T_m ice	T_h L-V	Homog. to	NaCl wt %	Bulk density
Coastal Belt Coast Ranges									
KW98-93		6	4	0.05	2.4	90.8	l	3.9	0.985
Central Belt Coast Ranges									
P&S1e	1a				0.1	82.9	l	0.2	0.972
P&S1e	1b					82.9	l		
P&S1e	1c				0.9	82.9	l	1.5	0.978
P&S1e	1d				0.9	86.4	l	1.5	0.976
P&S1e	1e				0.9	90.9	l	1.5	0.973
P&S1e	1f				0.9	86.4	l	1.5	0.976
P&S1e	2a				2.8	83.3	l	4.6	0.991
P&S1e	2b				2.8	83.2	l	4.6	0.992
P&S1e	2c				2.8	82.8	l	4.6	0.992
P&S1e	2d				2.8	87.8	l	4.6	0.989
P&S1e	2e				2.8	87.8	l	4.6	0.989
P&S1e	2f				2.2	88.7	l	3.6	0.984
P&S1e	2g				2.2	89.5	l	3.6	0.984
P&S1e	2h				2.0	87.2	l	3.3	0.983
P&S1e	2i					110.5	l		0.954
8398-4a	1b			0.05		177.5	l		0.893
8398-4a	1d	8	4	0.05		169.6	l		0.902
8398-4a	1e	12	12	0.05		200.1	l		0.865
8394-4a	1g			0.05		161.2	l		0.911
8394-4a	1h	20	20	0.06		167.6	l		0.904
8398-4a	2b	10	5	0.05	0.7	135.0	l	1.2	0.942
8398-4a	2c	8	3	0.05	2.4	135.0	l	3.9	0.957
8398-4a	2d	10	3	0.05		135.0	l		0.935
8398-4a	2e	6	4	0.10	0.3	135.0	l	0.5	0.938
8398-4a	3a	5	3	0.10	0.7	130.8	l	1.2	0.945
8398-4a	3b	4	4	0.05	1.6	130.8	l	2.6	0.953
8398-4a	3c	6	2	0.05	0.1	125.0	l	0.2	0.944
8398-4a	3d	2	2	0.05	0.1	130.8	l	0.2	0.940
8398-4a	3e	5	3	0.10	1.1	125.0	l	1.8	0.953
8398-4a	3f	4	2	0.05		115.4	l		0.951
AP8-3	1a	5	4	0.05	1.8	93.0	l	3.0	0.979
AP8-3	1b	2	2	0.05	0.0	95.7	l	0.0	0.964
AP8-3	1c	3	3	0.10	1.3	95.7	l	2.1	0.973
AP8-3	1d	4	3	0.05		98.7	l		0.962
AP8-3	1e - 3 incs	4	3	0.07		93.0	l		0.965
AP8-3	1f	5	4	0.05		80.8	l		0.972
AP8-3	1g	4	3	0.15	1.3	93.0	l	2.1	0.975
AP8-3	2a	3	3	0.05	0.0	82.8	l	0.0	0.971
AP8-3	2b	5	3	0.05	0.0	95.7	l	0.0	0.964
AP8-3	2c	12	6	0.10		82.8	l		0.971
AP8-3	2d	3	2	0.05	1.0	85.0	l	1.7	0.977
AP8-3	2e	3	2		2.5	91.4	l	4.1	0.985
AP8-3	2f	4	2	0.05	3.8	104.8	l	6.1	0.988
AP8-3	2g	3	2	0.05	0.1	76.6	l	0.2	0.975
AP8-3	3-a	5	5	0.10	0.9	108.1	v	1.5	0.963
Ap8-3	3-c	3	3	0.05		108.1	v		0.956
AP8-3	3b	6	2.5	0.05		120.9	v		0.946

Table continues

TABLE 4. *Continued*

System H_2O-NaCl-[KCl] Sample	Inclusion	Size (μm) L	W	Vol. frac. vap.	T_m ice	T_h L-V	Homog. to	NaCl wt %	Bulk density
	**Central Belt Coast Ranges (*continued*)**
AP8-3	4-a			0.05	0.8	122.5	l	1.3	0.952
AP8-3	4-b			0.05	0.8	97.0	l	1.3	0.969
AP8-3	4-c			0.05	0.8	97.0	l	1.3	0.969
AP8-3	4-d	6	4	0.10		118.2	l		0.948
AP8-3	4-e	4	3	0.05		117.4	l		0.949
	**Eastern Belt Coast Ranges**
CR8-2a	1a				2.8	69.3	l	4.6	0.998
CR8-2a	1b				0.4	75.8	l	0.7	0.977
CR8-2a	1c				0.0	95.7	l	0.0	0.964
CR8-2a	1d				0.4	75.8	l	0.7	0.977
CR8-2a	1e				2.8	95.7	l	4.6	0.985
CR8-2a	1f					95.7	l		0.964
CR8-2a	2a				0.0	88.8	l	0.0	0.968
CR8-2a	2b				0.2	84.3	l	0.3	0.972
CR8-2a	2c				0.2	88.8	l	0.3	0.969
CR8-2a	3-a	4	3	0.05		107.8	l		0.956
CR8-2a	3-c	4	3	0.05		96.8	l		
CR8-2a	3b	2	2	0.05		96.8	l		
8alder2	1a	12	12	0.05		178.4	l		0.892
8alder2	2a	15	8	0.1	0.0	97.5	l	0.0	0.963
8alder2	2b	12	6	0.05	0.0	109.6	l	0.0	0.955
8alder2	2c			0.055	0.4	86.9	l	0.7	0.972
8alder2	2d			0.05	0.2	100.2	l	0.3	0.962
8alder2	2e			0.05	0.4	81.8	l	0.7	0.974
8alder2	2f	15	8	0.05	1.7	153.2	l	2.8	0.936
8alder2	2g	3	3	0.05	2.3	122.7	l	3.8	0.965
8alder2	2h	6	6	0.05	1.7	122.7	l	2.8	0.960
8alder2	2i	25	10	0.05	3.4	103.7	l	5.5	0.986
8alder2	2j	7	7	0.05	2.1	103.7	l	3.4	0.975
8alder2	3a	4	2	0.05		94.4	l		0.964
8alder2	3b	5	2	0.05		110.2	l		0.954
8alder2	3c	3	2	0.05		94.4	l		0.964
8alder2	3d	8	3	0.05		94.4	l		0.964
8alder2	3e	4	3	0.05		85.7	l		0.969
8alder2	4a	4	3	0.05	0.4	132.4	l	0.7	0.941
8alder2	4b	8	5	0.05	2.0	126.1	l	3.3	0.960
8alder2	4c	4	4	0.05	0.8	122.8	l	1.3	0.952
8alder2	4d	10	8	0.05	0.8	146.8	l	1.3	0.933
8alder2	4e	5	4	0.05	0.8	167.3	l	1.3	0.914
8alder2	4f	8	3	0.05		122.8	l		0.945
8alder2	4g	6	5	0.05	0.8	126.1	l	1.3	0.950
8alder2	4h	3	2	0.05		132.4	l		0.937
8alder2	4i	4	3	0.05	0.8	132.4	l	1.3	0.945
8alder2	4j	6	6	0.05	0.8	126.1	l	1.3	0.950
8alder2	4k	4	3	0.05		125.7	l		0.943
8alder2	4l	4	3	0.05	0.8	126.1	l	1.3	0.950
AS2-3a	1a	6	3	0.05	0.1	82.7	l	0.2	0.972
AS2-3a	1b	6	5	0.05	0.1	82.7	l	0.2	0.972
AS2-3a	1c	2	2	0.05	0.1		l	0.2	

Table continues

TABLE 4. Continued

System H$_2$O-NaCl-[KCl] Sample	Inclusion	Size (μm) L	W	Vol. frac. vap.	T$_m$ ice	T$_h$ L-V	Homog. to	NaCl wt %	Bulk density
Eastern Belt Coast Ranges (continued)									
AS2-3a	2a	10	6	0.05		88.8	l		0.968
AS2-3a	2b	5	3	0.05		94.0	l		0.965
AS2-3a	2c	6	3	0.05	1.3	94.0	l	2.1	0.974
AS2-3a	2d	4	3	0.05	1.3	101.3	l	2.1	0.970
AS2-3a	2e	4	3	0.05	1.3	101.3	l	2.1	0.970
AS2-3a	2f	4	4	0.05	1.3	107.9	l	2.1	0.966
AS2-3a	2g	4	3	0.05		86.0	l		0.969
AS2-3a	2h	4	3	0.05	1.3	99.3	l	2.1	0.971
AS2-3a	2i	4	3	0.05	1.3	114.4	l	2.1	0.962
AS2-3a	2j	4	3	0.05	1.3	114.4	l	2.1	0.962
Diablo Range									
EP 10f chip1	1-a	5	4	0.05	0.2	78.7	l	0.3	0.975
EP10f	1-c	4	3	0.05	0.2	75.3	l	0.3	0.976
EP10f	1-d	4	3	0.05	0.2	80.6	l	0.3	0.974
EP10F chip1	b	6	3	0.05	0.2	78.7	l	0.3	0.975
WP-2a calcite	1a	5	4	0.05	1.1	135.4	l	1.8	0.945
WP-2a	1a			0.05	0.1	191.0	l	0.2	0.879
WP-2a calcite	1b	6	3	0.05	1.1	120.7	l	1.8	0.956
WP-2a	1b				0.1	191.0	l	0.2	0.879
WP-2a	1c			0.5		191.0	l		0.877
WP-2a	1d			0.5		191.0	l		0.877
WP-2a	1e			0.5		191.0	l		0.877
WP-2a	2a			0.1	1.4	111.4	l	2.3	0.965
WP-2a	2a	4	2	0.1	2.7	161.2	l	4.4	0.940
WP-2a	2b	5	3	0.1	2.7	161.2	l	4.4	0.940
WP-2a	2c	10	5	0.1	1.1	109.1	l	1.8	0.964
WP-2a	2d	2	2		1.5	109.1	l	2.5	0.967
WP-2a	2e	2	2			125.8	l		0.943
WP-2a	2f and g				1.4	137.4	l	2.3	0.946
WP-2a	2h					158.8	l		0.913
WP-2a	2i				1.5	150.4	l		
DPC-4c	2a	6	3	0.1	4.2	90.0	l	6.7	0.999
DPC-4c	2b	3	3	0.3		85.8	l		0.969
DPC-4c	2c	3	3	0.3		90.0	l		0.967
DPC-4c	2d	5	3	0.3	1.5	76.4	l	2.5	0.985
DPC-4c	2e	4	4	0.3	0.5	88.1	l	0.8	0.972
DPC-4c	2f	5	5	0.3	1.5	124.9	l	2.5	0.956
DPC-4c	2g	3	2	0.3	0.5	81.0	l	0.8	0.975
DPC-4c	2h	5	5	0.3		108.1	l		0.956
DPC-4c	2i	3	2	0.1		88.1	l		0.968
DPC-4c	2j	4	3	0.2	1.9	81.0	l	3.1	0.986

[1]Abbreviations: L = length; W = width; T$_m$ = melting temperature; T$_h$ = homogenization temperature; l = liquid; v = vapor.

seawater buffering seems highly unlikely, given that these rocks experienced much higher temperatures than the Coastal and Central belts, and the veins appear to represent several generations of pre-, syn-, and post-kinematic growth. Mafic rocks from the Eastern Belt have slightly lower $\delta^{18}O$ and could have been buffered by a seawater-like fluid at temperatures of 100 to 120°C. Veins from the Pacheco Pass area (Fig. 8D) show a wider range in $\delta^{18}O$ than those of the other Franciscan subunits. It is possible to explain any of these $\delta^{18}O$ values as a record of equilibration with a seawater-like fluid over a wide

TABLE 5. Results of Fluid-Inclusion Microthermometry of CH_4-Rich Fluids

Sample	Inclusion	Inc. Type	T_h	Density[1]
		Coastal Belt Coast Ranges		
KW98-46	1a	Isolated	−74.1	
	1b	Planar cluster	−86.3	0.23
	1c	Planar cluster	−86.3	0.23
	1d	Planar cluster	−86.3	0.23
	1e	Planar cluster	−86.3	0.23
	1f	Planar cluster	−86.3	0.23
	1g	Planar cluster	−86.3	0.23
	1h	Planar cluster	−86.3	0.23
	2a	Irreg. cluster	−84.5	0.23
	2b	Irreg. cluster	−72.6	
	2c	Irreg. cluster	−84.5	0.21
	2d	Irreg. cluster	−86.3	0.23
	2e	Planar cluster	−81.6	0.162
	2f	Planar cluster	−86.3	0.23
	2g	Planar cluster	−86.3	0.23
	2h	Planar cluster	−86.3	0.23
	2i	Planar cluster	−86.3	0.23
	2j	Planar cluster	−85.5	0.225
	2k	Isolated	−72.3	
	2l	Planar cluster	−85.5	0.225
	2m	Planar cluster	−85.5	0.225
	2n	Isolated	−72.3	0.26
	3a	Planar cluster	−88.1	0.26
	3b	Planar cluster	−88.1	0.26
	3c	Planar cluster	−88.1	0.26
	3d	Planar cluster	−88.1	0.26
	3e	Planar cluster	−88.1	0.26
	3f	Planar cluster	−88.1	0.26
	3g	Planar cluster	−88.1	0.26
	3h	Planar cluster	−88.1	0.26
	3i	Planar cluster	−88.1	0.26
	3j	Planar cluster	−88.1	0.33
	4a	Isolated	−109.3	0.26
	4b	Isolated	−70.6	
	4c	Planar cluster	−73.4	
	4d	Planar cluster	−73.4	
	4e	Planar cluster	−73.4	
	4f	Planar cluster	−73.4	
	4g	Isolated	−72.9	
	4h	Isolated	−72.9	
	4i	Isolated	−72.7	
	4j	Isolated	−72.7	

[1]T_h = homogenization temperature.
[2]Bulk densities derived from Mullis (1987), Figure 5.11.

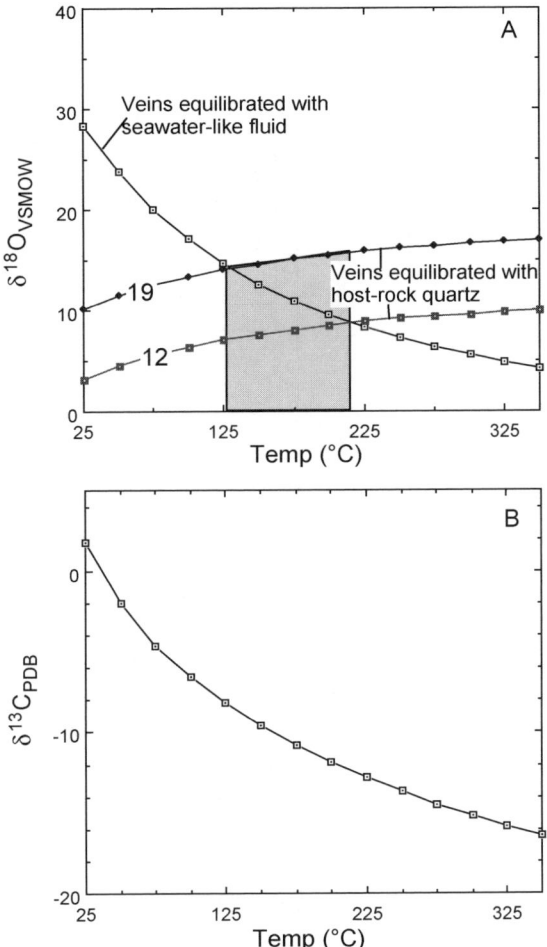

FIG. 7. Predicted isotopic compositons of $CaCO_3$ veins as a function of the temperature at which the veins are formed. A. $\delta^{18}O$ of veins formed from a seawater-like fluid ($\delta^{18}O = 0‰$) and veins formed by rock-buffered fluids. There are high and low estimates for the rock-buffered case, representing the range of likely $\delta^{18}O$ values of quartz in the graywackes (+12 to +19‰; see Magaritz and Taylor, 1976, Bebout, 1991a, 1991b; Grove and Bebout, 1995). Note the shaded region of temperature in which the models produce overlapping results. B. $\delta^{13}C$ of $CaCO_3$ veins formed in equilibrium with host-rock reduced C ($\delta^{13}C = -25‰$; from Sadofsky and Bebout, 2003).

range in the temperature of vein formation. However, the fact that the different textural generations within outcrops EP3 and EP4 appear to have overlapping ranges in $\delta^{18}O$ suggests that there were other factors in the formation of these veins.

Rock-buffering end member. Calculations of the rock-buffered end-member hypothesis of vein formation (Fig. 9) explain many of the isotopic compositions that are inconsistent with buffering of $\delta^{18}O$ by a seawater-like fluid. Many veins from the Coastal Belt have $\delta^{18}O$ and $\delta^{13}C$ in the range of rock-buffered possibilities for their likely metamorphic temperatures (up to 150°C). However, in taking the $\delta^{18}O$ and $\delta^{13}C$ values together (Fig. 9A), it is apparent that these veins were not entirely rock buffered with regard to isotopic composition. Most $\delta^{13}C$ values are consistent with rock buffering in the range of 75 to 125°C, whereas most veins have $\delta^{18}O$ values too high for rock buffering at that temperature. If C and O equilibrated at the same temperatures, it appears that most of the veins of the Coastal Belt have isotopic compositions that would

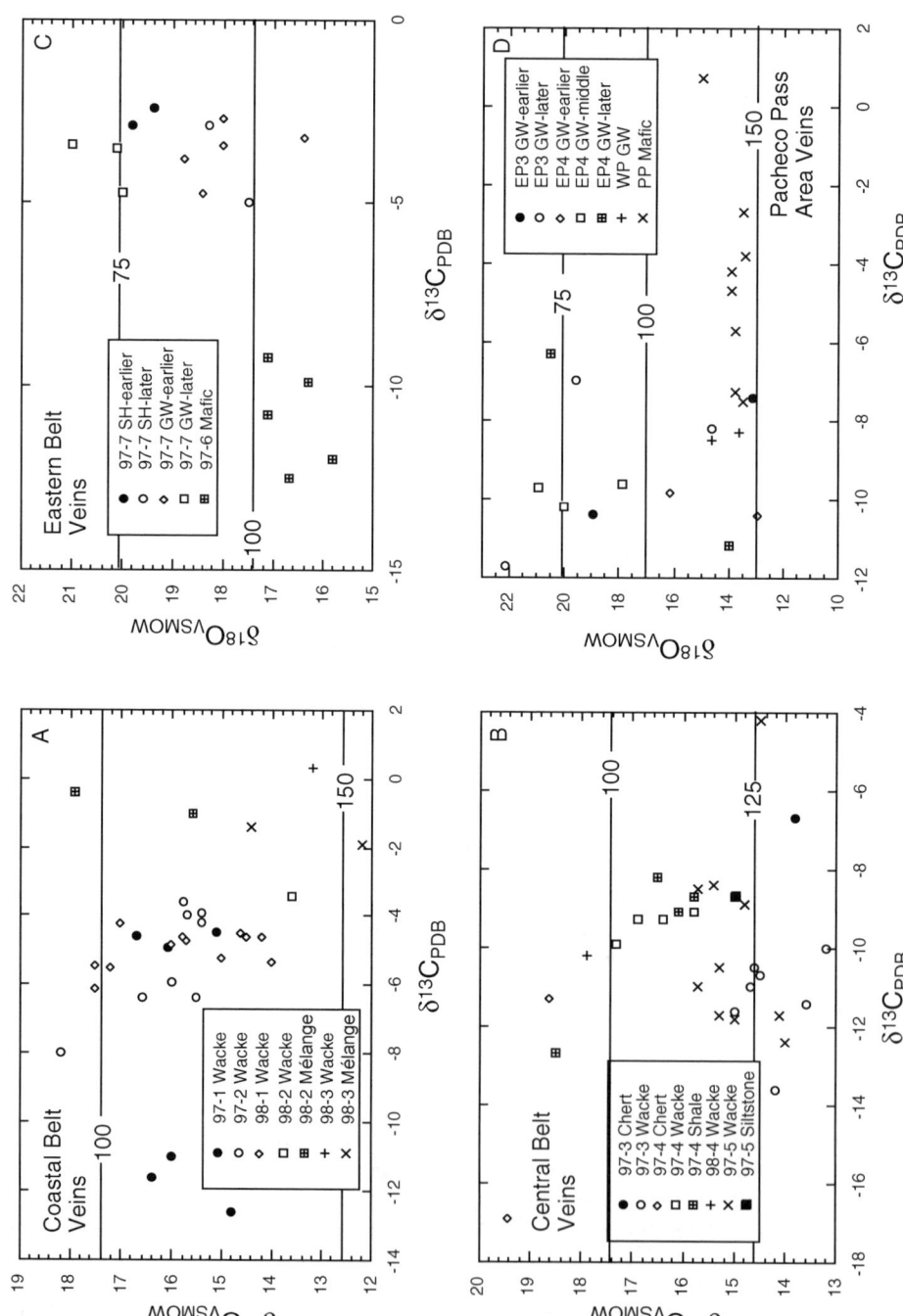

FIG. 8. C- and O-isotopic compositions of Franciscan $CaCO_3$ veins plotted by subterrane and outcrop. O-isotopic composition of veins predicted to form in equilibrium with a seawater-like fluid ($\delta^{18}O = 0‰$) are shown as a function of temperature in °C (horizontal lines; see relationships in Fig. 7).

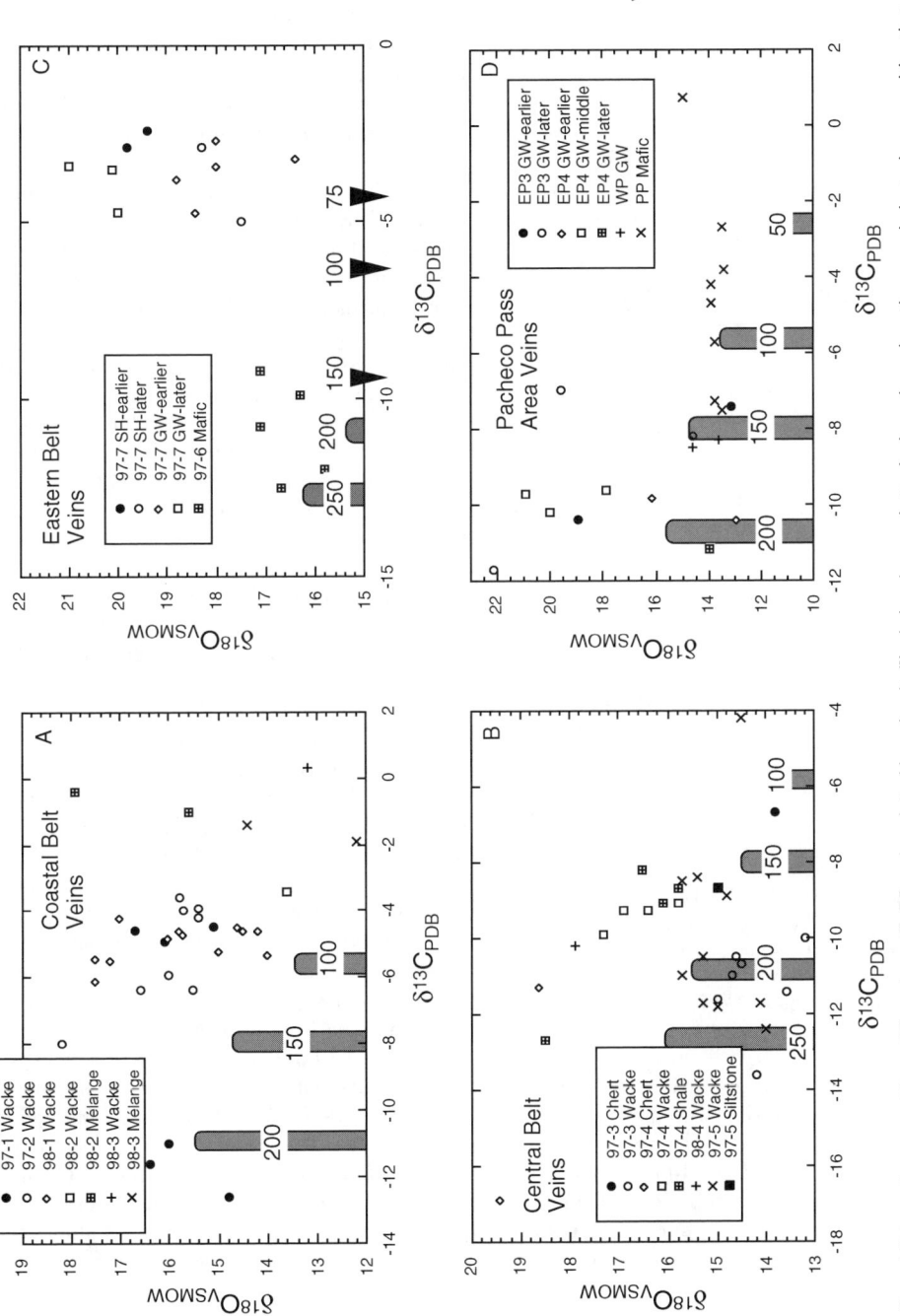

FIG. 9. C- and O-isotopic compositions of Franciscan CaCO$_3$ veins plotted by subunit. Shaded regions and attached numbers show the range in isotopic composition that would be expected under completely rock buffered conditions (i.e., equilibrated with clastic quartz and marine organic C) as a function of temperature in °C (see Fig. 7). Note that the horizontal and vertical scales differ among these diagrams.

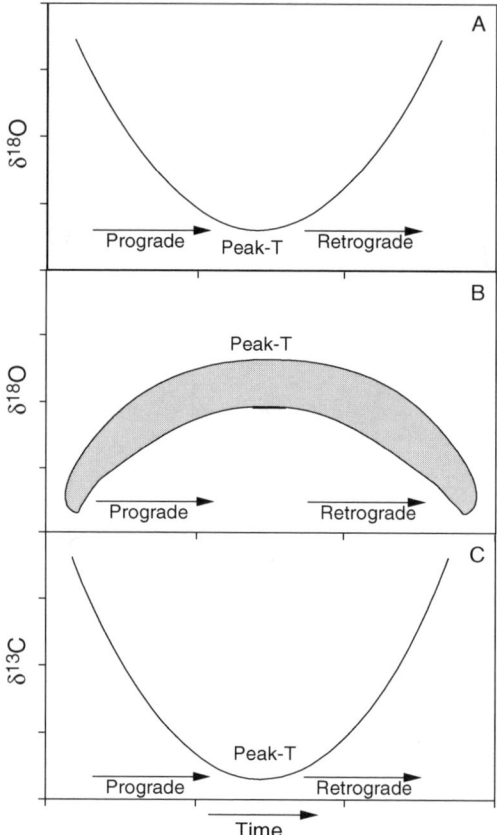

FIG. 10. Schematic representations of the changes in isotopic composition of calcite veins precipitated from an H_2O-rich fluid at differing temperatures, with constant buffering by host rocks of a fluid of constant composition. A. $\delta^{18}O$ of veins formed in equilibrium with a fluid of constant composition. B. $\delta^{18}O$ of calcite veins fomed in equilibrium with a host rock, with the shaded field representing a range in host-rock composition. C. $\delta^{13}C$ of calcite veins formed in equilibrium with graphite reservoir with constant $\delta^{13}C$.

represent $\delta^{13}C$ in equilibrium with the local host rocks and $\delta^{18}O$ in equilibrium with pore fluids (subducted seawater). The range in $\delta^{13}C$ (even at similar $\delta^{18}O$ values) could record equilibration of C isotopes in some of the veins at differing temperatures (i.e., see Figs. 10 and 11), thus suggesting that the variations in C- and O-isotopic compositions were not entirely coupled in these veins. Therefore, the question of whether the veins in these lowest-T rock were rock buffered or seawater buffered with respect to O remains equivocal because of the overlapping model $\delta^{18}O$ at these low temperatures and the inability to more precisely constrain vein formation temperatures.

Most veins from outcrops 97-3 and 97-5 in the Central Belt have C- and O-isotopic compositions that could have formed in equilibrium with the host rocks at a range of T (Fig. 9B). However, veins from outcrops 97-4 and 98-4, and a few of the samples from outcrop 97-5, have $\delta^{18}O$ values that are too high to have been formed by equilibration with the host rocks. These high $\delta^{18}O$ values could be explained by a process of up-dip (down-T) flow of fluids equilibrated with hotter rocks deeper in the subduction zone. Veins from the Eastern Belt are somewhat more difficult to explain (see Fig. 9C). Samples from mafic rocks of outcrop 97-6 have $\delta^{13}C$ values consistent with equilibrium at ~200 to 250°C. Samples of metasedimentary rocks from outcrop 97-7 have a range of $\delta^{18}O$ values that appear to be too high to represent equilibrium with clastic quartz, and may be high because of the abundance of shalier rocks in this area, or may represent up-dip flow of fluids that have equilibrated with similar rocks at higher-T. $\delta^{13}C$ values are also too high to have equilibrated with the host rocks at any reasonable temperature; thus, these increases in $\delta^{13}C$ could have been caused by participation of sedimentary/early diagenetic carbonate in the vein-forming fluids, or as with $\delta^{18}O$, the high $\delta^{13}C$ may be a product of fluid flow from higher-temperature areas of the subduction zone.

Many of the veins from the Pacheco Pass area have O-isotopic compositions that are similar to those expected under rock-buffered conditions, with some others deviating to higher $\delta^{18}O$ values (Fig. 9D). These higher $\delta^{18}O$ values are unlikely to have been caused by seawater effects, because the temperatures required for increasing $\delta^{18}O$ by seawater are too low for the $\delta^{13}C$ of these veins and fluid inclusion data from nearby veins. Therefore, these higher $\delta^{18}O$ values are likely to represent up-dip flow of fluids that formed at higher temperature.

Implications of the isotope data in relation to the models above. Superimposed on the complexity of the models discussed above is the fact that veins from an individual outcrop probably formed at different times during the heating and cooling histories of the rocks. Figure 10 shows a schematic representation of the variations in isotopic composition that veins would record, depending on when they were precipitated during the heating-cooling history of the rocks. These diagrams are presented only in schematic fashion because they are used to

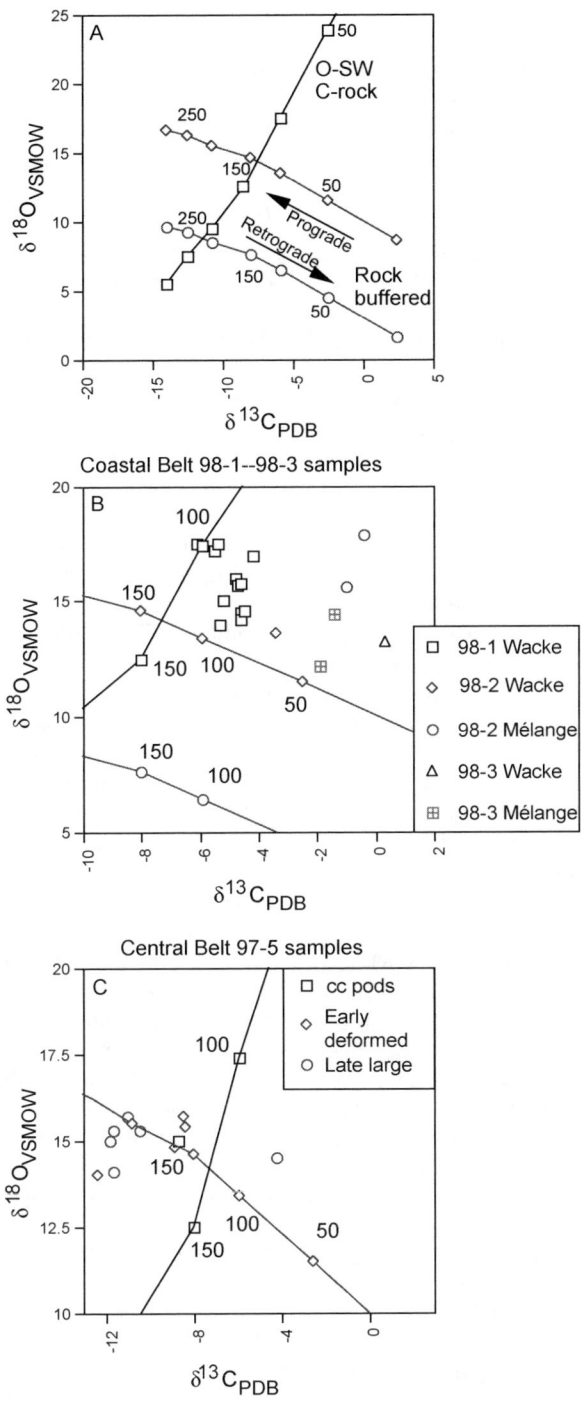

FIG. 11. A. Models of variations in vein $\delta^{18}O$ and $\delta^{13}C$ as a function of T (°C), with $\delta^{13}C$ buffered by host-rock organic C and $\delta^{18}O$ buffered by either seawater or host-rock quartz. Numbers indicate the temperature of the calculation (as in Fig. A-5), and high and low host-rock $\delta^{18}O$ are represented by parallel lines. B. Isotopic compositions of veins from sites 98-1 through 98-3 are plotted with model calculations. C. Isotopic compositions of veins from site 97-5 are plotted with the model calculations.

represent rocks that experienced a wide range in peak-T conditions. The veins that have isotopic compositions consistent with equilibration with seawater at lower temperatures may have formed earlier in the T-time history of the rocks than those that appear to be in equilibrium with the host rocks.

Modeled isotopic compositions and observed data for two subsets of the veins are shown in Figure 11. An example that appears to record differential retention of fluid signatures as a function of rock type or vein generation is in the Coastal Belt, where all samples of veins from the shaley-matrix mélange have $\delta^{13}C$ compositions higher than that of most of the veins in the graywacke units. Some of the later-stage veins in the graywacke units appear to be similar to those in the shaley matrix. This suggests that the veins preserved in the mélange may have formed at lower temperature than most of the veins in the graywacke units, and could indicate that the mélange preserves only the later-formed veins (earlier-formed veins in these lithologies are highly transposed and disrupted). The graywacke was perhaps stronger and preserved multiple generations of veins formed at or near peak-T and during prograde histories (see Fig. 11B; Wroblewski, 1999). Another example comparing model calculations with actual isotopic compositions for a specific outcrop is for site 97-5 (Fig. 11C) where it is apparent that almost all veins have isotopic compositions that are nearly entirely rock buffered at T of ~150°C, except for one of the late-stage veins that has a significantly higher $\delta^{13}C$.

Overall, a careful analysis of the C- and O-isotopic compositions of veins from the Coast and Diablo ranges of the Franciscan Complex in the context of these simple end member models shows significant variation in the type of fluid-rock interactions that produced these veins. Samples from the Coastal Belt have $\delta^{18}O$ that appears to be seawater buffered at T near the peak of metamorphism, and $\delta^{13}C$ that appears to have been rock buffered at a similar, but over a more variable, range of T. At the inferred peak temperature for this unit, aqueous fluid equilibrated with the rocks had $\delta^{18}O$ near that of seawater; thus seawater- and rock-buffered models cannot be distinguished (see discussion below). Veins from the Central Belt are more likely to record both $\delta^{18}O$ and $\delta^{13}C$ that are generally rock buffered at a range of T up to those estimated for peak metamorphism, with some departures to higher $\delta^{18}O$ that may be associated with up-dip fluid flow or nearby non-graywacke lithologies. Veins in metasedimentary rocks from the Eastern Belt, and many metasedimentary rocks from the Pacheco Pass area, have isotopic compositions that are too high for reasonable model calculations, suggesting that these rocks experienced infiltration by an externally derived fluid that participated in formation of these veins.

Discussion of fluid inclusion microthermometry

Salinity of the fluid inclusions is calculated directly from the observed freezing point depression (for fluid inclusions that melt at <0°C). Presumably the initial pore water in these sediments was seawater with a salinity of ~3 to 3.5 wt%. It is difficult to draw much of a conclusion from the one inclusion of the Coastal Belt that was measurable; however, this inclusion has a salinity of 3.9%, slightly higher than that of seawater. Fluid inclusions from the high-grade rocks have somewhat variable salinities. However, despite large and overlapping ranges, averages of salinity by metamorphic unit show a general decrease (or freshening) with increasing P-T conditions. This could be due to reduced influence of seawater and greater influence of dehydration waters at progressively greater depths (Fig. 4).

Homogenization temperatures of hydrous fluid inclusions represent minimum temperatures of trapping of fluids, and for most rocks, these temperatures are lower than those expected for peak metamorphism (Fig. 5). This temperature can be corrected for pressure (by the ideal gas law, inasmuch as an inclusion has a near-constant volume), and the resulting linear (in P-T space) limit of possible P-T conditions is referred to as an isochore. Isochores calculated for averages from each vein show that low-grade rocks preserve trapping of fluid inclusions at or near lithostatic pressure in the possible range of peak conditions (Fig. 6). However, many of the inclusions in high-grade rocks show isochores that are too high-T for a given P to reflect peak lithostatic conditions. These fluid inclusions may have been trapped during uplift and record conditions near peak T at lower than peak pressure (i.e., Moree, 1998).

Relationship to fluids observed at active convergent margins

Much recent work has been done to better understand the behavior of fluids in active forearcs, such as those at the Marianas, Barbados, and Cascadia margins, and to attempt to quantify and understand the release of fluids in the forearc. These studies have observed discharge of fluid at the décollement and in other fracture systems that is in many cases similar in composition to seawater. However, several distinct chemical and isotopic tracers sometimes vary from those fluids, including elevated $\delta^{18}O$, δD, NH_4^+, Ca, alkalinity, CH_4 and other hydrocarbons, Li, Rb, B and K, and lower Cl, Na and Sr (see relevant observations by Elderfield et al., 1990; Vrolijk et al., 1990; Kastner et al., 1995, 1997; Benton, 1997; Brown et al., 2001; Moore et al., 2001; Silver et al., 2000; Mottl et al., 2003, in press). These departures from local pore-water compositions, often associated with the "décollement zone" or other faults, have led the above-cited and others to suggest an external source for these fluids from greater depths. Suggestions for the source of these "fresher" waters have included dehydration waters (related to clay transitions, opal dehydration, or deeper metamorphism), water that interacts with clastic or pelagic sediments, and waters that have interacted with the oceanic crust (see Elderfield et al., 1990; Kastner et al. 1991; Silver et al., 2000; Brown et al., 2001; Mottl et al., 2003, in press).

Observations from this study suggest that fluids originating during devolatilization of the Franciscan Complex may be very similar to the "fresher" fluid component that is documented at sites of active accretion, and the processes recorded in these rocks are similar to those occurring at depth in active subduction zones (see Silver et al., 2000; Brown et al., 2001; Moore et al., 2001). Several lines of evidence indicate a connection between the shallower salinity anomalies, observed along the décollement and other shallow accretionary prism structures, and dehydration and fluid flow at greater depths; this evidence comes from isotopic compositions of the veins in these rocks, fluid inclusion microthermometry, and major- and trace-element geochemistry of rocks such as the Franciscan paleoaccretionary suite (see below).

There has likely been significant loss of H_2O from these rocks (see Sadofsky and Bebout, 2003) during diagenesis and early metamorphism, related to breakdown of clay minerals to stabilize low-grade white-micas (see Frey, 1987; also opal dehydration; Brown et al., 2001), and this fluid (and that generated at greater depths) would be likely to migrate upward in the accretionary complex or into/along zones of enhanced permeability (Silver et al., 2000). This dehydration fluid could have a range of $\delta^{18}O$ values, depending on the variations in source (clastic vs. pelagic sediment) and the temperature at which they exchange with the local rocks. However it is highly likely (see Fig. 12) that any fluid migrating up the décollement would have $\delta^{18}O$ that is similar, or slightly elevated, relative to the local pore waters, because deeper-sourced fluids would have equilibrated to some degree with the host rocks at elevated temperatures. The $\delta^{18}O$ of the fluids that produced veins in the lowest-grade rocks are somewhat variable, but most likely within $\pm \sim 3$‰ of seawater, like most of the fluids that are venting at sites of active accretion (Vrolijk et al., 1990; Kastner et al., 1991; Benton, 1997; Mottl et al., 2003, in press).

Elevations in CH_4 concentration are one of the most commonly cited pieces of evidence for an externally sourced fluid at active accretionary complexes (i.e., Moore, 1989; Vrolijk et al., 1990; Kastner et al., 1991). The lowest-grade rocks of this study, which should be most similar to rocks underneath the shallow décollement, contain both H_2O-rich and CH_4-rich fluids, suggesting that CH_4 was a significant mobile fluid component during vein production.

Increases in the concentration of NH_4^+ and B in décollement fluids are probably due to the mobility of the adsorbed components of B and N during low-T diagenesis, and by initial loss of structurally bound B-N during low-temperature devolatilization (Bebout and Nakamura, 2003; Sadofsky and Bebout, 2003). Finally, the lower concentrations of Cl and Na in the "fresher" décollement fluids (Kastner et al., 1991) correspond to the low salinity of most H_2O-rich fluid inclusions analyzed in veins from these rocks (with most inclusions being less saline than seawater; Fig. 4).

Comparison with other studies of fluids in paleo-accretionary complexes

High-P/T metamorphic rocks such as those in the Franciscan Complex in California are a direct time-integrated product of the evolving thermal regime of ancient accretionary wedges (Ernst, 1988). Detailed studies using fission tracks and thermal models point to potentially complex thermal histories for individual rock packets in such settings, in response to thermal relaxation due to plate motion reorganization, varying rates of exhumation, and other variables such as shear stress, thickness of the upper plate, and upper-plate radiogenic heat production (e.g., Dumitru, 1989; Royden, 1993; Tagami and Dumitru, 1996). Field-based and theoretical models of exhumation of most high-P/T metamorphic rocks in such settings in general invoke buoyant rise of subducted material (Cloos and

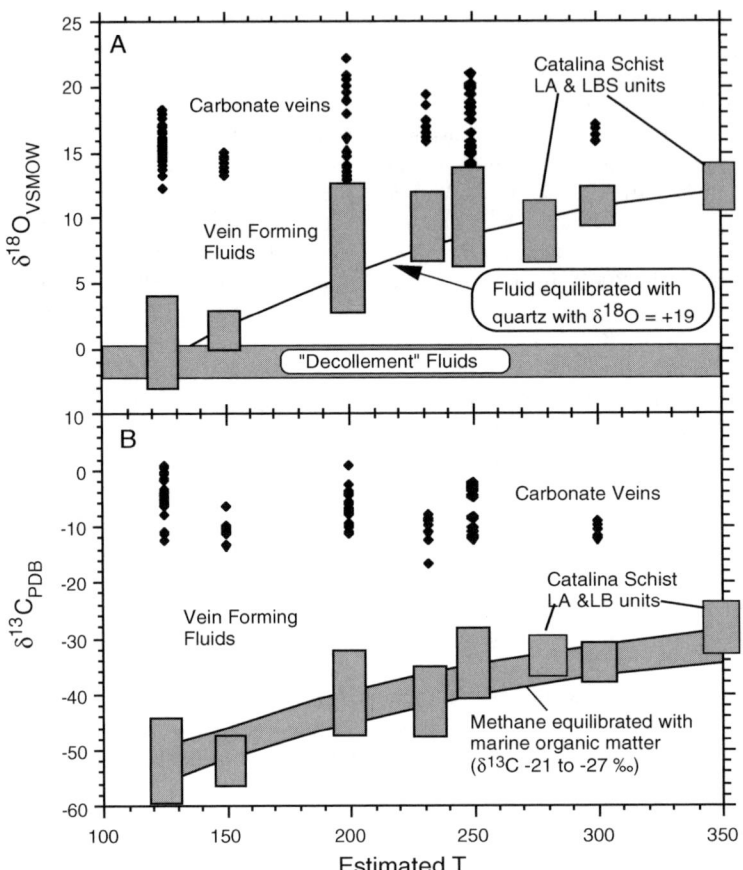

FIG. 12. A. $\delta^{18}O$ of $CaCO_3$ veins plotted vs. estimated peak-T that the rocks have experienced. Also plotted are estimates of $\delta^{18}O$ of fluids in equilibrium with the veins; $\delta^{18}O$ of fluids in equilibrium with quartz in host rocks; and $\delta^{18}O$ of fluids in active accretionary complexes (fractionation data from Friedman and O'Neil, 1977; Sharp and Kirschner, 1994). Note that the curve for equilibrium of the fluids with host-rocks is only for quartz $\delta^{18}O$ of +19‰, and quartz $\delta^{18}O$ in these rocks likely ranges from +12 to +19‰—thus, some veins are consistent with rock buffering (see similarity in fluid $\delta^{18}O$ calculated from $CaCO_3$ veins, in patterned boxes, with the curve for fluids equilibrated with quartz with $\delta^{18}O$ of +19‰), but many are higher in $\delta^{18}O$ than could be explained by equilibration with rock quartz with $\delta^{18}O$ of +12 to +19‰ (the curve for quartz $\delta^{18}O$ of +12‰ would parallel that for +19‰, but would be 7‰ lower on this figure). Interestingly, at lower T, rock-buffered $\delta^{18}O$ straddles that of seawater (0‰). B. $\delta^{13}C$ of $CaCO_3$ veins; most have $\delta^{13}C$ similar to that calculated in equilibrium with $C_{reduced}$ from these rocks (−27 to −21‰). The ranges below the raw $\delta^{13}C$ data are calculated CH_4 in equilibrium with the veins, and for CH_4 in equilibrium with host-rock $C_{reduced}$. Note calculated fluids in both (A) and (B) for veins in the two lowest-grade units of the Catalina Schist, for comparison (data and peak temperature estimates for these units are from Bebout, 1995).

Shreve, 1988; Ernst et al., 1997) or combinations of normal faulting during synsubduction extension, driven by underplating and/or maintenance of a critically tapered accretionary wedge (see Platt, 1986), and the effects of oblique subduction (Avé Lallement, 1996; Mann and Gordon, 1996).

One of the primary constraints in every model to explain the uplift and exposure of blueschist belts has been to account for the preservation of the low-T, high-P/T metamorphic mineral parageneses (particularly those containing aragonite, for which back-reaction to calcite during exhumation is expected to be rapid; Carlson and Rosenfeld, 1981). This, and $^{40}Ar/^{39}Ar$ data indicating extremely rapid cooling of such rocks (e.g., Grove and Bebout, 1995) or slower uplift/cooling in low geothermal gradients maintained by steady-state subduction (Underwood et al., 1987; Dumitru, 1989; Baldwin, 1996; Ernst and Peacock, 1996), constitute reasons to believe that the record of volatile contents/devolatilization history is retained. Forearc metamorphism at P-T conditions similar to those estimated by thermal models is confirmed by the presence of blueschist-metamorphosed clasts in forearc serpentinite seamounts in the Marianas subduction zone (Maekawa et al., 1993; Peacock, 1996).

A striking resemblance in structural character and lithology is observed between the low-grade parts (lawsonite-albite facies; Grove and Bebout, 1995) of the Catalina Schist and large parts of the Franciscan Complex and the Western Baja Terrane (Bailey et al., 1964; Ernst et al., 1970; Magaritz and Taylor, 1976; Sedlock, 1988a, 1988b; Bebout and Barton, 1993; comparison with Sanbagawa Complex by Ernst et al., 1970; also see Lewis et al., 2000). However, veining and fluid-rock interaction and chemical/isotopic composition have not previously been studied in sufficient detail in most of the other areas to make direct comparisons with the Catalina results. Bebout (1996) summarized the existing data relevant to consideration of fluid production and transfer and volatile cycling in Franciscan-like complexes (non-collisional; Ernst, 1988; Patrick and Day, 1995). Comparison of the Catalina Schist and other rocks representing 15-30 km depths of accretionary wedges (e.g., Western Baja Terrane, Mexico, the Eastern Belt of the Coast Ranges Franciscan Complex, California, and Franciscan rocks at Pacheco Pass, Diablo Range, California; Sedlock, 1988a; Blake et al., 1988; Jayko et al., 1986; Ernst, 1993) with the rocks representing shallower parts of circum-Pacific accretionary wedges (e.g., <15 km for Kodiak accretionary complex; see detailed field, fluid inclusion, and stable isotope studies by Vrolijk, 1987a, 1987b; Vrolijk et al., 1988; <12 km for the Central Belt melange matrix and the Coastal Belt of the Coast Ranges Franciscan Complex, California, and Franciscan melange matrix near San Simeon; Mashburn and Cloos, 1985; Blake et al., 1988) can ultimately provide a continuum in inferred depth-T over which to examine extents of devolatilization, fluid-rock interactions, and fluid isotopic compositions. A number of the similarities between our results and those presented in other studies of paleo-accretionary fluid-rock interactions have been noted above; however, there are a few particularly useful comparisons that should be emphasized here.

Vrolijk (1987a, 1987b) presented a detailed fluid inclusion and C-O isotope data set for paleoaccretionary rocks in the Kodiak Complex (representing 5–15 km depths in a accretionary prism; also see Vrolijk et al., 1988). In the isotopic study, apparent isotopic disequilibrium between veins and host rocks (see Fig. 12) similarly was attributed to sluggish reequilibration rates of fluid with minerals such as quartz, and infiltration by a deeper-sourced fluid was inferred to explain these relationships. As in the studies of the Franciscan (Mashburn, 1995; this study), CH_4 was identified in many of the fluid inclusions, and Vrolijk (1987b) used fluid inclusion data to infer temporal pressure fluctuations related to hydraulic fracturing and pore-fluid pressure variation.

Also notable is a detailed fluid inclusion and vein textural study undertaken by Mashburn and Cloos (1985; also see Mashburn, 1985) on shaley mélange matrix and tectonic blocks in the Franciscan exposures at San Simeon, California (Coastal Belt, representing subduction to up to 15–20 km depths; Cloos, 1983). These authors noted salinities (NaCl equivalent wt. %) mostly > 3%, higher than those obtained for aqueous fluid inclusions in this study (which are mostly more dilute than seawater), perhaps reflecting a larger pore/seawater fluid component in these rocks relative to those in the Coast Ranges. Many of the fluid inclusions they examined were CH_4-bearing, and the fluid inclusions were used to infer paleo-P-T and an elevated geothermal gradient inferred to represent upward advection of heat in rising fluids.

Lewis et al. (2000) and Lewis and Byrne (2003) likewise identified CH_4-bearing fluid inclusions, and they attempted to reconstruct intraprism fluid flow during prism thickening based on the fluid inclusions and other veining textures. These authors estimated paleoconditions, and inferred relatively high geothermal gradients affected by permeability constrasts within the Shimanto accretionary prism, and the possible proximity of an active ridge segment. Salinities of the aqueous fluid inclusions were in the range of 2–3 wt% NaCl equivalent.

Significance of observations in paleo-accretionary suites in considerations of forearc fluid fluxes (convergent margin volatiles cycling)

The studies of seafloor sediment proportions, together with information regarding the major- and trace-element and isotopic composition of seafloor pelagic and trench sediments (Ben Othman et al.,

1989; Plank and Ludden, 1992; Plank and Langmuir, 1998), afford assessment of chemical flux into subduction zones in sediments (Jarrard, 2003). The concentrations and, to a lesser extent, the stable isotope compositions of C, S, and N and H_2O contents have been characterized through analyses of deep-sea cores of pelagic and trench sediments (e.g., Patience et al., 1990; Taira et al., 1991; Li et al., 2003; Sadofsky and Bebout, 2004). The geochemical, petrologic, and structural state of the variably hydrothermally altered ocean crust entering trenches has also been considered and clarified through extensive study of deep-sea drill cores (Jarrard, 2003).

The overall mass-balancing of recycled material, including volatile components, is a rapidly evolving field (see recent studies of C-N recycling by Fischer et al., 2002; Hilton et al., 2002; Jarrard, 2003; Sadofsky and Bebout, 2004). Fyfe (1983), in one of the earlier qualitative assessments of volatile subduction, considering only structurally bound H_2O in hydrated ocean-floor crust (with 5% H_2O) and in pelagic sediments (assuming 10% H_2O), concluded that at the calculated rate of H_2O subduction using these H_2O contents, the entire ocean mass (1.4 × 10^{24} grams H_2O) would be subducted in a billion years (cf. DeVore, 1983; Kasting et al., 1993; see recent discussion by Jarrard, 2003). Many studies have concluded, based on such calculations, that massive return flow to the oceans or some other shallow crustal reservoir probably occurs (see Fyfe and McBirney, 1975; Fyfe, 1983; Ito et al., 1983; Fyfe and Kerrich, 1985; Peacock, 1990; Moran et al., 1992; Bebout, 1995, 1996). Rea and Ruff (1996) estimated that seawater (pore water) constitutes 40% of the mass initially subducted in sediments, and that 1.4×10^{15} grams of sediment and 0.9×10^{15} grams of seawater are subducted globally per year. This large amount of pore H_2O has generally been excluded from calculations of crust-mantle volatiles cycling, instead considering "structurally bound" volatiles, inasmuch as the primary goal of these studies has been to assess the magnitude of the subduction of materials past the accretionary prism into the deep mantle (e.g., Peacock, 1990; Moran et al., 1992; Bebout, 1995, 1996).

A notable recent exception is the extremely thorough evaluation of chemical, including H_2O fluxes, by Jarrard (2003)—in that study, the annual global flux of subducted H_2O was estimated at 1.83×10^{15} g/year, with this total H_2O input flux including sediment pore water (42%), sedimentary structurally bound H_2O (7%), igneous crustal pore water (18%), and igneous crustal structurally bound H_2O (33%). At shallow levels of accretionary complexes, it is likely that the mechanical expulsion of this pore H_2O dominates the overall fluid flux budget and, indeed, a reasonable estimation of the flux of this H_2O (42% of the initially subducted H_2O, according to calculations by Jarrard, 2003) out of shallow parts of accretionary prisms could be, as a rough approximation, obtained by simply assuming that *all* of the pore fluid initially subducted is expelled, probably in an extremely heterogeneous fashion over space and time (see discussions by Carson and Screaton, 1998; Moore and Vrolijk, 1992; Fisher, 1996). However, the extent to which pore fluid is retained in sedimentary rocks subducted to depths >5 km (i.e., not accreted) is largely unknown (see porosity data vs. depth for *accreted* sediments in Moore and Vrolijk, 1992). Cloos and Shreve (1988) assumed a pore-water content of ~20% in their subduction channel model. The study of volatile contents in the Franciscan Complex (the same exposures for which we present isotopic vein data in the present paper), Western Baja Terrane, and low-grade Catalina Schist paleo-accretionary suites, together representing 5–45 km depths, indicates that a large proportion of the structurally bound H_2O component is retained to 45 km depths, after some significant loss during clay-to-mica reactions at depths of <10 km (Bebout, 1991a; Bebout et al., 1999; Sadofsky and Bebout, 2003).

Remembering that there are large errors associated with many of the flux estimates (Bebout, 1995), it is apparent, in general, that the budgeted inputs of structurally bound volatiles into subduction zones are at least twice the magnitude of the known outputs of these volatile components in arcs and, in the case of H_2O, perhaps 10 times higher (Ito et al., 1983; Peacock, 1990; Moran et al., 1992; Bebout, 1995, 1996; Jarrard, 2003). This relationship has led previous workers to conclude that the remaining material is either added to and retained in deep crust and mantle reservoirs, or transported back toward the surface, perhaps along faults in forearc regions. This latter flux can be addressed by study of forearc metamorphic suites such as the Franciscan Complex, Western Baja Terrane, and the Catalina Schist (also see studies by O'Hara et al., 1997; Smith and Yardley, 1999; Breeding and Ague, 2002).

Investigation of the Franciscan Complex (this study and that of Mashburn, 1985), and also of similar other suites (e.g., Kodiak Complex—Vrolijk, 1987a, 1987b; Fisher, 1996, Fisher and Brantley, 1992, Fisher et al., 1995; Catalina Schist—Bebout, 1991b; Shimanto—Lewis et al., 2000, Lewis and Byrne, 2003) has indicated a complex interplay between differences in structural style at varying P-T conditions and among different lithologies, and the fluid pressure history in the rocks. Graywacke blocks preserve multiple generations of veins, believed in some cases to reflect fracturing and mineralization over a wide range of P-T conditions. In the lower-grade Coastal and Central belts, shalier domains preserve far fewer intact veins. This may reflect deformation of earlier-formed veins and/or the predominance of more diffuse infiltration in these more pervasively deformed mélange-like zones. In the higher-grade Eastern Belt, shaley domains preserve multiple generations of veins similar to those in graywackes.

Despite significant variation in both $\delta^{18}O$ and $\delta^{13}C$ at outcrop scale, C- and O-isotopic compositions of the veins in the Coast and Diablo Ranges show several trends that provide insight into the history of fluid-rock interactions. In the Coastal Belt, vein $\delta^{18}O$ may have been produced by fluids with values within ±3‰ of seawater at these low temperatures. $\delta^{13}C$ ranges from –12.6 to +0.8‰, with higher values for mélange (Wroblewski, 1999) and a few samples of metamafic rocks. Central Belt $\delta^{18}O$ is similar in magnitude and variability to the Coastal Belt, and $\delta^{13}C$ is more consistently negative, showing a strong influence of sedimentary organic matter. Eastern Belt vein $\delta^{18}O$ tends to be slightly higher than in the other subterranes, and $\delta^{13}C$ is intermediate between that of veins in the Coastal and Central belts. Veins from the Pacheco Pass area are isotopically similar to those in the rest of the Coast Ranges. $\delta^{18}O$ appears to be higher in texturally later-stage veins, and $\delta^{13}C$ is higher in metamafic rocks than in the metaclastic rocks. This may reflect reduced isotopic influence of organic C in the mafic lithologies.

Calculations aimed at evaluating extents of open- and closed-system isotopic behavior (above) show that vein-forming fluids were predominantly rock buffered with respect to both $\delta^{13}C$ and $\delta^{18}O$, complicating isotopic identification of the ultimate fluid sources (e.g., deeply subducted seawater or influx of fluids produced by devolatilization at greater depth). At the lower temperatures, by virtue of the increasing $\Delta^{18}O$ for rock-H_2O with decreasing T, calculated $\delta^{18}O$ values of rock-buffered H_2O approaches seawater (0‰). Thus, rock-buffered fluid (with multiple possible sources and initial $\delta^{18}O$) is indistinguishable from deeply subducted seawater. Some of the $\delta^{18}O$ values and a few of the $\delta^{13}C$ values of vein samples are too high to be explained by this rock-buffered model, and may record up-dip (down-T) flow of fluid that equilibrated with rocks at higher T and greater depth in the accretionary prism (Fig. 12). Carbon-isotope behavior among graywacke and mélange domains in the Coastal Belt may reflect varying degrees of open-system behavior during prograde and retrograde P-T paths among different textural domains. Preferential preservation of late-stage, lower-T veins in the mélange zones would be expected to have the observed higher $\delta^{13}C$, due to greater C-isotopic fractionations at lower T (Fig. 12).

Regarding the possibility of significant entrainment of pore waters in sediments to depth in subduction zones, the δD_{VSMOW} for fluids calculated in equilibrium with the low-grade units of the Franciscan Complex (Magaritz and Taylor, 1976) and the Catalina Schist (Bebout, 1991a, 1991b) are ~ –30 to +10‰, straddling the composition of seawater. The relative sizes of fluid and rock reservoirs would make the H-isotope system more likely to preserve seawater signatures and the O-isotope system more likely to be rock buffered (Magaritz and Taylor, 1976). δD values near seawater could reflect the isotopic influence of entrained seawater (δD ~ 0‰) in deeply subducting sediments or a complex hydration-dehydration process beginning on the seafloor; and releasing of H_2O during dehydration with seawater-like δD (Bebout, 1991a, 1991b). Fluid-inclusion H_2O from veins of this study is generally lower in salinity than seawater. This low salinity implies either mixing of deeply subducted seawater with a more "fresh" fluid, perhaps derived by devolatilization, or conceivably, a "freshening" process involving uptake of F and Cl by apatite (see Smith and Yardley, 1999).

Fluids that formed the lower-T veins have $\delta^{18}O$ spanning the range documented at venting sites in active accretionary complexes (Vrolijk et al., 1990; Kastner et al., 1991; Benton, 1997). Although the non-accreting Mariana arc-trench system differs from the extensive accretionary prism in Mesozoic western North America, we suggest that deeply accreted rocks studied here could lend insight into metamorphic reaction histories and fluid mobility in subducting sediments in such systems. The presence of calcite cement in lower-grade rocks of the Coastal and Central belts and absence in higher-grade rocks of the Eastern Belt and Diablo Range suggest that cements are lost to decarbonation reactions (some $CaCO_3$ may then be reprecipitated in veins). Similar relationships exist in the Catalina Schist, where lawsonite-albite–facies metasedimentary rocks contain finely disseminated $CaCO_3$ that is absent in higher-grade units (Bebout, 1995). Decarbonation and mobility of finely disseminated calcite at depths represented here have been inferred to contribute to high alkalinity and formation of carbonate chimneys in forearc serpentinite seamounts (Fryer et al., 1999). Mottl et al. (2003, in press) noted an increase in $\delta^{18}O$ of fluids emanating from serpentinite seamounts (from near 0‰ to near +4‰), as a function of increasing distance from the trench (and thus increasing depth to the subducting slab and increasing slab temperature). We suggest that the more "inboard" seamounts could sample fluids recharging from greater depths at greater flux rates preventing continual reequilibration with rocks along their paths.

Interestingly, Mottl et al. (2003, in press) noted elevations in LILE (K, Cs, Rb) and B in fluids emanating from greater depths in the Mariana forearc, and proposed that these increases reflect the greater solubilities of these elements in aqueous fluids produced at higher temperatures by prograde dehydration reactions in subducting crust and sediment—the fluids they inferred as having emanated from the greatest slab-sediment-slab depths are also the highest in $\delta^{18}O$. In a separate investigation of major and trace element (and volatiles concentrations) in the Coast and Diablo Range Franciscan metasedimentary rocks (and also for Western Baja Terrane lithologies), Sadofsky and Bebout (2003) noted a large decrease in concentrations of "structurally bound" H_2O, with increasing inferred depth of subduction, related to clay-to-mica reactions. Even B and Cs, known to be relatively mobile in metamorphic fluids at greater temperatures in subduction zones (e.g., Bebout et al., 1999) do not show noticeable decrease across metamorphic grade in the Coast and Diablo Range metasedimentary rocks (see peak P-T conditions in Fig. 1B). However, significant mobilization of these elements (and also Rb and K) could occur without leaving an obvious depletion in these lithologies, particularly given the obscuring effect of large variations in concentration within single grades. As another possibility, B and LILE could be delivered in aqueous fluids moving toward the surface, *through* the rocks we studied; as noted above, such fluids could, if moving at sufficient fluxes, inherit higher $\delta^{18}O$ from fluid-rock interactions at greater depths. Sadofsky and Bebout (2003) *did* report significant depletion of Sr in the higher-grade Coast Ranges units (Central and Eastern belts) and attributed this, and correlated Ca loss, with the decarbonation of abundant carbonate cement found only in the Coastal Belt rocks representing the shallowest levels (likely <5 km). Mottl et al. (in press) noted Sr enrichments in fluids emanating from nearer-trench seamounts that they sampled.

Finally, the information presented here for more deeply subducted rocks, combined with the information regarding complex, multiple sources of vein and cement generations at shallower levels of active accretionary prisms, points to complex C cycling in such systems. Carbon that is sequestered as veins and cements at shallow levels can be more deeply subducted. Ultimately, some of this C is returned to shallower levels when it is released by devolatilization, to mix with fluid C generated by shallow-level biogenic and lower-T abiotic processes (and may again be precipitated in veins). It is worth noting that, despite the production of significant amounts of CH_4 in sediment-rich forearcs (Mashburn, 1985; Vrolijk, 1987; Lewis et al., 2000; this study), examination of the concentrations of reduced C (the organic C fraction) in high-P/T metasedimentary suites indicates very little effect of this loss to diagenetic and metamorphic fluids on whole-rock reduced C contents. Much of this C fraction is probably subducted to subarc depths and beyond into the deeper mantle (Bebout, 1995; Sadofsky and Bebout, 2003).

Conclusions

Rocks of the Franciscan Complex exposed in the California Coast and Diablo Ranges contains veins that formed under a variety of conditions during metamorphism, and that likely reflect processes occurring at active subduction zones. Veins in the lowest-T rocks could have been buffered by either subducted seawater or fluids that were derived from—or produced elsewhere then equilibrated with—the host rocks. Models of the production of forearc fluid (e.g., Sadofsky and Bebout, 2003) suggest the mobility of dehydration fluids at P-T conditions lower than those experienced by the lowest-grade rocks (related to clay-to-mica reactions). Regardless of fluid sources, these veins appear to have been formed by fluids within ± 3‰ of seawater, a range that is similar to the fluids emerging at the décollement of active accretionary prisms (Vrolijk et al., 1990; Kastner et al., 1991; Benton, 1997). The rocks that have been subducted to the greater depths (Central and Eastern belts) appear to have been predominantly rock buffered with respect to both O- and C-isotopic composition. Although some individual veins were likely to have formed from seawater-buffered fluids, and the lowest-T veins have O-isotopic compositions that are somewhat equivocal in their source and may have been produced by subducted pore water, the general trend of rock buffering of vein isotopic compositions provides a reasonable fit to the data observed in this study (see Fig. 12). Some disequilibrium is related to infiltration by fluids from greater depths as well as sluggish equilibration rates of minerals in the rocks with externally derived fluids.

Interpretation of the data in this study and the combination of these interpretations with evidence from the Kodiak accretionary complex (i.e., Vrolijk, 1987a, 1987b), suggests that fluid-rock interactions in low-grade accretionary metamorphic rocks are characterized by several common features. Subducted sediments initially contain abundant pore water, and although much of this water is lost in the first few kilometers of subduction (i.e., ~50% to <10% in 4 km of depth, Moore and Vrolijk, 1992; these data are for accreted sediments), the extent of deeper pore-water entrainment in the subducting sediments remains poorly constrained. Pore-water entrainment to depths equal to or greater than those represented by the Coast Ranges Franciscan exposures could have profound implications for models of convergent margin cycling of volatiles. Many of the fluids that formed veins in these rocks appear to have been rock buffered at the relatively local scale and nearly in equilibrium with their immediate host rocks. However, fluids can be mobilized down-T (up-dip) to transfer heat (Vrolijk et al., 1988) or to potentially affect the isotopic composition of veins (this study), given sufficiently high fluxes. Fluids like those in equilibrium with veins in the lowest-grade Franciscan rocks, together with those released at somewhat shallower levels by clay reactions such as the smectite to illite reaction or opal dehydration (Moore et al., 2001; Brown et al., 2001), are the probable sources of "fresh" or "deeper-sourced" fluids observed at active accretionary complexes, inasmuch as these fluids have the appropriate $\delta^{18}O$, presence of CH_4, and low salinity (Vrolijk et al., 1990; Silver et al., 2000).

Acknowledgements

We are pleased to present this paper in tribute to the profound contributions Gary Ernst made to this and many other fields in the earth sciences. This work was funded by NSF (grant EAR-9805050 to GEB), with some additional funding from GSA (grant 6127-97 to SJS). We would like to thank Gary Ernst and A. Jayko for their assistance in planning and executing this study, and Mark Evans (University of Pittsburgh) for his help in obtaining the fluid inclusion data. Finally, we thank Mark Cloos for his extremely thoughtful and useful review of our manuscript.

REFERENCES

Avé Lallement, H. G., 1996, Displacement partitioning and arc-parallel extension: Example from the southeastern Caribbean plate margin, in Bebout, G. E., Scholl, D. W., Kirby, S. H., and Platt, J. P., eds., Subduction: Top to bottom: American Geophysical Union, Geophysical Monograph 96, p. 113–118.

Bailey, Edward H., Irwin, W. P., and Jones, D. L., 1964, Franciscan and related rocks, and their significance in the geology of western California: California Department of Mines Geological Bulletin, 183.

Baldwin, S. L., 1996, Contrasting P-T-t histories for blueschists from the Western Baja Terrane and the Aegean: effects of synsubduction exhumation and backarc extension, in Bebout, G. E., Scholl, D. W., Kirby, S. H., and Platt, J. P., eds., Subduction: Top to bottom: American Geophysical Union, Geophysical Monograph 96, 135–141.

Bebout G. E., 1991a, Field-based evidence for devolatilization in subduction zones: Implications for arc magmatism, Science, v. 251, p. 413–416.

Bebout, G. E., 1991b, Geometry and mechanisms of fluid flow at 15 to 45 kilometer depths of an early Cretaceous accretionary complex: Geophysical Research Letters, v. 18, p. 923–926.

Bebout, G. E., 1995, The impact of subduction-zone metamorphic processes on the mass-balance of mantle-ocean chemical exchange: Chemical Geology, 126, 191–218.

Bebout, G. E., 1996, Volatile transfer and recycling at convergent margins: Mass-balance and insights from high-P/T metamorphic rocks, in Bebout, G. E., Scholl, D. W., Kirby, S. H., and Platt, J. P., eds., Subduction: Top to bottom: American Geophysical Union, Geophysical Monograph, 96, p. 179–193.

Bebout, G. E., and Barton, M. D., 1993, Metasomatism during subduction: Products and possible paths in the Catalina Schist, California: Chemical Geology, v. 108, p. 61–92.

Bebout, G. E., and Nakamura, E., 2003, Record in metamorphic tourmalines of subduction zone devolatilization and boron cycling: Geology, v. 31, p. 407–410.

Bebout, G. E., Ryan, J. G., Leeman, W. P., and Bebout, A. E., 1999, Fractionation of trace elements during subduction-zone metamorphism: Impact of convergent margin thermal evolution: Earth and Planetary Science Letters, v. 171, p. 63–81.

Ben Othman, D., White, W. M., and Patchett, J., 1989, The geochemistry of marine sediments, island arc magma genesis, and crust-mantle recycling: Earth and Planetary Science Letters, v. 94, p. 1–21.

Benton, L., 1997, Origin and evolution of serpentine seamount fluids, Marianna and Izu-Bonin forearcs: Implications for the recycling of subducted material: Unpubl. Ph.D. dissertation, University of Tulsa.

Berner, R. A., and Lasaga, A. C., 1989, Modeling the geochemical carbon cycle: Scientific American, v. 260, p. 74–81.

Blake, M. C., Jr., Jayko, A. S., and McLaughlin, R. J., 1988, Metamorphic and tectonic evolution of the Franciscan Complex, Northern California, in Ernst, W. G., ed., Metamorphism and crustal evolution of the western United States (Rubey Volume VII): Englewood Cliffs, NJ, Prentice Hall, p. 1023–1060.

Bolhar, R., and Ring, U., 2001, Deformation history of the Yolla Bolly terrane at Leech Lake Mountain, Eastern belt, Franciscan subduction complex, California Coast Ranges: Geological Society of America Bulletin, v. 113, p. 181–195.

Breeding, C. M., and Ague, J. J., 2002, Slab-derived fluids and quartz-vein formation in an accretionary prism, Otago Schist, New Zealand: Geology, v. 30, p. 499–502.

Brown, K. M., Saffer, D. M., and Bekins, B. A., 2001, Smectite diagenesis, pore-water freshening and fluid flow at the toe of the Nankai wedge: Earth and Planetary Science Letters, v. 194, p. 97–109.

Brown, P. E., and Hagemann, S. G., 1995, The program MacFlinCor and its application to geobarometry in Archaean lode-gold deposits: Geochimica et Cosmochimica Acta, v. 59, p. 3943–3952.

Carlson, W. D., and Rosenfeld, J. W., 1981, Optical determination of topotactic aragonite-calcite growth kinetics: Metamorphic implications: Journal of Geology, v. 89, p. 615–638.

Carson, B., and Screaton, E., 1998, Fluid flow in accretionary prisms: Evidence for focused, time-variable discharge: Reviews of Geophysics, v. 36, p. 329–351.

Cloos, M., 1983, Comparative study of mélange matrix and metashales from the Franciscan subduction complex with the basal Great Valley Sequence, California: Journal of Geology, v. 91, p. 291–306.

Cloos, M., 1986, Blueschists in the Franciscan complex of California, petrotectonic constraints on uplift mechanisms: Geological Society of America Memoir 164, p. 77–93.

Cloos, M., and Shreve, R. L., 1988, Subduction-channel model of prism accretion, melange formation, sediment subduction, and subduction erosion at convergent plate margins: 2. Implications and discussion: Pure and Applied Geophysics, v. 128, p. 501–545.

Dalla Torre, M., Livi, K., Veblen, D., and Frey, M., 1996, White K-mica evolution from phengite to muscovite in shales and shale-matrix melange, Diablo Range, California: Contributions to Mineralogy and Petrology, v. 123, p. 390–405.

Deming, D., 1999, On the possible influence of extraterrestrial volatiles on Earth's climate and the origin of the oceans: Palaeaogeography, Palaeoclimatology, and Palaeoecology, v. 146, p. 33–51.

DeVore, G. W., 1983, Relations between subduction, slab heating, slab dehydration, and continental growth: Lithos, v. 16, p. 255–263.

Dumitru, T. A., 1989, Constraints of the uplift in the Franciscan subduction complex from apatite fission track analysis: Tectonics, v. 8, p. 197–220.

Elderfield, H., Kastner, M., and Martin, J. B., 1990, Compositions and sources of fluids in sediments of the Peru Subduction Zone: Journal of Geophysical Research, v. 95, p. 8819–8827.

Ernst, W. G., 1971, Petrologic reconnaissance of Franciscan metagreywacke from the Diablo Range, Central California Coast Ranges: Journal of Petrology, v. 12, p. 413–437.

Ernst, W. G., 1988, Tectonic history of subduction zones inferred from retrograde blueschist P-T paths, Geology, v. 16, p. 1081–1084.

Ernst, W. G., 1993, Metamorphism of Franciscan tectonostratigraphic assemblage, Pacheco Pass area, east-central Diablo Range, California Coast Ranges: Geological Society of America Bulletin, v. 105, p. 618–636.

Ernst, W. G., Maruyama, S. and Wallis, S., 1997, Buoyancy-driven, rapid exhumation of ultrahigh-pressure metamorphosed continental crust: Proceedings of the National Academy of Sciences, v. 94, p. 9532–9537.

Ernst, W. G., and Peacock, S. M., 1996, A thermotectonic model for preservation of ultrahigh-pressure phases in metamorphosed continental crust, in Bebout, G. E., Scholl, D. W., Kirby, S. H., and Platt, J. P., eds., Subduction: Top to bottom: American Geophysical Union, Geophysical Monograph 96, p. 171–178.

Ernst, W. G., Seki, Y., Onuki, H., and Gilbert, M. C., 1970, Comparative study of low-grade metamorphism in the California Coast Ranges and the Outer Metamorphic Belt of Japan: Geological Society of America Memoir 124, 276 p.

Evans, B., 1980, Phase relations of epidote-blueschists: Lithos, v. 25, p. 3–23.

Fischer, T. P., Hilton, D. R., Zimmer, M. M., Shaw, A. M., Sharp, Z. D., and Walker, J. A., 2002, Subduction and recycling of nitrogen along the Central American Margin: Science, v. 297, p. 1154–1157.

Fisher, D., 1996, Fabrics and veins in the forearc: A record of cyclic fluid flow at depths of <15 km, in Bebout, G. E., Scholl, D. W., Kirby, S. H., and Platt, J. P., eds., Subduction: Top to bottom: American Geophysical Union, Geophysical Monograph 96, p. 75–90.

Fisher, D. M., and Brantley, S. L., 1992, Models of quartz overgrowth and vein formation: Deformation and episodic fluid flow in an ancient subduction zone: Journal of Geophysical Research, v. 97, p. 20,043–20,061.

Fisher, D. M., Brantley, S. L., Everett, M., and Dzvonik, J., 1995, Cyclic fluid flow through a regionally extensive fracture network within the Kodiak accretionary prism: Journal of Geophysical Research, v. 100, p. 12,881–12,894.

Franck, S., Kossacki, K., and Bounama, C., 1999, Modelling the global carbon cycle for the past and future evolution of the earth system: Chemical Geology, v. 159, p. 305–317.

Frey, M., 1987, Very low-grade metamorphism of clastic sedimentary rocks, in Frey, M., ed., Low temperature metamorphism: Glasgow, UK, Blackie, p. 9–58.

Frey, M., De Capitani, C., and Liou, J. G., 1991, A new petrogenetic grid for low-grade metabasites: Journal of Metamorphic Geology, v. 9, p. 497–509.

Friedman, I., and O'Neil, J. R., 1977, Compilation of stable isotope fractionation factors of geochemical interest, in Data of geochemistry: U.S. Geological Survey Professional Paper 440-KK.

Fryer, P., Wheat, C., and Mottl, M., 1999, Mariana blueschist mud volcanism; implications for conditions within the subduction zone: Geology, v. 27, p. 103–106.

Fyfe, W. S., 1983, Subduction and the geochemical cycle: Tectonophysics, v. 99, p. 271–277.

Fyfe, W. S., and Kerrich, R., 1985, Fluids and thrusting: Chemical Geology, v. 49, p. 353–362.

Fyfe, W. S., and McBirney, A., 1975, Subduction and the structure of andesitic volcanic belts: American Journal of Science, v. 275-A, p. 285–297.

Gamo, T., Kastner, M., Berner, U., and Gieskes, J., 1993, Carbon isotope ratio of total inorganic carbon in pore waters associated with diagenesis of organic material at site 808, Nankai Trough, in Hill, I. A., Taira, A., Firth, J. V. et al., eds., Proceedings of the Ocean Drilling Program, Scientific Results, no. 131: College Station, TX, p. 259–163.

Gray, D. R., Gregory, R. T., and Durney, D. W., 1991, Rock-buffered fluid-rock interaction in deformed quartz-rich turbidite sequences, eastern Australia: Journal of Geophysical Research, v. 93, p. 4625–4656.

Grove, M., and Bebout, G. E., 1995, Cretaceous tectonic evolution of coastal southern California: Insights from the Catalina Schist, Tectonics, v. 14, p. 1290–1308.

Hilton, D. R., Fischer, T. P., and Marty, B., 2002, Noble gases and volatile recycling at subduction zones, in Porcelli, D., et al., eds., Noble Gases in Geochemistry and Cosmochemistry: Reviews of Mineralogy and Geochemistry, v. 47, p. 319–370.

Ito, E., Harris, D. M., and Anderson, A. T., Jr., 1983, Alteration of oceanic crust and geologic cycling of chlorine and water: Geochimica et Cosmochimica Acta, v. 47, p. 1613–1624.

Jarrard, R. D., 2003, Subduction fluxes of water, carbon dioxide, chlorine, and potassium: Geochemistry, Geophysics, Geosystems (G3), v. 5 (online).

Javoy, M., 1997, The major volatile elements of the Earth: Their origin, behavior, and fate: Geophysical Research Letters, v. 24, p. 177–180.

Javoy, M., 1998, The birth of the Earth's atmosphere: the behaviour and fate of its major elements: Chemical Geology, v. 147, p. 11–25.

Jayko, A., and Blake, M. C., 1989, Deformation of the eastern Franciscan Belt, northern California: Journal of Structural Geology, v. 11, p. 375–390.

Jayko, A. S., Blake, M .C., and Brothers, R. N., 1986, Blueschist metamorphism of the eastern Franciscan belt, northern California, in Blueschists and eclogites: Geological Society of America Memoir 164, p. 107–123.

Kasting, J. F., Eggler, D. H., and Raeburn, S. P., 1993, Mantle redox evolution and oxidation state of the Archaen atmosphere: Journal of Geology, v. 101, p. 245–257.

Kastner, M., Elderfield, H., and Martin, J. B., 1991, Fluids in convergent margins: What do we know about their composition, origin, role in diagenesis and importance for oceanic chemical fluxes?: Philosophical Transactions of the Royal Society of London, v. 335, p. 243–259.

Kastner, M., Sample, J. C., Whiticar, M., Hovland, M., Cragg, B. A., and Parkes, J. R., 1995, Relation between pore fluid chemistry and gas hydrates associated with bottom-simulating reflectors at the Cascadia Margin, sites 889 and 892, in Carson, B., Westbrook, G. K., Musgrave, R. J., and Suess, E., eds., Proceedings of the Ocean Drilling Program, Scientific Results, no. 146: College Station, TX, p. 375–384.

Kastner, M., Zheng, Y., Laier, T., Jenkins, W., and Ito, T., 1997, Geochemistry of fluids and flow regime in the décollement zone at the Northern Barbados Ridge, in Shipley, T. H., Ogawa, Y., Blum, P., and Bahr, J. M., eds., Proceedings of the Ocean Drilling Program, Scientific Results, no. 156: College Station, TX, p. 311–320.

Lewis, J. C., and Byrne, T. B., 2003, History of metamorphic fluids along outcrop-scale faults in a Paleogene accretionary prism, SW Japan: Implications for prism-scale hydrology: Geochemistry, Geophysics, Geosystems (G3), v. 4, no. 9 [9007, doi:10.1029/2002GC000359].

Lewis, J. C., Byrne, T. B., Pasteris, J. D., London, D., and Morgan, G. B., VI, 2000, Early Tertiary fluid flow and pressure-temperature conditions in the Shimanto accretionary complex of south-west Japan: Constraints from fluid inclusions: Journal of Metamorphic Geology, v. 18, p. 319–333.

Li, L., Sadofsky, S. J., and Bebout, G. E., 2003, Carbon and nitrogen input fluxes in subducting sediments at the Izu-Bonin and Central America convergent margins [abs.]: American Geophysical Union, Fall Meeting, San Francisco, CA.

Liou, J. G., Maruyama, S., and Cho, M., 1987, Very low-grade metamorphism of volcanic and volcaniclastic rocks—mineral assemblages and mineral facies, in Frey, M., ed., Low temperature metamorphism: Glasgow, UK, Blackie, p. 59–113.

Maekawa, H., Shozui, M., Ishii, T., Fryer, P., and Pearce, J. A., 1993, Blueschist metamorphism in an active subduction zone: Nature, v. 364, p. 520–523.

Magaritz, M., and Taylor, H., 1976, Oxygen, hydrogen and carbon isotope studies of the Franciscan formation, Coast Ranges, California: Geochimica et Cosmochimica Acta, v. 40, p. 215–234.

Mann, P., and Gordon, M. B., 1996, Tectonic uplift and exhumation of blueschist belts along transpressional strike-slip fault zones, in Bebout, G. E., Scholl, D. W., Kirby, S. H., and Platt, J. P., eds., Subduction: Top to bottom: American Geophysical Union, Geophysical Monograph 96, p. 143–154.

Maruyama, S., Liou, J. G., and Sasakura, Y., 1985, Low-temperature recrystallization of Franciscan greywackes from Pacheco Pass, California: Mineralogical Magazine, v. 49, p. 345–355.

Mashburn L., 1985, Mineralized veins in the Franciscan Melange and Cambria Slab trench-slope basin, near San Simeon, California: A fluid inclusion analysis with implications for dewatering subducting and accreted sediments: Unpubl. M.S. thesis, University of Texas.

Mashburn, L. E., and Cloos, M., 1985, Mineralized veins from Franciscan melange blocks and Cambria Slab trench slope basin sediments near San Simeon, California: A fluid inclusion study [abs.], Geological Society of America Abstracts with Program, v. 17, p. 63–64.

McCrea, J. M., 1950, The isotopic chemistry of carbonates and paleotemperatures scale: Journal of Chemical Physics, v. 18, p. 849–857.

Moore, G. F., Taira, A., Klaus, A., and 23 others, 2001, New insights into deformation and fluid flow processes in the Nankai Trough accretionary prism: Results of Ocean Drilling Program Leg 190: Geochemistry, Geophysics, Geosystems (G3), v. 2 [10.239/2001GC000166].

Moore, J. C., 1989, Tectonics and hydrogeology of accretionary prisms: Role of the décollement zone: Journal of Structural Geology, v. 11, p. 95–106.

Moore, J. C., and Vrolijk, P., 1992, Fluids in accretionary prisms: Reviews of Geophysics, v. 30, p. 113–135.

Moran, A. E., Sisson, V. B., and Leeman, W. P., 1992, Boron depletion during progressive metamorphism: Implications for subduction processes: Earth and Planetary Science Letters, v. 111, p. 331–349.

Moree, M., 1998, The behavior of retrograde fluids in high pressure settings: Implications for the petrology and geochemistry of subduction-related metabasic rocks from Catalina Island (California, USA) and Syros (Greece): Unpubl. Ph.D. dissertation, Vrije University, Amsterdam, Netherlands.

Mottl, M. J., Wheat, C. G., Fryer, P., Gharib, J., and Hulme, S., 2003, Chemistry of springs across the Mariana forearc shows progressive devolatilization of the subducting Pacific plate: EOS (Transactions of the American Geophysical Union), v. 84 [T32A-0909].

Mottl, M. J., Wheat, C. G., Fryer, P., Gharib, J., and Martin, J. B., in press, Chemistry of springs across the Mariana forearc shows progressive devolatilization of the subducting slab: Geochimica et Cosmochimica Acta.

Mullis J., 1987, Fluid inclusion studies during very low-grade metamorphism, in Frey, M., ed., Low temperature metamorphism: Glasgow, UK, Blackie, p. 162–199.

O'Hara, K. D., Guoyan, X., and Li, Z., 1997, Regional $\delta^{18}O$ gradients and fluid-rock interaction in the Altay accretionary complex, northwest China: Geology, v. 25, p. 443–446.

Patrick, B. E., and Day, H. W., 1995, Cordilleran high-pressure metamorphic terranes: Progress and problems: Journal of Metamorphic Geology, v. 13, p. 1–8.

Patience, R. L., Clayton, C. J., Kearsley, A. T., Rowland, S. J., Bishop, A. N., Rees, A. W. G., Bibby, K. G., and Hopper, A. C., 1990, An integrated biochemical, geochemical, and sedimentological study of organic diagenesis in sediments from Leg 112, in Suess, E., von Huene, R., et al., Proceedings of the Ocean Drilling Program, Scientific Results, no. 112: College Station, TX, p. 135–153.

Peacock, S. M., 1990, Fluid processes in subduction zones: Science, v. 248, p. 329–337.

Peacock, S., 1992, Blueschist-facies metamorphism, shear heating, and P-T-t paths in subduction shear zones: Journal of Geophysical Research, v. 97, p. 17,693–17,707.

Peacock, S., 1993, Large-scale hydration of the lithosphere above subducting slabs: Chemical Geology, v. 108, p. 49–59.

Peacock, S. M., 1996, Thermal and petrologic structure of subduction zones, in Bebout, G. E., Scholl, D. W., Kirby, S. H., and Platt, J. P., eds., Subduction: Top to bottom: American Geophysical Union, Geophysical Monograph 96, p. 119–133.

Plank, T., and Langmuir, C. H., 1998, The chemical composition of subducting sediment and its consequences for the crust and mantle: Chemical Geology, v. 145, p. 325–394.

Plank, T., and Ludden, J., 1992, Geochemistry of sediments in the Argo Abyssal Plain at ODP Site 765: A continental margin reference section for sediment recycling in subduction zones, *in* Proceedings of the Ocean Drilling Program, Scientific Results, no. 123: College Station, TX, p. 167–189.

Platt, J. P., 1986, Dynamics of orogenic wedges and the uplift of high-pressure metamorphic rocks: Geological Society of America Bulletin, v. 97, p. 1037–1053.

Rea, D. K., and Ruff, L. J., 1996, Composition and mass flux of sedimentary materials entering the world's subduction zones: Implications for global sediment budgets, great earthquakes, and volcanism: Earth and Planetary Science Letters, v. 140, p. 1–12.

Royden, L. H., 1993, The steady state thermal structure of eroding orogenic belts and accretionary prisms: Journal of Geophysical Research, v. 98, p. 4487–4507.

Sadofsky, S. J., and Bebout, G. E., 2001, Paleohydrogeology at 5- to 15-km depths of accretionary prisms: The Franciscan Complex, Geophysical Research Letters, v. 28, p. 2309–2312.

Sadofsky, S. J., and Bebout, G. E., 2003, Record of forearc-to-subarc devolatilization in low-T, high-P/T metasedimentary suites: Significance for models of convergent margin chemical cycling: Geochemistry, Geophysics, Geosystems (G3), v. 4, no. 4 [DOI 10.1029/2002GC000412, 23 April 2003].

Sadofsky, S. J., and Bebout, G. E., 2004, Nitrogen geochemistry of subducting sediments: New results from the Izu-Bonin-Mariana Margin and insights regarding global nitrogen subduction: Geochemistry, Geophysics, Geosystems (G3), v. 5, no. 3 [DOI 10.1029/2003GC000543, 06 March 2004].

Saffer, D. M., 2003, Pore pressure development and progressive dewatering in underthrust sediments at the Costa Rican subduction margin: Comparison with northern Barbados and Nankai, Journal of Geophysical Research, v. 108, p. 2261 [DOI: 10.1029/2002JB001787, 2003].

Saffer, D. M., Silver, E. A., Fisher, A. T., Tobin, H., and Moran, K., 2000, Inferred pore pressures at the Costa Rica subduction zone: Implications for dewatering processes: Earth and Planetary Science Letters, v. 177, p. 193–207.

Sample, J. C., 1996, Isotopic evidence from authigenic carbonates for rapid upward fluid flow in accretionary wedges: Geology, v. 24, p. 897–900.

Sample, J. C., and Reid, M. R., 1998, Contrasting hydrogeologic regimes along strike-slip and thrust faults in the Oregon convergent margin: Evidence from the chemistry of syntectonic carbonate cements and veins: Geological Society of America Bulletin, v. 110, p. 48–59.

Sample, J. C., Reid, M. R., Tobin, H. J., and Moore, J. C., 1993, Carbonate cements indicate channeled fluid flow along a zone of vertical faults at the deformation front of the Cascadia accretionary wedge (northwest U.S. coast): Geology, v. 21, p. 507–510.

Sedlock, R. L., 1988a, Tectonic setting of blueschist and island-arc terranes of west-central Baja California, Mexico: Geology, v. 16, p. 623–626.

Sedlock, R. L., 1988b, Metamorphic petrology of a high-pressure, low-temperature subduction complex in west-central Baja California, Mexico: Journal of Metamorphic Geology, v. 5, p. 205–233.

Sharp, Z. D., and Kirschner, D. L., 1994, Quartz-calcite oxygen isotope thermometry: A calibration based on natural isotopic variations: Geochimica et Cosmochimica Acta, v. 58, p. 4491–4501.

Silver, E., Kastner, M., Fisher, A., Morris, J., McIntosh, K., and Saffer, D., 2000, Fluid flow paths in the Middle America Trench and Costa Rica Margin: Geology, v. 28, p. 679–682.

Smith, M. P., and Yardley, B. W. D., 1999, Fluid evolution during metamorphism of the Otago Schist, New Zealand; (II), Influence of detrital apatite on fluid salinity: Journal of Metamorphic Geology, v. 17, p. 187–193.

Tagami, T., and Dumitru, T., 1996, Provenance and thermal history of the Franciscan accretionary complex: Journal of Geophysical Research, v. 101, p. 11,353–11,364.

Taira, A., Hill, I. A., Firth, J., and Shipboard Scientific Party, 1991, Proceedings of the Ocean Drilling Program, Initial Results, no. 131: Washington, DC, U. S. Government Printing Office, 434 p.

Terabayashi, M., and Maruyama, S., 1998, Large pressure gap between the Coastal and Central Franciscan Belts, northern and central California: Tectonophysics, v. 285, p. 87–101.

Terabayashi, M., Maruyama, S., and Liou, J. G., 1996, Thermobaric structure of the Franciscan Complex in the Pacheco Pass region, Diablo Range, California: Journal of Geology, v. 104, p. 617–636.

Underwood, M. B., O'Leary, J. D., and Strong, R. H., 1987, Contrasts in thermal maturity within terranes and across terrane boundaries of the Franciscan Complex, Northern California: Journal of Geology, v. 96, p. 399–415.

von Huene, R., and Scholl, D. W., 1991, Observations at convergent margins concerning sediment subduction, subduction erosion, and the growth of continental crust: Reviews of Geophysics, v. 29, p. 279–316.

von Huene, R., and Scholl, D. W., 1993, The return of sialic material to the mantle indicated by terrigenous material subducted at convergent margins: Tectonophysics, v. 219, p. 163–175.

Vrolijk, P. J., 1987a, Paleohydrology and fluid evolution of the Kodiak Accretionary Complex, Alaska: Unpubl. Ph.D. dissertation, University of California, Santa Cruz.

Vrolijk, P., 1987b, Tectonically-driven fluid flow in the Kodiak accretionary complex, Alaska: Geology, v. 15, p. 466–469.

Vrolijk, P., Chambers, S. R., Gieskes, J. M., and O'Neil, J. R., 1990, Stable isotope ratios of interstitial fluids from the Northern Barbados accretionary prism, ODP Leg 110, *in* Moore, J. C., and Mascle, A., eds., Proceedings of the Ocean Drilling Program, Scientific Results, no. 110: College Station, TX, p. 189–205.

Vrolijk, P., Myers, G., and Moore, J. C., 1988, Warm fluid migration along tectonic melanges in the Kodiak accretionary complex, Alaska: Journal of Geophysical Research, v. 93, p. 10,313–10,324.

Wroblewski, K. M., 1999, Fluid-rock interactions in a Tertiary paleo-accretionary complex, California: Unpubl. senior thesis, Lehigh University.

Zhang, Y., and Zindler, A., 1993, Distribution and evolution of carbon and nitrogen in Earth: Earth and Planetary Science Letters, v. 117, p. 331–345.

Aluminous Xenoliths in Miocene Andesite, Central California Coast Ranges: Magma-Crust Interaction in a Subduction-Transform Transitional Setting

ELLEN P. METZGER,[1]

Department of Geology, San Jose State University, San Jose, California 95192-0102

W. G. ERNST,

Department of Geological and Environmental Sciences, Stanford U6niversity, Stanford, California 94305-2115

AND DENNIS SORG

17737 Patricia Way, Grass Valley, California 95949

Abstract

Non-foliated, Al-rich inclusions in Miocene andesite dikes and plugs cropping out in the Dowdy Ranch area, central Diablo Range, record two stages of high-temperature metamorphism, as well as complex chemical interactions between the inclusions and their mafic magma host. Both xenoliths and andesitic host show marked compositional zoning and partial resorption of minerals. The first stage of subsolidus contact metamorphism produced andalusite + alkali-feldspar + Na-plagioclase + muscovite + biotite + quartz (± actinolite or epidote?) in xenoliths of the originally quartzofeldspathic wall rock. P-T conditions attending this event were approximately 600°C, ~2 kbar, with P_{fluid} approaching P_{total}. Capture of fragments of metamorphosed wall rock by ascending mafic magma resulted in partial fusion of the xenoliths, loss of silica and alkalis to the liquid, and calcium-enrichment in the restitic, increasingly aluminous inclusions. Second-stage neoblastic assemblages are sillimanite + hercynite-rich spinel + corundum + Ca-plagioclase ± Mg-rich biotite or phlogopite ± orthopyroxene. Physical conditions in the xenoliths accompanying decompression partial melting were approximately 700–750°C and ~1 kbar (with $P_{fluid} < P_{total}$). The presence of corundum and ZnO-poor spinel is consistent with a restitic residue produced by dehydration melting under high-temperature, silica-deficient conditions. The parageneses developed in the Al-rich inclusions were an upper crustal thermal response to Miocene passage of the Mendocino triple junction at this latitude.

Introduction

Miocene (~9.5–11.6 Ma; Nakata et al., 1993) basaltic to dacitic dikes and plugs in the Diablo Range host a diverse array of plutonic and metasedimentary xenoliths that provide a window into the underlying continental basement. The inclusions offer a unique opportunity to study complex thermal and chemical interactions between mafic magma and quartzofeldspathic crust in a tectonic setting that evolved from subduction to transform faulting. This study focuses on small (< 2 cm) aluminous xenoliths that display marked disequilibria textures. Mineral parageneses include andalusite porphyroblasts that are partially replaced by Ca-plagioclase, corundum, and sillimanite, and rimmed by granoblastic intergrowths of plagioclase + hercynite-rich spinel. Quartz is absent. Similar silica-deficient, aluminous xenoliths have been described in contact aureoles and within extrusive and intrusive igneous rocks around the world (e.g., van Bergen, 1983; van Bergen and Barton, 1984; Montel et al., 1986; Suarez et al., 1992; White et al., 2003; Johnson et al., 2004), and are generally interpreted as restitic products resulting from the dehydration melting of biotite. Textural relationships and mineral chemistry record two stages of high-temperature (T), low-pressure (P) metamorphism: (1) an early contact metamorphic event, followed by (2) incorporation in rising magma, additional heating, and partial melting. The enclosing andesite shows abundant textural and mineralogical evidence for the mixing of quartzofeldspathic wall rock with mafic magma, suggesting that partial fusion of crustal rocks played an impor-

---

[1]Corresponding author; email: metzger@geosun.sjsu.edu

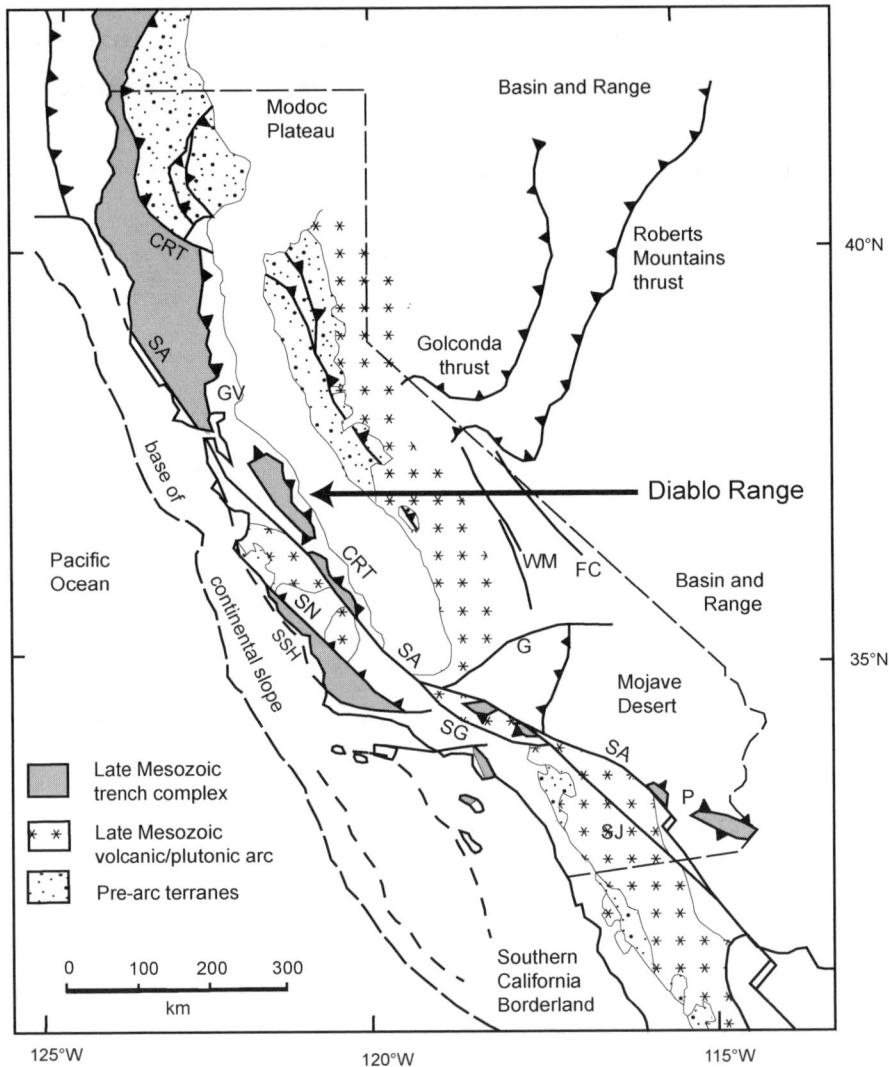

FIG. 1. Geologic index map of California and aspects of western Nevada, after Hamilton (1978). Location of Diablo Range indicated. Major fault abbreviations are: CRT = Coast Range thrust; FC = Furnace Creek; G = Garlock; SA = San Andreas; SG = San Gabriel; SJ = San Jacinto; SSH = San Simeon–Hosgri–San Gregorio; WM = White Mountain shear zone; GV = Great Valley equivalent terrane in the southern California borderland; P = Pelona-Orocopia terrane.

tant role in the chemical evolution of the andesite lavas.

Geologic Setting of the Diablo Range Volcanics

The Diablo Range Volcanics (DRV) are located within the San Jose 30' × 60' Quadrangle, west-central California (Wentworth et al., 1999; Figs. 1 and 2). Plugs and feeder dikes intrude the uplifted and eroded bedrock of the Jura-Cretaceous Franciscan Complex. No pyroclastic or flow deposits are associated with the DRV. Like the more extensive Quien Sabe volcanic field (QSF), which covers an area of ~260 km² approximately 20 km to the southeast, the DRV suite is part of a northwestward-younging sequence of Miocene and younger volcanic rocks resting on older lithologic units of

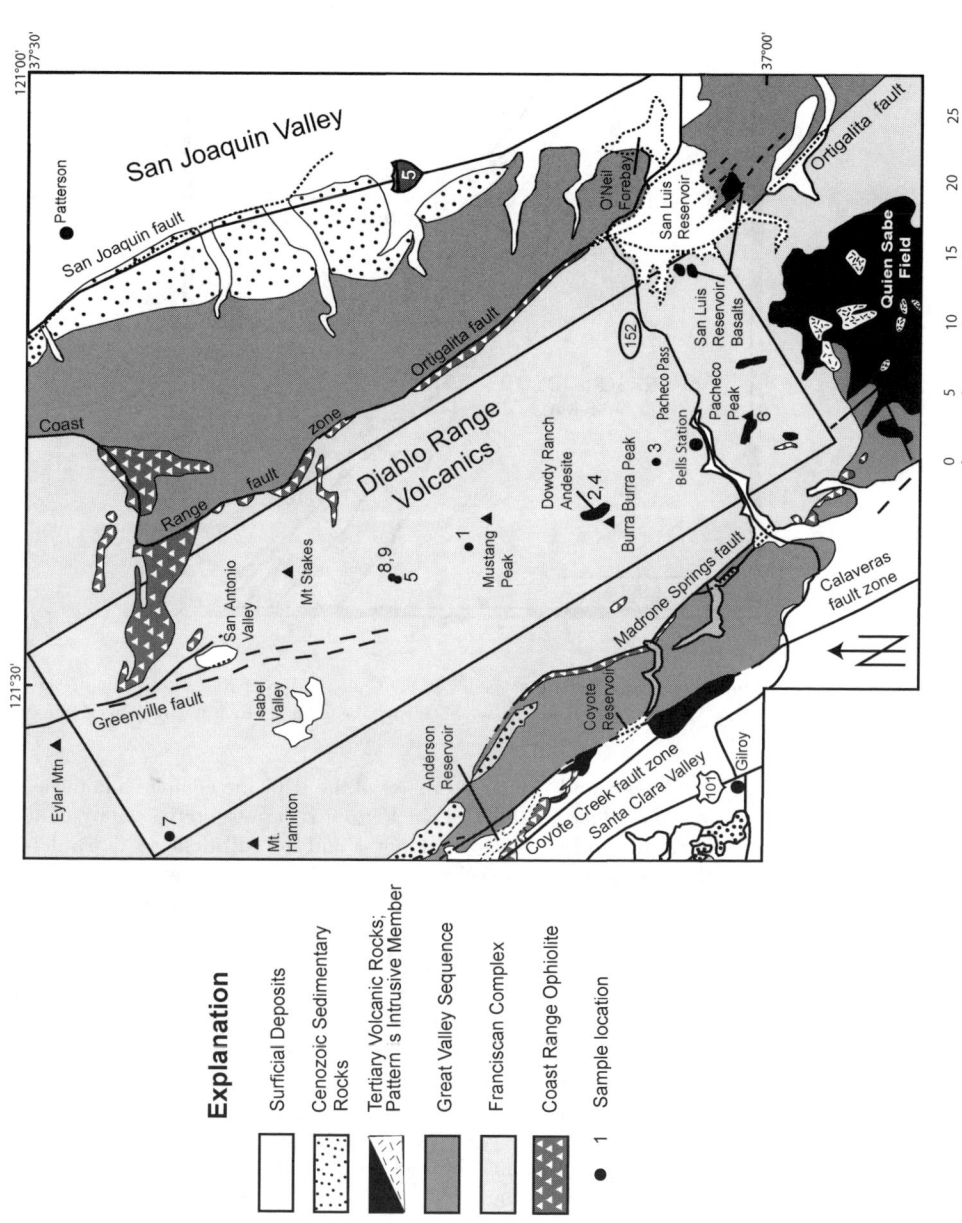

FIG. 2. Simplified geologic map of the study area. The Diablo Range Volcanics are shown inside the rectangle. Modified from Nakata et al. (1993).

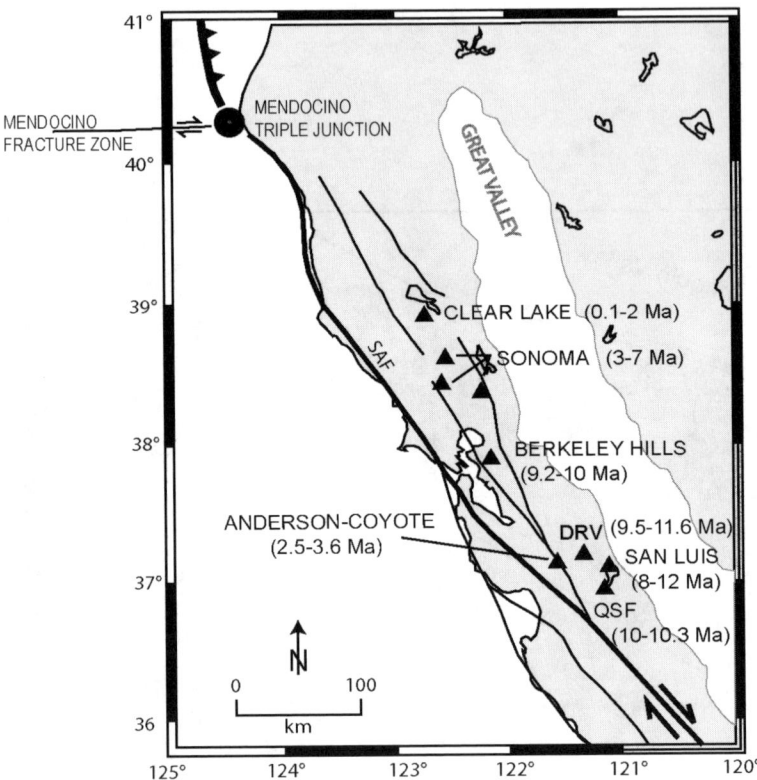

FIG. 3. Current position of northwest-younging Tertiary volcanics in the California Coast Ranges. Modified from Whitlock (2002) and Dickinson (1997). Ages for the Berkeley Hills and Quien Sabe fields are from Wakabayshi (1999).

the North American plate (Fig. 3; Johnson and O'Neil, 1984; Fox, et al., 1985; Cole and Basu, 1995). These volcanic rocks apparently were generated by northward passage of the Mendocino triple junction (MTJ), formation of the lithospheric slab window, and resultant infilling by asthenospheric mantle (Dickinson and Snyder; 1979; Wakabayashi, 1999).

Wakabayashi (1999) discussed uncertainties associated with geochronology of these Cenozoic volcanics, for which nearly all published dates are conventional K-Ar ages with typically large error ranges. Recent redating of some of the volcanic rocks using Ar-Ar step heating has led to significant refinement of previously obtained K-Ar ages. Figure 3 shows conventional K-Ar ages (see Dickinson, 1997 for sources) except for the Berkeley Hills and Quien Sabe fields, for which ages are by Ar-Ar step heating from Wakabayashi (1999).

We have obtained major and trace element data for 26 Diablo Range Volcanics. Bulk-rock chemical characteristics of the DRV are compared to those of the QSF in Figure 4; representative analyses are listed in Table 1 and the full data set is available from the first author. Bulk-rock major and selected trace element analyses for samples reported in this paper were determined by XRF and ICP-MS at the Geoanalytical Laboratory of Washington State University; Johnson et al. (1999) presented a discussion of the precision and accuracy of these methods. Additional data were obtained by XRF analysis at U.S. Geological Survey laboratories in Menlo Park, California and Denver, Colorado.

Analyses of the DRV range from basalt to dacite; andesite and basaltic andesite are the most abundant lava types (Fig. 4A). Sample PP5 (Table 1) is the Dowdy Ranch andesite. All but one analyzed sample in the DRV are subalkaline. The single alkaline basalt is undated, but instead may be related to younger (2.5–3.6 Ma) alkaline rocks that occur at Anderson-Coyote Reservoir (Nakata, 1977; Nakata et al., 1993; Jové and Coleman, 1998).

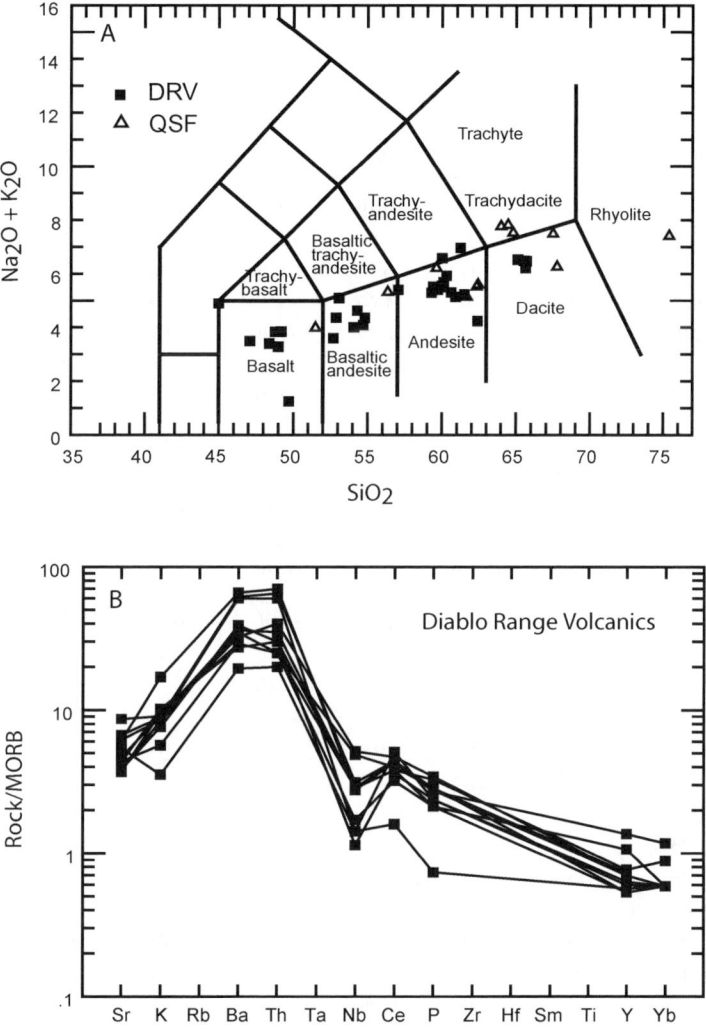

FIG. 4. A. Bulk-rock alkalis-silica plot of samples from the Diablo Range Volcanics (DRV) and the Quien Sabe field (QSF). B. MORB-normalized spider diagram for basalts and basaltic andesites from the DRV. Data for Quien Sabe Volcanics from Drinkwater et al. (1988). Normalizing values for MORB are from Pearce (1983).

Basalts of the DRV are enriched in large-ion lithophile elements (LILE) relative to high field-strength elements (HFSE), and possess a distinct negative Nb anomaly (Fig. 4B). Post mid-Miocene (< 14 Ma) basalts and basaltic andesite from 9 other localities in the northern Coast Ranges, including the Quien Sabe, San Luis Reservoir, Berkeley Hills and Clear Lake fields, display a similar enrichment in LILE and depletion in HFSE (Whitlock et al., 2001; Whitlock, 2002; Furlong and Schwartz, 2004). This trace element signature is typical of calc-alkaline magmatism at convergent margins (Gill, 1981), and suggests that the slab window was filled by the relict supra-slab mantle wedge to the east. Fluid-influenced partial melting of the slab wedge formed the DRV and other northern Coast Range volcanics (Whitlock et al., 2001; Whitlock, 2002; Furlong and Schwartz, 2004).

Two types of xenoliths occur in the DRV: (1) plutonic rocks ranging from granite to gabbro; and (2) a diverse array of metasedimentary rocks including quartzite, biotite schist, and silica-undersaturated

TABLE 1. Representative Compositions of the Diablo Range Volcanics[1]

Sample:	MC47	PP5A	PP35	PP5	MC7	PP12	EM3	MS01	MS03
Location in Figure 2:	1	2	3	4	5	6	7	8	9
Type:[2]	B	B	BA	A	A	A	TA	D	D
SiO_2	47.10	48.40	54.30	60.10	60.30	61.50	61.24	65.70	65.64
Al_2O_3	13.40	23.50	17.10	17.00	14.50	15.80	15.69	15.38	15.41
Fe_2O_3[3]	7.60	6.26	7.04	5.17	3.70	5.46	4.23	3.68	3.54
MgO	8.77	3.98	5.70	1.65	3.31	4.24	6.26	3.87	4.01
CaO	7.22	11.60	7.95	5.34	4.26	6.07	4.59	4.12	4.28
Na_2O	2.22	2.93	3.26	3.86	3.62	3.42	3.91	3.67	3.57
K_2O	1.27	0.47	1.36	1.83	2.31	1.81	3.05	2.80	2.65
TiO_2	0.70	1.29	1.16	0.81	0.41	0.69	0.55	0.46	0.46
P_2O_5	0.29	0.05	0.26	0.21	0.14	0.20	0.18	0.16	0.16
MnO	0.14	0.07	0.11	0.03	0.06	0.09	0.08	0.08	0.07
LOI	10.80	0.93	1.64	3.11	6.66	0.85	–	–	–
Total	99.51	99.48	99.88	99.11	99.27	100.13	99.77	99.91	99.79
Ba	1200	220	720	880	671	629	915	918	801
Ce	51	13	38	45	33	36	31	34	47
Co	42	24	33	17	15	19	–	–	–
Cr	370	12	190	180	182	160	235	177	192
Ga	15	18	19	19	17	18	19	18	19
La	31	8	23	27	21	23	20	19	22
Nb	5	6	10	9	7	9	7	10	10
Nd	27	11	23	23	16	19	–	–	–
Ni	230	15	110	31	34	38	109	32	30
Pb	–	4	–	–	–	–	22	20	20
Sr	740	710	560	430	707	449	605	630	661
Th	12	4	7	8	9	6	14	11	10
V	190	300	180	120	61	105	89	63	55
Y	18	17	32	27	13	15	11	17	13
Yb	2	2	2	3	2	2	–	–	–
Zn	71	37	61	72	42	59	46	52	54

[1]Major elements in wt%; trace elements in ppm.
[2]Abbreviations: B = basalt; BA = basaltic andesite; A = andesite; TA = trachyandesite; D = dacite.
[3]Total iron as Fe_2O_3.

aluminous xenoliths consisting of corundum, aluminosilicate, feldspars, and spinel. This study focuses on the low-Si, high-Al, non-foliated inclusions that occur as small xenoliths and xenocrysts in the Dowdy Ranch andesite exposed near Burra Burra Peak (Fig. 2).

Field relationships and K-Ar ages of vent bodies and feeder dikes in the Quien Sabe volcanic field (QSF) indicate that the initial focus of volcanism was located near the northern and northeastern margins of the field, and subsequently shifted southwestward from these older vents. We regard the

Dowdy Ranch xenolith-bearing andesites as representing a part of this somewhat older, more highly eroded, more mafic and less evolved intrusive and extrusive phase of the QSF (Sorg and Drinkwater, unpubl. data).

A similar suite of xenoliths also has been observed in several other DRV andesite dikes as well as at Pacheco Peak, approximately 6–8 km south of the Dowdy Ranch andesite location (Fig. 2). One of these dikes a short distance south of Pacheco Peak has been dated at 11.2 Ma; the vent body of Pacheco Peak was dated at 11.6 Ma and is the oldest DRV volcanic rock (Nakata et al., 1993). No aluminous or plutonic xenoliths have been found in the younger, less eroded and more highly evolved intrusive and extrusive rocks in the main part of the Quien Sabe volcanic field.

QSF rocks are transected by two sets of steeply dipping normal and reverse faults. The predominant fault set trends N40-50W, and has very steep northeasterly or southwesterly dips. A subordinate fault set trends approximately N60E with nearly vertical dips. The northwest-trending reverse faults postdate and truncate the NE-trending normal faults. Vent centers in the greater QSF form a broad NW-trending zone that indicates structural control. The same set of faults extends northwestward along the Diablo Range from the QSF through the area of the DRV rocks. NE-trending faults are probably related to an initial extensional tectonic regime that formed during passage of the MTJ through this latitude at approximately 10–15 Ma. The initial migration of magma into the upper crust and the formation of vent centers appear to have been largely controlled by these pre-volcanic bedrock faults. Several DRV andesite dikes northwest of the QSF were emplaced near and along some of these NE-trending extensional faults. The younger NW-trending faults may represent preexisting basement rock faults that were reactivated by the transpressive tectonic regime that was superimposed on the region after passage of the MTJ and development of the San Andreas transtensional-transpressional fault system (Atwater and Stock, 1998; Argus and Gordon, 2001; Sorg and Drinkwater, unpubl. data).

Petrography

Enclosing volcanic rocks

Diablo Range Volcanics are highly porphyritic with intersertal and hyalopilitic textures. The most common large crystals are intensely zoned glomeroporphyritic plagioclase phenocrysts set in a finer-grained mesostasis; the latter consists of lath-like plagioclase, clinopyroxene, and orthopyroxene, with lesser amounts of hornblende and biotite. Zoning is common in the plagioclase and also occurs in the pyroxene. DRV plagioclase phenocrysts are reversely zoned and exhibit abundant sieve textures and wormy, spongy, rounded margins. These disequilibrium textures are clear evidence for magma mixing. Coarse (up to ~ 3 mm) xenocrysts of strongly embayed quartz (Fig. 5A) are typically bordered by augite coronas. Quartz is apparently absent from the andesitic groundmass.

Al-rich xenoliths

Al-rich, silica-poor inclusions occur as xenoliths and pseudomorphed andalusite xenocrysts in the Dowdy Ranch andesite of the DRV. They are elongate or equant, angular, and less than 2 cm across. The presence of pseudomorphs, reaction rims, and coronas suggests a complex paragenetic history for these inclusions, which can be described in terms of three textural components: (1) partly replaced andalusite porphyroblasts; (2) spinel-plagioclase rims around the porphyroblasts; and (3) a sporadically developed, very fine grained granoblastic matrix consisting of feldspars and lesser quantities of spinel intergrown with biotite or phlogopite and/or orthopyroxene.

Blocky and elongate porphyroblasts of andalusite ranging in size from <1 mm to 1 cm are largely replaced by intergrowths of corundum, plagioclase, and sillimanite ± alkali feldspar ± spinel, along with accessory rutile, ilmenite, and apatite (Figs. 5B, 5C, and 5D). Corundum occurs in acicular bundles, with crystals up to a few mm in length. Fibrolitic sillimanite partially replaces andalusite in some porphyroblasts.

Dark rims around the porphyroblasts (Figs. 5B, 5C, and 5D) consist of intergrown plagioclase and spinel. The thickness of the rim varies from sample to sample, reflecting variable degrees of consumption of the original andalusite. Spinel is green to black and occurs as very small (a few tenths of a mm) anhedral to larger blocky euhedral crystals. In some cases, a multicrystal corona of clear plagioclase surrounds the spinel-rich rim (Fig. 5C).

The granoblastic matrix is a fine-grained mosaic of plagioclase and spinel ± alkali feldspar ± biotite or phlogopite and/or orthopyroxene (Figs. 5B and C). Ilmenite is a common accessory. Garnet and cordierite are both lacking. Quartz is absent from the

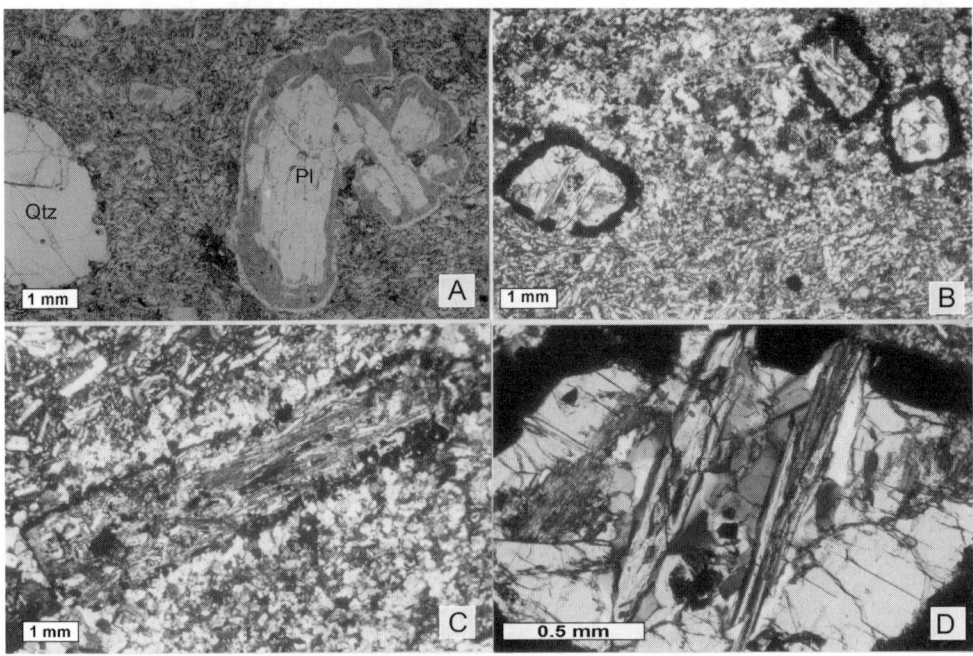

FIG. 5. Photomicrographs. A. Andesite with quartz xenocryst (Qtz) and glomeroporphyritic plagioclase (Pl) showing sieve structure. B. Al-rich inclusion showing partly pseudomorphed andalusite porphyroblasts with dark spinel-rich rims. Large acicular crystals in andalusite are corundum. Note granoblastic matrix at top of photo and volcanic host in the lower part. C. Al-rich porphyroblast partly replaced by fibrolite, plagioclase, spinel, and corundum and surrounded by a multi-crystal plagioclase corona and granoblastic matrix. D. Close-up of andalusite-bearing porphyroblast. Large acicular crystals are corundum. Fibrolitic sillimanite, blocky plagioclase, and spinel are also present.

aluminum-rich inclusions; it is a minor phase in the surrounding andesite, typically occurring as large, partially resorbed xenocrysts rimmed by fine-grained augite.

Four samples of aluminous xenolith-bearing Dowdy Ranch andesite were selected for detailed microprobe study. Each specimen contains several Al-rich inclusions. Based on relationships with the surrounding material, two types of Al-rich xenoliths are present. Type-1 inclusions (samples PP15 and PP20a) consist of partially pseudomorphed andalusite porphyroblasts encircled by a very fine grained granoblastic matrix composed primarily of plagioclase feldspar; in turn, this matrix is surrounded by the andesitic groundmass. Type-2 inclusions (samples PP32 and PP5h) are partially replaced andalusite porphyroblasts in direct contact with the volcanic groundmass. Glass is not present in either Al-rich inclusions or the andesitic mesostasis.

Analytical Procedure

Microprobe analyses were obtained using a JEOL 733 microprobe at Stanford University. Operating conditions were 15 keV accelerating voltage and 15 nA beam current with 20 seconds counting time. Standard analytical techniques were employed (Ernst, 1993).

Mineral Chemistry

Mineral compositions vary with textural setting. Table 2 provides a summary of mineral compositions within the porphyroblasts, along their rims, in the granoblastic matrix surrounding them, and in the volcanic host. Representative mineral analyses are presented in Tables 3–5.

Plagioclase

Representative feldspar analyses are listed in Table 3. Plagioclase is the most compositionally

TABLE 2. Mineral Compositions in Al-rich Inclusions and Volcanic Host Rocks

Mineral	Inside porphyroblast	Rim of porphyroblast	Granoblastic matrix	Volcanic host
		Type-1 inclusions		
Plagioclase feldspar	An17-37	An32-48	An34-60	Ave core: ~ An50
Alkali feldspar	Ab44-61An2Or37-54		Ab72An6Or22	–
Spinel	XFe0.66-0.73	XFe0.54-0.66	XFe0.63-0.67	–
		Type-2 inclusions		
Plagioclase	An26-37	An73-97	–	Ave core: ~An50
Alkali feldspar	Ab67An2Or31		–	
Spinel	XFe0.61-0.69	XFe0.56-0.67	–	

variable phase (An24–An97) in the Al-rich inclusions and host andesite; its chemistry is a function of texture (Table 2). Plagioclase phenocrysts in the andesite host exhibit normal, reverse, and oscillatory zoning; some exhibit a prominent calcic culmination (~An70) at the rim. Average core composition of plagioclase phenocrysts is ~An50, suggesting that the original liquid was basalt or basaltic andesite.

In Type-1 Al-rich inclusions, plagioclase is more sodic (An17–37) inside the porphyroblasts than in the porphyroblast-girdling spinel-rich rims (An32–48), or in the plagioclase coronas (An32–42) surrounding the rims (Tables 2 and 3). Plagioclase in the granoblastic matrix is the most calcic (An34–60) feldspar found in Type-1 inclusions. In Type-2 xenoliths, plagioclase compositions inside the inclusions range from An26 to A37; the plagioclase in spinel-rich rims in direct contact with volcanic rock is much more calcic (An73–97).

Alkali feldspar

Alkali feldspar ranges from anorthoclase to sanidine (Or22–54), and is present both within the porphyroblasts and in the surrounding granoblastic matrix (Table 3). The K content of the alkali feldspar in the partly pseudomorphed andalusite core is relatively high (Or37–54), whereas it is lower (~Or22) in the granoblastic matrix of Type-1 xenoliths.

Spinel

Representative spinel analyses are presented in Table 4. They are binary solid solutions of Mg-spinel and hercynite, with trace amounts of MnO (≤0.4 wt%) and ZnO (0.08–0.4 wt%). X_{Fe} ranges from 0.53 to 0.73; no significant variation of X_{Fe} or of Fe^{2+}/Fe^{3+} ratio related to textural setting is discernible.

Orthopyroxene

Orthopyroxene in the andesite is of variable composition (Wo01En60Fs39 to Wo03En78Fs19), and exhibits both normal and reverse zoning (Table 5). Orthopyroxene in the granoblastic matrix of Type-1 inclusions is of more uniform composition (~Wo03-4En64-68Fs28-33), but is similar to that in the volcanic groundmass.

Biotite and phlogopite

A small amount of biotite or phlogopite is present in the granoblastic matrix of Type-1 samples PP20 and PP15, and in the volcanic groundmass adjacent to Type-2 inclusions in sample PP5h. Because most of the mica is largely altered to retrograde chlorite and to prograde dehydration products, only a few microprobe analyses were obtained. Table 5 provides representative data. Mica is rich in TiO_2 (5.65–7.41 wt%); the highest Ti content occurs in biotite in the volcanic groundmass of sample PP5h, in close proximity to an aluminous inclusion. Values of X_{Fe} range from 0.26 to 0.49, and Al^{VI} contents range from 0.03 to 0.20 atoms per formula unit (11 oxygens).

Origin of the Al-rich Xenoliths

Possible protoliths

The phase assemblages and lack of foliation in the aluminous xenoliths suggest that they recrystal-

TABLE 3. Representative Compositions of Feldspars in Al-Rich Xenoliths

Sample: Setting:[1]	Plagioclase					Alkali feldspar			
	PP20a a	PP20a b	PP20a c	PP32 a	PP32 b	PP5h a	PP20a a	PP20a c	PP15 a
SiO_2	60.42	57.64	58.12	58.59	43.36	66.53	66.44	66.44	66.91
TiO_2	0.02	0.01	0.01	0.08	0.02	0.14	0.04	0.04	0.13
Al_2O_3	24.43	26.13	27.20	25.83	36.43	20.34	18.22	18.22	20.96
FeO	0.00	0.53	0.08	0.31	0.45	0.00	0.32	0.32	0.05
MnO	0.08	0.00	0.08	0.02	0.00	0.00	0.00	0.00	0.00
Cr_2O_3	0.04	0.07	0.00	0.00	0.02	0.01	0.00	0.00	0.02
MgO	0.00	0.02	0.02	0.00	0.01	0.04	0.00	0.00	0.00
CaO	5.37	7.38	8.14	7.75	19.57	0.34	0.35	0.35	0.39
Na_2O	8.01	7.18	6.62	6.88	0.32	7.61	5.03	5.03	7.02
K_2O	0.81	0.48	0.40	0.57	0.01	5.66	9.38	9.38	6.44
ZnO	0.14	0.00	0.00	n.d.[2]	n.d.	n.d.	n.d.	n.d.	n.d.
Total	99.32	99.44	100.68	100.04	100.19	100.69	99.78	99.78	101.91
	Cations based on 8 oxygens								
Si	2.71	2.60	2.58	2.62	2.01	2.95	3.01	3.01	2.93
Ti	0.00	0.00	0.00	0.00	0.00	0.00	0.00	0.00	0.00
Al	1.29	1.39	1.43	1.36	1.99	1.06	0.97	0.97	1.08
Fe	0.00	0.02	0.00	0.01	0.02	0.00	0.01	0.01	0.00
Mn	0.00	0.00	0.00	0.00	0.00	0.00	0.00	0.00	0.00
Cr	0.00	0.00	0.00	0.00	0.00	0.00	0.00	0.00	0.00
Mg	0.00	0.00	0.00	0.00	0.00	0.00	0.00	0.00	0.00
Ca	0.26	0.36	0.39	0.37	0.97	0.02	0.02	0.07	0.02
Na	0.70	0.63	0.57	0.60	0.03	0.44	0.44	0.73	0.60
K	0.05	0.03	0.02	0.03	0.00	0.54	0.54	0.22	0.36
Zn	0.00	0.00	0.00	n.d.[2]	n.d.	n.d	n.d.	n.d.	n.d.
Total	13.01	13.03	13.00	13.01	13.01	13.00	13.00	13.02	13.00
Ab	0.70	0.62	0.58	0.60	0.03	0.44	0.44	0.72	0.61
An	0.26	0.35	0.40	0.37	0.97	0.02	0.02	0.06	0.02
Or	0.05	0.03	0.02	0.03	0.00	0.54	0.54	0.22	0.37

[1] a = inside porphyroblast; b = rim of porphyroblast, c = granoblastic matrix around porphyroblast.
[2] n.d. = not determined.

lized during contact rather than regional metamorphism. The inclusions could represent metamorphosed pelite and/or graywacke of the Franciscan Complex basement that underlies the Dowdy Ranch andesite. In this case, the early-stage contact metamorphic event could well represent Miocene heating associated with development of a slab window in the wake of passage of the MTJ (Wakabayashi, 1996). Alternatively, the suite of non-foliated aluminous and associated schistose xenoliths (not studied in this report) might represent fragments of crystalline basement of Sierran affinity that were entrained in the rising mafic magma. In addition to the Al-rich and schistose inclusions, the Dowdy Ranch andesite encloses plutonic xenoliths, including diorite and gabbro. Jones et al. (1994) cited the mafic and intermediate plutonic and metasedimentary xenoliths in the Dowdy Ranch

TABLE 4. Representative Compositions of Spinels in Al-Rich Xenoliths

Sample:	PP20a	PP20a	PP20a	PP15	PP15	PP15	PP5h	PP5h	PP32	PP32
Setting:[1]	a	b	c	a	b	c	a	b	a	b
SiO_2	0.00	0.10	0.07	0.01	0.03	0.02	0.00	1.58	2.01	3.07
TiO_2	0.20	0.09	0.58	0.21	0.29	0.36	0.22	0.59	0.25	0.19
Al_2O_3	57.52	58.15	53.85	60.65	59.40	59.64	58.27	55.26	59.62	57.33
FeO	33.08	32.32	33.46	29.09	28.91	30.96	30.72	30.33	26.44	25.65
MnO	0.43	0.28	0.20	0.32	0.28	0.23	0.39	0.20	0.39	0.20
Cr_2O_3	0.08	0.02	1.00	0.08	0.09	0.05	0.24	0.04	0.05	0.02
MgO	7.62	9.12	10.76	9.57	10.40	8.76	8.24	10.72	10.07	10.15
CaO	0.02	0.09	0.07	0.00	0.01	0.01	0.00	0.10	0.16	1.59
Na_2O	0.00	0.02	0.01	0.00	0.00	0.00	0.00	0.24	0.22	0.05
ZnO	0.10	0.14	0.40	n.d.[2]	n.d.	n.d.	n.d.	n.d	n.d.	n.d.
Total	99.04	100.32	100.42	99.92	99.42	100.03	98.09	99.07	99.24	98.25
Cations based on 4 oxygens										
Si	0.00	0.00	0.00	0.00	0.00	0.00	0.00	0.04	0.05	0.08
Ti	0.00	0.00	0.01	0.00	0.01	0.01	0.00	0.01	0.01	0.00
Al	1.92	1.87	1.76	1.95	1.92	1.94	1.94	1.81	1.90	1.85
Fe^{+2}	0.68	0.72	0.74	0.61	0.59	0.65	0.66	0.59	0.60	0.59
Fe^{+3}[3]	0.10	0.00	0.01	0.05	0.08	0.06	0.07	0.11	0.00	0.00
Mn	0.01	0.01	0.00	0.01	0.01	0.01	0.01	0.00	0.01	0.00
Cr	0.00	0.00	0.02	0.00	0.00	0.00	0.01	0.00	0.00	0.00
Mg	0.32	0.37	0.44	0.39	0.43	0.36	0.35	0.45	0.41	0.41
Ca	0.00	0.00	0.00	0.00	0.00	0.00	0.00	0.00	0.00	0.05
Na	0.00	0.00	0.00	0.00	0.00	0.00	0.00	0.01	0.01	0.00
K	0.00	0.00	0.00	0.00	0.00	0.00	0.00	0.00	0.00	0.00
Zn	0.00	0.00	0.01	n.d.	n.d.	n.d.	n.d.	n.d.	n.d.	n.d.
X_{Fe}[4]	0.68	0.66	0.63	0.61	0.58	0.64	0.66	0.57	0.60	0.59

[1] a = inside porphyroblast; b = rim of porphyroblast; c = granoblastic matrix around porphyroblast.
[2] n.d. = not determined.
[3] Fe^{+3} calculated as described by Droop (1987).
[4] $X_{Fe} = Fe^{+2}/(Fe^{+2} + Mg)$.

andesite as evidence for Sierran-type crystalline rocks beneath the Franciscan (see also Wentworth et al., 1984; Wentworth and Zoback, 1990). They also noted the presence of metamorphic rocks of possible Sierran affinity preserved in Neogene thrust sheets as far west as Loma Prieta in the Santa Cruz Range and presented lower-crustal seismic velocity data appropriate for mafic to intermediate granitoid rocks. Additionally, andalusite- and sillimanite-bearing schists are known to occur in Mesozoic and Paleozoic metamorphic rocks exposed approximately 100–120 km due east in the Sierra Foothills near the towns of Mariposa and Chowchilla (Murdoch and Webb, 1966). However, in mapping the neighboring Pacheco Pass quadrangle ~10 km to the east of the DRV, Ernst (1993) described lenses of disaggregated ophiolitic lithologies within the Franciscan Complex but found no samples of Sierran basement, which in any case would be more mafic in its western extension than the Al-rich inclusions studied in this report.

Regardless of their Franciscan or Sierran affinity, the Al-rich inclusions represent continental crustal rocks that underwent at least two stages

TABLE 5. Representative Compositions of Micas and Pyroxenes in Al-rich Xenoliths and Volcanic Host

Sample:	PP20	PP15	PP5h	PP15		PP20a	PP5h	PP32	PP15	PP15
Setting[1]	c	c	d	c		c	d	d	c	d
	\multicolumn{4}{c}{Biotite and plagioclase}			\multicolumn{5}{c}{Orthopyroxene}
SiO_2	36.62	38.62	33.44	35.84		51.99	53.02	53.42	53.50	53.17
TiO_2	5.88	5.65	7.41	5.81		0.39	0.55	0.63	0.29	0.42
Al_2O_3	15.37	14.68	14.24	14.82		0.85	1.54	2.89	2.07	2.56
FeO^2	10.48	12.34	17.64	14.59		20.30	20.23	17.83	17.41	19.78
MnO	0.02	0.00	0.11	0.03		0.40	0.53	0.35	0.33	0.37
Cr_2O_3	0.00	0.06	0.04	0.00		0.04	0.06	0.06	0.00	0.01
MgO	16.47	15.27	10.21	13.57		22.19	22.08	22.26	23.69	22.81
CaO	0.08	0.06	0.16	0.02		1.70	1.35	2.35	1.79	2.09
Na_2O	1.33	0.81	0.86	0.70		0.03	0.02	0.18	0.01	0.01
K_2O	8.07	8.66	6.93	8.85		0.00	0.00	0.16	0.02	0.00
Total	94.33	96.16	91.05	94.22		97.91	99.37	100.13	99.11	100.23
	\multicolumn{4}{c}{Cations based on 11 oxygens}			\multicolumn{5}{c}{Cations based on 6 oxygens}
Si	2.70	2.81	2.65	2.71		1.97	1.97	1.95	1.97	1.94
Ti	0.33	0.31	0.44	0.33		0.01	0.02	0.02	0.01	0.01
Al	1.34	1.26	1.33	1.32		0.04	0.07	0.12	0.09	0.11
Fe	0.65	0.75	1.17	0.92		0.64	0.63	0.55	0.54	0.60
Mn	0.00	0.00	0.01	0.00		0.01	0.02	0.01	0.01	0.01
Cr	0.00	0.00	0.00	0.00		0.00	0.00	0.00	0.00	0.00
Mg	1.81	1.66	1.21	1.53		1.25	1.22	1.21	1.30	1.24
Ca	0.01	0.00	0.01	0.00		0.07	0.05	0.09	0.07	0.08
Na	0.19	0.11	0.13	0.10		0.00	0.00	0.01	0.00	0.00
K	0.76	0.80	0.70	0.85		0.00	0.00	0.01	0.00	0.00
Total	18.78	18.71	18.66	18.78		10.00	9.98	9.98	9.98	10.00
Al^{IV}	1.30	1.19	1.30	1.35		0.03	0.03	0.05	0.03	0.06
Al^{VI}	0.04	0.07	0.03	—		0.01	0.04	0.08	0.06	0.05
X_{Fe}^3	0.26	0.31	0.49	0.38		0.34	0.34	0.31	0.29	0.33
Mg#[4]	0.74	0.69	0.51	0.62	Wo	0.03	0.03	0.05	0.04	0.04
					En	0.64	0.64	0.66	0.68	0.65
					Fs	0.33	0.33	0.29	0.28	0.31

[1] c = granoblastic matrix around porphyroblast; d = volcanic host.
[2] All Fe as FeO.
[3] X_{Fe} = FeO/(FeO + MgO).
[4] Mg# = Mg/(Mg + Fe^{2+}).

of high-T, low-P metamorphism—early contact metamorphism—followed by later incorporation in a rising basaltic andesite melt that supplied additional heat and caused partial fusion of the xenoliths. Petrologic relationships are documented below.

Early contact metamorphism

The first high-T event (Stage 1) produced a quartzofeldspathic hornfels that probably also initially contained muscovite, biotite, and possibly a Ca-bearing phase (actinolite or epidote?). Continued prograde heating generated andalusite porphy-

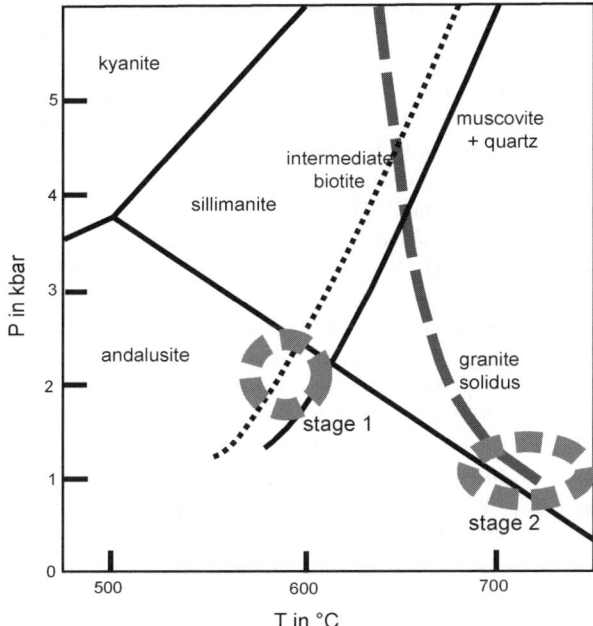

FIG. 6. Schematic phase equilibria for minerals and assemblages of aluminous xenoliths present in the Dowdy Ranch andesite. Al_2SiO_5 polymorphic relations after Holdaway (1971) and Kerrick (1990); breakdown of muscovite and quartz to aluminosilicate, K-feldspar and H_2O after Evans (1965) and Kerrick (1972); stability of intermediate Fe-Mg biotite solid solutions at fO_2 values defined by the Ni-NiO oxygen buffer after Wones and Eugster (1965); granite solidus (P_{H2O} = P_{total}) after Tuttle and Bowen (1958). Reflecting prograde devolatilization of initially hydrous mineral-bearing assemblages, aqueous fluid pressure is equated with total pressure in this diagram.

roblasts, alkali feldspar, and Na-rich intermediate plagioclase through a complex multivariant reaction involving the consumption of layer-lattice silicates + quartz ± a Ca-bearing phase. The resulting Al_2SiO_5 polymorph was rimmed by biotite that, interacting with associated phases, partially devolatilized at still higher temperatures, producing iron-rich spinel, Mg-rich biotite, and Ca-plagioclase. This stage of contact metamorphism took place within the andalusite P-T stability field, and at temperatures exceeding those of intermediate Fe-Mg biotite + muscovite (+ a Ca-phase?) under conditions exceeding ~600°C at pressures around 2 kbar. Generalized P-T phase relations are shown in Figure 6.

Contact metamorphism that developed the andalusite-bearing hornfels probably was caused by heating of Franciscan pelite or graywacke at moderate crustal depths of about 8 km. We correlate this thermal event with cessation of outboard subduction/refrigeration and rising temperatures accompanying the initial stages of slab window formation during passage of the Mendocino triple junction (Dickinson and Snyder, 1979). There may have been a delay between initial high-temperature contact metamorphism (due to asthenospheric upwelling and attendant devolatilization), and the arrival of mafic magma (Furlong, 1984; Liu and Furlong, 1992). Alternatively, the contact metamorphism might have transformed older Sierran quartzofeldspathic basement that may exist beneath the Franciscan. If so, such recrystallization likely would have been Mesozoic or even Paleozoic in age, and would have been unrelated to thermal perturbation associated with the Miocene transit of the MTJ (Dickinson and Snyder, 1979; Wakabayashi, 1996). In either case, however, breakdown of H_2O-bearing minerals would have ensured that contact metamorphism took place under conditions in which P_{fluid} approached P_{total}.

Incorporation in mafic magma

Rising basaltic and basaltic andesite melt captured fragments of the andalusite hornfels wall rock, leading to yet higher temperature metamorphism

(Stage 2), partial replacement of andalusite by corundum and Ca-plagioclase ± sillimanite ± alkali feldspar, and partial fusion attending decompression. Temperatures (Fig. 6) must have exceeded the granite minimum solidus at 700–750°C and ~1 kbar $P_{fluid} < P_{total}$. In this situation, proximity to the Earth's surface would be expected to result in aqueous fluid pressures approaching hydrostatic rather than lithostatic values. Furthermore, if conditions involved elevated fluid pressure, H_2O would strongly partition into the initial melt, further supporting the idea that P_{fluid} was less than that provided by the lithostatic load. Because of sluggish reaction rates, andalusite evidently persisted metastably into the sillimanite P-T field after partial melting of the inclusions began.

Aluminous xenoliths enclosed by the Dowdy Ranch andesite are remarkably similar to Al-rich inclusions reported by Suarez et al. (1992) in the calc-alkaline granitoids of the Hercynian belt of Spain. Suarez et al. interpreted the aluminous, silica-deficient assemblages as products of dehydration melting of biotite to form spinel in a reaction of the type: biotite + sillimanite (and/or andalusite) → hercynite-rich spinel + melt.

In this simplified formulation of the reaction, quartz, K-feldspar, Na-plagioclase, and H_2O preferentially would have entered the calc-alkaline melt. For the DRV xenoliths, the partial replacement of andalusite by sillimanite, and the possible breakdown of single-phase domains of muscovite to K-feldspar + corundum help explain the increase in Or-content of the coexisting alkali-feldspar as well as the formation of minor corundum in the former porphyroblasts. Thus, chemical gradients between magma and aluminous xenoliths led to movement of SiO_2, K_2O, and Na_2O from inclusions to the melt and the transfer of CaO from magma to the Al-rich inclusions. Such extraction of silica and alkalis from inclusions to the magma and the subsequent crystal-melt differentiation on cooling could have produced the slightly quartz-normative Dowdy Ranch andesite from initially more mafic basalt or basaltic andesite. Abundant textural and mineralogical evidence for mixing in the andesite is consistent with incorporation of quartzofeldspathic or pelitic crustal material into an initially more mafic magma.

The inclusions described by Suarez et al. (1992) lack cordierite but contain rare garnet; both minerals are absent from the aluminous xenoliths studied in the DRV. The absence of these two minerals, which are common restitic phases in pelitic rocks and graywackes that have undergone partial melting, suggests that spinel-forming reactions in inclusions in both the Hercynian granitoids and the Dowdy Ranch andesite took place at pressures intermediate between a lower pressure cordierite field and a higher pressure garnet field, consistent with the chemographic analysis by Montel et al (1986). Alternatively, for the Al-rich inclusions in the Diablo Range Volcanics, relatively higher oxidation states attending the near-surface dehydration reactions would have lowered the P-T stability ranges for intermediate Fe-Mg biotites, cordierites, and pyralspite garnets, and instead would have favored expansion of the field for spinel-hercynite solid solutions.

Some spinel + corundum assemblages in the Hercynian calcalkaline rocks and the Dowdy Ranch andesite are surrounded by plagioclase. Plagioclase grains along the rims of porphyroblasts and in granoblastic coronas around the spinel-rich rims are more An-rich than plagioclase grains inside the porphyroblasts (Tables 2 and 3), and could have formed by a reaction such as: biotite + Na-rich plagioclase → hercynite-rich spinel + An-rich plagioclase + melt (Suarez et. al., 1992; Cesare, 2000). Van Bergen and Barton (1984) reported a similar increase in An content of plagioclase from the core of xenoliths outward in aluminous, silica-deficient xenoliths enclosed in siliceous lavas of central Italy. The highly calcic plagioclase surrounding Type-2 inclusions in the Dowdy Ranch andesite reflects the instability of Fe-rich biotite and Ab component in the presence of siliceous liquid, and the crystallization of An-rich plagioclase and hercynitic spinel (van Bergen and Barton, 1984). At higher temperatures, progressively more magnesian biotite plus silica from the melt reacted to form orthopyroxene + K-rich alkali feldspar. The persistence of minor amounts of phlogopitic biotite at high temperature reflects its stabilization by high Ti, Mg, and Al contents.

The concentration of silica-deficient mineral assemblages around partially replaced andalusite porphyroblasts in the aluminous xenoliths within the Dowdy Ranch andesite is similar to textural relationships described by White et al. (2003) and Johnson et al. (2004). Using pseudosections in the $MnO-Na_2O-CaO-K_2O-FeO-MgO-Al_2O_3-SiO_2-H_2O-TiO_2$ model system, Johnson et al. showed that a close subsolidus textural association of andalusite and biotite led to a restricted silica-deficient domain focused around the andalusite porphyroblasts.

Within these inclusion domains, the breakdown of biotite, aluminosilicate, and quartz first produced a moat of cordierite around andalusite. As the reaction progressed, all remaining quartz was consumed, resulting in silica-deficient conditions and the production of spinel. The moat of cordierite surrounding andalusite porphyroblasts may be analogous to the plagioclase coronas that surround andalusite pseudomorphs in xenoliths in the Dowdy Ranch andesite. The formation of Ca-plagioclase instead of cordierite reflects pressures exceeding the stability of cordierite, or low-Mg bulk compositions of the xenoliths. White et al. (2003) used similar modeling of hornfels mineralogy to show that large, isolated andalusite porphyroblasts spatially controlled the development of quartz-absent, spinel-bearing domains produced by the silica-deficient breakdown of coexisting andalusite and biotite.

P-T conditions

Application of quantitative thermobarometry is hindered by the lack of suitable mineral assemblages and, more importantly, marked disequilibrium textures displayed by the aluminous xenoliths. The latter, however, allow the paragenetic sequence to be studied semiquantitatively. Because the original protoliths evidently consisted of moderately hydrous assemblages, contact and magma-induced recrystallization (at least in the early stages) would have resulted in the evolution of H_2O. Accordingly, simplified phase equilibria presented in Figure 6 are appropriate for conditions of P_{fluid} approaching P_{total}. As the mafic magma ascended and decompressed, volatile expulsion would have lowered the aqueous fluid pressure toward (or even below) hydrostatic values.

Mineral assemblages in the Al-rich inclusions are consistent with metamorphism under low-pressure, high-temperature pyroxene hornfels-facies conditions. The absence of both cordierite and garnet suggests low pressures (~1–3 kbar), Mg-poor wall rock bulk compositions, and possibly relatively high values of fO_2. Contact metamorphism resulted in the reaction of layer-lattice silicates + quartz to form andalusite porphyroblasts at ~600°C and 2 kbar (Evans, 1965; Kerrick, 1972). Continued heating and partial replacement of andalusite by fibrolitic sillimanite as the melt-entrained inclusions neared the Earth's surface suggest a temperature approaching or exceeding 700°C (Holdaway, 1971; Kerrick, 1990) at low pressures. The production of corundum, orthopyroxene, and calcic plagioclase accompanying partial fusion requires temperatures above about 700–750°C, at ~1 kbar (Winkler, 1974; Vielzeuf and Holloway, 1988). The Mg, Al, and Ti contents of biotite in the Al-rich inclusions also are consistent with pyroxene hornfels-facies conditions (Pattison and Tracy, 1991; Spear, 1993). Montel et al. (1986) showed that the presence of ZnO-poor spinel such as that observed in Al-rich inclusions in the Dowdy Ranch andesite reflects advanced partial melting due to breakdown of ferruginous biotite and aluminosilicate under silica-undersaturated conditions.

Estimates of pressure conditions for xenoliths hosted by the DRV are substantially lower than those reported by Stimac (1993) for crustal inclusions enclosed in andesite in the Clear Lake volcanic field to the northwest (Hearn et al., 1981; Fig. 3). In contrast, however, metasedimentary xenoliths at Clear Lake contain cordierite, suggesting that they were metamorphosed at somewhat different physical conditions or possessed more magnesian bulk compositions.

Discussion

Miocene volcanism in the central Diablo Range involved the production of calc-alkaline magma. The DRV rocks are typical of northward-younging extrusive igneous rocks found throughout the Coast Ranges on the North American side of the San Andreas transform system. The volcanics appear to represent mantle-derived basaltic liquids contaminated in varying degrees by assimilation of quartzo-feldspathic lithologies. The assimilated continental crustal components were either derived from the Franciscan Complex or from Sierran basement lithologies. Parageneses developed in the Al-rich inclusions of the DRV bear testament to increasing temperatures of recrystallization of the wall rock prior to capture and partial melting in the rising, differentiating liquid. The studied Al-rich xenoliths reflect a thermal response of the shallow crust to Miocene passage of the Mendocino triple junction.

Conclusions

Aluminous xenoliths within Miocene andesite exposed in the Diablo Range of west-central California were produced by two stages of metamorphism in concert with complex chemical and thermal interactions between the inclusions and their mafic magma host. The first stage of contact metamorphism produced the non-foliated mineral assemblage andalusite + alkali-feldspar + Na-plagioclase + muscovite + biotite + quartz (± actinolite or epidote?). Physical conditions during this thermal event were approximately 600°C, 2 kbar with P_{fluid} approaching P_{total}. This thermal event could be a result of the Miocene passage of the Mendocino triple junction. Capture of fragments of the originally quartzofeldspathic wall rock by an ascending mafic magma resulted in partial fusion of the xenoliths, loss of silica and alkalis to the liquid, and Ca-enrichment in the restitic Al-rich inclusions. Neoblastic assemblages include sillimanite + hercynite-rich spinel + corundum + Ca-plagioclase ± Mg-rich biotite or phlogopite. Physical conditions accompanying decompression partial melting in the xenoliths were approximately 700–750°C and ~1 kbar (with $P_{fluid} < P_{total}$). The presence of corundum and ZnO-poor spinel is consistent with a restitic residue produced by dehydration melting under high-temperature, silica-deficient conditions.

Acknowledgments

Support for this investigation was provided by a California State University Research Grant, Stanford University, and the U. S. Geological Survey. A preliminary draft of the manuscript was reviewed, constructively criticized, and improved by Carter Hearn, John Wakabayashi, and Chris Mattinson. To the above institutions and researchers, we extend our sincere thanks.

REFERENCES

Argus, D. F., and Gordon, R. G., 2001, Present tectonic motion across the Coast Ranges and San Andreas fault system in central California: Geological Society of America Bulletin, v. 113, p. 1580–1592.

Atwater, T., and Stock, J. M., 1998, Pacific-North America plate tectonics of the Neogene Southwestern United States: An update: International Geology Review., v. 40, p. 375–402.

Cesare, B., 2000, Incongruent melting of biotite to spinel in a quartz-free restite at El Joyazo (SE Spain): Textures and reaction characterization: Contributions to Mineralogy and Petrology, v. 139, p. 273–284.

Cole, R. B. and Basu, A. R., 1995, Nd-Sr isotopic geochemistry and tectonics of ridge subduction and middle Cenozoic volcanism in western California: Geological Society of America Bulletin, v. 107, p. 167–179.

Dickinson, W. R., 1997, Tectonic implications of Cenozoic volcanism in coastal California: Geological Society of America Bulletin, v. 109, p. 936–954.

Dickinson, W. R., and Snyder, W. S., 1979, Geometry of triple junctions related to San Andreas transform: Journal of Geophysical Research, v. 84, p. 561–572.

Drinkwater, J. L., Weigand, P. W., and Sorg, D. H., 1988, Petrographic and chemical data for the Quien Sabe volcanic rocks in the Mariposa Peak Quadrangle, central Diablo Range, California: U. S. Geological Survey Open-File Report 88-642, 32 p.

Droop, G. T. R., 1987, A general equation for estimating Fe^{3+} concentrations in ferromagnesian silicates and oxides from microprobe analyses, using stoichiometric criteria: Mineralogical Magazine, v. 51, p. 431–435.

Ernst, W. G., 1993, Metamorphism of Franciscan tectonostratigraphic assemblage, Pacheco Pass area, east-central Diablo Range, California Coast Ranges: Geological Society of America Bulletin, v. 105, p. 618–636.

Evans, B. W., 1965, Application of a reaction-rate method to the breakdown equilibria of muscovite and muscovite plus quartz: American Journal of Science, v. 263, p. 647–667.

Fox, K. F., Fleck, R. J., Curtis, G. H., and Meyer, C. E., 1985, Implications of the northwestwardly younger age of the volcanic rocks of west-central California: Geological Society of America Bulletin, v. 96, p. 647–654.

Furlong, K. P., 1984, Lithospheric behavior with triple junction migration: An example based on the Mendocino Triple Junction: Physics of The Earth and Planetary Interiors, v. 36, p. 213–223.

Furlong, K. P., and Schwartz, S. Y, 2004, Influence of the Mendocino Triple Junction on the tectonics of coastal California: Annual Review of Earth and Planetary Sciences, v. 32, p. 403–433.

Gill, J., 1981, Orogenic andesites and plate tectonics: New York, NY, Springer-Verlag, 390 p.

Hamilton, W., 1978, Mesozoic tectonics of the western U.S., in Howell, D. G., and McDougall, K., eds., Mesozoic paleogeography of the western United States: Los Angeles, CA, Society of Economic Paleontologists and Mineralogists, p. 33–70.

Hearn, B. C., Donnelly-Nolan, J. M., and Goff, F. E, 1981, The Clear Lake Volcanics: Tectonic setting and magma sources, in McLaughlin, R. J., and Donnelly-Nolan, J. M., eds., Research in the Geysers–Clear Lake geothermal area, northern California, U. S. Geological Survey Professional Paper 1141, p. 25–45.

Holdaway, M. J., 1971, Stability of andalusite and the aluminum silicate phase diagram: American Journal of Science, v. 271, p. 97–131.

Johnson, C. M., and O'Neil, J. R., 1984, Triple junction magmatism: A geochemical study of Neogene volcanic rocks in western California: Earth and Planetary Science Letters, v. 71, p. 241–262.

Johnson, D. M., Hooper, P. R., and Conrey, R. M., 1999, XRF analysis of rocks and minerals for major and trace elements on a single low dilution Li-tetraborate bead: Advances in X-ray Analysis, v. 41, p. 843–867.

Johnson, T., Brown, M., Gibson, R., and Wing, B., 2004, Spinel-cordierite symplectites replacing andalusite: Evidence for melt-assisted diapirism in the Bushveld complex, South Africa: Journal of Metamorphic Geology, v. 22, p. 529–545.

Jones, D. L., Graymer, R., Wang, C., McEvilly, T. V., and Lomax, A., 1994, Neogene transpressive evolution of the California Coast Ranges: Tectonics, v. 13, p. 561–574.

Jové, C. F., and R. G. Coleman, 1998, Extension and mantle upwelling within the San Andreas fault zone, San Francisco Bay area, California: Tectonics, v. 17, p. 883–890.

Kerrick, D. M., 1972, Experimental determination of muscovite + quartz stability with $PH_2O < Ptotal$: American Journal of Science, v. 272, p. 946–958.

Kerrick, D. M., 1990, The Al_2SiO_5 polymorphs: Mineralogical Society of America, Reviews in Mineralogy, v. 22, 406 p.

Liu, M., and Furlong, K. 1992, Cenozoic volcanism in the California Coast Ranges: Numerical solutions: Journal of Geophysical Research, v. 97, p. 4941–4951.

Montel, J. M., Weber, C., and Pichavant, M., 1986, Biotite-sillimanite-spinel assemblages in high-grade metamorphic rocks: Occurrence, chemographic analysis, and thermobarometric interest: Bulletin de Minéralogie, v. 109, p. 555–673.

Murdoch, J., and Webb, R. W., 1966, Minerals of California, Centennial Volume (1866–1966): California Division of Mines and Geology, Bulletin 189, 559 p.

Nakata, J. K., 1977, Distribution and petrology of the Anderson-Coyote Reservoir Volcanic Rocks, Unpubl. M.S. thesis, San Jose State University, San Jose, California, 105 p.

Nakata, J. K, Sorg, D. H., Russell, P. C., Meyer, C. E., Wooden, J., Lanphere, M. A., McLaughlin, R. J, Sarna-Wojcicki, A. M., Saburomaru, J. Y., Pringle, M. S., and Drinkwater, J., 1993, New radiometric ages and tephra correlations from the San Jose and the northeastern part of the Monterey 1:100,000 quadrangles, California: Isochron/West no. 60 (Dec. 1993), p. 19–32.

Pattison, D. R. M., and Tracy, R. J., 1991, Phase equilibria and thermobarometry of metapelites, *in* Kerrick, D. M., ed., Contact metamorphism: Mineralogical Society of America Reviews in Mineralogy, v. 25, p. 105–206.

Pearce, J. A., 1983, Role of sub-continental lithosphere in magma genesis at active continental margins, *in* Hawkesworth, C. J., and Norry, M. J., eds., Continental basalts and mantle xenoliths: Nantwich, UK, Shiva Publications, p. 230–249.

Spear, F. S., 1993, Metamorphic phase equilibria and pressure-temperature-time paths: Mineralogical Society of America, Monograph Series, 799 p.

Stimac, J. A., 1993, The origin and significance of high-grade metamorphic xenoliths, Clear Lake Volcanics, California, *in* Rytuba, J. J. ed., Active geothermal systems and gold-mercury deposits in the Sonoma–Clear Lake volcanic fields, California: Society of Economic Paleontologists and Mineralogists Guidebook 16, p. 171–189.

Suarez, O., Cuesta, A., Corretge, G., and Fernandez-Suarez, J., 1992, Spinel-bearing inclusions in calc-alkaline granitoids of the Cantabrian and west Asturian Leonese zones, Hercynian Iberian belt: Bulletin de la Société Géologique de France, v. 163, p. 611–623.

Tuttle, O. F., and Bowen, N. L., 1958, Origin of granite in the light of experimental studies in the system $NaAlSi_3O_8$-$KAlSi_3O_8$-SiO_2-H_2O: Geological Society of America Memoir 74, 153 p.

van Bergen, M. J., 1983, Polyphase metamorphic sedimentary xenoliths from Mt. Amiata volcanics (Central Italy): Evidence for a partially disrupted contact aureole: Geologische Rundschau, v. 72, p. 637–662.

van Bergen, M. J., and Barton, M., 1984, Complex interaction of aluminous metasedimentary xenoliths and siliceous magma: An example from Mt. Amiata (Central Italy): Contributions to Mineralogy and Petrology, v. 86, p. 374–385.

Vielzeuf, D., and Holloway, J. R., 1988, Experimental determination of the fluid-absent melting relations in the pelitic system: Contributions to Mineralogy and Petrology, v. 98, p. 257–276.

Wakabayashi, J., 1996, Tectono-metamorphic impact of a subduction-transform transition and implications for interpretation of orogenic belts: International Geology Review, v. 38, p. 979–994.

Wakabayashi, J., 1999, Distribution of displacement on and evolution of a young transform fault system: the northern San Andreas fault system, California: Tectonics, v. 18, p. 1245–1274.

Wentworth, C. M., Blake, M. C., Jr., Jones, S. L., Walter, A. W., and Zoback, M. D., 1984, Tectonic wedging associated with emplacement of the Franciscan assemblage, California Coast Ranges, *in* Blake, M. C., Jr., ed., Franciscan geology of northern California: Pacific Section, Society of Economic Paleontologists and Mineralogists, v. 43, p. 163–173.

Wentworth, C. M., Blake, M. C., Jr., McLaughlin, R. J., and Graymer, R. W., 1999, Preliminary geologic map of the San Jose 30 × 60-Minute Quadrangle, California: A digital database: U. S. Geological Survey Open-File Report 98-795.

Wentworth, C. M., and Zoback, M. D., 1990, Structure of the Coalinga area and thrust origin of the earthquake, *in* Rymer, M. J., and Ellsworth, W. L., eds., The Coalinga, California earthquake of May 2, 1983: U. S. Geological Survey Professional Paper 1487, p. 41–67.

White, R. W., Powell, R., and Clarke, G. L., 2003, Prograde metamorphic assemblage evolution during partial melting of metasedimentary rocks at low pressures: Migmatites from Mt. Stafford, central Australia: Journal of Petrology, v. 44, p. 1937–1960.

Whitlock, J. S., 2002, Evidence of a mantle wedge source for slab window volcanism in the northern California Coast Ranges: Unpubl. M.S. thesis, Pennsylvania State University, State College, Pennsylvania, 78 p.

Whitlock, J. S., Furlong, K. P., Lesher, C. E., and Furman, T., 2001, The Juan de Fuca slab-window and Coast Range Volcanics, California: Correlation between subducted slab age and mantle wedge geochemistry [abs.]: EOS (Transactions of the American Geophysical Union), v. 82, p. F1186.

Winkler, H. G. F., 1974, Petrogenesis of metamorphic rocks: New York, NY, Springer-Verlag, 320 p.

Wones, D. R., and Eugster, H. P., 1965, Stability of biotite: Experiment, theory, and application: American Mineralogist, v. 50, p. 1228–1272.

DOMINICAN REPUBLIC

UHP Magma Paragenesis, Garnet Peridotite, and Garnet Clinopyroxenite: An Example from the Dominican Republic

RICHARD N. ABBOTT, JR.,[1]

Department of Geology, Appalachian State University, Boone, North Carolina 28608

GRENVILLE DRAPER, AND SHANTANU KESHAV[2]

Department of Earth Sciences, Florida International University, Miami, FL 33199

Abstract

Spinel-bearing garnet peridotite and corundum-bearing variants in the Cuaba Gneiss of the Cretaceous Rio San Juan Complex show evidence for ultrahigh pressure (UHP) partial melting and magmatic fractionation (orthocumulate textures). The paragenesis involves the following sequence of assemblages (plus inferred melt) with declining T: (1) Grt + Ol + Spl + Cpx + Liq (partial melt assemblage); (2) Grt + Spl + Cpx + Liq (cumulate Cpx with interstitial Grt); (3) Grt + Spl + Crn + Cpx + Liq (pegmatite, Cpx with interstitial Grt and late Crn). Comparison with 3 GPa liquidus relationships in CMAS (Milholland and Presnall, 1998) and extrapolation to pressures above which sapphirine is not possible (> 3.4 GPa at ~1570°C) show that assemblage (1) is consistent with the equilibrium Spl + Cpx = Ol + Grt + Liq. Liquid fractionated from this assemblage crystallized in equilibrium with Grt + Cpx + Spl (2). Further fractionation resulted in the crystallization of Crn according to the equilibrium, Cpx + Grt + Crn = Spl + Liq (3). This last reaction is only possible at P > 3.4 GPa and T > 1550°C. The assemblages constrain a short but well-defined liquid line of descent. The inferred conditions of T are much higher than previous estimates that did not take melt into account. Previously estimated conditions (P = 2.8–3.4 GPa, T = 740–810°C) are presumed to reflect subsolidus reequilibration. Evidently, the rocks originated in the deepest part of the lithosphere, or shallowest part of the asthenosphere, and cooled more or less isobarically as they were delivered to the subduction zone, prior to ascent.

Introduction

THIS STUDY CONCERNS Alpine-type garnet peridotite, garnet clinopyroxenite, and a corundum-bearing variant of the latter, from the Dominican Republic (Abbott et al., 2001a, 2001b, 2003a, 2003b; Draper et al., 2002). Although peridotite and clinopyroxenite are relatively common in orogenic belts, garnet-bearing varieties are very rare, and only associated with ultrahigh pressure (UHP) or high pressure (HP)–ultrahigh temperature (UHT) metamorphic terranes. The rocks yield valuable insight into convergent plate–boundary tectonics, especially with regard to extreme conditions in a subduction zone, deep subduction of crust, the transfer of mantle rock to deep-subducted crust, and the mechanism by which such rocks are delivered to the surface.

The purpose of this contribution is to describe textural and mineralogical evidence for magmatic fractionation at UHP conditions, and to characterize a liquid line of descent in the CMAS system (CaO-MgO-Al_2O_3-SiO_2). The liquid line of descent is derived from the results of 3 GPa melting experiments in the CMAS system (Milholland and Presnall, 1998) and results of experiments on the stability of sapphirine (Ackermand et al., 1975).[3]

Geologic Setting

Hispaniola is one of the four large islands of the Greater Antilles (Fig. 1). Three of the islands (Jamaica, Hispaniola, and Puerto Rico) lie just south of the northern edge of the present-day Caribbean plate. The fourth and largest island (Cuba) lies just north of the boundary, on the North America

---

[1]Corresponding author; email: abbottrn@appstate.edu
[2]Shantanu Keshav is currently at Geophysical Laboratory, Carnegie Institution of Washington, Washington, DC 20015.

[3]Mineral and component abbreviations are consistent with Kretz (1983); liq = liquid (silicate melt).

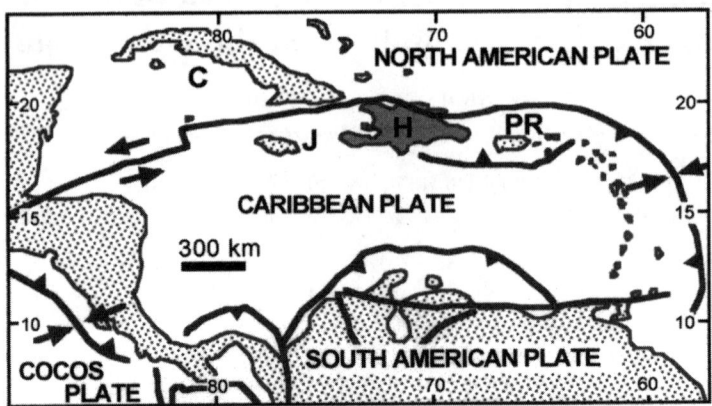

FIG. 1. Geologic provinces of the Caribbean. Islands of the Greater Antilles include Cuba (C), Jamaica (J), Hispaniola (H), and Puerto Rico (PR).

plate. The basements of Jamaica, Puerto Rico, Hispaniola, and the southern part of Cuba began as an Early Cretaceous, intra-oceanic island arc complex above a NE-dipping, Pacific-derived plate (Pindell and Barrett, 1990; Draper et al., 1996). A fundamental change in the plate boundary configuration occurred in the mid-Cretaceous (~120–100 Ma; Pindell and Barrett, 1990). The change involved displacement of the plate boundary from the Pacific side to the Atlantic side of the island arc complex and reversal in the polarity of subduction (Pindell and Barrett, 1990), thus establishing the present-day configuration (i.e., SW- to W-dipping subduction with the plate boundary on the Atlantic side of the island-arc complex). From the Late Cretaceous to the mid-Eocene, the intra-oceanic island arc was augmented in response to SW-directed subduction of oceanic parts of the North and South American plates beneath the northeastern and eastern edges of the Caribbean plate (Pindell and Barrett, 1990; Draper et al., 1996). The original island-arc system has since been modified by E-W, left-lateral, transcurrent tectonics that began in the mid-Eocene and continues today (Mann et al., 1990; Draper et al., 1996).

In Hispaniola (Fig. 2A), the northernmost mountain range, the Cordillera Septentrional, consists of Upper Eocene to Lower Miocene siliciclastic and carbonate sedimentary rocks, underlain by a metamorphic and igneous basement complex (Eberle et al., 1982; Lewis and Draper, 1990). Where the cover has been eroded away, the basement is exposed in a number of stratigraphic windows, or inliers (Fig. 2B). The largest inlier exposes the Rio San Juan Complex (Eberle et al., 1982; Draper and Nagle, 1991).

The Rio San Juan Complex (Draper and Nagle, 1991) is divided into distinct northern and southern parts (Fig. 2C), which were juxtaposed by faulting, probably in the Paleogene (Draper and Nagle, 1991; Draper et al., 1994). The northern half of the inlier consists of serpentinite and blueschist-eclogite mélange with serpentine matrix, faulted against fine-grained, coherent greenschist-blueschist–facies rocks. The HP/low temperature (LT) metamorphism of these rocks is interpreted as having resulted from SW-directed subduction in the Late Cretaceous (Draper and Nagle, 1991; Draper et al., 1994). The southern half of the Rio San Juan Complex consists of the Cuaba Gneiss and the Rio Boba Intrusive Complex. The Cuaba Gneiss is predominantly hornblende gneiss and hornblende schist. The common mineral assemblage is Hbl + Pl(andesine) + Qtz + Rt ± Grt ± Bt ± Ep. Draper and Nagle (1991) suggested a mafic protolith (basalt/diabase/gabbro) of oceanic crustal origin. Retrograded eclogite in the Cuaba Gneiss was first reported in 1998 (Abbott and Draper, 1998). Evidence for eclogite is in the form of Pl-Di (symplectite) + Grt, with greater or lesser amounts of hornblende, depending on the extent of retrograde hydration (Abbott and Draper, 1998, 2002). The retrograded eclogite occurs as mm- to dm-scale layers in otherwise symplectite-free hornblende gneiss. Garnet peridotite was first reported in 2001 (Abbott et al., 2001a). The garnet peridotite and more recently discovered garnet clinopyroxenite (Abbott et al., 2003a, 2003b), which are the subject

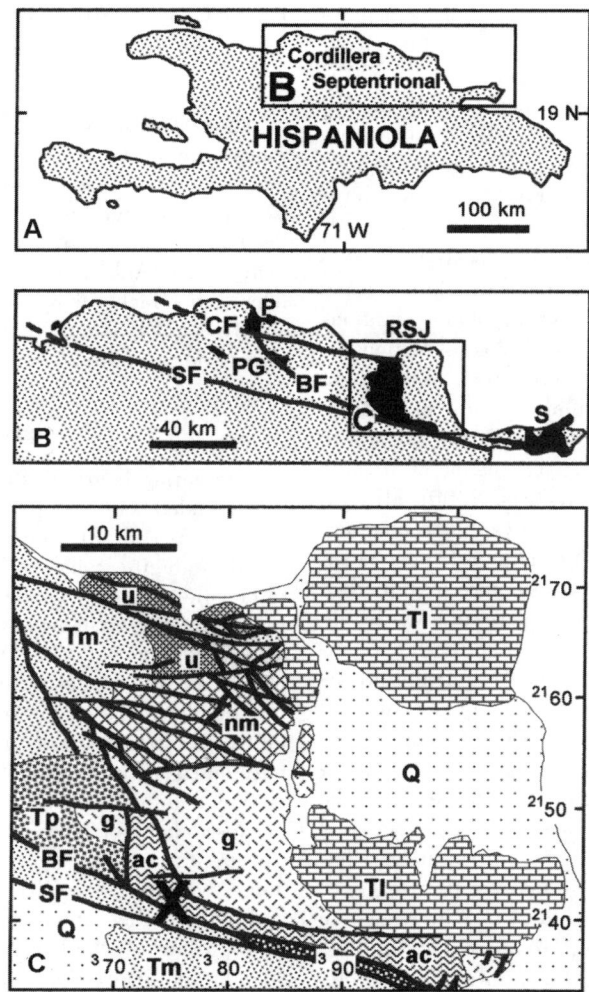

FIG. 2. Geologic Setting. A. Island of Hispaniola and location of Cordillera Septentrional. B. Major inliers of pre-Middle Eocene rocks: RSJ = Rio San Juan; S = Samana; P = Puerta Plata; PG = Pedro Garcia (PG). Faults: CF = Camu; BF = Bajabonico; SF = Septentrional. C. Simplified geologic map (Draper and Nagle, 1991). Rio San Juan inlier (Cretaceous): serpentinite (u), blueschist/eclogite (nm), Cuaba Gneiss (ac), Rio Bobo gabbro (g). Younger sediments and sedimentary rocks: Upper Eocene–Oligocene sedimentary rocks (Tm), Tertiary conglomerate (Tp), Neogene limestone (Tl), Quaternary alluvium and reef deposits (Q). "X" marks location of garnet-bearing ultramafic rocks.

of this contribution, are interpreted as minor constituents of the Cuaba Gneiss. The Cuaba Gneiss was intruded by gabbroic to quartz dioritic rocks of the Rio Boba Intrusive Complex (Draper and Nagle, 1991). The petrogenetic relationship (if any) between the HP/LT rocks in the northern half of the inlier and HP/UHP rocks of the Cuaba Gneiss remains unclear.

Previous P-T Estimates

Constraints on the maximum P-T conditions for Cuaba hornblende gneiss and hornblende schist are not particularly helpful in constraining the conditions for the associated garnet-bearing ultramafic rocks. Interpreting the retrograded eclogite as a predecessor to much, if not all, of the hornblende gneiss and schist, the original mineral assemblage (omphacite + garnet + quartz) indicates only that the maximum conditions are P > 1.2 GPa and T > 600°C (Abbott and Draper, 2002), consistent with HP conditions, but well below UHP conditions. Evidence for UHP conditions in the gneiss, schist, and retrograded eclogite (e.g., coesite, microdiamonds) has yet to be recognized.

When the garnet peridotite was first reported (Abbott et al., 2001a), neither corundum-bearing assemblages nor evidence for magmatic conditions had been recognized. The available observations — mineral assemblages, textures, rock association, tectonic setting, and preliminary mineral analyses (Table 1)—suggested that the garnet peridotite belonged in the low-P/T (UHT) regime of Medaris (1999; Brueckner and Medaris, 2000). Mineral analyses (Table 1) were used in various geothermobarometers to estimate P-T conditions (Abbott et al., 2001a). Variation in the compositions of individual minerals (e.g., garnet, core versus rim) was within expected limits of analytical error and within a standard deviation from the average. Despite this apparent chemical homogeneity, a need for more chemical analyses is certainly indicated. Cr-in-Cpx and en-in-Cpx thermobarometry suggested T > 900 °C and P > 1.2 GPa (Abbott et al., 2001a), consistent with the low-P/T (UHT) regime. However, these estimated conditions are questionable because of the very low Cr values in the Cpx (~0.001 Cr per 6-oxygen formula unit, Abbott et al., 2001a). With respect to Fe-Mg exchange, garnet, clinopyroxene, olivine, and hornblende were last in equilibrium at T = 600–660°C (Abbott et al., 2001a), consistent with retrograde, amphibolite-facies conditions.

Corundum was discovered in the garnet peridotite in cm-scale regions where the olivine is locally absent (Draper et al., 2002; Abbott et al., 2003a). The local assemblage, Cpx + Grt + Spl + Crn (+ retrograde hornblende), offered the first evidence of UHP conditions. The assemblage is unusual for the coexistence of Grt + Spl + Crn. According to the experiments on the stability of sapphirine (Ackermand et al., 1975), the assemblage Grt + Spl + Crn is only stable under UHP conditions (Ackermand et al., 1975). Equilibrium analysis (WEBINVEQ: Gordon, 1999) involving CFMAS components in relevant combinations of olivine, garnet, clinopyroxene, spinel, and corundum indicated the following UHP conditions (Draper et al., 2002; Abbott et al., 2003a):

P = 2.8 GPa, T = 810 °C, for Grt + Cpx + Spl + Crn,
and
P = 3.5 GPa, T = 740 °C, for Ol + Grt + Cpx + Spl.

These WEBINVEQ P-T conditions are very robust with regard to reasonable uncertainty in mineral chemistry. The low Al in the Cpx (0.11 Al per 6-oxygen formula unit) is consistent with the relatively low temperatures. More recent evidence for a magmatic origin (Abbott et al., 2003b, and see below), hence much higher temperatures, indicates that the WEBINVEQ P-T conditions relate to subsolidus, retrograde reequilibration.

Petrography

The garnet-bearing ultramafic rocks occur as well-rounded boulders up to 4 meters in diameter in Rio Cuevas (Fig. 3). Outcrops along the river are mostly eclogite. Smaller (<1 m) garnet-bearing ultramafic boulders have been discovered in neighboring drainage systems to the east of Rio Cuevas. The distribution of the boulders and reconnaissance of bedrock geology restrict the source to a narrow, NW-trending zone on the southwest slope of Loma de Quita Espuela. Efforts to find *in situ* exposures have failed so far because of the terrain, dense vegetation, and thick deposits of landslide debris. But, because other rock types (mainly retrograded eclogite) are exposed nearly continuously in bedrock along relevant parts of Rio Cuevas, its tributaries, and in dry gullies, and along smaller streams to the east, presumably the boulders did not come from a single, large body. More likely, the boulders came from many, small bodies of about the sizes of the boulders themselves.

Three texturally and mineralogically distinct lithologies were investigated, exemplified by samples DR03-10 (Fig. 4), DR03-5 (Fig. 5) and DR03-12 (Fig. 6). The rock types and mineral assemblages are as follows:

Garnet peridotite (DR03-10), **Cpx + Ol + Grt + Spl** + Mg-hornblende + serpentine,
Garnet clinopyroxenite (DR03-5), **Cpx + Grt + Spl** + hornblende,
Corundum-bearing garnet clinopyroxenite (DR03-12), **Cpx + Grt + Spl + Crn** + hornblende.

Accessory minerals include magnetite and pyrite. Unlike other occurrences of corundum-garnet–bearing ultramafic rocks (Kornprobst et al., 1990; Morishita et al., 2001; Zhang et al., 2004), all

TABLE 1. Mineral Analyses

	ASU[1]			FCAEM[2]			
	Garnet	Spinel	Corundum	Garnet	Olivine	Diopside	Hornblende
No. of analyses:	10	7	2	4	5	22	5
	wt%			wt%			
Na_2O	n.a.[3]	n.a.	n.a.	0.00	0.03	0.12	1.49
MgO	13.30	20.46	n.a.	9.47	40.12	15.49	16.48
Al_2O_3	23.29	68.40	99.73	22.57	0.03	2.47	11.40
SiO_2	41.17	n.a.	n.a.	38.86	37.62	50.90	46.95
K_2O	n.a.	n.a.	n.a.	0.00	0.01	0.00	0.01
CaO	11.09	n.a.	n.a.	8.57	0.01	23.05	12.48
TiO_2	0.02	0.10	n.a.	0.01	0.00	0.19	0.48
Cr_2O_3	0.09	0.09	0.03	0.00	0.02	0.10	0.56
MnO	0.47	0.18	n.a.	0.81	0.23	0.15	0.05
FeO[4]	10.59	10.78	n.a.	21.10	23.21	5.29	7.09
Fe_2O_3[5]	n.a.	n.a.	0.24	n.a.	n.a.	n.a.	n.a.
Total	100.00	100.00	100.00	101.40	101.27	97.76	96.99
	Cations p.f.u.			Cations p.f.u.			
Na	–	–	–	0.00	0.00	0.01	0.41
Mg	1.45	0.76	–	1.05	1.53	0.87	3.51
Al	2.00	2.01	1.996	1.99	0.00	0.13	1.92
Si	3.01	–	–	2.90	0.96	1.90	6.72
K	–	–	–	0.00	0.00	0.00	0.00
Ca	0.87	–	–	0.69	0.00	0.93	1.91
Ti	0.00	0.00	–	0.00	0.00	0.01	0.05
Cr	0.00	0.00	0.000	0.00	0.00	0.00	0.06
Mn	0.03	0.00	–	0.05	0.00	0.00	0.01
Fe^{2+}	0.65	0.23	–	1.11	0.42	0.10	0.85
Fe^{3+}	–	–	0.003	0.21	0.07	0.06	0.00
Total cations	8.00	3.00	1.999	8.00	3.00	4.00	15.45
Total O	12.01	4.00	3.000	12.00	4.00	6.00	23.00
Mg#	0.69	0.77	–	0.49	0.78	0.90	0.81
prp	0.48			0.36			
grs	0.29			0.24			
sps	0.01			0.02			
alm	0.21			0.38			
fo					0.78		
fa					0.22		
wo						0.49	
en						0.46	
fs						0.05	

[1]Electron microscope facility, Appalachian State University.
[2]Florida Center for Analytical Electron Microscopy, Florida International University (Abbott et al., 2001a).
[3]n.a. = not analyzed.
[4]Total Fe as FeO.
[5]Total Fe as Fe_2O_3.

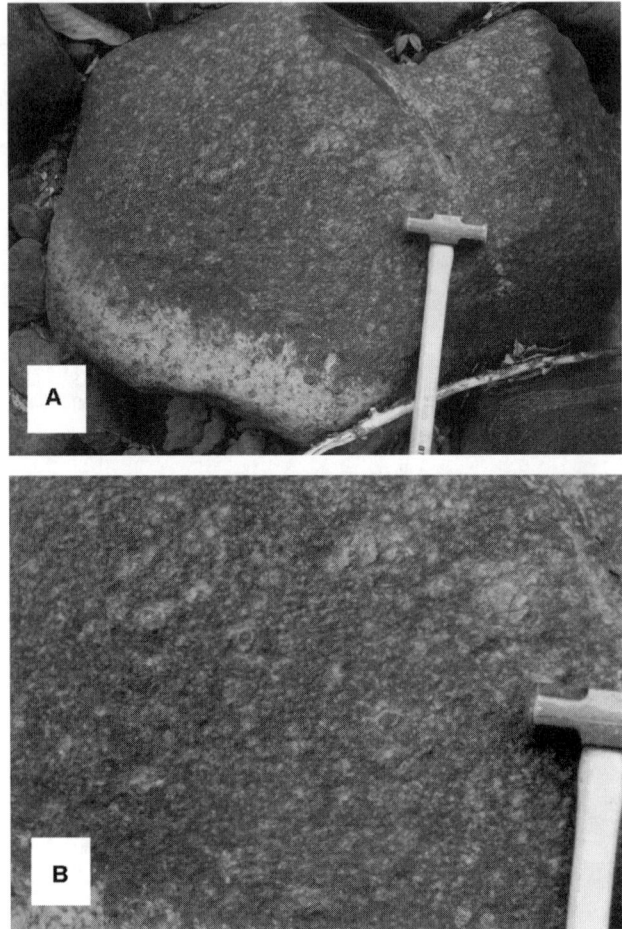

FIG. 3. A. Typical boulder of the garnet peridotite (~1.5 m). The essential mineralogy can be recognized even at this scale. Garnet forms subspherical phenocrysts up to 2 cm in diameter (grey). These are rimmed by light grey clinopyroxene and magnesiohornblende. The black matrix is partially serpentinized olivine. B. Close-up of the same boulder. Note the garnet phenocrysts with the "kelyphitic" rims of late (retrograde) magnesiohornblende. Spinel occurs as fine (<0.5 mm) inclusions in the garnet and also in the kelyphitic rims.

of the assemblages reported here are conspicuously devoid of plagioclase and sapphirine. Coexisting Grt + Spl + Crn in the DR03-12 assemblage has not been reported elsewhere, and by itself precludes sapphirine (Ackermand et al., 1975).

Garnet peridotite

Coarse, granoblastic garnet peridotite (DR03-10, Figs. 3 and 4) is by far the most common of the three lithologies. The rock is homogeneous in a given boulder (Fig. 3), and from one boulder to another. Four textural elements dominate the rock:

Garnet (up to 25%). Crystals of pink garnet, up to 2 cm in diameter, are the most conspicuous feature of the rock (Figs. 3 and 4). The crystals are anhedral, but crudely equidimensional. The garnet contains fine (<1.5 mm) inclusions of transparent, emerald-green spinel (Fig. 7A). Magnesiohornblende occupies cracks and fractures.

Kelyphitic magnesiohornblende (~20%). Magnesiohornblende forms light grey-green, so-called "kelyphite" rims on the garnet (Figs. 3 and 4). The kelyphite also occupies deep embayments in the garnet, and locally replaces garnet completely. In

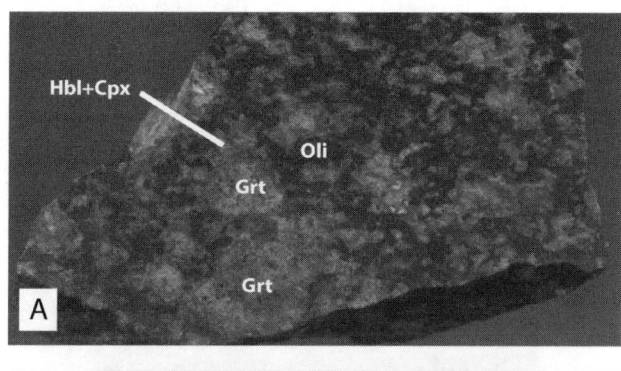

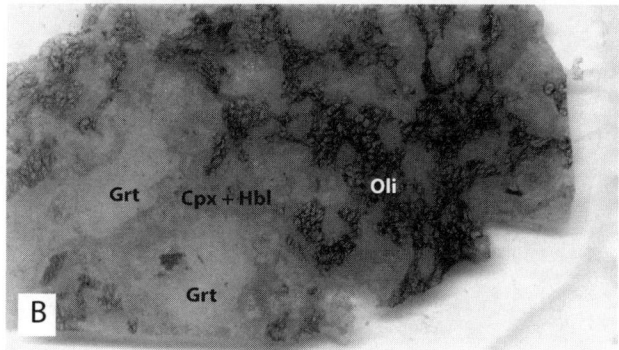

FIG. 4. Garnet peridotite (DR03-10): **Cpx + Ol + Grt + Spl** + Mg-hornblende + serpentine. A. Photograph of slab, width of field = 12.3 cm. B. Photomicrograph of a thin section, width of field = 3.7 cm. Garnet (Grt), partially serpentinized olivine (Oli, black), spinel (fine "black" flecks in Grt), clinopyroxene + late magnesiohornblende (light grey). The magnesiohornblende and serpentine are products of retrograde hydration. The original assemblage was Grt + Cpx + Ol + Spl.

thin section, the kelyphitic magnesiohornblende resolves into pale-green, fine- to medium-grained (1–2 mm) euhedral to subhedral crystals. Close to the garnet, spinel occurs in the kelyphite as fine (<0.5 mm), disseminated, mainly intragranular grains and clusters of grains.

Clinopyroxene (~15%). Clinopyroxene forms large (~6 mm), subhedral, generally rounded or ovate crystals that are clouded to a greater or lesser extent by minute (submicron-scale), crystallographically oriented, platy inclusions. The mineralogy of these inclusions is unknown (ilmenite?). Clinopyroxene does not contain inclusions of spinel. Crystals of clinopyroxene are rimmed by kelyphitic magnesiohornblende. Rare, less-rounded grains of clinopyroxene, with little or no kelyphite, occur within domains of partially serpentinized olivine. Enstatite was reported (Abbott et al., 2001a, 2001b) on the basis of optical properties of two or three grains in one thin section. Re-examination of the grains in light of subsequent theoretical considerations (Abbott et al., 2003a, 2003b) and examination of newer thin sections could not verify the earlier report. Enstatite is neither required by stoichiometry, nor is it observed.

Partially serpentinized olivine (~40%). On the surfaces of boulders and in hand specimen, partially serpentinized olivine forms a continuous, black matrix (Figs. 3 and 4A). In thin section (Fig. 4B), this matrix resolves into a fractured mosaic texture, with serpentine filling the late fractures. The olivine itself is free of inclusions.

Interpretation. Magnesiohornblende and serpentine are the products of retrograde hydration. The original assemblage was Grt + Cpx + Ol + Spl.

Garnet clinopyroxenite

Garnet clinopyroxenite (DR03-5, Fig. 5) is known from only six boulders. The mineral assemblage is distinguished by the absence of olivine and

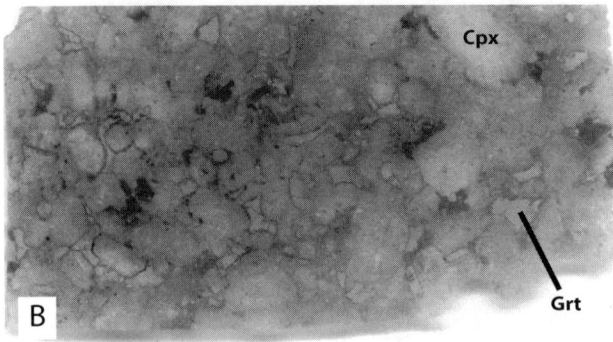

FIG. 5. Garnet clinopyroxenite (DR03-5): **Cpx + Grt + Spl** + Mg-hornblende. A. Photograph of slab, width of field = 12.3 cm. B. Photomicrograph of a thin section, width of field = 3.7 cm. In hand specimen (A): Orthocumulate clinopyroxene (grey, subhedral phenocrysts), interstitial garnet (light grey), late hornblende (black). In thin section (B): Orthocumulate clinopyroxene (Cpx, colorless, 80% replaced by colorless hornblende), rimmed by thin (0.1 mm) selvage of hornblende (black). Garnet (Grt) is interstitial (cuspate shapes). Spinel occurs as fine (<0.2 mm) inclusions in garnet. The hornblende is interpreted as the product of retrograde hydration. The original assemblage was Cpx + Grt + Spl.

corundum. Euhedral to subhedral crystals of clinopyroxene (5–10 mm) form an orthocumulate texture. Dark green Mg-hornblende forms thin (~0.1 mm) selvages on clinopyroxene and on clinopyroxene that has otherwise been partially to completely replaced by pale green to colorless hornblende. In this example (Fig. 5B), only 20 to 30 % of the original clinopyroxene remains. The dark hornblende selvages also serve to outline the unusual cuspate shapes of the interstitial garnet. Spinel occurs as fine (< 0.5 mm) inclusions in garnet. Opaque minerals are too weathered and too fine-grained to properly identify. Hornblende is interpreted as the product of retrograde hydration. The original assemblage was Cpx + Grt + Spl.

Corundum-bearing garnet clinopyroxenite

The corundum-bearing garnet clinopyroxenite (DR03-12, Fig. 6) occurs as cm- to dm-scale irregular, cross-cutting veins in three of the boulders of garnet clinopyroxenite. Large (up to 3 cm) euhedral to subhedral crystals of clinopyroxene form an orthocumulate texture, similar to the garnet clinopyroxenite (DR03-5) but much coarser and more distinct. Interstices are filled by garnet. Spinel and corundum occur as fine (<1 mm) inclusions in garnet (Figs. 7B and 7C). The corundum is partially replaced around its edges and along fine (<0.1 mm) fractures by unidentified, colorless, isotropic (amorphous?) products of late hydrothermal alteration or weathering (Figs. 7B and 7C). Hornblende forms thin (1 mm) selvages on the clinopyroxene. The hornblende is interpreted as the product of late, retrograde hydration. The original assemblage was Cpx + Grt + Spl + Crn.

Orthocumulate textures serve as prima-facie evidence of a magmatic origin. By inference, the granoblastic texture of the garnet peridotite is also

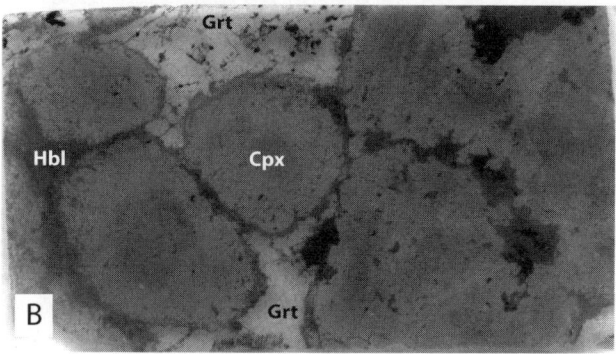

FIG. 6. Corundum-bearing garnet clinopyroxenite (DR03-12): **Cpx + Grt + Spl** + Crn + hornblende. A. Photograph of slab, width of field = 12.3 cm. B. Photomicrograph of a thin section, width of field = 3.7 cm. Orthocumulate clinopyroxene (Cpx, grey euhedral to subhedral phenocrysts), interstitial garnet (Grt, light grey). Spinel occurs as fine (<1 mm) inclusions in garnet. The fine, white inclusions in garnet are corundum. Hornblende forms thin (1 mm) selvages on phenocrysts of clinopyroxene. The hornblende is interpreted as the product of late, retrograde hydration. The original assemblage was Cpx + Grt + Spl + Crn.

magmatic, admittedly much altered by the products of late, subsolidus hydration. As such, textural relationships suggest that the original, magmatic assemblages were related by fractional crystallization involving a liquid (silicate melt) in equilibrium with Grt + Cpx + Ol + Spl (DR03-10), then Cpx + Grt + Spl (DR03-5), and finally Cpx + Grt + Spl + Crn (DR03-12). In the next section, the magmatic conditions (P, T) are established for the sequence of assemblages.

CMAS System, Liquidus Relationships

Figure 8 shows primary liquidus volumes for mineral phases in a portion of the CMAS system, $CaMgSi_2O_6$-$CaAl_2Si_2O_8$-Mg_2SiO_4-SiO_2, at P slightly > 3.0 GPa, as determined experimentally by Milholland and Presnall (1998). Liquidus equilibrium (1), Liq + Fo + Grt = Cpx + Spl, is stable at P > 3.0 GPa (ibid.). At the experimental conditions (3 GPa), the equilibrium can be balanced by linear algebraic methods:

$$1 \text{ Liq}(1) + 0.23 \text{ Fo} + 0.01 \text{ Grt} = 0.33 \text{ Spl} + 2.45 \text{ Cpx}, \quad (1)$$

using Grt, Cpx and Liq(1) from Milholland and Presnall (1998), Fo (Mg_2SiO_4), and Spl ($MgAl_2O_4$). The composition of the liquid is normalized to 10 cations.

According to experiments on the stability of sapphirine (Ackermand et al., 1975), with increasing P the primary volume for sapphirine (Spr) must shrink and finally disappear at about 3.4 GPa (Fig. 8, upper right inset). At P > 3.4 GPa, the equilibria represented by liquids at W, X, Y, and Z in Figure 8 (Milholland and Presnall, 1998) are metastable. The stable equilibrium at these higher pressures can be inferred from Schreinemakers analysis, (3) Liq + Spl = Grt + Cpx + Crn. Although this equilibrium is

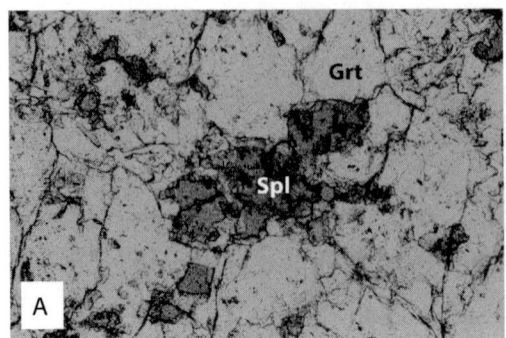
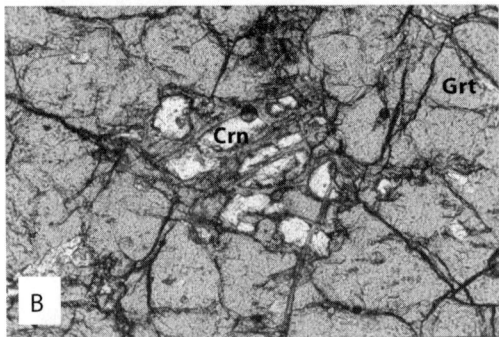
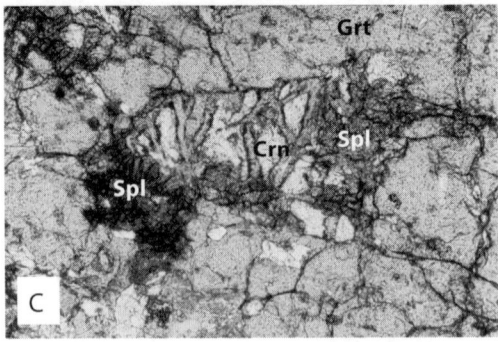

FIG. 7. Photomicrographs of inclusions of spinel and corundum in garnet, width of field of view, 1.3 mm. A. Spinel (Spl) inclusions in garnet (Grt), plane polarized light. Sample DR99-2; granoblastic garnet peridotite, **Cpx +Ol + Grt + Spl** (+ Hbl) (equivalent to DR03-10, Fig. 4). B. Corundum inclusion (Crn, bright), partially cross-polarized light. C. Inclusion of corundum (Crn, bright) + spinel (Spl) in garnet (Grt), partially cross-polarized light. Sample DR00-3; orthocumulate corundum-spinel-bearing garnet clinopyroxenite, **Cpx + Grt + Spl + Crn** (+ Hbl) (equivalent to DR03-12, Fig. 6). Corundum is partially replaced along fractures by unidentified, isotropic (amorphous?) products of hydrothermal alteration or weathering. High birefringence crystals near lower edge of (C) are hornblende.

metastable at the experimental conditions (3 GPa), it can still be balanced by linear algebraic methods for the metastable conditions,

$$1 \text{ Liq}(3) + 0.80 \text{ Spl} = 0.92 \text{ Grt} + 0.91 \text{ Cpx} + 0.91 \text{ Crn}. \quad (3)$$

The composition of Liq(3), normalized to 10 cations, was taken to be the average of the compositions of liquids W, X, Y and Z. This assumes that the metastable Liq(3) coexisting with Spl, Grt, Cpx and Crn is within the 3 GPa sapphirine volume, and close to the average of the compositions of liquids W, X, Y and Z. The compositions for W and X are reasonably well constrained by the experimental data, whereas the compositions for Y and Z are not very well constrained. Equilibrium (3) involves the composition of the most aluminous Cpx reported by Milholland and Presnall (1998), the most calcic Grt reported by Milholland and Presnall (1998), Crn (Al_2O_3), and Spl ($MgAl_2O_4$).

Invariant Point at ~3.4 GPa and ~1550°C

Equilibria involving Cpx, specifically W, X, Y, and (3), define a CMAS invariant point. In P-T space, the invariant point lies on the equilibrium Spr = Grt + Spl + Crn (Ackermand et al., 1975) at ~3.4 GPa and ~1550 °C. Figure 9 shows this equilibrium and the sequence of stable and metastable parts of the equilibria W, X, Y, and (3). The temperature for metastable equilibrium (3) at 3 GPa must be greater than T_X = 1540°C and less than T_Y = 1545°C. Also shown in Figure 9 is the locus of equilibrium (1). Liquids coexisting with just Grt, Cpx, and Spl are only possible in the shaded region. Presumed conditions represented by the rock samples are indicated by their labels in the figure. The garnet peridotite (DR03-10) corresponds to equilibrium (1) and the corundum-bearing garnet clinopyroxenite (DR03-12) corresponds to equilibrium (3). Conditions for the garnet clinopyroxenite (DR03-5) are only possible in the shaded region in Figure 9 — i.e., between the locus of equilibrium W and the locus of equilibrium (1) at P < 3.4 GPa, and between the locus of equilibrium (3) and the locus of equilibrium (1) at P > 3.4 GPa.

A'W'E' Projection and Effects of Other Components

Figure 10 projects the compositions of minerals and liquids from spinel onto the plane defined by A' = Crn/3 = Al_2O_3/3, W' = wo/3 = $CaSiO_3$/3, and E' = en/3 = $MSiO_3$/3 (M = Fe + Mg), such that all compositions are normalized to one oxygen.

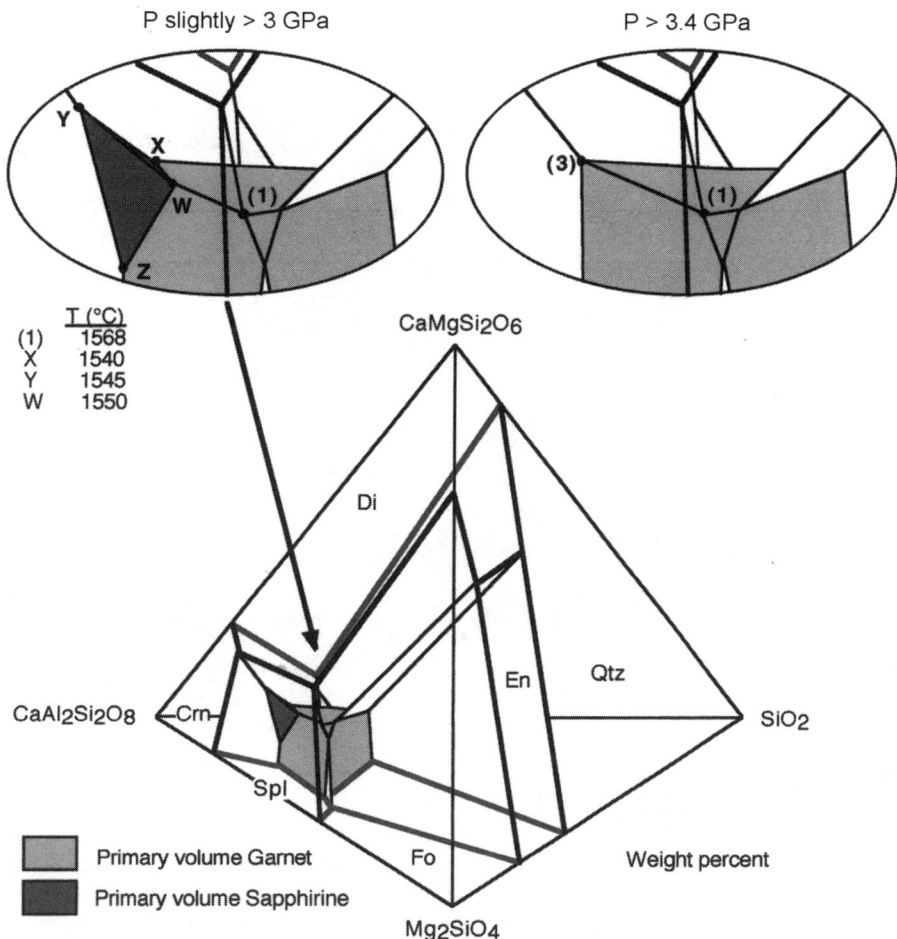

FIG. 8. Primary liquidus volumes for phases in a portion of the CMAS system at P slightly greater than 3 GPa (Milholland and Presnall, 1998). Primary liquidus volume for garnet and sapphirine are highlighted. Details of the sapphirine volume and part of the garnet volume are shown in the upper left inset, where univariant equilibria W, X, Y, Z, and (1) are indicated. With increasing pressure, the sapphirine volume shrinks and finally disappears at ~3.4 GPa. Upper right inset shows relevant details for P > 3.4 GPa. The line connecting (1) and (3) is the locus of liquids in equilibrium with Cpx, Grt, and Spl.

Analyses of 3 GPa Grt, Cpx, and Liq(1) are from Milholland and Presnall (1998). Liq(3) is the average of 3 GPa compositions for liquids at X, Y, W and Z (Milholland and Presnall, 1998). The hypothetical locus of Liq(2) in equilibrium with Grt, Cpx, and Spl is also shown. The average composition of N-MORB (Normal Mid-Ocean Ridge Basalt; Best and Christiansen, 2001) is shown for comparison.

Non-CMAS components, in particular Fe and Na, influence the projected compositions of Liq(3), Liq(1), and the locus of Liq(2) in ways that are difficult to quantify. Milholland and Presnall (1998) argued that addition of Na and Fe to the system does not change the general forms of these equilibria, and has little effect on their temperatures or pressures. Of the phases involved in the equilibria, only Cpx tolerates appreciable Na as the component $NaAlSi_2O_6$ (jadeite). Thus, adding Na to the system should expand the liquidus volume for Cpx, effectively displacing the locus of Liq(2), i.e., the

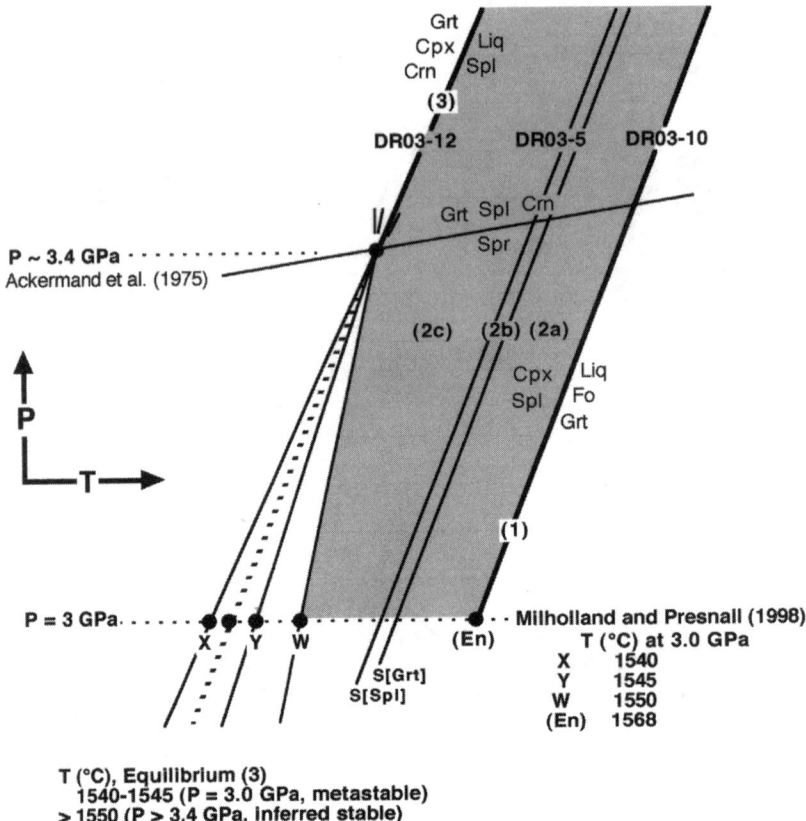

FIG. 9. P-T diagram, showing liquidus equilibria involving Cpx, extrapolated from Milholland and Presnall (1998), and equilibrium Sap = Grt + Spl + Crn (Ackermand et al., 1975). With increasing pressure, equilibria X, Y, and W converge at an invariant point on the equilibrium Spr = Grt + Spl + Crn, at ~3.4 GPa. At higher pressures the stable equilibrium is (3) Liq + Spl = Grt + Cpx + Crn. Liquids in equilibrium with Cpx + Grt + Spl are possible in the shaded region. Conditions for the three lithologies are indicated by their respective labels, DR03-12, DR03-5, and DR03-10.

segment (1)–(3), away from the W' (Fig. 10). On the other hand, the partitioning of Fe and Mg between garnet and Cpx (Pattison and Newton, 1989) is such that adding Fe to the system should expand the garnet volume, thus displacing the locus of Liq(2) closer to W'. The combined effects of adding both Na and Fe the system is therefore not entirely clear, and may be more or less neutral. Presumably, the effect would depend on the amounts and proportions of Na and Fe added to the system. All other factors considered equal, expansion of the primary volumes for both Grt and Cpx should lengthen the locus of Liq(2), creating a greater compositional disparity between Liq(1) and Liq(3).

Details

The nature of the equilibrium involving liquid (2) coexisting with Grt, Cpx, and Spl warrants closer examination. At 3 GPa, the equilibrium is limited at high temperature by equilibrium (1) (1568°C; Milholland and Presnall, 1998) and at low temperature by equilibrium (3) (1540–1545°C; Fig. 9). At 3 GPa, the range of temperature is not more than 23–28°C. Presumably, the temperature difference between equilibrium (1) and (3) would be about the same at higher pressures.

For practical purposes assuming that the locus of Liq(2) in equilibrium with Cpx, Grt, and Spl is

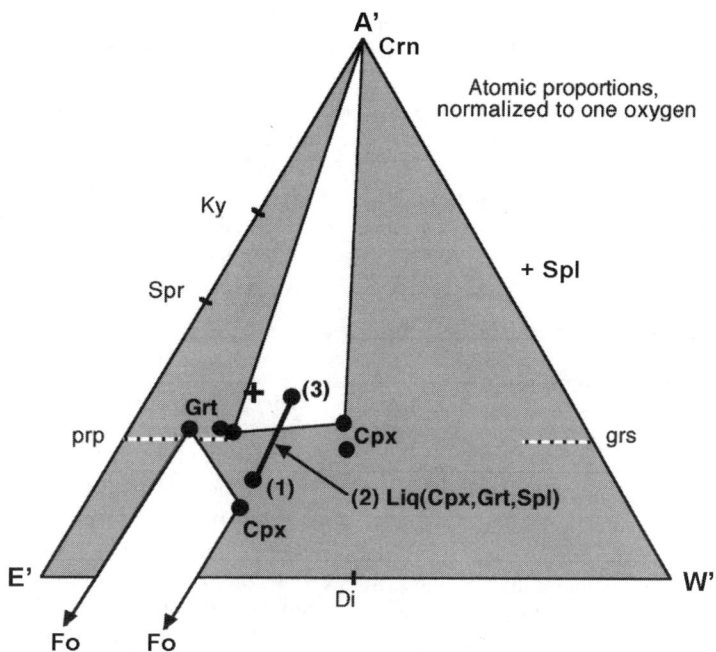

FIG. 10. A'W'E' projection from spinel onto the plane defined by A' = Crn/3, W' = wo/3, and E' = en/3. Analyses of Grt, Cpx, and Liq (1) are from Milholland and Presnall (1998). Liq (3) is the average of liquid compositions W, X, Y, and Z (Milholland and Presnall, 1998). Liq (1) is involved in the equilibrium Liq + Fo + Grt = Spl + Cpx. Liq (3) is involved in the equilibrium Liq + Spl = Grt + Cpx + Crn. Locus of liquids in equilibrium with Cpx, Grt, and Spl connects (1) and (3). Symbols: + = average N-MORB (Best and Christiansen, 2001).

straight (Fig. 10), such that the tangent to the locus is the line (1)–(3), then the nature of the equilibrium can be determined at the high-temperature end of the locus, i.e., at (1), in the following way:

1. Find the point of intersection, $P(1)$, of the line (1)–(3) and the plane defined by Grt(1), Cpx(1), and Spl(1) in equilibrium with Liq(1).

2. Express the composition $P(1)$ as a linear combination of Grt(1), Cpx(1) and Spl(1). This gives the nature of equilibrium (2) at the high-T end of the line (1)–(3),

$$1\, P(1) + 0.276\, \text{Grt}(1) = 1.197\, \text{Cpx}(1) + 0.079\, \text{Spl}(1), \quad (2a)$$

where the mineral formulae and composition $P(1)$ are normalized to one oxygen.

Likewise, for equilibrium (2) at the low-T end of the locus, i.e., at (3):

$$1\, P(3) + 0.099\, \text{Spl}(3) = 0.607\, \text{Cpx}(3) + 0.492\, \text{Grt}(3). \quad (2c)$$

Along the locus from (1) to (3) the role of Grt changes from reactant to product; hence the equilibrium passes through a singularity, S[Grt], where the coefficient of Grt is zero. Likewise, the role of Spl changes from product to reactant, hence another singularity, S[Spl], where the coefficient of Spl is zero. Assuming T and the coefficients of Grt, Cpx, and Spl vary linearly between (1) and (3), and taking T(3) = 1543°C, then T = 1559°C at S[Grt], and T = 1557°C at S[Spl]. In the 2°C interval between S[Grt] and S[Spl], the nature of the equilibrium is:

$$\text{Liq} = \text{Grt} + \text{Cpx} + \text{Spl}. \quad (2b)$$

The coefficients of Grt and Spl are very small in this interval, such that equilibrium (2b) is very close to Liq = Cpx. Hypothetical loci for S[Grt] and S[Spl] in P-T space are shown in Figure 6. The loci divide the shaded region into 3 zones according to the nature of equilibrium (2). The relationships are consistent with the growth of Cpx phenocrysts (2a, b) in samples DR03-5 and DR03-12, accumulation of the phenocrysts, and subsequent growth of Grt in the interstices (2c); in the case of DR03-12, before the growth of Crn (3).

Conclusions

Orthocumulate clinopyroxene textures and the observed mineral assemblages indicate a magmatic origin for the studied garnet peridotites and garnet clinopyroxenites. Hornblende and serpentine are products of late, retrograde hydration. The mineral assemblages are consistent with fractional crystallization at P > 3.4 GPa, in the range T = 1570–1540°C. The temperature is much higher than previous estimates that did not take into account a melt. Previously estimated conditions (T = 740–810°C, P = 2.8–3.4 GPa; Abbott et al., 2003a), based on mineral analyses (Table 1), are presumed to reflect subsolidus reequilibration. The magmatic paragenesis involves three important liquidus equilibria, with declining T:

$$Liq + Grt + Fo = Cpx + Spl; \qquad (1)$$

$$\text{Equilibrium involving Liq, Cpx, Spl, and Grt;} \qquad (2)$$

$$Liq + Spl = Grt + Cpx + Crn. \qquad (3)$$

Magmatic fractionation took place at a depth of approximately 105 km. The findings imply crystallization in the context of a magma chamber near the base of the lithosphere or within the uppermost part of the asthenosphere.

At present there is insufficient information to wholly constrain the tectonic evolution. We offer the following speculative scenario: (1) Ultramafic orthocumulate rocks formed during the Late Cretaceous at the base of the Caribbean lithosphere or in the shallowest part of the asthenosphere beneath the plate. (2) From mid-Cretaceous to mid-Eocene time, Atlantic lithosphere was being subducted beneath the Caribbean plate in an intra-oceanic arc setting. (3) We suggest that the host Cuaba Gneiss was Atlantic oceanic crust, metamorphosed in the subduction zone. At depth, the Cuaba Gneiss incorporated fragments of the overriding Caribbean lithosphere, including the ultramafic orthocumulates. (4) The mechanism of ascent remains problematic, inasmuch as the Cuaba Gneiss is relatively dense (mafic eclogite and amphibolite), and hence buoyancy cannot be invoked. We suggest that reverse-flow processes associated with subduction (Gerya and Yuen, 2003; Gerya et al., 2004) may be responsible for the exhumation of dense, deep-subducted Cuaba Gneiss to higher levels in the lithosphere.

Acknowledgments

The project is supported by National Science Foundation Grants EAR-8306145, EAR-8509542, and INT-0139536 to Draper and NSF Grants EAR-0111471 and INT-0139490 to Abbott. The research was supported by an Appalachian State University Research Grant to Abbott. Keshav was supported by a Dissertation Year Fellowship from the University Graduate School, FIU, and NSF grants to Gautam Sen, FIU. We greatly appreciate the thoughtful comments of the reviewers, D. C. Presnall, J. G. Liou, and W. G. Ernst.

REFERENCES

Abbott, R. N., Jr., and Draper, G., 1998, Retrograde eclogite in the Cuaba Amphibolite of the Rio San Juan Complex, northern Hispaniola: Paper presented at the 15th Caribbean Geological Conference, Kingston, Jamaica.

Abbott, R. N., Jr., and Draper, G., 2002, Retrograded eclogite in the Cuaba Amphibolite of the Rio San Juan Complex, northern Hispaniola, in Jackson, T. A., ed., Caribbean geology: Into the third millennium: Transactions of the 15th Caribbean Geological Conference: Kingston, Jamaica, The University of the West Indies Press, p. 97–108.

Abbott, R. N., Jr., Draper, G., and Keshav, S., 2001a, Garnet peridotite found in the Greater Antilles: EOS (Transactions of the American Geophysical Union), v. 82, no. 35, p. 381–388.

Abbott, R. N., Jr., Draper, G., and Keshav, S., 2001b, Alpine-type garnet peridotite in the Caribbean [abs.]: Geological Society of America Abstracts with Programs, v. 33, no. 6.

Abbott, R. N., Jr., Draper, G., and Keshav, S., 2003a, Tectonic implications of UHP garnet peridotite in the Cuaba Unit, Rio San Juan Complex, Dominican Republic [abs.]: Geological Society of America Abstracts with Programs, v. 35, no. 2.

Abbott, R. N., Jr., Draper, G., and Keshav, S., 2003b, UHP magmatic paragenesis, garnet peridotite, Cuaba Unit, Rio San Juan Complex, Dominican Republic [abs.]:

Geological Society of America Abstracts with Programs, v. 35, no. 6.

Ackermand, D., Seifert, F., and Schreyer, W., 1975, Instability of sapphirine at high pressures: Contributions to Mineralogy and Petrology, v. 50, p. 79–92.

Best, M. G., and Christiansen, E. H., 2001, Igneous Petrology: Oxford, UK, Blackwell, 458 p.

Brueckner, H. K., and Medaris, L. G., 2000, A general model for the intrusion and evolution of "mantle" garnet peridotites in high-pressure and ultra-high-pressure metamorphic terranes: Journal of Metamorphic Geology, v. 18, p. 123–133.

Draper, G., Abbott, R. N., Jr., and Keshav, S., 2002, Indication of UHP metamorphism in garnet peridotite, Cuaba Unit, Rio San Juan Complex, Dominican Republic [abs.]: Sixteenth Caribbean Geological Conference, Barbados.

Draper, G., Gutierrez, G., and Lewis, J. F., 1996, Thrust emplacement of the Hispaniola peridotite belt: Orogenic expression of the mid-Cretaceous Caribbean arc polarity reversal?: Geology, v. 24, p. 1143–1146.

Draper, G., Mann, P., and Lewis, J. F., 1994, Hispaniola, in Donovan, S. K., and Jackson, T. A., eds., Caribbean Geology: An introduction: Kingston, Jamaica, University of the West Indies Publishers Association, p. 129–150.

Draper, G., and Nagle, F., 1991, Geology, structure, and tectonic development of the Rio San Juan Complex, northern Dominican Republic, in Mann, P., Draper, G., and Lewis, J. F., eds., Geologic and tectonic development of the North American–Caribbean plate boundary in Hispaniola: Geological Society of America Special Paper 262, p. 77–95.

Eberle, W., Hirdes, W., Muff, R., and Pelaez, M., 1982, The geology of the Cordillera Septentrional (Dominican Republic), in Snow, W., Gil, N., Llinas, R., Rodriguez-Torres, R., and Tavares, I., eds., Transactions, 9th Caribbean Geological Conference, Santo Domingo, Dominican Republic, 1980, p. 619–632.

Gerya, T. V., and Yuen, D. A., 2003, Rayleigh-Taylor instabilities from hydration and melting propel "cold plumes" at subduction zones: Earth and Planetary Science Letters, v. 212, p. 47–62.

Gerya, T.V., Yuen, D. A., and Sevre, E. O. D., 2004, Dynamic causes for incipient magma chambers above slabs: Geology (in press).

Gordon, T., 1999, Generalized thermobarometry: WEBINVEQ with the TWQ 1.02 data base [http://ichor.geo.ucalgary.ca/~tmg/Webinveq/rgb95.html].

Kretz, R., 1983, Symbols for rock-forming minerals: American Mineralogist, v. 68, p. 277–279.

Kornprobst, J., Piboule, M., Roden, M., and Tabit, A., 1990, Corundum-bearing garnet clinopyroxenites at Beni Bousera (Morocco): Original plagioclase-rich gabbros recrystallized at depth within the mantle?: Journal of Petrology, v. 31, p. 717–745.

Lewis, J. F., and Draper, G., 1990, Geology and tectonic evolution of the northern Caribbean region, in Dengo, G., and Case, J. E., eds., The Caribbean Region: Boulder, CO, Geological Society of America, Geology of North America, v. H, p. 77–140.

Mann, P., Schubert, C., and Burke, K., 1990, Review of Caribbean neotectonics, in Dengo, G., and Case, J.E., eds., The Caribbean Region: Boulder, Colorado, Geological Society of America, Geology of North America, v. H, p. 307–338.

Medaris, L. G., Jr., 1999., Garnet peridotites in Eurasian high-pressure and ultrahigh-pressure terranes: A diversity of origins and thermal histories: International Geology Review., v. 41, p. 799–815.

Milholland, C. S., and Presnall, D. C., 1998, Liquidus phase relationships in the system $CaO-MgO-Al_2O_3-SiO_2$ at 3.0 GPa: The aluminous pyroxene thermal divide and high-pressure fractionation of picritic and komatiitic magmas: Journal of Petrology, v. 39, p. 3–27.

Morishita, T., Arai, S., and Gervilla, F., 2001, High-pressure aluminous mafic rocks from the Ronda peridotite massif, southern Spain: Significance of sapphirine- and corundum-bearing mineral assemblages: Lithos, v. 57, p. 143–161.

Pattison, D., and Newton, R. C., 1989, Reversed experimental calibration of the garnet-clinopyroxene Fe-Mg exchange thermometer: Contributions to Mineralogy and Petrology, v. 101, p. 87–103.

Pindell, J. L., and Barrett, S. F., 1990, Geological evolution of the Caribbean region; a plate tectonic perspective, in Dengo, G., and Case, J. E., eds., The Caribbean Region: Boulder, CO, Geological Society of America, Geology of North America, v. H, p. 405–432.

Zhang, R. Y., Liou, J. G., and Zheng, J. P., 2004, Ultra high-pressure corundum-bearing garnetite in garnet peridotite, Sulu terrane, China: Contributions to Mineralogy and Petrology, v. 147, p. 21–31.